Invertebrate Fossils

INVERTEBRATE FOSSILS

RAYMOND C. MOORE

Professor of Geology, University of Kansas

CECIL G. LALICKER

Professor of Geology, University of Kansas

ALFRED G. FISCHER

International Petroleum Company, Peru
Formerly Assistant Professor of Geology
University of Kansas

New York   Toronto   London

McGRAW-HILL BOOK COMPANY, INC.

1952

INVERTEBRATE FOSSILS

Library of Congress Catalog Card Number: 51-12632

Preface

Work on this textbook was begun in 1948, mainly in response to needs in teaching the general course in Invertebrate Paleontology offered at the University of Kansas. Here, as at many other institutions which have a curriculum of training in geological subjects, knowledge of the essential features pertaining to invertebrate fossils is judged to be an indispensable part of the preparation for almost any type of geological employment. A study of the organic remains preserved in rocks, from oldest to youngest and representing all sorts of environments, is no less fundamental for students of geology than the study of minerals, rocks, and geological processes. In addition, paleontological knowledge is nearly as valuable to the biologist as to the geologist, even though most students of zoology and botany receive little instruction concerning fossils. It seems beyond doubt that the study of paleontology has been handicapped by the existence of too few good textbooks, and accordingly, the descriptions, discussions, and illustrations offered in the book on *Invertebrate Fossils* are intended partly to fill this need.

The book is designed to serve as a text for students who are taking a first course in Invertebrate Paleontology, and because many majors in geology lack preparatory training in zoology, no such previous zoological study is assumed. Obviously, the completion of a course in biology or elementary zoology prior to enrollment in a course on invertebrate fossils is an advantage, but in our view it is not required. Naturally, attention is focused mainly on the nature of hard parts belonging to each group of invertebrates represented among

fossils, but we have endeavored to keep in mind that the objects of study are all remains of once living animals. It is important to learn enough about their soft structures, life histories, manner of reproduction, and mode of life to permit an understanding of them as living things. Although many deductions can be made from a study of the fossils themselves, most observations of soft-part anatomy depend on investigations of modern relatives. The same is true for interpretations of ecologic adaptations. Fortunately, all but a few kinds of fossil invertebrates are represented by similar living animals.

Invertebrate Fossils is a book intended to supplement instruction in the classroom and laboratory by providing a fairly complete description of each main group of fossil invertebrates, with illustrations of examples. It is planned as a textbook, not as a reference work. This means that the chapters undertake to explain in consecutive manner important features of structure, classification, reproduction and growth stages, characters of main divisions, evolutionary trends, and geological distribution. Condensed summaries of morphological features are given in illustrated form, accompanied by brief cross-indexed definitions of terms. Stratigraphical distribution is represented graphically, and in most chapters there are supplementary illustrations in which fossils are grouped by geologic systems. These latter are included with two main purposes: to serve as a background in the student's first general course, aiding the comprehension of the stratigraphical usefulness of invertebrate fossils; and to provide carefully selected material for more advanced study, in which attention may be given to distinguishing characters of the fossil assemblages. In an introductory study, it is appropriate that little or no effort should be expended on becoming acquainted with the names and many distinguishing characters of the genera illustrated, but as the student progresses, he may undertake to learn these, adding much to his knowledge by a study of the stratigraphical groups.

Arrangement of study by zoologically defined groups and emphasis on morphology are the soundest approaches to a proficiency in the practical uses of invertebrate paleontology. On the knowledge of significant structural characters of the fossils, classification essentially depends. A thorough acquaintance with morphological attributes also is necessary in differentiating youthful, mature, and old-age stages in the growth of individuals, in defining race-history relationships, and in recognizing evolutionary trends, all of which are important in taxonomy and in the stratigraphical use of fossil invertebrates. Hence, studies directed to understanding how preserved skeletal features are constructed, how they originate, and how they function furnish a firm foundation for interpreting them correctly and using them in stratigraphical work reliably. An effort to learn the identity of any fossil on the basis of its general appearance, without becoming able to discern and understand its distinguishing structural characters, is largely misdirected. Such labor is comparable to memorizing many illustrated items of a mail-order catalogue. One may gain familiarity with the items pictured, recite their names, and give correctly the page number on which they are listed, but if the items to be identified do not exactly match those of the catalogue and if they differ in association, all is strange. The ability to recognize and rightly name many "index fossils" does have value, but this needs to be supported by a thorough knowledge of the morphological attributes of the fossils, because

these relate them to others and at the same time distinguish them from others.

No previously published textbook on invertebrate paleontology brings together illustrations and definitions of morphological features belonging to each fossil group as comprehensively and compactly, and in our belief, as clearly and usably, as the student will find in *Invertebrate Fossils*. A desirable feature of the book also is the provision of lettering on nearly all illustrations, which makes them understandable without reference to the accompanying captions. Mostly, therefore, these captions are supplementary in nature. We think that teachers will approve the many features of graphic presentation contained in the book, and we hope that students will find them helpful. Special care has been taken to ensure accuracy.

This textbook is conservative in following a long-tested general plan of organization and in seeking to reflect the consensus of authoritative judgment on all the subjects treated. Sources are spread throughout the literature. They are so voluminous and generally well known that individual acknowledgments are omitted except in a few places. Many of the original sources are cited in bibliographies at the end of chapters. The book is original in many features of the text. Among these are some departures from classification previously employed for various groups. This pertains to coelenterates, brachiopods, pelecypods, cephalopods, arthropods, and crinoids. None of the changes are sweeping, for this would be inappropriate in a general textbook, and each is discussed in a manner permitting the teacher and student to make his own conclusions concerning them. For dentition of some of the clams, the commonly used system of symbols is explained, but because it is difficult and cumbersome, a simplified scheme of designation is introduced. Nearly all illustrations consist of line drawings made by us; a few were made by assistants, and some are copied without change from published sources. Each author prepared the figures in the chapters with which his name appears in the Table of Contents, planning the layout, making the drawings, and placing the legends. This task proved fully equal to that of writing the text and figure captions.

A difficult question in organizing the textbook on *Invertebrate Fossils* was how to determine the proportion of space for each division. Judgment concerning the relative importance of main groups is bound to differ. Professional paleontologists in the employ of geological surveys, museums, and commercial concerns such as oil companies put emphasis on groups with which they are especially concerned. Teachers of invertebrate paleontology vary in viewpoint according to the influence of their own preparation in the subject, interest in particular groups, judgment of future needs of their students, and other considerations. They come from institutions where courses in paleontology differ in emphasis, owing largely to geographic location. Schools in the Great Lakes region or New England tend to stress Paleozoic organisms at the expense of Mesozoic and Cenozoic groups, whereas those of the Gulf Coast and Pacific Border reverse such weight. Most teachers agree that specialization of this sort is undesirable, at least in the first course, but concentration on particular assemblages of fossils may be encouraged in advanced study.

Our effort to meet the problem of balance between various chapters led first to a survey of all available textbooks on invertebrate fossils, old and new, in English and other languages. This survey showed wide variation in the allotment of space to different groups. The pattern of the course offered at the University of Kansas was not a satisfactory guide, although it approximated what we thought was desirable. A tentative schedule was drawn up showing the estimated book pages, including text and illustrations, judged suitable for each chapter. Strict adherence to this schedule proved impractical. We cannot claim to have solved

ideally the distribution of space among respective chapters, but we feel confident that no group is greatly overweighted and that with the possible exception of conodonts (which in our view are not invertebrates), none is seriously underweighted.

Patently, this book is not suited as a text for a short survey course in paleontology. It is restricted to invertebrates. We judge, however, that by the selection of study assignments the book is adapted to serve in a flexible manner the requirements of most courses on invertebrate fossils. Such selection may be made in various ways, such as the omission of subdivisions of chapters or designated items of morphology and stratigraphical occurrence. The text itself has been written simply, in plain English, with an effort to avoid needless technical terminology; also, care has been exercised to avoid the introduction of unfamiliar terms without an accompanying explanation. As a result, even beginning students who have had no previous course work in zoological subjects should be able to understand readily all parts of the text. The references given following each chapter are selective rather than comprehensive; they are included for the convenience of the teacher and as a guide for interested students. They call attention to the most important supplementary and source materials.

Counsel and criticism have been given on various chapters, and for this assistance we are very grateful. Dr. W. Storrs Cole, of Cornell University, helped in reviewing larger post-Paleozoic Foraminifera; Dr. John W. Wells, also of Cornell University, and Dr. Erwin C. Stumm, of the University of Michigan, gave valuable suggestions based on an examination of the text and illustrations for the chapter on coelenterates; Dr. G. A. Cooper, Curator of Invertebrate Paleontology and Paleobotany at the U.S. National Museum, similarly read the chapter on brachiopods critically and contributed in several ways to its improvement; Dr. J. Brookes Knight, also of the U.S. National Museum, helped greatly in the chapter on gastropods; Dr. Norman D. Newell, of Columbia University and the American Museum of Natural History, reviewed the chapter on pelecypods; Dr. Hertha Sieverts-Doreck, of Stuttgart, Germany, and Dr. G. Ubaghs, of the Université de Liége, read the chapter on crinoids, making suggestions concerning morphological features and classification.

It is hardly necessary to state that none of these friends is responsible for the content of the chapters on which they gave aid, particularly in view of the fact that some of the suggestions were rejected. The authors themselves bear this responsibility. Readers are invited to send us corrections of any errors they may find, and criticisms or suggestions they may have to offer.

UTRECHT, THE NETHERLANDS
LAWRENCE, KANSAS
TALARA, PERU
    May, 1952

RAYMOND C. MOORE
CECIL G. LALICKER
ALFRED G. FISCHER

Contents

Preface v

1. THE NATURE OF FOSSILS (Moore, Fischer, Lalicker) 1

Requisites of fossilization 1
Types of preservation 1
   Unaltered soft parts—Unaltered hard parts
   —Altered hard parts—Traces of animals—
   Distortion and obliteration of fossils
Diversity of fossils 6
Value of fossils 7
   Stratigraphic indicators—Records of past
   forms of life—Significance in historical
   geology
Kinds of animals and names for them . . 9
   Concepts of species and genera—Scientific
   names of species and genera—Subdivisions
   of species—Subgenera—Categories of higher
   rank than genus—Principal divisions of
   animals

Adaptation to environment 16
   Factors in marine environments—Kinds of
   marine environments—Land environments—
   Facies fossils
Development of individual animals . . . 22
   Kinds of individuals and their associations—
   Beginning of the individual—Factors in-
   fluencing development—Nature of growth—
   Life history
Evolution of animals 28
   Evidence of evolution—Evolutionary changes
   in animal assemblages—Branching of species
   —Extinction of animals—Patterns of evolu-
   tion—Rates of evolution
Fossil-bearing rock divisions 35
References 38

2. FORAMINIFERA AND RADIOLARIA (Lalicker) 39

Foraminifera 39
   Living animal—Life history—Test—Ecol-
   ogy—Classification—Geological distribution
Larger Foraminifera 60
   Fusulinids—Camerinids—Orbitoidids—Dis-
   cocyclinids—Miogypsinids—Ecology
Radiolaria 72
   Skeletal structures—Physiological features

   —Classification—Geological distribution
   and importance
Flagellates 75
   Silicoflagellates — Coccoliths — Dinoflagel-
   lates
Tintinnids 76
References 76

3. SPONGES AND SPONGELIKE FOSSILS (Moore) 79

General characters 79
Reproduction 80
Fossil record 81
Structural features 81
   Soft parts—Skeleton
Classification 87

Calcareous sponges 89
Siliceous sponges 91
Spongelike organisms 93
   Pleosponges—Receptaculitids—Nidulites
References 97

4. COELENTERATES (Moore) 99

Characters of representative modern coelenterates 99
   Hydra, a simple hydrozoan—Obelia, a co-
   lonial hydrozoan—Metridium, an anthozoan
Classification 103

Hydrozoans 105
   Hydroids—Milleporids and stylasterids—
   Sphaeractiniids—Stromatoporids and labe-
   chiids
Scyphozoans 109

Alcyonarians 110
Tabulate corals 112
   Schizocorals—Thallocorals
Zoantharians 115
Rugose corals 118
   Structures illustrated by Streptelasma and
   Lambeophyllum—Structures illustrated by
   Heritschia—Structures illustrated by Synap-
   tophyllum—Structures illustrated by Hexa-

gonaria and Pachyphyllum—Main groups
and their geological occurrence—Evolutionary trends
Heterocorals. 142
Scleractinian corals. 143
   Morphological features—Main groups and
   evolutionary trends of scleractinians—Origin
   and geological history
References 153

5. BRYOZOANS (MOORE). 156

Anatomical features 156
Collection and study of bryozoans . . . 158
   Preparation of thin sections—Interpretation
   of thin sections
Classification 159

Skeletal features of main bryozoan groups 161
   Trepostome bryozoans—Cryptostome bryo-
   zoans—Ctenostome bryozoans—Cyclostome
   bryozoans—Cheilostome bryozoans
Geological distribution and importance. 193
References 195

6. BRACHIOPODS (MOORE). 197

Anatomical features 199
   Soft parts—Hard parts
Reproduction and larval growth. . . . 204
Inarticulate shells 205
   Morphological features
Articulate shells 207
   Morphological features
Shell growth 213
   Types of enlargement—Evolution of inter-
   areas
Shell form 216
Homeomorphy 217

Classification 220
   Definition of main divisions—Significance of
   shell structure—Summary of classification
Inarticulate brachiopods 221
   Atremates—Neotremates
Articulate brachiopods 226
   Palaeotremates—Orthids—Terebratulids—
   Pentamerids—Triplesiids—Rhynchonellids—
   Strophomenids—Spiriferids
Geological distribution 261
References 264

7. MOLLUSKS (MOORE) 268

Classification 269
Anatomical features 270
   Soft parts—Hard parts

Minor classes 271
   Chitons—Scaphopods
References 275

8. GASTROPODS (MOORE) 276

Soft parts 277
   Head—Viscera—Foot and mantle
Hard parts 279
   Noncoiled shells—Coiled shells
Classification 287
Evolutionary trends 290
Geological distribution 296
Amphigastropods 297
Prosobranchs 301

Archaeogastropods—Mesogastropods—Neo-
gastropods
Opisthobranchs 328
   Pleurocoels—Pteropods
Pulmonates. 329
   Basommatophorans—Stylommatophorans
Indirect paleontological evidence of
   gastropods 332
References 333

9. CEPHALOPODS (FISCHER) 335

Living cephalopods. 336
   Loligo, a squid—Nautilus—Octopus
Geologic importance 341
Classification 341
Functions and development of shell . . . 343

Nautiloid cephalopods 345
   Kinds of nautiloids—Volborthella—Elles-
   meroceroids—Michelinoceroids—Ascoceroids
   —Oncoceroids—Endoceroids—Actinoceroids
   —Discosoroids—Nautilids—Geological his-
   tory of nautiloids

CONTENTS

Ammonoid cephalopods 361
  *Goniatites—Variations of ammonoid shells
  —Geological history of ammonoids—Evolu-
  tionary trends—Life habits*
Belemnoids 388
  Megateuthis—Other belemnoids—Origin of

*belemnoids—Function of the belemnoid shell
—Geological history and evolutionary trends*
Sepioids 393
Teuthoids 394
Octopods 395
References 395

10. PELECYPODS (MOORE) 398

Anatomical features 400
  Soft parts—Shell
Modes of life 406
Reproduction and ontogeny 407
Classification 409
Palaeoconchs 412
Taxodonts 414
  Nuculacea—Arcacea
Schizodonts 418
  Trigoniacea—Cardiniacea
Isodonts 422
  Spondylacea—Anomiacea

Dysodonts 423
  *Pectinacea—Mytilacea—Pinnacea—Ostre-
  acea—Limacea—Dreissensiacea*
Heterodonts 434
  Cypricardiacea—Lucinacea—Cyrenacea
Pachyodonts 440
  Chamacea—Rudistacea
Desmodonts 442
  *Anatinacea—Myacea—Adesmacea—Mac-
  tracea—Solenacea—Poromyacea*
Evolutionary trends 447
Geological distribution and importance . 450
References 450

11. ANNELIDS AND OTHER WORMS (LALICKER, MOORE) 452

Morphological features of annelids . . . 452
  *External features—Digestive, circulatory,
  and respiratory systems—Excretory and
  nervous systems—Reproductive organs*
Geologic work of worms 453
Classification 454

Conulariids 458
  *Morphological features—Kinds of conu-
  lariids—Ecologic association—Zoological
  affinities*
References 462

12. ARTHROPODS (MOORE) 463

Skeletal features 464
  *Segmentation of body—Organization of body
  segments—Appendages*
Growth stages 467
Physiological structures 468
  *Digestive, circulatory, and respiratory sys-
  tems—Nervous system and sense organs*

Mode of life 469
Classification 470
Geological distribution 471
Pararthropods 472
References 474

13. TRILOBITES (MOORE) 475

Structure of main skeletal divisions . . . 478
  Cephalon—Thorax—Pygidium
Internal anatomy 484
Growth stages 484
  *Larval development—Attainment of adult
  characters*
Classification 486
Character of main trilobite groups . . . 488

*Protoparians—Proparians—Opisthoparians
—Hypoparians—Eodiscids—Agnostids*
Evolutionary trends of trilobites 497
  *Primitive characters—Facial sutures—Gla-
  bella—Eyes—Thorax—Pygidium—Spi-
  nosity—General conclusions*
Geological distribution 515
Minor trilobitomorphs 519
References 519

14. OSTRACODES AND OTHER CRUSTACEANS (LALICKER, MOORE) 521

Characters of crustaceans 521
  *Morphological features—Growth stages—
  Physiological features—Crustaceans as food*
Ostracodes 526

*Hard parts—Orientation of shell—Growth
stages—Appendages—Physiological features
—Reproduction—Mode of life—Classification—Geological distribution and importance*

Other crustaceans 541
   Branchiopods—Cirripeds—Malacostracans

References 550

15. CHELICERATES, MYRIAPODS, AND INSECTS (Moore) 552

Chelicerates. 552
   Distinguishing features—Merostomes (Xiphosurans, Eurypterids)—Arachnids—Pycnogonids
Myriapods 566

   Diplopods—Chilopods
Insects 567
   Primitive wingless insects—Winged insects
References 572

16. ECHINODERMS (Moore) 574

Kinds of echinoderms. 577
Modes of life 578
Larval development 578

Relationships 580
Geological distribution 581
References 581

17. PRIMIIVE ATTACHED ECHINODERMS (Moore) 582

Eocrinoids 583
Paracrinoids. 583
Carpoids 585
Edrioasteroids. 586
Cystoids 587
   Cystoids bearing pore rhombs—Cystoids bearing paired pores—Evolution of cystoids

Blastoids 594
   Structural features of Pentremites—Geological record and evolutionary trends of blastoids
References 601

18. CRINOIDS (Moore) 604

Modern crinoids. 604
   Soft parts—Hard parts—Orientation
Classification 613
Structures of fossil crinoids 614
   Calyx—Dorsal cup—Ray system—Column
Inadunate crinoids. 621
   Disparid inadunates—Hybocrinid inadunates—Cladid inadunates

Flexible crinoids. 626
Camerate crinoids 628
   General evolutionary trends—Diplobathrid camerates—Monobathrid camerates
Articulate crinoids. 635
   Stem-bearing forms—Free-swimming forms
References 650

19. HOLOTHUROIDS (Fischer) 653

Morphology. 653
   Soft parts—Skeleton
Life habits 656
Classification 657

Geological distribution 657
Relationships 658
References 658

20. STARFISHES (Fischer). 659

Classification of starfishes 659
Somasteroids 659
Asteroids 661
   Structure of a sea star—Mode of life—Geological distribution and evolution

Ophiuroids 665
   Structure—Mode of life—Geological distribution and evolution
References 674

21. ECHINOIDS (Fischer) 675

Regular echinoids 675
   Morphology—Orientation—Classification of regular echinoids—Structural variations of regular echinoids—Lepidocentroids—Melonechinoids—Cidaroids—Stirodonts—Aulodonts—Camarodonts

Irregular echinoids. 684
   Main divisions of irregular echinoids—Holectypoids—Cassiduloids—Spatangoids—Clypeastroids
Ecology and mode of life 703

CONTENTS

Geological history and importance. . . 709
   Relationships and origin—Development of
   orders—Geological importance
Primitive eleutherozoans of uncertain

affinities 712
   Bothriocidaroids—Ophiocystids
References 713

22. GRAPTOLITES AND PTEROBRANCHS (Fischer) 715

Relation of graptolites and pterobranchs 716
Rhabdopleura, a representative ptero-
branch. 716
Classification of protochordates 717

Graptolites 718
   Preservation and preparation—Dendroid
   graptolites—Tuboids, camaroids, and stolo-
   noids—Graptoloids
References 732

23. CONODONTS (Lalicker) 733

Sedimentary environment. 733
Classification 733

Orientation. 738
References 738

Index. 739

CHAPTER 1

The Nature of Fossils

Fossils are the remains or traces of animals or plants which have been preserved by natural causes in the earth's crust. The term fossil originally referred to anything dug from the earth. Since organisms living at the present time are not considered fossils when they become buried, many paleontologists arbitrarily exclude from classification as fossils all organisms which have been buried since the beginning of historic times. Most men of ancient and medieval times believed that fossils found in rocks were freaks of nature, or grew from the earth. It was demonstrated later that they are remains or evidences of former life, and that marine fossils found on land formerly lived in the sea and were raised to their present positions by upward movements of the land or relative lowering of sea level.

Fossils are unequally distributed in sedimentary rocks. They are abundant in some formations, but rare or absent in others. Fossils are common at certain localities and lacking elsewhere in the same formation. The places where they are common generally denote conditions of burial favorable for preservation of organic remains, such as protected bays along a former shore line. Some relatively thin fossil zones are persistent throughout large areas, making it possible to correlate formations in widely separated localities.

REQUISITES OF FOSSILIZATION

In the process of fossilization, preservation of the organic remains is a necessity, and this depends chiefly, though not entirely, upon two requisites: (1) quick burial in a protective medium and (2) some kind of hard parts, such as a shell or skeleton. Any condition unfavorable to life of bacteria hinders decay. Decomposition may be retarded or prevented by burial in soft muds or volcanic ash, by low temperatures or very dry air, by sea water, or by a covering of tar or resin.

Since preservation by a protective covering is of first importance in the process of fossilization, animals and plants which live in water have a much better chance of becoming fossils than terrestrial organisms. It is not surprising, therefore, that marine animals are much more common as fossils than those which lived on land.

TYPES OF PRESERVATION

Unaltered Soft Parts

If bacteria are excluded entirely from organisms, their soft parts, as well as skeletal structures, may be preserved. The best-known examples of complete preservation are remains of mammoths and rhinoceroses in the frozen tundra of Siberia. Natural mummies have been formed by the dry air of deserts, preserving some of the soft tissues and the hard parts.

Unaltered Hard Parts

Most invertebrates possess hard parts composed of calcium carbonate, calcium phosphate, silica, complex organic compounds, or combinations of these. Calcium carbonate occurs in the form of calcite or aragonite, and some shells contain both these minerals. Silica takes the noncrystalline, hydrated form (opal). These organically precipitated mineral substances generally are not chemically pure but contain admixtures of such elements as magnesium, strontium, manganese, iron, and sulfur.

Many shells and skeletons have been preserved in the rocks with no recognizable change except for removal of the less-stable organic matter. Examples are abundant shells of Cenozoic mollusks or Paleozoic brachiopods, which retain their original microstructure. Hard parts of insects have been preserved perfectly in amber (fossil resin), in Oligocene rocks of the Baltic region, and in Cretaceous rocks of Manitoba.

Calcitic Shells. Calcium carbonate in hexagonal-rhombohedral form (the mineral calcite) is one of the most widely used skeletal substances. It is found in certain flagellate protozoans, most Foraminifera, some sponges, stromatoporoids, many of the extinct corals, most bryozoans, most brachiopods, some mollusks and crustaceans, and all echinoderms. Unaltered or virtually unaltered skeletal remains composed of calcite are common in Paleozoic, Mesozoic, and Cenozoic rocks.

Aragonitic Shells. Calcium carbonate in orthorhombic form (the mineral aragonite) is secreted by most scleractinian corals and mollusks. Aragonitic shells commonly are preserved unaltered in Cenozoic clays, are less commonly found intact in Mesozoic rocks, and are very rarely found in rocks of Paleozoic age, for aragonite is comparatively unstable and tends to be removed in solution or recrystallized into calcite.

Phosphatic Shells. The shells of conulariids and all but a few groups of inarticulate brachiopods are composed mainly of tricalcium phosphate. This substance is also found in the armor of many arthropods, and composes the microscopic fossils

called conodonts. It is chemically resistant; unaltered phosphatic shells, or at least showing no perceptible signs of alteration, are found in rocks ranging from Cambrian to Recent.

Siliceous Shells. Amorphous, hydrous silica (the mineral opal, $SiO_2.nH_2O$) serves as skeletal substance for some flagellates, most radiolarians, and many sponges. Opal is comparatively unstable, and unaltered opaline skeletons are therefore largely restricted to rocks of Cenozoic age.

Resistant Organic Hard Parts. Certain organic compounds (complex molecules of carbon, hydrogen, oxygen, and other elements) are resistant to bacterial action, and are not readily altered. Chitin, for example, a substance found in arthropods and graptolites, may be hard physically and resistant chemically; it has been preserved intact where tightly locked up in impervious rock matrix.

Altered Hard Parts

Many fossils show varying sorts and degrees of alteration of their original structure. Such changes may affect physical structure, chemical composition, or both. They may consist only of slight rearrangement of molecules, or involve hardly noticeable removals, additions, or substitutions. If carried far enough, not a trace of the original chemical and physical structure will remain.

Carbonized. Solution and other chemical action under water commonly transforms the composition of the tissues of plants and animals to a thin film of carbon. The organic remains are then carbonized, and the process of making such fossils is termed **carbonization** (Fig. 1-1, 7). It is accomplished by a decrease in the volatile

---

Fig. 1-1. **Types of preservation of fossils.**

1, 2, Neogene pelecypod exhibiting an internal mold within an unaltered shell and boring made by a gastropod.

3, External mold of a Pennsylvanian brachiopod.

4, Cast from the external mold of 3.

5, Internal mold of a trilobite of Ordovician age.

6, External molds of a gastropod of Paleogene age.

7, Graptolite of Ordovician age, preserved by carbonization.

8, Fossil tracks on a slab of sandstone.

9, Silicified brachiopod of Permian age.

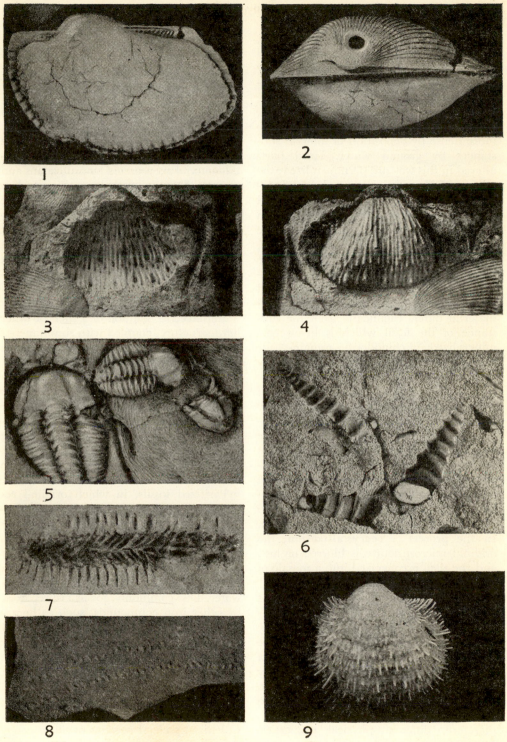

Fig. 1-1.

3

constituents, comprising the hydrogen, oxygen, and nitrogen content of original tissues. Fossils commonly preserved by carbonization include graptolites, arthropods, fishes, and plants.

Permineralized. Shells and bones, which are somewhat porous, may be made more dense by deposition of mineral substances by ground water. Hard parts altered in this way are permineralized; the process of alteration is termed **permineralization** or **petrifaction**. The added solids may have a chemical composition exactly similar to the hard parts of the fossil, as is very common in echinoderms, the skeletal parts of which are made more dense by addition of calcite in crystallographic continuity with the original hard parts. Commonly, however, the infiltrated material differs chemically from the substance of the fossil which is being permineralized. Echinoderm remains, for example, are commonly permineralized by the iron–potassium silicate, glauconite.

Recrystallized. The internal physical structure of some shells is changed by shuffling of molecules as the result of solution and reprecipitation. Commonly the molecules become arranged in crystalline aggregates. Hard parts which have undergone this sort of alteration in appreciable degree are said to be recrystallized, even though the original substance may be too fine-textured to show crystalline structure. In the process of **recrystallization,** the original microstructure is blurred or lost, and the shell is converted into a mosaic of interlocking crystals.

Recrystallization commonly maintains the original mineral composition. A brachiopod or foraminifer shell, originally composed of fibrous calcite, may simply be transformed into a nonfibrous shell of interlocking calcite grains. Also, it may involve change from one mineral to another of similar chemical composition but different molecular structure and, therefore, different physical properties. Most Mesozoic and Cenozoic shells origi-

nally composed of aragonite are now composed of calcite, owing to regrouping of the constituent molecules into more stable form as calcite. They are recrystallized. Outside of possible small volume changes, the external appearance of such fossils is retained.

Dehydrated and Crystallized. Large quantities of noncrystalline opaline skeletal material are secreted by protozoans and sponges. Opal is unstable, tending to lose its water; it then crystallizes into the minerals chalcedony or quartz. Not surprising, therefore, is the fact that most siliceous fossils now consist of chalcedony or quartz. Crystallization of such complex microscopic bodies as radiolarian tests generally has destroyed their original structure, partly or entirely.

Replaced. Solution of a shell or other hard structure, coupled with simultaneous deposition of some other mineral substance in the voids formed, leads to replacement. Change of this sort may also comprise substitution of one chemical ion for another in a mineral, as magnesium or iron for calcium. Whether the original microstructure is preserved in this process or not, the hard parts are described as replaced. Many kinds of minerals may replace others.

Pyritized fossils, in which original remains, generally calcareous, have been replaced by pyrite, are common in many formations. The iron sulfide is thought to be produced by interaction of iron present in the sediment, and sulfur formed from decaying organic matter and from sea water. Less commonly, fossils are replaced by other iron compounds, such as hematite, limonite, and glauconite. Hematitized fossils are abundant in some sedimentary iron ores of Ordovician and Jurassic age in central Europe and Silurian age in eastern North America.

Various carbonate minerals may replace noncarbonate materials, or each other. Calcitized skeletal remains of silica-secreting organisms are common in some rocks, for example, siliceous sponge spicules replaced by calcite (**calcitization**). Dolomite or other carbonates may replace originally calcitic or aragonitic shells (dolomitization). Changes of this sort, and much less common ones in which carbonates such as siderite, rhodochrosite, and others are the replacing mineral substance, may be grouped under the general term carbonatization.

Silica (chalcedony or quartz) has commonly replaced calcareous skeletons, in a process termed **silicification**. Many limestones contain vast quantities of beautifully silicified shells, which may be extracted by means of acid (Fig. 1-1, *9*).

Traces of Animals

Molds and Casts. Following burial of the hard parts of an organism, sedimentary materials are packed closely around them, and if there are cavities, these commonly are filled also by sediment. The impression of skeletal remains in adjoining rock constitutes a **mold.** It may be termed an **external mold** (Fig. 1-1, *3, 6*), if the shape of the outer sides of the hard parts is shown, or **internal mold** (Fig. 1-1, *1, 2, 5*), if the impression reveals the form and markings of inner surfaces. An internal mold is frequently called a **steinkern.**

Many shells and skeletons buried in sediment are removed by solution. This is particularly true of organic remains in permeable rocks, such as sandstones and porous limestones, which permit free circulation of water. Opaline material and aragonite are particularly liable to this type of destruction. Where skeletal parts are thus removed from the matrix, natural molds can be observed. Filling of the cavity by mineral matter or other substance forms a **cast** (Fig. 1-1, *4*).

Molds can be obtained artificially by removing shells from the rock mechanically, as by fracture along the contact of shell and matrix, or chemically, as by use of acid which dissolves the shell without attacking the matrix. Either natural or artificial molds provide material for making artificial casts, and these often are very useful in paleontological study.

Tracks, Trails, and Borings. Tracks and trails made by the feet, tails, and other portions of the bodies of animals are frequently preserved in muds (Fig. 1-1, *8*). Tracks made by dinosaurs are abundant in certain localities. Less commonly, impressions made by the soft bodies of jellyfish and other delicate organisms in fine sediment may be preserved. Burrows or tubes made by worms and other animals are common also. Gastropods bore round holes in clam and other shells in order to secure the soft parts as food (Fig. 1-1, *2*).

Coprolites. Fossil excrements of animals, termed coprolites, may be important fossils because many contain undigested remains of food. They also show the approximate shape of the anus of some animals.

Distortion and Obliteration of Fossils

The original shape of organic remains is not always preserved in fossils; it may be obliterated partially or completely. Fossils may be flattened by compaction of sediments and distorted or even completely destroyed by transformation of sediments into metamorphic and igneous rocks.

Innumerable shells are obliterated in the recrystallization of limestones. For example, in the coarse crystalline Burlington limestone, of Mississippian age, in Missouri, the only fossils commonly found at many outcrops are relatively large thick-shelled brachiopods, firmly constructed corals, and crinoids. Chert nodules in the same limestone, however, may yield a host of small delicate fossils, such as bryozoans, small thin-shelled brachiopods, and others. At the edge of the chert nodules, portions of shells which would normally project outside the nodules are absent. This signifies that the chert must have been emplaced before recrystallization of the surrounding limestone obliterated small delicate fossils in these parts of the formation. The unfossiliferous character of most dolomites also generally attests obliteration of fossils.

DIVERSITY OF FOSSILS

Remains of invertebrates and other animals preserved in rocks of the earth crust—not to mention plants—are extraordinarily numerous and varied. They are the essential working materials of paleontological study, and only a cursory inspection of fossil collections in most large universities and museums is sufficient to impress anyone with their richness. The task of cleaning them properly for study, of sorting and labeling them carefully as to stratigraphic and geographic source, of comparing them with one another and with most nearly similar modern forms of life in order to classify them, of investigating the significance of their characters in relation to environmental adaptations and evolutionary

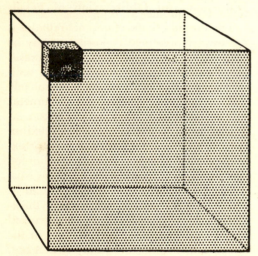

Fig. 1-2. **Diversity of living and extinct species of animals.** The small, dark-shaded cube represents the approximate number of living species of animals, estimated to be at least 1 million. The large cube depicts the probable order of magnitude of the total number of animal species on earth, past and present. These belong to successive populations distributed through more than 500 million years of geologic time. Only a small fraction of these enormously diverse kinds of animals are known from fossil remains.

trends, and of describing and illustrating them for publication in scientific literature—all this is enormous. Work by many generations of trained specialists in studies on fossils, together numbering thousands of individuals, has led to present knowledge, the printed record of which assembled in one place would fill a large library. Fossils are highly diverse, and man's acquaintance with them is considerable.

With the foregoing brief statement stressing amplitude of paleontological materials in mind, it is well to obtain a concept, if possible, of the actual magnitude of the fossil record, or more important, the magnitude of the fossil record which man can find and study.

All kinds of living animals are estimated to include approximately 1 million species. Whether the true number is somewhat larger or smaller has no appreciable bearing on an estimate of how many species, now extinct, have lived on the earth in past geologic time. Unquestionably, the beginning of animal life must far antedate earliest Cambrian time, even though the oldest invertebrate fossils in any considerable numbers are Cambrian. Thus, extinct animals which successively appeared and disappeared during more than 500 million years—possibly more than 1,000 million years—of earth history surely have an aggregate number greater than 100 times the modern 1 million species (Fig. 1-2).

Obviously, even a single specimen representing each extinct animal species never can be available for examination by man. The chief limiting factors in learning about past forms of life are as follows: (1) Vast numbers of species have no hard parts adapted for preservation as fossils. (2) Species which possess hard parts may not be fossilized, because nowhere were their remains suitably buried. (3) Fossils which existed at one time have been destroyed by erosion, by metamorphism, or by recrystallization. (4) Fossils exist today, but they occur only in deposits beyond reach of man (as beneath the sea or at very great depth on land). (5) Fossils occur in formations at or near the surface, but they are undiscovered because outcrops are lacking or very poor, or because the fossils are rare. (6) Fossils occur in well-exposed formations but in geologically unexplored parts of the world. (7) Fossils which have been found are too fragmentary or too poorly preserved to show definitive characters of the once-living animals that they represent. The combined effect of these, and several lesser hindrances, is to provide an extremely small sample of the past forms of animal life. Indeed, the sample is almost infinitesimal. Nevertheless, each effort to find and collect fossils has possible reward in uncovering a brand new, conceivably very exciting page in the paleontological record of life on earth.

VALUE OF FOSSILS

Remains of organisms preserved in rocks of the earth's crust have three chief kinds of value. These are (1) as stratigraphic indicators, for correlation of deposits containing them and for determination of relative geologic age; (2) as records of past forms of life, showing the course of evolutionary modifications of animals and plants during geologic time; and (3) as evidence of changing environments and geographic patterns during geological history.

Stratigraphic Indicators

Fossils are an indispensable means for classifying and correlating sedimentary formations distributed throughout the world, except very ancient (Pre-Cambrian) rocks in which fossils are virtually absent. Such classification and correlation have the utmost importance in economic geology, as in exploring for petroleum, coal, and many other mineral resources.

Arrangement of fossils in order of their geologic age requires, first, that the relative position of the fossil-bearing rock layers be observed. Where strata are not disturbed by folding, the oldest are lowermost and the youngest are topmost. Even where sedimentary layers stand on end or are overturned, physical criteria of various sorts generally permit firm conclusion as to original bottoms and tops of beds. The order of succession of fossil assemblages in each exposed rock section thus can be established. Wherever fossil assemblages in two or more sections can be matched, indicating approximate age equivalence, tie points are given for making a composite section which reaches stratigraphically from oldest to youngest rocks included in any of the joined sections (Fig. 1-3). When enough sections have been pieced together in this manner, the whole fossil-bearing part of the geologic column is covered and succession in age of the fossils is determined.

A problem encountered in correlating various rock successions is how to deal with differences in fossil assemblages which reflect dissimilarities of environment and existence of geographic barriers that prevent mingling of contemporaneous animals. Paleontological differences of this sort do not signify that one group of fossils is older than another. The problem can be solved by finding places where the unlike fossil-bearing deposits come together and by searching out every possible clue to the relative age of rock sequences in geographically isolated provinces. For example, if rocks in Montana and central Europe containing dissimilar fossils are underlain by Lower Cambrian and overlain by Upper Cambrian in both regions, they are reasonably correlated as Middle Cambrian and approximately equivalent in age. Most difficult and at present least known is correct correlation of many individual fossil assemblages located in widely distant regions, especially where these regions seem to have been separated during much of geologic time by geographic barriers. Thus, placement of some fossiliferous deposits of North America in terms of subdivisions of the European section is not determined.

Each geologic system is characterized broadly by distinguishing features of its fossils. Lesser stratigraphic divisions, such as series and stage, are similarly differentiated, although some of them vary regionally. Many rock successions can be divided into parts characterized by restricted occurrence in them of some species or genus, and these parts, called **zones,** then bear the name of the chosen fossil. Some rock divisions classed as zones do not contain all beds in which the selected zone fossil occurs but only those in which it is relatively abundant or in which it is associated with certain other fossils. Such divisions are really part zones, since they embrace only an arbitrarily defined fraction of strata included in the entire range of the chosen zone fossil. Rocks of this entire range are classed as a **biozone.** Because animals are influenced greatly by environment, fossils occur commonly or exclusively in a particular type of sedimentary rock, and if deposition of this rock shifts laterally in the course of geologic time, the fossils vary in age from place to place.

Records of Past Forms of Life

Few fossils completely show the form and structural features of once-living organisms, for only exceptionally are traces of soft parts preserved. Hard parts of animals are very abundant, however, in all Cenozoic, Mesozoic, and Paleozoic systems, and these furnish indisputable proof of the existence and nature of ancient animal life. If these fossils were unknown, the diversity of modern animals and many peculiarities of their distribution would be quite incomprehensible. Hypotheses to explain them could be formulated from study of living creatures, but there would be no adequate means for testing them. The fossils are actual records of successive past forms of life, distributed over the world in places where the organisms lived. To the extent we can find all significant parts of this voluminous record and correctly interpret them, we shall gain understanding of the "tree of life." This includes

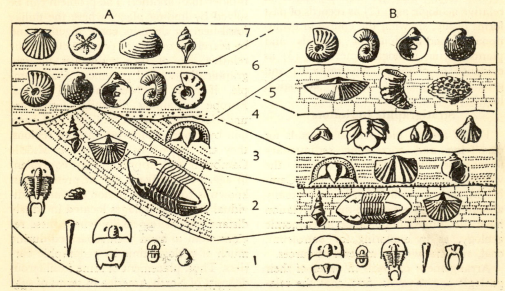

FIG. 1-3. **Fossils as stratigraphic indicators.** Diagrammatic sections *A* and *B* show successions of strata at two places about 200 miles apart and some of the invertebrate fossils in the various rock divisions. Observation of the sequence of the different assemblages furnishes basis for determining their relative geologic age, and comparison of fossils found in the two sections permits correlation of one with the other.

knowledge of what kinds of animals have existed, when they appeared and vanished, how and where they lived, and what sorts of environment they preferred. Especially, it consists of learning how the innumerable species and genera of animals have evolved, how some are descended from others, and how classificatory groups of varying rank should be defined. Taking account of all features of their nature and occurrence, fossils furnish evidence for determining these things.

Significance in Historical Geology

Study of the history of the earth, which is the domain of historical geology, so largely depends on fossils that without them we should be able to decipher only parts of the rock record and this on a regional basis at most. We could guess that some sedimentary formations were laid down in shallow seas which invaded continental areas, but without finding marine fossils we could not be sure. Study of rock structure in the Appalachian Mountain region would reveal that folding and faulting occurred there after deposition of thick coal-bearing strata named Pennsylvanian and before accumulation of red sandy strata called Newark, but without fossils, geologists have no way of fixing the time of Appalachian mountain building with respect to crustal deformation in the Rockies, Sierra Nevada, and Alps. Without fossils, they could not know whether oil-bearing strata in Louisiana correspond in age to oil-bearing rocks in western Texas, Wyoming, and the Middle East. Also, without fossils, the geologic date of igneous activity represented by diabase sheets in Newark strata of New Jersey could not be compared with that indicated by similar igneous rocks in bottom parts of the Grand Canyon or by thick basalts of the Columbia Plateau.

Fossils not only furnish evidence for determining the age of geologic events in different regions and for correlations of rocks from place to place throughout the world, but they are the foundation for correct interpretation of nearly all sorts of sedimentary environments, on land and in the sea. In paleontology, as in physical aspects of historical geology, the present is key to the past.

KINDS OF ANIMALS AND NAMES FOR THEM

No scientific training is needed for awareness of distinctions between innumerable kinds of animals which man finds all around him in any part of the world. A child begins to perceive such things even before he can talk, and among the earliest words learned are the names used in the language of his parents for the most common and important animals of the surroundings. Wherever the young human animal grows up—in an American, European, or Asiatic city, on a farm, in a tropical jungle, or in an Eskimo igloo— knowledge of living things about him expands and words for them increase in number. Three observations ·with respect to this universal acquaintance of man with the animal kingdom are noteworthy: (1) sharpness of discrimination of individual kinds of organisms varies between wide limits; (2) recognition of group relationships of animals is more or less vague; and (3) nomenclature is utterly diverse.

Concepts of Species and Genera

Scientific classification of animals is based essentially on conception of unit groups called **species,** which are assemblages of individuals having identity or near identity of form and anatomical features, except for sex differences, and measurable distinctness from other assemblages. Among living animals, one test of belonging to a given species is the capacity of individuals to interbreed with one another, but of course, the validity of species defined on the basis of fossils cannot be confirmed or challenged by evidence of this sort. Example of a species is the large cat called African lion, variously designated in different languages (as

Löwe, leeuw, léon, etc.), and another is the striped Asiatic cat called tiger (Tiger, tijger, tigris, tigre, etc.).

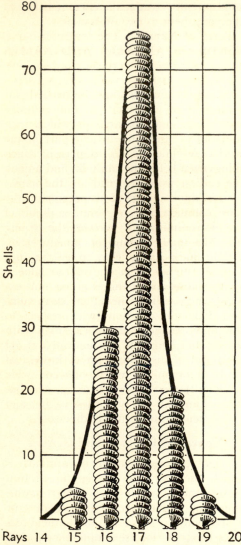

Rays 14   15   16   17   18   19   20

FIG. 1-4. **Variation among individuals comprising sample of a clam species.** The diagram shows specimens in a pailful of shells belonging to a common species of scallop (*Pecten*) collected on an Atlantic beach. These were divided into groups according to number of ribs (rays) on the shell. The diagram indicates that 56 per cent of them possess 17 rays and that 93 per cent have 16 to 18 rays. This is a normal curve of variation in populations belonging to a species. (*Modified from American Museum of Natural History.*)

In order to understand the concept of species, it is very necessary to take account of the fact that no two individuals are absolutely identical. Despite possession of common features, individuals belonging to any one species vary in size, shape, and many details of external and internal characters. If this were not so, we could not distinguish one dog from another or at a glance recognize the difference between red-haired, blue-eyed Sally Smith and brunette, black-eyed Jessie Jones. Thus, each species comprises a "population" ranging generally in magnitude from many thousands to unnumbered billions, in which a preponderant majority possess attributes closely grouped around a norm and a minority diverge appreciably from this norm (Fig. 1-4). Accordingly, we may expect that wherever two or more somewhat nearly related species are found associated in space or time, divergent minor fractions of each may so simulate one another as to be indistinguishable. Individuals belonging to these fractions often cannot be classified firmly.

Differentiation of species and definition of their characters have been undertaken by zoologists and paleontologists in two very dissimilar ways. A widely used old method is selection of a single specimen or small group of specimens, followed by investigation of this sample and classification of other individuals on the basis of comparison with the chosen standards or types. Such part-for-the-whole technique can serve satisfactorily at least for qualitative discrimination of species if the type or types come from the norm of the whole assemblage that really belongs together; it is obviously unsatisfactory if the type or types represent abnormal divergent minor fractions of the whole assemblage. The method gives no basis for measuring the nature and degree of convergence in characters properly assignable to the species or the nature and extent of divergence shown by variants properly classifiable in the species. A second method, which is the only accepted basis of modern zoological taxonomy, is study of populations, which means statistical analysis of a sample composed of many individuals. The larger the sample and the more completely any misleading factors in sampling are eliminated, the truer is the determination of significant features of the species, as well as the nature and range of variation from the norm (Fig. 1-4). Classification of individuals not included in the sample is made by comparison of their characters with data furnished by analysis of the whole assemblage represented in the sample.

After the existence of numerous species is recognized and their several collective distinguishing characters have been ascertained, the next step in classification of animals is association together of species having significant common attributes which are judged to denote genetic relationships. Such groups of interrelated species are called **genera** (sing. **genus**). Thus, the lion, tiger, puma, leopard, and several other species are grouped together as cats. The generic assemblage of cats clearly differs from another such assemblage consisting of dogs, among which are separate species (domestic dog, wolf, jackal, etc.). As noted subsequently, related genera are assembled in families, and these in higher-rank categories of increasing comprehensiveness, among which the most commonly used are termed order, class, and phylum.

Scientific Names of Species and Genera

Requirements. Scientific nomenclature of animals requires (1) that each species and genus found in the world shall have a name that is **independent of change,** such as pertains to common names used in many languages; (2) that each species and genus shall have **separate** names, duplicated by none which refer to some other species or genus; and (3) that **different** names shall not be applicable to any one species or genus. For these purposes International Rules of Zoological Nomenclature have been formulated and a Commission of zoologists drawn from various countries has been set up for the purpose of interpreting the Rules as may be needed and of suspending them in case confusion in nomenclature is judged likely to result from applying them to certain names.

Form of Names. The scientific name of a species invariably consists of two parts, each of which is a word having classic form, if not actually derived from the classic languages, Greek or Latin. Such nomenclature is binominal, and it dates from a monumental work on the animal kingdom published by the Swedish naturalist Linné in 1758. The first word is the name of the genus to which the species belongs, written with an initial capital letter; the second word is the so-called trivial name, according to present practice invariably written with an initial small letter. The combination of generic and trivial names constitutes a specific name. A distinction in the use of these component parts of a specific name is that the name of a genus may stand alone but the trivial name belonging to a species can never be used in an independent manner; always a generic name or abbreviation for it (provided ambiguity is avoided) must precede the trivial name. Thus, the trivial name referring to the species lion is *leo,* and the genus of cats to which this animal belongs is *Felis;* the name of the species is *Felis leo.* In referring to a number of species of cats, one may write *Felis leo, F. concolor* (puma), *F. pardus* (leopard), *F. catus* (wildcat), etc. Generic names have gender masculine, feminine, or neuter; and trivial names (unless nominative nouns) must agree with the generic name in gender. According to convention, the scientific names of animals are printed in type (generally italics) different from that of accompanying text, so that they may be discerned readily.

The names of genera and species must be words written in Roman letters, not Greek, Russian, Chinese, or some other non-Latin characters. Zoological nomenclature is designed to be expressed in a universally understood alphabet.

Homonyms. An important requisite

of the scientific names of animals, whether of genera or of species, is that none shall duplicate another. Obviously, confusion and error are unavoidable if identical names may refer to different kinds of animals. Therefore, each recognized genus must have a scientific name of its own, distinct from all others in the animal kingdom, and the same rule applies to each species. Homonyms (identical names applied to different things) are not allowed, and if, as has happened frequently, such names for genera or species appear in print, the later-published name is invalid. It must be replaced by a name that does not duplicate any already in use. Thus, the name *Craterophyllum*, proposed by one author in 1909, is the valid name of a fossil coral, whereas the same name published by other authors in 1911 and in 1931 for other fossils are junior homonyms, not allowed to stand; the last two genera were renamed, respectively, *Barbouria* in 1940 and *Cypellophyllum* in 1933. Concerning species, it must be remembered that the name consists of two parts, and it is this combination which cannot be applied to more than one species of animal. The same trivial name, however, may be used in combination with different generic names without making homonyms. For example, the brachiopods called *Plaesiomys subquadrata* and *Christiania subquadrata* are not homonyms.

Synonyms. Another requirement in scientific nomenclature of animals is that any one genus or species shall have only a single valid designation. Two or more different names for the same thing (synonyms) are less serious hindrances than the same name for different things, but objectionable. In removing synonyms from scientific literature, the junior ones must go, which means that the first-published valid name prevails. For example, *Monilopora* (1879) is a junior synonym of *Cladochonus* (1847), and accordingly must be suppressed.

Citation of Names. Because generic and specific names are extremely numerous and because avoidance of homonyms and synonyms is essential in zoological nomenclature, publication of these names in scientific literature commonly is accompanied by citation of the author of the name; also, the year of first publication of the name may be given. Such information is very useful in taxonomic study. Rules governing citation of authors' names provide that they follow the scientific name, without intervening punctuation of any kind, as *Nucula* Lamarck, *Felis leo* Linné. If the combination of generic and trivial names constituting the name of a species is altered, the author's name is enclosed in parentheses, as *Belemnitella americana* (Morton), the originally published name of this species being *Belemnites americanus* Morton. Changes of this sort are frequent, as when a very broadly defined genus is split up into more narrowly diagnosed new genera, or when study indicates that a species really belongs to some genus other than the one in which it was classed originally. If a genus is subdivided into new genera, the name of the old genus must be retained for one of the subdivisions, and the revised definition of it constitutes a restriction.

Fixation of Names. Firmness in application of names for genera and species of animals depends very largely on anchorage to entities called **types.** This means that an example belonging to the assemblage is chosen as inseparable signification in the use of the name. It is the primary name bearer. Because a genus is composed of one or more species, the type of a genus is one of its constituent species; and because a species is made up of individual specimens, the type of a species is a particular specimen. The type of a genus cannot be a specimen, for it comprises the entire assemblage of individuals belonging to the species which is type. Once established, types are not subject to change.

Type species of genera are fixed in various ways, according to specifications of International Rules of Zoological Nomenclature. The chief modes are by (1) original designation, in which an author declares choice of the type species in connection with first publication of the generic name; (2) subsequent designation, in which the author of a generic name failed to select a type species and a later author who first publishes choice from among species originally assigned to the genus fixes the type; (3) monotypy, in which only one species was originally included in the genus and it automatically becomes the type; and (4) tautonymy, in which one of the originally included species has a trivial name identical to the generic name, as *Bison bison.*

Type specimens of species may be chosen by the original author, or under various conditions determined by later workers. In former practice, two or more specimens frequently were designated as coequal types, termed **syntypes.** Experience has shown that these may represent taxonomically distinct kinds of animals, belonging not only to different species but even to different genera. Rules now provide that a species shall have a single type, called **holotype,** which stands alone as name bearer, although other specimens serving for characterization of the species may be associated with it; these latter specimens are classed as **paratypes.** If the holotype of a species is not preserved carefully in some suitable place where it can be examined by zoologists or paleontologists, it may be difficult or impossible to recognize the species. Type specimens, therefore, are extremely important objects.

Subdivisions of Species

In many animal assemblages classed as species, two or more subordinate groupings based on minor distinctions can be recognized, and because these seem to denote significantly closer relationship between individuals belonging to each segment than between units classed in different segments, there is need to take account of them in zoological taxonomy. Terms most generally used for subdivisions of species are **subspecies** and **variety,** but concepts of their meaning have differed, particularly as between neozoologists working on living animals and paleozoologists working on fossils. Except that the student of fossils has the advantage of studying assemblages of animals in which time, as well as dispersion in space, works changes, there should be no difference in essential viewpoints about infraspecific categories of animals.

A subspecies may be defined as an entire population or race within a species, which has become morphologically differentiated by reason of isolation during some period of time from other populations or races belonging to the species. Subspecies interbreed freely when brought together, either under natural or artificial conditions, thus indicating their close zoological relationship. They maintain identity only as conditions keep them apart. It follows that true subspecies are not found living together in the same geographic and ecologic setting, and they should not be associated in a single collection of fossils. Since time may operate to modify characters of the form and structure of animals in a manner exactly analogous to that effected by geographic isolation and change of environment, subspecies may be distinguished which differ slightly in geologic age. Indeed, the paleontologist generally has no means for discriminating subspecies which originated as the result of geographic and ecologic separation from those which were produced by evolution during a part of geologic time.

A variety is a distinctive morphological group of individuals within a population, differentiated from other parts of the population by possession of some character or combination of characters lacking in the others. Segregation of individuals belonging to such a group is artificial in that it consists of picking out from the graded morphological series of a whole population those units which happen to have particular attributes. Thus, a variety is not a sort of subspecies. Most zoologists do not recognize assemblages called varieties as useful in systematic classification, but many paleontologists have described and named subdivisions of species which they designate as varieties. Some of these probably are mere variants which should not be differentiated taxonomically, whereas others seem actually to be subspecies.

Individuals properly classifiable in a species or subspecies may be nearly identical or appreciably variable. In the study of fossils from a given rock formation, the test of belonging together in a taxonomic group is degree of identity, and in varying forms, completeness of intergradations (Fig. 1-5).

FIG. 1-5. **Intergradation of individuals belonging to a species.** The population comprising any species includes individuals which vary more or less in different features from the norm of the species as defined by characters belonging to a preponderant majority of individuals. The shells here figured represent a species of the gastropod *Cassidaria*, from Pliocene deposits of northern Italy, ×0.5; they come from a single collecting locality.

A subspecies may bear its own trivial name, placed next following the trivial name of the species to which it belongs, without any intervening term or punctuation. Thus, the subspecies of African wildcat is named *Felis ocreata maniculata*. The addition of a subspecific trivial name is not construed to change the fundamental binominal pattern of zoological nomenclature of species.

Subgenera

In classification of animals, provision has been made for putting together within a genus groups of species which are judged to have special characters in common and to be marked by distinctions between the groups. These assemblages of species are called **subgenera,** and names for them, like those for genera, are written with an initial capital letter. According to convention, subgeneric names are enclosed by parentheses when cited in the name of a species: as, *Felis* (*Panthera*) *pardus*, the leopard. Such a name is treated as binominal, for incorporation of the subgeneric term is considered merely optional and supplementary.

Categories of Higher Rank Than Genus

Groups of related genera are placed together in zoological classification as **families.** They are named by choosing one of the included genera as type of the family and adding the termination -idae to the root of the generic name. The family of the cats, comprising the genera *Felis*, *Lynx*, the saber-toothed fossil cats called *Dinictis*, *Smilodon*, and numerous others, is named from *Felis* as type genus and thus called Felidae. Family names, as well as names of all other categories of higher rank than genus, are printed in ordinary type, not in italics.

Families may be divided into **subfamilies** and grouped together in **superfamilies.** The Rules provide for use of the ending -inae for all subfamily assemblages, added to the root of names belonging to genera chosen as types of the subfamilies. Whenever a family is divided in this manner, one of the subfamilies which contains the type genus of the family must be named from this genus. For example, the family of foraminiferal protozoans called Miliolidae (type genus *Miliola*), is divided into subfamilies called Miliolinae, Cornuspirinae (type genus *Cornuspira*), and a number of others. A uniform mode of making names of superfamilies is not provided by the Rules, but generally they are based on the name of one of the constituent families with change of the terminal letters to -acea or other combination.

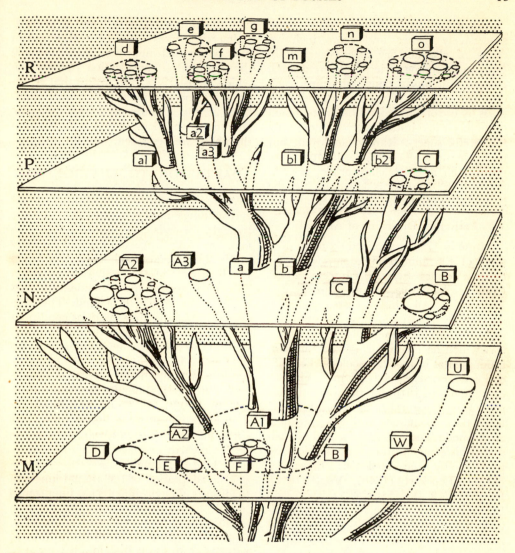

Fig. 1-6. **Small portion of the "tree of life."** The diagram represents a few main branches and numerous lesser ones, which correspond to zoological taxonomic categories called phyla, classes, orders, families, genera, and species. The planes M, N, P, R denote arbitarily chosen successive moments of geologic time, and for sake of simplicity in drawing, parts of branches extending above the different planes are omitted.

Letter symbols on plane M serve for identification of branches as follows: U, W, Porifera and Coelenterata (phyla); D, E, F, Echinoidea, Holothuroidea, Stelleroidea (classes), which together form the subphylum Eleutherozoa; B, Cystoidea (class), $A1$, Camerata (subclass), and $A2$, Inadunata, (subclass), all of which belong to the subphylum Pelmatozoa, $A1$, $A2$, and $A3$ (plane N) are main divisions of the class Crinoidea. The broken line on plane M encircles branches of the phylum Echinodermata.

Groups of branches represented on plane R indicate families containing varying numbers of genera and species (but most individual species are not shown). Families d, e, f, g belong to suborders of the camerate crinoid order Monobathrida (a, plane N), and other families, m, n, o, belong to the order Diplobathrida (b, plane N).

Above families (or superfamilies) come successively higher and more inclusive ranks of classificatory assemblages (Fig. 1-6). In ascending series, these are **orders, classes,** and **phyla** (sing. **phylum**), each of which may be divided into lesser-rank parts by use of the prefix sub-, or combined in higher-rank units under names using the prefix super-. Rarely employed, but useful in some parts of classification, are terms such as tribe, clan, grade, branch, and others. Highest is kingdom, for the kingdom Animalia includes all animals.

Principal Divisions of Animals

Linné in 1758 recognized six main divisions of the animal kingdom, which he called Vermes (worms), Insecta, Pisces (fishes), Amphibia, Aves (birds), and Mammalia. All kinds of creatures now collectively known as invertebrates were included under Vermes. Distinctions of the sort which have come to be recognized as fundamental in modern zoological classification were first grasped by Lamarck in 1809 and Cuvier in 1816, for the work of these great French naturalists established characters of the phyla named Mollusca, Annelida, Arthropoda, and Vertebrata, and of their main divisions. Subsequent studies, especially by the zoologists Ehrenberg (1838), Leuckart (1848), Vogt (1851), Haeckel (1874), Hatschek (1888), and Lankester (1900), have shown that definition of the principal divisions of animals needs to be based on consideration of groups of characters, rather than on any one sort of evidence consisting of anatomical or embryological facts. Among the chief characters having significance in classification are (1) grade of body construction, cellular or with organized tissues and organs; (2) type of body symmetry, primarily radial or bilateral; (3) presence or absence of a body cavity (coelom) distinct from the digestive tract, and mode of origin of such cavity; (4) presence or absence of an anus; (5) segmentation of the body or lack of it; and (6) nature of circulatory, respiratory, excretory, and nervous systems, if present. Using these bases, the animal kingdom may be divided into main parts shown in the outline of classification opposite, in which an asterisk (*) indicates that representatives occur among fossils, and numbers in the column at right designate chapters of this book in which members of the phyla are described.

ADAPTATION TO ENVIRONMENT

Animals are greatly influenced by the surroundings in which they live. In the struggle for existence, many species become specially fitted for particular modes of life. They obtain food, adapt themselves to all aspects of their environment, find ways of protection from enemies, and successfully reproduce their kind. They become established under various conditions of depth, temperature, and bottom conditions in the seas; become differentiated for floating freely or swimming; invade hypersaline, brackish, and fresh waters; achieve ability to move about on land; and even take to the air. Many burrow in sediment at the bottom of water bodies or in the earth on land. Virtually no place on or near the earth's surface is uninhabited by some kind of animals.

This adaptation of the animal kingdom to all sorts of environments has importance in paleontological study, for what is true of the present is equally true of the past. Knowledge of the mode of life of modern animals of all kinds therefore is helpful in understanding adaptations of ancient invertebrates and vertebrates, because most ecologic adaptations of animal groups seem to have been acquired early in their history and retained tenaciously. Thus, the fact that modern corals build reefs only in shallow warm seas, correlated with observation that Mesozoic and Paleozoic coral reefs occur chiefly in low latitudes and in deposits having signs of shallow marine origin, strongly suggests that the pre-Tertiary reef corals, like modern ones, thrived in warm shallow waters.

Classification of Animals

Character of Division	Phylum

I. *EOZOA (*subkingdom*), one-celled animals.......................   *Protozoa (2) ✓

II. *METAZOA (*subkingdom*), many-celled animals

   A. Agnotozoa (*branch*), simplest cellular animals, no endoderm........   Mesozoa

   B. *Parazoa (*branch*), cells not organized in tissues or organs.........   *Porifera (3) ✓

   C. *Eumetazoa (*branch*), cells organized in tissues or organs

      1. *Radiata (*subbranch*), primary radial symmetry, two tissue layers (ectoderm, endoderm), no open body space (coelom) additional to digestive tract, no anus

         a. *Coelenterata (*phylum*), tentacles around mouth, has stinging cells...   *Coelenterata (4) ✓

         b. Ctenophora (*phylum*), mouth not surrounded by tentacles, no stinging cells..   Ctenophora

      2. *Bilateria (*subbranch*), primary bilateral symmetry, three tissue layers (ectoderm, mesoderm, endoderm), anus generally present

         a. Acoelomata (*grade*), space between digestive tract and body wall filled with mesenchyme, no true coelom. Includes Platyhelminthes, Nemertina (*phyla*)

         b. Pseudocoelomata (*grade*), space between digestive tract and body wall not a true coelom. Includes Entoprocta, Aschelminthes (*phyla*)

         c. *Eucoelomata (*grade*), true body cavity (coelom) in addition to digestive tract

            (1) *Schizocoela (*subgrade*), coelom originates as space in mesoderm

               (*a*) *Bryozoa (*phylum*), unsegmented body, with tentacle-bearing structure (lophophore) for feeding, colonial...   *Bryozoa (5) ✓

               (*b*) *Brachiopoda (*phylum*), unsegmented body, with lophophore and bivalve shell.....................   *Brachiopoda (6) ✓

               (*c*) *Mollusca (*phylum*), unsegmented body, without lophophore....................................   *Mollusca (7–10) ✓

               (*d*) *Annelida (*phylum*), segmented body without segmented appendages.............................   *Annelida (11) ✓

               (*e*) *Arthropoda (*phylum*), segmented body with segmented appendages.............................   *Arthropoda (12–15) ✓

               (*f*) Phoronida, Sipunculoidea, Priapuloidea, Echiuroidea (*phyla*)

            (2) *Enterocoela (*subgrade*), coelom originates from digestive tract

               (*a*) *Echinodermata (*phylum*), secondary pentamerous radial symmetry, with water-vascular system.............   *Echinodermata (16–21) ✓

               (*b*) *Chaetognatha (*phylum*), without notochord, gill slits, or endoskeleton................................   *Chaetognatha (11)

               (*c*) *Hemichordata (*phylum*), without well-developed notochord but with gill slits or endoskeleton............   *Hemichordata (22)

               (*d*) *Chordata (*phylum*), notochord in embryo, adults with vertebral column or gill slits or both...............   *Chordata (23)

Factors in Marine Environments

Most invertebrate fossils are remains of animals which lived in the sea, and therefore it is most important to give attention to the conditions affecting life in marine environments. These conditions can be analyzed and classified in terms of physical and biological factors, which in large degree are interrelated. Based on combinations of environmental factors and adaptations of animals to them, several categories of ecologic settings can be discriminated.

Temperature. A chief controlling factor in distribution of marine animals is temperature, which in different parts of ocean areas ranges from a minimum of 28°F. in polar regions and at great depths, to a maximum of approximately 85°F. in shallow tropical waters. Seasonal range of temperature affects only surface waters, amounting to about 3.5°F. in equatorial regions and less than 10°F. in most other areas; seasonal change of temperature diminishes downward, and below a depth of 200 m. it is absent. Marine animals are adjusted to the temperature, cold or moderately warm, of the place in which they live, and because this temperature is nearly constant, they cannot tolerate appreciable deviation from it. Larvae are observed to be more susceptible to changes in temperature than adult animals, and if they are carried by currents from cool to warm waters, or vice versa, death generally results. Temperature acts as a barrier to the dispersal of marine animals in a manner comparable to a high mountain range as a hindrance to the spread of land animals.

Salinity. The concentration of dissolved salts in sea water is much more nearly constant than temperature. Although salinity does vary from place to place, chiefly by the effects of evaporation which increases the salt content, and by dilution resulting from rain water and runoff from lands, the proportions of different chemical compounds in sea water remain unchanged. Marine animals are thoroughly adapted to these conditions of salinity, and most of them quickly die if they are transferred to fresh water or placed in a concentrated brine. Some species are much more tolerant than others to salinity variations. For example, oysters and mussels, which abound in normal marine environment, are able to live near mouths of rivers where the salt content of the water around them is less than one third that of the oceans; also, the brine shrimp can live in almost any concentration of salt water, but thrives best in hypersaline environments.

Light Penetration. The extent to which light penetrates sea water is important to marine animals, because all of them depend directly or indirectly on plants for food, and the plants require light for growth. Floating microscopic plants of the photic zone are the chief source of food; others are green, brown, and red algae, most of which live at depths of less than 50 m. Light of short wave length, at the violet end of the spectrum, reaches downward in sea water much farther (maximum about 300 m.) than long light waves (10 m. or less), and rays of the sun which strike the water surface vertically in low latitudes penetrate much farther than oblique sun's rays in high latitudes.

Water Movements. Currents, flow of tides, and wave-produced movements, especially surf along coasts, are important factors in many environments.

Ocean currents affect the distribution of temperature, carry nutrients for plant growth and food for marine animals, and transport some of the animals themselves. Warm-water currents, such as the Gulf Stream, and cold-water currents, such as the Labrador Current, are mainly responsible for the nature of floating, swimming, and shallow-water bottom-dwelling animal populations in areas which they affect. Also, there are deep-seated slow circulatory movements of ocean water, for cold, relatively heavy surface waters of polar regions sink to the bottom, flow equatorward, and rise in low latitudes, thus carrying oxygen and food to inhabitants of great depths. Shore currents, which generally are constant in direction of movement, except as modified by storms, carry sediment, food materials, and larvae of marine animals from place to place.

Tides are responsible for once- or twice-daily changes of sea level along most sea borders, and locally, as in the Bay of Fundy in eastern Canada, this may be as much as 16 m. (50 ft.). Both the change of level and the to-and-fro movement of water due to tides greatly affect bottom inhabitants of the intertidal zone. Animals living in this environment must have efficient means to prevent drying out during times of exposure to the air. Many burrow in the mud or sand; others protect themselves by sealing in a supply of water to bathe their soft parts. Tide-zone animals have less time for feeding than those which live off shore, and they are subject to hazards of molestation by land animals (including man) when they are uncovered.

Surf-agitated waters can be inhabited only by animals which possess very firm means of attachment. Most marine organisms cannot stand the battering of breakers, and hence surf environment constitutes an ecologic setting to which only hardy, specially adapted marine animals are admitted.

Bottom Conditions. Marine animals other than swimmers and floaters are influenced by the nature of the sea bottom, on which they crawl about or burrow or live attached in fixed position. In places, the sea floor consists of hard rock, which affords opportunity for firm anchorage, but elsewhere it is covered with gravel, coarse to fine sand, clayey or calcareous mud, or very fine ooze. That these conditions influence bottom-dwelling animals importantly is shown by their abundance in some environments, coupled with their absence or near absence in others; also,

different kinds of marine animals are adapted for existence on unlike kinds of sea bottom.

Biologic Associations. Animals and plants are important factors of environment, both because they may supply food and because they may make life easier by furnishing protection or more difficult by providing competition and enemies. Seaweeds provide a refuge for many animals, especially in the zone between low and high tides, where they provide moisture and hiding places for crustaceans and various other creatures. Competition for space is a chief problem of numerous clams, snails, barnacles, and other invertebrates of the near-shore parts of seas. Close association of various organisms has led to growth together of different species in a manner beneficial to the participants (**symbiosis**), or in a manner helpful to one without hurt to its host (**commensalism**), or in a manner detrimental to the host (**parasitism**). Biologic associations commonly are not less important than physical conditions as factors of environment.

Kinds of Marine Environments

A twofold classification may be applied to include all types of life surroundings in oceans and their borders. These are (1) bottom environments, termed **benthonic,** and (2) water environments, termed **nektoplanktonic** (Fig. 1-7) or pelagic. Animals which live on the bottom of shallow or deep seas are collectively termed **benthos;** those capable of moving about are classed as **vagile** (wandering) benthos, and those fastened in fixed position are known as **sessile** (seated) benthos. Swimming animals are termed **nektonic,** and floating forms are called **planktonic;** hence, the nektoplanktonic environment includes all animals which move freely through the water or at its surface.

Benthonic Environments. Bottom-dwelling marine animals may be grouped in four general assemblages according to the depth of water in which they live (Fig.

1-7). (1) Between the limits of high and low tides is the benthonic environment called **littoral;** daily or twice-daily fluctuations of sea level amount to only a few feet along most coasts, but the width of the belt alternately covered and uncovered by sea water may be more than a mile. (2) From low-tide mark to a depth of 200 m., which approximately coincides with the outer edge of the continental shelf in all parts of the world, and which marks the lower limit of effective penetration of light rays, is the **neritic** environment; it is divided somewhat arbitrarily into an inner neritic belt, extending to a maximum depth of 50 m., and an outer neritic belt, between 50 and 200 m. (3) The conti-

nental slope, ranging in depth from 200 to 2,000 m., is known as the **bathyal** environment; bottom life is much reduced in this region. (4) Deep ocean bottoms, below 2,000 m., are classed as **abyssal** environment.

A preponderant part of all known marine fossils represents the neritic benthonic environment.

Nektoplanktonic Environments. Swimming and floating animals are much less clearly and sharply grouped in subdivisions than benthonic animals. Two main categories are discriminated, however, and each of these is divisible into two. (1) Surface and near-surface animals of the oceans and their borders, mostly living

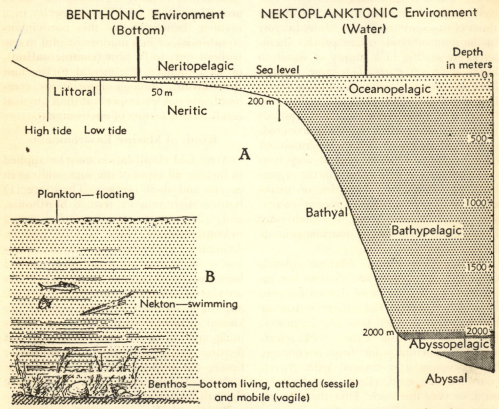

BENTHONIC Environment (Bottom)   **NEKTOPLANKTONIC** Environment (Water)

FIG. 1-7. **Environments of marine animals.** The diagrams show classification of environments recognized by the Committee on Marine Ecology and Paleocology, National Research Council. Simplest and most important is differentiation of bottom habitats (benthonic) and water habitats (nektoplanktonic), which includes all above-bottom modes of life. Both are subdivided according to depth below sea level.

99% of sea life in surface areas.

at depths of less than 200 m., are grouped in (a) the so-called **neritopelagic** environment, comprising shallow waters above the continental shelf areas, and (b) the **oceanopelagic** environment, including near-surface parts of open ocean basins. In these parts of the sea live perhaps 99 per cent of all nektonic and planktonic marine animals. (2) Deep waters include (a) the **bathypelagic** environment, between depths of 200 and 2,000 m., and (b) the **abyssopelagic** environment, below 2,000 m. A few sorts of peculiarly modified fishes are chief inhabitants of these waters.

Land Environments

Ecologists recognize numerous major and minor types of environments on land, the differences between them consisting chiefly of conditions controlled by climate and topography. Classification of these environments has little meaning for study of invertebrate fossils, and hence may be neglected. Most invertebrates of the land which are found preserved in Tertiary and older sedimentary deposits are animals adapted for life in fresh water, as in ponds, lakes, and streams. The environment in which they lived is indicated partly by physical characters of the rocks containing them and partly by the nature of the fossils themselves, including zoological affinities to known types of nonmarine animals and absence of marine organisms.

Air-breathing invertebrates, adapted to dry-land environments, include insects, spiders, scorpions, and other kinds of jointed-leg animals (arthropods), and the pulmonate snails. These have acquired the ability to thrive in almost all sorts of ecologic surroundings, even deserts and high mountains.

Facies Fossils

The term facies is used in stratigraphy to designate areally segregated sedimentary deposits which have significant characters differentiating them from contiguous strata of equivalent or partly equivalent age.

Thus, rocks which are divisible into two or more facies comprise diverse kinds of contemporaneous sedimentation in different areas. A facies intergrades or interfingers with neighboring ones, and at some places rocks of one facies may rest on another.

Because animals generally are adapted to particular modes of life and environments most favorable for them, different facies of rock formations almost invariably are characterized by more or less unlike assemblages of fossils (faunas). It is natural to distinguish such environment-controlled dissimilar assemblages as fossils of facies, or more simply, **facies fossils.** Examples of facies fossils are so numerous in all parts of the geologic column and in all regions that it is more difficult to cite species which lack impress of facies (environment) than those which do. From this comes the untrue but near-true dictum that "all fossils are facies fossils," meaning that virtually all reflect environments to which they are specially suited or restricted. Thus, in Middle Ordovician rocks of the Appalachian region we find limestones containing many brachiopods and bryozoans, which interfinger eastward with graptolite-bearing black shale which lacks the brachiopods and bryozoans. This and many similar occurrences express the influence of environment.

Study of the kinds of fossils occurring in some successions of sedimentary deposits has shown numerous specimens of a distinctive marine invertebrate in certain rock layers, whereas none are found in adjacent strata until a zone is reached 200 ft. or so higher in the section; here are many shells identical to the distinctive fossils previously noted. They are called a **recurrent fauna,** because the assemblage makes a new appearance after seeming extinction in the region. Observation of the fossil-bearing rocks indicates that the two zones under discussion are closely similar in lithologic features and different from associated strata. The temporary disappearance of the recurrent fauna in the area of study undoubtedly is correlated with change in sedimentary environment, and we may infer that during this absence the invertebrates lived on in other places where favorable environment prevailed, returning when conditions were propitious.

DEVELOPMENT OF INDIVIDUAL ANIMALS

Living matter which can exist independently or as a more or less interdependent member of a community is termed an **individual.** Some animals, like the amoeba, are organisms composed of a microscopic single cell; at the other extreme are huge animals like the whale, formed of billions of cells and weighing many tons.

The first step in a paleontological investigation is study of the remains of individual animals or groups of animals. Such samples provide means for recognizing different species and for understanding evolutionary changes which have occurred with lapse of geologic time. A species may be represented by several kinds of individuals, each of which passes through various forms in its life history. It is needful, therefore, to survey the nature of differentiation of individuals into kinds and to notice the modes of their development.

Kinds of Individuals and Their Associations

Among some species of animals, each individual carries on all the life functions, and therefore, each closely resembles others in form. There are also species in which two or more kinds of individuals perform different functions, and accordingly these differ in form and structure. Most animals, including man, are represented by males and females (**dimorphism**), specialized for different roles in reproduction. Many species of invertebrates, like the social ants, comprise three or more kinds of individuals (**polymorphism**), distinct in form and function.

In some invertebrate groups, individuals are organized together in **colonies,** and the independence of individual members is then partly or largely lost. The individual polyps of coral colonies may remain separate, may be connected by common tissues, or may fuse. Development of colonies commonly is associated with specialization of individuals. Such floating coelenterate colonies as the Portuguese man-of-war and the genus *Velella* (Fig. 1-8) contain individuals specialized for catching food, others devoted entirely to digestion, and still others which carry on reproduc-

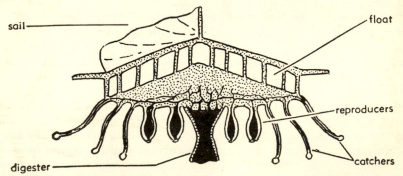

Fig. 1-8. **Colony of specialized individuals.** The coelenterate genus *Velella* comprises floating colonies made up of a number of different kinds of individuals. The upper surface of such a colony is composed of a chambered gas float, topped by a sail, which catches the wind and carries the colony along. The lower surface is composed of three types of individuals: catchers (dactylozooids), which, obtain food as well as ward off enemies by means of poison darts; a digester (gastrozooid) in the middle which serves as stomach for the entire colony; and reproducers (gonozooids), which bud off new little Velellas.

tion; none of these individuals could live alone.

Individuals of different species, generally belonging to different phyla, may live together in close association (**consortium**). As already noted in discussing adaptation to environment, this association may be classified as symbiotic, commensal, or parasitic.

Beginning of the Individual

Individuals arise by reproduction, which is accomplished in two chief ways. In **asexual reproduction,** a parent produces one or more offspring by budding or by dividing itself into pieces, each of which grows into a new animal. **Sexual reproduction** involves the union of two parents

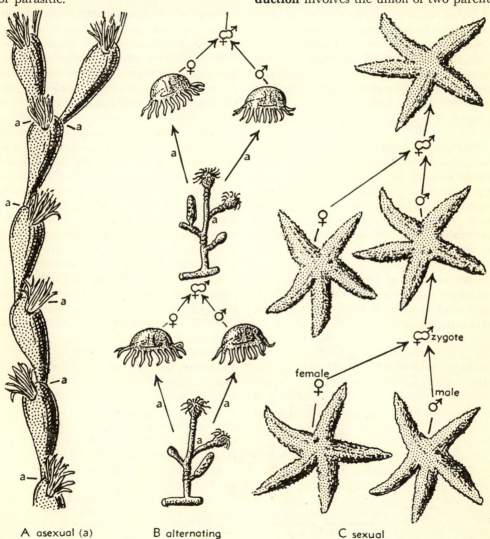

A asexual (a)   B alternating   C sexual

FIG. 1-9. **Types of reproduction.** Asexual reproduction, illustrated by the bryozoan (*A*), leads to essentially identical offspring. Sexual reproduction involves sexual union of male and female cells. It is illustrated by the starfish (*C*). The offspring shares characters of both parents. Many kinds of invertebrates, such as hydroid coelenterates (*B*), combine these two methods and alternate between asexual and sexual reproduction.

or of their reproductive cells (egg and sperm) to form a single cell (zygote), which develops into an offspring. Some species reproduce only asexually, but in most invertebrates, reproduction is either sexual, or alternately sexual and asexual (Fig. 1-9).

The life span of individuals is limited, ranging from a few hours at one extreme to 100 years or more at the other. Therefore, reproduction is essential for the maintenance of a species through a time longer than the individual life spans. Reproduction also gives rise to variation which has brought about diversification of the animal kingdom.

Factors Influencing Development

Heredity. Everyone knows that children resemble their parents and that puppies born of purebred dachshunds grow up into dachshunds, rather than becoming collies, or cows, or sparrows. This transmission of characters from parents to offspring is termed heredity. The science which deals with this process is called **genetics.**

The concept of heredity incorporates judgment that each of the microscopic reproductive cells of man, for example, contains all factors which make up a human being, complete to such details as eye color, nose shape, tendency toward baldness, and sense of taste. These and innumerable other characters are transmitted by postulated genetic units termed **genes,** too small to be seen, but contained in visible bands, called **chromosomes,** of the reproductive cells.

When cells multiply, they split in two in such manner that each of their chromosomes divides lengthwise, so as to furnish the two resulting cells with identical genetic attributes. Individuals arising from asexual reproduction are genetically identical to the parent, therefore.

In sexual reproduction, the offspring derives half of its chromosomes from its father and half from its mother. It partakes of both paternal and maternal characters. As a consequent of sexual reproduction, individuals are united in **populations**—groups which interbreed as generations pass, and within which there is a constant shuffling and recombination of genes in new individuals.

New characters, such as differences in size, color, proportions, or number of appendages, are introduced into asexual lineages or into sexually reproduced populations by changes (**mutations**) in genes and chromosomes. Gene mutations can be studied only in their effects on offspring. They are common in nature and may be induced artificially by radioactivity. Changes at the chromosome level are due to accidental overlapping or other disarrangement of chromosomes in the process of cell division, and may, like gene mutations, have a profound effect upon the nature of the **mutant** (offspring).

Many mutations upset development of offspring, resulting in death. Others have a less drastic effect, but result in crippling. Still others are not deleterious, and they may be advantageous to the new animal. Mutations are inheritable; consequently, interbreeding of mutants with normal individuals may result in the spread of mutations through a whole population. Without doubt, variability of populations has been acquired in this fashion.

Environment. Heredity can endow an organism only with potential size, potential structure, and potential abilities. These hereditary factors collectively define the so-called **genotype** of the animal. Proper growth and development can occur, however, only in a suitable environment. The resulting structure of an organism, partly hereditary and partly influenced by environment, constitutes what is termed a **phenotype.** Dwarf faunas, frequently encountered in the fossil record, indicate phenotypic stunting of individuals by unfavorable environment. Experiments have shown that genotypically similar reef corals will develop delicate lacy fronds in quiet water but produce massive low-growing colonies in surf.

The paleontologist cannot transplant his fossils to different places in order to watch their reaction to environment, and he cannot breed his specimens to study their genetics. He is restricted, therefore, to studying phenotypes, in which segregation of genotype and environmentally induced characters is inferential at best.

Nature of Growth

The ability to grow is one of the most distinctive properties of living matter. Many invertebrates grow throughout their life history; others cease to grow at some stage. Since all structures of organisms are produced by growth, it follows that differences in homologous structures are largely due to differences in growth. For example, the structural diversity between a dachshund and a greyhound is an expression of differential growth rates of the various bones in the skeleton. Figure 1-10 illustrates four crabs of strikingly unlike appearance, which closer examination shows are basically quite similar, the differences in shape being merely distortions due to differences in growth rates of various parts.

Because parts of an individual grow at different rates, proportions of these parts change during life history. In man, all parts of the body grow from babyhood until full size is reached, but from birth onward, the head grows less rapidly than the body and limbs; as a result, the head is *relatively* larger in a baby than in an adult.

Similar relations exist in invertebrates. Figure 1-11 shows the measurements of total length and length of the jaws (mandibles) of stag beetles belonging to a single species. Among the smallest beetles, the jaws account for about one-fifth of the total length, whereas in the largest they account for half of it. This is not a true growth series, for the beetles do not grow from small to large, but spring fully developed from the larva; nevertheless, it is akin to growth in exhibiting a graded series within which proportions vary with total size.

We may conclude that organisms are subject to a law of relative growth. So long as growth continues, whether rapid or slow, organs will develop at predetermined **relative rates.** These relative growth rates vary little within a species, for otherwise form would be startlingly diverse. Two related species may differ from each other in relative growth rates of one or more characters, or in the extent to which growth occurs. Of two groups of closely

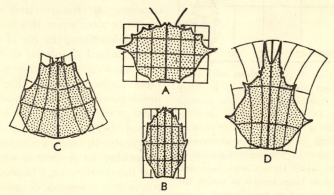

Fig. 1-10. **Differences in structure due to variable growth gradients.** The four crabs illustrated (*Geryon*, *A*; *Corystes*, *B*; *Paralomis*, *C*; *Chorinus*, *D*) seem at first glance to be quite dissimilar. Closer analysis shows that they are basically akin, being composed of similar elements, and that they differ in proportions. These differences may be expressed by a system of coordinates, constructed in squares for *A*, and variously distorted in the other genera. (*After D'Arcy Thompson.*)

related animals having similar growth rates, if individuals of group A grow to twice the total size shown in group B, the two assemblages will show strikingly different proportions.

The marvelous symmetry shown by coiled structures in nature, such as the shells of foraminifers, shells of mollusks, and horns of sheep, results from differential growth of the inner and outer sides of these structures. The relative growth rates are generally maintained throughout life, so that the spirals formed are mathematically definable. They belong to the class of logarithmic or equiangular spirals, in which the angle between radius and tangent drawn through any point on the curve is constant (Fig. 1-12).

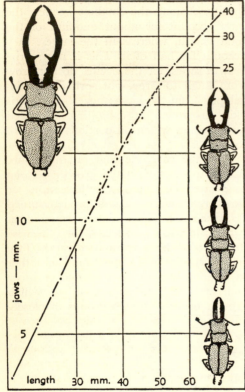

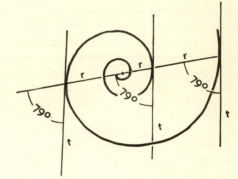

FIG. 1-12. **Equiangular spiral of Nautilus.** The spiral shell of the cephalopod *Nautilus* expands at a constant geometric rate, in such fashion that any radius (*r*) forms an angle of 79 deg. with a tangent (*t*).

Life History

The sequence of events from inception to death of an individual comprises its life history, or **ontogeny.** In many protozoans, which reproduce chiefly by splitting, ontogeny is characterized chiefly by increase in size; when this has reached certain proportions, the life of the individual is terminated simply by division into new ones. Among many other protozoans and most higher animals, life histories are much more complex. In the higher forms, such as the starfish (Fig. 1-13), the individual begins life as a single cell. It progresses through a single-walled sphere (**blastula**) to a double-walled **gastrula,** and thence through a procession of further stages into a mature organism. In insects, for example, this development may take years, and the organism may spend much time in one or more **larval** stages which differ radically from the adult in form and habits.

FIG. 1-11. **Variation of proportions with size.** The total length and jaw length of a great many individuals of the stag beetle *Cyclommatus tarandus* were measured by Dudich. The beetles range in size from about 20 to 75 mm., jaws from 4 to 35 mm. A statistical plot of the measurements yields an almost straight line on a logarithmic base, leading from small beetles with jaws one fifth of total length to large beetles in which the jaws account for one-half of total length. Conclusion: jaw size is a function of body size but increases at a much greater rate. Similar relationships exist between all the parts of animal bodies, which are linked to each other by relative rates of growth. (*After Huxley.*)

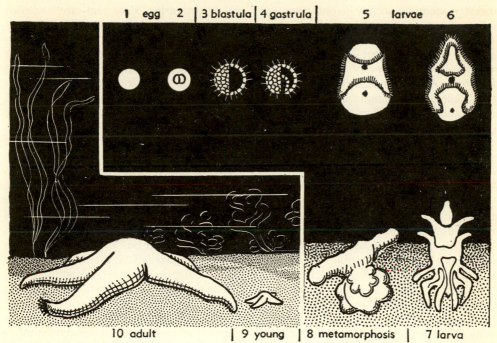

1 egg **2** | **3** blastula | **4** gastrula | **5** larvae **6**

10 adult | **9** young | **8** metamorphosis | **7** larva

FIG. 1-13. **Life history of a starfish.** A starfish begins life as (1) a single-celled fertilized *egg*, which passes into (2) multicellular stages and develops into (3) a *blastula*. Invagination of one side forms (4) a double-walled *gastrula*, which develops into a feeding and growing *larva*. Larval stages are (5) the free-swimming *dipleurula* (common to most echinoderms), (6) the free-swimming *bipinnaria* (restricted to star-fishes), and (7) the attached *brachiolaria*. (8) Metamorphosis changes the larva radically into (9) a baby starfish, which grows into (10) an adult.

The life histories of animals may be divided generally into five stages: **embryonic,** growth before the organism reaches independence; **larval (nepionic),** independent early development, in which structure and function differ markedly from those of the adult; **neanic,** close to the adult in structure but small in size; **adult (ephebic),** normal in size and able to reproduce; and **old age (gerontic).** Embryos and larvae are rarely encountered as fossils, because most lack hard parts or their delicate skeletons are very minute. Most fossils represent the pre-adult (neanic) and adult stages. Gerontic specimens are generally rare, for few animals survive to a ripe old age; the majority are killed at or before the height of life.

The life history of higher organisms passes through stages which recall to some extent the adult organization of lower phyla. The single-celled egg recalls the protozoan; the blastula, a protozoan colony (*Volvox*); and the gastrula, the coelenterates. The embryos of mammals and birds develop gill arches and gill slits, well supplied with blood. These fishlike structures evidently serve no useful purpose, and are resorbed during further development. The heart develops from a two-chambered (fishlike) to a three-chambered (amphibian) condition before reaching the final four-chambered stage. Thus, an individual mammal passes through stages which are comparable in a general way to structures of protozoans, coelenterates, higher invertebrates, fishes, and amphibians. There is much evidence that during geologic time mammals have indeed developed in this manner. Thus, traces of ancestry seem to persist in ontogeny, an observation summarized in the **biogenetic law** (also known as Haeckel's law or the law of **palingenesis**): that **ontogeny recapitulates phylogeny** (individual life history repeats race history) (Figs. 4-23, 9-24, 9-26). This recapitulation is highly incomplete, at best. Furthermore, special features may develop in the larval stages—features which are necessary for the larva to survive in its environment, and which bear no relation to any ancestors. Stages 7 and 8 of Fig. 1-13 are of this type.

In general, closely related animals show similar types of development, and ontogeny may be used as a clue to relationships. The barnacle is a good example; although the adult animal bears very little resemblance to crustaceans, its development shows beyond doubt that it is a highly aberrant member of this class. On the other hand, differences in development do not necessarily disprove relationships. Of closely related species among starfishes or sea urchins, one may undergo a normal type of development, including a series of free-drifting larval stages, whereas another passes its development within the egg and on the mother, and short-cuts development by omitting the larval state. Certain salamanders of the genus *Ambystoma* may reach sexual maturity in the larval state, and may thus omit the normal ephebic stage of development, though they will attain it if the proper environment is provided. This omission of normal adult characters, and maturation in what are normally earlier stages, is termed **neoteny** or **paedomorphism.**

We may conclude that ontogenetic development may furnish important clues to relationships and descent, but that, like any other character of an organism, it must be used with discretion.

Paleontologists are interested primarily in the development of preservable parts of an animal, namely, the skeleton. Skeletal formation generally does not begin until larval (nepionic) or pre-adult (neanic) stages are reached, but commonly furnishes extensive data on the neanic, ephebic, and gerontic development of fossil species.

Among some groups of organisms (foraminifers, corals, and mollusks, for example), each individual retains within its skeleton a record of its development. Others, such as the trilobites, periodically shed the skeleton and secrete a new one, large enough for the growing body. In still others, such as echinoderms and mammals, the skeletal elements persist throughout life but are changed in size and shape. Therefore, it is possible to reconstruct neanic and ephebic development of a coral from a single adult specimen, whereas in trilobites this can be done only by finding a large series of individual skeletons representing various stages of development.

The deciphering of skeletal development is very important to the paleontologist, for a number of reasons. (1) Successive growth stages of some animals have been found to be so dissimilar that before discovery of true relationships they were classified as different species, some of which were assigned to different genera, families, or even phyla. (2) Development of the skeleton tells more about relationships than does the adult portion of the skeleton alone; for example, some adult rugose corals may resemble closely certain hexacorals, but they are distinguished readily by the manner in which partitions are added during life history. (3) Development of the skeleton may pass through stages shown by adult or juvenile stages of ancestors and may thus aid in deciphering evolution; an example of this is shown in Fig. 9-26.

Colonies may show stages in development corresponding to stages in the ontogeny of an individual; this is termed **astogeny.** It is displayed by some bryozoans, corals, and many graptolites. The factors underlying astogenic changes are obscure.

EVOLUTION OF ANIMALS

The theory of organic evolution holds that life on earth has developed gradually, from simple to complex, and from few ancestors to the present multitude of diverse organisms, all by change and branching of species. Alternate theories are (1) that the earth was devoid of all life until the kinds of plants and animals now living were created; (2) that during past time many species have been created one after another, and when species were exterminated, they were replaced by new creations. The first alternate hypothesis is entirely at variance with the findings of geology and paleontology, which indicate that life goes far back in the earth's past and has changed very greatly with the lapse of time. The second hypothesis fails to account for fossil records of gradual transition from one species into another during geologic time. Instead of constituting independent creations, we find that each species has been developed from an ancestral one by a series of interacting processes, which operate in shaping life today just as they have done in the past. Some of these processes are familiar to everyone; some are not so familiar but can be demonstrated by experiment or studied with the microscope; and some are known to us only by their effects.

Concepts of organic evolution are stated in writings of the ancient Greeks, but the theory was first clearly outlined in scientific form by Lamarck (1744–1829), who also wrote the first great treatise on invertebrate animals. Lamarck was greatly impressed with the perfection of organic adaptation to environment, and devised a theory as to how this might have come about. He postulated that animals are able to transmit to their offspring such characters as they have acquired during their own lifetime (**Lamarckism,** the theory of inheritance of acquired characteristics). This means that the son of a blacksmith, for example, can be expected to have more powerful biceps than the son of a lawyer, because the blacksmith constantly labors with his arms. The legs of wading birds and the necks of giraffes grew long because each of many generations stretched these structures by wading in deep water or by reaching for high branches. Throughout a long period of time, the use of organs was thought to strengthen and improve them, whereas disuse was believed to lead to their shrinkage and disappearance. Modern genetics has not substantiated this simple and ingenious hypothesis.

Charles Darwin's *Origin of Species* (1859) was the first work in which the theory of evolution was backed up with convincing evidence and the description of a working mechanism. This book established the theory in all fields of biology. Subsequent work has substantiated all the essential features of Darwin's views, although modifying and refining them in many ways. Especially, the rise of genetics had led to a much greater insight into the basic processes of evolution.

Evidence of Evolution

Various fields of biology yield evidence or proof of gradual development of the animal kingdom. The structure of the system of **classification** offers strong support. Although devised by Linné, who believed that species, or at least genera, were individually created, the treelike arrangement of graded categories of plant and animal divisions from smallest to largest defines relationships which are intelligible only as an expression of common descent (Fig. 1-6).

Study of **comparative anatomy** furnishes evidence supplementing that given by botanical and zoological classification. The bodies of all sea urchins, or of all mammals, are built on the same general plan. A bear's skeleton contains the same basic elements as those in the frame of a horse or a cow, and such similarities are most reasonably explained as the resultant of common descent.

Not only are main organs belonging to these animals very much alike, but **vestigial structures** may show relationship. The appendix of man is the vestige of a digestive structure which is functional in other mammals. The useless eyes of blind cave-dwelling animals, such as fishes, amphibians, and crayfishes, have no reason for existence except as vestigial organs derived from seeing ancestors which lived in light. If these animals were created especially for a cave-dwelling life, they should have no eyes at all.

The relatively new science of **comparative physiology** has discovered that supposed relationships based on structural similarities are likewise indicated by correspondence in the composition of hemoglobin and other protein complexes of the body fluids.

Embryology also supports the theory of evolution, for in the course of life history all higher animals pass through a sequence of stages which correspond to lower forms of life, paralleling the supposed line of descent.

Geographic distribution of animals indicates gradual development of life, because dissimilarities between faunas of different areas are largely correlated with the length of time during which they have been isolated from one another. The peculiar mammalian fauna of Australia, for example, composed largely of marsupials, is evidently a result of Australia's separation from Asia in Cretaceous or early Cenozoic time. Evolution on a subspecific level is indicated by tree snails of some tropical islands in the Pacific, in which trees grow only in the valleys; here, each valley has its own subspecies of snails, morphologically distinct from those of adjacent valleys, because mixing and interbreeding of the various populations is prevented by dry, treeless divides between the valleys. This situation is most readily explained by postulating descent of the various subspecies from a single ancestral form which spread widely during a time of continuous forest cover. Presumably, the subspecies evolved from local populations which became isolated by the development of arid treeless divides.

The **fossil record** shows indisputable proof of change and gradual development of life through time. Although lack of knowledge is immeasurably greater than knowledge, many lineages among fossils of various groups have been firmly established. These demonstrate the transformation of one species or genus into another and thus constitute documentary evidence of gradual evolution.

Man himself has brought about evolution on a subspecific scale, in having developed the many diverse races of domesticated plants and animals, by rigorous selection and **controlled breeding.**

The science of **genetics** has discovered some of the ways and means by which variation, the fundamental basis of evolution, may come about.

In summation, nearly a dozen lines of evidence, each measurably independent of all others, convergently support the theory of evolution. None conflicts. Therefore, it is not surprising that biologists of the present day agree in recognizing evolution of life as a fact, rather than a theory. As yet very incompletely known, however, is the actual course of evolution in developing the tree of life, the mechanism of evolution, and the factors that affect evolution. Indeed, no one has yet discovered what life really is, why organisms grow, reproduce, die, and show the various manifestations which differentiate them from quartz crystals in granite or pebbles in a stream bed.

Evolutionary Changes in Animal Assemblages

All natural assemblages of animals—the fauna of part of a shallow sea or of an area on land, and the population comprising a species or subspecies—are subject to evolutionary change. The amount of change depends on the potency of factors causing evolution and defining evolutionary rate, and the duration of time in which all processes have been at work. For a brief survey of this subject, a single representative example is chosen: several thousand land snails belonging to a single species, living in a mountain valley.

Variation. Like all populations, the group of mountain-valley land snails is variable. Some individuals have thick, heavy shells, others thin, fragile ones; some grow larger than others; some are striped, others tend toward a speckled pattern; some are able to get along with less moisture than others; some are timid, others are bold. These and countless other differences are hereditary, because of differences in genes which are exchanged throughout the population. In addition to this preexisting diversity, mutations occur, introducing new variations into the population.

Struggle for Existence. The snails lead precarious lives in a world full of danger. Prolonged drought may desiccate and kill them; predatory animals may eat large numbers of them; parasites and diseases may take a toll. Thus, influenced by physical and biologic factors of their environment, the snails continuously wage a struggle for existence.

Natural Selection. Among varying individuals of the snail population, some are more likely to survive than others. Hardy types may last through droughts while others perish; conspicuously colored snails may be found and eaten by enemies, while camouflaged individuals are likely to remain unnoticed. The individuals best fitted to cope with all the features of their environment are those most likely to stay alive long enough to perpetuate the population, passing on advantageous attributes to their offspring. The less fit will be eliminated.

New mutants constantly make an appearance within the population, but of these, many die in embryonic stages, and many others which survive simply duplicate variations already existing in the population. A few are likely to differ in one or more features from any of their associates, and if, as is likely, these are less well fitted to the environment than normal individuals, they will be eliminated by natural selection. Occasionally, a mutation may appear which possesses a new advantageous factor, and if such an individual manages to survive and reproduce, its offspring may spread the advantageous variation. In the course of many generations, the new factor may be disseminated throughout the population.

Influence of Environment. If environmental conditions of the mountain valley remain unchanged for long periods of time, the snail population will become thoroughly adjusted to them. The amount of variation will tend to be restricted, and the nature of the population will change only very slowly, by rare appearance of some new advantageous mutations.

If the environment alters as a result of lowering the mountain range by erosion, or as a consequent of climatic changes, accompanied by introduction of new plants and animals and by elimination of others, the old snail population would find itself out of step with the environment. Former valueless or detrimental variations may now be highly advantageous, and characters formerly beneficial may now lead to destruction. Selection will therefore change the population, the rate and extent of this change depending on the rate and degree of environmental change, toleration of the snails for the changed conditions, and variability of the population. Variability is a prime factor, for under greatly altered conditions a species may be destroyed completely, if it cannot furnish a few individuals capable of surviving.

In this manner, populations evolve through time (Fig. 1-14). More or less continuous fossilization has recorded sequences of evolutionary stages in successive beds of sediment; several examples of this sort are described in the following chapters. The change through several thousand years, represented by inches or feet of rock, may be so gradual that alteration can be detected only by statistical methods. Continued for millions of years, evolution is likely to produce organisms different enough from their ancestors to warrant

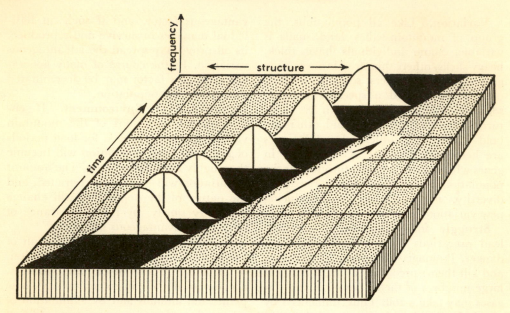

FIG. 1-14. **Statistical comparison of successive populations of an evolving strain.** Populations are represented by frequency curves showing structural variation. Successive populations show a shift in average values, as well as in extremes. Two populations which are close to each other in time overlap structurally to a large extent, but greater lengths of time cause such change (evolution) that an overlap in structural characters disappears.

classifying them in different species, genera, or families. Where the fossil record is relatively complete, the gradational nature of evolutionary transformation is likely to make the choice of boundary lines between successive species an arbitrary matter.

Branching of Species

The processes of evolution which have been described lead to changes in a population or species, but do not explain branching of a single species into two or more descendent species. This branching may also be illustrated by the snails of a mountain valley (Fig. 1-15).

Members of this snail population are designated as species *A* (illustrated by the lowest snail in Fig. 1-15). During an unusually warm summer, a few members of species *A* migrate over the divide into an adjacent valley, not previously invaded by the species. Although environment in the neighboring valley differs somewhat from that to which they were accustomed, the pioneers manage to survive and multiply, forming a new population. They are not joined by other emigrants from the old valley.

The pioneers are unlikely to be perfectly average representatives of species *A* and are certain to lack many of the genes current in it. On this basis alone, they and their descendants differ statistically from *A*. As a result of environmental differences which cause the selection of different factors in the two valleys, the populations or lineages come to differ increasingly from one another in the course of time. At first, they are nearly identical populations, but later the structural differences become sufficient to warrant taxonomic distinction (subspecies *x* in the old valley and subspecies *y* in the other, Fig. 1-15).

Let us now assume that conditions

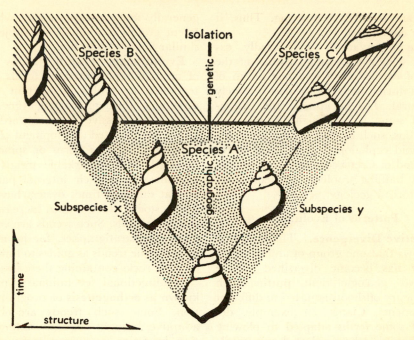

FIG. 1-15. **Evolution and branching of species.** An ancestral population, the average of which is represented by the snail at the bottom, becomes geographically split into two populations, each of which evolves and diverges (lineage 1 and lineage 2). Change leads first to morphologically different forms still capable of interbreeding and maintained as separate units only by virtue of the intervening geographic barrier (subspecies *x* and *y*). Eventually differences become so great that the two populations will not fuse by interbreeding though they may mingle in the same area. They are then considered as separate species (*B* and *C*).

change so as to permit each population freely to invade the territory of the other. If this occurs before evolutionary differentiation is far advanced, subspecies *x* and *y* will interbreed so as to fuse into a single population. If divergent evolution of the lineages has been carried far enough to make separate species (*B*, *C*, Fig. 1-15), such fusion will not occur. Individuals might be physically unable to mate, inclined to mate at different seasons, and otherwise not interested in each other. Even if individuals did undertake to interbreed, differences in the structure of their chromosomes would be likely to prevent development of viable offspring, or could result in sterility of any hybrids formed. Thus, geographic isolation, coupled with

divergent evolution, leads to branching of species.

Extinction of Animals

The pattern of organic evolution shows not only transformation and branching of species, but also their extinction. Species, genera, families, and even whole orders and classes of the animal kingdom have been exterminated in the geologic past; some now are on the verge of elimination. Extinction is part of the pattern of competition and elimination of the less fit. A species or larger unit, like an individual, survives only so long as it is able to make adjustment to changes in its environment; it is in constant danger of extinction from disease-producing organisms, parasites, predators, and competitors, all of which are themselves evolving. Thus, it may be eliminated.

These observations undoubtedly express a widely applicable truth; yet the cause of extinction of any one group of animals during the geologic past is largely conjectural. The abruptness and magnitude of some paleontologically recorded extinctions, such as that of the dinosaurs, ammonoid cephalopods, and other stocks at the end of Cretaceous time, are among the most baffling problems of paleontology and historical geology.

Patterns of Evolution

Adaptive Divergence. The fossil record shows that one group of animals after another has become diversified in the course of geologic time, putting forth evolutionary offshoots adjusted to different environments. Clams, for example, developed some forms adapted to plowing through mud, others adjusted to a sessile life on rocks, still others became suited for burrowing deeply into mud or boring into wood or rock, and a few developed the ability to swim. This evolutionary diversification is termed **adaptive divergence** or radiation.

Adaptive Convergence. The paleontologic record is full of examples comprising two or more distinct groups of animals which show evolution in the same direction—that is, leading to similar forms. For example, corals, certain brachiopods, and the rudistid clams developed very similar conical shells in adaptation to an attached life on reefs; ammonoid cephalopods and fusulinid foraminifers both show a trend from simple shell partitions to complex wavy ones. Such similarity of trends in evolutionary development is termed **parallelism,** and similarity in the form of the products of evolution is termed **convergence.** Two convergent species, so similar as hardly to be distinguished on superficial characters, are called **homeomorphs.** Parallelism and convergence

generally make an appearance when different groups of animals are subjected to similar forces of selection.

Evolutionary Trends. Many groups of animals represented by fossils show a tendency to evolve in certain definite directions during geologic time. Some show a progressive trend toward larger size or accentuated development of special structures, such as horns or spines. Ammonoids show remarkable trends in the complication of their internal partitions, and at different times oysters have given rise to stocks which tended increasingly to coil their shells. Such trends are very useful to the stratigrapher, for he may use stages in the trends as guides to time placement of rocks containing these fossils.

Unidirectional evolutionary trend is known as **orthogenesis** or program evolution. Some such trends are obviously adaptive, brought about by natural selection in fitting organisms more perfectly either to stable environments or to environments which change progressively in a certain direction.

Some trends are expressed by structures which may have little or no selective value. These can develop by reason of being linked to other characters which are primary elements of evolutionary change. For example, the development of mandibles in stag beetles (Fig. 1-11) is linked to body size, for no direct adaptive function seems to be served by enlargement of the jaws beyond moderate size, whereas evolutionary selection for greater body size may be correlated with the presence of overlarge jaws. Presumably, nonadaptive "overdevelopment" among other animals, such as antlers in deer, tusks in elephants, and spines of many invertebrates, may express similar effects.

Irreversibility. Evolution never plays its record in reverse. This observation, first expressed by Dollo, is accordingly known as **Dollo's law.** Some terrestrial reptiles and mammals, which are descendants of semiaquatic amphibians and aquatic fishes, have become established as inhabitants of the sea (ichthyosaurs, seals, whales); but in doing so, none of them have undergone backward evolution so as again to become amphibians and fishes. Though fishlike in form, they retain reptilian and mammalian characteristics of breathing air, circulation of blood, and skeletal structure. When a group of organisms once loses an organ or skeletal part, these can never be regained by descendants, although functions of the lost parts may be served in another way. This rule harmonizes with the concepts that mutations arise by chance and that possibility for variation is nearly infinite. Thus, evolution seems never to follow either identical forward or backward courses.

Rates of Evolution

Evolutionary rates, measured in the duration of species and genera, vary greatly. The spirally coiled foraminifer *Ammodiscus* (Silurian to Recent) has persisted for more than 300 million years, whereas many other genera of protozoans and metazoans represent transitory stages in evolutionary lines which persisted 5 million years or less before being superseded by much altered descendants. Lineages of short duration are of special interest to the stratigrapher, for if their remains are abundantly preserved in rock strata, and if they are distributed widely, they furnish the most useful tools for the age correlation of sedimentary deposits (Fig. 1-3).

The fossil record indicates that evolutionary rates are by no means uniform either in different stocks or in the same stock at different times. Some organisms exhibit such extremely sluggish evolutionary modification that they seem to stand still, whereas others undergo amazing sorts of morphological change with geological abruptness. Also, paleontological study of various lineages indicates that periods of slow, gradual adaptive change alternate with "explosive" spurts of evolution. These accelerated bursts develop descendants which differ markedly from their ancestors in form and function. The lack of known fossil connecting links between various orders and classes of the animal kingdom suggests that their origin may lie in this type of accelerated evolution or abrupt divergence.

FOSSIL-BEARING ROCK DIVISIONS

Fossil-bearing rocks in all parts of the world are classed according to placement in geologic systems. In many areas, they are assigned also to lesser-rank stratigraphic divisions, such as series and stage. The occurrence of invertebrate fossils illustrated in this book commonly is recorded in terms of subdivisions of systems, both because the information given is more precise and because some of the lesser-rank divisions are divergently treated in systemic classification. For example, Tremadocian rocks of Europe are included in the Cambrian System by many geologists but placed at the base of Ordovician by others; similarly, Downtonian is assigned varyingly to uppermost Silurian or lowermost Devonian.

A tabulation of stratigraphic divisions recognized in this book is given in Figs. 1-16 and 1-17.

GENERAL			NORTH AMERICA	EUROPE
CENOZOIC	NEOGENE	Pleistocene (contains Recent)		
		Pliocene		
		Miocene		PONTIAN
				SARMATIAN
				TORTONIAN
				HELVETIAN
				BURDIGALIAN
				AQUITANIAN
	PALEOGENE	Oligocene		CHATTIAN
				RUPELIAN
				TONGRIAN
		Eocene	JACKSONIAN	LUDIAN
				BARTONIAN
			CLAIBORNIAN	AUVERSIAN
				LUTETIAN
			WILCOXIAN	CUISIAN
				YPRESIAN
		Paleocene	MIDWAYAN	THANETIAN
				MONTIAN
MESOZOIC	CRETACEOUS	UPPER CRETACEOUS	LARAMIAN	DANIAN
			MONTANAN	MAESTRICHTIAN
				SENONIAN: CAMPANIAN
			COLORADOAN	SENONIAN: SANTONIAN
				SENONIAN: CONIACIAN
				TURONIAN
			DAKOTAN	CENOMANIAN
			WASHITAN	
		LOWER CRETACEOUS	FREDERICKSBURGIAN	ALBIAN
			TRINITIAN	APTIAN
				NEOCOMIAN
	JURASSIC	UPPER JURASSIC		PORTLANDIAN
				KIMMERIDGIAN
				OXFORDIAN
				CALLOVIAN
		MIDDLE JURASSIC		BATHONIAN
				BAJOCIAN
		LOWER JURASSIC		TOARCIAN
				PLIENSBACHIAN
				SINEMURIAN
				HETTANGIAN
	TRIASSIC	UPPER TRIASSIC		RHAETIAN
				NORIAN
				KARNIAN
		MIDDLE TRIASSIC		LADINIAN
				ANISIAN
		LOWER TRIASSIC		SCYTHIAN

FIG. 1-16. **Fossil-bearing divisions of Mesozoic and Cenozoic rocks.**

	GENERAL		NORTH AMERICA	EUROPE
PALEOZOIC	PERMIAN	UPPER PERMIAN	OCHOAN	CHIDERUAN
			GUADALUPIAN	KAZANIAN
				KUNGURIAN
		LOWER PERMIAN	LEONARDIAN	ARTINSKIAN
			WOLFCAMPIAN —	SAKMARIAN
	PENNSYL-VANIAN	UPPER PENNSYLVANIAN	VIRGILIAN	URALIAN
			MISSOURIAN	GSHELIAN
		MIDDLE PENNSYLVANIAN	DESMOINESIAN	MOSCOVIAN
			ATOKAN	
		LOWER PENNSYLVANIAN	MORROWAN	
			SPRINGERAN	NAMURIAN
	MISSISSIP-PIAN	UPPER MISSISSIPPIAN	CHESTERAN	VISÉAN
			MERAMECIAN	
		LOWER MISSISSIPPIAN	OSAGIAN	TOURNAISIAN
			KINDERHOOKIAN	ETROEUNGTIAN
	DEVONIAN	UPPER DEVONIAN	CONEWANGOAN	FAMENNIAN
			CASSADAGAN	
			CHEMUNGIAN	FRASNIAN
			FINGERLAKESIAN	
		MIDDLE DEVONIAN	TAGHANICAN	GIVETIAN
			TIOUGHNIOGAN	
			CAZENOVIAN	EIFELIAN
		LOWER DEVONIAN	ONESQUETHAWAN	COBLENZIAN
			DEERPARKIAN	
			HELDERBERGIAN	GEDINNIAN
	SILURIAN	UPPER SILURIAN (CAYUGAN)	KEYSERAN	DOWTONIAN
			TONOLOWAYAN	LUDLOVIAN
			SALINAN	
		MIDDLE SILURIAN (NIAGARAN)	LOCKPORTIAN	
			CLIFTONIAN	WENLOCKIAN
			CLINTONIAN	LLANDOVERIAN
		LOWER SILURIAN	ALEXANDRIAN	
	ORDOVICIAN	UPPER ORDOVICIAN (CINCINNATIAN)	RICHMONDIAN	ASHGILLIAN
			MAYSVILLIAN	CARADOCIAN
			EDENIAN	
		MIDDLE ORDOVICIAN (MOHAWKIAN)	TRENTONIAN	
			BLACKRIVERAN	
			CHAZYAN	LLANDEILIAN
		LOWER ORDOVICIAN	CANADIAN	SKIDDAVIAN
				TREMADOCIAN
	CAMBRIAN	UPPER CAMBRIAN (CROIXIAN)	TREMPEALEAUAN	(LINGULA FLAGS)
			FRANCONIAN	
			DRESBACHIAN	
		MIDDLE CAMBRIAN	ALBERTAN	MENEVIAN
				SOLVAN
		LOWER CAMBRIAN	WAUCOBAN	COMLEYAN

FIG. 1-17. Fossil-bearing divisions of Paleozoic rocks.

REFERENCES

CAMP, C. L., & HANNA, G. D. (1937) *Methods in paleontology:* University of California Press, Berkeley, pp. 1–153, figs. 1–58. Good discussion on methods of working with different groups of fossils.

DACQUÉ, E. (1921) *Vergleichende biologische Formenkunde der fossilen niederen Tiere:* Borntraeger, Berlin, pp. 1–777, figs. 1–345 (German). Contains a wealth of material on functional and ecologic interpretation of fossil invertebrates.

DEECKE, W. (1923) *Die Fossilisation:* Borntraeger, Berlin, pp. 1–216, no illus. (German). Discusses types of fossilization and changes in fossils in rocks.

DOBZHANSKY, T. (1937) *Genetics and the origin of species:* Columbia University Press, New York, pp. 1–446, figs. 1–23.

DUNBAR, C. O. (1949) *Historical geology:* Wiley, New York, pp. 1–567, figs. 1–350. Contains descriptions of fossil groups and chief features of their geologic record.

GOLDRING, W. (1950) *Handbook of paleontology for beginners and amateurs,* Part 1, *The fossils:* New York State Museum, Albany, Handbook 9, 2d ed., pp. 1–394, figs. 1–97. Gives a good discussion of fossils and types of preservation (pp. 17–64), with descriptions of phyla and bibliography.

HUXLEY, J. S. (1932) *Problems in relative growth:* Dial Press, New York, pp. 1–276, figs. 1–104.

HYMAN, L. H. (1940) *The Invertebrates: Protozoa through Ctenophora:* McGraw-Hill, New York, pp. 1–726, figs. 1–221. Discusses classification of animals (pp. 22–43) and basic zoological concepts (pp. 248–283).

JEPSEN, G. L., MAYR, E., & SIMPSON, G. G. (1949) *Genetics, paleontology, and evolution:* Princeton University Press, Princeton, pp. 1–474. A symposium containing 23 papers on various subjects relating to the title.

LULL, R. S. (1947) *Organic evolution:* Macmillan, New York, 2d ed., pp. 1–744, figs. 1–265. An excellent comprehensive treatment, well illustrated.

MAYR, E. (1942) *Systematics and the origin of species:* Columbia University Press, New York, pp. 1–334. A lucid account of modern evolutionary theory.

MOORE, R. C. (1949) *Introduction to historical geology:* McGraw-Hill, New York, pp. 1–582, figs. 1–364. Describes and illustrates main groups of invertebrate fossils and discusses their evolution.

RAYMOND, P. E. (1939) *Prehistoric life:* Harvard University Press, Cambridge, pp. 1–324, figs. 1–156. Discusses nature of fossils, types of preservation, and methods of collecting fossils (pp. 3–18).

RICHTER, R. (1948) *Einführung in die zoologische Nomenklatur:* Verlag Kramer, Frankfurt a. M., pp. 1–252 (German). Excellent exposition of the rules of zoological nomenclature.

SCHENCK, E. T., & McMASTERS, J. H. (1936) *Procedure in taxonomy:* Stanford University Press, Stanford University, pp. 1–72. Discusses nomenclature applied to fossils and gives summary of international rules.

SCHINDEWOLF, O. H. (1950) *Grundfragen der Paläontologie:* Schweizerbart, Stuttgart, pp. 1–506, figs. 1–332, pls. 1–32 (German). A lengthy treatment of evolution is illustrated by many examples (invertebrate and vertebrate).

SCOTT, W. B. (1917) *The theory of evolution:* Macmillan, New York, pp. 1–183. An excellent summary of the evidence for evolution.

SHIMER, H. W. (1933) *An introduction to the study of fossils:* Macmillan, New York, 3d ed., pp. 1–496, figs. 1–175. Discusses types of fossilization (pp. 1–30).

SIMPSON, G. G. (1944) *Tempo and mode in evolution:* Columbia University Press, New York, pp. 1–237, figs. 1–36. A most penetrating discussion of evolution from the paleontological standpoint.

SWINNERTON, H. H. (1946) *Outlines of palaeontology:* E. Arnold & Co., London, pp. 1–393, figs. 1–368. A general text, with emphasis on principles and many examples of evolutionary series.

THOMPSON, D'ARCY W. (1942) *On growth and form:* Cambridge, London, and Macmillan, New York, pp. 1–1116, figs. 1–554. A fascinating, lucidly written excursion into the mathematics of growth and organic structure.

CHAPTER 2

Foraminifera and Radiolaria

The structurally simplest members of the animal kingdom represented by fossils are Foraminifera and Radiolaria. They belong to the phylum of one-celled organisms called Protozoa, which includes the amoeba and a host of soft-bodied microscopic creatures unknown in the paleontologic record. Whether some of them are actually animals or plants is doubtful, and accordingly the term Protista has been proposed (Haeckel) to embrace all one-celled organisms.

Protozoans are divided into four main groups, designated as classes. These are named (1) Flagellata (or Mastigophora), characterized by the fixed shape of the cell wall and the presence of one or more long whiplike flexible projections (flagellae) used for locomotion; (2) Sarcodina (or Rhizopoda), distinguished by a changeable cell form which is associated with mobile extensions of the body (pseudopodia); (3) Sporozoa, parasitic protozoans having a fixed cell wall but no hard parts; and (4) Ciliata (or Infusoria), mainly differentiated by abundant short threadlike processes (cilia) which cover the exterior of the definitely shaped cell wall. The first two of these classes contain divisions which possess hard parts capable of preservation. The most important by far is the order Foraminifera, most of which secrete a calcareous test around the soft substance of the single cell, and next is the order Radiolaria, which generally bear an internal delicately complex skeleton of silica. Both of these groups belong to the class Sarcodina. The only other protozoans known as fossils are minor divisions of the Flagellata and Ciliata.

FORAMINIFERA

Foraminifera are mostly very small. The overwhelming majority have a diameter of less than 1 mm., but some attain dimensions of 100 mm. across the shell. Many important geologic formations in the world consist largely of their remains. They are common in shallow marine waters, and floating forms are widely distributed in the near-surface parts of the oceans. The sands of many tropical beaches are made up almost entirely of foraminiferal tests.

The Foraminifera are among the most useful fossils in stratigraphic work because they are widely distributed in marine sediments in the geologic column, especially in late Paleozoic, Mesozoic, and Cenozoic deposits. Many foraminiferal stocks exhibit definite evolutionary trends which are toward a more simple type of test in some groups but toward a more complex one in others.

Although studies on Foraminifera embrace a period of nearly 400 years, the most important work on them has been done during the past century. Their special importance in stratigraphic zonation of rocks penetrated in deep borings has become recognized in the course of exploration for petroleum since about 1915.

Living Animal

A test is secreted by all except a few of the simplest foraminifers. It consists of a complex organic compound (chitin), of cemented arenaceous or other particles, or of calcite. The living substance is termed **cytoplasm,** differentiated into **ectoplasm** on the outside, and **endoplasm** on the inside. The endoplasm contains the **nucleus** of the cell, or perhaps several nuclei. Parts of the cytoplasm may extend outside the test through its aperture or apertures, or through tiny pores in the wall. In feeding, the cytoplasm may completely enclose the test, sending out long, threadlike, extensions called **pseudopodia.** Material of the cytoplasm is continually in motion, some flowing out into the pseudopodia and some flowing back into the test. The inward-moving flow carries with it tiny particles of organic matter caught by the pseudopodia, thus supplying the animal with food. The pseudopodia also aid in locomotion by creeping.

Life History

Successive generations of a foraminifer are produced partly by subdivision of the body substance of fully grown individuals (asexual reproduction) and partly by development of young from the union of male and female elements released by different adults (sexual reproduction). The asexual type of generation is prevalent, but periodically it is replaced by sexual reproduction. The alternation of sexual and asexual generations was first recognized in studying fossil Foraminifera. It was noted that shells of nearly identical sort, seemingly representing the same species, comprise two types: a **megalospheric form** (Fig. 2-1), characterized by the small size of the test and the relatively large initial chamber (**proloculus**); and a **microspheric form,** distinguished by the large size of the test and the very small proloculus.

The complete life cycle has been observed in several living species. The microspheric form, which reproduces asexually (Fig. 2-2), has a large number of nuclei scattered throughout the cytoplasm. The nuclei increase in number by simple division during growth. When the adult stage is reached, the cytoplasm leaves the inside of the test and divides into many rounded masses, each of which contains a nucleus. Then, the asexually produced units develop tests of megalospheric forms. These individuals reproduce sexually. On attaining adult growth, the single nucleus in the cytoplasm of each splits into a large number of tiny parts. The minute nucleus segments leave the test, carrying with them some cytoplasm, and become flagellated zoospores (gametes) which conjugate. The sexually produced new individuals develop tests of microspheric forms, thus completing the life cycle of two generations. The complete life cycle extends over a period of as much as two years.

The ratio of megalospheric to microspheric forms varies from 2 to 1 in some genera to as much as 30 to 1 in others. Megalospheric forms are fewer in genera which conjugate in cysts where fertilization is most complete.

The size of the proloculus of megalospheric forms varies considerably in certain species. This variation in size has been suggested to signify the existence of more than two forms of a species, but probably this is not so. Several megalospheric generations may intervene between microspheric generations, each represented by individuals having a proloculus of different size.

The biogenetic law, which states that the life history of an individual at least partly reflects the evolution of its race, is supported by features of the microspheric and megalospheric shells of many foraminiferal species. The microspheric test indicates ancestral features because these

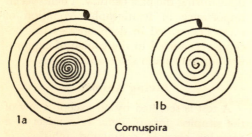

Cornuspira

microspheric form   megalospheric form

Fig. 2-1. **Two forms of test belonging to a single species of foraminifers.** The microspheric shell has embryonic coils at its center which are lacking in the megalospheric test.

are developed in the first-formed portion of the test. The trend of development of a genus may be shown in the megalospheric form, since its last-built chambers in the adult stage are more advanced than corresponding chambers of microspheric forms. The relations between different genera may be determined from a study of the evolutionary trends manifested by microspheric and megalospheric forms belonging to each.

Test

Tests of Foraminifera consist of different materials. Some are composed of chitinous matter, some of agglutinated particles cemented together, and some—the vast majority—of calcareous material. Siliceous tests are reported, but they are rare, and it is possible that all are secondary in origin.

Chitinous Test. A thin, flexible, transparent test composed of complex organic substance (chitin) is believed to be the most primitive type. Some very simple genera have a covering of this sort. Also, an inner layer of chitin is observed to characterize the test of various genera, such as *Textularia*, which have arenaceous walls, and of some genera which have an outer layer of calcareous material.

Arenaceous or Agglutinated Test. An arenaceous or agglutinated test consists of foreign particles bound together with cement. The foreign particles may consist of quartz sand, calcareous grains, mica flakes, sponge spicules, and tests of other Foraminifera. Many species are very selec-

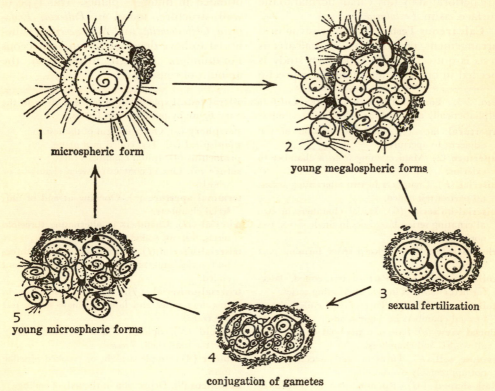

1 microspheric form
2 young megalospheric forms
3 sexual fertilization
4 conjugation of gametes
5 young microspheric forms

FIG. 2-2. **Life cycle of Spirillina vivipara.** The diagram illustrates alternation in generation of a foraminifer. A fully grown microspheric individual (1) by *asexually* subdividing produces many small megalospheric individuals (2); *sexual* fertilization of two grown megalospheric individuals (3) by exchange of nuclear elements is followed by subdivisions forming nucleus-bearing gametes, two of which coalesce (4) and grow into microspheric forms (5, 1).

tive in choosing materials for construction of the test, some using only fine quartz grains, some rejecting everything except sponge spicules, and so on. Certain genera, like *Textularia*, build their test of quartz grains in cold-water areas and of calcareous grains in tropical waters. The cementing material is commonly calcareous, although ferruginous cement is used by some genera. In certain tests, the cementing material, which may be secreted by the cytoplasm, can hardly be seen. The amount of cement generally is greater in tests composed of calcareous fragments, but complete gradation seems to exist between tests consisting mostly of cement and those having a minimum amount of cement. The arenaceous tests of some genera are perforated by tiny canals normal to the surface, as in *Textularia*.

Calcareous Tests. The microstructural arrangement of materials in calcareous tests is quite variable, and much study is needed in order to determine their precise features. Classification of calcareous Foraminifera in groups having perforate and imperforate tests, respectively, does not seem significant, because perforations are found in arenaceous tests and they occur in many calcareous tests which have been called imperforate.

Many genera have fibrous tests made up of tiny crystals of calcite which have their *c*-axes normal to the outer surface. Such forms as *Lagena*, *Nodosaria*, *Bolivina*, and *Epistomina*, characterized by transparent or translucent tests, commonly called **hyaline,** have this type of wall structure. The walls are perforate.

Many other genera, some having hyaline tests, possess granular calcareous walls. The granules are of the same size and are oriented in different planes. This type of wall structure, found in *Pullenia*, *Endothyra*, *Chilostomella*, and *Monogenerina*, is believed to have developed from arenaceous Foraminifera through increase in the amount of cement.

Fig. 2-3. **Form and structural features of foraminiferal tests.** Explanation of terms is given in the alphabetically arranged list, which is cross-indexed to the figure by numbers and letters.

apertural face (7). Flattened portion of test adjacent to aperture.

aperture (2). Main opening from a chamber to exterior.

biserial (*F*). Chambers in two alternating series, no part of test coiled.

biserial-uniserial (*G*). Initial chambers in two alternating series, later ones in single series, test not coiled.

chamber (1). Wall-enclosed space forming part of test.

coiled-biserial (*E*). Part of test coiled, later-formed chambers in two alternating series.

coiled-uniserial (*C*). Part of test coiled, later-formed chambers in single series.

dorsal view (8). Side of a trochoid test showing all whorls and chambers.

evolute coil (*K*). Turns of shell in contact but not appreciably overlapping.

flask-shaped (*M*). Rounded outline joined to more or less extended neck with aperture at its end.

involute coil (*L*). Outer whorls of shell strongly overlapping inner ones.

periphery (3). Outer margin of the test.

planispiral (*B*). Test coiled in a plane.

proloculus (4). Initial chamber of test.

suture (6). Line of contact between chambers or whorls.

terminal aperture (5). Opening at end of uniserial chambers.

triserial (*H*). Chambers in alternating threefold series, test not coiled.

triserial-biserial (*I*). Chambers initially in threefold series and later in two series, test not coiled.

triserial-uniserial (*J*). Chambers initially in threefold series and later in single series, test not coiled.

trochoid (*N*). Spirally coiled in conical form, like the snail shell, *Trochus*.

tubular (*A*). Simple straight or crooked pipelike form.

umbilicus (9). Depression in the axis of coiling.

uniserial (*D*). Chambers in a single series, test not coiled.

ventral view (10). Side of a trochoid test showing only chambers of last whorl.

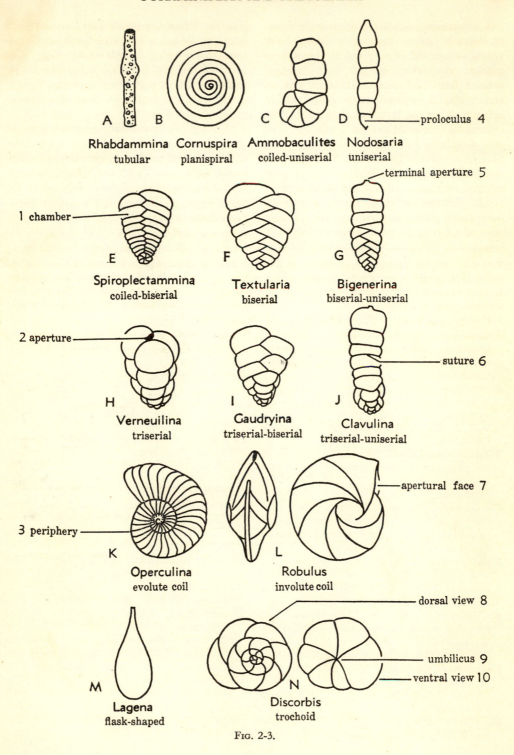

A Rhabdammina
tubular

B Cornuspira
planispiral

C Ammobaculites
coiled-uniserial

D Nodosaria
uniserial

proloculus 4

terminal aperture 5

1 chamber

E Spiroplectammina
coiled-biserial

F Textularia
biserial

G Bigenerina
biserial-uniserial

2 aperture

H Verneuilina
triserial

I Gaudryina
triserial-biserial

J Clavulina
triserial-uniserial

suture 6

3 periphery

K Operculina
evolute coil

L Robulus
involute coil

apertural face 7

dorsal view 8

M Lagena
flask-shaped

N Discorbis
trochoid

umbilicus 9

ventral view 10

FIG. 2-3.

Calcareous, so-called imperforate Foraminifera of porcelaneous appearance, have a test composed of tiny calcite crystals of uniform size which do not have constant orientation in all genera. In some genera, the crystals are oblique to the outer wall in some parts of the test, but they exhibit no consistent orientation in others. The megalospheric forms of some porcelaneous genera are perforate.

Morphology of Test. The fundamental unit of the test is the **chamber** (*1*) (italic figures and letters in the text on morphology refer to Fig. 2-3), which consists of a cavity and the surrounding wall.

The first chamber formed in Foraminifera is the **proloculus** (*4*), and the line of contact between chambers is a **suture** (*6*). The form of the test depends upon the shape and arrangement of the chambers. Principal forms are found in tests of different composition. For example, genera such as *Textularia* (*F*) and *Bolivina* have the same biserial arrangement of chambers, but they differ in composition of the test, the first being arenaceous and the second calcareous. Also, different genera may have the same form but dissimilar types of aperture. An **aperture** (*2*) is the opening (other than a pore in the wall) from a chamber to the

FIG. 2-4.   Representative Ordovician and Silurian Foraminifera. The oldest-known Foraminifera are from Ordovician rocks (12–14). The genera are mostly simple arenaceous tubular and globular forms belonging to the families Astrorhizidae and Saccamminidae, some of which have living representatives. Silurian Foraminifera (1–11) are simple arenaceous forms, which consist mainly of straight and coiled tubular tests, some of which have more than one aperture.

Ammodiscus Reuss, Silurian–Recent. The arenaceous test is planispirally coiled; aperture at end of tube. *A. exsertus* Cushman (7, ×65), Silurian, Oklahoma.

Bathysiphon M. Sars, Silurian–Recent. The test is cylindrical and may be curved; sponge spicules commonly form part of the arenaceous wall. *B. curvus* Moreman (11, ×35), Silurian, Oklahoma.

Bifurcammina Ireland, Silurian. The last coil of the planispiral arenaceous test is divided into two tubes. *B. parallela* Ireland (6, ×60), Oklahoma.

Colonammina Moreman, Silurian. The arenaceous attached test is planoconvex and elliptical in outline; it bears a single aperture. *C. verruca* Moreman (10, ×110), Oklahoma.

Hyperammina H. B. Brady, Silurian–Recent. The elongate arenaceous test has an aperture at end of the tube. *H. harrisi* Ireland (4, ×40), Silurian, Oklahoma.

Kerionammina Moreman, Ordovician. The irregular, flattened test composed of sand grains bears apertures at ends of arms. *K. favus* Moreman (13, ×40), Viola limestone, Oklahoma.

Lagenammina Rhumbler, Silurian–Recent. The arenaceous flask-shaped test has a rounded aperture. *L. stilla* Moreman (8, ×100), Silurian, Oklahoma.

Marsipella Norman, Ordovician–Recent. The test is elongate, cylindrical, composed of sand grains and sponge spicules. *M. aggregata* Moreman (12, ×45), Ordovician (Viola), Oklahoma.

Psammosphaera Schulze, Silurian–Recent. The globular test has a chitinous layer beneath the outer covering of sand grains and other fragments; aperture indefinite. *P. gigantea* Dunn (5, ×50), Silurian, Tennessee.

Rhabdammina M. Sars, Ordovician–Recent. The branching cylindrical tubes of sand grains have apertures at ends of tubes. *R. trifurcata* Moreman (14, ×40), Ordovician (Viola), Oklahoma.

Saccammina M. Sars, Silurian–Recent. The globular arenaceous test has a simple aperture on a short neck. *S. moremani* Ireland (1, ×40), Silurian, Oklahoma.

Tholosina Rhumbler, Silurian–Recent. The hemispherical arenaceous test is attached and has small apertures above the base. *T. convexa* Moreman (2, ×50), Silurian, Oklahoma.

Thurammina H. B. Brady, Silurian–Recent. The simple spherical test has several apertures at ends of short protuberances, and the arenaceous wall has an inner chitinous layer. *T. subspherica* Moreman (9, ×30), Silurian, Oklahoma.

Webbinella Rhumbler, Silurian–Recent. The attached test is circular and planoconvex; wall composed of fine sand grains. *W. coronata* Ireland (3, ×50), Silurian, Oklahoma.

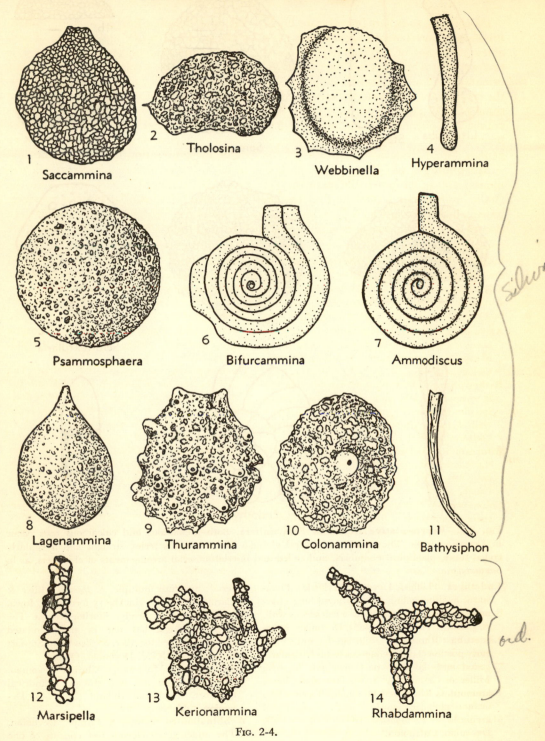

1 Saccammina
2 Tholosina
3 Webbinella
4 Hyperammina
5 Psammosphaera
6 Bifurcammina
7 Ammodiscus
8 Lagenammina
9 Thurammina
10 Colonammina
11 Bathysiphon
12 Marsipella
13 Kerionammina
14 Rhabdammina

Fig. 2-4.

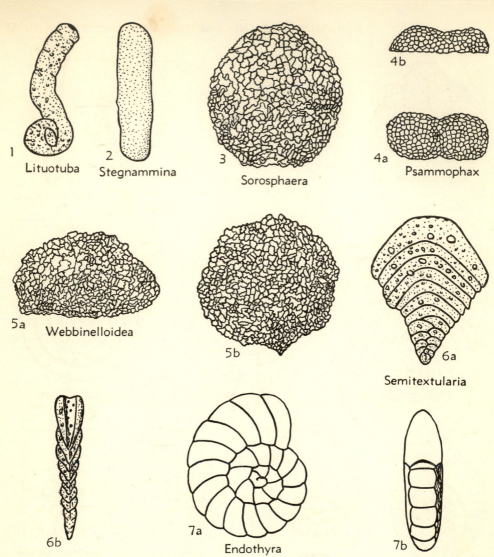

1 Lituotuba **2** Stegnammina **3** Sorosphaera **4b** **4a** Psammophax

5a Webbinelloidea **5b** **6a** Semitextularia

6b **7a** Endothyra **7b**

FIG. 2-5. **Representative Devonian Foraminifera.** Both arenaceous and calcareous tests occur in Devonian rocks. They consist of several globular chambers in a series, straight and irregularly coiled tubes, planispiral chambers, and coiled and biserial-uniserial arrangements of chambers, as in *Semitextularia*.

Endothyra Phillips, Devonian–Triassic. Planispiral evolute coiling is a distinguishing character of this genus, which has a calcareous wall. *E. gallowayi* Thomas (7, ×85), Devonian, Iowa.

Lituotuba Rhumbler, Silurian–Recent. The early portion of the arenaceous test is irregularly coiled and the later part uncoiled. *L. dubia* Miller & Carmer (1, ×90), Devonian, Iowa.

Psammophax Rhumbler, Devonian–Recent. Two or more globular chambers occur in series; wall arenaceous. *P. bipartita* Ireland (4, ×100), Devonian, Oklahoma.

Semitextularia Miller & Carmer, Devonian. The compressed arenaceous test is coiled in early portion, then biserial, and finally uni-

serial; aperture multiple. *S. thomasi* Miller & Carmer (6, ×85), Hackberry formation, Iowa.

Sorosphaera H. B. Brady, Silurian–Recent. The arenaceous test has one or more inflated chambers joined together. *S. columbiensis* Stewart & Lampe (3, ×75), Devonian, Ohio.

Stegnammina Moreman, Silurian–Devonian. The arenaceous test is a straight cylindrical chamber. *S. elongata* Ireland (2, ×60), Devonian, Oklahoma.

Webbinelloidea Stewart & Lampe, Devonian. The attached arenaceous test consists of one or more plano-convex chambers. *W. hemispherica* Stewart & Lampe (5, ×55), Ohio.

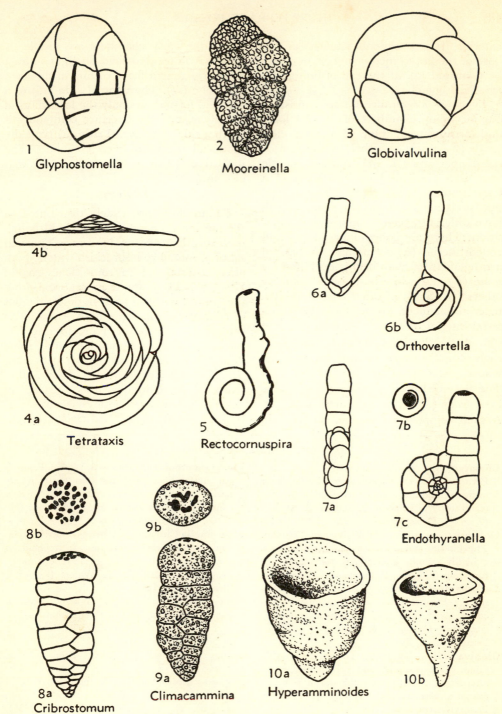

1 Glyphostomella

2 Mooreinella

3 Globivalvulina

4b

6a

6b Orthovertella

4a Tetrataxis

5 Rectocornuspira

7a

7b

7c Endothyranella

8b

9b

8a Cribrostomum

9a Climacammina

10a Hyperamminoides

10b

FIG. 2-6. **Representative Pennsylvanian Foraminifera.** In addition to Fusulinidae, which are treated separately, Pennsylvanian rocks have yielded a variety of arenaceous and calcareous Foraminifera. Several genera have chambers in a biserial and uniserial arrangement, as in *Climacammina*. Several others have the chambers in a trochoid arrangement, as in *Tetrataxis*. Many of the smaller Foraminifera of the Pennsylvanian are also found in Permian rocks, but several are restricted to the Pennsylvanian. (*Continued on next page.*)

exterior. It is commonly situated at the base of the last-formed chamber, but may be **terminal** (5), at the end of uniserially arranged chambers.

Types of foraminiferal tests consist of one or many chambers of various form and arrangement. One of the simplest forms is a single chamber having one or more apertures (A, M). Another simple form is a planispiral test consisting of a long tubular chamber (B). The uniserial test (D) is common to several genera which have different apertures and wall composition. There are coiled-uniserial (C) and coiled-biserial (E) forms also. A biserial arrangement of chambers is exhibited by some (F) and a biserial-uniserial plan by others (G). Shells of other types exhibit a triserial chamber arrangement (H), triserial-biserial (I), and triserial-uniserial (J). A tightly coiled form in which the early chambers are covered by the last whorl is known as an involute coil (L). The flattened portion below the aperture in *Robulus* (L) is the **apertural face** (7). A shell composed of a loose coil (K), in which the early chambers are visible, is called evolute. Trochoid shells, common to several genera, have coiling in the form of a cone (N). The **ventral side** (10) of a trochoid form shows only the last whorl of chambers, and the aperture commonly is on that side. The **dorsal side** (8) typically shows all the whorls and chambers. The **umbilicus** (9) is a depression in the axis of coiling.

Ecology

Foraminifera are widely distributed in the marine waters of the world. Although most of them occur in seas of normal salinity, some live in brackish waters, both along coasts and in some lakes, as in Hungary. They are even reported from the ground water of wells in desert regions of central Asia and northern Africa. The central Asia fauna of living foraminifers includes several Miliolidae and some doubtful forms called *Lagena, Discorbis, Textularia, Globigerina,* and *Nodosaria,* all of which have chitinous tests, instead of normal calcareous or arenaceous ones. Several genera (such as *Quinqueloculina,*

(Fig. 2-6 continued from preceding page.)

Climacammina H. B. Brady, Pennsylvanian–Permian. The early portion is biserial, later uniserial, and the aperture irregularly sievelike (cribrate). *C. cylindrica* Cushman & Waters (9, ×20), Pennsylvanian, Texas.

Cribrostomum Möller, Pennsylvanian–Permian. The biserial and uniserial cylindrical test has a highly cribrate aperture. *C. marblense* H. J. Plummer (8, ×15), Pennsylvanian, Texas.

Endothyranella Galloway & Harlton, Mississippian–Pennsylvanian. Early chambers close-coiled, later ones in a rectilinear series. *E. armstrongi* Cushman & Waters (7, ×50), Pennsylvanian, Texas.

Globivalvulina Schubert, Pennsylvanian–Permian. Test trochoid and subglobular, with alternating chambers on the dorsal side. *G. biserialis* Cushman & Waters (3, ×100), Pennsylvanian, Texas.

Glyphostomella Cushman & Waters, Pennsylvanian. The involute test is nearly symmetrical and has apertures at the base of the chambers. *G. triloculina* (Cushman & Waters) (1, ×35), Texas.

Hyperamminoides Cushman & Waters, Pennsylvanian–Permian. The test has an elongate, flaring second chamber marked by growth constrictions. *H. expansus* H. J. Plummer (10, ×60), Pennsylvanian, Texas.

Mooreinella Cushman & Waters, Pennsylvanian. The coarsely arenaceous test is trochoid in early stages and biserial in later chambers. *M. biserialis* Cushman & Waters (2, ×55), Strawn group, Texas.

Orthovertella Cushman & Waters, Pennsylvanian–Permian. The early portion of the undivided calcareous test is coiled in varying planes, later portion uncoiled. *O. protea* Cushman & Waters (6, ×85), Pennsylvanian, Texas.

Rectocornuspira Warthin, Pennsylvanian. Early portion planispiral, and adult portion uncoiled, straight. *R. lituiformis* Warthin (5, ×35), Oklahoma.

Tetrataxis Ehrenberg, Pennsylvanian–Permian. The conical-shaped test has elongate, crescentic chambers. *T. scutella* Cushman & Waters (4, ×60), Pennsylvanian, Texas.

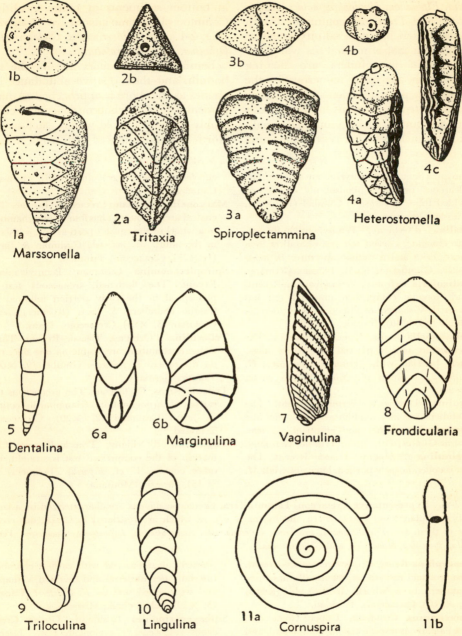

Fig. 2-7. **Representative Triassic, Jurassic, and Cretaceous Foraminifera.** A modern foraminiferal fauna appears first during the Triassic period (5, 9–11). Most genera have living representatives, and almost all are members of the families, Astrorhizidae, Lituolidae, Miliolidae, and Lagenidae. Genera of the families Lagenidae and Polymorphinidae are the most common in Jurassic faunas (6–8). In general, a marked increase in surface ornamentation and in differentiation of species is observed in Jurassic Foraminifera, which differ mainly from Cretaceous forms in absence of planktonic species. A few Cretaceous arenaceous foraminifers are illustrated here (1–4). (*Continued on next page.*)

Rotalia, Elphidium, Haplophragmoides, Ammo-baculites, and *Textularia*) commonly live in brackish waters having a salinity as low as one thirtieth that of normal sea water.

Planktonic foraminifers are floating forms which are most numerous in the near-surface parts of open oceans. They are distributed by currents, and their tests fall to the bottom in large numbers, making up as much as 80 per cent of the fauna in bottom sediments at many localities. Common planktonic genera are *Globigerina* (Fig. 2-13, *6*), *Globigerinoides, Globigerinella, Orbulina,* and *Globorotalia.*

Benthonic foraminifers live on the sea bottom, and their relative abundance increases as the shore is approached. A high percentage of arenaceous genera commonly is found in cold waters, such as the North Sea, as compared with tropical seas.

(Fig. 2-7 continued from preceding page.)

Cornuspira Schultze, Pennsylvanian–Recent. The calcareous test is a planispirally coiled undivided tube. *C. pachygyra* Gümbel (11, ×95), Triassic, Germany.

Dentalina d'Orbigny, Pennsylvanian–Recent. The elongate, curved test has chambers in a linear series and a radiate aperture. *D. transmontana* Gümbel (5, ×65), Triassic, Germany.

Frondicularia Defrance, Pennsylvanian–Recent. The elongate compressed uniserial test has inverted V-shaped chambers. *F.* cf. *franconia* Gümbel (8, ×40), Jurassic, Canada.

Heterostomella Reuss, Upper Cretaceous. The arenaceous test is triserial in earliest stage, later biserial with terminal aperture. *H. austinana* Cushman (4, ×60), Austin chalk, Texas.

Lingulina d'Orbigny, Permian–Recent. The chambers are in a rectilinear series, and the aperture is terminal and elliptical. *L. intumescens* Gümbel (10, ×65), Triassic, Germany.

Marginulina d'Orbigny, Triassic–Recent. The test is coiled in early portion, later uniserial. *M.* cf. *limata* (Schwager) (6, ×75), Jurassic, Canada.

Marssonella Cushman, Cretaceous–Eocene. The conical arenaceous test has four or five chambers to a whorl in the earliest portion and is biserial in the adult. *M. indentata* Cushman & Jarvis (1, ×25), Cretaceous, Trinidad.

Spiroplectammina Cushman, Pennsylvanian–Recent. The flattened arenaceous test is planispiral in the early portion followed by biserial chambers. *S. laevis* (Roemer) *cretosa* Cushman (3, ×75), Cretaceous, Texas.

Triloculina d'Orbigny, Triassic–Recent. Three elongate chambers are visible on one side, two on the other. *T. raibliana* Gümbel (9, ×60), Triassic, Germany.

Tritaxia Reuss, Cretaceous. The arenaceous test is triserial throughout and triangular in section. *T. jarvisi* Cushman (2, ×20), Cretaceous, Trinidad.

Vaginulina d'Orbigny, Triassic–Recent. One margin of the compressed test is straight, the other convex. *V.* cf. *disparilis* (Terquem) (7, ×15), Jurassic, Montana.

FIG. 2-8. **Representative Cretaceous Foraminifera.** Cretaceous faunas are characterized by a large number of planktonic Foraminifera and by many genera which are restricted to Cretaceous rocks. Some of them are: *Flabellammina, Frankeina, Heterostomella, Arenobulimina, Kyphopyxa, Eouvigerina, Ventilabrella, Stensiöina, Schackiona,* and *Globotruncana.*

Allomorphina Reuss, Upper Cretaceous–Recent. The trochoid calcareous test usually has three chambers to a whorl. *A. trochoides* (Reuss) (6, ×110), Cretaceous, Trinidad.

Globotruncana Cushman, Upper Cretaceous. The calcareous test is trochoid, compressed, and with a keel on margins. *G. canaliculata* (Reuss) (1, ×75), Annona chalk, Texas.

Gümbelitria Cushman, Upper Cretaceous–Eocene. The triserial, calcareous test has globular chambers. *G. cretacea* Cushman (4, ×190), Cretaceous, Texas.

Kyphopyxa Cushman, Upper Cretaceous. Compressed calcareous test with earliest chambers biserial, later uniserial and inverted V-shaped and with raised sutures. *K. christneri* (Carsey) (3, ×30), Selma chalk, Mississippi.

Siphogenerinoides Cushman, Upper Cretaceous–Paleocene. The elongate test is biserial in the earliest portion, later uniserial. *S. parva* Cushman (2, ×35), Cretaceous, Venezuela.

Stensiöina Brotzen, Cretaceous. The test is trochoid, flattened dorsally and convex ventrally. *S. excolata* (Cushman) (5, ×70), Cretaceous, Mexico.

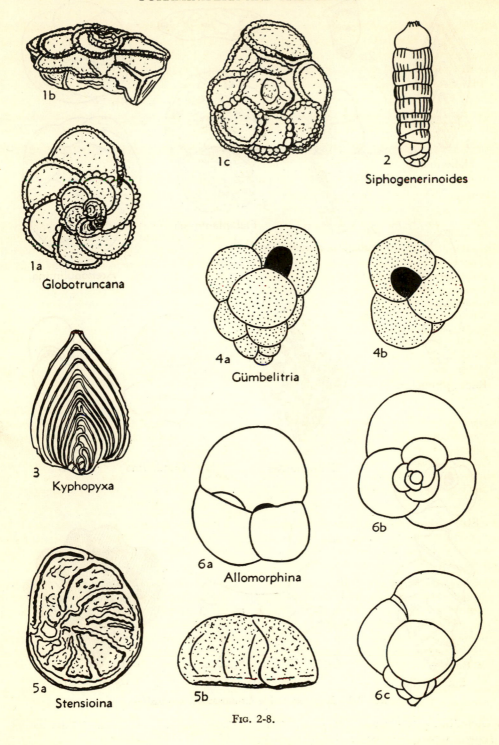

1b

1c

2
Siphogenerinoides

1a
Globotruncana

4a

4b
Gümbelitria

3
Kyphopyxa

6b

5a
Stensioina

6a
Allomorphina

5b

6c

Fig. 2-8.

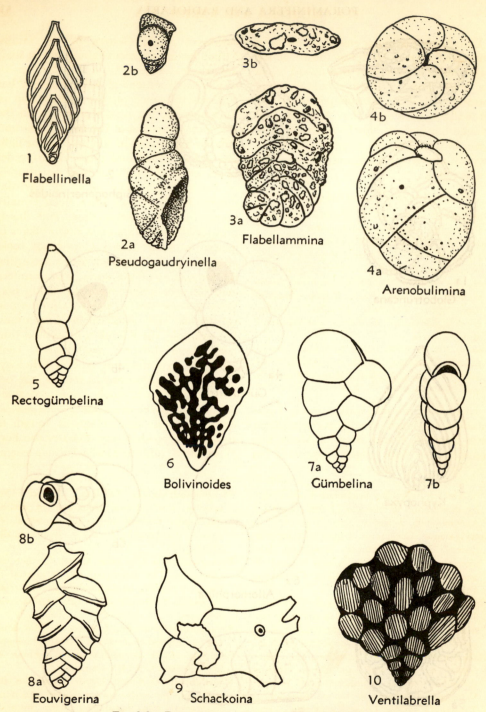

FIG. 2-9. **Representative Cretaceous Foraminifera.**

Arenobulimina Cushman, Cretaceous. The spirally arranged arenaceous test is triserial in early portion, later portion with more than three chambers. *A. americana* Cushman (4, ×65), Cretaceous, Arkansas.

Bolivinoides Cushman, Upper Cretaceous. The

52

Distribution of living genera seems to be controlled largely by temperature of the water, which decreases downward from sea level. Four distinctive faunal zones, controlled largely by temperature, are recognized both on the Atlantic and Pacific borders of North America. The zones of the Atlantic slope and some of their characteristic genera of Foraminifera are as follows:

Temperature-controlled Foraminiferal Zones

Zone *1*, 0 to 5 m., 0 to 27°C.; abundant *Elphidium, Rotalia, Quinqueloculina,* and *Eggerella.*

Zone *2*, 15 to 90 m., 3 to 16°C.; *Cibicides, Proteonina, Elphidium, Guttulina, Eponides, Bulimina, Quinqueloculina,* and *Triloculina* are common genera.

Zone *3*, 90 to 300 m., 9 to 13°C.; *Gaudryina, Pseudoclavulina, Massilina, Pyrgo, Robulus, Marginulina, Nonion, Nonionella, Virgulina, Gyroidina, Discorbis, Eponides, Epistomina, Cassidulinoides,* and *Textularia.*

Zone *4*, 300 to 1,000 m., 5 to 8°C.; *Listerella, Bulimina, Gyroidina, Nonion, Angulogerina, Uvigerina, Cassidulina, Bolivina, Valvulina, Karreriella,* and *Pseudoglandulina.*

In shallow tropical waters, many Foraminifera are large, having average diameters between 5 and 20 mm. Many live commensally with algae, especially in coral-reef areas. Genera of the family Miliolidae are especially abundant in shallow tropical waters.

The environment of sedimentation has a direct influence on the distribution of Foraminifera, the type of material on the sea floor being one of the most important factors. Variations in the amount of light and the chemical and physical properties of sea water also influence the distribution of Foraminifera. The effect of environmental features must have been equally strong in past geologic time, and this is illustrated by variations in assemblages of genera found in Lower Cretaceous rocks of northern Texas, which differ according to the nature of the rocks. Foraminifera generally are absent from the sandy portions of these strata but are abundant in calcareous and argillaceous formations. For example, the Kiamichi formation contains three different assemblages, each restricted to a different type of lithology. The *Virgulina-Trochammina* assemblage occurs only in dark, finely laminated shale of the Red River area; the *Flabellammina-Spiroplectammina* assemblage is found in brown to gray shale and marl of the Fort Worth area; and the *Lituola-Palmula* assem-

---

(*Fig. 2-9 continued.*)

biserial test has surface ornamented with irregular depressions. *B. decorata* (Jones) (6, ×55), Annona chalk, Texas.

Eouvigerina Cushman, Upper Cretaceous–Eocene. Earliest chambers planispiral, later biserial, and last ones irregularly triserial. *E. americana* Cushman (8, ×125), Cretaceous, Texas.

Flabellammina Cushman, Jurassic–Cretaceous. The arenaceous test is coiled in the early portion followed by uniserial chambers. *F. saratogaensis* Cushman (3, ×30), Cretaceous, Arkansas.

Flabellinella Schubert, Cretaceous. The earliest portion is coiled with one margin straight; the adult portion is uniserial with inverted V-shaped chambers. *F. delicata* Loeblich & Tappan (1, ×35), Lower Cretaceous, Texas.

Gümbelina Egger, Cretaceous–Oligocene. The calcareous biserial test has globular chambers. *G. globulosa* (Ehrenberg) (7, ×155), Cretaceous, Tennessee.

Pseudogaudryinella Cushman, Upper Cretaceous. Triserial and triangular in earliest stage, later biserial, and finally uniserial. *P. capitosa* (Cushman) (2, ×15), Selma chalk, Mississippi.

Rectogümbelina Cushman, Upper Cretaceous–Oligocene. The calcareous test is biserial in the early portion, later chambers are uniserial. *R. cretacea* Cushman (5, ×125), Cretaceous, Arkansas.

Schackoina Thalmann, Cretaceous. The adult test has inflated chambers in a single plane and tubular spines. *S. multispinata* (Cushman & Wickenden) (9, ×145), Canada.

Ventilabrella Cushman, Upper Cretaceous. Biserial in the early portion, later becoming multiserial and fan-shaped with multiple apertures. *V. eggeri* Cushman (10, ×55), Taylor marl, Texas.

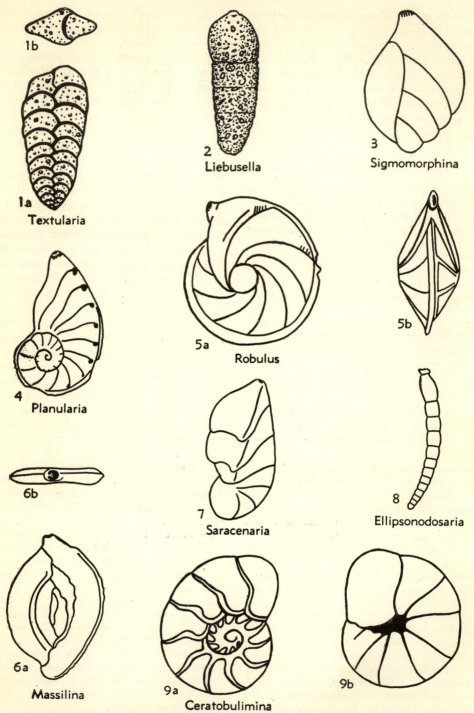

1b

1a
Textularia

2
Liebusella

3
Sigmomorphina

4
Planularia

5a
Robulus

5b

6b

7
Saracenaria

8
Ellipsonodosaria

6a
Massilina

9a
Ceratobulimina

9b

FIG. 2-10. **Representative Eocene Foraminifera.** Eocene faunas are marked by an increase in the number of modern genera. *Hantkenina, Cribrohantkenina,* and *Coleites* are restricted to Eocene rocks. (*Continued on page 56.*)

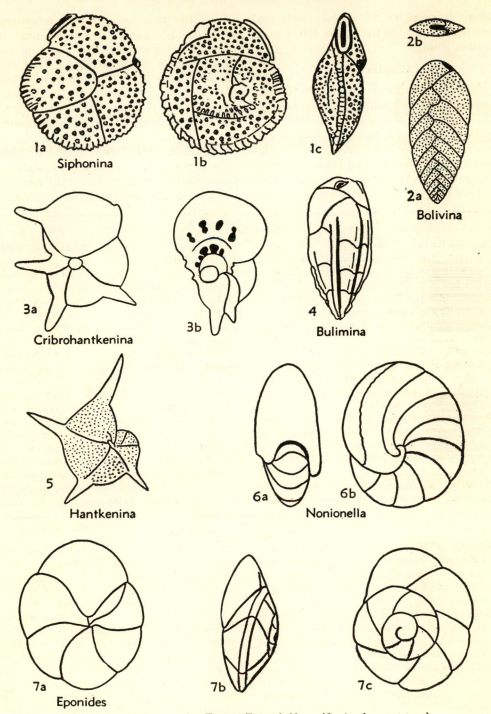

1a Siphonina
1b
1c
2b
2a Bolivina
3a Cribrohantkenina
3b
4 Bulimina
5 Hantkenina
6a 6b Nonionella
7a Eponides
7b
7c

FIG. 2-11. **Representative Eocene Foraminifera.** (*Continued on next page.*)

blage is restricted to yellow marls and white limestones of the Brazos River area, farther south.

Classification

The Foraminifera have been grouped in approximately 50 families on the basis of (1) phylogeny of the order, as shown by the fossil record; (2) ontogeny of individual genera, as shown by microspheric and megalospheric forms; and (3) morphology of the shell and physiology of living forms.

It is believed that chitinous forms constitute the most primitive Foraminifera, and that arenaceous types are next in order of advancement, for many of the latter have a chitinous lining. The calcareous forms are judged to have developed from those with arenaceous shells by gradual increase in the amount of calcareous cement, gradual loss of agglutinated particles, and increase of secreted calcareous particles. This is well shown by *Cymbalopora* which has arenaceous early chambers consisting of fragments held together by calcareous cement and deposited on a

---

(Fig. 2-10 continued from page 54.)

Ceratobulimina Toula, Upper Cretaceous–Recent. The close-coiled trochoid test has a polished wall and a ventral aperture. *C. eximia* (Rzehak) (9, ×90), Eocene. Texas.

Ellipsonodosaria A. Silvestri, Cretaceous–Tertiary. The uniserial, elongate, rounded test has a subelliptical aperture. *E. cocoaensis* (Cushman) (8, ×25), Eocene, Alabama.

Liebusella Cushman, Eocene–Recent. Four or five chambers to a whorl in the earliest stage and a uniserial adult portion. *L. byramensis* (Cushman) var. *turgida* Cushman (2, ×25), Eocene, South Carolina.

Massilina Schlumberger, Cretaceous–Recent. The elongate chambers are in a single plane in adult stage, margin sharp or rounded. *M. pratti* Cushman & Ellisor (6, ×60), Eocene, Texas.

Planularia d'Orbigny, Cretaceous–Recent. The flattened coiled test tends to become uncoiled

in the adult. *P. ouachitaensis* Wallace (4, ×35), Eocene, Alabama.

Robulus Montfort, Jurassic–Recent. The involute symmetrical test has a radiate aperture at the margin. *R. arcuatostriatus* (Hantken) *carolinianus*, Cushman (5, ×20), Eocene, Alabama.

Saracenaria Defrance, Jurassic–Recent. Early chambers close-coiled, later uncoiled and triangular in section. *S. arcuata* (d'Orbigny) *hantkeni* Cushman (7, ×50), Eocene, South Carolina.

Sigmomorphina Cushman & Ozawa, Cretaceous–Recent. The chambers of the test added between angles of 144 and 180 deg. and a radiate aperture. *S. vaughani* Cushman & Ozawa (3, ×30), Eocene, South Carolina.

Textularia Defrance, Pennsylvanian–Recent. The elongate, compressed arenaceous test is biserial throughout. *T. hockleyensis* Cushman & Applin (1, ×15), Eocene, Texas.

(Fig. 2-11 continued from preceding page.)

Bolivina d'Orbigny, Cretaceous–Recent. The compressed calcareous test is biserial throughout with aperture at base of chamber. *B. jacksonensis* Cushman & Applin (2, ×40), Eocene, Texas.

Bulimina d'Orbigny, Jurassic–Recent. The calcareous test is triserial, and the aperture is loop-shaped. *B. jacksonensis* Cushman (4, ×50), Eocene, Alabama.

Cribrohantkenina Thalmann, Eocene. Close-coiled test has acicular spine on most chambers and a multiple aperture. *C. mccordi* (Howe & Wallace) (3, ×40), Cocoa sand, Alabama.

Eponides Montfort, Jurassic–Recent. The trochoid test is usually biconvex with the aperture

at base of chamber. *E. minima* Cushman (7, ×150), Eocene, South Carolina.

Hantkenina Cushman, Eocene–lowest Oligocene. A prominent spine occurs at the anterior angle of each chamber; a single aperture. *H. alabamensis* Cushman (5, ×40), Eocene, Alabama.

Nonionella Cushman, Cretaceous–Recent. The involute test is unsymmetrical; aperture at base of the chamber. *N. cockfieldensis* Cushman & Ellisor (6, ×100), Eocene, Texas.

Siphonina Reuss, Cretaceous–Recent. The biconvex trochoid test has an aperture at end of a short neck on the margin. *S. jacksonensis* Cushman & Applin (1, ×100), Eocene, Mississippi.

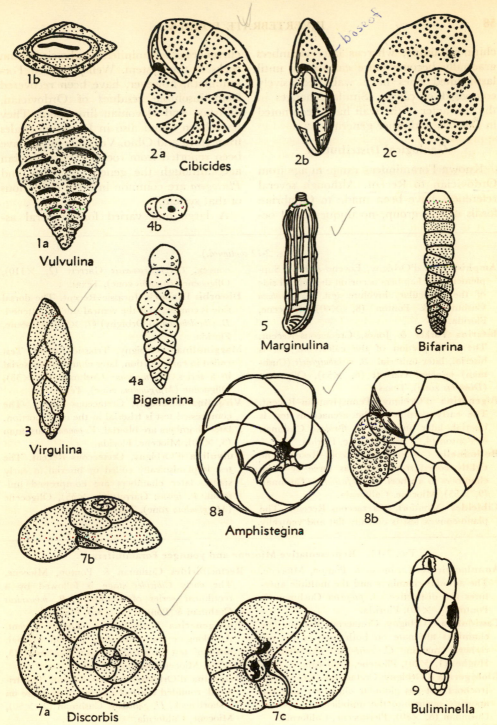

Fig. 2-12. **Representative Oligocene and Miocene Foraminifera.** Oligocene Foraminifera (1, 2, 5, 6) are similar to those of Eocene age. Distinctive faunas occur in areas, such as the Gulf Coast, which has the faunas of the Anahuac formation (Texas), the *Discorbis*, *Heterostegina*, and *Marginulina* zones. Several genera are restricted to Miocene rocks (3, 4, 7–9), among them *Rectocibicides*, *Annulocibicides*, *Eponidella*, and *Virgulinella*. The percentage of living species in Miocene rocks is less than 60 per cent. (*Continued on next page.*)

57

chitinous lining, whereas later chambers gradually become more calcareous, until last-formed ones have walls exclusively composed of calcite. Similar changes in the character of the wall have been noted in *Dorothia* and other genera.

Geological Distribution

Known Foraminifera range in age from Ordovician to Recent. Although several references have been made to Cambrian fossils of this group, no unquestioned oc-

currences of foraminifers can be cited from rocks of this system. Well-preserved Foraminifera, however, have been recovered from insoluble residues of Ordovician, Silurian, and Devonian limestones. They have been found also in Devonian shales from Iowa and Ohio. Very few forms have been reported from rocks of Mississippian age, although the genera *Endothyra* and *Plectogyra* are common in some formations of that system.

A large and varied foraminiferal as-

(Fig. 2-12 continued.)

Amphistegina d'Orbigny, Eocene–Recent. Supplementary chambers occur on the ventral side of the lenticular, involute test. *A. floridana* Cushman & Ponton (8, ×40), Miocene, Florida.

Bifarina Parker & Jones, Cretaceous–Recent. The early portion of the calcareous test is biserial, later uniserial. *B. vicksburgensis* (Cushman) *monsouri* Garrett (6, ×55), Oligocene (*Discorbis* zone), Texas.

Bigenerina d'Orbigny, Pennsylvanian–Recent. The early portion of the arenaceous test is biserial, later uniserial. *B. floridana* Cushman & Ponton (4, ×30), Miocene, Florida.

Buliminella Cushman, Cretaceous–Recent. The tightly coiled spiral test has three or more chambers to a whorl. *B. subfusiformis* Cushman (9, ×55), Miocene, California.

Cibicides Montfort, Cretaceous–Recent. The plano-convex test is dorsally flat and ventrally

convex. *C. jeffersonensis* Garrett (2, ×110), Oligocene (*Discorbis* zone), Texas.

Discorbis Lamarck, Jurassic–Recent. The dorsal side is convex, and the ventral side is flattened. *D. vilardeboana* (d'Orbigny) (7, ×90), Miocene, Florida.

Marginulina d'Orbigny, Triassic–Recent. Test coiled in early portion, later chambers uniserial in a series. *M. mexicana* Cushman (5, ×35), Oligocene (*Marginulina* zone), Texas.

Virgulina d'Orbigny, Cretaceous–Recent. The compressed test is triserial in the early portion, later chambers are biserial. *V. pontoni* Cushman (3, ×80), Miocene, Florida.

Vulvulina d'Orbigny, Cretaceous–Recent. The test is planispirally coiled or biserial in early stages, later chambers are compressed uniserial. *V. ignava* Garrett (1, ×35), Oligocene (*Marginulina* zone), Texas.

FIG. 2-13.   **Representative Miocene and younger Foraminifera.**

Annulocibicides Cushman & Ponton, Miocene. The annular chambers and the multiple apertures are distinctive. *A. projectus* Cushman & Ponton (7, ×40), Florida.

Cassidulina d'Orbigny, Cretaceous–Recent. The chambers alternate on both sides of the peripheral margin. *C. californica* Cushman & Hughes (1, ×20), Pliocene, California.

Globigerina d'Orbigny, Cretaceous–Recent. The trochoid test has globular chambers, and the aperture opens into the umbilicus. *G. apertura* Cushman (6, ×80), Pleistocene, California.

Gyroidina d'Orbigny, Cretaceous–Recent. The trochoid test is convex ventrally and flattened dorsally. *G. soldanii* d'Orbigny (8, ×65), Recent, Atlantic Ocean.

Rectocibicides Cushman & Ponton, Miocene. The early *Cibicides* stage is followed by a rectilinear series of chambers. *R. miocenicus* Cushman & Ponton (5, ×35), Florida.

Siphogenerina Schlumberger, Eocene–Recent. An elongate test with early portion triserial, most of test uniserial. *S. reedi* Cushman (3, ×35), Miocene, California.

Uvigerina d'Orbigny, Eocene–Recent. The triserial, rounded test has a terminal aperture on a short neck. *U. gallowayi* Cushman (2, ×50), Miocene, California.

Valvulineria Cushman, Cretaceous–Recent. Trochoid test close-coiled, and commonly with a plate over umbilicus. *V. californica* Cushman (4, ×40), Miocene, California.

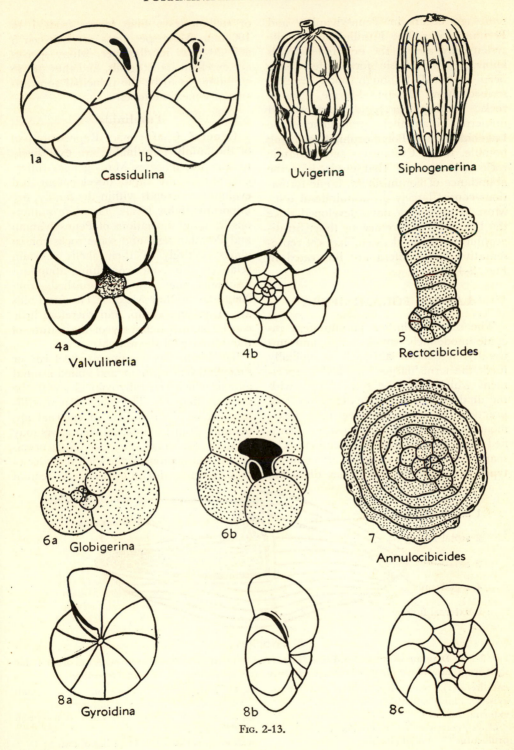

1a 1b
Cassidulina

2
Uvigerina

3
Siphogenerina

4a
Valvulineria

4b

5
Rectocibicides

6a Globigerina

6b

7
Annulocibicides

8a Gyroidina

8b

8c

FIG. 2-13.

semblage occurs in Pennsylvanian and Permian rocks. The Fusulinidae of those systems are among the best index fossils known. Few Triassic foraminifers have been collected, although some are described from Europe. Marine Jurassic rocks, on the other hand, have yielded numerous Foraminifera, prominence of the Lagenidae and Polymorphinidae being notable. Many Cretaceous and Cenozoic rocks are especially characterized by the abundance of Foraminifera. Some formations consist largely of foraminiferal tests. Most living genera have developed since the beginning of Cretaceous time. Stratigraphically grouped assemblages of representative Foraminifera are illustrated in Figs. 2-4 to 2-13.

LARGER FORAMINIFERA

The so-called larger Foraminifera include moderately small, large, and unusually large forms, although as originally used, the term "larger" applied to specimens which could be seen readily with the unaided eye. A better definition from practical viewpoints embraces those specimens, large or small, which can be identified only by means of thin sections. The shape of the test varies, although common types have fusiform, lenticular, discoidal,

or stellate tests. Size ranges from 1 to 100 mm. Most specimens are between 5 and 20 mm. in diameter. Many species are excellent index fossils, and they are so abundant as to be the principal constituents of some limestones.

Fusulinids

General Characters. Representatives of the family Fusulinidae are exclusively marine invertebrates. Many genera, differentiated by the rapid development and evolutionary trends within the family, are important index fossils. They are numerous in many formations of Pennsylvanian and Permian age, and some rocks consist almost entirely of their shells. Certain relatively thin zones containing abundant fusulinids may be distinguished from Nebraska to New Mexico. The fusulinids are an extinct group, and therefore little can be known definitely of the nature of the soft parts of the animal.

The shells are commonly fusiform or subcylindrical in shape and coiled around an axis which generally coincides with the greatest diameter. The shells of many different genera are similar in external appearance, but the internal characters may be quite dissimilar. Therefore, it is necessary to cut sections of the shell in order to make correct identifications. A section

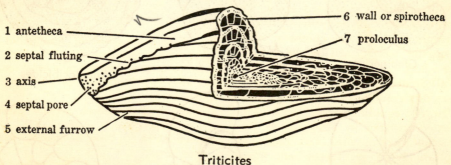

1 antetheca
2 septal fluting
3 axis
4 septal pore
5 external furrow
6 wall or spirotheca
7 proloculus

Triticites

Fig. 2-14. **Diagram of a fusulinid test (Triticites) showing structural features.** A quadrant of the shell is cut away along planes of sagittal and axial sections so as to show internal structures.

antetheca (1). Front wall of last-formied volution.
axis (3). Line about which test is coiled.
external furrow (5). Axial depression on exterior of test, corresponding in position to a septum.
proloculus (7). Initial chamber.

septal fluting (2). Corrugations of a septum which decrease in intensity toward top.
septal pore (4). Minute perforation of a septum or the antetheca.
wall or spirotheca (6). Covering of test.

parallel to the axis of coiling is termed an **axial section,** and one at right angles to the axis is a **sagittal section.** A section parallel to the axis of coiling but not through the initial chamber is a **tangential section.**

The minute subspherical initial chamber (proloculus) is at the center of the shell (Fig. 2-14, 7). It varies in diameter from 20 to 750 microns. The outer wall (**spirotheca,** Fig. 2-14, 6), coils about the proloculus in an expanding spiral. Elongate thin plates parallel to the axis, called **septa,** subdivide the volutions into elongate, narrow chambers which extend from one end to the other. The shell increases in size by secretion at the outer margin of the shell, forming new chambers. The front wall of the last-formed chamber (**antetheca,** Fig. 2-14, 1) becomes a septum when an additional chamber is added. The external surface of the shell is divided by shallow closely spaced meridional grooves (**external furrows,** Fig. 2-14, 5) which mark the tops of the septa.

Since the last-formed septum commonly lacks an aperture, communication with the exterior of the shell is through numerous small openings called **septal pores** (Fig. 2-14, 4). A centrally located low basal slitlike opening, the **tunnel,** penetrates all septa except those of the last volution. Probably it was formed by resorption of parts of previously deposited septa. The tunnel is bordered on each side by secondary ridges of calcite, the **chomata,** which rest on the floor formed by the next interior whorl.

Wall Structure. The wall is composed of very tiny granular calcite crystals which are nearly equidimensional and firmly cemented together by clear calcite. The wall structure of the shell is complex. Two important types are recognized, fusinellid and schwagerinid, named from genera in which each type is well developed.

The **fusulinellid** type of wall consists of four layers in the inner volutions and two or three layers in the outermost volutions. The external thin dark layer is the **tectum** (Fig. 2-15, 3, 6). Bounding the tectum on

the inside is a transparent layer, the **diaphanotheca** (Fig. 2-15, 4). The outer and inner layers are the tectoria, the outer one the **upper tectorium,** and the inner the **lower tectorium** (Fig. 2-15, 2, 5). The tectoria are somewhat thinner than the diaphanotheca and are intermediate in appearance between the tectum and the diaphanotheca (Fig. 2-16). The tectoria are secondary in origin and are deposited as a lining on all sides of the chambers of the inner volutions.

The last-formed volution consists of only tectum and diaphanotheca, but chambers of the early part of this volution are lined

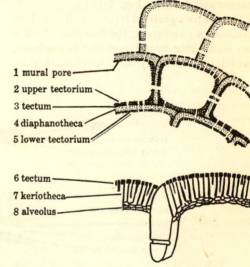

1 mural pore
2 upper tectorium
3 tectum
4 diaphanotheca
5 lower tectorium

6 tectum
7 keriotheca
8 alveolus

Fig. 2-15. **Comparison of fusulinellid and schwagerinid wall structures.** The upper diagram shows outer parts of a fusulinellid test, and the lower represents part of a schwagerinid spirotheca.

alveolus (8). Deep prismatic or cylindrical cavity in keriotheca.
diaphanotheca (4). Clear, transparent, relatively thick layer.
keriotheca (7). Inner, relatively thick layer which has deep prismatic or cylindrical alveoli.
lower tectorium (5). Secondary dark layer on the inside of wall of inner volutions.
mural pore (1). Minute perforation of a wall.
tectum (3, 6). Thin dark layer of spirotheca.
upper tectorium (2). Secondary layer on outside of wall of inner volutions.

with tectoria. The wall in this position consists of three layers: tectum, diaphanotheca, and lower tectorium. In earlier volutions there are four layers: upper tectorium, tectum, diaphanotheca, and lower tectorium. All layers are pierced by tiny pores known as **mural pores** (Fig. 2-15, *7*).

The fusulinellid type of wall structure, which is primitive, characterizes all genera of Early and Middle Pennsylvanian time (Morrowan–Desmoinesian, Fig. 2-19) with certain modifications. In *Profusulinella*, the diaphanotheca is absent, and accordingly, only three layers are found in the wall of early volutions. In the genera *Wedekindellina*, *Fusulinella*, and *Fusulina*, the four layers are distinct (Fig. 2-16).

The **schwagerinid** type of wall structure consists of two layers: the **tectum**, which is an outer thin layer; and the **keriotheca,** which is a relatively thick layer marked by transverse dark lines. These dark lines are the walls of **alveoli,** which are deep cell-like cylindrical or prismatic cavities or pits (Fig. 2-15, *8*). Both the tectum and keriotheca are pierced by tiny mural pores.

The schwagerinid type of wall is characteristic of several Late Pennsylvanian (Missourian and Virgilian) and Permian genera, common ones being *Triticites*, *Schwagerina*, *Pseudoschwagerina*, *Paraschwagerina*, *Parafusulina*, and *Polydiexodina* (Figs. 2-16, 2-17, 2-19, 2-20).

Septa. The septa are partitions between chambers of the fusulinid shell. In primitive genera they have a plane surface, but in more advanced forms the septa become wavy or fluted along the lower margin (Fig. 2-17). The fluting of two adjacent septa is opposed, that is, backwardly directed folds in one correspond in position to forwardly directed folds in the

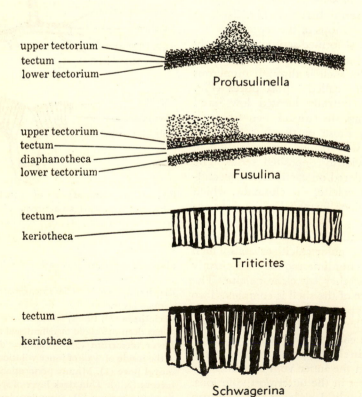

upper tectorium
tectum
lower tectorium

Profusulinella

upper tectorium
tectum
diaphanotheca
lower tectorium

Fusulina

tectum
keriotheca

Triticites

tectum
keriotheca

Schwagerina

FIG. 2-16.   **The wall structure of Profusulinella, Fusulina, Triticites, and Schwagerina.**

other. The tips of opposed fluted septa overlap in *Parafusulina* and some other genera, so as to form low passageways between chambers. These are called **cuniculi** (Fig. 2-17).

In early genera, such as *Fusulinella* and *Triticites*, the septa are fluted near the poles and in the equatorial portion. In later genera, as *Fusulina* and *Schwagerina*,

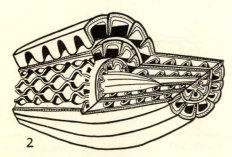

1

Fusulinella

2

Triticites

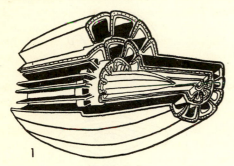

3

Parafusulina

FIG. 2-17. **Diagrammatic sections of Fusulinella, Triticites, and Parafusulina.** The drawings show characters of the septa which are simple in *Fusulinella*, and more complex in *Triticites* and *Parafusulina*.

the septa are highly fluted throughout the shell. In axial sections the septal fluting appears as a series of loops.

Ecology. The environments in which the fusulinids lived is interpreted from the study of the rocks in which they are found. As indicated especially by their occurrence in deposits of cyclic nature, formed during advance and retreat of shallow seas on the continental platform, fusulinids chiefly lived in a clear-water marine environment far from shore. They characterize rock strata formed during maximum marine invasions, being most abundant in shallow-water deposits in far offshore locations. They are not present in brackish-water near-shore sediments or in evaporites.

Evolutionary Trends. Changes in the structural characters of the fusulinid shell were very rapid during Pennsylvanian and Permian time. Similar evolutionary trends occurred at approximately the same time throughout widely separated parts of the world, which makes these fossils particularly useful in correlation.

Evolutionary trends among fusulinids include change of shape, increase of size, complication of shell-wall structure, fluting of septa, and development of special structures such as chomata, septula, axial fillings, and parachomata in certain groups.

The shape of the shell changed from discoidal to spherical, spindle-shaped, or subcylindrical (Fig. 2-18). The earliest-known genus (*Millerella*) has a discoidal form, the shortest diameter coinciding with the axis of coiling. In more advanced genera, the greatest diameter coincides with the axis of coiling.

The shell increased in size from a fraction of a millimeter in late Mississippian time (*Millerella*) to as much as 60 mm. in late Permian time in the genera *Polydiexodina* and *Parafusulina* (Fig. 2-20, 2, 3).

The trend in the evolution of the wall was the development of a thicker primary wall and a reduction in the number of wall layers. The primary wall of the earliest genus *Millerella* is a single thin dense layer, the tectum. This was covered above and below by secondary layers, the tectoria. This three-layered wall is characteristic also of *Pseudostaffella* and *Profusulinella* and developed into the four-layered wall of *Fusulinella* and *Fusulina*, which exhibit typical fusulinellid wall structure, by the addition of the diaphanotheca below the tectum (Fig. 2-16). In some advanced species of *Fusulina*, the diaphanotheca is relatively thick, and it contains pores and alveoli, as in *Triticites*. The two-layered wall, representing the schwagerinid wall structure, probably developed from *Fusulinella*. It consists of an outer tectum and a thick keriotheca containing alveoli. The keriotheca increased in thickness, and the alveoli became coarser in *Schwagerina* and later genera, as compared with the earlier *Triticites*. In *Verbeekina*, the wall consists of a single dense layer (Fig. 2-21).

The septa and antetheca of primitive genera are plane and unfluted throughout the shell. They are also unfluted in some later genera, such as *Verbeekina* and *Ozawaina*. The development of septal fluting occurred at different rates in the different genera, but was quite rapid in *Fusulina*, *Schwagerina*, *Parafusulina*, and *Polydiexodina* (Fig. 2-17).

The chomata of the earliest fusulinids are small, but they became more massive in *Fusulinella* and *Triticites*, then less massive in genera such as *Fusulina* and *Schwagerina*.

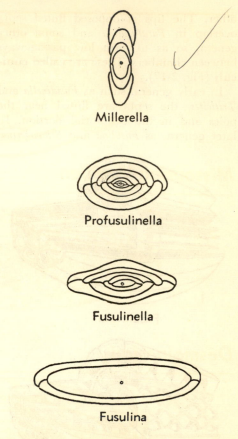

Millerella

Profusulinella

Fusulinella

Fusulina

FIG. 2-18. **Development of shell shape in the subfamily Fusulininae.** The primitive genus *Millerella* has a discoid form, whereas more advanced genera illustrated exhibit trend toward an elongate spindle-shaped outline.

FIG. 2-19. **Representative Pennsylvanian fusulinids.** Most Pennsylvanian fusulinids have the fusulinellid type of wall structure. Common genera include *Fusulinella*, *Fusulina*, *Wedekindellina*, and *Triticites*. Forms here illustrated characterize divisions of the Pennsylvanian succession as indicated.

Fusulina Fischer, late Middle Pennsylvanian (Desmoinesian). The fusiform test has the fusulinellid-type wall and deeply fluted septa throughout the test. *F. acme* Dunbar & Henbest (2, ×15), Illinois.

Fusulinella Möller, early Middle Pennsylvanian (upper Atokan and lower Desmoinesian). The elongate test has the fusulinellid-type wall, plane septa except near poles, a narrow tunnel, and massive chomata. *F. fugax* Thompson (3, ×25), New Mexico.

Millerella Thompson, Upper Mississippian–Upper Pennsylvanian. The very small discoidal test has a short axis, thin wall, and narrow tunnel. *M. marblensis* Thompson (4, ×55), Pennsylvanian, Arkansas.

Triticites Girty, Upper Pennsylvanian (Missourian and Virgilian) and Lower Permian (Wolfcampian). The fusiform test has a schwagerinid-type wall, fluted septa in the middle and plane septa in the outer portion. *T. cullomensis* Dunbar & Condra (1, ×10), Virgilian, Kansas.

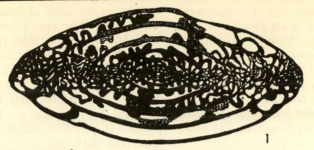

Missourian-Virgilian   1
                                    Triticites

Desmoinesian   2
                                    Fusulina

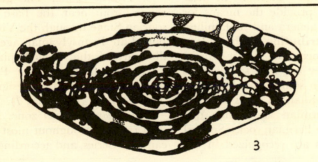

Atokan   3
                                    Fusulinella

                                    4
                                    Millerella

Morrowan

FIG. 2-19.

Some giant fusulinids which have plane septa exhibit another sort of evolutionary development. This is an increase in the complexity of the septula, which are ridges extending downward from the lower surface of the wall so as to subdivide the chambers partially. The septula are short, broad, and transverse in *Cancellina* but develop into complex transverse and axial partitions in *Yabeina*, *Sumatrina*, and other genera (Fig. 2-20, *1*).

Most subfamilies of the Fusulinidae exhibit deposits of dense calcite in the axial portions of the shells. Among late members of the Fusulininae and Schwagerininae, development of heavy axial fillings occurs as the chomata are reduced.

Classification. The Fusulinidae have been subdivided into several subfamilies and approximately 50 genera. The characters which are most useful in the classification of the group include the structure of the wall, size and shape of the shell, degree of fluting of the septa, and nature of the tunnel and chomata. Variations in these characters mainly provide the means for the recognition of genera and subfamilies.

Geological Distribution. The oldest-known fusulinids occur in Upper Mississippian rocks. They are very small forms but are common in some beds. Fusulinid shells are most abundant in rocks of Pennsylvanian and Permian age. They disappear in very late Permian rocks. Several stratigraphic zones are recognized in the midcontinent region of the United States mainly on the basis of fusulinids. In upward order, these are defined as follows:

1. *Millerella* zone. The primitive fusulinid *Millerella* ranges from Chesteran (Upper Mississippian) to Virgilian (Upper Pennsylvanian). It is most common in Morrowan rocks, of Early Pennsylvanian age, and this portion of the Pennsylvanian System is therefore called the *Millerella* zone (Fig. 2-19, *4*).

2. *Profusulinella* zone. Lower Atokan rocks, of early Middle Pennsylvanian age, are characterized by *Profusulinella*, which is confined to this part of the section.

3. *Fusulinella* zone. The genus *Fusulinella* is restricted to the upper part of the Atokan and the lower part of the Desmoinesian Stages of Middle Pennsylvanian age (Fig. 2-19, *3*).

4. *Fusulina* zone. Rocks of the Desmoinesian Stage, of Middle Pennsylvanian age, are referred to the *Fusulina* zone (Fig. 2-19, *2*).

5. *Triticites* zone. The genus *Triticites* ranges from near the base of the Missourian Stage through the Virgilian Stage, both Late Pennsylvanian age, to the middle part of the Wolfcampian Series, of Early Permian age. The Pennsylvanian part of this section is designated as the *Triticites* zone (Fig. 2-19, *1*).

6. *Pseudoschwagerina* zone. *Pseudoschwagerina* ranges throughout most of the Wolfcampian Series, and accordingly the Lower

---

FIG. 2-20. **Representative Permian fusulinids.** Most Permian fusulinids have the schwagerinid-type wall structure and fluted septa throughout. Common genera include *Schwagerina*, *Pseudoschwagerina*, *Parafusulina*, and *Polydiexodina*. The genera illustrated define Permian zones.

Parafusulina Dunbar & Skinner, Permian (Leonardian and Guadalupian). The elongate test has a schwagerinid-type wall and overlapping opposed folds of the fluted septa forming cuniculi. *P. bösei* Dunbar & Skinner (3, ×10), Guadalupian (Word formation), Texas.

Polydiexodina Dunbar & Skinner, Upper Permian (Upper Guadalupian). A very elongate tightly coiled test with accessory paired tunnels on opposite sides of the center. *P.*

capitanensis Dunbar & Skinner (2, ×7), Capitan limestone, Texas.

Pseudoschwagerina Dunbar & Skinner, Lower Permian (Wolfcampian). The subspherical test has closely coiled inner volutions followed by widely spaced volutions. *P. uddeni* (Beede & Kniker) (4, ×7), Wolfcampian, Texas.

Sumatrina Volz, Upper Permian, Orient. The elongate test has a thin wall composed of a single layer; septa are plane, and several septula occur between each pair of septa. *S. longissima* Deprat (1, ×10), Indo-China.

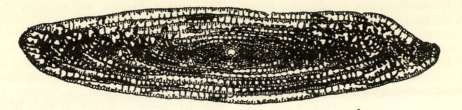

Upper Permian

1

Sumatrina

Guadalupian

2

Polydiexodina

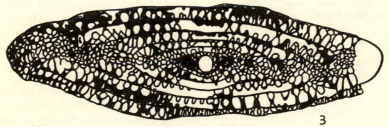

Leonardian

3

Parafusulina

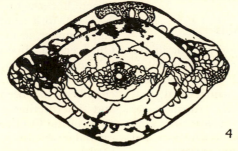

Wolfcampian

4

Pseudoschwagerina

FIG. 2-20.

Permian is called the *Pseudoschwagerina* zone (Fig. 2-20, *4*).

7. *Parafusulina* zone. *Parafusulina* ranges from the base of the Leonardian Series to the middle portion of the Guadalupian Series of the Permian System (Fig. 2-20, *3*).

8. *Polydiexodina* zone. The upper part of the Guadalupian Series contains abundant specimens of *Polydiexodina*, and accordingly these Permian rocks are assigned to the *Polydiexodina* zone (Fig. 2-20, *2*).

9. Other zones. The *Verbeekina-Neoschwagerina* and *Yabeina* zones are recognized in eastern Asia (Fig. 2-21). The *Yabeina* zone is stratigraphically the highest of the fusulinid zones.

Verbeekina

Fig. 2-21. **Verbeekina, a Permian fusulinid.** This genus has a spheroidal test, plane septa, and wall consisting of tectum and a thin keriotheca. The species illustrated is *V. heimi* Thompson & Foster (×8), from China.

Camerinids

Foraminifera of the family Camerinidae (by suspension of rules called Nummulitidae) are characterized by a generally planispiral bilaterally symmetrical test, which is involute in early growth stages and commonly evolute in later stages. A secondary skeleton and complex canal system are developed in some genera. Important representatives of the family include *Camerina* (by suspension, *Nummulites*),

Paleocene to Oligocene (Fig. 2-22, *4*), *Operculinoides*, Tertiary (Fig. 2-22, *3*), and *Heterostegina*, Eocene to Recent (Fig. 2-22, *8*).

Orbitoidids

The test of foraminifers belonging to the family Orbitoididae is large, discoidal, saddle-shaped (selliform), or stellate. Initial chambers are coiled, at least in the microspheric form. In the megalospheric form, the initial chambers may be bilocular or multilocular, and they may or may not be arranged in a coil. They are followed by rhombical, arcuate, spatulate, or hexagonal equatorial chambers which connect with one another by means of tubular openings called **stolons** (Fig. 2-23, *1*, *2*). Layers of lateral chambers occur on both sides of the equatorial chambers in an irregularly overlapping manner or in regular layers. No canal system, such as characterizes camerinid shells, is observed in this family.

The initial chambers are particularly useful in classifying megalospheric orbitoidid shells. The size and position relationships of these chambers show a distinct evolutionary trend in proceeding from older to younger Cenozoic specimens. They vary from a spiral arrangement around the initial chamber in *Lepidocyclina* (*Polylepidina*), middle and upper Eocene (Fig. 2-23, *6a*), to equal bilocular chambers in *L.* (*Lepidocyclina*), upper Eocene to upper Oligocene (Fig. 2-23, *6b*), to a partially embracing second chamber in *L.* (*Nephrolepidina*), upper Eocene to lower Miocene (Fig. 2-23, *6c*), to a completely embracing second chamber, as in *L.* (*Eulepidina*), middle Oligocene to Miocene (Fig. 2-23, *6d*).

The development of the periembryonic chambers, which surround the initial chambers, has been used in classification and interpreting phylogenetic development of the Orbitoididae. Three developmental series are recognized: (1) uniserial, (2) biserial, and (3) quadriserial. The importance of these in classification is seriously questioned, however, since it has

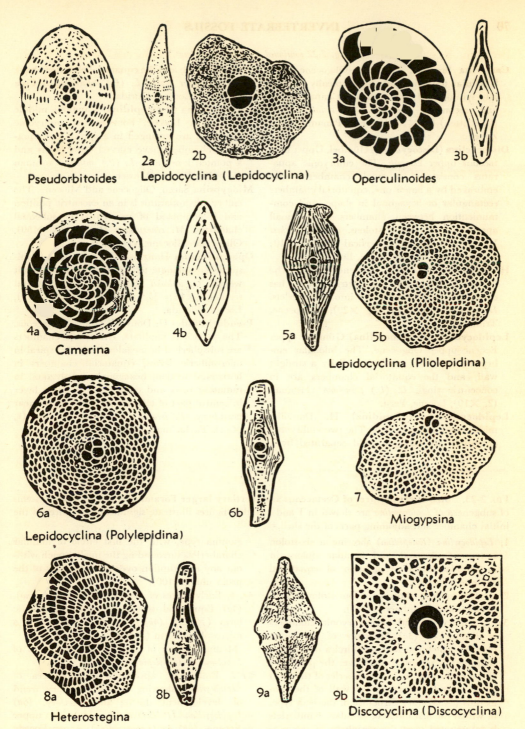

FIG. 2-22. **Representative Cretaceous and Tertiary larger Foraminifera.** Several genera of the larger Foraminifera, including *Orbitoides*, *Lepidorbitoides*, and *Pseudorbitoides*, are restricted to Cretaceous rocks. The most important Eocene genera include *Discocyclina*, *Lepidocyclina*, and *Camerina*. The common genera in Oligocene rocks are *Heterostegina*, *Miogypsina*, *Miogypsinoides*, and *Lepidocyclina*. (*Continued on next page.*)

(*Fig. 2-22 continued from preceding page.*)

Camerina Brugière, Paleocene–Oligocene. The lenticular test is involute and typically bilaterally symmetrical; the wall has a canal system and secondary skeleton. *C. striatoreticulata* (L. Rutten) (4, ×10), Eocene, Panama Canal Zone.

Discocyclina (Discocyclina) Gümbel, Upper Cretaceous–upper Eocene. The embryonic apparatus consists of a small chamber partly embraced by a larger one, equatorial chambers rectangular or hexagonal in shape, and communication between chambers is by small apertures and annular stolons. *D. (D.) barkeri* Vaughan & Cole (9a, vertical section, ×10; 9b, horizontal section, ×25), Eocene, Cuba.

Heterostegina d'Orbigny, Eocene–Recent. The flattened, bilaterally symmetrical test has chambers which are divided into chamberlets. *H. antillea* Cushman (8, ×20), Oligocene, Trinidad.

Lepidocyclina (Lepidocyclina) Gümbel, upper Eocene–upper Oligocene. The bilocular embryonic chambers are separated by a straight wall, and the equatorial chambers are in concentric rings. *L. (L.) peruviana* Cushman (2, ×10), Eocene, Peru.

Lepidocyclina (Pliolepidina) H. Douvillé, middle and upper Eocene. The two embryonic chambers are subequal and separated by a straight wall, and the equatorial chambers have curved outer walls. *L. (P.) cedarkeysensis* Cole (5, ×20), Eocene, Florida.

Lepidocyclina (Polylepidina) Vaughan, middle and upper Eocene. Four to ten embryonic chambers are arranged in a spiral; the equatorial chambers have curved outer walls and a pointed inner end. *L. (P.) antillea* Cushman (6, ×20), Eocene, Mexico.

Miogypsina Sacco, Oligocene and Miocene. The embryonic apparatus is in an excentric position and is composed of two equal or subequal chambers. *M. venezuelana* Hodson (7, ×50), Oligocene–Miocene, Venezuela.

Operculinoides Hanzawa, Tertiary. The planispiral, complanate test is involute in the adult whorls. *O. wilcoxii* (Heilprin) (3a, horizontal section, ×15; 3b, vertical section, ×10), Eocene, Florida.

Pseudorbitoides H. Douvillé, Upper Cretaceous. The surface is papillate, embryonic chambers are subspherical in megalospheric and spiral in microspheric forms, equatorial chambers in horizontal section, hexagonal or polygonal in radiating rows and comprising a single layer in central part of test, two or three layers near periphery. *P. israelski* Vaughan & Cole (1, ×25), Taylor marl, Louisiana.

FIG. 2-23.   **Structural features of Cretaceous and Tertiary larger Foraminifera.** The stolon systems of subgenera of *Lepidocyclina* are shown in 1 and 2. Other figures illustrate significant structures of the initial chamber and adjoining parts of the shell.

1, *Lepidocyclina (Eulepidina)* showing a six-stolon system of diagonal and annular stolons in a Canada-balsam preparation of equatorial chambers (×100).

2, *L. (Pliolepidina)* showing a four-stolon system of equatorial chambers (×100).

3a, c, Embryonic and periembryonic chambers of *Discocyclina*. (3a) Structure of the alpha type, in which the first two circles of annular chamberlets are interrupted over the periphery of the proloculus, and a chamberlet of the third circle extends inward to the wall of the proloculus (×35). (3b) Structure of the beta type, in which the first circle of annular chamberlets is interrupted over the periphery of the proloculus, and one or more chamberlets of the second circle extends inward to the wall of the proloculus (×100). (3c) Structure of the

gamma type, with the first circle of annular chamberlets surrounding the nucleoconch without any interruption over the periphery of the proloculus (×100).

4a, b, Early stages of *Lepidocyclina (Lepidocyclina)*. (4a) Equatorial section of a megalospheric form (×125). (4b) Equatorial section of a microspheric form (×500).

5, Median section of nucleoconch and spire of *Lepidocyclina (Polylepidina)* (×60).

6a–d, Embryonic apparatus of subgenera of *Lepidocyclina*, showing the evolutionary trend of development during the Tertiary. (6a) *Lepidocyclina (Polylepidina)*, middle to upper Eocene. (6b) *L. (Lepidocyclina)*, upper Eocene to upper Oligocene. (6c) *L. (Nephrolepidina)*, upper Eocene to lower Miocene. (6d) *L. (Eulepidina)*, middle Oligocene to Miocene.

1

Lepidocyclina (Eulepidina)

2

Lepidocyclina (Pliolepidina)

3a

α type

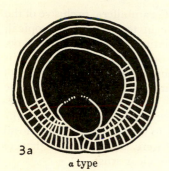

3b   Discocyclina
β type

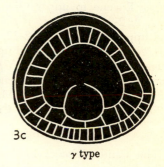

3c

γ type

4a   Lepidocyclina (Lepidocyclina)   4b

5 Lepidocyclina
(Polylepidina)

6a
Lepidocyclina
(Polylepidina)

6b
Lepidocyclina
(Lepidocyclina)

6c
Lepidocyclina
(Nephrolepidina)

6d
Lepidocyclina
(Eulepidina)

FIG. 2-23.

been shown that different specimens of a single species from the same locality may exhibit all three types of structure.

The stolon systems of orbitoidids are interesting, and probably they show some evolutionary trends; but because it is too difficult to make preparations showing them, they seem to have little practical value. The number of stolons in the wall of each chamber increases during phylogenetic development of the family.

Important genera and subgenera in this family include *Pseudorbitoides*, Upper Cretaceous (Fig. 2-22, *1*); *Lepidocyclina ss.*, upper Eocene to upper Oligocene (Figs. 2-22, *2*; 2-23, *4*); *L.* (*Polylepidina*), middle and upper Eocene (Figs. 2-22, *6*; 2-23, *5*); *L.* (*Pliolepidina*), middle and upper Eocene (Figs. 2-22, *5*; 2-23, *2*); *L.* (*Nephrolepidina*), upper Eocene to lower Miocene, *L.* (*Eulepidina*), middle Oligocene to Miocene (Fig. 2-23, *1*); and *Lepidorbitoides*, Cretaceous.

Discocyclinids

The family Discocyclinidae is characterized by rectangular to faintly hexagonal equatorial chambers and an intraseptal and intramural canal system. Members of this group differ from orbitoidids in these characters.

The types of embryonic and periembryonic chambers of the megalospheric generation are important in the classification of the discocyclinids (Fig. 2-23, *3*). The embryonic chambers consist of a subspherical proloculus, which may be partially or completely embraced by an outer chamber, as in *Discocyclina* (*Discocyclina*), or separated by a straight wall as in *D.* (*Asterocyclina*).

Three kinds of stolons—annular, vertical, and radial—occur in the equatorial layer outside the initial chambers, and oblique, tangential, radial, and vertical stolons are found in the lateral chambers.

Important genera of this family are *Discocyclina*, Upper Cretaceous to upper Eocene (Fig. 2-22, *9*); *Pseudophragmina*, Paleocene (?) and Eocene; and *D.* (*Asterocyclina*), middle and upper Eocene.

Miogypsinids

The Miogypsinidae include short-ranged genera which are small and have spiral and interseptal canals, a net of canals in the chamber walls, and generally lozenge-shaped equatorial chambers. Some species of *Miogypsina* develop hexagonal chambers. This genus (Fig. 2-22, *7*), which occurs in Oligocene and Miocene rocks, and *Miogypsinoides*, Oligocene and middle Miocene, are the most important genera in the family.

Ecology

The larger Foraminifera, exclusive of the fusulinids, live in temperate, subtropical, and tropical latitudes, and commonly occur in rocks containing coral-reef deposits and calcareous algae. Living genera of larger Foraminifera live in shallow marine waters of the tropics or subtropics. Therefore, it is reasonable to judge that the environment to which the larger fossil Foraminifera were adapted was that of warm, shallow, clear seas.

RADIOLARIA

The Radiolaria are a large group of marine Rhizopoda. They are commonly pelagic and live mainly in the surface waters. They have a wide distribution, especially in the warmer seas, being most abundant in the central Pacific and Indian oceans. Their shells fall to the bottom, forming a distinctive deposit known as radiolarian ooze, which occurs at depths ranging from 2,300 to 4,500 fathoms. Radiolaria are so abundant in some formations of Devonian, Jurassic, and Tertiary ages that they are the principal constituents of rock layers.

The Radiolaria are distinguished in having a shell in the form of a perforated membranous **central capsule** (*7;* italic numbers in the text describing radiolarians refer to Fig. 2-24) and generally, in addition, **a skeleton of silica or of strontium sulfate.**

Skeletal Structures

The skeleton of the Radiolaria, which is mainly external, is commonly formed of silica, but in one group it is made up of strontium sulfate. The membranous central capsule is the only skeleton in some genera. The siliceous skeleton typically consists of an intricate latticework having a globular, discoid, stellate, or conical shape, bearing spines (9) or lacking them. Some have several layers of skeletal silica, one inside the other (1, 3, 5).

Although colonial forms are rare among Radiolaria, they do occur in some genera by repeated divisions of the central capsule.

Physiological Features

The body is commonly spherical but may be bilaterally or radially symmetrical. The central capsule (7), a single or double perforated membrane of mucinoid or pseudochitinous material, lies embedded in the protoplasm, dividing it into two regions, called **extracapsular** (2) and **intracapsular** (6). Reproduction occurs in the intracapsular region. In the extracapsular region, food is assimilated and waste products are excreted. The intracapsular protoplasm contains a large nucleus (4), or in some radiolarians, many nuclei.

Digestive System. The Radiolaria live on microplankton, consisting of diatoms, copepods, and protozoans. This food is collected by stiff, threadlike **pseudopodia** (8), which pass it into the central part of the body where it is digested by food vacuoles. Radiolaria can live in the absence of such food by using the nutrition products of algae (zooxanthellae) living symbiotically with them. The Radiolaria furnish the zooxanthellae with nitrogenous waste matter and carbon dioxide, and in return, the zooxanthellae supply oxygen, fats, and carbohydrates to their host.

Reproduction. The life cycle of radiolarians is not well known. Reproduction is

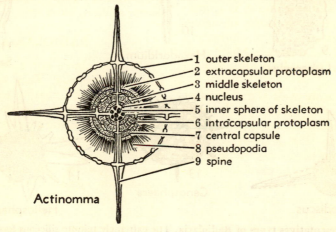

1 outer skeleton
2 extracapsular protoplasm
3 middle skeleton
4 nucleus
5 inner sphere of skeleton
6 intracapsular protoplasm
7 central capsule
8 pseudopodia
9 spine

Actinomma

Fig. 2-24. **Morphology of a radiolarian.** Parts are defined in the alphabetical list, cross-indexed to the figure by numbers.

central capsule (7). Perforated chitinoid internal skeleton.
extracapsular protoplasm (2). Cell materials outside the central capsule.
inner sphere of skeleton (5). Innermost part of test.
intracapsular protoplasm (6). Cell materials inside central capsule.

middle skeleton (3). Part of test between innermost and outermost spheres.
nucleus (4). Rounded mass of protoplasm enclosed in a delicate membrane.
outer skeleton (1). Outermost part of test.
pseudopodia (8). Projections of cell substance in form of delicate radiating threads.
spine (9). Stiff sharp spicule.

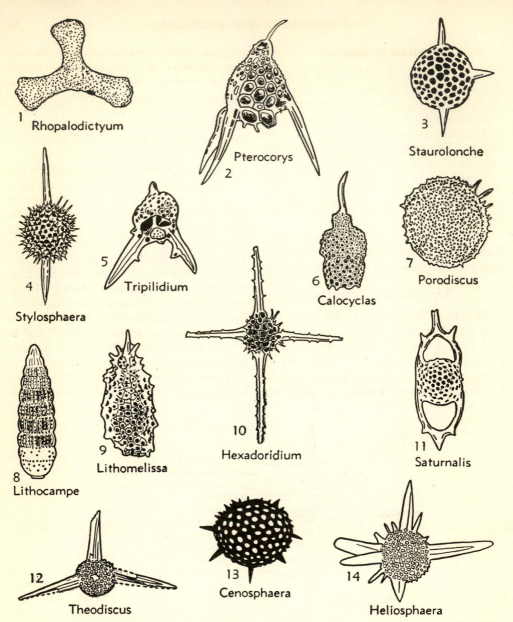

1 Rhopalodictyum

2 Pterocorys

3 Staurolonche

4 Stylosphaera

5 Tripilidium

6 Calocyclas

7 Porodiscus

8 Lithocampe

9 Lithomelissa

10 Hexadoridium

11 Saturnalis

12 Theodiscus

13 Cenosphaera

14 Heliosphaera

FIG. 2-25. **Representatives types of Radiolaria.** The extremely minute siliceous fossils of this group range from Pre-Cambrian to late Cenozoic.

Calocyclas Haeckel, Tertiary–Recent. A simple corona with no ribs in the shell wall. *C. (Calocyclella) semipolita semipolita* Clark & Campbell (6, ×150), Eocene, California.

Cenosphaera Ehrenberg, Paleozoic–Recent. The simple, single lattice sphere has subangular, unequal pores. *C. affinis* Hinde (13, ×200), Devonian, England.

Heliosphaera Haeckel, Paleozoic–Recent. The simple lattice sphere is covered by large and small spines. *H. clavata* Hinde (14, ×160), Devonian, England.

Hexadoridium Haeckel, Mesozoic–Recent. A spongy spherical shell with two latticed medullary shells in its center bears six equal-length spines. *H. magnificum* Campbell & Clark (10, ×75), Cretaceous, California.

Lithocampe Ehrenberg, Mesozoic–Recent. An ovate spindle-shaped shell. *L. (Lithocampanula) andersoni* Campbell & Clark (8, ×85), Cretaceous, California.

by multiple or binary fission or by budding. In multiple fission, the intracapsular protoplasm divides into a swarm of small units, each of which becomes a new radiolarian. Further development is unknown.

Classification

The Radiolaria are classified on composition of the skeleton, variation in perforations of the central capsule, number and arrangement of spines, and pattern of the skeleton. An outline of a classification is as follows:

Main Divisions of Radiolaria

Radiolaria (*order*), marine Sarcodina having a central capsule and emitting fine, radially directed pseudopodia; most kinds have a siliceous skeleton. Pre-Cambrian–Recent.

Actipylea (*suborder*), skeleton composed of strontium sulfate and made up of radiating spicules; central capsule perforated finely and uniformly. Recent.

Peripylea (*suborder*), skeleton siliceous; central capsule perforated finely and uniformly. Lower Paleozoic–Recent.

Monopylea (*suborder*), skeleton siliceous; central capsule bears a porous plate at one pole, forming the base of an inwardly directed cone; not uniformly perforated. Lower Paleozoic–Recent.

Tripylea (*suborder*), skeleton siliceous; central

capsule provided with a main aperture (astropyle) and two accessory apertures (parapyles). Recent.

Geological Distribution and Importance

Radiolarians have been reported from rocks ranging in age from Pre-Cambrian to Recent and are an important constituent of many siliceous rocks, such as quartzites. They are common in Pre-Cambrian quartzites interbedded with gneiss in northwestern France, and in Devonian cherts in Texas and California. Mississippian quartzites in Europe have yielded many species. Radiolaria are abundant in great thicknesses of Jurassic rocks in California. They have also been found in Cretaceous and Tertiary rocks in many parts of the world. Approximately 800 genera of these protozoans, fossil and Recent, have been described (Fig. 2-25).

FLAGELLATES

Some organisms assigned to the class Flagellata combine characteristics of primitive plants and protozoans. For example, they may contain chlorophyll, the green coloring matter of plants. Representatives of this group are relatively unimportant as fossils, although in some deposits their re-

---

(Fig. 2-25 continued.)

Lithomelissa Ehrenberg, Mesozoic–Recent. Three solid spines occur on sides and one or more horns at tip. *L.* (*Sethomelissa*) *armata* Campbell & Clark (9, ×200), Cretaceous, California.

Porodiscus Haeckel, Mesozoic–Recent. A simple circular disk made up of several rings. *P.* (*Trematodiscus*) *charlestonensis* Clark & Campbell (7, ×150), Eocene, California.

Pterocorys Haeckel, Tertiary–Recent. Characterized by three simple, lateral wings on sides. *P.* (*Pterocyrtidium*) *splendens* Campbell & Clark (2, ×150), Miocene, California.

Rhopalodictyum Ehrenberg, Tertiary–Recent. Three spongy arms radiating from central disk. *R.* (*Rhopalodictyum*) *malagaense* Campbell & Clark (1, ×150), Miocene, California.

Saturnalis Haeckel, Mesozoic–Recent. Two concentric lattice spheres and two opposite

spines. *S. lateralis* Campbell & Clark (11, ×125), Cretaceous, California.

Staurolonche Haeckel, Tertiary–Recent. Two concentric lattice spheres and four simple spines. *S.* (*Staurolonchantha*) *aculeata* Campbell & Clark (3, ×150), Miocene, California.

Stylosphaera Ehrenberg, Mesozoic–Recent. Two concentric lattice shells and equal-sized pores. *S.* (*Stylosphaerella*) *datura* Clark & Campbell (4, ×120), Eocene, California.

Theodiscus Haeckel, Paleozoic–Recent. A simple latticed disk, with three marginal radial spines. *T. hastatus* Hinde (12, ×125), Devonian, England.

Tripilidium Haeckel, Tertiary–Recent. Characterized by three simple or branched terminal feet and an apical horn. *T.* (*Tristylocorys*) *clavipes advena* Clark & Campbell (5, ×150), Eocene, California.

mains are very abundant. Three divisions of flagellates call for notice: silicoflagellates, coccoliths, and dinoflagellates.

Silicoflagellates

Silicoflagellates are tiny (0.05 to 0.1 mm. in diameter) marine flagellates, which are widely distributed in plankton of present oceans. They have a siliceous skeleton consisting of a simple system of arcs, spines, and thin rings. They are a division of an order called Chrysomonadina.

Fossil silicoflagellates are reported from Cretaceous and Cenozoic rocks of California and several other places. They are most common in rocks which contain abundant diatoms.

Coccoliths

Another group of protozoans belonging in the order which contains silicoflagellates comprises very minute calcareous planktonic organisms, classed as members of the family Coccolithophoridae. Each animal consists of a spherical cell, which is covered by the microscopic bodies of calcite called coccoliths. These have several forms. Most common are perforate or imperforate disks, star-shaped bodies, and paired disks which have a central tubular connection and bear a single perforation. The diameter of the disks ranges from 0.002 to 0.01 mm.

Coccoliths have been reported from rocks as old as Upper Cambrian. They are especially common in some Cretaceous and Tertiary formations such as chalks and limestones. Locally, they are abundant enough to be termed rock builders.

Dinoflagellates

Dinoflagellates are marine planktonic protozoans, which comprise an order of the Flagellata. Some are naked, but others have a thick wall of cellulose, which may be spiny and divided by furrows into angular plates. From time to time they locally multiply into dense swarms that give surface sea water a brownish hue over many square miles, killing large numbers of fishes, apparently by clogging their gills. Some kinds are responsible for phosphorescence of ocean waters.

Fossil dinoflagellates have been reported from Upper Jurassic rocks of France and Germany, in Upper Cretaceous flints of Germany, and from Tertiary rocks.

TINTINNIDS

The class of protozoans called Ciliata (or Infusoria) contains numerous kinds of cilia-bearing one-celled animals, but only one small group of them is known to be represented by fossils. This is the family Tintinnidae, which includes abundant marine planktonic forms and a few freshwater species. The conical or trumpet-shaped cell, generally less than 0.1 mm. in length, is surrounded by a gelatinous or pseudochitinous protective envelope, termed lorica. The body can be extended outward from its cover, movement being effected by cilia and bristles which surround the mouth. The lorica may incorporate mineral grains, which give it an agglutinated appearance, and in favorable circumstances may be preserved as a fossil.

The known geologic distribution of tintinnids is from Jurassic to Recent. Chief records are from Upper Jurassic and Cretaceous rocks of the Alps, Carpathians, and Spain.

REFERENCES

Foraminifera

BARKER, R. W. (1939) *Species of the foraminiferal family Camerinidae in the Tertiary and Cretaceous of Mexico:* U.S. Nat. Mus. Proc., vol. 86, pp. 305–330, pls. 11–22. Contains descriptions and illustrations of Camerinidae from Mexico.

BRADY, H. B. (1884) *Report on the Foraminifera dredged by H.M.S. Challenger during the years 1873–1876:* Rept. Sci. Results Voyage of H.M.S. Challenger, Zoology, vol. 9, text vol., pp. 1–814, 1 vol., pls. 1–115. A very complete and well-illustrated monograph of Recent Foraminifera of the world; the largest publication on the Foraminifera.

CUSHMAN, J. A. (1923) *The Foraminifera of the Vicksburg group:* U.S. Geol. Survey Prof. Paper 133, pp. 11–71, pls. 1–8. Descriptions and illustrations of Oligocene Foraminifera from Mississippi and Alabama.

——— (1935) *Upper Eocene Foraminifera of the southeastern United States:* Same, Prof. Paper 181, pp. 1–88, pls. 1–23. An excellent, well-illustrated monograph of Eocene Foraminifera.

——— (1946) *Upper Cretaceous Foraminifera of the Gulf Coastal region of the United States and adjacent areas:* Same, Prof. Paper 206, pp. 1–241, pls. 1–66. Includes illustrations, descriptions, and range charts of Upper Cretaceous Foraminifera.

——— (1948) *Foraminifera, their classification and economic use:* Harvard University Press, Cambridge, 4th ed., pp. 1–605, pls. 1–55. An excellent, well-illustrated general text on morphology, classification, and geologic and geographic distribution of Foraminifera.

——— & ELLISOR, A. C. (1945) *The foraminiferal fauna of the Anahuac formation:* Jour. Paleontology, vol. 19, pp. 545–572, pls. 71–78. Contains illustrations and descriptions of Foraminifera from subsurface beds of Oligocene age on the Gulf Coast.

——— & PONTON, G. M. (1932) *Foraminifera of the Upper, Middle, and part of the Lower Miocene of Florida:* Florida Geol. Survey Bull. 9, pp. 1–147, pls. 1–17. Illustrates and describes Foraminifera of the Miocene of Florida.

——— & WATERS, J. A. (1930) *Foraminifera of the Cisco group of Texas:* Texas Univ. Bull. 3019, pp. 22–81, pls. 2–12. A well-illustrated paper on the smaller Foraminifera of the Upper Pennsylvanian.

DUNBAR, C. O., & CONDRA, G. E. (1927) *The Fusulinidae of the Pennsylvanian system in Nebraska:* Nebraska Geol. Survey, ser. 2, Bull. 2, pp. 1–135, pls. 1–15. Includes sections on morphology and classification and illustrations and descriptions of Fusulinidae from the Mid-Continent region.

——— & HENBEST, L. G. (1942) *Pennsylvanian Fusulinidae of Illinois:* Illinois Geol. Survey Bull. 67, pp. 1–218, pls. 1–23. The morphology of the test and methods of study are important parts of this paper, which illustrates and describes Pennsylvanian Fusulinidae from Illinois.

——— & SKINNER, J. W. (1937) *Permian Fusulinidae of Texas:* Texas Univ. Bull. 3701, pp. 517–825, pls. 42–81. A well-illustrated and detailed study of Permian Fusulinidae of Texas.

ELLIS, B. F. & MESSINA, A. R. (1940 *et seq.*) *Catalogue of Foraminifera:* American Museum of Natural History, New York, 45 volumes, looseleaf. A catalogue of foraminiferal species with one or more pages for each species giving (1) copy of original figure, (2) reference to type description and figure, (3) original description, (4) stratigraphic occurrence, (5) type locality, and (6) depository of type specimen.

GALLOWAY, J. J. (1933) *A manual of Foraminifera:* Principia Press, Inc., Bloomington, pp. 1–483, pls. 1–42. A textbook on Foraminifera which includes classification, morphology, geographic and stratigraphic distribution, and illustrations and descriptions of genera (out of print).

GARRETT, J. B., JR. (1939) *Some Middle Tertiary smaller Foraminifera from subsurface beds of Jefferson County, Texas:* Jour. Paleontology, vol. 13, pp. 575–579, pls. 65, 66. Contains descriptions and illustrations of Foraminifera of Oligocene age from the Gulf Coast.

GLAESSNER, M. F. (1947) *Principles of Micropaleontology:* Wiley, New York, pp. 1–296, pls. 1–14, figs. 1–64. A general textbook on micropaleontology which includes a detailed section on Foraminifera and brief sections on other groups such as Ostracoda, Radiolaria, diatoms, and conodonts. Contains an excellent section on collecting and studying microfossils.

IRELAND, H. A. (1939) *Devonian and Silurian Foraminifera from Oklahoma:* Jour. Paleontology, vol. 13, pp. 190–202, figs. A1–36, B1–39. Contains illustrations and figures of Foraminifera from insoluble residues of limestones.

KLEINPELL, R. M. (1938) *Miocene stratigraphy of California:* American Association of Petroleum Geologists, Tulsa, pp. 1–450, pls. 1–22, figs. 1–13. Includes a section on stratigraphic distribution of the Foraminifera and illustrations and descriptions.

MOREMAN, W. L. (1930) *Arenaceous Foraminifera from Ordovician and Silurian limestones of Oklahoma:* Jour. Paleontology, vol. 4, pp. 42–59, pls. 5–7. Arenaceous Foraminifera from insoluble residues are described and illustrated. Describes oldest unquestioned Foraminifera.

MYERS, E. H. (1936) *The life-cycle of Spirillina vivipara Ehrenberg, with notes on morphogenesis, systematics and distribution of the Foraminifera;* Jour. Royal Micr. Soc., vol. 56, pp. 120–146, pls. 1–3, An outstanding study of the life history of living Foraminifera.

NATLAND, M. L. (1933) *The temperature and depth distribution of some Recent and fossil Foraminifera in the southern California region:* Bull. Scripps Inst. Oceanography Tech. Ser., vol. 3, pp.

225–230, figs. Gives the foraminiferal faunas from bottom samples at different depths and temperatures.

ORBIGNY, A. D. D' (1846) *Foraminifères fossiles du Bassin tertiare de Vienne:* Paris, pp. 1–312, pls. 1–21. Contains descriptions and illustrations of Miocene Foraminifera of Austria.

PARKER, F. L. (1948) *Foraminifera of the continental shelf from the Gulf of Maine to Maryland:* Bull. Mus. Comp. Zool., vol. 100, no 2, pp. 213–241, pls. 1–7. Demonstrates that the living Foraminifera collected in bottom samples are in four distinctive faunal zones controlled largely by temperature.

SHERBORN, C. D. (1893, 1896) *An index to the genera and species of the Foraminifera:* Smithsonian Misc. Coll., Publ. 856. A valuable index to the Foraminifera published before 1896.

THOMPSON, M. L. (1948) *Studies of American fusulinids:* Kansas Univ. Paleont. Contr., Protozoa, Art. 1, pp. 1–184, pls. 1–38. This excellent paper contains sections on morphology, classification, distribution and illustrations, and descriptions of Pennsylvanian and Permian genera of the world and fusulinid faunas from the Pennsylvanian of New Mexico.

VAUGHAN, T. W. (1933) *The biogeographic relations of the orbitoid Foraminifera:* Nat. Acad. Sci. Proc., vol. 19, pp. 922–938. Discusses the environment of living larger Foraminifera and speculates on the environment of fossil forms.

——— (1945) *American Paleocene and Eocene larger Foraminifera:* Geol. Soc. America Mem. 9, 1–175, pls. 1–46. Discusses structure of American Discocyclinidae and describes and illustrates larger Foraminifera of Paleocene and Eocene ages in America.

——— & COLE, W. S. (1941) *Preliminary report on the Cretaceous and Tertiary Larger Foraminifera of Trinidad, British West Indies:* Same, Spec. Paper 30, pp. 1–137, pls. 1–46. A well illustrated paper in which the larger Foraminifera of Trinidad are described. Some stolon systems are illustrated.

——— & ——— (1948) *The families Orbitoididae, Discocyclinidae, and Miogypsinidae:* in Cushman, J. A., Foraminifera, their classification and economic use, Harvard University Press,

Cambridge, pp. 347–376. The classification and stratigraphic and geographic distribution are given, and important species are illustrated and described.

Radiolaria

CAMPBELL, A. S., & CLARK, B. L. (1944) *Miocene radiolarian faunas from southern California:* Geol. Soc. America Spec. Paper 51, pp. 1–76, pls. 1–7. Describes and illustrates a well-preserved fauna from California.

——— & ——— (1944a) *Radiolaria from Upper Cretaceous of Middle California:* same, Spec. Paper 57, pp. 1–61, pls. 1–8. A rich Upper Cretaceous fauna is described and figured.

CLARK, B. L. & CAMPBELL, A. S. (1945) *Radiolaria from the Kreyenhagen formation near Los Banos, California:* Geol. Soc. America Mem. 10, pp. 1–66, pls. 1–7. Contains figures and descriptions of Eocene radiolarians.

HAECKEL, E. (1887) *Report on the Radiolaria collected by H.M.S. Challenger during the years 1873–1876:* Rep. Sci. Results Voyage of H.M.S. Challenger, Zoology, vol. 18, pp. 1–1893, pls. 1–140. An excellent monograph on Recent Radiolaria containing a section on morphology.

HINDE, G. J. (1899) *On the Radiolaria in the Devonian rocks of New South Wales:* Geol. Soc. London Quart. Jour., vol. 55, pp. 38–64, pls. 8, 9. Contains descriptions and figures of well-preserved radiolarians.

Flagellates and Tintinnids

COLOM, G. (1948) *Fossil tintinnids: loricated Infusoria of the order of the Oligotricha:* Jour. Paleontology, vol. 22, pp. 233–263, pls. 33–35, text figs. An excellent account of the morphology of the tintinnids. Many Jurassic and Cretaceous species are described and illustrated.

GLAESSNER, M. F. (1947) *Principles of Micropalaeontology:* Wiley, New York, pp. 17–18. Describes some of the flagellates which occur as fossils.

HANNA, G. D. (1931) *Diatoms and silicoflagellates of the Kreyenhagen shale:* California Dept. Nat. Res., Div. Mines, vol. 27, no. 2, pp. 187–201, pls. A–E. Eocene diatoms and silicoflagellates are described and figured.

CHAPTER 3

Sponges and Spongelike Fossils

The sponges are aquatic, dominantly marine invertebrates, which are ranked next above protozoans in classification, because they are composed of many cells having specialized functions but lacking organization into definite tissue. Attainment of a cellular grade of construction distinguishes them from the protozoans, and absence of a tissue grade, found in coelenterates and all higher invertebrates, separates them from these. Sponges now are assigned to a phylum of their own, called Porifera (pore-bearing). Grouped with the sponges in this chapter are assemblages of spongelike fossils which have doubtful classificatory status. They are important and thus deserve study, but because none of them can be placed surely among true sponges, they are treated separately.

General Characters

The form and size of sponges vary exceedingly. Many grow as solitary individuals which have globular, cylindrical, conical, or irregular shape. Their dimensions range from those of a pinhead to a height or width of more than 1 m. Colonial sponges are common; they are mostly irregular branching forms which may attain a diameter of 2 m. Some sponges, both solitary and colonial, comprise thin incrustations on foreign objects such as rock or shell. A few kinds of modern sponges are illustrated in Fig. 3-1.

The sponges have no internal organs, nervous tissue, or circulatory and digestive systems, such as occur in higher invertebrates. There is neither mouth nor anus. Food particles and oxygen are brought to cells of the body interior by water which is introduced through myriads of external apertures, termed **dermal pores.** Outlet of this water is by way of larger openings, called **oscula.** The interior of the sponge consists simply of an open space, the **cloaca** or spongocoel (sponge hollow), or it comprises a branching system of canals. Nearly all sponges possess an internal skeleton of separate or joined elements, which are calcareous, siliceous, or composed of a horny organic substance, called spongin. The individual skeletal elements, termed **spicules,** if not united, are scattered about when soft parts decay after death of the animal. On the other hand, if they are knit firmly together, the form of the sponge is preserved by the skeleton. Destruction of the organic substance of horny sponge skeletons happens readily and quickly, however, and solution may obliterate the calcareous and siliceous types.

All adult sponges live in an attached position. They have no power of locomotion, although some are carried from place to place by other organisms to which they may be fixed. For example, crabs are known to nip off pieces of sponge and hold them on their backs until the sponge becomes anchored. By continued growth, the sponge eventually covers most of the crab, protecting it from molestation, for sponges are avoided by other animals; probably this is explained by the relatively inedible nature of their spicular skeleton and by the offensive odor given off by many.

The great majority of sponges inhabit shallow seas, in tropical, temperate, and arctic waters, but siliceous types are mostly restricted to depths between 500 and 1,000 m., some occurring on the ocean bottom

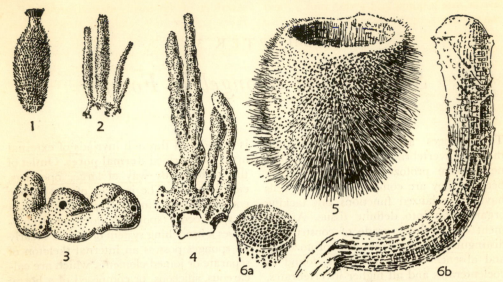

FIG. 3-1. Modern sponges belonging to classes represented by fossils. The sponges grow as solitary individuals or in colonies, some having porous, nearly smooth exterior surfaces and some bearing many projecting spicules, which produce a bristly to hairy appearance. Except for one rather minor family, sponges are exclusively marine. (*After Hyman and Dendy, not to scale.*)

1, 2, Calcareous sponges (class Calcispongea), distinguished by their one- to four-rayed skeletal spicules of calcite. (1) *Sycon*, a solitary vaselike form. (2) *Leucosolenia*, a colonial sponge composed of slender cylindrical individuals.

3, 4, Siliceous sponges having other than six-rayed spicules (class Demospongea). (3) *Chondrilla*, which has tetractinellid spicules.

(4) *Haliclona*, a monactinellid sponge.

5, 6, Siliceous sponges having six-rayed spicules (class Hyalospongea). (5) *Staurocalyptus*, a deep-water, bowl-shaped glass sponge, which has abundant long projecting spicules. (6a, b) *Euplectella*, the "Venus's-flower-basket," showing sievelike enclosure of the osculum and side view of a specimen.

5,000 m. below the surface. One family, containing about 50 species, lives in fresh water, being found in ponds, lakes, and streams. These like the marine forms, have world-wide distribution. The greatest number and variety of sponges commonly are found where firm fastening is provided, as on rocky bottom, but some kinds can fix themselves on muddy sediment by rootlike expansions of the base. They are animals of quiet waters, for if sediment is agitated by waves or currents, it tends to clog their pores. A few are able to live on surf-beaten rocks and attached to oysters within the zone of high and low tides, where they are periodically exposed to air and sunlight. These are decidedly exceptional.

Numerous animals of the sea use sponges as a shelter, living part or nearly all of their existence inside a sponge. This is especially true of marine worms and various arthropods. Some of the siliceous lace sponges, of the type shown in Fig. 3-1, 6a, b, are found to be inhabited by a pair of shrimps, which at their mature size cannot escape through any of the openings in the sponge skeleton; it is reported that such imprisoned shrimps are a desired wedding gift in Japan, symbolizing a marriage that endures until death.

Reproduction

All sponges are reproduced sexually by union of eggs and sperm, produced by specialized cells of the parents. The fertilized egg develops into a free-swimming larva, which settles to the bottom after a few or many hours and grows into a form characteristic of the species.

Virtually all sponges can reproduce or regenerate asexually also. Reproduction consists of the making of a new individual, and among sponges this may be accomplished asexually in two ways: by branching of a parent, as in the development of a colony; or by liberation of buds (called gemmules), which become separated from the parent and eventually (even after freezing and some drying) grow into new sponges. Regeneration is the development of a fully formed individual from a fragment of some previous sponge growth. This is a very general attribute of the Porifera, correlated with their low grade of body organization. A piece of sponge consisting of only one kind of cells cannot regenerate, but if different cell types needed for growth are contained in the fragment, it will proceed to make a whole sponge. As shown by experiment, cells broken apart by pressing a sponge through fine silk cloth are able to reunite and grow into several new sponges. Thus, the lowly invertebrates belonging in this group are tenacious of life, and they tend to be spread wherever environments are suitable.

Fossil Record

The sponges have a paleontological record which is not exceeded in length by any other animals. Several kinds of siliceous sponges occur in Cambrian rocks, including the Lower Cambrian, and spicules are recorded from Pre-Cambrian formations of northwestern France. The siliceous sponges are numerous in many Paleozoic and younger deposits, but calcareous forms are recorded only from Devonian to Recent. They are most abundant in some Jurassic, Cretaceous, and early Tertiary rocks. Horny sponges are not known as fossils.

STRUCTURAL FEATURES

The elements of sponge anatomy about which one needs to know for an understanding of the fossils classed among Porifera are few and relatively simple. They comprise the nature of cells and their arrangement with respect to water passageways in the sponge body, types of skeletal spicules, and organization of the spicules.

Soft Parts

Two main categories of cells in the body of sponges can be discriminated. One of these comprises so-called choanocytes or collar cells, which bear a mobile, whiplike flagellum, guarded by a cylindrical wall— the collar. These cells are identical in nature to some protozoans. In sponges, they serve the dual functions of producing water currents by motion of their flagella and of withdrawing food particles from the water for digestion. The other category of cells lacks digestive functions. It includes especially those which form the outer portions of the body wall, a sort of epiderm. Some of these are specialized for the secretion of skeletal spicules, and a few serve other purposes, such as reproduction. All of this second group of cells are dependent on the collar cells for sustenance.

Three general types of body structure are recognized. The ascon type, represented by the living calcareous sponge, *Leucosolenia*, constitutes the simplest known sort (Fig. 3-2, *1*). It has a saclike form, with collar cells lining the central cavity and epiderm cells forming the exterior. Spicules of calcite strengthen the body wall, which is perforated in many places by pores. Water enters the sponge through these pores, yields its contained food particles to the collar cells, and flows outward through a large rounded vent (osculum) at the top of the sponge. Water movement is slow because vibrations of the flagella belonging to the collar cells are not effectively concentrated.

More complex than ascon sponges are those of the sycon type, represented in simplest form by the genus *Sycetta* (Fig. 3-2, *2*) and in more advanced stage by *Sycon* (Fig. 3-2, *3*). In both, convolution of the walls gives origin to chambers lined with collar cells, and in these chambers a more concentrated flow of water can be effected. *Sycon* differs from *Sycetta* in a partial blocking of the inward folds of the exterior so as to make enclosed vestibules for the inflowing water. They are radial canals, which connect by small passageways with the collar-cell chambers. Outlets from the chambers lead to the central sponge cavity or cloaca and thence to the osculum at the summit of the sponge.

A third sort of sponge, the **leucon type,** is represented by *Leuconia* and other genera in which the collar-cell chambers become subdivided (Fig. 3-2, *4–6*) and the pathways of water movement more circuitous. Stages in complication may be distinguished on the basis of the degree of separation of the collar-cell chambers and the length of the inlet and outlet canals serving them. Excurrent water is collected in channels, which correspond to tributaries of a river in joining together as a

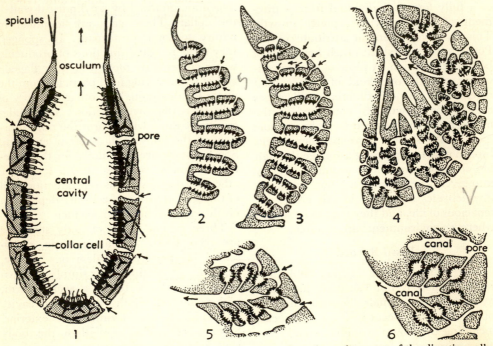

FIG. 3-2. **Types of sponge structures.** Diagrammatic sections show placement of the digestive collar cells (black), which control water movement by means of their whiplike flagella; direction of water currents is indicated by arrows. (*Modified from Hyman.*)

1, Ascon type, the simplest known sort of sponge. It has a vaselike form, in which the outer wall bears scattered spicules and an inner layer of collar cells lines the central cavity. Openings in the wall made by perforated cells admit water which escapes through the vent (osculum) at the top.

2, 3, Sycon type, distinguished by folding of the walls to form separate chambers lined by collar cells. Primitively (2), the infolded parts of the outer wall constitute depressions open to the exterior, but in advanced forms (3), these

depressions are concealed and access to them is by means of dermal pores.

4-6, Leucon type, characterized by subdivision of the collar-cell chambers. Least specialized arrangement (4) is recognized by the nearly confluent relations of the chambers; a more advanced type (5) is distinguished by separation of the chambers, which have very short inlets from the incurrent canals; a most advanced form (6) has lengthened inlets to the chambers, which thus are individually isolated in the sponge wall.

trunk stream leading to the osculum. Sponges of this type lack a widely open central cavity.

The siliceous sponges differ from the calcareous forms of the three types just described in lacking a distinct epiderm and in having an internal fibrous network containing regularly arranged collar-cell chambers of the sycon or leucon type. Some of these sponges, also, have no central cavity or system of branching excurrent channels. Their body has the form of a fanlike expansion standing upright, or of a broad thin-walled bowl, or part of it spreads laterally like an umbrella (Fig. 3-9, 4, 5). An osculum is lacking, since water entering on one side of the sponge departs through many small openings on the other side. The canal system of such types is uncomplicated.

Skeleton

Only calcareous and siliceous hard parts of sponges have paleontological importance, and so attention may be confined to them. Sponges having skeletons of these kinds are very numerous, but no sponge has an internal support consisting of both calcareous and siliceous spicules. It has either one sort or the other. Moreover, the nature of the hard parts is a chief feature in classifying sponges.

Spicules of sponges, whether composed of calcite or silica, are of two general sorts: relatively large ones (megascleres), which constitute the main skeletal framework; and minor ones (microscleres), which are irregularly distributed as accessory skeletal elements in the body. Both types are classified in groups according to the number of their axes, and within each group are many varieties. Examples are illustrated in Fig. 3-3, to which the numbers cited in the following paragraphs on spicules refer unless indicated otherwise.

Monaxons. Calcareous or siliceous spicules which grow in one or both directions along a single axis are termed monaxons. Seemingly, there is not much room for variation in an architectural element of such simplicity, but actually more than two dozen main kinds are separately named. Some are straight (1a–g), others curved (1h–j), and several bear terminal expansions of different sorts (1c, j–k). Monaxons may be distributed in regular or irregular orientation through the sponge walls, and they tend to cluster as projecting bristles around openings in the wall. At the borders of the osculum, they may be exceptionally long, forming a brushlike marginal fringe.

Triaxons. A very important type of spicule in many siliceous sponges but absent in calcareous forms is the triaxon (5a–g). It consists essentially of three axes which cross one another at right angles. Growth along these axes produces a six-rayed spicule, which may be very regular (5b, f), or it may be distinguished by elongation and special modification of some rays (5a, c). A few types are peculiarly specialized (5d, e). Fusion of evenly spaced, uniformly oriented triaxons produces a very regular cribwork (5g), which is a characteristic feature of the skeleton of many siliceous sponges.

Tetraxons. Spicules having four axes not in the same plane, which diverge from a common point, are termed tetraxons. They occur both in calcareous and siliceous sponges. The rays may be more or less equal (2a), but generally one ray is considerably elongated and others are reduced (2c–e). Also, some rays may be suppressed wholly (2f), and it is presumed that the most common sort of calcareous spicule (triradiate type, 2b, 6, 7) originates in this way. Special sorts of double-headed siliceous spicules (amphidisks, 2g, h) are classed among tetraxons.

Polyaxons. Siliceous spicules which have several equal rays diverging from a point are known as polyaxons (3a–c).

Desmas. A special kind of siliceous spicule, characterized by seeming lawlessness of structural plan, is termed desma (4a, b). Divergent main members of the spicule commonly bear spiny or warty excrescences growing in all directions. Actually, the various shapes of desmas are derived from monaxons and tetraxons by irregular growth of silica in lumps and branches. Spicules of this type join together to form a reticulate sort of skeleton, called **lithistid** (8), which distinguishes one rather important group of siliceous sponges.

Organization of Spicules. Whether disconnected or bound firmly together, the spicules of most sponges have rather definite placement. They are not haphazard in arrangement. Some siliceous sponges have only one type of spicule, monaxon or triaxon, but a majority possess two or more types, which are chiefly a combination of monaxons and tetraxons. These facts lack paleontological significance in so far as loose spicules in the sponge body are concerned, for if they occur as fossils, they are scattered and mixed indiscriminately. Spicules which

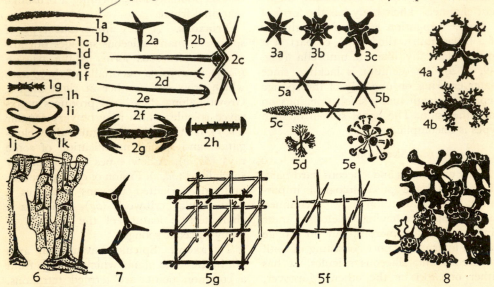

Fig. 3-3. Sponge spicules. The form of calcareous and siliceous skeletal elements is so highly varied that classification is difficult; yet the scores of named kinds can be assembled in a few groups. Examples of each are illustrated (not to scale).

1*a–k*, Monaxons, produced by growth along a single straight or curved axis. They are common in calcisponges and the siliceous demosponges.

2*a–h*, Tetraxons, consisting typically of rays projected in four directions not in the same plane. They are abundant in some of the demosponges and are judged to include the dominant triradiate type of calcisponge spicule, in which one ray is suppressed. A tetraxon type of spicule found in some glass sponges (hyalosponges) is shown in 2*g*.

3*a–c*, Polyaxons, which occur in some demosponges.

4*a, b*, Desmas, very irregular siliceous spicules derived from monaxons and tetraxons. They characterize types of demosponges called lithistids.

5*a–g*, Triaxons, consisting typically of three axes

of growth crossing one another at right angles. (5*a–e*) Individual spicules, partly of aberrant form. (5*f*) Six-rayed triaxons which are individually distinct but arranged in a regular lattice. (5*g*) Latticework formed by fused spicules arranged as in 5*f*. Triaxon spicules characterize the hyalosponges and are restricted to them.

6, Section of part of the wall of *Grantia*, a calcareous sponge, showing arrangement of spicules, exterior of sponge at top. (*After Dendy.*)

7, Adjoined positions of triradiate tetraxons of a calcareous sponge; they may be cemented to form a firm skeleton (compare 5*f, g*).

8, Part of skeleton of a lithistid sponge, consisting of fused desmas.

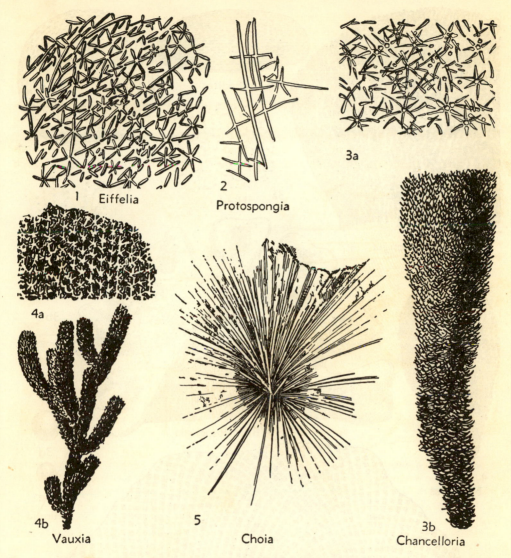

FIG. 3-4. **Cambrian siliceous sponges.** Except *Choia*, in which the spicules consist of radiating long monaxons, indicating classification among the Demospongea, the fossils illustrated are hexactinellid types, belonging to the Hyalospongea.

Chancelloria Walcott, Middle to Upper Cambrian. Elongate tubular sponge having coarse six-rayed spicules. *C. eros* Walcott (3a, spicules, ×6; 3b, side view, ×1), Middle Cambrian (Burgess), British Columbia.

Choia Walcott, Cambrian. *C. carteri* Walcott (5, ×2), Middle Cambrian (Burgess), British Columbia.

Eiffielia Walcott, Middle Cambrian. Subglobular sponge with spicules forming a close meshwork.

E. globosa Walcott (1, spicules, ×3), Burgess, British Columbia.

Protospongia Salter, Lower to Middle Cambrian. Globular sponge with long-rayed spicules. *P. fenestrata* Salter (2, spicules, ×3), Middle Cambrian (Burgess), British Columbia.

Vauxia Walcott, Middle Cambrian. Slender branching sponge, thin-walled. *V. gracilenta* Walcott (4a, part showing spicules, ×6; 4b, colony, ×2), Burgess, British Columbia.

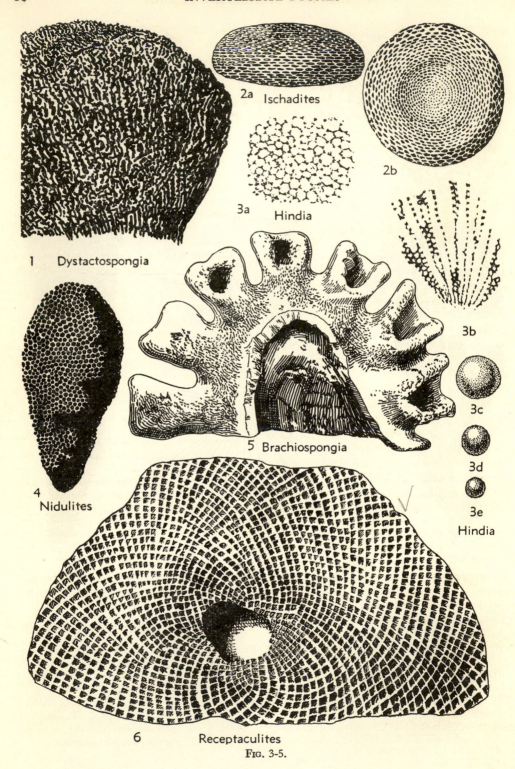

1 Dystactospongia

2a Ischadites

2b

3a Hindia

3b

3c

3d

3e

Hindia

4 Nidulites

5 Brachiospongia

6 Receptaculites

Fig. 3-5.

unite to make a firm skeleton are quite another matter, for they preserve a record of body shape and commonly show distribution of dermal pores, oscula, canals of various sorts, and internal chambers. These are identifiable in terms of species and genera, which may be stratigraphically useful when found as fossils. Paleontological interest in sponges, therefore, is largely conditioned by the extent to which skeletal elements are organized together so as to yield information on form and structure of the whole organism.

CLASSIFICATION

Division of the sponges into natural major and minor groups is by no means easy. Likewise difficult is an answer to the questions of the origin of the recognized main types. Virtual identity of the collar cells to types of protozoans included in the class Flagellata, some of which attain a colonial mode of growth hardly separable from primitive sponges, like *Proterospongia*, strongly indicates that the sponges are descendants from protozoan ancestors. One may ask, however, whether contrasted kinds of sponges having calcareous, siliceous, or organic skeletal structures are branches derived from a single primordial kind of sponge, or independently developed. Classification now generally adopted by zoologists is fairly well suited to paleontological needs because it is largely based on characters of hard parts. This is outlined in the following tabulation. Not incorporated in the phylum Porifera, but added as a sort of appendix, are categories of spongelike organisms which have doubtful taxonomic placement.

Divisions of Sponges and Spongelike Organisms

Porifera (*phylum*), solitary or colonial animals having cellular grade of construction, with body which bears many pores, canals, and chambers, but no mouth or internal organs; calcareous, siliceous, or horny skeletal support generally present. Pre-Cambrian–Recent.

Calcispongea (*class*), skeleton composed of discrete or united calcareous spicules. Devonian–Recent.

Homocoelida (*order*), structure of ascon type. Recent.

Heterocoelida (*order*), structure of sycon or leucon type. Devonian–Recent.

Hyalospongea (*class*), skeleton composed of discrete or united siliceous spicules of triaxon type; the glass sponges. Cambrian–Recent.

Hexasterophorida (*order*), lack double-headed (amphidisk) spicules. Cambrian–Recent.

Amphidiscophora (*order*), have amphidisks. Recent.

Demospongea (*class*), skeleton of siliceous or horny spicules, or both; no triaxons. Pre-Cambrian–Recent.

Monaxonida (*order*), spicules consist of siliceous monaxons, with or without horny fibers. Cambrian–Recent.

FIG. 3-5. **Representative Ordovician sponges.** Among fossils here illustrated, *Dystactospongia* and *Hindia* are siliceous sponges belonging among the Demospongea, but classificatory position of others is uncertain.

Brachiospongia Marsh, Ordovician. A siliceous sponge characterized by the lobate lateral projections. *B. digitata* (Owen) (5, approximately half of a specimen, ×1), Middle Ordovician, Kentucky.

Dystactospongia Ulrich, Ordovician. A massive siliceous sponge with dense radiating structure. *D. minor* Ulrich & Everett (1, part of specimen, ×1), Middle Ordovician, Tennessee.

Hindia Duncan, Ordovician–Mississippian. A lithistid siliceous sponge of spherical form having radiate structure. *H. parva* Ulrich (3a, b, transverse and longitudinal sections, enlarged; 3c, specimens, ×1), Middle Ordovician, Minnesota.

Ischadites Lonsdale, Ordovician–Devonian. Skeleton calcareous. *I. iowensis* (Owen) (2a, b, side and top views, ×1), Middle Ordovician, Minnesota.

Nidulites Salter, Ordovician. Calcareous skeleton with small hexagonal surface pits. *N. pyriformis* Bassler (4, side view, ×1), Middle Ordovician, Pennsylvania.

Receptaculites de Blainville, Ordovician–Devonian. Broad discoid calcareous fossil resembling center of sunflower. *R. oweni* Hall (6, incomplete mold, ×1), Middle Ordovician, Minnesota.

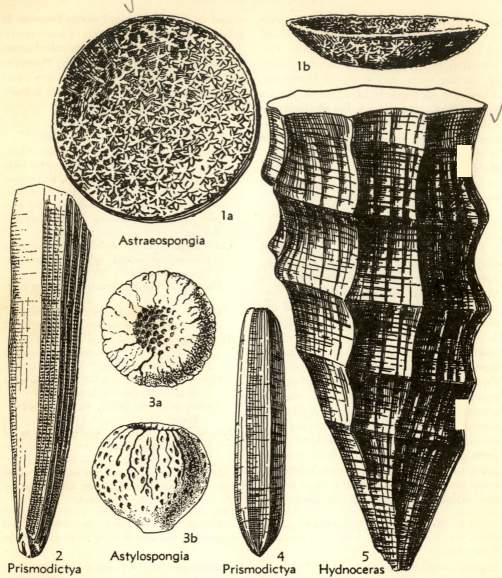

Fig. 3-6. Representative Silurian and Devonian sponges. The Silurian forms (1, 3) are common Niagaran fossils; the others are Upper Devonian glass sponges. All are siliceous.

Astraeospongia Roemer, Silurian. This sponge, of uncertain placement in classification, is distinguished by its low saucer-like form and prominent stellate spicules. *A. meniscus* (Roemer) (1*a*, *b*, top and side views, ×1), Niagaran, Tennessee.

Astylospongia Roemer, ?Ordovician, Silurian. A globular lithistid type of demosponge. *A. praemorsa* (Roemer) (3*a*, *b*, top and side views, ×1), Niagaran, Tennessee.

Hydnoceras Conrad, Devonian–Mississippian. An eight-sided conical hexactinellid sponge having regularly disposed nodose elevations. *H. tuberosum* Conrad (5, side view, ×1), Upper Devonian (Chemung), New York.

Prismodictya Hall & Clarke, Devonian–Mississippian. Eight-sided, like *Hydroceras*, but slender pillar-like forms, lacking nodes. *P. prismatica* (Hall) (2, ×1); *P. telum* (Hall) (4, ×1); both Upper Devonian (Chemung), New York.

Tetraxonida (*order*), spicules consist of siliceous tetraxons or desmas, or spicules lacking. Pre-Cambrian–Recent.

Ceratosida (*order*), spicules horny, none of siliceous type. Recent.

Spongelike organisms, classification doubtful.

Pleospongea (*class*), calcareous single- or double-walled, saucer-shaped to cylindrical porous fossils, which commonly bear internal radial, horizontal, and vesicular structures. Divided into 5 subclasses and 11 orders. Lower and Middle Cambrian.

Receptaculitida (*class*), calcareous subglobular or saucer-shaped fossils consisting of regular, closely spaced plates or pillars which expand at their outer extremities to form an even pavement; interior hollow. Ordovician–Devonian.

Nidulitida (*class*), calcareous hollow ovoid fossils with outer wall composed of small, regular honeycomb cells which are open toward the exterior. Middle Ordovician.

CALCAREOUS SPONGES

Fossils belonging to the Calcispongea are of two readily distinguished types: thin-walled chambered sponges having relatively simple, short canals for admission of water to the chambers; and thick-walled sponges characterized by comparatively long, very irregular canals. Both are included in the Heterocoelida, which have sycon or leucon arrangements of collar-cell chambers.

The first group, classed in the suborder Sycones, is illustrated by the genera *Girtyocoelia, Girtycoelia, Maeandrostia, Amblysiphonella,* and *Cystauletes* (Fig. 3-8, *1–4, 6, 9*), of Pennsylvanian age, and *Barroisia* (Fig. 3-9, *1*), found in Cretaceous rocks. These fossils consist of a cylindrical column of superposed chambers, with or without a central cloacal passageway leading to an osculum at the summit. In some genera, the chambers are globular and the connection between them is so narrowed that the sponge resembles a short string of round beads. Others are nearly smooth-sided. *Barroisia* differs from the Pennsylvanian genera in its colonial mode of growth, having many closely spaced

branches. The Sycones range from Pennsylvanian to Recent.

The thick-walled second group of calcareous sponges belongs in the suborder Pharetrones. They include solitary and colonial types of growth, which generally are larger and rougher externally than the

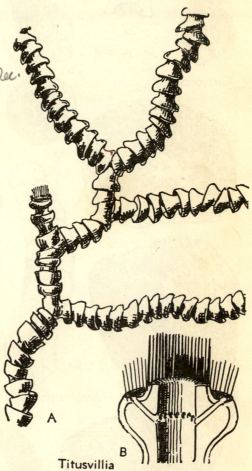

Titusvillia

Fig. 3-7. **An early Mississippian colonial glass sponge.** This species, called *Titusvillia drakei* Caster, consists of numerous varyingly oriented cuplike nodes distributed in series along branches. The skeleton is composed of three meshworks of hexactinellid spicules, among which the innermost lines a cloacal cavity leading to outlets (oscula) at tips of branches. (*A*) Part of colony, ×0.5. (*B*) Restoration showing a terminal node in section, ×4. Cussewago beds, northwestern Pennsylvania.

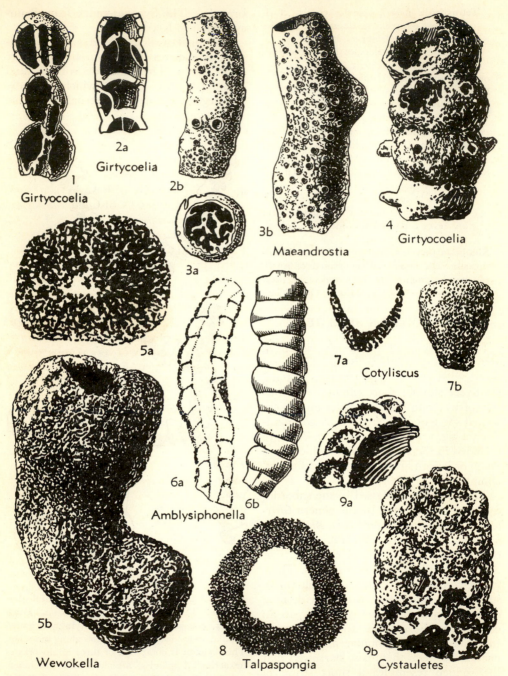

1 Girtyocoelia

2a Girtycoelia

2b

3a

3b Maeandrostia

4 Girtyocoelia

5a

6a

6b Amblysiphonella

7a

Cotyliscus

7b

9a

5b

Wewokella

8 Talpaspongia

9b Cystauletes

FIG. 3-8. **Representative Mississippian, Pennsylvanian, and Permian sponges.** All are classed among Calcispongea.

Amblysiphonella Steinmann, Pennsylvanian. A series of chambers penetrated by an axial tube; walls bear pores. *A. prosseri* Clarke (6a, b, longitudinal section and side view, ×1),

Upper Pennsylvanian (Virgilian), Nebraska. **Cotyliscus** R. H. King, Mississippian. A small, cup-shaped sponge with walls penetrated by short meandering canals. *C. ewersi* King

Sycones. Some kinds have a clearly differentiated central cavity, whereas others do not. Illustrated examples of pharetronid sponges include *Cotyliscus* (Fig. 3-8, *7*), from Mississippian rocks; *Wewokella* (Fig. 3-8, *5*), from the Pennsylvanian; *Talpaspongia* (Fig. 3-8, *8*), of Permian age; and *Stellispongia* and *Corynella* (Fig. 3-9, *2, 3*), from the Jurassic. Some of these sponges are very useful guide fossils. The group of Pharetrones is best developed in Mesozoic parts of the column, but it ranges from Devonian to Recent.

SILICEOUS SPONGES

The classes Hyalospongea and Demospongea are represented by very numerous fossils, distinguished by the siliceous nature of the skeleton. Both groups have a very long geologic range, for the first extends from Cambrian to Recent and the second is represented in Pre-Cambrian rocks, as well as younger formations down to those being made at the present day. Sponges characterized by discrete, unconnected spicules occur in each of the two classes, but these are unimportant paleontologically in comparison with those having skeletons which show the form of the whole sponge.

Representatives of the Hyalospongea, or glass sponges, from Cambrian rocks include *Eiffelia*, *Protospongia*, *Chancelloria*, and *Vauxia* (Fig. 3-4, *1–4*), which are mostly thin-walled tubular sponges growing as solitary individuals or colonial branching forms. Their six-rayed spicules distinguish them. An unusual sponge which has a wall of moderate thickness is the Middle Ordovician *Brachiospongia* (Fig. 3-5, *5*). It spreads out laterally with projecting blunt fingers on all sides. A distinctive Silurian sponge, *Astraeospongia* (Fig. 3-6, *1*), of saucer-shaped outline, is doubtfully classed among the Hyalospongea, because its six-rayed spicules are not disposed along axes crossing one another at right angles but form a regular star with button-like prominences on either side at the center. Some Devonian rocks are especially noted for their content of glass sponges which have a skeletal framework of very regular quadrate meshes. *Prismodictya* and *Hydnoceras* (Fig. 3-6, *2, 4, 5*) are examples, some of which are smooth-sided and some marked by nodose elevations. A colonial hyalosponge of early Mississippian age is *Titus-*

(Fig. 3-8 continued.)

(*7a, b*, section and side view, ×5), Lower Mississippian, Texas.

Cystauletes R. H. King, Pennsylvanian. Porous-walled chambers surround a large enclosed axial cavity. *C. mammilosus* King (*9a*, broken fragment showing chambers and wall of inner cavity; *9b*, exterior, ×2), Desmoinesian, Oklahoma.

Girtycoelia R. H. King, Pennsylvanian. Cylindrical branches with coarsely porous walls, internally composed of superposed chambers. *G. typica* King (*2a, b*, longitudinal section and side view, ×1), Upper Pennsylvanian (Missourian), Texas.

Girtyocoelia Cossmann, Pennsylvanian–Permian. Succession of spheroidal chambers which bear spoutlike incurrent openings; perforated by an axial cavity similar to that of *Amblysiphonella*. (Formerly called *Heterocoelia* Girty, 1908, but this is an invalid junior homonym. *Girtycoelia*

and *Girtyocoelia* are distinct, although the names are nearly identical.) *G. beedei* (Girty) (*1*, section, ×3), Upper Pennsylvanian (Missourian), Kansas. *G. dunbari* King (*4*, side view, ×2), Permian (Leonardian), Texas.

Maeandrostia Girty, Pennsylvanian. Thick-walled branching stem, bearing a large axial cavity. *M. kansasensis* Girty (*3a, b*, transverse section and side view, ×1), Upper Pennsylvanian (Missourian), Texas.

Talpaspongia R. H. King, Lower Permian. Cylindrical sponge with fine irregular canals penetrating the wall. *T. clavata* King (*8*, transverse section of this useful guide fossil, ×2), Wichita group, Texas.

Wewokella Girty, Pennsylvanian. Irregular cylindrical, thick-walled sponge with coarse meandering canals. *W. contorta* King (*5a, b*, transverse section and side view, ×2), Upper Pennsylvanian (Missourian), Texas.

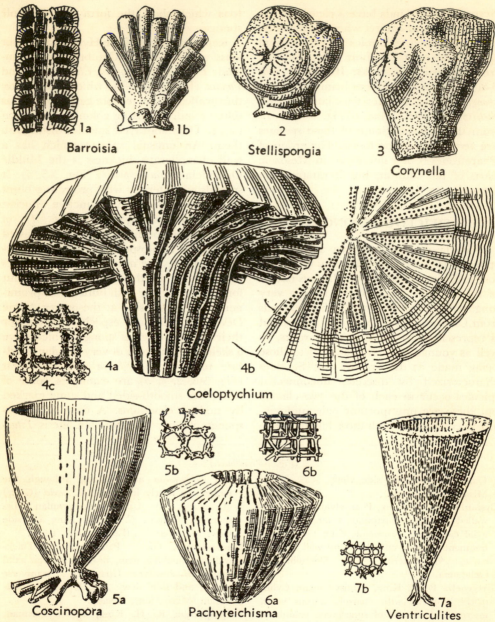

1a 1b Barroisia

2 Stellispongia

3 Corynella

4a 4c 4b Coeloptychium

5b 6b

5a Coscinopora

6a Pachyteichisma

7b 7a Ventriculites

FIG. 3-9. **Representative types of Mesozoic sponges.** Fossils shown in the upper row (1–3) are calcareous forms, and the others are siliceous.

Barroisia Steinmann, Cretaceous. A colonial sponge in which the cylindrical branches consist of superposed chambers surrounding an axial cloaca; simple passageways through the walls. *B. anastomans* (Mantell) (1*a*, longitudinal section of a branch, ×3; 1*b*, colony, ×1), Lower Cretaceous, England.

Coeloptychium Goldfuss, Upper Cretaceous. A mushroom-shaped siliceous sponge characterized by strongly fluted walls with inlets (ostia) confined to the convex mid-lines of the ribs; upper surface porous, lacking a well-defined osculum. *C. agaricoides* Goldfuss (4*a*, *b*, side view and part of upper surface, ×1; 4*c*, crib-

villia (Fig. 3-7), which has branches composed of variously directed cuplike nodes. Although the branches are only a little more than a half inch in diameter, the colony may exceed 15 in. in height and width. Mesozoic representatives of this class are common, and they include several interesting types, most of which have a very regular skeletal framework resembling the steel columns and beams of a skyscraper. The genus *Coeloptychium* (Fig. 3-9, *4*) is specialized in having the shape of an umbrella or mushroom, with prominent radial ribs on the undersurface. These bear openings for inward-moving water. An osculum is lacking, for escape of the water is provided by very numerous small apertures on the upper surface of the sponge. Thin-walled bowl- or funnel-shaped sponges which similarly lack an osculum are *Coscinopora* and *Ventriculites* (Fig. 3-9, *5, 7*). Water travels through the walls of these sponges from the outer to inner sides and thence upward through the widely open top. An unusually thick-walled hyalosponge is *Pachyteichisma* (Fig. 3-9, *6*), which has a top-shaped form with longitudinal corrugations.

The Demospongea include many siliceous fossil sponges, most of them characterized by very irregular desma types of spicules joined together in a reticulate wall of considerable thickness. These are the lithistid sponges. Examples are *Hindia* (Fig. 3-5, *3*), ranging from Ordovician to Mississippian, which has the form of a ball with radial structure; *Dystactospongia* (Fig. 3-5, *1*), a massive, coarse Ordovician sponge

with irregular long radial canals; and *Astylospongia* (Fig. 3-6, *3*), from Middle Silurian rocks, which has a pear-shaped outline with a crater-like depression at the top. A monaxon-type sponge from the Cambrian, distinguished by unusually long spicules which project in all directions, is *Choia* (Fig. 3-4, *5*).

SPONGELIKE ORGANISMS

Pleosponges

Although some authors judge that the fossils grouped under the name Pleospongea are an extinct type of true sponge, as is implied by the name (meaning full sponge), others express great doubt that they belong among Porifera. They are treated here as a class of uncertain taxonomic status. Inasmuch as their calcareous skeleton includes radial walls and they are common in some Cambrian rocks, it has been suggested that pleosponges are really a primordial type of coral and thus may comprise the ancestors of the oldest-known undoubted corals, found in Ordovician rocks. Such interpretation seems to be far from reasonable. Surely, they resemble sponges more than corals.

The pleosponges have world-wide distribution in Lower and Middle Cambrian deposits but are unknown in younger rocks. Specimens range in diameter from 1.5 to 60 mm. and have a maximum height of 100 mm. (4 in.). Average size is approximately 15 by 30 mm. Although small, their abundance in some places is sufficient to form thick reefs, and in Australia pleo-

---

(Fig. 3-9 continued.)

work of hexactinellid spicules, ×40), Germany.

Corynella Zittel, Triassic–Cretaceous. One of the thick-walled calcareous sponges having long irregular canals; osculum bordered by branching furrows. *C. quenstedti* Zittel (*3*, ×1), Upper Jurassic, Germany.

Coscinopora Goldfuss, Cretaceous. A goblet-shaped, thin-walled siliceous sponge anchored by spreading roots. *C. infundibuliformis* Goldfuss (*5a*, side view, ×0.5; *5b*, spicules, ×8).

Pachyteichisma Zittel, Upper Jurassic. Top-shaped, thick-walled siliceous sponge with furrowed surface. *P. carteri* Zittel (*6a*, side view, ×0.5; *6b*, spicules, ×8), Germany.

Stellispongia d'Orbigny, Triassic–Jurassic. Differs from *Corynella* in spicules and mode of growth. *S. glomerata* (Quenstedt) (*2*, ×1), Upper Jurassic, Germany.

Ventriculites Mantell, Cretaceous. Thin reticulate wall longitudinally convoluted. *V. striatus* Smith (*7a*, side view, ×0.5; *7b*, spicules, ×8).

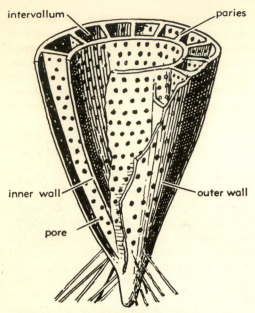

FIG. 3-10. **Structural features of pleosponges.**
These fossils, confined to Lower and Middle Cambrian rocks, have a porous calcareous skeleton.
Most of them have an inner and outer wall
separated by a space (intervallum) which contains radial walls (parietes, sing. paries).

sponge-rich limestone can be traced along
the outcrop for a distance of at least 400
miles. The pleosponges include some 80
described genera, which are distributed in
16 families and 11 orders.

Morphological Features. Most pleosponges are cup- or vase-shaped, but some
have the form of saucers or tall cylinders.
Conical specimens may have a slender
projection at the lower point (spitz). The
skeleton consists essentially of a thin **outer
wall,** perforated by very numerous **pores,**
and of internal structures which vary somewhat in different genera (Fig. 3-10). The
simplest pleosponges, however, possess no
skeletal elements inside the porous wall.
Others typically have an **inner wall** parallel to the outer one, both bearing many
pores. The space (**intervallum**) between
the walls is divided into compartments by
radial partitions (**parietes,** sing. **paries**),
which are perforated by pores. These are
the most important and characteristic

structural features, but in some pleosponges horizontal platforms, crossbars,
or irregular vesicles may occur in the
intervallum. The central cavity is empty,
except for the presence of vesicles in the
narrow lower part in some genera. The
top is broadly open.

Representative Types. Illustrations of
a few Lower Cambrian pleosponges are
given in Fig. 3-11. *Archaeocyathus* (*1*) is a
slender cylindrical fossil characterized by
irregularity of structures between the outer
and inner walls; it has world-wide distribution. A conical to cylindrical pleo-
sponge, which has a finely porous outer
wall and simple, regular partitions of the
intervallum, is *Ajacicyathus* (*2*), which occurs in North America, Australia, Asia,
and Europe. *Ethmophyllum* (*3*), which has
the same distribution as *Ajacicyathus,* is distinguished by its complex inner wall, composed of vesicles perforated by oblique
canals. A world-wide Lower Cambrian
genus is *Pycnoidocyathus* (*4*), which has
many slightly uneven partitions of the
space between the outer and inner walls.

Receptaculitids

Specially puzzling fossils are those collectively called receptaculitids, found most
commonly in Ordovician rocks but ranging upward into Devonian and possibly
younger formations. Two characteristic
genera of this group are *Ischadites* and
Receptaculites (Figs. 3-5, *2, 6;* 3-12). They
are round bowl-shaped-to-globose objects
which bear a distinctive pattern of intersecting spiral lines that define rhombic
interspaces. In most specimens, the lines
are an external expression of calcareous
walls which extend inward to a uniform
depth and spaces between the walls are
rounded passageways extending from outer
to inner surface. The interior of globose
specimens is hollow. Broad discoid shapes
exhibited by some Middle Ordovician
species of *Receptaculites,* which may attain
a diameter of 50 cm., suggest the central
part of a huge sunflower, and accordingly,
they are popularly known as "sunflower

corals." Assuredly, they are not corals, however.

Specimens of the sort just described are misleading in that they actually are molds, consisting of a limestone matrix surrounding the skeletal elements of the fossil organism, now dissolved away. This is demonstrated both by study of thin sections of such molds and by discovery of other specimens having the original hard parts preserved. They consist of calcite. The rounded passageways of the mold correspond to cylindrical pillars of the calcareous skeleton. The pillars widen at

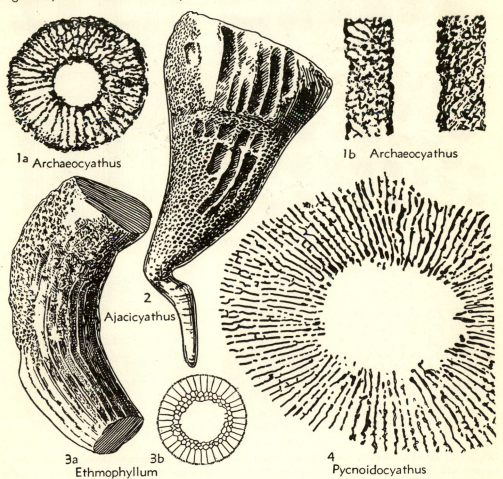

1a Archaeocyathus

1b Archaeocyathus

2 Ajacicyathus

3a   3b
Ethmophyllum

4
Pycnoidocyathus

FIG. 3-11. **Cambrian pleosponges.** This group of fossils, known only from Cambrian formations and world-wide in distribution, is distinguished by radial compartmentation of space around a central open cavity. At many places they build reefs. The illustrated genera are all restricted to Lower Cambrian.

Ajacicyathus Bedford. The apical part of this fossil consists of a simple conical tube, termed spitz. *A. nevadensis* Okulitch (2, ×2), Nevada.

Archaeocyathus Billings. The outer and inner walls are porous. *A. atlanticus* Billings (1a, b, transverse and longitudinal sections, ×4), Labrador.

Ethmophyllum Meek. A subcylindrical fossil having regular radial walls in outer portion and oblique canals adjoining central cavity. *E. whitneyi* Meek (3a, b, side view and transverse section, ×3), Nevada.

Pycnoidocyathus Taylor. The numerous radial walls are connected by crossbars. *P. occidentalis* (Raymond) (4, transverse section, ×1), Nevada.

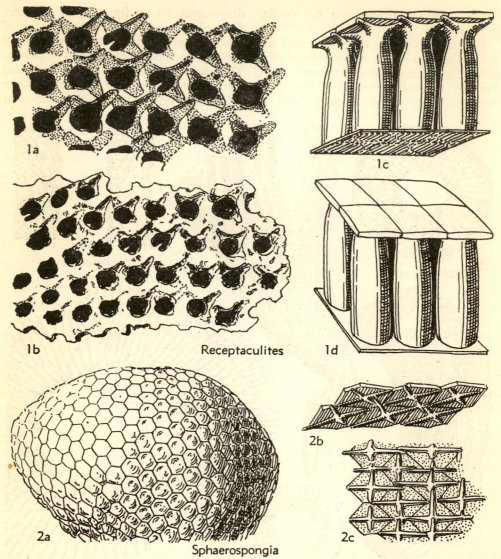

Receptaculites

Sphaerospongia

Fig. 3-12. **Receptaculitids.** Fossils of this group, known from Ordovician to Devonian, have a calcareous skeleton consisting of plates or pillar-like elements surrounding a central cavity. Externally, the individual units fit together to form a smooth pavement of quincunxial pattern, formed by intersecting spirals.

1a–d, *Receptaculites oweni* Hall, Middle Ordovician, Missouri; (1a, b) exterior surface of fossil mold, showing round cavities which correspond to pillars, and impressions of ridges on underside of the rhombic expanded ends of the pillars, ×4 and ×3; (1c, d) reconstructed small portion of the skeleton viewed obliquely from below and from above, ×4.

2a–c, *Sphaerospongia tesselata* Phillips, Middle Devonian, England; (2a) exterior of a specimen showing regular hexagonal plates, ×0.7; (2b,c) inside of plates, viewed obliquely in manner corresponding to 1c, and normal to the inner surface, showing four-rayed spicule-like ridges, ×2.

their outer extremities into rhomb-shaped flat-surfaced ends, arranged so that all join neatly together in the manner of a pavement (Fig. 3-12, *1c, d*). The lines of juncture between the paving blocks are intersecting spirals, exactly like the pattern of the sunflower coral. The underside of the expanded summit portion of each pillar is strongly ridged in four directions running to corners of the rhomb, and points of the ridges are extended in such manner as to help lock the pillar with its neighbors (Fig. 3-12, *1a, b*). Some paleontologists have thought that a layer of hexagonal plates covered the rhomb-topped pillars, but this is not true. There are no such plates. The Devonian genus *Sphaerospongia*, however, has a skeleton of closely fitting hexagonal plates which on their underside bear four ridges diverging at right angles, like those of *Receptaculites* (Fig. 3-12, *2a–c*). Although *Sphaerospongia* lacks pillars, it clearly belongs among the Receptaculitida. *Receptaculites* and *Ischadites* have stout-pillared skeletons.

The genera are not sponges, nor really are they spongelike. The pillars have no resemblance to coarse spicules, and there are neither inlet pores nor canals for the flow of water. No osculum exists. Excluded from the Porifera, they now have no classificatory home.

Nidulites

A Middle Ordovician fossil which is very useful as a guide in identifying and correlating a zone in which they occur, especially in deposits of differing facies in the Appalachian region, is named *Nidulites* (Fig. 3-5, *4*). It has a calcareous skeleton consisting of a thin reticulate wall surrounding a large central hollow. Specimens have an ovoid outline, with one diameter approximately twice as large as the other, typically measuring 50 by 25 mm. One extremity tends to be evenly rounded like the transverse circumference, whereas the other is somewhat pointed. Thus, *Nidulites* is more or less pear-shaped. The surface network consists of hexagonal cells of even size, about 1 mm. in width, and nearly the same in depth. It is like a very fine honeycomb with no cells covered over.

Resemblance of *Nidulites* to a sponge is limited to the shape of the fossil, which means little, and the nature of the surface, which seems to show many closely spaced pores. Actually, the small hollows between cell walls have closed bottoms; they are not openings into canals of sponge type. *Nidulites* may be a calcareous alga, but this is merely a guess. It is not actually a sponge.

REFERENCES

CASTER, K. E. (1939) *Siliceous sponges from Mississippian and Devonian strata of the Penn-York embayment:* Jour. Paleontology, vol. 13, pp. 1–20, pls. 1–4, figs. 1–8.

CLARKE, J. M. (1920) *The great glass-sponge colonies of the Devonian:* Jour. Geology, vol. 28, pp. 25–37, figs. 1–19.

EASTMAN, C. R., & ZITTEL, K. A. (1913) *Textbook of palaeontology:* Macmillan & Co., Ltd., London, 2d ed., vol. 1, pp. 1–839, figs. 1–1593 (sponges, pp. 46–74, figs. 46–93).

HALL, J., & CLARKE, J. M. (1898–1899) *Palaeozoic reticulate sponges constituting the family Dictyospongidae:* New York State Geologist, 15th Ann. Rept., vol. 2, pp. 743–984, pls. 1–47, figs. 1–17; 16th Ann. Rept., pp. 343–436, pls. 48–70,

figs. 18–45; also New York State Mus. Mem. 2, pp. 1–350, pls. 1–70, figs 1–45. A comprehensive, well-illustrated work on Devonian glass sponges.

HINDE, G. J. (1887–1893) *British fossil sponges:* Palaeont. Soc. London Mon., pp. 1–264, pls. 1–19, figs. 1–6.

HYMAN, L. H. (1940) *The invertebrates: Protozoa through Ctenophora:* McGraw-Hill, New York, pp. 1–726, figs. 1–221 (sponges, pp. 284–364, figs. 77–105). An excellent description of structures of modern sponges.

KING, R. H. (1932) *Pennsylvanian sponge fauna from Texas:* Texas Univ. Bull. 3201, pp. 75–85, pls. 7–8.

———— (1938) *Pennsylvanian sponges of north-*

central *Texas:* Jour. Paleontology, vol. 12, pp. 498–504, figs. 1–14.

———— (1943) *New Carboniferous and Permian sponges:* Kansas Geol. Survey Bull. 47, pp. 1–36, pls. 1–3, figs. 1–2.

MacGinitie, G. E., & MacGinitie, N. (1949) *Natural history of marine animals:* McGraw-Hill, New York, pp. 1–473, figs. 1–282 (sponges, pp. 108–116).

Okulitch, V. J. (1943) *North American Pleospongia:* Geol. Soc. America Spec. Paper 48, pp. 1–112, pls. 1–18, figs. 1–19. Contains bibliography.

Walcott, C. D. (1920) *Middle Cambrian Spongiae:* Smithsonian Misc. Coll., vol. 67, pp. 261–364.

Weller, J. M. (1930) *Siliceous sponges of Pennsylvanian age from Illinois and Indiana:* Jour. Paleontology, vol. 4, pp. 233–251, pls. 15–20.

Winchell, N. H., & Schuchert, C. (1895) *Sponges from Lower Silurian [Ordovician] of Minnesota:* Minnesota Geol. Survey, vol. 3, pt. 1, pp. 55–80, pl. F.

CHAPTER 4

Coelenterates

The coelenterates are aquatic invertebrates of highly varied form which are the most simply organized animals having well-developed body tissues. Except for a very few kinds adapted for existence in fresh waters, they are all inhabitants of the sea. A majority grow together as colonies, but solitary individuals occur also. Many are attached throughout life to part of the sea bottom, whereas others, including both individuals and colonial assemblages, float or swim about freely. Chief groups are the corals, sea anemones, jellyfishes, and plantlike colonies of hydroids. Many of these animals are very brightly colored, others are like delicately tinted flowers, and some have a translucent colorless appearance.

The word coelenterate (*coel*, hollow; *enteron*, gut) was originally applied to the sponges and ctenophores (so-called comb jellies), as well as to hydrozoans, scyphozoans, and anthozoans, which are included in the phylum, but sponges and ctenophores now are recognized as separate phyla. Because the coelenterates as restricted are characterized by the possession of stinging cells, the name Cnidaria (*knide*, nettle) is employed for them by some zoologists who reject the term Coelenterata because of its too-inclusive original application. It is preferable to retain this older name, both because of very widespread long usage and because it refers fittingly to the simple body organization of these invertebrates.

The chief characteristic of the coelenterates is their radial symmetry and the presence of a two-layered body wall surrounding a space to which the mouth gives both entrance and exit. This is a saclike digestive cavity or enteron, which in some forms is divided by radial partitions, but in others wholly lacks any compartmentation. There are no respiratory or excretory organs and no central nervous system or circulatory system; an anus is lacking. Many kinds of coelenterates lack hard parts, but others, such as the corals, are able to secrete calcium carbonate and to form skeletal structures which are readily capable of fossilization. Naturally, only the types of coelenterates which possess hard parts have importance in the paleontological record. These include two main divisions and several subdivisions of the coelenterates, however.

The size of most individual coelenterate animals is small. Those attached, whether solitary or growing together as colonies, are termed polyps. The unattached jellyfish types, which commonly have a fringed umbrella-like form, are known as medusae. The individual polyps range in diameter and length from less than 1 mm. to as much as 1 m. (some anemones). Most medusae have a diameter of 10 to 50 mm. A few attain the extraordinary diameter of 2 m., and at least one form is reported to have tentacles 10 m. long. Among coelenterates having calcareous hard parts, some solitary corals attain a diameter of more than 500 mm. (20 in.) and others an equally large total length. Colonial structures, however, may exceed 2 m. in height and diameter.

CHARACTERS OF REPRESENTATIVE MODERN COELENTERATES

Significant structural features and mode of growth of coelenterates in general may be studied advantageously by observation of three common modern genera. These are *Hydra*, which grows as a small solitary individual; *Obelia*, which has a colonial mode of growth; and *Metridium*, a common sea anemone, which is noncolonial. The first two belong to the major division of coelenterates called Hydrozoa, and the third to the class Anthozoa, which includes the corals.

Hydra, a Simple Hydrozoan

Hydra is one of the simplest of tissue-bearing animals, which has a flexible body of slender cylindrical form, generally 1 in. or less in length (Fig. 4-1). One end of the

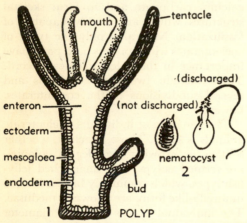

FIG. 4-1. **Simple hydrozoan polyp and stinging cells.** The common fresh-water coelenterate known as *Hydra* (1) has a saclike body formed by two layers of cell-built tissue, with noncellular gelatinous substance between them. Tentacles surround the mouth, which opens directly into an undivided digestive cavity. Stinging cells (2) occur in the body wall.

body is closed, forming a basal disk for attachment to foreign objects. The opposite end bears a small opening which is the mouth. It leads directly to the digestive tract comprising all of the interior part of the body. There is no throat or gullet, and the body interior is not subdivided in any way. Surrounding the mouth are 6 to 10 slender tentacles arranged in radial manner, each consisting of a hollow extension of the body cavity, like the finger of a

glove. The wall of the body and tentacles is composed of two layers of cells, separated by a somewhat gelatinous noncellular substance, termed mesogloea. The thin outer cell layer is termed ectoderm, and it has protective and sensory functions. The thick inner layer is termed endoderm and serves mainly for digestion. Within both outer and inner tissue layers of the body wall, but most abundantly in the ectoderm, are muscular cells. Some of these are arranged for longitudinal contraction, so as to shorten the body, whereas others contract transversely, so as to reduce the diameter of the body and extend it. The tentacles are moved readily in a similar manner.

A distinctive feature of *Hydra*, and of coelenterates in general, is the presence at many places in the body of stinging cells, termed nematocysts (Fig. 4-1, *2*). These are minute rounded capsules filled with fluid and containing a coiled tubular thread, which by action of a trigger-like mechanism at an exposed edge of the capsule, may be discharged suddenly so as to pierce the skin of small animals or coil tightly around the body parts of animals for aid in capturing them as prey. Many of the nematocysts bear thorny barbs, and their fluid has a paralyzing action on tissues of victims coming in contact with the stinging cells. Nematocysts are most abundant along the tentacles. When they have been discharged, their tubular threads cannot be withdrawn into the capsule, and they are then replaced by newly formed nematocysts.

Hydra lives in the fresh water of lakes, ponds, and streams. Normally it is attached by its basal disk, but it can move about also by a gliding or tumbling movement, or with the aid of a gas bubble secreted in mucus, it can float at the surface. The food of the *Hydra* consists mainly of minute crustaceans and insect larvae which are swallowed whole, even though some of them may exceed the *Hydra* itself in size.

New *Hydra* polyps are produced either by asexual budding or by sexual union of egg and sperm. A bud consists of a projection of the body wall of the parent, somewhere between the basal disk and the circle of tentacles. When the bud is sufficiently extended, a mouth opening and a circle of blunt tentacles appear at its outer end, the base constricts until separation is effected, and the small independent *Hydra* is set free. Both egg and sperm may be produced by the same individual in temporary structures (gonads) on the sides of the body. When an egg is fertilized by union with a sperm, division of cells leads to development of ectoderm and endoderm layers which make a small *Hydra*, without passing through a larval stage.

Because *Hydra*, like many other hydrozoans, lacks skeletal structures, it is not adapted for fossilization.

Obelia, a Colonial Hydrozoan

The vast majority of coelenterates are animals of shallow seas. Most of them are colonial. *Obelia* is a typical, relatively simple representative of the colonial hydrozoans, which merits study because it furnishes an example of colonial growth

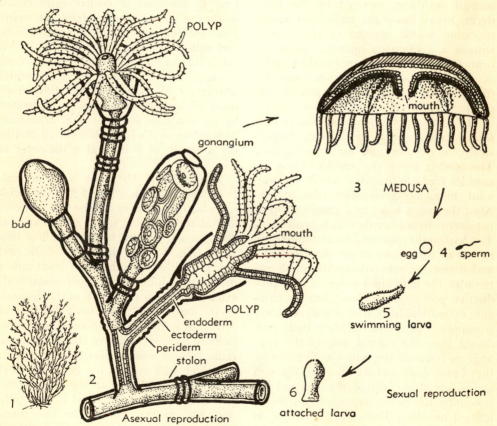

FIG. 4-2. **A colonial hydrozoan with alternation of generations.** The marine hydrozoan *Obelia*, shown natural size (1) and part of the colony, enlarged (2), consists of feeding polyps and specialized reproductive polyps, joined by hollow stems. All individuals, including young free-swimming medusae, are asexually produced. An adult medusa (3) liberates eggs or sperm (4), which on uniting develop into a larva (5) that eventually becomes attached (6) and grows into the initial polyp of a new colony.

containing two distinct types of polyps and because it well illustrates alternation in sexual and asexual regeneration.

A colony of *Obelia* comprises a soft tuft of threadlike branching stems which bear hundreds of microscopic polyps (Fig. 4-2). Such a colony may have a height and width of approximately one inch. Although the branches are flexible, the colony is attached in a fixed position by a rootlike base; thus, the polyps cannot move from place to place, like *Hydra*. The walls of the polyps and of the branches which connect them consist of ectoderm and endoderm, but in addition, there is an outer translucent chitinous covering termed periderm, secreted by the ectoderm. Space inside the endoderm, corresponding to the enteron of *Hydra*, constitutes a continuous opening along the branches which is connected to the interior of each individual polyp. Accordingly, digested food products may be passed from one part of the colony to another.

The two types of polyps observed in *Obelia* colonies are specialized, respectively, for feeding and for reproduction. The feeding polyp (hydranth) closely resembles *Hydra* in form but has 20 or more solid tentacles surrounding the mouth. Also, the polyp is set in a vaselike cup of periderm (hydrotheca), which protects it. The hydranths capture minute animals by means of their nematocysts and tentacles and digest the prey within the small enteron of the individual polyp. The reproductive polyp (gonangium) has an elongate cylindrical form with a wall of nearly transparent periderm. Minute buds produced from the central axis of the gonangium resemble tiny saucers. Eventually, they break away and are liberated through an orifice at the tip of the gonangium, becoming free-swimming small jellyfishes, termed medusae. These have a mouth on the underside, in the central part of the concavity, and bear tentacles around the periphery of their bell or umbrella (Fig. 4-2, *3*). Each individual bears gonads which are capable of producing eggs or sperm, but not both.

Both the medusae and the colonial polyps of *Obelia* are classed among hydrozoans. They lack calcareous hard parts and are not suited to fossilization. Some hydrozoans, such as *Hydra*, lack a medusa stage, whereas in others, the polyp generation is reduced or lacking; these are represented predominantly or exclusively by medusae. In the class of coelenterates called Scyphozoa, medusae or jellyfishes greatly predominate and the polyp stage may be lacking.

Metridium, an Anthozoan

Marine polyps of flower-like form, which include the sea anemones and corals, belong to the class of coelenterates called Anthozoa. In this assemblage are many genera which grow as solitary individuals, and still more numerous are those which exhibit a colonial mode of growth. None of them have a medusa stage. Essential features of the polyps, which differ from those of hydrozoans, are well illustrated by the common sea anemone, *Metridium*. This grows as a solitary individual attaining a height of 8 in. and a diameter of 4 or 5 in.

The body of *Metridium* is cylindrical. It is provided with a broad flat basal disk by which it adheres to the sea bottom, and at the opposite, upper end is an oral disk, mostly covered by short hollow tentacles (Fig. 4-3). At the center of the oral disk is a slitlike mouth which leads into a moderately elongate gullet (stomodaeum), extending half or two thirds of the distance to the basal disk. The body interior of *Metridium* is divided into compartments by radially arranged walls of tissue. All these walls, termed **mesenteries,** are attached to the outer wall and to the floor and oral disk. Some are joined to the sides of the gullet, whereas others extend inward only part way to the gullet. The free inner margins of each mesentery are thickened and convoluted, and they may bear long threadlike processes near the floor of the enteron. These mesenterial structures contain numerous nematocysts and gland cells. On one side or other of each mesentery, but arranged in systematic manner, are thick bands of muscle fibers which run longitudinally between the basal and oral disks (Fig. 4-3, *2–3*). Contraction of these muscles serves to pull the anemone down to a small fraction of its size when extended. According to the disposition of the muscle bands, the mesenteries are seen to be arranged in couples. The two couples which join opposite extremities of the slit-like gullet carry muscle bands on the sides facing away from each other, whereas other couples have muscle bands on facing sides. Therefore, those which join the narrowly rounded extremities of the gullet are termed **directive mesenteries.** It is clear that the slitlike form of the gullet and the arrangement of muscles on the mesenteries define a bilateral symmetry which modifies the otherwise dominant radial symmetry. This is an important character which aids in understanding the skeletal features of anthozoans having hard parts.

Survey of the structural characters of *Metridium* emphasizes two chief differences of anthozoans as compared with hydrozoans. These are the presence of a gullet and division of the body interior into radial compartments. On the other hand, the structure of the body wall, consisting of ectoderm, mesogloea, and endoderm, and the presence of abundant nematocysts are characters in common. The facts that individuals of *Metridium* have separate sexes and that a medusa stage is lacking are also worthy of mention.

CLASSIFICATION

Coelenterates are grouped in three classes, each containing several orders. In paleontological importance, these divisions are very uneven. None of the medusa-stage representatives of the class Hydrozoa or jellyfishes included in the class Scyphozoa secrete hard parts, and consequently, fossil records of such coelenterates consist only of rare impressions preserved in sedimentary rock. The Anthozoa far outrank other coelenterates from the standpoint of paleontological importance, but a few kinds of hydrozoans also have calcareous hard parts and are preserved as fossils. The following outline of classification shows the main divisions which are recognized among the coelenterates, those having some measure of paleontological importance being marked by an asterisk (*). Omitted from this tabulation of divisions is the extinct group of organ-

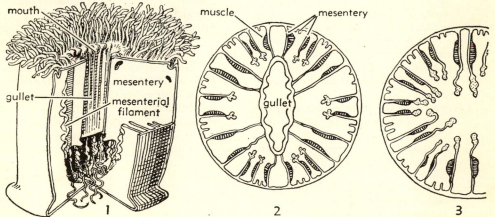

Fig. 4-3. **An anthozoan polyp; the common sea anemone.** A sectional view of a small *Metridium* (1), natural size, shows the slitlike mouth surrounded by tentacles, the gullet, and radial divisions of the digestive cavity, called mesenteries. Transverse sections of the body, which lacks hard parts, show structure at mid-length of the gullet (2) and just below the gullet (3). Muscle bands run vertically (longitudinally) on infacing sides of mesentery pairs, except those termed directive mesenteries located at narrow extremities of the gullet.

isms called Conulariida, which is represented by many fossils, classed as jellyfish-like coelenterates by some paleontologists. The conulariids are discussed in Chap. 11 of this book.

Classification of Coelenterates

*Hydrozoa (*class*), solitary or colonial, gullet lacking. Cambrian–Recent.

   Hydroida (*order*), polypoid generation well developed, some colonial forms with hard parts, distinct stolonal structures.

   Milleporida (*order*), dimorphic colonial polyps which secrete calcareous hard parts. Cretaceous–Recent.

   Stylasterida (*order*), like millepores but differing in nature and arrangement of dimorphic polyps. Cretaceous–Recent.

   Sphaeractiniida (*order*), colonial polyps which build spheroidal concentrically laminated calcareous remains. Permian–Cretaceous.

   *Stromatoporida (*order*), colonial polyps which secrete laminated calcareous deposits with pillars between laminae. Cambrian–Cretaceous.

   *Labechiida (*order*), colonial polyps which build cylindrical vesiculose colonies with scattered pillars. Ordovician–Permian.

   Siphonophorida (*order*), chiefly medusoid coelenterates attached to a stem or disk. Devonian–Recent.

   Trachylida (*order*), medusae provided with a shelf extending inwardly from the margin. Recent.

Scyphozoa (*class*), the true medusae or jellyfishes; no shelf projecting inward from margin. Cambrian–Recent.

   Carybdeida (*order*), cuboidal forms with four tentacles or groups of tentacles. Jurassic–Recent.

   Coronatida (*order*), scalloped margin separated from central bell by furrow. Cambrian–Recent.

   Semaeostomida (*order*), corners of mouth prolonged in four frilled lobes. Jurassic–Recent.

   Rhizostomida (*order*), oral lobes fused, containing many small mouths, tentacles lacking. Cambrian–Recent.

   Lucernariida (*order*), attached by aboral stalk. Recent.

*Anthozoa (*class*), polypoid generations only, possess gullet and mesenteries, solitary or colonial. Ordovician–Recent.

Alcyonaria (*subclass*), polyps have eight pinnate tentacles, colonial. Triassic–Recent.

   Pennatulacea (*order*), sea pens; very long axial polyp and many lateral polyps, dimorphic. Triassic–Recent.

   Coenothecalia (*order*), secrete massive skeleton containing cylindrical cavities for polyps. Cretaceous–Recent.

   Alcyonacea (*order*), lower part of polyp bodies fused in fleshy mass, skeleton of separate calcareous spicules. Cretaceous–Recent.

   Gorgonacea (*order*), sea fans and sea feathers; short polyps borne on sides of calcareous colonial axis. Cretaceous–Recent.

   Stolonifera (*order*), polyps connected by basal stolons or mat, calcareous spicules or tubes. Recent.

   Telestacea (*order*), very long axial polyps bearing lateral polyps as branches. Recent.

*Tabulata (*subclass*), extinct colonial corals characterized by presence of tabulae but weak or absent septa. Ordovician–Jurassic.

   *Schizocoralla (*order*), polypoid tubes multiply by fission. Ordovician–Jurassic.

   *Thallocoralla (*order*), polypoid tubes increase by lateral budding. Ordovician–Permian.

*Zoantharia (*subclass*), corals and sea anemones, tentacles simple, six primary pairs of mesenteries, solitary or colonial, with or without skeleton but not consisting of loose spicules. Ordovician–Recent.

   *Rugosa (*order*), tetracorals, characterized by serial order of appearance of septa in quadrants, tending to produce bilateral symmetry. Ordovician–Permian.

   Heterocorallia (*order*), like Rugosa but having a different plan of septal arrangement. Mississippian.

   *Scleractinia (*order*), hexacorals, characterized by introduction of septa in cycles with regular hexameral symmetry. Triassic–Recent.

   Actiniaria (*order*), sea anemones, solitary, lack hard parts. Recent.

   Zoanthidea (*order*), chiefly distinguished by features of mesenteries, no hard parts. Recent.

   Antipatharia (*order*), black corals, colonial, with horny axial skeleton. Recent.

   Cerianthharia (*order*), long, solitary, anemone-like forms. Recent.

HYDROZOANS

Although seven of the eight orders recognized among hydrozoans are represented among described fossils, most of these divisions have small paleontological importance. Some of them are decidedly noteworthy as rock builders, but their skeletal structures are mostly not enough varied and distinctive to permit differentiation of species which are useful in stratigraphic correlation or age determinations. A few, such as impressions of medusae doubtfully classed as belonging to the Siphonophorida, are so rare as to have interest only from the standpoint of indicating the existence of very ancient probable relatives of this modern group of hydrozoans. All recorded fossil hydrozoans are marine.

Hydroids

This group includes the fresh-water hydras of many kinds, widely distributed throughout the world, and a host of marine forms. They comprise a majority of all kinds of hydrozoans. Many genera, like *Obelia*, which belongs to the Hydroida, are colonial in the polypary generation. Commonly, such colonies are largely encased in a chitinous covering of periderm, and some of them secrete calcium carbonate. This deposit may be in the form of a thin incrusting layer on a shell or other foreign surface to which the colony is attached, or it may build moderately thick laminated masses of irregular form. A thin layer of fleshy tissue covers the surface of the calcareous deposits, and from this common colonial tissue the individual polyps rise (Fig. 4-4, *2a*).

Three genera from Ordovician rocks of North America and Europe and others from Silurian and Jurassic strata have been described on the basis of chitinous colonial skeletons somewhat resembling those of *Obelia* and other modern hydroids. They do not resemble graptolites, with which some of them are associated, and are judged to belong among hydroid hydrozoans because of the way in which the

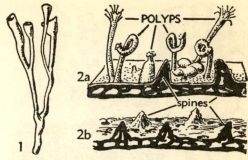

Palaeotuba **Hydractinia**

FIG. 4-4. **Fossil and modern Hydroida.** Colonial polyps belonging to this branch of the hydrozoans are polymorphic, some being adapted for functions other than feeding. Chitinous or calcareous skeletal deposits may be preserved as fossils. (1) *Palaeotuba dendroidea* Eisenack, Ordovician, Germany, ×20. (*2a*) A modern species of *Hydractinia*, showing skeletal deposit and polymorphic polyps; (*2b*) *Hydractinia pliocaena* Allman; Pliocene, Italy; both much enlarged.

polypary cups (hydrothecae) are joined to slender, irregularly branching stems. One of them is illustrated in Fig. 4-4, *1*. Calcareous colonial growths referred to the Hydroida include Cretaceous and Tertiary fossils, among which some are species classed as belonging to *Hydractinia*, a living hydroid (Fig. 4-4, *2*). This genus forms thin incrusting growths of cellular construction, characterized by projecting spines. Approximately 10 genera of these calcareous hydrozoans have been described.

Milleporids and Stylasterids

Colonial hydrozoans belonging to the orders Milleporida and Stylasterida secrete massive calcareous structures of varied form which may attain a height and diameter of more than two feet. They range from subhemispherical, lumpy growths to delicately frilled and fluted walls, leaflike expansions, and closely crowded slender branches, like staghorn coral (Fig. 4-5). They commonly grow in association with other corals, and together with calcareous algae, contribute significantly to the building of coral reefs in warm shallow seas in many places.

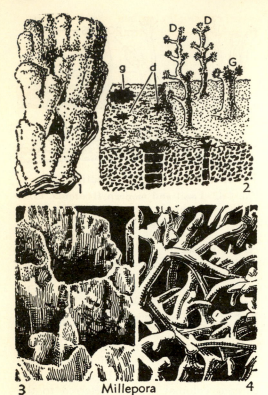

FIG. 4-5. Hydrozoans belonging to the order Milleporida. The millepores are colonial coelenterates in which two types of polyps, gastrozooids (*G*) and dactylozooids (*D*), are interconnected by a mat of tubular tissue; the colony secretes a calcareous skeleton, in which gastropores (*g*) and dactylopores (*d*) mark places of lodgment of the dimorphic polyps. (1) *Millepora* sp., part of a colony, natural size; (2) part of same, much enlarged. (3) *M. squarrosa* Lamarck, Recent, Atlantic, part of colony, ×1. (4) *M. intricata* Edwards, Recent, South Pacific, part of colony, ×1. (3, 4, *from Boschma.*)

The skeletal structure in both of these groups of calcareous hydrozoans consists of finely porous deposits with innumerable interconnecting minute passageways, penetrated by relatively straight-sided tubes of larger size with axes normal to the surface of the colony. The tubes are circular or stellate in cross section, and they are intersected by transverse partitions, called tabulae. Two distinct sizes of the tabulated tubes are observed, a large set which pro-

vides lodgment for short feeding polyps, termed gastrozooids, and a small set which is occupied by elongate, slender mouthless polyps, call dactylozooids. The latter are equipped with numerous stinging cells and serve a protective and food-capturing function. Both types of polyps can be withdrawn completely into their tubes, respectively designated as **gastropores** and **dactylopores** (Fig. 4-5, 2). The individual polyps are joined laterally by a thin layer of colonial fleshy tissue which covers the skeletal deposits between the pores.

Colonies of *Millepora* commonly are white, yellowish, or flesh-colored, whereas *Stylaster* and other genera of the stylasterids mostly are reddish to purplish. The chief difference in skeletal features consists in the more distinctly stellate outline of polypary tubes among the stylasterids, their arrangement of dactylopores around the gastropores in more systematic patterns, and generally more delicate branching colonial form. Both groups are recognized in Tertiary deposits and are recorded also in the Cretaceous.

Sphaeractiniids

The sphaeractiniids are a Permian-to-Cretaceous group of hydrozoans, characterized chiefly by the subglobular shape of their colonial calcareous skeleton and its concentrically laminated structure (Fig. 4-6). Fine tubular passageways connect

FIG. 4-6. Fossil hydrozoans of the order Sphaeractiniida. This group builds concentrically laminated calcareous skeletons. (1) *Sphaeractinia diceratina* Steinmann, Upper Jurassic, Germany, ×1. (2) *Parkeria sphaerica* Carpenter, Lower Cretaceous, England, ×1.

the adjoining layers, and in most genera irregularly distributed pillars radiate in all directions from the center. These fossils range from the size of a small marble to approximately that of a baseball.

Stromatoporids and Labechiids

Calcareous colonial structures which are assigned to orders called Stromatoporida and Labechiida are appropriately studied together because they are dominantly Paleozoic assemblages having some features of organization in common and because both are extinct groups which conceivably do not belong among hydrozoans or even among coelenterates. Various authors have suggested classification of them with algae, sponges, foraminiferal protozoans, and bryozoans, but almost surely they cannot be associated with any of these. Possibly, one or both of them represent an otherwise unknown phylum of invertebrates, distinct from any living animals, but this seems unlikely. The consensus of opinion by specialists in studies of them is that they should be ranged among hydrozoans.

Stromatoporids. The stromatoporids build incrusting or massive, relatively dense calcareous laminated deposits which are mostly several inches in diameter and thickness. Some growths belonging to a single colony attain thickness of three or four feet and a width exceeding six feet. Especially in some Silurian and Devonian rocks, they are locally very abundant and build reefs, either in conjunction with corals or largely by themselves. As observed in the field or by inspection of a hand specimen in the laboratory, most stromatoporids show a tendency to split in layers parallel to the surface of the colony (Fig. 4-7, *1*). The fossils break less readily and evenly in directions transverse to these laminae. The surface of a colony or any of its layers may appear smooth and nearly plane or gently curved, devoid of any markings; or low to rather prominent monticules may be distributed evenly or irregularly over such surfaces. Also, readily visible without the use of a lens,

clusters of small channels may be present, diverging radially from a center and branching toward their outer terminations. They are termed **astrorhizae** (*astro*, star; *rhiza*, root) because of their stellate pattern. Where they occur in colonies which bear monticules, the centers of the astrorhizae are invariably located on summits of the small elevations.

For discrimination of genera and species of stromatoporids, thin sections cut transversely and parallel to the laminae are requisite. These show the existence of closely spaced microlaminae—very many within one of the sheets into which a specimen ordinarily breaks—and walls or rodlike pillars approximately at right angles to them (Fig. 4-7, *2–7*). In some genera, the pillars are continuous through several laminae, but in others there is no such continuity. The calcareous tissue of the colonial skeleton is minutely porous in some genera (*Stromatoporella*, Fig. 4-7, *7*), but in others it is dense. Relatively large round tubes (generally called caunopores) transect the laminae of some stromatoporids, and these may denote places of lodgment of special polyps, such as the gastrozooids of milleporids and stylasterids, but some such tubes evidently belong to tabulate corals with which the stromatoporids are found intergrown. The recorded range of stromatoporids is from Cambrian to Cretaceous, but they are very uncommon in upper Paleozoic and most Mesozoic rocks.

Labechiids. The Labechiida differ from Stromatoporida in having a generally cylindrical growth form of the colony and in lacking well-defined laminae. Some of them (*Labechia, Stromatocerium*, Fig. 4-7, *8, 9*) have rodlike structures running transverse to the surface, but other parts of the skeleton consist of rounded vesicles arranged roughly in rows with the convex walls of the vesicles on the side nearest the surface of the colony. *Beatricea* and *Cryptophragmus*, which are useful guide fossils in some Ordovician rocks, are distinguished by an axial region of large blister-

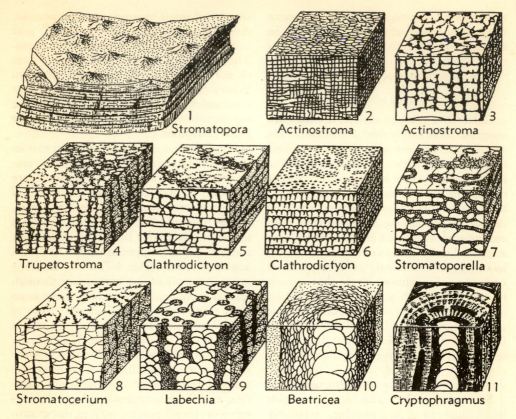

FIG. 4-7. **Fossil hydrozoans of the orders Stromatoporida and Labechiida.** The stromatoporids (1–7) build massive, laminated calcareous skeletal deposits which are especially common as fossils in Silurian and Devonian rocks, whereas labechiids (8–11) have generally cylindrical vesiculose hard parts and are most common as Ordovician fossils.

1, *Stromatopora* Goldfuss, Silurian–Jurassic. *S. concentrica* Goldfuss, Middle Devonian, Germany, ×1. This specimen shows astrorhiza-bearing monticules on the surface and indicates tendency to split into sheets.

2, 3, *Actinostroma* Nicholson, Silurian–Jurassic. (2) *A. whiteavesi niagarense* Parks, Niagaran, Ontario, ×7; polished block, showing structure transverse and tangential to laminae, pillars penetrating a number of laminae. (3) *A. clathratum* Nicholson, Devonian, England, ×7.

4, *Trupetostroma* Parks, Devonian. *T. warreni* Parks, Middle Devonian, northwestern Canada, ×7; pillars and laminae porous.

5, 6, *Clathrodictyon* Nicholson & Murie, Silurian–Devonian. (5) *C. laxum* Nicholson, Lower Devonian, Ohio, ×7; vertical elements irregular. (6) *C. striatellum* (d'Orbigny), Niagaran, New York, ×7; distinctive regular structure.

7, *Stromatoporella* Nicholson, Silurian–Devonian. *S. tuberculata* Nicholson & Murie, Lower Devonian, New York, ×7; thick, porous laminae.

8, *Stromatocerium* Hall, Ordovician. *S. huronense* Billings, Upper Ordovician, Ohio, ×7; a labechiid which simulates stromatoporids in mode of growth.

9, *Labechia* Edwards & Haime, Ordovician–Silurian. *L. conferta* Lonsdale, Middle Silurian, England, ×7.

10, *Beatricea* Billings, Ordovician. *B. nodulosa* Billings, Upper Ordovician, Kentucky, ×1; distinguished by cylindrical growth and large axial hemispherical chambers.

11, *Cryptophragmus* Raymond, Ordovician. *C. antiquatus* Raymond, Middle Ordovician, Virginia, ×0.7; resembles *Beatricea* but has finer cystose sheath penetrated by transverse openings.

like vesicles (Fig. 4-7, *10, 11*). Genera of the Labechiida range from Ordovician to Permian.

SCYPHOZOANS

The jellyfishes or scyphozoans lack hard parts, and consequently preservation of them as fossils can occur only as the somewhat fortuitous result of impressing their form on sediments, or as in the exceptional set of conditions represented by the Middle Cambrian Burgess shale of western Canada, by leaving traces of soft parts. Four out of the five recognized orders of scyphozoans seem to be represented by fossils, some of which occur in Cambrian rocks. In addition, there are fossil jellyfishes which cannot be placed definitely in groups of living scyphozoans. All modern jellyfishes are characterized by strongly developed tetramerous symmetry.

The group called Carybdeida is distinguished by the nearly cube-shaped form of the bell. The top and four sides are flattened, and the outline of the transverse sections is square. Modern carybdeids are strong, graceful swimmers which subsist mostly on fish, and some are moderately large, attaining a height or width of 10 in. They are noted for the virulence of their stinging cells, to which they owe their common name of sea wasps; they are regarded as the most dangerous of coelenterates, greatly feared by swimmers on eastern Asiatic coasts. A Jurassic fossil carybdeid is illustrated in Fig. 4-8, *1*.

The Coronatida possess a conical or dome-shaped rounded bell which is characterized by a circular groove separating

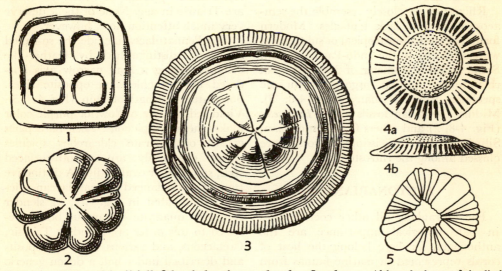

FIG. 4-8. **Types of fossil jellyfishes belonging to the class Scyphozoa.** Although these soft-bodied coelenterates are an important zoological division, they are rare as fossils. Three of the five orders are represented among specimens here figured, however.

1, *Medusina* Haeckel, Jurassic. *M. quadrata* Haeckel, Upper Jurassic, Germany, ×0.4; a cuboidal jellyfish belonging to the Carybdeida.

2, *Brooksella* Walcott, Cambrian. *B. alternata* Walcott, Middle Cambrian, Alabama, ×0.7; classed among Rhizostomida.

3, *Rhizostomites* Haeckel, ?Triassic–Jurassic. *R. admirandus* Haeckel, Upper Jurassic, Germany,

×0.12; this rhizostomid, here partly restored, had a diameter of 16 in.

4, *Camptostroma* Ruedemann, Lower Cambrian. *C. roddyi* Ruedemann, Pennsylvania, ×0.7; (4a, b) aboral and side views, restored.

5, *Peytoia* Walcott, Middle Cambrian. *P. nathorsti* Walcott, Burgess shale, British Columbia, ×0.7; oral view; ordinal assignment of this scyphozoan is doubtful.

the crown from marginal areas. Typically, they inhabit rather deep ocean waters, but some kinds are common in shallow seas, as off the coast of Florida. If the fossil named *Camptostroma* (Fig. 4-8, *4*), from Lower Cambrian rocks of southeastern Pennsylvania, is rightly assigned to this division of the scyphozoans, the order ranks as one of the two longest ranging branches of the class. The pitted markings of the crown of this fossil and its radially furrowed periphery make resemblance to a sunflower.

The only kinds of scyphozoan commonly seen in temperate regions belong to the order Semaeostomida. They have a low saucer- or bowl-shaped bell with scalloped margin. Jellyfishes of this type have been found in the Solnhofen limestone, Jurassic, of southern Germany but are not known otherwise as fossils.

Rhizostomida closely resemble the semaeostomids but lack tentacles. Modern forms live in shallow tropical or subtropical waters, chiefly in the Indo-Pacific region. Some attain a diameter of 32 in. A fossil rhizostomid, half as large, is a Jurassic specimen from Germany (Fig. 4-8, *3*). Middle Cambrian fossils called *Brooksella* (Fig. 4-8, *2*) and others from Ordovician, Silurian, and Mississippian rocks are classed among the rhizostomids.

ALCYONARIANS

Far outranking all other coelenterates in paleontological importance are the anthozoans, for here belong the host of corals represented in marine faunas from early Ordovician to the present day. The calcareous skeletal remains of corals are a quantitatively significant component of many rock formations, and many of them are useful guides in stratigraphic correlation and as geologic age indicators. The anthozoans are distinguished by a radial compartmentation of the digestive cavity, a slitlike mouth, the possession of a gullet, and the entire absence of a medusoid stage. They include solitary individuals but are

mostly colonial. They are exclusively marine. Although a majority secrete hard parts, several divisions are wholly soft-bodied and thus not suited for fossilization.

The subclass Alcyonaria comprises anthozoans having eight pinnately branched tentacles and eight mesenteries joined to the gullet at their inner margins. They are exclusively colonial, but the polyps are not directly joined with one another, as in many colonial corals, for they interconnect only by tubular passageways (stolons), generally located in a laterally extending mat (Fig. 4-9). Newly formed polyps arise at various places along these outgrowing stolonal tubes. The colonies are supported by a horny or calcareous skeleton or by a combination of calcareous spicules and horny connective tissue. Four of the six orders recognized among alcyonarians are represented by fossils, of which the oldest are Triassic in age. None of them deserve very much attention here, however.

The Pennatulacea include the so-called sea pens, distinguished by feather-like colonies having a long narrow axis bordered by branchlets on opposite sides. Some attain a length of 40 in. (Fig. 4-9, *3*). Vivid red, orange, yellow, or purple color makes them conspicuous. Their hard parts consist of separate calcareous spicules which commonly are smooth-surfaced rods, spindles, or oval bodies. A distinctive type of three-flanged pennatulacean spicule is illustrated in Fig. 4-9, *4;* slender round or quadrate calcareous bodies referred to this order are found in Triassic, Cretaceous, and various Tertiary deposits and described under half a dozen generic names (Fig. 4-9, *5*).

The order Coenothecalia contains three genera, all of which occur as fossils in Cretaceous or Tertiary formations, and one—the blue coral, *Heliopora*—persists as a modern reef builder in the Indo-Pacific region (Fig. 4-9, *7*). The massive calcareous skeleton of *Heliopora*, stained blue, seemingly by iron salts, contains rounded tubular openings approximately 1 mm. in diameter in which the polyps are lodged.

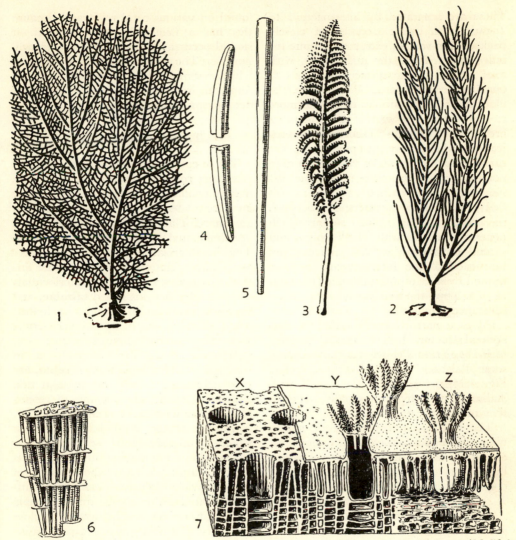

FIG. 4-9. **Types of alcyonarian anthozoans.** With one exception (5), all are Recent forms. (*Modified from Hyman and various sources.*)

1, 2, *Gorgonia*, belonging to the order Gorgonacea; (1) the common sea fan; (2) sea-whip type; secretes calcareous spicules; approximately ×0.1.

3, 4, *Pennatula*, a sea pen; (3) colony showing lateral branches containing rows of small polyps arranged on opposite sides of the axis, approximately ×0.2; (4) a three-flanged calcareous spicule.

5, *Graphularia desertorum* Zittel, Eocene, Libya, ×0.7; calcareous axial skeleton of a penntulacean.

6, *Tubipora*, calcareous skeleton of an organ-pipe coral, belonging to the order Stolonifera, ×1; not known as fossil.

7, *Heliopora*, polyps and skeletal deposit of the blue coral, much enlarged; (*X*) part of corallum with soft parts of colony removed, showing large rounded tubes occupied by the polyps; (*Y*) a polyp and adjacent part of fleshy colonial mat in section, showing blind glove-finger extensions of the mat; (*Z*) skeleton removed from one of the polyps and adjoining part of colonial mat, but represented in section below.

These are separated by interspaces 1 to 2 mm. wide, which consist of closely packed, thin-walled, extremely minute tubules disposed like the main tubes, with axes normal to the surface. Both sets of openings are intersected by transverse tabulae, those of the tubules being the more closely spaced. The walls of the large tubes are indented by 12 to 25 short inward projections, which are termed pseudosepta because their number and arrangement does not conform to the eight mesenterial divisions of the polyp body. A thin mat of fleshy tissue covers the interspaces between the polyps; it bears tubular stolons which penetrate the side walls of the polyps and extend downward as blind narrow sacs into each of the interpolypary tubules, limited by their topmost tabulae (Fig. 4-9, 7*Y*, *Z*). Thus, the living polyps and fleshy substance common to the colony are confined to a narrow surficial fringe of the skeletal structure. Because the skeletal form of *Heliopora* rather closely resembles that of some Paleozoic corals (*Heliolites*, *Lyellia*, Fig. 4-16, *3*, *6*) which are classed with Tabulata, this division actually may belong among alcyonarians.

The so-called horny corals or gorgonians, belonging to the order Gorgonacea, comprise the sea fans and sea whips, which build very graceful, somewhat flexible colonies with delicate coloration (Fig. 4-9, *1*, *2*). Some of them attain a height of 10 ft. Although their horny skeletal support is perishable, many of them secrete calcareous spicules and these may be preserved as fossils. The red precious coral, *Corallium*, is a gorgonian which is distinguished by having solid calcareous axes composed of spiniform spicules bonded by finely fibrous calcite. It is known from Cretaceous and Tertiary rocks but is rare. Remains of other genera of the order have similar distribution.

The Stolonifera is one of the most simply organized divisions of alcyonarians, in which the polyps arise singly from a basal fleshy mat containing stolons. It includes the red organ-pipe coral (Fig. 4-9, *6*),

found on various coral reefs, and because this has a firm calcareous skeleton, it should occur as a fossil. None are known, however. The remaining order, Telestacea, which resembles the gorgonians, also has some calcareous skeletal deposits, but these are lacking among fossils.

TABULATE CORALS

The so-called tabulate corals are a group of extinct colonial anthozoans, which are mainly confined to Paleozoic rocks. They are here treated as correlative in rank with alcyonarians and zoantharians, and thus are designated as the subclass Tabulata. The characterization "so-called tabulates" seems appropriate, inasmuch as other anthozoans possess many of the interior platforms or diaphragms called **tabulae,** and they are common in colonies of hydrozoans, such as milleporids and stylasterids, also. Furthermore, among rugose and scleractinian corals are some forms in which the radial walls, termed **septa,** are extremely short, as they are in tabulates, if present at all. These points of resemblance have no real importance, however, because characters of tabulates, taken as a whole, readily and clearly set them apart. Not overlooked in making this statement is the difficulty in deciding whether some of the Ordovician genera should be identified as tabulates or classed among simple colonial rugose corals. Likewise, the close similarity of *Heliolites*, an accepted tabulate coral, and *Heliopora*, which is unquestionably an alcyonarian, is recognized. The tabulates are a group of unknown origin or exact zoological relationships, which well may include ancestors of zoantharian and alcyonarian coelenterates.

Essential structural characters of the tabulates are rather easily apprehended. The hard parts of a tabulate polyp consist of a polygonal, elliptical, or circular tube provided with imperforate or perforate walls and bearing many or few tabulae; short septa or septa-like spines may be present or absent. The coral tubes may grow (1) in approximately parallel position, closely packed together so that each impinges on its neighbors, (2) loosely packed so that tubes are joined to one another only by stolonal cross tubes, (3) united laterally with a neighbor along parts of the sides without compression of the walls, or (4) one budding from another in such manner that the tubes have divergent positions. The walls of tubes belonging to some genera are perforated by pores which provide intercommunication. Mostly, however, such openings are absent.

Two main groups of tabulates may be differentiated, and these are defined as orders. The Schizocoralla (*schizo*, divide) are tabulates in which new individuals are introduced by fission or apparent splitting of a parent, which branches upward into two or more separate units. The Thallocoralla (*thallo*, sprout) reproduce by budding, commonly from the side of the parent. To this group the bulk of widely known tabulates belongs.

Colonies of tabulate corals range in size from a greatest dimension of a few millimeters to masses fully 1 m. in height or width. Representatives of the subclass are distributed from Lower Ordovician to Jurassic.

Schizocorals

The oldest known genus of schizocorals, and indeed, the oldest discovered anthozoan (unless the Middle Cambrian supposed bryozoan *Archaeotrypa* is a tabulate coral), is *Lichenaria* (Fig. 4-10, *5*). The colonial fossil consists of subhemispherical masses 4 in. or less in height and width, composed of imperforate polygonal tabula-bearing tubes, which in different species have a diameter of 0.7 to 2 mm. Septa are lacking. Species and individual specimens are most numerous in Middle Ordovician rocks (Chazyan to Trentonian), but at least two species are described from near the base of Lower Ordovician rocks.

Another Ordovician genus (*Tetradium*, Fig. 4-10, *3*), which is a typical schizocoral, is represented by more than three dozen described species and subspecies. The colo-

nies are varied in shape and size, but all species are characterized by irregularly polygonal to rounded polypary tubes, which have imperforate walls and mostly bear four distinct septa. The tubes range in diameter from 0.4 to 1.7 mm. in different species. Tabulae are numerous in some forms but rare or lacking in others. Multiplication is effected by union of the four septa of the parent so as to form part of the walls of four new individuals, which for a space lack septa of their own. This is a simple method of fission that produces four out of one.

Chaetetes is a very common, stratigraphically useful fossil in some Lower and Middle Pennsylvanian marine rocks, and reported to be abundant in Lower Carboniferous deposits of Europe (Fig. 4-21, *1*, *2*). It has thin-walled, imperforate polygonal tubes which have an average diameter of only 0.2 to 0.5 mm. Colonial growths vary greatly in shape and size, some attaining a height or width of 4 ft. Septa are rare or lacking, but tabulae may be numerous. The tubes increase in numbers by fission. Although several features serve to put *Chaetetes* in a category of its own, it is judged to belong among the tabulate corals, and because multiplication of individuals is by fission, it is assigned to the Schizocoralla. The genus is lacking in Upper Pennsylvanian and Permian strata of North American but is identified elsewhere in rocks as young as Jurassic.

Thallocorals

Tabulate corals which reproduce primarily by lateral budding or growing from stolonal branches given off by parent polyps are grouped in the order Thallocoralla. It comprises four families, of which two include colonies of cylindrical or trumpet-shaped tubes with few lateral contacts (Syringoporidae, Auloporidae), another made up of rounded tubes joined at their sides in a manner giving the cross sections the appearance of a chain (Halysitidae), and a fourth consisting of massive growths of polygonal tubes (Favositidae).

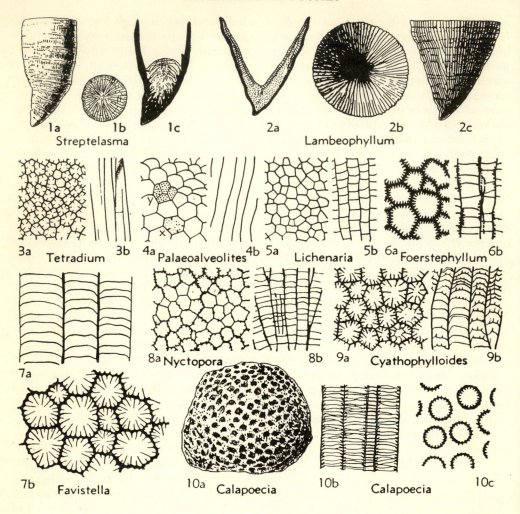

FIG. 4-10. **Representative Ordovician rugose and tabulate corals.** Forms classed among Rugosa include 1, 2, 7, 9; the others are included in the Tabulata.

Calapoecia Billings, Ordovician–Silurian. *C. canadensis* Billings (10*a*, corallum, ×1; 10*b, c*, sections, ×2), Upper Ordovician, Quebec.

Cyathophylloides Dybowski, Upper Ordovician–Middle Silurian. *C. ulrichi* Bassler (9*a, b*, sections, ×3), Upper Ordovician, Minnesota.

Favistella Dana, Middle and Upper Ordovician. *F. alveolata* (Goldfuss) (7*a, b*, sections, ×1), Middle Ordovician, Tennessee.

Foerstephyllum Bassler, Middle and Upper Ordovician. *F. halli* (Nicholson) (6*a, b*, sections, ×2), Blackriveran, Tennessee.

Lambeophyllum Okulitch, Middle Ordovician. *L. profundum* (Conrad) (2*a–c*, longitudinal section, calycal and side views, ×1), Blackriveran, New York.

Lichenaria Winchell & Schuchert, Lower and Middle Ordovician. *L. heroensis* (Raymond) (5*a, b*, sections, ×3), Chazyan, Vermont.

Nyctopora Nicholson, Middle and Upper Ordovician, *N. billingsi* Nicholson (8*a, b*, sections, ×3), Trentonian, Kentucky.

Palaeoalveolites Okulitch, Middle Ordovician. *P. carterensis* (Bassler) (4*a, b*, sections, ×3), Blackriveran, Tennessee.

Streptelasma Hall, Ordovician–Devonian. *S. corniculum* Hall (1*a–c*, side view, transverse and longitudinal sections, ×0.7), Middle Ordovician, New York.

Tetradium Dana, Ordovician. *T. fibratum* Safford (3*a, b*, sections, ×3), Trentonian, Tennessee.

Auloporids. Simplest in structure seemingly are the auloporids, which range from Ordovician to Permian. They have short, trumpet-shaped coralites, one growing outward from near the base of another or from some other point along the side of the parent. In *Aulopora* (Fig. 4-19, *1*), the colony spreads out over some foreign surface, to which each individual is cemented along the side opposite its aperture; this genus has curved tabulae but only a trace of septa. In other genera, members grow erect above a basal circlet of individuals which establish anchorage around a crinoid stem or similar support, and the apertures face in opposite directions (Figs. 4-20, *10*; 4-21, *4*); or colonies are distinguished by an outgrowth of newly budded individuals in whorls (Fig. 4-19, *2*).

Syringoporids. This group of tabulates also is distributed from Ordovician to Permian. It includes several genera, based on mode of colonial growth, nature of tabulae, and other features. Faint septal markings occur in some, but mostly they are lacking. The skeletal tubes of *Syringopora* (Fig. 4-11, *8*) grow upward in approximately parallel position, with stolonal connections at various levels. Colonies are common in many formations, some of them attaining a height and width of 2 ft. or more.

Halysitids. The halysitids are Ordovician and Silurian tabulates, most of which are readily identified by their chain-like mode of growth. The coral tubes of *Halysites* (Fig. 4-11, *7*) vary in shape among different species; generally they have an elliptical cross section, with the long axis coincident to the trend of the chain, but units having perfectly circular or quadrate cross sections are well known. In some chain corals, all members of a colony have uniform size, with or without distinct short septa, but in others a very small tube occurs between each pair of large ones. The small individuals are quadrangular or subtriangular in transverse shape and are characterized by especially close spacing of the tabulae. Halysitid species typically have 12 septa in each tube.

Favositids. The massive thallocorals, grouped in the family Favositidae, make first appearance in the Ordovician; they are abundant in many Silurian and Devonian formations, and persist to the Upper Permian. In colonies of most genera, close packing of individuals gives them a polygonal transverse section, but some (Figs. 4-10, *10*; 4-11, *3, 6*; 4-19, *3*), exhibit relatively large tubes of subcircular section, separated by small polygonal tubes or by irregular colonial structures. A characteristic feature of the "honeycomb coral," *Favosites* (Fig. 4-11, *7*; 4-19, *6*), and numerous other favositids is the presence of rounded openings, called **mural pores,** in the walls. They are clearly visible in weathered or broken specimens which reveal the sides of tubes, and they may be evident in transverse sections (Fig. 4-19, *5a*). The passageways denote interconnection of tissues belonging to the living polyps, but it seems doubtful indeed that the pores represent abortive buds for production of new individuals, as a few authors have suggested. Some genera possess irregular wall pores (Fig. 4-10, *10*), whereas others lack them entirely (Fig. 4-11, *3, 6*).

Among typical favositids, colonial form and size of the individual tubes varies greatly (Figs. 4-11, *4, 5, 9*; 4-21, *3, 5*). Some have exceptionally numerous tabulae, but none of them show more than vestiges of septa. Many colonies exhibit a well-developed covering of the base, termed holotheca. An unusual feature seen in some Paleozoic favositids (Fig. 4-21, *5*) is differentiation of a thin-walled interior part of the colony and a very dense peripheral region, like the immature and mature portions of some bryozoan colonies. Many favositids are useful guide fossils in stratigraphic correlation.

ZOANTHARIANS

The subclass Zoantharia mainly consists of corals, but it includes also the sea anemones and some other orders which lack hard parts. Among zoantharians

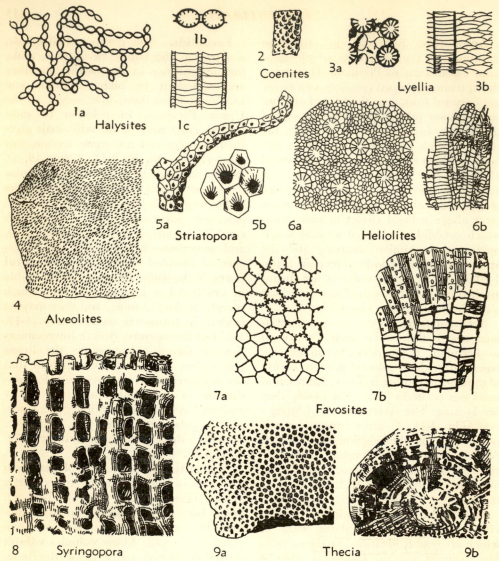

FIG. 4-11. **Representative Silurian tabulate corals.**

Alveolites Lamarck, Silurian–Devonian. *A. un-dosus* Miller (4a, ×1), Niagaran, Kentucky.

Coenites Eichwald, Silurian–Devonian. *C. seriata* (Hall) (2, ×1), Niagaran, New York.

Favosites Lamarck, Ordovician–Permian. *F. favosus* (Goldfuss) (7a, top view of part of corallum, showing polygonal corallites, many with marginally fluted tabulae; 7b, side view, showing natural section, below, and walls of corallites with mural pores, above; both ×1), Niagaran, Michigan.

Halysites Fischer, Upper Ordovician–Silurian. *H. catenularia* (Linné) (1a, top view of part of corallum, ×1; 1b, c, sections, ×3), Niagaran, Indiana.

Heliolites Dana, Ordovician–Devonian. *H. inter-stinctus* (Linné) (6a, b, sections, ×3), Niagaran, Tennessee.

Lyellia Edwards & Haime, Silurian. *L. americana* Edwards & Haime (3a, b, sections, ×2), Niagaran, Indiana.

Striatopora Hall, Silurian–Pennsylvanian. *S. flexuosa* Hall (5a, part of corallum, ×1; 5b, calyces of a few corallites, ×3), Niagaran, New York.

Syringopora Goldfuss, Silurian–Pennsylvanian. *S. verticillata* Goldfuss (8, side view of corallum, ×1), Niagaran, Michigan.

Thecia Edwards & Haime, Silurian–Devonian. *T. minor* Rominger (9a, top of corallum; 9b, base of corallum, showing holotheca: both ×1), Niagaran, Michigan.

which secrete a calcareous skeleton, none have loose spicules, like many alcyonarians, which are scattered when the polyps die. As previously noted, this division of the coelenterates is characterized by entire absence of a medusa stage in generation and by internal features of body organization, including especially the radial partitions called mesenteries and a gullet (Fig. 4-12).

Basically and most simply, the skeleton of a coral comprises three elements: a **basal disk,** which is the first-formed structure of the very young polyp; radial partitions, called **septa;** and an outer wall, generally termed **epitheca.** The basal disk may be so overshadowed by later-built hard parts as to disappear from view, but it exists, and may be evident in the mature coral. The septa invariably are located between mesenterial partitions of the enteric cavity, not coincident with them,

and they extend inward from the periphery, reaching part or all of the distance to the center. They are secreted by inwardly folded segments of the ectoderm. The outer wall similarly is formed by parts of the ectoderm, with or without an overfold (**edge zone**) at the growing lip of the wall. Study of various corals shows the need to recognize modifications in this generalized description, but it is important to observe that all skeletal secretion in every coral is ectodermal. Thus, without regard to complexity of skeletal structure, the coral polyp sits in or on the hard parts which it builds. Calcium carbonate is not secreted so as to permeate the soft tissues.

The Zoantharia are confined to marine waters and are most abundant in warm shallow seas. Some grow as solitary individuals, attached firmly or loosely embedded in soft sediment, but most are colonial. The latter are especially impor-

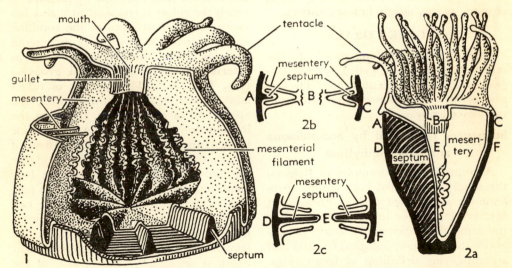

FIG. 4-12. **Zoantharian coral polyps showing relation of soft parts to skeleton.** The drawings are diagrammatic representations of modern corals, those at right modified from Vaughan & Wells (1943).

1, Solitary coral having soft parts essentially the same as those of *Metridium* (Fig. 4-3) except for inward foldings of the base where parts of the ectoderm build septa by secretion of calcium carbonate; the septa are thus constructed of material deposited on their opposite sides.

2, *Flabellum*, a simple scleractinian coral, approxi-

mately ×1. (2a) Longitudinal section of polyp and its skeleton, cut through one of the septa (at left) and between septa (at right); the septum is composed of upwardly inclined curved rows of calcite fibers, called trabeculae. (2b) Part of transverse section through points marked A, B, C in 2a. (2c) Part of transverse section through points marked D, E, F in 2a.

tant as reef builders. Reef corals live only between sea level and a depth of approximately 90 m., but reef building is confined to depths less than 50 m. and active building to 15 m. or less. Solitary cup corals mostly live above the 1,000-m. depth contour, but a few are found sparingly to 5,870 m. Modern coral reefs are chiefly developed in two regions: the Indo-Pacific area, extending from the east coast of Africa northeastward to Hawaii and the Paumotu Islands; and the Caribbean–West Indies area, reaching as far as Bermuda. The largest and one of the most noted reefs is the Great Barrier Reef, off the northeastern coast of Australia, which locally attains a width of 80 miles and has a length of 1,200 miles. Some South Pacific coral atolls are as much as 50 miles across. Ancient coral reefs are found especially in some Silurian and Devonian formations, and in Upper Triassic, Jurassic, Cretaceous, and middle to late Tertiary rocks of many places, mostly in low latitudes.

RUGOSE CORALS

Paleozoic corals belong mostly to the order of zoantharians called Rugosa (*rugosa*, wrinkled), or Tetracoralla. The rugose corals exhibit many important structures excellently.

Structures Illustrated by Streptelasma and Lambeophyllum

As representative examples of Rugosa, the simply constructed horn corals, *Streptelasma* and *Lambeophyllum*, of Ordovician age, are chosen for initial notice, and at the same time reference is made to diagrams showing morphological features and accompanying definition of terms (Figs. 4-14, 4-15).

Growth Form. Solitary corals, such as those just named, are designated as **simple corals,** in contrast to colonies termed **compound** corals (Fig. 4-14, *9, 47*), but the skeleton of either a solitary or colonial polyp is called a **corallite** (Fig. 4-14, *8, 49*). According to adopted descriptive terms, the shape of corallites belonging to

our selected genera is **trochoid** or **turbinate** (Fig. 4-15, *8, 27*). Other commonly used designations for shape of solitary corallites include **discoid, patellate, cylindrical, calceoloid, ceratoid, pyramidal,** and **scolecoid** (Fig. 4-15, *9–14, 28*). In some corals it is noteworthy that the shape alters distinctly from stage to stage in growth. The proximal, initially formed part of a corallite is its **apical end** (Fig. 4-14, *10*), and this becomes increasingly blunt as expansion in width matches or exceeds elongation by upward growth.

Outer Wall. The rounded outer wall (**epitheca**) of corallites belonging to *Streptelasma* and *Lambeophyllum* shows two sorts of markings: irregular wrinkles formed by slight constrictions or thickened parts of the wall, which conform to the upper edge of the corallite and represent successive positions of this margin during growth (therefore termed **growth lines**); and longitudinal ridges divided by furrows, of which the ridges correspond to spaces between the internal septa of the corallite, and the furrows to juncture of these septa with the outer wall (hence defined as **interseptal ridges** and **septal grooves**) (Fig. 4-14, *4–7*). The number of ridges and grooves increases upward from the apical end of the corallite, reflecting the introduction of new septa in the interior (**thecarium,** Fig. 4-14, *40*). The pattern of this increase is not uniform around the corallite but shows arrangement in four quadrants. A spread-out map of this pattern is given in Fig. 4-13, *5*, and one aspect of it appears in the side view of *Lambeophyllum*, shown in Fig. 4-10, *2c*. The tetrameral plan of septal arrangement which is denoted by distribution of the interseptal ridges and septal grooves, and which is confirmed by transverse sections of the corallite at various levels, is a characteristic feature of the Rugosa, even though many of them seem to have no sign of fourfold grouping of the septa when the upper edges of these radial structures in a mature specimen are viewed (Fig. 4-10, *1b, 2b*).

Septa. At the summit of corallites of *Streptelasma* and *Lambeophyllum* is a depression known as the **calyx** (Fig. 4-14, *1*). It is formed by the accordant upper (distal) edges of the septa, which are considerably lower at the upper limit of their **axial edges** in the central part of the thecarium than at their **peripheral edges** adjoining the epitheca (Figs. 4-10, *1c, 2a, b;* 4-15, *22, 23*). The septa of fully grown specimens of these corals, as seen in the calyx, are clearly divisible into two orders on the basis of size, particularly the extent to which they reach inward. The longest septa, which have axial edges at or near the center, are termed **major septa** or **metasepta,** whereas the short ones are **minor septa** or **tertiary septa** (Fig. 4-14, *11, 12, 26, 27*). The latter group make an appearance only in the mature parts of the corallite, as can be proved by their absence in transverse sections somewhat below the base of the calyx or by the downward tracing of the septal grooves corresponding to them on the exterior of the epitheca.

A—alar septum   C—cardinal septum   K—counter septum   KL—counter-lateral septum

FIG. 4-13. **Septal patterns of rugose and scleractinian corals.**

The four series of transverse sections of corallites diagrammatically indicate the successive growth stages of (1) a rugose coral having well-defined tetracoral attributes, because septa are grouped in distinct quadrants, with prominent cardinal and alar fossulae; (2) a rugose coral having perfectly radial arrangement of septa at maturity, with quadrants not defined; (3) an early Mesozoic type of scleractinian coral in which two of the "sextants" between protosepta (stippled areas) are smaller than others, initial stages hypothetical; and (4) a typical scleractinian having equal sextant divisions. The mode of insertion of septa during growth of each series is worthy of special notice.

Fan-shaped diagrams at lower left are spread-out maps of the arrangement of septal grooves on the exterior of a rugose coral (5) and of septal edges at the periphery of a scleractinian coral (6). The obliquely pinnate arrangement of lines seen in 5 contrasts with radial regularity shown in 6.

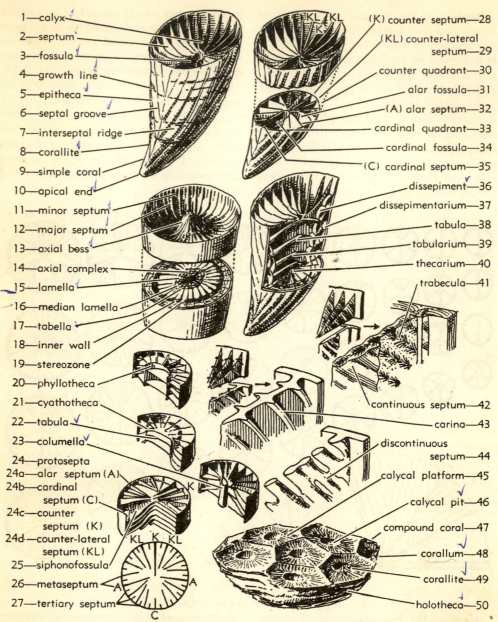

1—calyx
2—septum
3—fossula
4—growth line
5—epitheca
6—septal groove
7—interseptal ridge
8—corallite
9—simple coral
10—apical end
11—minor septum
12—major septum
13—axial boss
14—axial complex
15—lamella
16—median lamella
17—tabella
18—inner wall
19—stereozone
20—phyllotheca
21—cyathotheca
22—tabula
23—columella
24—protosepta
24a—alar septum (A)
24b—cardinal septum (C)
24c—counter septum (K)
24d—counter-lateral septum (KL)
25—siphonofossula
26—metaseptum
27—tertiary septum

(K) counter septum—28
(KL) counter-lateral septum—29
counter quadrant—30
alar fossula—31
(A) alar septum—32
cardinal quadrant—33
cardinal fossula—34
(C) cardinal septum—35
dissepiment—36
dissepimentarium—37
tabula—38
tabularium—39
thecarium—40
trabecula—41
continuous septum—42
carina—43
discontinuous septum—44
calycal platform—45
calycal pit—46
compound coral—47
corallum—48
corallite—49
holotheca—50

Fig. 4-14. **Morphologic terms commonly applied to rugose corals.** The various terms are explained briefly in the accompanying alphabetically arranged list, cross-indexed to the figure by numbers.

alar fossula (31). Gap in calyx in position of an alar septum or adjoining it.

alar septum (24a, 32). Protoseptum located about midway between cardinal and counter septa, generally identifiable by pattern of inserted new septa which join it pinnately on the counter side.

apical end (10). Pointed proximal extremity of a corallite where growth begins.

axial boss (13). Central prominence in calyx formed by an axial structure.

axial complex (14). Differentiated structure in central part of thecarium formed by twisted

(Continued on next page.)

(Fig. 4-14 continued.)

inner edges of septa, by tabellae and lamellae, or by abruptly deflected parts of tabulae.

calycal pit (46). Localized depression in central part of calyx.

calycal platform (45). Flat or gently sloping part of calycal floor.

calyx (1). Bowl-shaped depression at summit of a corallite, chiefly formed by upper edges of septa.

cardinal fossula (34). Gap in calyx in position of the cardinal septum; may be open, when edges of adjoining septa do not join, or closed, when they unite.

cardinal quadrant (33). Part of thecarium between cardinal septum and either of alar septa.

cardinal septum (24*b*, 35). Protoseptum which is adjoined pinnately on both sides by newly inserted septa.

carina (43). Small flangelike projection on side of septum, formed by thickened trabecula.

columella (23). Relatively solid axial structure.

compound coral (47). Skeleton of a colonial coral.

continuous septum (42). Radial wall of corallite formed of laterally adjoined trabeculae, or of uninterrupted fibers or laminar tissue.

corallite (8, 49). Skeleton formed by an individual coral polyp, whether solitary or forming part of a colony.

corallum (48). Skeletal deposit of a coral colony.

counter-lateral septum (24*d*, 29). Protoseptum adjacent to counter septum on either side.

counter quadrant (30). Part of thecarium between counter septum and either of alar septa.

counter septum (24*c*, 28). Protoseptum opposite cardinal septum.

cyathotheca (21). Inner wall formed by sharp deflections and union of tabulae.

discontinuous septum (44). Radial wall formed of trabeculae which are not solidly joined together, or wall which is interrupted longitudinally or peripherally.

dissepiment (36). Vesicle typically occurring in marginal parts of thecarium, with convexly curved wall toward axis of corallite.

dissepimentarium (37). Part of thecarium occupied by dissepiments.

epitheca (5). Outer wall of corallite.

fossula (3). Gap or depression in floor of calyx.

growth line (4). Marking on epitheca, such as slight ridge or depression parallel to margin of calyx, defining a former position of this edge.

holotheca (50). Generally wrinkled lamina deposited by colonial corals to cover base of corallum.

inner wall (18). Partition approximately parallel to epitheca, which may develop in various ways inside thecarium.

interseptal ridge (7). Longitudinal elevation on exterior of epitheca corresponding in position to space between two septa.

lamella (15). Radially disposed wall in central part of corallite forming part of axial complex; may or may not be joined to a septum.

major septum (12). Any protoseptum or metaseptum, distinguished from minor septa by greater length.

median lamella (16). Wall in axial complex in plane of counter and cardinal septa.

metaseptum (26). Any major septum other than a protoseptum.

minor septum (11). Generally short, secondarily introduced septum belonging to a cycle which makes appearance nearly simultaneously between major septa.

phyllotheca (20). Inner wall formed by abrupt bending and union of axial edges of septa.

protoseptum (24). One of the first-formed six septa.

septal groove (6). Longitudinal furrow on exterior of epitheca in position of a septum.

septum (2). Radial wall generally extending from peripheral edge of corallite partly or entirely to its axis, but in some species not extended to the periphery.

simple coral (9). Equivalent to solitary; not colonial.

siphonofossula (25). Gap in calyx formed by localized abrupt depression of tabulae adjacent to largely suppressed cardinal septum.

stereozone (19). Thin or thick band of relatively dense calcareous deposits which is formed in various ways within thecarium.

tabella (17). Steeply inclined tabula-like lamina in central part of thecarium, generally forming part of an axial complex.

tabula (22, 38). Convex, plane, concave, or irregular lamina extending across thecarium transversely, or developed only in central part of thecarium.

tabularium (39). Part of corallite occupied by tabulae.

tertiary septum (27). Equivalent to minor septum in classification which designates protosepta as primary and metasepta as secondary.

thecarium (40). Interior of a corallite, consisting of entire space enclosed by epitheca, or equivalent to such space if external wall is lacking.

trabecula (41). Row of calcite fiber bundles comprising a main structural element of septa.

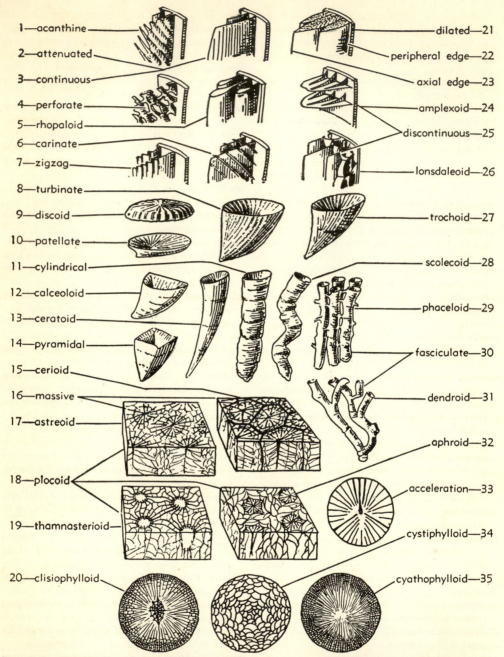

1—acanthine
2—attenuated
3—continuous
4—perforate
5—rhopaloid
6—carinate
7—zigzag
8—turbinate
9—discoid
10—patellate
11—cylindrical
12—calceoloid
13—ceratoid
14—pyramidal
15—cerioid
16—massive
17—astreoid
18—plocoid
19—thamnasterioid
20—clisiophylloid

dilated—21
peripheral edge—22
axial edge—23
amplexoid—24
discontinuous—25
lonsdaleoid—26
trochoid—27
scolecoid—28
phaceloid—29
fasciculate—30
dendroid—31
aphroid—32
acceleration—33
cystiphylloid—34
cyathophylloid—35

FIG. 4-15. **Descriptive terms commonly applied to rugose corals.** The alphabetically arranged brief explanations on facing page are cross-indexed to the figure by numbers.

acanthine (1). Spinose; type of septal edge in which trabeculae are not laterally joined together.

acceleration (33). More rapid introduction of septa in some quadrants than in others.

amplexoid (24). Characterized by very short septa, which commonly are interrupted longitudinally, as in *Amplexus*.

aphroid (32). Structure of a massive corallum in which absence of epithecal walls between

The major septa extend down a considerable part or all of the way to the apical end, and study of them establishes that, in turn, they are classifiable in two groups: a primary set of so-called **protosepta,** which are first-formed of all, and a secondary set, which includes all other metasepta, introduced as growth proceeds (Figs. 4-13, *1;* 4-14, *24).* Cross sections near the initially formed tip of the corallite show the identity and placement of the protosepta, and they serve also to demonstrate a fundamental bilateral symmetry of the septal pattern. First in order of appearance are two septa on opposite sides of the very young corallite; they are, respectively, named **cardinal septum** and counter-cardinal septum, but the latter is more conveniently shortened to **counter septum.** Next come two septa from nearly middle parts of the sides, and these are designated as **alar septa** because they have

---

(Fig. 4-15 continued.)

corallites is accompanied by occurrence of a dissepimental zone that separates septate parts of adjoining corallites.

astreoid (17). Massive corallum in which absence of epithecal walls between corallites is accompanied by near contact of long straight septa belonging to adjoining corallites.

attenuated (2). Relatively long and thin; characterizes some septa.

axial edge (23). Margin of septum toward center of corallite.

calceoloid (12). Shaped like the toe of a slipper, as in *Calceola.*

carinate (6). Bearing keels or flanges; a character of septa having laterally widened trabeculae.

ceratoid (13). Hornlike, having slender conical form, with apical angle approximately 20 deg. or less.

cerioid (15). Massive corallum in which corallites possess epithecal walls between them.

clisiophylloid (20). Having an axial structure which in transverse section resembles a spider web, as in *Clisiophyllum.*

continuous (3). Uninterrupted, imperforate; characterizes solidly built septa.

cyathophylloid (35). Having many long, thin, evenly radial septa and no axial structure, as in *Cyathophyllum.*

cylindrical (11). Subcircular, parallel-sided corallite.

cystiphylloid (34). Having entire thecarium filled by dissepiments and septa virtually absent, as in *Cystiphyllum.*

dendroid (31). Corallum composed of cylindrical corallites which branch irregularly in treelike manner.

dilated (21). Appreciably thickened; commonly characterizes septa.

discoid (9). Button-shaped.

discontinuous (25). Interrupted; characterizes septa in which trabeculae are not well joined

(perforate type), peripheral edges do not reach epitheca (lonsdaleoid type), and longitudinal interruptions may appear (amplexoid type).

fasciculate (30). Bundled; refers to any compound coral in which neighboring corallites are widely enough spaced to avoid mutual interference.

lonsdaleoid (26). Characterized by septa which disappear peripherally in a dissepimental zone, without reaching epitheca, as in *Lonsdaleia.*

massive (16). Refers to any compound coral in which adjoining corallites press against each other so as to change their shape.

patellate (10). Extremely low conical, having an apical angle of 120 deg. or more.

perforate (4). Containing holes; refers to septa in which spaces between trabeculae are incompletely filled by calcareous deposits.

peripheral edge (22). Margin of septum toward epitheca.

phaceloid (29). Fasciculate corallum having subparallel corallites.

plocoid (18). Massive corallum in which epithecal walls between corallites are lacking.

pyramidal (14). Shaped like a pyramid, with angular edges.

rhopaloid (5). Club-shaped; refers to septa which are dilated along their axial edges.

scolecoid (28). Wormlike; refers to cylindrical corallites which bend crookedly.

thamnasterioid (19). Massive corallum in which absence of epithecal walls between corallites is accompanied by confluent, interconnecting septa, curved as in lines of a magnetic field.

trochoid (27). Moderately steep conical, having an apical angle of approximately 40 deg.

turbinate (8). Low conical, having an apical angle of approximately 70 deg.

zigzag (7). Angularly crooked; characterizes axial portions of some septa.

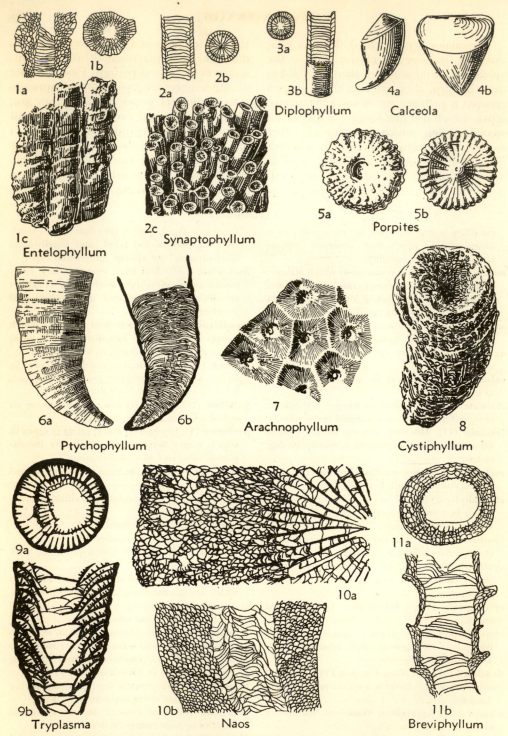

1b

1a

2b

2a

3a

3b
Diplophyllum

4a

4b
Calceola

1c
Entelophyllum

2c Synaptophyllum

5a

5b
Porpites

6a

6b

Ptychophyllum

7
Arachnophyllum

8
Cystiphyllum

9a

10a

11a

9b
Tryplasma

10b Naos

11b
Breviphyllum

FIG. 4-16. **Representative Silurian rugose corals.**

Arachnophyllum Dana, Silurian. *A. pentagonum* (Goldfuss) (7, ×1), Niagaran, Quebec.

Breviphyllum Stumm, Silurian–Devonian. *B. cliftonense* (Amsden) (11*a, b,* sections of specimen

(*Continued on next page.*)

124

a winglike position. Third is a pair of septa adjacent to the counter septum, disposed on either side; these are called **counter-lateral septa** (Figs 4-13, *1;* 4-14, *24a–d, 28, 29, 32, 35*). After the six protosepta have been introduced, the secondary major septa are added according to a definite plan: first pairs on the cardinal side of the counter-laterals and alars; then, second pairs on the cardinal side of the first two pairs; and so on with third, fourth, and other sets of pairs. This explains the tetrameral organization of hard parts which especially characterizes the Rugosa. There are four quadrants: two **cardinal quadrants,** comprising areas between the cardinal and alar septa; and two **counter quadrants,** comprising areas between the counter and alar septa (Fig. 4-14, *30, 33*).

Septa in the mature part of corallites belonging to *Streptelasma* and *Lambeophyllum* are evenly disposed, but in many other genera, such as *Zaphrenthis, Siphonophrentis* (Fig. 4-17, *7a, 9b*), *Hadrophyllum* (Fig. 4-18, *2b*), *Homalophyllum, Triplophyllites, Hapsiphyllum,* and *Dipterophyllum* (Fig. 4-20, *2a, 4b, 6b, 8b*), the cardinal septum is much shortened or it may even vanish distally. The gap thus produced in the floor of the calyx, supplemented by arrangement of septa in the cardinal quadrants, forms a more or less prominent depression, which is termed the **cardinal fossula** (Fig. 4-14, *3, 34*). **Alar fossulae** may be present on the counter side of the alar septa (Figs.

4-13, *1;* 4-17, *11;* 4-20, *4b*), and a counter fossula occurs rarely in position of the counter septum. Without doubt, this bilateral symmetry in arrangement of septa reflects placement of the directive mesenteries and the related pattern of soft-part radial walls of the corals.

Microscopic examination of the septal structure in *Streptelasma* and *Lambeophyllum* shows evenly distributed, closely packed fine calcite fibers disposed obliquely or nearly normally to the sides of the septa, diverging from the center line. The septa of most corallites, however, reveal organization of fibers in definite bundles, generally arranged in rows which slope upward and inward from the peripheral edge of the septum, and they may form ridges on the sides of the septa. Such rows of fiber bundles are called **trabeculae** (Fig. 4-14, *41*). Knowledge of this structural feature serves for the understanding of many peculiarities of septa which have importance in the classification of the corals. Among types of septa which particularly reflect trabecular features are those called **continuous, discontinuous, perforate, acanthine,** and **carinate** (Figs. 4-14, *42–44;* 4-15, *1, 4, 6*).

Tabulae. Last among structural characters illustrated by *Streptelasma* (but seemingly lacking in *Lambeophyllum*) are thin, gently arched platforms, which extend across the thecarium beneath the floor of the calyx. They are called **tabulae** (Figs.

(Fig. 4-16 continued.)

showing regeneration or spasmodic growth, ×2), Niagaran, Tennessee.

Calceola Lamarck, Silurian–Pennsylvanian. *C. tennesseensis* Rominger (*4a, b,* side views showing operculum, ×1), Niagaran, Tennessee.

Cystiphyllum Lonsdale, Silurian. *C. niagarense* (Hall) (*8,* ×1), Niagaran, Michigan.

Diplophyllum Hall, Silurian–Devonian. *D. caespitosum* Hall (*3a, b,* corallite of a phaceloid colony, ×1), Niagaran, Ontario.

Entelophyllum Wedekind, Silurian. *E. rugosum* (Smith) (*1a–c,* sections and part of phaceloid corallum, ×1), Niagaran, Ontario.

Naos Lang, Silurian–Devonian. *N. sewellensis* Amsden (*10a,* part of transverse section, ×4;

10b, longitudinal section, ×2), Niagaran, Tennessee.

Porpites Schlotheim, Silurian–Cretaceous. *P. rotuloides* (Hall) (*5a, b,* calycal and basal views, ×2), Niagaran, New York.

Ptychophyllum Edwards & Haime, Silurian–Devonian. *P. stokesi* Edwards & Haime (*6a, b,* ×0.5), Niagaran, Ontario.

Synaptophyllum Simpson, Silurian–Devonian. *S. multicaule* (Hall) (*2a, b,* sections, ×2; *2c,* corallum, ×1), Niagaran, Michigan.

Tryplasma Lonsdale, Silurian–Mississippian. Distinguished by short, discontinuous acanthine septa. *T. brownsportensis* (Amsden) (*9a, b,* sections, ×2), Niagaran, Tennessee.

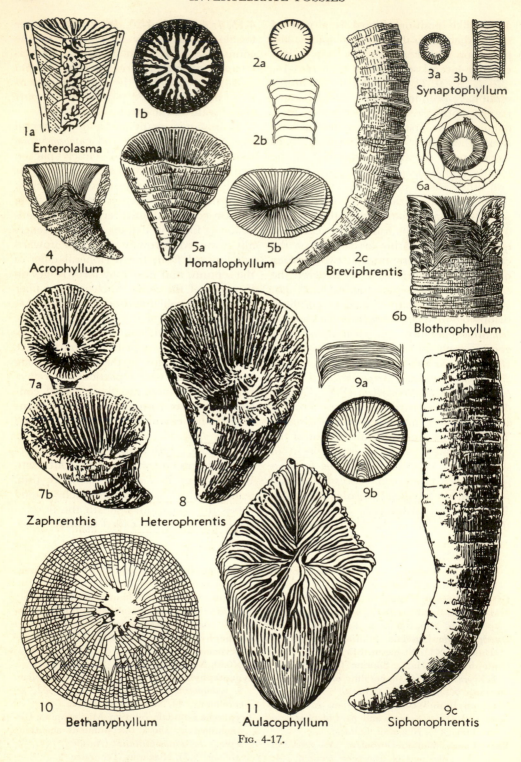

1a Enterolasma
1b
2a
2b
3a 3b Synaptophyllum
4 Acrophyllum
5a
5b
2c Breviphrentis
6a
6b Blothrophyllum
7a
7b
Zaphrenthis
8
Heterophrentis
9a
9b
9c Siphonophrentis
10 Bethanyphyllum
11 Aulacophyllum

Fig. 4-17.

4-10, *1c;* 4-14, *22).* In other rugose corals, they may bend downward in the central region, or be nearly plane. They extend through the interseptal spaces, serving as a floor for the polyp, which no longer occupies the apical part of the corallite. Among corals having short septa, tabulae may be the only sort of structural element in the axial part of the corallite (Figs. 4-16, *9, 11;* 4-17, *2;* 4-18, *4;* 4-21, *7).*

Structures Illustrated by Heritschia

The Lower Permian genus *Heritschia* (Fig. 4-22, *1)* is suitable for representation of some structural features common in rugose corals but not observed in *Streptelasma* or *Lambeophyllum. Heritschia* has long cylindrical corallites which externally bear strongly marked septal grooves and interseptal ridges. It has many long, evenly disposed septa, like the Ordovician fossils previously studied, but they are mostly very attenuated and flexuous, showing a tendency to lose their identity near the periphery. Septa which are impersistent or disappear toward margins of a corallite are termed **lonsdaleoid** (Fig. 4-15, *26).* Inner parts of the septa of *Heritschia* are thickened in a well-defined zone, and such a feature is termed a **stereozone;** it appears in many corals, but in some it is produced by thickening of other structural elements (Fig. 4-14, *19).*

Axial Structure. A striking character of *Heritschia* is the structure of its axial region. This is indicated externally by a sharply projecting prominence at the center of the calyx, constituting a spine or **axial boss** (Figs. 4-14, *13;* 4-22, *1d).* Transverse and longitudinal sections (Fig. 4-22, *1b,* c) show that the axial structure is mainly composed of concentric subvertical walls which in part may be upturned central sections of the tabulae, but they seem to be more numerous than the tabulae and are not traceable into them. Consequently, these parts of the **axial complex** are named **tabellae** (Fig. 4-14, *14, 17).* Other elements in the axial structure of *Heritschia* are radially directed vertical walls, which include extensions of the septa, especially of the counter septum; these elements are called **lamellae,** the one coinciding with the greatest width of the axial complex and connecting with the counter septum being known as the **median lamella** (Fig. 4-14, *15, 16).* A comparatively solid axial structure in which

---

FIG. 4-17.   **Representative Devonian rugose corals.**

Acrophyllum Thomson & Nicholson, Devonian. *A. oneidaense* (Billings) (4, side view, with top part in section, ×0.5), Lower Devonian, Ontario.

Aulacophyllum Edwards & Haime, Silurian–Devonian. *A. sulcatum* (d'Orbigny) (11, a weathered corallite, ×1), Lower Devonian, Indiana.

Bethanyphyllum Stumm, Devonian. *B. robustum* (Hall) (10, section, ×1), Middle Devonian, Ohio.

Blothrophyllum Billings, Silurian–Devonian. *B. decorticatum* Billings (6a, transverse section; 6b, side view with upper part sectioned; both ×0.5), Lower Devonian, Ontario.

Breviphrentis Stumm, Silurian–Mississippian. *B. yandelli* (Edwards & Haime) (2a–c, sections and side view of corallite, ×1), Lower Devonian, Ohio.

Enterolasma Simpson, Silurian–Devonian. *E.*

strictum (Hall) (1a, b, corallite cut in half longitudinal and transverse section, ×1.5), Lower Devonian, New York.

Heterophrentis Billings, Devonian. *H. prolifica* (Billings) (8, ×1), Middle Devonian, Michigan.

Homalophyllum Simpson, Devonian–Mississippian. *H. ungula* (Rominger) (5a, b, side and calycal views of corallite, which grows in a flattened form, ×1), Lower Devonian, Indiana.

Siphonophrentis O'Connell, Devonian. *S. gigantea* (Lesueur) (9a–c, sections and side view of corallite, ×0.5), Lower Devonian, Ohio.

Synaptophyllum Simpson, Silurian–Devonian. *S. arundinaceum* (Billings) (3a, b, sections, ×1), Lower Devonian, Ontario.

Zaphrenthis Rafinesque & Clifford, Devonian. *Z. phrygia* (Rafinesque & Clifford) (7a, b, corallites showing denticulate, carinate septa ×1), Lower Devonian, Kentucky.

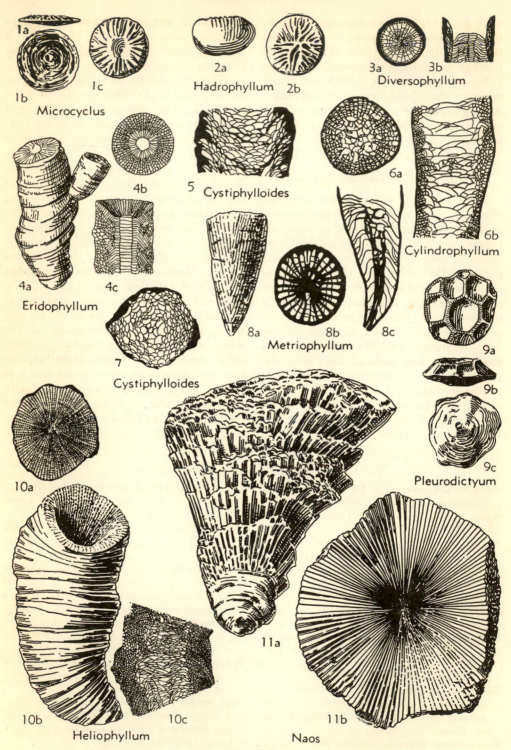

1a
1c
1b
Microcyclus
2a
Hadrophyllum   2b
3a   3b
Diversophyllum
4b
5 Cystiphylloides
6a
6b
Cylindrophyllum
4a   4c
Eridophyllum
7   Cystiphylloides
8a
8b
Metriophyllum   8c
9a
9b
Pleurodictyum   9c
10a
11a
10b   10c
Heliophyllum   Naos
11b

FIG. 4-18.

diverse structural elements cannot be observed is commonly called a **columella** (Figs. 4-14, *23;* 4-22, *2e, 4c, d*). Corals having an axial structure composed of tabellae and lamellae, with spider-web cross section, are described as **clisiophylloid,** from the genus *Clisiophyllum,* which shows these features excellently (Fig. 4-15, *20*).

Dissepimentarium. Peripheral parts of the thecarium of *Heritschia* corallites contain innumerable vesicles bounded by convexly rounded walls on the sides facing toward the inner and upper parts of the coral interior. These are called **dissepiments** and the area in which they are concentrated constitutes a **dissepimentarium** (Figs. 4-14, *36, 37;* 4-22, *1a–c*). The dissepiments are built between the outer parts of the septa, but in some corals where well-defined septa are lacking, they may be the dominant element of the entire thecarium (Figs. 4-16, *8;* 4-18, *5, 7*). Corallites which exhibit unusual prominence of dissepiments may be described as **cystiphylloid** (Fig. 4-15, *34*).

Structures Illustrated by Synaptophyllum

Synaptophyllum is a Silurian and Devonian coral which may be chosen as an example of a large number of colonial rugose genera (Figs. 4-16, *2;* 4-17, *3;* 4-19, *8*). The internal structure of the corallites is much more simple than that of some which resemble it externally. The colonies of *Synaptophyllum* consist of many cylindrical individuals, growing parallel to one another and rather closely spaced. The collective skeletal remains of the entire colony constitute a **corallum,** and a corallum in which neighboring corallites are sufficiently separated to permit retention of their cylindrical form is called **fasciculate** (Figs. 4-14, *48;* 4-15, *30*). Two types of fasciculate compound corals are recognized; **dendroid,** in which treelike branching and divergent attitude of the corallites are prominent; and **phaceloid,** in which bifurcations are inconspicuous and the corallites grow in subparallel position (Fig. 4-15, *29, 31*). *Synaptophyllum* is a phaceloid coral.

Sections of *Synaptophyllum* show regularly distributed septa, closely spaced, nearly flat tabulae, and a narrow dissepimentarium, which forms an **inner wall** (Figs. 4-14, *18;* 4-16, *2a, b;* 4-17, *3a, b*).

Increase of Corallites. Multiplication of individuals, on which expansion of compound coralla obviously depends, is a feature calling for notice. In colonies of

---

FIG. 4-18. **Representative Devonian corals.** All forms illustrated are Rugosa except *9a–c,* which is a tabulate coral.

Cylindrophyllum Simpson, Devonian. *C. panicum* (Winchell) (*6a, b,* sections, ×2), Middle Devonian, Michigan.

Cystiphylloides Chapman, Devonian. *C. conifollis* (Hall) (*5,* section, ×2), Middle Devonian, New York. *C. americanus* (Edwards & Haime) (*7,* section, ×1), Lower Devonian, Ohio.

Diversophyllum Sloss, Devonian. *D. traversense* (Winchell) (*3a, b,* sections, ×1), Middle Devonian, Michigan.

Eridophyllum Edwards & Haime, Devonian. *E. archiaci* Billings (*4a,* side view of corallite; *4b, c,* sections; all ×1), Middle Devonian, Michigan.

Hadrophyllum Edwards & Haime, Devonian–Mississippian. *H. orbignyi* Edwards & Haime (*2a, b,* side and calycal views of corallite, ×1), Middle Devonian, Kentucky.

Heliophyllum Dana, Devonian. *H. halli* Edwards & Haime (*10a–c,* corallite and sections, ×1), Middle Devonian, Michigan.

Metriophyllum Edwards & Haime, Devonian. *M. rectum* (Hall) (*8a–c,* corallite and sections, ×1.5), Middle Devonian, New York.

Microcyclus Meek & Worthen, Devonian–Mississippian. *M. discus* Meek & Worthen (*1a–c,* side, basal, and top views of corallite, ×1), Middle Devonian, Illinois.

Naos Lang, Silurian–Devonian. *N. magnificus* (Billings) (*11a, b,* side and calycal views of weathered specimen, ×1), Lower Devonian, Michigan.

Pleurodictyum Goldfuss, Devonian–Pennsylvanian. *P. lenticulare* (Hall) (*9a–c,* top, side, and basal views of corallum, ×1), Lower Devonian, New York.

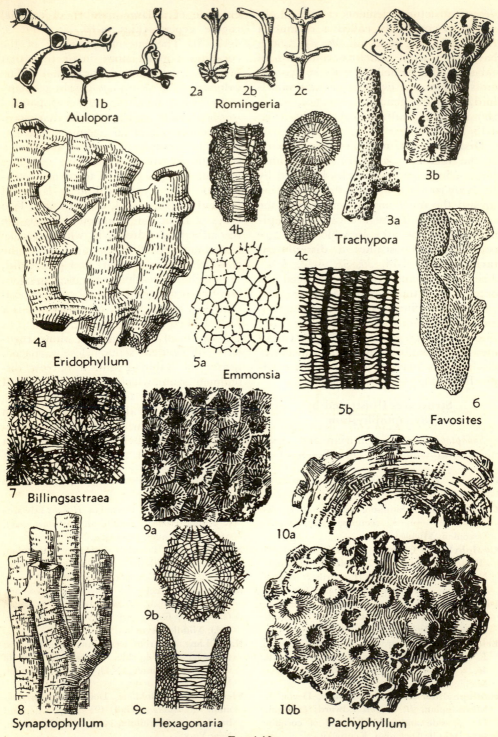

1a   1b
Aulopora

2a   2b   2c
Romingeria

3b

3a
Trachypora

4b

4c

4a
Eridophyllum

5a
Emmonsia

5b

6
Favosites

7 Billingsastraea

9a

10a

8
Synaptophyllum

9b

9c
Hexagonaria

10b
Pachyphyllum

Fig. 4-19.

Synaptophyllum, one corallite here and there springs from the side of another; this is known as **lateral increase** (Fig. 4-19, *8*). Obviously, however, such budding cannot represent divergence of a daughter polyp from the side of its parent at some point below the latter's calyx, for such parts of the parent corallite are solidly walled in by epitheca and they are already vacated by the polyp which rests on its last-formed tabula. Budding takes place, then, at the margin of the calyx, and subsequently the old and new individual grow upward side by side.

In some corals, both compound and simple, the parent itself evidently ceases to grow when one or more new individuals rise from the axial or peripheral parts of its calyx, or skeletal structures formed by parent and offspring cannot be discriminated clearly. Unless there is multiplication of corallites, **regeneration** may be difficult to distinguish from what is called **rejuvenescence,** which is indicated by new growth after seemingly complete or nearly complete cessation of old growth. An example of rejuvenescence, or possibly of spasmodic growth interrupted by rest periods, is the Silurian coral illustrated in Fig. 4-16, *11*.

Structures Illustrated by Hexagonaria and Pachyphyllum

Relations of Corallites. Characteristic Devonian compound corals which build **massive** coralla are *Hexagonaria* and *Pachyphyllum* (Figs. 4-15, *16;* 4-19, *9, 10*). The corallites are so closely packed that each continuously abuts its neighbors on all sides, and accordingly their shape in cross section is polygonal rather than circular. In *Hexagonaria*, epithecal walls separate adjoining corallites, a structure called **cerioid,** whereas no such walls occur in *Pachyphyllum*, a character termed **plocoid** (Fig. 4-15, *15, 18*). Plocoid corals may be classed as **astreoid,** if the septa are relatively straight and those of neighboring corallites nearly meet; **thamnasterioid,** if the septa of adjoining corallites are confluent and curved so as to resemble lines of a magnetic field; and **aphroid,** if the septa are short and adjacent corallites are separated by a dissepimental zone (Fig. 4-15, *17, 19, 32*). Specimens of *Pachyphyllum* commonly exhibit astreoid structure, but some are aphroid.

Calyces. A few surface features of the corallites deserve notice. The calyces of *Hexagonaria* have a relatively broad outer

---

FIG. 4-19.   **Representative Devonian corals.** Colonial rugose corals include 4, 7–10; the others are tabulates.

Aulopora Goldfuss, Silurian–Pennsylvanian. *A. elleri* M. A. Fenton (1*a–b*, corallites ×2 and ×1), Middle Devonian, New York.

Billingsastraea Grabau, Devonian. *B. billingsi nevadensis* Stumm (7, section, ×1), Middle Devonian, Nevada.

Emmonsia Edwards & Haime, Devonian. *E. emmonsi* (Hall) (5*a, b*, sections, ×1), Lower Devonian, New York.

Eridophyllum Edwards & Haime, Devonian. *E. seriale* Edwards & Haime (4*a–c*, corallites and sections, ×1), Lower Devonian, Kentucky.

Favosites Lamarck, Ordovician–Permian. *F. limitaris* Rominger (6, side view of corallum, ×1), Lower Devonian, Indiana.

Hexagonaria Gürich, Devonian. *H. percarinata*

Sloss (9*a*, surface of corallum, ×1; 9*b, c*, sections, ×2), Middle Devonian, Michigan.

Pachyphyllum Edwards & Haime, Devonian. *P. woodmani* (White) (10*a, b*, base of corallum showing holotheca, and top view, ×1), Upper Devonian, Iowa.

Romingeria Nicholson, Silurian–Devonian. *R. umbellifera* (Billings) (2*a–c*, parts of coralla, showing individuals budded in whorls, ×1), Lower Devonian, Michigan.

Synaptophyllum Simpson, Silurian–Devonian. *S. arundinaceum* (Billings) (8, part of corallum, ×1), Lower Devonian, Ontario.

Trachypora Edwards & Haime, Devonian–Pennsylvanian. *T. ornata* Rominger (3*a*, part of corallum, ×1; 3*b*, portion ×2.5), Middle Devonian, New York.

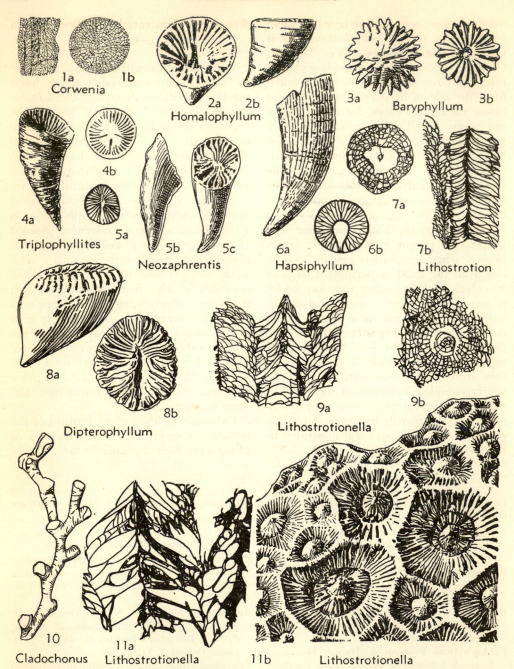

1a 1b
Corwenia

2a 2b
Homalophyllum

3a Baryphyllum **3b**

4b

4a
Triplophyllites **5a**

5b 5c
Neozaphrentis

6a 6b
Hapsiphyllum

7a

7b
Lithostrotion

8a

8b
Dipterophyllum

9a

9b
Lithostrotionella

10 **11a** **11b**
Cladochonus Lithostrotionella   Lithostrotionella

FIG. 4-20. **Representative Mississippian corals.** All forms illustrated but 10, which is a tabulate, are rugose corals.

Baryphyllum Edwards & Haime, Mississippian. *B. verneuilianum* Edwards & Haime (3a, b, basal and top views of corallite, ×1), Lower Mississippian, Tennessee.

Cladochonus McCoy, Devonian–Pennsylvanian.

C. beecheri (Grabau) (10, part of corallum, ×1), Lower Mississippian, Missouri.

Corwenia Smith & Ryder, Mississippian. *C. rugosa* (McCoy) (1a, b, sections, ×2), Lower Carboniferous, Wales.

(*Continued on next page.*)

floor, which is defined as the **calycal plat-form,** and a circular, abruptly depressed center, which is termed the **calycal pit** (Figs. 4-14, *45, 46;* 4-19, *9*). The calyces of *Pachyphyllum* are everted in such manner as almost to represent a cast made from the mold of *Hexagonaria* calyces. They resemble low volcanic cones, each with a broad crater at its summit. This is a rather unusual structure in the Rugosa.

Holotheca. The lower side of coralla belonging to *Hexagonaria* and *Pachyphyllum* is covered by a wrinkled dense calcareous lamina, which is secreted by confluent ectodermal tissue of polyps adjoining it. This protecting layer is called **holotheca** (Figs. 4-14, *50;* 4-19, *10a*).

Main Groups and Their Geological Occurrence

At least three main groups are recognizable among the rugose corals, each distinguished by its combination of structural characters and general mode of growth. Because family assemblages of closely related genera are placed wholly within one group or another, the main divisions may be viewed as superfamilies or suborders, but if they are so designated, it is important to observe that all components of one may not be derivatives of a single ancestral stock. If a group contains units of diverse origin, it is polyphyletic. The rugose coral divisions now being considered are called the suborders Streptelasmacea, Stauriacea,

and Cystiphyllacea. A chief basis for separation of these suborders is the presence or absence of dissepiments and, secondarily, features of the septa. Mode of growth broadly characterizes the groups but is not diagnostic.

Characters of Main Groups. The Streptelasmacea are solitary corals which lack dissepiments and possess strongly developed septa, grouped in distinct quadrants. This simple definition is accurate as applied to all but a few genera, and therefore it is useful. Among streptelasmacean corals, a few dissepiments do occur in certain genera, although not in some where they are reported to exist, probably on the basis of erroneous interpretation of transverse sections. Portions of tabulae can be misidentified very easily in such sections, but distinction between tabulae and dissepiments is clearly evident in most longitudinal sections. Colonial growth is rare but not excluded from the Streptelasmacea, and accentuation of septal grouping in cardinal and counter quadrants is not universal. These qualifications, however, do not modify our definition importantly.

The Stauriacea comprise a host of rugose corals in which dissepiments generally are abundant and septa well developed but not grouped in quadrants; colonial growth predominates greatly. This statement also requires minor emendation, for some Ordovician genera belonging here lack dissepiments. Also, septal quadrants

---

(Fig. 4-20 continued.)

Dipterophyllum Roemer, Mississippian. *D. glans* (White) (8*a, b,* side and calycal views of corallite, ×2), Lower Mississippian, Missouri.

Hapsiphyllum Simpson, Mississippian–Pennsylvanian. *H. calcariforme* (Hall) (6*a, b,* side view of corallite and transverse section, ×1.5), Upper Mississippian, Indiana.

Homalophyllum Simpson, Devonian–Mississippian. *H. calceolum* (White & Whitfield) (2*a, b,* corallite, ×2), Lower Mississippian, Missouri.

Lithostrotion Fleming, Mississippian–Pennsylvanian. *L. whitneyi* Meek (7*a, b,* sections, ×1.5), Upper Mississippian, Montana.

Lithostrotionella Yabe & Hayasaka, Mississippian. *L. castelnaui* Hayasaka (9*a, b,* sections, ×1.5), Upper Mississippian, Kentucky. *L. hemisphaerica* Hayasaka (11*a, b,* section and surface of corallum, ×2), Upper Mississippian, Missouri.

Neozaphrentis Grove, Silurian–Pennsylvanian. *N. tenella* (Miller) (5*a–c,* section and two views of corallites, ×1.5), Lower Mississippian, Missouri.

Triplophyllites Easton, Mississippian. *T. spinulosus* (Edwards & Haime) (4*a, b,* corallite and section, ×1), Lower Mississippian, Illinois.

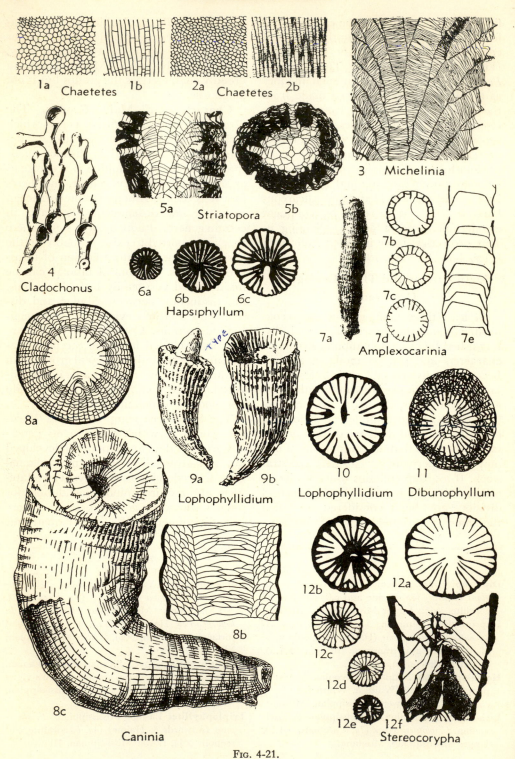

1a Chaetetes 1b   2a Chaetetes 2b

3 Michelinia

4
Cladochonus

5a Striatopora 5b

6a 6b 6c
Hapsiphyllum

7a 7b 7c 7d 7e
Amplexocarinia

8a

9a 9b
Lophophyllidium

10
Lophophyllidium

11
Dibunophyllum

8b

12b 12a

12c

12d

12e 12f

8c

Caninia

Stereocorypha

FIG. 4-21.

are clearly evident in *Aulacophyllum*, for example. Otherwise, discrimination of stauriacean corals is easy.

The Cystiphyllacea are distinguished by an extraordinary profusion of dissepiments, coupled with a reduced vertical continuity of septa to the point of near disappearance; both colonial and solitary growth are common.

Streptelasmacea. Among Middle and Upper Ordovician corals, *Streptelasma* and *Lambeophyllum* (Fig. 4-10, *1*, *2*), which have been studied, are good examples of this group and are prominent. Silurian rocks contain several genera of small solitary corals containing a more or less complex axial structure. Others (Fig. 4-16, *9*) lack such a structure, being distinguished by much shortened septa and a prominent tabularium.

Devonian strata yield a large variety of streptelasmacean corals. One important group is known as zaphrenthids (Figs. 4-17, *2*, *5*, *7–9*; 4-18, *8*, *10*), characterized by numerous long septa and a well-developed cardinal fossula. Some attain large size, and they are excellent examples of the so-called horn corals or cup corals. *Zaphrenthis* (Fig. 4-17, *7*), which has distinctly denticulate, carinate septa, is very probably an ancestor of the similarly characterized *Heliophyllum* (Fig. 4-18, *10*), as well as *Caninia* and other genera of the Stauriacea. A descendant of *Zaphrenthis* having very short septa throughout the long-extended mature region is illustrated in Fig. 4-17, *2*. All these genera are characterized by upwardly arched tabulae, which extend from wall to wall of the thecarium; they lack dissepiments.

Very unlike most corals are the discoid, button-like Devonian-to-Pennsylvanian members of the family Hadrophyllidae (Figs. 4-18, *1*, *2*; 4-20, *3*, *8*). Similarly shaped corals occur in Silurian rocks (Fig. 4-16, *5*), but are assigned to another family because of the difference in structure and arrangement of the septa. Also, discoid scleractinian corals are distributed from Triassic to Recent (Figs. 4-28, *5*; 4-29, *1*, *7*; 4-30, *1*, *6*, *13*, *15*). This growth form, accordingly, is not unique, nor does it distin-

---

FIG. 4-21. Representative Pennsylvanian corals. Rugosa include the solitary corallites shown in *6–12*, and the others are Tabulata.

Amplexocarinia Soschkina, Pennsylvanian–Permian. *A. corrugata* (Mather) (*7a*, a corallite, ×1.3; *7b–e*, sections, ×2), Lower Pennsylvanian (Morrowan), Oklahoma.

Caninia Michelin, Devonian–Pennsylvanian. *C. torquia* (Owen) (*8a–c*, a corallite and sections, ×1), Upper Pennsylvanian (Virgilian), Nebraska.

Chaetetes Fischer, Ordovician–Jurassic. *C. milleporaceus* Edwards & Haime (*1a*, *b*, sections, ×3), Middle Pennsylvanian (Desmoinesian), Kansas. *C. eximius* Moore & Jeffords (*2a*, *b*, sections, ×3.3), Lower Pennsylvanian (Morrowan), Oklahoma.

Cladochonus McCoy, Devonian–Pennsylvanian. *C. texasensis* Moore & Jeffords (*4*, part of corallum, ×0.7), Middle Pennsylvanian (Atokan), Texas.

Dibunophyllum Thomson & Nicholson, Mississippian–Pennsylvanian. *D. valeriae* Newell (*11*, section, ×1.5), Upper Pennsylvanian (Missourian), Kansas.

Hapsiphyllum Simpson, Mississippian–Pennsylvanian. *H. crassiseptum* Moore & Jeffords (*6a–c*, sections, ×2), Lower Pennsylvanian (Morrowan), Oklahoma.

Lophophyllidium Grabau, Pennsylvanian–Permian. *L. proliferum* (McChesney) (*9a*, corallite with edge of calyx broken, showing the prominent columella; *9b*, another specimen; both ×1.5), Upper Pennsylvanian (Missourian), Illinois. *L. conoideum* Moore & Jeffords (*10*, section, ×2), Middle Pennsylvanian (Atokan), Texas.

Michelinia De Koninck, Devonian–Permian. *M. referta* Moore & Jeffords (*3*, section, ×1), Middle Pennsylvanian (Atokan), Texas.

Stereocorypha Moore & Jeffords, Pennsylvanian. *S. annectans* Moore & Jeffords (*12a–f*, sections, ×2), Middle Pennsylvanian (Atokan), Texas.

Striatopora Hall, Silurian–Pennsylvanian. *S. oklahomaensis* (Snider) (*5a*, *b*, sections, ×1), Lower Pennsylvanian (Morrowan), Arkansas.

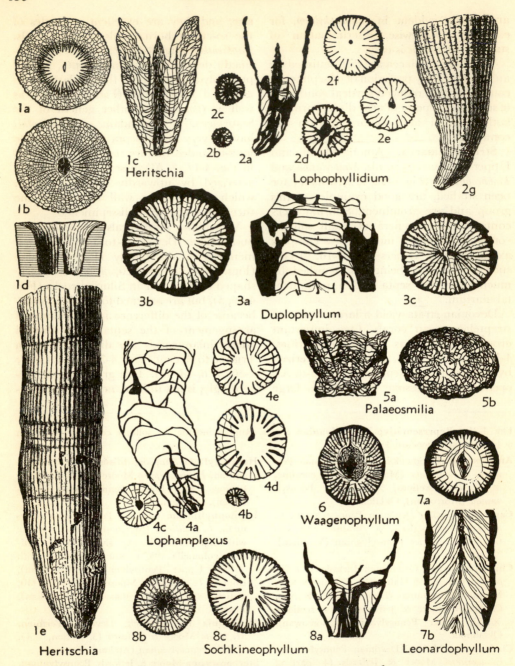

FIG. 4-22. **Representative Permian rugose corals.**

Duplophyllum Koker, Permian. *D. septarugosum* Moore & Jeffords (3a–c, sections, ×2), Lower Permian (Leonardian), Texas.

Heritschia Moore & Jeffords, Lower Permian. *H. girtyi* Moore & Jeffords (1a–e, sections and a corallite of the phaceloid corallum, ×2), Kansas.

Leonardophyllum Moore & Jeffords, Lower Permian. *L. distinctum* Moore & Jeffords (7a, b, sections, ×2), Leonardian, Texas.

(*Continued on next page.*)

guish any particular taxonomic group. It seems to be an ecologic adaptation, for modern discoid corals are chiefly found lying on sandy bottoms.

Mississippian streptelasmaceans, in addition to the discoid forms just noted, include numerous genera and species of zaphrenthids (Fig. 4-20, *2, 4–6*). Some especially well show an inner wall of the type termed **phyllotheca** (Fig. 4-14, *20*), formed by an abrupt bending and union of the inner extremities of the septa, the central cavity thus formed being open toward the much-shortened cardinal septum so as to make a prominent fossula.

Most common of the Pennsylvanian and Permian Streptelasmacea is *Lophophyllidium* (Figs. 4-21, *9, 10;* 4-22, *2*), which has a prominent **columella** (Fig. 4-14, *23*), formed by thickening of the axial edge of the counter septum. Numerous species have been described. Other genera are characterized by the club-shaped (**rhopaloid**) (Fig. 4-15, *5*) form of the septa in transverse section (Fig. 4-22, *8*); by the shortening of the septa in mature parts of the corallite (Figs. 4-21, *7, 8;* 4-22, *4*); and by the clear differentiation of protosepta and fossulae (Figs. 4-21, *6, 10, 12;* 4-22, *2, 4, 8*). Strong **acceleration** (Fig. 4-15, *33*) of the counter quadrants is an important feature well shown by some of these corals (Fig. 4-21, *6, 10, 12*).

Stauriacea. The beginning of the stauriacean group, which probably are ancestors, or very like the ancestors of Streptelasmacea, as well as other Stauriacea, is seen in Ordovician compound corals which bear well-developed septa and tabulae, but no dissepiments (Fig. 4-10, *7, 9*).

Silurian rocks, which contain the oldest well-defined coral reefs, are rich in stau-riaceans, both solitary and compound. Among horn corals illustrated here, some are distinguished by their many long, thin, axially twisted septa and closely spaced tabulae (Fig. 4-16, *6*); others have well-defined septa in the axial part of the corallite, but they become lost in crossing the wide zone of dissepiments (Fig. 4-16, *10*); and some have short septa, nearly plane tabulae, and narrow dissepimentarium (Fig. 4-16, *11*). One distinctive form (Fig. 4-16, *9*) has discontinuous strongly acanthine (Fig. 4-15, *1, 25*) septa which are very short. Three types of phaceloid Silurian corals and a typical massive corallum are shown in Fig. 4-16, *1–3, 7*.

The peak in variety and number of rugose coral genera is found in the Devonian, and stauriaceans considerably predominate over the two other main groups. Reef deposits occur in many areas, among them the richly productive petroleum reservoir rocks at Leduc and elsewhere in western Canada. Representative Devonian "simple" stauriacean corals, which are not at all simple in their internal structures, include relatively small forms with a narrow zone of dissepiments and more or less strongly arched tabulae (Figs. 4-17, *4;* 4-18, *3*) and a variety of robust horn corals characterized by prominence of dissepiments (Figs. 4-17, *6, 10, 11;* 4-18, *10, 11*). In most of these corals, the septa are long and very numerous. Compound coralla here illustrated include a few phaceloid types, which have cylindrical corallites, and three massive forms. In the phaceloid group are some distinguished by irregularity of the tabularium and dissepimentarium (Fig. 4-18, *6*) and others in which the structure of the tabularium and dissepimentarium is fairly regular, with or

---

(*Fig. 4-22 continued.*)

Lophamplexus Moore & Jeffords, Lower Permian. *L. eliasi* Moore & Jeffords (4*a–e*, sections, ×2), Kansas.

Lophophyllidium Grabau, Pennsylvanian–Permian. *L. dunbari* Moore & Jeffords (2*a–g*, sections and a corallite, ×2), Lower Permian, Kansas.

Palaeosmilia Edwards & Haime, Mississippian–Permian. *P. schucherti* Heritsch (5*a, b,* sections, ×1), Lower Permian, Texas.

Sochkineophyllum Grabau, Pennsylvanian–Permian. *S. mirabile* Moore & Jeffords (8*a–c,* sections, ×2), Lower Permian, Kansas.

Waagenophyllum Hayasaka, Middle and Upper Permian. *W. texanum* Heritsch (6, section, ×1.7), Upper Permian (Guadalupian), Texas.

without the presence of an inner wall (Figs. 4-17, *3;* 4-18, *4;* 4-19, *4, 8*). The massive stauriaceans include *Hexagonaria* and *Pachyphyllum*, already noticed, and a genus like *Hexagonaria* except for the lack of walls between the corallites (Fig. 4-19, *7, 9, 10*).

Mississippian coral faunas contain several sorts of stauriaceans, both simple and compound, but the genera are almost entirely different from those known in the Devonian. The Lower Carboniferous rocks of northwestern Europe have yielded a larger number of genera and species of these corals than is yet reported from North America in equivalent deposits, but this may partly represent more intensive study of these fossils on the eastern side of the Atlantic. The British section belonging to this part of the column is largely zoned in terms of corals; also, prolific faunas have been described from Russia. Characteristic stauriacean corals of Upper Mississippian rocks in North America, which include both phaceloid and massive types of colonies, are the closely similar *Lithostrotion* and *Lithostrotionella* (Fig. 4-20, *7, 9, 11*). In these genera, the tabularium is well differentiated, and a sharp upward projection of these platforms at the center produces an axial boss.

Pennsylvanian corals of this group are less numerous than streptelasmaceans. A most common type is a large simple coral, called *Caninia* (Fig. 4-21, *8*), in which the evenly distributed septa do not reach the axis. The cardinal septum can be identified in transverse sections, but other protosepta are not differentiated except by study of the apical region. In some other genera, long septa are developed in association with a spider-web type of axial structure (Fig. 4-21, *11*).

Only three stauriacean genera of Permian age, among many which are known, are shown in Fig. 4-22, *1, 5, 6*. They are characterized by the prominence of their axial structures.

Cystiphyllacea. From a numerical standpoint, the Cystiphyllacea are relatively unimportant. They include only a few genera, and all these are confined to Silurian and Devonian rocks. Notice is merited chiefly from the standpoint of their relation to evolutionary trends. They seem to be an end product in the tendency to develop vesiculose tissue inside the corallite walls, and concurrently with this, such universal coral structures as the septa dwindle and in some forms vanish. Reduction of the septa is a tendency of evolution seen in other groups of corals. In this shortening, they tend also to become developed in an axial direction only on the upper, distal side of successive tabulae. Among Cystiphyllacea, this tendency results in virtual disintegration of septa through fragmentation.

An alternative explanation of the short, discontinuous septa of cystiphyllacean corals, which postulates that these structures are incipient, rather than degenerate—signifying retention in mature growth stages of a character belonging to earliest youth (paedogenesis or neoteny)—seems to be wholly discordant with the extraordinary development of dissepimental structures.

Cystiphyllum is a genus to which both Silurian and Devonian corals formerly were assigned. Study of specimens from these rocks, however, has shown that the older ones possess septal crests of acanthine type, whereas the Devonian forms show mere septal striae. These latter are now separated under the name *Cystiphylloides* (Fig. 4-18, *5, 7*). A Silurian *Cystiphyllum* is shown in Fig. 4-16, *8*. Derivatives of different stauriacean groups probably are included among the cystiphyllacean rugose corals. If this is true, the suborder is polyphyletic. It incorporates various end products of evolution which trend in directions of increasing the dissepiments and reducing the septa.

Evolutionary Trends

Comparative study of many rugose coral groups which indicate sequences in development permits recognition of their evolutionary trends. Segregation of groups best adapted for this sort of analysis is made by observing those which possess certain common structural elements and which also supply evidence of genetic interrelationship by their growth stages. In addition, placement of the compared fossils in the geologic time scale must be taken into account.

A summary of evolutionary changes which can be defined in diverse coral stocks emphasizes two directions of trend: (1) increasing complexity of structural organization, and (2) simplification of structure. These together define ascending and descending portions of an evolutionary curve, which has a rounded peak representing maximum diversification and complication. Such a curve can be discerned by paleontologists in the evolution of many other groups of organisms, and accordingly, the pattern seems to have the attributes of a "law" of life.

Among rugose corals, the first-formed and simplest structure is the flat or gently rounded calcareous platform built by ectoderm of the base. Then come the septa, produced by deposits on either side of their mid-line by upward and inward foldings of the ectoderm. These features, which are basic elements of the rugose coral skeleton, are observed to develop in ontogeny, and they suggest that the primordial type of corallite is a discoid or patellate solitary form of small size, having perhaps only four or six septa (protosepta). The placement and order of appearance of the septa reflect the arrangement of directive mesenteries and other mesenteries. The addition of a few metasepta in serial order in each quadrant, defining in simple manner the characteristic tetrameral arrangement of rugose septa, constitutes an advance on the ascending limb of the evolutionary curve.

The shape of solitary corallites is postulated to progress during evolution from the primitive discoid form to a low cup-shaped outline (turbinate), and through moderately steep conical (trochoid) and elongate horn shapes (ceratoid) to cylindrical. Such a tendency is indicated in various stocks.

Septa may be expected to increase in numbers, in extent toward the axis, and in regularity of radial arrangement. In fact, this does characterize early evolution. In different lineages it is found associated with an axial structure, developed in one way or another by inner extremities of the septa. Reversal of these trends is seen in later evolution, however, for septa become shorter and shorter, leaving a central space occupied only by tabulae, as in *Caninia* and *Amplexus*, or severing connection with an axial structure that persists, as in *Clisiophyllum*. Such a tendency is defined as caninoid, amplexoid, or clisiophylloid, depending on the form which is approached. Caninoid corallites have peripherally shortened but longitudinally continuous septa and a prominent dissepimentarium, whereas amplexoid corallites exhibit extreme shortening and longitudinal discontinuity of the septa and absence of dissepiments. These corals represent similar trends but in different stocks. Shortening of septa may occur in another way—by attenuation and disappearance of peripheral parts, so that the septa cannot be traced to contact with the epitheca. Dissepimental vesicles occupy peripheral parts of such corals exclusively, and the evolutionary trend toward a structure of this sort is termed lonsdaleoid. It is less common than shortening of the septa on the axial side, but is not rare. Last to be mentioned is another modification of the septa which consists in localized or extensive thickening. Dilation confined to axial parts produces so-called rhopaloid septa, with club-shaped transverse section. Thickening of intermediate sections of the septa gives rise to one type of stereozone, and this is common. A few genera, mostly in Pennsylvanian and Permian rocks, have septa in which all parts are so dilated that virtually no space remains between them.

Tabulae and dissepiments are structures of rugose corals which reflect evolutionary trends. A primitive condition of tabulae may be differentiated readily from specialized ones, for primitive types should resemble the initial calcareous floor of the corallite in being plane or gently concave

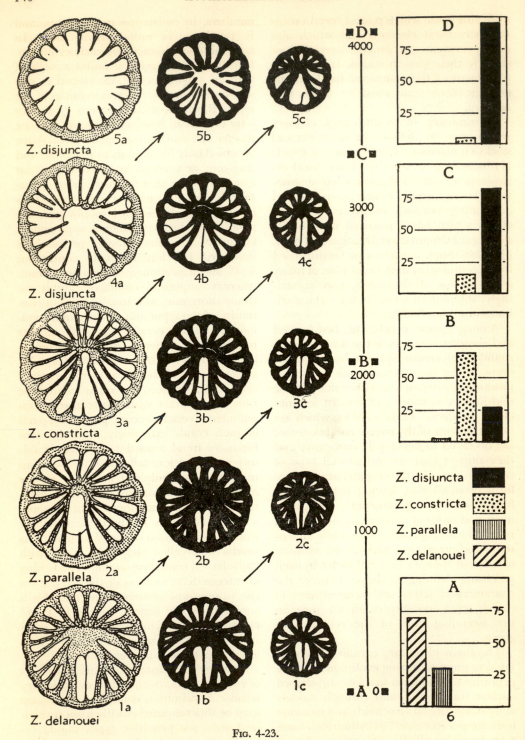

Z. disjuncta
5a 5b 5c

Z. disjuncta
4a 4b 4c

Z. constricta
3a 3b 3c

Z. parallela
2a 2b 2c

Z. delanouei
1a 1b 1c

D
4000
C
3000
B
2000
1000
A 0

D
75 50 25

C
75 50 25

B
75 50 25

Z. disjuncta
Z. constricta
Z. parallela
Z. delanouei

A
75 50 25
6

FIG. 4-23.

and in extending to the peripheral wall of the skeleton on all sides. Modifications of tabulae consist chiefly in a reduction of their lateral span until they are confined to a narrow axial part of the corallite, and in deflection of their attitude, generally by upward arching. Whether numbers and spacing of tabulae define significant evolutionary trends or merely reflect rates of growth is doubtful. Some rugose corals possess extremely abundant, closely crowded tabulae, whereas others have few or none at all.

Dissepiments are entitled to front rank as indicators of evolutionary change in the Rugosa. Their presence or absence, relative prominence, and to some extent, their form and size are useful in classification, in spite of the fact that independent development of them in different stocks contributes to production of homeomorphs. The primitive, geologically oldest corallites lack dissepiments, and this is true both for solitary and colonial types. Many highly specialized corals also lack dissepiments, but clearly there is prevalent tendency in evolution to introduce them. Stauriacea and Cystiphyllacea are characterized by all but universal occurrence of numerous dissepiments, and among corallites of the latter, they fill the thecarium. Recognizing that culmination in development of dissepiments occurs in the probably polyphyletic Cystiphyllacea, it is surprising to find

no corals of this type in post-Devonian rocks. Even if descendants of Devonian genera such as *Cystiphylloides* failed to survive, homeomorphic equivalents derived from Mississippian or Pennsylvanian *Caninia*-like genera seemingly should be found. None are known.

Colonial rugose corals are mainly, if not wholly, evolved from solitary forms. Some genera exhibit both modes of growth, and it is probable that an increase in the asexual mode of reproduction accounts for the observed advance in colonial compound types of Paleozoic corals as compared with solitary types. With the advent of numerous colonial corals, reef building became common in suitable environments, and this dates from Silurian time. Evolution in colonial organization may be seen in progression from (1) dendroid fasciculate growths to (2) phaceloid fasciculate corals, and from these to (3) massive colonies having cerioid structure (with epithecal walls persisting between individuals); most advanced are (4) plocoid corals, in which intervening walls have vanished. In the last-mentioned group, astreoid colonies are evidently less highly organized than thamnasterioid and aphroid types, which are specialized in divergent directions.

A classic example of evolutionary change in a coral assemblage is furnished by a succession of zaphrenthid populations col-

FIG. 4-23. **Evolution of a group of Mississippian zaphrenthids.** Analysis of collections made from four zones (A, B, C, D) in Lower Carboniferous rocks of Scotland shows evolutionary changes of the coral named *Zaphrentoides delanouei* during the time represented by the accumulation of approximately 4,000 ft. of strata.

Graphs at the right (6) indicate composition of collections, in per cent, specimens from "A" horizon (110 individuals) dominantly (69 per cent) include *Z. delanouei* (1a–c, black figures represent growth stages); these are associated with individuals referable to *Z. parallela* (30 per cent), which exhibits a *Z. delanouei* growth stage (2b) in its development, and one specimen classifiable as *Z. constricta* (3a–c), which passes through both *Z. delanouei* (3c) and *Z. parallela* (3b) growth stages. Collections at "B" horizon (679 specimens) contain only a trace of per-

sisting *Z. delanouei*, a few *Z. parallela* (3 per cent), a preponderant assemblage of *Z. constricta* (69 per cent), the remainder belonging to a product of evolution called *Z. disjuncta* (28 per cent), which exhibits growth stages typical of older species (4b, c). Collections at "C" (315 specimens) and "D" (20 specimens) also show changes, for *Z. disjuncta* strongly predominates. Thus, the trend of evolution is indicated both by composition of successive faunas and by growth stages (data from Carruthers, 1910).

lected mainly at four stratigraphic levels distributed through 4,000 ft. of Lower Carboniferous rocks in Scotland. As represented graphically in Fig. 4-23, the lowest collection was found to consist dominantly of the species now known as *Zaphrentoides delanouei*, associated with specimens which pass through a *Z. delanouei* stage during growth of the corallite but are so different at maturity as to warrant classifying them under another name, *Z. parallela*. Geologically younger populations include an insignificant remnant of these two species, which are replaced by others having *Z. delanouei* and *Z. parallela* stages successively developed in their ontogeny. Thus, an evolutionary succession of forms is demonstrated. Descendent types are chiefly characterized by changes in the fossulae and marked shortening of the septa.

The subject of evolutionary trends of corals, which has been reviewed in condensed form, remains very inadequately treated unless mention is made of one important point. This is the inequality or unevenness of progress in evolutionary change affecting different coral structures in various groups. Modification of some characters in one group of corals is rapid, while change in others lags. Evolution in another group is likewise uneven, but structures affected chiefly may differ from those in the first. Thus, complexity is great and comparisons are difficult.

HETEROCORALS

A minor branch of the late Paleozoic corals differs so fundamentally in structure from the Rugosa as to warrant separation in an independent order. It is called Heterocorallia. Although this division of the Zoantharia comprises only two described genera with short stratigraphic range, being confined to part of the Viséan stage (equivalent to much but not all of the Upper Mississippian of North America), representatives of both corals have been found in Scotland and Germany and as far east as Japan. Chief interest in them is directed to understanding their structure and how they originated.

The two genera included in the order of heterocorals are named *Hexaphyllia* and *Heterophyllia*. Both have slender cylindrical corallites, 5 to 15 mm. in diameter, estimated to attain a total height of more than 50 cm. (Fig. 4-24, *1a*, *2a*). They pose difficult questions as to their mode of life. The simplest septal arrangement, observed in cross sections of *Hexaphyllia* and approached in young stages of *Heterophyllia*, consists of four septa which meet to form a right-angled cross, two of the opposed septa bifurcating peripherally and the other two being undivided (Fig. 4-24, *1b*, *2b*). During growth of *Heterophyllia*, the initially unforked septa develop bifurcations and subordinate septa are introduced between each pair of limbs of the Y-shaped primary septa. This arrangement produces four fossulae in quadrants between the initial septa, where no new septa are developed.

The heterocoral septal arrangement, although quite unlike that of the Rugosa, can be explained as an abruptly introduced new pattern produced by divergence at a juvenile stage of rugose development when only the first four protosepta are present (Fig. 4-24, *3b*). If, instead of forming two more (counter-lateral) protosepta and proceeding with construction of metasepta in regular succession, as is normal in the Rugosa (Fig. 4-24, *4a–c*), a genetic line should strike off in a new direction so as to invent the heterocoral septal pattern (Fig. 4-24, *5a–c*), a radically different group of corals would originate suddenly. The fact that it is not possible to change the rugose pattern into that of the heterocorals by gradual rearrangement of septal growth, coupled with the abrupt appearance and widespread distribution of the new stock, leads to judgment that the introduction of *Hexaphyllia* and *Heterophyllia* was by such a jump. Abrupt evolutionary branching of this sort out of an embryonic or early juvenile growth stage of another organic stock is termed **proterogenesis**.

It may well be the mode of origin of many major animal groups, which branched off from others without intermediate growth forms. If this is true, missing links are missing because they never existed.

SCLERACTINIAN CORALS

The order Scleractinia includes all kinds of solitary and colonial modern corals, exclusive of some alcyonarians and hydrozoans, which loosely are called corals also. They have a fundamental hexameral organization, as is shown clearly by the septal pattern of many genera and quite as well indicated by the six pairs of protomesenteries in those which lack evident sixfold arrangement of skeletal parts. Relationship to the Rugosa may be seen in these characters. The scleractinians differ from rugose corals chiefly in the mode of addition of new septa, which appear in regular cycles distributed evenly in the sextants defined by the protosepta. Fossil scleractinians are common in many Mesozoic and Cenozoic rocks, the oldest known being found in Middle Triassic.

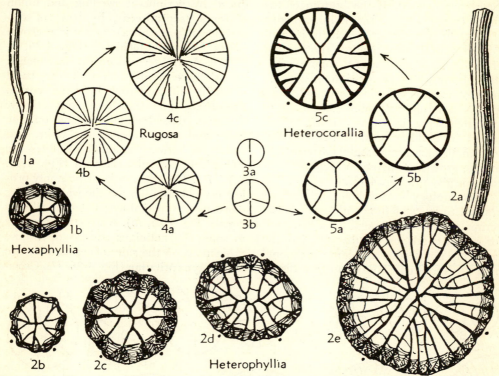

FIG. 4-24. **Characters of heterocorals.** These are a distinctive but short-lived branch which developed in Late Mississippian time. Earliest growth stages correspond to those of rugose corals, indicating derivation of the group from this stock. (*In part, modified from Schindewolf.*)

1a, b, *Hexaphyllia* Stuckenberg, Upper Mississippian. (1a) Corallite of *H. mirabilis* (Duncan) from Germany, ×1. (1b) Transverse section of same, ×8.

2a–e, *Heterophyllia* McCoy, Upper Mississippian. (2a Corallite of *H. reducta* Schindewolf, from Germany, ×1. (2b) Section of very early growth stage of *H.* cf. *parva*, ×10. (2c, d) Sections of young and mature *H. reducta*, ×8 and ×5. (2e) Sections of mature *H. grandis* McCoy, ×6.

3a, b, Early growth stages common to Rugosa and Heterocorallia.

4a–c, Later stages of Rugosa.

5a–c, Later stages of Heterocorallia; fossular areas extending to center of corallite marked by dots.

Scleractinian modern corals

Morphological Features

Study of growth stages of modern corals reveals that the mesenterial partitions, between pairs of which the septa form, make an appearance in a certain sequence. They are grouped in cycles, with that of the protomesenteries first, but those of each cycle are not produced simultaneously. The progession in development of these structures in one of the scleractinians is shown in Fig. 4-25. Both bilateral symmetry and polarity of the plane of symmetry are defined. If the bottom of the circles in Fig. 4-25 (at the peripheral edge of protomesenteries p2), which is considered ventral, corresponds to the position of the cardinal septum of a rugose coral, then the order of progression in appearance of metamesenteries (m1, m2, etc.) follows the rule governing placement of new metasepta in rugose corals, which results in building toward the cardinal septum. Comparison of septal patterns during the growth of rugose and scleractinian corals is given in Fig. 4-13.

Terminology. Morphological terms applied to parts of the corallites of Scleractinia mostly correspond to those used for rugose corals, except that many of the latter are not needed and some additional ones are required. Thus septa are distinguished as **entosepta** if they are located inside a mesenterial pair, and as **exosepta** if they are located outside such a pair, which means between two neighboring pairs (Figs. 4-26, *5, 6;* 4-27). This has importance in connection with understanding the structural plan of isolated pillars, called **pali,** and peripheral bifurcations of some septa in the juvenile part of corallites or persisting to the adult stage, as in genera which display what is known as the **Pourtalès plan** of septa (Figs. 4-26, *4, 21;* 4-27). Scleractinian corals may possess a comparatively dense side wall produced by peripheral swelling and union of septa (**septotheca**), by rows of dissepiments (**paratheca**), or by intergrown spines (**synapticulotheca**) (Fig. 4-26, *1, 13, 20*). A wall termed **epitheca,** which is essentially a continuation of the basal plate, is differentiated in some corals. **Endotheca** and **exotheca,** respectively, comprise dissepimental regions on the inner and outer sides of the walls, and structures between colonial corallites are collectively termed **peritheca** (Fig. 4-26, *8, 17, 18*). The septa rather commonly project peripherally beyond the wall, forming more or less prominent external ribs or **costae** (Fig. 4-26, *15*). A structure peculiar to some scleractinians consists of spinose projections on the sides of septa; they are named **synapticulae** (Fig. 4-26, *11*). Cross-

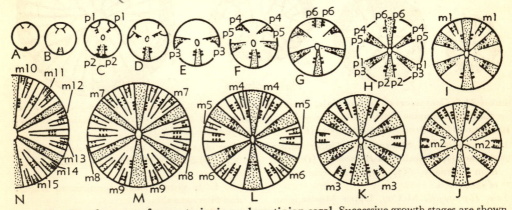

Fig. 4-25.   **Development of mesenteries in a scleractinian coral.** Successive growth stages are shown in diagrams *A–N.* The position and order of appearance of the protomesenteries (*p1–p6*) are indicated in diagrams *A–H.* The first metamesenteries (*ml*) introduced are shown in stage *I* and others in succeeding diagrams. (*Based on studies by Krempf and Duerden.*)

bars between septa may be formed by them.

Reproduction. The polyps belonging to scleractinian genera have separate sexes or are hermaphroditic, but some are sterile. Larvae (planulae) generally develop within the enteron of the parent until they attain a length of 1 to 3 mm. and are ready to swim after they have been ejected through the parent's mouth. The swimming stage may last for weeks, but commonly it is only a few days. Then the planulae become attached and begin development toward maturity. Colonial species reproduce asexually, as well as sexually, and the mode of budding has importance in determining the nature of the colonial skeleton. Accordingly, specialists in the study of this group of corals must be able to discriminate the various modes of budding which have importance in classifying genera.

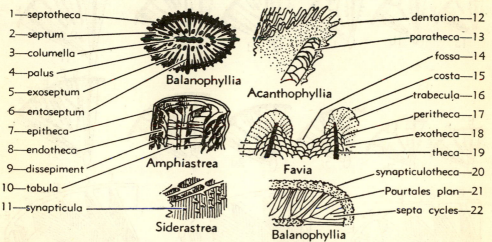

FIG. 4-26. **Morphologic features of scleractinian corallites.** Alphabetically arranged explanation of terms is cross-indexed to the figure by numbers.

columella (3). Solid or spongy axial structure.

costa (15). External projection of septum beyond corallite wall (theca), forming a rib.

dentation (12). Toothlike projection on edge of septum.

dissepiment (9). Vesicle having convex wall facing more or less upward; like similar structure in Rugosa.

endotheca (8). Dissepimental and tabular structures inside thecal or epithecal wall of corallite.

entoseptum (6). Septum formed in space (entocoel) between mesenteries of the same pair.

epitheca (7). External wall in some corallites, consisting essentially of extension of basal plate.

exoseptum (5). Septum formed in space (exocoel) between adjacent mesenteries of different pairs.

exotheca (18). Dissepimental structures outside thecal wall.

fossa (14). Depression in bottom of calyx.

palus (4). Rodlike structure in central part of corallite, derived from an exoseptum, to which it is attached basally, but in calyx it is opposite an entoseptum.

paratheca (13). Corallite wall formed of dissepimental tissue.

peritheca (17). Dissepimental and other structures between corallites of a compound coral.

Pourtalès plan (21). Arrangement of peripherally branching septa in such corals as *Balanophyllia*.

septa cycles (22). Arrangement of septa according to their order of appearance in successive groups.

septotheca (1). Corallite wall formed by peripheral thickening and union of septa.

septum (2). Radial wall of corallite.

synapticula (11). Lateral spinelike projection on side of septum.

synapticulotheca (20). Porous outer wall of corallite formed by union of numerous synapticulae.

tabula (10). Approximately horizontal platform extending across all or part of corallite interior.

theca (19). Dense wall at or near margin of corallite, distinct in origin from epitheca.

trabecula (16). Row of calcite fiber bundles, generally curved obliquely upward in septum.

Main Groups and Evolutionary Trends of Scleractinians

The Scleractinia are now recognized to be divisible into five suborders, which are based primarily on structural features of the septa. All but one contain numerous genera of solitary corals and equally abundant genera of colonial habit.

Astrocoeniida. The presumably most primitive scleractinians belong in this group, which consists almost exclusively of colonial forms. They have small corallites with rudimentary laminar septa composed of a few simple trabeculae. Examples are *Astrocoenia*, which grows in massive cerioid colonies, and *Stylina*, which contains massive plocoid colonies with colu-

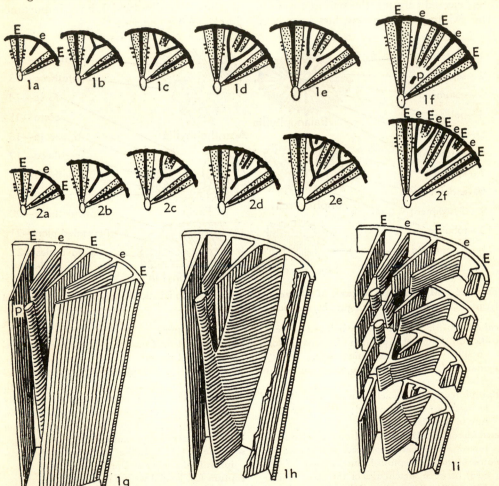

FIG. 4-27. **Relations of entosepta and exosepta in scleractinian corals and origin of pali.** The upper row of diagrams (1*a–f*) shows successive growth stages in one of the sextants of a corallite. Septa developed in entocoels (between mesenteries of the same pairs), marked by stippled pattern, are entosepta (**E**). Septa introduced in exocoels (between mesenteries of different pairs) are exosepta (**e**). During growth, the exoseptum bifurcates peripherally and an entoseptum appears between the branches, and when the axial part of the forked exoseptum becomes isolated (1*e, f*), forming a palus (**p**), it has position opposite the introduced entoseptum. These features are illustrated in perspective in drawings 1*g, h,* and *i.* Diagrams 2*a–f* represent similar relations of entosepta and exosepta leading to development of the so-called Pourtalès plan, **in** which bifurcations of exosepta persist.

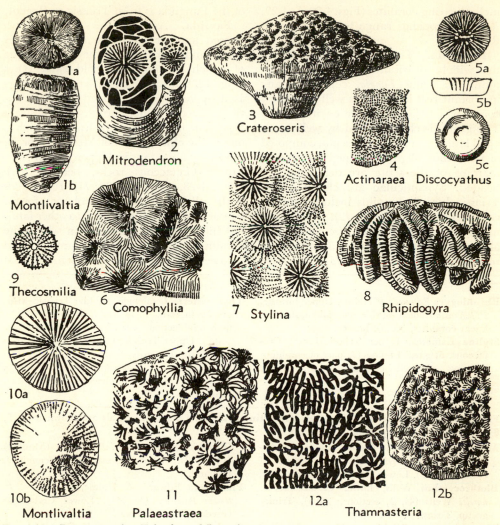

FIG. 4-28. **Representative Triassic and Jurassic scleractinian corals.** Solitary types are illustrated in 1, 5, 10; others are compound.

Actinaraea d'Orbigny, Upper Jurassic. *A. granulata* (Münster) (4, part of surface of corallum, calycal pits distinct, ×2), Kimmeridgian, Germany.

Comophyllia d'Orbigny, Upper Jurassic. *C. polymorpha* (Koby) (6, part of corallum, ×1.5), Portugal.

Crateroseris Tomes, Upper Jurassic. *C. fungiformis* Tomes (3, side view of corallum, ×0.7), England.

Discocyathus Edwards & Haime, Middle Jurassic–Middle Cretaceous. *D. eudesii* Edwards & Haime (5a–c, top, section, and base of corallite, ×1), Middle Jurassic (Bajocian), France.

Mitrodendron Quenstedt, Upper Jurassic. *M. schaferi* (Ogilvie) (2, part of corallum, showing vesicular area surrounding corallites, ×0.7) Portlandian, Czechoslovakia.

Montlivaltia Lamouroux, Middle Triassic–Cretaceous. *M. nattheimensis* Milaschewitsch (1a, b, top and side views of corallite, ×0.3), Upper Jurassic, Germany. *M. norica* Frech (10a, b, top and base of corallite, ×1), Upper Triassic, Alps.

Palaeastraea Kühn, Middle and Upper Triassic. *P. decussata* (Reuss) (11, surface of corallum, ×2), Upper Triassic, California.

(Continued on next page.)

mella-bearing corallites (Figs. 4-28, *7;* 4-29, *10*). Astrocoeniids range from Triassic to Recent.

Fungiida. This suborder, named from the widely distributed flat-based mushroom-like genus *Fungia*, contains many important reef builders, as well as solitary corals which live in other environments. The group is characterized by well-developed laminar or fenestrate septa, composed of numerous, slightly inclined trabeculae; margins of the septa are denticulate or beaded. Synapticulae are characteristic. Triassic-Jurassic fungiids include colonial corals (Fig. 4-28, *3, 4, 6, 12*), which have massive plocoid coralla. Discoid fungiids are common in Cretaceous and Tertiary rocks (Figs. 4-29, *7;* 4-30, *1, 15*). The range of the Fungiida is from Triassic to Recent.

Faviida. Among the Faviida are classed another large assemblage of reef corals and numerous solitary forms, some of which are deep- or cold-water types (Fig. 4-29, *13*). The suborder is characterized by very well-developed laminar denticulate septa of nonporous, nonfenestrate type. The family Montlivaltiidae, which is one of the chief groups of Mesozoic corals, especially abundant in Upper Jurassic and Lower Cretaceous rocks, belongs here (Figs. 4-28, *1, 9, 11;* 4-29, *11*). It includes both solitary corals and colonial forms. Cretaceous faviids belonging to other families include plocoid genera, in which the corallites have a spongy columella, and colonies having a dendroid or meandroid

---

(Fig. 4-28 continued.)

Rhipidogyra Edwards & Haime, Upper Jurassic. *R. percrassa* Étallon (8, side of corallum, showing linear corallites, ×0.3), France.

Stylina Lamarck, Upper Triassic–Lower Cretaceous. *S. girodi* Étallon (7, part of corallum, showing elevated calyces, ×1.5), Upper Jurassic, France.

Thamnasteria Lesauvage, Middle Triassic–Middle Cretaceous. *T. rectilamellosa* Winkler (12a, tangential section, ×2; 12b, surface of corallum, ×0.7), Upper Triassic, California and Austria.

Thecosmilia Edwards & Haime, Middle Triassic–Cretaceous. *T. fenestrata* (Reuss) (9, calyx of corallite belonging to phaceloid colony, ×2), Alaska.

Fig. 4-29. **Representative Cretaceous scleractinian corals.** Compound corals are shown in 3, 5, 8–11; others are solitary forms.

Astrocoenia Edwards & Haime, Upper Triassic–Recent. *A. whitneyi* Wells (10, surface of corallum, ×3), Lower Cretaceous, Texas.

Blothrocyathus Wells, Lower Cretaceous. *B. harrisi* Wells (4a, b, sections, ×0.5), Trinity group, Texas.

Dimorphastrea d'Orbigny, Upper Jurassic–Middle Cretaceous. *D. crassisepta* d'Orbigny (11, part of corallum, ×0.3), Lower Cretaceous, France.

Diploastrea Matthai, Cretaceous–Recent. *D. harrisi* Wells (9, part of corallum, ×2), Lower Cretaceous, Texas.

Micrabacia Edwards & Haime, Cretaceous–Recent. *M. rotatilis* Stephenson (7a–c, side, base, and top of corallite, ×2), Upper Cretaceous, Maryland.

Montastrea de Blainville, Upper Jurassic–Recent. *M. roemeriana* (Wells) (8a, b, corallum, ×3, ×1), Lower Cretaceous, Texas.

Parasmilia Edwards & Haime, Lower Cretaceous–Recent. *P. austinensis* Roemer (2a, b, top and side views, ×1), Lower Cretaceous (Fredericksburg), Texas. *P. centralis* (Mantell) (14, ×2), Upper Cretaceous (Senonian), England.

Pleurocora Edwards & Haime, Cretaceous. *P. texana* Roemer (3, ×1), Lower Cretaceous (Fredericksburg), Texas.

Stephanosmilia de Fromentel, Cretaceous. *S. perlata* de Fromentel (12a, b, top and side views, ×3), Aptian, France.

Strotogyra Wells, Upper Cretaceous–Eocene. *S. undulata* (Reuss) (5, top of corallum, ×1), Upper Cretaceous (Turonian), Austria.

Tiarasmilia Wells, Lower Cretaceous. *T. casteri* Wells (6, top of corallite, ×1), Trinitian, Texas.

Trochocyathus Edwards & Haime, Middle Jurassic–Recent. *T. scottianus* (1a–c, side, base, and top of corallite, ×3), Lower Cretaceous (Washita), Texas.

Trochosmilia Edwards & Haime, Cretaceous–Miocene. *T. didymophyla* Felix (13a, b, top and side views of corallite, ×0.7), Upper Cretaceous (Turonian), Austria.

type of growth (Fig. 4-29, *3, 5, 8*). Tertiary representatives of this group include massive coralla and ramose colonies (Fig. 4-30, *2, 3, 9, 16*). The Faviida range from Triassic to Recent.

Caryophylliida. The suborder of corals called Caryophylliida is distinguished by the single fan-shaped arrangement of trabeculae in their laminar, nonporous septa, which have smooth edges. Jurassic

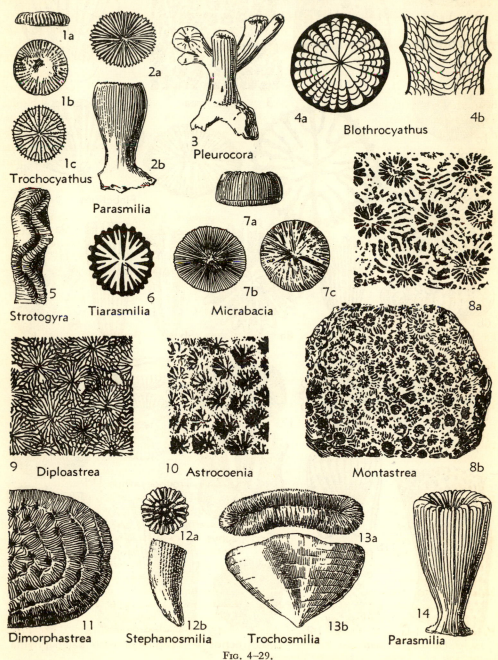

1a
1b
1c
Trochocyathus

2a
2b
Parasmilia

3
Pleurocora

4a
4b
Blothrocyathus

5
Strotogyra

6
Tiarasmilia

7a
7b
7c
Micrabacia

8a
8b

9 Diploastrea

10 Astrocoenia

Montastrea

11
Dimorphastrea

12a
12b
Stephanosmilia

13a
13b
Trochosmilia

14
Parasmilia

Fig. 4–29.

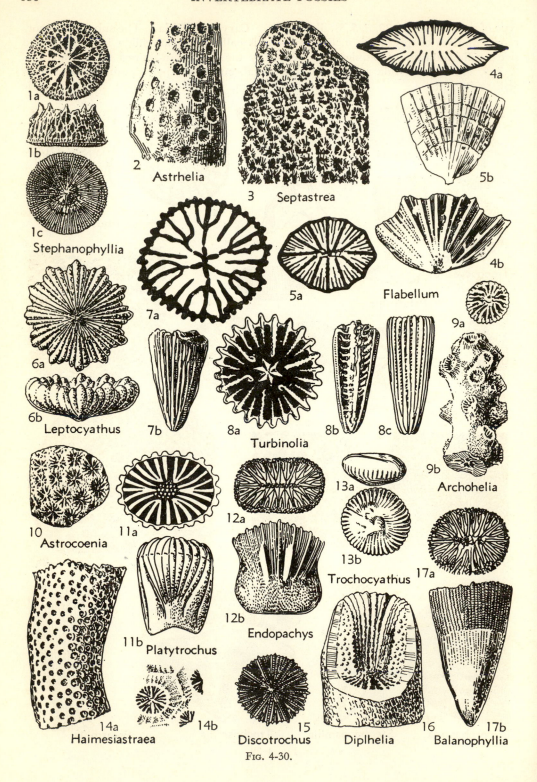

1a

1b

1c
Stephanophyllia

2
Astrhelia

3   Septastrea

4a

5b

5a

Flabellum

4b

7a

6a

6b
Leptocyathus

7b

8a
Turbinolia

8b

8c

9a

9b
Archohelia

10
Astrocoenia

11a

12a

13a

13b

Trochocyathus

17a

11b
Platytrochus

12b
Endopachys

14a

14b

Haimesiastraea

15
Discotrochus

16
Diplhelia

17b
Balanophyllia

Fig. 4-30.

genera here illustrated include a button-shaped coral and colonial types, one of which has a broad vesicular zone at the periphery of corallites, and another in which the corallites grow in a laterally confluent manner (Fig. 4-28, *2*, *5*, *8*). Cretaceous examples figured are all solitary forms (Fig. 4-29, *1*, *2*, *4*, *6*, *12*, *14*). *Blothrocyathus*, which has two distinct endothecal zones, composed, respectively, of tabulae and dissepiments, resembles some stauriacean rugose corals. Tertiary caryophylliids are abundant and much varied. They include colonial types (Fig. 4-30, *14*), conical solitary corals with strongly defined costae (Fig. 4-30, *7*, *8*), wedge-shaped solitary forms with rounded or pointed calycal edges (Fig. 4-30, *4*, *5*, *11*), and discoid types (Fig. 4-30, *6*, *13*). The Caryophylliida range from Jurassic to Recent.

Dendrophylliida. This is a relatively small suborder of scleractinians, which has septal structure like that of the caryophylliids except for the irregular porous and much thickened nature of these parts in some forms. Commonly, the corallites are enclosed by a porous synapticulotheca. Characteristic representatives are solitary corals (Fig. 4-30, *12*, *17*) which have branched Pourtalès plan septa, but colonial types occur also. The suborder is known to be distributed from Late Cretaceous to Recent. Only a few genera are reef builders.

Evolutionary Trends. According to studies by Vaughan & Wells (1943), a chief evolutionary modification observed in the scleractinian corals is the development of the edge zone and interpolypary soft tissue (coenosaec) permitting a great increase of colonial growth; this is accompanied by the development of greater com-

---

FIG. 4-30. **Representative Cenozoic scleractinian corals.** All are solitary corals except 2, 3, 9–10, 14, and 16.

Archohelia Vaughan, Middle Cretaceous–Pliocene. *A. vicksburgensis* (Conrad) (9a, calyx, ×2; 9b, part of corallum, ×1), Oligocene, Mississippi.

Astrhelia Edwards & Haime, Miocene. *A. palmata* (Goldfuss) (2, corallum, ×1), Maryland.

Astrocoenia Edwards & Haime, Upper Triassic–Recent. *A. decaturensis* Vaughan (10, corallum ×3), Oligocene, Georgia.

Balanophyllia Wood, Eocene–Recent. *B. elaborata* (Conrad) (17a, b, top and side views of corallite, ×2), Eocene, Maryland.

Diplhelia Edwards & Haime, Eocene. *D. papillosa* Edwards & Haime (16, corallite with part of calyx wall cut away, ×4), Eocene, England.

Discotrochus Edwards & Haime, Eocene. *D. orbignianus* Edwards & Haime (15, calycal view, ×2), Louisiana.

Endopachys Lonsdale, Eocene–Recent. *E. maclurii* (Lea) (12a, b, top and side views of corallite, ×1), Eocene, Mississippi.

Flabellum Lesson, Eocene–Recent. *F. redmondianum* Gabb (4a, b, top and side views, ×2), Paleocene, California. *F. cuneiforme* Lonsdale (5a, b, top and side views, ×1), Upper Eocene, Texas.

Haimesiastraea Vaughan, Eocene. *H. conferta* Vaughan (14a, part of corallum, ×1; 14b, corallite, ×5), Alabama.

Leptocyathus Edwards & Haime, Cretaceous–Eocene. *L. elegans* Edwards & Haime (6a, b, base and side views, ×2), Lower Eocene, England.

Platytrochus Edwards & Haime, Upper Cretaceous–Recent. *P. stokesi* (Lea) (11a, b, top and side views of corallite, ×3), Eocene, South Carolina.

Septastrea d'Orbigny, Miocene–Pliocene. *S. marylandica* (Conrad) (3, ×1), Miocene, Maryland.

Stephanophyllia Michelin, Eocene–Recent. *S. imperialis* Michelin (1a–c, top, side, and basal views, ×0.7), Miocene, Austria.

Trochocyathus Edwards & Haime, Middle Jurassic–Recent. *T. californianus* Vaughan (13a, b, side and top views of corallite, ×3), Eocene, California.

Turbinolia Lamarck, Eocene–Recent. *T. dickersonia* Nomland (7a, calyx, ×10; 7b, side view, ×5), Eocene, California. *T. pharetra* Lea (8a, calyx, ×10; 8b, corallite broken longitudinally to show interior, ×5; 8c, exterior, ×5), Eocene, Alabama.

pactness of the entire corallum. Structural changes of the corallite wall are seen to trend toward replacement of the primitive epitheca by a septotheca or paratheca, and in some, replacement of septotheca by a synapticulotheca. Inner parts of corallites also exhibit distinct evolutionary trends. Especially they include divergent sorts of changes in the septa, such as increased compactness or increased porosity, and in some a reduction of the septa to small remnants consisting of spines or ridges. Most successful of modern reef-building corals are families (Acroporidae, Poritidae) of astrocoeniids and fungiids which are distinguished by lightness of their porous skeletal structures, a feature that is correlated with their rapid growth. These groups include approximately as many species as all other scleractinians put together.

Origin and Geological History

Much has been written about the relationship of the scleractinians to older corals and the question of the ancestry of the post-Paleozoic corals. Two main postulates have been advanced: (1) that all corals have a common origin in unknown Cambrian or older primitive anthozoans, the Rugosa comprising a branch which culminated and died before Triassic time, and the Scleractinia constituting another branch which persisted without acquiring capacity to secrete hard parts until several kinds began in Triassic time to form a skeleton of calcium carbonate; and (2) that scleractinians are descendants of rugose corals.

The fact that sea anemones (Actiniaria), which lack hard parts, are nearly identical to scleractinians except for the absence of a skeleton, may be cited in favor of the first hypothesis, for the anemones almost surely are not derived from scleractinians. Also, most prominent among oldest known Scleractinia, in Middle Triassic rocks, are the astrocoeniids, which are least like rugose corals; this is somewhat negative evidence supporting the first hypothesis.

The fact that some late Paleozoic Rugosa show the introduction of metasepta between the counter and counter-lateral protosepta (a section of the rugose corallite which normally remains suppressed as regards insertion of new septa) supports the second hypothesis. The additional fact that some Triassic scleractinians have two adjoining spaces between protosepta, which contain fewer metasepta than the four other such spaces (Fig. 4-13, 3), also supports the second hypothesis, as does the order of development of mesentery pairs in modern scleractinians (Fig. 4-25).

Early Triassic coral faunas, as yet unknown, may help to bridge the gap in the coral record which now exists. The Middle Triassic coral faunas of Germany and southern Europe include many reef-type genera, but there are no true reefs. In Late Triassic time, however, reef building by corals can be recorded not only in southern Europe but eastward to Malaysia, as well as in California and Alaska. Genera and species increased greatly in numbers.

Early Jurassic coral assemblages are very similar to the Triassic, but no important reef building occurred until Middle Jurassic and again in Late Jurassic time. In both these epochs, coral faunas and reef building were extensive in areas northward to the British Isles and Japan.

Although corals are abundant at many places in Lower Cretaceous rocks, and although reef building occurred in lower latitudes throughout the world during part of Early Cretaceous time, the middle part of the Cretaceous lacks numerous corals. The Upper Cretaceous contains many coral reefs, however.

Solitary corals are common in shallow-water deposits of Early Tertiary age in the Gulf Coastal region, the Paris and London basins, and various other places, but reef corals are restricted to parts of the Tethyan belt, including the West Indies. From the Miocene epoch to modern time, corals have thrived, especially in the Caribbean and Indo-Pacific regions.

REFERENCES

General

HYMAN, L. H. (1940) *The invertebrates: Protozoa through Ctenophora:* McGraw-Hill, New York, pp. i–xii, 1–726, figs. 1–221 (coelenterates, pp. 365–661, figs. 106–208). The best general work on living forms, although scleractinians are slighted; accompanied by selected bibliography.

MACGINITIE, G. E., & MACGINITIE, N. (1949) *Natural history of marine animals:* McGraw-Hill, New York, pp. 1–473, figs. 1–282 (coelenterates, pp. 117–143, figs. 15–32).

SHIMER, H. W., & SHROCK, R. R. (1944) *Index fossils of North America:* Wiley, New York, pp. i–ix, 1–837, pls. 1–303 (coelenterates, pp. 58–122, pls. 18–46; includes graptolites).

SWINNERTON, H. H. (1947) *Outlines of palaeontology:* E. Arnold & Co., London, 3d ed., pp. 1–393, figs. 1–368 (coelenterates, pp. 29–68, figs. 27–60). Good on corals, with discussion of zaphrenthid evolution.

VAUGHAN, T. W. (1913) *Cnidaria:* in Eastman, C. R., & Zittel, K. A., Textbook of palaeontology, Macmillan & Co., Ltd., London, 2d ed., vol. 1, pp. 74–124, figs. 94–192.

Hydrozoans

KUHN, O. (1939) *Hydrozoa:* Handbuch der Paläozoologie, Borntraeger, Berlin, Lief. 5, Bd. 2A, pp. A1–A68, figs. 1–96. Morphology, classification, and descriptions of fossil hydrozoans, including stromatoporoids.

NICHOLSON, H. A. (1886–1892) *A monograph of the British stromatoporoids:* Palaeont. Soc. Mon., pp. 1–234, pls. 1–29. A classic work on this group of fossils.

PARKS, W. A. (1907) *Stromatoporoids of the Guelph formation in Ontario:* Toronto Univ. Studies, Geol. ser., vol. 4, pp. 1–40, pls. 1–6. Silurian.

——— (1908) *Niagaran stromatoporoids:* Same, vol. 5, pp. 1–68, pls. 7–15.

——— (1909) *Silurian stromatoporoids of America:* Same, vol. 6, pp. 1–52, pls. 16–20.

——— (1910) *Ordovician stromatoporoids of America:* Same, vol. 7, pp. 1–52, pls. 21–25.

——— (1936) *Devonian stromatoporoids of North America:* Same, vol. 33, pp. 1–125, pls. 1–19.

Scyphozoans

KIESLING, A. (1939) *Scyphozoa:* Handbuch der Paläozoologie, Borntraeger, Berlin, Lief. 5, Bd. 2A, pp. A69–A109, figs. 1–42. Describes known types of fossil jellyfishes.

WALCOTT, C. D. (1898) *Fossil medusae:* U.S. Geol. Survey Mon. 30, pp. 1–201, pls.

Anthozoans

AMSDEN, T. W. (1949) *Stratigraphy and paleontology of the Brownsport formation (Silurian) of western Tennessee:* Peabody Mus. Nat. History Bull. 5, pp. 1–138, pls. 1–34, figs. 1–29. Contains descriptions and illustrations of Niagaran corals.

BASSLER, R. S. (1937) *The Paleozoic rugose coral family Palaeocyclidae:* Jour. Paleontology, vol. 11, pp. 189–201, pls. 30–32.

——— (1950) *Faunal lists and descriptions of Paleozoic corals:* Geol. Soc. America Mem. 44, pp. 1–315, pls. 1–20. Compilation of recorded occurrence of species of fossil corals from entire world, with tables of correlation for stratigraphic divisions.

BUSCH, D. A. (1941) *An ontogenetic study of some rugose corals from the Hamilton of western New York:* Jour. Paleontology, vol. 15, pp. 392–411, figs. 1–73.

CARRUTHERS, R. G. (1910) *On the evolution of Zaphrentis delanouei in Lower Carboniferous times:* Geol. Soc. London Quart. Jour., vol. 66, pp. 523–538, pls. 36–37.

CUMINGS, E. R., & SHROCK, R. R. (1927) *The Silurian coral reefs of northern Indiana and their associated strata:* Indiana Acad. Sci. Proc., vol. 36, pp. 71–85, figs. 1–4.

DAVIS, W. J. (1887) *Kentucky fossil corals, part 2:* Kentucky Geol. Survey, pp. i–xiii, pls. 1–139 (part 1, comprising text, never published). Excellent illustrations, mostly of Devonian fossils.

DUNCAN, P. M. (1866–1873) *A monograph of the British fossil corals:* Palaeont. Soc. Mon., pt. 1, pp. 1–66, pls. 1–10; pt. 2, pp. 1–46, pls. 1–15; pt. 3, pp. 1–24, pls. 1–7; pt. 4; pp. 1–73, pls. 1–17. Excellent work on Mesozoic and Cenozoic corals of Great Britain.

DURHAM, J. W. (1942) *Eocene and Oligocene coral faunas of Washington:* Jour. Paleontology, vol. 16, pp. 84–104, pls. 15–17, fig. 1.

EASTON, W. H. (1944) *Corals from the Chouteau and related formations of the Mississippi Valley region:* Illinois Geol. Survey Rept. Inv. 97, pp. 1–62, pls. 1–17. Lower Mississippian rugose and tabulate corals; contains morphologic definitions.

——— (1944a) *Revision of Campophyllum in North America:* Jour. Paleontology, vol. 18, pp. 119–

132, pl. 22, figs. 1–4. Late Paleozoic stauriacean corals.

EDWARDS, H. M., & HAIME, J. (1850–1854) *A monograph of the British fossil corals:* Palaeont. Soc. Mon., pp. 1–299, pls. 1–72. Important work on Paleozoic corals.

────── (1851) *Monographie des polypiers fossiles des terrains paléozoiques:* Mus. nat. histoire nat. Arch., vol. 5, pp. 1–502, pls. 1–20 (French). A leading work on rugose and tabulate corals.

────── (1857–1860) *Histoire naturelle des coralliaires:* Paris, vol. 1, pp. 1–326; vol. 2, pp. 1–633; vol. 3, pp. 1–560; atlas, pls. 1–31 (French). Ranked by some specialists as the outstanding report of all time on corals.

FELIX, J. (1903) *Die Anthozoen der Gosauschichten in den Ostalpen:* Palaeontographica, vol. 49, pp. 163–359, pls. 17–25, figs. 1–67 (German). Classic report on Cretaceous corals.

FROMENTEL, E. DE (1862–1887) *Terrains crétacés, Zoophytes:* Paléontologie française, vol. 7, pp. 1–624, pls. 1–192 (French). Classic report on Cretaceous corals.

────── & FERRY, H. B. A. T. DE (1865–1869) *Terrains jurassiques, Zoophytes:* Paléontologie française, pp. 1–240, pls. 1–60 (French). A chief work on Jurassic corals.

GERTH, H. (1931) *Palaeontologie von Niederlandisch Oost-Indien, Coelenterata:* Leidsche Geol. Meded., vol. 5, pp. 387–445 (German). Permian corals.

GRABAU, A. W. (1922–1928) *Palaeozoic corals of China, Part 1, Tetraseptata;* Palaeontologia Sinica, ser. B, vol. 2, fasc. 1, pp. 1–70, pl. 1, figs. 1–73; fasc. 2, pp. 1–151, pls. 1–6, figs. 1–22.

GROVE, B. H. (1934–1935) *Studies in Paleozoic corals:* Am. Midland Naturalist, vol. 15, pp. 97–137; vol. 16, pp. 337–378.

HALL, J. (1887) *Corals and Bryozoa:* New York Geol. Survey, Paleontology, vol. 6, pp. i–xxvi, 1–298, pls. 1–66.

HILL, D. (1935) *British terminology for rugose corals:* Geol. Mag., vol. 72, pp. 481–519, figs. 1–21. A well-organized description of rugose coral morphology.

────── (1936) *The British Silurian rugose corals with acanthine septa:* Royal Soc. London Phil. Trans., ser. B, vol. 206, pp. 189–217, pls. 29–30.

────── (1938–1941) *A monograph on the Carboniferous rugose corals of Scotland:* Palaeont. Soc. Mon., pp. 1–213, pls. 1–11.

HUANG, T. K., (1932) *Permian corals of southern China:* Palaeontologia Sinica, ser. B, vol. 8, fasc. 2, pp. 1–163, pls. 1–16.

JEFFORDS, R. M. (1942) *Lophophyllid corals from Lower Pennsylvanian rocks of Kansas and Okla-*

homa: Kansas Geol. Survey Bull, 41, pp. 185–260, pls. 1–8.

JONES, O. A., & HILL, D. (1940) *The Heliolitidae of Australia, with a discussion of the morphology and systematic position of the family:* Royal Soc. Queensland Proc., vol. 51, pp. 183–215, pls. 6–11.

LANG, W. D. (1917) *Homeomorphy in fossil corals:* Geol. Assoc. Proc., vol. 28, pp. 85–94.

────── (1923) *Trends in British Carboniferous corals:* Same, vol. 34, pp. 120–136, figs. 1–24.

────── (1938) *Some further considerations on trends in corals:* Same, vol. 49, pp. 148–159, pl. 7, figs. 25–28.

──────, SMITH, S., & THOMAS, H. D. (1940) *Index of Palaeozoic coral genera:* British Mus. Nat. History, London, pp. 1–231. An indispensable reference for taxonomic research on Rugosa and Tabulata. Contains no morphological descriptions but includes large bibliography.

MOORE, R. C. & JEFFORDS, R. M. (1941) *New Permian corals from Kansas, Oklahoma, and Texas:* Kansas Geol. Survey Bull. 38, pp. 65–120, pls. 1–8. Contains large bibliography.

────── & ────── (1945) *Description of Lower Pennsylvanian corals from Texas and adjacent States:* Texas Univ. Bull. 4401, pp. 63–208, pl. 14, figs. 1–213. A varied and representative assemblage of Pennsylvanian rugose and tabulate corals.

OKULITCH, V. J., (1935) *Tetradidae—a revision of the genus Tetradium:* Royal Soc. Canada Proc. and Trans., 3d ser., vol. 29, pp. 49–74, pls. 1–2.

ROBINSON, W. I. (1917) *The relationship of the Tetracoralla to the Hexacoralla:* Connecticut Acad. Arts and Sci. Trans., vol. 21, pp. 145–200, pl. 1, figs. 1–7.

ROMINGER, C. L. (1876) *Fossil corals:* Michigan Geol. Survey, vol. 3, pt. 2, pp. 1–161, pls. 1–55. Important reference on Devonian corals.

SANFORD, W. G. (1939) *A review of the families of tetracorals:* Am. Jour. Sci., vol. 237, pp. 295–323, 401–423, figs. 1–16. Contains useful descriptions of morphologic features of rugose corals.

SCHINDEWOLF, O. H. (1930) *Über die Symmetrie-Verhältnisse der Steinkorallen:* Palaeont. Zeitschr., vol. 12, pp. 214–263, figs. 1–60 (German).

────── (1938) *Zur Kenntnis der Gattung Zaphrentis (Anthoa. Tetracorall.) und der sogenannten Zaphrentiden des Karbons:* Preuss. geol. Landesanstalt Jahrb. (1937), Bd. 58, pp. 439–454, pls. 44–45 (German).

────── (1940) *"Konvergenzen" bei Korallen und bei Ammoneen:* Fortschr. Geologie u. Palaeontologie, Bd. 12, Heft 41, pp. 289–492, pl. 1, figs. 1–33

(German).

——— (1941) *Zur Kenntnis der Heterophylliden, einer eigentümlichen paläozoischen Korallengruppe:* Palaeont. Zeitschr., Bd. 22, pp. 213–306, pls. 9–16, figs. 1–54 (German). Describes and interprets the order of heterocorals.

SLOSS, L. L. (1939) *Devonian rugose corals from the Traverse beds of Michigan:* Jour. Paleontology, vol. 13, pp. 52–73, pls. 9–12.

SMITH, S. (1945) *Upper Devonian corals of the Mackenzie River region, Canada:* Geol. Soc. America Spec. Paper 59, pp. 1–126, pls. 1–35. Contains definition of morphologic features of Rugosa and Tabulata.

STEWART, G. A. (1938) *Middle Devonian corals of Ohio:* Geol. Soc. America Spec. Paper 8, pp. 1–120, pls. 1–20.

STUMM, E. C. (1937) *The Lower Devonian tetracorals of the Nevada limestone:* Jour. Paleontology, vol. 11, pp. 423–443, pls. 53–55.

——— (1949) *Revision of the families and genera of the Devonian tetracorals:* Geol. Soc. America Mem. 40, pp. 1–92, pls. 1–25. Especially valuable because of concise definitions of rugose genera and illustration of many type species.

VAUGHAN, A. (1905) *The palaeontological sequence in the Carboniferous limestone of the Bristol area:* Geol. Soc. London Quart. Jour., vol. 62, pp. 275–323, pls. 29–30. Describes zonation of British rocks of Mississippian age, largely in terms of corals.

VAUGHAN, T. W. (1900) *The Eocene and Oligocene coral faunas of the United States:* U.S. Geol. Survey Mon. 39, pp. 1–263, pls. 1–24.

——— (1901) *Corals:* Maryland Geol. Survey, Eocene, pp. 222–232, pl. 61.

——— (1904) *Corals:* Same, Miocene, pp. 438–447, pls. 122–129.

——— (1919) *Fossil corals from Central America, Cuba, and Puerto Rico, with an account of the American Tertiary, Pleistocene, and Recent coral reefs:* U.S. Nat. Mus. Bull. 103, pp. 189–524, pls. 68–152.

——— & POPENOE, W. P. (1935) *The coral fauna of the Midway Eocene of Texas:* Texas Univ. Bull. 3301, pp. 325–343, pls. 3–4.

——— & WELLS, J. W. (1943) *Revision of the suborders, families, and genera of the Scleractinia:* Geol. Soc. America Spec. Paper 44, pp. 1–363, pls. 1–51, figs. 1–39. A chief source of information on this group of corals, containing description of anatomy, skeletal morphology, reproductions, ecology, distribution, evolution, and summary of taxonomy, accompanied by bibliography of 1,024 titles.

WANG, H. C. (1950) *A revision of the Zoantharia Rugosa in the light of their minute skeletal structures:* Royal Soc. London Phil. Trans., ser. B, no. 611, vol. 234, pp. 175–246, pls. 4–9, figs. 76–79. An effort to apply study of septal structure to classification of rugose corals.

WELLS, J. W. (1932) *Corals from the Trinity group of the Comanchean of central Texas:* Jour. Paleontology, vol. 6, pp. 225–256, pls. 30–39.

——— (1933) *Corals of the Cretaceous of the Atlantic and Gulf Coastal Plains and western interior of the United States:* Bull. Am. Paleontology, vol. 18, pp. 85–288, pls. 14–28.

CHAPTER 5

Bryozoans

The invertebrates called bryozoans are aquatic animals, of which the great majority live in the sea and only a few kinds inhabit fresh waters. They are tiny creatures, averaging less than 1 mm. in length, but invariably they grow together in colonies which commonly have dimensions of an inch or more. Some colonies of living species are as much as 30 cm. (12 in.) in diameter, and a few known fossil forms are at least 60 cm. (2 ft.) across (Fig. 5-1).

The name Bryozoa, which signifies moss animals (*bryon*, moss; *zoon*, animal), is derived from the tufted, mosslike growths of some modern bryozoan colonies. British zoologists prefer to call these animals Polyzoa (meaning many animals), a name which alludes to the numerous individuals united together in the colonies. Although Polyzoa is a slightly older term than Bryozoa, it is not used generally.

Bryozoans are found in present-day seas in all latitudes and at depths ranging downward to at least 5,500 m. (18,000 ft.). They are most abundant in shallow seas of temperate and tropical zones. The colonies grow on the sea bottom, attached to some foreign object, cemented by their base or anchored by small rootlike appendages, or lying loose on the sea floor. Many colonies are thin, delicate incrustations growing on stones, shells, or seaweeds. Some types, which lack hard parts, can crawl slowly by movement of the colonial base.

The bryozoans seem mostly to live in clear water rather than turbid, muddy water. They prefer shallow seas, and many live fairly close to the shore. Some are found in waters which are constantly agitated by waves and strong currents, but areas of the sea bottom made up chiefly of moving sands are not a favorable environment for bryozoans. The colonies of these animals are most numerous on rocky bottom and in places where shells of other invertebrates afford places for attachment. Their food supply consists almost exclusively of microscopic floating organisms, such as diatoms and radiolarians.

ANATOMICAL FEATURES

The individual bryozoan animal (**zooid**) is considerably more advanced in body organization than the coelenterates, which it resembles in having tentacles and a tubular or a saclike body form. The tentacles are joined at their base to a fleshy ring (**lophophore**), which surrounds the mouth. The chief distinguishing structure within the body of the bryozoan is the possession of a U-shaped digestive tract which consists of an esophagus, a stomach, and an intestine (Fig. 5-2). The anus is located near the mouth on the upper surface of the animal, just outside the circlet of tentacles. Not included in this characterization is a small group of somewhat similar animals (Entoprocta) which have the anus within the circle of tentacles. Although classed as bryozoans by many authors, these creatures are now treated as an independent phylum.

Surrounding the digestive tube of the bryozoans is a fluid-filled body cavity which contains reproductive elements. A nerve ganglion occurs in a space between the mouth and anus. Most bryozoans are bisexual in that they possess both eggs (ova) and sperm, the former being produced generally in the upper part of the

FIG. 5-1. **A colony of the modern cheilostome Bugula.** This bryozoan grows in bushy tufts 2 or 3 in. high. Four rows of zooecial chambers, which are chitinous and somewhat flexible, form each branch.

body and the latter in a lower region. The bryozoans have neither heart nor vascular system, but they possess numerous white blood cells (leucocytes) which move about in the body cavity. Several sets of muscles serve for protrusion and retraction of the animal from the membranous or calcareous skeleton which is secreted around it. Like many coelenterates, the hard parts of most bryozoans are composed of calcium carbonate.

Fertilized eggs produced by a bryozoan are lodged within the skeletal wall or in a special chamber (**ovicell**), where they develop until ready to emerge as free-swimming larvae. When set free, the larvae move about for a few hours by rhythmical vibration of many short cilia and may be carried far by currents. Eventually they settle to the bottom and become attached. Each larva develops into the first adult individual (**ancestrula**) of a new colony. Asexual budding of this individual produces new animals which are attached to

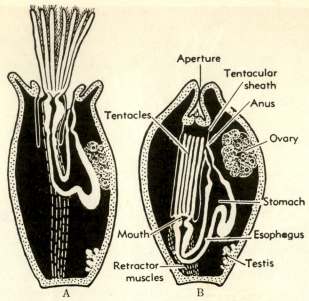

FIG. 5-2. **Anatomy of a living bryozoan species.** The body wall of this form, *Alcyonidium albidum* Alder, is membranous and flexible; hard parts are lacking. The drawings show the animal in longitudinal section, (*A*) with the zooid extended, and (*B*) retracted.

it, and similar divisions of successively formed zooids build the colony. Thus, all individuals belonging to any bryozoan colony, except the parent animal, are products of budding.

COLLECTION AND STUDY OF BRYOZOANS

Fossil bryozoans are extremely abundant in many marine formations, chiefly shales and limestones, from Ordovician to Recent. Specimens imbedded in limestone are not readily separated from the matrix, but those in shaly limestone and shale are commonly weather-free. They may be obtained in large numbers by brushing or scraping together the weathered specimens at an outcrop and packing them in a sack with disintegrated shale. Such collections can be washed in the laboratory. When necessary, the shale may be broken down by alternate thorough soaking and drying.

Specimens in which adherent clay obscures the surface characters may be cleaned by the use of caustic potash,

preferably in pellet form. Deliquescence of the potash when it is placed on the fossil loosens the clay so that it may be brushed off easily, but in order to stop chemical action, it is necessary to neutralize the specimen carefully by washing in water containing a little hydrochloric or acetic acid. Removal of clay may be effected also by placing the specimens in a saturated solution of Glauber's salt, which in crystallizing loosens the clay.

Bryozoan colonies embedded in limestone are hard to obtain as free specimens. The calcareous matrix adheres so tightly to the fossils that when one breaks the rock the colonies are broken also. Although such fragments do not show surface characters very well, generic and specific identification can be made ordinarily in quite satisfactory manner by properly oriented thin sections.

Many post-Paleozoic deposits are sufficiently unconsolidated to permit the collection of bryozoans by simple screening and washing. Weathered outcrops where the matrix is partly decomposed are likely to yield the best specimens. The matrix and fossils together should be collected in quantity and prepared for sorting in the laboratory.

Preparation of Thin Sections

The task of cutting and mounting thin sections of bryozoans is neither very difficult nor time-consuming, but experience is necessary in order to make sections of requisite thinness. The materials needed are coarse and fine abrasive, a piece of plate glass or grinding lap, slides and cover glasses, and a substance such as Canada balsam or a plastic cement by which the smoothly ground specimen may be cemented to the glass. A pair of medium-sized nippers is useful for cutting chips in the desired direction from the bryozoan colony. The best abrasive is carborundum in the form of a powder or grinding stone.

As a preparation for grinding, the specimen is oriented in the desired manner and mounted on a glass slide by cementing it with heated balsam or plastic. When cool, the specimen is ground with a suitable abrasive in a plane parallel to the glass. By heating the cement, the specimen may be turned over so as to grind the other side until the slice is thin enough to transmit light readily and reveal the structural features clearly. The final grinding needs to be done carefully on very fine carborundum or a hone. The slide is completed by heating cement on a cover glass and inverting it over the ground section.

Thin sections needed for the identification of bryozoans include one parallel to the tubes and another transverse to the tubes, parallel to the surface of the colony. These are termed longitudinal and tangential sections, respectively.

Interpretation of Thin Sections

Structural features of bryozoans which have importance in classification and which may be studied in thin sections particularly include the nature of the walls built by the individual bryozoan animals. In some, the walls of adjacent tubes seem to be coalesced, whereas they are clearly distinct in others. The first-mentioned type of wall structure is termed **amalgamate**, and the second is called **integrate**.

Other features are calcareous projections or partitions within the walled areas. The relation of thin sections to bryozoan structures is illustrated in Fig. 5-3.

CLASSIFICATION

Bryozoans have been classed with brachiopods in a phylum, called Molluscoidea, because both groups have somewhat similar internal body organization and possess a tentacle-bearing lophophore around the mouth. Prevailing modern judgment, however, takes account of numerous divergences of these animals, and treats them as independent phyla. The solitary mode of growth of brachiopods and the bivalve nature of their hard parts contrast greatly with the microscopic skeletal chambers of the bryozoan animals growing together as a colony.

Bryozoans are divided into two very unequal groups. The smaller of these includes fresh-water forms. They lack hard parts and are characterized by tentacles arranged in a horseshoe shape and by a lip which overhangs the mouth. These are assigned to the class Phylactolaemata (meaning protected gullet).

Remaining bryozoans, so preponderant in numbers as to include nearly all, mostly have hard parts, a circular row of tentacles around the mouth, and no lip. Virtually all are marine. Paleontological study of bryozoans is wholly devoted to this class, called Gymnolaemata (meaning uncovered gullet).

The Gymnolaemata are divided into five orders, of which two (Trepostomata, Cryptostomata) are composed of fossils known only in Paleozoic rocks. Two orders (Ctenostomata, Cyclostomata) are distributed from Ordovician to Recent. The fifth order (Cheilostomata) has no known Paleozoic representatives but is abundant in Mesozoic and Cenozoic strata. It is

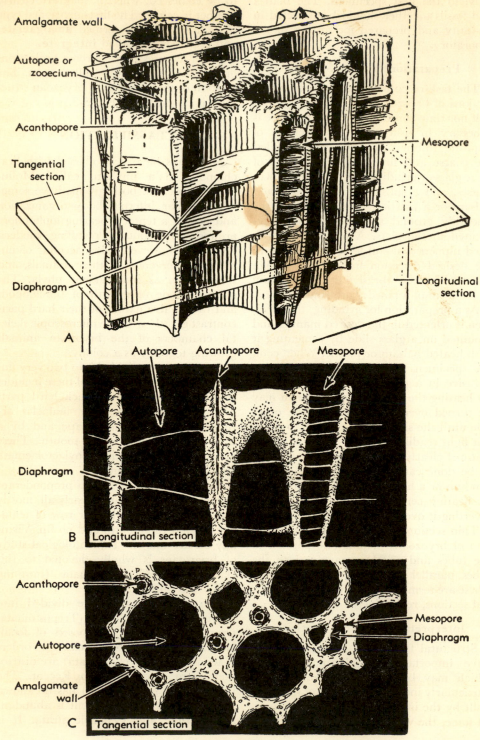

Amalgamate wall

Autopore or
zooecium

Acanthopore

Tangential
section

Mesopore

Diaphragm

Longitudinal
section

A

Autopore   Acanthopore   Mesopore

Diaphragm

B Longitudinal section

Acanthopore

Mesopore

Autopore

Diaphragm

Amalgamate
wall

C Tangential section

FIG. 5-3.

strongly dominant in present-day seas and includes the largest number of species known in any order. These divisions are indicated in the following tabulation.

ntoprocta
Ectoprocta

Main Divisions of Bryozoans

Phylactolaemata (*class*), fresh-water bryozoans having a horseshoe-shaped loop of tentacles around the mouth, which is protected by an overhanging lip. Recent (not known as fossils).

Gymnolaemata (*class*), almost exclusively marine bryozoans having a circular row of tentacles around the mouth. Ordovician–Recent.

matrypa
onticuloporella
nstellare
rasopora
allopara

Trepostomata (*order*), animals enclosed by a long curved calcareous tube, generally intersected by partitions. Immature and mature parts of the colonial structure are distinct. The name signifies turned mouth. Ordovician–Permian.

rchemede

Cryptostomata (*order*), animals enclosed in a relatively short calcareous tube, walls near the periphery of the colony much thickened. The name means hidden mouth. Ordovician–Permian.

Cyclotrypa
neekopara

Cyclostomata (*order*), animals enclosed in a calcareous tubular chamber having a lidless circular aperture. The name means circular mouth. Ordovician–Recent.

Ctenostomata (*order*), individual animals enclosed in a gelatinous chamber; processes resembling the teeth of a comb close the aperture when the tentacles are retracted. The name means comb mouth. Ordovician–Recent.

Cheilostomata (*order*), animals enclosed in a short saclike calcareous or chitinous chamber provided with a hinged chitinous lid (operculum) which closes the aperture when the tentacles are retracted. The name means rimmed mouth. Jurassic–Recent.

SKELETAL FEATURES OF MAIN BRYOZOAN GROUPS

Study of the most important skeletal features of bryozoans may be organized best according to the orders which are recognized in this group of invertebrates. Although we know nothing of the soft parts of animals belonging to the Trepostomata, it seems desirable to begin our survey with this dominant group of older Paleozoic bryozoans. The main structural features of their skeletal remains are illustrated in Fig. 5-4 with explanation of the terms; italic numbers and letters in the text describing trepostomes refer to parts of this figure.

Trepostome Bryozoans

As examples of this important group, the common Ordovician genera *Dekayella* and *Prasopora* well illustrate the characteristic features. Both have relatively massive calcareous colonies (**zoaria,** *1*) in which the long cylindrical or polygonal tubes (**autopores** or **zooecia,** *6, 14*) are the dominant structural elements. In both, also, there is clear differentiation of an inner or axial region, in which the tubes have thin walls, and an outer zone, which is characterized by thicker walls and more complicated structure. The thin-walled inner part of the colony, which is made up of the early-formed, relatively rapid-growing portions of the autopore tubes, is called the **immature region** (*13*). The thick-walled outer part of the colony, which represents a very slow-growing or static condition of the tubes, is called the **mature region** (*12*). These features especially distinguish the trepostomes.

Dekayella. Colonies belonging to this genus are coarse branches of irregular form having an average thickness of somewhat less than 1 in. (*G*).

A longitudinal section parallel to the autopore tubes shows that in the interior region these trend roughly parallel to the axis of colonial growth, whereas in the exterior portion, the tubes change direction

---

Fig. 5-3. **Structural features of bryozoans shown by thin sections.** (*A*) Perspective drawing of a small fragment of a calcareous bryozoan colony (*Dekayella*, an Ordovician trepostome) showing positions of thin slices cut longitudinally and transversely with respect to the tubes; (*B*) longitudinal thin section; (*C*) tangential section. The structural elements indicated are illustrated and explained in Fig. 5-4.

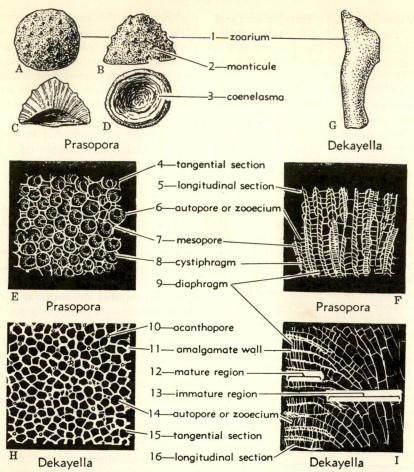

FIG. 5-4. **Structural features of trepostome bryozoans.** (*A–F*) *Prasopora conoidea* Ulrich, Middle Ordovician (Trentonian), Minnesota; (*A–D*) top, side, vertical section, and bottom views of zoarium, ×1; (*E, F*) tangential and longitudinal sections, ×10. (*G–I*) *Dekayella praenuntia* Ulrich, Middle Ordovician (Trentonian), Minnesota; (*G*) side view of zoarium, ×1; (*H, I*) tangential and longitudinal sections, ×10. Terminology is explained below.

acanthopore (10). Minute tube parallel to auto-pores in mature region of colony, commonly projecting at surface as a short spine.

amalgamate wall (11). Coalesced wall between adjacent autopores, between autopores and mesopores, or between mesopores.

autopore (6, 14). Relatively large tube or chamber occupied by one of the main zooids of the colony; also termed zooecium.

coenelasma (3). Calcareous lamina, generally wrinkled, secreted at the base of a colony.

cystiphragm (8). Curved thin wall forming a vesicle along the side of an autopore; commonly in a uniformly oriented vertical series within the autopore.

diaphragm. (9). Thin platform extending transversely across an autopore or mesopore.

immature region (13). Interior part of colonial structure containing the proximal (initial) part of autopores, characterized by thin-walled tubes having widely spaced diaphragms; area of budding of autopores.

longitudinal section (5, 16). Thin slice of colony cut parallel to autopore tubes.

mature region (12). Peripheral part of colony containing the distal (terminal) part of autopores, commonly characterized by thickened walls and by presence of mesopores, acanthopores, and more closely crowded diaphragms.

mesopore (7). Tube parallel to autopores, comso as to intersect the surface nearly at right angles (*I*). As shown by the beginnings of the tubes, the individual zooids which secrete them originate by budding from previously formed members of the colony. As each animal builds a calcareous wall around itself, it moves gradually upward, abandoning the first-formed part of its tube. At rather widely spaced intervals, the autopore tubes are blocked off by cross partitions (**diaphragms,** *9*), and each such partition represents a temporary floor secreted by the zooid. The diaphragms, therefore, are built in succession, each representing a temporary halt in the building of the tubes. It follows that the portion of any tube below the last-formed diaphragm has been vacated by the zooid and is not occupied by living tissue. The diaphragms of *Dekayella* are complete in that they extend entirely across the tubes.

Some other trepostome bryozoans have incomplete diaphragms with a central perforation or reaching only part way across the tube. The zoological significance of such structures is uncertain. Obviously, the soft parts of a bryozoan could extend through the aperture of such incomplete diaphragms. A few trepostomes have no diaphragms at all (Figs. 5-5, *3*; 5-16, *7*).

In *Dekayella*, the part of the autopores near the surface of the colony is characterized by much more numerous and closely spaced diaphragms than in the immature region (*I*). Also, the walls are noticeably thicker here. Both of these characters denote a slowing down in the outward building of the tubes, which indicates the attainment of maturity in these parts of the colony.

Longitudinal sections of *Dekayella* commonly show some tubes in the mature region which are distinctly more slender than the average. Also, these have more closely crowded diaphragms than the normal-sized tubes (*I*). Such small tubes are termed **mesopores** (*7*), for clearly they are unlike the large ones called autopores (pores or tubes standing by themselves, that is, main ones). As among modern bryozoan colonies, which contain specialized individuals associated with the normal animals, the trepostomes evidently included more than a single type of zooid. This differentiation, or polymorphism, is common among colonial invertebrates.

A tangential section of the *Dekayella* colony (*H*) intersects the autopores and mesopores approximately at right angles and thus serves to show the transverse shape and size of the tubes. They have a distinctly polygonal form. The mesopores are identifiable by their small diameter. Both autopores and mesopores, which the thin section happens to cut between diaphragms, appear to be open, for only clear crystalline calcite generally is seen as a filling of the tubes. Wherever a diaphragm occurs in the plane of the section, however, the autopore or mesopore containing it appears cloudy, because the diaphragm interferes with the passage of light. In addition to these features, tangential sections of *Dekayella* show scattered tiny, thick-walled tubes, which are located generally at angles between two or three adjoining autopores or at junctions of the walls of autopores and mesopores. They are called **acanthopores** (*acanthos*, spine, *10*) because they project above the general surface of the colony as minute spines. Acanthopores are confined to the mature regions of the colony and undoubtedly represent another specialized type of zooid. Their functions,

---

(*Fig. 5-4 continued.*)

monly smaller and more angular in section than autopores and containing more numerous diaphragms; presumably occupied by a specialized zooid.

monticule (*2*). Localized elevation of colonial surface, commonly formed by a cluster of mesopores surrounded by larger-than-average autopores; generally monticules are regularly spaced.

tangential section (*4, 15*). Thin slice of colony cut transverse to autopores near surface of colony.

zoarium (*1*). Entire bryozoan colonial skeleton.

zooecium (*6, 14*). Same as autopore.

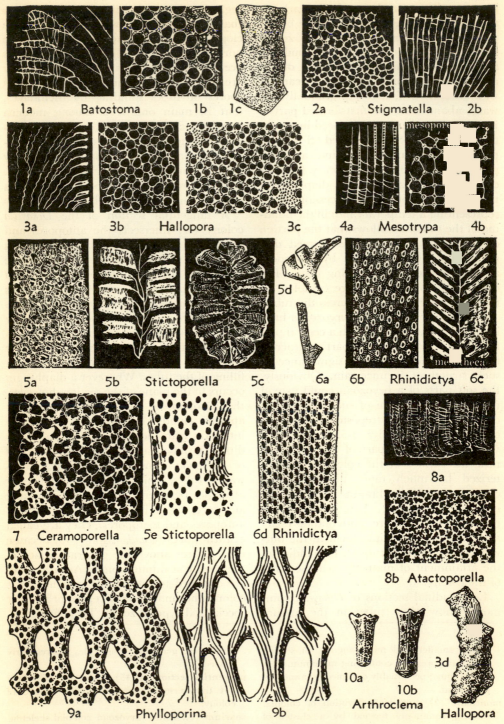

FIG. 5-5. **Representative Ordovician bryozoans.** Enlarged ×10 except as otherwise indicated. Trepostomes include 1–4, 8, 9; cryptostomes, 5, 6, 10; and cyclostomes, 7. (*Continued on next page.*)

however, like those of the mesopore zooids, are quite unknown.

Prasopora. The calcareous structures which comprise the skeletal deposits of colonies of *Prasopora* are massive growths, mostly subhemispherical, like a gumdrop (*A–D*). At the base of the colony is a concentrically wrinked thin plate or sheet (**coenelasma:** *coen*, common, belonging to the whole colony; *elasma*, sheet or lamina, *3*). Tubes secreted by the individual zooids grow upward from this base, curving so as to reach the surface approximately at right angles. Structures shown by thin sections generally are comparable to those of *Dekayella* except for a greater abundance of mesopores and the occurrence of many strongly curved partitions (**cystiphragms,** *8*), which are associated with the straight diaphragms of the autopores. These features are best shown in longitudinal sections of *Prasopora* (*F*), but they are recognizable also in tangential sections (*E*). In the latter, the mesopores are identified by their small size and polygonal outline, in contrast to the moderately large autopore tubes, which are nearly circular. The arched walls of cystiphragms are intersected obliquely by tangential sections, and thus they appear as indistinctly bounded curved lines within the autopores. They tend to have a somewhat constant orientation because they occur typically on the same side of the tubes (*E*).

The surface of some colonies of *Prasopora* is marked by regularly spaced, small elevations called **monticules** (*2*). Examination with a lens shows that these local prominences are formed by a special grouping of cells. The central area is composed wholly of mesopores, and around this are autopores which are distinctly larger than average size. The biological significance of these regularly spaced little communities of mesopores and large autopores within the colony is unknown. Many other bryozoans exhibit the same sort of localized differentiation of the tubes, but

---

(Fig. 5-5 continued.)

Arthroclema Billings, Ordovician. Zoarium consists of articulated segments. *A. cornutum* Ulrich (10*a*, *b*, side views of two segments), Middle Ordovician, Minnesota.

Atactoporella Ulrich, Ordovician–Silurian. Numerous mesopores and acanthopores, walls of autopores indented. *A. typicalis* Ulrich (8*a*, *b*, longitudinal and tangential sections), Middle Ordovician, Minnesota.

Batostoma Ulrich, Ordovician–Silurian. Acanthopores relatively prominent. *B. fertile* Ulrich (1*a*, *b*, longitudinal and tangential sections; 1*c*, side view of zoarium showing monticules, ×1), Middle Ordovician, Minnesota.

Ceramoporella Ulrich, Ordovician–Silurian. Incrusting colonies having short autopores, lunaria forming hoods over apertures, mesopores numerous. *C. inclusa* Ulrich (7, tangential section), Middle Ordovician, Minnesota.

Hallopora Bassler, Ordovician–Devonian. *H. crenulata* (Ulrich) (3*a*, *b*, longitudinal and tangential sections; 3*c*, surface showing two maculae; 3*d*, side view of zoarium, ×1), Middle Ordovician, Minnesota.

Mesotrypa Ulrich, Ordovician–Silurian. Large acanthopores at angles between autopores, few mesopores. *M. angularis* Ulrich & Bassler (4*a*, *b*,

longitudinal and tangential sections), Middle Ordovician, Kentucky.

Phylloporina Ulrich, Ordovician–Silurian. Zoarium a network of anastomosing branches bearing apertures on one side. Absence of hemisepta and other characters distinguish this bryozoan as a lacy trepostome. *P. sublaxa* Ulrich (9*a*, *b*, obverse and reverse sides), Middle Ordovician, Minnesota.

Rhinidictya Ulrich, Ordovician–Silurian. A very widely distributed ribbon-like bifoliate form. *R. mutabilis* Ulrich (6*a*, part of colony viewed normal to plane of mesotheca, ×1; 6*b*, *c*, tangential and longitudinal sections; 6*d*, surface), Middle Ordovician, Minnesota.

Stictoporella Ulrich, Ordovician. Mature very thick-walled region unusually prominent; colony bifoliate. *S. angularis* Ulrich (5*a–c*, tangential, longitudinal, and transverse sections; 5*d*, side view of zoarium, ×1; 5*e*, surface), Middle Ordovician, Minnesota.

Stigmatella Ulrich & Bassler, Upper Ordovician–Silurian. Small polygonal autopores, numerous acanthopores, few mesopores. *S. spinosa* Ulrich & Bassler (2*a*, *b*, transverse and longitudinal sections), Upper Ordovician, Indiana.

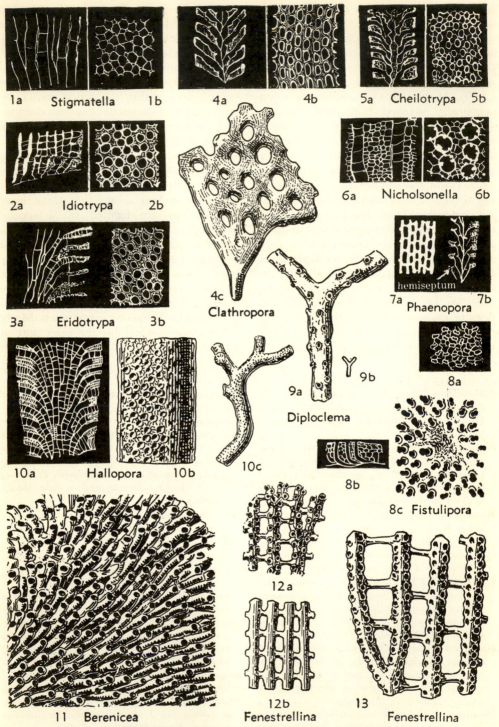

1a Stigmatella 1b
4a 4b
5a Cheilotrypa 5b
2a Idiotrypa 2b
6a Nicholsonella 6b
3a Eridotrypa 3b
4c Clathropora
7a Phaenopora 7b
hemiseptum
8a
10a Hallopora 10b
9a Diploclema 9b
10c
8b
8c Fistulipora
11 Berenicea
12a
12b Fenestrellina
13 Fenestrellina

FIG. 5-6. **Representative Silurian bryozoans.** Enlarged ×10 unless otherwise indicated. Trepostomes include 1–3, 6, 10; cryptostomes, 4, 7, 12, 13; cyclostomes, 5, 8, 9, 11. (*Continued on next page.*)

surface expression of the spotlike areas is in the form of a slight depression rather than an elevation, or is in a plane even with the surface of the colony. The depressed or even-surface spots are termed **maculae.** The size, shape, spacing, and relative prominence of monticules and maculae are important specific characters.

Both *Prasopora* and *Dekayella* belong to the group of trepostome bryozoans which is characterized by having fused or amalgamate walls (suborder Amalgamina). In this group the boundaries of adjacent tubes are not discernible. Another division (suborder Integrina) is distinguished by a sharply defined separation of the walls of adjoining tubes. In tangential sections, a dark-colored divisional line is seen where the walls of two tubes come together.

Cryptostome Bryozoans

The chief distinguishing character of the cryptostome bryozoans is shortness of the zooecial tubes or autopores. Like trepostomes, they have a thin-walled immature region and a thicker-walled mature region, but the contrast between these two parts of the colonial structure is distinctly greater in the cryptostomes. The distal (near-surface) part of each autopore comprises a **vestibule** (Fig. 5-7, *7*), which extends inward to a projection from the wall (**hemiseptum,** Fig. 5-7, *10*) or to the point of abrupt bending of the tube. The bottom of the vestibule is inferred to mark the location of the true aperture and its operculum (probably chitinous, as among cheilostomes)—hence the designation of

---

(Fig. 5-6 continued.)

Berenicea Lamouroux, Ordovician–Recent. Thin incrusting colony consisting of a single layer of parallel autopores. *B. consimilis* (Lonsdale) (11, portion of obverse side), Middle Silurian, Gotland, Sweden.

Cheilotrypa Ulrich, Silurian–Mississippian. Cylindrical branching colonies having an axial concavity. *C. ostiolata* (Hall) (5*a, b,* longitudinal and tangential sections), Middle Silurian, New York.

Clathropora Hall, Silurian–Devonian. Anastomosing bifoliate branches, autopore apertures in regular rows. *C. frondosa* Hall (4*a, b,* longitudinal and tangential sections; 4*c,* side view of colony, ×1), Middle Silurian, New York.

Diploclema Ulrich, Silurian. Slender branching colonies; circular apertures of autopores very widely spaced. *D. sparsum* (Hall) (9*a,* side view of colony, 9*b,* same, ×1), Middle Silurian, New York.

Eridotrypa Ulrich, Ordovician–Devonian. Slender branching colonies, autopores thick-walled, mesopores common. *E. similis* Bassler (3*a, b,* longitudinal and tangential sections), Middle Silurian, Ontario.

Fenestrellina d'Orbigny, Silurian–Permian. Lace bryozoans having two rows of autopore apertures on obverse face of branches, none on dissepiments. *F. cribrosa* (Hall) (12*a, b,* obverse and reverse); *F. elegans* (Hall) (13, obverse); both Middle Silurian, New York.

Fistulipora McCoy, Silurian–Permian. Rounded autopore tubes separated by vesicular coenosteum, lunaria more or less well developed. *F. tuberculosa* (Hall) (8*a, b,* tangential and longitudinal sections; 8*c,* portion of surface showing a macula), Middle Silurian, New York.

Hallopora Bassler, Ordovician-Devonian. Mesopores prominent. *H. elegantula* (Hall) (10*a, b,* longitudinal section and external view; 10*c,* part of colony, ×1), Middle Silurian, New York.

Idiotrypa Ulrich, Silurian. Incrusting colonies, short cylindrical dense-walled autopores separated by tissue containing small acanthopores; diaphragms evenly spaced. *I. punctata* (Hall) (2*a, b,* longitudinal and tangential sections), Middle Silurian, Indiana.

Nicholsonella Ulrich, Ordovician–Silurian. Walls of rounded autopores indented by acanthopores; coenosteum prominent between autopores. *N. florida* (Hall) (6*a, b,* longitudinal and tangential sections), Middle Silurian, New York.

Phaenopora Hall, Silurian. Bifoliate laminar colonies, hemisepta prominent. *P. ensiformis* Hall (7*a, b,* tangential and longitudinal sections), Middle Silurian, Ontario.

Stigmatella Ulrich & Bassler, Upper Ordovician–Silurian. *S. globata* Bassler (1*a, b,* longitudinal and tangential sections), Middle Silurian, New York.

cryptostome, which means hidden mouth.

Some representatives of this order have mesopores and acanthopores, which suggest relationship with the trepostomes, but in the mature region of most cryptostome bryozoans, spaces between the zooecial tubes are filled with irregular minute vesicles or a nearly solid calcareous deposit. These interzooecial structures cannot have been made by specialized zooids, such as occupied the tubes called mesopores and acanthopores. Rather, they are deposits formed by living tissue of the whole colony, belonging to no particular individual members, and thus may be designated as **coenosteum,** which means common skeletal substance (Fig. 5-8, 8).

Four genera of representative crypto-

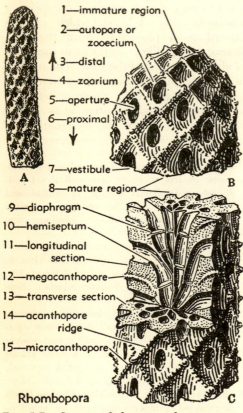

1—immature region
2—autopore or zooecium
↑ 3—distal
4—zoarium
5—aperture
6—proximal
↓
7—vestibule
8—mature region
9—diaphragm
10—hemiseptum
11—longitudinal section
12—megacanthopore
13—transverse section
14—acanthopore ridge
15—micracanthopore

A   B   C

Rhombopora

FIG. 5-7. **Structural features of cryptostome bryozoans—Rhombopora.** (*A*) Side view of a colony of *Rhombopora constans* Moore, Upper Pennsylvanian (Virgilian), north-central Texas, ×10. (*B*) Growing tip (distal extremity) of this species, ×40. (*C*) Transversely and longitudinally cut part of colony, showing internal features, ×40.

acanthopore ridge (14). Narrow elevation formed by acanthopores (megacanthopores and micracanthopores) in interzooecial areas.

aperture (5). Opening of an autopore or zooecial tube. Although this term is commonly used for the outer opening, which is visible from the exterior, it is employed also for the opening at the base of the vestibule, which constitutes the "hidden mouth" of cryptostomes.

autopore (2). One of main tubes of colony; also termed zooecium.

diaphragm (9). Thin platform extending transversely across an autopore.

distal (3). Direction away from initial autopore (ancestrula) of a bryozoan colony, corresponding to direction of growth of successively budded autopores.

hemiseptum (10). Platform extending part way across an autopore, designated as superior if it projects from proximal wall and as inferior if it projects from distal wall of autopore; marks base of vestibule in some cryptostomes and possibly furnished support for a hinged lid (operculum) for closure of tube.

immature region (1). Thin-walled axial part of colony, which reaches the surface only at growing tip of a branch.

longitudinal section (11). A cut parallel to autopore tubes.

mature region (8). Peripheral part of colony, characterized by greatly thickened interzooecial walls and by presence of acanthopores.

megacanthopore (12). Unusually large, prominent acanthopore.

micracanthopore (15). Relatively small acanthopore.

proximal (6). Direction toward initial autopore (ancestrula) of a bryozoan colony, opposite to direction of growth of successively budded autopores.

transverse section (13). Slice or cut normal to direction of colony growth, intersecting mature and immature regions.

vestibule (7). Outer part of autopore tube which runs approximately normal to surface of colony; inner limit of vestibule is marked in some cryptostomes by a hemiseptum.

zoarium (4). Hard parts of the whole bryozoan colony.

zooecium (2). Same as autopore.

stome bryozoans are selected for the purpose of making an acquaintance with the important structural features of this group: *Rhombopora* comprises slender, cylindrical colonies which have twiglike branches or are unbranched; *Sulcoretepora* typically has the form of a narrow ribbon, which may branch in a Y-shaped manner like a forked

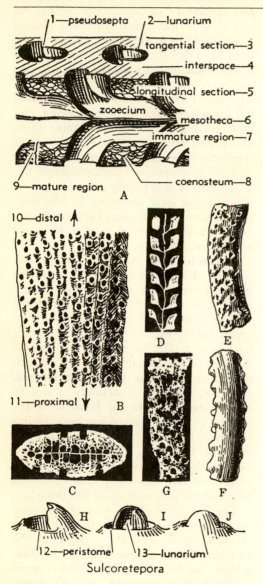

FIG. 5-8. **Structural features of cryptostome bryozoans—Sulcoretepora.** (*A*) Diagrammatic much enlarged view of a colony of *Sulcoretepora* showing structure in longitudinal section and portion of tangentially planed peripheral region of colony. (*B-D*) *Sulcoretepora obliqua* McNair, Middle Devonian (Cazenovian), northern Michi-

gan; (*B*) external view normal to plane of mesotheca; (*C*) transverse section showing mesotheca, autopores in immature region, and dense coenosteum in mature region; (*D*) longitudinal section; all ×10. (*E-G*) *Sulcoretepora formosa* Moore, Upper Pennsylvanian (Virgilian), north-central Texas; (*E*) external view normal to plane of mesotheca; (*F*) external view of edge, in plane of mesotheca; (*G*) tangential section showing autopores at various depths within colony and intervening coenosteum; all ×10. (*H-J*) Much enlarged views of a hood-shaped lunarium from side, distal, and proximal directions.

coenosteum (8). Vesicular or dense skeletal tissue between autopores in mature region, secreted by interconnecting soft parts of whole colony rather than by individual zooids.

distal (10). In the direction of colonial growth.

immature region (7). Thin-walled part of colony next to mesotheca of *Sulcoretepora*, characterized by autopores growing parallel to plane of mesotheca.

interspace (4). Area at surface of a colony between adjacent apertures of autopores.

longitudinal section (5). Slice or cut of colony parallel to direction of growth of autopores.

lunarium (2, 13). Thick-walled posterior (proximal) part of autopores belonging to some bryozoan colonies, commonly more strongly curved transversely than distal wall and having edges which may project as pseudosepta into autopores; surficial portion of some projects as a hood over aperture.

mature region (9). Peripheral part of colony characterized by interspaces between autopores filled by coenosteum.

mesotheca (6). Median lamina of bifoliate colonies, composed of two coenelasma layers which form respective bases of oppositely growing halves of colony.

peristome (12). Rim surrounding an autopore aperture.

proximal (11). Direction toward beginning of colonial growth.

pseudosepta (1) Projections of edges of a lunarium forming longitudinal ridges on walls of an autopore.

tangential section (3). Thin slice or cut of colony near surface of colony, transverse to autopores.

highway, different parts of the colony being in the same plane; *Fenestrellina* and *Archimedes* are characterized by a netlike form of slender branches joined by crossbars. *Archimedes* is further distinguished by the presence of a spiral axial support for the colony.

Rhombopora. The autopores of this genus (Fig. 5-7) have regularly distributed elliptical openings arranged in obliquely intersecting rows all around the cylindrical branch. The walls of the mature region are much thickened. Around the openings they produce upward sloping areas which culminate in ridges of granules or short spines. These latter are the surface expression of regular rows of closely spaced acanthopores. The intersecting arrangement of the rows gives rise to the characteristic rhomb-shaped outline of each autopore area in *Rhombopora*. Two distinct sizes of acanthopores commonly are recognizable. A single very large one (**megacanthopore,** from *mega*, big) is located at the proximal extremity of the ridge which girdles each opening, that is, on the side toward the base of the colony (Fig. 5-7, *12*). The other acanthopores, which are very numerous, are small (**micracanthopores,** from *micro*, small) (Fig. 5-7, *15*). They project at the surface only moderately.

Mesopores occur between the zooecial tubes of the mature region of some other cryptostomes, but there are none in *Rhombopora*.

The immature region of colonies of *Rhombopora*, as shown clearly both by longitudinal and transverse sections, consists of very thin walls. The tubes are narrowest in their beginning portions, which run nearly parallel to the axis of the branch. Their diameter increases gradually as they diverge to the point of abrupt bending into the mature region, where their attitude is approximately at right angles to the surface of the colony. Immature parts of the autopores commonly contain a few widely spaced diaphragms. In the mature region, near the elliptical opening, one or two half diaphragms

hemisepta, Fig. 5-7, *10*) project from the wall. These are typical structures of *Rhombopora* and several other cryptostomes.

The growing end of a colony of *Rhombopora* is bluntly rounded (Fig. 5-7*B*). An end view of the branch in this region is comparable to a transverse section of the branch, in that the walls between the tubes in the axial region are very thin and the outline of the tubes is strongly polygonal. The fact that the branch has a normal full diameter close to the growing end indicates that elongation of the tubes virtually ceases after maturity is reached, as represented by the thickening of the walls and the development of acanthopores.

Rhombopora is typically a late Paleozoic genus, being most abundant in Mississippian, Pennsylvanian, and Permian strata. It is recorded, however, from rocks as old as Ordovician.

Sulcoretepora. A distinguishing feature of this genus, which is seen also in numerous other cryptostomes, is the ribbon-like form of the colony and the arrangement of the short autopores (zooecial tubes) in a double series growing back to back (Fig. 5-8*A, C, D*). This double layer, or bifoliate mode of growth, is developed from a median lamina (**mesotheca,** Fig. 5-8, *6*). This is actually a double lamina, consisting of the basal layer belonging to half of the colony, built against the basal layer of the other half. It comprises the floor from which the tubes grow outward in two directions. This structural plan is clearly shown by transverse and longitudinal thin sections of the branch.

The autopores of *Sulcoretepora* first grow nearly parallel to the mesotheca (Fig. 5-8, *A*) then bend abruptly outward to the surface. The immature region is thinwalled, as in trepostomes. The mature region is thicker walled, and the **interspaces** (Fig. 5-8, *4*) between the tubes are filled with small irregular vesicles or solid calcareous material (stereome), secreted not by the individual zooids but by common tissue of the colony. This filling is termed **coenosteum** (Fig. 5-8, *8*).

The apertures of the autopores are surrounded by a narrow rim (**peristome,** Fig. 5-8, *12*). Some species of *Sulcoretepora* have evenly rounded or elliptical apertures, but more commonly, species belonging to this genus possess indented apertures, with an accompanying hoodlike elevation of part of the peristome. This form of aperture is associated with a structure called **lunarium** (Fig. 5-8, *2, 13*), which occurs in the mature part of the autopore tubes. It consists of a thickened wall which curves transversely with distinctly smaller radius than that of the remainder of the tube, and the edges of the strongly curved thickened part of the wall may project into the tube. Such projections form longitudinal ridges called **pseudosepta** (Fig. 5-8, *1*). The lunaria invariably occur on the side of the autopore tubes which is nearest to one of the margins of the colony.

Sulcoretepora is common in many Paleozoic marine formations from Devonian to Permian.

Fenestrellina. Colonies of the lace bryozoans, which include several different types, grow in fan-shaped or funnel-like expansions, cemented to some foreign object by calcareous tissue or rootlike extensions near the base of the colony (Fig. 5-9; italic letters and numbers throughout the succeeding text on cryptostome bryozoans refer to this figure). The main parts of the colonial structure consist of nearly parallel slender branches, joined to one another at intervals by crossbars (**dissepiments**) (*5*). The openings of the autopores are all situated on one side of the network. This front surface is designated as the **obverse side** (*10E*) of the colony, and the nonporiferous back surface is termed the **reverse side** (*7G*). In *Fenestrellina*, the obverse face of the colony is on the inner side of funnel-shaped growths (*C*), but in some other genera of similar form, the apertures may be on the outer side.

A distinguishing feature of *Fenestrellina* is the arrangement of the zooecial apertures in two rows along each branch (*A,*

E), in contrast to various other genera having more than two rows. The mid-line of each branch commonly is raised in the form of a keel (**carina,** *11*), and in some species this ridge bears evenly spaced nodes or **spines** (*2*), which represent **acanthopores** (*6*). The size and spacing of the zooecial apertures and of nodes along the keel are features employed in defining species. Also useful is the presence or absence of a distinct rim (**peristome,** *13*) around the zooecial aperture. The dissepiments lack autopores. The openings (**fenestrules,** *1*) formed by adjacent branches and dissepiments vary considerably in size and shape. The dimensions of the fenestrules, branches, and dissepiments (commonly expressed by the number occurring in the space of 5 or 10 mm.) are important in the differentiation of species. Although surface characters and dimensions are sufficiently distinctive to serve for identification of most species of *Fenestrellina*, tangential sections are also useful because they show the varying shape in outline and arrangement of the zooecial tubes within the branches (*A,* lower left). The calcareous tissue composing the dissepiments, as well as supports for the colony and some parts of the branches, constitutes **coenosteum** (*14*), secreted by extrazooecial cells of the colony.

The reverse side of colonies belonging to *Fenestrellina* may be smooth, striated, or marked by fine granules (*G*). As seen from this side, the dissepiments may be flush with the branches or appear strongly depressed. Such features must be noted in the description of species.

Fenestrellina is first seen in mid-Silurian rocks. The genus is abundant in many Devonian, Mississippian, Pennsylvanian, and Permian marine deposits, and is world-wide in distribution (Figs. 5-6, 5-11, 5-12, 5-14).

Archimedes. Special types of colonial support developed by some lace bryozoans are a basis for defining genera. One of the most remarkable of these is the spiral *axis* (*9*) of the bryozoans called *Archimedes* (*B,*

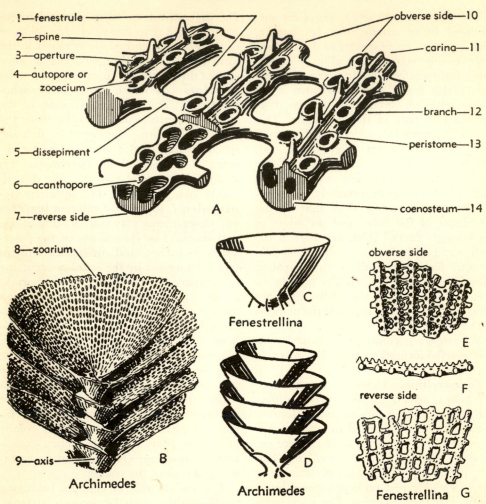

1—fenestrule
2—spine
3—aperture
4—autopore or zooecium
5—dissepiment
6—acanthopore
7—reverse side
8—zoarium
9—axis

obverse side—10
carina—11
branch—12
peristome—13
coenosteum—14

A

Fenestrellina

C

Archimedes

B

obverse side

E

F

reverse side

Fenestrellina   G

Archimedes

D

FIG. 5-9. **Structural features of cryptostome bryozoans—Fenestrellina and Archimedes.** (A) Enlarged oblique view of part of the lacy network which characterizes the zoarial fronds of *Fenestrellina* and *Archimedes*, branches transversely and tangentially sectioned at front of the drawing. (B) Part of the zoarium of *Archimedes wortheni* (Hall), ×1, Mississippian, Illinois, showing screwlike axis of the colony and spirally arranged network attached to the central support. (C) Diagram of the funnel-like zoarial form of some species of *Fenestrellina*, showing extensions for support at the base. (D) Diagram of the spirally arranged fronds of the *Archimedes* zoarium. (E–G) *Fenestrellina pectinis* Moore, ×10, Upper Pennsylvanian (Virgilian), north-central Texas; obverse, edge, and reverse sides of part of a colony.

acanthopore (6). Slender dense-walled tube which projects as a node or spine at surface; arranged along mid-part of branches of many fenestrate bryozoans.

aperture (3). Opening of zooecium at surface.

autopore (4). Chamber occupied by an individual zooid, very short in fenestrate bryozoans and having ovoid, pentagonal, or hexagonal outline as seen in tangential sections; also called zooecium.

axis (9). Solid calcareous support of *Archimedes*.

branch (12). Slender rodlike structure of fenestrate colonies extending outward radially from base, place of attachment, or point of bifurcation; it bears zooecial chambers, which have apertures on one side of branch only.

carina (11). Median keel running along side of branches which bears zooecial apertures.

coenosteum (14). Dense or vesicular calcareous tissue formed by interzooecial cells of bryozoan

(Continued on next page.)

D). *Fenestrellina*-like networks grew out from the flanges of this axis. Thus, *Archimedes* differs from *Fenestrellina* essentially in the spiral structure of the colony and in its solid screwlike central support.

Colonies of *Archimedes* showing parts of the spreading fronds attached to the solid spiral axis can be found in many places. Generally, however, the delicate lacelike parts of the colony are broken away from the axis and the latter, which is much more sturdy, is found alone (Fig. 5-12, *4*). Since the lacelike parts of *Archimedes* differ in no way from *Fenestrellina*, it is difficult to identify such networks if they are found separated from the spiral axis. Some isolated fragments of lacy bryozoans classed as *Fenestrellina* may actually belong to *Archimedes*, because only the spiral axis distinguishes these two genera. Thickness and spacing of the spiral flanges on the axis of *Archimedes* are characters useful in discriminating species. The lacy parts belonging to many species is unknown.

A recent suggestion that the screwlike axis of *Archimedes* was produced by lime-secreting algae which grew in symbiotic union with fenestrellinid bryozoans does not accord with constancy of characters of the spiral axes. Also, microscopic structure of the axes corresponds to that of thickened parts of the lacework branches and to the branching roots of *Fenestrellina*. Accordingly, the screwlike axis of *Archimedes* seems to be the product of secretion by extra-zooecial cells of the colony, and its material may be termed coenosteum.

Archimedes is abundant in some Lower and Upper Mississippian formations, and the genus belongs essentially to this part of the geologic column. A few species have been found in Pennsylvanian rocks of western North America and in Permian deposits of Russia (Fig. 5-16).

Ctenostome Bryozoans

The comb-mouth bryozoans, or cteno-stomes, are characterized by the growth of individuals which bud from internodes of a tubular stem (**stolon**), which is attached throughout its length to a shell or some other foreign object (Fig. 5-10).

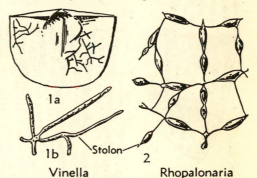

Vinella   Rhopalonaria

FIG. 5-10. **Fossil ctenostome bryozoans.** (1*a*, *b*) *Vinella repens* Ulrich, Middle Ordovician (Black-riveran), Minnesota; (1*a*) stolons of colonies attached to interior of brachiopod shell, ×1; (1*b*) portion of radiating stolons showing regularly spaced minute pores which mark places of attachment of bryozoan individuals, ×10. (2) *Rhopalonaria venosa* Ulrich, Upper Ordovician (Cincinnatian), southern Ohio; spindle-shaped shallow excavations in shell on which this ctenostome colony grew, with threadlike stolonal connections, ×10.

In an unexplained manner, possibly by chemical solution of the calcareous substance of shells which furnish the base, some ctenostome stolons, such as *Rhopa-*

(Fig. 5-9 continued.)

colony, rather than by any individual zooid.

dissepiment (5). Crossbar connecting branches of a lacy bryozoan.

fenestrule (1). Opening framed by adjacent branches and dissepiments of a lacy bryozoan.

obverse side (10). Surface of colony which bears autopore apertures.

peristome (13). Rim partly or entirely encircling an autopore aperture.

reverse side (7). Surface of a colony lacking autopore apertures.

spine (2). Surficial projection formed by an acanthopore on obverse side of a lacy bryozoan branch, generally located on a carina.

zoarium (8). Hard parts of entire bryozoan colony.

zooecium (4). Same as autopore.

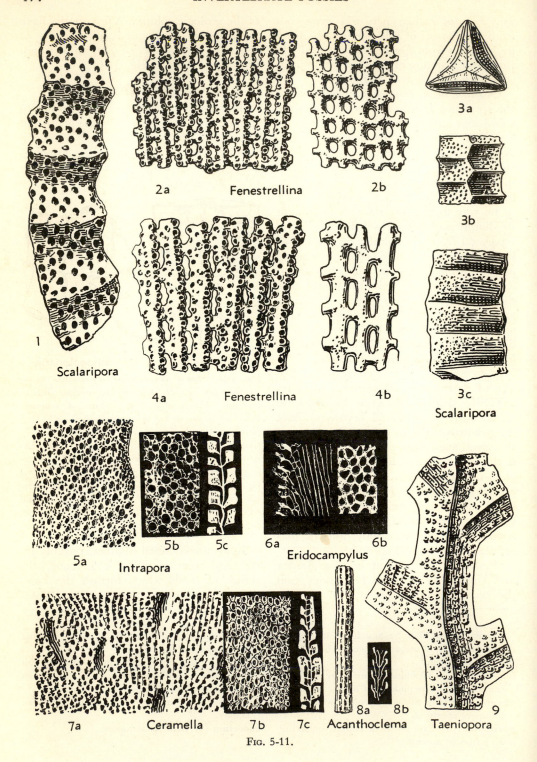

1 Scalaripora

2a Fenestrellina 2b

3a

3b

3c Scalaripora

4a Fenestrellina 4b

5a 5b 5c Intrapora

6a Eridocampylus 6b

7a Ceramella 7b 7c Acanthoclema 8a 8b Taeniopora 9

Fig. 5-11.

lonaria (Fig. 5-10, *2*), become imbedded in their support. Depressions ascribable to these bryozoans are seen on some fossil shells. They constitute the only record of ctenostome bryozoans in various Paleozoic formations.

The stolon, which commonly is thread-like, gives off cylindrical stalks, each of which dilates at its end into a bryozoan zooid. The aperture is closed by a membranous lid (operculum), which bears toothlike projections resembling a comb. Although the stolon of ctenostomes may be calcified and thus preserved as a fossil, no other parts of these organisms are known to secrete calcium carbonate.

The known range of this group of bryozoans is from Ordovician to Recent.

Vinella. Representative of the ctenostomes is the genus *Vinella*, known from Middle Ordovician rocks (Fig. 5-10, *1a, b*) and recorded also from Pennsylvanian strata of the central United States. The species here illustrated consists of very slender, cylindrical stolons attached to a brachiopod shell. The stolons radiate from centers, and at intervals along them are small apertures which presumably mark the location of zooids.

Because preservable characters of this group of bryozoans are not highly varied, distinctive, or abundant, the fossils are relatively unimportant.

Cyclostome Bryozoans

The skeleton of cyclostome bryozoans consists of simple calcareous tubes which generally lack transverse partitions and which have a plain rounded aperture, not closed by an operculum. The walls of the tubes are thin and minutely porous. The shape of the colonial structures varies widely, although the types of growth belonging to any one species are fairly constant.

Stomatopora. One of the simplest types of colonies is *Stomatopora*, which consists of tubular autopores attached to some foreign support (Fig. 5-13, *4*). They grow in a series, which may branch. Each autopore has a single round aperture and comprises the skeletal deposit of one zooid.

Colonies of other genera closely resemble *Stomatopora* in the form of autopores, but instead of a single row of cells, they form sheetlike expansions attached to a shell (Fig. 5-6, *11*) or are unattached, growing from a basal lamina (coenelasma). Colonies

---

FIG. 5-11. **Representative Devonian bryozoans.** Enlarged ×10 except as otherwise indicated. All are cryptostomes except 6, which is a treptostome.

Acanthoclema Hall, Devonian–Mississippian. Slender cylindrical branches, autopores in longitudinal rows separated by ridges. *A. ohioense* McNair (8*a, b*, side view of exterior and longitudinal section), Middle Devonian, northern Ohio.

Ceramella Hall & Simpson, Devonian. Laminar bifoliate colonies, prominent depressed maculae, autopore apertures surrounded by strong peristome. *C. casei* McNair (7*a–c*, exterior, tangential, and longitudinal sections), Middle Devonian, northern Michigan.

Eridocampylus Duncan, Middle Devonian. Hooklike incomplete diaphragms, called heterophragms, occur in the mature region. *E. laxatus* Duncan (6*a, b*, longitudinal and tangential sections), northern Michigan.

Fenestrellina d'Orbigny, Silurian–Permian. *F. regularis* McNair (2*a, b*, obverse and reverse

sides); *F. rockportensis* McNair (4*a, b*, obverse and reverse sides); both Middle Devonian, northern Michigan.

Intrapora Hall, Devonian–Mississippian. Bifoliate, interspaces between autopores filled by coenosteum. *I. puteolata* Hall (5*a*, exterior view; 5*b, c*, tangential and longitudinal sections), Middle Devonian, northern Michigan.

Scalaripora Hall, Devonian. Branches triangular in section bearing transverse ridges, autopore apertures on all faces. *S. separata* Ulrich (1, side view), Middle Devonian, northern Michigan. *S. scalariformis* Hall (3*a–c*, end and two side views), Lower Devonian, Ohio.

Taeniopora Nicholson, Devonian. Bifoliate bifurcating branches having strong median ridges and autopore apertures on both sides. *T. exigua* Hall (9, side view), Middle Devonian, New York.

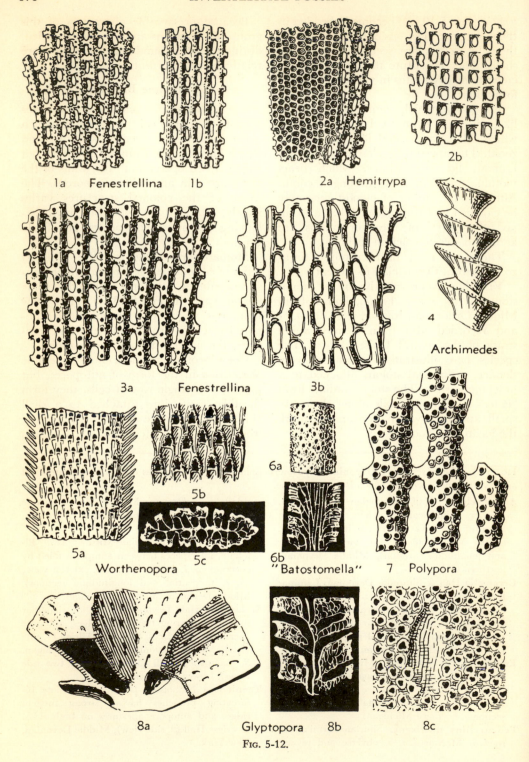

1a   Fenestrellina   1b   2a   Hemitrypa   2b

3a   Fenestrellina   3b   4   Archimedes

5a   5c   6a   6b   7   Polypora
Worthenopora   5b   "Batostomella"

8a   Glyptopora   8b   8c

FIG. 5-12.

of cylindrical twiglike form have apertures, like some of the cryptostomes, distributed on all parts of the exposed surface (Figs. 5-17, *1–6, 8, 9;* **5-18,** *5*). Other branching forms, especially those of flattened cross section, tend to have apertures on one face only, which then may be designated as the obverse (frontal) side (Figs. 5-18, *1–4, 6;* 5-19, *1, 5*). Subhemispherical, globular, or irregularly massive growths are common in some types of cyclostomes. Many of the structural features of this group of bryozoans resemble skeletal parts of trepostomes and cryptostomes (Fig. 5-15).

Fistulipora. One of the most common Paleozoic cyclostome genera is *Fistulipora*, which typically forms massive irregular colonial growths. Some attain horizontal and vertical dimensions of 15 in. or more. Colonies of other species are thin, sheetlike expansions.

The rounded apertures of the autopores are sufficiently large to be readily visible to the naked eye, but vesicular colonial tissue (coenosteum, Fig. 5-13*A*), which separates the autopore tubes, cannot be observed readily without the use of a lens. A distinguishing feature of this genus is the separation of autopores from neighbors all around by interspaces occupied by coenosteum.

Another feature of *Fistulipora* is the lunarium, which commonly projects into the zooecial space and has a more sharply curved surface than other parts of the tube (Fig. 5-13, *1a*). The lunaria are readily observed at the surface. In some species, they appear as elevated hoods, partly projecting over the apertures. They are directed away from regularly spaced monticules or maculae composed of coenosteum.

Longitudinal sections of *Fistulipora* show diaphragms distributed at intervals in the autopore tubes. The irregular vesicles of coenosteum between the tubes are much smaller in diameter than the autopores

---

FIG. 5-12. **Representative Mississippian bryozoans.** Enlarged ×10 except as indicated otherwise. All are cryptostomes except 6, which is a trepostome.

Archimedes Hall, Mississippian–Permian. Distinguished by solid screwlike axis which supports *Fenestrellina*-like lacy expansions. *A. wortheni* (Hall) (4, side view of part of axis, ×1), Lower Mississippian (Osagian), western Illinois.

Batostomella Ulrich, Ordovician. Cylindrical bifurcating branches with small acanthopores and numerous mesopores. Although called *Batostomella*, the Mississippian fossils differ from the type species of this genus. "*B.*" *spinulosa* Ulrich (6*a, b*, side view and longitudinal section), Upper Mississippian (Chesteran), Kentucky.

Fenestrellina d'Orbigny, Silurian–Permian. *F. serratula* (Ulrich) (1*a, b*, obverse and reverse sides); *F. rudis* (Ulrich) (3*a, b*, obverse and reverse sides); both Lower Mississippian (Osagian), southeastern Iowa and western Illinois.

Glyptopora Ulrich, Mississippian–Pennsylvanian. Bifoliate laminar expansions, bifurcating at intervals, surface marked by elongate depressed maculae, coenosteum between auto-pores, which commonly bear lunaria. *G. elegans* Prout (8*a*, side view of colony, ×1; 8*b, c*, longitudinal section and detail of surface showing a macula), Lower Mississippian (Osagian), western Illinois.

Hemitrypa Phillips, Silurian–Mississippian. Like *Fenestrellina* except for having a fine network superstructure on obverse side, supported by spines along the carinae. *H. proutana* Ulrich (2*a, b*, obverse and reverse sides), Lower Mississippian (Osagian), western Illinois.

Polypora McCoy, Ordovician–Permian. Like *Fenestrellina* but coarser and branches have three to eight rows of autopore apertures. *P. simulatrix* Ulrich (7, obverse side of fragment), Lower Mississippian (Osagian), southeastern Iowa.

Worthenopora Ulrich, Mississippian. Narrow bifoliate branches, rhomboidal autopores in regular rows, proximal side of apertures truncate. *W. spinosa* Ulrich (5*a*, side view of part of colony; 5*b, c*, detail of exterior and transverse section, ×20), Lower Mississippian (Osagian), southeastern Iowa.

(Fig. 5-13, *1b*). Interzooecial spaces near the surface tend to become dense by deposit of calcium carbonate within the coenosteum.

Fistulipora is questionably identified in Ordovician rocks but is represented by characteristic species in Silurian formations. It is most common in later Paleozoic marine deposits, including Permian.

Entalophora. Among geologically young cyclostomes is *Entalophora*, which ranges from Jurassic to Recent and is common in early Tertiary deposits of the Gulf Coast region (Fig. 5-13, *3a, b*). Colonies of this genus are branching growths which have somewhat widely spaced cir-

cular apertures projecting distinctly above the general surface of the colony. Transverse sections show that the colony consists of a very simple grouping of long autopore tubes. There are neither mesopores nor diaphragms, and no other special type of cell is observed.

Spiropora. Another genus, which resembles *Entalophora* in the general simplicity of its colonial structure, is *Spiropora*, Jurassic to Recent (Fig. 5-13, *2a, b*). A distinguishing feature is the arrangement of the apertures in a spiral series around the branch. The walls between the autopores can be discerned externally near the apertures, but the immature parts of the tubes

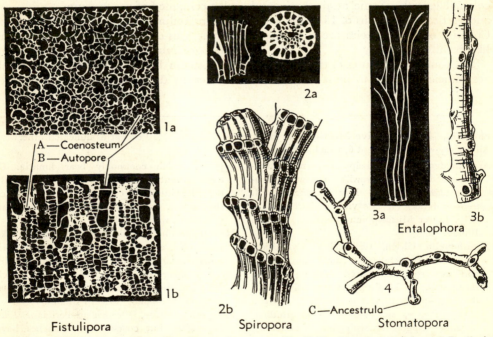

Fig. 5–13. **Representative types of cyclostome bryozoans.** (*1a, b*) *Fistulipora proiecta* (Moore & Dudley), Lower Permian (Leonardian), western Texas; tangential and longitudinal sections, ×10. (*2a, b*) *Spiropora majuscula* Canu & Bassler, Eocene (Jacksonian), South Carolina; (*2a*) longitudinal and transverse sections; (*2b*) external view showing spirally arranged rows of apertures; all ×10. (*3a, b*) *Entalophora proboscidea* Milne-Edwards, Oligocene, Alabama; longitudinal section and external view, ×10. (4) *Stomatopora parvipora* Canu & Bassler, Eocene (Jacksonian), Mississippi, ×20.

ancestrula (*C*). First-formed autopore of colony, which develops from a sexually produced larva and by asexual budding gives rise to other autopores of colony.

autopore (*B*). Tube or chamber occupied by

one of main zooids of colony; also called zooecium.

coenosteum (*A*). Vesicular tissue between autopores, formed by colony in common.

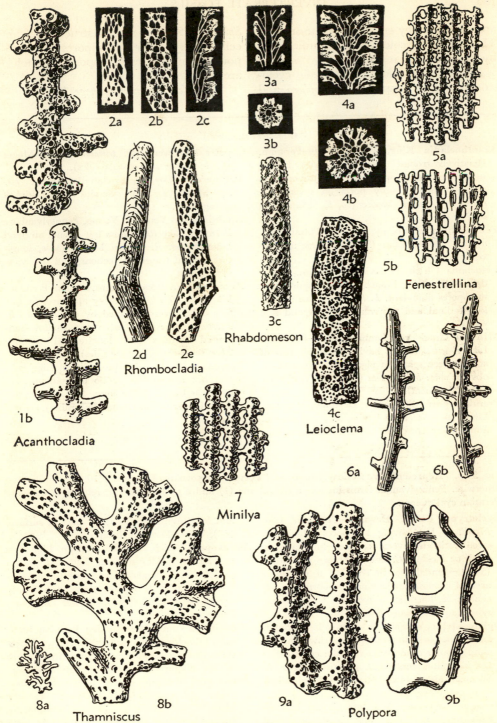

1a

1b

Acanthocladia

2a 2b 2c

2d 2e
Rhombocladia

3a

3b

3c
Rhabdomeson

4a

4b

4c
Leioclema

5a

5b
Fenestrellina

6a 6b

7
Minilya

8a 8b
Thamniscus

9a

9b
Polypora

FIG. 5-14. **Representative Pennsylvanian bryozoans.** Enlarged ×10 and from Upper Pennsylvanian of north-central Texas unless indicated otherwise. All are cryptostomes except 4, which is a trepostome.

(Continued on next page.)

are concealed by cells nearer the surface. This part of the colony can be studied only by means of thin section.

Not all Mesozoic and Cenozoic cyclostomes are so simple as *Entalophora* and *Spiropora*. Several special structures (cancelli, dactylethrae, tergopores, nematopores, firmatopores, mesopores, vacuoles) are recorded.

(*Text continued on page 186.*)

(*Fig. 5-14 continued.*)

Acanthocladia King, Pennsylvanian–Permian. Relatively robust pinnate colonies having peristome-circled apertures of autopores on obverse and accessory pores on reverse. *A. ciscoensis irregularis* Moore (1*a*, *b*, obverse and reverse sides).

Fenestrellina d'Orbigny, Silurian–Permian. *F. mimica raymondi* Elias (5*a*, *b*, obverse and reverse sides).

Leioclema Ulrich, Ordovician–Permian. Cylindrical bifurcating branches, acanthopores and mesopores abundant. *L. hirsutum* Moore (4*a–c*, longitudinal and transverse sections and side view).

Minilya Crockford, Pennsylvanian–Permian. Like *Fenestrellina* but having a double row of acanthopore spines in zigzag arrangement on obverse side. *M. binodata* (Condra) (7, obverse side of fragment).

Penniretepora d'Orbigny, Devonian–Permian. Like *Acanthocladia* but having only two rows of autopore apertures on obverse and no accessory pores on reverse. *P. trilineata texana* Moore (6*a*, *b*, reverse and obverse sides).

Polypora McCoy, Ordovician–Permian. *P. hirsuta* Moore (9*a*, *b*, obverse and reverse sides of a fragment).

Rhabdomeson Young & Young, Pennsylvanian–Permian. Like *Rhombopora* but having an axial hollow tube. *R. decorum* Moore (3*a–c*, longitudinal and transverse sections, and side view of fragment).

Rhombocladia Rogers, Pennsylvanian–Permian. Like *Rhombopora* but bearing apertures on one face only, reverse covered by coenelasma. *R. delicata* Rogers (2*a–c*, deep and shallow tangential sections and longitudinal section; 2*d*, *e*, reverse and obverse sides).

Thamniscus King, Silurian–Permian. Like *Polypora* but having few or no dissepiments. *T. octonarius* Ulrich (8*a*, colony, ×1; 8*b*, obverse side), Upper Pennsylvanian (Virgilian), central Kansas.

FIG. 5-15. **Representative Pennsylvanian and Permian bryozoans.** Enlarged ×10 unless indicated otherwise. Pennsylvanian forms include a trepostome (1) and a cyclostome (3); remaining ones are Permian cyclostomes.

Cyclotrypa Ulrich, Devonian–Permian. Like *Fistulipora* but having autopores which are circular in transverse section, lunaria very weak or lacking. *C. candida* Moore & Dudley (3*a*, tangential section showing a macula—area of coenosteum lacking autopores—in left center; 3*b*, longitudinal section), Upper Pennsylvanian (Virgilian), Kansas.

Dybowskiella Waagen & Wentzel, Permian. Like *Fistulipora*, but autopores grow from an axial tubular coenelasma and the colony thus has a hollow center. *D. grandis* Waagen & Wentzel (2*a*, *b*, tangential and longitudinal sections), Upper Permian, India.

Hexagonella Waagen & Wentzel, Permian. Like *Fistulipora* but bifoliate and having polygonal ridges on the surface surrounding maculae. *H. ramosa* Waagen & Wentzel (5*a*, *b*, tangential and longitudinal sections; 5*c*, part of surface, ×5), Upper Permian, India.

Meekopora Ulrich, Silurian–Permian. Like *Hexagonella* but lacking surficial ridges. *M. prosseri* Ulrich (4*a*, *b*, tangential and longitudinal sections; 4*c*, part of surface showing a faintly depressed macula, ×5), Lower Permian, Nebraska.

Tabulipora Young, Mississippian–Permian. Laminar, branching, or massive; walls of autopores thickened in beadlike manner in mature region, diaphragms centrally perforated; surface marked by maculae or monticules; mesopores and acanthopores common. *T. cava* Moore (1*a*, *b*, part of exterior showing macula (upper left) and tangential section; 1*c*, part of zoarium, ×5), Upper Pennsylvanian (Virgilian), north-central Texas.

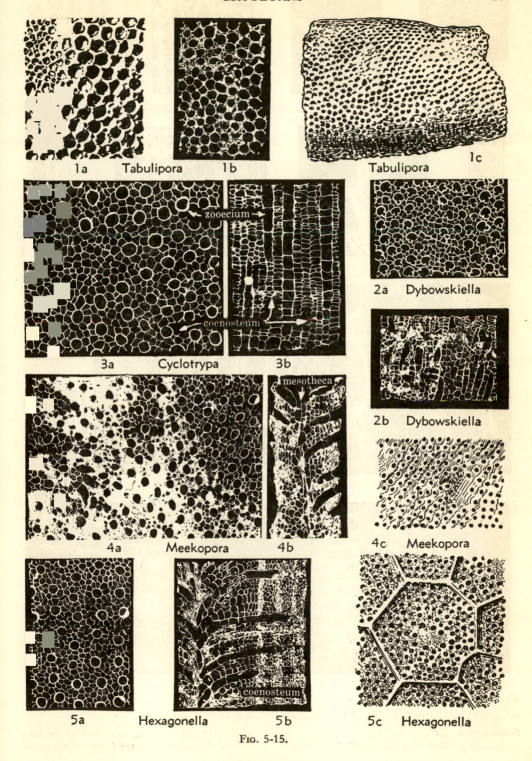

1a   Tabulipora   1b   Tabulipora   1c

3a   Cyclotrypa   3b

zooecium

coenosteum

2a   Dybowskiella

2b   Dybowskiella

4a   Meekopora   4b

mesotheca

4c   Meekopora

5a   Hexagonella   5b   5c   Hexagonella

coenosteum

Fig. 5-15.

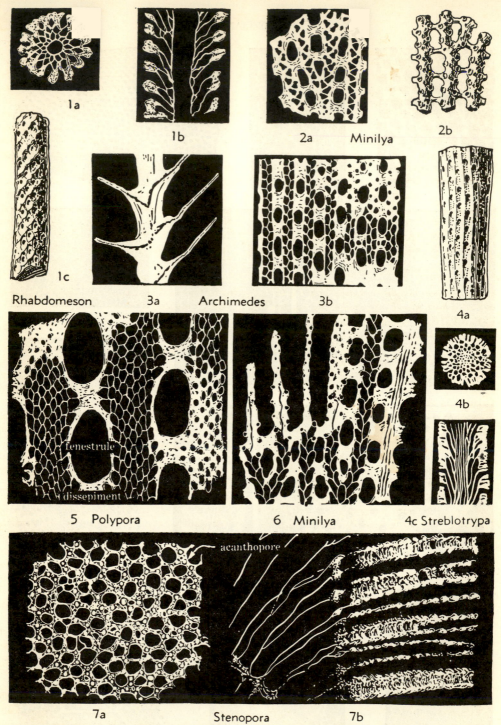

1a

1b

2a Minilya

2b

1c

Rhabdomeson **3a** Archimedes **3b** **4a**

4b

fenestrule

dissepiment

5 Polypora **6** Minilya **4c** Streblotrypa

acanthopore

7a Stenopora **7b**

Fig. 5-16. **Representative Permian bryozoans.** Enlarged ×10 unless indicated otherwise. A trepostome is illustrated in 7; the remaining forms are cryptostomes.

Archimedes Hall, Mississippian–Permian. *A. stuckenbergi* Nikiforova (3*a*, longitudinal section of central axis, showing extensions of the lacy part of the colony from the flanges of the spiral

(*Continued on next page.*)

(*Fig. 5-16 continued.*)

screw, ×1; *3b*, tangential section of branches and dissepiments forming part of the lacy network), Lower Permian, Russia.

Minilya Crockford, Pennsylvanian–Permian. *M. duplaris* Crockford (*2a, b*, tangential section and obverse view of a fragment), Permian, western Australia; *M. permiana* (Stuckenberg) (6, tangential section showing below and at right double rows of alternating autopores, and in upper left sections of acanthopore spines disposed in zigzag), Lower Permian, Russia.

Polypora McCoy, Ordovician–Permian. *P. cyclopora* Eichwald (5, tangential section showing several rows of hexagonal autopores), Lower Permian, Russia.

Rhabdomeson Young & Young, Pennsylvanian–Permian. *R. mammillatum* (Bretnall) (*1a, b*, transverse and longitudinal sections; *1c*, external view, ×5), Permian, Australia.

Stenopora Lonsdale, Permian. Like *Tabulipora* but having few or no diaphragms, those present not being centrally perforated. *S. pustulosa* Crockford (*7a, b*, tangential and longitudinal sections). Tasmania.

Streblotrypa Ulrich, Devonian–Permian. Slender cylindrical bifurcating branches having autopore apertures in longitudinal rows with several intervening mesopores. *S. marmionensis* Etheridge (*4a–c*, external view and transverse and longitudinal sections), Permian, western Australia.

FIG. 5-17.   (*See next page.*) **Representative Cretaceous and Eocene bryozoans,** Enlarged ×10 unless indicated otherwise. Cretaceous cyclostomes include 1–6, 8, 9; Eocene, cheilostomes, 7, 10–12.

Aplousina Canu & Bassler, Cretaceous–Recent. Colony encrusting, zooecia large, separated by furrow. *A. contumax* Canu & Bassler (*10a, b*, a few zooecia showing slitlike distal openings called septulae, ×10 and ×20), Eocene (Wilcoxian), Vincentown, New Jersey.

Cardioecia Canu & Bassler, Lower Cretaceous. Bifoliate, irregularly placed round apertures, broad gently arched ovicells. *C. neocomiensis* (d'Orbigny) (*6a*, exterior showing two broad smooth ovicell areas; *6b–c*, transverse and longitudinal sections), Switzerland.

Diacanthopora Lang, Cretaceous–Eocene. Prominent ribs (costules) on frontal wall, triangular avicularia between zooecia. *D. convexa* Canu & Bassler (*7a, b*, several zooecia, ×10 and ×20), Eocene (Wilcoxian), Vincentown, New Jersey.

Diplotresis Canu & Bassler, Cretaceous–Eocene. Two prominent pores in frontal wall, ovicells numerous, long pointed avicularia. *D. sparsiporosa* Ulrich & Bassler (*11a, b*, showing bulbous ovicells on distal side of apertures and long pointed avicularia, ×20 and ×10), Eocene (Wilcoxian), Vincentown, New Jersey.

Entalophora Lamouroux, Jurassic–Recent. Relatively large simple tubes internally; widely spaced, commonly projecting apertures. *E. icaunensis* d'Orbigny (9, external view), Lower Cretaceous, Switzerland.

Laterocavea d'Orbigny, Lower Cretaceous. Subcylindrical branches having an axial tube and characterized by abundant mesopores in peripheral region. *L. dutempleana* d'Orbigny

(*1a–c*, transverse and longitudinal sections, and exterior), Aptian, England.

Leiosoecia Canu & Bassler, Cretaceous. Cylindrical branches, autopore apertures polygonal, mesopores few, large. *L. grandipora* Canu & Bassler (*3a, b*, longitudinal section and exterior), Lower Cretaceous, Switzerland.

Meliceritites Roemer, Cretaceous. Cylindrical branches, triangular apertures in transverse rows, very long slender tubes in axial region. *M. transversa* Canu & Bassler (*4a–c*, longitudinal and transverse sections, and exterior), Lower Cretaceous, England.

Multicrescis d'Orbigny, Lower Cretaceous–Miocene. Colonies subcylindrical, hollow, or encrusting; mesopores numerous. *M. pulchella* Canu & Bassler (*2a, b*, longitudinal section and exterior), Lower Cretaceous, Switzerland.

Nematifera Canu & Bassler, Cretaceous. Compressed bifoliate branches. *N. reticulata* (d'Orbigny) (*5a–c*, transverse and longitudinal sections, and exterior), Lower Cretaceous, Switzerland.

Siphodictyum Lonsdale, Cretaceous. Slender cylindrical branches, very thick-walled peripherally. *S. gracile* Lonsdale (8, exterior, transverse, and longitudinal sections), Lower Cretaceous, England.

Stichocados Marsson, Cretaceous–Eocene. Unilamellar colony, large pores in frontal wall, avicularia on each side of apertures. *S. compositus* Lang (*12a, b*, several typical zooecia, ×10 and ×20), Eocene (Wilcoxian), Vincentown, New Jersey.

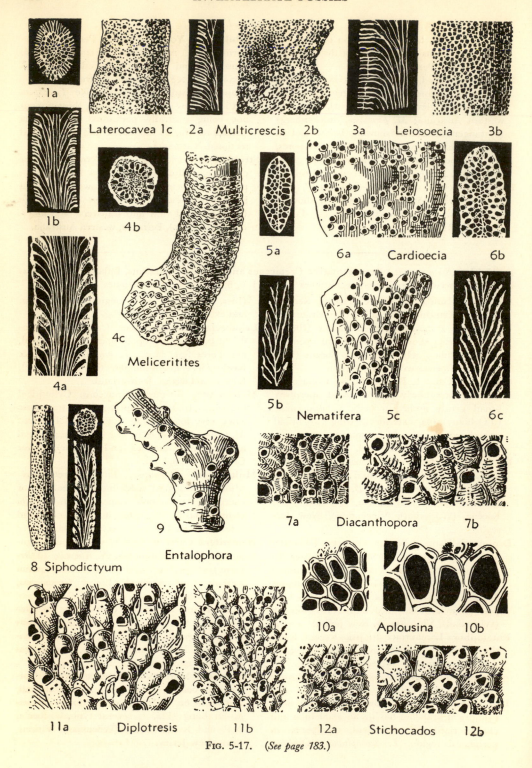

1a Laterocavea 1c 2a Multicrescis 2b 3a Leiosoecia 3b

1b 4b

4c

5a 6a Cardioecia 6b

Meliceritites

5b Nematifera 5c 6c

4a

7a Diacanthopora 7b

9

Entalophora

8 Siphodictyum

10a Aplousina 10b

11a Diplotresis 11b 12a Stichocados 12b

FIG. 5-17. (See page 183.)

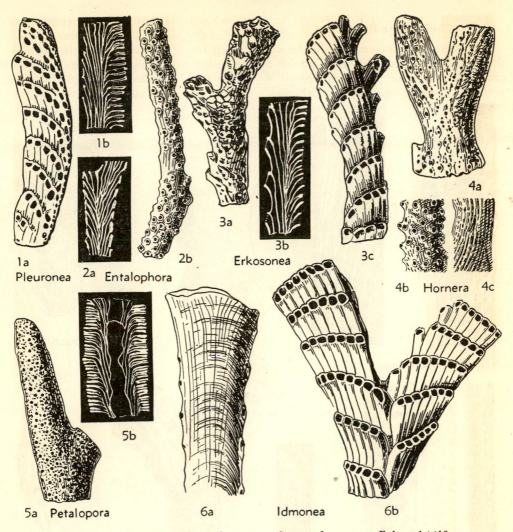

1a Pleuronea   2a Entalophora

1b   2b   3a   3b Erkosonea   3c   4a   4b Hornera 4c

5a Petalopora   5b   6a   Idmonea   6b

Fig. 5-18. **Representative Paleogene cyclostome bryozoans.** Enlarged ×10.

Entalophora Lamouroux, Jurassic–Recent. *E. cylindrica* (Canu & Bassler) (2a, b, longitudinal section and external view), Eocene (Jacksonian), North Carolina.

Erkosonea Canu & Bassler, Eocene. Contains closed pores (dactylethrae) on reverse side of colony. *E. semota* Canu & Bassler (3a–c, reverse side, longitudinal section, and obverse side), Jacksonian, Mississippi.

Hornera Lamouroux, Eocene–Recent. Apertures on obverse side only, slitlike pores (vacuoles) on both obverse and reverse sides. *H. jacksonica* Canu & Bassler (4a–c, two frontal views and reverse side), Eocene (Jacksonian), North Carolina.

Idmonea Lamouroux, Jurassic–Recent. Apertures on obverse side only, reverse plain. *I. magna* Canu & Bassler (6a, b, reverse and obverse sides), Eocene (Jacksonian), Georgia.

Petalopora Lonsdale, Cretaceous–Paleocene. Hollow cylindrical branches, numerous mesopores between autopore apertures. *P. consimilis* (Ulrich) (5a, b, external view and longitudinal section), Paleocene (Midwayan), Arkansas.

Pleuronea Canu & Bassler, Paleocene–Pliocene. Grouped curving lines of apertures on one side of colony, and specialized cells (tergopores) on opposite side. *P. subpertusa* Canu & Bassler (1a, b, exterior and longitudinal section), Eocene (Jacksonian), Mississippi.

Cheilostome Bryozoans

The cheilostome bryozoans are characterized by closure of the aperture by a chitinous operculum when the animal is withdrawn into its membranous, chitinous, or calcareous chamber. Representatives of this group are the most numerous and varied of all bryozoans, and they exhibit an amazing diversity of structural features. Some have considerable complexity. The colonies are delicate, and many are ex-

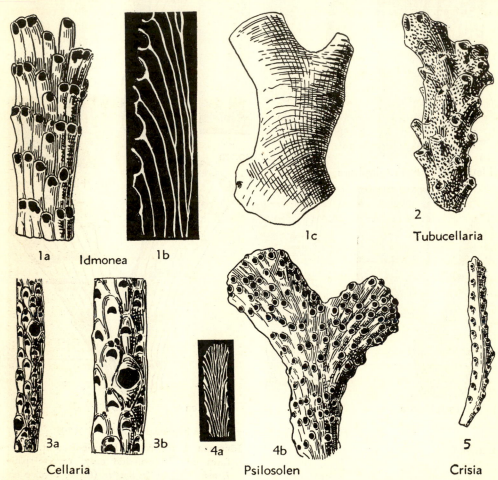

1a Idmonea 1b 1c 2 Tubucellaria

3a Cellaria 3b ·4a 4b Psilosolen 5 Crisia

FIG. 5-19. **Representative late Neogene bryozoans.** Enlarged ×10 unless indicated otherwise. Cyclostomes include 1, 4, 5; cheilostomes, 2, 3. All are from Pleistocene deposits of California.

Cellaria Ellis & Solander, Eocene–Recent. Articulated cylindrical branches, ovicell apertures subcircular, on distal side of crescentic zooecial apertures. *C. mandibulata* Hincks (3a, b, part of a branch, ×10 and ×20; the very large round apertures are specialized avicularia).

Crisia Lamouroux, Eocene–Recent. Slender branches of biserially arranged zooecia. *C. serrata* Gabb & Horn (5, obverse side).

Idmonea Lamouroux, Jurassic–Recent. *I. cali-fornica* d'Orbigny (1a–c, obverse side, longitudinal section, and reverse side), a characteristic and very common species.

Psilosolen Canu & Bassler, Pleistocene–Recent. Very simple branching cylindrical cyclostomes composed of autopore tubes only. *P. capitiferax* Canu & Bassler (4a, longitudinal section, ×5; 4b, side view).

Tubucellaria d'Orbigny, Eocene–Recent. *T. punctulata* Gabb & Horn (2, side view).

ceptional in the beauty of their markings and regularity of their exquisite design. Undoubtedly this group represents the highest type of development among the bryozoans. The cheilostomes, which probably are derived from cryptostomes, are known from Jurassic to the present time. They are exclusively marine and predominate strongly over other types of living bryozoans.

Structure of a Typical Living Cheilostome, Bugula. We may advantageously begin a survey of the cheilostomes by studying one of the most common living representatives of the group, known as *Bugula* (Fig. 5-1). This genus lacks a calcareous skeleton, but the chitinous hard parts may be found along the coasts of almost any part of the world. The colonies grow in bushy tufts several inches high and during life are attached to rocks or other objects on the sea bottom. When examined under a lens, *Bugula* is seen to be made up of narrow branching stems, along which the zooecial chambers are arranged in four rows. Near the distal end of each chamber, opposite the point of its attachment, is a broad crescentic aperture with a blunt spine on each side (Fig. 5-20).

Near the aperture on the outer side of many zooecia is a short projection which resembles the head of a bird, a resemblance which is increased by the manner in which the beak is opened and closed. This appendage, known as the **avicularium** (Fig. 5-20), is a characteristic structure of the cheilostomes. Although it may vary considerably from the bird-head form of *Bugula*, it is found on or between the zooecial chambers of most genera. It is not calcified and therefore cannot be preserved in the fossil state, but its place of attachment is readily determined. Like mesopores and acanthopores of Paleozoic genera, which are judged to have been formed by specialized zooids, the avicularia are structures built by modified individuals of the cheilostome colonies.

An individual zooid of the *Bugula* colony consists essentially of a U-shaped digestive tract, surrounded by fluid of the body cavity (Fig. 5-20*C*, *D*). Fourteen long ciliated tentacles are attached to a circular ring (lophophore) around the mouth. The tentacles are probably sensory in function and may serve also for respiration. The mouth and tentacles may be protruded (Fig. 5-20*B*, *D*) or withdrawn into the body chamber (Fig. 5-20*A*, *C*) by the action of muscles attached to the flexible walls of the chamber.

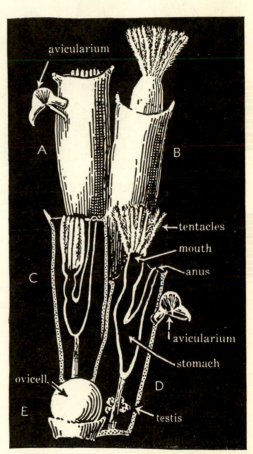

FIG. 5-20. **Structural features of Bugula.** (*A*) External view of a zooecium (autopore) and attached "bird's-head" avicularium, with animal in retracted position; (*B*) adjoining zooecium with bryozoan extended; (*C*) sectional view of zooecium and retracted animal; (*D*) sectional view of zooecium and extended animal; (*E*) ovicell belonging to another zooecium. Much enlarged.

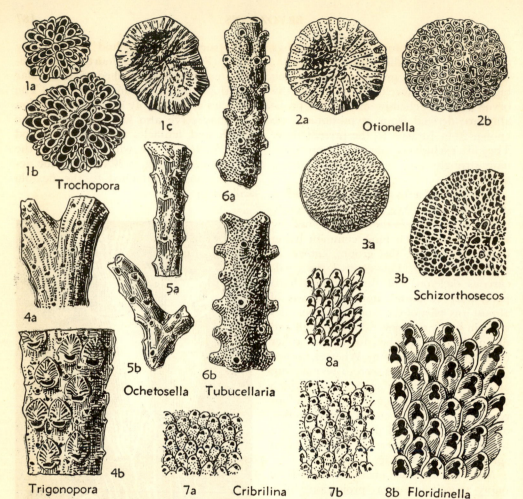

1a, 1b, 1c Trochopora
2a, 2b Otionella
3a, 3b Schizorthosecos
4a, 4b Trigonopora
5a, 5b Ochetosella
6a, 6b Tubucellaria
7a, 7b Cribrilina
8a, 8b Floridinella

FIG. 5-21. **Representative Paleogene cheilostome bryozoans.** Enlarged ×10 unless indicated otherwise.

Cribrilina Gray, Eocene–Recent. *C. verrucosa* Canu & Bassler (7a, b, typical specimens showing avicularia adjoining apertures and pores in frontal walls of zooecia), Paleocene (Midwayan), Arkansas.

Floridinella Canu & Bassler, Cretaceous–Recent. Distinguished by shape of apertures. *F. vicksburgica* Canu & Bassler (8a, b, part of colony, ×10 and ×20), Oligocene, Alabama.

Ochetosella Canu & Bassler, Eocene. Resembles *Trigonopora*. *O. jacksonica* Canu & Bassler (5a, b, two colonies), Jacksonian, North Carolina.

Otionella Canu & Bassler, Cretaceous–Eocene. Resembles *Trochopora*. *O. perforata* Canu & Bassler (2a, b, reverse and obverse sides), Eocene (Jacksonian), Mississippi.

Schizorthosecos Canu & Bassler, Eocene. Discoid colonies, zooecial apertures separated by small cells (mostly avicularia). *S. interstitea* (Lea)

(3a, b, obverse view, ×5 and ×10), Claibornian, Alabama, a guide form.

Trigonopora Maplestone, Eocene–Recent. Small arcuate apertures, frontal wall of zooecia pierced by numerous pores. *T. monilifera* (Milne-Edwards) (4a, colony showing normal zooecia; 4b, part of colony showing nine ovicell-bearing zooecia), Oligocene, Alabama.

Trochopora d'Orbigny, Eocene–Miocene. Unilamellar disks, zooecial apertures in radial rows separated by bases of vibracula. *T. bouei* (Lea) (1a–c, two colonies from obverse side and one from reverse side), Eocene (Claibornian), Alabama, an abundant guide fossil.

Tubucellaria d'Orbigny, Eocene–Recent. Colony consists of articulated branches, frontal walls of zooecia bear many pores. *T. vicksburgica* Canu & Bassler (6a, b, two representative specimens), Oligocene, Alabama.

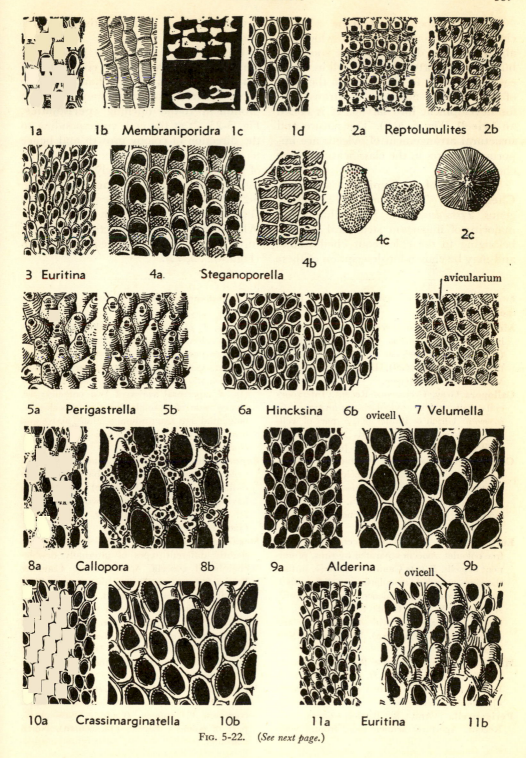

1a   1b   Membraniporidra   1c   1d   2a   Reptolunulites   2b

3 Euritina   4a.   Steganoporella   4b   4c   2c

5a   Perigastrella   5b   6a   Hincksina   6b   ovicell   7 Velumella   avicularium

8a   Callopora   8b   9a   Alderina   9b   ovicell

10a   Crassimarginatella   10b   11a   Euritina   11b

FIG. 5-22.   (*See next page.*)

Near the aperture of some zooecial chambers is a round structure, which has about the same diameter as the zooecium but is much shorter. This is a cell in which eggs, after being fertilized within the body of the bryozoan adjacent, are transferred for development into larvae. It is called an **ovicell** (Fig. 5-20E). The nature and structural arrangement of ovicells are important features in the classification of the cheilostomes.

Structure of a Typical Calcareous Cheilostome, Tessardoma. The living genus, *Tessardoma* (Fig. 5-23), illustrates a majority of important structural features belonging to the calcareous cheilostomes and may be chosen for description of these structures. Fossil representatives are not known.

The colony of *Tessardoma* grows in a slender branch which is composed of parallel rows of zooecial chambers. The exposed surface of each chamber has a prominent round opening (**peristomice,** Fig. 5-23, *2*) at the end, corresponding to the direction of colony growth and hence termed distal. This is not the true aperture of the chamber but merely an opening into a tubular vestibule (**peristome,** Fig. 5-23, *10*) which leads to the true **aperture** (Fig. 5-23, *1*) located within the chamber. When the animal is retracted, this aperture is closed by a hinged **operculum** (Fig. 5-23, *5, 12*).

FIG. 5-22. (*See page 189.*) **Representative Paleogene cheilostome bryozoans.** Enlarged ×10 unless indicated otherwise.

Alderina Norman, Cretaceous–Recent. Encrusting, similar to *Euritina*. *A. rustica* d'Orbigny (9a, b, showing ovicells and zooecia lacking them, ×10 and ×20), Eocene (Wilcoxian), Vincentown, New Jersey.

Callopora Gray, Cretaceous–Recent. Interzooecial walls thick, avicularia numerous. *C. jerseyensis* Ulrich & Bassler (8a, b, part of well-preserved colony, ×10 and ×20), Eocene (Wilcoxian), Vincentown, New Jersey.

Crassimarginatella Canu, Cretaceous–Recent. Encrusting, distinguished by keeled ovicells and scattered avicularia. *C. intermedia* Canu & Bassler (10a, b, regular zooecia, a few having ovicells, ×10 and ×20), Eocene (Wilcoxian), Vincentown, New Jersey.

Euritina Canu, Cretaceous–Eocene. Bilamellar free colonies, zooecia separated by rims, prominent ovicells. *E. tecta* Canu & Bassler (3, surface showing rounded avicularia—smaller cells—between zooecia), Paleocene (Midwayan), Arkansas; *E. torta* Gabb & Horn (11a, b, typical zooecia, many with ovicells, ×10 and ×20), Eocene (Wilcoxian), Vincentown, New Jersey.

Hincksina Norman, Eocene–Recent. Bilamellar colonies, large apertures. *H. jacksonica* Canu & Bassler (6a, b, parts of two colonies), Eocene (Jacksonian), Georgia.

Perigastrella Canu & Bassler, Cretaceous–Recent. Apertures semicircular, frontal walls porous. *P. plana* Canu & Bassler (5a, b, parts of two colonies), Oligocene, Alabama.

Membraniporidra Canu & Bassler, Cretaceous–Recent. Colonies bilamellar, zooecia rimmed by thick walls. *M. spissimuralis* Canu & Bassler (1a, d, apertural views of two colonies; 1b, reverse side of a single layer of zooecia; 1c, tangential and longitudinal sections, above and below, respectively), Eocene (Jacksonian), North Carolina.

Reptolunulites d'Orbigny, Cretaceous–Recent. Colonies conical, concave at base, zooecia in radial rows. *R. distans* (Lonsdale) (2a, b, zooecia having narrow avicularia between apertures; 2c, basal membrane of colony, ×1), Eocene (Jacksonian), North Carolina.

Steganoporella Smitt, Eocene–Recent. Two slightly different types of internally double-chambered zooecia. *S. vicksburgica* Canu & Bassler (4a, surface showing two types of large ovicelled zooecia distributed among normal ones, characterized by smaller apertures, which lack ovicells; 4b, part of colony with roofs of zooecia removed so as to show interzooecial and intrazooecial walls, the latter recognized by the vertical tube within them; 4c, colonies ×1), Oligocene, Alabama.

Velumella Canu & Bassler, Cretaceous–Recent. Zooecia surrounded by rims, elongate avicularia common. *V. fusiformis* (Canu & Bassler) (7, part of colony), Eocene (Jacksonian), North Carolina.

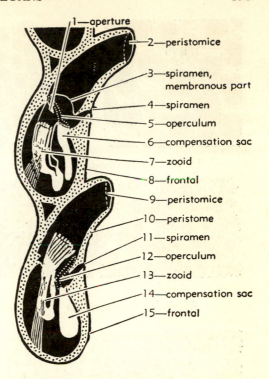

Fig. 5-23. **Diagram showing structural features of a modern calcareous cheilostome.** The drawing shows zooecial chambers belonging to *Tessardoma gracile* Sars. (*A*) Sectional view of zooecium showing retracted animal and operculum closed; (*B*) similar view of an adjoining zooecium showing animal partly extended and aperture open; much enlarged.

1—aperture
2—peristomice
3—spiramen, membranous part
4—spiramen
5—operculum
6—compensation sac
7—zooid
8—frontal
9—peristomice
10—peristome
11—spiramen
12—operculum
13—zooid
14—compensation sac
15—frontal

aperture (1). Opening which is closed by operculum.

compensation sac (6, 14). Flexible hydrostatic organ which by taking water from outside of zooecial chamber causes displacement of zooid through aperture.

frontal (8, 15). Portion of zooecial chamber which contains openings leading to its interior.

operculum (5, 12). Membranous covering of aperture and of opening to compensation sac.

peristome (10). Calcareous rim or tube surrounding aperture.

peristomice (2, 9). Orifice of an elongate tube-like peristome.

spiramen (3, 4, 11). Opening through frontal wall for passage of water to compensation sac.

zooid (7, 13). Bryozoan animal.

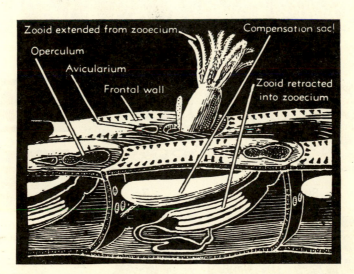

Zooid extended from zooecium
Compensation sac
Operculum
Avicularium
Frontal wall
Zooid retracted into zooecium

Fig. 5-24. **Diagram of cheilostome bryozoans belonging to the suborder Ascophorina.** These are provided with a hydrostatic organ (*compensation sac*) as a means of pushing the zooid outward through the aperture of the zooecial chamber. When the animal is retracted, the aperture and opening to the compensation sac are closed by a membranous lid (*operculum*), which in this bryozoan is located in the roof portion (*frontal wall*) of the zooecium. The operculum is supported on pivots near its middle. Attachment places for avicularia are shown. Much enlarged.

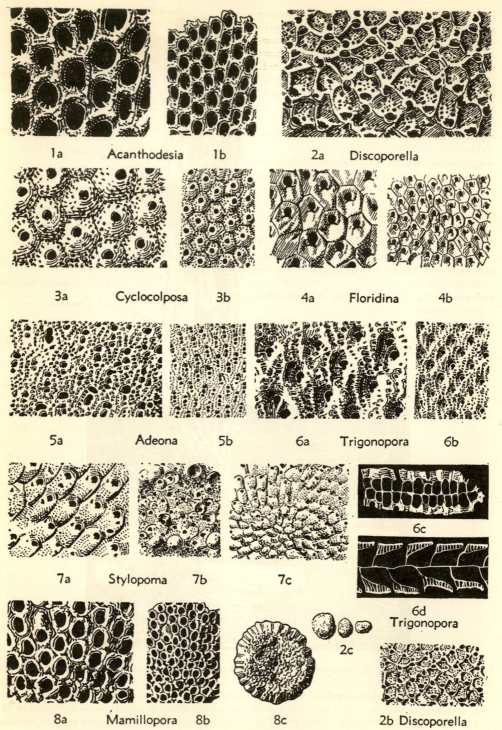

1a   Acanthodesia   1b   2a   Discoporella

3a   Cyclocolposa   3b   4a   Floridina   4b

5a   Adeona   5b   6a   Trigonopora   6b

6c

6d
Trigonopora

7a   Stylopoma   7b   7c

2c

8a   Mamillopora   8b   8c   2b Discoporella

FIG. 5-25. **Representative Neogene cheilostome bryozoans.** Enlarged ×10 unless indicated otherwise. (*Continued on next page.*)

Protrusion of a *Tessardoma* zooid from its rigid calcareous chamber is accomplished by a very different mechanism from that of *Bugula*. Within the zooecium is a flexible hydrostatic organ (**compensation sac,** Figs. 5-23, *6, 14;* 5-24) which by receiving water from the outside and discharging it can compensate the volume of the zooid body in the extended or retracted position. When the hydrostatic organ is filled with water, the zooid is displaced upward through the aperture. When the animal by its muscles pulls back into the chamber, water is expelled from the hydrostatic organ. Communication of this organ with the outside is through an aperture (**spiramen,** Fig. 5-23, *4, 11*) or a special pore in the upper (frontal) wall. Bryozoans which possess this somewhat complex mechanism are placed in the suborder Ascophorina (*ascon,* sac; *phora,* bearing). *Bugula* (Fig. 5-1), which does not require this device, belongs in the suborder Anascina (meaning no sac). Some calcareous cheilostomes belonging to the latter group, however, have an external hydrostatic system in which a cavity (**hypostege**) in the outer wall operates in a manner similar to the internal sac of the Ascophorina.

Around the margins of the zooecial chamber roof in *Tessardoma* and other genera of the Ascophorina are radially disposed low ridges (**costules**), and between these are pores (**areolae**) which penetrate the superficial layer of calcareous material of the cell wall (Fig. 5-24). Finally, there are small places of attachment, each marked by an articulating crossbar, to which avicularia (similar to the "bird's-head" organ of *Bugula*) are attached. Another appendage (**vibraculum**) on the outer side of the zooecium has the form of a long whip lash. The exact functions of these organs, which are common in the cheilostomes, is not known.

Fossil Cheilostomes.   The abundance, variety, and excellent preservation of cheilostome bryozoans in many marine deposits of Jurassic and later age make these fossils exceptionally important. They are useful in paleontological zonation and correlation (Figs. 5-17, 5-21, 5-22, 5-25).

GEOLOGICAL DISTRIBUTION AND IMPORTANCE

Bryozoans are virtually unknown from rocks older than Middle Ordovician, but the abundance and variety of Ordovician forms indicate that they must have had a considerable antecedent existence during which they became so differentiated (Fig.

(Fig. 5-25 continued.)

Acanthodesia Canu & Bassler, Eocene–Recent. Bilamellar colonies, zooecial rim denticulate. *A. savarti bifoliata* Ulrich & Bassler (1*a, b,* surface ×20 and ×10), Miocene, Maryland.

Adeona Lamouroux, Oligocene–Recent. Frontal walls bear numerous small pores and two larger openings, zooecial aperture and orifice (ascopore) of compensation sac. *A. heckeli* (Reuss) (5*a, b,* surface ×20 and ×10), Miocene, North Carolina.

Cyclocolposa Canu & Bassler, Miocene–Pliocene. Apertures rounded, large pores in frontal walls. *C. perforata* Canu & Bassler (3*a, b,* surface ×20 and ×10), Pliocene, South Carolina.

Discoporella d'Orbigny, Miocene–Recent. Ridges between zooecia, frontal walls porous. *D. umbellata* (Defrance) (2*a, b,* surface ×20 and ×10; 2*c,* three colonies, ×1), Miocene, Santo Domingo.

Floridina Jullien, Cretaceous–Recent. Distinguished by apertures and polygonal zooecia. *F. regularis* Canu & Bassler (4*a, b,* surface ×20 and ×10), Miocene, North Carolina.

Mamillopora Smitt, Eocene–Recent. Discoid colonies, some avicularia. *M. tuberosa* Canu & Bassler (8*a, b,* surface showing one ovicelled zooecium ×20 and ×10; 8*c,* a colony, ×1), Miocene, Santo Domingo.

Stylopoma Levinsen, Miocene–Recent. Frontal wall bears fine pores, small avicularia near zooecial apertures, spherical ovicells. *S. spongites* (Pallas) (7*a, c,* surface of typical colony, ×20 and ×10; 7*b,* specimen bearing several ovicells), Miocene, Virginia.

Trigonopora Maplestone, Eocene–Recent. *T. auriculata* (Canu & Bassler) (6*a, b,* surface ×20 and ×10; 6*c, d,* transverse and longitudinal sections), Pliocene, Florida.

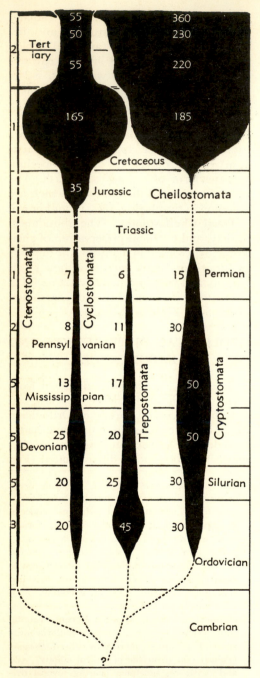

55
50
55
165

360
230
220
185

Cretaceous

35 Jurassic   Cheilostomata

Triassic

Ctenostomata	Cyclostomata			Permian
7	6		15	Permian
8	11		30	Pennsylvanian
13	17	Trepostomata	50	Cryptostomata — Mississippian
25	20		50	Devonian
20	25		30	Silurian
20	45		30	Ordovician

Cambrian

?

FIG. 5-26. **Geologic distribution of bryozoans.** The figures indicate the approximate number of described genera belonging to the respective orders in each geological division.

5-26). A supposed trepostome bryozoan (*Archaeotrypa*), from Upper Cambrian rocks of western Canada, may prove to be a coral. Only one or two Lower Ordovician (Canadian) bryozoans are recorded, but in many Middle and Upper Ordovician strata the remains of bryozoans are so abundant as to form a large part of the rock. All orders are represented except the cheilostomes. The greatly predominant group is the Trepostomata—the "stony bryozoans," which are so called because the colonies are compact stony structures. Many have branching form, some are hemispherical or globular, and some grow in extended sheets. Several species build large irregularly shaped massive colonies.

Silurian and Devonian bryozoans are distinguished mainly by an abundance of slender branching colonies and delicate lacy types. The massive trepostomes declined in importance. These trends are more pronounced in Mississippian, Pennsylvanian, and Permian strata, in which the most common bryozoan colonies are lacelike networks and slender branches. A majority of these latter forms belong to the order Cryptostomata.

The cheilostomes first appear in Jurassic rocks and are very abundant in Jurassic, Cretaceous, and Cenozoic formations of many areas. The variety of colonial growths and especially the range of structures shown by the individual bryozoan skeletons is almost endless.

In nearly all rocks which contain fossil bryozoans, recognized genera and species are found to have short stratigraphic range, and many of them have wide geographic distribution. Hence, they are useful guide fossils for the correlation of the formations which contain them. The fact that special techniques are required in order to determine their structural features adequately and the use in paleontological literature of a somewhat complicated terminology are factors which have deterred study of the group. These barriers, however, are by no means sufficient to warrant neglect of the bryozoans.

REFERENCES

BASSLER, R. S. (1906) *The bryozoan fauna of the Rochester shale:* U.S. Geol. Survey Bull. 292, pp. 1–136, pls. 1–31. Almost the only available publication on mid-Silurian bryozoans in North America; well-preserved fossils which can be collected at many places in these rocks need research.

—— (1911) *Bryozoa of the Middle Devonic of Wisconsin:* Wisconsin Geol. Survey Bull. 21, pp. 49–67, pls. 5–11.

—— (1911a) *The early Paleozoic Bryozoa of the Baltic provinces:* U.S. Nat. Mus. Bull. vol. 77, pp. 1–382, pls. 1–13, figs. 1–226. A well-illustrated, very useful report on European bryozoans, mostly resembling American forms of corresponding age.

—— (1922) *The Bryozoa or moss animals:* Smithsonian Inst. Ann. Rept. 1920, pp. 339–380, pls. 1–4, figs. 1–13. A useful description of morphology.

—— (1927) *Bryozoa:* in Twenhofel, W. H., Geology of Anticosti Island, Canada Geol. Survey Mem. 154, pp. 143–168, pls. 5–14. Describes Upper Ordovician and Lower Silurian forms.

—— (1929) *The Permian Bryozoa of Timor:* Palæontologie von Timor, Lief. 16, Abh. 28, pp. 36–90, pls. 225–247. The fauna consists chiefly of cryptostomes.

—— (1935) *Bryozoa:* Fossilium Catalogus, pt. 67, pp. 1–229. Gives outline of classification of living and fossil genera, with type species of each genus and references to literature.

—— (1939) *The Hederelloidea, a suborder of Paleozoic cyclostomatous Bryozoa:* U.S. Nat. Mus. Proc., vol. 87, pp. 25–91, pls. 1–16, figs. 1–7.

CANU, F., & BASSLER, R. S. (1920) *North American Early Tertiary Bryozoa:* U.S. Nat. Mus. Bull. 106, pp. 1–878, pls. 1–162, figs. 1–279. Although burdened by complex terminology which is inadequately explained, this is an indispensible source of information on American cheilostomes (99 genera, 482 species) and cyclostomes (61 genera, 200 species) from early Cenozoic rocks, chiefly in Atlantic and Gulf regions. Stratigraphic distribution is summarized in faunal tables.

—— & —— (1923) *North American Later Tertiary and Quaternary Bryozoa:* Same, Bull. 125, pp. 1–302, pls. 1–47, figs. 1–38. A continuation of studies published in Bull. 106. A valuable bibliography of world literature on Bryozoa

from 1898 to 1923 is given (pp. 209–243) with descriptions of cheilostomes (109 genera, 234 species), cyclostomes (31 genera, 26 species), and ctenostomes (2 genera, 7 species).

—— & —— (1926) *Studies on the cyclostomatous Bryoza:* U.S. Nat. Mus. Proc., vol. 67, art. 21, pp. 1–124, pls. 1–31, figs. 1–46. Describes many Lower Cretaceous fossils from England and Switzerland, showing nature of Mesozoic cyclostomes.

—— & —— (1933) *The bryozoan fauna of the Vincentown limesand:* U.S. Nat. Mus. Bull. 165, pp. 1–108, pls. 1–21, fig. 1. Judged by the authors to be the only known rich late Cretaceous bryozoan assemblage, it is now classed as Eocene. It occurs in New Jersey.

CONDRA, G. E. (1903) *The Coal Measure Bryozoa of Nebraska:* Nebraska Geol. Survey Bull. 2, pp. 1–162, pls. 1–21. Although inadequate, this is the only comprehensive paper on Pennsylvanian bryozoans.

—— & ELIAS, M. K. (1944) *Study and revision of Archimedes (Hall):* Geol. Soc. America Spec. Paper 53, pp. 1–243, pls. 1–41, figs. 1–6. A detailed study of late Paleozoic fenestrate genera, presenting the thesis that *Archimedes* and some other forms are composite fossils produced by intergrowth of bryozoans and algae.

CROCKFORD, J. (1943) *Permian Bryozoa of eastern Australia:* Royal Soc. New South Wales Jour. and Proc., vol. 76, pp. 258–267, pl. 15. This and other papers by Crockford are chief sources of information on late Paleozoic bryozoans of the southwestern Pacific region.

—— (1944) *Bryozoa from the Permian of western Australia:* Linnean Soc. New South Wales Proc., vol. 69, pp. 139–175, pls. 4–5, figs. A–C, 1–49.

—— (1945) *Stenoporoids from the Permian of New South Wales and Tasmania:* Same, vol. 70, pp. 9–24, pls. 1–3, figs. 1–25.

—— (1947) *Bryozoa from the Lower Carboniferous of New South Wales and Queensland:* Same, vol. 72, pp. 1–48, pls. 1–6.

CUMINGS, E. R. (1904) *Development of some Paleozoic Bryozoa:* Am. Jour. Sci., 4th ser. vol. 17, pp. 49–79, figs. 1–83. Describes stages in colonial development of *Fenestrellina* and other genera.

—— & GALLOWAY, J. J. (1913) *Stratigraphy and paleontology of the Tanners Creek section of the Cincinnatian series of Indiana:* Indiana Geol. Survey Ann. Rept. 37, pp. 353–478, pls. 1–20.

A useful paper which brings together information on the prolific Upper Ordovician bryozoans of the Ohio Valley region.

——— & ——— (1915) *Studies of the morphology and histology of the Trepostomata or monticuliporoids:* Geol. Soc. America Bull., vol. 26, pp. 349–374, pls. 10–15.

DEISS, C. F. (1932) *Description and stratigraphic correlation of the Fenestellidae from the Devonian of Michigan:* Michigan Univ., Mus. Paleontology Contr., vol. 3, pp. 233–275, pls. 1–14.

DUNCAN, H. (1939) *Trepostomatous Bryozoa from the Traverse group of Michigan:* Michigan Univ., Mus. Paleontology Contr., vol. 5, pp. 171–270, pls. 1–16, fig. 1. Describes well-preserved Middle Devonian bryozoans (19 genera, 74 species).

ELIAS, M. K. (1937) *Stratigraphic significance of some late Paleozoic fenestrate bryozoans:* Jour. Paleontology, vol. 11, pp. 306–334, figs. 1–3. Gives conclusions on evolutionary trends.

LOEBLICH, A. R., JR. (1942) *Bryozoa from the Ordovician Bromide formation, Oklahoma:* Jour. Paleontology, vol. 16, pp. 413–436, pls. 61–64.

MACFARLAN, A. C. (1942) *Chester Bryozoa of Illinois and western Kentucky:* Jour. Paleontology, vol. 16, pp. 437–458, pls. 65–68.

McGUIRT, J. H. (1941) *Louisiana Tertiary Bryozoa:* Louisiana Geol. Survey Bull. 21, pp. 1–177, pls. 1–31.

McNAIR, A. H. (1937) *Cryptostomatous Bryozoa from the Middle Devonian Traverse Group of Michigan:* Michigan Univ., Mus. Paleontology Contr., vol. 5, pp. 103–170, pls. 1–14, fig. 1.

MOORE, R. C. (1929) *A bryozoan faunule from the upper Graham formation, Pennsylvanian, of north-central Texas:* Jour. Paleontology, vol. 3, pp. 1–27, 121–156, pls. 1–3, 15–18, figs. 1–5. A representative assemblage of Upper Pennsylvanian bryozoans.

SIMPSON, G. B. (1895) *Handbook of the genera of the North American Palaeozoic Bryozoa:* New York State Geol. Ann. Rept. 14, pp. 407–669, pls. A–E, 1–25, figs. 1–222. A well-illustrated but not entirely reliable summary of characters belonging to many genera.

ULRICH, E. O. (1890) *Palaeozoic Bryozoa:* Illinois Geol. Survey, vol. 8, pp. 285–688, pls. 29–78, figs. 1–15. A classic work which is the best of its period and still an invaluable source of accurately reported morphology of bryozoans.

——— (1895) *Lower Silurian [Ordovician] Bryozoa of Minnesota:* Minnesota Geol. Survey, vol. 3, pt. 1, pp. 96–332, pls. 1–28, figs. 1–20. An important report, especially on trepostomes. (Author's separate, 1893.)

——— & BASSLER, R. S. (1904) *A revision of the Palaeozoic Bryozoa,* pt. 1, *On genera and species of Ctenostomata:* Smithsonian Misc. Coll., vol. 45, pp. 256–294, pls. 65–68, figs. 32–33; pt. 2, *On genera and species of Trepostomata:* Same, vol. 47, pp. 15–55, pls. 6–14.

CHAPTER 6

Brachiopods

One of the chief divisions of invertebrate fossils consists of the Brachiopoda. They rank high in terms of usefulness for stratigraphic correlation, value for age determination, diversity of morphologic features, and richness of materials offered for study of evolution. Members of the group are distributed throughout the entire geologic column, exclusive of Pre-Cambrian rocks, and especially in some Paleozoic and Mesozoic deposits their well-preserved shells are extraordinarily abundant.

Brachiopods are marine invertebrates in which the soft parts are enclosed by two deposits of shell, termed valves. Thus, superficially, they resemble the bivalve mollusks called pelecypods or clams. Brachiopod shells are readily distinguished from clams, however, by inequality of the two valves in size and shape, and by equilateral symmetry of each valve. A longitudinal median plane through either valve divides it into two halves, in which one is the mirror image of the other. The name brachiopod (*brachio*, arm; *pod*, foot) refers to internal paired appendages (brachia or lophophore), which at first were assumed to function for locomotion, like the foot of mollusks.

The brachiopods are chiefly inhabitants of the shallow sea bottom. Except during larval existence, none can swim about, nor in adult life can they move from place to place on the sea bottom like many clams. They live in fixed position and thus are classed as sessile benthos. They may be anchored by a stemlike attachment, termed pedicle, which is pushed downward into sediment, as in *Lingula* (Fig. 6-1), or fastened to some foreign object, such as another shell. Some brachiopods are fixed by cementation of the shell or by projecting spines, and others live without anchorage. A large majority of living brachiopods are found between the strand line and a depth of 100 fathoms. A few species live in deep waters along the continental slopes. Also, a few are able to live in the brackish waters of estuaries, and they may be found between the zone of high and low tides, especially in tropical

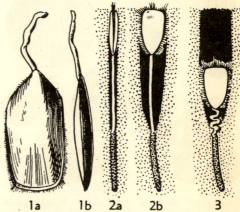

FIG. 6-1. Lingula, a modern inarticulate brachiopod. The thin, partly translucent shell has a brownish to greenish color and horny appearance. It consists mainly of calcium phosphate but contains some chitin and calcium sulfate. The two similarly shaped and nearly equal-sized valves are held together only by muscles. Hairlike bristles (setae) project outward from the margins, and a muscular stalk (pedicle) extends through a gap beneath the beaks.

1a, b, *Lingula hawaiiensis*, view normal to one of the valves and an edge view, ×1.

2a, b, Edge and normal views of *Lingula* in feeding position in its burrow, the pedicle extended (reduced scale). The pedicle is anchored by its terminal part, encased in mucus-cemented sand grains.

3, Normal view of *Lingula* in retracted position, with the shell drawn to the bottom of its burrow.

waters. No brachiopods are known to have lived in fresh water. *Lingula* commonly makes a vertical burrow as much as 12 in. in depth, in which it lives attached by its pedicle to mucus-cemented sand on the

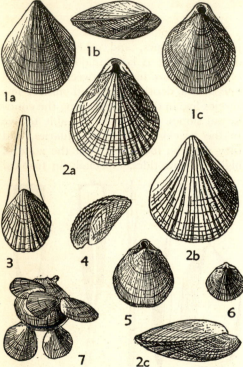

1b

1a

1c

2a

3   4   2b

5   6

7   2c

FIG. 6-2. **Modern articulate brachiopods.** The shells in this group are composed of calcium carbonate. Like most brachiopods, each valve is divisible along its longitudinal mid-line into symmetrical left and right halves, but one valve is distinctly larger than the other. The hinge line is transverse to the shell axis and located just below the beaks. The aperture for protrusion of the pedicle occupies the beak region of the larger valve. All the shells, except 3, belong to the genus *Terebratulina*, mostly natural size. (1a–c) *T. hawaiiensis*, normal views from opposite sides and edge view; (2a–c) *T. cailletti*, a somewhat coarsely ribbed species (×2); (4) *T. reevei*, side view of shell in normal position of growth; (5) *T. floridana*, normal view from side of the smaller valve; (6) *T. filosa*, a small shell; (7) *T.* sp., five small shells in position of growth, attached by means of their short pedicles to a larger brachiopod shell; (3) *Terebrirostra* sp., distinguished by great elongation of the beak region of the larger valve.

bottom and sides of the burrow (Fig. 6-1, 2a, b, 3). By extension of the pedicle, two thirds of the shell may extend above the surface; but it disappears instantly when the animal is disturbed. The burrows occur on mud flats exposed at low water.

The shells of modern brachipods range in greatest length or width from less than $\frac{1}{4}$ in. (5 mm.) to slightly more than 3 in. (80 mm.). Most fossil forms fall within this range, but some adults have a shell less than 1 mm. in diameter, and a few attain a width of approximately 375 mm. (15 in.). Such gigantic forms, however, are decidedly exceptional.

Two main types are recognized: mostly diminutive forms having valves held together without hingement (**inarticulates,** Fig. 6-1), and small to large forms having valves which bear teeth and sockets for articulation along edges hinged together (**articulates,** Fig. 6-2).

The food of brachiopods consists of microscopic material which is drawn to the interior of the slightly open valves by water currents induced by movements of cilia on the arms or brachia. The margins of *Lingula* are fringed by bristles (setae) which are brought together so as to form three funnels for the passage of water, two at the side for incurrent movement and one in the middle for excurrent flow.

Oldest-known brachiopods appear near the base of Lower Cambrian rocks. Without doubt, these invertebrates originated in Pre-Cambrian time, for more than 500 species of Cambrian brachiopods have been described. Ordovician rocks contain a vastly larger and more varied assemblage of brachiopods, and throughout Paleozoic time, this group of invertebrates held a very important, if not always dominant, position in the faunas of the shallow seas. Many stocks died out before the beginning of the Mesozoic Era, although orders which persisted are abundantly represented in some post-Paleozoic deposits. There are over 200 species of living brachiopods.

ANATOMICAL FEATURES

Soft Parts

The brachiopod animal consists of soft parts, which carry on essential body functions, and hard parts, which are mainly external and serve for protection. Before studying the characters of the shells, it is appropriate to survey the soft parts, because these build the shell and determine its internal and external characters.

Soft parts of a brachiopod (Fig. 6-3) consist essentially of the body covering (mantle), a digestive tract, visceral organs, various muscles, and tentacle-bearing appendages, termed brachia. The mantle lines the inside of both valves and contains cells which deposit the mineral substance of the shell. Among many living brachiopods (and undoubtedly certain groups of fossils), innumerable fine rodlike extensions of the mantle project outward into the shell, which has a correspondingly perforated or punctate structure. The mantle of other modern brachiopods and many fossil forms lacks such extensions. These have an impunctate or falsely punctate (pseudopunctate) shell. On the inside, the mantle encloses an area occupied by visceral organs and muscles, and it also forms the lining of a space open to sea water, termed the mantle cavity. The visceral organs and muscles are located in the posterior part of the shell, farthest from the growing margins. This is near the apices or beaks of the valves, and among brachiopods in which the valves are joined by hingement, the viscera and muscles are adjacent to the hinge line. The mantle cavity occupies the middle and anterior part of the shell interior, farther from the beaks and hinge line.

Digestive Tract and Internal Organs. The mouth, located at the rear of the mantle cavity, is the entrance to an esophagus which leads to the stomach. Living inarticulate brachiopods have a long intestine beyond the stomach, leading to an anus, but all modern hinged calcareous forms have a short intestine which ends blindly. Waste products are excreted through the mouth. Adjoining the stomach is a large digestive gland or liver, which furnishes most of the digestive juices. A genital gland is located near the liver, and connecting tubes provide for the discharge of its products into the mantle cavity. There is no heart, but circulation of fluid which fills the body cavity is provided by a branching system of passageways leading into the mantle and extending outward to its margins. These avenues are termed pallial sinuses. Their position may be impressed on the interior of the shell as **pallial markings** (Fig. 6-12). A nerve surrounds the esophagus and sends branches to the mantle and various parts of the body. Sense organs are mostly lacking, but experiments indicate that some brachiopods are sensitive to light and to touch.

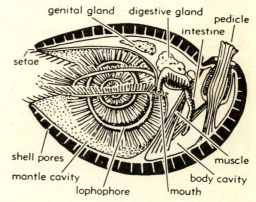

genital gland   digestive gland
pedicle
intestine
setae
shell pores
mantle cavity
lophophore
mouth
muscle
body cavity

FIG. 6-3. **Anatomical features of a modern articulate brachiopod.** The diagrammatic median longitudinal section represents one of the terebratulids, such as illustrated in Fig. 6-2. The shell (solid black) is perforated by small tubes or pores (puncta), occupied by extensions of the body wall (mantle), which reach almost to the outer surface. The interior of the shell is divided into two main parts: a body cavity enclosed by the mantle, containing the digestive tract, various glands, and muscles; and a mantle cavity, bathed by sea water. The mantle cavity is mostly filled by the food-gathering apparatus called lophophore, which consists generally of a pair of cilia-bearing "arms" or brachia—hence the name brachipod.

Pedicle. The pedicle is a muscular stalk which serves as the attachment organ of the brachiopods. Generally, it is short and stout, the part protruded from the shell being protected by a brownish tough hornlike cover (Fig. 6-4). Such a pedicle commonly has a fibrous expanded base of attachment. Its inner part passes through a sort of capsule, just inside the shell

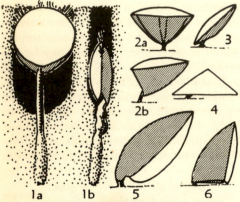

FIG. 6-4. **Pedicle attachments of brachiopods.** The diagrams show the attitude of shells attached by pedicles. The valves may be held in upright position (1a, b, 6), obliquely inclined (3, 5), or subhorizontally (2a, b, 4). The valve to which the pedicle is attached internally—hence called the pedicle valve—is indicated by a ruled pattern.

1a, b, *Dicellomus*, an inarticulate (atremate type) which probably lived like modern *Lingula*.

2a, b, *Acrotreta*, an inarticulate (neotremate type) in which the pedicle emerges from a rounded opening between the apex of the pedicle valve and the nearest shell margin.

3, *Schizambon*, an inarticulate (neotremate type) in which the pedicle opening is located between the apex of the pedicle valve and the shell margin.

4, *Orbiculoidea*, an inarticulate (neotremate type) having a nearly plane pedicle valve of circular outline in which the pedicle issues from a slit located between the centrally placed apex and margin of the valve.

5, *Composita*, an articulate (spiriferid type) in which the pedicle opening is a circular foramen in the beak of the pedicle valve.

6, *Tritoechia*, an articulate (orthid type) somewhat resembling *Composita* in location of the pedicle aperture but having an extended hinge line and probably an extremely short pedicle.

orifice, and by muscle fibers it is attached to the floor of one of the valves, which accordingly is designated as the **pedicle valve.**

The pedicle of *Lingula* and of some other modern brachiopods may be very long, as much as nine times the greatest length of the shell (Fig. 6-1). The elongate pedicle is flexible and capable of movement. A central tubular space which connects with the visceral cavity is filled by body fluid, and when the pedicle is muscularly contracted, its contained fluid is forced into the brachiopod body. The pedicle is extended by pushing the fluid back into the axial tube. In this manner, the shell of *Lingula* can be moved up or down within its burrow.

One of the deep-sea brachiopods (*Chlidonophora*), which has a pedicle twice as long as the shell, shows the terminal part of the pedicle split into numerous fine fibers which entangle foraminifer shells, thus obtaining anchorage. It is known that the pedicle of many living and fossil brachiopods becomes atrophied at an early stage, in which case the shell simply lies loose on the sea bottom, or it may be held in position by spines or by cementation of the beak of the pedicle valve to rock or to some other shell.

Muscles. A more or less complex system of muscles occupies part of the visceral cavity. They are joined to various parts of the floor of the valves in such manner as to produce scars, which may be prominent features of the shell interior and useful in the classification of the brachiopods. Accordingly, further attention to muscles will be given in the notice of shell characters. Brachiopod muscles serve for opening and closing the valves, and for movement of the shell on the pedicle (Fig. 6-5).

Lophophore. The lophophore is an appendage extending from the vicinity of the mouth into the mantle cavity. It consists of a simply lobed disk or of two elongate brachia arranged in folds or spirals. On its edge, the lophophore bears long tentacle-like cirri, at the base of which is a groove leading to the mouth (Fig. 6-3). The cirri bear numerous fine cilia, which vibrate so as to set up water currents that bathe the lophophore and carry food particles toward the mouth. They also function to produce an outward moving current along the median axis. There seems to be no evidence that cirri of the lophophore serve a respiratory function as has been supposed. Study indicates that respiration is principally effected by means of the mantle.

The lophophore, or brachia, is supported by attachment to the valve opposite the one which carries muscular attachment of the pedicle. The valve which bears the brachia therefore is termed the **brachial valve.** Some types of brachiopod lophophores and markings made by them on interior surfaces of the valves are illustrated in Fig. 6-6.

Hard Parts

Wide diversity of appearance characterizes the shells of brachiopods. Some are perfectly circular or regularly elliptical in outline, whereas others have angulated margins. The shell may be considerably extended along the longitudinal axis or transverse to it, and either valve may be gently to strongly convex or concave, or it may be plane. The contact between the valves may lie in a plane, but in different parts of some shells this junction deviates considerably from a plane. The exterior of the valves may be smooth, marked by concentric growth lines, covered by spines, roughened by shingle-like lamellae, or traversed by radially disposed ridges and grooves. All these readily visible external characters are useful in differentiating genera, but study indicates that many internal features of the shell are even more important for classification.

As already noted, two main types of brachiopod shells are distinguished. These are based partly on composition and general form but essentially on the nature of articulation between the valves. One division comprises the inarticulates which lack hingement, the valves being held together only by muscles attached to their interiors. The other division comprises the articulates, which are joined along a distinct hinge line where teeth and sockets are developed. Such articulation provides for permanent contact of the valves along one margin, while permitting other edges to be moved apart and brought together again.

FIG. 6-5. **Mechanism for opening and closing the brachiopod shell.** The drawings show a longitudinally bisected terebratulid shell with half of the hinge line and muscle system. The larger (pedicle) valve is placed in conventional orientation below the smaller (brachial) valve. A rounded projection (tooth) on the hinge line of the pedicle valve articulates with a depression (socket) on the hinge line of the brachial valve, thus forming a fulcrum. The valves are closed by muscles (adductors) attached to opposite portions of their interiors (stippled muscle in *A*). The valves are opened by muscles running from the floor of the pedicle valve to a projection (cardinal process) at the beak of the brachial valve, between the pair of hinge teeth; contraction of this muscle (stippled in *B*) pulls the front and sides of the two valves apart.

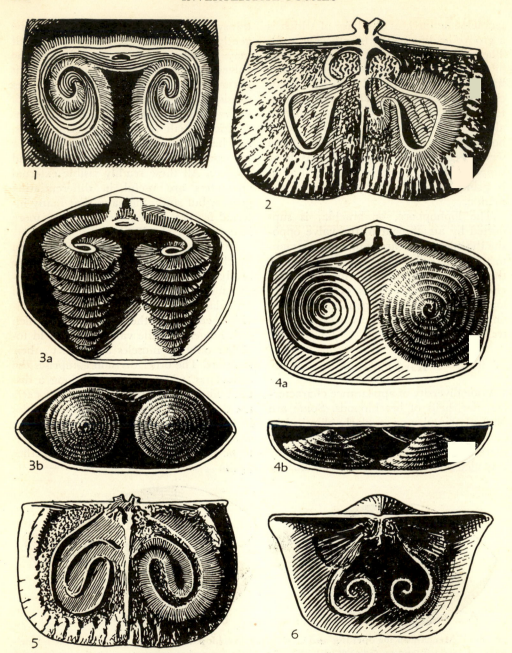

Fig. 6-6. **Lophophore patterns of some modern and fossil brachiopods.** The arrangement of lophophores in living brachiopods is observed readily, whereas that of many fossil forms is much less certainly known. The terebratulids, which greatly predominate among modern articulate brachiopods, have a simple or reflexed calcareous loop as support for the lophophore, but the spirally coiled lophophores of modern rhynchonellid articulates (3a, b) and of inarticulates, such as *Lingula* (1), lack calcareous supports. A large group of Paleozoic brachiopods (Spiriferida) is characterized by spirally coiled internal calcified structures attached to the brachial valve for support of the lophophores, but other

(*Continued on next page.*)

Shell Composition and Structure.
Brachiopod shells are classifiable in two
groups on the basis of their chemical composition. Most inarticulate brachiopods
have hard parts composed predominantly
of calcium phosphate and chitinous organic matter, associated with minor
amounts of calcium sulfate, calcium carbonate (in the form of calcite), and magnesium carbonate. Such shells are termed
chitinophosphatic. The hard parts of all
articulates and of a few inarticulates are
composed almost exclusively of calcium
carbonate. This group is called **calcareous.**
A third type of shell probably distinguished
Pre-Cambrian brachiopods, although it is
unknown in any fossil or in a living species, except in the tiny embryonic hard
parts called **protegulum.** The nature of
this initially formed shell suggests that
earliest, most primitive brachiopods in the
adult state had a chitinous covering which
lacked the admixture of appreciable
amounts of calcium phosphate or other
inorganic salts.

Structurally, two classes of chitinophosphatic shells are recognized, and three of
calcareous shells. One type of chitinophosphatic shell is represented by *Lingula,* in
which thin phosphatic laminae alternate
with chitinous ones. In *Discinisca* and various other genera, however, the calcium
phosphate and chitinous material are uniformly admixed.

The three structural types of calcareous

shells are impunctate, punctate, and pseudopunctate (Fig. 6-7). In each, there are
two discernible shell layers: an outer, relatively thin, finely laminated layer, and an
inner, relatively thick layer composed of
obliquely inclined calcareous prisms or
fibers. In **impunctate shells,** these layers
are solid (Fig. 6-7*A*). In **punctate shells,**
the inner fibrous layer is perforated by
fine tubes or pores, extending from the
interior almost to the outer surface (Fig.
6-7*C*). Each pore, termed a punctum, is
somewhat expanded in the lamellar layer
and connected with the surface by several
very minute tubules. As seen in tangential
sections, the puncta are cavities having
rounded outlines. They may be somewhat
uniformly distributed through the shell or
arranged in rows. **Pseudopunctate shells**
lack pores, but the fibrous layer contains
rodlike bodies of calcite, the ends of which
commonly form projections on the shell
interior. Because the rodlike bodies tend
to be dissolved more readily in weathering
than the surrounding fibrous material,
rounded cavities resembling puncta may
be produced. Thus, the shell is falsely
punctate.

Shell structure is an important basis of
classification among brachiopods because
study shows that each type is constantly
associated with other significant structures
and general shell form within groups. Evidence seems to indicate, however, that the
nature of shell composition and structure

(Fig. 6-6 continued.)

groups show the arrangement of lophophores by impressions on the shell interior, or there is no trace of
this structure.

1, *Lingula anatina* Lamarck, Recent; an inarticulate (much enlarged).

2, *Dictyoclostus americanus* Dunbar & Condra,
Pennsylvanian, Kansas; interior of a brachial
valve, showing (at right) inferred appearance of
cilia-bearing lophophore ($\times$1).

3a, b, *Tegulorhyncha nigricans* (Sowerby), Recent,
New Zealand; a modern rhynchonellid articulate, showing interior of brachial valve with
lophophore and shell in transverse section
viewed from the front (much enlarged).

4a, b, *Davidsonia* sp., Devonian, England; interior

of brachial valve showing (4a) spiral impression
of lophophore on floor of valve and restoration
(at right), and (4b) inferred appearance of
lophophore in transverse section of shell (much
enlarged).

5, *Leptaenisca* sp., Devonian, Oklahoma; interior
of brachial valve showing lophophore ridge and
restoration ($\times$3).

6, *Gigantoproductus* sp., Lower Carboniferous
(Viséan), England; interior of pedicle valve,
showing spiral impressions of lophophore
(reduced).

is not a fundamental character for differentiation of phylogenetically divergent stocks. It is true that pseudopunctate shells are confined to a single order of articulates (Strophomenida), but there are both punctate and impunctate divisions in at least three seemingly natural main groups as defined by other shell characters (Orthida, Rhynchonellida, Spiriferida). Other main divisions of the articulates are either exclusively impunctate or exclusively punctate.

Orientation and Nomenclature of Valves. In the discussion of soft parts, we have noted that the valve to which the pedicle is attached appropriately is termed the pedicle valve and that which bears the brachia is known as the brachial valve. The relation of these soft parts to the shell are constant, and because of their special morphological significance, they are best chosen as the basis of nomenclature. According to a long-established presumption, the brachial valve occupies a dorsal position with respect to the animal and the pedicle valve a ventral position. Recent work on embryological development of brachiopods (Percival, 1944) shows that the reverse of the conventionally applied orientation of the valves belongs to early growth stages of some forms at least. This means that in the embryonic brachiopod the pedicle valve may be morphologically dorsal and the brachial valve actually ventral. In view of this observation, it is preferable to use the names pedicle valve and brachial valve rather than names denoting position. Nevertheless, when it is needful or desirable to use terms for position, it seems best to retain dorsal and ventral in the sense so firmly established in literature, that is, dorsal corresponding to the position of the brachial valve and ventral, to the position of the pedicle valve.

REPRODUCTION AND LARVAL GROWTH

Brachiopods are reproduced from eggs which are fertilized by sperm. The sexes are separate. Fertilization takes place in

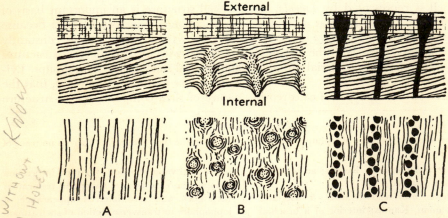

External

Internal

A   B   C

Fig. 6-7. **Types of calcareous shell structure in brachiopods.** Three main types are recognized: (A) impunctate, (B) pseudopunctate, (C) punctate. Each of these reveals two distinct shell layers, as shown by the longitudinal sections in the upper part of the figure. A relatively thin outer lamellar layer is readily differentiated from an inner fibrous layer, composed of inclined calcareous fibers which slope toward the lamellar layer in directions approaching the shell margins. Punctate shells are traversed by small tubes or pores which extend from the shell interior almost to the outer surface; as seen in tangential sections (lower figures), the pores appear as round openings. The pseudopunctate type of shell contains solid rods of structureless calcite in part of the fibrous layer.

the open sea water or within the mantle cavity; and in some brachiopods, the larvae develop in brood pouches formed by a fold of the mantle inside the shell of the female.

Developmental stages vary in different genera, but commonly there is a free-swimming stage of 10 or 12 days, during which the larva moves about freely by means of cilia on its surface. Three segments are developed, a head region, median portion, and caudal or foot segment, which later may become modified to form the pedicle. Minute shell plates develop on the inner side of lobes which grow downward along the median segment (Fig. 6-8A). Subsequently, they turn upward to embrace the head segment (Fig. 6-8B, C), the larva settles down and becomes attached by the caudal segment or embryonic pedicle. The diminutive covering valves constitute the protegulum. This embryonic shell develops on the outer side of the median part of the larva in some brachiopods and then is not turned over during growth. As the animal grows, the shell increases in size, mainly by deposits along the lateral and anterior margins.

INARTICULATE SHELLS

Morphology of inarticulate brachiopod shells is illustrated in Fig. 6-9, to which italic numbers in the text refer unless indicated otherwise. Inarticulates have a pedicle valve (7, 18) and a brachial valve (14, 17), but positive identification of them is not always easy. Inarticulate pedicle valves lack hinge teeth, and some contain no perceptible notch or opening for passage of the pedicle. The pedicle valve commonly is larger than the brachial, but the reverse may be true. The inarticulate brachial valve lacks dental sockets and marks of lophophore attachment. In spite of all this, the mode of shell attachment, placement of the pedicle opening which generally is present, and internal features, such as muscle scars, afford definite evidence for identification of the valves.

Morphological Features

Inarticulate shells may be divided into equal halves by a longitudinal plane which intersects the marginal or excentric **apex** (1, 10); the apex coincides with the protegulum, marking the point of initial shell growth. Growth lines parallel to the shell margins are concentric around the apex. Because shell growth is dominantly in anterior and lateral directions from the apex, the posterior edge (22) of a valve is defined by its relation to the apex. In shells having a pedicle notch or **foramen** (35, 39), the posterior margin is the edge of the shell nearest to this opening. The anterior edge (9) is opposite to the posterior.

In many inarticulate brachiopod shells, the posterior portions of one or both valves are distinguished by difference in slope or convexity from the remainder of the outer surface. Such areas are termed **interareas** (24). They may slope posteriorly from the apex of the valve or in reverse manner turned forward to the shell margin. Lateral

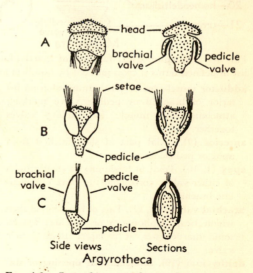

FIG. 6-8. **Larval stages of a modern brachiopod.** The drawings represent three stages in the embryonic development of *Argyrotheca*, a terebratulid; soft parts stippled; short projections indicate cilia and longer ones, setae; shell in sectional views at right is solid black (*b*, brachial; *p*, pedicle).

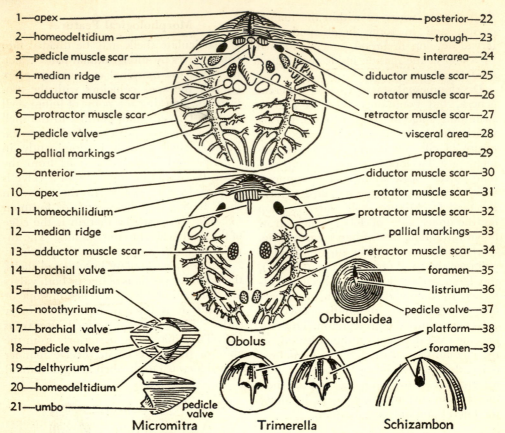

1—apex — posterior—22
2—homeodeltidium — trough—23
3—pedicle muscle scar — interarea—24
4—median ridge — diductor muscle scar—25
5—adductor muscle scar — rotator muscle scar—26
6—protractor muscle scar — retractor muscle scar—27
7—pedicle valve — visceral area—28
8—pallial markings — proparea—29
9—anterior — diductor muscle scar—30
10—apex — rotator muscle scar—31
11—homeochilidium — protractor muscle scar—32
12—median ridge — pallial markings—33
13—adductor muscle scar — retractor muscle scar—34
14—brachial valve — foramen—35
15—homeochilidium — listrium—36
16—notothyrium — pedicle valve—37
17—brachial valve — platform—38
18—pedicle valve — foramen—39
19—delthyrium
20—homeodeltidium
21—umbo

Orbiculoidea

Obolus

pedicle valve

Micromitra   Trimerella   Schizambon

FIG. 6-9. **Morphological features of inarticulate brachiopod shells.** The alphabetically arranged list of defined terms is cross-indexed by numbers to parts of the figure.

adductor muscle scar (5, 13). Impression on interior of brachial or pedicle valve marking attachment of muscle for pulling valves together.

anterior (9). Axial part of shell farthest from apex or pedicle foramen.

apex (1, 10). Point of beginning of shell growth of either valve; corresponds to beak of articulate brachiopods.

brachial valve (14, 17). Part of brachiopod shell which bears lophophore or brachia; among some inarticulates, it is larger than pedicle valve.

delthyrium (19). Subtriangular opening or slitlike notch beneath apex of pedicle valve, for passage of pedicle.

diductor muscle scar (25, 30). Impression on posterior part of interior of either valve for attachment of muscles for opening valves.

foramen (35, 39). Completely enclosed rounded or slitlike opening in pedicle valve for passage of pedicle.

homeochilidium (11, 15). Convex shell structure restricting notothyrium.

homeodeltidium (2, 20). Convex shell structure restricting the delthyrium.

interarea (24). Differentiated tract on posterior part of either valve.

listrium (36). Shell deposit partly restricting area of pedicle foramen.

median ridge (4, 12). Linear elevation along mid-line of interior of either valve.

notothyrium (16). Subtriangular opening beneath apex of brachial valve.

pallial markings (8, 33). Branching impressions on interior of either valve formed by fluid-filled passageways in mantle.

pedicle muscle scar (3). Impression on interior of pedicle valve made by attachment of muscle which pulls pedicle.

(Continued on next page.)

parts of the interarea may be differentiated as triangular tracts termed **propareas** (*29*), and the median, generally somewhat convex part is termed **homeodeltidium** (*2, 20*) on the pedicle valve and **homeochilidium** (*11, 15*) on the brachial valve. Although growth lines of the interarea generally are somewhat deflected in crossing these median parts, they are uninterrupted. The homeodeltidium may bear a longitudinal **trough** (*23*), denoting indentation of the shell by an external part of the pedicle. An unclosed indentation of the posterior edge of the pedicle valve for passage of the pedicle is termed **delthyrium** (*19*), and if an adjacent corresponding notch occurs in the brachial valve, it is called **notothyrium** (*16*). A rounded or slitlike opening for the pedicle which is enclosed entirely by substance of the pedicle valve is called a foramen (*35, 39*), and a shell deposit which partly closes the foramen of some inarticulates is termed **listrium** (*36*).

Internal features of the shells of inarticulates consist mainly of **muscle scars** (*3, 5, 6, 13, 25–27, 30–32, 34*), **pallial markings** (*8, 33*), a **median ridge** (*4, 12*), and depression in the **visceral area** (*28*). In some calcareous-shell inarticulates, muscles are attached to raised **platforms** (*38*) of the valve interiors.

ARTICULATE SHELLS

Morphological Features

Articulate brachiopod shells are distinguished from inarticulates by the possession of a hinge line transverse to the median axis of the valves (*50;* italic numbers in the text describing articulate shells refer to Fig. 6-10). It may be straight or curved, and laterally much extended or very short. The hinge line coincides with the posterior margin (*23*) of each valve, but generally it does not mark the posterior extremity (*21*) of the shell. This extremity is almost invariably at a pointed projection of the pedicle valve called its **beak** (*16*) or at an arched part (**umbo,** *18*) of the valve near the beak. The beak corresponds to the apex of inarticulate brachiopod valves. Each valve has a beak, although that of the brachial valve may project little or not at all above the hinge line. The beaks mark the beginning of shell growth, as shown by concentric arrangement of **growth lines** (*10*) around them and by the radially disposed ribs (**costae,** *63*) or shell corrugations (**plications,** *19, 69, 70;* **fold,** *15;* **sulcus,** *20*) which generally diverge from them.

Application of directional terms, such as **dorsal** (*1*), **ventral** (*5*), **posterior** (*6*), and

---

(Fig. 6-9 continued.)

pedicle valve (*7, 18, 37*). Part of brachiopod shell to which pedicle is attached; opposite brachial valve.

platform (*38*). Raised shell structure on interior of either valve for muscle attachments.

posterior (*22*). Part of shell occupied by viscera, including area nearest to pedicle opening; posterior extremity designated in figure.

proparea (*29*). Differentiated portion of shell exterior adjoining posterior margin and bordering delthyrium and homeodeltidium or notothyrium and homeochilidium.

protractor muscle scar (*6, 32*). Impression on interior of either valve made by attachment of muscles which serve for slight sliding movement of valves.

retractor muscle scar (*27, 34*). Impression on interior of either valve made by muscles which move valves longitudinally.

rotator muscle scar (*26, 31*). Impression on interior of either valve made by muscles which serve for slight twisting of valves.

trough (*23*). Furrow on posterior part of pedicle valve beneath apex which provides space for pedicle.

umbo (*21*). Prominent convexity of either valve adjacent to apex.

visceral area (*28*). Depression in posterior part of interior of pedicle valve, indicating placement of viscera.

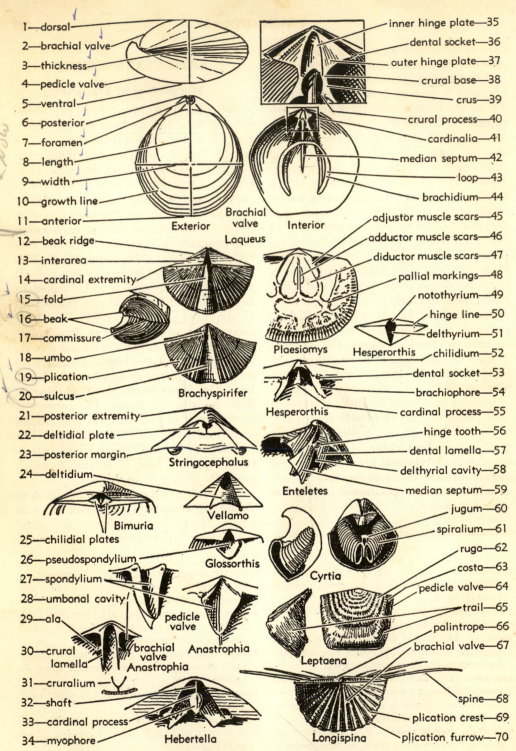

1—dorsal
2—brachial valve
3—thickness
4—pedicle valve
5—ventral
6—posterior
7—foramen
8—length
9—width
10—growth line
11—anterior
12—beak ridge
13—interarea
14—cardinal extremity
15—fold
16—beak
17—commissure
18—umbo
19—plication
20—sulcus
21—posterior extremity
22—deltidial plate
23—posterior margin
24—deltidium
25—chilidial plates
26—pseudospondylium
27—spondylium
28—umbonal cavity
29—ala
30—crural lamella
31—crurialium
32—shaft
33—cardinal process
34—myophore

inner hinge plate—35
dental socket—36
outer hinge plate—37
crural base—38
crus—39
crural process—40
cardinalia—41
median septum—42
loop—43
brachidium—44
adjustor muscle scars—45
adductor muscle scars—46
diductor muscle scars—47
pallial markings—48
notothyrium—49
hinge line—50
delthyrium—51
chilidium—52
dental socket—53
brachiophore—54
cardinal process—55
hinge tooth—56
dental lamella—57
delthyrial cavity—58
median septum—59
jugum—60
spiralium—61
ruga—62
costa—63
pedicle valve—64
trail—65
palintrope—66
brachial valve—67
spine—68
plication crest—69
plication furrow—70

Exterior   Brachial valve   Interior
Laqueus
Brachyspirifer
Stringocephalus
Bimuria
Glossorthis
Vellamo
pedicle valve
brachial valve
Anastrophia
Anastrophia
Hebertella
Plaesiomys
Hesperorthis
Hesperorthis
Enteletes
Cyrtia
Leptaena
Longispina

FIG. 6-10. **Morphological features of articulate brachiopod shells.** Terms are defined in the alphabetically arranged list which is cross-indexed by numbers to parts of this figure.

(*Continued on next page.*)

208

(Fig. 6-10 continued.)

adductor muscle scar (46). Impression on interior of either valve marking attachment of muscles for pulling valves together.

adjustor muscle scar (45). Impression on interior of pedicle valve marking attachment of pedicle muscle.

ala (29). Lateral flange on outer side of crural lamellae.

anterior (11). Part of shell adjacent to mid-line at extremity opposite beak.

beak (16). Generally pointed extremity of shell which marks beginning of shell growth; radial costae diverge from it, and successive growth lines are progressively distant from it.

beak ridge (12). Generally angulated line extending from beak to lateral extremity of cardinal area.

brachial valve (2, 67). One of two main parts of brachipod shell; bears supports for lophophore or brachia, and by convention is defined as dorsal in position.

brachidium (44). Simple or complex calcareous support for lophophore or brachia; chief types are loop and spiralium.

brachiophore (54). Short, generally stout projections from beak region of brachial interior at edge of notothyrium, for attachment of lophophore; corresponds to crural base of more advanced brachiopods.

cardinal extremity (14). Lateral terminus of hinge line.

cardinal process (33, 55). Projection at or near beak of brachial valve for attachment of diductor muscles which function for opening valves.

cardinalia (41). Collective term for structures in beak region of brachial valve functioning for articulation, muscle attachment, and brachial support.

chilidial plate (25). Shell deposit at side of notothyrium, serving partly to close this opening of brachial valve.

chilidium (52). Single plate extending across notothyrium, partly or entirely closing it.

commissure (17). Line of lateral and anterior contact of closed valves.

costa (63). Riblike thickening of shell extending generally from beak region to anterior or lateral margins.

crura (39). pl. of crus.

crural base (38). Projection from hinge plate of brachial valve at edge of notothyrium for attachment of one of crura.

crural lamella (30). One of two short septa adjoining cavity below notothyrium in beak region of brachial valve, extending from inner edge of outer hinge plate to floor of valve; in some brachiopods, the two crural lamellae unite above floor of valve, forming spoon-shaped structure supported by one or more septa.

crural process (40). Point near base of crus directed obliquely inward and toward opposite valve.

cruralium (31). Spoon-shaped structure in beak region of brachial valve formed by union of crural lamellae, supported by one or more septa.

crus (pl. **crura**) (39). Proximal part of calcified brachial support, attached to hinge plate in beak region of brachial interior.

delthyrial cavity (58). Space beneath beak of pedicle valve, generally bounded laterally by dental lamellae.

delthyrium (51). Notch beneath beak of pedicle valve for passage of pedicle; extends to hinge line, medially bisecting pedicle interarea.

deltidial plate (22). Shell deposit at side of delthyrium, serving to constrict it or with adjoining deltidial plate, to close it.

deltidium (24). Undivided shell covering of delthyrium (exclusive of fused deltidial plates).

dental lamella (57). Oblique or vertical septum extending from floor of pedicle valve to edge of delthyrium, supporting a hinge tooth; dental lamellae divide space beneath beak into three compartments, of which middle one is delthyrial cavity.

dental socket (36, 53). Depression on hinge plate of brachial valve for reception of hinge tooth of pedicle valve.

diductor muscle scar (47). Impression on interior of brachial valve marking attachment of muscle for opening valves.

dorsal (1). According to convention, direction away from position of pedicle valve toward opposite valve; dorsal valve is thus equivalent to brachial valve.

fold (15). Major rounded elevation of shell along longitudinal mid-line, affecting both outer and inner shell surfaces; generally on brachial valve.

foramen (7). Rounded opening at or near beak of pedicle valve for passage of pedicle.

growth line (10). Marking on shell surface parallel to valve margin, indicating former position of this margin.

(Continued on next page.)

(Fig. 6-10 continued.)

hinge line (50). Edge of shell where the two valves articulate.

hinge plate (35, 37). Divided or undivided platform in beak region of brachial interior, generally joined to dental sockets and crural bases; may be divided into inner and outer plates.

hinge tooth (56). Projection on hinge line of pedicle valve which fits dental socket of brachial valve, serving as pivot in articulation.

inner hinge plate (35). Subhorizontal small plate extending medially from crural base.

interarea (13). Plane or longitudinally curved surface between beak and posterior margin of either valve, any transverse line across it being straight or nearly straight.

jugum (60). Simple or complex connection between halves of a brachidium.

length (8). Linear distance in plane of bilateral symmetry from anterior extremity to beak or posterior margin of shell, whichever is greater; term is applicable to shell as a whole or to either valve separately.

loop (43). Simply curved or doubly bent nonspiral type of brachidium.

median septum (42, 59). Calcareous wall built up along part of mid-line of a valve interior.

myophore (34). Muscle attachment on tip of cardinal process; also in some shells, a platform developed above floor of a valve for muscle attachments.

notothyrium (49). Triangular opening at middle of interarea of a brachial valve, furnishing space (with delthyrium) for passage of pedicle.

outer hinge plate (37). Part of hinge plate extending laterally outward from crural base.

palintrope (66). Recurved or turned-back posterior portion of either valve; interarea developed on palintrope faces opposite valve.

pallial marking (48). Sinuous branching impression on parts of shell interior outside muscle scars, formed by fluid-filled passageways of mantle (pallial sinuses), which connect with body cavity in the posterior part of shell.

pedicle valve (4, 64). One of two main parts of brachiopod shell which bears attachment of pedicle; by convention, defined as ventral in position.

plication (19). Corrugation affecting both outer and inner shell surfaces, disposed radially with respect to beak; smaller in amplitude than fold and may be located on any part of shell except palintropes.

plication crest (69). Convex upper part of a plication.

plication furrow (70). Concave lower part of a plication.

posterior (6). Part of shell occupied by viscera; generally marked externally by pedicle opening and development of a palintrope.

posterior extremity (21). Part of valve or shell farthest back of anterior margin; in most articulates this is not identical with posterior margin, which is located forward of beak.

posterior margin (23). Growing edge of valve farthest back from anterior margin; among shells with a palintrope, posterior margin lies forward of posterior extremity.

pseudospondylium (26). Shell deposit in delthyrial cavity which may simulate a true spondylium.

ruga (62). Concentric shell corrugation which affects both outer and inner surfaces.

shaft (32). Axis of cardinal process.

spine (68). Long or short, straight or curved, solid or hollow projection of the shell surface.

spiralium (61). One of the pair of spiral brachidia on interior of some brachial valves (pl. spiralia).

spondylium (27). Curved platform for muscle attachment in beak region of pedicle valve, formed by convergence and union of the two dental plates; termed *sessile spondylium* if median part rests on floor of valve, *spondylium simplex* if base rests on a median septum, and *spondylium duplex* if base is supported by a pair of septa.

sulcus (20). Major rounded depression of shell along longitudinal mid-line affecting both outer and inner surfaces; generally on pedicle valve.

thickness (3). Linear distance from farthest opposite points on surface of the two valves.

trail (65). Anterior prolongation of some brachiopod shells, generally at strong angle to general plane of posterior portion of valves.

umbo (18). Convex portion of valve adjacent to beak.

umbonal cavity (28). Space on inside of either valve between a dental or crural lamella and adjacent postero-lateral shell wall.

ventral (5). According to convention, direction away from position of brachial valve toward opposite valve; ventral valve is thus equivalent to pedicle valve.

width (9). Linear distance between farthest opposite points on lateral margins of a valve or shell.

anterior (*11*), to articulate brachiopod shells is illustrated in Fig. 6-9. Likewise, the commonly used dimensional terms **thickness** (*3*), **length** (*8*), and **width** (*9*) are represented.

Posterior Part of Valves. Between the beak and the hinge line of one or both valves of many articulates is a plane or longitudinally curved smooth tract termed **interarea** (*13*). It extends laterally to the limits of the hinge line (**cardinal extremities**, *14*) and is differentiated from other parts of the valve exterior by a more or less angulated line known as the **beak ridge** (*12*). Directly beneath the beak, the interarea of either valve may be interrupted by a triangular open space, termed **delthyrium** (*51*) in the pedicle valve and **notothyrium** (*49*) in the brachial valve. The delthyrium provides space for passage of the pedicle. It may remain open throughout the growth of the shell or become closed in varying degree by shell substance deposited over it. Such covering of the delthyrium is called a **deltidium** (*24*) if it consists of a single plate formed as a unit, or **deltidial plates** (*22*) if constriction or enclosure consists of two separate shell elements that grow from the sides of the delthyrium. Different terms for covering structures of the delthyrium (pseudodeltidium, henidium, xenidium) which are based on presumed differences in origin, are not distinguished here, all being classed as a deltidium. A single plate enclosing or extending across the apical part of the notothyrium is called a **chilidium** (*52*), and separate plates in this area are termed **chilidial plates** (*25*). A rounded opening for passage of the pedicle which may be left in the deltidium, between deltidial plates, or in the beak region of the pedicle valve, is termed a **foramen** (*7*). Development of the deltidium and deltidial plates in different articulate genera is indicated diagrammatically in Fig. 6-11.

Pedicle Valve Interior. Internal features of articulate shells are varied and important in classification. Chief struc-

tures of the pedicle valve are the **hinge teeth** (*56*), vertical or oblique **dental lamellae** (*57*) in the beak region, **muscle scars** (*45–47*), and **pallial markings** (*48*). The teeth occur along the hinge line at angles of the delthyrium, and in many shells they are buttressed by dental lamellae extending to the floor of the valve below them. The space between dental lamellae, extending backward to the beak, is known as the **delthyrial cavity** (*58*).

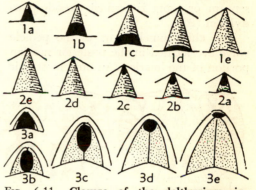

FIG. 6-11. **Closure of the delthyrium in articulate brachiopods.** The triangular notch beneath the beak of the pedicle valve, called delthyrium, may remain widely open for passage of the pedicle throughout all growth stages, but in many genera it is constricted by shell growth forming a single, transversely arched or flat wall (deltidium) or consisting of a pair of laterally extended walls (deltidial plates) which may meet.
1*a–e*, Deltidium (stippled) extended downward from the apex of the delthyrium during successive growth stages, eventually closing the space in some forms but leaving an unclosed area near the hinge line in others.
2*a–e*, Deltidium (stippled) developed across the lower part of the delthyrial space so as to leave a rounded apical foramen, which may become reduced in size during growth, and with atrophy of the pedicle, disappear.
3*a–e*, Deltidial plates (stippled) formed during growth stages of a terebratulid articulate. In a young shell the plates may be a pair of inconspicuous flanges along sides of the delthyrium, but in a mature shell only a round foramen is left along the mid-line of delthyrial space (3*d*), or by continued growth of the deltidial plates and resorption of shell in the beak region, the delthyrial space may be closed completely (3*e*).

In some pedicle valves, a **median septum** (59) of varying length and height extends forward from the beak along the mid-line. In some articulate brachiopods, a spoon-shaped structure, termed **spondylium** (27), is formed in the pedicle beak by convergence and coalescence of the dental lamellae. It rests on the floor of the valve or is supported by one, two, or three septa. A somewhat similar but differently formed structure which is not supported by a septum or septa is termed **pseudospondylium** (26).

The chief muscle scars of the pedicle valve are located in the posterior median portion of the interior. An outermost pair of impressions are scars of adjustor muscles (45) which function in moving the shell on the pedicle. Next inside them are the considerably larger scars made by the diductor muscles (47) for opening the valves. The innermost pair of scars belongs to the adductor muscles (46) which pull the valves together. Pallial markings show general correspondence in pattern among representatives of each major group or order of brachiopods, but they are not so useful as other internal features of the shells for distinction of genera (Fig. 6-12).

Brachial Valve Interior. Chief interior features of the brachial valve are those of the beak region related to articu-

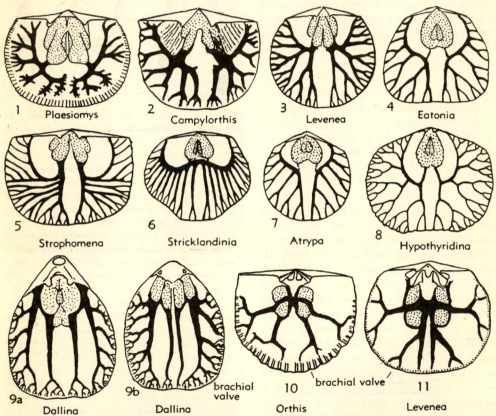

FIG. 6-12. **Types of pallial markings on brachiopod shells.** Canals beneath the mantle comprise avenues of circulation, and they may be impressed on the interior of the shell. The pattern of these markings is essentially constant in different genera and, where preserved, is useful in classification. Muscle-scar areas are stippled. All the shells here illustrated except 9b, 10, and 11 are pedicle valves. Orthid shells include 1–3, 10, 11; rhynchonellids, 4, 8; strophomenids, 5; pentamerids, 6; spiriferids, 7; and terebratulids, 9. Ordovician forms include 1, 2, 5, 10; Silurian, 3, 6, 11; Devonian, 4, 7, 8; Recent, 9.

lation, muscle scars, septa on the floor of the valve, and most important, structures which function for support of the brachia. At margins of the notothyrium or borne by outer parts of a **hinge plate** (*35, 37*) are the **dental sockets** (*36*), shaped to receive the teeth of the pedicle valve. The hinge plate may be medially divided or undivided. At the beak is a projection termed **cardinal process** (*33*), which in some brachiopods consists of a distinct **shaft** (*32*) and a bulbous end called **myophore** (*34*) for the attachment of the diductor muscles. The other end of these muscles is attached to the floor of the pedicle valve, and contraction of the muscles pulls the cardinal process so as to lift the anterior parts of the brachial valve, thus opening the shell (Fig. 6-5). Muscle scars on the floor of the brachial valve belong to the adductor muscles.

Structures for support of the brachia or lophophore are somewhat varied, and they may be complex and delicate. A primitive structure of this sort, confined to some of the early articulates, is termed **brachiophore** (*54*). It consists of short, relatively thick calcareous projections from the position of the dental sockets, directed obliquely forward and in some shells extended downward to the shell floor. Similar, much more delicate projections, termed **crura** (sing. **crus**, *39*), are a more advanced type of support for the basal part of the brachia. They may be relatively short, long and pointed, or extended into a continuous calcareous ribbon having a looped or spiral pattern. These extensions, which serve for rather complete support of the brachia, are termed **brachidia** (*44*), and the two main types are the **loop** (*43*) and **spiralium** (*61*). A variously shaped crossbar between the spiralia is called **jugum** (*60*); it has considerable importance in classifying some groups of spire-bearing brachiopods. Attachment of the crura, or **crural base** (*38*), is between the dental sockets at the edge of the hinge plate, and an angular or spinelike inward projection of the crura

near their base is termed **crural process** (*40*). Comparable to dental lamellae in the pedicle valve are **crural lamellae** (*30*) in the brachial valve. These are erect or obliquely inclined shell walls extending to the floor of the brachial valve in the position of the crura and serving to support them. Collectively, the varied structures of the brachial interior in the beak region are designated as **cardinalia** (*41*); and characters of the cardinalia are especially useful in classifying all articulate brachiopods.

SHELL GROWTH

Types of Enlargement

The initially formed part of each brachiopod valve is a very minute chitinous cover, generally of semicircular outline, called **protegulum** (Fig. 6-13, *1–3*). As shell material is deposited beneath it and along part or all of its margins, the protegulum becomes located at the apex or tip of the beak of the growing shell. It is bordered by concentric growth lines.

According to the manner in which additional laminae of shell substance are added to the protegulum, three modes of growth are differentiated. One of these, termed **hemiperipheral**, is characterized by additions of shell along the lateral and anterior margins but not along the posterior edge (Fig. 6-13, *1a, b*). A brachiopod valve so constructed becomes elongated anteriorly, and the apex or beak is at the posterior margin of the shell.

A second mode of growth, called **holoperipheral**, is distinguished by extension of the valve on all sides of the protegulum. The apex or beak thus comes to have an excentric or central location in the partly developed and fully grown shell (Fig. 6-13, *2a, b*).

The third type is known as **mixoperipheral**. Shell accretion proceeds on all sides of the protegulum, corresponding to holoperipheral growth, but the shell is not enlarged in a backward direction because growth on the posterior side of the valve is turned forward from the position of the protegulum toward the plane of commissure of the valve. The apex or beak thus has a marginal position, constituting a posterior extremity which lies farther back than the growing posterior margin of the valve. This inflected posterior portion of the valve is termed a **palintrope** (*palin,* reverse; *trope,* turned). The outer surface of a palintrope forms an interarea, but it should be noted that interareas include posterior portions of valves which slope backward from the apex or beak, as well as palintrope surfaces which slope in reverse direction forward. The most important feature for notice in studying the many shapes developed by brachiopod valves is the manner in which growth proceeds from the protegulum.

Evolution of Interareas

Normal growth of nearly all brachiopods includes some addition of shell along the posterior margin, but increment in forward and lateral directions predominates greatly, except in a few inarticulates. Among many articulates, both valves have interareas formed by shell enlargement on the posterior side of the protegulum; but in many others, it is confined to the pedicle valve.

Interarea Angle. Inclination of the interarea of either valve has importance in classifying brachiopod shells and in the

---

FIG. 6-13.   **Growth of brachiopod shells.** The diagrams illustrate features of shell shape which depend on amount and placement of shell accretions along various parts of the valve margins. The upper figures (1–3) are applicable both to inarticulate and articulate shell growth; remaining figures illustrate characters belonging mainly to articulate brachiopods.

1*a, b,* Normal and sectional views of a very young shell enlarged by **hemiperipheral growth** (greatly magnified). Part of the shell shown in black indicates the embryonic valve (protegulum, *p*). Shell growth, as indicated by alternating blank and stippled bands, is confined to lateral and anterior shell margins.

2*a, b,* Normal and sectional views of a very young shell enlarged by **holoperipheral growth;** accretion of shell substance proceeds along all margins (greatly magnified).

3*a, b,* Normal and sectional views of a very young shell enlarged by **mixoperipheral growth;** this corresponds to holoperipheral growth except that, along the posterior side, the shell is strongly reflexed, forming a palintrope (greatly magnified).

4*a, b,* Normal and side views of pedicle valve of *Retrorsirostra,* showing holoperipheral growth and development of interarea (*ia*) inclined backward from the origin of growth at the protegulum (*p*), making an **acute angle** with the plane of commissure. Arrows indicate direction of shell growth.

5*a, b,* Normal and side views of pedicle valve of *Tenticospirifer* (letters and symbols as in 4*a, b*). This shell has a high **right-angle** interarea developed in a plane normal to the commissure.

6*a–e,* Cross sections of pedicle valves showing types of interareas, defined by angles between them and the plane of commissure; (6*a*) acute angle (less than 90 deg.); (6*b*) right angle (90 deg.); (6*c*) obtuse angle (between 90 and 180 deg.); (6*d*) straight angle (180 deg.); (6*e*) reflex angle (more than 180 deg.).

7*a, b,* Normal and side views of *Vellamo,* which exhibits mixoperipheral shell growth, giving rise to an **obtuse-angle** interarea that slopes forward from the protegulum.

8*a, b,* Normal and side views of *Glossorthis,* illustrating mixoperipheral growth in which the **straight-angle** interarea comes to lie in the plane of commissure.

9*a, b,* Normal and side views of *Cyrtonotella,* illustrating mixoperipheral growth in which the **reflex-angle** interarea inclines toward the floor of the valve.

10–14, Diagrams illustrating position of the interarea at different growth stages (indicated by dotted lines numbered 1–5). Study of these drawings makes clear that the angle between interarea and commissure may change progressively during shell growth. Thus, *Glossorthis* (11) passes from acute-angle through right-angle and obtuse-angle positions of the interarea during growth before attaining straight-angle condition in the adult stage.

15, 16, Graphs showing changes in attitude of the interarea during growth of brachiopods shown in Figs. 10–14.

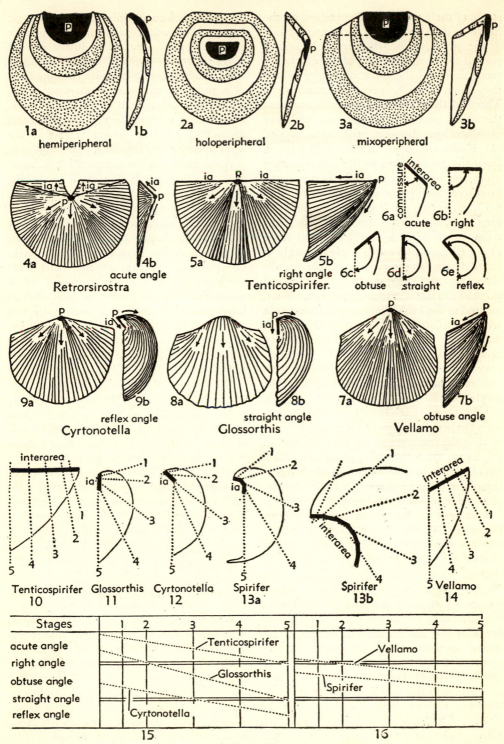

FIG. 6-13.

study of their evolution. Dependent on the mode of shell growth on the posterior side of the protegulum, whether backward or turned forward, five types of interareas are recognized on the basis of angles which they form with the plane of commissure. Three of these are chiefly significant, and the other two are only narrow boundary classifications between the main types. (1) Backward slope in the shell growth from the protegulum to the valve margin gives rise to an **acute angle** between the interarea and plane of commissure (Fig. 6-13, *4a*, *b*, *6a*). (2) If growth of the interarea from the beak is directed toward the commissure with neither backward slope nor inclination forward, a **right angle** is formed between the interarea and commissure (Fig. 6-13, *5a*, *b*, *6b*). (3) Forward slope of the interarea from the protegulum, located at the beak, makes an **obtuse angle** with the plane of commissure (Fig. 6-13, *6c*, *9a*, *b*), and this is most common in articulate brachiopods. (4) The interarea may grow straight forward, in line with the commissure; this makes a **straight angle** of 180 deg. (Fig. 6-13, *6d*, *8a*, *b*). (5) In some brachiopods, the beak projects beyond the plane of valve commissure so that the interarea slopes forward and at the same time inward toward the valve which bears it. The interarea then meets the plane of commissure in a **reflex angle,** of more than 180 deg. (Fig. 6-13, *6e*, *7a*, *b*).

Changes in Interarea Angle during Growth. The interarea angle is generally not the same during successive stages of shell growth, and this significant fact is shown diagrammatically in Fig. 6-13, *10–16*. Important also is the constant nature of change, for shift is always in the same direction, from acute-angle interareas as the initial (most primitive) condition to reflex-angle interareas as the terminal (most specialized) condition. No one brachiopod, however, is known to extend through this whole range. Many pass through two or three interarea angles. Thus, *Glossorthis* successively exhibits acute-angle, right-angle, and obtuse-angle

interareas before attaining a straight-angle interarea in adult growth. The pedicle valve of *Retrorsirostra* has an interarea that never develops beyond the acute-angle stage, and hence it may be distinguished as a more archaic shell than *Cyrtonotella*, which in youth has an obtuse-angle interarea and at maturity a reflex-angle interarea. Both of these genera belong to the same suborder (Orthacea).

A somewhat different condition of interarea development is illustrated by *Spirifer* (Fig. 6-13, *13a*, *b*, *16*). In this genus, the extremely young shell has an obtuse-angle interarea, and the same is true of a fully grown shell, despite the fact that anterior growth from the beak so exceeds that of the posterior portion of the valve that the shell becomes strongly curved longitudinally. The interarea is also curved longitudinally, and it is owing to this curvature that the obtuse-angle condition of the interarea is maintained. Otherwise, *Spirifer* would develop a highly reflex-angle interarea.

We may now generalize in saying that the slope characters of the interarea largely reflect differential growth in anterior and posterior directions along the median axis of the brachiopod valve, particularly where this differential growth involves shell curvature. The tendency in evolution is to produce strongly arched shells, with such excess of growth at the anterior margin as compared with the posterior that interareas are changed in inclination, tending to be overturned. At least in some stocks, the interarea tends to disappear.

SHELL FORM

Brachiopod shells have diverse shapes. The simplest, as illustrated by many inarticulates, have valves of subequal convexity and subcircular or elliptical outline. Among articulates especially, distinguished by hingement of the valves, several stocks show a tendency to become transversely widened so as to develop valves having a semicircular or semielliptical outline. The greatest width of the shell may be along the hinge line. An extreme is marked by such shells as *Mucrospirifer mucronatus* (Fig. 6-37, *3a–c*) from the Middle Devonian, but many others approach this shape. Other shells become very narrow transversely and longitudinally extended. The length of these shells, which have a very short hinge line, may be two or three times the maximum width.

The shape of shells in longitudinal profile of the valves is important and gives a basis for characterization, as illustrated in Fig. 6-14. The shape of one valve with respect to the other must be taken into account; and by convention in nomenclature, the shape term applicable to the brachial valve precedes that referring to the pedicle valve. Thus, a concavo-convex shell, such as that of *Rafinesquina* (Fig. 6-14D), has a concave brachial valve and convex pedicle valve, whereas a convexi-concave shell, like *Valcourea* (Fig. 6-14G), has a convex brachial valve and concave pedicle valve.

HOMEOMORPHY

Among brachiopods, the condition known as homeomorphy is common. This consists of such striking external resemblance between shells belonging to different genera that, without careful study, one may be mistaken very readily for the other. Not only do we find shells belonging to different groups which have similar shapes, but their patterns of ornamentation may be nearly identical. Examination of the internal structure may show that these differ radically, or the shell substance of one is impunctate, whereas the other is perforated by numerous minute puncta. Other sorts of evidence may establish true relationships, indicating wide classificatory separation in spite of resemblances.

The British paleontologist Buckman (1901–1908) has called attention to several striking examples of homeomorphy among brachiopods which are common in Jurassic rocks of England. Punctate shells be-

longing to the group called terebratulids contain at least five independent developments of the same character. Thus, it is easy to confuse species belonging to different genera and even to different families. Buckman recognized two categories of homeomorphy. One of these refers to closely similar but unrelated shells having identical or nearly identical occurrence in geologic time. Such contemporaneous or

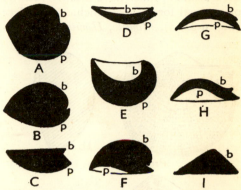

Fig. 6-14. **Diagrams showing characteristic shell form of brachiopods.** The drawings represent shells sectioned along the longitudinal midline, with pedicle valve (*p*) below and brachial valve (*b*) above. In some of the diagrams (*D–H*), the edge of one of the valves appears beyond the plane of the median section.

A, Biconvex shell, brachial valve less convex than pedicle valve (*Gypidula*).

B, Biconvex shell, brachial valve more convex than pedicle valve (*Atrypa*).

C, Plano-convex shell, brachial valve plane, pedicle valve convex (*Hesperorthis*).

D, Concavo-convex shell, brachial valve gently concave, pedicle valve gently convex (*Rafinesquina*).

E, Concavo-convex shell, like *D* but valves much more strongly curved (*Dictyoclostus*).

F, Convexi-concave shell, brachial valve strongly convex, pedicle valve concave anteriorly (*Hebertella*).

G, Convexi-concave shell, brachial valve gently convex, pedicle valve gently concave (*Valcourea*).

H, Resupinate shell, brachial valve chiefly convex but concave near hinge line; pedicle valve chiefly concave but convex near hinge line (*Strophomena*).

I, Convexi-plane shell, brachial valve conical, pedicle valve plane (*Orbiculoidea*).

near-contemporaneous forms he designated as **isochronous homeomorphs.** Many of the Jurassic examples are of this type. The other category includes two or more shells of different geologic age, in which one simulates another. These are termed **heterochronous homeomorphs.**

Examples of Homeomorphy. Schuchert & Cooper (1932) call attention to at least eight distinct orthid genera, distributed from Lower Ordovician to Middle Devonian, which have essential identity of external form. These include *Pionodema* (Fig. 6-22, *5*), *Mimella, Hemipronites, Dole-*

roides (Fig. 6-20, *5*), *Deltatreta, Finkelnburgia* (Fig. 6-20, *8*), *Schizophorella,* and *Schizophoria* (Fig. 6-23, *5, 7*). Two of these have punctate shells, and others are impunctate; all have important internal differences. They are classed among five separate families.

Two striking examples of homeomorphy are shown in Fig. 6-15. The first involves the Middle Ordovician genus *Productorthis,* belonging to the order Orthida, and Upper Pennsylvanian representatives of the genus *Dictyoclostus,* belonging to the order Strophomenida. These shells have a rather

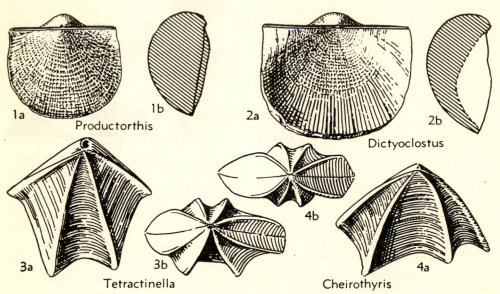

1a
1b
Productorthis
2a
2b
Dictyoclostus
3a
3b
Tetractinella
4b
4a
Cheirothyris

FIG. 6-15. **Homeomorphy of brachiopod shells.** Simulation of one shell by another which is actually very dissimilar constitutes homeomorphy, and resemblances of this sort are relatively numerous among brachiopods. The near identity of appearance may be such as to mislead anyone who fails to take account of significant distinctions, such as shell structure and nature of internal features. The homeomorphs not only belong to different genera, but they may be classed in different families, or even in different orders. They may be contemporaneous or widely separated in geologic age.

1a, b, Productorthis eminens (Pander), brachial view and median longitudinal section showing shell form, Middle Ordovician, Russia, ×2. This impunctate shell belongs in the family Orthidae of the suborder Orthacea and order Orthida.

2a, b, Dictyoclostus americanus Dunbar & Condra, brachial view and median section, Upper Pennsylvanian, Kansas, ×0.7. This is a pseudopunctate shell belonging in the family Productidae, suborder Productacea, order Stropho-

menida. *Productorthis* and *Dictyoclostus* are homeomorphs.

3a, b, Tetractinella trigonella (Schlotheim), brachial and posterior views, ×2; an impunctate spire-bearing brachiopod of the suborder Rostrospiracea, order Spiriferida; Middle Triassic, Italy.

4a, b, Cheirothyris fleuriausa (d'Orbigny), brachial and posterior views, ×2; a punctate loop-bearing shell of the order Terebratulida; Upper Jurassic, Germany.

closely similar profile and outline. The hinge line is extended, and the interareas are wide but very short, nearly lacking. Even the surface markings of radial costae crossed by concentric lines closely resemble one another. Yet the stocks to which these shells belong must have been quite distinct throughout virtually all of Paleozoic time. *Productorthis* has an impunctate shell and orthid internal structure, whereas *Dictyoclostus* has a pseudopunctate shell and productacean internal characters.

Tetractinella is a Middle Triassic genus having a shell which so closely resembles that of *Cheirothyris*, from Upper Jurassic rocks, that examples of one have been mistaken as belonging to the other (Fig. 6-15, *3, 4*). *Tetractinella* has an impunctate shell and a spirally coiled brachidium. It belongs to the suborder Rostrospiracea of the order Spiriferida. *Cheirothyris*, however, has a punctate shell and a looped brachidium. It is properly classed in the order Terebratulida. Thus, in spite of their resemblance, these two shells could hardly be farther apart in classification. They constitute remarkable examples of homeomorphy.

CLASSIFICATION

Definition of Main Divisions

For nearly 100 years, division of the brachiopods into two main groups has been recognized. Different names for them have been employed, some based on the presence or absence of an anus, others on differences in embryological development and the nature of the pedicle, and still others on the manner in which the two valves are held together. Yet the content of major units in any of these classifications is almost identical. The first-proposed names were based on the nature of union of the valves, whether unhinged or hinged—Lyopomata, meaning loose valves, and Arthropomata, meaning articulated valves. Somewhat different names having the same significance were introduced by Huxley in 1869 and have come

into almost universal use. They are Inarticulata, which includes brachiopods having the valves held together by muscles only, and Articulata, which comprises brachiopods having the valves joined together along a hinge line. The first group is composed predominantly of chitinophosphatic shells, whereas the second is made up exclusively of shells composed of calcium carbonate.

Since studies of embryological development of brachiopod shells by Beecher in 1891, names proposed by him, based on the nature of the pedicle opening, have been used very generally. These are Atremata and Neotremata, which consist of inarticulate brachiopods; and Protremata and Telotremata, which are composed of articulate brachiopods. To these, a minor group called Palaeotremata (Thomson, 1927) has been added; it contains shells classed as rudimentary articulates.

Significance of Shell Structure

Brachiopod studies in recent years, especially by Cooper (1944), throw such doubt on the validity of the divisions called Protremata and Telotremata as to warrant their rejection, because characters presumed to distinguish each of them have been found in the other. Work by Kozlowski (1929), Cooper, and others has emphasized the importance of shell microstructure as a character having far-reaching importance in classification. For example, all articulate brachiopods having pseudopunctate shell structure possess other attributes which point to close genetic affinity, and they are judged to be associated properly together as an order (Strophomenida). On the other hand, various groups of articulates which have punctate shell structure differ so radically in other features that they cannot reasonably be associated as a closely related stock having common ancestry.

Taking account of all known shell characters, we may recognize four distinct groups of punctate articulates, of which three are judged (on the basis of numerous characters) to have close affinities with different impunctate stocks. If this judgment is sound, it means that the presence of puncta is a secondary, rather than primary, basis of taxonomic grouping. A corollary of this conclusion is that punctate shell structure must have been evolved independently at different times among somewhat widely separate branches of the articulate brachiopods. There seems to be enormously greater probability of this than that all punctate-shelled brachiopods represent a single genetic group, containing assemblages which in amazing detail duplicate combinations of characters belonging to divergent impunctate stocks. Accordingly, genera of punctate-shelled brachiopods, which clearly belong together in the group called Dalmanellacea, are placed close to the impunctate group called Orthacea, because only the shell structure constitutes a significant difference. Even more striking is the morphological near identity of the punctate-shelled Rhynchoporacea and the impunctate-shelled Rhynchonellacea (Figs. 6-28, 6-29).

Summary of Classification

The classification adopted in this book, with the accompanying notation of distinguishing characters and geologic distribution, is shown in the following table. A graphic summary of distribution is given in Fig. 6-40.

Main Divisions of Brachiopods

Brachiopoda (*phylum*), bilaterally symmetrical inequivalve marine bivalves, especially characterized by possession of a food-gathering apparatus termed lophophore. Lower Cambrian–Recent.

Inarticulata (*class*), valves unhinged, lacking teeth and sockets, shell generally chitinophosphatic. Lower Cambrian–Recent.

Atremata (*order*), opening for pedicle shared by both valves. Lower Cambrian–Recent.

Lingulacea (*suborder*), chitinophosphatic shell thickened at posterior edge, trough below apex of pedicle valve. Lower Cambrian–Recent.

Trimerellacea (*suborder*), calcareous shell

with muscle areas on platforms. Middle Ordovician–Upper Silurian.

Neotremata (*order*), opening for pedicle confined to pedicle valve or lacking. Lower Cambrian–Recent.

Paterinacea (*suborder*), apex of pedicle valve at posterior extremity, homeodeltidium present. Lower and Middle Cambrian.

Siphonotretacea (*suborder*), like Paterinacea but lacking a homeodeltidium, pedicle aperture at apex or in front of it. Lower Cambrian–Middle Ordovician.

Acrotretacea (*suborder*), apex of pedicle valve generally in front of posterior extremity, homeodeltidium lacking. Lower Cambrian–Upper Ordovician.

Discinacea (*suborder*), pedicle aperture slitlike, brachial valve conical. Middle Ordovician–Recent.

Craniacea (*suborder*), calcareous shell, pedicle valve generally attached by cementation, brachial valve conical. Middle Ordovician–Recent.

Articulata (*class*), valves hinged, calcareous, generally bearing well-defined teeth and sockets. Lower Cambrian–Recent.

Palaeotremata (*order*), lack well-developed teeth and sockets. Lower Cambrian.

Orthida (*order*), subcircular to semielliptical, generally biconvex shells with radial ribs, having brachiophores and simple or lobate cardinal process. Lower Cambrian–Upper Permian.

Orthacea (*suborder*), shell impunctate. Lower Cambrian–Lower Devonian.

Dalmanellacea (*suborder*), shell punctate. Middle Ordovician—Upper Permian.

Terebratulida (*order*), shell punctate, hinge line very short, bear a looped brachidium. Upper Silurian–Recent.

Pentamerida (*order*), biconvex impunctate shells having a short hinge line, open delthyrium, and generally a well-defined spondylium (the aberrant genus *Enantiosphen* has a loop). Middle Cambrian–Upper Devonian.

Syntrophiacea (*suborder*), generally smooth shells, muscle scars in brachial valve not enclosed by crural lamellae. Middle Cambrian–Lower Devonian.

Pentameracea (*suborder*), smooth or plicate shells, muscle scars in brachial valve enclosed by crural lamellae. Upper Ordovician–Upper Devonian.

Triplesiida (*order*), biconvex impunctate shells having a moderately short hinge line, forked cardinal process, and flat deltidium. Middle Ordovician–Middle Silurian.

Rhynchonellida (*order*), biconvex, generally strongly plicate shells having a very short hinge line and prominent beaks. Middle Ordovician–Recent.

Rhynchonellacea (*suborder*), shell impunctate. Middle Ordovician–Recent.

Rhynchoporacea (*suborder*), shell punctate. Mississippian–Permian.

Strophomenida (*order*), pseudopunctate shells having a wide hinge line, one valve generally concave, surface costate. Lower Ordovician–Recent.

Strophomenacea (*suborder*), interarea well developed on one or both valves, with deltidium and chilidium; pedicle foramen very minute or lacking. Lower Ordovician–Recent.

Productacea (*suborder*), interareas reduced or lacking, spines along posterior margin or distributed over shell surface. Upper Ordovician–Upper Permian.

Spiriferida (*order*), shells containing a spiral brachidium. Middle Ordovician–Jurassic.

Atrypacea (*suborder*), impunctate shells mostly having a very short hinge line and spiralia not directed toward cardinal extremities. Middle Ordovician–Lower Mississippian.

Spiriferacea (*suborder*), impunctate shells mostly having an extended hinge line, surface marked by costae or plicae, and spiralia directed toward cardinal extremities. Middle Silurian–Jurassic.

Rostrospiracea (*suborder*), impunctate shells having a short hinge line and lacking costae or plicae. Middle Silurian–Jurassic.

Punctospiracea (*suborder*), punctate shells containing a spiral brachidium. Upper Silurian–Jurassic.

INARTICULATE BRACHIOPODS

Atremates

One of the oldest, most primitive, and at the same time conservative brachiopod stocks is that of order called the atremates. Most of them—in fact all except a short-lived branch (Trimerellacea) of Ordovician–Silurian age—have a chitinophosphatic shell. The valves are held in apposed position only by muscles, and although small inward projections of the posterior shell margin may occur at the borders of the homeodeltidium on the pedicle valve, there are no hinge teeth or sockets. They are typical inarticulates. The pedicle emerges from a gape or shallow notch at the posterior margin of the valves. This feature, together with the marginal position of the apex on either valve and the development of palintropes, characterizes or distinguishes the Atremata.

The valves of most atremates have a subcircular to elongate elliptical outline and a very gentle, even convexity. The exterior generally is featureless except for the presence of growth lines, or on some shells, regularly spaced concentric fine ridges. Muscle scars are the chief features of the interior. Figure 6-16 shows the arrangement of muscle impressions on the pedicle and brachial valves of *Lingula*, together with a diagrammatic representation of the muscles in position as attached to the pedicle valve. By means of the different pairs of muscles, the valves may be shifted laterally or longitudinally to some extent. Contraction of the diductor muscle in the posterior part of the valves serves to open the anterior part of the shell slightly, for uncontracted muscles in the median region serve as a pivot. When the main pair of these central muscles (adductors) are contracted, the valves are pulled together. Muscle impressions similar to those of *Lingula* are found on the interior of various fossil atremate shells.

Lingulacea. Shells of this suborder (Fig. 6-17) are small, mostly less than 0.5 in. in length or width, although the valves of *Lingula* may attain a length of 2 in.; several species of Cambrian Lingulacea have a length of nearly an inch. Development of a prominent palintrope on the pedicle valve, and generally also on the brachial valve, characterizes the shells. The central part of the pedicle palintrope is depressed in a trough for the accommodation of an emergent part of the pedicle.

The Lingulacea range from Lower Cambrian to Recent, representatives being known in all parts of the geologic column (Figs. 6-4, 6-17). Shells are found in normal marine deposits, but they are much more common in shaly beds probably laid down in poorly oxygenated brackish waters, ill-suited for most marine invertebrates. In this setting, the group has maintained itself with virtually no change in external form down to the present day. In most Paleozoic and younger deposits, the Lingulacea are useful as indicators of the environmental conditions to which they are adapted, rather than as fossils helpful in stratigraphic correlation. In Cambrian rocks, they are an important component of the brachiopod faunas.

Trimerellacea. Moderately thick calcareous shells, which attain a width of 2 in. and a length of more than 3 in., are classed as a branch of the atremates called Trimerellacea (Fig. 6-17). Both the shell composition and unusual size remove them from consideration as typical Atremata; but the absence of any sign of valve articulation

gives sanction to classifying them among the inarticulates. Also, features of their palintropes are nearly identical with those of Lingulacea. So far as known, this group of shells ranges from Middle Ordovician to Upper Silurian rocks.

The chief distinguishing feature of the Trimerellacea is the presence of fairly broad calcareous platforms raised above the floor of both pedicle and brachial valves. The platforms are supported by septa, and they serve as a base of the attachment for muscles.

Neotremates

The order of inarticulates called Neotremata is characterized by emergence of the pedicle from a notch or foramen confined to the pedicle valve (Fig. 6-18). Some genera, however, in which the pedicle of adults is atrophied, lack a pedicle opening in the shell. Like the atremates, all divisions of the neotremates, except one, have chitinophosphatic shells. The Craniacea possess shells composed of calcium carbonate.

The neotremates are a relatively unimportant assemblage of brachiopods, which nevertheless are represented by fos-

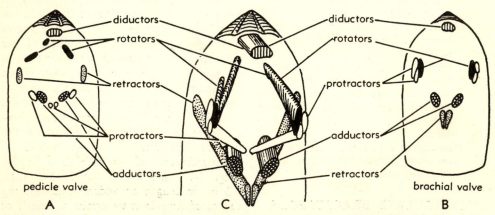

Fig. 6-16. **Musculature of an inarticulate brachiopod.** The illustrated form is *Lingula*, a genus of the Atremata (not to scale).

A, Interior of pedicle valve showing position of muscle scars.

B, Interior of brachial valve showing position of muscle scars.

C, Oblique view of pedicle interior showing the various muscles as they would appear with the brachial valve removed.

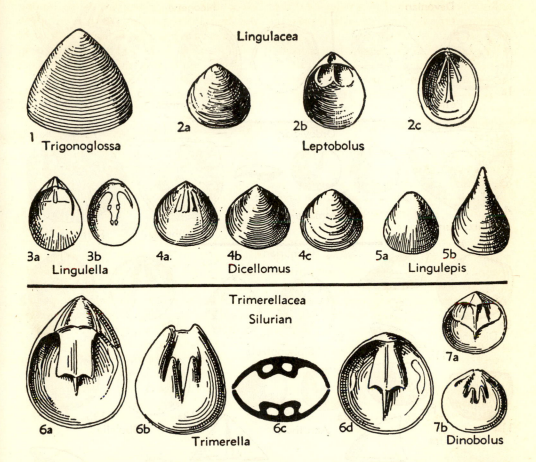

FIG. 6-17. **Representative atremate inarticulate brachiopods.** The geologic range of this group, characterized by generally small size and mostly by calcium phosphate shell composition, is from Lower Cambrian to Recent.

Dicellomus Hall, Upper Cambrian. Moderately thick shell with marginal apices; propareas and trough on pedicle valve. *D. politus* (Hall) (4a, pedicle interior; 4b, brachial valve; 4c, pedicle valve; ×3), Upper Cambrian, Wisconsin.

Dinobolus Hall, Middle Silurian. Calcareous shell, resembles *Trimerella*. *D. conradi* Hall (7a, pedicle interior; 7b, internal mold, pedicle valve; ×1), Middle Silurian, Iowa.

Leptobolus Hall, Middle and Upper Ordovician. Distinct propareas and trough on pedicle valve. *L. lepis* Hall (2a, b, pedicle exterior and interior; 2c, brachial interior; ×2), Trentonian, Ohio.

Lingulella Salter, Lower Cambrian–Lower Ordovician. Resembles *Lingula*. *L. similis* Walcott (3a, b, pedicle and brachial interiors, ×4), Upper Cambrian, South Dakota.

Lingulepis Hall, Upper Cambrian. Pedicle valve produced posteriorly. *L. pinnaformis* (Owen) (5a, b, brachial and pedicle valves, ×1), Wisconsin.

Trigonoglossa Dunbar & Condra, Mississippian–Pennsylvanian. Flat subtriangular shell bearing regular concentric fine ridges. *T. nebrascensis* (Meek) (1, pedicle valve, ×1), Virgilian, Nebraska.

Trimerella Billings, Middle Silurian. Thickened calcareous shell with internal platforms. *T. ohioensis* Meek (6a, b, pedicle valve interior and internal mold; 6c, transverse section through platforms, pedicle valve below; ×0.5), Niagaran, Ohio.

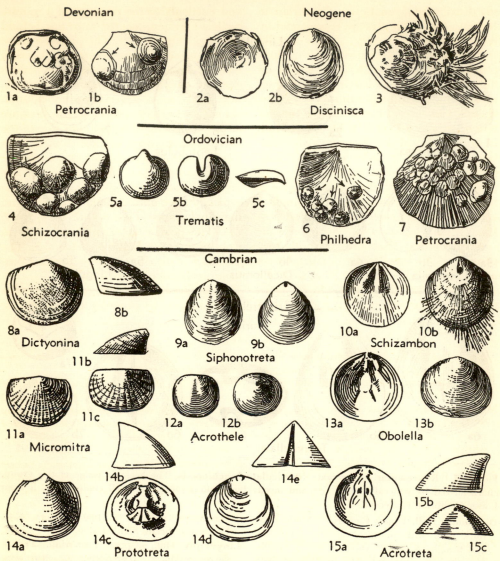

FIG. 6-18. **Representative neotremate inarticulate brachiopods.** This group of mostly phosphatic shells is distinguished by restriction of the pedicle opening to the pedicle valve or by its disappearance; commonly the brachial valve is conical. Paterinacea includes 8, 11; Siphonotretacea, 9, 10, 13; Acrotretacea, 12, 14, 15; Discinacea, 2–5; and Craniacea, 1, 6, 7.

Acrothele Linnarsson, Middle Cambrian. Differs from *Acrotreta* chiefly in its lower pedicle valve and larger size. *A. coriacea* Linnarsson (12*a, b,* brachial and pedicle valves, ×2), Sweden.

Acrotreta Kutorga, Lower to Upper Cambrian. The conical pedicle valve has a small apical foramen and distinct interarea. *A. idahoensis* Walcott (15*a,* brachial interior; 15*b, c,* side and posterior views of pedicle valve; ×8), Upper Cambrian, Idaho.

Dictyonina Cooper, Lower and Middle Cambrian. Surface marked by obliquely intersecting rows of minute pits. *D. pannula* (White) (8*a, b,* apical and side views of pedicle valve, ×7), Middle Cambrian, Nevada.

Discinisca Dall, Tertiary–Recent. Apices subcentral; pedicle valve flat or concave, with elongate foramen. *D. sparselineata* Dall (3,

(Continued on next page.)

sils distributed from Lower Cambrian to Recent. Two suborders are represented by living species.

Paterinacea. Rounded chitinophosphatic shells, mostly having maximum dimensions of 0.1 to 0.3 in. and characterized by a distinct palintrope on the pedicle valve, are classed among the Paterinacea (Fig. 6-18, *8, 11*). The delthyrium is largely closed by a homeodeltidium. Growth of the pedicle valve is mixoperipheral, as shown by presence of a palintrope. That of the brachial valve is hemiperipheral. Known representatives are confined to Lower and Middle Cambrian rocks.

Acrotretacea. Shells of this suborder correspond to the Paterinacea in general outline and very small size (Fig. 6-18, *12, 14, 15*). They are distinguished by the more distinctly conical form of the pedicle valve and particularly the presence of a minute pedicle foramen, located at or just behind the apex (Fig. 6-4, *2*). The posterior slope of the pedicle valve commonly bears a distinct trough; but because the pedicle does not emerge at the base of this trough, next to the valve margin, the indentation cannot be interpreted to denote pressure of the pedicle against the outer side of the valve. Acrotretacea are distributed from Lower Cambrian to Upper Ordovician.

Siphonotretacea. Both valves of the genera belonging to the Siphonotretacea typically have obtuse-angle interareas and marginal apices (Figs. 6-4, *3;* 6-18, *9, 10, 13*). The pedicle foramen is apical or in front of the apex of the pedicle valve. The suborder is distributed from Lower Cambrian to Middle Ordovician.

Discinacea. Both valves of the genera belonging to the suborder Discinacea are

---

(Fig. 6-18 continued.)

brachial valve, showing prominent setae along anterior margin, ×3), Recent, Japan; *D. lugubris* (Conrad) (*2a, b*, pedicle interior and exterior, ×1), Miocene, Maryland.

Micromitra Meek, Lower to Upper Cambrian. Surface marked by concentric lines and low radial ridges. *M. sculptilis* (Meek) (*11a, b*, pedicle valve; *11c*, brachial valve; ×4), Middle Cambrian, Montana.

Obolella Billings, Lower Cambrian. Valves low oval, nearly equal; minute foramen near apex of pedicle valve. *O. chromatica* Billings (*13a*, brachial interior; *13b*, pedicle exterior; ×3), Labrador.

Petrocrania Raymond, Ordovician–?Permian. Pedicle valve flat, attached to some foreign shell and concealed by low conical brachial valve. *P. hamiltoniae* (Hall) (*1a*, brachial interior, ×2; *1b*, two specimens attached to a *Stropheodonta* shell, showing brachial valves, ×0.7), Middle Devonian, Michigan; *P. scabiosa* (Hall) (7, some 20 specimens attached to a *Rafinesquina* shell, the brachial valves ridged in a manner reflecting costae of the host shell, ×0.7), Upper Ordovician, Indiana.

Philhedra Koken, Ordovician–?Permian. Like *Petrocrania*, but brachial valve marked by radial costae. *P. laelia* (Hall) (6, shells attached to a *Rafinesquina* brachial valve, ×0.7), Upper Ordovician, Kentucky.

Prototreta Bell, Middle Cambrian. Pedicle valve high conical, with distinct propareas and trough; brachial valve interior has median septum which expands anteriorly into a divided plate. *P. trapeza* Bell (*14a, b, e*, pedicle valve; *14c, d*, brachial valve; ×8), Montana.

Schizambon Walcott, Upper Cambrian. Pedicle foramen and groove anterior to apex of pedicle valve. *S. typicalis* Walcott (*10a*, brachial interior; *10b*, pedicle exterior; ×4), Nevada.

Schizocrania Hall & Whitfield, Ordovician–Lower Devonian. Radially costate brachial valve overlaps attached flat pedicle valve. *S. filosa* (Hall) (4, specimens attached to a *Rafinesquina* shell, ×1), Upper Ordovician, Ohio.

Siphonotreta de Verneuil, Upper Cambrian–?Lower Ordovician. Oblique apical foramen on pedicle valve. *S. tertia* (Walcott) (*9a, b*, brachial and pedicle valves, ×3), Upper Cambrian, Alberta.

Trematis Sharpe, Middle and Upper Ordovician. Deep pedicle fissure extending from apex to posterior edge of pedicle valve. *T. millipunctata* Hall (*5a, b*, brachial and pedicle valves; *5c*, side view of shell, pedicle valve below; ×1), Upper Ordovician, Indiana.

rounded, discoid to conical shells, which develop by holoperipheral growth (Fig. 6-18, *2–5*). The pedicle opening is a deep narrow notch indenting the posterior margin of the pedicle valve, or an enclosed narrow slit located behind the apex (Fig. 6-4, *4*). It may be constricted by a plate, termed listrium. The shell substance is dominantly chitinophosphatic, but some genera have external calcareous shell laminae. The group ranges from Middle Ordovician to the present.

Craniacea. Members of the suborder Craniacea have subcircular shells composed of calcium carbonate (Fig. 6-18, *1, 6, 7*). In the adult stage, they lack a pedicle opening but are cemented by the exterior of the pedicle valve to a foreign surface, generally another shell. Commonly, the host is a brachiopod, but it may be a coral, bryozoan colony, crinoid stem, or mollusk.

Longitudinal sections of some attached shells (*Philhedra*) demonstrate reduction of the pedicle valve, the brachial valve being attached by flangelike marginal areas which precisely fit irregularities of the underlying surface. Some species seem to choose a particular species of articulate brachiopod as host, for no specimens have been found attached otherwise. A margin of the attached craniacean shell may coincide with the edge of the valve of its host, but it is never found extending onto two adjoined valves, as it might if the craniacean grew on an empty shell.

A few members of this group (*Petrocrania*) have the peculiarity of reflecting rather perfectly in the markings of their brachial valve the surface of the host shell to which the craniacean is attached. Such adherent neotremates are inconspicuous. The range of Craniacea is from Middle Ordovician to Recent.

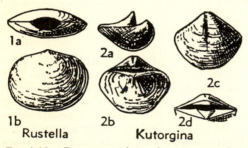

1a

2a

2c

1b   2b   2d

Rustella   **Kutorgina**

FIG. 6-19. **Representative paleotremate articulate brachiopods.** These are calcareous shells having only imperfectly developed articulation but a distinct hinge line. They are known only from Lower Cambrian strata.

Kutorgina Billings, Lower Cambrian. May belong among Orthacea. Pedicle valve has prominent beak which curves over interarea of brachial valve; delthyrium very wide. *K. cingulata* (Billings) (*2a*, side view of shell, pedicle valve below; *2b, c*, brachial and pedicle views; *2d*, interareas, showing partial closure of delthyrium by wide, narrow deltidium and notothyrial space of brachial valve covered by chilidium; ×1), Vermont.

Rustella Walcott, Lower Cambrian. Shell biconvex, broad delthyrium on pedicle valve, interareas indistinct or lacking on brachial valve. *R. edsoni* Walcott (*1a, b*, posterior and pedicle views, ×1), Vermont.

ARTICULATE BRACHIOPODS

Palaeotremates

The most primitive articulate brachiopods consist of Lower Cambrian calcareous-shelled forms assigned to the order Palaeotremata (Fig. 6-19). The valves are joined along what seems to be a somewhat extended hinge line; and one or both valves bear an interarea. Distinct hinge teeth and sockets are not observed, however. Beneath the beak on the posterior side of the pedicle valve is a transverse shell deposit, which constricts an otherwise broadly open space corresponding to the delthyrium.

Only one genus, *Rustella*, is assigned definitely to the order Palaeotremata. Its brachial valve lacks an interarea. Another brachiopod, *Kutorgina*, is a doubtful member of the order; although generally classified here, it may actually be a primitive orthid. The brachial valve of *Kutorgina* has a relatively prominent interarea and a plate, equivalent to the chilidium of other articulates, which closes the space beneath the beak.

Orthids

The order called Orthida is a very important assemblage of Paleozoic brachiopods, which includes the oldest well-developed articulates (Figs. 6-20 to 6-23.) The shells are rounded to semielliptical in outline, and the outer surface of both valves is marked by fine to coarse radial costae. The hinge line is generally wide, so that commonly it coincides with greatest width of the shell. In some genera, however, it is of intermediate width or relatively short, and such hinge lines are associated with shells having rounded outlines. A characteristic feature of the orthids is the development of a distinct interarea on both the pedicle and brachial valves. They may be indented by open notches (delthyrium, notothyrium), or these spaces beneath the beak may be more or less completely closed by transverse plates (deltidium, chilidium). Most genera have biconvex shells, but some are concavo-convex, and a few convexi-concave. The interareas of both valves are predominantly obtuse angled. Muscle scars are prominent on the valve interiors; and in general, structural features of the interior are important in differentiating genera.

Two main divisions of the orthids are recognized, one having an impunctate shell structure and the other having a punctate structure. Of these, the impunctate forms (Orthacea) are an antecedent group, ranging from Lower Cambrian to Middle Devonian. Orthids possessing punctate shells are grouped in the suborder Dalmanellacea. Undoubtedly, these were derived from the impunctate group. Their known stratigraphic range is from Middle Ordovician to Upper Permian.

Orthacea. The oldest well-developed articulate brachiopod is a member of the suborder Orthacea which occurs in Lower and Middle Cambrian rocks, respectively, in the Appalachian and Cordilleran geosynclines (Fig. 6-20, *10a, b*). It is a small subquadrate shell, which has obtuse-angle interareas on the two valves and fine radiating costae on the surfaces extending laterally and forward from the beaks. Space for passage of the pedicle is furnished by indentations of the posterior side of both valves. The presence of a deltidium and chilidium reduces this space.

Upper Cambrian genera are characterized by a broad, obtuse-angle interarea on the pedicle valve, with a convex deltidium or open delthyrium, and a pseudospondylium in the pedicle valve (Fig. 6-20, *11, 13*). A simple rodlike cardinal process may be observed beneath the beaks of the brachial valves.

A great expansion of the Orthacea occurred in Ordovician times (Figs. 6-20, 6-21, 6-40). As compared with 6 known genera of Cambrian Orthacea, Lower Ordovician rocks contain 21 described genera and at least 147 species. Middle and Upper Ordovician Orthacea are much more numerous but many are not yet described. The richness of Ordovician brachiopod faunas is suggested by the fact that a report by G. A. Cooper on Middle Ordovician brachiopods contains more than 250 plates; it includes not only Orthacea, but other groups present in these rocks.

Some Lower Ordovician orthacean shells (Fig. 6-20, *7*) closely resemble the punctate shells of *Dalmanella*, but they are distinguished by impunctate shell structure and internal features. *Finkelnburgia* (Fig. 6-20, *8*), which contains 23 described species, is characterized by the shape and markings of its shell, the presence of a pseudospondylium in the pedicle valve, and crural lamellae, which converge toward the floor of the brachial valve beneath strong brachiophores. *Tritoechia*, which is one of the most widely distributed genera, contains 14 described species (Fig. 6-20, *9*). Both valves have obtuse-angle interareas, that of the pedicle valve being much more prominent. The rounded foramen at the summit of the convex deltidium and deep delthyrial cavity between the strong dental lamellae of the pedicle valve and the chilidial plates and simple cardinal process of the brachial valve are features of the genus.

Genera and species of Orthacea are extremely numerous in Middle and Upper Ordovician rocks (Figs. 6-20, *1, 2, 4;* 6-21, *2–9*). Several important types are finely costate biconvex shells, and less common are the convexi-concave shells, which externally resemble members of the Strophomenacea. Some genera are coarsely costate, with a plano-convex shell and prominent, nearly right-angle interareas on both valves, or with a convexi-concave shell bearing small obtuse-angle interareas. An important genus of biconvex shells (*Platystrophia*), common in some Upper

---

FIG. 6-20.   **Representative Cambrian–Middle Ordovician orthid articulate brachiopods belonging to the suborder Orthacea.** This group contains impunctate shells which commonly have well-developed interareas on both valves. Internal calcareous supports for lophophores are lacking.

Apheoorthis Ulrich & Cooper, Upper Cambrian–Lower Ordovician. Pedicle valve has unclosed delthyrium and a pseudospondylium; surface marked by bundled costae. *A. lineocosta* (Walcott) (12a, pedicle valve, ×1; 12b, c, interior of brachial and pedicle valves, ×3), Upper Cambrian, Colorado.

Billingsella Hall, Upper Cambrian. Broad interarea on pedicle valve; delthyrium largely closed by convex deltidium, with pedicle foramen at its tip, adjoining beak. *B. corrugata* Ulrich & Cooper (11a, pedicle interior, ×3; 11b, brachial valve, ×1), Oklahoma; *B. perfecta* Ulrich & Cooper (13a, b, pedicle and brachial interiors, ×2; 13c, d, pedicle and brachial exteriors, ×1), Montana.

Dinorthis Hall & Clarke, Middle and Upper Ordovician. Convexi-concave shell with rounded outline and strong costae. *D. pectinella* (2, pedicle valve, ×1), Trentonian, Kentucky.

Diparelasma Ulrich & Cooper, Lower Ordovician. Biconvex shell with subcircular outline; pseudospondylium in pedicle valve. *D. elegantulum* (Butts) (3a–c, brachial, posterior, and side views, ×1), Alabama.

Doleroides Cooper, Middle Ordovician, Biconvex, finely costate shell. *D. gibbosus* (Billings) (5a, b, posterior and pedicle views, ×1.5), New York.

Finkelnburgia Walcott, Upper Cambrian–Lower Ordovician. Gently biconvex finely costate shells of semicircular outline; pseudospondylium in pedicle valve. *F. virginica* Ulrich & Cooper (8a, b, pedicle and brachial interiors; 8c, d, brachial and pedicle exteriors; ×1), Canadian, Virginia.

Hesperorthis Schuchert & Cooper, Middle Ordovician–Middle Silurian. Plano-convex coarsely ribbed shell, large interarea on pedicle valve; delthyrium open except for closure by small deltidium adjoining beak. *H. tricenaria* (Conrad) (1a–d, posterior, side, brachial interior, and brachial exterior views, ×1), Middle Ordovician, New York.

Nanorthis Ulrich & Cooper, Lower Ordovician. Biconvex shell with rounded outline and bundled costae; delthyrium open; homeomorphic with the punctate *Dalmanella*. *N. hamburgensis* (Walcott) (7a, 7c, brachial and pedicle interiors, ×4; 7b, brachial exterior, ×2), Nevada.

Nisusia Walcott, Lower and Middle Cambrian. Oldest well-developed articulate brachiopod; gently biconvex, finely costate shell having subequally large interareas on the valves; delthyrium and notothyrium partly closed by deltidium and chilidium. *N. montanaensis* Bell (10a, b, posterior and brachial views, ×1), Middle Cambrian, Montana.

Orthambonites Pander, Lower and Middle Ordovician. Biconvex somewhat coarsely costate shell; brachial valve has prominent, simple cardinal process. *O. eucharis* Ulrich & Cooper (6a, b, brachial exterior and interior, ×2, ×4), Lower Ordovician, Nevada.

Tritoechia Ulrich & Cooper, Lower Ordovician. Pedicle valve has large interarea and convex deltidium with apical foramen; costae crowded and commonly hollow, but the shell is not punctate. *T. typica* (Ulrich) (9a, pedicle exterior, showing openings into hollow costae; 9b, pedicle interarea; 9c, brachial interarea, showing chilidial plates and cardinal process; ×2), Oklahoma.

Valcourea Raymond, Middle Ordovician. Convexi-concave shell; delthyrium generally covered by deltidium; quadrate muscle-scar area in pedicle valve. *V. strophomenoides* (Raymond) (4a, b, brachial view of shell and pedicle interior, ×2), Chazyan, New York.

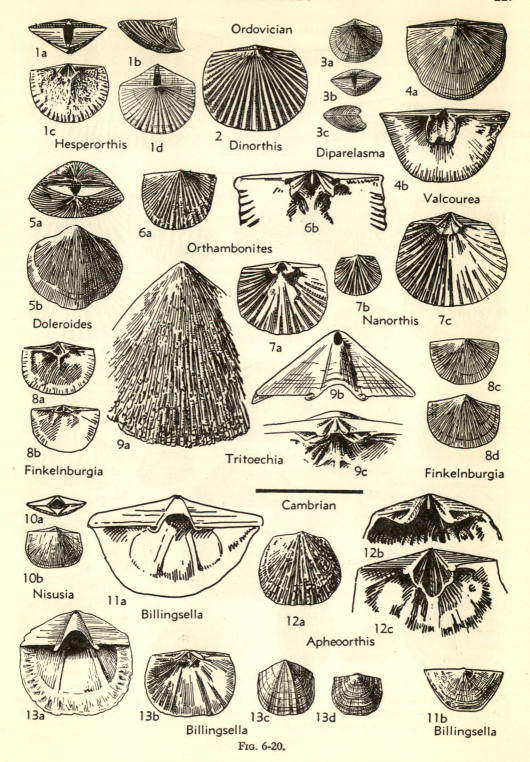

Ordovician

1a 1b 1c Hesperorthis 1d 2 Dinorthis

3a 3b 3c Diparelasma

4a 4b Valcourea

5a 5b Doleroides 6a Orthambonites 6b

7a 7b Nanorthis 7c

8a 8b Finkelnburgia 9a Tritoechia 9b 9c 8c 8d Finkelnburgia

Cambrian

10a 10b Nisusia 11a Billingsella 12a Apheoorthis 12b 12c

13a 13b Billingsella 13c 13d 11b Billingsella

Fig. 6-20.

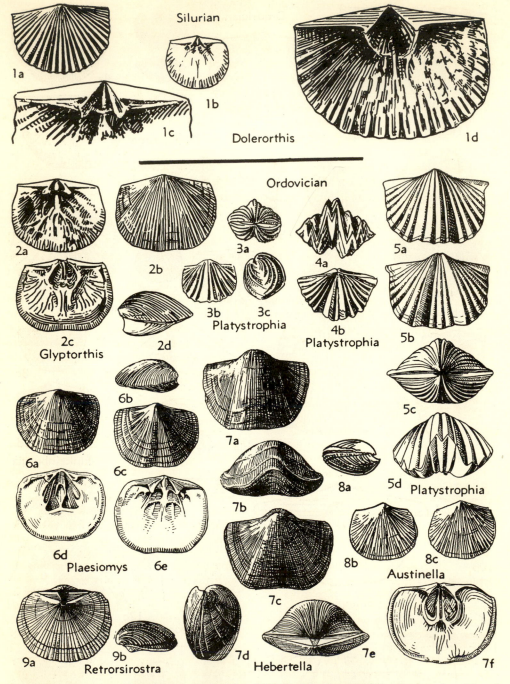

Fig. 6-21. Representative Upper Ordovician–Middle Silurian orthid articulate brachiopods belonging to the suborder Orthacea. A remnant of the orthacean stock persisted into Devonian time, but no post-Silurian forms are illustrated here.

(*Continued on next page.*)

Ordovician rocks, is characterized by wide hinge line and coarse angular plicae, which interlock along the anterior margins.

Silurian rocks contain several Upper Ordovician orthacean genera and a few new forms, one of which is a coarsely costate shell with a widely open delthyrium and typical rodlike cardinal process (Fig. 6-21, 1). Some orthaceans continue into the Devonian, but they are not known higher (Fig. 6-40).

Dalmanellacea. This important suborder, ranging from Middle Ordovician to the end of the Paleozoic, may be described simply and briefly as orthids having a punctate shell structure. Most of them have biconvex shells of rounded outline which bear fine radiating costae (Figs. 6-22, 6-23).

Typical Ordovician species of this group have an open delthyrium and prominent muscle scars in the pedicle valve and generally strong crural lamellae in the brachial valve (Fig. 6-22, 4–6).

The most common Silurian examples have nearly circular outlines and narrow obtuse-angle interareas on both valves (Fig. 6-22, 1, 3). One genus, which is well known in North America and Europe, is characterized by its small shell strongly constricted along the median axis and bearing a relatively large obtuse-angle interarea on the pedicle valve (Fig. 6-22, 2).

Devonian members of the Dalmanellacea are numerous and important. Several belong to long-ranging genera which have very finely costate shells of similar appearance, with large muscle scars on the interior of the valves (Fig. 6-23, 4, 5, 7). A useful guide fossil, restricted to the Devonian, is *Tropidoleptus*, distinguished by its concavo-convex shell, which bears broad low costae (Fig. 6-23, 6).

Among post-Devonian Dalmanellacea, one of the most distinctive genera, is *Enteletes* which first appears in Upper Pennsylvanian rocks (Fig. 6-23, 1). It has a globose shell with very short hinge line and small obtuse-angle interareas. The surface bears fine radial costae, and the marginal areas are coarsely plicate. Internal features include prominent dental and crural lamellae and a high median septum in the pedicle valve.

(Fig. 6-21 continued.)

Austinella Foerste, Upper Ordovician. Shell strongly biconvex; quadrate muscle area in pedicle valve. *A. kankakensis* (McChesney) (8*a–c*, side, pedicle, and brachial views, ×0.7), Illinois.

Dolerorthis Schuchert & Cooper, Lower and Middle Silurian. Convexi-plane coarsely costate shell; interior like *Hesperorthis*. *D. flabellites* (Foerste) (1*a*, pedicle exterior, ×1; 1*b, c*, brachial interior, ×0.7, ×2; 1*d*, pedicle interior, ×2), Niagaran, Ohio.

Glyptorthis Foerste, Middle Ordovician–Lower Silurian. Like *Dolerorthis* but shell biconvex and more finely costate. *G. insculpta* (Hall) (2*a, b*, brachial interior and exterior; 2*c*, pedicle interior; 2*d*, side view; ×1), Upper Ordovician (Richmondian), Ohio.

Hebertella Hall & Clarke, Middle and Upper Ordovician. Shell moderately large, pedicle valve nearly flat and sulcate, brachial valve distinctly convex. *H. sinuata* (Hall) (7*a–e*, brachial, anterior, pedicle, side, and posterior views; 7*f*, pedicle interior; ×0.7), Upper Ordovician, Kentucky.

Plaesiomys Hall & Clarke, Middle and Upper Ordovician. Biconvex finely costate shell; prominent muscle scars on interior. *P. subquadrata* (Hall) (6*a–c*, brachial, side, and pedicle views; 6*d, e*, pedicle and brachial interiors; ×0.7), Upper Ordovician, Ohio.

Platystrophia King, Middle Ordovician–Middle Silurian. Biconvex, strongly plicate shells, generally widest along hinge line; sulcus prominent on pedicle valve and fold on brachial valve. *P. crassa* James (3*a–c*, posterior, brachial, and side views, ×0.7), Upper Ordovician, Kentucky; *P. cypha* James (4*a, b*, anterior and brachial views, ×0.7), Upper Ordovician, Indiana; *P. ponderosa* Foerste (5*a–d*, brachial, pedicle, posterior, and anterior views, ×0.7), Upper Ordovician, Ohio.

Retrorsirostra Schuchert & Cooper, Upper Ordovician. Convexi-concave shell mainly distinguished by the strongly acute-angle interarea of the pedicle valve. *R. carleyi* (Hall) (9*a, b*, pedicle and side views, ×0.7), Kentucky.

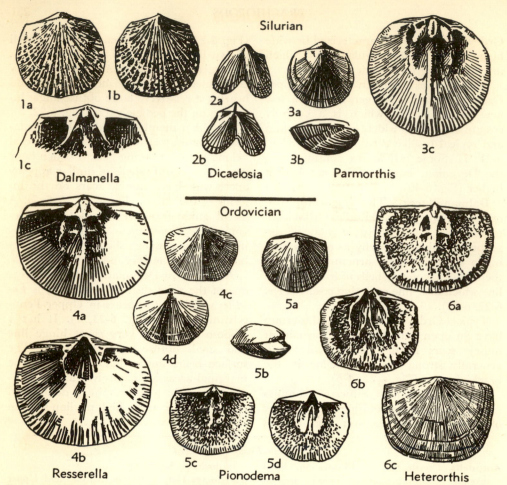

Silurian

1a 1b

1c

Dalmanella

2a

2b

Dicaelosia

3a

3b

Parmorthis

3c

Ordovician

4c

4a

4d

5a

5b

6a

6b

4b

5c

5d

6c

Resserella   Pionodema   Heterorthis

FIG. 6-22. **Representative Ordovician and Silurian orthid articulate brachiopods belonging to the suborder Dalmanellacea.** This group corresponds closely in appearance and general structural features to the Orthacea but differs in having punctate shells. Dalmanellacean genera are distributed from Middle Ordovician to Upper Permian.

Dalmanella Hall & Clarke, Lower Silurian. Biconvex, rather coarsely costate shells with rounded outline; cardinal process lobed. *D. edgewoodensis* Savage (1*a*, *b*, brachial and pedicle views, ×2; 1*c*, brachial interior, ×4), Illinois.

Dicaelosia King, Upper Ordovician–Lower Devonian. Small shells having a distinctive median constriction. *D. biloba* (Linné) (2*a*, *b*, pedicle and brachial views, ×1), Niagaran, Indiana.

Heterorthis Hall & Clarke, Middle Ordovician. Moderately large plano-convex, finely costate shells; strong paired muscle scars on pedicle interior. *H. clytie* (Hall) (6*a*, *c*, brachial interior and exterior; 6*b*, pedicle interior; ×1), Trentonian, Kentucky.

Parmorthis Schuchert & Cooper, Middle Silurian. Biconvex to plano-convex, finely costate shell; pedicle interior has deep delthyrial cavity and stout dental lamellae. *P. waldronensis* (Foerste) (3*a*, *b*, brachial and side views, ×1; 3*c*, brachial interior, ×2), Indiana.

Pionodema Foerste, Middle Ordovician–Lower Silurian. Semielliptical biconvex, finely costate shells; strong crural lamellae and crural bases in brachial interior. *P. subaequata* (Conrad) (5*a*, *b*, brachial and side views; 5*c*, *d*, brachial and pedicle interiors; ×1), Middle Ordovician, Minnesota.

Resserella Bancroft, Middle Ordovician–Lower Silurian. Pedicle valve more convex than brachial, surface finely costate; internal features distinctive. *R. meeki* (Miller) (4*a*–*b*, brachial and pedicle interiors, ×2; 4*c*, *d*, pedicle and brachial views, ×1), Upper Ordovician, Indiana.

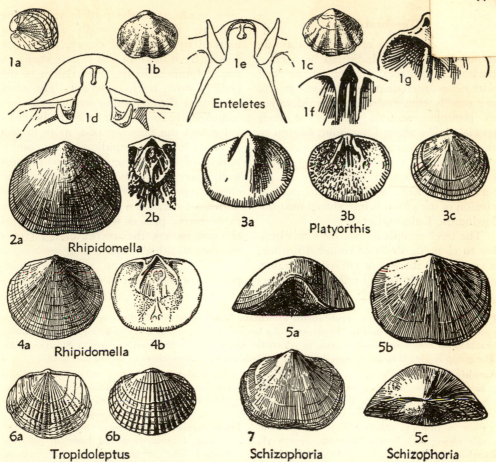

FIG. 6-23. **Representative Devonian–Pennsylvanian orthid articulate brachiopods belonging to the suborder Dalmanellacea.** The shells illustrated are geologically younger members of the group shown in Fig. 6-22.

Enteletes Fischer, Pennsylvanian–Permian. Coarsely plicate, subglobular shell ornamented by fine radial costate; thin dental lamellae and median septum in pedicle valve. *E. hemiplicatus* (Hall) (1*a–c*, side, brachial, and pedicle views, ×1; 1*d, e*, brachial interior, view normal to commissure and tipped backward, showing cardinal process, curved crural processes, and crural lamellae, ×5; 1*f, g*, pedicle interior, ×2), Upper Pennsylvanian (Missourian), Kansas.

Platyorthis Schuchert & Cooper, Silurian–Devonian. Plano-convex shells with subcircular outline; large muscle area in pedicle valve. *P. planoconvexa* (Hall) (3*a*, pedicle internal mold; 3*b*, brachial interior; 3*c*, brachial view; ×1), Lower Devonian, New York.

Rhipidomella Oehlert, Middle Silurian–Permian. Compressed biconvex, finely costate

shells with subcircular outline; large muscle area in pedicle valve. *R. penelope* (Hall) (4*a, b*, pedicle exterior and interior, ×0.7), Middle Devonian, New York; *R. oweni* Hall & Clarke (2*a, b*, pedicle exterior and interior, ×1), Lower Mississippian, Kentucky.

Schizophoria King, Middle Silurian–Permian. Brachial valve more convex than pedicle, finely costate. *S. australis* Kindle (5*a–c*, anterior, pedicle, and posterior views, ×0.7), Upper Devonian, New Mexico; *S. nevadensis* Merriam (7, pedicle view, ×0.7), Lower Devonian, Nevada.

Tropidoleptus Hall, Devonian. Concavo-convex shells with elliptical outline, bearing broad low costae. *T. carinatus* (Conrad) (6*a, b*, brachial and pedicle views, ×0.7), Middle Devonian, New York.

Terebratulids

The origin of the order Terebratulida is not at all clearly indicated by a comparison of structural features of the shell with other antecedent groups. Some groups are readily eliminated from consideration as possible ancestors because of their specialization along lines quite foreign to those of the terebratulids. The dalmanellacean branch of the orthids is judged most probably to include progenitors of the terebratulid line; and accordingly, discussion of the group is introduced next following the Dalmanellacea.

The terebratulids mostly have subcircular to elongate shells of rounded outline, a very short, straight or curved hinge line, and a small obtuse- to reflex-angle inter-

area, confined to the pedicle valve. The punctate shell is generally smooth or marked by fine radial costae, but a few genera have more or less plicate shells. A round foramen in the beak region of the pedicle valve is characteristic. The most important distinguishing features of terebratulids are the cardinalia in the brachial valve and the presence of a more or less complex looped brachidium. This combination of features sets the terebratulids well apart from other brachiopods. The oldest representatives appear in Upper Silurian beds; from this beginning, they range to the present day. In modern brachiopod faunas, the terebratulids far outnumber all other types.

Terebratulids are an important component of Devonian brachiopod assem-

---

Fig. 6-24. **Representative terebratulid articulate brachiopods.** This division is characterized by punctate shell structure and by the presence of a looped brachidium. It includes a majority of the living brachiopods and many fossil genera, of which the oldest known are from uppermost Silurian rocks. The group is judged to have been developed probably from dalmanellacean orthid ancestors.

Amphigenia Hall, Devonian. Elongate nearly smooth shells having a spondylium in the pedicle valve. *A. elongata* (Vanuxem) (12a, d, spondylium in pedicle valve and divided hinge plate in brachial valve; 12b, c, brachial and side views; ×1), Lower Devonian, New York.

Beachia Hall & Clarke, Lower Devonian. Valves equally convex; brachidium distinctive. *B. suessana* (Hall) (8a, c, section and brachial interior views showing brachidium; 8b, d, side and brachial views of exterior; ×0.7), New York.

Choristothyris Cooper, Upper Cretaceous. Subcircular biconvex shell with coarse plications and moderately large foramen. *C. plicata* (Say) (3, brachial view, ×1), New Jersey.

Cranaena Hall & Clarke, Middle Devonian–Mississippian. Short, strongly biconvex smooth shell; pedicle beak incurved. *C. romingeri* (Hall) (6a, b, side and brachial views, ×1), Middle Devonian, New York.

Cryptonella Hall, Lower and Middle Devonian. Elongate smooth shell, valves equally convex. *C. rectirostra* (Hall) (10a, b, brachial and side views, ×1), Middle Devonian, New York.

Dielasma King, Mississippian–Permian. Elongate gently biconvex smooth shell; dental lamellae in pedicle valve. *D. illinoisense* Weller (7a, b,

side and brachial views, ×1), Upper Mississippian (Chesteran), Illinois; *D. bovidens* (Morton) (5a–c, side, brachial, and pedicle views, ×0.7), Upper Pennsylvanian, Texas.

Etymothyris Cloud, Lower and Middle Devonian. Prominent dental lamellae converge toward floor of pedicle valve. *E. gaspensis* (Clarke) (9a, b, brachial and side views, ×0.7), Lower Devonian, Quebec.

Heterelasma Girty, Permian. Small shell; brachidium is a reflexed loop. *H.* sp. (4a, b, brachial and side views, ×1; 4c, brachial interior showing loop, ×4), Texas.

Kingena Davidson, Lower Cretaceous. Oval outline, pedicle beak incurved. *K. wacoensis* (Roemer) (1a, b, side and brachial views, ×1), Texas.

Plectoconcha Cooper, Triassic. Strongly biconvex shell, anteriorly costate; dental lamellae lacking. *P. aequiplicata* (Gabb) (2a, c, brachial and side views, ×1; 2b, brachial interior, ×2), California.

Rensselaeria Hall, Lower Devonian. Moderately large, narrowly elongate shell, surface costate. *R. marylandica* (Hall) (11a, b, brachial interior and section showing brachidium, ×0.7), Maryland; *R. elongata* (Conrad) (13a, b, side and brachial views, ×0.7), Tennessee.

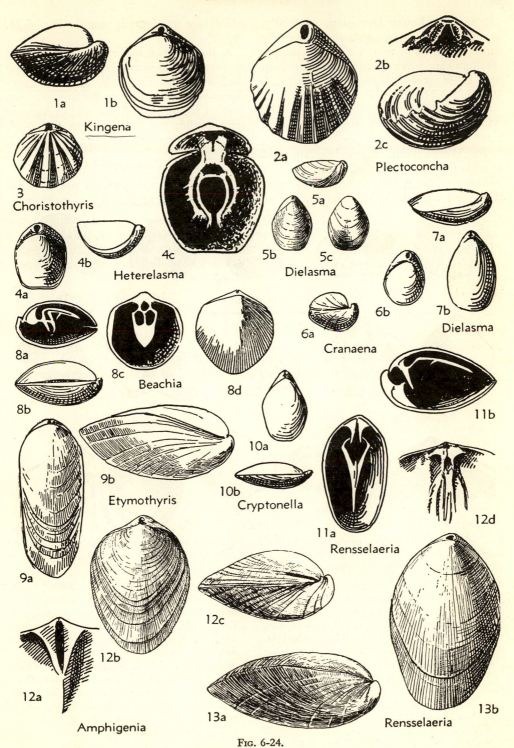

1a 1b
Kingena

2a
2b
2c
Plectoconcha

3
Choristothyris

4a
4b
4c
Heterelasma

5a
5b 5c
Dielasma

6a
6b
Cranaena

7a
7b
Dielasma

8a
8b
8c
8d
Beachia

9a
9b
Etymothyris

10a
10b
Cryptonella

11a
11b
Rensselaeria

12a
12b
12c
12d
Amphigenia

13a
13b
Rensselaeria

Fig. 6-24.

blages. The Lower Devonian, especially, is characterized by several moderately large, elongate genera, which are mainly distinguished by the characters of the loop and other internal features (Fig. 6-24, *8, 9, 11–13*). The exterior of these shells is marked by faint to distinct, fine costae. Much smaller, smooth-shelled Devonian terebratulids also occur (Fig. 6-24, *6, 10*).

In Mississippian, Pennsylvanian, and Permian rocks, terebratulids are a minor element of faunas; but there are several described smooth-shelled genera, all of which are moderately small (Fig. 6-24, *4, 5, 7*).

A large expansion of the terebratulids occurred in Mesozoic time, and despite decline in the Cenozoic Era, they predominate among living brachiopods (Fig. 6-40). Examples of Triassic and Cretaceous terebratulids are illustrated in Fig. 6-24, *1–3*, and common modern types in Fig. 6-2.

Pentamerids

The pentamerids are an order of biconvex impunctate-shelled brachiopods,

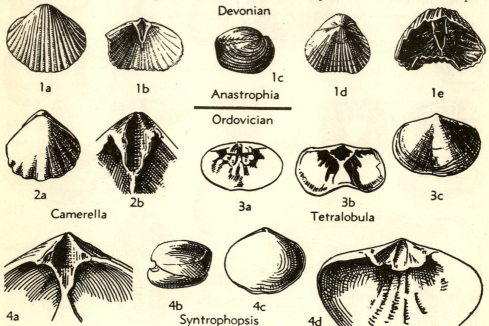

FIG. 6-25. **Representative pentamerid articulate brachiopods belonging to the suborder Syntrophiacea.** Shells classed in this group range from Upper Cambrian to Lower Devonian. They are biconvex, have impunctate shell structure, and generally possess a spondylium.

Anastrophia Hall, Silurian–Devonian. Costate subglobular shell; prominent spondylium in pedicle valve and crural lamellae in brachial valve. *A. verneuili* (Hall) (1*a, c, d*, pedicle, side, and brachial views; 1*b*, pedicle interior; 1*e*, interior of beak region of both valves, showing spondylium, below, and crural lamellae, above; ×0.7), Lower Devonian, Tennessee.

Camerella Billings, Lower and Middle Ordovician. Shell anteriorly costate, large spondylium in pedicle valve. *C. plicata* (Schuchert & Cooper) (2*a*, brachial view, ×1; 2*b*, pedicle interior, ×4), Middle Ordovician, Tennessee.

Syntrophopsis Ulrich & Cooper, Lower Ordovician. Shell wider than long, smooth; spondylium well developed. *S. magna* Ulrich & Cooper (4*a, d*, pedicle and brachial interiors, ×2; 4*b, c*, side and brachial views, ×1), Arkansas.

Tetralobula Ulrich & Cooper, Lower Ordovician. Shell short, wide, with fold and sulcus; brachial interior contains quadripartite muscle platform. *T. delicatula* Ulrich & Cooper (3*a, b*, brachial and pedicle interiors; 3*c*, pedicle exterior; ×2), Alabama.

which undoubtedly developed in Middle Cambrian time or earlier from the Orthacea. Small obtuse-angle interareas are developed on both valves behind the short hinge line, in some shells nearly concealed by the beaks or actually reduced to point of disappearance. Openings beneath the beaks (delthyrium, notothyrium), wherever observed, are uncovered. One of the chief distinguishing characters of the group is a well-developed spondylium in the pedicle valve. A similar structure in the brachial valve, made by convergence and union of crural lamellae, is present in some shells but not in others. The order ranges from Middle Cambrian through the Devonian (Fig. 6-40).

Syntrophiacea. The more primitive and older pentamerids are grouped in the suborder Syntrophiacea (Fig. 6-25), represented by 4 described Upper Cambrian genera (one of which occurs also in Middle Cambrian rocks) and at least 11 species. Rapid expansion in the Lower Ordovician is indicated by the presence in these rocks of some 17 genera and 86 species. Many of these genera and some additional ones are found in Middle and Upper Ordovician rocks, of which a few range upward to Middle Devonian. The group is most abundant in the Lower Ordovician and is specially characteristic of this division.

Syntrophiacean brachiopods have a less well-developed spondylium generally than that observed in the Pentameracea, from which they are most clearly distinguished by the characters of the brachial interior. Crural lamellae are weakly to fairly well developed in the brachial valve of Syntrophiacea, but in none of them are muscle scars on the floor of the valve enclosed by these plates. Important Ordovician genera have smooth to strongly costate shells, which generally are wider than long (Fig. 6-25, *2–4*). Some Siluro-Devonian shells have plicate valves, a well-developed spondylium in the pedicle valve, and long crural lamellae in the brachial valve (Fig. 6-25, *1*).

Pentameracea. Silurian and Devonian pentamerids, which mostly are distinctly larger than the syntrophiaceans, belong to the suborder Pentameracea (Fig. 6-26). Elongated shells are more common than transverse forms, and plication of all or part of the valves characterizes many species. Some, however, like *Pentamerus*, have smooth shells. Especially prominent is the large spondylium inside the pedicle valve. Some genera are distinguished by prominence of the pedicle beak which arches over the opposite valve, and they differ from most brachiopods in having a sulcus on the brachial valve and a fold on the pedicle valve (Fig. 6-26, *1, 2, 4, 5*).

Triplesiids

A small group of Ordovician and Silurian smooth-shelled or plicated brachiopods is set apart from others as an independent order called Triplesiida (Fig. 6-27). They have a short hinge line and normally have a small obtuse-angle interarea on each valve. The biconvex shell has a rounded outline. A fold and sulcus are well developed. The smooth or costate shell is impunctate. Internal features of the triplesiids do not permit classifying them with any other of the major brachiopod groups. The most significant structures are in the brachial valve, which has short projecting brachiophores or crura, and a very prominent bifurcate cardinal process. There is no sign of a spiral or other sort of brachidium.

Rhynchonellids

A very distinct group of small- to medium-sized, biconvex plicated shells comprises the order Rhynchonellida (Figs. 6-28, 6-29). They have a subtriangular to rounded outline, and many are decidedly globose. The hinge line is very short, and because the pointed pedicle beak generally is prominent, the shells are termed rostrate (*rostrum*, beak). A preponderant majority of these brachiopods have impunctate shells, but a division which is distinguished by very well-defined punc-

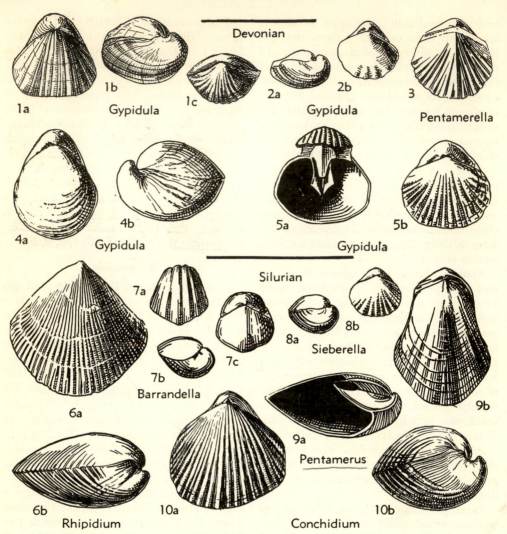

FIG. 6-26. **Representative pentamerid articulate brachiopods belonging to the suborder Penta-meracea.** Internal structure of shells in this group is similar to that of the Syntrophiacea but more advanced. Known pentameracean brachiopods are confined to Silurian and Devonian rocks.

Barrandella Hall & Clarke, Middle Silurian. Subglobular small shells having a fold and sulcus; crural lamellae in the brachial interior unite with a median septum. *B. fornicata* (Hall) (7a–c, pedicle, brachial, and side views, ×1), New York.

Conchidium Hisinger, Middle and Upper Silurian. Moderately large costate shells which are like *Pentamerus* internally. *C.* sp. (10a, b, brachial and side views, ×0.7), Tennessee.

Gypidula Hall, Middle Silurian–Upper Devonian. Strongly biconvex shells with subcircular to elongate outline, distinctly costate or nearly

smooth; pedicle beak strongly incurved. *G. coeymanensis* Schuchert (1a–c, brachial, side, and posterior views, ×0.7), Lower Devonian, New York; *G. comis* (Owen) (2a, b, side and brachial views, ×0.7), Middle Devonian, Iowa; *G. pseudogaleata* (Hall) (4a, b, brachial and side views, ×0.7), Lower Devonian, New York; *G. romingeri* Hal & Clarke (5a, interior showing underside of spondylium, above, and crural lamellae joined to floor of brachial valve, below; 5b, brachial view, ×0.7), Middle Devonian, Michigan.

Pentamerella Hall, Middle and Upper Devon-

(Continued on next page.)

tate shell structure is set apart as a sub-order (Rhynchoporacea). Diminutive in-terareas are present on one or both valves in some genera but generally cannot be discerned. This group of brachiopods lacks a calcareous brachidium of looped or spiral form, but among modern rhynchonellids are unsupported anteriorly directed spiral lophophores (Fig. 6-6, *3a*, *b*). The rhyncho-nellids range from Middle Ordovician to the present time. They are most abundant as fossils in Mesozoic rocks.

Rhynchonellacea. Impunctate rhyn-chonellids, classed as the suborder Rhyn-chonellacea, have the general characters already described. The essential feature is lack of puncta in the shell.

Among numerous Middle and Upper Ordovician rhynchonellids (Fig. 6-28, *13–17*), genera are distinguished partly on shell form but more on internal features, such as muscle areas, the presence or ab-sence of dental and crural lamellae, crural processes, and similar features. Compari-son of the posterior portion of the interior of valves shows readily perceived differ-ences. Deltidial plates largely close the delthyrium of some shells. The Ordovician rhynchonellids include useful guide fossils.

Among Silurian genera are shells char-acterized by the attenuated, sharply pointed posterior part of the shell (Fig. 6-28, *9–12*).

Devonian rocks include a large variety of rhynchonellids, among which most spe-cies have short stratigraphic range and some unusually wide geographic distribu-tion. For instance, closely similar or identi-cal species of *Hypothyridina* (Figs. 6-11, *8;* 6-28, *3a–c*) are found just below and above

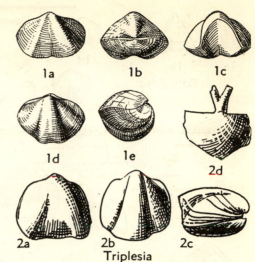

Triplesia

FIG. 6-27. **Representative triplesiid articulate brachiopods.** These have an impunctate shell. The pedicle valve bears a flat deltidium, and the brachial valve has a forked cardinal process. The group is confined to Ordovician and Silurian rocks.

Triplesia Hall, Middle Ordovician–Middle Si-lurian. Shell biconvex, prominent fold and sulcus; hinge line short *T. cuspidata* Clark (*1a–e*, pedicle, posterior, anterior, brachial, and side views, ×0.7), Middle Ordovician, New York; *T. ortoni* (Meek) (*2a–c*, pedicle, brachial, and side views, ×0.7; *2d*, cardinal process from exterior, ×1.5), Lower Silurian, Ohio.

the Middle Devonian–Upper Devonian boundary in many parts of the world. Several distinctive types of rhynchonellids occur in Lower Devonian rocks (Fig. 6-28, *2, 4, 5, 7*), and others which range into Mississippian deposits are found in the Upper Devonian (Fig. 6-28, *1, 6*).

(Fig. 6-26 continued.)

ian. Sulcus on pedicle valve, fold on dorsal valve; interior as in *Gypidula. P. arata* (Conrad) (3, brachial view, ×1), Lower Devonian, Ohio.

Pentamerus Sowerby, Middle Silurian. Large, smooth, elongate shell; large spondylium and long crural plates. *P. laevis* Sowerby (*9a*, section view, pedicle valve below, showing spondylium supported by septum; *9b*, brachial view; ×0.5), Illinois.

Rhipidium Schuchert & Cooper, Middle Si-lurian. Differs from *Pentamerus* in being radially costate. *R. knappi* (Hall & Whitfield) (*6a*, *b*, brachial and side views, ×0.7), Tennessee.

Sieberella Oehlert, Silurian–Devonian. Differs from *Gypidula* in nature of crural lamellae, which unite with a median septum. *S. roemeri* Hall & Clarke (*8a*, *b*, side and brachial views, ×0.7), Middle Silurian, Oklahoma.

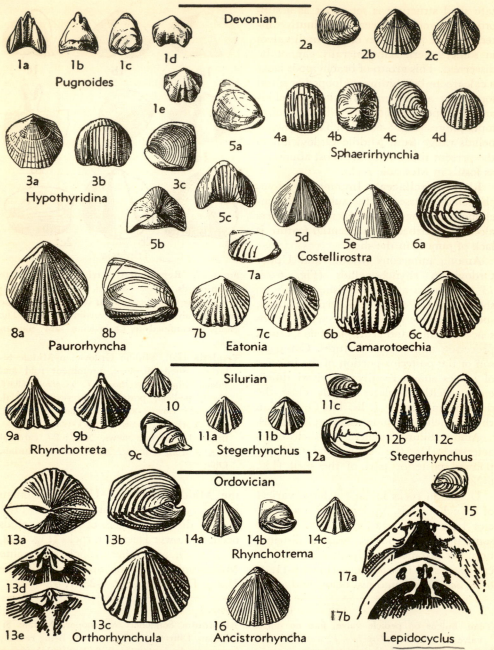

FIG. 6-28.   **Representative Ordovician–Devonian rhynchonellid articulate brachiopods belonging to the suborder Rhynchonellacea.** This group has impunctate shells with a narrow hinge line and relatively prominent beaks. Rounded or angular unbranched costae and the presence of a prominent fold and sulcus are characteristic. Stratigraphic range is from Middle Ordovician to Recent.

Ancistrorhyncha Ulrich & Cooper, Middle Ordovician. Costae relatively fine. *A. costata* Ulrich & Cooper (16, brachial view, ×2), Tennessee.

Camarotoechia Hall & Clarke, Devonian–Mississippian. Pedicle interior has strong dental

(Continued on next page.)

Rhynchonellids are very abundant in some Mississippian formations but a distinctly subordinate element in brachiopod faunas of other deposits (Fig. 6-29, 4-7). In oölitic limestones particularly, rhynchonellids may be almost the only type of brachiopods found, and their shells may be very numerous. They preponderate also in some shaly deposits, especially black shales which lack a diversified marine fauna. These observations indicate that at least some of the rhynchonellids were well adapted to environments in which other brachiopods could not get along well.

Pennsylvanian and Permian rhynchonellids are mostly minor constituents of brachiopod faunas from these rocks. Individuals of some species (Fig. 6-29, 3) locally exceed all other brachiopods, however. There are several kinds of Permian rhynchonellids, one of the larger of which is *Stenoscisma* (Fig. 6-29, 1), distinguished by a prominent spondylium in the pedicle valve and by unusual internal features of the brachial valve.

Rhynchoporacea. As shown by illustrated species of *Rhynchopora* (Fig. 6-29, 8, 9), these shells are indistinguishable externally from some genera of rhynchonellids, and they have also virtually identical internal structures. The shell of *Rhynchopora* is perforated by numerous puncta, however, and this furnishes fully sufficient basis for putting them in a separate category from other members of the Rhynchonellida. The known range of Rhyncho-

---

(Fig. 6-28 continued.)

lamellae. *C. congregata* (Conrad) (6*a–c*, side, anterior, and brachial views, ×1), Middle Devonian, Pennsylvania.

Costellirostra Cooper, Lower Devonian. Differs from *Eatonia* in costae, cardinal process, and muscle areas. *C. tennesseensis* (Dunbar) (5*a–e*, side, posterior, anterior, pedicle, and brachial views, ×0.7), Tennessee.

Eatonia Hall, Lower Devonian. Dental lamellae absent; brachial interior has median septum and myophore. *E. medialis* (Hall) (7*a–c*, side, pedicle, and brachial views, ×0.7), New York.

Hypothyridina Buckman, Middle and Upper Devonian. Cuboid to globular shell, anterior valve margins strongly sinuate; an important guide fossil. *H. venustula* (Hall) (3*a–c*, brachial, anterior, and side views, ×0.7), Middle Devonian, New York.

Lepidocyclus Wang, Upper Ordovician. Like *Rhynchotrema* except for internal features. *L.* sp. (17*a, b*, pedicle and brachial interiors, ×2).

Orthorhynchula Hall & Clarke, Middle and Upper Ordovician. Shell has small interarea. *O. linneyi* (James) (13*a–c*, posterior, side, and brachial views, ×1; 13*d, e*, brachial interiors, ×2), Upper Ordovician, Ohio.

Paurorhyncha Cooper, Upper Devonian. Prominent fold and sulcus; brachial hinge plate divided and supported by median septum. *P.* sp. (8*a, b*, brachial and side views, ×0.7), New Mexico.

Pugnoides Weller, Upper Devonian–Upper Mississippian. Plications weak or lacking near beaks; fold and sulcus prominent. *P. calvini* Fenton & Fenton (1*a–e*, anterior, posterior, side, pedicle, and brachial views, ×0.7), Upper Devonian, Iowa.

Rhynchotrema Hall, Middle and Upper Ordovician. Concave deltidial plates on pedicle valve; divided hinge plate and thick median septum in brachial valve. *R. dentatum* Hall (14*a–c*, pedicle, side, and brachial views, ×0.7), Upper Ordovician (Richmondian), Indiana; *R. argenturbicum* (White) (15, side view, ×0.7), Upper Ordovician, New Mexico.

Rhynchotreta Hall, Silurian. Subtriangular shell with pointed beaks; long dental lamellae in pedicle valve. *R. americana* Hall (9*a–c*, pedicle, brachial, and side views, ×1), Middle Silurian, Kentucky.

Sphaerirhynchia Cooper & Muir-Wood, Middle Silurian–Lower Devonian. Subglobular shell distinguished by large oval muscle scar in pedicle valve and small cruralium in brachial valve. *S. nucleolatus* (Hall) (2*a–c*, side, brachial, and pedicle views, ×0.7), Lower Devonian, Oklahoma; *S. ventricosus* (Hall) (4*a–d*, anterior, posterior, side, and brachial views, ×0.7), Lower Devonian, New York.

Stegerhynchus Foerste, Middle Silurian. Like *Camarotoechia* except in having a thin lamellar cardinal process. *S. indianense* (Hall) (10, pedicle view, ×0.7), Indiana; *S. whitei* (Hall) (11*a–c*, pedicle, brachial, and side views, ×0.7), Indiana; *S. acinus* (Hall) (12*a–c*, side, pedicle, and brachial views, ×1), Indiana.

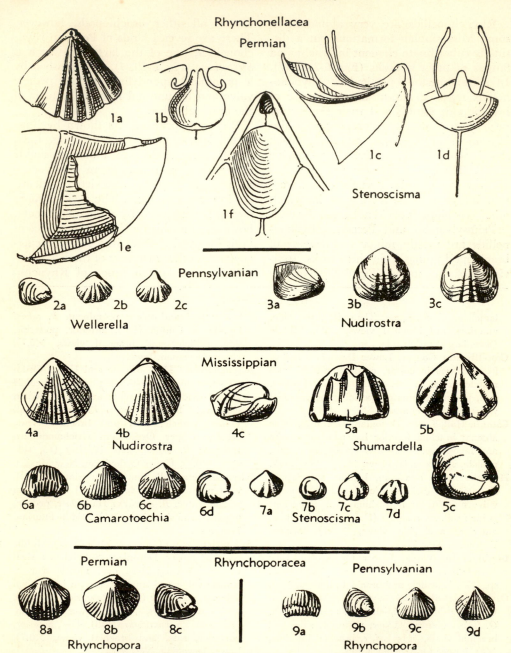

FIG. 6-29. **Representative Mississippian–Permian rhynchonellid articulate brachiopods belonging to the suborders Rhynchonellacea and Rhynchoporacea.** Shells shown in the bottom row (8*a–c*, 9*a–d*) have punctate structure, which distinguishes the Rhynchoporacea. The others are impunctate shells and are geologically younger members of the group illustrated in Fig. 6-28.

Rhynchonellacea

Camarotoechia Hall & Clarke, Devonian–Mississippian. *C. mutata* (Hall) (6*a–d*, anterior, brachial, pedicle, and side views, ×1), Upper Mississippian (Meramecian), Indiana.

Nudirostra Cooper & Muir-Wood, Middle De-

(Continued on next page.)

poracea is from Mississippian to Permian (Fig. 6-40).

Strophomenids

The chief distinguishing feature of the Strophomenida, which sets this order apart from all other brachiopods, is the pseudo-punctate structure of the shells (Fig. 6-7B). In addition, this important group is rather easily distinguished on the basis of external form (Figs. 6-30 to 6-34). Generally, the hinge line is extended so that it equals the greatest width. Commonly, but by no means prevailingly, the shell width exceeds the length. A characteristic feature, also, is that one valve is convex and the other concave. The surface is marked by radial costae or spines or both. The strophomenids range from Lower Ordovician to the present day but are mainly a Paleozoic assemblage. Undoubtedly, they are derivatives of the early Orthacea.

Strophomenacea. The oldest and longest ranging division of the Strophomenida is the suborder Strophomenacea. Genera of this group are distinguished from the somewhat more specialized, and in part distinctly aberrant, division called Productacea in having well-defined inter-areas on each valve and in lacking spinose projections. Most species have a short stratigraphic range, and many of them are important guide fossils.

Described Lower Ordovician representatives of Strophomenacea include some 14 species assigned to 8 genera. Among these, some occur also in Middle Ordovician rocks (Fig. 6-30, *1, 3, 10*). The expansion of this stock in Middle Ordovician time is remarkable. Not only are genera and species very numerous, but individuals belonging to the group occur by the millions in many Middle and Upper Ordovician formations. Most shells are rather small, finely costate, concavo-convex types (Fig. 6-30, *3–7, 9, 10*), but some are relatively large (Fig. 6-30, *8*). Genera are distinguished chiefly by internal features, of which the most important are markings on the brachial valve. Probably, as inferred from studies by Kozlowski (1929), the irregularities of the brachial valve floor reflect placement of the brachia or lophophore. At any rate, they are constantly distinguishing features. Such impressions, denoting looped and spiral patterns of the brachia, are found among strophomenid brachiopods (Fig. 6-6, *4, 5*). A few species of Ordovician Strophome-

---

(Fig. 6-29 continued.)

vonian–Pennsylvanian. Relatively large, weakly plicate shells. *N. carbonifera* (Girty) (4*a–c*, pedicle, brachial, and side views, ×0.7), Upper Mississippian, Arkansas; *N. rockymontana* (Marcou) (3*a–c*, side, pedicle, and brachial views, ×0.7), Upper Pennsylvanian (Missourian), Oklahoma.

Shumardella Weller, Mississippian. Weakly plicate shells distinguished from *Nudirostra* by internal features. *S. missouriensis* (Shumard) (5*a–c*, anterior, pedicle, and side views, ×1), Lower Mississippian, Missouri.

Stenoscisma Conrad, Ordovician-Permian. A spondylium supported by a low median septum occurs in the pedicle valve; the brachial valve bears a shallow spoon-shaped structure on a high median septum and long, delicate crural processes. *S. venustum* (Girty) (1*a*, brachial view, ×1; 1*b–d*, interior of brachial beak, approximately ×5; 1*e–f*, spondylium, approximately

×5), Permian, Texas; *S. explanatum* (McChesney) (7*a–d*, pedicle, side, brachial, and anterior views, ×1), Upper Mississippian (Chesteran), Illinois.

Wellerella Dunbar & Condra, Pennsylvanian–Permian. Like *Pugnoides* externally, but brachial valve has undivided hinge plate. *W. osagensis* (Swallow) (2*a–c*, side, brachial, and pedicle views, ×1), Upper Pennsylvanian, Kansas.

Rhynchoporacea

Rhynchopora King, Mississippian–Permian. Almost identical externally and internally to *Camarotoechia*, except for perforation of the shell by many coarse puncta. *R. taylori* Girty (8*a–c*, pedicle, brachial, and side views, ×1), Permian, Texas; *R. magnicosta* Mather (9*a–d*, anterior, side, brachial, and pedicle views, ×0.7), Lower Pennsylvanian (Morrowan), Arkansas.

nacea differ from the majority in having a convexi-concave shell (Fig. 6-30, *2, 4, 11*). They are abundant in some Middle and Upper Ordovician formations.

Most Silurian representatives of the Strophomenacea externally resemble the Ordovician concavo-convex shells, but some have gently convex, nearly flat pedicle and brachial valves (Fig. 6-31, *10, 11*).

Strophomenacea are a prominent constituent of many Devonian brachiopod assemblages. Most of them are finely costate concavo-convex shells of small to large size (Fig. 6-31, *1, 4, 5, 8, 9*), among which are some with finely denticulate hinge margins. Biconvex and convexi-concave shells are also common, and many are characterized by large muscle scars (Fig. 6-31, *2, 6, 7*).

Mississippian Strophomenacea include the last representatives of the long-ranging *Leptaena* (Fig. 6-31, *3*) and other genera. Except in a few formations, this element of the brachiopod fauna is not especially important.

Pennsylvanian and Permian Strophomenacea include shells characterized by rounded outline, fine costae, and prominent pedicle interarea; in some, the surfaces of both valves are marked by radial corrugations as well as costae (Fig. 6-32, *1, 3, 4*).

Undoubtedly, most specialized among all strophomenacean genera is *Leptodus*,

---

FIG. 6-30. **Representative Ordovician strophomenid articulate brachiopods belonging to the suborder Strophomenacea.** The strophomenids are differentiated from all other groups in having pseudopunctate shell structure. The Strophomenacea mostly have greatest width along the hinge line and possess one convex valve and one concave valve. A deltidium and chilidium generally are well developed. The group ranges from Lower Ordovician to Recent, but is most prominent in older Paleozoic rocks.

Bimuria Ulrich & Cooper, Middle Ordovician. Small concavo-convex shells; floor of brachial valve bears two long septa closely adjoining a thin median septum. *B. superba* Ulrich & Cooper (6*a*, brachial view, ×1; 6*b*, brachial interior, ×2), Virginia.

Christiania Hall & Clarke, Middle and Upper Ordovician. Small quadrate to elongate, concavo-convex shells; brachial interior bears four prominent longitudinal ridges. *C. subquadrata* (Hall) (5*a*, pedicle interior, ×2; 5*b, c*, posterior and normal views of brachial interior, ×2; 5*d*, brachial view, ×1), Middle Ordovician, Tennessee.

Leptellina Ulrich & Cooper, Lower and Middle Ordovician. Subquadrate concavo-convex shell; prominent median ridge on floor of brachial valve. *L. tennesseensis* Ulrich & Cooper (1*a–c*, brachial pedicle, and posterior brachial interior views, ×2), Virginia.

Öpikina Salmon, Middle Ordovician. Small concavo-convex shells; brachial interior bears two longitudinal ridges on each side of median septum. *Ö. septata* Salmon (7*a*, brachial view, ×1; 7*b*, brachial interior, ×2), Tennessee.

Rafinesquina Hall & Clarke, Middle and Upper Ordovician. Relatively large concavo-convex finely costate shell; prominent muscle area in pedicle valve and short median ridge in brachial valve. *R. loxorhytis* (Meek) (8*a, b*, brachial and pedicle views, ×0.7; 8*c*, pedicle interior, ×1), Upper Ordovician (Richmondian), Indiana.

Sowerbyella Jones, Lower to Upper Ordovician. Medium small concavo-convex shell with uneven-sized fine costae; brachial interior has divergent longitudinal ridges but no median septum. *S. punctostriata* (Mather) (3*a–c*, pedicle and brachial interiors, brachial view, ×1; 3*d*, brachial exterior, ×2), Trentonian, New York; *S. clarksvillensis* (Foerste) (10, brachial interior, ×2), Richmondian, Ohio.

Sowerbyites Teichert, Middle Ordovician. Like *Sowerbyella* externally but brachial interior has median septum flanked by septum on each side. *S. triseptatus* Willard (9*a, c*, pedicle and brachial interiors, ×2; 9*b*, brachial view, ×1), Pennsylvania.

Strophomena de Blainville, Middle and Upper Ordovician. Convexi-concave shell which internally is like *Rafinesquina*. *S. planoconvexa* Hall (2*a–c*, side, pedicle, and brachial views, ×0.7); *S. nutans* (Meek) (4, pedicle view, ×0.7); *S. neglecta* James (11*a–c*, pedicle interior, pedicle, and brachial views, ×0.7); all Upper Ordovician, Ohio.

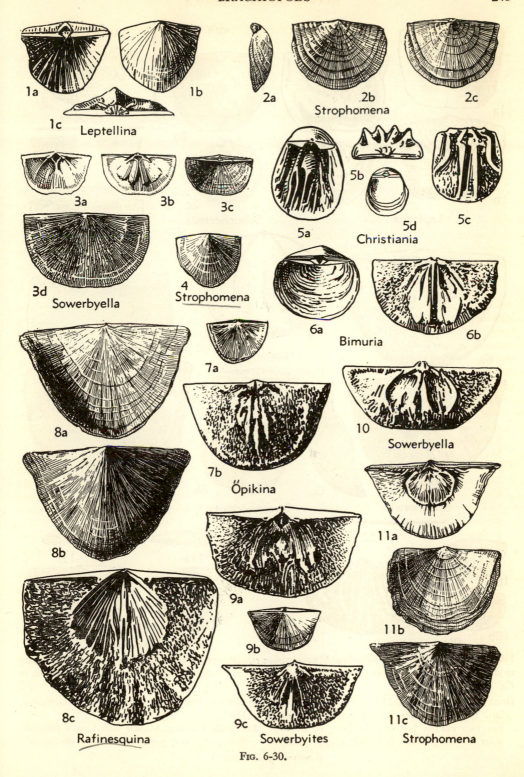

1a 1b 1c Leptellina

2a 2b 2c Strophomena

3a 3b 3c 3d Sowerbyella

4 Strophomena

5a 5b 5c 5d Christiania

6a Bimuria 6b

7a 7b Öpikina

8a 8b 8c Rafinesquina

9a 9b 9c Sowerbyites

10 Sowerbyella

11a 11b 11c Strophomena

Fig. 6-30.

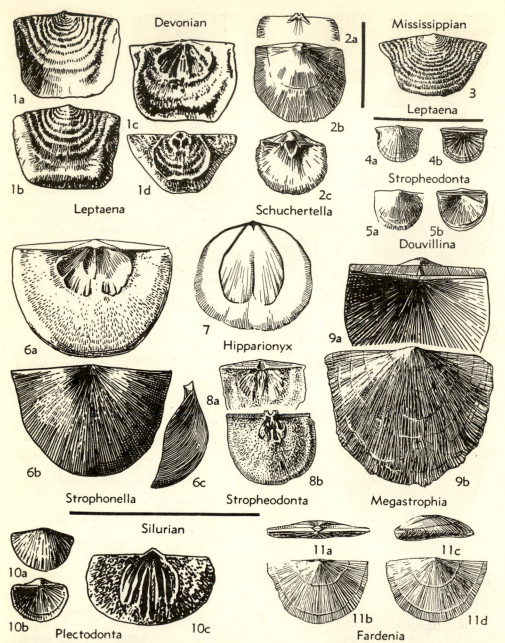

FIG. 6-31. **Representative Silurian–Mississippian strophomenid articulate brachiopods belonging to the suborder Strophomenacea.** This assemblage is continuation of the Ordovician group shown in Fig. 6-30.

Douvillina Oehlert, Middle and Upper Devonian. Concavo-convex shells; brachial interior has two curved lamellae in front of muscle scars. *D. inaequistriata* (Conrad) (5a, b, pedicle and brachial views, ✕0.7), Middle Devonian, Michigan.

Fardenia Lamont, Middle and Upper Silurian. Both valves very gently convex; has distinguish-

(*Continued on next page.*)

of Permian age (Fig. 6-32, 2). The pedicle valve somewhat resembles an oyster, and like the oyster, this brachiopod is attached by cementation of its pedicle valve to other shells. The interior of the pedicle valve bears a low median ridge, and on either side of it are grooves and ridges arranged normal to the median ridge. The brachial valve is extremely thin, and it is flat or somewhat concave, fitting as a lid over the pedicle valve. Slits extend laterally from the more solid central part, arranged in such manner that each slit overlies one of the ridges of the pedicle interior. Probably the unusual features of both valves are related to the pattern of a pair of long, convoluted brachia. The occurrence of *Leptodus* in Permian rocks of western Texas, but not in deposits of equivalent age in north central Texas, Kansas, and Nebraska, offers a problem because shallow seas connected these regions. Perhaps this genus was adapted to life on the flanks of reefs, such as existed in the Permian sea of western Texas but not in the other areas mentioned, or distribution may be influenced by temperature of the waters, or other environmental factors.

Productacea. The strophomenid branch called the suborder Productacea is characterized by concavo-convex shells accompanied by development of spines, either along the posterior margin, or distributed more or less abundantly over other parts of the shell surface (Figs. 6-33, 6-34). Interareas, although present in some shells, are not conspicuous features of any. The hinge line is generally extended so as to equal or nearly equal the greatest width of the shell. A prominent feature of many genera is the unusual convexity of the pedicle valve, accompanied by prolongation of the anterior portion of both valves so as to produce the feature called trail. In the region of the trails, the two valves are nearly or actually in contact when the valves are closed; and growth of the shell in this manner adds virtually nothing to the space of the shell interior.

Earliest representatives of the Productacea are found in Upper Ordovician rocks, but the group did not attain importance

---

(Fig. 6-31 continued.)

ing internal features. *F. subplana* (Conrad) (11*a–d*, posterior, brachial, side, and pedicle views, ×0.7), Middle Silurian, Indiana.

Hipparionyx Vanuxem, Lower Devonian. Biconvex shell with circular outline; very large muscle scar in pedicle valve. *H. proximus* Vanuxem (7, mold of pedicle interior, ×0.7), Lower Devonian, New York.

Leptaena Dalman, Middle Ordovician–Mississippian. Concavo-convex shell with abruptly bent anterior and lateral portions; surface marked by fine costae and concentric rugae. *L. rhomboidalis* Wilckens (1*a, b*, brachial and pedicle views; 1*c, d*, pedicle and brachial interiors; ×1), Middle Devonian, Michigan; *L. analoga* (Phillips) (3, pedicle view, ×0.7), Lower Mississippian, Missouri.

Megastrophia Caster, Middle Devonian. Large deeply concavo-convex shell with denticles along entire hinge line. *M. concava* (Hall) (9*a, b*, brachial and pedicle views, ×0.7), Pennsylvania.

Plectodonta Kozlowski, Middle Silurian. Like *Sowerbyella* externally, but brachial interior

bears three curved ridges on each side of low median septum. *P. transversalis* (Dalman) (10*a, b*, pedicle and brachial views, ×1; 10*c*, brachial interior, ×2), New York.

Schuchertella Girty, Lower Devonian–Permian. Shell irregular, biconvex to convexi-concave; no dental lamellae; cardinal and brachial processes fused. *S. woolworthana* (Hall) (2*a, c*, brachial and pedicle interiors; 2*b*, brachial view; ×0.7), Lower Devonian, New York.

Stropheodonta Hall, Devonian. Small to medium-sized concavo-convex shell with denticles along all of hinge line; large muscle scars. *S. erratica* Winchell (4*a, b*, pedicle and brachial views, ×0.7), Middle Devonian, Michigan; *S. demissa* (Conrad) (8*a, b*, pedicle and brachial interiors, ×0.7), Middle Devonian, New York.

Strophonella Hall, Middle Silurian–Lower Devonian. Large convexi-concave shell like *Strophomena* but has partly denticulate hinge line. *S. ampla* (Hall) (6*a*, pedicle interior; 6*b, c*, brachial and side views, ×0.7), Lower Devonian, New York.

until Late Devonian time. Marine deposits of Mississippian, Pennsylvanian, and Permian age throughout the world are especially characterized by the importance of this group of brachiopods; but no Mesozoic survivors are known.

The oldest and also longest-ranging segment of the Productacea is the stock consisting of *Chonetes* and closely allied genera, which range from Middle Silurian to Upper Permian (Figs. 6-33, *6–11;* 6-34, *5, 7*). These have shallow concavo-convex shells which externally closely resemble small Ordovician Strophomenacea. They are distinguished by their pseudopunctate shell structure, by spines along the posterior margin of the pedicle valve, pointing obliquely outward from the beak ridge

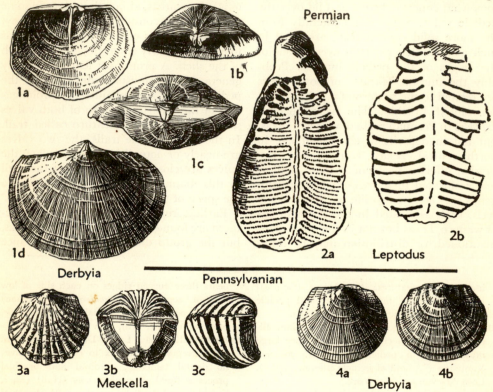

Fig. 6-32. **Representative Pennsylvanian–Permian strophomenid articulate brachiopods belonging to the suborder Strophomenacea.** Older fossils classed in this group are illustrated in Figs. 6-30 and 6-31.

Derbyia Waagen, Pennsylvanian–Permian. Shell somewhat irregular, biconvex or having pedicle valve nearly plane; pedicle interior has strong median septum but no dental lamellae. *D. cymbula* Hall & Clarke (1*a*, pedicle interior; 1*b*, posterior view of brachial valve showing forked cardinal process; 1*c, d*, posterior and brachial views; ×0.7), Lower Permian, Kansas; *D. crassa* (Meek & Hayden) (4*a, b*, brachial and pedicle views, ×0.7), Upper Pennsylvanian (Virgilian), Kansas.

Leptodus Kayser, Permian. Large elongate, concavo-convex shell, attached by cementation of posterior part of pedicle valve, producing irregularity of form; pedicle interior marked by long, low median ridge and many lateral grooves; brachial valve very thin, with slits located above ridges on pedicle interior. *L. americanus* (Girty) (2*a, b*, pedicle interior and brachial exterior, ×0.7), Texas.

Meekella White & St. John, Pennsylvanian–Permian. Medium-sized biconvex, pedicle beak commonly twisted; long thin dental lamellae and forked cardinal process. *M. striatocostata* (Cox) (3*a–c*, brachial, posterior, and side views, ×0.7), Upper Pennsylvanian, Missouri.

at edge of the pedicle interarea, and by internal features. Some Devonian chonetids have unusually large spines and coarse costae; others are distinguished by denticles on the hinge margin and strong convexity of the pedicle valve. Mississippian, Pennsylvanian, and Permian rocks all contain useful guide fossils belonging to this group. One of the best known is *Mesolobus*, distinguished by longitudinal corrugations of the mid-portion of the valves; it is confined to Middle Pennsylvanian rocks (Fig. 6-34, 7).

The group of shells called productids differs from chonetids in having much greater convexity of the pedicle valve and generally a more concave brachial valve; also, small or large spines of varying number are distributed over the surface (Figs. 6-33, 1-5; 6-34, 1, 2, 4, 9-14). Relatively few species occur in Devonian rocks, but they are very numerous in younger Paleozoic formations nearly to uppermost Permian. Two main shell types are distinguished: in one, the surface is marked chiefly by longitudinal fine to coarse ribs, with few spines; in the other, spines are very numerous and costae are lacking or inconspicuous. Some members of the rib-marked group have evenly spaced concentric corrugations on the posterior part of both valves so as to produce a reticulate pattern; the interior of the brachial valve bears looped ridges which probably are impressions of the lophophore (Figs. 6-6, 2, 6; 6-33, 1; 6-34, 4, 13).

One of the Mississippian costate productids (*Gigantoproductus*, Fig. 6-6, 6) is the largest brachiopod known, attaining a width of approximately 12 in. Among noncostate productids which have shells covered by abundant obliquely disposed spines, some genera are characterized by unevenness in the arrangement of the spines, others by very regular concentric rows of spines, and a Permian genus (*Waagenoconcha*) has spines arranged in curved, regularly intersecting oblique rows (Figs. 6-33, 2; 6-34, 1, 2, 10-12, 14).

One of the strangest types of produc-

tacean brachiopods is the Permian genus called *Prorichthofenia* (Fig. 6-34, 3). Superficially, the pedicle valve resembles a horn coral which is held in an upright position by its anchorage of outspread spines. The brachial valve is a lidlike structure which articulates with the pedicle valve at a level well below the shell edge. Near the margins of each valve on their inner side, spines are arranged in alternating position so that when the valves are partly open, they guard the entrance to the shell interior. Some species even have a lacelike calcareous grillwork that arches over the brachial valve, and in such species, the spines on the inner sides of the pedicle and brachial valves are lacking. Individuals of *Prorichthofenia* commonly grow together in clusters; but because one is not budded from another, they constitute a community rather than a colony.

Spiriferids

The order of brachiopods called Spiriferida, which is characterized mainly by the spiral form of the brachidium, is a very large and highly varied assemblage (Figs. 6-35 to 6-39). The great majority of spiriferids have an impunctate shell, and these are mostly confined to rocks of Paleozoic age, ranging from Middle Ordovician to Upper Permian. Several kinds of spiriferids, however, survived into middle Mesozoic time (Fig. 6-40).

Nearly all spiriferids have a biconvex shell, but a very few genera have one valve which is plane or concave. The hinge line is wide or narrow, and the outline of the shell in views normal to the plane of commissure ranges from highly transverse forms with angular extremities, through elliptical and subcircular shapes with smoothly curved lateral margins, to elongate shells with prominent beaks. Shells marked by radial costae or plicae greatly predominate over smooth forms. A few genera are characterized by the presence of concentric lamellae, and in some shells these lamellae bear denticulate margins or have spinose extensions. Prominent

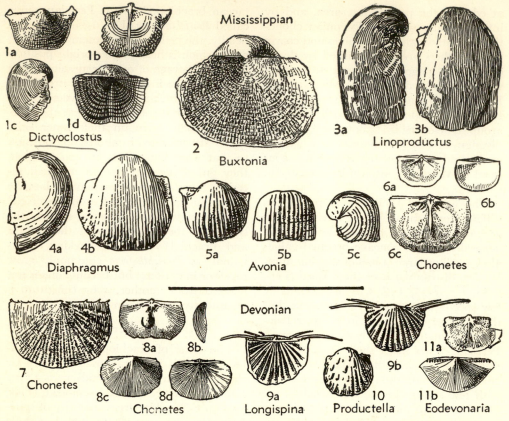

FIG. 6-33. **Representative Devonian–Mississippian strophomenid articulate brachiopods belonging to the suborder Productacea.** This group consists mostly of strongly concavo-convex spine-bearing shells. They are distributed from Silurian to Upper Permian but are most abundant in post-Devonian rocks.

Avonia Thomas, Mississippian–Permian. Small to medium-sized shells; pedicle valve anteriorly costate and bearing scattered erect spines; brachial valve lacking spines. *A. oklahomensis* Snider (5a–c, pedicle, anterior, and side views, ×0.7), Upper Mississippian, Oklahoma.

Buxtonia Thomas, Mississippian–Permian. Moderately large shells having oblique spines on both valves. *B. semicircularis* Sutton & Wagner (2, brachial view, ×0.7), Upper Mississippian (Chesteran), Kentucky.

Chonetes Fischer, Middle Silurian–Permian. Generally small, shallow concavo-convex shells, with short obliquely outward pointing spines along posterior edge of pedicle valve. *C. oklahomensis* Snider (6a, b, pedicle interior and brachial views, ×0.7; 6c, brachial interior, ×1.5), Upper Mississippian, Oklahoma; *C. aurora* Hall (7, pedicle view, ×3), Middle Devonian, New York; *C. coronatus* (Conrad)

(8a–d, brachial interior, side, pedicle, and brachial views, ×0.7), Middle Devonian, New York.

Diaphragmus Girty, Upper Mississippian. Brachial valve contains a transverse partition. *D. cestriensis* (Worthen) (4a, b, side and pedicle views, ×1), Chesteran, Kentucky.

Dictyoclostus Muir-Wood, Mississippian–Permian. Medium to large costate shells bearing rugae in umbonal region and scattered erect spines. *D. inflatus* (McChesney) (1a–d, posterior, brachial interior, side and brachial views, ×0.7), Upper Mississippian (Chesteran), Illinois.

Eodevonaria Breger, Lower and Middle Devonian. Generally like *Chonetes*, but pedicle valve much more convex; denticles along hinge line. *E. arcuata* (Hall) (11a, b, brachial interior, posterior view of pedicle valve, ×0.7), Lower Devonian, New York.

(*Continued on next page.*)

granules or short spines are scattered over the surface of some spiriferids. Commonly the pedicle valve bears a distinct interarea, and it may be very prominent. Unlike groups such as the orthids and strophomenids, the brachial valve does not have a perceptible interarea. The triangular delthyrial space of the pedicle valve may be constricted by deltidial plates growing inward from side of the delthyrium, or in some genera by a deltidium (Fig. 6-11). Some genera of spiriferids have a very small pedicle interarea, or none may be visible.

As shown in the tabular summary of brachiopod classification, the Spiriferida include three impunctate suborders, named Atrypacea, Spiriferacea, and Rostrospiracea. The fourth suborder, composed of punctate shells, is known as Punctospiracea.

Atrypacea. The oldest spiriferids, which occur in Ordovician rocks, and others related to them ranging upward to Lower Mississippian, are included in the Atrypacea (Fig. 6-35). They have a short hinge line and very inconspicuous interarea on the pedicle valve. All have a rounded outline; but in longitudinal profile, the shells range from biconvex, which is the most common form, to convexiplane, plano-convex, and concavo-convex. A small minority of the Atrypacea have smooth shells, others being rather finely costate or having medium to coarse plications. Concentric growth lines or lamellar outgrowths of shell distinguish some genera, such as *Atrypa*.

Internally, the posterior part of the pedicle valve generally bears large muscle scars, and some shells have clear pallial markings (Fig. 6-12, 7). Several genera have well-developed dental lamellae, whereas others lack these structures. In the brachial valve, strong crural processes are directly continuous with the initial large coil of each spiralium. A distinguishing feature of the Atrypacea is the manner in which the beginnings of the spiralia bend outward so as to enclose between them the remainder of the spiral cones. The axis of these cones may be parallel to the hinge line or approximately at right angles to this direction, and apices of the cones may be directed outward, inward, or toward the middle part of the floor of the brachial valve, as in *Atrypa* (Fig. 6-35, 1d, e).

Ordovician Atrypacea are moderately small biconvex shells having fine radial plications and an inconspicuous interarea on the pedicle valve (Fig. 6-35, 8, 9). The coils of the spiralia are directed inward, the crossbar (jugum) connecting the outermost coil of the spires being located posteriorly in some and variably in others.

In Silurian rocks are smooth-shelled Atrypacea in which the spires inside the outermost loop are pointed outward or turned at right angles toward the floor of the brachial valve, as in *Atrypa* (Fig. 6-35, 10, 11). Long-ranging *Atrypa reticularis* is common in some Silurian formations but is more abundant in the Devonian.

Various species of *Atrypa* are the most common representatives of the Atrypacea in Devonian deposits (Fig. 6-35, 1–4). Most are finely costate, but some have coarse ribs crossed by rough lamellae. In all, the brachial valve is much more convex than the pedicle valve, which tends to

---

(Fig. 6-33 continued.)

Linoproductus Chao, Mississippian–Permian. Shell distinguished by long trail and fine, commonly somewhat flexuous costae. *L. ovatus* (Hall) (3a, b, side and pedicle views, ×1), Lower Mississippian (Osagian), Iowa.

Longispina Cooper, Lower and Middle Devonian. Distinguished by coarse costae and very long oblique spines at posterior edge of pedicle valve. *L. emmetensis* (Winchell) (9a, b, brachial and pedicle views, ×1), Middle Devonian, Michigan.

Productella Hall, Devonian–Mississippian. Pedicle valve bears a short interarea and scattered long oblique spines. *P. spinulicosta* Hall (10, pedicle view, ×1), Middle Devonian, Michigan.

be nearly flat and in some individuals somewhat concave in the anterior region. The base of the spiral cones lies closest to the pedicle valve, and apices of the many-coiled spires are directed toward the posterior middle portion of the brachial valve. The jugum is located posteriorly, not far from the crural processes, which connect the spire with the hinge plate of the brachial valve. A number of small plicate genera differ mostly from *Atrypa* in being wider than long and in having a plano-convex to concavo-convex shell profile (Fig. 6-35, *5–7*). They are distinguished both by external and internal characters.

Some species of short stratigraphic range and fairly wide geographic distribution are valuable guide fossils.

An interesting peculiarity of occurrence distinguishes some species of *Atrypa* in Devonian formations. This consists in the presence locally of hundreds of pedicle valves, without any associated brachial valves—not even a few loose ones scattered among the others. Concentrations of pedicle valves in this manner could be effected only by separation of the valves after death of the brachiopods and sorting of the valves by water movement, such as may be produced on a shallow sea bottom by waves

---

Fig. 6-34.   **Representative Pennsylvanian–Permian strophomenid articulate brachiopods belonging to the suborder Productacea.** Older members of this group are illustrated in Fig. 6-33.

Buxtonia Thomas, Mississippian–Permian. *B. peruviana* (d'Orbigny) (2, brachial view, ×0.7), Permian, Texas.

Chonetes Fischer, Middle Silurian–Permian. *C. granulifer* Owen (6*a, c*, pedicle and brachial interiors; 6*b*, brachial view; ×0.7), Upper Pennsylvanian, Kansas.

Chonetina Krotow, Pennsylvanian. Strong sulcus on pedicle valve and fold on brachial valve. *C. flemingi* (Norwood & Pratten) (5*a, b*, pedicle and brachial views, ×0.7), Upper Pennsylvanian (Missourian), Missouri.

Dictyoclostus Muir-Wood, Mississippian–Permian. *D. americanus* Dunbar & Condra (4*a, c–e*, posterior, brachial, pedicle, and side views; 4*b*, brachial interior, ×0.7), Upper Pennsylvanian, Kansas.

Echinoconchus Weller, Mississippian–Permian. Both valves marked by regular concentric rows of short, oblique spines. *E. semipunctatus* (Shepard) (12*a, b*, brachial and pedicle views, ×0.7), Upper Pennsylvanian, Kansas.

Heteralosia King, Devonian–Permian. Beak of pedicle valve bears scar of attachment; oblique spines on pedicle valve but not on brachial. *H. slocomi* King (10*a, b*, brachial and pedicle views, ×1), Pennsylvanian, Texas.

Juresania Fredericks, Pennsylvanian–Permian. Both valves bear two sizes of oblique spines which are less clearly arranged in concentric bands than in *Echinoconchus*. *J. nebrascensis* (Owen) (11*a*, brachial interior; 11*b, c*, brachial and side views; ×0.7), Upper Pennsylvanian, Nebraska; *J. symmetrica* (McChesney) (14*a, b*,

pedicle and brachial views, ×0.7), Upper Pennsylvanian, Kansas.

Linoproductus Chao, Mississippian–Permian. *L. oklahomae* Dunbar & Condra (9*a*, pedicle view; 9*b*, mold of brachial exterior, ×0.7), Upper Pennsylvanian (Missourian), Oklahoma.

Marginifera Waagen, Mississippian–Permian. Small to medium-sized shell; interior of brachial valve has marginal ridge. *M. muricatina* Dunbar & Condra (8*a, b*, pedicle and brachial views, ×0.7), Middle Pennsylvanian (Desmoinesian), Missouri; *M. lasallensis* (Worthen) (13*a, b*, pedicle and brachial views, ×0.7), Upper Pennsylvanian, Texas.

Mesolobus Dunbar & Conrad, Middle Pennsylvanian. Pedicle valve bears a median fold bordered by narrow sulci, and brachial valve has a corresponding shallow sulcus bordered by ridges. *M. mesolobus* (Norwood & Pratten) (7*a, b*, pedicle and brachial views, ×1), Desmoinesian, Texas.

Prorichthofenia King, Permian. Pedicle valve deep conical, anchored by irregular spines; brachial valve flat, located well within margin of pedicle valve; commonly grows in clusters. *P. uddeni* (Böse) (3*a*, oblique view of brachial valve in position of normal articulation; 3*b*, brachial interior; 3*c*, side view of pedicle valve; ×0.7), Texas.

Waagenoconcha Chao, Permian. Surface covered by many small oblique spines arranged in intersecting rows. *W. montpelierensis* Girty (1*a, b*, pedicle and brachial views, ×1), Idaho.

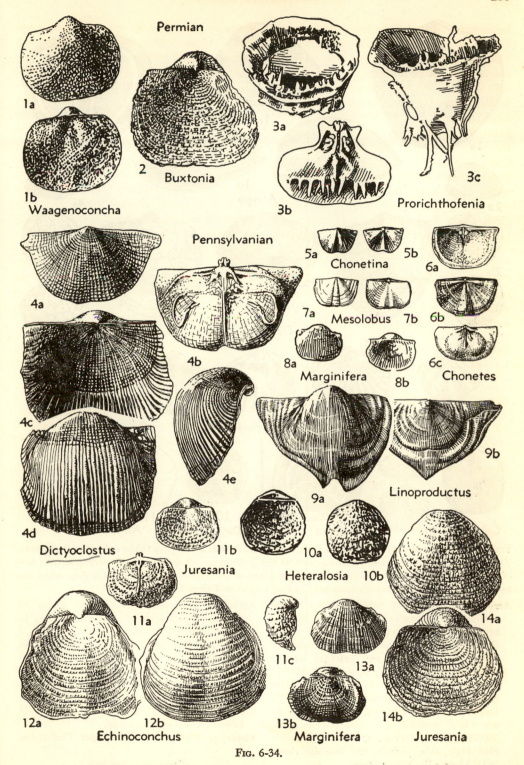

Permian

1a

1b
Waagenoconcha

2
Buxtonia

3a

3b

3c
Prorichthofenia

Pennsylvanian

4a

4b

4c

4d
Dictyoclostus

4e

5a 5b
Chonetina

7a Mesolobus 7b

8a

Marginifera 8b

6a

6b

6c
Chonetes

9a

9b
Linoproductus

11b
Juresania

11a

10a

Heteralosia 10b

11c

13a

13b
Marginifera

14a

14b
Juresania

12a

12b
Echinoconchus

Fig. 6-34.

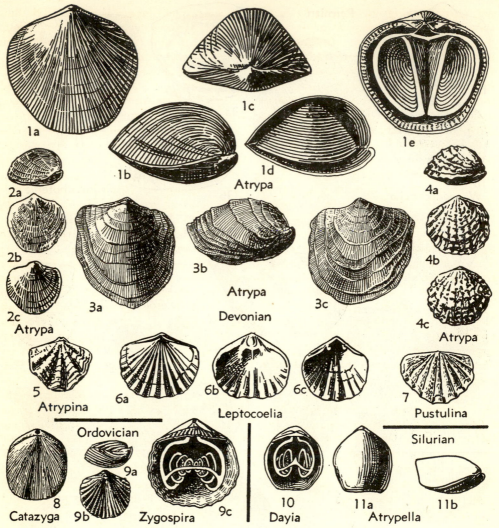

FIG. 6-35. Representative spiriferid articulate brachiopods belonging to the suborder Atrypacea.
These fossils have spiral calcareous supports of the lophophore, characteristically arranged with axes
of the spiral cones approximately normal to the plane of commissure. The group is distinguished also
by a short hinge line. It ranges from Middle Ordovician to Lower Mississippian.

Atrypa Dalman, Middle Silurian–Lower Mississippian. Pedicle valve flat or gently convex;
brachial valve strongly convex. *A. reticularis*
(Linné) (1*a–c*, brachial, side, and posterior
views; 1*d*, *e*, section and brachial interior
views showing spiralia in position; ×1), Middle
Devonian, New York; *A. devoniana* Webster
(2*a–c*, side, pedicle, and brachial views, ×0.7),
Upper Devonian, Iowa; *A. independensis* Webster (3*a–c*, brachial, side, and pedicle views,
×0.7), Middle Devonian, Iowa; *A. bellula*
Hall (4*a–c*, side, pedicle, and brachial views,
×1), Middle Devonian, Iowa.

Atrypella Kozlowski, Middle and Upper Silurian. Smooth-surfaced subequal valves; internally like *Atrypa*. *A. shrocki* Cooper (11*a*, *b*,
side and brachial views, ×1), Niagaran,
Indiana.

Atrypina Hall & Clarke, Middle Silurian–Lower
Devonian. Coarsely plicate, concentrically
lamellose; spiralia as in *Atrypa* but having few
coils. *A. imbricata* (Hall) (5, brachial view, ×2),
Lower Devonian, Tennessee.

Catazyga Hall & Clarke, Upper Ordovician.

(Continued on next page.)

and currents. It is not difficult to understand how such movements might transport the strongly convex brachial valves readily, but be unable effectively to shift the nearly flat pedicle valves, partly buried in sediment. Thus, the latter came to be a sort of lag concentrate. Of course, segregation of individual valves in this manner is an exception, rather than the rule.

Spiriferacea. This suborder of spire-bearing brachiopods may be said to constitute the most typical division of the Spiriferida, because it contains the genus *Spirifer*. Also, in most shells, the spiral brachidium is remarkably developed, the cone-shaped coils on either side of the median plane being formed of many turns. The impunctate shell has a wide hinge line, so that commonly the cardinal extremities coincide with the position of the greatest shell width (Figs. 6-36, 6-37). Nearly all shells bear costae or plications, and they are almost invariably marked by a sulcus on the pedicle valve and a fold on the brachial valve. A plane or longitudinally curved, generally obtuse-angle interarea occurs on the pedicle valve. Directly beneath the beak, the interarea is interrupted by the delthyrium, which may be partly closed by a deltidium. Internal structures are varied, and in some shells they are important in generic diagnosis. Generally, there are large muscle scars, stout dental lamellae, and close-coiled, outwardly pointed spiralia. The complete or incomplete jugum is simple. The Spiriferacea range from Middle Silurian to Jurassic, having a peak development in variety and numbers in the Devonian.

Silurian Spiriferacea are uncommon. Generally, the valves bear a fold and sulcus, and in addition, they are marked by very fine radiating lines and concentric growth lines (Fig. 6-36, *16*, *17*). The pedicle valve of some shells has a low curved interarea, whereas in others it has an elevated, nearly plane right-angle interarea. The delthyrium is closed by a convex deltidium which may have a rounded foramen near its apex.

In Devonian rocks, the Spiriferacea are a prominent brachiopod assemblage. They comprise two main groups and a third very minor one, defined mainly on the features of the shell exterior. (1) In the first group are shells having a nonplicated fold and sulcus (Fig. 6-36, *4*, *5*, *9*, *15*). Such brachiopods are rather easily distinguished on the basis of external features but are separated also by internal characters. Middle Devonian genera, some of which range into the Upper Devonian, have distinctive features, such as granules distributed over the surface, spinose concentric lamellae or fine concentric markings (Fig. 6-36, *1*, *3*, *8*, *10*, *12*). (2) A second group of Devonian Spiriferacea is characterized by the occurrence of plications on the fold and sulcus. Some have relatively coarse even plications and others medium to fine plications (Fig. 6-36,

(*Fig. 6-35 continued.*)

Finely plicate biconvex shells with spiral coils of brachidium directed inward; jugum posterior. *C. headi* (Billings) (8, brachial view, ×1), Indiana.

Dayia Davidson, Silurian. Shell smooth; coils of spiralia directed outward, jugum anterior. *D. navicula* (Sowerby) (10, interior view from the brachial side, ×2), Upper Silurian, England.

Leptocoelia Hall, Lower and Middle Devonian. Concavo-convex plicate shell. *L. flabellites* (Conrad) (6a–c, brachial view, brachial and pedicle interiors, ×1), Lower Devonian, New York.

Pustulina Cooper, Middle Devonian. Plano-convex plicate shell having pustulose exterior. *P. pustulosa* (Hall) (7, brachial valve, ×2), New York.

Zygospira Hall, Middle Ordovician–Lower Silurian. Resembles *Catazyga* but more coarsely plicate and position of jugum variable. *Z. modesta* (Say) (9a, b, side and brachial views, ×2; 9c, interior view from brachial side, ×4), Upper Ordovician, Kentucky.

2, 6, 11, 13, 14). (3) A third group includes diminutive nonplicated shells which closely approach a plano-convex form (Fig. 6-36, 7). They are abundant in some shaly deposits but rare in limestones.

Mississippian Spiriferacea differ from those of the Devonian in the general lack of species which have a nonplicated fold and sulcus. The genus *Spirifer*, as now restricted, is essentially a Mississippian

Fig. 6-36. **Representative Silurian and Devonian spiriferid articulate brachiopods belonging to the suborder Spiriferacea.** This group, which ranges from Middle Silurian to Upper Permian, is characterized by wide hinge line and laterally directed spiralia. The shell structure is impunctate.

Ambocoelia Hall, Devonian–Mississippian. Pedicle valve strongly convex with prominent incurved beak; brachial valve very gently convex. *A. umbonata* (Conrad) (7*a*, brachial interior, ×2; 7*b*, brachial view, ×1), Middle Devonian, New York.

Brachyspirifer Wedekind, Middle Devonian. Fold and sulcus smooth; short median septum in brachial valve. *B. audaculus* (Conrad) (12, brachial view, ×0.7), New York.

Brevispirifer Cooper, Lower and Middle Devonian. Like *Mucrospirifer* but valve narrow. *B. gregarius* (Clapp) (5*a*, *b*, side and brachial views, ×0.7), Lower Devonian, Ohio.

Costispirifer Cooper, Lower Devonian. Fold and sulcus costate. *C. arenosus* (Conrad) (11*a*, *b*, brachial and pedicle views, ×0.7), Maryland.

Cyrtia Dalman, Middle Silurian. Pedicle valve pyramidal, brachial valve gently convex; apically perforate deltidium. *C. exporrecta* (Wahlenberg) (17*a-c*, posterior, side, and brachial views, ×0.7), Indiana.

Cyrtospirifer Nalivkin, Upper Devonian. Medium-sized, costate fold and sulcus; dental lamellae strong. *C. whitneyi* (Hall) (14*a-c*, pedicle, brachial, and side views, ×0.7), Iowa.

Delthyris Dalman, Middle Silurian–Lower Devonian. Hinge line wide, surface plicate, lamellose; long dental lamellae and high median septum in pedicle valve. *D. perlamellosus* (Hall) (9, brachial view, ×0.7), Lower Devonian, New York.

Elytha Fredericks, Devonian. Surface bears low plications and concentric rows of fine spines. *E. fimbriata* (Conrad) (1*a-c*, brachial, pedicle, and side views, ×0.7), Middle Devonian, New York.

Eospirifer Schuchert, Middle Silurian–Lower Devonian. Surface bears fine radiating costae and may be faintly plicate also. *E. radiatus* (Sowerby) (16*a*, *b*, brachial and side views, ×0.7), Middle Silurian, New York.

Fimbrispirifer Cooper, Middle Devonian. Rounded lateral extremities, costate fold and sulcus; surface bears concentric lamellae terminating in small spines. *F. venustus* (Hall) (13, brachial view, ×0.7), Middle Devonian, Ohio.

Metaplasia Hall & Clarke, Lower Devonian. Small coarsely plicate shell with strong fold and sulcus. *M. pyxidata* (Hall) (4*a*, *b*, brachial view and interior, ×0.7), New York.

Mucrospirifer Grabau, Middle and Upper Devonian. Hinge line extended, fold and sulcus nonplicate. *M. mucronatus* (Conrad) (3*a-c*, posterior, pedicle, and brachial views, ×0.7), New York; *M. thedfordensis* Shimer & Grabau (3*d*, *e*, brachial and pedicle views, ×0.7), Ontario; *M. consobrinus* (d'Orbigny) (3*f*, *g*, brachial and pedicle views, ×0.7), New York; all Middle Devonian.

Paraspirifer Wedekind, Lower and Middle Devonian. Very prominent noncostate keel-like fold and deep sulcus. *P. acuminatus* (Conrad) (15*a-c*, anterior, side, and brachial views, ×0.7), Lower Devonian, Maryland.

Platyrachella Fenton & Fenton, Middle and Upper Devonian. Fold and sulcus noncostate, surface covered by radial lines of fine granules. *P. oweni* (Hall) (8*a*, *b*, side and brachial views, ×0.7), Middle Devonian, Kentucky.

Spinocyrtia Fredericks, Middle Devonian. Like *Platyrachella;* surface entirely covered with tear-shaped granules. *S. granulosa* (Conrad) (10*a*, *b*, posterior and brachial views, ×0.7; 10*c*, part of surface enlarged), Pennsylvania.

Tenticospirifer Tien, Upper Devonian. Small shell, fold and sulcus costate, pedicle valve with high interarea. *T. cyrtiniformis* (Hall & Whitfield) (2*a-d*, pedicle, brachial, posterior, and side views, ×0.7), Iowa.

Theodossia Nalivkin, Upper Devonian. Shell outline rounded, fold and sulcus weakly defined, entire surface finely and evenly costate. *T. hungerfordi* (Hall) (6*a*, *b*, brachial and side views, ×0.7), Iowa.

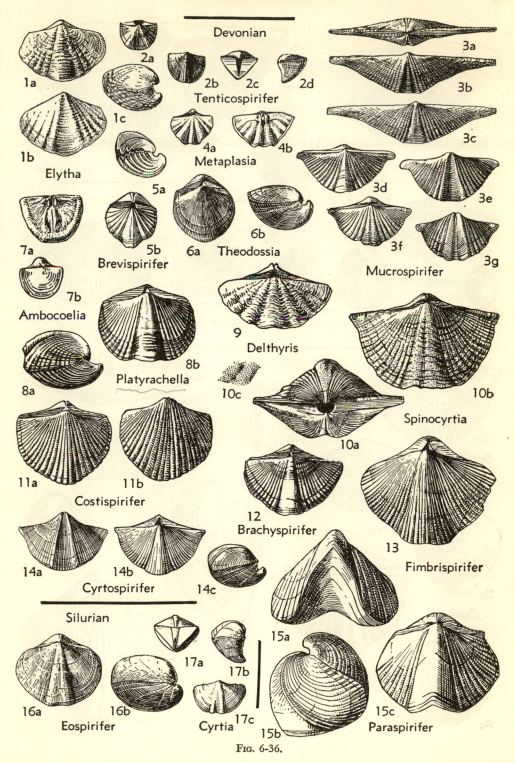

Devonian

1a 1b 1c **Elytha**

2a 2b 2c 2d **Tenticospirifer**

4a 4b **Metaplasia**

3a 3b 3c 3d 3e 3f 3g **Mucrospirifer**

5a 5b **Brevispirifer**

6a 6b **Theodossia**

7a 7b **Ambocoelia**

8a 8b **Platyrachella**

9 **Delthyris**

10a 10b 10c **Spinocyrtia**

11a 11b **Costispirifer**

12 **Brachyspirifer**

13 **Fimbrispirifer**

14a 14b 14c **Cyrtospirifer**

Silurian

16a 16b **Eospirifer**

17a 17b 17c **Cyrtia**

15a 15b 15c **Paraspirifer**

Fig. 6-36.

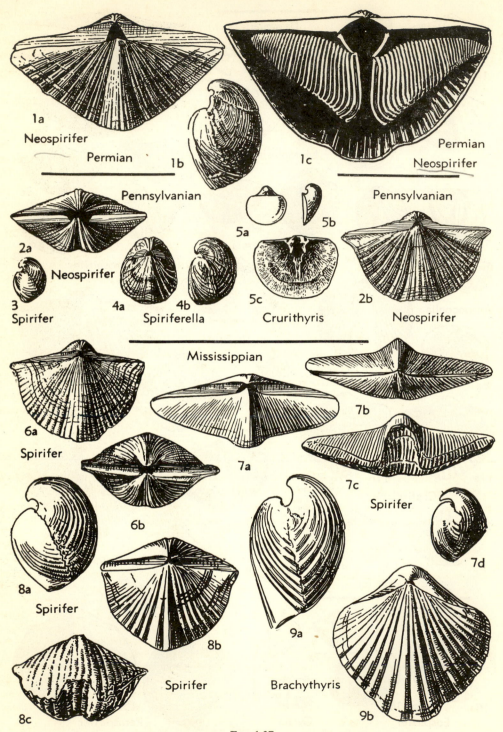

1a
Neospirifer
Permian   1b   1c   Permian
                                                                        Neospirifer

Pennsylvanian   5a   5b   Pennsylvanian

2a
        Neospirifer   5c
3   4a   4b   2b
Spirifer   Spiriferella   Crurithyris   Neospirifer

Mississippian   7b

6a
Spirifer   7a   7c
                    6b   Spirifer

8a   7d
Spirifer   8b
                    Spirifer   9a

8c   Brachythyris
                                                              9b

FIG. 6-37.

genus, but it ranges into the Lower Pennsylvanian (Fig. 6-37, *3, 6–8*). Some species are characterized by great width, narrow plications, and a sharp carinate fold; others are short-hinged and more coarsely ribbed. A common genus, related to *Spirifer*, has a laterally rounded shell outline and broad low plications (Fig. 6-37, *9*).

Pennsylvanian and Permian Spiriferacea are characterized especially by bifurcation and bundling of plications on the shell surface (Fig. 6-37, *1, 2, 4*).

Rostrospiracea. The suborder of spiriferids called Rostrospiracea generally is distinguished by nonplicate, impunctate shells of rounded outline, having prominent beaks but no perceptible interarea on the pedicle valve (Fig. 6-38). Except for their spiral brachidium, directed laterally outward, they have almost nothing in common with the Spiriferacea. Externally, they resemble a few of the Atrypacea which have a smooth shell, but these differ in the nature of the brachidium. A characteristic feature of the rostrospiracean brachidium is the more or less complicated nature of the jugum, which may bear long pointed processes (Fig. 6-38, *7a*) or loops like scissor handles. Stratigraphic range of the Rostrospiracea is from Middle Ordovician to Jurassic. They are most abundant in Silurian and Devonian rocks but are common in the Mississippian, and some genera are represented by multitudes of individuals in post-Mississippian rocks.

Several important and characteristic Silurian genera belong in this group (Fig. 6-38, *7–10*). Devonian forms are common (Fig. 6-38, *4–6*). Especially numerous in some Middle and Upper Devonian rocks is a transverse, elliptical shell (*Athyris*) which has closely spaced concentric growth lines and lamellae; it ranges upward into the Lower Mississippian.

One of the most abundant rostrospiraceans in some Mississippian and Pennsylvanian rocks is *Composita* (Fig. 6-38, *1, 2*). Faunules composed mainly of this brachiopod denote an environment to which it was particularly adapted and one not favored by most others.

Punctospiracea. Spire-bearing brachiopods which have punctate shell structures are included in the suborder Punctospiracea (Fig. 6-39). As with other groups of punctate shells, such as terebratulids and dalmanellaceans, the presence of puncta constitutes a basis which is presumed to be

---

Fig. 6-37. **Representative Mississippian–Permian spiriferid articulate brachiopods belonging to the suborder Spiriferacea.** Older fossils of this group are illustrated in Fig. 6-36.

Brachythyris McCoy, Mississippian. Shell outline rounded, broad low plications on flanks and fold and sulcus. *B. subcardiformis* (Hall) (*9a, b*, side and brachial views, ×1), Upper Mississippian, Illinois.

Crurithyris George, Devonian–Permian. Brachial valve nearly flat, pedicle valve strongly convex, with incurved beak; hinge line short. *C. planoconvexa* (Shumard) (*5a, b*, brachial and side views, ×1; *5c*, brachial interior, ×3), Middle Pennsylvanian, Missouri.

Neospirifer Fredericks, Pennsylvanian–Permian. Plications bundled, both on flanks and on fold and sulcus. *N. condor* (d'Orbigny) (*1a, b*, brachial and side views, ×0.7; *1c*, brachial interior showing spiralia, ×1), Lower Permian, Texas; *N. cameratus* (Morton) (*2a, b*, posterior and brachial views, ×0.7), Middle Pennsylvanian, Ohio.

Spirifer Sowerby, Mississippian–Middle Pennsylvanian. Completely plicate shells with wide hinge line, plications mostly undivided. *S. rockymontanus* Marcou (*3*, side view, ×0.7), Middle Pennsylvanian, Oklahoma; *S. arkansanus* Girty (*6a, b*, brachial and posterior views, ×0.7), Upper Mississippian, Oklahoma; *S. lateralis* Hall (*7a–d*, brachial, posterior, anterior, and side views, ×1), Upper Mississippian (Meramecian), Indiana; *S. keokuk* Hall (*8a–c*, side, brachial, and anterior views, ×1), Lower Mississippian (Osagian), Iowa.

Spiriferella Tschernyschew, Pennsylvanian. Subquadrate, thick shell with strong dental lamellae. *S. texanus* (Meek) (*4a, b*, brachial and side views, ×0.7), Upper Pennsylvanian, Texas.

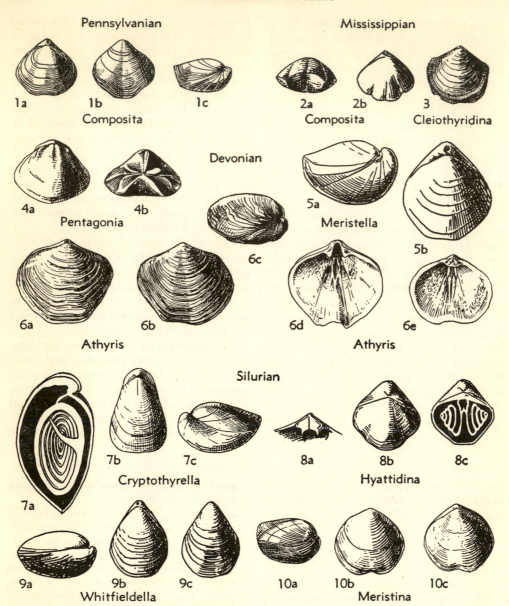

FIG. 6-38. **Representative spiriferid articulate brachiopods belonging to the suborder Rostro-spiracea.** Shells of this group are impunctate and very short-hinged, lacking a distinct interarea; they have a rounded or elongate outline and lack costae or plicae. Stratigraphic range is from Middle Silurian to Upper Permian.

Athyris McCoy, Middle Devonian–Lower Mississippian. Subequally biconvex shell with rounded outline, generally wider than long; surface lamellose. *A. spiriferoides* (Eaton) (6a–c, brachial, pedicle, and side views; 6d, e, pedicle and brachial interiors, ×0.7), Middle Devonian, New York.

Cleiothyridina Buckman, Mississippian–Permian. Like *Athyris* but lamellae have spinose projections and internal structure differs. *C. sublamellosa* (Hall) (3, brachial view, ×0.7), Upper Mississippian (Chesteran), Illinois.

Composita Brown, Mississippian–Permian. Smooth shell with more or less distinct fold

(*Continued on next page.*)

fully adequate for separating them from other spire bearers. On the other hand, diversity in form of shell and, to some extent, divergence of internal characters suggest that the Punctospiracea may be polyphyletic. That is to say, the group may embrace derivatives of more than one ancestral stock. Rostrate forms, such as represented by various genera distributed from Silurian to Permian (Fig. 6-39, *2, 5, 6, 8–12*), are similar in many ways. In spite of internal differences, such as the presence or absence of dental lamellae, features of the cardinalia, and details of the spiralia, they are clearly related.

Obviously different from the shells just cited are punctate spire-bearing brachiopods having a wide hinge line and a generally prominent interarea on the pedicle valve (Fig. 6-39, *1, 3, 4, 7, 9*). A spondylium occurs in the pedicle valve of some of these genera but is absent in others. The delthyrium of most wide-hinged Punctospiracea is open, but in a few it is closed by a convex, apically perforate deltidium. The shells of this group possibly are derivatives of one or more components of the

Spiriferacea, from which they are distinguished by the acquisition of a punctate shell structure. A good deal of study is needed in order to establish satisfactorily the features of classification and phylogeny of these brachiopods.

GEOLOGICAL DISTRIBUTION

Graphic summary of the geological distribution of each main division of the brachiopods is given in Fig. 6-40. At the left of the diagram, separated from divisions at right by a broken line, are narrow lines representing the occurrence of inarticulate brachiopod groups. Although these dominate in Cambrian formations, studies so far published indicate that they are numerically and stratigraphically unimportant. Some types, both among atremates and neotremates are very conservative, long-ranging brachiopods, which seem to have changed very little since their first appearance in the paleontological record, in Cambrian or Ordovician strata. Actually, this may be a very misleading appraisal, based on lack of knowledge, for recent etching of lower Paleozoic calcare-

(Fig. 6-38 continued.)

and sulcus; round foramen in pedicle valve beak. Very common in many Pennsylvanian and Permian beds. *C. subtilita* (Shepard) (*1a–c*, brachial, pedicle, and side views, ×0.7), Upper Pennsylvanian, Kansas; *C. trinuclea* (Hall) (*2a, b*, anterior and brachial views), Upper Mississippian (Chesteran), Illinois.

Cryptothyrella Cooper, Lower Silurian. Elongate shell, pedicle beak small, incurved, bearing minute foramen. *C. cylindrica* (Hall) (*7a*, section showing spiralium, ×1; *7b, c*, brachial and side views, ×0.7), Ohio.

Hyattidina Schuchert, Middle Silurian. Subpentagonal shell outline; median and lateral sulci on pedicle valve with corresponding folds on brachial valve; spiralium loosely coiled. *H. congesta* (Conrad) (*8a*, divided hinge plate of brachial valve, ×2; *8b*, brachial view, ×0.7; *8c*, same, with part of shell removed to show spiralium, ×0.7), New York.

Meristella Hall, Upper Silurian–Middle Devonian. Oval smooth shell, prominent incurved

beak on pedicle valve, dental lamellae strong. *M. nasuta* (Conrad) (*5a, b*, side and brachial views, ×0.7), Lower Devonian, New York.

Meristina Hall, Middle Silurian. Closely similar to *Meristella* externally and internally but distinguished by form of jugum. *M. maria* (Hall) (*10a–c*, side, pedicle, and brachial views, ×0.7), Indiana.

Pentagonia Cozzens, Lower and Middle Devonian. Distinguished mainly by angulated plication of shell along lines bounding shallow sulci; interior like *Meristella*. *P. bisulcata* (Hall) (*4a, b*, brachial and posterior views, ×0.7), Middle Devonian, Ontario.

Whitfieldella Hall & Clarke, Lower Silurian–Lower Devonian. Smooth elongate oval shell with round foramen in pedicle beak; concave hinge plate in brachial valve supported by median septum. *W. nitida* (Hall) (*9a–c*, side, brachial, and pedicle views, ×1), Middle Silurian, Indiana.

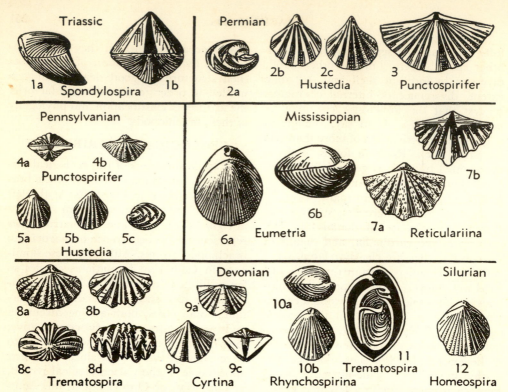

FIG. 6-39. **Representative spiriferid articulate brachiopods belonging to the suborder Puncto-spiracea.** These spire-bearing shells differ from others in having punctate structure of the calcareous exterior. They are distributed from Middle Silurian through later Paleozoic and Triassic formations.

Cyrtina Davidson, Middle Silurian–Lower Mississippian. Small to medium-sized, pedicle valve subpyramidal; a large foramen occurs near apex of the convex deltidium. *C. hamiltonensis* (Hall) (9*a–c*, brachial, pedicle, and posterior views, ×0.7), Middle Devonian, Michigan.

Eumetria Hall, Mississippian. Elongate oval costate shell with round foramen in pedicle valve beak. *E. vera* (Hall) (6*a, b*, brachial and side views, ×1), Upper Mississippian (Chesteran), Illinois.

Homeospira Hall & Clarke, Middle Silurian. Elongate oval costate shell having a high median septum in brachial valve. *H. evax* (Hall) (12, brachial view, ×0.7), Indiana.

Hustedia Hall & Clarke, Mississippian–Permian. Small costate shells with prominent beak on pedicle valve; dental lamellae lacking. *H. hessensis* King (2*a–c*, side, brachial, and pedicle views, ×1), Permian, Texas; *H. mormoni* (Marcou) (5*a–c*, brachial, pedicle, and side views, ×1), Upper Pennsylvanian, Kansas.

Punctospirifer North, Mississippian–Permian. Wide-hinged small shells having nonplicate fold and sulcus; surface imbricate or pustulose, *P. pulcher* (Meek) (3, brachial view, ×0.7). Permian, Wyoming; *P. kentuckiensis* (Shumard)

(4*a, b*, posterior and brachial views, ×0.7), Pennsylvanian, Texas.

Reticulariina Fredericks, Upper Mississippian. Like *Punctospirifer* but surface covered by spines. *R. spinosa* (Norwood & Pratten) (7*a, b*, brachial view and pedicle interior, ×1), Chesteran, Illinois.

Rhynchospirina Schuchert & LeVene, Upper Silurian–Lower Devonian. Elongate oval shell with low costae; stout crural bases and median septum in brachial valve. *R. formosa* (Hall) (10*a, b*, side and brachial views, ×1), Lower Devonian, New York.

Spondylospira Cooper, Triassic. Costate fold and sulcus; high interarea on pedicle valve and spondylium inside it. *S. alia* (Hall & Whitfield) (1*a, b*, side and brachial view, showing high pedicle interarea, ×0.7), Upper Triassic, Nevada.

Trematospira Hall, Middle Silurian–Middle Devonian. Strongly costate shell wider than long; circular foramen in pedicle beak. *T. gibbosa* (Hall) (8*a–d*, brachial, pedicle, posterior, and anterior views, ×1), Middle Devonian, New York; *T. multistriata* (Hall) (11, section showing spiralium and jugum, ×1), Lower Devonian, Tennessee.

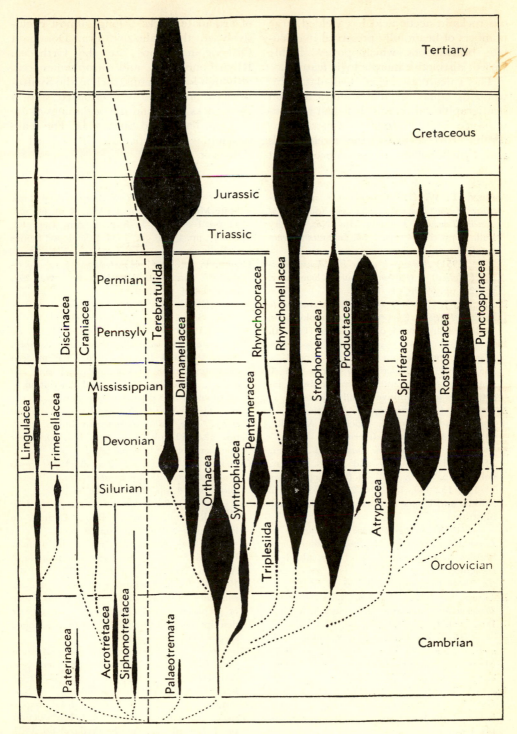

Fig. 6-40. **Geological distribution of brachiopod groups.** Inarticulate brachiopods are indicated at the left of the broken line and articulate groups at the right.

ous rocks in acetic acid has yielded large numbers of beautifully preserved inarticulate brachiopods which possess highly varied, distinctive morphologic features.

The calcareous-shelled articulate brachiopods equal the inarticulates in total stratigraphic range, but their first real importance is in Ordovician deposits. Excepting two divisions, the terebratulids and rhynchonellids, the great development of articulate stocks belongs almost exclusively to the Paleozoic Era. Dominant Ordovician groups are the Orthacea, Rhynchonellacea, and Strophomenacea, although lesser groups, such as the Syntrophiacea, are not to be overlooked; and in these rocks are found the beginnings of the Dalmanellacea and Spiriferida. The main development of articulates is confined to Paleozoic rocks.

REFERENCES

AMSDEN, T. W. (1949) *Stratigraphy and paleontology of the Brownsport formation (Silurian) of western Tennessee:* Peabody Mus. Nat. History Bull. 5, pp. 1–134, pls. 1–34. Includes characteristic brachiopods.

—— (1951) *Brachiopods of the Henryhouse formation (Silurian) of Oklahoma:* Jour. Paleontology, vol. 25, pp. 69–96, pls. 15–20.

BARRANDE, J. (1879) *Systeme silurien du centre de la Bohême, Partie 1:* Recherches paléontologiques, Praha, vol. 5, pp. 1–226, pls. 1–153 (French).

BELL, W. C. (1941) *Cambrian brachiopods from Montana:* Jour. Paleontology, vol. 15, pp. 193–255, pls. 28–37. Describes 45 species and 20 genera of Middle and Upper Cambrian inarticulates and primitive orthid articulates.

BITTNER, A. (1890) *Brachiopoden der Alpinen Trias:* Kaiserlich-Königl. geol. Reichanst. Abh., Bd. 14, pp. 1–325, pls. 1–41 (German).

BRANSON, E. B. (1924) *The Devonian of Missouri:* Missouri Bur. Geology and Mines, 2d ser., vol. 17, pp. 1–279, pls. 1–71 (brachiopods, pp. 71–110, 132–149, 182–204, 228–252). Typical Middle and Upper Devonian assemblages well illustrated.

—— (1938) *Stratigraphy and palentology of the Lower Mississippian of Missouri:* Missouri Univ. Studies, vol. 13, no. 3, pp. 1–205, pls. 1–20 (brachiopods, pp. 21–76); no. 4, pp. 1–242, pls. 21–48 (brachiopods, pp. 15–31, 50–52).

BUCKMAN, S. S. (1901) *Homeomorphy among Jurassic Brachiopoda:* Cotteswold Naturalists Field Club Proc., vol. 13, pp. 231–290, pls. 12–13.

—— (1917) *The Brachiopoda of the Namyan beds, northern Shan States, Burma:* Geol. Survey India Mem. Palaeontologia Indica, n.s., vol. 3, Mem. 2, pp. 1–254, pls. 1–21. Jurassic.

CHAO, Y. T. (1927–28) *Productidae of China:* China Geol. Survey, Palaeontologia Sinica, ser. B, vol. 5, fasc. 2, pp. 1–244, pls. 1–16; fasc 3, pp. 1–81, pls. 1–6.

CLARKE, J. M. (1908–09) *Early Devonic history of New York and eastern North America:* New York State Mus. Mon. 9, pt. 1 (Ann. Rept. 60, vol. 4), pp. 1–366, pls. 1–48; pt. 2 (Ann. Rept. 62, vol. 4), pp. 1–250, pls. 1–34. Brachiopod faunas well illustrated.

CLOUD, P. E., JR. (1941) *Homeomorphy, and a remarkable illustration:* Am. Jour. Sci., vol. 239, pp. 899–904, pl. 1.

—— (1941) *Color patterns in Devonian terebratuloids:* Same, vol. 239, pp. 905–907, pl. 1.

—— (1942) *Terebratuloid Brachiopoda of the Silurian and Devonian:* Geol. Soc. America Spec. Paper 38, pp. 1–182, pls. 1–26. Chief modern study on earliest terebratuloids including 260 species, 31 genera, with careful description of morphology and discussion of evolution.

—— (1948) *Some problems and patterns of evolution exemplified by fossil invertebrates:* Evolution, vol. 2, pp. 322–350, pls. 1–4, figs. 1–4. Describes homeomorphy of brachiopods.

COOPER, G. A. (1930) *The brachiopod genus Pionodema and its homeomorphs:* Jour. Paleontology, vol. 4, pp. 369–382, pls. 35–37.

—— (1937) *Brachiopod ecology and paleoecology:* Nat. Research Council Rept. Comm. on Paleoecology, 1936–1937, pp. 26–53.

—— (1944) *Phylum Brachiopoda:* in Shimer, H. W., & Shrock, R. R., Index Fossils of North America, Wiley, New York, pp. 277–365, pls. 105–143. An indispensible reference containing summary of morphologic features, discussion ot preparation and classification, and concise descriptions of important American brachiopod species, with illustrations.

CRICKMAY, C. H. (1933) *Attempt to zone the American Jurassic on the basis of its brachiopods:* Geol. Soc. America Bull., vol. 44, pp. 877–893, pls. 19–22. Rhynchonellids and terebratulids.

DAVIDSON, T. (1851–1886) *British fossil Brachiopoda:* Palaeont. Soc. Mon., vol. 1, Introduction, pp. 1–136, pls. 1–9; pt. 1 (Tertiary) pp. 1–23, pls. 1–2; pt. 2 (Cretaceous), pp. 1–117, pls.

1–12; pt. 3 (Jurassic) pp. 1–100; pls. 1–18; append., pp. 1–30, pl. A. Same, vol. 2, pt. 4 (Permian), pp. 1–51, pls. 1–4; pt. 5 (Carboniferous), pp. 1–280, pls. 1–55. Same, vol. 3, pt. 6 (Devonian), pp. 1–131, pls. 1–20; pt. 7 (Silurian), pp. 1–397, pls. 1–37. Same, vol. 4, (Carb.-Rec. Suppl.), pp. 1–383, pls. 1–42. Same, vol. 5 (Sil.-Dev. Suppl.), pp. 1–476, pls. 1–21. Same, vol. 6 (bibliog.), pp. 1–163.

——— (1880) *Brachiopoda dredged by H.M.S. Challenger during the years 1873–1876:* Rept. Sci. Results Voyage of H.M.S. Challenger, Zoology, vol. 1, pt. 1, pp. 1–67, pls. 1–4.

——— (1886–1888) *Monograph of Recent Brachiopoda—Part 1:* Linnean Soc. London Trans., vol. 4, pt. 1, pp. 1–248, pls. 1–30, figs. 1–24.

DUNBAR, C. O. (1920) *New species of Devonian fossils from western Tennessee:* Connecticut Acad. Arts and Sci., vol. 23, pp. 109–158, pls. 1–5.

———, & CONDRA, G. E. (1932) *Brachiopoda of the Pennsylvanian System in Nebraska:* Nebraska Geol. Survey, 2d. ser., Bull. 5 pp., 1–377, pls. 1–44. A valuable report which is the only comprehensive presentation of Pennsylvanian brachiopods in North America.

FENTON, C. L., & FENTON, M. A. (1924) *Stratigraphy and fauna of the Hackberry stage of Upper Devonian:* Michigan Univ. Mus. Geol., Contrib. vol. 1, pp. 1–260, pls. 1–45. A varied, well-preserved brachiopod assemblage from Iowa.

GEMMELLARO, G. G. (1898–1899) *La fauna dei calcari con Fusulina della valle del fiume Sosio nella Provincia di Palermo:* Michele Amenta, Palermo, fasc. 4, pp. 231–338, pls. 25–36, figs. 1–46 (Italian). A classic work on Permian brachiopods.

GIRTY, G. H. (1909) *The Guadalupian fauna:* U.S. Geol. Survey Prof. Paper 58, pp. 1–651, pls. 1–31. Figures many Permian brachiopods.

——— (1911) *Fauna of the Moorefield shale of Arkansas:* Same, Bull. 439, pp. 1–148, pls. 1–15. Upper Mississippian brachiopods.

——— (1915) *Fauna of the Wewoka formation of Oklahoma:* Same, Bull. 544, pp. 1–353, pls. 1–34. Middle Pennsylvanian brachiopods.

——— (1920) *Carboniferous and Triassic faunas:* Same, Prof. Paper 111, pp. 641–657, pls. 52–57.

HALL, J. (1867) *Fossil Brachiopoda of Upper Helderberg, Hamilton, Portage, and Chemung groups:* New York Geol. Survey, Paleontology, vol. 4, pp. 1–428, pls. 1–63. Classic report on Devonian brachiopods.

——— (1879) *Fauna of the Niagara group in central Indiana:* New York State Mus. Ann. Rept. 28, pp. 99–203, pls. 3–34. First description of many important Silurian species.

——— (1882) *Descriptions of the species of fossils found in the Niagara group at Waldron, Indiana:* Indiana Dept. Geol. Nat. History Ann. Rept. 11, pp. 217–414, pls. 1–66. Numerous brachiopods.

———, & CLARKE, J. M. (1892–1894) *An introduction to the study of the genera of Paleozoic Brachiopoda:* New York Geol. Survey, Paleontology, vol. 8, pt. 1, pp. 1–367, 40 pls., figs. 1–39; pt. 2, pp. 1–394, pls. 21–84, figs. 40–251. The largest, most comprehensive work yet published on American brachiopods; contains different text and many more figures than the parallel work intended as a book for students.

——— & ——— (1892–1894) *An introduction to the study of the Brachiopoda:* New York State Geol. Ann. Rept. 11, pp. 133–300, pls. 1–22, figs. 1–286 (same, New York State Mus. Ann. Rept. 45, pp. 449–616); Ann. Rept. 13, pp. 749–1015, pls. 23–54, figs. 287–669; same, Ann. Rept. 47, pp. 945–1137, illus.).

———, & WHITFIELD, R. P. (1872) *Descriptions of invertebrate fossils, mainly from the Silurian system:* Ohio Geol. Survey, Palaeontology, vol. 2, pt. 2, pp. 65–179, pls. 1–13. Includes numerous brachiopods.

JONES, O. T. (1928) *Plectambonites and some allied genera:* Great Britain Geol. Survey, Palaeontology, vol. 1, pt. 5, pp. 367–527, pls. 21–25. Important study of Ordovician and Silurian strophomenids.

KING, R. E. (1931) *Permian fauna of the Glass Mountains:* Texas Univ. Bull. 3042, pp. 1–245, pls. 1–44. Includes only brachiopods; belongs with Girty's Guadalupian report as chief source of information on Permian forms.

KING, R. H. (1938) *New Chonetidae and Productidae from Pennsylvanian and Permian strata of north-central Texas:* Jour. Paleontology, vol. 12, pp. 257–279, pls. 36–39. Excellent figures.

KOZLOWSKI, R. (1914) *Les brachiopodes du Carbonifére superieur de Bolivie:* Annales de paléontologie, vol. 9, pp. 1–100 (French).

——— (1929) *Les brachiopodes gothlandiens de la Podolie Polonaise:* Palaeontologia Polonica, vol. 1, pp. 1–254, pls. 1–12, figs. 1–95 (French). Contains important contributions to brachiopod morphology and excellent description of Silurian species from the eastern Baltic region.

MEEK, F. B. (1873) *Descriptions of invertebrate fossils of the Silurian and Devonian Systems:* Ohio Geol. Survey, vol. 1, pt. 2, pp. 1–243, pls. 1–23. Descriptions and figures of many important brachiopod species.

MUIR-WOOD, H. M. (1928) *The British Carbon*

iferous Producti II—Productus (sensu stricto) and longispinus groups: Great Britain Geol. Survey Mem., Palaeontology, vol. 3, pt. 1, pp. 1–217, pls. 1–12, figs. 1–35.

——— (1934) On the internal structure of some Mesozoic Brachiopoda: Royal Soc. London Philos. Trans., ser. B. vol. 223, pp. 511–567, pls. 62–63, figs. 1–14.

——— (1936) The Brachiopoda of the Fuller's Earth: Palaeont. Soc. Mon. 1935, pt. 1, pp. 1–144, pls. 1–5. Jurassic rhynchonellids and terebratulids from England.

PAECKELMANN, W. (1930–1931) Die fauna des deutschen Unterkarbons, Die Brachiopoden: Preuss. geol. Landesanstalt, Heft 122, pp. 143–326, pls. 9–24; Heft 136, pp. 1–352, pls. 1–41, figs. 1–14 (German).

PERCIVAL, E. (1944) A contribution to the life-history of the brachiopod Terebratella inconspicua Sowerby: Royal Soc. New Zealand Trans., vol. 74, pp. 1–23. Establishes dorsal position of pedicle valve and ventral position of brachial valve in early development.

RAYMOND, P. E. (1911) Brachiopoda and Ostracoda of the Chazy: Carnegie Mus. Annals, vol. 7, pp. 215–259, pls. 33–36, figs. 1–26.

SAHNI, M. R. (1929) A monograph of the Terebratulidae of the British chalk: Palaeont. Soc. Mon. 1927, pp. 1–62, pls. 1–10.

ST. JOSEPH, J. K. S. (1938) Pentameracea of the Oslo region: Norsk geol. tidssk., vol. 17, pp. 225–336, pls. 1–8.

SALMON, E. S. (1942) Mohawkian Rafinesquinae: Jour. Paelontology, vol. 16, pp. 564–603, pls. 85–87. Good paper on morphology and classification of important Ordovician strophomenids.

SAVAGE, T. E. (1913) Stratigraphy and Paleontology of the Alexandrian series in Illinois and Missouri: Illinois Geol. Survey Bull. 23, pp. 67–160, pls. 3–9. Early Silurian brachiopods.

SCHUCHERT, C. (1897) A synopsis of American fossil Brachiopoda: U.S. Geol. Survey Bull. 87, pp. 1–464. Important although largely superceded.

——— (1913) Brachiopoda: in Eastman, C. R., & Zittel, K. A., Textbook of palaeontology, Macmillan & Co., Ltd., London, 2d ed., vol. 1, pp. 355–420, figs. 526–636. Classifies articulates in orders Protremata and Telotremata; useful selected bibliography of foreign literature.

———, & LEVENE, C. M. (1929) Brachiopoda: Fossilium Catalogus, pars 42, pp. 1–140. Lists all described genera and gives classification.

———, & COOPER, G. A. (1932) Brachiopod genera of the suborders Orthoidea and Pentameroidea: Peabody Mus. Nat. History Mon. 4, pt. 1, pp. 1–270, pls. 1–29, figs. 1–36. A thorough study utilizing shell structure and internal features for discrimination of numerous genera; good description of morphology.

STAINBROOK, M. A. (1938–1943) [Brachiopods] from the Cedar Valley beds of Iowa: Jour. Paleontology, (Atrypa, Stropheodonta), vol. 12, pp. 229–256, pls. 30–35; (Terebratulacea), vol. 15, pp. 42–45, pls. 7–8, figs., 1–10; (Inarticulata, Rhynchonellacea, Rostrospiracea), vol. 16, pp. 604–619, pls. 88–89, figs. 1–6; (Strophomenacea), vol 17, pp. 39–59, pls. 6–7; (Spiriferacea), vol. 17, pp. 417–450, pls. 65–70, figs. 1–14; (Pentameracea) Am. Midland Naturalist, vol. 19, pp. 723–739, pls. 1–2, figs. 1–8; (Orthids), vol. 23, pp. 482–492, figs. 1–37; (Elytha), vol. 24, pp. 414–420, figs. 1–29.

——— (1945) Brachiopoda of the Independence shale of Iowa: Geol. Soc. America Mem. 14, pp. 1–74, pls. 1–6, figs. 1–2. Devonian.

——— (1947) Brachiopoda of the Percha shale of New Mexico and Arizona: Jour. Paleontology, vol. 21, pp. 297–328, pls. 44–47.

SUTTON, A. H. (1938) Taxonomy of the Mississippian Productidae: Jour. Paleontology, vol. 12, pp. 537–569, pls. 62–66. Morphology and classification of 16 genera.

THOMAS, I. (1910) The British Carboniferous Orthotetinae: Geol. Survey Great Britain Mem., Paleontology, vol. 1, pt. 2, pp. 83–134. Describes strophomenids.

——— (1914) The British Carboniferous Producti—Genera Pustula and Overtonia: Same, vol. 1, pt. 4, pp. 197–366.

THOMSON, J. A. (1927) Brachiopod morphology and genera (Recent and Tertiary): New Zealand Board Sci. and Art (Wellington, N.Z.) Manual 7, pp. 1–338, pls. 1–2, figs. 1–103. One of the best available texts on brachiopods but barely mentions pre-Cenozoic forms.

TSCHERNYSCHEW, T. (1902) Die obercarbonischen Brachiopoden des Ural und des Timan: Mém. comité géol., vol. 16, no. 2, Lief. 1, pp. 1–432 (Russian), 433–749 (German), figs. 1–85, Lief 2, pls. 1–63.

ULRICH, E. O., & COOPER, G. A. (1938) Ozarkian and Canadian Brachiopoda: Geol. Soc. America Spec. Paper 13, pp. 1–323, pls. 1–57, figs. 1–14. Chief source of information on brachiopods from rocks generally classed as uppermost Cambrian and Lower Ordovician; introductory pages (1–22) give useful general discussion.

WALCOTT, C. D. (1884) *Paleontology of the Eureka District, Nevada:* U.S. Geol. Survey Mon. 8, pp. 1–298, pls. 1–24. Includes Cambrian to Carboniferous brachiopods.

———— (1912) *Cambrian Brachiopoda:* Same, Mon. 51, pp. 1–872, pls. 1–104. Treats genera and species of the entire world.

WANG, Y. (1949) *Maquoketa Brachiopoda of Iowa:* Geol. Soc. America Mem. 42, pp. 1–55, pls. 1–12. Excellent illustrations of an Upper Ordovician fauna.

WELLER, S. (1914) *Mississippian Brachiopoda of the Mississippi Valley basin:* Illinois Geol. Survey Mon. 1, pp. 1–508, pls. 1–83. Best available work on this group.

WINCHELL, N. H., & SCHUCHERT, C. (1895) *Lower Silurian [Ordovician] Brachiopoda of Minnesota:* Minnesota Geol. Survey, vol. 3 pt. 1, pp. 333–474, pls. 30–34, figs. 21–34.

CHAPTER 7

Mollusks

One of the main groups of invertebrates, especially from the standpoint of paleontological study, is the assemblage contained in the phylum Mollusca. Here belong the almost endless variety of marine, fresh-water, and terrestrial snails; oysters, mussels, scallops, and many other sorts of clams living in the sea and in streams, lakes, and ponds; squids, cuttlefishes, octopuses, the pearly nautilus, and a host of extinct related forms, all restricted to a marine habitat; and the lesser groups called chitons (coat-of-mail shells) and scaphopods (tusk shells), which live on the sea bottom. Among the 60,000 or more living species of mollusks are forms ranging in adult size from snails less than 0.5 mm. in length to the giant squids of Atlantic waters, which attain a length of 16 m. (53 ft.). This cephalopod, with body 20 ft. long and tentacles which reach nearly 35 ft., is much the largest invertebrate known. Some mollusks are active swimmers, some float or drift passively about as planktonic organisms, some burrow into mud or sand, bore into wood or rock, or attach themselves solidly to almost any firm foundation, but most crawl about by means of a muscular organ of locomotion, called the foot.

The name Mollusca (*mollusca*, soft-bodied) was originally employed for soft-bodied invertebrates such as slugs, sea squirts (tunicates), and others which lack hard parts. Shell-bearing animals were excluded. Linné (1758) thus differentiated between groups having hard parts and those which lack them. The foundation of modern classification was established by the French naturalist, Cuvier, who in 1795 defined the fundamental similarities in structure of soft-bodied members of the phylum and many shell-bearing forms, and divided mollusks into three classes called Gastropoda, Cephalopoda, and Acephala (clams). Subsequently, brachiopods, cirripeds (barnacles), and some types of worms were grouped for many years among the mollusks but now are removed.

The chief distinguishing features of mollusks are (1) complete or nearly complete absence of segmentation; (2) normally elongate, bilaterally symmetrical organization of body; (3) enclosure of the viscera by a body wall, of which the lower part (foot) is modified for locomotion and the upper part (mantle) hangs down as a fold enclosing the free space between it and the body; (4) concentration of sensory structures in a head (except pelecypods); (5) special characters of the digestive and nervous systems; and (6) distinctive aspects of larval stages. The mantle contains cells which secrete the shell. In all mollusks, the shell consists mainly of calcium carbonate, partly or wholly in the form of calcite, or partly or wholly in the form of aragonite. Normally, the shell is external, but it may become completely internal. These main common denominators and many lesser ones clearly differentiate the mollusks as a group of related invertebrates which stand well apart from most others. Whether the Mollusca originated from an annelid source, characterized by segmented body structure, or otherwise, is quite uncertain, but their common features and the divergent trends of molluscan main divisions have been established since the beginning of the fossil record.

CLASSIFICATION

The Mollusca are divided into five classes on the basis of well-defined differences in general form, character of the shell (wherever present), and mode of life (Fig. 7-1). Two classes are distinguished by the broad, flat nature of the foot in all but a few kinds of specialized members: these are the chitons (Amphineura) and snails (Gastropoda). The chitons have hard parts consisting of eight separate calcareous pieces which articulate with one another, or they lack a skeleton. The snails are univalves; their shell consists of a simple cone with somewhat broadly flaring sides or of a spirally coiled tube which varies greatly in cross section in different species but invariably widens toward the aperture. Like some chitons, there are snails which have no shell; these are commonly known as slugs. Other univalves among the mollusks include the classes called Cephalopoda (nautilus and allied forms) and Scaphopoda (tusk shells). Some fossil cephalopods have a straight or slightly curved elongate conical shell, but most representatives of this class have planispirally coiled shells. A distinguishing feature of cephalopod shells is the presence

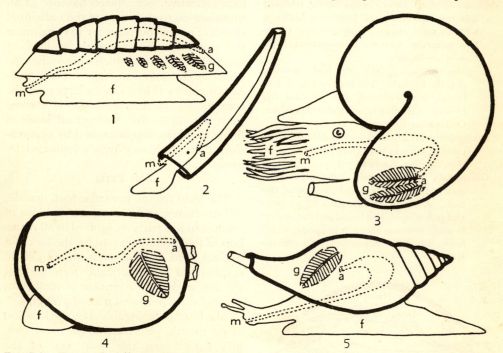

FIG. 7-1. **Types of mollusks.** The diagrams show characteristic forms of the shell (heavy lines) in each of the five molluscan classes and location of the digestive tract and gills (a, anus; f, foot; g, gill; m, mouth).

1, Amphineura (chitons), shell segmented in form of eight transverse pieces; foot broad and flat; gills and anus in posterior position.

2, Scaphopoda (tusk shells) shell elongate tubular, open at both ends; foot conical; respiration by the mantle; no gills.

3, Cephalopoda (nautilus), shell planispirally coiled; soft parts in living chamber next to shell aperture; head-foot bears numerous tentacles; gills and anus in posterior position.

4, Pelecypoda (clams), distinguished by the two-valved nature of shell, hinged dorsally; foot laterally compressed; gills and anus in posterior position.

5, Gastropoda (snails), shell generally coiled in a conical spire; foot mostly broad and flat; gills and anus in anterior position as result of torsion (but primitively or secondarily in posterior position in many forms).

of internal partitions which make chambers. Some kinds have solid calcareous hard parts inside the body, and some lack any sort of shell. The clams, which belong to the class Pelecypoda, are bivalves. They have two shells which generally are equal in size and mirror image in form with respect to each other, with hingement of the valves at the dorsal edge of the animal.

Definition of main characters of the divisions of mollusks is given in the following outline of classification.

Main Divisions of Mollusca

Mollusca (*phylum*), invertebrates having typically bilateral symmetry of body organization, without segmentation; body covering (mantle) generally secretes a calcareous shell. Cambrian–Recent.

Amphineura (*class*), elongate body covered dorsally by eight-piece shell or naked; foot broad, head reduced, gills posterior but in some extending forward also along sides. Ordovician–Recent.

Scaphopoda (*class*), shell and mantle slenderly tubular, open at both ends, foot conical, no gills. Silurian–Recent. tusk shells

Gastropoda (*class*), body generally asymmetrical in spirally coiled shell; head distinct, with one or two pairs of tentacles and pair of eyes; foot broad and flat, except in some nektonic forms (pteropods) which have winglike expansions. Cambrian–Recent.

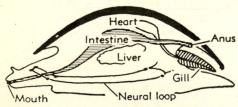

FIG. 7-2. **Generalized ancestral mollusk.** The shell (black) and main soft parts belonging to a hypothetical archaic type of mollusk are shown in median longitudinal section. Such a form has structural features which suit the ancestor of all molluscan classes. A simple digestive tract extends from front to rear, where the anus adjoins a pair of feather-shaped gills in the mantle cavity. From ganglia in the head, nerve cords form a ventrally located loop.

Cephalopoda (*class*), shell external, internal, or none; large head with eyes, horny jaws, and many tentacles, fused with foot. Cambrian–Recent.

Pelecypoda (*class*), shell mostly consisting of bilaterally symmetrical valves hinged dorsally, with ligament; foot generally hatchet-shaped, pointed; head lacking; gills posterior. Cambrian–Recent.

ANATOMICAL FEATURES

Despite the wide range in body form and mode of life, the mollusks are remarkably alike in essential anatomical characters. It is desirable to survey these features in a collective, generalized manner as an introduction to the study of the individual classes. For this purpose, we may examine the structure of a synthetic mollusk of simplest body plan, which approaches the probable common ancestor of all molluscan classes (Fig. 7-2). Evidence for the construction of this prototype is derived from a comparative study of all kinds of living mollusks, supplemented by observations of evolutionary trends indicated by fossils.

Soft Parts

The head region, muscular foot, mantle, gill structures, and viscera are divisions of molluscan soft parts recognized in all members of the phylum, except for the lack of a distinct head in pelecypods and of gills in scaphopods. It is reasonable to judge that the unspecialized progenitor of molluscan invertebrates possessed a head and gills.

The head is distinguished by the location of the mouth, placement of the nerve ganglia constituting the front end of the nervous system, and position of the sense organs such as the tentacles and eyes. Crawling mollusks move in the direction of the head, and accordingly the head region is defined as anterior. The opposite extremity is posterior.

The foot is a broad fleshy mass composed chiefly of muscle fibers but containing also cavities into which blood (generally colorless) may be forced. Typically, the foot has a flat sole.

Dorsal and lateral parts of the molluscan body are covered by the mantle, which is a layer of tissue containing cells that secrete the shell. Between marginal parts of the mantle and the body, especially at the rear of the mollusk, is an open space termed the mantle cavity. Owing to torsion of the body in most snails, the mantle cavity comes to have an anterior position in these mollusks, but primitively this open space and the gills which it contains are placed posteriorly. The anus opens into it.

Respiration of mollusks normally is by means of gills, of which the common, unspecialized type consists of a feather-like structure of minute tubes covered with cilia. The prototype mollusk almost certainly was provided with a pair of such gills in the mantle cavity at the rear end of the body.

The viscera comprise internal organs for digestion, excretion, and circulation. A digestive tract extends backward from the mouth through the median and dorsal parts of the body, terminating at the anus. A somewhat expanded part of the tract corresponds to a stomach, and this is adjoined by the chief digestive organ, the so-called liver. A heart, which is three-chambered in most mollusks, is located dorsally, and it connects by vessels with the gills. Kidneys are present.

A nervous system consists of the cerebral ganglia and various others which are joined by a cord of nerve fibers in the form of a loop. In most mollusks, as in the primitive synthetic form here described, this loop is a simple circle, but in many gastropods which have undergone torsion, the loop is twisted into a figure-eight pattern.

Hard Parts

The generalized ancestral mollusk may be pictured as having a calcareous shell, because nearly all mollusks bear such shells. Modern species which lack hard parts are surely not primitive in this respect, but specialized, for the evolutionary tendency toward reduction of the shell is seen in numerous molluscan stocks. The simplest type of shell is a single shallow inverted bowl covering the dorsal side of the body, and probably this univalve structure of the hard parts distinguished earliest invertebrates classifiable as mollusks.

MINOR CLASSES

The minor groups of chitons and scaphopods are described in this chapter because they cannot be treated conveniently elsewhere and because the chitons have special significance for study of molluscan evolution. They are judged to be more primitive than any other group of living mollusks.

Chitons

General Features. The chitons (Amphineura) are sluggish crawlers on the sea bottom, living mostly in comparatively shallow water. They have an evenly rounded elliptical outline and thickness along the mid-line which is roughly about one fourth of their greatest width (Fig. 7-3). The body is bilaterally symmetrical. No head is visible from the dorsal side, and there are no tentacles or other projections reaching beyond the periphery. A marginal band of uniform width, but wider in some species than in others, commonly is differentiated from the central portion of the back. This band, known as the **girdle,** belongs with the soft parts of the animal, although its covering may be studded with needle-like or scalelike calcareous spicules. It is formed by the mantle and is quite flexible. The surface of the girdle may be smooth or shaggy. The central part of the back consists of shell, which is not a single arched oval plate but a series of eight articulating pieces, with joints between them running transverse to the axis of the body. If disturbed, the chiton can roll up like a pill bug, bending at each joint between the dorsal cross plates. The surface of the plates generally is somewhat sculptured or roughened. In a few chitons, the dorsal shell may be concealed entirely by extension of the mantle over it, and in such members of the group, the girdle is not clearly differentiated.

The underside of a chiton is mostly a broad flat muscular foot, like that of a snail. Along the border, a groove may be visible, for the mantle cavity opens in this position. At one end is the head, which is identified merely by a constriction separating it from the foot and by the presence of the mouth opening. The anus is at the opposite extremity (Fig. 7-3, *1b, c*).

An average-sized adult chiton is an inch or two in length, and its width is one half to two thirds of the length. The largest known specimens attain a length of 13 in., but many full-grown individuals are less than a half inch long. In color, they may be black, brick-red, brown, greenish, or otherwise tinted.

These mollusks are exclusively herbivorous, eating seaweed or subsisting on plant debris sucked up along the sea floor. The sexes are separate. Eggs are laid in jelly-like strings or masses, similar to those of many gastropods, and the larvae have a free-swimming stage before they settle to the bottom and begin to develop the form of their parents.

The description just given applies to chitons, but not to shell-less invertebrates of wormlike form which have been classed with the chitons as members of the Amphineura. These soft-bodied "worms," called Aplacophora, are doubtfully placed among mollusks, and because they do not occur in the fossil record, we may ignore them. For chitons alone, the term Loricata is sometimes employed.

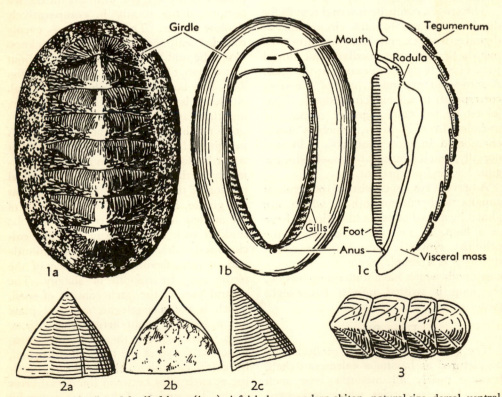

FIG. 7-3.　**Modern and fossil chitons** (*1a–c*). A fairly large modern chiton, natural size, dorsal, ventral, and median longitudinal section (diagrammatic). (*2a–c*) Dorsal, ventral, and side views of terminal valve of *Priscochiton canadensis* (Billings) (×2), Middle Ordovician (Blackriveran), Ontario. (3) Anterior valves of *Gryphochiton priscus* (Münster) (×1), Lower Mississippian (Tournaisian), Belgium.

Soft Parts. The main features of soft-parts anatomy of the chitons are more important in paleontological study than the characters of the hard parts. This seems anomalous indeed, until we explain that a comparison of the internal organization of the chitons with the structure of gastropods and other mollusks throws considerable light on the evolutionary significance of the symmetrical or asymmetrical arrangement of the body parts. The shell of chitons furnishes no help in this direction, and accordingly, analysis of its special features merits effort only in proportion to the importance of the group as fossils. This is not very great.

The soft parts of a chiton are essentially those of the simplified, perfectly symmetrical prototype mollusk, which has posterior placement of the respiratory apparatus and anus. The chiton does not quite meet these specifications, because the gills, although symmetrically paired, are multiplied in number, evidently by proliferation from an original single pair. Some chitons have two gills on each side of the anus, but in others the number ranges to 40 on each side, filling most of the anteriorly extended mantle cavities which stretch like grooves along the sides of the body to the head. Whether few or many gills are present, one pair dominates in size, and because this pair is located at or near the rear end of the mantle cavity, it may be interpreted as the primitive initial pair, corresponding to the single pair in primitive snails. The gill structure also is the same, having double rows of leaflets.

The alimentary tract is a tube which extends from the mouth, on the underside of the head, to the anus, located on the mid-line of the body at the rear. The chiton has a hard rasplike mouth structure composed of many horny teeth borne on a flexible tough ribbon. This is called a radula, and generally it is like those of gastropods. The amphineuran heart contains two auricles. There are two kidneys.

We cannot be wrong in thinking that the anatomy of the chitons closely approaches the internal organization of the ancient primitive snails, which are characterized by bilateral symmetry of form. Chitons are specialized in adaptation of their body for longitudinal bending, but otherwise they have persisted throughout geologic time with extremely little change. The archaic symmetrical gastropods, however, vanished early in Mesozoic time.

Hard Parts. The discrete shell elements of chitons comprise the eight main pieces somewhat inaptly called **valves.** They occupy the central part of the back. In addition, innumerable minute calcareous spicules are borne by the girdle. Fossil chitons having all eight main skeletal parts in natural union with one another are known, but more commonly, the pieces are disarticulated. The spicules should be found among washings of shallow-water marine sediment, associated with various microfossils, but so far they have not been distinguished.

The valves are two-layered. An external rough-surfaced layer, which is perforated by coarse and fine pores, is termed **tegmentum** (*tegmen*, roof). The inner layer has a very different structure and appearance, for it is a dense porcelaneous laminated element. Because it may project shingle-fashion in joining with the plates next in series, this layer is called **articulamentum.** Front and rear valves are readily distinguished from intermediate ones by their shape (Fig. 7-3).

Paleozoic chitons are distinguished from later ones by the weak development or absence of imbrication of the valves. They fit against one another instead of overlapping. Mesozoic and Cenozoic chitons, like modern kinds, have more or less prominent extensions of the lower layer as an aid in articulation. The geologic range of this class is from Ordovician to Recent. All together, about 100 species of fossil chitons have been described.

Scaphopods

The scaphopods (Fig. 7-4) hold unquestioned status as members of the Mollusca in good standing, but if they were known only as fossils, it is almost certain that they would be put on an *incertae sedis* (uncertain classification) list. No other mollusks have relatively long, slender tubes which are open at both ends. The popular name, elephant's-tusk shells, is appropriate as a suggestion of the gently curved, tapering shape, but is distinctly a misnomer as regards size. Most living scaphopods have a maximum length of only 80 mm. (3 in.) and a diameter at the larger end of little more than 6 mm. The total length of adults belonging to some genera is only 5 mm. However, a fossil scaphopod from Pennsylvanian rocks of Texas attains a diameter of 40 mm., and the length of a complete individual probably is 60 cm. (2 ft.) or more.

General Features. The scaphopods are exclusively marine animals, which live from low-tide limits outward into fairly deep water, burrowing a short distance in the sediment of the sea bottom. In normal position, they do not lie flat but stand steeply upright, the wide end of the tube

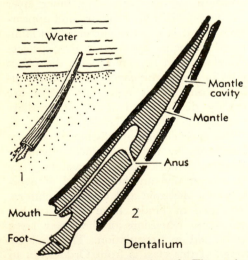

Fig. 7-4. **A modern scaphopod.** The entire animal (×1) in natural living position, the anterior part of the shell buried in sediment of the sea floor, and the posterior part projecting into the water. At right is a diagrammatic longitudinal section (enlarged) showing the head, foot, digestive tract, and mantle cavity.

buried and the narrow end projecting into the water (Fig. 7-4, *1*). They are reported to feed on Foraminifera, young pelecypods, and similar animals, being carnivores rather than plant eaters. The sexes of scaphopods are separate, and eggs are laid singly.

Soft Parts. Description of the anatomy of scaphopods may be based on the genus which is most common both among fossils and living forms. This is known as *Dentalium*.

The soft parts of *Dentalium* consist of a poorly developed head, a conical foot, a visceral mass, and the mantle (Fig. 7-4, *2*). All may be enclosed within the tubular shell, but ordinarily the foot and sensory appendages of the head project from the larger, down-pointed opening of the shell. The head is a two-lobed cylindrical mass which contains the mouth, and it bears long slender appendages which are sense organs and which serve in capturing small organisms used as food. The foot is capable of considerable extension, and its tip is effective in digging. The viscera include a simple digestive tract, liver, heart, and kidneys; there are no gills. The mouth parts contain a radula, which serves to shred food into particles for passage through the esophagus to the stomach. Waste is delivered by an intestine to the anus, which opens into the mantle cavity near the mid-length of the shell. Blood is oxygenated along the inner surface of the mantle, water for respiration being conducted inward from the small opening of the shell which projects upward. The mantle cavity extends throughout the length of the shell, opening at the large end as well as at the small one. As a whole, the soft parts are bilaterally symmetrical, the head and visceral mass being located in a dorsal position and the mantle cavity, ventral.

Shell. The shell of *Dentalium* is circular in cross section. It is slightly curved longitudinally, and it tapers evenly throughout its length. The large end is designated as anterior, because the head is near the large opening, and the small end is posterior. The exterior of the shell commonly bears fine growth lines, which run transversely around the shell, and more or less prominent longitudinal costae. There is no operculum.

Known fossil scaphopods range from Silurian to Pleistocene. They are not very abundant, except locally, and because simple tapered tubular shells can exhibit few sorts of difference other than size and surface markings, the scaphopods are not very useful as index fossils. Their remains are most abundant in Cenozoic deposits.

REFERENCES

BORRADAILE, L. A., *et al.* (1948) *The Invertebrata:* Cambridge, London, pp. 1–725, figs. 1–483 (mollusks, pp. 543–605, figs. 372–414). An authoritative, compact description of main characters.

HENDERSON, J. B. (1920) *A monograph of the East North American scaphopod mollusks:* U.S. Nat. Mus. Bull. 111, pp. 1–130, pls. 1–20.

PELSENEER, P. (1906) *Mollusca:* in Lankester, E. R., Treatise on zoology, A. & C. Black, London, pp. 1–355, figs. 1–301 (Amphineura, pp. 40–65, figs. 23–43; Scaphopoda, pp. 197–204, figs. 181–186).

PILSBRY, H. A. (1913) *Scaphopoda, Amphineura:* in Eastman, C. R., & Zittel, K. A., Textbook of palaeontology, Macmillan & Co., Ltd., London, 2d. ed, vol. 1, pp. 508–513, figs. 837–842.

ROBSON, G. C. (1929) *Scaphopoda:* Encyclopaedia Britannica, 14th ed., vol. 20, p. 51, figs. 1–4. A good summary.

CHAPTER 8

Gastropods

Gastropods are one of the main divisions of the phylum Mollusca. They include animals which bear a coiled or uncoiled calcareous shell and others, called slugs, which have no hard parts. Originally, they were exclusively marine, but especially in Mesozoic and Cenozoic time, large numbers of them became adapted for life in fresh waters and in the air, enabling them to invade dry land. A large majority of the group, however, has maintained existence in the sea. Taken together, recent and fossil species of gastropods considerably outnumber all other species of mollusks combined. The average size of shells belonging to the group, commonly called snails, is approximately 25 mm. (1 in.) in length or diameter, but fully grown adults of different kinds range from less than 0.5 mm. to approximately 60 cm. (2 ft.).

As illustrated by the common land snail, *Helix*, and shallow-water marine gastropod, *Buccinum* (Fig. 8-1, *1*, *5*), the animal is seen to be an elongate flat-soled organism which carries a spirally coiled shell on its back. The head is identified both by observing the creature's direction of movement and by the presence at one extremity of the body of paired tentacles, which are placed in the vicinity of the mouth, generally above it. They may bear eyes. The muscular sole or foot, on which the snail crawls, is not associated closely with the stomach, located inside the shell, and hence the name gastropod (*gastro*, stomach; *pod*, foot) is not really very appropriate. The tip of the spiral shell points backward, and the opening into the largest, last-formed turn of the shell is in a forward position, directed downward. The "back" of the animal extends into

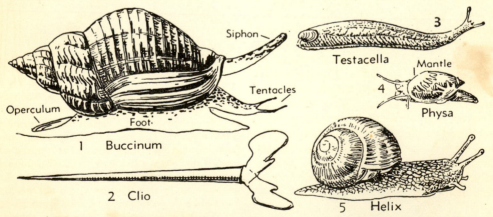

FIG. 8-1. **Types of living gastropods.** All except *Clio* (2), which is a pelagic swimming snail, crawl about on their broad muscular foot, carrying the shell with its apex pointed obliquely or directly toward the rear. Most snails can pull themselves inside their shell, and many are equipped with a horny or calcareous lid (operculum) which fits over the aperture when the animal is inside. *Buccinum* (1) is a shallow-water marine gastropod. *Testacella* (3) and *Helix* (5) are terrestrial air breathers. *Physa* (4) is a fresh-water snail, which, like many marine forms, may extend the shell depositing mantle over the exterior of the shell ($\times$1, except 2, $\times$2).

this opening to and beyond the area of muscle attachment, where the soft parts are firmly joined to the shell interior. Some snails can pull the movable soft parts into the shell and close the opening with a horny or calcareous lid, called operculum (Fig. 8-1, *7*), but others cannot do this (for example, *Testacella*, Fig. 8-1, *3*, which carries a mere vestige of shell at its rear).

The gastropods chiefly live on shallow sea bottoms, but some have been dredged from ocean depths of more than 3 miles. Others swim in near-surface waters of the open oceans far from land. Terrestrial snails are the only mollusks which have acquired lungs, and thus have been able to move out of water bodies. They can climb trees and ascend mountains to an elevation of 18,000 ft. above sea level. Most snails feed on various sorts of vegetation, such as abundant marine algae or leaves of land plants. Some are adapted to eating flesh, either that of dead animals which they chance to find or soft parts of invertebrates which they kill, as by boring a hole through the shell of the victim (Fig. 1-1). A few live commensally with other invertebrates, partly buried in a sponge or attached over the anal vent of a crinoid, feeding mainly on this animal's refuse. A few are parasites. Most sessile forms, like firmly cemented snail colonies (*Vermetus*) resembling an organ-pipe coral, subsist on microscopic planktonic organisms carried to them in the sea water. On the whole, gastropods are inactive animals, well characterized by the adjective sluggish, a word derived from slug. They depend on their shell and on their retiring habits to protect them from enemies. Many can burrow in mud or sand, protected by devices which prevent overfouling of their gills by sediment.

SOFT PARTS

Paleontological study of the gastropods first requires examination of their most significant anatomical features. The soft parts of a gastropod are divisible into three

main sections: head, visceral mass, and integument, which includes mantle and foot.

Head

The head is a mobile anterior part of the body which bears the mouth, eyes, and one or two pairs of sensory tentacles.

The mouth is simply the opening at the front or underside of the head which leads to a mouth cavity and thence to the esophagus. The mouth cavity has distinctive peculiarities which have been employed in classification of living gastropods, but because these structures are missing in fossils, they cannot be used directly in classifying them. Beginning just inside the mouth is a long, muscularly movable rasping mechanism, which is composed of many minute horny teeth arranged in transverse rows on a tough flexible ribbon. This structure, termed **radula,** functions for tearing food into particles (Fig. 8-2). In addition, most

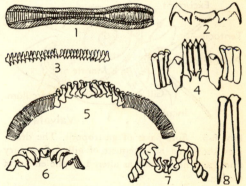

FIG. 8-2. **Types of gastropod radulas.** The tiny horny teeth, arranged in transverse rows on a flexible cartilaginous base, are an effective rasping mechanism which is carried in protrusible front parts of the alimentary tract. The pattern of teeth in the rows is a character useful for definition of main groups of living snails, but because the teeth are not preserved in fossils, it is not directly applicable to classification of fossil gastropods. (1) Whole radula of a fresh-water snail, *Paludina* (Taenioglossa). (2–8) Teeth of a single crossrow of marine snails and a chiton; (2) *Nassa* (Rachiglossa); (3) *Ianthina* (Taenioglossa); (4) *Patella* (Docoglossa); (5) *Trochus* (Rhipidoglossa); (6) *Cypraea* (Taenioglossa); (7) *Trachydermon* (Amphineura); (8) *Conus* (Toxiglossa).

gastropods have a pair of laterally placed horny jaws adjoining the radula, or the jaws may be fused together so as to operate as a unit with the radula. The mouth parts of many snails can be so deeply infolded as to leave a surficial opening which is not the true mouth. The inbent walls can be pushed outward far enough to bring the tip of the radula even with the outer sur-

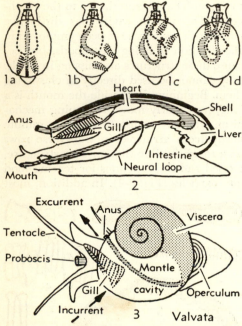

FIG. 8-3. **Soft parts of gastropods.** The drawings show the arrangement of gills, alimentary tract, and nerve cord as altered by the effects of torsion.

1a–d, Dorsal views of the shell and soft parts of hypothetical ancestral gastropod (1a) and inferred steps in twisting the body (1b–d) so as to change location of gills and anus from posterior to anterior and produce a figure-eight pattern of the neural loop. This torsion occurs within a few days during early ontogeny. (*Modified from Robert.*)

2, Median longitudinal section of a prosobranch gastropod, showing typical arrangement of soft parts.

3, Dorsal view of a modern marine prosobranch, *Valvata piscinalis*, showing asymmetry of shell and soft parts. The direction of water movement which aerates the single gill is indicated (Yonge).

face or considerably beyond it, like the extended finger of a glove. This structural feature of the mouth region is called a **proboscis.** It can be introverted or everted at will.

The eyes are borne at the tips of tentacles of some gastropods, but in others they are situated near the base of the tentacles. In all but some very primitive forms, they have a crystalline lens, and all have a retina of pigmented, light-sensitive cells.

The tentacles are not segmented structures, as in insects, but are merely slender extensions of the body wall which contain soft tissue and nerve fibers. They can be retracted by muscles.

Viscera

The viscera comprise parts of the digestive tract behind the mouth cavity, a relatively large liver, kidney, heart, circulatory passageways, reproductive organs, and nervous system (Fig. 8-3). This division of the body overlies the foot and extends backward so as to fill the shell cavity, which may be a very long, narrow, spirally twisted space. Thus, the viscera may be extended in a decidedly wormlike form.

The anterior part of the digestive tract is a long tubular esophagus, which near the mouth contains the opening of a salivary gland. The secretion of this gland in a few genera is virulently poisonous. If a live *Conus* is held in the hand and its radula makes a puncture in the skin, the entrance of a minute amount of this snail's venom into a man's body causes quick death. Beyond the esophagus is a thin-walled stomach, which may be lined with hard cuticle. It serves no digestive function but is merely for storage of food until the liver comes into play, converting the food to uses of the body. The kidney aids withdrawal of waste. The heart pumps blood, which is mostly a colorless fluid but may be red from hemoglobin or blue from hemocyanin.

In living gastropods, reproductive organs consist of a single, unpaired gland. The sexes are separate, except for all land snails (pulmonates) and a few hermaphroditic genera of other groups.

The nervous system is made up of ganglia, which are concentrated clusters of nerve cells in the head and various other parts of the body, and nerve cords which connect the ganglia and lead to sensory organs. The placement of the main nerve cords has bearing on the classification of the snails, for among those which have primary or secondarily developed body symmetry, the cords form a single loop, whereas torsion of the body which is expressed by the characteristic asymmetry of most gastropods is represented by a figure-eight pattern of the nerve cords (Fig. 8-3, *1a–d*).

Foot and Mantle

The foot is a fleshy, highly muscular part of the body, which ordinarily has a broadly flattened sole on which the snail crawls. It may be modified, however, into winglike swimming structures (Fig. 8-1, *2*).

The mantle comprises the portion of the body covering which secretes the shell and which hangs downward so as to enclose between itself and the body a respiratory chamber. This mantle cavity is a distinctive feature of the mollusks. Located primitively at the rear (Figs. 7-2; 8-3, *1a*), it has attained a forward position in most snails (Fig. 8-3, *1d, 2, 3*) but is secondarily reversed to a posterior placement in some. Outgrowths of the mantle called **ctenidia** (*ctenos*, comb) are feather-like gills which project into the mantle cavity and are the means of respiration in most gastropods. Water currents are produced by cilia on the surface of the gills. Normally paired, the ctenidia are represented in many gastropods by only a single one (Fig. 8-3, *3*), the other having disappeared; or both may vanish, being replaced by a modified structure of the mantle surface as a means of respiration. Gastropods which lack gills include some marine and fresh-water forms and the air-breathing pulmonates.

An important feature of respiration among aquatic gastropods is the avoidance of fouling the gills by fine sediment. To some extent, the ctenidial filaments are cleaned by the action of the cilia which line them, but this mechanism cannot serve if the water is too turbid. Accordingly, various ways of meeting the problem of getting reasonably clear water into the mantle cavity have been developed. One device is a sensitive "mud-smeller" (**osphradium**), located near the point where incoming water reaches the gill. Experiments with living snails show that when a limit of silt tolerance is reached, the mantle cavity closes, evidently in response to stimulus from this sense organ, for if osphradia are removed, there is no such reaction to the muddy water. When the water clears, the mantle cavity reopens. Snails are thus able to avoid mud, and many choose to live only on clean rocky bottom. A method of cleaning the gills, which supplements the action of cilia, is a spasmodic opening and closing of the mantle cavity in such manner as to produce pulsatory currents that clear the silt away. The most effective protection for the gills against mud, however, is an anterior prolongation of the mantle in the form of an inhalant tube or snout, called a **siphon.** By pulling in clean water through this intake, marine snails belonging to several progressive families are able to travel over muddy bottom, and some burrow in soft sediment of the sea floor.

HARD PARTS

The calcareous shell, such as is secreted by a preponderant majority of gastropods, is the structure on which paleontological study of the group must be based almost exclusively. Naturally this is so, because other hard parts, including the teeth of the radula, are not adapted to preservation. Horny opercula are rarely found as fossils, and those formed of calcium carbonate commonly are separated from shells to which they belong. Most of them, moreover, are not distinctive enough to be very useful in identifying species. Many trails and other marks on bedding planes of stratified rocks undoubtedly represent the work of gastropods. Of course, these are not classifiable as skeletal remains, and they are almost negligible paleontologically. Fortunately, most gastropod shells exhibit features of form and structure which can be used in differentiating species and genera, and in many deposits—even including wind-borne dust (loess)—shells of these mollusks may be extremely abundant. Also, because some environments favorable for snails are ill suited for various other invertebrates, faunas are found which consist almost exclusively of gastropods.

Shells of gastropods may be divided broadly into two groups: those which exhibit little or no sign of coiling, and those which are partly or very distinctly coiled. Such division facilitates description but is not usable as a basis for the definition of major phylogenetic branches, because each of the four classes of snails (amphigastropods, prosobranchs, opisthobranchs, pulmonates) contains noncoiled and coiled types of shells.

Noncoiled Shells

The most common sort of noncoiled gastropod shell has the form of an asymmetrical cone, elliptical or subcircular at the base and generally somewhat more steeply sloping on one side than the other. The apex is more or less excentric. The outer surface may be marked simply by growth lines parallel to the rounded lower margin of the shell, by concentric folds or wrinkles, by radial ridges and grooves, or by combinations of these. The inside is smooth, except possibly for scars of muscle attachment.

Some cap-shaped noncoiled shells are distinguished by the presence of a marginal indentation, which may have the form of a deep narrow **slit** extending well toward the apex. Such indentation is constant in its position and is a structural feature of importance. It marks the outlet of the digestive tract (anus), but because of torsion that has altered the internal structure of the gastropods occupying these shells,

the side which bears the anus is anterior, rather than posterior. The apex of most simple open-based cone-shaped shells is imperforate, but in the keyhole limpet and similar shells, an elliptical opening occurs at the peak of the cone. It is termed **apical aperture**, in order to distinguish it from the true **aperture**, which is the large opening at the base of the shell. The vent at the top serves for the escape of water from the mantle cavity after traversing the gills, and for the anus.

A few gastropod shells consist of nondescript flat rounded plates which may be devoid of surface markings. Commonly these are internal. Other shells, particularly of free-swimming marine snails of the group called pteropods, may represent an extreme type of very narrow elongate cone, which is virtually a tube (Fig. 8-1, 2). Such shells are never perforate at the apex and thus are distinguishable from the tubular cone-shaped shells of scaphopods (Fig. 7-4).

A majority of the noncoiled shells here discussed are classed zoologically with groups having coiled shells, because the earliest-formed parts, constituting the embryonic shell, are found to show coiling. Generally, this delicate, very minute initial shell is lost before the adult growth stage is reached, and in any case, it is so inconspicuous as not to be evident without hunting for it with a lens.

The nomenclature applied to the noncoiled cone-shaped shells is not burdened by technical terms. Descriptive words such as base, apex, sides, aperture, apical aperture, slit, and muscle scars, supplemented by characterization of external ornament, are sufficient.

Coiled Shells

Gastropods generally possess coiled shells, and in most, the plan of coiling gives rise to or is associated with pronounced asymmetry of the body. This is shown by the torsion of soft parts in which the anus and gills have become shifted from their primitive position at the rear of the body to one in front. The gill and kidney on one side of the body tend to diminish in size and may disappear entirely. A few gastropods have curved shells which define only part of a single 360-deg. volution. Although such shells can be said to exhibit only a tendency toward coiling, either incipient or vestigial, they may be associated with those having a distinctly coiled form.

Types of Coiling. Snail shells typically have a spirally coiled structure, and the spiral is invariably one or other of two types: planispiral and conispiral. No snail shell exhibits the form of spiral coil represented by an ordinary spring or wire wound around a cylinder at more or less constant angle.

A simplest and seemingly most primitive type of gastropod coiling is that in which the mid-line of the curved tubular shell lies entirely in a single plane (Fig. 8-4, *1a–c*). This is **planispiral.** The initial turns are close to the point of origin, and the later ones are progressively farther out from the center, following the course of a logarithmic spiral (Fig. 1-12), but the mid-line does not swerve from a plane. The half of a shell lying on one side of the plane of coiling is the mirror image of the other half (Fig. 8-4, *1c*). Such coiling is illustrated by *Bellerophon* (Fig. 8-13, *2a–c*), and it is the normal type of coiling among cephalopods, but few living gastropods have planispiral shells.

Another, much more common pattern of spiral coiling in snail shells is distinguished by deviation from a plane. A component of shift away from a plane produces a form like that of a wire wound around a cone from the apex to its base (Fig. 8-4, *3a*). Such a spirally wound wire, representing the mid-line of the coiled gastropod shell, defines a **conispiral** form. Nearly all snails have gently to steeply sloping conispiral patterns.

A mathematical concept which matches a zoological fact in the gastropods is a conispiral form having negative elevation —in other words, an inverted cone, with the apex directed downward (Fig. 8-4, *4a*).

Some snail shells thus coil in a depressed cone and are designated as **hyperstrophic** (*hyper*, ultra; *stroph*, turn). The normal type of conispiral shell is termed **orthostrophic** (*ortho*, erect).

Some conispiral shells are so flat that the mid-line of all turns of the coil may lie nearly or exactly in a plane. Such shells are not planispiral, because one side is not the mirror image of the other. They may be described as **pseudoplanispiral** (Fig. 8-4, *2a–c*).

The **axis of coiling,** in all types, is an imaginary line drawn through the origin of the spiral in such direction that successive points outward along the spiral mid-line of the coiled shell have a gradually (logarithmically) increased distance from any selected point on the axis. It is the line around which the shell seems to coil (*a* in Fig. 8-4). In planispiral shells, the axis is normal to the plane of coiling; in conispiral shells, it runs from the apex to the center of the base of the cone.

Orientation. Coiled gastropod shells are oriented in a consistent manner for purposes of description and illustration. Planispiral shells ordinarily are placed with the plane of coiling in a vertical position and the axis horizontal. Conispiral shells are placed with the axis of coiling vertical and (except in most French publications) with the origin of coiling (apex of the shell) uppermost. Although some features of shell morphology are defined by pointing the aperture toward the observer, convention does not require this. For various purposes, other placement of the shell may be employed.

Planispiral shells, like that of *Bellerophon* (Fig. 8-13), presumably were carried by the snails which inhabited them with the plane of coiling vertical and the aperture directed downward. Generally, paleontologists have guessed that the downward pointing aperture was also forward, because this is the normal position of conispiral shells carried by living snails. We judge this orientation erroneous as applied to planispiral shells, because it denotes torsion (so as to bring the gills into anterior position). Such torsion could hardly have occurred without disturbing bilateral symmetry. The outer part of the aperture, farthest from the center of coiling, is interpreted as the most posterior and an opposite point toward the head of the animal, the most anterior. This problem of orientation is discussed later in the description of the snails called amphigastropods.

Conispiral shells normally are borne by living snails with the aperture directed forward and downward and the point of the coiled shell backward (Fig. 8-1, *1, 4, 5*). Thus, anterior means in the direction of the aperture, as used for these shells, and posterior is toward the point of the cone. As such shells are oriented in illustrations,

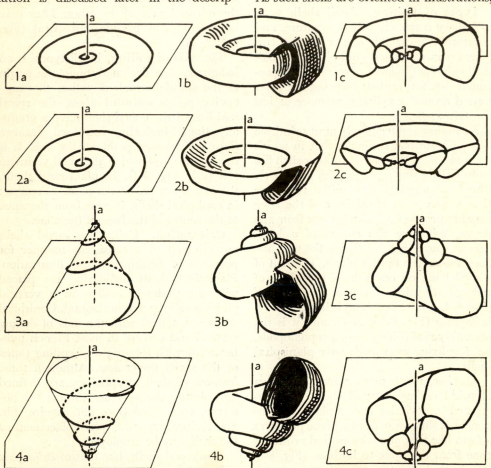

Fig. 8-4. Types of coiling of gastropod shells. The spiral lines in the four drawings at the left represent the center lines of the coiled whorls, which lie in a plane (1*a*, 2*a*) or along the surface of a cone (3*a*, 4*a*). The other drawings show complete shells and sectioned shells corresponding to the geometrical figures at left. Lines marked *a* are the axes of coiling.

1*a*, *b*, Planispiral shell, having center line of whorls in a plane and divided by this plane into symmetrical halves.

2*a–c*, Pseudoplanispiral shell, having center line of whorls in or near a plane but not divided symmetrically by this plane.

3*a–c*, Conispiral shell, orthostrophic type, having center line of whorls coiled around an erect cone.

4*a–c*, Conispiral shell, hyperstrophic type, having center line of whorls coiled around an inverted cone.

the posterior extremity is uppermost and the anterior part lowermost.

Parts of Shell. Each single complete turn of the shell, consisting of a 360-deg. volution, is termed a **whorl** (*17;* italic numbers accompanying names for parts of snail shells refer to Fig. 8-5). The shell as a whole consists of a **body whorl** (*2*), which comprises one complete volution extending backward from the **aperture** (*3*), and the **spire** (*1*), which includes all other whorls to the **apex** (*14*). The **base** (*4*) is the extremity opposite to the apex. A constricted anterior part of the body whorl of various shells is differentiated as the **neck** (*7*).

Junctions of the whorls, as seen on the exterior of the shell, are known as **sutures** (*16*); they may have various expression, such as flush, impressed, or channeled, and their inclination (**dip**, *18*) may be a significant shell character. The transverse section of whorls is expressed by the **whorl profile** (*41*), which may be quite different from the shape of the terminus of the body whorl (**labral outline**, *39*). The sloping surface of a whorl next below a suture is termed **ramp** (*44*), and a horizontal projection in this position, **shelf** (*29*). An angulation between the ramp or shelf and the surface next to it constitutes a **shoulder** (*47*), or if it is sharply accentuated, a **carina** (*30*). A groove on the surface of a whorl, parallel to its coiling, is called a **channel**. The exterior of the shell farthest removed from the axis of coiling is termed **periphery** (*6*), and lines tangent to various whorls, converging toward the apex, define the **spiral angle** (*10*). Some paleontologists judge that the so-called **pleural angle** subtended by lines on opposite sides of the shell, tangent to the two last-formed whorls, is more useful in the definition of species; this angle may differ appreciably from the spiral angle.

All shells in which the inner sides of the whorls lie outside the axis of coiling possess a narrow or wide open space within the encircling whorls. This is an **umbilicus** (*42*). Shells in which the inner edges of the whorls touch the axis of coiling have a solid rodlike center, which is termed **columella** (*31*). It may be smooth or bear spirally arranged ridges (**columellar folds**, *33*). Shells having a narrow umbilicus are said to possess a **perforate columella**.

The embryonic part (**nucleus**, *34*) of a shell, located at the origin of coiling, frequently is distinguishable by differences in the form and surface marking of the whorls, and in some species, by divergence in attitude of the axis of coiling. It comprises one to four whorls. Change in position of the axis of coiling, which may be abrupt, is **heterostrophy** (*35*) (*hetero*, different; *strophy*, turning).

The aperture has parts which are important in the description of the gastropod shell and in the definition of genera. The margin of the aperture constitutes the **peristome** (*37*). The side of the aperture next to the axis of coiling is termed **inner lip** (*20*), and in many shells this is divisible into a posterior part (**parietal lip**, *24*) controlled in shape by the adjoining whorl, and an anterior part (**columellar lip**, *25*) beyond this whorl. A smooth-surfaced shell deposit made by the mantle along the inner lip, covering the outer surface of the next-to-last whorl, is termed **inductura** (*36*), and a localized thickened part of the inductura outside the aperture, which may partly or entirely seal the umbilicus, is the **callus** (*38*). The outer side of the aperture, away from the shell, is the **outer lip** (*21*). Inductural deposits may be formed also on this side of the aperture, if the shell-secreting mantle turns back beyond the outer lip.

At the anterior extremity of the aperture, where outer and inner lips meet, many advanced types of gastropods bear a groove (**siphonal notch**, *9*) which may be extended along a shell outgrowth (**canal**, *12*) that holds the siphon. This takes water into the mantle cavity. A shell corrugation marked by curved growth lines which define the shift in position of the siphonal notch is termed **basal fasciole** *8*). An excurrent passageway is marked at the posterior edge of the aperture in some gastropods by a notch or groove (**gutter**, *19*). This also may indicate the position of the anus.

Other snails, including especially many Paleozoic planispiral and conispiral genera, have an indentation of varying depth in the outer lip which is inferred to have served the same function as the gutter—

a passageway for waste from the anus and for water after it had bathed the gills. This indentation is called a **sinus** (*49*), if it is comparatively broad, and a **slit** (*26*), if it is narrow, parallel-sided, and more or less deep. In a few fossil snails, the slit is excessively deep, reaching more than halfway around the body whorl. The extremity of the slit, on the side away from the aper-

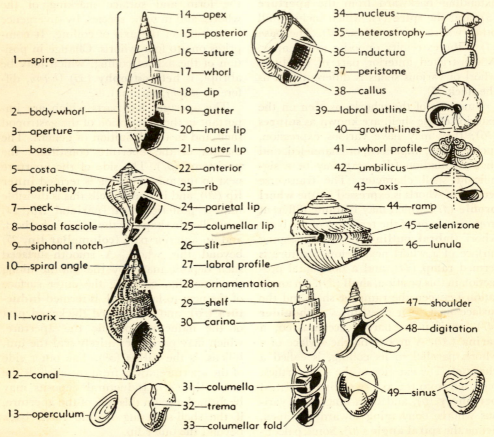

FIG. 8-5. **Structural features of gastropods.** The illustrated parts of the shell are cross-indexed to alphabetically arranged explanations of terms.

anterior (22). Direction toward head of gastropod, in spiral shells toward aperture.

aperture (3). Opening of shell, out of which part of gastropod body may extend.

apex (14). Point of beginning of shell growth, at tip of spire or cone.

axis (43). Imaginary center line around which a shell coils.

basal fasciole (8). Tract of more or less strongly inflected growth lines on the base of a shell,

marking the location of a siphonal notch.

base (4). Part of shell surface at extremity opposite apex.

body whorl (2). Last-formed single complete loop of a spiral shell.

callus (38). Generally thickened deposit of shell substance on anterior part of inner lip of aperture, partly or wholly covering umbilicus; it is part of inductura.

(Continued on next page.)

(*Fig. 8-5 continued.*)

canal (12). Semitubular anterior extension of aperture, enclosing siphon; at least slightly open along side, not closed like a pipe.

carina (30). Spiral keel on exterior of whorl, generally at edge of shelf.

columella (31). Solid or perforate pillar formed by inner walls of a conispiral shell.

columellar fold (33). Spiral elevation on columella produced by localized thickening of shell.

columellar lip (25). Part of inner lip of aperture adjoining columella.

costa (5). Coarse threadlike thickening of shell running spirally or axially.

digitation (48). Finger-like outward projection of outer lip of aperture.

dip (18). Deviation of suture from a plane normal to the shell axis.

growth line (40). Marking on shell parallel to apertural margin, denoting a former position of aperture.

gutter (19). Groove or canal at posterior extremity of aperture, in some gastropods marking location of anal outlet.

heterostrophy (35). Abrupt change in type of coiling between nucleus and later-formed part of shell.

inductura (36). Layer of lamellar shell material along inner lip of aperture or extending over shell surface beyond outer lip, characterized by smooth surface; includes callus.

inner lip (20). Margin of aperture adjacent to the next-to-last whorl; may include parietal lip and columellar lip.

labral outline (39). Shape of outer lip in view normal to aperture.

labral profile (27). Shape of outer lip in view parallel to edge of aperture.

lunula (46). Crescentic growth line on selenizone.

neck (7). Constricted anterior part of body whorl of some gastropods, exclusive of canal.

nucleus (34). Embryonic gastropod shell, commonly consisting of one to four whorls.

operculum (13). Horny or horny-calcareous plate carried on posterior part of foot, used to close aperture when gastropod withdraws into its shell.

ornamentation (28). Raised or depressed markings of shell surface other than growth lines.

outer lip (21). Edge of aperture on side away from next-to-last whorl.

parietal lip (24). Part of inner lip which adjoins next-to-last whorl.

periphery (6). Part of whorl farthest from shell axis.

peristome (37). Margin of whole aperture.

posterior (15). Direction backward from head of gastropod, in spiral shells toward apex.

ramp (44). Sloping surface of a whorl next below a suture.

rib (23). Well-marked linear elevation of shell surface, larger and broader than a costa.

selenizone (45). Sharply defined band parallel to coiling of whorls, which bears crescentic growth lines denoting a notch or slit in outer lip.

shelf (29). Subhorizontal part of whorl surface next to a suture, bordered on side toward periphery of whorl by a sharp angulation or by a carina.

shoulder (47). Salient angulation of a whorl parallel to coiling.

sinus (49). Reentrant in outer lip with nonparallel sides.

siphonal notch (9). Reentrant at junction of outer and columellar lips occupied by siphon.

slit (26). More or less deep, parallel-sided reentrant in outer lip, which gives rise to a selenizone.

spire (1). Coiled gastropod shell exclusive of body whorl.

spiral angle (10). Angle formed by lines tangent to two or more whorls on opposite sides of shell; inasmuch as lines tangent to all whorls of the spire may define a curve, spiral angle is commonly determined by drawing straight-line tangents to lowermost whorls of spire.

suture (16). Spiral line of junction between surfaces of any two whorls; includes external sutures on outer side of shell and umbilical sutures within umbilicus.

trema (pl. *tremata*, 32). Perforation of shell, generally formed by periodic closure of a slit, but occurring also at apex of some cap-shaped shells.

umbilicus (42). Central cavity of a shell formed by walls on inner sides of whorls; most common is basal umbilicus of orthostrophic conispiral shells but also included are apical umbilicus of convolute and hyperstrophic shells and lateral umbilici of planispiral (isotrophic) shells.

varix (11). Ridge, flange, or row of spines parallel to growth lines and marking modification of shell at former position of aperture.

whorl (17). Single complete loop of a spiral shell.

whorl profile (41). Transverse contour of surface of a whorl in a plane intersecting the axis of coiling; differs from labral profile and labral outline.

ture, is narrowly rounded, and as the shell margin is built forward the shape of the base of the slit is recorded by successive sharp-curved growth lines (**lunulae,** *46*). The belt of lunulae, which commonly is defined by bordering ridges, constitutes the **selenizone** (*45*) or so-called **slit band.** In some shells, like that of the modern abalone (*Haliotis*), the slit is discontinuous, and unsealed openings (*tremata,* sing. *trema,* *32*) are left behind; these function as outlets for water and waste from the shell interior.

The outer lip of some gastropods carries projections, those which extend inward being called **teeth,** and periodically formed outward extensions being known as **digitations** (*48*). A few species are distinguished by remarkably long and slender digitations, which denote unusual narrow and attenuated temporary protrusions of the shell-secreting mantle in building them.

Surface features of snail shells commonly include pigmentation which is highly varied as to hue and pattern, and sculptural markings, which vie with color in making the shells of these mollusks the most-sought objects of seashore collectors. The pigment, which forms part of the **ornamentation** (*28*), is seldom preserved in fossils, although numerous shells with clearly defined color patterns can be found in some formations as old as Paleozoic; the surface sculpture, which generally is well preserved, mainly consists of **growth lines** (*40*), linear elevations (**costae,** *5; **ribs,** 23*) parallel or transverse to the direction of coiling, linear depressions similarly disposed, and regularly or irregularly distributed granules, tubercles, spines, pits, or combinations of almost any of these.

The configuration of the aperture, which corresponds closely to that of the whorl, is represented by the characters of the **labral profile** (*27*) and **labral outline** (*39*). On parts of the body whorl and spire behind the aperture, this may be shown distinctly by the growth lines, and in some shells also by flangelike ridges called **varices** (*39*). A varix marks a former location of the aperture where a pause in forward growth of the shell was accompanied by construction of special apertural features. These may include digitations which have to be cut away by resorption when growth of the body whorl brings the functional aperture around to the position of the abandoned one. Varices on successive whorls may be aligned, indicating rather remarkable regularity in pulsatory building of the shell, or they may be offset, indicating unevenly distributed pauses in growth.

Shell Structure. All but a minor part of the shell of most gastropods and the entire shell of others is composed of calcium carbonate. Mineralogically, this consists of calcite or aragonite, or both. Shells built largely or wholly of calcite are more likely to retain features necessary for identification of species than those having a good deal of aragonite, because aragonite is very much less stable than calcite. The inner, smooth-polished or iridescent pearly layer of many gastropods is constructed of thinly laminated aragonite, and the outer enamel-like layer of some shells is similarly formed. Parts of shells belonging to different species also show layers of varying thickness which are composed of calcite, some with lamination parallel to the shell surface and some with fine fibrous structure, the fibers running normal to the surface.

Outside the calcareous shell is a thin, horny layer (**periostracum**), composed of a complex organic substance called conchiolin. It is a somewhat elastic tough material which protects the underlying surface. The conchiolin layer is lacking on shells which are covered by turned-back extensions of the mantle.

Although brachiopods and other invertebrates in many fossil collections from Paleozoic and Mesozoic rocks are well preserved, gastropods in the same collections may be indeterminable, even generically. They are internal molds (**steinkerns**) which do not indicate outer form of the shells and furnish no information about surface markings. Unless good external molds can be found, the gastropod element of such faunas must remain very little known. This type of preservation (or lack of it) indicates a difference in constitution of the gastropod shells, as compared with the other invertebrates. Fortunately, poor preservation is by no means universal. Rocks of all ages from Ordovician to Recent have yielded very numerous excellent fossil snails. Especially useful and promising are silicified shells which show minute features of the surface and form of the shell, including internal parts, although the layered construction of the shell substance is not observable. Exceptionally good preservation of shells composed of calcite furnishes supplementary material for study by the paleontologist.

Form of Shell. Some features of the form of gastropod shells have been incorporated in reviewing their parts. Additional characters are important in classifying and describing this group of fossils.

Shells of low conical form which lack coiled whorls commonly are designated as **patelliform** (*20;* italic numbers accompanying the terms for shell form refer to Fig. 8-6), because of their resemblance to the cap-shaped *Patella*.

Coiled shells, as already noted, can be divided into groups according to their symmetry or asymmetry: planispiral shells having bilateral symmetry and conispiral and pseudoplanispiral shells which lack symmetry.

The asymmetrical coiled shells can be divided into two groups on the basis of the direction of their coiling: right-handed or **dextral** (*1*), and left-handed or **sinistral** (*10*), defined on the basis of the position of the aperture when held facing an observer with the apex of the shell pointed upward. Each of these two groups, in turn, contains two divisions: **orthostrophic** (*2*) shells, which spiral downward from the apex, and **hyperstrophic** (*19*) shells, which spiral upward (negative cones).

Both symmetrical and asymmetrical coiled shells are classifiable according to the tightness of their coiling: **evolute** (*4*), whorls not in contact; **advolute** (*5*), whorls barely in contact; **involute** (*12*), outer whorls moderately embracing next inner ones; and **convolute** (*14*), outer whorls completely embracing and concealing inner ones.

Conispiral shells are variously characterized according to the uniform or progressively changing profile of their spiral angles, as **conical** (*22*), **conoidal** (*24*), or **extraconical** (*17*); according to the increase of height with relation to the number of component whorls, as **multispiral** (*7*) or **paucispiral** (*16*); or according to the shape of the spire and body whorl, as **discoidal** (*3*), **pupaeform** (*6*), **turbiniform** (*8*), **ovoid** (*15*), **turreted** (*18*), **trochiform** (*21*), **biconical** (*23*), **fusiform** (*25*), **obconical** (*26*), and others. The shape of the aperture, whether having a continuous peristome (**holostomatous,** *9*) or interrupted by a siphonal notch or canal (**siphonostomatous,** *27*), is the basis for classifying shells.

The presence or absence of an umbilicus and, among umbilicate shells, the open, partially closed, or completely sealed nature of the base are expressed by the descriptive terms **phaneromphalous, hemiomphalous, cryptomphalous,** and **anomphalous** (Fig. 8-7, *1–4*). Some shells in which part of the whorls rise above the apex (origin of coiling) are provided with an **apical umbilicus,** and such a hollow may occur both in orthostrophic shells (Fig. 8-7, *5*) and hyperstrophic ones (Fig. 8-6, *19*).

CLASSIFICATION

Division of the very large and highly varied host of gastropods into natural groups is a difficult task. In order to accomplish it satisfactorily, forms belonging to each main evolutionary branch must be recognized and segregated. They should be characterized by having certain significant structures, inherited in common. But what are these significant structures?

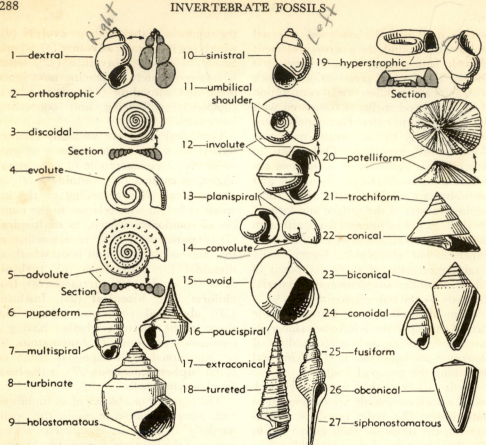

FIG. 8-6. **Form of gastropod shells.** Most commonly used descriptive terms applied to snail shells are illustrated with accompanying cross-indexed explanations.

advolute (5). Whorls in contact but not distinctly embracing.

biconical (23). Spire and base both having moderately elevated conical form.

conical (22). Spire and body whorl having evenly confluent sides which form a straight-walled cone, base flattened.

conoidal (24). Spire and body whorl forming a cone which is distinctly steeper-sided near the base than at the apex.

convolute (14). Outer whorls so deeply embracing inner ones that latter are nearly or quite invisible externally.

dextral (1). Right-handed; in apertural view with apex directed upward, aperture is on right side of shell; in apical view, shell coils in a clockwise direction.

discoidal (3). Having shape of a wheel or disk; spire extremely low conical, flat, or shallowly concave, and base also having one of these shapes.

evolute (4). Loose-coiled, whorls not in contact.

extraconical (17). Spire and body whorl forming a cone distinctly steeper-sided near apex than toward base.

fusiform (25). Spindle-shaped; largest in middle and sharp-pointed at both extremities.

holostomatous (9). Peristome continuous, not interrupted anteriorly by siphonal notch or canal.

hyperstrophic (19). Spire depressed instead of elevated; a hyperstrophic dextral shell may be identical in form to an orthostrophic sinistral shell, and a hyperstrophic sinistral shell likewise may correspond to an orthostrophic dextral shell; identification is based on organization of soft parts.

involute (12). Outer whorls slightly to strongly embracing inner whorls but not completely.

multispiral (7). Spire composed of numerous whorls.

obconical (26). Reversed-cone shape, base strongly conical and spire nearly flat.

orthostrophic (2). Spire slightly to strongly

(Continued on next page.)

Classification which has come to be adopted generally by zoologists is based on interpretation of the anatomy of soft parts, with little attention to the paleontological record, for of course, the nature of the gills, heart, digestive tract, and other internal structures of long-extinct gastropod genera cannot be discerned by study of the shell. Fortunately, placement of most fossil groups in the classification based on soft parts can be established reasonably well, because these groups contain living representatives. On the other hand, we must take account of important assemblages which have left no known descendants. It is not yet possible to trace evolutionary differentiation of the gastropods from the study of shell features shown by fossils and living forms, but advance in paleontological knowledge promises to introduce changes in definition of major taxonomic divisions among the snails. For example, the noncoiled or partially coiled gastropods of the superfamily Tryblidiacea, and perhaps also the planispirally coiled shells of the Bellerophontacea, stand distinctly apart from conispirally coiled forms. They include some of the oldest known genera, found in Cambrian rocks, and without indication of trends toward conispiral modification, they persisted throughout Paleozoic time. Seemingly, this is not a minor unit assignable to one of the classes defined on soft-parts anat-omy. Certain orthostrophic conispiral gastropods (raphistomatids) and hyperstrophic ones (macluritids, and less certainly, euomphalids) are other ancient stocks, first known in the Cambrian, which may be independent lines. It is appropriate here only to make note of these questions. The outline of classification which follows incorporates no previously unpublished features.

Main Divisions of the Gastropoda

Amphigastropoda (*class*), shells symmetrical, non-coiled or planispirally coiled; mantle cavity and gills located in posterior, primitive position (*amphi*, on both sides, referring to symmetry). Cambrian–Permian, ?Triassic.

Prosobranchia (*class*), cap-shaped or conispiral shell with mantle cavity and gills in anterior position; neural loop twisted in a figure of eight (*proso*, forward; *branchia*, gills). Cambrian-Recent.

Archaeogastropoda (*order*), primitively having two subequal gills provided with double rows of leaflets but advanced forms with only one such gill (*archaeo*, ancient). Cambrian–Recent.

Mesogastropoda (*order*), having only one gill provided with a single row of leaflets (*meso*, intermediate). Pennsylvanian–Recent.

Neogastropoda (*order*), gill structure like that of Mesogastropoda but having a siphonostomatous aperture, typically provided with a well-developed canal (*neo*, new, recent). Cretaceous–Recent.

Opisthobranchia (*class*), shell reduced in size, commonly internal or absent; mantle cavity

---

(*Fig. 8-6 continued.*)

elevated; as measured by form of whorl center lines, this embraces shells in which upper sides of whorls are all tangent to a single plane.

ovoid (15). Egg-shaped, apical and basal parts of shell somewhat evenly rounded.

patelliform (20). Low cap-shaped, like shell of *Patella*, noncoiled.

paucispiral (16). Spire composed of very few whorls.

planispiral (13). Coiling in a plane, with part of shell on one side of a plane the mirror image of other.

pupaeform (6). Elevated ovoid, like shell of *Pupa*, in which late-formed whorls have decreasing radii of curvature.

sinistral (10). Left-handed; in apertural view with apex directed upward, aperture is on left side of shell; in apical view, shell coils counter-clockwise.

siphonostomatous (27). Peristome discontinuous, interrupted anteriorly by a siphonal notch or canal.

trochiform (21). Sides of shell evenly conical, base flat, like shell of *Trochus*.

turbinate (8). Top-shaped, generally with rounded base.

turreted (18). Very high-spired shell with flat or gently rounded base, like shell of *Turritella*.

umbilical shoulder (11). Angulation of whorls at margin of umbilicus and within it.

and gill, where present, in rear position as result of twisting back from prosobranch condition, gill commonly absent, being replaced by respiratory structure developed in the mantle or entire outer surface; neural loop not crossed in figure eight (*opistho*, backward). Pennsylvanian–Recent.

Pleurocoela (*order*), shell, mantle cavity, and gill present. Mississippian–Recent.

Pteropoda (*order*), slender conical shell present or absent; lack distinct head, foot modified as paired winglike fins for swimming; gills present (*ptero*, wing; *pod*, foot). ?Cambrian–?Permian; Cretaceous–Recent.

Sacoglossa (*order*), shell lacking except in larval stage; no gills; unknown as fossils. Recent.

Pulmonata (*class*), mostly shell-bearing but lacking an operculum; neural cord not in form of figure eight; mantle cavity modified as an air-breathing lung (*pulmo*, lung). Pennsylvanian–Recent.

Basommatophora (*order*), fresh-water pulmonates having eyes at base of posterior tentacles (*bas*, base; *ommato*, eye; *phora*, carry). Jurassic–Recent.

Stylommatophora (*order*), terrestrial pulmonates having eyes at tip of posterior tentacles (*stylo*, stalk). Pennsylvanian–Recent.

EVOLUTIONARY TRENDS

The most evident evolutionary changes of gastropods are those related to torsion of the body and shell, and next to this, but independent of it, are modifications of respiratory structures for breathing air, and changes of the foot, which serve to adapt some of these invertebrates for a free-swimming mode of life. Many divergent sorts of gradual alteration in shell form and surface markings are found among all main groups, with or without obvious indication of their correlation with soft-part structures or mode of living.

Most primitive of all known snails are representatives of the amphigastropods which belong to the Tryblidiacea. They have a noncoiled cap-shaped shell or one that coils planispirally in a forward direction, and the arrangement of muscle scars indicates that the mantle cavity and gills were in the rear part of the shell—as in

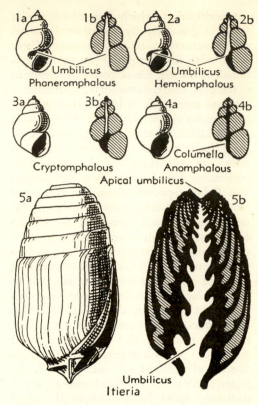

FIG. 8-7. **Structures of the axial region of gastropod shells.** The diagrams illustrate four types of structure by apertural views and sections of shells; these depend on the presence or absence of an open space in the position of the shell axis and its partial or complete closure by callus deposits (black in sectional views).

1a, b, Shells having a wide-open umbilicus are termed phaneromphalous (*phanero*, evident; *omphalus*, navel).

2a, b, In hemiomphalous (*hemi*, half) shells, opening of the umbilicus is partly closed by a callus.

3a, b, Complete closure of the umbilicus by callus deposits distinguishes shells called cryptomphalous (*crypto*, hidden).

4a, b, Shells lacking an umbilicus are termed anomphalous (*a*, not).

5a, b, A complexly built umbilicate shell, *Itieria cabanetiana* (d'Orbigny) (×0.3), from Upper Jurassic rocks of France. The strongly involute whorls, which are septate in their anterior portions, surround a large umbilicus having unusually developed shoulders. The concavity at the shell summit is an apical umbilicus.

the generalized ancestral type of mollusk (Fig. 7-2) and in the living chitons and pelecypods (Fig. 7-1). Accentuation of forward planispiral coiling (as in the Devonian *Cyrtonella*) is an evolutionary step toward the involute and convolute bellerophontid shell types, and it is possible that the symmetrical-shelled Bellerophontacea also belong among amphigastropods.

Giving attention to some soft-part structures and the radula, we may judge that gills provided with a double row of leaflets, like those of the archaeogastropod prosobranchs, are more primitive than the single-row type. The double-row type belongs to gastropods having other unspecialized anatomical features, and living kinds possess shells which most closely resemble those predominating in Paleozoic assemblages, whereas the single-row gill is associated with advanced organization of other soft parts and is borne by snails having shells such as predominate in post-Paleozoic rocks but are not known in Paleozoic formations. This signifies evolution of gill structure which partly underlies differentiation of the archaeogastropods from the mesogastropods and neogastropods.

Probably antedating the oldest known prosobranchs and belonging to the evolutionary stage in which the gills and anus were in a posterior position, is the development of the silt-sensing organs called osphradia. These are distinctive structures of all gill-bearing bottom-dwelling gastropods, and they have special importance because of the pocket-like nature of the mantle cavity which impedes water circulation. Combined with protection furnished by osphradia, a somewhat complicated intestinal mechanism was evolved by the archaeogastropods (studied in living representatives), which serves to consolidate digestive waste into compact pellets, too heavy to be carried by inhalant water currents to the gills. This device is not required by gastropods which lack a mantle cavity and gills, and in evolution it has disappeared.

The profound effects of torsion in twisting the soft parts of the rear to a position in the front are judged to be marks of pre-Paleozoic evolution, for Cambrian coni-

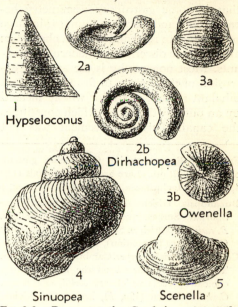

FIG. 8-8. **Representative Cambrian gastropods.** Amphigastropods include 1 and 5 (Tryblidiacea) and possibly 3 (Bellerophontacea). Others are archaeogastropods (Pleurotomariacea). Natural size unless indicated otherwise.

Dirhachopea Ulrich & Bridge, Cambrian. Whorls have a quadrate cross section; the body whorl is somewhat evolute. *D. normalis* Ulrich & Bridge (2a, b, oblique lower and apical views), Upper Cambrian, Missouri.

Hypseloconus Berkey, Upper Cambrian–Ordovician. A high uncoiled shell. *H. elongatus* Berkey (1, side view), Upper Cambrian (Dresbachian), Wisconsin.

Owenella Ulrich & Scofield, Upper Cambrian–Ordovician. A convolute bellerophontid. *O. antiquata* (Whitfield) (3a, b, top and side views), Upper Cambrian (Trempealeauan), Wisconsin.

Scenella Billings, Cambrian–Devonian. A cap-shaped shell with curved apex, probably forward, as in other tryblidiids. *S. reticulata* Billings (5, oblique view from side, ×2), Lower Cambrian, Newfoundland.

Sinuopea Ulrich, Upper Cambrian–Ordovician. The mid-part of the outer lip is deeply sinuate. *S. sweeti* (Whitfield) (4, ×2), Upper Cambrian (Trempealeauan), Wisconsin.

spiral snails have shells which differ little from various living prosobranchs (Fig. 8-8). Once established, the altered placement of soft parts remained stable, except that in some gastropods asymmetry became more and more pronounced; in large part these are recorded by shell characters and they are discerned in modern descendants. Most noteworthy is the anterior extension of the mantle to form an efficient inhalant siphon, partly or wholly incased by an outgrowth of shell which constitutes the canal. This improvement was not achieved by archaeogastropods or a majority of the mesogastropods, but progressive groups of the latter (Strombacea, Cerithiacea, Tonnacea, some Cypraeacea) and most of the neogastropods illustrate it excellently.

Although the radula cannot be used as an aid in the paleontological study of gastropods, the correlation of its structural types with shell groups, which can be recognized in work on fossils, permits conclusions on some features of its evolution. The type of radula (rhipidoglossate, Fig.

FIG. 8-9. Representative Ordovician gastropods. The fossils shown in 10 and 13 are Bellerophontacea; all others are archaeogastropods (Pleurotomariacea, 1–3, 5, 8, 11, 14–15; Euomphalacea, 4, 6, 7, 9; Macluritacea, 12; Trochonematacea, 16). All natural size unless indicated otherwise.

Bucanopsis Ulrich & Scofield, Ordovician–Silurian. Surface marked by fine spiral ribs. *B. carinifera* Ulrich & Scofield (13, ×2), Middle Ordovician (Trentonian), Kentucky.

Ceratopea Ulrich, Upper Cambrian–Ordovician. This genus is defined on the basis of thick conical opercula which are useful zone fossils. *C. keithi* Ulrich (12a, b), Lower Ordovician, Virginia.

Ecculiomphalus Portlock, Ordovician–Silurian. Distinguished by evolute coiling and cross section of whorls. *E. triangulus* Whitfield (9a, b), Lower Ordovician, Minnesota.

Eotomaria Ulrich & Scofield, Ordovician–Silurian. Slit band prominent. *E. supracingulata* (Billings) (15), Middle Ordovician, eastern Canada.

Helicotoma Salter, Ordovician. *H. tennesseensis* Safford (14a, b, side and umbilical views), Middle Ordovician, Kentucky.

Lecanospira Butts, Ordovician. This shell has advolute whorls, a flat base, and very distinctly depressed apex; it is thus classed as hyperstrophic. *L. compacta* Salter (7, oblique view of top), Lower Ordovician, Quebec.

Liospira Ulrich & Scofield, Ordovician. A low-spired shell having even sides and a peripheral selenizone. *L. micula* (Hall) (2a, b, oblique views of top and base, ×3), Upper Ordovician (Maquoketa); Wisconsin.

Lytospira Koken, Ordovician–Silurian. Shell very evolute, with an angular sinus, whorl angulated on the inner side of its curvature. *L. subrotunda* (Ulrich & Scofield) (6a, b), Middle Ordovician, Minnesota.

Ophileta Vanuxem, Ordovician. Low-spired broadly phaneromphalous shell having advolute quadrangular whorls and a narrow obscure selenizone. *O. complanata* Vanuxem (1a–c, apical, side, and oblique basal views). Lower Ordovician, New York.

Orospira Butts, Ordovician. Distinguished by shape, aperture, and sculpture. *O. bigranosa* Butts (11a, b, ×2), Lower Ordovician, Missouri.

Ozarkispira Walcott, Ordovician. Rather closely resembles *Ophileta* but differs in whorl cross section and absence of selenizone. *O. leo* Walcott (4a, b, apical and oblique basal views, ×4), Lower Ordovician, Alberta.

Phragmolites Conrad, Ordovician–Silurian. A planispiral shell characterized by oblique transverse frills and raised selenizone. *P. obliqua* (Ulrich & Scofield) (10a, b, ×2), Middle Ordovician, Minnesota.

Raphistoma Hall, Ordovician–Silurian. Top very low conical, basal part produced. *R. striatum* (Emmons) (8a, b), Middle Ordovician (Chazyan), New York.

Raphistomina Ulrich & Scofield, Lower Cambrian–Silurian. More discoidal than *Raphistoma*. *R. lapicida* (Salter) (3a, b, oblique apical and basal views, ×2), Middle Ordovician (Blackriveran), Quebec.

Schizopea Butts, Upper Cambrian–Ordovician. Discoidal shell having subangulated periphery. *S. typica* Ulrich & Bridge (5a, b), Lower Ordovician, Missouri.

Trochonema Salter, Ordovician–Devonian. Shell turreted, whorls angular in section, no slit. *T. umbilicatum* (Hall) (16, ×2), Middle Ordovician (Trentonian), New York.

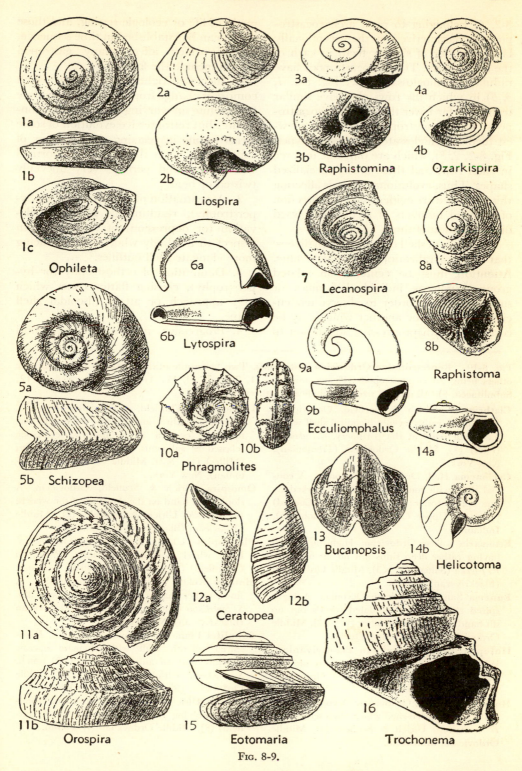

1a
1b
1c Ophileta
2a
2b Liospira
3a
3b Raphistomina
4a
4b Ozarkispira
5a
5b Schizopea
6a
6b Lytospira
7 Lecanospira
8a
8b Raphistoma
9a
9b Ecculiomphalus
10a
10b Phragmolites
11a
11b Orospira
12a
12b Ceratopea
13 Bucanopsis
14a
14b Helicotoma
15 Eotomaria
16 Trochonema

Fig. 8-9.

8-2, *5*) belonging to most archaeogastropods is characterized by the exceptionally large number of similar teeth in each of the many rows. The mesogastropods have radulas (taenioglossate type, Fig. 8-2, *1, 3, 6*) in which the rows also have similar teeth, but far fewer than in archaeogastropods. The neogastropod radulas have fewest teeth (rachiglossate, toxiglossate types, Fig. 8-2, *2, 8*), which are only one to three in each row, and these are of specialized shapes. Thus, evolutionary trends affecting this structure are evidently in the direction of reduced numbers of teeth and toward differentiation of them.

Evolution of shell characters is so diverse that few generalizations have much value. Attention must be centered on related groups, such as individual families or superfamilies, in order to observe tendencies of one sort or another which may be correlated with progressive adaptation to mode of life or ecologic setting, and those which can be established in time sequence. Trends which affect many gastropod groups include the following:

1. Increased tightness of coiling, which comprises change from evolute or advolute to involute in varying degree, and convolute; but retrogressive evolution of some forms results in relaxing the tightness of coiling to a point that leaves last-formed whorls free, and perhaps very unevenly twisted or bent.

2. Accentuation of orthostrophy (or hyperstrophy), resulting in increased elevation (or depression) of the apex with respect to the body whorl and producing more elongate shell outlines.

3. Diminution of orthostrophy (or hyperstrophy), causing flatter cones which may approach or attain discoidal shell shape.

4. Progressive modification of whorl

FIG. 8-10. **Representative Ordovician gastropods.** Two bellerophontids (3, 8) are illustrated, the others being archaeogastropods (Pleurotomariacea, 1, 2, 4, 7, 10, 11; Trochonematacea, 6, 9, 12; Subulitacea, 5). All natural size unless indicated otherwise.

Clathrospira Ulrich & Scofield, Ordovician–Silurian. A rather evenly biconical shell which has a slit band at the periphery. *C. subconica* (Hall) (7), Middle Ordovician (Trentonian), New York.

Cyclonema Hall, Ordovician–Devonian. A conical shell with rounded periphery, marked by fine spiral costae; no slit. *C. bilix* (Conrad) (9, ×2), Upper Ordovician (Richmondian), Indiana.

Ectomaria Koken, Ordovician. High-spired shell having rather deep sinus but no slit or band. *E. pagoda* (Salter) (1, ×3), Middle Ordovician (Blackriveran), Quebec.

Eunema Salter, Ordovician–Devonian. A high-spired shell having angulated whorls but no slit band. *E. strigillatum* Salter (12, ×2), Middle Ordovician (Blackriveran), Quebec.

Holopea Hall, Ordovician–Pennsylvanian. Whorls well rounded, oblique growth lines. *H. symmetrica* Hall (6, ×3), Middle Ordovician (Trentonian), New York.

Hormotoma Salter, Ordovician–Devonian. Moderately high-spired shell which has a selenizone. *H. trentonensis* Ulrich & Scofield (4), Middle Ordovician, Minnesota.

Lophospira Whitfield, Ordovician–Devonian. Whorls bear three angulations, the middle one marking location of a narrow selenizone (classed as subgenus of *Loxoplocus*). *L. milleri* (Miller) (11, ×2), Middle Ordovician (Trentonian), New York.

Omospira Ulrich & Scofield, Ordovician. A sloping slit band on the shoulder of the whorls. *O. laticincta* Ulrich & Scofield (2), Middle Ordovician (Blackriveran), Tennessee.

Plethospira Ulrich & Scofield, Ordovician. Body whorl elongated anteriorly. *P. cassina* (Whitfield) (10), Lower Ordovician, Vermont.

Sinuites Koken, Ordovician–Pennsylvanian. A sinus, rather than a slit, indents the outer lip. *S. cancellatus* (Hall) (3a, b), Middle Ordovician (Blackriveran), Minnesota.

Subulites Emmons, Ordovician–Silurian. High-spired smooth shell with elongate narrow aperture and an anterior siphonal (?) notch. *S. regularis* Ulrich & Scofield (5), Middle Ordovician (Blackriveran), Kentucky.

Tetranota Ulrich & Scofield, Ordovician. Prominent revolving ridges. *T. wisconsinensis* (Whitfield) (8), Middle Ordovician, Wisconsin.

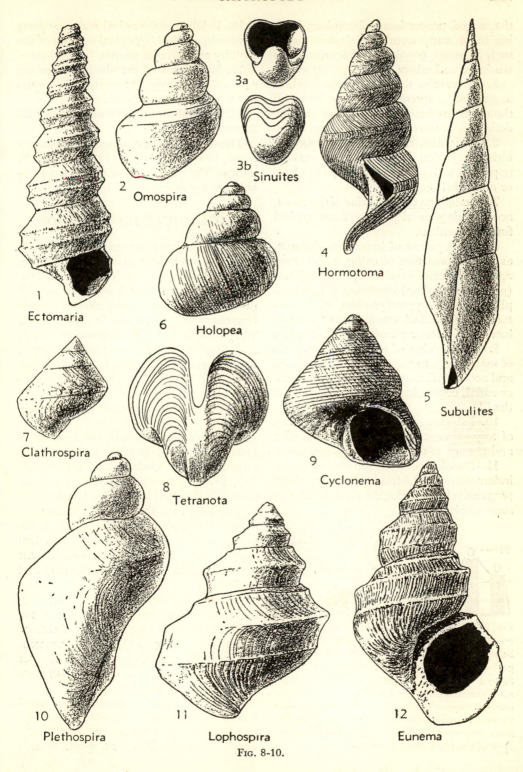

1 Ectomaria

2 Omospira

3a

3b Sinuites

4 Hormotoma

5 Subulites

6 Holopea

7 Clathrospira

8 Tetranota

9 Cyclonema

10 Plethospira

11 Lophospira

12 Eunema

Fig. 8-10.

shape and proportional dimensions leading to or away from special shell forms, such as ovoid, conoidal, extraconical, pupaeform, and others.

5. Increase or decrease of upper-whorl angulation expressed by characters of shoulders, carinae, and ramps or shelves, defining shape trends of turbiniform shells.

6. Flattening, rounding, or strong anterior projection of the shell base, making approach to typical trochiform, biconical, or obconical shapes.

7. Elongation or shortening of the canal, accentuating or departing from typical fusiform outlines.

8. Development of inductural deposits, as in the formation of callus, which may establish a trend from an open umbilicate (phaneromphalous) condition to hemiomphalous and cryptomphalous, or by extension of the inductura, so as to cover increased parts or all of the shell exterior.

9. Modification of apertural characters of many sorts, as expressed in the position and accentuation of the slit, siphonal notch or canal, columellar folds, digitations, and the like.

10. Increase or decrease in prominence of any element of surface ornamentation and change of their form or arrangement.

11. Development of spirally arranged indentations on the inside of whorls, which progressively constrict the shell interior of some stocks.

12. Infilling of the apical region or parts of whorls, or blocking apical sections of the shell by septa across whorls.

13. Reduction of the shell to a minor external or internal remnant, and in some stocks to its disappearance.

14. Loss of the operculum, as in pulmonates.

These, and many other evolutionary trends in various groups of the gastropods, indicate the complexity of changes observable among these mollusks.

GEOLOGICAL DISTRIBUTION

The geological range of the Gastropoda is from Lower Cambrian to the present. They are not common in Cambrian faunas, but at least nine families of amphigastropods and archaeogastropod prosobranchs are represented in this system. They include noncoiled cap-shaped forms, and planispiral, orthostrophic conispiral, and hyperstrophic conispiral shells (Fig. 8-8). The beginning of the amphigastropods and all main divisions of the early prosobranchs is identified in Cambrian strata.

A great enlargement in the number of known gastropod genera and species is recorded from Ordovician rocks (Figs. 8-9, 8-10). These belong to families previously introduced in the Cambrian and to 14 additional families. Nearly all these are confined to Paleozoic rocks. Eight additional families of archaeogastropods first appear in Silurian strata, two in the Devonian, one in the Mississippian, and three in the Permian.

The mesogastropods range through Mesozoic rocks but increase in the Cenozoic, whereas neogastropods are confined to Cretaceous and post-Cretaceous deposits. All but a few superfamily groups of these two orders are more abundant today than at any time in the past.

Opisthobranch gastropods are recorded from Mississippian to Recent. If the narrow conical hyolithids are pteropods, which is very doubtful, the opisthobranch

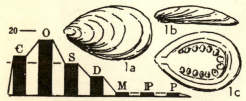

FIG. 8-11. **Geologic distribution of amphigastropods.** The graph shows number of genera belonging to the superfamily **Tryblidiacea** by geologic systems (Cambrian to Permian). The figured example of this group is *Tryblidium reticulatum* Lindström, ×0.7, from Middle Silurian rocks of Sweden (1*a–c*, apical, side, and apertural views). The horseshoe-shaped pattern of muscle scars (1*c*) opens anteriorly, indicating that the apex of the low shell tips forward.

group is as old as Cambrian. The opisthobranchs are far outranked by prosobranchs in paleontological importance.

The pulmonate gastropods are late arrivals, as might be expected from their evolutionary advancement in being adapted to live on land. A few forms belonging to this class seem to be represented among Pennsylvanian fossils, but the main geological record is in the Cenozoic. Modern faunas include nearly 1,000 genera of pulmonates. Thus, they seem to be at the peak of their development.

AMPHIGASTROPODS

The amphigastropods include the superfamily Tryblidiacea and possibly also the Bellerophontacea, which together contain nine families. Both appear in Cambrian rocks but attain maximum development a little later in early Paleozoic time. They persist into or slightly beyond the Permian.

Tryblidiids. The tryblidiids include shells ranging from rather high conical noncoiled shapes to very low, partly coiled forms (Fig. 8-8, *1*, *5*). All are symmetrical, and the apex of the cone or tip of the coiled shells is pointed forward. This is indicated by orientation of the shell in relation to a horseshoe-shaped pattern of paired muscle scars on the interior, which opens in what must be the direction of the head's protrusion. There is no sign of torsion that disturbs bilateral symmetry, and we may conclude rather confidently that a mantle cavity which contained a pair of gills and the anus was located in the rear part of the shell. As already noted, this is inferred to be the most archaic condition. Tryblidiacean gastropods are Cambrian to Devonian (Fig. 8-11).

Bellerophontids. The bellerophontids are more abundant in genera, species, and individuals than the tryblidiids, and they are distinctly more advanced in evolution of the shell. Some have an evolute, incompletely coiled shell form, whereas others are advolute, slightly to strongly involute, or distinctly convolute. The advolute and

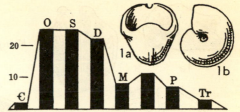

Fig. 8-12. **Geologic distribution of bellerophontid gastropods.** The graph indicates number of genera belonging to the superfamily **Bellerophontacea** by geologic systems (Cambrian to Triassic). This important group of planispiral shells is judged to belong to the class Amphigastropoda but is not certainly assignable there. 1*a*, *b*, *Bellerophon insculptus* De Koninck, ×0.5, from Upper Mississippian (Viséan) rocks of Belgium.

involute shells have a more or less open umbilicus on each side, but the convolute types lack umbilici and only the body whorl is visible from the exterior. A slit generally is present at the mid-point of the outer lip, and a selenizone can be traced along the mid-line of the last-formed whorl by the sharply curved pattern of growth lines (lunulae) and commonly by the slightly raised or depressed surface along the band. Different species exhibit a considerable range of spiral and transverse markings. The occurrence of both together produces a reticulate pattern. As represented by numbers of genera, peak development of the bellerophontids is in the Ordovician and Silurian, but the Devonian is only a little behind these systems (Fig. 8-12). Representative members of this group are illustrated in Figs. 8-9, 8-10, 8-13, 8-17, 8-19, and 8-21.

An unusually interesting late Paleozoic bellerophontid snail is the Pennsylvanian genus *Knightites* (Fig. 8-13, *1a–e*). The shell of this genus is mainly distinguished from others of the superfamily by the presence of paired tubelike snouts on either side of the slit. As growth of the shell proceeds, one pair of snouts is left behind and a new pair is built at the apertural margin. Evidently these projections of the shell are analogous to the siphon of highly devel-

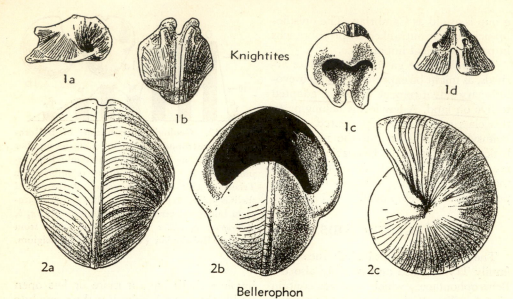

Knightites

1a 1b 1c 1d

2a 2b 2c

Bellerophon

FIG. 8-13. Pennsylvanian bellerophontid gastropods. *Knightites* is characterized by its spirally arranged fine costae, flaring aperture, and especially the paired tubular snouts which project on either side of the slit. The inferred function of the snouts was for inflow of water to bathe the gills, outlets from the mantle cavity being probably at sides of the shell near the umbilici. The anus presumably opened through the slit. Figures 1a–d are various views of the shell of *K. multicornutus* Moore, ×1, from Upper Pennsylvanian (Virgilian) rocks of Kansas. Figures 2a–c show *Bellerophon graphicus* Moore, ×2, from the same rocks.

oped prosobranchs, such as the neogastropods, serving as inhalant passageways for water going to the pair of gills. Outlet of the water used in respiration probably was at the sides of the aperture near the position of the umbilici. The anus doubtless discharged waste matter at the base of the slit.

As first interpreted, the head of *Knightites* was supposed to be located, as in prosobranchs, beneath and in front of the slit (Fig. 8-14, *3a, b*), which is exactly opposite to the primitive orientation of mollusks and to that judged to characterize the

FIG. 8-14. Orientation of bellerophontid gastropods. The diagrams are designed to show comparison of amphigastropod (*2a, b*) and prosobranch (*3a, b*) orientations of *Bellerophon* with *Tryblidium* (*1a, b*), an undoubted amphigastropod, *Pleurotomaria* (*4a, b*), an undoubted prosobranch having a prominent slit, and a planispirally coiled cephalopod (*5a, b*). The location of the anus is indicated by *a* and of gills by *g*. The left-hand drawings are side views, and those on the right are dorsal views.

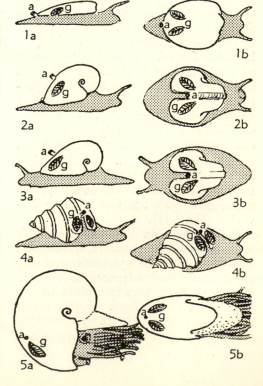

1a 1b 2a 2b 3a 3b 4a 4b 5a 5b

trylidiids (Fig. 8-14, *1a, b*). According to alternative interpretation, the head and tail of *Knightites*, and of the Bellerophon-

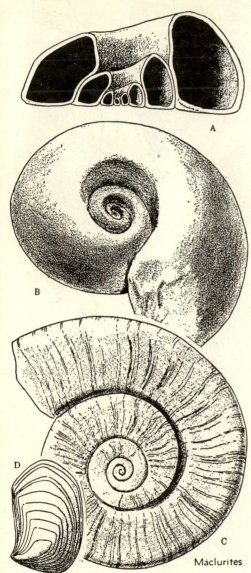

A

B

D

C

Maclurites

FIG. 8-15. **A lower Paleozoic hyperstrophic archaeogastropod.** The fossil is *Maclurites magnus* LeSueur, ×1, type species of the genus, from the Middle Ordovician (Lenoir limestone) of Tennessee. (*A*) Cross section showing apical umbilicus and form of the whorls. (*B*) Oblique apical view. (*C*) Basal view. (*D*) Operculum, which shows counterclockwise spiral twisting of the growth lines, the pattern invariably belonging to dextrally coiled shells that have spirally coiled opercula.

tacea generally, is reversed (Fig. 8-14, *2a, b*), putting these as in the amphigastropods. A compelling reason for this reorientation is the complete symmetry of the planispiral bellerophontid shell, for it is difficult to understand how torsion, such as characterizes the prosobranchs, could interchange the position of front and rear soft parts without affecting the shell. The pleurotomarian gastropods, which seem closely related to bellerophontids in possessing a well-defined slit, clearly are prosobranchs; and in the process of twisting body parts, strong asymmetry of shell characters is introduced (Fig. 8-14, *4a, b*). Coiled cephalopods, on the other hand, which have the anus and gills in a posterior position, as in primitive mollusks, have symmetrical planispirally coiled shells (Fig. 8-14, *5a, b*). In spite of these considerations, J. Brookes Knight, a leading specialist in studies of Paleozoic gastropods, summarily rejects classification of the Bellerophontacea among amphigastropods, judging that they are prosobranchs. His main reason is the essential identity of the bellerophontid slit with that of Pleurotomariacea; also given weight is the difference in muscle-scar patterns of trylidiids and bellerophontids. Although the sum of evidence seems to support assignment of

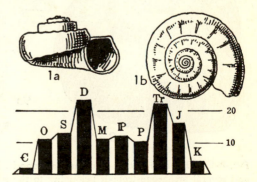

FIG. 8-16. **Geologic distribution of archaeogastropods.** The graph indicates number of genera belonging to the superfamily **Euomphalacea** by geologic systems (Cambrian to Cretaceous). The figured example (*1a, b*) is *Straparolus* (*Euomphalus*) *tuberosus* De Koninck, ×0.7, Lower Mississippian (Tournaisian), Belgium.

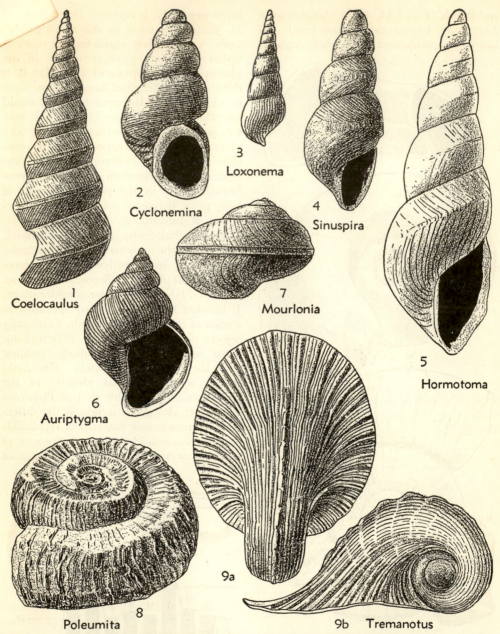

FIG. 8-17. **Representative Silurian gastropods.** The illustrated fossils include a bellerophontid (9) and several archaeogastropods belonging to the superfamilies Pleurotomariacea (1, 4, 5, 7), Euomphalacea (8), Trochonematacea (2), Subulitacea (6), and Loxonematacea (3). Natural size except as indicated otherwise.

Auriptygma Perner, Silurian. *A. fortior* Barrande (6, ×2), Middle Silurian, Czechoslovakia.

Coelocaulus Oehlert, Ordovician–Devonian. A high-spired umbilicate shell with distinct slit band. *C. macrospira* (Hall) (1), Middle Silurian, New York.

Cyclonemina Perner, Silurian. Like *Cyclonema* (Fig. 8-10, 9) but higher spired. *C. delicatula*

(*Continued on next page.*)

the Bellerophontacea to the amphigastropods, rather than prosobranchs, the opposed view which has been cited signifies that the question is not settled.

PROSOBRANCHS

The prosobranchs comprise the vast majority of Paleozoic gastropods, and they are an important constituent of post-Paleozoic faunas. Living genera are assigned to two orders, based primarily on the structure of the gills and heart: (1) the Aspidobranchia (*aspido*, shield; *branchia*, gills) have one or two gills which bear double rows of respiratory leaflets, and the heart contains two auricles, which gives rise to an alternative name for the order, Diotocardia (*di*, two; *oto*, auricle; *cardia*, heart); and (2) the Pectinibranchia (*pectini*, comb) have only one gill which bears a single row of leaflets, and the heart contains but one auricle, which is the basis for an alternative name, Monotocardia. The aspidobranch snails are coextensive with the Archaeogastropoda, and the pectinibranch snails include both Mesogastropoda and Neogastropoda. This class is also known as Streptoneura (*strepto*, twisted; *neura*, nerve cord), in reference to the figure-eight pattern of the nerve cord.

Archaeogastropods

Living archaeogastropods (or aspidobranchs) are divided into two suborders on the basis of the structure of the radula: Rhipidoglossa (*rhipido*, fan; *glossa*, tongue),

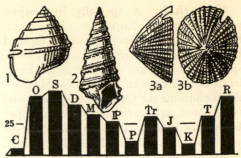

FIG. 8-18. Geologic distribution of archaeogastropods. The graph indicates number of genera belonging to the superfamily **Pleurotomariacea** (Cambrian to Recent). Figured examples are (1) *Schizolopha textilis* Ulrich, ×0.7, Middle Ordovician, Tennessee; (2) *Murchisonia turbinata* (Schlotheim), ×0.7, Middle Devonian, Germany; (3a, b) *Loxotoma neocomiensis* (d'Orbigny), ×1.3, Lower Cretaceous, France; side and apical views, showing slit, but not the spirally coiled embryonic shell (nucleus) at apex.

in which the radula has rows of very numerous teeth that diverge like ribs of a fan (Fig. 8-2, 5); and Docoglossa (*doco*, bar), in which the radula has rows made up of a few strong teeth (Fig. 8-2, 4). These suborders can be recognized fairly well among fossils by their shell characters, but because the radulas are not preserved and many fossil genera cannot be classified definitely in terms of the suborders just noted, the archaeogastropods are grouped in 10 superfamilies, which contain 59 families. We may examine briefly the nature and distribution of the superfamilies, because this is the simplest way of studying the enormously varied assemblage.

(*Fig. 8-17 continued.*)

(Lindström) (2, ×2), Middle Silurian, Sweden.

Hormotoma Salter, Ordovician–Silurian. *H. whiteavesi* Clarke & Ruedemann (5), Middle Silurian, New York.

Loxonema Phillips, Ordovician–Mississippian. Slender high-spired shell-bearing flexuous axial costae. *L. leda* Hall (3), Middle Silurian, Indiana.

Mourlonia Foerste, Silurian. Shell moderately low rounded conical, slit band distinct. *M. filitexta* (Foerste) (7, ×4), Middle Silurian (Clinton), Ohio.

Poleumita Clarke & Ruedemann, Silurian–Devonian. Low-spired, rough-surfaced advolute shell. *P. discors* (Sowerby) (8, oblique apical view), Middle Silurian, England.

Sinuspira Perner, Silurian. *S. tenera* Perner (4, ×2), Middle Silurian, Czechoslovakia.

Tremanotus Hall, Ordovician–Devonian. Spirally costate bellerophontid having a wide bell-shaped aperture and row of tremata in position of the slit band. *T. alpheus* Hall (9a, b), Middle Silurian, New York.

Macluritids. A specially interesting, paleontologically important, but short-lived archaeogastropod group is typified by the genus *Maclurites* (Fig. 8-15), some species of which are excellent guide fossils in Ordovician rocks. The family Macluritidae contains six Ordovician genera and two (based on opercula only) from Upper Cambrian rocks, but no Silurian or younger forms. Most of the shells are large. They are very low conispiral coils having advolute or slightly involute whorls. One side of the shell is nearly flat, and the opposite one bears a concavity formed by the wide umbilicus. If the umbilicate side is directed downward and the flat side upward, the shell is interpreted as sinistral, for the aperture when held toward the observer is on the left. Actually, *Maclurites* is a dextral shell which has a depressed spire, instead of an elevated one. The orientation with the flat side uppermost is wrong, because this is the base of the shell. It is hyperstrophic. Proof of this is furnished by the operculum (Fig. 8-15D), which happens to be calcareous and hence preservable as a fossil—commonly separated from the coiled whorls, but also found in position, fitting against the aperture. The growth lines of opercula may be arranged concentrically or, as in *Maclurites*, in a spiral pattern, and because the direction of twist of the spiral markings on the operculum of a dextral snail is invariably counterclockwise (whereas that of a sinistral shell is opposite), this character establishes the true nature of coiling of the shell, of whatever shape, to which it is attached. The origin and classificatory significance of the early Paleozoic hyperstrophic macluritids are unresolved paleontological problems.

FIG. 8-19. **Representative Devonian gastropods.** The illustrations show two bellerophontids (6, 7) and archaeogastropods (Pleurotomariacea, 3, 5, 13; Euomphalacea, 2, 8; Trochonematacea, 1, 9, 10, 14; Subulitacea, 15; Loxonematacea, 4; Neritacea, 11, 12). Natural size except as indicated otherwise.

Bellerophon Montfort, Ordovician–Triassic. Globular convolute shell. *B. vasulites* Montfort (7, the type species, ×2), Middle Devonian, Germany.

Bembexia Oehlert, Ordovician–Pennsylvanian. Rounded aperture, slit band well developed. *B. larteti* (Munier-Chalmas) (13, ×2), Lower Devonian, France.

Isonema Meek & Worthen, Devonian. Ovoid anomphalous shell having base of columella covered by inductura. *I. depressum* Meek & Worthen (14, ×2), Lower Devonian, Ohio.

Loxonema Phillips, Ordovician–Mississippian. *L. hamiltoniae* Hall (4), Middle Devonian, New York.

Murchisonia d'Archiac & de Verneuil, Ordovician–Permian. High-spired shell having rounded or straight-sided whorls which bear a slit band. *M. bachelieri* Rouault (3, ×2), Lower Devonian, France; *M. turbinata* (Schlotheim) (5), Middle Devonian, Germany.

Philoxene Kayser, Devonian. *P. laevis* (d'Archiac & de Verneuil) (2), Middle Devonian, Germany.

Platyostoma Conrad, Silurian–Devonian. Body whorl and aperture very large. *P. ventricosum* Conrad (10), Lower Devonian, New York.

Ptychospirina Perner, Silurian–Devonian. Aperture obliquely elongate. *P. varians* (Hall) (9), Lower Devonian, New York.

Scalitina Spriestersbach, Devonian. Whorls bear an angular shoulder. *S. montana* Spriestersbach (12, ×0.5), Middle Devonian, Germany.

Straparolus Montfort, Devonian–Permian. Low-spired smooth shell having round aperture. *S. cyclostomus* (Hall) (8), Middle Devonian, Iowa.

Strobeus De Koninck, Silurian–Permian. Nearly smooth-sided tall shell having slightly impressed sutures and a low fold on columella. *S. arculatus* (Schlotheim) (15), Middle Devonian, Germany.

Strophostylus Hall, Silurian–Devonian. Ovoid shell having very inconspicuous spire; columella stout, bearing a prominent fold. *S. andrewsi* Hall (1), Lower Devonian, New York.

Tropidodiscus Meek, Ordovician–Pennsylvanian. Laterally compressed bellerophontid having sharp periphery and large umbilicus. *T. curvilineatus* (Conrad) (6a, b), Lower Devonian, New York.

Turbonopsis Grabau & Shimer, Devonian. A large, low-spired shell having a peripheral carina. *T. shumardi* (Hall) (11), Lower Devonian, Kentucky.

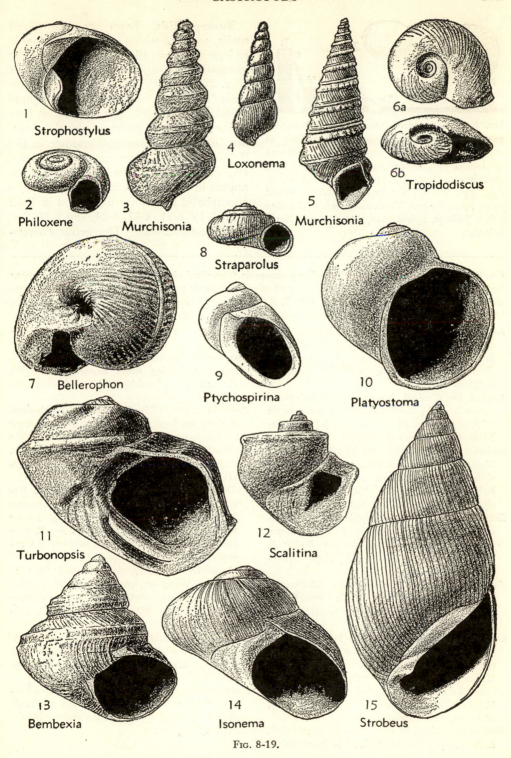

1 Strophostylus

2 Philoxene

3 Murchisonia

4 Loxonema

5 Murchisonia

6a

6b Tropidodiscus

7 Bellerophon

8 Straparolus

9 Ptychospirina

10 Platyostoma

11 Turbonopsis

12 Scalitina

13 Bembexia

14 Isonema

15 Strobeus

FIG. 8-19.

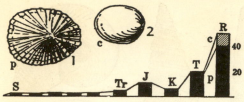

FIG. 8-20. **Geologic distribution of archaeo-gastropods.** The graph indicates number of genera belonging to the superfamilies **Patellacea** (*p*, black) and **Cocculinacea** (*c*, oblique-ruled) (Silurian to Recent). The examples shown in the figure are (1) *Patella coerulea* Linné, apical view, ×0.7, Recent, Mediterranean; (2) *Cocculina galeola* Jeffreys, apical view, ×3, Recent, North Atlantic.

Euomphalids. Low- to flat-spired snails, some of which may be classifiable as hyperstrophic, like the macluritids, are grouped in the superfamily Euomphalacea (Fig. 8-16). They are very abundant in some Paleozoic formations and include useful index fossils. The group, as now defined, has two peaks of development, one in the Devonian and another in the Triassic, but these very possibly are actually distinct stocks. The euomphalids disappear in the Cretaceous. Examples of the euomphalid group are illustrated in Figs. 8-17, 8-19, 8-21, 8-22, 8-24, and 8-26.

Pleurotomariids. The genus *Pleuroto-*

FIG. 8-21. **Representative Mississippian gastropods.** The planispiral shell shown in 13*a, b* has been variously interpreted as a bellerophontid and discoid pleurotomariid. Other illustrated fossils are archaeogastropods (Pleurotomariacea, 1–3, 7, 9–12; Euomphalacea, 6, 15; Trochonematacea, 4, 5; Subulitacea, 14; Neritacea, 8). Natural size except as indicated otherwise.

Angyomphalus Cossmann, Mississippian. Low-spired hemiomphalous, narrow slit band, a row of nodes adjoining sutures. *A. radians* (De Koninck) (3*a, b*, ×2), Lower Mississippian, Belgium.

Baylea De Koninck, Devonian, Mississippian–Pennsylvanian. Turbinate shell, selenizone on edge of shoulder. *B. yvanii* Léveillé (9), Upper Mississippian (Viséan), Belgium.

Bulimorpha Whitfield, Mississippian. Smooth high-spired shell having elongate aperture, siphonal (?) notch, no columellar fold. *B. bulimiformis* (Hall) (14, ×4), Upper Mississippian (Meramecian), Indiana.

Eotrochus Whitfield, Mississippian. Very smooth-sided conical shell having narrow oblique aperture, base concave, sutures flush. *E. tenuimarginatus* (Miller) (4, ×3), Upper Mississippian (Meramecian), Indiana.

Foordella Longstaff, Mississippian. Moderately high-spired shell having rounded whorls and prominent slit band. *F. hibernica* Longstaff (10), Mississippian, Ireland.

Gosseletina Bayle, Mississippian–Triassic. Whorls round, having inconspicuous slit band on upper part. *G. callosa* (De Koninck) (12), Upper Mississippian (Viséan), Belgium.

Mourlonia De Koninck, Silurian–Pennsylvanian. Slit very deep. *M. carinata* (Sowerby) (11), Mississippian, England.

Naticopsis McCoy, Silurian–Triassic. Ovoid shell having large erect aperture. *N. carleyana* (Hall) (8), Upper Mississippian (Meramecian), Indiana.

Platyceras Conrad, Silurian–Permian. Liberty-cap shell having small asymmetrical spire. *P. vetustum* (Sowerby) (5), Mississippian, Ireland.

Porcellia Léveillé, Silurian–Mississippian. Slightly involute planispiral shell ornamented by fine network of costae and coarse nodelike plications of sides; peripheral slit deep, narrow. *P. puzo* Léveillé (13*a, b*), Lower Mississippian (Tournaisian), Belgium.

Rhineoderma De Koninck, Mississippian. Revolving costae prominent, slit band present. *R. radula* (De Koninck) (1, ×2), Lower Mississippian (Tournaisian), Belgium; *R. wortheni* (Hall) (2), Upper Mississippian (Meramecian), Indiana.

Schizostoma Bronn, Devonian–Triassic. A subgenus of *Straparolus*, characterized by angular whorls. *Straparolus* (*Schizostoma*) *catillus* (Martin) (15*a, b*), Upper Mississippian (Viséan), Belgium.

Stegocoelia Donald, Mississippian–Permian. Small *Murchisonia*-like shell. *S. illinoiensis* (Weller) (7), Upper Mississippian (Meramecian), Illinois.

Straparolus Montfort, Devonian–Permian. Discoid or low spiral, round-whorled shells. *S. dionysii* Montfort (6, type of the genus), Mississippian, Belgium.

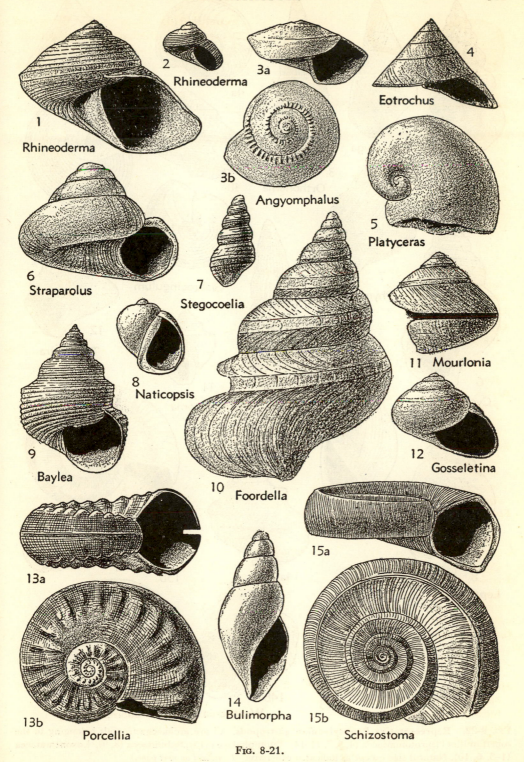

1 Rhineoderma
2 Rhineoderma
3a
3b Angyomphalus
4 Eotrochus
5 Platyceras
6 Straparolus
7 Stegocoelia
8 Naticopsis
9 Baylea
10 Foordella
11 Mourlonia
12 Gosseletina
13a
13b Porcellia
14 Bulimorpha
15a
15b Schizostoma

Fig. 8-21.

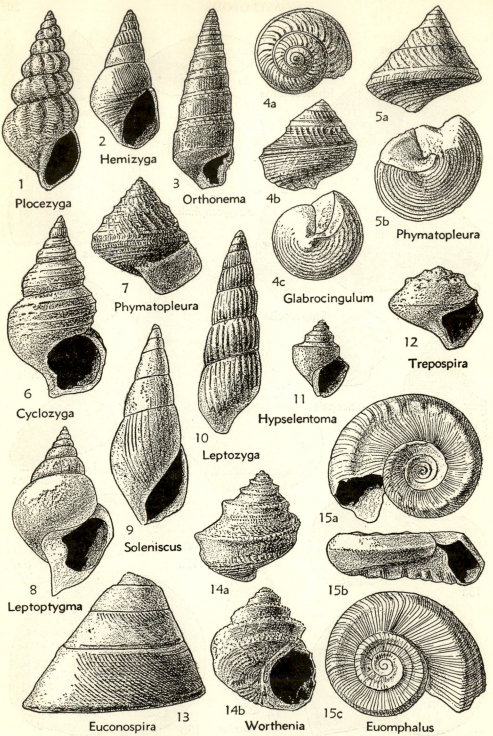

FIG. 8-22. **Representative Pennsylvanian gastropods.** All are archaeogastropods belonging to the superfamilies Pleurotomariacea (4, 5, 7, 11–14), Euomphalacea (15), Subulitacea (8, 9), Loxonematacea (1–3, 6, 10). Natural size except as indicated otherwise. (*Continued on next page.*)

maria has a conispiral shell characterized by its moderately wide spiral angle, fairly even sides, and somewhat flattened base, but its most important feature is a prominent slit, accompanied by a well-defined selenizone or slit band. The selenizone ex-

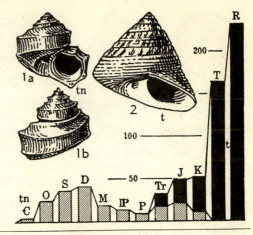

FIG. 8-23. **Geologic distribution of archaeogastropods.** The graph indicates number of genera belonging to the superfamilies **Trochonematacea** (*tn*, oblique-ruled) and **Trochacea** (*t*, black) (Cambrian to Recent). The figured examples are (*1a, b*) *Trochonema umbilicatum* (Hall), ×0.7, Middle Ordovician, Quebec; (*2*) *Trochus erythraeus* Brocchi, ×0.7, Recent, Red Sea.

(Fig. 8-22 continued.)

Cyclozyga Knight, Pennsylvanian. The rounded whorls of this diminutive shell carry revolving costae, but the nucleus has transverse costae. *C. mirabilis* Knight (6, type of the genus, ×20), Middle Pennsylvanian (Desmoinesian), Missouri.

Euconospira Ulrich & Scofield, Devonian–Pennsylvanian. Smooth-sided conical shell having slightly impressed sutures, selenizone just above suture, surface marked by fine revolving and oblique costae. *E. turbiniformis* (Meek & Worthen) (13, ×2), Upper Pennsylvanian (Missourian), Illinois.

Euomphalus Sowerby, ?Silurian–Permian. Subgenus of *Straparolus* characterized by carinae on upper angulation of whorls and wide umbilicus. *Straparolus (Euomphalus) pernodosus* Meek & Worthen (15*a–c*, basal, side, and apical views), Middle Pennsylvanian (Desmoinesian), Illinois.

Glabrocingulum Thomas, Mississippian–Pennsylvanian. A narrow slit band occurs along the shoulder of whorls. *G. grayvillense* (Norwood & Pratten) (4*a–c*, apical, side, and basal views), Middle Pennsylvanian (Desmoinesian), Oklahoma.

Hemizyga Girty, Pennsylvanian. A small shell having whorls marked by axial costae and by revolving ones on base. *H. elegans* Girty (2, ×5), Middle Pennsylvanian (Desmoinesian), Missouri.

Hypselentoma Weller, ?Devonian, Mississippian–Permian. *H. perhumerosa* (Meek) (11, ×2), Pennsylvanian, Nebraska.

Leptoptygma Knight, Pennsylvanian. Anterior end of aperture extended but not as siphonal canal. *L. virgatum* (Knight) (8, ×4), Middle Pennsylvanian (Desmoinesian), Missouri.

Leptozyga Knight, Pennsylvanian. Minute high-spired shell marked by axial costae on whorls. *L. minuta* (Knight) (10, ×20), Middle Pennsylvanian (Desmoinesian), Missouri.

Orthonema Meek & Worthen, Pennsylvanian. Straight-sided, high-spired shell, slight shoulders on the whorls. *O. salteri* (Meek & Worthen) (3, ×4), Middle Pennsylvanian (Desmoinesian), Illinois.

Phymatopleura Girty, Pennsylvanian. Small conical shells ornamented by oblique and revolving costae. *P. brazoensis* (Shumard) (5*a, b*, side and basal views), Upper Pennsylvanian (Missourian and Virgilian); *P. nodosa* (Girty) (7, ×4), Middle Pennsylvanian (Desmoinesian); both widespread, North America.

Plocezyga Knight, Pennsylvanian. Small high-spired shell marked by axial folds on whorl sides and fine revolving costae. *P. corona* (Knight) (1, ×10), Middle Pennsylvanian (Desmoinesian), Missouri.

Soleniscus Meek & Worthen, Mississippian–Pennsylvanian. High-spired smooth shell, siphonal (?) notch and columellar fold. *S. typicus* Meek & Worthen (9, type of the genus, ×2), Upper Pennsylvanian (Missourian), Illinois.

Trepospira Ulrich & Scofield, Devonian–Pennsylvanian. Low-spired smooth shell, except for tubercles adjoining sutures. *T. sphaerulata* (Conrad) (12, ×2), Pennsylvanian, Pennsylvania.

Worthenia De Koninck, Devonian–Triassic. Strongly turbinate shell, shoulder of whorls produced as a carina which carries a narrow slit band and crenulated selenizone. *W. tabulata* (Conrad) (14*a, b*), Upper Pennsylvanian (Conemaugh), Pennsylvania.

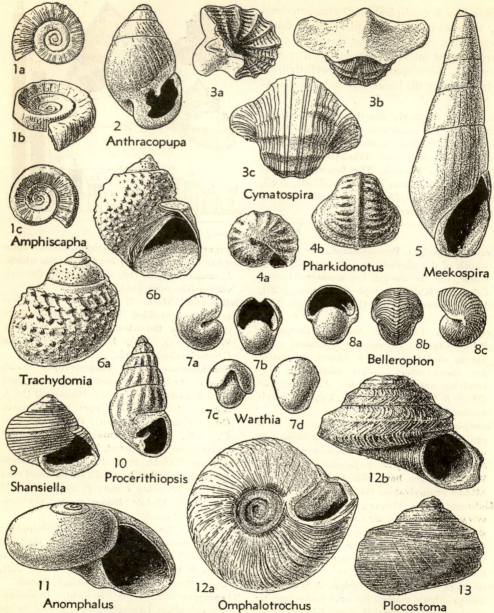

FIG. 8-24. Representative Pennsylvanian and Permian gastropods. Bellerophontids include 3, 4, 7, 8; a pulmonate, oldest of the air-breathing terrestrial snails, is shown in 2; others are archaeogastropods, distributed in the superfamilies Pleurotomariacea (9, 13), Euomphalacea (1, 12), Trochonematacea (11), Subulitacea (5), Loxonematacea (10), and Neritacea (6). Natural size except as indicated otherwise.

Amphiscapha Knight, Pennsylvanian. This shell, classed as a subgenus of *Straparolus*, is distinguished by its flat base, carinate whorls, and concave apical surface. *Straparolus* (*Amphi-* *scapha*) *reedsi* Knight (1*a*–*c*, basal, oblique apical, and apical views), Middle Pennsylvanian (Desmoinesian), Missouri.

Anomphalus Meek & Worthen, Devonian–

(Continued on next page.)

tends around the outer part of the body whorl and commonly can be discerned on spiral whorls near the sutures between whorls. Many Ordovician (Figs. 8-9, 8-10) and younger genera, some very low-spired and some high-spired, are associated with *Pleurotomaria* in a superfamily which ranges from Late Cambrian to Recent (Fig. 8-18), the largest number of genera have been described from Silurian rocks (Fig. 8-17). Modern snails classed among the Pleurotomariacea are characterized chiefly by slit-bearing, cap-shaped shells (Fig. 8-18, *3a, b*). Other examples of the Pleurotomariacea not already cited are forms shown in Figs. 8-8, 8-19, 8-21, 8-22, 8-24, 8-26, 8-35, and 8-39.

Patellids and Cocculinids. These are cap-shaped shells, most of which lack a sign of coiling in the adult. Included among archaeogastropods because of their soft-part anatomy, they are moderately common as fossils in Mesozoic and Cenozoic deposits but are not recorded from

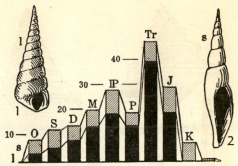

Fig. 8-25. Geologic distribution of archaeogastropods. The graph indicates number of genera belonging to the superfamilies **Loxonematacea** (*l*, black) and **Subulitacea** (*s*, oblique-ruled) (Ordovician to Tertiary). The figured examples are (1) *Loxonema exaltata* (Roemer), ×0.3, Middle Devonian, Germany; (2) *Subulites regularis* Ulrich & Scofield, ×0.5, Middle Ordovician, Kentucky.

Paleozoic rocks. Living genera are more than twice as numerous as those genera known in Tertiary or Mesozoic beds (Fig. 8-20).

(Fig. 8-24 continued.)

Pennsylvanian. A smooth flatly ovoid small shell which has either an open umbilicus or thick callus deposits closing it. *A. rotulus* Meek & Worthen (11, ×10), Middle Pennsylvanian (Desmoinesian), Illinois.

Anthracopupa Whitfield, Pennsylvanian. Aperture bears toothed indentations. *A. ohioensis* Whitfield (2, type of the genus, ×10), Upper Pennsylvanian, Ohio.

Bellerophon Montfort, Ordovician–Triassic. Globose convolute shell having surface marked by transverse ribs. *B. regularis* (Waagen) (8*a–c*), Permian, India.

Cymatospira Knight, Pennsylvanian. A bellerophontid having transverse plications of the shell and revolving costae, widely expanded aperture at final growth stage; inductural deposits prominent on inner lip of aperture. *C. montfortiana* (Norwood & Pratten) (3*a–c*), Middle Pennsylvanian (Desmoinesian), Oklahoma.

Meekospira Ulrich & Scofield, Ordovician–Permian. High-spired smooth shell having nearly flush sutures. *M. peracuta* (Meek & Worthen) (5, ×2), Upper Pennsylvanian (Missourian), Illinois.

Omphalotrochus Meek, Devonian–Permian. Distinguished by conical form and by outer lip with sinus above and protrusion below. *O. whitneyi* (Meek) (12*a, b*, basal and side views), Permian, California.

Pharkidonotus Girty, Pennsylvanian. Slit band elevated, transverse corrugations of whorls. *P. percarinatus* (Conrad) (4*a, b*), Pennsylvanian, Pennsylvania.

Plocostoma Gemmellaro, Permian. *P. neumayri* (Gemmellaro) (13, ×2), Permian, Sicily.

Procerithiopsis Mansuy, Permian. Small axially ribbed high-spired shell. *P. ambigua* Mansuy (10, ×4), Permian, Indo-China.

Shansiella Yin, Permian. Rotund shell which bears revolving low ribs, shallow slit, selenizone just above sutures. *S. planicosta* (Girty) (9, ×5), Lower Permian, Texas.

Trachydomia Meek & Worthen, Pennsylvanian–Permian. Surface nodose, neritacean lip. *T. nodosa* (Meek & Worthen) (6*a, b*, ×2), Middle Pennsylvanian (Desmoinesian), Missouri.

Warthia Waagen, Permian. Smooth surface, outer lip deeply sinuate. *W. polita* Waagen (7*a–d*, ×2), Permian, India.

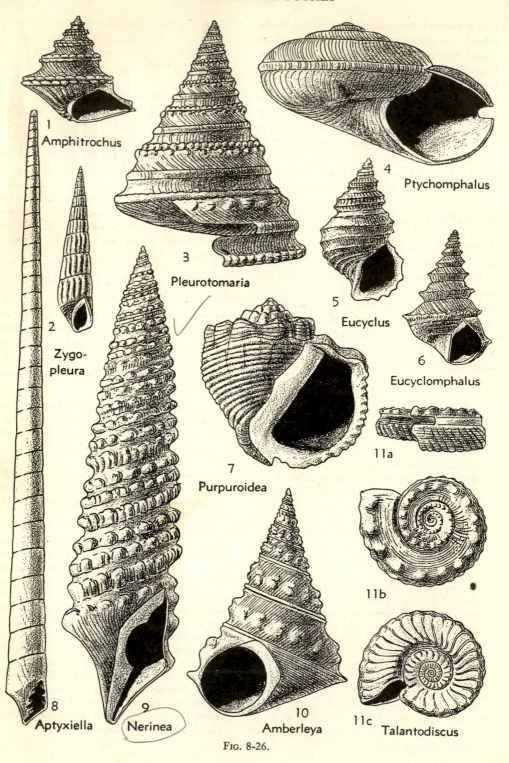

1 Amphitrochus

2 Zygo-
pleura

3 Pleurotomaria

4 Ptychomphalus

5 Eucyclus

6 Eucyclomphalus

7 Purpuroidea

8 Aptyxiella

9 Nerinea

10 Amberleya

11a

11b

11c Talantodiscus

Fig. 8-26.

Trochonematids and Trochids. Shells which are assigned to superfamilies called Trochonematacea and Trochacea represent a broadly defined stock, which probably was derived polyphyletically from pleurotomariid ancestors. Trochonematids are common in Devonian and other Paleozoic systems (Figs. 8-9, 8-10, 8-17, 8-19, 8-21, 8-24). The genera mostly have a pleurotomarian shape but lack a slit and slit band. The first-mentioned group is mainly Paleozoic but ranges to the close of Mesozoic (Figs. 8-23, 8-26, 8-28), whereas the second group is exclusively post-Paleozoic and mostly Cenozoic (Fig. 8-23). The younger superfamily is judged to comprise descendants of the older. Together, they include several hundred genera and many abundant, widely distributed species. Representatives of the Trochacea are illustrated in Figs. 8-26, 8-28, 8-34, 8-35, 8-39, and 8-40.

Loxonematids and Subulitids. These are somewhat similar but not evidently

related assemblages which generally are distinguished by their high narrow spire. Both make an appearance in Ordovician rocks and persist through much or all of

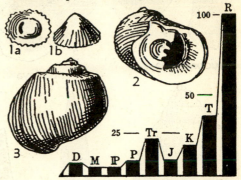

FIG. 8-27. **Geologic distribution ot archaeogastropods.** The graph indicates number of genera belonging to the superfamily **Neritacea** (Devonian to Recent). The figured examples are (1a, b) *Paleolus plicatus* Sowerby, apertural and side views, ×1.3, Middle Jurassic, England; (2) *Nerita peloronta* Linné, ×0.7, Recent, West Indies; (3) *Trachynerita quadrata* Stoppani, ×0.7, Middle Triassic, Italy.

FIG. 8-26. **Representative Jurassic gastropods.** Archaeogastropods include representatives of the superfamilies Pleurotomariacea (3, 4, 11), Euomphalacea (6), Trochonematacea (5, 10), Trochacea (1), and Loxonematacea (2). Mesogastropods include members of the Littorinacea (7) and Nerineacea (8, 9). Natural size except as indicated otherwise.

Amberleya Morris & Lycett, ?Triassic, Jurassic–Cretaceous. Whorls bear nodes; sutures dip steeply. *A. bertheloti* (d'Orbigny) (10), Upper Jurassic, France.

Amphitrochus Cossmann, Triassic–Cretaceous. Periphery of whorls marked by double row of nodes. *A. duplicatus* (Sowerby) (1, ×2), Middle Jurassic, Spain.

Aptyxiella Fischer, Jurassic–Cretaceous. A remarkably high-spired slender shell having internal indentations of the whorls. *A. implicata* (d'Orbigny) (8), Middle Jurassic, France.

Eucyclomphalus Ammon, Jurassic. An unusually high-spired euomphalid having a broadly open umbilicus. *E. cupido* (d'Orbigny) (6, ×3), Upper Jurassic, Spain.

Eucyclus Deslongchamps, Triassic–Jurassic. *E. ornatus* (Sowerby) (5), Middle Jurassic, England.

Nerinea Deshayes, Jurassic–Cretaceous. Very high-spired, smooth or ornamented shells having interior of whorls indented by projec-

tions. *N. trinodosa* Voltz (9), Upper Jurassic, France.

Pleurotomaria Sowerby, Triassic–Cretaceous, ?Eocene. Conical shell having prominent slit and slit band. *P. bitorquata* Deslongchamps (3), Upper Jurassic, Spain.

Ptychomphalus Agassiz, Pennsylvanian–Jurassic, ?Cretaceous. Low-spired, round-whorled shell having a well-defined slit and slit band. *P. heliciformis* (Deslongchamps) (4), Upper Jurassic, Spain.

Purpuroidea Lycett, Triassic–Cretaceous. A very robust, thick-shelled littorinid. *P. maureausea* (Buvignier) (7, ×0.5), Upper Jurassic, France.

Talantodiscus Fischer, Jurassic. A discoidal pleurotomariid, *T. mirabilis* (Deslongchamps) (11a–c, apical, side, and basal views, ×0.5), Lower Jurassic, Spain.

Zygopleura Koken, Silurian–Jurassic. High-spired axially ribbed shell. *Z. periniana* (d'Orbigny) (2), Lower Jurassic, France.

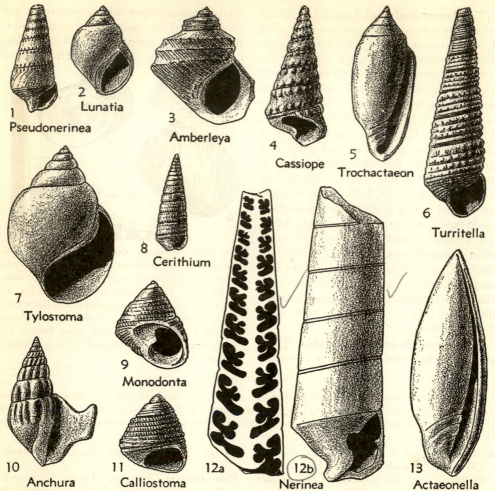

FIG. 8-28. **Representative Lower Cretaceous gastropods.** Archaeogastropods include species of Trochonematacea (3) and Trochacea (9, 11). Excepting two opisthobranchs (5, 13), the others are mesogastropods belonging to the superfamilies Cerithiacea (4, 6, 8), Nerineacea (1, 12), Strombacea (10), and Naticacea (2, 7). Natural size except as indicated otherwise.

Actaeonella d'Orbigny, Cretaceous. A thick-shelled spindle-shaped opisthobranch which has a smooth surface, slitlike aperture, and three columellar folds. *A. dolium* Roemer (13), Fredericksburgian, Texas.

Amberleya Morris & Lycett, ?Triassic, Jurassic–Cretaceous. *A. mudgeana* (Meek) (3, ×2), Fredericksburgian, Kansas.

Anchura Conrad, Cretaceous. The outer lip is prominently extended. *A. mudgeana* White (10), Washitan, Texas.

Calliostoma Swainson, Cretaceous–Recent. A spirally ornamented conical shell having a rounded aperture. *C. cragini* Stanton (11, ×2), Fredericksburgian, Kansas.

Cassiope Coquand, Cretaceous. High-spired shell ornamented by rows of tubercles. *C. branneri* (Hill) (4), Trinitian, Texas.

Cerithium Bruguière, Cretaceous–Recent. High-spired shell, small aperture notched anteriorly. *C. kickapooense* Stanton (8, ×2), Fredericksburgian, Texas.

Lunatia Gray, Cretaceous–Recent. *L. cragini* Stanton (2), Fredericksburgian, Kansas.

Monodonta Lamarck, Cretaceous–Recent. Like *Calliostoma* but distinguished by aperture and fold on columella. *M. bartonensis* Stanton (9), Fredericksburgian, Texas.

Nerinea Deshayes, Jurassic–Cretaceous. A distinctive genus. *N. riograndensis* Stanton (12a, b,

(Continued on next page.)

the Mesozoic (Figs. 8-10, 8-17, 8-19, 8-21, 8-22, 8-24, 8-26). The subulitids are the longer-lasting superfamily, the number of described genera in each system being surprisingly near the same. The loxonematids show a gradual increase of genera to a peak in the Triassic, followed by a quick decline (Fig. 8-25).

Neritids. The superfamily Neritacea consists mostly of ovoid shells but contains some cap-shaped forms. Deposits of callus on the inner lip and other special features of the aperture particularly distinguish many genera, especially the later ones. The group is a minor constituent of Paleozoic gastropod faunas, although large numbers of individuals belonging to a single species occur in some formations (Figs. 8-19, 8-21, 8-24). The number of neritid genera has increased since Permian time to a present-day maximum (Fig. 8-27).

Mesogastropods

The Mesogastropoda include considerably more than half of the order, defined by characters of the gill and heart, which is called Pectinibranchia (*pectini*, comb) because of the single row of gill leaflets, or Monotocardia because of the single auricle of the heart. Living mesogastropods are all classed on the basis of their radula structure in the suborder Taenioglossa (*taenio*, band), which typically has seven teeth in each row. The fossil mesogastropods, defined by shell characters, are arranged in 14 superfamilies which contain 86 families. If we neglect a half-dozen genera from Pennsylvanian and Permian rocks, which rather doubtfully are placed with the

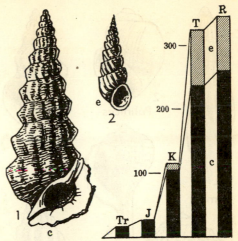

FIG. 8-29. **Geologic distribution of mesogastropods.** The graph indicates number of genera belonging to the superfamilies **Cerithiacea** (*c*, black) and **Epitoniacea** (*e*, oblique-ruled) (Triassic to Recent). The figured examples are (1) *Cerithium nodulosum* Bruguière, ×0.7, Recent, West Indies; (2) *Epitonium gradatum* (Hinds), ×0.7, Recent, southwestern Pacific.

mesogastropods, this order is wholly post-Paleozoic in distribution. It includes most known Mesozoic snails, but the number of Cenozoic genera is far greater than those of Mesozoic age.

Cerithiids.   One of the leading superfamily groups is called Cerithiacea, characterized by the generally very numerous whorls and tall, turreted form of the shell. The aperture of many, but not all, is siphonostomatous (Fig. 8-29, *1*), although a long canal is not developed. Among genera which are specially numerous and important as fossils are *Cerithium* (Cretaceous–Recent), including well-ornamented

---

(*Fig. 8-28 continued.*)

longitudinal section and side view), Washitan, Texas.

Pseudonerinea de Loriol, Jurassic–Cretaceous. Differs internally from *Nerinea*. *P. proctori* Cragin (1), Fredericksburgian, Texas.

Trochactaeon Meek, Cretaceous. An opisthobranch which differs from *Actaeonella* in visibility of the spire. *T. cumminsi* Stanton (5), Washitan, Texas.

Turritella Lamarck, Cretaceous–Recent, worldwide. Very high-spired, columella smooth. *T. seriatim-granulata* Roemer (6), Fredericksburgian, Texas.

Tylostoma Sharpe, Jurassic–Cretaceous. *T. elevatum* (Shumard) (7), Fredericksburgian, Texas.

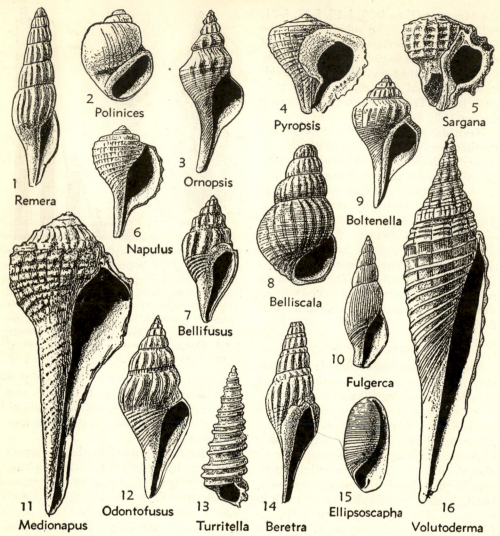

1 Remera
2 Polinices
3 Ornopsis
4 Pyropsis
5 Sargana
6 Napulus
7 Bellifusus
8 Belliscala
9 Boltenella
10 Fulgerca
11 Medionapus
12 Odontofusus
13 Turritella
14 Beretra
15 Ellipsoscapha
16 Volutoderma

FIG. 8-30. **Representative Upper Cretaceous gastropods.** All are prosobranchs except 15, which is an opisthobranch. Mesogastropods include members of the superfamilies Cerithiacea (13), Epitoniacea (8), and Naticacea (2). Neogastropods shown belong to the Muricacea (5), Buccinacea (1, 3, 7, 9, 12), Volutacea (4, 6, 11, 16), and Conacea (10, 14). Natural size and from the Navarro group of Texas except as indicated otherwise.

Bellifusus Stephenson, Cretaceous. Spindle-shaped shell having aperture longer than height of spire. *B. robustus* Stephenson (7).

Belliscala Stephenson, Cretaceous. Plump rounded whorls, axially ribbed and spirally costate. *B. forsheyi* (Shumard) (8, ×2).

Beretra Stephenson, Cretaceous. Distinguished by shape, sculpture, and aperture. *B. firma* Stephenson (14, ×3).

Boltenella Wade, Cretaceous. *B. excellens* Wade (9), Ripley, Tennessee.

Ellipsoscapha Stephenson, Cretaceous. An elongate ellipsoidal spirally costate opisthobranch. *E. striatella* (Shumard) (15).

Fulgerca Stephenson, Cretaceous. Axially fine costate slender fusiform shell. *F. venusta* Stephenson (10, ×3).

Medionapus Stephenson, Cretaceous. Canal very long and slender. *M. elongatus* Stephenson (11).

Napulus Stephenson, Cretaceous. Distinguished by shape and ornamentation. *N. tuberculatus* Stephenson (6, ×2).

(*Continued on next page.*)

high-spired shells, and *Turritella* (also Cretaceous–Recent but chiefly Tertiary), containing a host of exceptionally tall, slender, many-whorled shells (Figs. 8-28, 8-30, 8-35, 8-37, 8-39, 8-40, 8-43). The aberrant cerithiid genus, *Vermicularia*, starts growth like *Turritella* but soon builds its narrow shell tube irregularly in almost any direction. Another member of the superfamily is *Vermetus*, which is attached by the apical part of its shell in a fixed position with the aperture pointed upward; individuals of some species grow closely packed together in the form of gastropod "reefs." *Vermicularia* ranges from Cretaceous to Recent, and *Vermetus* from Pliocene to Recent. More than 300 cerithiid genera are found in Tertiary rocks, and a still larger number occurs in living marine faunas (Fig. 8-29).

Epitoniids. The superfamily Epitoniacea somewhat resembles the cerithiids, but the shells have round holostomatous apertures and the whorls commonly are marked by oblique frill-like expansions (Figs. 8-30, 8-32, 8-35, 8-39). Many species have very distinctive, rather delicate shells. They are smaller on the average than the cerithiids, and they are less numerous. A few genera occur in Cretaceous rocks, but they are mainly Tertiary and Recent (Fig. 8-29).

Strombids. The Strombacea are an important superfamily of marine mesogastropods, which is represented by some of the larger, more colorful shells belonging to the genus *Strombus* found on Florida and other sea beaches. The exterior gener-

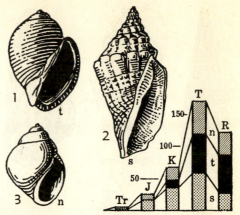

Fig. 8-31. **Geologic distribution of mesogastropods.** The graph indicates number of genera belonging to the superfamilies **Strombacea** (*s*, stippled), **Tonnacea** (*t*, black), and **Naticacea** (*n*, oblique-ruled) (Triassic to Recent). The figured examples are (1) *Parvitonnia perselecta* Iredale, ×0.7, Recent, Australia (tonnid); (2) *Canarium urceum* (Linné), ×0.7, Recent, southwestern Pacific (strombid); (3) *Billiemia diblasii* (Gemmellaro), ×0.7, Upper Jurassic, Sicily (naticid).

ally bears axial and revolving ribs and nodes. The spire is moderately elevated in turret form, and the aperture is long, with flaring lips in the adult. The group is first recognized in the Triassic, but numerous, distinctive fossils belong to Jurassic, Cretaceous, and Tertiary formations (Fig. 8-31, *s*). Some of these, such as *Aporrhais*, *Drepanocheilus*, and *Anchura*, from the Jurassic and Cretaceous, have very striking digitations of the adult outer lip, which

---

(Fig. 8-30 continued.)

Odontofusus Whitfield, Cretaceous. Spire higher than in *Bellifusus*. *O. curvicostatus* Wade (12), Ripley, Tennessee.

Ornopsis Wade, Cretaceous. More slender than *Boltenella*. *O. pulchra* Stephenson (3).

Polinices Montfort, Cretaceous–Recent. Ovoid shell having callus deposits which partly or entirely seals the umbilicus. *P. rectilabrum* (Conrad) (2).

Pyropsis Conrad, Cretaceous. Spire low, outer lip flaring. *P. lanhami* Stephenson (4).

Remera Stephenson, Cretaceous. Somewhat resembles *Beretra* but differs in axial ribs and aperture. *R. microstriata* Stephenson (1, ×2).

Sargana Stephenson, Cretaceous. Like *Pyropsis* but has different ornamentation and apertural characters. *S. stantoni* (Weller) (5).

Turritella Lamarck, Cretaceous–Recent. *T. bilira* Stephenson (13, a distinctive species, ×2).

Volutoderma Gabb, Cretaceous. A graceful slender shell having a long narrow aperture. *V. appressa* Wade (16), Ripley, Tennessee.

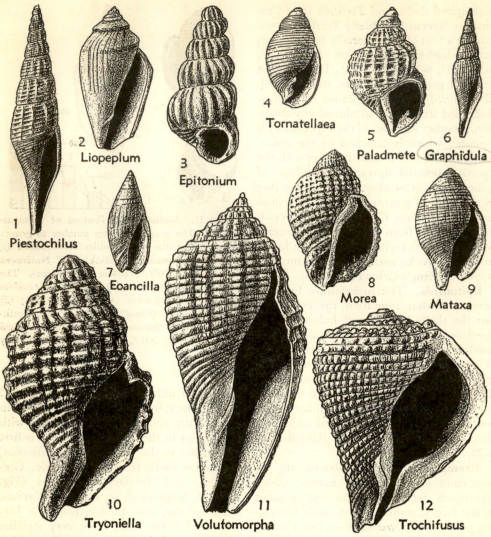

FIG. 8-32. **Representative Upper Cretaceous gastropods.** Except an opisthobranch (4) and a meso-gastropod prosobranch (Epitoniacea, 3), all fossils here shown are neogastropods. They belong to the superfamilies Muricacea (8) and Volutacea (1, 2, 5–7, 9–12). Natural size and from the Navarro group of Texas except as indicated otherwise.

Eoancilla Stephenson, Cretaceous. Resembles the modern *Oliva*. *E. acutula* Stephenson (7, ×3).

Epitonium Bolten, Cretaceous–Recent. Strong axial ribs, round aperture. *E. bexarense* Stephenson (3).

Graphidula Stephenson, Cretaceous. Very slender fusiform shell, finely costate axially. *G. terebriformis* Stephenson (6).

Liopeplum Dall, Cretaceous. Somewhat resembles *Conus* but has quite different columella. *L. leioderma longum* Stephenson (2).

Mataxa Wade, Cretaceous. *M. valida* Stephenson (9, ×3).

Morea Conrad, Cretaceous. Elongate egg-shaped, sculptured. *M. marylandica bella* Stephenson (8, ×2).

Paladmete Gardner, Cretaceous. Strongly axial ribs and spiral costae. *P. cancellaria* (Conrad) (5, ×2), Ripley, Tennessee.

Piestochilus Meek, Cretaceous. Very slender shell having spire longer than aperture. *P. pergracilis* Wade (1), Ripley, Tennessee.

(Continued on next page.)

make them distinguishable at a glance (Figs. 8-28, 8-34, 8-35). Even more unique in its display of long curved digitations is *Pterocera*, a modern strombid, which is known as a fossil only in some Pleistocene deposits.

Tonnids. Late-appearing, highly organized marine mesogastropods which resemble the strombids in many ways are grouped in the superfamily Tonnacea (Figs. 8-31, *t;* 8-34). Most of them have a strongly siphonostomatous aperture, and they are rather strongly sculptured. The shells are medium to large, with the body whorl predominating greatly over the spire. Many are very graceful. Several genera are characterized by the strong development of varices, which tend to be aligned on the whorls. The group is represented by 11 Cretaceous genera, but there are nearly six times as many in the Tertiary, and they are equally abundant at the present time.

Naticids. A conservative assemblage of ovoid to globose marine snails having nearly smooth shells which mostly are only an inch or two in greatest diameter belong to the Naticacea. Many of them bear prominent callus deposits which modify the configuration of the inner lip and tend to close the umbilicus. Hemiomphalous and cryptomphalous genera are numerous. Fossil naticids are distributed from Triassic through the Pleistocene, and there are many living genera (Fig. 8-31, *n*). The shells are extremely common in some Tertiary formations. Some of the most important genera are *Gyrodes* and *Tylostoma*, of Cretaceous age, and *Amaurellina, Polinices,* and *Neverita* in various Tertiary rocks. A spirally ribbed naticid having almost no visible spire is *Sinum;* it is fairly common in

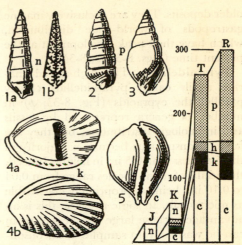

FIG. 8-33. **Geologic distribution of mesogastropods.** The graph indicates number of genera belonging to the superfamilies **Cypraeacea** (*c*, white), **Pyramidellacea** (*p*, stippled), **Calyptraeacea** (*k*, black), **Hipponicacea** (*h*, obliqueruled), and **Nerineacea** (*n*, white) (Jurassic to Recent). The figured examples are (1) *Nerinea defrancei* (Deslongchamps), ×0.7, Middle Jurassic, France; (2) *Pyramidella jamaicensis* Dall, ×7, Miocene, Jamaica; (3) *P. ventricosa* Guérin, ×0.7, Recent, Philippines; (4*a, b*) *Crepidula fornicata* (Linné), apertural and apical views, ×0.5, Recent, Mediterranean; (5) *Cypraeorbis sphaeroides* (Conrad), ×0.7, Oligocene, Mississippi.

some Eocene, Miocene, and Pliocene strata and can be found along modern seashores. Illustrated naticids appear in Figs. 8-28, 8-30, 8-34, 8-39, and 8-40.

Cypraeids. The brilliantly polished, mostly very smooth-surfaced "cowry shells" which bear no visible spire as adults and which have a long, narrow apertural slit on one side are members of the Cypraeacea. They occur as fossils in Cretaceous and Tertiary rocks but have not been found in

---

(Fig. 8-32 continued.)

Tornatellaea Conrad, Jurassic–Miocene. An opisthobranch. *T. scatesi* Stephenson (4, ×3).
Trochifusus Gabb, Cretaceous. Distinctive shape and sculpture. *T. perornatus* Wade (12), Ripley, Tennessee.

Tryoniella Stephenson, Cretaceous. Differs from *Morea* in development of canal. *T. valida* (Stephenson) (10).
Volutomorpha Gabb, Cretaceous. Tall sculptured shell having long evenly curved outer lip. *V. mutabilis* Wade (11), Ripley, Tennessee.

older deposits. They are exclusively marine gastropods of world-wide distribution, which have their peak development at the present time (Figs. 8-33, c; 8-34).

Pyramidellids. The high-spired marine snails called pyramidellids closely parallel the cypraeids (Fig. 8-33, p) in number of genera represented by fossils and in geologic distribution, but they are far less conspicuous because a majority of them are less than 0.5 in. in height. Indeed, the adults of many species range from 0.04 to 0.10 in., being appropriately classifiable as microfossils. In some respects, this is an advantage because large numbers of them can be washed from samples of unconsolidated sediment containing them. Pyramidellids are included in Figs. 8-37 and 8-39.

Calyptraeids and Hipponicids. These are two quantitatively less prominent assemblages of marine mesogastro-pods, but they are structurally distinctive, especially in the development of internal platforms for muscle attachment. Also, a few genera, such as *Calyptraea* and *Crepidula*, the latter commonly known as the "slipper shell," are represented by many fossils. Distribution of the two mentioned genera, like that of the superfamilies, is from Cretaceous to Recent (Figs. 8-33, h, k; 8-37; 8-39; 8-40).

Nerineids. The superfamily Nerineacea includes Jurassic and Cretaceous gastropods, many of which draw attention on account of their exceptional form, for the height of the shell may be more than twenty times the diameter of the body whorl. The spiral angle of such a shell is so small that the sides of the spire are nearly parallel. The external surfaces of the whorls tend to be evenly confluent. Some members of the group have a more

---

FIG. 8-34. **Representative Paleogene gastropods.** An archaeogastropod is shown in 14 (Trochacea.) Mesogastropods include Strombacea (7, 11), Cypraeacea (13), Naticacea (2, 4–6), and Tonnacea (15, 17). Neogastropods belong to the Muricacea (8), Volutacea (1, 16), and Conacea (12). Opisthobranchs are represented by 9 and 10, and a scaphopod by 3. Natural size and from Jacksonian Eocene deposits of Mississippi except as indicated otherwise.

Actaeon Montfort, Cretaceous–Recent. A moderately high-spired opisthobranch having a few spiral grooves. *A. idoneus* Conrad (10, ×4).

Athleta Conrad, Cretaceous–Recent. Spinose nodes along shoulder of whorls. *A. petrosa* (Conrad) (16).

Cassidaria Eocene–Recent. Inductural deposits of inner lip extended; outer lip reflected. *C. millsapsi* Sullivan & Gardner (17a, b).

Conus Linné, ?Cretaceous, Eocene–Recent. Biconical to obconical shell, aperture a long slit. *C. sauridens* Conrad (12).

Cypraedia Swainson, Cretaceous–Oligocene. Spire not externally visible, surface reticulate. *C. fenestralis* Conrad (13).

Dentalium Linné, Eocene–Recent. Scaphopod, with tapering tubular shell open at both ends, longitudinally costate. *D. danvillense* Harris & Palmer (3, ×2), Louisiana.

Ectinochilus Cossmann, Eocene–Oligocene. Sides of whorls axially costate, base spirally grooved. *E. laqueatum* (Conrad) (7, 11), Alabama.

Ficus Bolten, Paleocene–Recent. Rounded whorls, expanded aperture, reticulate ornamentation.

F. filia (Meyer) (15, ×2).

Globularia Swainson, Jurassic–Recent. Distinguished by shape of whorls and aperture. *G. morgani* (Jordan) (5).

Murex Linné, Eocene–Recent. Moderately elevated spire, shoulders of whorls spinose. *M. vanuxemi* Conrad (8, ×2).

Polinices Montfort, Cretaceous–Recent. Elongate ovoid shell, callus near base of umbilicus. *P. eminulus* (Conrad) (2, ×2), Alabama.

Pseudoliva Swainson, Cretaceous–Recent. Pear-shaped shell, aperture bearing siphonal notch, columella bent anteriorly. *P. vetusta perspectiva* Conrad (1), Louisiana.

Sinum Bolten, ?Cretaceous, Eocene–Recent. Shell ovoid, aperture widely flaring, body whorl large. *S. danvillense* Harris & Palmer (4, 6, ×2), Louisiana.

Solariella Wood, ?Triassic, Cretaceous–Recent. Low-spired. *S. cancellata jacksoniana* Harris & Palmer (14, ×5).

Tornatellaea Conrad, Jurassic–Miocene. Columella bears spiral folds. *T. lata* (Conrad) (9), Alabama.

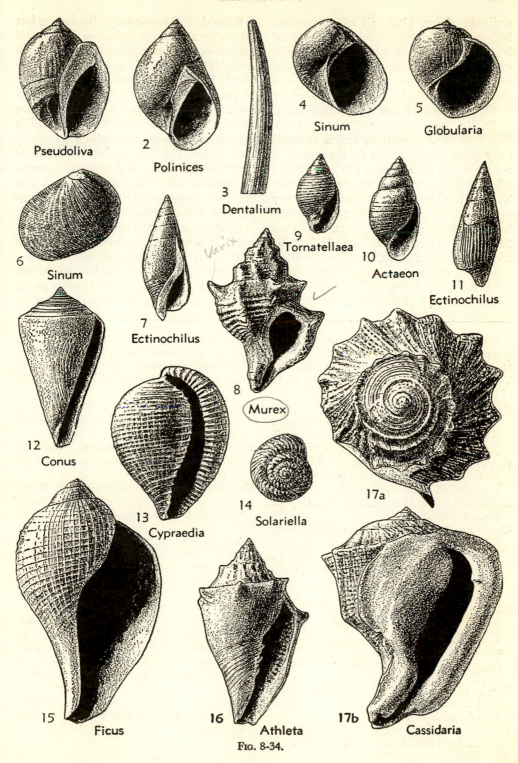

1 Pseudoliva

2 Polinices

3 Dentalium

4 Sinum

5 Globularia

6 Sinum

7 Ectinochilus

8 Murex

9 Tornatellaea

10 Actaeon

11 Ectinochilus

12 Conus

13 Cypraedia

14 Solariella

15 Ficus

16 Athleta

17a

17b Cassidaria

Varix

Fig. 8-34.

ordinary shape, although still somewhat high-spired. Ornamentation of the exterior is much less common than a nearly smooth outer surface. The most distinctive feature of the nerineids, however, is internal and can be seen only in broken shells or by making sections. When cut in half longitudinally, the inner walls of the whorls are found to bear projections which run spirally from the aperture to the apex, or nearly to it. Some shells are umbilicate, whereas others have a strong columella. The spiral ridges occur in both, and they produce varying cross sections of the whorl interiors from genus to genus. Although 21 genera of these fossils occur in Jurassic rocks and 25 in the Cretaceous, neither antecedents nor descendants are known (Fig. 8-33, *n*). A few nerineids are illustrated in Figs. 8-26 and 8-28.

Rissoids. A numerically important but otherwise rather unimpressive division of the mesogastropods comprises the Rissoacea. They include a few fresh-water forms, but most of them live in the sea. The shells are small, a large majority having a height of 0.5 to 1 in. They have slightly rounded conical outlines and a holostomatous aperture which generally lacks a rim. The shell surface is smooth or moderately sculptured. A few genera occur in Jurassic and Cretaceous rocks. The main development of the group is Tertiary and Recent (Figs. 8-36, 8-43).

Cyclophorids, Valvatids, and Littorinids. Superfamilies composed mainly of fresh-water mesogastropods, but including some which are at home in brackish water or live along coasts in the zone between high and low tides, are grouped together

---

FIG. 8-35. Representative Paleogene gastropods. Archaeogastropods include members of the Pleurotomariacea (13) and Trochacea (2); mesogastropods are represented by Epitoniacea (11), Cerithiacea (14–16), and Strombacea (6). Neogastropod superfamilies include Buccinacea (4, 5, 9), Volutacea (1, 3, 7, 10, 12), and Conacea (8, 17, 18). Natural size and from Eocene (Jacksonian) deposits of Mississippi except as indicated otherwise.

Architectonica Bolten, Cretaceous–Recent. Moderately low-spired conical ornamented shell having open umbilicus. *A. trilirata* (Conrad) (16*a–c*, side, basal, and apical views, ×2).

Calyptraphorus Conrad, Cretaceous–Eocene. Inductural deposits largely cover spire, aperture expanded. *C. velatus stamineus* (Conrad) (6*a, b*).

Caricella Conrad, Cretaceous–Eocene. *C. polita* Conrad (12).

Cirsotrema Mörch, Eocene–Recent. High-spired axially ribbed shell, round aperture. *C. nassulum creolum* Harris & Palmer (11, ×2), Louisiana.

Coronia Gregorio, Paleocene–Oligocene. High-spired, short canal. *C. genitiva* (Casey) (17, ×4), Louisiana.

Diodora Gray, Cretaceous–Recent. A keyhole limpet. *D. tenebrosa veatchi* Harris & Palmer (13, ×4), Louisiana.

Fusimitra Conrad, Cretaceous–Eocene. Evenly fusiform shell, aperture narrow, slitlike. *F. conquisita* (Conrad) (1).

Harpa Walch, Eocene–Recent. Axially ribbed shell, aperture somewhat expanded. *H. jacksonensis* Harris (10).

Lapparia Conrad, Paleocene–Eocene. Stout columella bears folds. *L. dumosa exigua* Palmer (7).

Latirus Montfort, ?Cretaceous, Eocene–Recent. Fusiform axially ribbed, canal well developed. *L. leaensis* Harris (5, ×2).

Levifusus Conrad, ?Cretaceous, Paleocene–Eocene. Long slender canal. *L. moodianus* Cooke (4, ×2).

Mazzalina Conrad, Cretaceous–Eocene. Nearly smooth shell, with spiral grooves on neck of body whorl. *M. inaurata oweni* (Dall) (9, ×2), Arkansas.

Scobinella Conrad, Eocene–Miocene. Elongate aperture less than height of spire. *S. louisianae* Harris (18, ×3), Louisiana.

Solariella Wood, ?Triassic, Cretaceous–Recent. *S. cancellata jacksoniana* Harris & Palmer (2*a, b*, side and basal views, ×5).

Terebra Bruguière, Eocene–Recent. Slender high-spired conacean. *T. jacksonensis* Cooke (8, ×2).

Tritonoatractus Cossmann, Eocene. Periphery of whorls bears nodose ribs, canal long. *T. pearlensis montgomeryensis* (Vaughan) (3, ×2).

Turritella Lamarck, Cretaceous–Recent. High-spired slender cerithiid. *T. arenicola* (Conrad) (14, ×2); *T. clevelandia* Harris (15, ×2), Arkansas.

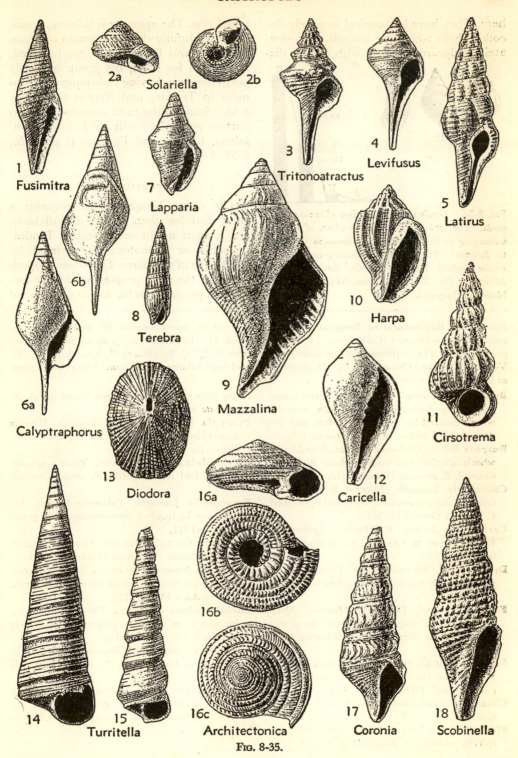

1 Fusimitra
2a Solariella 2b
7 Lapparia
8 Terebra
3 Tritonoatractus
4 Levifusus
5 Latirus
10 Harpa
6b
6a
Calyptraphorus
9 Mazzalina
13 Diodora
16a
12 Caricella
11 Cirsotrema
14 15 Turritella
16b
16c Architectonica
17 Coronia
18 Scobinella

FIG. 8-35.

here. They have low conical to nearly flat coiled shells which are smooth or moderately sculptured, mostly with axially dis-

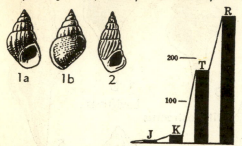

FIG. 8-36.   **Geologic distribution of mesogastropods.** The graph indicates number of genera belonging to the superfamily **Rissoacea** (Jurassic to Recent). The figured examples are (1a, b) *Rissoa violacea* Desmarest, ×3, Recent, North Atlantic; (2) *Rissoina dilomus* Woodring, ×3, Miocene, Jamaica.

posed ribs. The aperture is holostomatous. Some doubtfully classified genera of Pennsylvanian and Permian age are included, but except for these, the group is wholly post-Paleozoic, having maximum development in Tertiary and Recent time (Fig. 8-38). Some of the most common and important genera of fossils are *Littorina*, *Campeloma*, *Valvata*, and *Viviparus* (Figs. 8-26, 8-39, 8-40, 8-43).

Neogastropods

The order Neogastropoda contains a good deal less than half of the division, also classed as an order, called Pectinibranchia or Monotocardia, defined by characters of soft parts. The neogastropods comprise the more progressive, highly specialized pectinibranchs, and as represented

---

FIG. 8-37.   **Representative Neogene gastropods.** Except an opisthobranch (9) and three mesogastropod prosobranchs (Cerithiacea, 14; Pyramidellacea, 11; Hipponicacea, 7), all are neogastropod prosobranchs. They represent the superfamilies Muricacea (1, 2, 13), Buccinacea (5, 6, 8, 10, 12, 15, 18), Volutacea (3), and Conacea (4, 16, 17). Natural size and from Miocene formations of Maryland except as indicated otherwise.

Buccinofusus Conrad, Cretaceous–Recent. Large, moderately high-spired shell with rounded whorls marked by axial corrugations and two sizes of spiral costae. *B. parilis* Conrad (15).

Busycon Bolten, Oligocene–Recent. Shoulders of whorls angular and nodose; shells dextral or sinistral. *B. coronatus* (Conrad) (12, ×0.5).

Cancellaria Lamarck, Miocene–Recent. Axial corrugations pronounced, apertural lips ribbed. *C. alternata* Conrad (3, ×2).

Cerithiopsis Forbes & Hanley, Cretaceous–Recent. Close to *Cerithium*. *C. subulata* (Montagu) (14, ×5).

Drillia Gray, Eocene–Recent. Lower part of whorls axially corrugated. *D. calvertensis* Martin (17, ×2).

Ecphora Conrad, Cretaceous–Miocene. Spiral corrugations T-shaped in section, like a rail. *E. quadricostata* (Say) (13, a well-known index fossil).

Fossarus Philippi, ?Cretaceous, Eocene–Recent. Shaped like *Cancellaria* but not axially corrugated. *F. dalli* (Whitfield) (7, ×2).

Lirosoma Conrad, Cretaceous–Miocene. *L. sulcosa* Conrad (6).

Mangilia Risso, Eocene–Recent. Like *Drillia* but spirally ribbed. *M. patuxentia* Martin (4, ×4).

Nassa Bolten, Miocene–Recent. Short canal. *N. trivittata* Say (8, ×2).

Pterygia Bolten, Miocene–Recent. Whorls marked by spiral grooves. *P. calvertensis* (Martin) (10, ×2).

Ptychosalpinx Gill, Miocene. Whorls spirally ribbed, fold on columella. *P. altilis* (Conrad) (5).

Pyramidella Lamarck, Paleocene–Recent. Diminutive high-spired mesogastropod. *P. granulata* (Lea) (11, ×8).

Scalaspira Conrad, Miocene. Coarse reticulate ornamentation. *S. strumosa* Conrad (2, ×2).

Siphonalia Adams, ?Cretaceous, Eocene–Recent. Like *Buccinofusus* but axial corrugations weak. *S. marylandica* Martin (18).

Surcula Adams & Adams, Cretaceous–Recent. Periphery of whorls projecting, obliquely nodose. *S. rotifera* Conrad (16, ×2).

Urosalpinx Stimpson, Eocene–Recent. Like *Buccinofusus* but smaller. *U. rusticus* (Conrad) (1).

Volvula Adams, Eocene–Recent. Elongate ovoid, spire concealed, spiral grooves except on middle of body whorl. *V. iota patuxentia* Martin (9, ×5).

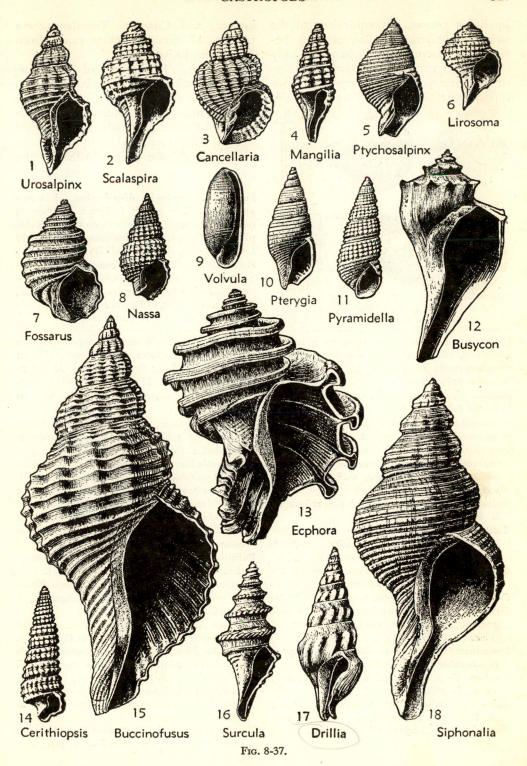

1 Urosalpinx
2 Scalaspira
3 Cancellaria
4 Mangilia
5 Ptychosalpinx
6 Lirosoma
7 Fossarus
8 Nassa
9 Volvula
10 Pterygia
11 Pyramidella
12 Busycon
13 Ecphora
14 Cerithiopsis
15 Buccinofusus
16 Surcula
17 Drillia
18 Siphonalia

FIG. 8-37.

by living genera, are equivalent to the suborder Stenoglossa (*steno*, narrow), because they have a radula in which only one to three teeth occur in each row. Fossil neogastropods are grouped in four superfamilies: Muricacea, Buccinacea, Vo-

lutacea, and Conacea. Together, these contain 17 families. They range from Cretaceous to Recent.

Muricids. This group of neogastropods is distinguished especially by the prominence of the canal, which extends forward from the aperture, and by the absence of columellar folds. The surface is strongly sculptured, and many shells bear knobby or spinose varices. Although the

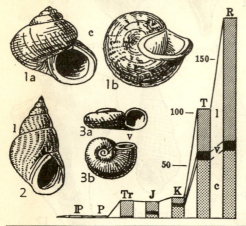

Fig. 8-38. **Geologic distribution of mesogastropods.** The graph indicates number of genera belonging to the superfamilies **Cyclophoracea** (*c*, stippled), **Valvatacea** (*v*, black), and **Littorinacea** (*l*, oblique-ruled) (Pennsylvanian to Recent). The figured examples are (1*a, b*) *Cyclophorus validus* (Sowerby), ×0.7, Recent, Philippines; (2) *Littorinopsis angulifera* (Lamarck), ×0.7, Recent, Africa; (3*a, b*) *Valvata cristata* Müller, side and basal views, ×3, Recent, Europe.

Fig. 8-39. **Representative Neogene gastropods.** An opisthobranch (3) and scaphopod (10) are illustrated, others being prosobranchs. Archaeogastropods (Pleurotomariacea, 4; Trochacea, 11) and mesogastropods (Littorinacea, 15; Cerithiacea, 6, 20; Epitoniacea, 1, 2; Pyramidellacea, 9; Calyptraeacea, 5; Naticacea, 18) slightly outnumber neogastropods (Muricacea, 17; Buccinacea, 14; Volutacea, 12, 13, 19; Conacea, 7, 8, 16). Natural size and from Miocene deposits of Maryland except as indicated otherwise.

Actaeon Montfort, Cretaceous–Recent. A spirally grooved opisthobranch. *A. ovoides* Conrad (3, ×3).

Architectonica Bolten, Cretaceous–Recent. *A. trilineata* (Conrad) (20*a–c*, side, basal, and apical views, ×2).

Bulliopsis Conrad, Miocene. Folds on columella. *B. integra* Conrad (14).

Calliostoma Swainson, Cretaceous–Recent. Conical, spirally ornamented shell, aperture rounded. *C. philanthropus* (Conrad) (11, ×2).

Conus Linné, ?Cretaceous, Eocene–Recent. Biconical shell. *C. diluvianus* Green (16).

Crepidula Lamarck, Cretaceous–Recent. "Slipper shell" having large oval aperture and internal platform for muscle attachment. *C. fornicata* (Linné) (5*a, b*).

Dentalium Linné, Eocene–Recent. Scaphopod. *D. alternatum* Say (10, ×2).

Diodora Gray, Cretaceous–Recent. A keyhole limpet. *D. alticosta* (Conrad) (4*a, b*).

Epitonium Bolten, Cretaceous–Recent. Strong axial ribs, round aperture. *E. calvertense* (Martin) (1, ×2); *E. expansum* (Conrad) (2).

Littorina Férussac, Paleocene–Recent. Ovoid smooth or spirally costate shells, aperture rounded. *L. irrorata* (Say) (15, ×2).

Marginella Lamarck, Eocene–Recent. Columella bears four folds. *M. calvertensis* Martin (12, ×3).

Melanella Bowdich, Eocene–Recent. High-spired smooth shell. *M. eborea* (Conrad) (9, ×4).

Oliva Martyn, Eocene–Recent. Pointed ovoid shell having long narrow aperture, folds on columella. *O. litterata* Lamarck (13, ×2).

Polinices Montfort, Cretaceous–Recent. Ovoid shell, open umbilicus. *P. heros* (Say) (18).

Scaphella Swainson, Cretaceous–Recent. Nodes on spiral whorls, four folds in columella. *S. typa* (Conrad) (19).

Terebra Bruguière, Eocene–Recent. Slender high-spired conacean. *T. unilineata* Conrad (7).

Tritonalia Fleming, Oligocene–Recent. Whorls corrugated axially and costate spirally. *T. topangensis* (Arnold) (17), Miocene, California.

Turris Bolten, Eocene–Recent. High-spired shell, aperture siphonostomatous. *T. communis* (Conrad) (8, ×2).

Turritella Lamarck, Cretaceous–Recent. High-spired cerithiid. *T. variabilis* Conrad (6).

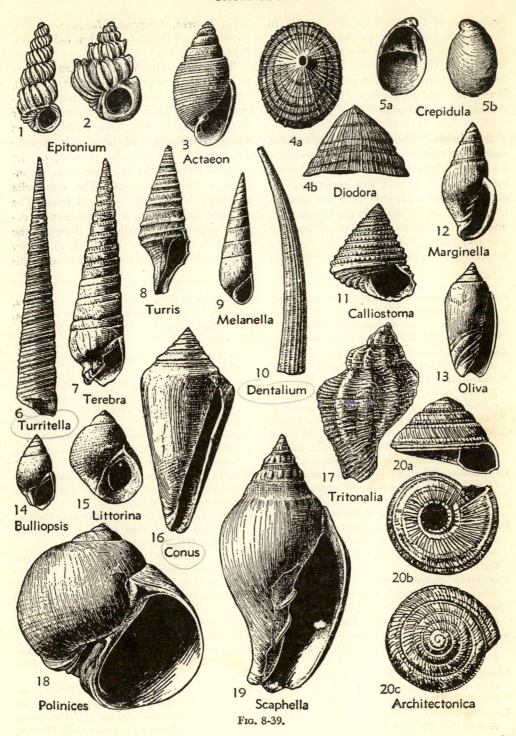

1 Epitonium
2
3 Actaeon
4a
4b Diodora
5a
Crepidula 5b
6 Turritella
7 Terebra
8 Turris
9 Melanella
10 Dentalium
11 Calliostoma
12 Marginella
13 Oliva
14 Bulliopsis
15 Littorina
16 Conus
17 Tritonalia
18 Polinices
19 Scaphella
20a
20b
20c Architectonica

FIG. 8-39.

beginning of the muricids is observed in Cretaceous rocks, the preponderant majority of known fossils comes from Tertiary formations (Fig. 8-41, *m*). Among more important genera of the superfamily, some here illustrated are *Ecphora*, *Forreria*, *Morea*, *Murex*, and *Tritonalia* (Figs. 8-30, 8-32, 8-34, 8-37, 8-39, 8-40).

Buccinids. The shells of Buccinacea mostly are spindle-shaped and less strongly sculptured than the muricids. They have a long or short canal, and generally, the columella lacks folds. Genera are numerous, for 43 are recorded in the Cretaceous, and there are more than 240 of Tertiary age and among living forms (Fig. 8-41, *b*). Representative fossil forms include *Buccino-*

fusus, *Busycon*, *Levifusus*, *Mazzalina*, *Nassa*, *Neptunea*, and *Odontofusus* (Figs. 8-30, 8-35, 8-37, 8-39, 8-40).

Volutids. The Volutacea are characterized by egg-shaped to fusiform shells, which mostly are little sculptured, although some bear fairly strong axial or spiral ribs and a few have knobby spines along the shoulders of the whorls. A well-developed canal may be present, or the anterior edge of the aperture merely contains a siphonal notch. The columella generally is marked by spiral folds. The geologic distribution and numbers of volutid genera are similar to those of the buccinids and conids (Fig. 8-41, *v*). Many common living marine snails belong in this

---

Fig. 8-40. Representative Neogene gastropods. The group includes archaeogastropods (Trochacea, 1, 8), mesogastropods (Littorinacea, 16; Cerithiacea, 4, 6, 7, 15, 18; Epitoniacea, 12; Calyptraeacea, 20; Naticacea, 5), and neogastropods (Muricacea, 9, 10, 19; Buccinacea, 2, 3, 11, 13, 17, 22; Volutacea, 14; Conacea, 21). Natural size except as indicated otherwise.

Amphissa Adams & Adams, Oligocene–Recent. *A. versicolor* Dall (13, ×2), Pliocene, California.

Barbarofusus Grabau & Shimer, Pliocene–Recent. Spirally costate whorls are axially corrugated, canal moderately elongate. *B. arnoldi* (Cossmann) (3), Pleistocene, California.

Bittium Leach, Paleocene–Recent. A small cerithiid. *B. asperum* (Gabb) (4, ×2), Pleistocene, (15, ×2) Pliocene, both California.

Calicantharus Clark, Eocene–Pleistocene. Whorls of spire axially corrugated, all whorls spirally ribbed, distinct basal fasciole. *C. fortis* (Carpenter) (11), Pleistocene, California.

Calliostoma Swainson, Cretaceous–Recent. Spire of intermediate height, revolving costae. *C. coalingaensis* Arnold (8, ×2), Pleistocene, California.

Cantharus Bolten, Eocene–Recent. *C. fortis* (Carpenter) (22, ×⅔), Pliocene, California.

Clathrodrillia Dall, Oligocene–Recent. *C. mercedensis* Martin (21), Pliocene, California.

Crepidula Lamarck, Cretaceous–Recent. Slipper shell. *C. princeps* Conrad (20, ×0.5), Pliocene, California.

Forreria Jousseaume, Miocene–Recent. A muricid which lacks revolving ribs. *F. magister* (Nomland) (19, ×0.5), Pliocene, California.

Littorina Férussac, Paleocene–Recent. *L. mariana* Arnold (16), Pliocene, California.

Nassarius Duméril, Miocene–Recent. Strong

folds on columella. *N. californianus* Conrad (17), Pliocene, California.

Neptunea Bolten, Eocene–Recent. Spirally costate whorls have distinct shoulder, moderately tall shell. *N. tabulata* (Baird) (2), Pleistocene, California.

Neverita Risso, Eocene–Recent. Like *Polinices*, callus prominent. *N. reclusiana* (Deshayes) (5), Pleistocene, California.

Nucella Bolten, Miocene–Recent. Smooth siphonostomatous shell. *N. etchegoinensis* (Arnold) (9), Pleistocene, California.

Olivella Swainson, Cretaceous–Recent. Aperture somewhat shorter than in *Oliva*. *O. biplicata* Sowerby (14), Pliocene, California.

Opalia Adams & Adams, Miocene–Recent. High-spired, axially ribbed shell. *O. varicostata* Stearns (12), Pliocene, California.

Tritonalia Fleming, Oligocene–Recent. Turreted shell, spirally costate, axially corrugated, canal well developed. *T. squamulifera* (Carpenter) (10), Pleistocene.

Turcica Adams, Miocene–Recent. Aperture rounded, fold on columella. *T. caffea brevis* Stewart (1), Pleistocene.

Turritella Lamarck, Cretaceous–Recent. *T. jewettii* Carpenter (6, ×¾), Pleistocene, California. *T. pedroensis* Applin (7), Pleistocene, California. *T. cooperi* Carpenter (18), Pliocene, California.

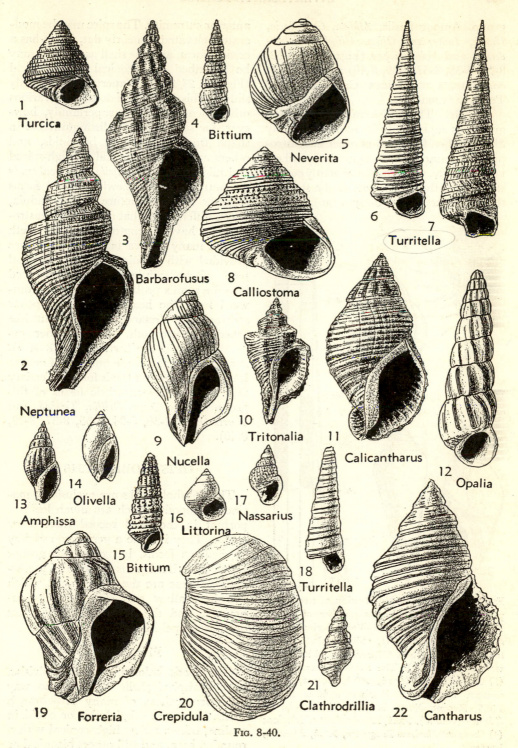

1 Turcica
4 Bittium
5 Neverita
3 Barbarofusus
8 Calliostoma
6 7 Turritella
2 Neptunea
9 Nucella
10 Tritonalia
11 Calicantharus
12 Opalia
13 Amphissa
14 Olivella
15 Bittium
16 Littorina
17 Nassarius
18 Turritella
19 Forreria
20 Crepidula
21 Clathrodrillia
22 Cantharus

FIG. 8-40.

group. Among fossils, *Athleta*, *Cancellaria*, *Harpa*, *Liopeplum*, *Oliva*, *Olivella*, *Volutoderma*, and *Volutomorpha* (Figs. 8-30, 8-32, 8-34, 8-35, 8-37, 8-39, 8-40) are important constituents of various Cretaceous and Tertiary faunas.

Conids. The shell form of *Conus*, which gives this superfamily its name, is readily distinguished from that of other neogastropods, because the elongate body whorl tapers evenly, with straight or gently curved sides, from the widest part of the shell, at the shoulder of the body whorl, to the

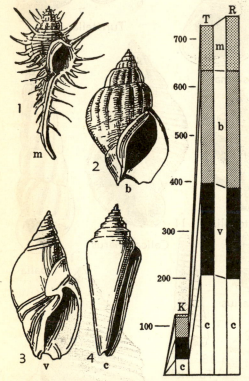

anterior extremity. The spire may be moderately elevated or nearly flat, but it has a conical form, and the shell as a whole may be described as biconical or obconical (Fig. 8-6, *23*, *26*). The aperture is decidedly slitlike. This description does not fit various other members of the superfamily, however, for some are not distinguishable in shape from volutids or buccinids, and others are nearly identical to high-spired cerithiids, such as *Turritella*. The real basis for putting these somewhat varied forms together is the structure of the radula, which differs from that of other neogastropods in having lengthened sharp teeth that in many genera bear barbs and are provided with a poison duct. Individual teeth of the radula may be introduced into the snail's proboscis and pushed outward from the head. The conids occur sparingly in Cretaceous strata, but like other neogastropods, their number expands greatly in the Tertiary (Fig. 8-41, *c*). A slight decrease in the number of genera is noted, comparing totals for the Tertiary and Recent. Characteristic fossils of this group include *Conus*, *Drillia*, *Terebra*, and *Turris* (Figs. 8-30, 8-34, 8-35, 8-37, 8-39, 8-40).

OPISTHOBRANCHS

The opisthobranchs are exclusively marine gastropods, which are much less important in the fossil record than other classes. They include a very large variety of shell-less forms, but only two groups which have hard parts capable of preservation. These are the Pleurocoela, which live in shallow seas, like most prosobranchs, and the Pteropoda, which are open-ocean pelagic snails.

Pleurocoels

Gastropods belonging to this division have conispiral shells which in no way differ from those of many prosobranch snails. Some are subglobular or ovoid in outline, others rather high conical with a rounded base, and still others biconical or

FIG. 8-41. **Geologic distribution of neogastropods.** The graph indicates number of genera belonging to the superfamilies **Muricacea** (*m*, stippled), **Buccinacea** (*b*, oblique-ruled), **Volutacea** (*v*, black), and **Conacea** (*c*, white) (Cretaceous to Recent). The figured examples are (1) *Murex tribulus* Linné, ×0.3, Recent, southwestern Pacific; (2) *Buccinum undatum* Linné, ×0.5, Recent, North Sea; (3) *Ancilla glabrata* (Linné) (volutid), ×0.5, Recent, Gulf of Mexico; (4) *Conus antediluvianus* Bruguière, ×0.5, Pliocene, France.

obconical. A few have thick shells. Some are so delicate that they are very ill-suited for preservation. As a whole, the order Pleurocoela contains 4 superfamilies, divided into 22 families. A few genera, such as the fairly common and widely distributed *Actaeonina*, are recorded from Mississippian to Recent. The large majority of described forms are restricted to Mesozoic and Cenozoic rocks, however. Representative members of the group are shown in accompanying illustrations (Figs. 8-28, *5, 13;* 8-30, *15;* 8-32, *4;* 8-34, *9–10;* 8-37, *9;* 8-39, *3*).

Pteropods

The body of the wing-footed swimming gastropods, called pteropods, is generally elongated and bilaterally symmetrical. Only part of this group is provided with a shell, for many—indeed, a majority—are naked. Both kinds swarm in many portions of the open sea, swimming some distance below the surface during the day and rising to the surface at nightfall. The calcareous covering of shelled forms is very thin and may be transparent. They are mostly very small, less than a half inch in length, and shaped like a very narrow straight-sided cone, but a few are spirally coiled. Some have flattened shells provided with lateral keels, and the aperture may be covered by a thin operculum. Abundance of pteropod shells in deep-sea deposits of parts of the ocean floor is the basis for calling them pteropod ooze.

A widely distributed pteropod-like fossil in Cambrian rocks, and recorded at many places in Ordovician and Silurian strata, is termed *Hyolithes*. The genus occurs sparingly in younger Paleozoic deposits as high as Permian (Fig. 8-42). The shell has an elliptical or subtriangular cross section and narrows from the aperture to a sharp point. The surface is smooth or marked by fine longitudinal striations. An operculum fits over the aperture. That this fossil is really a gastropod has been doubted, for conceivably it belongs to some other, entirely extinct group of inverte-brates. Fossils which are identified certainly as pteropods range from Cretaceous through Pleistocene, and these are a good deal smaller than most specimens of *Hyolithes*. Discovery of a Middle Cambrian *Hyolithes* in the Burgess shale of western Canada, which not only has the operculum joined to the aperture but shows two symmetrical impressions projecting laterally in the position of the paired wing foot of the pteropods, suggests that this ancient fossil may belong among the pteropods (Fig. 8-42, *1*). If this is true, the range of the order is Cambrian to Recent. The Burgess shale fossils do not at all prove the pteropod affinities of *Hyolithes*, however.

PULMONATES

Next to the Prosobranchs, pulmonate gastropods are most numerous, for they include not less than 7,000 living species and some 800 fossil species. These are distributed among approximately 1,000 gen-

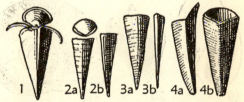

FIG. 8-42. Cambrian, Ordovician, and Devonian hyolithids. All are classed as belonging to the genus *Hyolithes* Eichwald, Lower Cambrian to Permian, North America and Europe. These fossils are grouped very doubtfully with the pteropods.

1, *H. carinatus* Matthew, Middle Cambrian (Burgess shale), British Columbia; a specimen having the operculum in position and showing impressions of paired extensions in position of swimming appendages of modern pteropods (×1.3). The pteropod "wings" entirely lack hard parts, however.

2a, b, *H. billingsi* Walcott, Lower Cambrian, Nevada; broad and narrow sides, with an operculum (×2).

3a, b, *H. baconi* Whitfield, Middle Ordovician, Wisconsin; broad and narrow sides (×0.7).

4a, b, *H. neapolis* Clarke, Upper Devonian, New York; narrow and broad sides (×2).

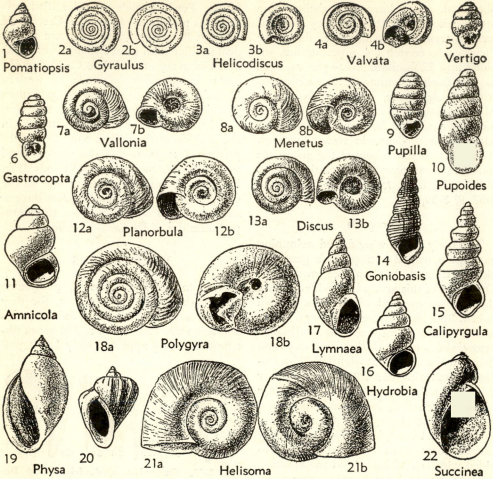

FIG. 8-43. Representative fresh-water and terrestrial gastropods of Pleistocene age. Except as indicated otherwise, all are from deposits of the Great Plains region of Texas, Oklahoma, Kansas, Nebraska, and Iowa. All are pulmonates except 4, 11, 14, and 16, which are mesogastropod prosobranchs. Aquatic pulmonates (Basommatophora) include 2, 3, 8, 12, 13, 15, 17, 19–21; land snails, (Stylommatophora) include 1, 5–7, 9, 10, 18, 22.

Amnicola Gould & Haldeman, Cretaceous–Recent. Moderately elevated smooth shell, round aperture; a rissoid. *A. longiqua* Gould (11, a zone fossil of Tulare beds, ×5), California.

Calipyrgula Pilsbry, Pleistocene. High-spired, whorls slightly angulated. *C. carinifera* Pilsbry (15, ×10), California.

Discus Fitzinger, Cretaceous–Recent. Discoidal umbilicate shell, round aperture. *D. cronkhitei* (Newcomb) (13a, b, apical and basal views, ×3).

Gastrocopta Wollaston, Oligocene–Recent. Elongate ovoid pupaeform shell, aperture toothed. *G. cristata* (Pilsbry & Vanatta) (6, ×6).

Goniobasis Lea, Cretaceous–Recent. High-spired, spirally costate cerithiid shell. *G. kettlemanensis woodringi* Pilsbry (14, ×1), California.

Gyraulus Charpentier, Paleocene–Recent. An advolute discoid shell. *G. pattersoni* Baker (2a, b, apical and basal views, ×3).

Helicodiscus Morse, Pleistocene–Recent. The broadly umbilicate base is concave. *H. parallelus* (Say) (3a, b, apical and basal views, ×3).

Helisoma Swainson, Pliocene–Recent. Discoid shell having rapidly expanding whorls. *H. trivolvis* (Say) (21a, b, basal and apical views, ×2).

(Continued on next page.)

era, which are grouped in 50 families. Although their total recorded stratigraphic range is from Pennsylvanian to Recent, fossil forms are preponderantly from Tertiary deposits. They are adapted to freshwater and terrestrial modes of living, and hence supplement marine invertebrates in furnishing paleontological materials for zonation and correlation of the rock column, inasmuch as they occur where marine shells are lacking. In coastal regions, as along the Atlantic, Gulf, and Pacific borders of the United States, interfingered marine and nonmarine deposits permit determination of equivalent points on the two scales, for one can be calibrated in terms of the other. Also, pulmonates are washed into the sea by rivers and become intermingled locally with marine shells. Interior regions, far removed from the sea, are less easily studied, but here also, sequences of nonmarine gastropod assemblages can be differentiated and dated in terms of their stratigraphic position with respect to vertebrates and floras occurring in the same continental deposits, or in terms of nonmarine molluscan faunas of

coastal regions with which they may be correlated. Pleistocene stratigraphy of the nonglaciated part of the Great Plains has been worked out reliably in recent years mainly because it has been found possible to differentiate and correlate widely distributed pulmonate gastropod faunas in the sediments. Moreover, the Pleistocene stratigraphic units of this region can be dated in terms of the glacial succession of the upper Mississippi Valley, because fossil pulmonates occur in the interglacial deposits.

Basommatophorans

This group of aquatic pulmonates is invariably provided with shells, which range in shape from moderately elevated cones with rounded bases to discoidal forms, all classifiable as conispiral. A few left-handed genera, such as *Physa* (Fig. 8-43, *19*, *20*), are very common, but most shells are dextral. The aperture is holostomatous, and the inner lip is smooth or bears spiral folds. Callus deposits near the umbilical opening, if present, are inconspicuous or lacking, and accordingly most shells are

---

(Fig. 8-43 continued.)

Hydrobia Hartmann, Jurassic–Recent. Moderately high-spired smooth littorinid shell, aperture rounded. *H. andersoni* (Arnold) (16, ×5), California.

Lymnaea Lamarck, Jurassic–Recent. High-spired smooth shell, aperture elongate oval. *L. caperata* Say (17, ×3).

Menetus Adams & Adams, Paleocene–Recent. Like *Helisoma* but aperture more oblique and umbilicus deeper. *M. pearlettei* Leonard (8*a, b,* apical and basal views, ×3).

Physa Draparnaud, Jurassic–Recent. Left-handed moderately elevated smooth or axially ribbed shells, aperture having reflected lips. *P. anatina* Lea (19, ×3), Great Plains; *P. wattsi* Arnold (20, ×2), California.

Planorbula Haldeman, Pleistocene–Recent. Closely resembles *Menetus*. *P. vulcanata occidentalis* Leonard (12*a, b,* apical and basal views, ×3).

Polygyra Say, Paleocene–Recent. Whorls involute, only body whorl visible at the base, aperture indented. *P. texasiana* (Moricand) (18*a, b,* apical and basal views, ×3).

Pomatiopsis Tryon, Oligocene–Recent. Moderately elevated smooth shell, aperture round; a rissoid. *P. cincinnatiensis* (Lea) (1, ×3).

Pupilla Leach, Oligocene–Recent. Ovoid pupaeform shell, aperture indented by teeth. *P. muscorum* (Linné) (9, ×6).

Pupoides Pfeiffer, Oligocene–Recent. Like *Pupilla* but taller and having fewer whorls. *P. albilabris* (Adams) (10, ×6).

Succinea Draparnaud, Eocene–Recent. Elongate ovoid paucispiral shell, aperture large. *S. grosvenori* Lea (22, ×3).

Vallonia Risso, Paleocene–Recent. Discoid shell having moderately involute whorls, oblique round aperture. *V. gracilicosta* Reinhardt (7*a, b,* apical and basal views, ×6).

Valvata Müller, Jurassic–Recent. Whorls angulated. *V. tricarinata* Say (4*a, b,* apical and basal views, ×3).

Vertigo Müller, Eocene–Recent. Relatively low, few-whorled pupaeform shell. *V. ovata* Say (5, ×6).

either phaneromphalous or anomphalous, the latter having a columella. Genera represented by fossil shells are identifiable as belonging to the Basommatophora either on the basis of studying the soft parts of living representatives, or by comparison of the shell with that of some living member of the order. Representative examples, which include some of the more common genera, are illustrated in Fig. 8-43.

Stylommatophorans

The land snails include many shell-less forms, and these we may neglect. Others have thin to moderately thick calcareous conispiral shells, which range from very low-spired to steep-sided, high-spired forms. A majority have many whorls, generally but not invariably more than among basommatophorans. They are small to medium in size, averaging about 0.5 in. in length or width. The aperture, which invariably has a rather evenly rounded outer lip, may have a thin peristome or bear a thickened rim. Toothlike projections of the inner lip, and in some shells of the outer lip also, modify the appearance of the aperture. Such features especially characterize genera of the Pupidae, and largely because of the close resemblance of some Pennsylvanian nonmarine gastropods to modern *Pupa*, these Paleozoic fossils are judged to belong among the Stylommatophorans. Typical fossil forms belonging to this group are illustrated in Fig. 8-24 and 8-43.

INDIRECT PALEONTOLOGICAL EVIDENCE OF GASTROPODS

In connection with the study of gastropod remains preserved in rocks, primarily consisting of their mineralogically altered or unaltered hard parts, and secondarily of external or internal impressions left when the shell is removed by solution, it is appropriate to notice traces of activities by these animals which are classifiable as "indirect fossils." These include trails on bedding planes of stratified rocks, similar to those observable today on exposed tidal flats or on the bottom of shallow bays along the coast. That rather distinctive sorts of markings can be produced by gastropods is readily demonstrated by finding the makers of trails at work. It is rarely possible, however, to establish such a connection between fossil snails and trails which they may have made, and even where this can be done, the information gained is not very important. It is sufficient to know that gastropods are responsible for some features of this sort.

Carnivorous gastropods may leave a record of their presence by the neatly drilled circular holes in the shells of victims. A few other invertebrates can excavate the calcium carbonate of the shell belonging to another, generally as a means of finding a protected living site, but only snails obtain food by going right through the protecting cover of a shell so as to reach the soft parts. Many fossil shells, mostly clams and snails in Tertiary rocks, bear a single round hole—one is enough—which furnishes a mute record of the cause of death. Identity of the gastropod species to which the killer belonged, however, is likely to be quite indeterminable. Mesozoic and Paleozoic bored shells of this sort are rare, but at least one Ordovician brachiopod from the Cincinnati region having such a drilled hole has been reported. This is not enough to be convincing. On the other hand, many Pennsylvanian and Permian crinoids (but not other invertebrates) from the midcontinent region are disfigured by circular borings of uniform diameter (about 4 mm.). They are all round-bottomed craters, rather than holes penetrating the interior of the crinoid cup, and thus they represent unsuccessful efforts to get through the relatively thick plates, if that was why they were cut. Perhaps it is a calumny to ascribe these borings to a gastropod, or can it represent dawning ambition? As many as 15 "dry holes" in random location on a single crinoid cup have been observed!

REFERENCES

BORRADAILE, L. A., et al. (1948) *The Invertebrata:* Cambridge, London, pp. 1–725, figs. 1–483 (gastropods, pp. 550–573, figs. 378–390).

BOWLES, E. (1939) *Eocene and Paleocene Turritellidae of the Atlantic and Gulf Coastal Plain of North America:* Jour. Paleontology, vol. 13, pp. 267–336, pls. 31–34.

CLARK, W. B. (1906) *Mollusca:* in Shattuck, G. B., et al., Pliocene and Pleistocene (of Maryland), Maryland Geol. Survey, Pliocene and Pleistocene, pp. 176–210, pls. 42–65 (gastropods, pp. 176–192, pls. 42–51).

——— & MARTIN, G. C. (1901) *Mollusca:* in Clark, W. B., et al., The Eocene deposits of Maryland, Maryland Geol. Survey, Eocene, pp. 122–203, pls. 17–57 (gastropods, pp. 123–158, pls. 20–29).

DALL, W. H. (1909) *Contributions to the Tertiary paleontology of the Pacific Coast, I, The Miocene of Astoria and Coos Bay, Oregon:* U.S. Geol. Survey Prof. Paper 59, pp. 1–278, pls. 1–23. Includes gastropods.

——— (1915) *A monograph of the molluscan fauna of the Orthaulax pugnax zone of the Oligocene of Tampa, Florida:* U.S. Nat. Mus. Bull., vol. 90, pp. 1–173.

DAVIES, A. M. (1935) *Tertiary faunas:* Murby, London, pp. 1–406, figs. 1–565 (gastropods, pp. 209–347, figs. 283–542). Excellent summary of younger gastropods.

DURHAM, J. W. (1937) *Epitoniidae from Mesozoic and Cenozoic of the west coast of North America:* Jour. Paleontology, vol. 11, pp. 479–512, pls. 56–57.

GARDNER, J. A. (1937) *Molluscan fauna of the Alum Bluff group of Florida:* U.S. Geol. Survey Prof. Paper 142, pp. 251–435, pls. 37–48.

GRANT, U. S., IV, & GALE, H. R. (1931) *Pliocene and Pleistocene Mollusca of California:* San Diego Soc. Nat. History Mon. 1, pp. 1–1036, pls. 1–32.

HARRIS, G. D. (1892) *The Tertiary geology of southern Arkansas:* Arkansas Geol. Survey, vol. 2, pp. 1–207, pls. 1–7. Contains descriptions and figures of Eocene gastropods.

KNIGHT, J. B. (1930–1934) *The gastropods of the St. Louis, Missouri, Pennsylvanian outlier:* Jour. Paleontology, vol. 4, Supp., pp. 1–88, pls. 1–5; vol. 5, pp. 1–15, 177–229, pls. 1–2, 21–27; vol. 6, pp. 189–202, pls. 27–28; vol. 7, pp. 30–58, 359–392, pls. 8–12, 40–46; vol. 8, pp. 139–166, 433–447, pls. 20–26, 56–57.

——— (1941) *Paleozoic gastropod genotypes:* Geol. Soc. America Spec. Paper 32, pp. 1–510, pls.

1–96, figs. 1–32. A study of type specimens of the type species, world wide in scope.

KONINCK, L. G. DE (1881–1883) Faune du calcaire carbonifère de la Belgique, Gastéropodes: Mus. histoire nat. Belgique, Ann., ser. paléontologie, pt. 3, pp. 1–170, pls. 1–21 (1881); pt. 4, pp. 1–240, pls. 1–36 (1883). Describes many genera and species of Mississippian age.

LEONARD, A. B. (1950) *A Yarmouthian molluscan fauna in the Midcontinent region of the United States:* Kansas Univ. Paleon. Contr., Mollusca, art. 3, pp. 1–48, pls. 1–6, figs. 1–4. Describes a Pleistocene gastropod assemblage, mainly pulmonates.

LONGSTAFF, J. (1926) *A revision of the British Carboniferous Murchisonidae:* Geol. Soc. London Quart. Jour., vol. 82, pp. 526–555, pls. 25–27.

——— (1933) *A revision of the British Carboniferous members of the family Loxonematidae:* Same, vol. 89, pp. 97–124, pls. 7–12.

MACGINITIE, G. E., & MACGINITIE, N. (1949) *Natural history of marine animals:* McGraw-Hill, New York, pp. 1–473, figs. 1–282 (gastropods, pp. 352–385, figs. 198–239). Many interesting observations on living shallow-water marine snails.

MANSFIELD, W. C. (1930) *Miocene gastropods and scaphopods of the Choctawhatchie formation of Florida:* Florida Geol. Survey Bull. 3, pp. 1–142, pls. 1–21. A well-preserved assemblage of Tertiary fossils.

——— (1937) *Mollusks of the Tampa and Suwannee limestones of Florida:* Florida Geol. Survey Bull. 15, pp. 1–334, pls. 1–21, (gastropods, pp. 63–187, pls. 1–10). Oligocene and lower Miocene fossils.

MARTIN, G. C. (1904) *Gastropoda:* in Clark, W. B., et. al., Miocene deposits of Maryland: Maryland Geol. Survey, Miocene, pp. 131–270, pls. 40–63.

MERRIAM, C. W. (1941) *Fossil turritellas from the Pacific Coast region of North America:* California Univ. Dept. Geol. Sci., Bull., vol. 26, pp. 1–214, pls. 1–41.

MORET, L. (1948) *Manuel de paléontologie animale:* Masson et Cie., Paris, pp. 1–745, figs. 1–274 (gastropods, pp. 417–454, figs. 155–168) (French). A well-organized description of the group.

PALMER, K. V. (1937) *Claibornian Scaphopoda, Gastropoda, and dibranchiate Cephalopoda of the southern United States:* Bull. Am. Paleontology, vol. 7, pt. 32, pp. 1–730, pls. 1–90.

——— (1947) *Univalves:* in Harris, G. D., &

Palmer, K. V., The Mollusca of the Jackson Eocene of the Mississippi Embayment (Sabine River to the Alabama River), Bull. Am. Paleontology, vol. 30, no. 117, pt. 2, pp. 209–563, pls. 26–65.

Pelseneer, P. (1906) *Mollusca:* in Lankester, E. R., Treatise on zoology, A. & C. Black, London, pp. 1–355, figs. 1–301 (gastropods, pp. 66–196, figs. 44–180). An important standard work on general morphology.

Perner, J. (1903–1911) *Gastéropodes:* in Barrande, J., Système silurien du centre de la Bohême, vol. 1, tome 4, pp. 1–164, pls. 1–89, figs. 1–111 (Patellidae, Bellerophontidae); tome 2, pp. 1–380, pls. 90–175, figs. 1–153 (Pleurotomariidae); tome 3, pp. 1–390, pls. 176–247, figs. 1–59 (Capulidae).

Perry, L. M. (1940) *Marine shells of the southwest coast of Florida:* Bull. Am. Paleontology, vol. 26, no. 95, pp. 1–260, pls. 1–39 (gastropods, pp. 93–178, pls. 21–39).

Pilsbry, H. A. (1913) *Gastropoda:* in Eastman, C. R., Zittel, K. A., Textbook of palaeontology, Macmillan & Co., Ltd., London, 2d ed., vol. 1, pp. 514–583, figs. 843–1097.

Shimer, H. W., & Shrock, R. R. (1944) *Index fossils of North America:* Wiley, New York, pp. 1–837, pls. 1–303 (gastropods, pp. 433–521, pls. 174–215).

Stanton, T. W. (1947) *Studies of some Comanche pelecypods and gastropods:* U.S. Geol. Survey Prof. Paper 211, pp. 1–256, pls. 1–67.

Stephenson, L. W. (1941) *The larger invertebrate fossils of the Navarro group of Texas:* Texas Univ. Pub. 4101, pp. 1–641, pls. 1–95. Includes numerous Upper Cretaceous gastropods.

Ulrich, E. O., & Scofield, W. H. (1897) *The Lower Silurian [Ordovician] gastropoda of Minnesota:* Minnesota Geol. Survey, Paleontology, vol. 3, pt. 2, pp. 813–1081, pls. 61–82.

——— & Bridge, J. (1931) *Gastropoda:* in Bridge J., Geology of the Eminence and Cardareva quadrangles: Missouri Bur. Geology and Mines (2), vol. 24, pp. 186–207, pls. 18–22. Describes Cambrian and Lower Ordovician gastropods.

Vokes, H. E. (1939) *Molluscan fauna of the Domengine and Arroyo Hondo formations of the California Eocene:* New York Acad. Sci. Annals, vol. 38, pp. 1–246, pls. 1–22.

Wade, B. (1926) *The fauna of the Ripley formation of Coon Creek, Tennessee:* U.S. Geol. Survey Prof. Paper 137, pp. 1–272, pls. 1–72. A remarkably preserved Upper Cretaceous gastropod fauna.

Wenz, W. (1938–1944) *Gastropoda:* Handbuch der Paläozoologie, Borntraeger, Berlin, Teil 1 (1938), pp. 1–240, figs. 1–471; Teil 2 (1938), pp. 241–480, figs, 472–1235; Teil 3 (1939), pp. 481–720, figs. 1236–2083; Teil 4 (1940), pp. 721–960, figs. 2084–2787; Teil 5 (1941), pp. 961–1200, figs. 2788–3416; Teil 6 (1943), pp. 1201–1506, figs. 3417–4211; Teil 7 (1944), pp. 1507–1639. Treats amphigastropods and prosobranchs, describing all known genera of fossils. An extremely valuable work which can be consulted in only a few libraries in North America because nearly the entire stock of later parts was destroyed during the war.

Whitfield, R. P. (1892) *Gastropoda and Cephalopoda of the Raritan clays and greensand marls of New Jersey:* U.S. Geol. Survey Mon. 18, pp. 1–402, pls. 1–50, figs., 1–2. (Same, New Jersey Geol. Survey, Paleontology of New Jersey, vol. 2).

Woodring, W. P. (1928) *Miocene mollusks from Bowden, Jamaica, II, Gastropods:* Carnegie Inst. Washington Pub. 385, pp. 1–564, pls. 1–40.

———, Bramlette, M. N., & Kew, W. S. W. (1946) *Geology and paleontology of Palos Verdes Hills, California:* U.S. Geol. Survey Prof. Paper 207, pp. 1–145, pls. 1–37. Illustrates leading Miocene, Pliocene, and Pleistocene gastropods.

Yonge, C. M. (1946) *The pallial organs in the aspidobranch Gastropoda and their evolution throughout the Mollusca:* Royal Soc. London Philos. Trans., ser. b, vol. 232, no. 591, pp. 443–518, pl. 18, figs. 1–40. Important studies of respiratory structures of marine snails and their evolutionary significance.

CHAPTER 9

Cephalopods

Most of the animals discussed in the preceding portions of this book are sessile or relatively slow-moving invertebrates, crawling on the bottom or drifting about at the mercy of waves and currents. Not so the group we are about to investigate, which includes such creatures as the squids, the pearly nautilus, and the octopus (Fig. 9-1). Like fishes, the cephalopods are equipped with highly developed eyes and other sense organs. They have an efficient method of locomotion and live largely as predators. Unlike fishes, they possess such features as prehensile tentacles, commonly studded with sucker disks,

may emit clouds of inky fluid when angered or frightened, and change color according to their psychologic state. Such "anomalies" have earned the octopus, for example, the common name of "devil fish." Cephalopods are exclusively marine. They include the largest invertebrates known, attaining a length (including tentacles) of approximately 16 m. (53 ft.).

Cephalopods are among the most highly developed invertebrates. None are known from fresh water, but have been an important element of marine fauna since Ordovician time. They are unsurpassed as guide fossils in Paleozoic and Mesozoic rocks.

FIG. 9-1.   **Modern cephalopods.** *Nautilus* in center foreground, ×0.3; a school of squids in the background, an octopus at lower left.

LIVING CEPHALOPODS

Loligo, a Squid

In order to provide a basis for visualizing fossil forms, we shall first examine some modern cephalopods and their ways of life.

Loligo, the common squid, may be familiar to persons who live on the seacoast. This cephalopod can be found on ice in fish markets catering to Mediterranean or oriental customers. *Loligo* (illustrated in Fig. 9-2, to which the following italic numbers refer) is a streamlined creature, generally about 30 cm. (1 ft.) in length, divided into a long, tapering **body** and a rounded **head** (*2*). The head carries a pair of large eyes (*3*), which, like our own, possess a lens. The mouth, located at the anterior end, is surrounded by 10 muscular tapering **arms** (*1*) studded with sucker disks. Two of these (tentacles, *10*)

are much longer than the others, and their suckers are concentrated in "hands" at the ends. The tentacles are flung out to grasp prey and to pull it to the mouth, holding it there until is is killed and consumed by the beaklike jaws.

The body wall consists of a tough, muscular tissue, the **mantle.** This covers the **mantle cavity** (*8*), on the ventral side of the animal, and the elongate **visceral mass** (*4*), which contains most of the vital organs. The mantle cavity houses a pair of **gills** (*7*), and also furnishes the chief means of locomotion. Water is drawn into it through a slit behind the head, aerating the gills. The mantle then contracts, exerting pressure on the contents of the mantle cavity. The intake slit, equipped with a one-way valve, automatically closes, and the water is ejected through a nozzle (**hyponome,** *9*), located at the ventral

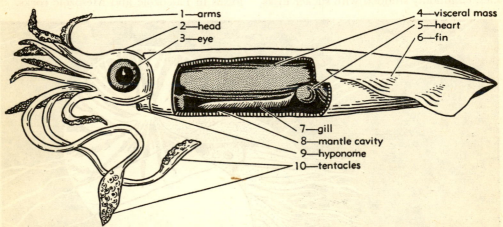

FIG. 9-2. Structure of Loligo. The drawing shows a partial dissection of the common squid, ×0.5. Main parts of the body are defined in the following list cross-indexed to the figure by numbers.

arms (1). Appendages surrounding mouth, bearing sucker disks.

eye (3). Large visual organ, equipped with lens, located on side of head.

fin (6). Horizontal extension of mantle, used for slow locomotion and for steering.

gill (7). Fringed respiratory organ (one of a pair), suspended in mantle cavity.

head (2). Anterior part of animal, containing brain, mouth, eyes, and bearing arms and tentacles.

heart (5). Contractile organ (one of a pair) which pumps blood through the body.

hyponome (9). Nozzle through which water is ejected from mantle cavity; used for rapid jet propulsion.

mantle cavity (8). Large space between ventral mantle and visceral mass, connecting with outside through a slit and the hyponome.

tentacles (10). Two of the appendages surrounding mouth; longer than the eight arms.

visceral mass (4). Organs, chiefly digestive and reproductive, located above and behind mantle cavity.

margin. The expulsion of water may be very forceful, making a jet which serves to propel the animal rapidly backward from the hyponome aperture. The direction of movement is controlled by moving the hyponome into different positions; by curving it to the rear, *Loligo* may shoot the jet backward and consequently move along headfirst. Maximum speed is attained when the jet is directed forward, and the animal then shoots along tailfirst. The effectiveness of this method of locomotion is attested by the fact that one of the squids, a relative of *Loligo*, may reach sufficient speed to leap from the water and fall into boats. For delicate steering, as well as for hovering in a given position or moving along very slowly, *Loligo* relies less on its jet motor than on its **fins** (6).

Above and behind the mantle cavity lies the visceral mass, which is overlain in turn by a long horny blade or "pen," a rudimentary internal skeleton.

Like many other cephalopods, *Loligo* has the ability to change color and pattern kaleidoscopically, before the eyes of an astonished observer.

Loligo is a gregarious animal, traveling coastal areas or the open ocean in large schools. Its life is largely spent suspended in the water, searching for food, mainly crustaceans, and avoiding enemies. For escape, it relies on its speed but when closely pressed, it can discharge from its ink gland a cloud of brownish sepia to baffle the pursuer. Enemies include the larger fishes and man, who catches squids for bait as well as for the dinner table.

Loligo is geologically significant, because its structures and habits tell us about those of various long-extinct cephalopods. It belongs to a group known as dibranchiates, which, among other things, are characterized by internal or rudimentary shells. They left an abundant fossil record in the Mesozoic, as described in the last portion of this chapter.

Nautilus

In the southwestern Pacific, from the Fiji Islands to the Straits of Malacca,

lives the pearly nautilus (Fig. 9-1, 9-3), famed, though somewhat misrepresented, in Oliver Wendell Holmes's poem of "Gulfs enchanted, where the siren sings and coral reefs lie bare." The genus *Nautilus* is the sole survivor of a host of shell-covered cephalopods, which populated Paleozoic and Mesozoic seas.

Body. The body of *Nautilus* consists of a **visceral mass** and **mantle cavity** (*16;* italic numbers in the text on morphologic features of *Nautilus* refer to Fig. 9-3) enclosed within a sac-shaped mantle, and a well-defined **head.** The latter bears about 90 **tentacles** (*18*) devoid of sucker disks, and a pair of **eyes** which are built like a pinhole camera, lacking a lens. The **mouth,** which opens inside the circle of tentacles, is equipped with a **radula** and two horny **jaws** (*12*), shaped like a parrot's beak. Above the head lies a thick, tough, fleshy fold—the **hood** (*11*). Ordinarily, the visceral mass and mantle cavity lie within the shell, and the head and hood protrude from the **aperture** (*10*). When danger approaches, the head may be pulled into the shell and the aperture closed by the hood. As in *Loligo,* water is drawn into the mantle cavity and expelled through the **hyponome** (*17*), but *Nautilus* possesses two pairs of **gills** (*14*), and two kidneys instead of one. It lacks an ink gland. From the rear of the visceral mass, the **siphon,** a long fleshy strand well supplied with blood vessels, extends backward.

The nature of the eyes and circulatory system and the lack of suckers and ink gland suggest that *Nautilus* is less advanced in evolution than the squids.

Outer Shell. The shell is composed of calcium carbonate, in the form of aragonite, and an admixture of organic matter. The outer shell may be conceived as a sort of bent cone, tightly coiled in a plane and expanding at such rate that the outer coils, or **whorls,** cover the inner ones. The shell expands at a constant geometric rate so that a tangent drawn to the periphery of the shell at any point makes a uniform angle with the radius drawn from this

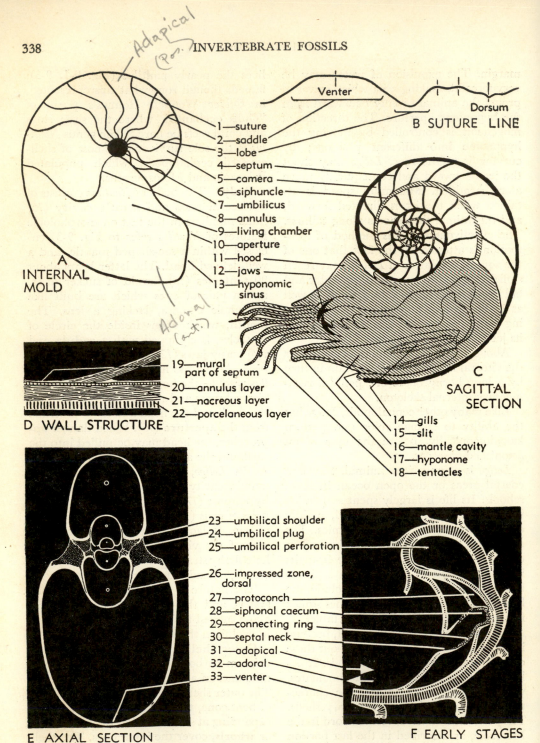

Adapical (Pos.)

Adoral (ant.)

1—suture
2—saddle
3—lobe
4—septum
5—camera
6—siphuncle
7—umbilicus
8—annulus
9—living chamber
10—aperture
11—hood
12—jaws
13—hyponomic sinus

A INTERNAL MOLD

Venter — Dorsum

B SUTURE LINE

C SAGITTAL SECTION

14—gills
15—slit
16—mantle cavity
17—hyponome
18—tentacles

19—mural part of septum
20—annulus layer
21—nacreous layer
22—porcelaneous layer

D WALL STRUCTURE

23—umbilical shoulder
24—umbilical plug
25—umbilical perforation
26—impressed zone, dorsal
27—protoconch
28—siphonal caecum
29—connecting ring
30—septal neck
31—adapical
32—adoral
33—venter

E AXIAL SECTION

F EARLY STAGES

Fig. 9-3. **Structure of the pearly Nautilus.** Various morphologic features are shown (*A*) as they appear on an internal mold (the most common way in which fossil nautiloids are found preserved) and (*B–F*) in various sections. The illustrated structures are defined in the following alphabetical list.

(Continued on next page.)

point to the center of the coil. In *Nautilus*, this angle is about 79 deg. Such a spiral is termed an equiangular or logarithmic spiral (Fig. 1-12). Because of the overlap of early coils, each whorl has a horseshoe-shaped cross section. The whorls do not close tightly around the axis of coiling, but leave an **umbilicus** (7), which may remain open, or may be closed by a calcareous deposit termed **umbilical plug** (24).

The outer shell is threefold, divided into an outer **porcelaneous layer** (22) or ostracum, made of aragonite prisms; a middle **nacreous layer** (21), composed of thin alternating sheets of aragonite and organic matter; and an inner thin, clear, prismatic **annulus layer** (20) of aragonite, secreted by muscles which attach the animal to the shell along a line termed the **annulus** (8). The porcelaneous layer is deposited only at the outer (ventral) and lateral margins of the aperture, not at the inner (dorsal) margin where the aperture is bounded by the preceding whorl of the shell, and where a thin black layer of organic matter is formed instead. Intermittent periods of nondeposition leave their mark on the porcelaneous layer in the form of **growth lines,** from which it is possible to determine the shape of the aperture at any growth stage. The aperture has a shallow embayment, the **hyponomic sinus** (13) on the ventral side.

The exterior shows a striking color

(Fig. 9-3 continued.)

adapical (31). Direction toward apex of shell.
adoral (32). Direction toward aperture of shell.
annulus (8). Line along which animal is attached to wall of living chamber.
annulus layer (20). Layer of shell secreted at annulus.
aperture (10). Opening of shell at outer end of living chamber.
camera (5). Gas chamber in shell.
connecting ring (29). Delicate ring forming wall of siphuncle from one septum to the next.
gills (14). Respiratory organs (four) suspended in mantle cavity.
hood (11). Tough, fleshy operculum located above head, for closure of aperture when head is retracted into shell.
hyponome (17). Nozzle through which water is expelled from mantle cavity.
hyponomic sinus (13). Embayment in ventral margin of aperture, giving hyponome freedom of action.
impressed zone (26). Concave depression along dorsal (inner) side of whorls, where they embrace the next-older whorl.
jaws (12). Pair of horny structures forming a parrot-like beak.
living chamber (9). Large cavity between last-formed septum and shell aperture.
lobe (3). Flexure of suture line toward rear (adapically).
mantle cavity (16). Space adjoining visceral mass; contains gills and connects with outside through slit and hyponome.

mural part of septum (19). Wedgelike extension of each septum along shell wall.
nacreous layer (21). Pearly middle layer of shell.
porcelaneous layer (22). Prismatic outer layer of shell.
protoconch (27). Initial chamber of shell.
saddle (2). Forward (adoral) flexure of suture line.
septal neck (30). Funnel-like extension of septum along siphuncle.
septum (4). Curved partition of aragonite, dividing shell into chambers.
siphonal caecum (28). Saclike end of siphuncle in protoconch.
siphuncle (6). Tube which leads from living chamber to protoconch, piercing septa near middle.
slit (15). Opening for intake of water into mantle cavity.
suture (1). Line of junction of a septum with walls of shell.
tentacles (18). Small appendages, devoid of sucker disks, located around mouth.
umbilical perforation (25). Opening from side to side through shell in position of axis of coiling.
umbilical plug (24). Calcareous deposit which fills umbilicus.
umbilical shoulder (23). Rim of umbilicus.
umbilicus (7). Open space, in position of axis of coiling, which is left when successive whorls do not reach this axis.
venter (33). Part of whorl farthest from axis of coiling.

pattern of flamelike, yellowish brown or orange stripes on a cream-colored background (Fig. 9-1). The stripes are produced at the aperture and are formed mainly during adolescence, so that a mature individual shows stripes on the portions of the shell facing upward only, the the remainder of the shell being cream-colored. This is protective coloration similar to that of many fishes which have color markings on the dorsal side and a plain, light-colored belly. They are thus inconspicuous, both when seen from above against the varicolored patterned bottom, and from below against the monotone light-colored water surface.

In contrast to the porcelaneous layer, the nacreous layer is built not only near the margin of the aperture, but continues to be secreted by the mantle lining the walls of the **living chamber** (9). It covers both the inner surface of the porcelaneous layer and the dorsal side of the living chamber, where it is laid on the previous whorl.

Interior of the Shell. The shell is divided into chambers (**camerae,** 5) by partitions called **septa** (4). An adult *Nautilus* shell consists of 33 to 36 chambers, filled with gases similar in composition to the atmosphere, but slightly higher in nitrogen content. The space between the last-formed septum and the aperture is much longer than that of the closed chambers, occupying one-third to one-half of the last whorl. It is called the **living chamber** (9). Except for a minor tubular organ extending backward through the shell, the living chamber is the only portion of the shell which, at any given stage, is occupied by the animal.

The septa, composed of nacreous aragonite, are strongly concave toward the living chamber—a direction called **adoral** (32), in contrast to **adapical** (31)—and become tangential to the outer shell wall where they meet it. The **suture** (1), which is a line formed by junction of a septum with the outer wall, has much importance in classifying fossil nautiloids. Deflections of the suture toward the aperture are termed

saddles (2), and those away from the aperture are called **lobes** (3).

A tube, the **siphuncle** (6), passes through the mid-regions of chambers and septa. It is formed by short funnel-like backward extensions of the septa, called **septal necks** (30), and of delicate **connecting rings** (29) (composed chiefly of organic matter), which connect each septal neck with the previous one. During life, the siphuncle contains the **siphon,** an organ which is thought to function in regulating gas pressure in the chambers.

Sexual Dimorphism. The sexes are separate, but there is only a slight differentiation in the shells, males being slightly broader than females.

Habits. *Nautilus* lives around reefs, ranging downward to about 600 m. Because all specimens captured alive have died soon afterward, we know little of its actual mode of life and nothing about its development. *Nautilus* uses its tentacles for temporary anchorage to the sea bottom but not for crawling, as has been supposed. It is a good swimmer, propelling itself like the squids by jetting water out of the hyponome. Greater knowledge of the habits and life history of *Nautilus* undoubtedly would contribute to reliable interpretation of its many fossil relatives.

Distribution. The shells of *Nautilus* are found far beyond areas in which this cephalopod lives, for when the animal dies and the soft parts disappear, the shell is buoyed up by the gas in the chambers. Being lighter than sea water, it rises to the surface and is carried by currents until waterlogged or washed up on some shore.

Function of the Shell. The shell serves as protection, and the gas chambers within it are a hydrostatic device by means of which the weight of the shell is compensated and the specific gravity of the entire organism brought near that of water. Some early workers believed that *Nautilus* could vary its specific gravity by flooding its chambers with water, thus being able to rise or sink at will. Detailed anatomical studies, however, have failed to reveal any such water-flooding mechanism. A more feasible method of changing specific gravity has been suggested as follows: The body of the animal is attached to the shell along two sinuous lines at the side, but not to the concave face of the last septum. A gas-filled space between this septum and the body could be altered by muscular contraction so as to compress or rarefy the gas, and thus raise or lower the specific gravity.

Octopus

Another type of cephalopod, common along modern coasts, is the group of octopods or devil fishes. The octopus (Fig. 9-1) possesses a small sack-shaped body, a large head which carries two keen eyes, and a circle of eight tapering tentacles, webbed at the base and studded with sucker disks, surrounding the mouth. The tentacles are amazingly sensitive and agile, their whip-like tips in continual motion, twitching, coiling, and uncoiling. Dark shades of brown or green come and go, various color patterns are developed and disappear, and the skin may be smooth at one moment or puckered and warty in the next. The tentacle spread of different species ranges from less than 1 m. to 3 m.

Ordinarily the octopus lies waiting in a rock cave or special lair constructed of stones, immobile except for the twitching ends of its tentacles and the pulsing of its "breathing slit." This aperture of the mantle cavity alternately opens as water is drawn in and closes as it is expelled through the hyponome. The approach of a potential dinner, such as a crab or fish, causes excitement. The octopus blushes darkly, and begins to "breathe" water at a rapid rate, shooting a forceful stream out of the hyponome, staying in place only because it is anchored by its sucker disks. Suddenly, as the suckers are released, the cephalopod shoots from its lair, body end first. Steering with tentacles and hyponome, the octopus executes a rapid about-face when within striking distance, wraps its tentacles around the victim, and carries the prey back to its stronghold.

Eye structure, a single pair of gills, and the presence of an ink gland indicate that the octopus is more closely related to squids than to the pearly nautilus.

GEOLOGIC IMPORTANCE

Many lines of cephalopods give evidence of rapid evolution, and their shells are widely distributed over the globe. The group called ammonoids are particularly good index fossils, and they furnish some of the finest examples of straight-line evolution and of stratigraphic zonation. The wide dispersal of individual genera and species may be accounted for partly by a swimming mode of life, which is not dependent on water depth or bottom conditions. Also by analogy to *Nautilus*, it is possible that empty shells, rising to the surface after death of the animal, may have been distributed posthumously by winds and currents.

CLASSIFICATION

Although cephalopods have been much studied and are represented by some of the best documented evolutionary sequences known, the classification of this group is far from satisfactory. Recent cephalopods fall into two main orders: naked forms (coleoids or dibranchiates), provided with one pair of gills, few arms, highly developed eyes, and an ink sac; and the nautilus (a tetrabranchiate), which has an external shell, two pairs of gills, many arms, relatively primitive eyes, and no ink sac. Paleontologists long have separated external-shelled cephalopods into two groups—the **nautiloids,** which embrace all forms having simple septa and a few offshoots having fluted septa, and the **ammonoids** (extinct), which are characterized by fluted septa. Some workers have combined ammonoids and nautiloids in a division called Tetrabranchiata (four-gilled). Others have held that because the number of gills possessed by ammonoids is unknown, this group should be classed as an independent unit having equal rank with nautiloids (four-gilled) and dibranchiates (two-gilled).

In the course of nautiloid studies, classification was first based mainly on external shape—the system mainly used by Barrande in his description of the Ordovician and Silurian nautiloids of Bohemia.

Later, Hyatt devised a classification in which major categories were based largely on the nature of the siphuncle. Five groups were recognized: (1) Holochoanites, characterized by greatly elongate septal necks, extending through one entire chamber or more; (2) Mixochoanites (= ascoceroids), possessing shells in which an immature part contains a tubular siphuncle and a mature part composed of strangely modified chambers contains a beaded siphuncle, the immature part being shed at maturity; (3) Schistochoanites, said to possess septal necks which are split on the ventral side; (4) Orthochoanites, characterized by tubular siphuncles composed of short septal necks and long, delicate connecting rings; (5) Cyrtochoanites, possessing siphuncles that are beaded, that is, contracted at the septa and expanded within the chambers.

This classification has been used widely, but studies since Hyatt's day have revealed an increasing number of defects. Characterization of the Schistochoanites was based on a misconception of internal siphonal deposits as septal necks. The Cyrtochoanites have been found to include a very diverse assemblage of which some are closely related to "Holochoanites" and others to "Orthochoanites." Thin-section studies indicate that the elongate septal necks which are supposed to distinguish the Holochoanites occur only in a minority of them, others having short septal necks and heavily calcified connecting rings that resemble septal necks.

The work of recent years has shown that the nautiloids in the broadly defined old sense are exceedingly diverse, composed of several widely separated divisions. Flower & Kummel (1950) classify the nautiloids in 75 families and 14 orders.

The classification adopted in this book is somewhat more compact. It is shown in the following table and in Figs. 9-5 and 9-23.

Main Divisions of Cephalopods

Cephalopoda (*class* of phylum Mollusca).
  Nautiloidea (*subclass*), possessing external shells divided by generally simple septa turned back along the siphuncle. Cambrian–Recent.
    Ellesmeroceroida (*order*), having curved shells, tubular siphuncles, short septal necks, and mostly thick connecting rings (includes Bassleroceratida of some authors). Cambrian–Ordovician.
    Michelinoceroida (*order*), shells straight to slightly curved, siphuncles tubular to beaded (cyrtochoanitic), composed of septal necks and delicate (largely uncalcified) connecting rings. Ordovician–Triassic.
    Ascoceroida (*order*), slightly curved shells, which at maturity have gas chambers located above living chamber, and shed normal, youthful part of shell. Ordovician–Silurian.
    Oncoceroida (*order*), strongly curved to straight shells, commonly short and blunt; siphuncle tubular in early stages, beaded at maturity, may contain longitudinal blades or ridges (actinosiphonate deposits). Ordovician–Devonian.
    Endoceroida (*order*), shells mainly straight, having large tubular spihuncles containing conical deposits (endocones). Ordovician.
    Actinoceroida (*order*), shells mainly straight. The siphuncle, which swells greatly in each chamber, is filled with calcareous deposits. These are separated from thick connecting rings by cavities (perispatia), and penetrated by central and radial canals. Ordovician–Pennsylvanian.
    Discosoroida (*order*), curved shells having beaded (cyrtochoanitic) siphuncles; connecting rings may be thick and the siphuncle filled with calcareous rings and cones. Ordovician-Devonian.
    Nautilida (*order*), coiled shells. (Includes the orders Tarphyceratida, Barrandeoceratida, Solenochilida, Rutoceratida, Centroceratida, and Nautilida of Flower & Kummel.) Ordovician–Recent.

Ammonoidea (*subclass*), having external(?) shell divided by fluted septa, which in early forms are turned back, in advanced forms turned forward, at the siphuncle. Devonian –Cretaceous.

Ammonitida (*order*), Devonian–Cretaceous.

Dibranchiata (*subclass*), having internal shells or lacking skeleton. Mississippian–Recent.

Belemnoida (*order*), having a chambered shell (phragmocone) surrounded by a counterweight (rostrum). Mississippian–Eocene.

Sepioida (*order*), shell consisting of a flat shield containing remnants of chambers; 10 arms. Jurassic–Recent.

Teuthoida (*order*), shell consisting of a horny "pen"; 10 arms. Jurassic–Recent.

Octopoida (*order*), no shell, eight arms. Cretaceous–Recent.

FUNCTIONS AND DEVELOPMENT OF SHELLS

In the Cambrian, numerous kinds of invertebrates grew protective shells, among them the mollusks. The latter had become differentiated at this time into three groups of diverse life habits: relatively passive bottom-dwelling filter feeders, which developed into the pelecypods; crawling carnivorous or herbivorous gastropods; and swimming predatory cephalopods. Whereas a bottom dweller can support a strong heavy shell, which offers a maximum of protection, a swimmer is limited to a light, delicate shell which gives little protection, unless attainment of buoyancy compensates for the weight of the increased shell thickness. The cephalopods developed along the latter line by installing buoyant gas chambers in the shell, separated by walls (septa) and connected with the body by a tube, the siphuncle. The buoyant chambers were located in the apical part of the conical shell, whereas the weight of the soft parts and the shell surrounding them was concentrated near the aperture. In consequence of this, the earliest cephalopods lived probably with the apex up and the aperture and head down. Several groups independently, in different ways

and at different times, shifted the centers of buoyancy and weight concentration into more nearly coincident positions. This allowed the body to lie on its ventral side and made for better streamlining and superior maneuverability. *Nautilus* and most ammonoids have coiled shells (Fig. 9-4, *6*), in which the body lies horizontally below the coil, the whole being streamlined in the direction of the animal's rear and of the general movement in swimming. Various straight-shelled cephalopods partly or completely filled the apical air chambers (Fig. 9-4, *2*, *3*) and the apical portions of the siphuncle (Fig. 9-4, *3*) with shell matter, thus distributing the weight so that the buoyant gas-filled chambers were between the living chamber and the counterweighted apex. In other straight forms, the siphuncle was greatly enlarged and located at the ventral side (Fig. 9-4, *4*). The living chamber was small, and most of the visceral mass apparently came to lie in the siphuncle below the gas chambers. In addition, these forms also filled the apical end of the siphuncle with counterweighting calcareous deposits.

In others (Fig. 9-4, *5*), the air chambers of the adult shell were extended forward over the dorsal side of the living chamber, and the earlier (more normal) portion of the shell was then broken off. In belemnoids (Fig. 9-4, *1*), which had an internal skeleton, a counterweight analogous to that of many nautiloids was deposited, but on the outside, rather than the inside, of the conical chambered shell. Finally, some of the Mesozoic cephalopods apparently developed to a stage in which mobility and the ink screen were more important factors in survival than the passive defense of a cumbersome armor. In these, the chambers (and the counterweight around them) were progressively reduced until only a remnant in the apertural region persisted. This is the condition shown by the cuttlefish *Sepia* and, in more advanced stages, by squids of the *Loligo* type (Fig. 9-2).

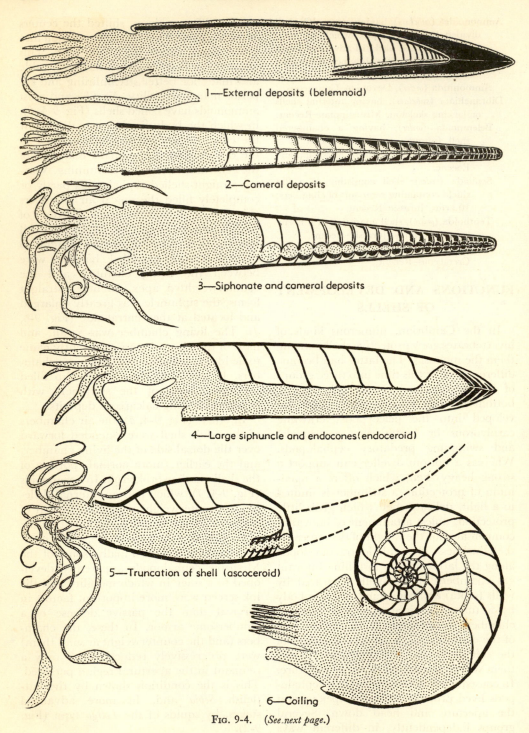

1—External deposits (belemnoid)

2—Cameral deposits

3—Siphonate and cameral deposits

4—Large siphuncle and endocones (endoceroid)

5—Truncation of shell (ascoceroid)

6—Coiling

FIG. 9-4. (See next page.)

NAUTILOID CEPHALOPODS

The term nautiloid, signifying like *Nautilus*, does not apply very clearly to the heterogeneous assemblage of shell-bearing cephalopods which are assigned to the subclass. A common feature of all, however, is the relative simplicity of the junctions between septa and the outer shell—that is, the sutures. The form and internal structures of the shell are quite diverse. Probably the Nautiloidea include the root stocks of all cephalopods, the ammonoids and dibranchiates merely representing particularly successful offshoots (Fig. 9-5). Among modern cephalopods, only one genus, the comparatively primitive *Nautilus*, is referred to the nautiloids.

Kinds of Nautiloids

Shell Form. Many fossil nautiloids differ greatly from modern *Nautilus* in the form of their shells. Some are perfectly straight, slender cylinders which taper very gradually toward the apical end. Others are short, relatively broad shells having straight or gently curved sides that converge strongly at the apex. Arc-shaped and loosely coiled forms are intermediate between the simple straight shells and the planispirally coiled shells having the successive whorls in contact. Also, shells in which the whorls barely touch their neighbors may be distinguished from more tightly coiled shells, in which outer whorls are concave on their inner sides, extending

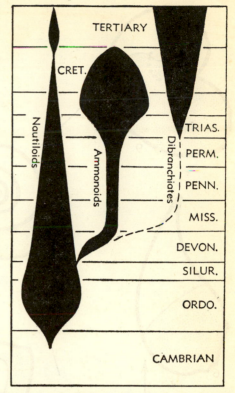

FIG. 9-5. **Geological distribution of main divisions of cephalopods.**

partly or entirely around the exterior of inner whorls. Finally, some nautiloids exhibit a conispiral type of coiling, in which the mid-line of the whorls has the course of a wire wrapped around a cone from its apex to the base. These several shapes and terms applied to them in study of nautiloids are indicated in Fig. 9-6.

FIG. 9-4. **Means of balancing the cephalopod shell.** Gas-filled chambers are shown in white, calcareous deposits in black, and soft parts by stippling. Different groups of cephalopods found various ways of counterbalancing the buoyant effect of gas chambers in the shell.

1, Belemnoids (shell internal) have solid calcareous guard (rostrum) surrounding chambered portion of the shell.

2–5, Straight-shelled nautiloids (shell external) attained balance by weighting parts of the shell or by placing gas chambers above the visceral mass; (2) type with apical chambers containing cameral deposits; (3) type with parts of siphuncle filled by siphonate deposits; (4) endoceroid type, with gas chambers above large siphuncle (probably occupied by visceral mass) and with endocone deposits at tip of siphuncle; (5) ascoceroid type, with gas chambers over living chamber and with early-formed part of shell discarded.

6, Coiled-shell nautiloid (shell external), with buoyant gas chambers above visceral mass as result of planispiral coiling; a similar arrangement characterizes nearly all ammonoids.

Fig. 9-6.   (See next page.)

Volborthella

The earliest cephalopod-like shells belong to the genus *Volborthella* (Fig. 9-7, *1a, b*), known from Lower Cambrian rocks of Europe and North America. These tiny, slender, conical shells have a central structure which looks like a siphuncle and possess closely spaced septa. Unfortunately, no well-preserved specimens have been found, and details of structure are therefore not known.

The occurrence of *Volborthella* poses a problem. No cephalopod-like shells have been recognized in Middle Cambrian rocks, and the primitive nautiloids which make their appearance in Upper Cambrian beds are curved forms (ellesmeroceroids). Straight shells which might be relatives of *Volborthella* do not appear until well above the base of the Ordovician column. This precocious appearance of the genus suggests that it is not ancestral to later shell-bearing cephalopods but an early armored offshoot of the stock which later gave rise to ellesmeroceroids.

Ellesmeroceroids

Members of the order Ellesmeroceroida have small shells, curved or straight, in which the siphuncle is located near the margin (Figs. 9-7, *2a, b;* 9-16, *3*). The septa generally are closely crowded. Septal necks are short. All the earlier ellesmeroceroids have thickly calcified connecting rings, but in advanced forms the connecting ring may be thin and tenuous, as in most other orders of nautiloids. The chambers of ellesmeroceroids do not contain secondary deposits of shell matter, but the siphuncle may show transverse partitions (**diaphragms**).

Ellesmeroceroids seem to be the root stock from which most of the nautiloid orders were derived. They are the only cephalopods known from Upper Cambrian rocks, and they range through the entire Ordovician.

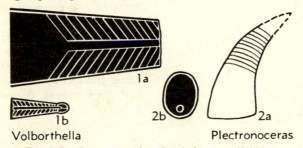

Volborthella   Plectronoceras

FIG. 9-7. **Representative Cambrian nautiloids.**

Volborthella Schmidt, Lower and Middle Cambrian. A small conical shell which possesses septa and what appears to be a siphuncle. Therefore, many but not all paleontologists regard it as a cephalopod. *V. tenuis* Schmidt (1*a, b,* diagrammatic, from Schindewolf, ×5). **Plectronoceras** Ulrich & Foerste, Upper Cambrian. Ellesmeroceroid. This fossil, definitely classed as a cephalopod, has an orthochoanitic siphuncle. *P. cambria* Walcott (2*a, b,* ×3).

FIG. 9-6. **Variations in form of cephalopod shells.** Earliest nautiloids had slightly curved shells, and from these a wide variety of architectural styles developed (indicated by arrows). Commonly used terms for the chief types are listed.

1, *Cyrtocone.* Curved, slender shell.
2, *Orthocone.* Straight, slender shell.
3, *Lituiticone.* Shell like *Lituites,* coiled in early stage, straight at maturity.
4, 6, *Brevicone.* Short, blunt shell.
5, *Ascocone.* Like *Ascoceras,* having a slender, curved early stage and a short, blunt mature stage in which gas chambers overlie living chamber.
7, *Gyrocone.* Loosely coiled shell, like *Gyroceras.*
8, *Advolute shell.* Coiled, whorls touching.
9, *Convolute shell.* Outer whorls embracing inner ones (also called nauticone, after *Nautilus*).
10, *Conispiral shell.* Coiled like a screw (also termed trochoid).

Michelinoceroids

Michelinoceratidae.   Straight-shelled nautiloids are well represented by *Micheli-noceras*, a genus which, along with others,

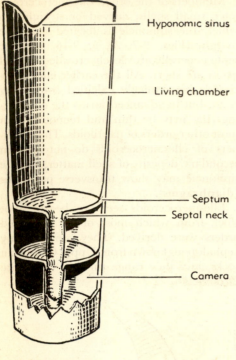

- Hyponomic sinus
- Living chamber
- Septum
- Septal neck
- Camera

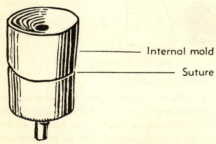

- Internal mold
- Suture

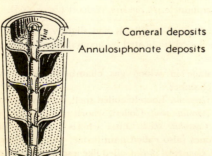

- Cameral deposits
- Annulosiphonate deposits

is found in the older literature under the name *Orthoceras*.

The shell of *Michelinoceras* (Fig. 9-8) is constructed much like that of *Nautilus*, but instead of being coiled, it is straight (orthoceraconic), growing into a long, slender cone of circular cross section. The chambered part (**phragmocone**) behind the living chamber is subdivided by simple watch-glass-shaped septa. The siphuncle is central and consists of short, cylindrical necks and long delicate, cylindrical connecting rings. As in most other long, slender straight nautiloids, the apex of the shell is weighted by calcium carbonate consisting of cameral deposits on the chamber walls and siphonal deposits in the siphuncle.

In exceptionally well-preserved shells, the **cameral deposits** may be seen to consist of alternating lamellae of aragonite and organic matter (Fig. 9-8). They completely fill the early chambers of a mature shell but are less prominent toward the aperture and are entirely lacking near the living chamber. Deposits may be differentiated into **episeptal deposits** (*epi*, upon), lining the rear wall (septum) of the chamber; **hyposeptal deposits** (*hypo*, under), lining the front wall (septum) of a chamber; and **mural deposits** (*mura*, wall), lining the sides (Fig. 9-9, *1–3*). Mural deposits and episeptal deposits commonly grade into each other, whereas the less extensively developed hyposeptal deposits are invariably distinguishable.

Cameral deposits generally are developed more strongly on the ventral side than on the dorsal.

A word of caution is in order. The

FIG. 9-8. Structure of Michelinoceras. The shell of this genus has the form of a straight, slender cone. Except for the shape, its structure agrees closely with that of *Nautilus*, and reference may be made to Fig. 9-3 for definition of most terms. Two structural features of *Michelinoceras* are not found in *Nautilus*. These are *cameral deposits*, consisting of calcareous secretions within the chambers, and *annulosiphonate deposits*, comprising calcareous rings formed in the siphuncle.

chambers of cephalopods commonly are not filled by matrix such as encloses the shell, but contain inorganic precipitates, generally calcite, deposited in the cavities after burial. Such filling, like true cameral deposits, tends to grow inward from the chamber walls. The following criteria permit discrimination between cameral deposits and secondary inorganic precipitates. Inorganic substances tend to cover all interior parts of a chamber uniformly, or they are distributed according to the position in which the shell was buried. Cameral deposits without exception are developed most strongly in the adapical and central regions of the shell, disappearing toward the living chamber. Generally they are thicker on the ventral side than on the dorsal, regardless of the position of the shell in the rock, and show in transverse section a striking bilateral symmetry. For a given growth stage of any one species, cameral deposits are essentially constant, although those of the apical region and nearer the living chamber may differ greatly. Well-preserved specimens show that the cameral deposits are built of alternating clear and organic-stained laminae.

The cameral deposits must have been secreted by tissues which persisted in the chambers after the visceral mass of the animal had moved ahead, building septa at its rear. The shape and size of these tissues are unknown, but it seems likely that they merely lined the chamber walls. But how could these tissues receive nourishment, and how were their waste products removed? Probably the siphuncle and the largely organic connecting rings performed these functions.

The calcareous additions to the siphuncle of *Michelinoceras* consist of so-called **annulosiphonate deposits,** which form rings in the siphuncle near the septal necks (Fig. 9-11, *1*) and join with the cameral deposits. Like the latter, the siphonal deposits are best developed in the apical region of the shell, and where chambers are filled, adjacent thick siphonal deposits commonly form an uninterrupted

lining in the siphuncle, which may leave open only a tiny tube. This observation is in harmony with the theory that a main function of the siphuncle was supplying blood to the cameral tissues. If this is true, the siphuncle was not necessary to chambers completely filled with deposits.

Michelinoceras and its close relatives, distinguished mainly on the basis of surface ornamentation, range from Ordovician to Triassic.

Pseudorthoceratidae. The genus *Pseudorthoceras* (Fig. 9-20, *6*) and similar nautiloids may resemble *Michelinoceras* from the exterior, but differ in internal structure. The siphuncle may be orthochoanitic in the chambers of the slightly curved apical tip, but it passes adorally through an intermediate (suborthochoanitic) stage into a beaded (cyrtochoanitic) condition (Fig. 9-10). The siphuncle is constricted where it passes through a septum and expanded in the chambers, coincident with flaring of the septal necks and bulging of the connecting rings. Because the shapes

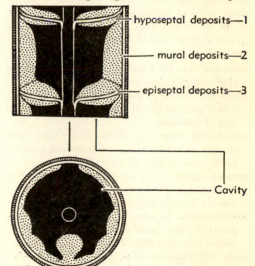

FIG. 9-9. **Cameral deposits.** Many of the straight-shelled nautiloids secreted calcareous deposits on the walls of their gas chambers, partly or entirely filling them. These deposits are of three types: (1) *hyposeptal*, on surface of septum forming front wall; (2) *mural*, on lateral walls of chamber; (3) *episeptal*, on septum forming rear wall.

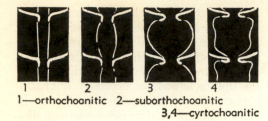

1—orthochoanitic 2—suborthochoanitic
3,4—cyrtochoanitic

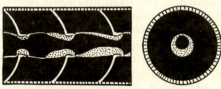

adnation area—5

neck—6

connecting ring—7

brim—8

FIG. 9-10. **Nautiloid siphuncles.** The siphuncles of all nautiloids are composed of *septal necks* and *connecting rings*. The upper diagrams illustrate three types of siphuncles, defined by characters of the necks and rings. The lower diagram is a section through half of a siphuncle and chambers adjoining it.

adnation area (5). Contact between a connecting ring and preceding septum.

brim (8). Recurved portion of a cyrtochoanitic septal neck, measured transverse to longitudinal axis of siphuncle.

connecting ring (7). Thin-walled tube with straight or bulging sides which extends between perforations of two adjacent septa.

cyrtochoanitic (3, 4). Type of siphuncle which is expanded in the chambers and contracted in passing through a septum.

holochoanitic. Tubular siphuncle essentially composed of long septal necks, merely lined with connecting rings (Fig. 9-14).

neck (6). Extension of septum along siphuncle; its length is measured parallel to longitudinal axis of siphuncle.

orthochoanitic (1). Tubular siphuncle composed of short septal necks and intervening connecting rings.

suborthochoanitic (2). Type of siphuncle intermediate between orthochoanitic and cyrtochoanitic.

and proportions of these features are important in classification, a standard terminology has come into use. Measurements of the septal neck are given as the length of the **neck** (Fig. 9-10, *6*), measured parallel to the longitudinal axis of the shell, and the width of the **brim** (Fig. 9-10, *8*), measured at right angles. Additional characters are the width of the area of **adnation** (Fig. 9-10, *5*), along which the rear portion of the connecting ring is in contact with the preceding septal neck, and the shape of the connecting ring (Fig. 9-10, *3*, *4*)

Pseudorthoceroids have both cameral and annulosiphonate deposits (Fig. 9-11, *1*). The pseudorthoceroids appeared in Ordovician time and persisted into the Permian. Examples are shown in Figs. 9-17, *4*, and 9-20, *2*, *3*, *6*.

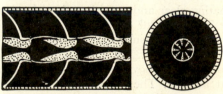

1—annulosiphonate deposits

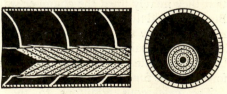

2—actinosiphonate deposits

3—endocones

FIG. 9-11. **Siphonal deposits.** Siphonal deposits are calcareous accumulations within the siphuncle.

1, **annulosiphonate deposits.** Calcareous rings inside siphuncle, generally thicker on ventral side.

2, **actinosiphonate deposits.** Radially arranged longitudinal blades inside siphuncle.

3, **endocones.** Apically pointed conical layers which fill siphuncle, building forward from rear.

Ascoceroids

The Silurian genus *Ascoceras* (Fig. 9-12, 5) achieved gravitational balance of the shell, not by counterweighting the tip, but by dropping off the early-formed portions of the shell at maturity. This abandoned part of the shell is a slightly curved, slender cone, subdivided by simple, transverse septa and having an orthochoanitic siphuncle near the ventral margin. The architecture of the shell changes abruptly at a certain growth stage. The shell expands, the siphuncle becomes cyrtochoanitic, and the septa are extended forward on the dorsal side, so as to place the main portion of the gas chambers above, rather than behind, the living chamber. At maturity, the aperture becomes contracted and the early part of the shell is broken off behind a certain septum, termed the septum of truncation. The break in the siphuncle is sealed by a callus. Derivation of *Ascoceras* and its relatives from more normal types of nautiloids can be traced, various stages in the sequence being represented by Middle and Upper Ordovician genera. The ascoceroids are judged to have descended from forms having slightly curved suborthochoanitic shells, such as

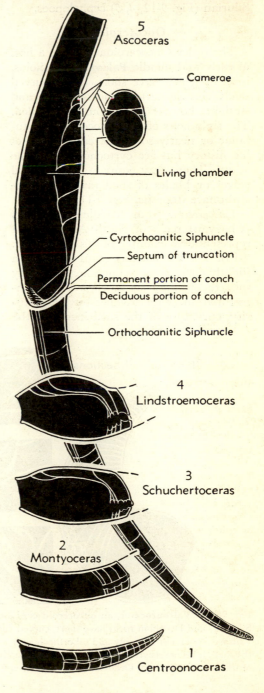

5
Ascoceras

Camerae

Living chamber

Cyrtochoanitic Siphuncle

Septum of truncation

Permanent portion of conch

Deciduous portion of conch

Orthochoanitic Siphuncle

4
Lindstroemoceras

3
Schuchertoceras

2
Montyoceras

1
Centroonoceras

FIG. 9-12. **Development of Ascoceroids.** An evolutionary sequence of nautiloids leads from simple, straight or gently curved forms to highly modified shells of Ordovician and Silurian age, classed in the family Ascoceratidae. Among advanced members, the early, simple part of the shell is shed at maturity; in the remaining portion the gas chambers are located partly or entirely above the living chamber, and the near-ventral siphuncle is cyrtochoanitic.

1, *Centroonoceras* Kobayashi, Lower Ordovician, a slightly curved, suborthochoanitic form, probably near or in the line of ascoceroid ancestry.

2, *Montyoceras* Flower, lower Middle Ordovician of Vermont; retains normal septa, but truncates the early part of the shell at maturity.

3, *Schuchertoceras* Miller, Middle and Upper Ordovician; a true ascoceroid, with mature siphuncle entirely cyrtochoanitic, and its last two septa extended forward over the living chamber.

4, *Lindstroemoceras* Miller, Silurian of Scandinavia; contains three "ascoceroid" septa.

5, *Ascoceras* Barrande, Silurian of Bohemia; contains as many as seven "ascoceroid" septa.

occur in Lower Ordovician rocks (Fig. 9-12, *1*). Genera of this group include Middle Ordovician (Fig. 9-12, *2, 3*), Upper Ordovician (Fig. 9-17, *3, 5*), and Silurian (Fig. 9-12, *4, 5*) cephalopods.

Oncoceroids

The order Oncoceroida includes a host of early and middle Paleozoic nautiloids, most of which have short, blunt, curved shells (brevicones). Some are long and straight, however, and some are coiled. The siphuncles of the oncoceroids are tubular or nearly so in the early stages of life history but are cyrtochoanitic in the adult. They may contain longitudinal ridges or blades of shell matter (**actinosiphonate deposits,** Fig. 9-11, *2*).

Valcouroceras from Middle Ordovician (Chazyan) rocks of North America is a representative example of this group (Fig. 9-13). The shell of this genus begins as a slightly curved cone, containing an empty suborthochoanitic siphuncle near the ventral margin. With advance toward maturity, curvature of the shell increases, the

siphuncle gradually becomes cyrtochoanitic, and actinosiphonate deposits are formed.

Many oncoceroids at maturity have bulbous living chambers but greatly constricted apertures (Fig. 9-18, *4a, b*). Such shells almost certainly were oriented with the apex up, the living chamber down.

Endoceroids

Among the most important groups of Ordovician nautiloids are the endoceroids. The genus *Endoceras* (Fig. 9-14) has a straight, nearly cylindrical shell, containing a large cylindrical subcentral or nearly ventral siphuncle. Not only is the siphuncle of much greater caliber than that of nautiloids previously discussed, but its structure is fundamentally different. Each septal neck is greatly elongated, reaching almost to the previous septum. The connecting rings are more strongly calcified and thicker; they line the inside of the siphuncle, thus making a double wall. Unlike most other nautiloids, in which the siphuncle is a delicate structure, that of

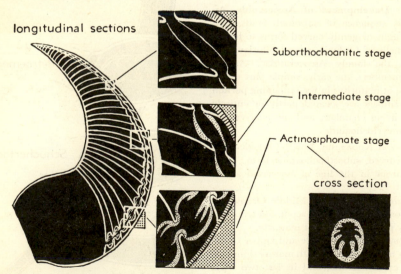

FIG. 9-13. **Valcouroceras, an oncoceroid cephalopod.** Many oncoceroids, common in Ordovician, Silurian, and Devonian rocks have short, curved shells and actinosiphonate deposits. The siphuncle of *Valcouroceras* Flower is empty and suborthochoanitic in the early stages, but in more mature portions of the shell, it becomes cyrtochoanitic and contains actinosiphonate blades, which probably are outgrowths of the connecting rings. Ordovician (Chazyan), eastern North America.

Endoceras is the strongest portion of the shell, and commonly is the only part preserved.

The chambers do not contain cameral deposits, but the apex of *Endoceras* is strongly weighted by distinctive siphonal deposits termed **endocones.** These are calcareous hollow cones, open toward the mouth, and they are stacked in the siphuncle like paper cups in a dispenser. A thin tube, the endosiphotube, extends through the apices of all the endocones (Fig. 9-14).

Many endoceroids lack a double-walled siphuncle, the septal necks being short; in all of them, however, the connecting rings are thick and strongly calcified. Many are breviconic (Fig. 9-16, *5*), but long cylindrical shells of nearly uniform diameter also are known (Fig. 9-16, *7*). In some, the siphuncle is located at the ventral margin and fills the entire apical portion of the shell (Fig. 9-4, *4*). Mid-Ordovician endoceroids are among the largest known nautiloids, some attaining a length of 4 m. (13 ft.).

The comparatively small size of the endoceroid living chamber and the exceedingly capacious nature of the siphuncle suggest that much of the visceral mass of the animal may have been housed within the siphuncle.

The endoceroids appeared in earliest Ordovician time, passed their climax in the middle of this period, and became extinct at its end.

Actinoceroids

The distinctive, highly specialized group called actinoceroids is exemplified by the genus *Actinoceras* (Fig. 9-15). It is characterized by a large, nearly cylindrical shell, which contains an unusually large siphuncle, as in the endoceroids. Unlike endoceroids, however, the siphuncle of *Actinoceras* is strongly cyrtochoanitic, composed of comparatively short, strongly recurved septal necks possessing a broad brim, and greatly inflated connecting rings (Fig. 9-10, *7, 8*). The siphuncle is so heavily

loaded with annulosiphonate deposits that in the central and rear (adapical) segments only small spaces are left open. These comprise a **central canal, radial canals,** and a peripheral space (**perispatium**) be-

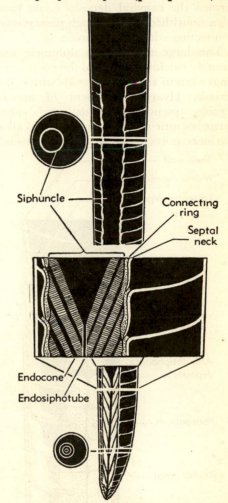

Siphuncle
Connecting ring
Septal neck
Endocone
Endosiphotube

Fig. 9-14. **Structure of endoceroids.** This group of nautiloids, containing *Endoceras* and its relatives, is characterized by the large, strongly calcified siphuncle, containing *endocones.* The apices of the endocones are open, forming an *endosiphotube.* The most advanced members of the group, such as *Endoceras*, possess *holochoanitic* siphuncles, in which septal necks extend back to the previous septum. The thick connecting rings line the interior of the siphuncle. *Endoceras* Hall is known from the Ordovician of Europe and North America.

tween the deposits and connecting rings (Fig. 9-15). The function of this canal system was not understood until its relation to cameral deposits in the chambers was noted. Tissues within the chambers which formed the cameral deposits could have been nourished only through passageways connecting with the siphuncle.

The large size of the siphuncle and strongly calcified nature of the connecting rings seem to ally actinoceroids with endoceroids. Hyatt's placement of actinoceroids, pseudorthoceroids, and other cyrtochoanitic groups in a suborder called Cyrtochoanites is judged to be artificial,

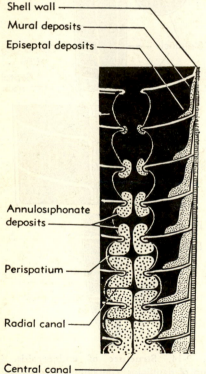

Shell wall
Mural deposits
Episeptal deposits
Annulosiphonate deposits
Perispatium
Radial canal
Central canal

Fig. 9-15. Structural features of actinoceroids. A diagrammatic section through part of a shell of the genus *Actinoceras* Bronn shows the structures distinctive of the actinoceroids. Thick *annulosiphonate* deposits obstruct the siphuncle to such an extent that only three sorts of openings remain: a *central canal*, *radial canals*, and *perispatia* (spaces between annulosiphonate rings and connecting rings). The siphuncle is strongly cyrtochoanitic, and the septal necks are sharply recurved.

inasmuch as the cyrtochoanitic structures seem to have been developed independently by diverse groups of nautiloids. Such a tendency constitutes convergence.

The earliest known actinoceroids occur in mid-Ordovician rocks. The group reached its climax during the Ordovician and Silurian periods, but representatives occur in Mississippian (Figs. 9-20, *1*; 9-23) rocks.

Discosoroids

The discosoroids are a small group of curved nautiloids differing from oncoceroids in having the siphuncle beaded (cyrtochoanitic) throughout ontogeny. The internal structure is highly varied. Some genera have thick-walled calcified connecting rings; others have delicate ones. Some contain rings of secondary shell matter in the siphuncle (annulosiphonate deposits), whereas others possess endocones like the endoceroids. The order ranges from Middle Ordovician into the Devonian.

Nautilids

Coiled nautiloid shells (with the exception of some coiled oncoceroids) are placed in the order Nautilida (see Fig. 9-23). Such assignment is judged to be arbitrary, as coiled nautiloids probably include distinct lineages. Early forms, having thick, calcified connecting rings, are derived from similarly constructed ellesmeroceroids. Later forms, characterized by delicate connecting rings, seem to be derived from advanced ellesmeroceroids, but in part they may be descendants of oncoceroids.

Types of Coiling. Nautilids exhibit a wide range in form of coiling, from open (evolute) planispirals (Fig. 9-6, *7*) to tight planispirals (Fig. 9-6, *8, 9*). Some of these have whorls barely touching (advolute). In others the outer whorls partly embrace the next inner ones (involute), and in still others, like *Nautilus* (Fig. 9-3), the outer whorls completely conceal all the inner ones (convolute). There are also conispirals (Fig. 9-6, *10*), which do not coil in a plane. Coiling generally is in the form

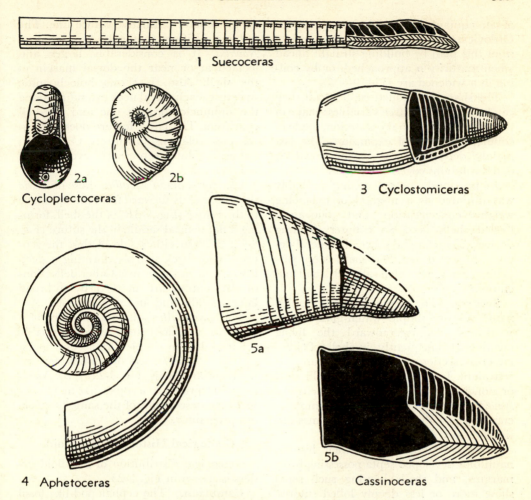

1 Suecoceras

Cycloplectoceras
2a 2b

3 Cyclostomiceras

5a

4 Aphetoceras

5b

Cassinoceras

FIG. 9-16. **Representative Lower Ordovician nautiloids.**

Aphetoceras Hyatt, upper Canadian. Nautilid. The early whorls of the shell may be in contact, but later ones invariably are free. The siphuncle seems to be orthochoanitic. *A. americanum* Hyatt (4, ✕0.4), Newfoundland.

Cassinoceras Ulrich & Foerste, upper Canadian. Endoceroid. A diagrammatic longitudinal section (5*b*) of this unusually short-shelled endoceroid shows the large siphuncle filled with endocones. An external view (5*a*) shows a specimen which has lost the apical chambers but retains apical parts of the siphuncle. *C. grande* Ulrich and Foerste (5*a*, *b*, ✕0.2), Cassin limestone, Vermont.

Cycloplectoceras Ulrich, Foerste, Miller, & Furnish, upper Canadian. Nautilid. The tightly coiled shells are annulated and growth

lines show a deep ventral sinus. *C. miseri* Ulrich, Foerste, Miller, & Furnish (2*a*, *b*, ✕2), Arkansas and Oklahoma.

Cyclostomiceras Hyatt, upper Canadian. Ellesmeroceroid. The straight, short shell has a nearventral siphuncle which contains short septal necks and thickly calcified connecting rings, suggesting relationship to the endoceroids. *C. cassinense* (Whitfield) (3, ✕0.7), Cassin limestone, Vermont.

Suecoceras Holm, Lower Ordovician. Endoceroid. This genus includes long, slender shells having a slightly expanded apical tip which curves slightly toward the venter. *S. barrandei* (Dewitz) (1, ✕0.7, apical region longitudinally sectioned), Sweden.

of an equiangular spiral, but numerous Ordovician and Silurian genera deviate from this plan by building straight ahead when maturity is approached, or by coiling in a widened arc (Fig. 9-6, *3*).

The shape of a coiled cephalopod shell can be reduced to three variables: shape of the whorls as seen in cross section, angle of coiling, and rate of expansion. The last mainly determines the extent to which outer volutions cover inner ones.

As previously noted, coiling probably was initiated as a method of balancing weight concentration and buoyancy. Coiled shells need no counterweighting devices. Therefore, absence of cameral deposits in coiled nautiloid shells is not surprising. Siphuncles generally are small and empty.

Sutures. The septa of most coiled nautiloids are simple watch-glass-shaped partitions, concave toward the living chamber. Consequently, in shells of circular cross section the projected suture line is straight. It is curved in shells of elliptical or noncircular cross section. In shells like *Nautilus*, which have more complexly curved septa, the suture line becomes distinctly wavy.

At several times during their history, nautiloids developed septa possessing wavy margins, and shells bearing such septa show more or less deeply lobed sutures. Derived from nautiloids and making an appearance in the Devonian is the group called **ammonoids,** characterized by complexly sutured shells. These developed along several lines in later Paleozoic time, culminated in the Mesozoic era, and disappeared before Cenozoic time. The conservative nautiloids persisted. In the Triassic, and again in the Tertiary, they produced offshoots which are characterized by wavy septa, but none of these became diversified like the ammonoids. An example is the Tertiary genus *Aturia.*

The shell of *Aturia* (Fig. 9-22, *2*) is similar in general appearance to that of *Nautilus*. The margins of the septa are strongly fluted, however, producing a deeply lobate, distinctive suture line. The siphuncle differs markedly from that of *Nautilus*. It is considerably larger and located at or near the dorsal margin of the shell. Also, it shows holochoanitic structure, each septal neck extending along the siphuncle through one and one-half chambers. The latter feature does not signify close relationship with the Ordovician *Endoceras*, but represents evolutionary parallelism.

Like the septa of *Nautilus*, those of *Aturia* slope concavely forward from the siphuncle to the venter (Fig. 9-3*B*) of the shell, forming a flat ventral saddle in the suture (Fig. 9-22, *2c*). On either side of this, the septum bends back sharply, forming a deep lateral lobe in the suture. Other deflections of the septum are marked by a broad lateral saddle and an umbilical lobe.

The genus *Aturia* is end member of a Cenozoic nautiloid stock, which probably originated from simple Cretaceous ancestors (Fig. 9-21, *1*), and developed through intermediate forms (Fig. 9-22, *1, 3*). *Nautilus* is interpreted as a somewhat more conservative descendant of the same Cretaceous progenitor.

Geological History of Nautiloids

Geological distribution of nautiloid orders is shown in Fig. 9-23.

Cambrian. The cephalopod-like fossil *Volborthella* lived in Early Cambrian time. No cephalopods have been recognized in Middle Cambrian rocks, but ellesmeroceroids appeared in the Late Cambrian (Fig. 9-7).

Ordovician. The Ordovician Period was the time of rapid and extreme differentiation in nautiloids; most of the orders arose at this time, springing directly or indirectly from the ellesmeroceroids. The ellesmeroceroids, endoceroids, and nautilids with thickly calcified connecting rings (Fig. 9-16) dominate Early Ordovician faunas. Siphuncles were largely limited to tubular form, and siphonal deposits include diaphragms and endocones. Chambers appear to have been empty.

Great development of internal structures (Fig. 9-17) is seen in the Middle and Upper Ordovician nautiloids. Cyrtochoanitic siphuncles containing calcareous rings and complex canal systems become prominent with the rise of the actinoceroids, which first appear in the Middle Ordovician.

These nautiloids are characterized also by cameral deposits. Concurrently the michelinoceroids developed annulosiphonate and cameral deposits, but oncoceroids developed actinosiphonate structure. The ascoceroids developed their peculiar mature shell, which sheds the early slender por-

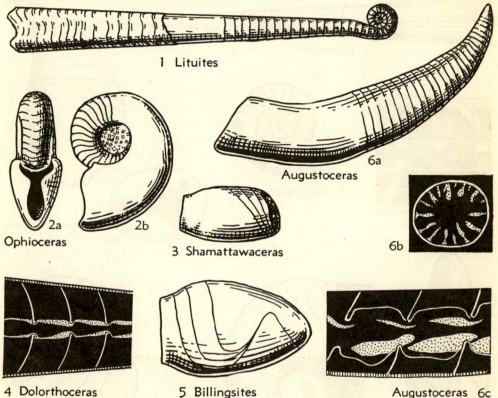

1 Lituites

Ophioceras 2a 2b

3 Shamattawaceras

6a
Augustoceras

6b

4 Dolorthoceras   5 Billingsites   Augustoceras 6c

FIG. 9-17. **Representative Middle and Upper Ordovician nautiloids.**

Augustoceras Flower, Cincinnatian. Oncoceroid. This close relative of *Valcouroceras* (Fig. 9-13) has a siphuncle which contains actinosiphonate deposits. *A. shideleri* Flower (6*a*, side view, ×0.7; 6*b, c*, longitudinal and transverse sections of siphuncle, ×10), Kentucky.

Billingsites Hyatt, Cincinnatian. Ascoceroid. The mature shell consists of three gas chambers which extend out over the living chamber. Related nautiloids are illustrated in Fig. 9-12. *B. canadensis* (Billings) (5, side view of mature shell, ×0.7), Anticosti Island.

Dolorthoceras Miller, Ordovician–Pennsylvanian. Michelinoceroid. A pseudorthoceratid genus having an orthochoanitic siphuncle in earliest stages, but cyrtochoanitic in later ones.

The siphuncle is partly filled by annulosiphonate deposits. *D. sociale* (Hall) (4, longitudinal section, ×5), Cincinnatian, Iowa.

Lituites Breyn, Ordovician. Nautilid. Shell growth begins in a tight coil but later becomes straight. *L. litmus* Montford (1, ×0.3), Baltic region.

Ophioceras Barrande, Ordovician. Nautilid. A tightly coiled shell which at maturity has a contracted aperture. *O. nackholmensis* Kjerulf (2*a, b*, ×0.7), Norway.

Shamattawaceras Foerste & Savage, Upper Ordovician. Ascoceroid. This primitive ascoceroid may be ancestral to *Billingsites*. *S. ascoceroides* Foerste & Savage (3, ×0.3), Manitoba.

tion. Endoceroids set the record for nautiloid size, attaining a length of up to 4 m. (13 ft.). Coiled nautiloids (order Nautilida) from the Middle Ordovician to the beginning of the Cenozoic Era are characterized by long, delicate connecting rings.

Silurian. Endoceroids and ellesmeroceroids died out at the end of the Ordovician, but the other orders continued on into Silurian time (Fig. 9-18). The ascoceroids reached their climax, but became extinct before the close of the period. Per-

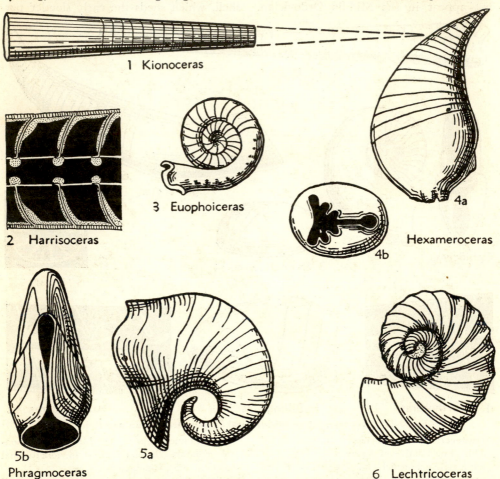

1 Kionoceras

2 Harrisoceras

3 Euophoiceras

4a

4b

Hexameroceras

5b
Phragmoceras

5a

6 Lechtricoceras

FIG. 9-18. **Representative Silurian nautiloids.**

Euophiceras Miller, Middle Silurian. Nautilid. The mode of growth resembles that of *Lituites*, but unlike this genus, the aperture is contracted at maturity. *E. simplex* Barrande (3, ×1).

Harrisoceras Flower, Middle Silurian. Michelinoceroid. This relative of *Michelinoceras* has an unusually large siphuncle containing annulosiphonate deposits. *H. orthoceroides* Flower (2, ×1), Niagaran, Illinois.

Hexameroceras Hyatt, Silurian. Oncoceroid. An intensely constricted lobate aperture and an actinosiphonate siphuncle characterize this genus. *H. hertzeri* (Hall & Whitfield) (4a, b, ×0.7), Niagaran, Ohio.

Kionoceras Hyatt, Ordovician–Devonian. Michelinoceroid. A straight, orthochoanitic shell bearing longitudinal ribs. *K. cancellatum* (Hall) (1, ×0.7), Niagaran, New York.

Lechtricoceras Foerste, Silurian. Nautilid. A transversely ribbed nautiloid which coils in a slightly trochoid spire. *L. desplainense* (McChesney) (6, ×0.7), Niagaran, Wisconsin.

Phragmoceras Sowerby, Silurian. Oncoceroid. This coiled relative of *Hexameroceras* has the venter on the inside of the coil, and an actinosiphonate siphuncle near the inner margin. *P. broderipi* Barrande (5a, b, ×0.3), Czechoslovakia.

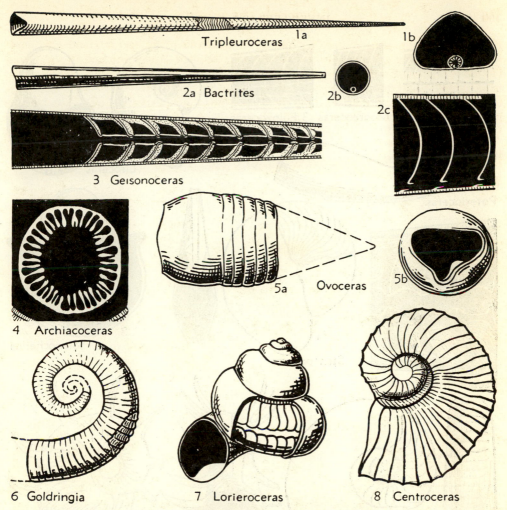

FIG. 9-19. **Representative Devonian nautiloids.**

Archiacoceras Foerste, Devonian. Oncoceroid. A curved shell with prominent actinosiphonate deposits. *A. subventricum* d'Archiac & de Verneuil (4, cross section of siphuncle, ×1.5).

Bactrites Sandberger, Ordovician–Permian. Michelinoceroid. This genus, commonly classified among the ammonoids, has structural features like those of *Michelinoceras*, but the protoconch is slightly swollen, the siphuncle is located at the ventral margin, and cameral or siphonal deposits are lacking. *Bactrites* is here classed as a nautiloid; it probably is ancestral to both ammonoids and belemnoids. (2a, diagrammatic side view; 2b, longitudinal section of a Devonian specimen from Michigan; 2c, diagrammatic cross section.)

Centroceras Hyatt, Devonian–Pennsylvanian. Nautilid. The planispiral shell shows flattened sides and a quadrangular cross section. *C. marcellense* (Vanuxem) (8, ×0.7), Middle Devonian, New York.

Geisonoceras Hyatt, Ordovician–Devonian. Michelinoceroid. This shell resembles *Michelinoceras* but is ornamented by transverse bands. *G. teicherti* Flower (3, longitudinal section, ×0.5), Middle Devonian, New York.

Goldringia Flower, Devonian. Nautilid. This loosely spiral shell possesses a large, ventral siphuncle. *G. trivolve* (Conrad) (6, ×0.3), Middle Devonian, New York.

Lorieroceras Foerste, Devonian. Probably oncoceroid. This genus has a conispiral shell; its marginal siphuncle contains actinosiphonate deposits. *L. lorieri* (Barrande) (7, ×0.7).

Ovoceras Flower, Devonian. Oncoceroid. A breviconic shell possessing a three-lobed aperture and a large cyrtochoanitic siphuncle. *O. oviforme* (Hall) (5a, b, ×0.7), New York.

Tripleuroceras Hyatt, Silurian–Devonian. Oncoceroid. A straight-shelled form characterized by its subtriangular cross section. *T. triangulare* Foerste (1a, b), Devonian, Germany.

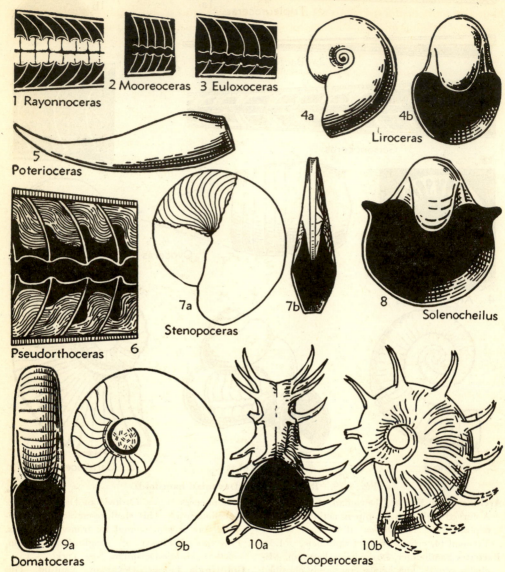

FIG. 9-20. **Representative Upper Paleozoic nautiloids.**

Cooperoceras Miller, Permian. Nautilid. Remarkable on account of its prominent spines. *C. texanum* Miller (10*a*, *b*, ×0.3), Leonard and Word groups, Texas.

Domatoceras Hyatt, Pennsylvanian and Permian. Nautilid. Venter and sides of this coiled form are flattened. *D. umbilicatum* Hyatt (9*a*, *b*, ×0.3), Middle Pennsylvanian, Kansas.

Euloxoceras Miller, Dunbar, & Condra, Pennsylvanian. Michelinoceroid. One of the most advanced member of the family Pseudorthoceratidae, having constricted connecting rings.

E. greeni Miller, Dunbar & Condra (3, longitudinal section, cameral and siphonal deposits not shown, ×2.5).

Liroceras Teichert, Mississippian–Permian. Nautilid. The rather low whorls of this deeply involute shell show a kidney-shaped cross section. *L. liratum* Girty (4*a*, *b*), Pennsylvanian, midcontinent region.

Mooreoceras Miller, Dunbar, & Condra, Devonian–Pennsylvanian. Michelinoceroid. This member of the pseudorthoceratid family has a straight shell and a subcentral cyrtochoanitic

(*Continued on next page.*)

haps the most distinctive feature of many Silurian cephalopod faunas is the prevalence of stubby oncoceroids having greatly constricted, commonly lobate apertures. Coiled forms include both planispiral and conispiral types of growth.

Devonian. Oncoceroids continued to flourish in the Devonian Period and reached the climax of actinosiphonate structure. Michelinoceroids were diversified; the genus *Bactrites*, referred to this order, possibly gave rise at this time to the subclass Ammonoidea. Coiled nautilids split into a number of lines which persisted on into the Mesozoic (Fig. 9-19).

Mississippian, Pennsylvanian, Permian. The michelinoceroids, a few actinoceroids, and several nautilid lineages were the only nautiloids to survive into the Mississippian (Fig. 9-20). Actinoceroids reached great size in this period, but are not common, and they seem to have died out early in Pennsylvanian time. Not much is known about the diversity of upper Paleozoic michelinoceroids, for variation chiefly affects internal features which have received little study. Many representatives of the Nautilida contrast with earlier members of the order in having highly involute form.

Mesozoic. Only *Michelinoceras* and convolute, coiled nautilids reached Triassic time, and the former did not survive this period (Fig. 9-21). The coiled forms of the Triassic include some which resemble the early ammonoids in pattern of wavy sutures, but only simply septate nautilids are found in the Cretaceous.

Cenozoic. Tertiary nautilids are derived from simple, coiled Cretaceous forms (Fig. 9-22). They developed along two main lines: one, characterized by highly fluted septa and a dorsal siphuncle of holochoanitic structure, produced *Aturia* (Paleocene to Miocene); the other, characterized by somewhat sinuous septa, led to the modern *Nautilus*. Miocene nautilids were fairly widespread, but at the present time, when *Nautilus* is the sole surviving genus, the order has become confined to a small area in the southwestern Pacific.

AMMONOID CEPHALOPODS

Ammonoids are mostly coiled cephalopods characterized by marginal siphuncles and fluted septa. This stock, derived from Early Devonian nautiloids, became highly diversified in structure and remarkably prolific in species and genera. It reached peak development during the Mesozoic era but vanished entirely before the beginning of Cenozoic time.

The ammonoids came from one of two possible ancestors. They may be derived from coiled nautiloids which developed fluted septa and a marginal siphuncle, or they may be descended from the genus *Bactrites* (Fig. 9-19, *2a–c*), a straight form possessing a marginal siphuncle. The second alternative requires coiling of the shell

---

(Fig. 9-20 continued.)

siphuncle. *M. tuba* (Girty) (2, longitudinal section). Middle Pennsylvanian, Nebraska.

Poterioceras McCoy, Ordovician–Pennsylvanian. Oncoceroid. A short, stout nautiloid which at maturity develops a constricted subtriangular aperture. *P. fusiforme* Sowerby (5, ×0.4), Mississippian, England.

Pseudorthoceras Girty, Devonian–Permian. Michelinoceroid. Type genus of the Pseudorthoceratidae; chambers near tip are filled with cameral deposits, siphuncle is partly filled with calcareous rings. *P. knoxense* (McChesney) (6, ×2), Pennsylvanian, North America.

Rayonnoceras Croneis, Mississippian–Pennsylvanian. Actinoceroid. Last of the actinoceroids. *R. solidiforme* Croneis (1, longitudinal section, ×1), Upper Mississippian (Fayetteville shale), Arkansas.

Solenocheilus Meek & Worthen, Mississippian–Permian. Nautilid. The subglobular shell is highly involute. *S. springeri* (White & St. John) (8, ×0.3), Upper Pennsylvanian, Iowa.

Stenopoceras Hyatt, Pennsylvanian–Permian. Nautilid. This strongly discoidal shell has a flat venter. *S. dumblei* (Hyatt) (7a, b, ×0.3), Lower Permian, Kansas.

and fluting of the septa, but seems to be most plausible, since the loosely coiled Devonian genus *Anetoceras* has characters of a connecting link, being intermediate between *Bactrites* and the true ammonoids.

Goniatites

As a typical example of a simple ammonoid, we may study the genus *Goniatites* (Fig. 9-24), widespread in Mississippian rocks of North America and Europe.

Superficially, the strongly involute shell of *Goniatites* resembles *Nautilus*, although it is much smaller and proportionately broader. Many of the same terms are used in the description of both. The **umbilicus** (*4; italic numbers in the text describing *Goniatites* refer to Fig. 9-24, unless indicated otherwise*) is small but deep and open. The **siphuncle** (*22*), unlike that of *Nautilus*, is located at the ventral (outer) margin of the shell. The septa are strongly fluted, and the shell shows periodic constrictions.

Septa. One of the most distinctive features of *Goniatites* is the fluting of its septa, shown by the **suture line** (*8–15*), or seen in a specimen broken so as to

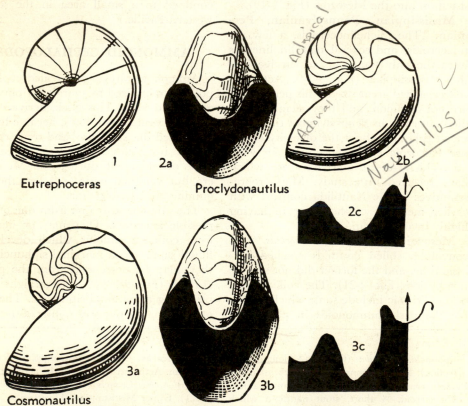

Eutrephoceras

Proclydonautilus

Cosmonautilus

Fig. 9-21.   **Representative Mesozoic nautiloids.**

Cosmonautilus Hyatt & Smith, Triassic. Nautilid. The shell has flattened sides and a highly sinuous suture. *C. dilleri* Hyatt & Smith (*3a–c*, ×0.7), Upper Triassic (*Tropites subbullatus* zone), California.

Eutrephoceras Hyatt, Cretaceous–Eocene. Nautilid. This genus includes many species formerly referred to *Nautilus* from which it differs in possessing a nearly straight suture. *E. dekayi* (Morton), (1, ×0.7), Upper Cretaceous, North America.

Proclydonautilus Mojsisovics, Triassic. Nautilid. The shell has a more rounded cross section and less sinuous suture than *Cosmonautilus*. *C. triadicus* Mojsisovics (*2a–c*, ×0.7), Upper Triassic, California.

expose the entire surface of one of the septa in the outer whorls (Fig. 9-25). The cross section of the whorl is crescent-shaped; hence, the septum is also crescentic in general outline. Instead of being simply concave, as in *Nautilus*, its marginal parts are folded into protruding arches (**saddles**) and receding troughs (**lobes**).

These folds flatten out away from the suture. A deep ventral lobe (*11*) is split by a small ventral saddle (*13*) in the position of the siphuncle. On the dorsal side is a corresponding but smaller dorsal lobe. The ventral and dorsal lobes are flanked on each side by narrow first lateral saddles, and beyond these, by deep, sharply pointed

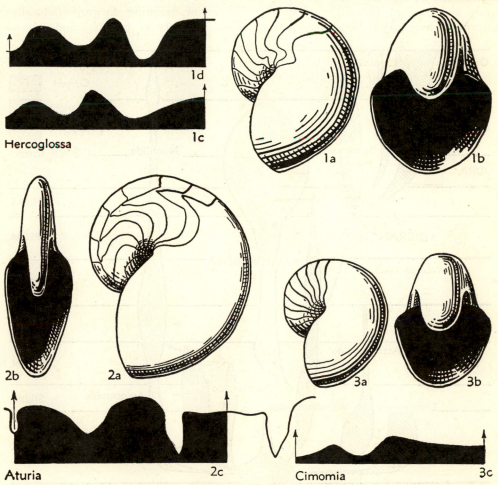

FIG. 9-22.   **Representative Tertiary nautiloids.**

Aturia Bronn, Paleocene–Miocene. Nautilid. The most complex of Tertiary nautiloids. *A vanuxemi* Conrad (2*a–c*, ×0.3), Eocene, New Jersey.

Cimomia Conrad, Paleocene–Oligocene. Nautilid. This genus is intermediate between *Eutrephoceras* (Fig. 9-21, *1*) and *Hercoglossa C. haughti* (Olsson) (3*a*, *b*, ×1), Eocene, Peru;

C. vincenti Miller, (3*c*, suture), Paleocene, Congo.

Hercoglossa Conrad, ?Cretaceous–Eocene. Nautilid. The suture is intermediate between those of *Cimomia* and *Aturia*. *C. harrisi* Miller & Thompson (1*a–c*, ×1) Paleocene, Trinidad; *C. orbiculata* (Tuomey) (1*d*), Paleocene, Alabama.

reentrants, which are lateral lobes. Then come broad second lateral saddles and umbilical lobes (8–15; Fig. 9-25).

The suture is best studied on internal molds; hence, it is necessary to peel off part of the shell from specimens preserved with the wall intact. The external (ventral and lateral) part of the suture generally is sufficient for identification. In order to study the internal (dorsal) part of the suture, it is necessary to break loose part of

a whorl and thus expose its dorsal surface.

For purposes of careful analysis and publication, it is customary to project the suture line on a plane, which may be done by tracing the suture under a camera lucida, while the ammonoid is rotated. On such projected suture lines, the venter (12) (and dorsum, if the dorsal suture is included in the diagram) is indicated by an arrow pointing forward (adorally). The boundary of external and internal

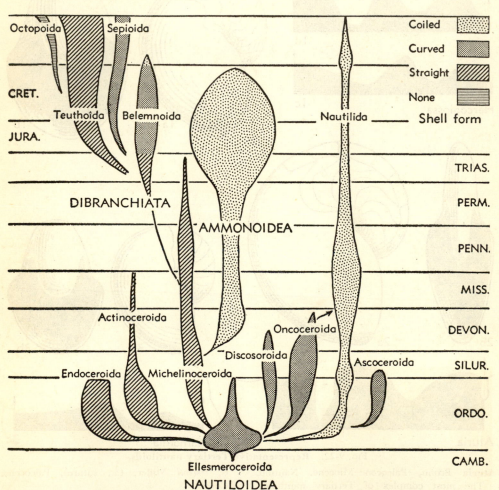

FIG. 9-23. **Geological distribution of cephalopod orders.** A great evolutionary outburst in Early and Middle Ordovician time formed most of the nautiloid orders. Ammonoids and belemnoids split later from the michelinoceroid line. Sepioids are derived from belemnoids, but the ancestry of teuthoids and octopoids is not clear. The Nautilida, here used to include nearly all coiled nautiloids, are an arbitrary unit embracing at least two lineages—some derived directly from the ellesmeroceroids, others from the oncoceroids.

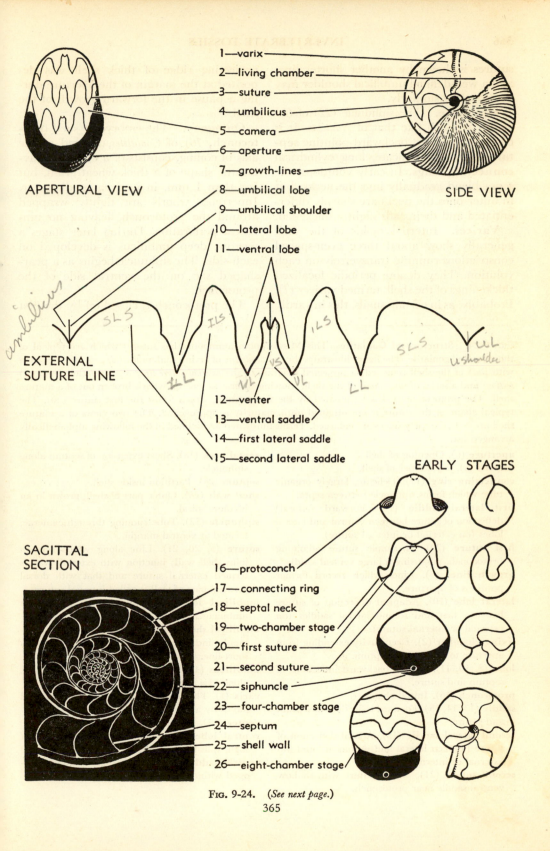

APERTURAL VIEW

SIDE VIEW

1—varix
2—living chamber
3—suture
4—umbilicus
5—camera
6—aperture
7—growth-lines
8—umbilical lobe
9—umbilical shoulder
10—lateral lobe
11—ventral lobe

EXTERNAL
SUTURE LINE

umbilicus

SLS ILS VS ILS SLS UL
LL VL VL LL u shoulder
ILL

12—venter
13—ventral saddle
14—first lateral saddle
15—second lateral saddle

EARLY STAGES

SAGITTAL
SECTION

16—protoconch
17—connecting ring
18—septal neck
19—two-chamber stage
20—first suture
21—second suture
22—siphuncle
23—four-chamber stage
24—septum
25—shell wall
26—eight-chamber stage

FIG. 9-24.   (*See next page.*)

sutures is shown by another shorter line, and a well-defined umbilical shoulder may be indicated similarly.

Siphuncle. The siphuncle (*22*) of *Goniatites* is much like that of *Nautilus*. It is composed of short backward-pointing **septal necks,** and delicate, long cylindrical **connecting rings.** In early volutions, the septa blend gradually into the necks, but in outer ones the necks are sharply differentiated and their ends slightly thickened.

Varices. Internal molds of the shell generally show about three even-spaced constrictions running transversely on each volution. They denote periodic localized thickenings of the shell, termed varices (*7*). Probably, as in certain snails, this inwardly

projecting ridge of thick shell was deposited at the margin of the aperture during a pause in the forward growth of the shell.

Ontogeny. The embryonic shell (**protoconch,** *16*) of *Goniatites*, located at the axis of coiling, consists of a chamber having the shape of a thick wheat grain, but less than 1 mm. in its greatest diameter. Innermost whorls are tightly wrapped around the protoconch, leaving no umbilical perforation. During later stages a narrow, deep umbilicus is developed on each side. The siphuncle begins as a pear-shaped sac on the ventral side of the protoconch.

The protoconch is closed off by the first

FIG. 9-24.   **Structure of Goniatites.** This Mississippian ammonoid has sutures which are typical of those termed goniatite. The figure illustrates the *apertural view* of an internal mold; *side view* of a specimen with part of the shell removed; a projected diagram of the *external suture line;* a *sagittal* (longitudinal) *section;* and a series of *early stages* of the shell, showing sutures as seen on whorls broken out of a mature shell. The protoconch (16) is terminated by the first septum, which forms the first suture (20). The typical shape of the suture is not attained until maturity is approached. The two views of a mature shell are ×1; the early stages are enlarged. Terms illustrated are defined in the following alphabetically arranged list.

aperture (6). Opening of shell.

camera (5). A chamber of shell.

connecting ring (17). Delicate, largely organic tube which forms siphuncle between septa.

first lateral saddle (14). Forward (adoral) deflection of suture between ventral and lateral lobes (on external or internal suture).

first suture (20). A simple suture adjoining protoconch, possessing a large ventral saddle.

growth lines (7). Lines which record former positions of aperture.

lateral lobe (10). Adapical deflection of suture between first and second lateral saddles (on external or internal suture).

living chamber (2). Deep cavity extending from aperture to last-formed septum.

lobe (8, 10, 11). Any of backward inflections of septum and suture.

protoconch (16). Initial chamber.

saddle (13, 14, 15). Forward inflections of septum and suture.

second lateral saddle (15). Adoral deflection of suture between lateral and umbilical lobes (on external or internal suture).

second suture (21). Simple suture with shallow ventral saddle near protoconch.

septal neck (18). Short extension of septum along siphuncle.

septum (24). Partition inside shell.

shell wall (25). Outer part of shell, grown in an involute spiral.

siphuncle (22). Tube running through camerae, located at ventral margin.

suture (3, 20, 21). Line along which septum joins shell wall; junction with external wall is termed external suture and that with dorsal internal part of wall is called internal suture.

umbilical lobe (8). Sutural lobe located at umbilicus.

umbilical shoulder (9). Rim of umbilicus, commonly indicated on a suture diagram by a short line.

umbilicus (4). Depression on each side of shell, along axis of coiling.

varix (1). Thickening of shell marked on internal molds by a transverse groove.

venter (12). Periphery of shell.

ventral lobe (11). Large lobe on periphery of shell, split by a smaller saddle.

ventral saddle (13). Small, sharp saddle developed within ventral lobe.

true septum (suture, *20*), which has a broadly rounded ventral saddle. The next following septum bears a well-defined lobe in the ventral saddle (*21*). A faint undulation which appears on the flanks of this saddle becomes more prominent in succeeding septa, developing into the lateral lobe (*10*) between the first and second lateral saddles (*14, 15*). The small saddle which splits the ventral lobe is limited to the outermost mature whorls.

Function of Septal Fluting. The fluting of ammonoid septa must have been useful to the animals which produced them. It is inconceivable that a structure of no value in natural selection should characterize thousands of ammonoid species, persist some 300 million years, reach a peak of complexity coincidently with greatest abundance and diversification of the ammonoids, and make an appearance also in specialized nautiloids. Unfortunately, no cephalopods possessing fluted septa are alive today, and the habits of their nearest living relative, the nautilus, are not well known. Therefore, it is not surprising that interpretations of the function of septal fluting vary.

A seeming advantage of the fluting is its buttressing effect. The area of contact between the septum and shell wall in a typical ammonoid is notably larger than in *Nautilus*, but in addition, the series of staggered arches of the ammonoid septum greatly increase the load-bearing capacity of the shell. Given two shells of equal size and shape, one possessing simple septa and the other fluted ones, the latter will be much the stronger if the walls are equal in thickness, or it may have equal strength if its walls are thinner. Many ammonoid shells were exceedingly delicate.

Computation indicates that if a *Nautilus* shell with a pressure of one atmosphere in its chambers is progressively submerged, the shell will collapse at a depth of 345 m. Yet the *Challenger* expedition dredged *Nautilus* from a depth of 549 m. Living specimens have been seen near the surface. Therefore it is certain that *Nautilus* is able to adjust the pressure within its shell to its surroundings. Probably the rate at which pressure adjustments can be made is limited, and the possession of a stronger shell might have permitted ammonoids to descend and rise more rapidly than *Nautilus*.

Possibly septal fluting contributed to the efficiency of the ammonoid's hydrostatic mechanism. In the discussion of *Nautilus*

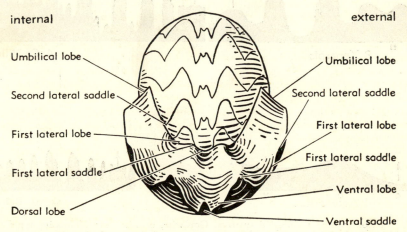

internal   external

Umbilical lobe   Umbilical lobe

Second lateral saddle   Second lateral saddle

First lateral lobe   First lateral lobe

First lateral saddle   First lateral saddle

Dorsal lobe   Ventral lobe

                                                           Ventral saddle

Fig. 9-25. **Septum of Goniatites.** The wavy surface of a nearly mature septum is shown as it appears in front view, the living chamber being broken away. The terms are defined in an alphabetically arranged explanation of Fig. 9-24 (first lateral lobe, same as lateral lobe; dorsal lobe is an adapical inflection on internal suture corresponding in position to the ventral lobe on the external suture).

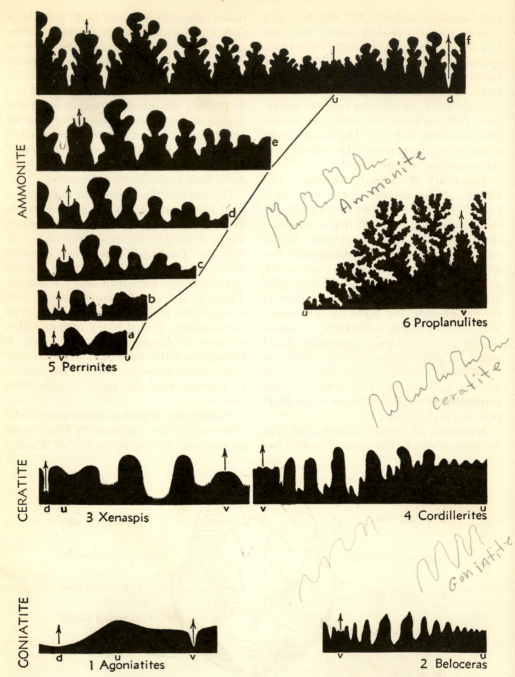

Fig. 9-26. Comparison of ammonoid sutures. Goniatite, ceratite, and ammonite types of sutures are shown, as well as ontogeny of the suture of *Perrinites*, a Permian ammonite, which illustrates gradation from the simple early stages to the complex adult form, through stages closely similar to the adult sutures of ancestral forms. The letters beneath the drawings of sutures signify venter (*v*), umbilical shoulder (*u*), and dorsum (*d*). (*Continued on next page.*)

it was pointed out that this animal probably possesses a gaseous space between the end of the body and the last-formed septum, and that muscular contraction of the body, by enlarging or decreasing the volume of this space, may regulate specific gravity within narrow limits, causing the animal to rise or sink slowly at will. A fluted septum at the rear of the living chamber might allow more efficient control of specific gravity, for the contraction of muscles attached to the saddles of the septal margins could pull the rear body wall forward, increasing the gas space, whereas muscles attached to lobes could pull the body like a piston toward the septum, compressing the gas much more strongly than *Nautilus* can do. Thus, complexity of septal fluting may be related to the ability of the ammonoids to change specific gravity, and consequently the rate of ascent or descent in the sea.

Variations of Ammonoid Shells

Other ammonoids differ from *Goniatites* chiefly in the character of the sutures, position and character of the siphuncle, general shape, and surface ornamentation.

Sutures. The suture of *Goniatites* and its contemporaries is comparatively simple. Some have fewer and others more lobes and saddles, but these individual deflections are undivided. Sutures of this comparatively simple type are called **goniatite** sutures (Fig. 9-26, *1, 2*). They characterize Paleozoic ammonoids and occur in a genus of Triassic age and one from the Cretaceous.

In Mississippian time, ammonoids appeared which have lobes subdivided into small second-order lobes and saddles. Such sutures, which persisted through Upper Paleozoic into Triassic time and reappeared in the Cretaceous, are termed the **ceratite** type (Fig. 9-26, *3, 4*).

At the beginning of Permian time, a third suture pattern, termed the **ammonite** type, was introduced. Both lobes and saddles are subdivided into smaller second-order lobes and saddles, and in advanced forms third-order serrations are developed, forming an exceedingly complex suture (Fig. 9-26, *5, 6*). Ammonite sutures persisted until extinction of the Ammonoidea at the end of Cretaceous time.

Ceratite and ammonite sutures, like those of *Goniatites*, develop gradually during the ontogeny of any one individual (Fig. 9-26, *5*), and earliest sutures of each belong to the goniatite type. Submature sutures of advanced ammonoid genera commonly, although not invariably, are closely similar to submature or mature sutures of genera which are their ances-

---

(Fig. 9-26 continued.)

1, Complete suture of *Agoniatites*, Devonian, showing a simple type of goniatite suture, which consists of a sharply incised ventral lobe, a very broad lateral saddle, and a broad, shallow dorsal lobe.

2, External suture of *Beloceras*, Devonian, showing an advanced type of goniatite suture, which has many sharply defined lobes and saddles, and a subdivided ventral lobe.

3, Complete suture of *Xenaspis*, Permian–Triassic, showing smoothly rounded saddles and serrated lobes characteristic of the ceratite suture.

4, External suture of *Cordillerites*, Triassic, showing a ceratite suture which differs from that of *Xenaspis* in the much coarser serration of the lobes.

5, Ontogeny of the suture of *Perrinites*, Permian, showing external sutures of juvenile and adolescent stages (*a–e*), and complete suture of adult (*f*). Early stages are goniatite types (*a–c*) and later ones ammonite type (*d–f*). (*a*) Sixth suture at *Goniatites* stage of development (Fig. 9-24), ×15. (*b*) Tenth suture at *Neoshumardites* stage of development, shell diameter, 2.6 mm., ×12. (*c*) Sixteenth suture, at *Shumardites* stage of development, shell diameter, 3.7 mm., ×12. (*d*) Nineteenth suture, at *Peritrochia* stage of development, shell diameter, 4.5 mm., ×12. (*e*) Late adolescent suture, at *Properrinites* stage of development (Fig. 9-32, 1), shell diameter, 7 mm., ×8. (*f*) Mature suture, shell diameter, 54 mm., ×1.6.

6, External suture of *Proplanulites*, Jurassic, showing a highly dissected ammonite suture.

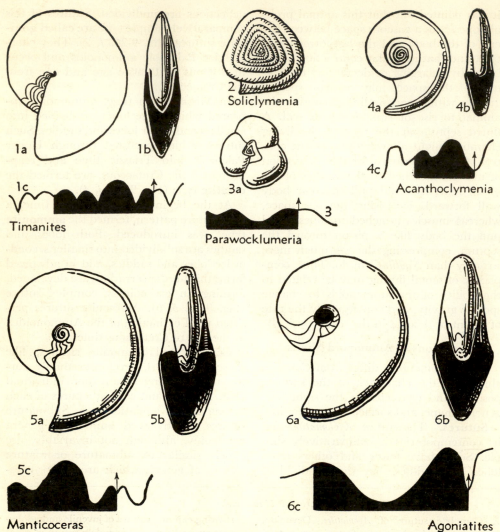

Fig. 9-27. **Representative Devonian ammonoids.** All Devonian ammonoids have goniatite sutures. Members of the family Clymenidae include 2–4.

Acanthoclymenia Hyatt, Upper Devonian. This comparatively evolute clymenid shows a well-developed ventral lobe in the suture. *A. neapolitana* (Clarke) (4a–c, ×2), Naples formation, New York.

Agoniatites Meek, Devonian. One of the simplest of the ammonoids, possessing a suture which, except for a well-marked ventral lobe, is more simple than that of *Nautilus*. *A. vanuxemi* (Hall) (6a–c, ×0.2), Middle Devonian, New York.

Manticoceras Hyatt, Middle and Upper Devonian. *M. sinuosum* (Hall) (5a–c, ×0.5), Upper Devonian, New York.

Parawocklumeria Schindewolf, Upper Devonian.

The shell is coiled in a three-lobed manner rather than in a normal equiangular spiral, the suture has a broad ventral saddle. *P. patens* Schindewolf (3a, b, ×1).

Soliclymenia Schindewolf, Upper Devonian. A peculiar clymenid, advolute and coiled in triangular manner. *S. paradoxa* (Muenster) (2, ×1.5), Germany.

Timanites Mojsisovics, Devonian. Complete convolution and a suture divided into five pairs of lobes mark this as one of the more highly developed among Devonian goniatites. *T. occidentalis* Miller & Warren (1a–c, ×0.5), Upper Devonian, Alberta.

tors. Thus, the suture of *Perrinites* (Fig. 9-26, *5*) passes through a stage which duplicates the adult suture of its ancestor *Properrinites*, furnishing a fine example of palingenesis ("ontogeny recapitulates phylogeny").

Siphuncles. Mid-Paleozoic ammonoids have septal necks pointing back toward the protoconch, a condition termed **retrosiphonate** (Fig. 9-29). In late Paleozoic and Mesozoic ammonoids, this structure is mostly confined to immature stages, and forward-pointing septal necks appear in submature whorls. This condition is described as **prosiphonate** (Fig. 9-29).

Near the living chamber of mature shells the septal necks may disappear entirely.

Most Devonian ammonoids have siphuncles like that of *Goniatites*, which is ventral throughout. One Upper Devonian family, the Clymenidae, however, has a dorsal siphuncle. Among many Upper Paleozoic and Mesozoic ammonoids, the siphuncle begins near the dorsal side or in the center of the early whorls, then shifts to a ventral position where it remains.

Shell Form. Most ammonoids, like nautiloids, are planispirally coiled and exhibit all degrees of involution of the whorls from evolute to advolute, involute, and

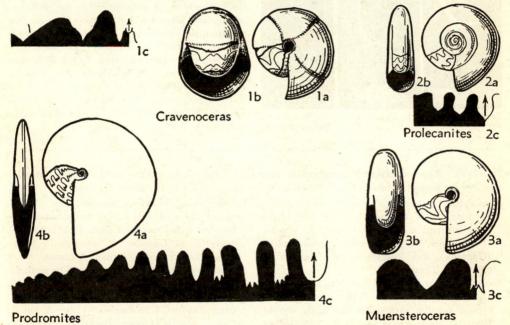

Cravenoceras

Prolecanites

Muensteroceras

Prodromites

Fig. 9-28. **Representative Mississippian ammonoids.** Except for *Prodromites*, which has a ceratitic suture, Mississippian ammonoids are characterized by goniatitic sutures. The shells generally are smooth and rounded.

Cravenoceras Bisat, Mississippian. The shell strongly resembles that of *Goniatites* but is differentiated by the shape of the saddles of its suture. *C. hesperium* Miller & Furnish, (*1a–c*, ×1), Upper Mississippian (Meramecian), Nevada.

Muensteroceras Hyatt, Mississippian. The sutures are simpler than those of *Prolecanites*. *M. parallelum* (Hall) (*3a–c*, ×0.5), Lower Mississippian (Kinderhookian), Indiana.

Prodromites Smith & Weller, Mississippian. A large discoidal genus, which shows the oldest known ceratite suture. *P. gorbyi* Miller (*4a–c*, ×0.5), Lower Mississippian (Kinderhookian), Indiana, Iowa, and Missouri.

Prolecanites Mojsisovics, Lower Mississippian. This unspecialized goniatite is thought to be the ancestor of important Pennsylvanian and Permian stocks (*Pronorites, Uddenites, Medlicottia*). *P. gurleyi* Smith (*2a–c*, ×0.7), Lower Mississippian (Kinderhookian), Missouri.

convolute. Some Mesozoic shells are coiled in the early stages only, becoming straight or U-shaped in mature life. A few built conispiral shells and some (*Nipponites*, Fig. 9-42, *8*) lost all semblance of symmetry.

The whorls also vary widely in the shape of their cross section. Among many shapes are subcircular, quadrate, crescentic, and V shapes (Figs. 9-27 to 9-42).

Most Paleozoic ammonoid shells are unornamented. Many Mesozoic forms, on the other hand, are highly ornate, bearing transverse ribs and protuberances, which range from gentle swellings to nodes and spines. The ribs may divide into two or more branches in their course from the umbilical region toward the venter. Many ammonoid genera are characterized by a sharp ridge or keel along the venter, or this part of the shell may be grooved.

Aperture. As among nautiloids, well-preserved ammonoid shells carry in their

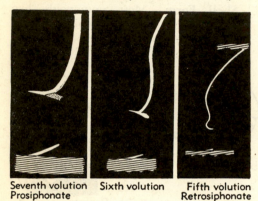

Seventh volution   Sixth volution   Fifth volution
Prosiphonate   Retrosiphonate

FIG. 9-29. **Ontogeny of the ammonoid siphuncle.** The Permian genus *Pseudogastrioceras* illustrates ontogenetic changes which occur in the siphuncles of many ammonoids. In the early part of the shell, illustrated by a section from the fifth volution, the siphuncle is *retrosiphonate*, much like that of a nautiloid, with septal neck (pointing to the rear) and connecting ring. In the sixth volution, a septal collar is added, pointing forward. In the seventh volution, the septum is bent forward (*prosiphonate*), but a secondary deposit forms a backward projection in place of the septal neck. (*After Miller & Unklesbay.*)

FIG. 9-30. **Representative Pennsylvanian ammonoids.** Goniatite sutures predominate, but some ceratites are present, and *Uddenites* is transitional from the ceratite to the ammonite type.

Eoasianites Ruzencev, Lower Pennsylvanian–Permian. Shell characterized by low broad volutions and a goniatite suture. *E. modestus* (Böse) (*8a–c*, ×2), uppermost Pennsylvanian (Gaptank) and lowermost Permian (Wolfcampian), Texas.

Eothalassoceras Miller & Furnish, Middle and Upper Pennsylvanian. This genus possess typical ceratite sutures having finely serrate lobes. *P. kingorum* (Miller) 4*a–c*, ×2), Upper Pennsylvanian (Virgilian), Texas.

Gastrioceras Hyatt, Pennsylvanian. A broadly rounded, smooth goniatite. *G. listeri* (Martin) (*2a–c*, ×1.5) described from rocks of Early Pennsylvanian age in England, but also occurs in the Atoka and Smithwick shales of Texas, Oklahoma, and Arkansas.

Gonioloboceras Hyatt, Pennsylvanian. The discoidal shell has simple but distinctive, deeply undulating goniatite sutures. *G. goniolobus* Meek (*5a–c*, ×0.5), Upper Pennsylvanian, New Mexico.

Pronorites Mojsisovics, Pennsylvanian. An involute discoidal shell having complex goniatite sutures; it is descended from *Prolecanites*, and ancestral to *Uddenites*. *P. arkansasensis* Smith (*6a–c*), Lower Pennsylvanian, Arkansas.

Schistoceras Hyatt, Pennsylvanian. The goniatite suture of this large, Upper Pennsylvanian ammonoid is characterized by a bottle-shaped ventral saddle. *S. missouriense* (Miller & Faber) (*7a–c*, ×0.7), Upper Pennsylvanian (Missourian), Missouri.

Shumardites Smith, Pennsylvanian. Globose ammonoid having a suture which contains several split lobes, and therefore primitively ceratitic. *S. uddeni* (Böse) (*3a–c*, ×1.5), Upper Pennsylvanian (*Uddenites* zone), Texas.

Uddenites Böse, Upper Pennsylvanian. This genus is a descendant of *Pronorites*, from which it is distinguished by a grooved venter and distinctively subdivided lateral saddles; its suture is not typical of the goniatite, ceratite, or ammonite patterns. *U. schucherti* Böse (*1a–c*, ×1), Graham group, Texas.

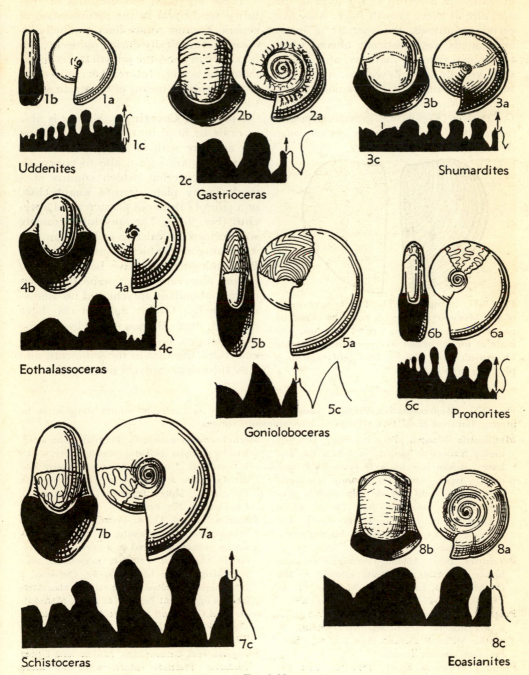

Uddenites

Gastrioceras

Shumardites

Eothalassoceras

Gonioloboceras

Pronorites

Schistoceras

Eoasianites

Fig. 9-30.

growth lines a record of the form of the aperture at every growth stage. Also, like certain nautiloids, the aperture of many ammonoids was drastically altered when the animal reached maturity, after which shell growth virtually ceased. The apertures are contracted, and among some, such as *Normannites* (Fig. 9-37, *3*) and *Oecoptychius* (Fig. 9-38, *2*), peculiar lateral outgrowths, termed ears or auricles, are developed.

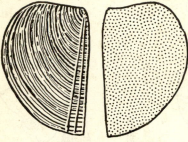

FIG. 9-31. Aptychus. The apertures of many of the Mesozoic ammonoids could be closed by paired calcareous plates, termed aptychus. The plate at the left shows its inner surface, marked by growth lines; that at the right presents a punctate outer surface.

These apertural modifications at maturity are helpful in the identification of genera. Because suture lines gradually increase in complexity during ontogeny, it is necessary to know the growth stages represented by a given suture. Apertural characters furnish a means of recognizing mature shells.

Apertural Covering. *Nautilus* is able to draw its head into the shell and close the aperture with its tough hood. Some ammonoids are known also to possess an apertural covering, which consists of a single horny plate, termed **anaptychus**, or a pair of calcareous plates termed **aptychus** (Fig. 9-31). The aptychus seemingly was formed of calcite, rather than less stable aragonite, which composed the shell. Paleontologists long have been intrigued by the observation that some limestones contain abundant aptychi and no ammonoid shells. Could the aptychi have dropped from ammonoid shells drifting over the area, or were shells and aptychi embedded together in the sediments, and only the calcitic aptychi preserved?

FIG. 9-32. Representative Permian ammonoids. Goniatite and ceratite sutures continue to be present, but most of the best index forms have ammonite sutures.

Medlicottia Waagen, Permian. Shell possesses highly distinctive sutures, in which the first lateral saddle is extensively subdivided (ammonite type), whereas the many remaining saddles are undivided or only slightly subdivided. *M. burckhardti* Böse (*2a–c*, ×1.5), Middle Permian (*Waagenoceras* zone), Las Delicias, Mexico, also found in equivalent rocks of Texas.

Perrinites Böse, Middle Permian. Possesses highly distinctive ammonite sutures. *P. hilli* (Smith) (*4a, b,* Las Delicias, Mexico, ×1, *4c,* Texas). Fig. 9-26 shows ontogeny of sutures.

Properrinites Elias, Lower Permian. This genus is ancestor of the Leonardian *Perrinites*. *P. mooreae* Miller & Furnish (*1a–c,* ×1.5), Lower Permian, western Texas.

Pseudogastrioceras Spath, Permian. This descendant of the Pennsylvanian *Gastrioceras* differs from the latter in showing spiral ornamentation. *P. beedei* (Plummer & Scott) (*3a–c,* ×2), Middle Permian, Texas.

Stacheoceras Gemmellaro, Permian. Characterized by globular shell and ceratite suture, which is composed of many saddles and mostly subdivided lobes. *S. toumanskaya* Miller & Furnish (*5a–c,* ×1), Upper Permian (*Timorites* zone), Las Delicias, Mexico.

Timorites Haniel, Upper Permian. Differs from *Waagenoceras* in possessing a more complex suture pattern arranged in a curve. It characterizes the latest Permian ammonoid zone recognized in the United States. *Cyclolobus,* a more highly developed close relative characterizes the uppermost Permian of India, Madagascar, and Greenland. *T. uddeni* Miller & Furnish (*7a–c,* ×0.3), Bell Canyon formation, western Texas.

Waagenoceras Gemmellaro, Middle and Upper Permian. Possesses sutures somewhat more complex than are those of *Perrinites*. *W. guadalupense* Girty (*6a–c,* ×1), Upper Permian (Guadalupian), Texas.

1b 1a

1c

Properrinites

2b 2a

3b 3a

1

3c

Pseudogastrioceras

2c Medlicottia

4b 4a

5b 5a

5c Stacheoceras

4c

Perrinites

6c

7c

6b 6a

Waagenoceras

7a 7b

Timorites

Fig. 9-32.

Geological History of Ammonoids

Devonian. All Devonian ammonoids have goniatite sutures. Two major groups occur—those in which the siphuncle is in the normal ventral position throughout ontogeny, and those (family Clymenidae) in which the siphuncle is located at the dorsal margin of the whorls. Some have peculiar triangle-shaped initial coils (Fig. 9-27).

Mississippian. The clymenids did not persist into the Mississippian, but many genera of goniatitic aspect occur in rocks of this period. Especially in northwestern

Europe, they include some of the most important fossils for stratigraphic zonation and correlation. The first ammonoids having ceratite sutures made their appearance (Fig. 9-28).

Pennsylvanian. Ammonoids characterized by goniatite sutures continued to be the most common, but ceratite sutures are also found in the Pennsylvanian. The Upper Pennsylvanian *Uddenites* possessed a suture which is transitional toward the ammonite type. Although several narrow discoid shells occur, a majority of the Pennsylvanian genera have broadly rounded venters (Fig. 9-30).

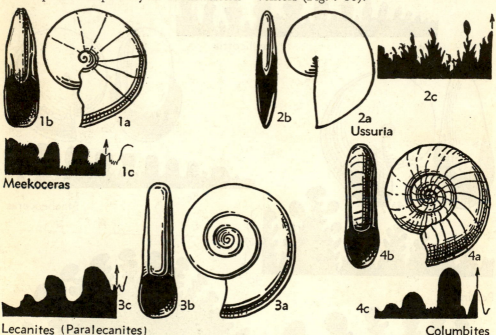

Meekoceras

Ussuria

Lecanites (Paralecanites)

Columbites

FIG. 9-33. Representative Lower Triassic (Scythian) ammonoids. Among Lower Triassic ammonoids, sutures of ammonite type are relatively less common than in the Permian, and finely serrate ceratites predominate. Shells are smooth or show simple ribs. All have restricted range, and most are geographically widespread.

Columbites Hyatt & Smith, Triassic. Advolute shell having simple ceratite sutures. *C. parisianus* Hyatt & Smith (4a–c, ×0.7), Lower Triassic (*Columbites* zone), Idaho.

Lecanites Mojsisovics, Triassic. One of the few Triassic genera which possess goniatite sutures. *L. (Paralecanites) arnoldi* Hyatt & Smith (3a–c, ×0.7), Lower Triassic (*Meekoceras* zone), California and Idaho.

Meekoceras Hyatt, Lower Triassic. Shell somewhat more involute than that of *Columbites*; its suture is more advanced. *M. gracilitate* White (1a–c, ×0.7), *Meekoceras* zone, Idaho and California.

Ussuria Diener, Triassic. One of the few Lower Triassic ammonites. *U. waageni* Hyatt & Smith (2a, b), Lower Triassic (*Meekoceras* zone), Idaho. *U. iwanowi* (2c), eastern Siberia.

Permian. Ammonoids possessing goniatite sutures continued to flourish in Permian time, but this period is characterized especially by development of complex ammonite sutures in several stocks (Fig. 9-32).

Triassic. The goniatite type of suture virtually disappeared in the Triassic. While ammonite sutures distinguish many genera, the faunas of Lower and Middle Triassic deposits were dominated by ceratite ammonoids. The Lower Triassic ammonoids, like Paleozoic forms, are mostly smooth-shelled (Fig. 9-33), but in Middle and Upper Triassic rocks many genera are found which are complexly ornamented by ribs, nodes, and keels (Figs. 9-34, 9-35). The earliest examples of uncoiled and conispiral (trochoid) shells among ammonoids are known from Upper Triassic formations (Fig. 9-35, 2).

Jurassic. The ammonoids of Jurassic age all seem to have descended from a small number of conservative Triassic forms. Only shells having ammonite sutures are known, and highly ornamented shells are common. No loosely coiled or conispiral forms are known from the Lower Jurassic (Fig. 9-36), but some shells of this type appear in mid-Jurassic rocks (Fig.

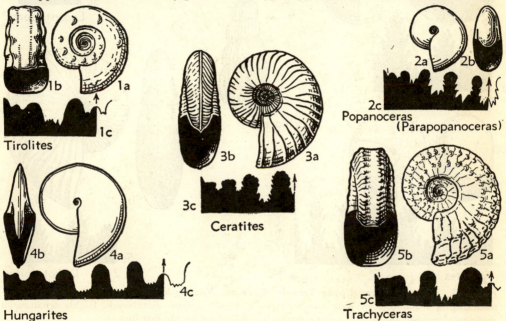

Fig. 9-34. **Representative Middle Triassic (Anisian and Ladinian) ammonoids.** Ceratite sutures predominate over ammonite types. The mid-Triassic ammonoid faunas differ from older ones chiefly in having more ornate shells, covered with ribs or tubercles.

Ceratites Haan, Middle Triassic. Primitive members have a typical ceratite suture, but more advanced species show sutures of the ammonite type. *C. (Gymotoceras) blakei* Gabb (3a–c, ×0.7), Nevada.

Hungarites Mojsisovics, Middle Triassic. Highly involute, sharp-keeled form having a finely serrate ceratite suture. *H. yatesi* Hyatt & Smith (4a–c, ×0.7), California.

Popanoceras Hyatt, Permian–Triassic. Possesses an ammonite suture. *P. (Parapopanoceras) haugi*

Hyatt & Smith (2a–c, ×0.7), Middle Triassic, California.

Tirolites Mojsisovics, Middle Triassic. Primitive ceratite which probably is ancestor of *Ceratites*. *T. pacificus* Hyatt & Smith (1a–c, ×0.7), Anisian, Nevada.

Trachyceras Laube, Middle and Upper Triassic. Strongly ribbed and noded. *T. (Anolcites) meeki* Mojsisovics (5a–c, ×0.7), Ladinian, Nevada; also known from the Alps.

9-37). The Upper Jurassic contains numerous genera which are not known in earlier rocks; among these are distinctively ornamented shells and forms bearing prominent auricles at the aperture (Fig. 9-38).

Cretaceous. Great diversity of ammonoid shells characterizes the Cretaceous system. The range of Early Cretaceous shells extends from involute to loosely spiraled and U-shaped ones (Fig. 9-39). Straight, conispiral, and irregularly contorted shells were added in the middle and late portions of the period (Figs. 9-40 to 9-42). All Lower Cretaceous ammonoids have ammonite sutures, though some of these show evidence of simplification from their ancestors. By Middle and Upper Cretaceous time, some ammonoids had

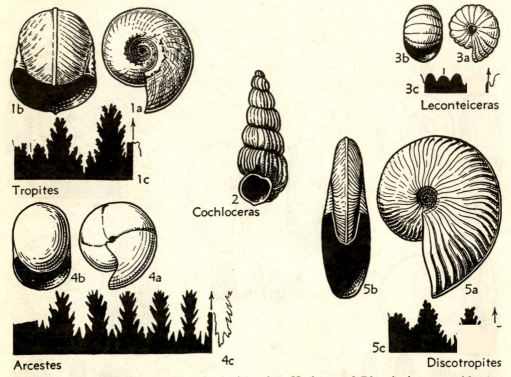

FIG. 9-35. **Representative Upper Triassic (Karnian, Norian, and Rhaetian) ammonoids.** Ammonite sutures dominate over the simpler types in Upper Triassic ammonoids. Highly ornamented shells are common, and uncoiled and conispiral shells appear for the first time.

Arcestes Suess, Upper Triassic. The suture is deeply dissected; the living chamber extends through more than one volution. *A. pacificus* Hyatt and Smith (4a–c, ×0.7), *Tropites subbullatus* zone, California.

Cochloceras Hauer, Upper Triassic. An aberrant ammonoid characterized by a strongly conispiral shell and goniatite suture. *C. fischeri* Hauer (2), Alps.

Discotropites Hyatt & Smith, Upper Triassic. Involute ammonoid ornamented by a strong keel and bifurcating ribs. *D. laurae* Mojsisovics (5a–c, ×0.7), *Tropites subbullatus* zone, California; also known from the Alps.

Leconteiceras Smith, Upper Triassic. Small globose shell ornamented by deep transverse furrows; the suture is unusual for an ammonoid of this age in that it is almost goniatite in nature, showing only incipient subdivision of lobes. *L. californicus* Hyatt & Smith (3a–c, ×1), *Tropites subbullatus* zone, California.

Tropites Hauer, Upper Triassic. One of the best-known index genera, marked by low, broad volutions and a well-defined keel. *T. subbullatus* Hauer, (1a–c, ×0.7), Hosselkus limestone, California; also known from the Mediterranean region and the Himalayas.

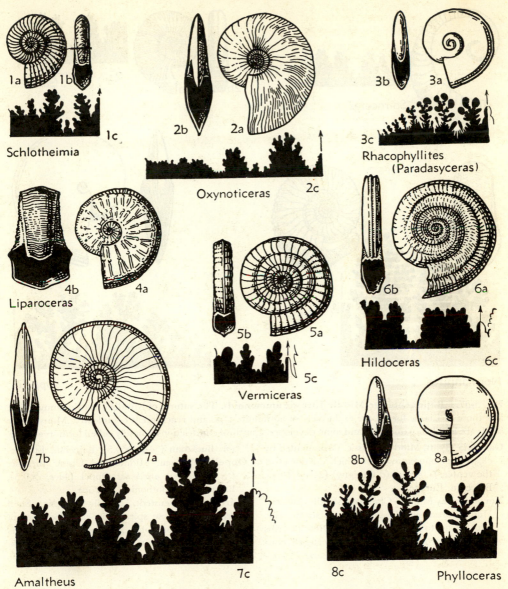

FIG. 9-36. **Representative Lower Jurassic (Liassic) ammonoids.** No goniatite or ceratite sutures are known from the Lias, but there is great diversity of ammonite sutures, ornamentation, and shell shape. No conispiral or unrolled forms have been discovered.

Amaltheus Montfort, Liassic. Distinctively ornamented with a corded keel, which projects beyond the aperture. *A. margaritatus* Montfort (7a–c, ×0.2), Domerian, Europe.

Hildoceras Hyatt, Liassic. The aperture of this shell is flanked by lateral crests, which leave their trace in the growth lines of the shell. Also distinctive is the keel bordered by grooves.

H. bifrons Bruguière (6a–c, ×0.3), Toarcian, western Europe.

Liparoceras Hyatt, Liassic. Highly ornamented form bearing bifurcating ribs and rows of tubercles. *L. henleyi* Sowerby (4a, b, ×0.3), Charmouthian, Europe.

Oxynoticeras Hyatt, Liassic. Characterized by

(*Continued on next page.*)

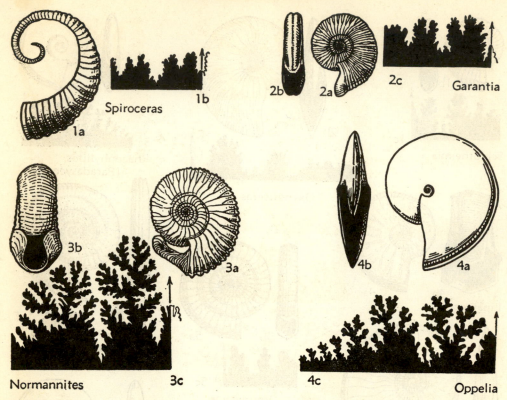

Spiroceras 1a 1b 2b 2a 2c Garantia

Normannites 3b 3a 3c 4b 4a 4c Oppelia

FIG. 9-37. Representative Middle Jurassic ammonoids. The sutures of Middle Jurassic ammonoids, like those of Liassic forms, are exclusively of ammonite type and reach great complexity. Many of the shells are highly ornamented, and some developed apertural constrictions. Loosely coiled forms reappear.

Garantia Hyatt, Middle Jurassic. Ornamented by closely set bifurcating ribs which do not cross the venter. *G. garanti* (d'Orbigny) (2a–c, ×0.2), Bajocian, Europe.

Normannites Munier–Chalmas, Middle Jurassic. Highly ornamented shell having strong ribs which lead to ventro-lateral tubercles; the ribs bifurcate across the venter. The aperture is constricted at maturity by two lateral ears. *N. orbi-*

gnyi Buckman (3a–c, ×0.5), Bajocian, Europe.

Oppelia Waagen, Middle and Upper Jurassic. *O. (Oxycerites) aspidoides* Oppel (4a–c, ×0.3), Bajocian and Bathonian, Europe.

Spiroceras Quenstedt, Middle Jurassic. Shell grows in a loose spiral, the early portions of which lie in one plane, whereas later portions become somewhat conispiral *S. bifurcatum* Quenstedt (1a, b, ×0.5), Bajocian, Europe.

(Fig. 9-36 continued.)

its discoidal shape, sharp-edged venter, and suture composed of many short lobes and saddles. *O. oxynotum* (Quenstedt) (2a–c, ×0.5), Sinemurian, Europe.

Phylloceras Suess, Triassic–Cretaceous. A close relative and descendent of *Rhacophyllites*. *P. heterophyllum* (Sowerby) (8a–c, ×0.3), Upper Liassic, Europe.

Rhacophyllites (*Paradasyceras*) Zittel, Triassic-

Liassic. Less involute and having a simpler suture than its descendant *Phylloceras*. *R. vermosense* Herbich (3a–c), Liassic, Europe.

Schlotheimia Bayle, Liassic. *S. angulata* Schlüter (1a–c, ×0.5), Europe.

Vermiceras Hyatt, Liassic. An advolute, discoidal shell, which has simple ribs and a keel. *S. spiratissimum* (Quenstedt) (5a–c, ×0.5), Sinemurian, central and southern Europe.

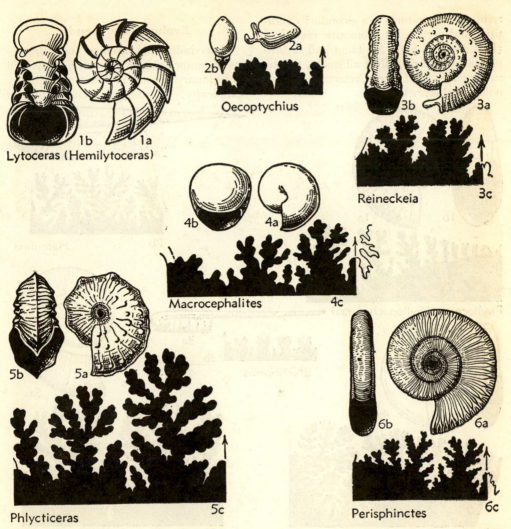

FIG. 9-38. **Representative Upper Jurassic ammonoids.** Upper Jurassic ammonoids continued to differentiate in ornamentation and shape, without deviating far from the closed spiral type of growth. All forms illustrated are European.

Lytoceras Suess, Lower Jurassic–Cretaceous. Strongly developed varices. *L. (Hemilytoceras) immane* Oppel (1*a, b,* ×0.7), Tithonian.

Macrocephalites Zittel, Middle and Upper Jurassic. Broad, highly convolute ammonoid which has a simple, somewhat constricted aperture. *M. (Kheraiceras) cosmopolita* Parona & Bonarelli (4*a–c,* ×0.3), Callovian.

Oecoptychius Neumayr, Middle and Upper Jurassic. An aberrant small shell, the last (mature) volution of which is subtriangular in shape; the aperture is surrounded by an extended ventral lip and a pair of curved

lateral ears. *O. refractus* Reinecke (2*a–c,* ×0.5), Callovian.

Perisphinctes Waagen, Upper Jurassic. An advolute shell ornamented by bifid and trifid ribs. *P. tiziani* Oppel (6*a–c,* ×0.5), Rauracian.

Phlycticeras Hyatt, Upper Jurassic. Highly ornate, its suture is very deeply dissected. *P. pustulatum* Reinecke (5*a–c,* ×0.5), Callovian.

Reineckeia Bayle, Middle and Upper Jurassic. A row of large, widely spaced tubercles occurs along each side; ribs extend toward the venter but do not cross it. *R. anceps* Bayle (3*a–c,* ×0.3), Callovian.

reduced their sutures to a secondarily ceratitic, and even nearly goniatitic condition
(Figs. 9-41, *1;* 9-42, *5*). The trend of simplification did not affect all stocks, for the
sutures of many Upper Cretaceous ammonoids are highly complex.

Evolutionary Trends

Many individual families and subfamilies of ammonoids exhibit clearly marked
evolutionary trends in change of shape,
complexity of suture, and elaboration of

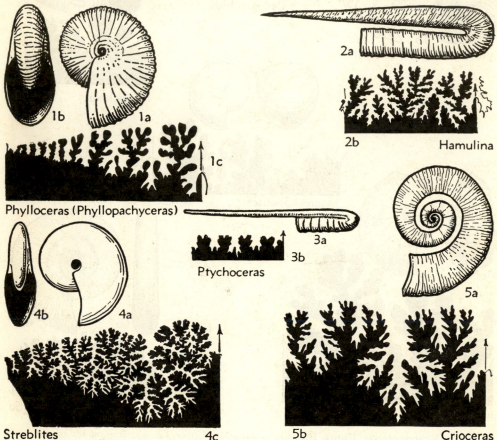

FIG. 9-39.   **Representative Lower Cretaceous (Neocomian) ammonoids.** In Early Cretaceous time
the diversity of ammonoids reached a new high—all gradations from highly involute forms to loose
spirals and U-shaped shells are present; sutures, all ammonitic, vary from exceedingly dissected to
some which are simplified.

Crioceras d'Orbigny, Lower Cretaceous. Shell
coiled in a loose spiral. *C. duvali* Léveillé
(5*a, b,* ×0.2), Hauterivian and Barremian,
French Alps.

Hamulina d'Orbigny, Lower Cretaceous. A
straight-chambered shell which, at maturity
turns abruptly 180 deg. to form a straight
mature living chamber. *H. astieri* d'Orbigny
(2*a, b,* ×0.2), Barremian, Europe.

Phylloceras Suess, Triassic–Cretaceous. A Liassic
species is shown in Fig. 9-36 *P.* (*Phyllopachyceras*)

infundibulum d'Orbigny (1*a–c,* ×0.5), Hauterivian and Barremian, Europe.

Ptychoceras d'Orbigny, Lower and Middle
Cretaceous (Albian). Resembles *Hamulina* in
form of growth, but differs in having a simplified
ammonitic suture. *P. emerici* d'Orbigny (3*a, b,*
×0.3), Barremian, Europe.

Streblites Hyatt, Upper Jurassic–Lower Cretaceous. Smooth, discoidal shell having extraordinarily dissected sutures. *S.* (*Uhligerites*)
kraffti Uhlig (4*a–c,* ×0.5), Tithonian and
Berriasian, Himalayas.

ornamentation. Examples from two evolutionary series are shown in Fig. 9-32. *Properrinites* (*1*), found in Wolfcampian rocks, is directly ancestral to *Perrinites* (*4*), characteristic of the overlying Leonardian series. A similar example, *Waagenoceras* (*6*), of Wordian and lower Capitanian age (subdivisions of the Guadalupian series), developed into the Capitanian *Timorites* (*7*) and to the even more complex *Cyclolobus* of uppermost Permian rocks. Such trends have been exceptionally useful to the stratigrapher, as well as in deciphering ammonoid relationships.

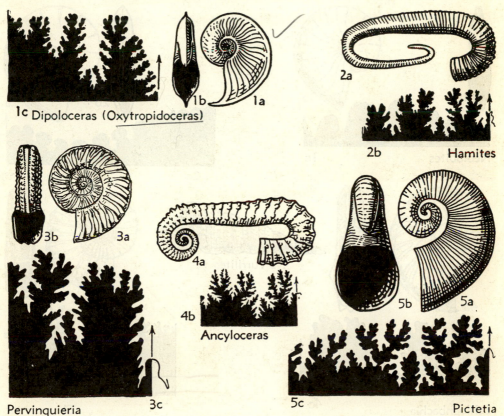

FIG. 9-40. Middle Cretaceous (Aptian and Albian) ammonoids. This late Mesozoic part of the column, approximately equivalent to Comanchean rocks of the United States, contains an ammonite fauna which shows great variation in shape and includes many highly ornate forms. Uncoiled shells flourished as never before. Many sutures show a new trend, toward simplification, which leads to ceratite sutures among some Aptian and Albian forms.

Ancyloceras d'Orbigny, Upper Jurassic–Middle Cretaceous. Shell begins as an open spiral, then grows straight, and finally turns 180 deg. to form the straight living chamber of the mature shell. *A. matheroni* d'Orbigny (*4a, b,* ×0.05), Aptian, Europe.

Dipoloceras Stieler, Albian. Characterized by a sharp flangelike keel. *D.* (*Oxytropidoceras*) *roissyi* d'Orbigny (*1a–c,* ×0.3), Albian, Europe.

Hamites Parkinson, Lower Cretaceous. Shell grows in repeated U-shaped loops, in one plane. *H. alternatus* Sowerby (*2a, b,* ×0.5), Gault, England.

Pervinquièria Böhm, Albian. One of the more common genera, in Europe and North America. *P. inflata* Sowerby (*3a–c,* ×0.3), Albian, Europe.

Pictetia Uhlig, Lower and Middle Cretaceous. Loosely coiled spiral shell having complex sutures. *P. astieri* d'Orbigny (*5a–c,* ×0.5), Gault, England.

One notable trend is the over-all increase in size. Most Paleozoic ammonoids are smaller than a golf ball, although a few reach the diameter of a saucer. Saucer-sized ammonoids are not uncommon in the Triassic, and larger forms occur. Jurassic ammonoids show greater average size, as well as a greater maximum. The Cretaceous has yielded the largest known ammonoids, which are discoidal forms 2 m. (6.5 ft.) in diameter. Small ammonoids continued to exist along with the larger ones throughout the Mesozoic.

Increasing complexity of sutures consti-

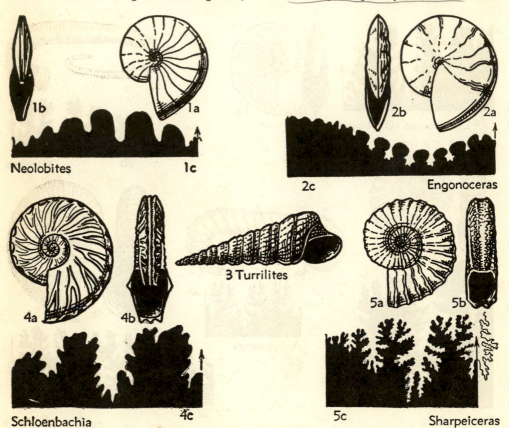

Fig. 9-41. **Representative Middle Cretaceous (Cenomanian) ammonoids.** The Cenomanian, which includes the lower part of the Upper Cretaceous or Gulfian of the United States, is especially noteworthy because of its content of numerous ammonoids which have simplified sutures.

Engonoceras Neumayr & Uhlig, Middle Cretaceous. A "pseudoceratite" having a distinctive suture composed of many bluntly inflated saddles and toothed lobes. *E. thomasi* Pervinquière (2a–c, ×0.3), Cenomanian, North Africa.

Neolobites Fischer, Cenomanian. Unique among Cretaceous ammonoids in having carried reduction of the suture to the goniatite stage. *N. vibrayi* d'Orbigny (1a–c, ×0.3), Mediterranean region.

Schloenbachia Neumayr, Cenomanian–Turonian. The shell has strong bifid or trifid ribs and possesses a simplified ammonitic suture. *S. varians* Sowerby (4a–c, ×0.5), Cenomanian, Europe.

Sharpeiceras Hyatt, Cenomanian. A highly ornate ammonite retaining a complex suture. *S. schlüteri* Hyatt (5a–c, ×0.3), Europe.

Turrilites Lamarck, Albian–Cenomanian. A high conispiral shell which has a complex ammonitic suture. *T. costatus* Lamarck (3, ×0.5), Cenomanian, France.

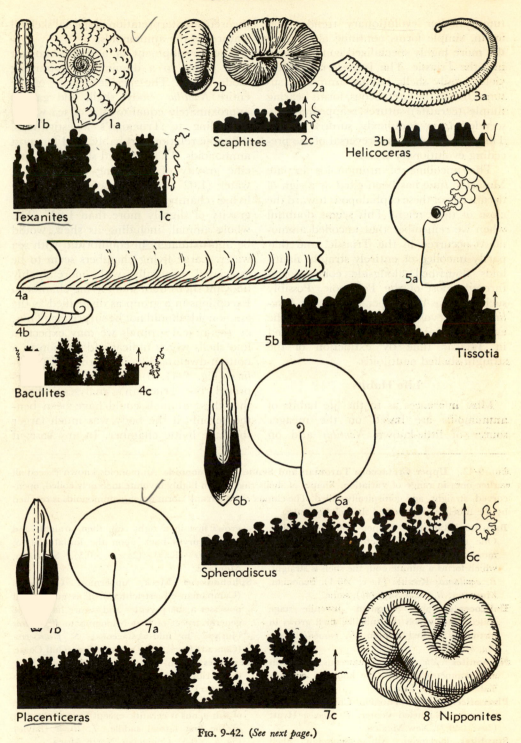

FIG. 9-42. (*See next page.*)

tutes another evolutionary trend. Here again, simple forms continue along with the more highly specialized complex ones into the Triassic. The Jurassic rocks have yielded only shells having ammonite sutures, but in the Cretaceous, forms bearing simple (ceratite) sutures reappeared, descended from complexly sutured stocks. This simplification is a reversal of the preceding evolutionary trend.

The uncoiling of ammonoids in late Mesozoic time has been cited as a sign of decadence of these cephalopods toward the close of their reign. This seems doubtful when we remember that uncoiled ammonoids occurred in the Triassic, and that partly uncoiled or entirely straight nautiloids flourished alongside coiled forms through most of the Paleozoic. Possibly such forms as the Upper Cretaceous *Baculites* were evolved in adaptation to the environmental domain which was vacated in Triassic time by extinction of the straight-shelled nautiloids.

Life Habits

Most inferences as to the life habits of ammonoids are based on the meager source of little-known *Nautilus* and on theoretical interpretation of special skeletal features of the ammonoids.

The development of a light, buoyant shell points to a dominantly swimming mode of life. The specific gravity of the entire *Nautilus* (body, shell, and gas) is approximately equal to that of sea water. According to Trueman's investigations, the same relations probably held for most ammonoids. The gas-filled shell has a specific gravity of something less than sea water (1.027), whereas soft parts in the living chamber should have a specific gravity of slightly more than 1.027. The whole animal, including its shell, would be approximately in equilibrium with sea water. Large living chambers seem to be associated generally with more highly buoyant shells than small living chambers. Exceptions in a group as diversified as the ammonoids should not be surprising. Thus, on theoretical grounds we may expect to find shells which indicate adaptation to a bottom-dwelling life. The trochoid *Turrilites* (Fig. 9-41, 3) has been thus interpreted, but Trueman's analyses indicate that these animals could have been benthonic only if the body was much larger than the living chamber. In any case, it

---

FIG. 9-42. **Upper Cretaceous Turonian and Senonian ammonoids.** Ammonoids known exceed all earlier ones in range of variation. Shapes of shells range from highly involute to loosely coiled, open-curved, straight, and conispirally coiled. The climax of "aberrant" form among ammonoids is reached by the tubelike, intertwined shell of *Nipponites*.

Baculites Lamarck, Upper Albian–Senonian. One of the most distinctive Cretaceous ammonoid genera. Except for the earliest portion, which forms a minute coil, the shell is straight. *B. aquilaensis* Reeside (4a–c, ×0.1), Senonian, Montana; *B. ovatus* Say, (4b), same.

Helicoceras Yabe, Senonian. Juvenile stage coiled, after which the tubelike shell grows in tortuously twisted fashion. *N. mirabilis* Yabe (3a, b, ×0.5).

Nipponites Yabe, Upper Cretaceous. Coiled in an asymmetrical, wormlike knot. *N. mirabilis* Yabe (8, ×0.5), Japan.

Placenticeras Meek, Senonian. Characterized by a flat or channeled venter. *P. planum* Hyatt (7a–c, ×0.2), New Mexico.

Scaphites Parkinson, Albian–Senonian. Shell grows first in a tight coil, then straight, and finally curves back upon the initial coil. *S. hippocrepis* (Kay) (2a–c, ×0.5), Senonian, Montana.

Sphenodiscus Meek, uppermost Cretaceous (Campanian–Maestrichtian). This guide fossil possesses a sharp venter and suture having the general aspect of that belonging to *Phylloceras* (Jurassic to mid-Cretaceous). *S. pleurisepius* (Conrad) (6a–c, ×0.7), Navarroan, Gulf Coast.

Texanites Spath, Senonian. Shell covered by thick ribs and blunt tubercles. *T. texanum* (Roemer) (1a–c, ×1), Austin chalk, Texas.

Tissotia Douvillé, lower Senonian. The suture of this genus is ceratitic except for a slight split in the first lateral saddle. *T. tissoti* (Bayle) (5a, b, ×0.3), Coniacian, North Africa.

seems probable that very irregular shells like *Nipponites* (Fig. 9-42, *8*) were bottom dwellers, and they may have grown in fixed position.

Most ammonoids probably were good swimmers, but their diversity of form suggests a wide variety of adaptations. Some may have spent part of their life on the bottom, others may have remained suspended in the water, drifting passively about with currents, and still others—the forms characterized by smooth, discoidal, streamlined shells—doubtless were energetic swimmers.

Fig. 9-43. **Structure of Megateuthis, a representative belemnoid.** The skeleton, less the apex of the epirostrum, is shown in partially cut-away side view and in cross section. Parts are explained in the aphabetical list, cross-indexed to the figure by numbers.

apical line (9). Minute tube or line which marks axis of rostrum.

asymptotic zones (2). Growth traces of sides of pro-ostracum on surface of conch; termed asymptotic because growth lines, more or less transverse on dorsum and venter, curve toward parallelism with the longitudinal axis.

clear layer (5). Unpigmented lamina of rostrum, which alternates with dark layers.

dark layer (6). Pigmented lamina in rostrum, which alternates with clear layers.

dorso-lateral furrow (11). Groove in dorso-lateral position on apical portion of rostrum; one of a pair.

embryonic rostrum (8). Beginning stage of rostrum, which encloses apex of phragmocone.

epirostrum (12). Long spear-shaped structure deposited over end of rostrum, and differing from it in shape and composition; central portion poorly calcified.

phragmocone (3). Chambered portion of shell.

pro-ostracum (1). Adoral projection of dorsal side of phragmocone, forming a protecting shield over visceral mass of the animal.

rostrum (4). Massive deposit of fibrous calcite, sometimes called guard, enclosing much of phragmocone and extending beyond its apex.

siphuncle (7). Tube located on ventral margin of phragmocone.

ventro-lateral furrow (10). Groove in ventro-lateral position on apical portion of rostrum; one of a pair.

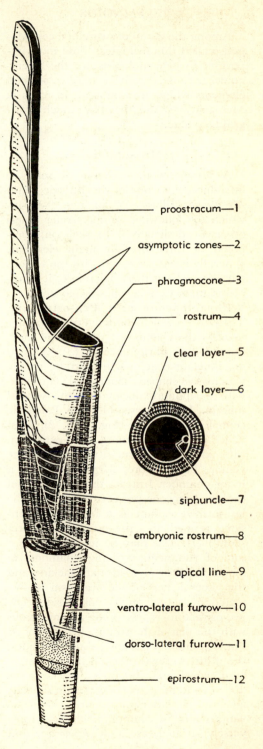

proostracum—1
asymptotic zones—2
phragmocone—3
rostrum—4
clear layer—5
dark layer—6
siphuncle—7
embryonic rostrum—8
apical line—9
ventro-lateral furrow—10
dorso-lateral furrow—11
epirostrum—12

BELEMNOIDS

Among the group of two-gilled cephalopods called dibranchiates, those most important to paleontologists are the belemnoids, because they have massive skeletal elements which are abundant in rocks of Mesozoic age. Also, they have much stratigraphic value.

Megateuthis

A survey of the belemnoids may be introduced by study of the Jurassic genus *Megateuthis* (illustrated in Fig. 9-43, to which the following numbers refer unless otherwise designated). Its skeleton, composed of three or four main elements, generally is not preserved with the various parts joined together. Therefore, it is appropriate to examine these separately.

Phragmocone and Pro-ostracum. The portion which will seem most familiar to us is the chambered part of the shell, termed the **phragmocone** (*3*). This is an elongate cone, partitioned by simple watch-glass-shaped septa and equipped with a siphuncle located at the ventral margin. It thus resembles the Paleozoic cephalopod *Bactrites* (Fig. 9-19, *2*), from which it differs by a more rapid expansion rate and a slight curvature toward the venter. Furthermore, *Megateuthis* lacks a living chamber. Instead, the dorsal part of the phragmocone projects far beyond the last-formed septum as a bladelike extension, termed the **pro-ostracum** ("foreshell") (*1*). This may be interpreted as a sort of rudimentary living chamber, in which the floor and sides are lacking and the visceral mass of the organism is protected by a roof on its dorsal side. The pro-ostracum is essentially homologous with the pen of the squids. Because of its blunt frontal margin and subparallel sides, the distinct growth lines of the pro-ostracum, which may be traced back to the apex of the phragmocone, divide the surface of the latter into four fields. The ventral and two lateral fields are transversely striated, as in *Michelinoceras* and

Bactrites. High up on the sides, the growth lines curve sharply forward, becoming hyperbolic (tangential) to the longitudinal axis of the conch, thus forming a pair of **asymptotic zones** (*a*, not; *symptotic*, falling together) (*2*). These mark former positions of lateral margins of the pro-ostracum. Between the asymptotic zones on the dorsal side are transverse lines having a gentle convexity forward; these growth lines denote successive positions of the front of the pro-ostracum. These markings make it possible to reconstruct the shape of the pro-ostracum from pieces of phragmocone, although the pro-ostracum itself is missing.

The phragmocone and pro-ostracum are exceedingly delicate structures. The former generally is crushed during compaction of the enclosing sediments unless protected by the massive guard or rostrum.

Rostrum. The most massive element of the belemnoid skeleton is the rostrum, a structure which seems to have no equivalent in the shells of nautiloids and ammonoids. It is the part most frequently preserved, and accordingly classification of belemnoids is based mainly on its variations. The rostrum of *Megateuthis* (*4*) is a subcylindrical structure, the rear of which tapers to a conical apex. The anterior two thirds of a mature rostrum encloses the phragmocone. The deep cavity in which the phragmocone is embedded is known as the **alveole.** The rostrum is composed of fibers of calcite, oriented at right angles to the surface. In addition to this radial structure, there is also a concentric structure of growth layers, suggestive of the growth rings of a tree trunk. Dark growth layers (*6*), containing a considerable admixture of organic matter, alternate with clear growth layers composed of nearly pure calcite. These may be seen both in transverse and longitudinal sections. They show the size and shape of the rostrum at each stage of growth. The fact that a mature rostrum carries a complete record of its ontogeny is an aid to classification of belemnoids.

From the tip of the phragmocone (base

of alveole) to the pointed extremity of the rostrum is a fine line which defines its axis, for calcite fibers radiate out from it on all sides to the surface. It is called the **apical line** (9). In *Megateuthis*, the line is approximately central; but in many other genera, it is excentric. The position of the apical line is useful in classification. For careful studies of the interior of belemnoids, it is necessary to prepare polished sections or thin sections, but the general structure of the interior and the position of the apical line may be determined readily by splitting the rostrum longitudinally. Owing to the fibrous structure of the rostrum, it nearly always breaks through the apical line.

The rostrum of *Megateuthis* is distinguished by a number of external longitudinal furrows located near the apex. *M. quinquesulcatus* carries five such grooves: one ventral, two ventro-lateral (*10*), and two dorso-lateral (*11*). In other species of the genus, the ventral and ventro-lateral grooves may be absent, and a dorsal groove may occur. There are no grooves in the mid-region or in the adoral portions of the rostrum.

Epirostrum. Like a few other Jurassic belemnoids, adult specimens of *Megateuthis* possess an additional skeletal structure, termed the **epirostrum** (*12*). This is a long, gently tapering rod extending backward (adapically) from the rostrum. Its hollow anterior part encloses the tip of the rostrum, in the same manner as the latter bears the phragmocone. The epirostrum differs from the rostrum both in shape and internal structure. It lacks differentiation of clear and dark growth layers. The interior commonly is poorly calcified and structureless, but the peripheral region is a hard calcareous cortex.

A mature skeleton, including epirostrum and pro-ostracum, has a maximum width of a few centimeters, but its length may reach 2 m.

Nonskeletal Features. Nonskeletal parts of belemnoid animals have been found preserved as fossils. These show that the arms, like those of many modern squids, were equipped with rows of strongly curved, sharp chitinoid hooklets (termed onychites) used for grasping and holding soft-bodied prey. None of these have been identified as belonging to *Megateuthis*, but probably this genus possessed them.

Exceptionally, arm hooks are found preserved in their original position in double rows, each pair indicating an arm. The occurrence of only eight double rows together in some fossils has been interpreted to indicate that belemnoids were eight-armed cephalopods, but impressions of 10 arms also are found. Accordingly, at least some belemnoids, like certain modern squids, probably bore hooks on eight arms and sucker disks on two.

Jurassic rocks of Europe contain some amazingly preserved belemnoids in which the arm hooks are well shown and the contents of the ink sac occur as a mass of blackish, fine-textured material, lying under the pro-ostracum.

Ontogeny. The skeleton of *Megateuthis*, like that of ammonoids and nautiloids, started with a protoconch, or first chamber. A short, blunt cone, comprising the earliest stage of the rostrum, was deposited over the rear wall of the globular protoconch, and new chambers were added in the other (adoral) direction, building the phragmocone. As the latter grew forward, successive conical layers were added to the rostrum, each extending the apex toward the rear and overlapping farther on the phragmocone toward the front. At a late stage in development of the skeleton, growth of the rostrum seemingly ceased, and the epirostrum was developed.

That the entire skeleton was internal is proved by the fact that all layers, from aragonitic bands at the frontal margin of the pro-ostracum to growth layers of the rostrum, were laid on from the outside.

Other Belemnoids

Most belemnoids of Permian and Triassic age (Fig. 9-44, *15, 16, 18*) resemble

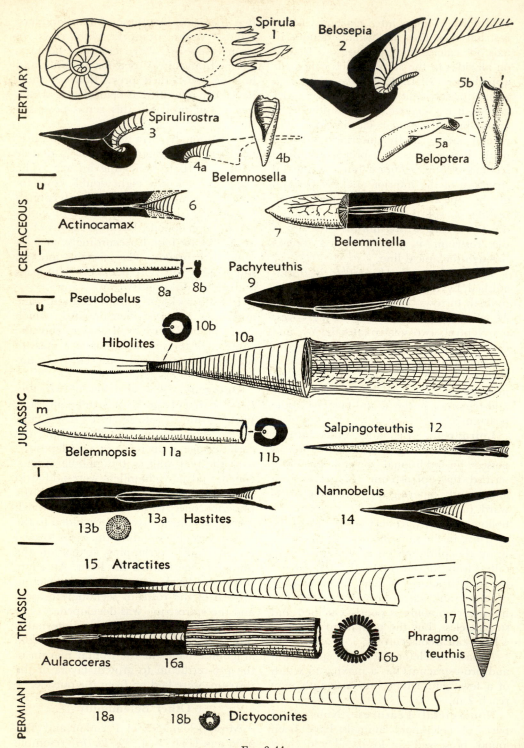

TERTIARY

Spirula
1

Belosepia
2

5b

Spirulirostra
3

4b

5a
Beloptera

4a

Belemnosella

CRETACEOUS

u

6

Actinocamax

7

Belemnitella

l

Pachyteuthis
9

Pseudobelus   8a   8b

u

Hibolites   10b

10a

JURASSIC

m

Belemnopsis   11a   11b

Salpingoteuthis   12

l

13b   13a   Hastites

Nannobelus
14

15   Atractites

TRIASSIC

17
Phragmo
teuthis

Aulacoceras   16a   16b

PERMIAN

18a   18b   Dictyoconites

Fig. 9-44.

Bactrites (Fig. 9-19, *2*) even more closely than does *Megateuthis*, as the phragmocones of these early forms expand at a smaller angle and possess longer chambers. Also, they lack any curvature and project far beyond the rostrum. The phragmocone of *Megateuthis*, which is relatively small and largely hidden in the alveole, is typical of most Jurassic and Cretaceous belemnoids.

The rostra of belemnoids vary from short conical sheaths over the phragmocone (Fig. 9-44, *14, 17*) to long, slender, club-shaped structures (Fig. 9-44, *10, 13*). They are variously shaped in cross section. Some forms have rostra ornamented by longitudinal ribs (Fig. 9-44, *16, 18*). In others, the rostrum is covered by branching grooves, representing impressions of blood

---

FIG. 9-44. Representative belemnoids and sepioids. In longitudinal sections, the rostrum is shown in black, the embryonic rostrum is outlined in white, and the epirostrum (12) is stippled. The pre-Tertiary genera illustrated are belemnoids, and Tertiary ones are sepioids. Approximately ×0.5 except as indicated otherwise.

Actinocamax Miller, Upper Cretaceous. Differs from other belemnoids in the abruptly terminated adoral end of the rostrum, instead of having a deep concavity for reception of the phragmocone. *A. granulatus* de Blainville (6), Upper Cretaceous, Europe.

Atractites Guembel, Triassic–Lower Jurassic. Resembles *Dictyoconites* but lacks ribs and furrows on the rostrum. *A. haueri* Diener (15), Triassic, Alps.

Aulacoceras Hauer, Upper Triassic. A rather blunt rostrum ornamented by coarse ribs. *A. timorense* Wanner (16*a, b*), Timor.

Belemnitella d'Orbigny, Upper Cretaceous. Rostrum possesses a short alveolar slit, a pair of dorso-lateral furrows, and impressions of blood vessels. *B. mucronata* Schloenbach (7), Maestrichtian, Europe.

Belemnopsis Bayle, Jurassic. Rostrum shows a deep ventral slit extending from the alveolar portion toward the apex. *B. bessinus* (d'Orbigny) (11*a, b*), Middle Jurassic, Europe.

Belemnosella Naef, Eocene. The phragmocone is slightly curved, and the rostral edges are bluntly rounded. *B. americana* (Mayer & Aldrich) (4*a, b*), Wautubbee marl, Mississippi.

Beloptera de Blainville, Eocene. Characterized by a blunt, club-shaped rostrum having edges which flare into short wings. *B. longa* Naef (5*a, b*), Eocene, Germany.

Belosepia Voltz, Eocene. A strikingly *Sepia*-like shell, but having open chambers and a short, tubular siphuncle. *B. sepioides* de Blainville (2), Eocene, Europe.

Dictyoconites Mojsisovics, Triassic. Long, straight, slender phragmocone and slender rostrum bearing two pairs of dorso-lateral furrows, as well as many fine longitudinal ribs.

D. groenlandicus Rosenkrantz (18*a, b*, ×1), Upper Permian, Greenland.

Hastites Mayer, Lower and Middle Jurassic. Rostrum club-shaped throughout growth. *H. clavatus* (Schlotheim) (13*a, b*), Liassic.

Hibolites Mayer Eymar, Middle Jurassic–Lower Cretaceous. The alveolar part of the rostrum is slit on the venter. *H. semisulcatus* (Münster) (10*a, b*), Upper Jurassic, Bavaria.

Nannobelus Pavlow, Lower Jurassic. Rostrum conical throughout growth. *N. acutus* (Miller) (14), Europe.

Pachyteuthis Bayle, Upper Jurassic–Lower Cretaceous. Possesses a slender juvenile rostrum, becomes blunt in older stabes; growth of the rostrum highly eccentric. *P. densus* (Meek) (9), Upper Jurassic (Sundance), Rocky Mountain region.

Phragmoteuthis Mojsisovics, Triassic. Characterized by crowded septa and a rudimentary rostrum which forms only a thin sheath over the phragmocone. *P. bisinuatus* (Bronn) (17), Wengener beds, Alps.

Pseudobelus de Blainville, Lower Cretaceous. Laterally compressed rostrum bears a dorsal furrow and pair of lateral furrows. *P. bipartitus* de Blainville (8*a, b*), Neocomian, Europe.

Salpingoteuthis Lissajous, Middle Jurassic. Conical rostrum abruptly succeeded by a long epirostrum. *S. acuarius* (Quenstedt) (12), Europe.

Spirula Lamarck, Miocene–Recent. The phragmocone is loosely coiled, and the rostrum wanting. *S. spirula* Hoyle (1), Recent.

Spirulirostra d'Orbigny, Miocene. The strongly curved phragmocone has tip covered by a pointed rostrum. *S. bellardii* d'Orbigny (3), Italy.

vessels (Fig. 9-44, 7). Many rostra bear longitudinal grooves or slits on the outer surface, and these features have been widely used in classification. The grooves may occur in pairs (dorso-lateral or ventro-lateral), or may be single (dorsal or ventral), restricted to the apical region (*Megateuthis*) or present in the adoral alveolar region (*Belemnitella*, Fig. 9-44, 7).

The rostra of Chitinoteuthidae (Fig. 9-45), recognized to date only from the Lower Jurassic of Germany, are uncalcified, chitinoid, nearly cylindrical rods.

On the basis of their early growth form, rostra are divisible into two types: those which begin as short cones, and those which begin as long, slender, or even needle-like structures.

Origin of Belemnoids

The most primitive belemnoids known seem to be the Jurassic Chitinoteuthidae (Fig. 9-45), which have a noncalcareous rostrum composed of organic matter. If these forms lacked a rostrum, they would be classified among straight-shelled nautiloids, possibly related to *Bactrites* (Fig. 9-19, 2a–c), which is known from Ordovician to Permian. Belemnoids, which are geologically older than the chitinoteuthids, as from Triassic, Permian, and perhaps even more ancient rocks, are more advanced in structure. This suggests possibility that mid-Paleozoic cephalopods like *Bactrites* connect the externally-shelled nautiloids and the internal-shelled belemnoids. The transition from one type of skeleton to the other offers no great problem, for many modern snails demonstrate various stages in the process. Some of these have shells which are intermittently covered by flaps of the mantle and others bear shells permanently overgrown by mantle tissues.

Function of the Belemnoid Shell

The shell of belemnoids evidently served a number of functions. The pro-ostracum protected the underlying visceral mass, provided internal support, and offered a basis for muscle attachment. The large phragmocones of early belemnoids (Fig. 9-44, 10, 15, 18) no doubt served an important hydrostatic function enabling the relatively large animals to suspend their weight in the water without undue muscular effort. The rostrum served as a counterweight, balancing the main part of the body, including the arms, located on the opposite side of the buoyant phragmocone. It also served as protection for the delicate tip of the phragmocone. Belemnoids, like squids, doubtless traveled backward at good rates of speed. Fractures and dislocations in rostra, healed over by later growth layers, record traffic accidents in the Mesozoic seas.

The function of the epirostrum (Figs. 9-43, 12; 9-44, 12), found only in two Jurassic genera, is quite conjectural. If it was a counterweight, it must have been developed in response to some special growth enlargement of the animal at maturity, which possibly may be correlated with reproduction.

The general organization of belemnoids indicates that very probably they resembled modern squids both in appearance and mode of life. Remains of true squids are found in rocks as old as Jurassic.

Geological History and Evolutionary Trends

A single belemnoid has been reported from Mississippian rocks of Oklahoma, and one species is abundant in the Permian of Greenland. Belemnoids characterized by straight, slender phragmocones projecting far beyond the rostrum were widespread in the Triassic, and lived also in Early Jurassic time where they are

FIG. 9-45. **Chitinoteuthis** Müller-Stoll. This illustrates the structure of the simplest known group of belemnoids, from Lower Jurassic rocks of Germany. The phragmocone is *Bactrites*-like; the rostrum is noncalcareous, and is preserved as a carbon film.

associated with forms possessing delicate, curved phragmocones which expand at a larger angle.

Belemnoids reached a climax in abundance of individuals, as well as in number of genera and species, during Jurassic and Cretaceous time. The phragmocone belonging to many of these belemnoids was relatively small and largely confined to the alveole. Forms having epirostra have been found only in Jurassic rocks.

The latest known belemnoids occur in the Eocene. Their shells are slender and their rostra longitudinally striate. They resemble the Permian and Triassic *Dictyoconites* and thus seem to be anachronisms.

SEPIOIDS

The sepioid cephalopods are named from the modern genus *Sepia*, the cuttlefish, which is common in warmer and temperate seas. Sepioids are represented by fossils in Tertiary rocks.

Sepia. In basic construction *Sepia* (Figs. 9-46, 9-47) resembles the squid *Loligo*. It differs chiefly in being flattened dorso-ventrally, and in possessing a markedly different skeleton. The shell (Fig. 9-47), familiar as the cuttlebone commonly supplied to canaries, is an internal, oval plate, which overlies and shields the visceral mass of the animal. It consists of several parts. A thin, dense **dorsal shield** terminates at the rear in a short spike, the **rostrum.** Below the shield is a thick, spongy mass of aragonite. In section, this is seen to be a modified **phragmocone,** composed of slanted, closely crowded septa and secondary septula, which are connected by many small pillars. Only the dorsum of the phragmocone is well developed. At the apical end, a small rim on the sides and venter represents the much-reduced sides and venter of the phragmocone, and the conical open space between them is a remnant of the siphuncle. A projection from the ventral side of the rostrum offers attachment for the tough

muscular mantle tissues which cover the underside of the animal.

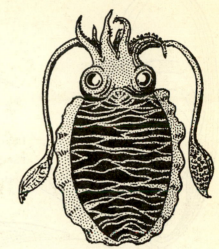

Fig. 9-46. **Sepia** Linné. Dorsal view of the common cuttlefish, ×0.3.

As might be expected from its flattened shape, *Sepia* is primarily a bottom dweller and spends much of its life lying in ambush for smaller animals. It can swim rapidly by jetting water, and is adept also at hovering and moving slowly forward or backward by wavelike motions of fleshy extensions of the body sides (Fig. 9-45). When pursued, it shoots forth a cloud of brownish inklike fluid termed sepia.

Spirula. Another, quite different animal, which is classified among sepioids, is the genus *Spirula* (Fig. 9-44, *1*). It is a small, free-drifting cephalopod of the deep sea, which possesses an internal coiled, chambered shell.

Fossil Sepioids. The remains of Tertiary dibranchiates are largely those of sepioids. These show progressive evolutionary trends: reduction in the size of the rostrum, increase in the curvature of the phragmocone (leading to the coiled *Spirula*), and flattening of the shell (leading to *Sepia* and its allies). The Tertiary sepioids have been judged to be descendants of the belemnoids. Discovery of a sepioid-like shell in Jurassic rocks of Cuba indicates that cephalopods of this group are

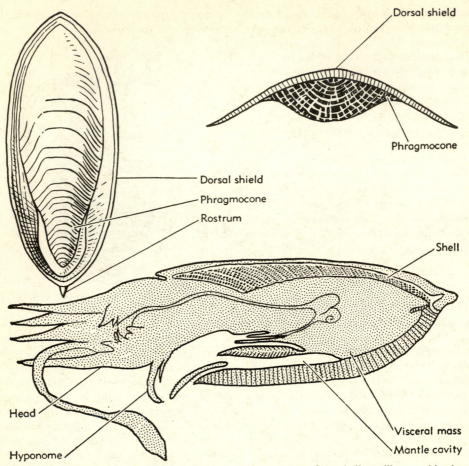

Fig. 9-47. **Structure of Sepia.** A ventral view and cross section of the shell are illustrated in the two upper drawings. The relationship of shell to body is shown by the lower drawing.

more ancient than was supposed, but it still is possible that the sepioids are derivatives of an early belemnoid stock.

TEUTHOIDS

The teuthoids include the true squids, of which one common representative (*Loligo*) has already been discussed. A majority of living cephalopod species are squids. They are exclusively swimming (nektonic) animals, and occur in all the oceans, from the surface to abyssal depths. Like deep-sea fishes, the deep-sea squids are studded with lights.

The skeletons of a number of teuthoid

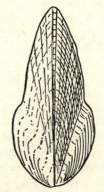

Fig. 9-48. **Beloteuthis.** The horny "pen" of a fossil squid (teuthoid), *Beloteuthis subquadrata* Münster, from Lower Jurassic rocks of Germany, ×0.1.

genera are found in Jurassic and Cretaceous rocks. They consist of horny "pens," which are perhaps analogous to the proostracum of belemnoids (Fig. 9-48).

Probably the teuthoids are an offshoot of the belemnoids which became differentiated in Triassic or earlier periods. It is also conceivable that they were derived independently from other shell-bearing cephalopods, or from some group of shell-less cephalopods of which we know nothing.

OCTOPOIDS

The common octopus has been discussed. Owing to the absence of skeletal structures, little is known about the geologic history of octopoids. The earliest known octopoid is an impression found in Upper Cretaceous rocks of Lebanon (Fig. 9-49).

The female of the nektonic genus *Argonauta* produces an exceedingly delicate calcareous brood pouch for transport of its eggs (Fig. 9-50). The exterior of this brood pouch strikingly resembles the shell of some Mesozoic ammonoids, but the interior differs in lacking septa and siphuncle. Fossil *Argonauta* shells have been found in the Pliocene.

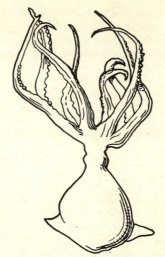

FIG. 9-49. **Palaeoctopus** Woodward, the oldest known octopoid, from Upper Cretaceous rocks of the Lebanon Mountains in Palestine. Short lateral fins occur at the rear of the body.

FIG. 9-50. **Argonauta** Linné. This planispiral, involute shell is secreted by the female of the swimming octopoid genus *Argonauta*. Pliocene of Italy.

REFERENCES

ARKELL, W. J. (1933) *The Jurassic system of Great Britain:* Oxford, New York, 1933, pp. 1–681, pls. 1–41, figs. 1–97. Includes much information on zonation by means of ammonoids.

BARRANDE, J. (1866–1877) *Système Silurien du centre de la Bohême,* partie 1, vol. 2, *Cephalopodes:* texte 1 (1867) pp. 1–712, texte 2 (1870) pp. 1–263, texte 3 (1874) pp. 1–804, texte 5 (1877) pp. 743–1505, pls. 108–244 (1866), 351–460 (1870) (French). The most voluminous work written on nautiloid cephalopods.

BÖHMERS, J. C. A. (1936) *Bau und Struktur von Schale und Sipho bei permischen Ammonoidea:* Drukkerij Universitas, Apeldoorn, Netherlands, pp. 1–125, pls. 1–2 (German). A detailed study of internal structure.

BRANCO, W. (1879–1880) *Beiträge zur Entwicklungsgeschichte der fossilen Cephalopoden.* Theil 1: *Die Ammoniten:* Palaeontographica, vol. 26,

pp. 19–50, pls. 4–13 (1879). Theil 2: *Die Goniatiten, Clymenien, Nautiliden, Belemnitiden und Spiruliden, nebst Nachtrag zu Theil 1:* Same vol. 27, pp. 17–81, pls. 4–11 (German). Ontogenetic studies, particularly of suture development.

FLOWER, R. H. (1939) *Study of the Pseudorthoceratidae:* Palaeontographica Americana, vol. 2, pp. 1–198, pls. 1–9, figs. 1–22.

——— (1941) *Notes on structure and phylogeny of eurysiphonate cephalopods:* Same, vol. 3, pp. 1–56, pls. 1–3, figs. 1–3. Revision of early nautiloids based on thin-section study of siphuncles.

——— (1941) *Development of the Mixochoanites:* Jour. Paleontology, vol. 15, pp. 524–548, pls. 76–77, figs. 1–32. A study of ascoceroid evolution.

——— (1946) *Ordovician cephalopods of the Cincinnati region, part 1:* Bull. Am. Paleontology, no. 116, pp. 1–656, pls. 1–50, figs. 1–22. Deals with nautiloid morphology in general and with description of a great nautiloid fauna, exclusive of trocholitids, michelinoceroids, endoceroids, and ascoceroids.

——— (1947) *Holochoanites are endoceroids:* Ohio Jour. Sci., vol. 47, pp. 155–172, figs. 1–3.

——— & KUMMEL, B. (1950) *A classification of the Nautiloidea:* Jour. Paleontology, vol. 24, pp. 604–616, fig. 1. Recognizes 75 families and 14 orders of nautiloids.

FOERSTE, A. F. (1926) *Actinosiphonate, trochoceroid, and other cephalopods:* Denison Univ. Sci. Lab., Bull., vol. 21, pp. 285–384, pls. 22–53. A revision of these peculiar nautiloids.

HYATT, A. (1883) *Genera of fossil cephalopods:* Boston Soc. Nat. History Proc., vol. 22, pp. 253–338. An early attempt at a comprehensive classification of cephalopods. Many of the generic names of nautiloids and ammonoids date back to this paper.

——— (1900) *Cephalopoda:* in Zittel, K. A., Textbook of paleontology (ed. by Eastman, C. R.) Macmillan & Co., Ltd., London, vol. 1, pp. 583–689, figs. 1098–1336.

MILLER, A. K. (1932) *Devonian ammonoids of North America:* Geol. Soc. America Spec. Paper 14, pp. 1–262, pls. 1–38, figs. 1–41.

——— (1947) *Tertiary nautiloids of the Americas:* Geol. Soc. America Mem. 23, pp. 1–234, pls. 1–100, figs. 1–30.

———, DUNBAR, C., & CONDRA, G. E. (1933) *The nautiloid cephalopods of the Pennsylvanian System in the mid-Continent region:* Nebraska Geol. Survey Bull. 9, ser. 2, pp. 1–240, pls. 1–24, figs. 1–32.

——— & FURNISH, W. M. (1940) *Permian ammonoids of the Guadalupe Mountain region and adjacent areas of Texas:* Geol. Soc. America Spec. Paper 26, pp. 1–242, pls. 1–44, figs. 1–59. Description of the most extensive Permian ammonoid faunas known from North America.

MOJSISOVICS VON MOJSVAR, E. (1882) *Die Cephalopoden der mediterranen Triasprovinz:* Abh. geol. Reichsanstalt Wien, vol. 10, pp. 1–317, pls. 1–94 (German). A monograph of the great Alpine Triassic ammonoid faunas.

MÜLLER-STOLL, H. (1936) *Beiträge zur Anatomie der Belemnoidea:* Nova Acta Leopoldina, n.s., vol. 4, no. 20, pp. 159–226, pls. I–XIII, 57–69, figs. 1–5 (in German).

NAEF, A. (1922) *Die fossilen Tintenfische:* Carl Fischer, Jena, pp. 1–322, figs. 1–100. The most comprehensive work on fossil dibranchiates.

PLUMMER, F. B., & SCOTT, G. (1937) *Upper Paleozoic ammonites in Texas:* The Geology of Texas, vol. 3, Texas Univ. Bull. 3701, pp. 1–516, pls. 1–41.

REESIDE, J. B. (1927) *The cephalopods of the Eagle sandstone and related formations in the Western Interior of the United States:* U.S. Geol. Survey Prof. Paper 151, pp. 1–40, pls. 1–45. One of the more comprehensive works on American Upper Cretaceous ammonoids.

ROMAN, F. (1938) *Les ammonites jurassiques et crétacées. Essai de genera:* Masson et Cie, Paris, pp. 1–554, pls. 1–53, figs. 1–496 (French). A most important summary of Jurassic and Cretaceous ammonoids, in which all genera (in a broad sense) are illustrated and described. An extensive bibliography is included.

SCHINDEWOLF, O. H. (1937) *Zur Stratigraphie und Paläontologie der Wocklumer Schichten (Oberdevon):* Preuss. geol. Landesanstalt Abh., no. 178, pp. 1–132, pls. 1–4, figs. 1–26. A study of clymenid ammonoids, including some triangularly coiled forms.

SMITH, J. P. (1914) *Mid-Triassic invertebrate faunas of North America:* U.S. Geol. Survey Prof. Paper 83, pp. 1–254, pls. 1–99.

——— (1927) *Upper Triassic marine invertebrate faunas of North America:* Same, Prof. Paper 141, pp. 1–262, pls. 1–111.

——— (1932) *Lower Triassic ammonoids of North America:* Same, Prof. Paper 167, pp. 1–111, pls. 1–81.

SPATH, L. F. (1933) *The evolution of the Cephalopoda:* Biol. Rev. vol. 8, pp. 418–462. Spath has also written important monographs on Jurassic and Cretaceous ammonite faunas.

TEICHERT, C. (1933) *Der Bau der actinoceroiden Cephalopoden:* Palaeontographica, vol. 78, pp. 111–230, pls. 8–15. A monograph of the actinoceroids (German).

TRUEMAN, A. E. (1941) *The ammonite body chamber, with special reference to the buoyancy and mode of life of the living ammonite:* Geol. Soc. London Quart. Jour., vol. 96, pp. 339–383, figs. 1–17.

ULRICH, E. O., FOERSTE, A. F., & MILLER, A. K. (1943) *Ozarkian and Canadian cephalopods, part 2: Brevicones:* Geol. Soc. America Spec. Paper 49, pp. 1–240, pls. 1–70, figs. 1–15.

———, ———, ——— & FURNISH, W. M. (1942) *Ozarkian and Canadian cephalopods, part 1: Nautilicones:* Same, Spec. Paper 37, pp. 1–157, pls. 1–57, figs. 1–23.

———, ———, ———, & UNKLESBAY, A. G. (1944) *Ozarkian and Canadian cephalopods, part 3: Longicones and summary:* Same, Spec. Paper 58, pp. 1–226, pls. 1–68, figs. 1–9. These three monographs by Ulrich, Foerste, Miller, *et al.*, represent a comprehensive study of Lower Ordovician cephalopod faunas.

WEDEKIND, R. (1918) *Die Genera der Palaeoammonoidea* (*Goniatiten*)*:* Palaeontographica, vol. 62, pp. 85–184, figs. 1–54 (German). A summary of goniatite genera.

——— (1935) *Einführung in die Grundlagen der historischen Geologie.* Vol 1, *Die Ammoniten-, Trilobiten-, und Brachiopodenzeit:* Enke, Stuttgart, pp. 1–109, pls. 1–24, figs. 1–17 (German). Contains much information on ammonoid evolution, especially the development of sutures and ornamentation.

CHAPTER 10

Pelecypods

The pelecypods are a division of mollusks which stands rather well apart from others in having a skeletal covering that consists of two calcareous valves. These are joined by a hinge on the dorsal side of the animal along a line paralleling the front-to-rear axis. Thus, one valve encloses the right side and the other the left side of the pelecypod. Other shell-bearing mollusks are all univalves (gastropods, cephalopods, scaphopods), except for the small group of chitons, which have several calcareous plates arranged transversely to the axis of the animal. Like the chitons and some of the oldest unspecialized gastropods, pelecypods are primitive in having gills located in a posterior mantle cavity. Also, the pelecypods are primitive in their general bilateral symmetry and relatively simple digestive, circulatory, and nervous systems, but they are specialized in lacking a dis-

tinct head and in the elaboration of gill structure found in most of them.

The name pelecypod signifies "hatchet foot," in reference to the shape of the fleshy muscular antero-ventral part of the body—the so-called foot—which may project outward when the valves are open. The group is also known as lamellibranchs, in allusion to the common sheetlike or lamellar form of the gills. The name clam may be applied to all pelecypods. Many fresh-water and a few marine clams (for example, *Mytilus*) are called mussels. Scallops and oysters are special sorts of marine clams.

Most pelecypods are bottom-dwelling aquatic invertebrates, among which the vast majority live in shallow marine waters. Unlike pulmonate gastropods, none are air breathers. Although a few pelecypods, such as the scallop, *Pecten*, and the so-called file shell, *Lima* (Fig. 10-1, *1*, *3*), can propel

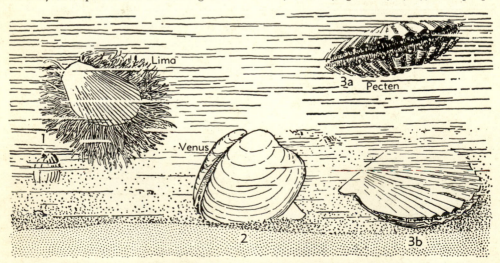

FIG. 10-1. **Mobile types of pelecypods.** Most so-called "normal" clams, like *Venus* (2), slowly crawl about on the floor of the water body in which they live, using the foot as an organ of locomotion. The beaks point upward and forward. A few pelecypods, such as *Lima* (1) and *Pecten* (3), are able to swim.

themselves rapidly through the water for short distances, nearly all others are sluggishly moving bottom dwellers, or they live firmly fixed in one place. Those that crawl about or burrow into bottom sediments are able to move by extending the foot outward and forward between the slightly opened valves, expanding the tip of the foot in such a way as to gain anchorage and then draw the body along by contracting the foot muscles. In this way, many clams can dig rapidly, and some (such as *Macoma*, Figs. 10-2, *3;* 10-27, *10*) use their shell as an aid in cutting into mud or sand by moving the body back and forth sideways. Some clams can travel slowly with nearly all the shell above the bottom sediment, the dorsal edge uppermost (Fig. 10-1, *2*); or they can burrow into the sediment in any direction so that the shell is varyingly oriented and entirely concealed (Fig. 10-2, *1–3*). Many burrowers do not move about but live in a fixed position, generally with the front end pointed downward and the rear upward. Oysters (Fig. 10-3, *4*) and several other sorts of clams also live in a fixed

position but do not bury themselves in sand or mud. They are attached by the exterior of one of the valves, which may be firmly cemented to rock or shells on the sea bottom; or they may fasten themselves tightly in place by means of a horny secretion (termed **byssus**) made by the foot. The byssus consists of a short solid cylinder of tough ligamentous threads or of long silky fibers, which extend outward from between the valves (Fig. 10-3, *2, 3*). Some clams can bore deeply into wood or solid rock, so as to live in self-inflicted imprisonment (Fig. 10-2, *4*).

The pelecypods range in size from adults barely 1 mm. in length to fossil forms having a shell nearly 1 m. (3.3 ft.) wide and 1.5 m. (5 ft.) long (*Haploscapha,* Upper Cretaceous, Kansas). The largest modern clam is *Tridacna,* of the South Pacific and Indian Oceans (Fig. 10-3, *1*). It secretes a coarsely corrugated shell as much as 90 cm. (3 ft.) in length and weighing more than 600 lb., although the soft parts of such a large clam are reported to weigh only 25 lb. Most pelecypods are an inch or two in length and width.

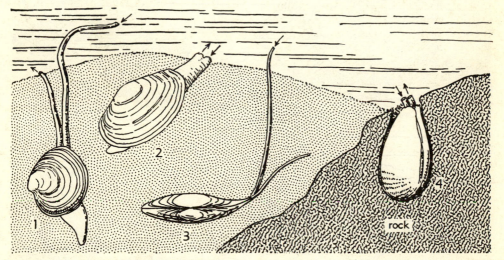

Fig. 10-2. **Burrowing types of pelecypods.** A majority of this group of clams burrows in the soft mud, silt, or sand beneath shallow seas or on the floor of fresh-water bodies. They may have long siphons which are separate from one another, as in *Scrobicularia* (1) and *Macoma* (3), or somewhat short and thick siphons bound together as in *Mactra* (2). A rock borer is illustrated by *Pholas* (4). The incurrent and excurrent siphons are marked by arrows pointing in the direction of the water movement (not drawn to scale).

ANATOMICAL FEATURES

Soft Parts

It is appropriate first to examine the organization of the soft parts of pelecypods because commonly these are reflected in the features of the shell. Also, they have importance in the classification of this group of mollusks.

The body of most clams is laterally compressed and elongated from front to rear (Fig. 10-4, *1–3*). It is enclosed by a membranous layer termed the **mantle** (Fig. 10-4, *4*), which contains cells that function in secreting the calcareous substance of the shell. The two halves of the mantle are joined together except along parts of the anterior, ventral, and posterior margins; an antero-ventral opening allows protrusion of the fleshy muscular organ termed the **foot.** Posterior openings lead into the mantle cavity which contains

the **gills.** The posterior part of the mantle may be more or less extended in the form of tubes called **siphons.** One of these nearest the ventral margin of the shell serves for the inflow of water containing oxygen and microscopic food particles. The other, nearest the dorsal margin, serves for the outflow of deoxygenated water containing body waste products (Fig. 10-4, *3, 4*). In some clams (such as *Lima* and *Pecten*, Fig. 10-1, *1, 3*), the edges of the mantle may project as filaments around all of the shell margin except the hinge line.

The foot forms the median ventral and anterior part of the body. It can be extended by forcing blood into cavities within it. Retraction of the foot is effected by evacuation of the blood spaces and by the action of muscles which generally have independent attachment to the interior of the shell. Other muscles, which run transverse to the body axis from

FIG. 10-3. **Sedentary types of pelecypods.** Many pelecypods remain fixed in one location throughout most of their life, anchoring themselves by adhesive threads (byssus) which may be hardened by calcification, or by cementation of the shell to foreign objects. Sedentary clams illustrated in this drawing include *Tridacna* (1), the slipper clam *Mytilus* (2), a long-eared *Pteria* (3), a group of oysters, *Ostrea* (4), and the spiny *Spondylus* (5).

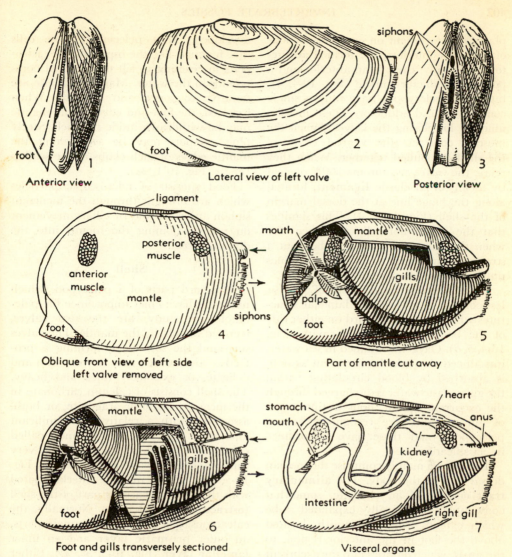

FIG. 10-4. Anatomical features of a pelecypod. The form illustrated is the common fresh-water mussel *Anodonta*, approximately natural size. The top figures show the animal in its shell, as in life, and the lower ones show various features of internal anatomy.

1, Front view of shell showing bilateral symmetry of the two valves, which normally are slightly open, permitting the foot to protrude.

2, Side view from the left, showing foot extended in the anterior ventral region and the siphons projecting at the rear.

3, Posterior view, showing the larger incurrent siphon, below, and the smaller excurrent siphon, above.

4, Oblique view with one valve removed, showing ligament and internal soft parts. Ends of the two adductor muscles, which run transversely from attachment scars on the inner sides of

the two valves, are seen. The soft parts are enclosed by a membranous covering, called the mantle, which functions also in secreting the shell.

5, View with mantle partly removed to show one of the two pairs of gills; the mouth, bordered by palps, which convey microscopic food particles to it; and part of the foot.

6, View similar to 5, but with central section of foot and gills removed to reveal their structural arrangement.

7, Diagrammatic representation of the alimentary tract and associated internal organs.

places of attachment on opposite portions of the valve interiors, are termed **adductors.** Normally, there is an anterior and a posterior adductor muscle (Fig. 10-4, *4*), but many pelecypods have only a single centrally placed adductor. These muscles function for closing the valves. When the muscles contract, the ventral edges of the shell are pulled together. When they relax, the valves are automatically opened by means of an elastic **ligament,** located along the hinge line at the dorsal margin of the shell (Fig. 10-4, *4*). This signifies that the valves normally are open; for when the adductors contract, they put a strain on the ligament, which continues until muscular effort ceases.

Gills of the pelecypod consist of more or less complicated leaflike or lamellar structures which hang downward on either side of the body in the mantle cavity (Fig. 10-4, *5, 6*). Oxygen contained in sea water, introduced through the incurrent siphon, is absorbed by blood circulating within the gills; waste water is discharged through the excurrent siphon. Four distinct types of gill structure are recognized among pelecypods, and these have importance in classification.

The viscera mainly comprise the median dorsal part of the body. The **alimentary tract** consists of an elongate, somewhat convoluted and locally expanded tube which extends from the **mouth,** located above the foot in the anterior region, to the **anus,** which empties into the excurrent siphon in the dorsal posterior region (Fig. 10-4, *7*). A digestive gland surrounds the stomach portion of the alimentary tract, and there is also a kidney, which functions as an excretory organ. The **circulatory system** consists of a three-chambered heart and blood vessels which connect with blood cavities in the foot, supply capillaries in the mantle, and provide for aeration of blood in the gills. The **nervous system** consists of a looped arrangement of nerve cords connecting ganglia located near the mouth, in the foot, and in the visceral region. Sense organs are poorly

developed. Most pelecypods have cells that function like the osphradia of gastropods in reacting to silt or chemical compounds in the water. Many clams also are sensitive to touch and to light. When irritated, the foot and edges of the mantle may be withdrawn inside the shell. Steely blue eyespots may be seen along the mantle edges of such clams as the scallop (*Pecten*, Fig. 10-1, *3a*).

Food consists of microscopic particles which are brought through the incurrent siphon and conveyed by ciliary movement on palps adjoining the mouth into the digestive tract.

Shell

The hard parts of a pelecypod, which naturally have chief importance for paleontological study, are the two valves, secreted mainly by the mantle. The valves surround the soft parts, forming a protective cover which may be very thin and delicate, or extremely thick and heavy. The shell consists of calcium carbonate in the form of calcite or aragonite, or both, associated with a relatively insignificant amount of dark organic substance, called conchiolin. The conchiolin forms a very thin outermost cover (**periostracum,** Fig. 10-5, *33*), and it may be interlaminated with the main calcareous part of the shell (**ostracum,** Fig. 10-5, *36*). Normally, the calcareous shell consists of two main parts, an outer **prismatic layer** and an inner **lamellar layer** (Fig. 10-5, *34, 35*). Either of these parts may dominate or be developed to the exclusion of the other. The minute calcite prisms of the outer ostracum are arranged normal to the shell surface. Where this layer is thick, as in some Cretaceous clams, like *Inoceramus* (Fig. 10-19, *15*), it may have a decidedly fibrous appearance; and in weathering, the shell may disintegrate into fine needle-like fragments. The lamellar layer consists of thin sheets of calcite or aragonite, in some shells intergrown with microscopically thin laminae of conchiolin. Very thin uniform lamellae of aragonite produce what is called the nacreous structure which has a pearly luster. This is seen in the shells of many living species and may be preserved in fossils. Still another shell element, which is significant, although quantitatively rather unimportant, consists of calcite deposited in the areas of muscle attachment. This shell material, which is not secreted by the mantle, is distinct from the lamellar layer and is termed **hypostracum.**

Morphological Features. The form of pelecypod shells varies widely; but except in highly specialized types, one may readily distinguish the **hinge line** (9; italic numbers in the text on morphological features refer to Fig. 10-5) from other shell margins, which in life are slightly separated from one another. The **ventral** (7) margin is opposite the hinge line, the **posterior** (29) at the extremity where the siphons protrude, and the **anterior** (5) opposite the posterior, generally at the end of the shell toward which the **beak** (2) points. Along the hinge line and generally externally visible is the **ligament area** (26), and in various shells are differentiated tracts termed **lunule** (10), **escutcheon** (11), and **cardinal area** (40). When the shell is viewed with the anterior end (generally with beaks pointed away from the observer), the **right valve** (20) is on the right-hand side and the **left valve** (43) on the left-hand side. Except for internal features such as **hinge teeth** (14), one valve is normally a mirror image of the other. Among some shells, such as the scallops, however, which have nearly identical anterior and posterior portions, identification of right and left valves requires notice of such things as asymmetry of the extensions along the hinge line called **auricles** (17), location of the **byssal notch** (18), **byssal sinus** (42), and **auricular sulcus** (44), all of which are anterior, or the position of **muscle scars** (19) on the posterior side of the shell interior in order to determine orientation. External features which are readily visible on many pelecypod shells are the **umbo**

(8), comprising an area of strong convexity which may extend in ridgelike manner from the beak; **growth lines** (6), which run concentrically parallel to free margins of the shells; and radially disposed ribs, formed by localized thickening or corrugations of the shell (**plicae,** 21; **costae,** 22). Characters of the shell commonly used in description include **length** (4), measured between the anterior and posterior extremities; **height** (3), measured from the dorsal to ventral margins; and **thickness** (12), consisting of the greatest dimension normal to the plane of union between the two valves (called **plane of commissure,** 13). When pulled together, the valves of some clams do not meet all around their margins, for an opening (**gape,** 47) is left for the passage of the siphons which cannot be withdrawn into the shell. This type of shell chiefly characterizes those having a burrowing or boring mode of life.

The first formed embryonic shell, called **prodissoconch** (37), can be distinguished at the tip of the beak of many pelecypods. Growth lines on the surface of the shell formed during juvenile, intermediate, and adult stages are parallel to margins of the prodissoconch.

Features of the interior of pelecypod shells may be grouped according to their location along the hinge line or in other parts of the interior. Structures include a varyingly prominent **hinge plate** (27), which bears teeth and sockets for articulation of the valves—**cardinal teeth** (28) and **sockets** (16) directly below the beaks and **lateral teeth** (15) and sockets along the hinge plate in front and behind the beaks. Among some clams, the hinge plate bears a variable number of small similar teeth which are not differentiated into cardinals and laterals. Also, the ligament may be partially or entirely internal; and its elastic substance is thus compressed instead of being put under tension when the valves are closed by the action of the muscles. This internal ligament, called **resilium** (23), rests in a

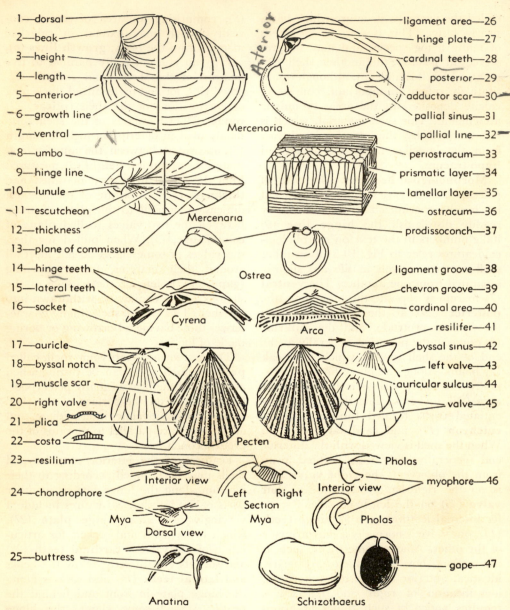

1—dorsal
2—beak
3—height
4—length
5—anterior
6—growth line
7—ventral
8—umbo
9—hinge line
10—lunule
11—escutcheon
12—thickness
13—plane of commissure
14—hinge teeth
15—lateral teeth
16—socket
17—auricle
18—byssal notch
19—muscle scar
20—right valve
21—plica
22—costa
23—resilium
24—chondrophore
25—buttress

26—ligament area
27—hinge plate
28—cardinal teeth
29—posterior
30—adductor scar
31—pallial sinus
32—pallial line
33—periostracum
34—prismatic layer
35—lamellar layer
36—ostracum
37—prodissoconch
38—ligament groove
39—chevron groove
40—cardinal area
41—resilifer
42—byssal sinus
43—left valve
44—auricular sulcus
45—valve
46—myophore
47—gape

Mercenaria
Mercenaria
Ostrea
Cyrena
Arca
Pecten
Pholas
Interior view
Left Right
Section
Mya
Mya
Dorsal view
Anatina
Interior view
Pholas
Schizothaerus
Anterior

Fig. 10-5. **Morphological features of pelecypod shells.** The illustrated parts are defined under alphabetically arranged terms, accompanied by numbers as cross index.

adductor scar (30). Impression on inside of valve made by attachment of muscle which functions for closure of valve.

anterior (5). Part of shell containing mouth; beaks of most pelecypods point forward.

auricle (17). Forward or backward projection of shell along hinge line in some pelecypods; also called ear.

auricular sulcus (44). Groove on shell exterior separating auricle from remainder of valve.

beak (2). More or less sharp-pointed projection at the initial point of shell growth, located along or above hinge line.

buttress (25). Ridge on inner surface of a valve which serves as support for part of hinge.

byssal notch (18). Indentation on anterior edge of some shells for protrusion of threadlike attachment called byssus; most common in

(Continued on next page.)

(Fig. 10-5 continued.)

pectinoid shell on right valve, which is lowermost, allowing protrusion of the small foot without opening valve widely.

byssal sinus (42). Indentation beneath auricle of left valve of pectinoid shells on anterior margin.

cardinal area (40). Plane or curved surface between beak and hinge line, generally distinguished from remainder of valve exterior by sharply angulated border.

cardinal teeth (28). Projections vertical or oblique to hinge line directly beneath or closely adjacent to beak; they fit into sockets of opposite valve.

chevron groove (39). Narrow depressions on cardinal area having an inverted V shape, marking ligament attachments.

chondrophore (24). Relatively prominent internal spoon-shaped structure, which holds an internal ligament (resilium).

costa (22). Radial ridge on shell surface formed by thickening of shell.

dorsal (1). Direction toward part of shell containing hinge line.

escutcheon (11). Depressed plane or curved area along hinge line behind beak, corresponding to posterior part of cardinal area.

gape (47). Anterior or posterior space between edges of valves when ventral margins are in contact.

growth line (6). More or less obscure concentric lines parallel to shell margin, marking successive advances of edge of shell.

height (3). Distance from dorsal to ventral margin measured normal to length.

hinge line (9). Edge of valve along dorsal margin which is in permanent contact with opposite valve.

hinge plate (27). Internal surface adjacent to hinge line along which hinge teeth project.

hinge teeth (14). Projections from hinge plate for articulation of valves.

lamellar layer (35). Generally innermost part of pelecypod shell, consisting of microscopically thin sheets of calcite or aragonite separated by layers of conchiolin.

lateral teeth (15). Projections from hinge plate nearly parallel to hinge line, situated in front or behind cardinal teeth.

left valve (43). Shell on left side of the anteroposterior axis; among pelecypods which characteristically lie on one side, the left valve is typically uppermost in some (pectinoids) but lowermost in others (oysters and many pachyodonts).

length (4). Distance from anterior to posterior margin at farthest points or measured parallel to hinge line.

ligament area (26). Portion of surface along hinge line to which ligament is attached.

ligament groove (38). Linear depression in cardinal area or ligament area marking attachment of ligament fibers.

lunule (10). Depressed plane or curved area along hinge line in front of beak, equivalent to anterior part of cardinal area.

muscle scar (19). Generally depressed (less commonly raised) area on inner surface of shell, marking attachment place of muscle.

myophore (46). Plate or rodlike structure on inside of shell for attachment of muscle.

ostracum (36). Calcareous structure composing all of pelecypod shell except thin outer conchiolin layer (periostracum).

pallial line (32). Linear depression on inside of pelecypod shell along ventral side, marking inner margin of thickened mantle edges.

pallial sinus (31). Inward deflection of posterior part of pallial line, defining space for retraction of siphons. *where foot comes out*

periostracum (33). Outer thin layer of conchiolin, which in many pelecypods covers calcareous ostracum.

plane of commissure (13). Surface approximately coinciding with valve margins.

plica (21). Radially disposed ribs formed by fold that involves entire thickness of shell.

posterior (29). Direction or part of shell toward position of anus and siphonal opening; in most pelecypods opposite to inclination of beak.

prismatic layer (34). Outer part of ostracum in many pelecypods, consisting of closely spaced polygonal prisms of calcite.

prodissoconch (37). Earliest-formed part of shell; generally preserved at tip of beak.

resilifer (41). Portion of hinge plate which bears internal ligament (resilium); generally a simple shallow pit.

resilium (23). Part of ligament below level of valve margins and under compression instead of tension.

right valve (20). Shell on right side of anteroposterior axis; among pelecypods which lie on their sides, it is generally lowermost in pectinoids and uppermost in oysters and pachyodonts.

(Continued on next page.)

shallow depression, called **resilifer** (*41*), or in an enlarged, somewhat spoon-shaped structure called **chondrophore** (*24*). Other structures beneath the beaks are **myophores** (*46*), for the attachment of muscles, and **buttresses** (*25*), for strengthening the shell. The scars of muscle attachment (adductors, *30*, for pulling the valves together) may be discerned on the anterior and posterior parts of the shell interior, or among clams having only a single such muscle in an intermediate position. A narrow line on the valve interior running parallel to the ventral margin, observed in many clam shells, denotes the outer edge of firm attachment of the mantle to the shell. It is called the **pallial line** (*32*). Parts of the mantle outside this line are not attached to the shell. A posterior indentation of the pallial line, observed in some pelecypods, denotes a space for at least partial retraction of the siphons. This indentation is termed the **pallial sinus** (*31*).

Descriptive Terms for Pelecypods. Among many descriptive terms which are needed for characterization of pelecypod shells and their parts, the more common and important ones are illustrated in Fig. 10-6 (to which all italic numbers accompanying the descriptive terms refer). Features of symmetry or asymmetry are expressed by the terms equilateral (*3*), inequilateral (*6*), equivalve (*2*), and inequivalve (*9*). The shape of shells is expressed by such words as compressed (*1*), rhomboidal (*5*), produced (*7*), orbicular (*20*), quadrate (*23*), mytiliform (*24*), trigonal (*25*), carinate (*27*), rostrate (*29*), truncate (*30*), gaping (*31*), and alate (*32*). Shells in which the beaks turn forward are called prosogyral (*8*), and this is the general rule; but in some they point backward, a condition termed opisthogyral (*33*). The long axis of pelecypod shells, particularly among pectinoid types (scallops) is expressed by the terms prosocline (*10*), pointing forward; acline (*13*), neutral; and opisthocline (*14*), pointing backward.

Special characters, such as the possession of winglike expansions along the hinge line and the occurrence of an opening through the shell for the purpose of byssal attachment, are indicated by the terms auriculate (*11*) and perforate (*45*).

The nature of surface ornamentation may be costate (*12*), multicostate (*21*), plicate (*22*), concentric (*28*), divaricate (*47*), and cancellate (*48*).

According to the nature of teeth concerned with the hingement of the valves, pelecypods may be described as isodont (*15*), schizodont (*16*), edentate (*17*), taxodont (*34*), dysodont (*39*), desmodont (*40*), heterodont (*41*), and pachyodont (*42*). The nature of the ligament structures, whether external, internal, or combined external and internal, and whether the ligament extends both in front and behind the beak or is restricted to areas behind the beak is indicated by the terms alivincular (*18*), duplivincular (*19*), opisthodetic (*37*), multivincular (*38*), amphidetic (*43*), and parivincular (*44*). According to the number and relative size of the adductor muscle impressions on the interior of the valve, the shells may be described as dimyarian (*35*), isomyarian (*36*), and monomyarian (*46*).

MODES OF LIFE

Three main divisions of pelecypods may be made according to the manner in which

---

(Fig. 10-5 continued.)

socket (*16*). Depression in hinge plate for reception of a hinge tooth of opposite valve.

thickness (*12*). Maximum dimension of pelecypod shell measured normal to plane of commissure.

umbo (*8*). Very strongly convex part of valve adjacent to beak.

valve (*45*). Part of shell lying on either side of hinge line.

ventral (*7*). Direction or part of shell lying opposite the hinge line; generally located lowermost in pelecypods which move about freely.

they live. The French paleontologist Dou-villé concluded that differentiation of this sort is useful as the chief basis for classifying these mollusks. On the basis of ecologic adaptation, he recognized (1) so-called normal or active pelecypods, which are accustomed to crawl about on the sea bottom or are able to swim above the sea floor (Fig. 10-1); (2) burrowing or boring clams, which dig into unconsolidated sediment or solid rock or wood which is covered by the sea (Fig. 10-2); and (3) a sedentary group, which lives in a fixed position on the sea bottom, attached by a byssus or cemented by the shell exterior (Fig. 10-3).

Classification according to ecologic adaptation, or mode of life, is complicated by the observation that many pelecypods are attached by a byssus during part of their existence, but later become active free-moving animals, as illustrated by taxodonts of the Arcacea group and by many scallops (Pectinacea); also, some sedentary pelecypods, such as the Chamacea and Rudistacea, are shown by anatomical features to be descendants of "normal" free-moving heterodont ancestral stocks. Thus, classification of pelecypods on the basis of ecologic adaptation seems to accord only partly with morphologic differentiation.

Whether active or sedentary, the functions of respiration and food gathering are much the same in all clams. Incurrent water is moved over the gills by rhythmic motions of cilia, which project through a covering of mucus that serves to strain out all food and sedimentary particles. The mucus is carried along to fleshy appendages (palps) adjoining the mouth, which partly sort out indigestible material from food that is passed to the mouth.

Many pelecypods have assumed a mode of life that entails considerable torsion of the soft parts. They lie on their side, some having the left valve lowermost and others with the right valve down, but in any such recumbent genus, orientation of the valves is constant. Oysters are attached with the left valve below the right, whereas the scallops have the right valve lowermost. In *Tridacna*, the shell is attached with the hinge line down and the soft parts are turned 180 deg., so that the dorsal region is uppermost.

REPRODUCTION AND ONTOGENY

Among practically all pelecypods, the sexes are separate, although in some species, a given individual may be male at one time and later a female. The eggs and sperm are discharged into the sea water through the excurrent siphon. The fertilized eggs develop into free-swimming veliger larvae, which begin secretion of a shell and settle to the sea bottom. Some marine clams produce eggs in enormous numbers, as many as half a billion by one individual in a single season. A few types of marine pelecypods retain their eggs in a brood pouch until they hatch. Among some fresh-water clams (Unionidae), eggs are carried to a space above the gills of the female where they are fertilized by sperm drawn in through the incurrent siphon; the larvae are ejected through the excurrent siphon. In order to survive, they must become attached to fishes, where they grow parasitically until they drop off and begin independent life on the water bottom.

The embryonic shell (prodissoconch, Fig. 10-5, *37*) consists of two minute equivalve calcareous coverings, which form the tips of the beaks of mature shells. Growth lines formed by successive accretions of shell substance are parallel to the margins of the prodissoconch. As the shell increases in size, the position of the muscle attachments on the interior shifts gradually away from the hinge line, and this is recorded in many shells by a "trail" of hypostracum, which is a special sort of calcareous deposit in the area of muscle attachment. Juvenile characters of the hinge teeth and ligament may be preserved similarly in the adult shell, serving as clues to ancestry.

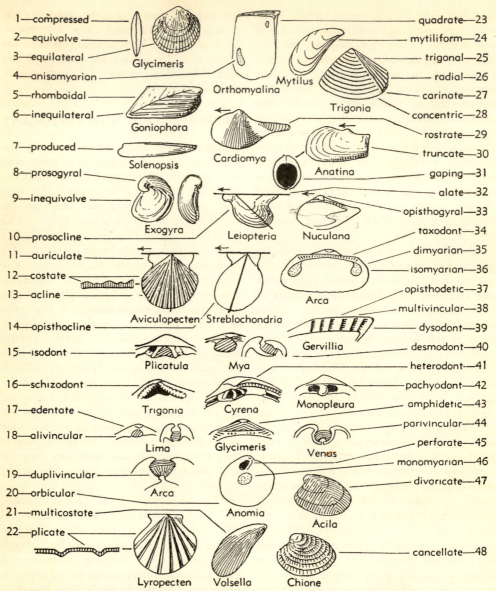

1—compressed
2—equivalve
3—equilateral
4—anisomyarian
5—rhomboidal
6—inequilateral
7—produced
8—prosogyral
9—inequivalve
10—prosocline
11—auriculate
12—costate
13—acline
14—opisthocline
15—isodont
16—schizodont
17—edentate
18—alivincular
19—duplivincular
20—orbicular
21—multicostate
22—plicate

Glycimeris
Orthomyalina
Mytilus
Trigonia
Goniophora
Cardiomya
Solenopsis
Anatina
Exogyra
Leiopteria
Nuculana
Aviculopecten Streblochondria
Arca
Plicatula
Mya
Gervillia
Trigonia
Cyrena
Monopleura
Lima
Glycimeris
Venus
Arca
Anomia
Acila
Lyropecten
Volsella
Chione

quadrate—23
mytiliform—24
trigonal—25
radial—26
carinate—27
concentric—28
rostrate—29
truncate—30
gaping—31
alate—32
opisthogyral—33
taxodont—34
dimyarian—35
isomyarian—36
opisthodetic—37
multivincular—38
dysodont—39
desmodont—40
heterodont—41
pachyodont—42
amphidetic—43
parivincular—44
perforate—45
monomyarian—46
divaricate—47
cancellate—48

FIG. 10-6. **Descriptive terms applied to pelecypod shells.** The terms alphabetically arranged below are briefly defined and cross-indexed by numbers to drawings in the figure.

acline (13). Shell having neither forward nor backward obliquity, mid-line of umbo being normal to hinge line.

alate (32). Shell characterized by possession of wings or auricles.

alivincular (18). Type of external ligament having greatest length transverse to plane of commissure.

amphidetic (43). Ligament located along hinge line both in front and behind beak.

anisomyarian (4). Adductor muscle scars conspicuously unequal.

auriculate (11). Shell possessing auricles; equivalent to alate.

cancellate (48). Shell surface marked by subequal concentric and radial markings.

carinate (27). Shell surface marked by sharpangled edge extending outward from beak.

compressed (1). Transversely flattened shell having small thickness.

(Continued on next page.)

CLASSIFICATION

The most significant structural character for use in the classification of living clams is judged by zoologists to be the nature of the gills. Most simple, and accordingly primitive, is a leaflike respiratory structure, which closely resembles the gills of gastropods and chitons. Paired gills of this sort, termed **protobranch** (Fig. 10-7, *1*), hang downward into the mantle cavities of clams such as *Solemya* (Fig. 10-8, *1*), which otherwise has the generalized characters of the group called palaeoconchs. They occur also in *Nucula* (Fig. 10-9, *10*), which is one of the taxodonts, characterized by numerous small teeth and sockets arranged in a row along the hinge plate.

A next more advanced type of gill, termed **filibranch** (Fig. 10-7, *2*), comprises parallel rows of filaments which hang downward in the mantle cavities but have

(Fig. 10-6 continued.)

concentric (28). Shell surface marked by ridges parallel to shell margin.

costate (12). Shell bearing radial ribs formed by localized thickening.

desmodont (40). Type of shell characterized mainly by prominence of internal ligament.

dimyarian (35). Valves having two adductor scars, whether equal or unequal.

divaricate (47). Shell surface marked by two sets of parallel lines which meet at a distinct angle.

duplivincular (19). Ligament composed partly of fibrous (compressional) tissue and partly of lamellar (tensional) tissue.

dysodont (39). Shells mainly characterized by absence or near absence of hinge teeth and narrow external ligament.

edentate (17). Lacking hinge teeth.

equilateral (3). Anterior and posterior halves of valve subequal and nearly symmetrical.

equivalve (2). Right and left valves subequal and comprising mirror images of one another except for hinge structures.

gaping (31). Part of valve margins not in contact when other parts are pulled tightly together.

heterodont (41). Characterized by hinge teeth of distinct type—cardinals beneath beak, and laterals in front or behind or both.

inequilateral (6). Anterior and posterior parts of valve unequal and lacking symmetry.

inequivalve (9). Opposite valves dissimilar in size or shape or both.

isodont (15). Characterized by two subequal prominent hinge teeth on one valve and corresponding sockets in the other.

isomyarian (36). Having two adductor muscles of approximately equal size.

monomyarian (46). Having only one adductor muscle, originally posterior but tending to be central in position.

multicostate (21). Surface marked by costae which increase by intercalation or bifurcation.

multivincular (38). Ligament chiefly formed by successive bands of fibrous (compressional) tissues.

mytiliform (24). Slipper-shaped, like the genus *Mytilus*.

opisthocline (14). Shell having backward obliquity, approach along mid-line to beak pointing backward.

opisthodetic (37). External ligament located behind beaks.

opisthogyral (33). Beaks turned backward instead of projecting forward.

orbicular (20). Shell subcircular in outline.

pachyodont (42). Thickened specialized teeth, typically developed in coral-like rudistids.

parivincular (44). Ligament having long axis parallel to hinge line, consisting mainly of lamellar (tensional) tissue.

perforate (45). Valve (right) characterized by rounded opening for passage of byssus.

plicate (22). Shell radially folded to form ribs.

produced (7). Shell much elongated in one direction.

prosocline (10). Having forward obliquity, approach to beak along mid-line of shell inclined forward.

prosogyral (8). Beaks directed forward.

quadrate (23). Shell rectangular in outline.

radial (26). Surface marked by costae or plicae diverging from beak.

rhomboidal (5). Shell outline rhomb-shaped.

rostrate (29). Having prominent beaks.

schizodont (16). Having prominent diverging or bifurcate hinge teeth.

taxodont (34). Characterized by more or less numerous subequal hinge teeth, generally arranged in a row.

trigonal (25). Shell outline subtriangular.

truncate (30). Edge of shell, generally posterior, having a chopped-off appearance.

their terminal portions bent upward. Cilia on the surface of these doubled filaments move rhythmically so as to produce water currents that bathe the gills. Filibranch gills occur in taxodont clams such as the superfamily Arcacea (Fig. 10-9, *1, 4, 5, 7, 9, 11*) and various toothless clams (dysodonts) such as the scallops (Fig. 10-16, *6–14*), slipper clams and their allies (Fig. 10-16, *1–5*), and others.

More advanced is the **eulamellibranch** type of gill structure (Fig. 10-7, *3*), in

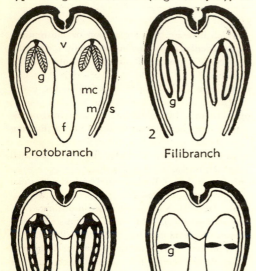

Protobranch   Filibranch

Eulamellibranch   Septibranch

FIG. 10-7. **Types of gill structure in pelecypods.** Four categories of gill structure are recognized among pelecypods, and this is the main basis of classification employed by many zoologists. These are illustrated by transverse diagrammatic sections through the mid-portion of the body, showing pairs of gills in the mantle cavities on each side of the foot. Most primitive is the (1) protobranch type, characterized by small leaflike gills resembling those of many gastropods, chitons, and cephalopods. By far the largest number of clams belong to the (2) filibranch and (3) eulamellibranch types, which have definitely advanced gill structure. Most specialized is a small group having (4) septibranch gills. (*f*, foot; *g*, gill; *m*, mantle; *mc*, mantle cavity; *s*, shell; *v*, viscera).

which the reflexed lamellae of the gills are laterally joined by interfilamentary union which forms enclosed spaces. Such gills characterize some toothless clams (dysodonts, Fig. 10-18), and most types having specialized tooth structure along the hinge line (heterodonts, Fig. 10-22), as well as the majority of burrowing clams (desmodonts, Fig. 10-26).

One small group of burrowing clams (Poromyacea) has profoundly modified gills of the type called **septibranch** (Fig. 10-7, *4*). Among these, the gill lamellae form a sort of partition that divides the mantle cavity into upper and lower sections.

Thus, according to gill structure, zoologists recognize four orders among the pelecypods: Protobranchia, Filibranchia, Eulamellibranchia, and Septibranchia.

Although we may recognize the importance of gill structure in differentiating various groups of pelecypods, divisions defined on this basis do not necessarily comprise natural taxonomic units. Other structural features are important, and they must be taken into account in studying the evolutionary differentiation of the clam assemblage. Especially important to the paleontologist are characters which may be discerned from study of the hard parts. These include valve hingement—both dentition and nature of the ligament—and musculature, as shown by attachment scars on the interior of valves. Study of divergence in adaptation to different modes of life may also be significant in determining natural classification. The threefold grouping of pelecypods into so-called normal, burrowing, and sedentary types does not serve to define orders and suborders because division in this way lacks correspondence with distinctions of structural characters, such as nature of gills and valve hingement.

The classification of pelecypods which is adopted here depends on a combination of structural features, among which characters of hingement rank first, but the nature of the gills, musculature, and structures resulting from adaptation to particular environments are also given weight. It is proper also to take account of the distribution of assemblages with respect to occurrence in geologic time. We may note that specialized types of clams are missing in the early parts of the paleontological record, whereas more or less highly modified forms make late geological appearance. On the other hand, we must conclude that the pelecypods are an unusually slow-changing group of invertebrates, among which main lines begin very early and are very persistent. Most divisions range from early Paleozoic to Recent (Fig. 10-29). This means that, by and large, the clams are an unusually conservative, slowly evolving group of organisms.

Main Divisions of Pelecypods

Pelecypods (*class*), aquatic bivalve mollusks. Ordovician–Recent.

Prionodesmacea (*subclass*), pelecypods having prismatic and nacreous shell structure; mantle lobes separated and siphons poorly developed; hinge teeth lacking or little specialized; mostly normal or sedentary, few burrowing. Ordovician–Recent.

Palaeoconcha (*order*), protobranch gills, hinge teeth lacking or poorly defined, subequal adductors. Ordovician–Recent.

Taxodonta (*order*), hinge teeth small, numerous, similar. Ordovician–Recent.

Nuculacea (*suborder*), taxodonts having protobranch gills. Ordovician–Recent.

Arcacea (*suborder*), taxodonts having filibranch gills. Ordovician–Recent.

Schizodonta (*order*), filibranch gills, hinge teeth few and distinct, diverging from beneath beak. Ordovician–Recent.

Trigoniacea (*suborder*), shell trigonal, marine. Ordovician–Recent.

Cardiniacea (*suborder*), shell rounded to elongate, mostly fresh water. Silurian–Recent.

Isodonta (*order*), filibranch gills, a pair of symmetrically placed teeth on each valve. Triassic–Recent.

Spondylacea (*suborder*), inequivalve spiny shells. Triassic–Recent.

Anomiacea (*suborder*), thin-shelled, lower valve commonly perforate. Jurassic–Recent.

Dysodonta (*order*), hinge teeth weak or lacking. Ordovician–Recent.

Mytilacea (*suborder*), filibranch gills, slipper shells, myalinids and allied forms. Ordovician–Recent.

Pectinacea (*suborder*), filibranch gills, auriculate shells; the scallops. Ordovician–Recent.

Pinnacea (*suborder*), eulamellibranch gills, elongate thin shells, terminal beaks. Devonian–Recent.

Ostreacea (*suborder*), eulamellibranch gills, the oysters, monomyarian. Triassic–Recent.

Limacea (*suborder*), eulamellibranch gills, resemble scallops, some free-swimming Pennsylvanian–Recent.

Dreissensiacea (*suborder*), like mytilids, but have eulamellibranch gills. Eocene–Recent.

Teleodesmacea (*subclass*), pelecypods having porcelaneous and in part nacreous structure; all eulamellibranch (except Porodesmacea); mantle lobes generally connected and siphons well developed; hinge teeth specialized in form of distinct cardinals and laterals, accompanied by external ligament, or in varying degree obsolete and accompanied by internal ligament; normal, sedentary, and burrowing or boring. Ordovician–Recent.

Heterodonta (*order*), hinge teeth include well-developed cardinals. Silurian–Recent.

Cypricardiacea (*suborder*), primitive generalized forms. Silurian–Recent.

Lucinacea (*suborder*), distinguished by arrangement of cardinal teeth. Silurian–Recent.

Cyrenacea (*suborder*), distinguished by arrangement of cardinal teeth. Triassic–Recent.

Pachyodonta (*order*), mostly thick-shelled sedentary specialized clams, highly inequivalve, thick teeth. Jurassic–Recent.

Chamacea (*suborder*), retain signs of heterodont derivation. Jurassic–Recent.

Rudistacea (suborder), include many robust coral-like forms. Cretaceous.

Desmodonta (*order*), hinge teeth weak or lacking, resilifers (or chondrophores) for reception of internal ligament; burrowers and borers. Ordovician–Recent.

Solenacea (*suborder*), elongate narrow razor clams. Cretaceous–Recent.

Myacea (*suborder*), elongate siphons, shell gaping posteriorly. Triassic–Recent.

Mactracea (*suborder*), retains cardinal teeth, siphons prominent. Cretaceous–Recent.

Adesmacea (*suborder*), includes rock and wood borers. Jurassic–Recent.

Ensiphonacea (*suborder*), shell degenerate, siphons enclosed in calcareous tube. Cretaceous–Recent.

Anatinacea (*suborder*), hinge teeth and internal ligament weakly developed. Ordovician–Recent.

Poromyacea (*suborder*), septibranch gill structure, borers. Jurassic–Recent.

PALAEOCONCHS

A predominantly Paleozoic assemblage of rather thin-shelled pelecypods, which have ill-developed hinge characters, is classed as palaeoconchs. They are unspecialized in having equivalve form. Some are rounded in outline and others very elongate (Fig. 10-8, *2*, *3*). The narrow ligament extends both in front and behind the beak (amphidetic). Like modern *Solemya* (Fig. 10-8, *1*), which is placed in this group, extinct genera of palaeoconchs are judged to have had protobranch gills, which clearly are the simplest and least specialized means of respiration found among pelecypods. On the interior of the valves are two adductor muscle scars, located in high anterior and posterior positions. There is no distinct pallial line and no pallial sinus. Judging from the form of the shell and the mode of occurrence of these clams in many Paleozoic deposits, some palaeoconchs probably were able to crawl about on the sea bottom in the manner of so-called normal active modern clams. Others (Figs. 10-8, *4*; 10-11, *19*) probably burrowed in the soft sediment. *Solemya* is an unusual clam in that the anterior part of the shell exceeds the posterior in length, and the axis of obliquity points backward instead of forward.

Genera grouped among the palaeoconchs are most numerous in Silurian and Devonian strata. Only a very few range into Mesozoic and younger rocks. An Ordovician palaeoconch is illustrated in Fig. 10-10, *7*, several Devonian representatives of this group in Fig. 10-11, and Pennsylvanian forms in Fig. 10-13. *Chaenomya* (Fig. 10-13, *15*) is a palaeoconch which evidently had a burrowing mode of life, as indicated by the wide gape at the posterior extremity of the shell.

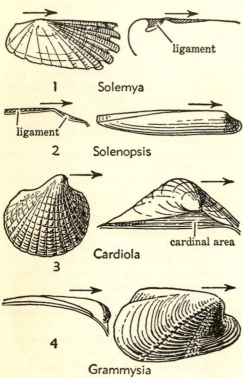

FIG. 10-8. **Representative palaeoconch pelecypods.** This group is characterized by generally primitive features and lack of specialized structures. The ligament, which is chiefly external, typically extends along parts of the hinge both in front and behind the beak. Hinge teeth are weak or lacking. Abundant in Paleozoic rocks, palaeoconchs also include living forms. Arrows point to anterior part of shell. Natural size except as indicated otherwise.

1, *Solemya australis* Lamarck, Recent, Australia.

2, *Solenopsis pelagica* Goldfuss, Middle Devonian, Germany.

3, *Cardiola cornucopiae* Goldfuss, Middle Devonian, Germany.

4, *Grammysia nodocostata* Hall, Middle Devonian, New York (×0.7).

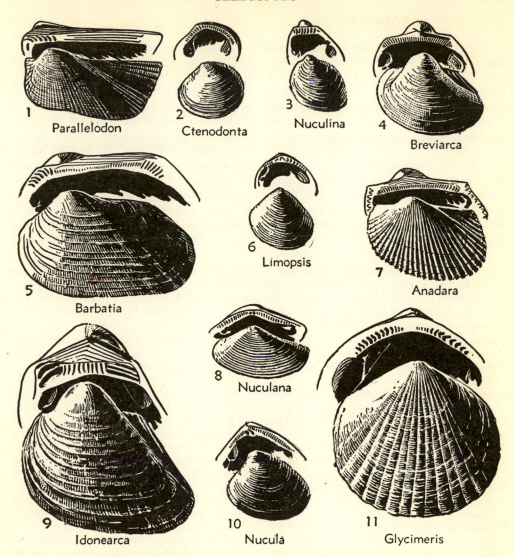

FIG. 10-9. **Representative taxodont pelecypods.** The presence of more or less numerous small teeth along the hinge line distinguishes this group, part of which has the most primitive (protobranch) type of gill structure found among the clams and part a more advanced (filibranch) type. They range from Ordovician to the present. External views are all of left valves, and others showing hinge teeth are of right valves. Natural size.

1, *Parallelodon hirsonensis* Morris & Lycett, Jurassic, England.

2, *Ctenodonta pectunculoides* Hall, Upper Ordovician, Ohio.

3, *Nuculina ovalis* (Wood), Miocene, Austria.

4, *Breviarca haddonfieldensis* Stephenson, Upper Cretaceous, New Jersey.

5, *Arca* (*Barbatia*) *micronema* (Meek), Upper Cretaceous, Wyoming.

6, *Limopsis aurita* Brocchi, Pliocene, Italy.

7, *Arca* (*Anadara*) *diluvii* Lamarck, Pliocene, Italy.

8, *Nuculana deshayesiana* Duchesne, Oligocene, Belgium.

9, *Cucullaea* (*Idonearca*) *vulgaris* Morton, Upper Cretaceous, Maryland.

10, *Nucula nucleus* Linné, Miocene, Austria.

11, *Glycimeris subovata* (Say), Miocene, Maryland.

TAXODONTS

The order of pelecypods termed Taxodonta is chiefly distinguished by the presence of numerous subequal teeth along the hinge plate (Fig. 10-9). These teeth may be short projections approximately normal to the hinge line, forming a closely spaced row on parts of the hinge plate both in front of the beak and behind it (Fig. 10-9, 2–8, 10, 11). Some members of this group also have a few elongate teeth running more or less nearly parallel to the hinge line (Fig. 10-9, 1, 9). The taxodont clams include two suborders, both of which range from Ordovician to Recent.

Nuculacea

Taxodonts which are mostly of diminutive size and characterized by the possession of protobranch gill structure belong to the suborder Nuculacea (Fig. 10-9, 2, 3, 8, 10). Typical genera have rounded to somewhat elongate outlines and equivalve shells. They are so-called normal, active pelecypods, which do not live in an attached position but are able to crawl about by means of the foot projecting from the antero-ventral part of the shell. The exterior is smooth or marked by closely spaced fine concentric lines. The outer shell layer is prismatic, and the inner one,

---

Fig. 10-10. **Representative Ordovician and Silurian pelecypods.** The beginnings of most main groups are found in this part of the geologic column. Among the illustrated species, are palaeoconchs (7), taxodonts (4, 8, 11, 13), schizodonts (10, 18), dysodonts (1–3, 5, 6, 12, 14, 17), heterodonts (15), and desmodonts (9, 16). Ordovician fossils include 1–13 and Silurian fossils 14–18. Natural size except as indicated otherwise. L, left valve; R, right valve.

Ambonychia Hall, Ordovician. A primitive pectinoid having fine radial costae. *A. bellistriata* Hall (12a, L; 12b, anterior view), Trentonian, New York.

Byssonychia Ulrich, Ordovician. Resembles *Ambonychia* but has coarser ribs and more terminal beaks. *B. radiata* (Hall) (5, L), Cincinnatian, Ohio.

Cleidophorus Hall, Ordovician–Devonian. A nuculid taxodont. *C. planulatus* (Conrad) (8, L), Cincinnatian, Ohio.

Colpomya Ulrich, Ordovician–Silurian. A mytilid dysodont. *C. constricta* Ulrich (1, R), Trentonian, New York.

Ctenodonta Salter, Ordovician–Silurian. A robust nuculid showing typical taxodont dentition. *C. gibberula* Salter (11a, L; 11b, R), Blackriveran, Ontario.

Cypricardinia Hall, Silurian–Mississippian. Classed among earliest heterodonts. *C. arata* Hall (15, R), Niagaran, Indiana.

Cyrtodonta Billings, Ordovician–Silurian. An early arcid taxodont. *C. grandis* (Ulrich) (13, R, ×0.5), Trentonian, Wisconsin.

Goniophora Phillips, Silurian–Devonian. The umbonal ridge is sharply angular; a mytilid. *G. bellula* Billings (14, R), Upper Silurian, Nova Scotia.

Leiopteria Hall, Silurian–Mississippian. A pectinoid dysodont. *L. subplana* (Hall) (17, L), Niagaran, New York.

Lyrodesma Conrad, Ordovician–Silurian. Oldest schizodont. *L. major* (Ulrich) (10a, b, R), Richmondian, Minnesota.

Megalomus Hall, Silurian. A robust thickshelled clam which has coarse ill-formed teeth. *M. canadensis* Hall (18, R, ×0.5), Niagaran, Ontario.

Modiolopsis Hall, Ordovician. A mytilid dysodont. *M. concentrica* Hall & Whitfield (3, L), Richmondian, Indiana.

Newsomella Foerste, Silurian. *N. ulrichi* Foerste (16, R), Niagaran, Tennessee.

Orthodesma Hall & Whitfield, Ordovician. One of the palaeoconch clams. *O. rectum* Hall & Whitfield (7, R), Cincinnatian, Indiana.

Pterinea Goldfuss, Ordovician–Pennsylvanian. An early pectinoid dysodont, characterized by fine concentric markings and extended hinge line. *P. demissa* (Conrad) (2, L), Cincinnatian, Ohio.

Rhytimya Ulrich, Ordovician. Classed among Anatinacea, desmodonts. *R. producta* Ulrich (9, R), Cincinnatian, Ohio.

Vanuxemia Billings, Ordovician. An arcid taxodont having teeth subparallel to hinge. *V. hayniana* (Safford) (4a, b, R), Trentonian, Tennessee.

Whiteavesia Ulrich, Ordovician. Resembles *Colpomya*. *W. cincinnatiensis* (Hall & Whitfield) (6, L), Trentonian, Ohio.

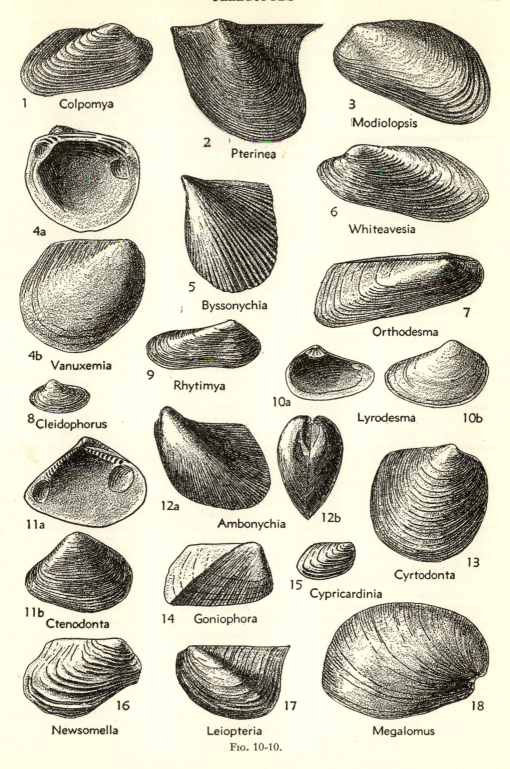

1 Colpomya

2 Pterinea

3 Modiolopsis

4a

4b Vanuxemia

5 Byssonychia

6 Whiteavesia

7 Orthodesma

8 Cleidophorus

9 Rhytimya

10a

Lyrodesma 10b

11a

12a Ambonychia 12b

13 Cyrtodonta

11b Ctenodonta

14 Goniophora

15 Cypricardinia

16 Newsomella

17 Leiopteria

18 Megalomus

FIG. 10-10.

nacreous or porcelaneous. Two subequal adductor muscle scars are seen near the anterior and posterior extremities of the hinge line. An unindented pallial line is observed in most genera, but a few have a small pallial sinus, denoting retractability of the siphon. Some forms are characterized by backwardly directed beaks (opisthogyral) and by more or less sharp-pointed posterior extremities of the valves

(Figs. 10-9, *8;* 10-11, *13;* 10-15, *3;* 10-23, *3*). *Ctenodonta* (Figs. 10-9, *2;* 10-10, *11*) is an Ordovician nuculid taxodont. Devonian representatives of the Nuculacea are illustrated in Fig. 10-11, *6, 13, 16;* Mississippian, Fig. 10-13, *1;* Pennsylvanian, Fig. 10-13, *6;* Permian, Fig. 10-15, *3, 4;* Cretaceous, Fig. 10-20, *9;* Paleogene, Fig. 10-23, *1–4;* and Neogene, Fig. 10-25, *9,* and Fig. 10-27, *4.* The genus *Acila* (Figs.

Fig. 10-11.   **Representative Devonian pelecypods.** This assemblage generally resembles that belonging to Ordovician and Silurian time. Among forms here illustrated, all from New York, are several palaeoconchs (7, 10, 12, 17, 19), some taxodonts (6, 9, 13, 16), a schizodont (23), dysodonts (1, 3–5, 11, 20–22), heterodonts (14, 15, 18), and desmodonts (2, 8). Natural size and from Middle Devonian except as indicated otherwise. L, left valve; R, right valve.

Amnigenia Hall, Devonian. A fresh-water clam which has obscure teeth. *A. catskillensis* (Vanuxem) (23, L, ×0.5).

Buchiola Barrande, Silurian–Devonian. Shell marked by wide flat plications; classed among palaeoconchs. *B. speciosa* (Hall) (10, attached valves, L below, R above, ×6), Upper Devonian.

Conocardium Bronn, Ordovician–Permian. This long-ranging pelecypod has very distinctive form; doubtfully placed with pectinoid dysodonts. *C. cuneus trigonale* Hall (1*a*, R; 1*b*, ventral; 1*c*, dorsal), Lower Devonian.

Cornellites Williams, Ordovician–Pennsylvanian. A typical Paleozoic pectinoid dysodont. *C. flabella* (Conrad) (4, L, ×0.5).

Cypricardella Hall, Devonian–Mississippian. A primitive heterodont. *C. bellastriata* (Conrad) (15*a*, R, cardinal teeth, enlarged; 15*b*, L).

Cypricardinia Hall, Silurian–Mississippian. Early heterodont. *C. indenta* Conrad (18, R).

Goniophora Phillips, Silurian–Devonian. A mytilid dysodont having a subrhombic outline and angular umbonal ridge. *G. hamiltonensis* (Hall) (20, R).

Grammysia Verneuil, Silurian–Mississippian. A palaeoconch. *G. bisulcata* (Conrad) (19, R, ×0.5).

Leiopteria Hall, Silurian–Mississippian. A pectinoid dysodont. *L. rafinesquei* Hall (3, L, ×0.5).

Leptodesma Hall, Silurian–Mississippian. A pectinoid dysodont having an extended posterior auricle. *L. rogersi* Hall (11, L).

Limoptera Hall, Devonian. A pectinoid dysodont. *L. macroptera* (Conrad) (5, 1, ×0.5).

Lunulicardium Münster, Devonian–Mississippian. A subtriangular pectinoid dysodont having prominent terminal beaks and anterior byssal notch, *L. curtum* Hall (22, R).

Lyriopecten Hall, Devonian. Large pectinoid dysodont. *L. tricostatum* (Vanuxem) (21, L, ×0.5), Upper Devonian.

Nuculana Link, Silurian–Recent. A nuculid taxodont which is rounded anteriorly, sharply pointed posteriorly, and has backward-turned beaks. *N. diversa* (Hall) (13, L, ×2), Upper Devonian.

Nuculoidea Williams & Breger, Devonian. A nuculid taxodont. *N. lirata* (Conrad) (6, L).

Orthonota Conrad, Silurian–Devonian. An elongate palaeoconch. *O. undulata* Conrad (7, L, ×0.5).

Palaeoneilo Hall, Devonian–Triassic. Large nuculid taxodont. *P. constricta* (Conrad) (16, L).

Paracyclas Hall, Devonian. Moderately large orbicular heterodont. *P. elliptica* Hall (14, L).

Parallelodon Meek, Devonian–Tertiary. A straight-hinged arcid taxodont having teeth parallel to hinge. *P. chemungensis* (Hall) (9, L), Upper Devonian.

Pararca Hall, Devonian. A palaeoconch. *P. erecta* Hall (17, L), Upper Devonian.

Pholadella Hall, Devonian. An early desmodont. *P. radiata* (Conrad) (2, R).

Solemya Lamarck, Devonian–Recent. A palaeoconch. *S. vetusta* Meek (12*a*, R; 12*b*, dorsal), Lower Devonian.

Sphenotus Hall, Devonian–Mississippian. A desmodont. *S. contractus* (Hall) (8, R), Upper Devonian.

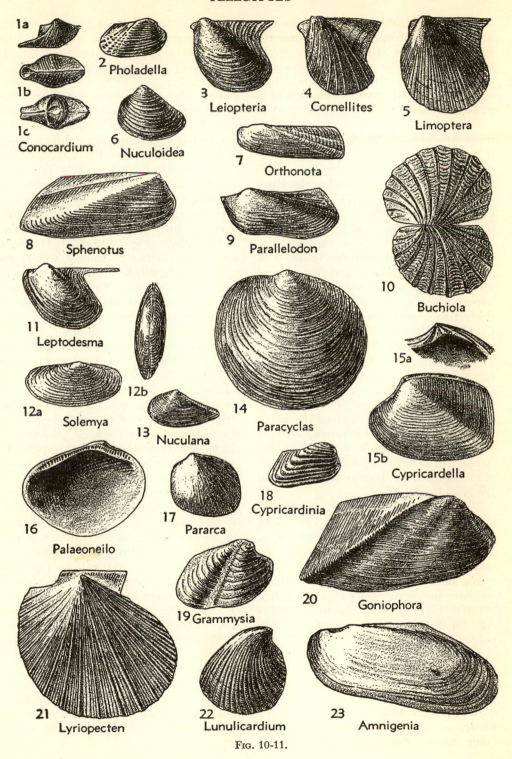

1a
1b
1c Conocardium
2 Pholadella
3 Leiopteria
4 Cornellites
5 Limoptera
6 Nuculoidea
7 Orthonota
8 Sphenotus
9 Parallelodon
10 Buchiola
11 Leptodesma
12a
12b
Solemya
13 Nuculana
14 Paracyclas
15a
15b Cypricardella
16 Palaeoneilo
17 Pararca
18 Cypricardinia
19 Grammysia
20 Goniophora
21 Lyriopecten
22 Lunulicardium
23 Amnigenia

Fig. 10-11.

10-23, *2;* 10-25, *9*), which is common in some Cenozoic deposits and in modern pelecypod faunas, is characterized by its divaricate (Fig. 10-6, *47*) surface ornamentations consisting of parallel ridges, which meet in the mid-portion of the shell so as to form an acute angle.

Arcacea

Taxodont pelecypods belonging to the suborder Arcacea generally are much larger than the nuculaceans just described (Fig. 10-9, *1, 4, 5, 7, 9, 11*). Most of them also are characterized by a longer, straighter hinge line and by the presence of a small to fairly large cardinal area between the hinge and the beaks. This area consists of a plane or slightly curved surface, which is differentiated from adjoining parts of the shell exterior by a sharp, somewhat angulated boundary. Commonly it bears ligament grooves in the form of broadly flaring chevrons (Fig. 10-5, *38–40*). Thus, the ligament is external; and because it occurs both in front and behind the beak, it is termed amphidetic. The gill structure of the arcacean taxodonts differs from that of the nuculacean group in being filibranch instead of protobranch (Fig. 10-7, *1, 2*). Thus, the Arcacea evidently comprise a distinct stock of taxodonts which is more advanced than the Nuculacea.

The Arcacea are marine pelecypods which live most abundantly in shallow water. At least during early growth stages, they do not move about freely like the nuculaceans but are attached by means of byssus threads; and thus, they are classed by some among sedentary clams. Fixation of this sort is not long enough to have influence on shell shape, however, for the two valves are equal in size.

Ordovician arcacean taxodonts are illustrated in Fig. 10-10, *4, 13*, and a Devonian form in Fig. 10-11, *9*. Mesozoic and Cenozoic representatives of the suborder are shown in Figs. 10-20, *2, 5, 8;* 10-23, *13, 14, 18;* and 10-27, *6, 7*. Many important guide fossils, especially in Cretaceous, Paleogene, and Neogene deposits, belong to this group.

SCHIZODONTS

The order of schizodonts comprise clams which have a filibranch type of gill structure, like the Arcacea among taxodonts, but they have a distinctively different sort of dentition, which consists of more or less prominent teeth and sockets diverging from beneath the beaks. This arrangement of teeth, which gives the group its name (*schizo*, divided; *odous*, tooth), is typically shown by such shells as *Trigonia*, *Schizodus*, and *Myophoria* (Fig. 10-12, *1–3*). The right valve contains two prominent widely diverging teeth which meet or nearly meet just below the beak. The opposite valve has a prominent bifurcate or Y-shaped tooth that fits between the two large ones of the right valve; in addition, the left valve may carry narrow teeth adjoining the hinge line, beyond the sockets which receive the teeth of the right valve. In *Schizodus* and *Myophoria*, the sides of the teeth are smooth or nearly so; but in *Trigonia*, the sides of the teeth bear long gently curved, evenly spaced grooves and ridges which interlock with corresponding corrugations on teeth of the opposite valve. This rather complicated hinge structure so fits together that the two valves of a *Trigonia* shell can hardly be separated without breaking some of the corrugations. In life, the shells are held together also by a narrow external ligament, located entirely on the posterior side of the beak (opisthodetic). On the inside of the shell are two subequal muscle scars belonging to the anterior and posterior adductors. They are close to the hinge line on the dorsal part of the valve. The shell consists of an outer prismatic layer and a more or less prominent inner nacreous layer. The inner side of some well-preserved *Trigonia* shells of Cretaceous age still have a well-defined pearly sheen, thus corresponding to many modern clam shells found along a sea beach.

Trigoniacea

The suborder Trigoniacea, to which *Trigonia*, *Schizodus*, and *Myophoria* belong, is one of two main divisions of the schizodonts. In addition to characters of dentition, most shells of the group have a more or less well-defined trigonal outline. This is emphasized by the oblique truncation of the posterior margin and by the relative prominence of the umbonal ridge, extending from the beak to the postero-ventral angle. This line is somewhat sharply carinate in some species of *Trigonia* and *Myophoria*. Surface markings of the shells of *Trigonia* commonly are distinctive. They variously consist of concentric ridges or obliquely disposed subradial nodose ribs, which cover the median and anterior parts of the valve, whereas surface markings of the posterior area, between the umbonal ridge and the hinge line, are altogether different. Thus, many species are readily differentiated.

Earliest representative of the Trigoniacea is *Lyrodesma* (Fig. 10-10, *10*), found in Ordovician and Silurian rocks. Other genera occur in Devonian and later Paleozoic formations. *Myophoria* (Fig. 10-17, *4*) is a characteristic Triassic genus. *Trigonia* (Figs. 10-17, *8a, b;* 10-20, *3*) is most abundant in Jurassic and Cretaceous rocks, but it is rare in Cenozoic strata. Some living pelecypods are very closely allied to *Trigonia* and are possibly assignable to this genus.

Cardiniacea

Essentially confined to fresh or brackish waters are filibranch pelecypods which differ rather widely in external and internal appearance from the Trigoniacea. This suborder, called Cardiniacea, is classed among the schizodonts because the characters of the hinge teeth are judged most closely to approach the schizodont pattern. The common fresh-water mussel *Unio* (Fig. 10-12, *4*), which has essentially world-wide distribution in modern lakes and streams, exhibits somewhat variable dentition, in

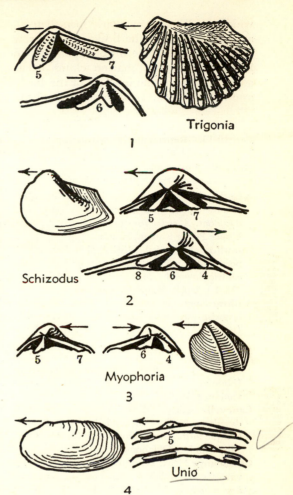

Trigonia

1

Schizodus

2

Myophoria

3

Unio

4

Fig. 10-12. **Representative schizodont pelecypods.** The divergent subequal hinge teeth beneath the beak, which in some groups assume characters of lateral teeth, characterize the schizodonts. They have filibranch gill structure and range from Ordovician to the present. Arrows point to anterior part of shell. Numbers adjoining hinge teeth are for identification of homologous elements, odd-numbered on the right valve and even-numbered on the left. Not to scale.

1, *Trigonia castrovillensis* Stephenson, Upper Cretaceous, Texas.

2, *Schizodus wheeleri* Swallow, Pennsylvanian, Kansas.

3, *Myophoria vulgaris* Schlotheim, Middle Triassic, Germany.

4, *Unio pictorum* Linné, Recent, Europe.

which small to large, poorly formed teeth are found beneath the beaks, and narrow widely divergent teeth corresponding to laterals of various marine clams occur along the hinge line in front of the beaks and behind them. Teeth beneath the beak are termed pseudocardinals, and those running parallel to the hinge line are called pseudolaterals. It is difficult to see in them resemblance to the dentition of *Schizodus* or *Trigonia;* but like the Trigoniacea, *Unio* and its many allies has an opisthodetic

FIG. 10-13. **Representative Mississippian and Pennsylvanian pelecypods.** Dysodont types (2, 4, 7, 8, 11, 13, 14, 17, 19–24) are dominant. Among other groups are palaeoconchs (9, 12, 15), taxodonts (1, 6, 10), schizodonts (18), and primitive heterodonts (3, 5, 16). Mississippian fossils include 1–5 and 8; the others are Pennsylvanian. Natural size except as indicated otherwise. L, left valve; R, right valve.

Acanthopecten Girty, Mississippian–Permian. Spinose costae distinguish this pectinoid dysodont. *A. carboniferus* (Stevens) (23, L, ×2), Upper Pennsylvanian, Kansas.

Annuliconcha Newell, Mississippian–Permian. A pectinoid dysodont having prominent concentric ridges. *A. interlineatus* (Meek & Worthen) (21, L), Upper Pennsylvanian, Kansas.

Anthracomya Salter, Pennsylvanian. A fresh or brackish-water schizodont. *A. elongata* (Dawson) (18, R, ×2), Lower Pennsylvanian, Nova Scotia.

Aviculopecten McCoy, Silurian–Permian. A long-ranging pectinoid dysodont. *A. occidentalis* (Shumard) (14a, R; 14b, L), Upper Pennsylvanian, Missouri.

Caneyella Girty, Mississippian. A pectinoid dysodont having terminal beaks and very oblique form. *C. richardsoni* Girty (2, R), Upper Mississippian, Oklahoma.

Chaenomya Meek, Pennsylvanian–Permian. The valves gape widely at the posterior edge. *C. leavenworthensis* Meek & Hayden (15a, dorsal; 15b, L; ×2), Upper Pennsylvanian, Kansas.

Clinopistha Meek & Worthen, Devonian–Pennsylvanian. A palaeoconch. *C. radiata* Hall (12, L), Upper Pennsylvanian, Missouri.

Cypricardella Hall, Devonian–Mississippian. A primitive heterodont. *C. oblongata* Hall (3, R, ×2), Meramecian, Indiana.

Cypricardinia Hall. Silurian–Mississippian. A primitive heterodont. *C. consimilis* Hall (5, R), Lower Mississippian, Ohio.

Edmondia De Koninck, Devonian–Pennsylvanian. A palaeoconch. *E. aspinwallensis* Meek (9a, R; 9b, dorsal; ×0.5), Upper Pennsylvanian, Nebraska.

Fasciculiconcha Newell, Pennsylvania. A pectinoid dysodont. *F. providencensis* (Cox) (22, L, ×0.5), Middle Pennsylvanian, Kentucky.

Lima Brugière, Pennsylvanian–Recent. A eulamellibranch dysodont. *L. retifera* Shumard (7, L), Upper Pennsylvanian, Kansas.

Monopteria Meek & Worthen, Pennsylvanian. A pectinoid dysodont. *M. longispina* (Cox) (19, L, ×0.5), Upper Pennsylvanian, Kansas.

Myalina De Koninck, Mississippian–Permian. A relatively thick-shelled mytilid dysodont which lacks auricles. *M. keokuk* Worthen (4, L, ×0.5), Osagian, Illinois. *M. wyomingensis* (Lea) (11a, R, ×1; 11b, L, ×0.5), Upper Pennsylvanian, Illinois.

Mytilarca Hall, Silurian–Mississippian. A strongly oblique mytilid dysodont. *M. fibristriata* White & Whitfield (8, L), Lower Mississippian, Ohio.

Naiadites Dawson, Pennsylvanian. A brackish or fresh-water mytilid dysodont. *N. carbonarius* Dawson (20, L), Lower Pennsylvanian, Nova Scotia.

Nuculopsis Girty, Pennsylvanian (subgenus of *Nucula*). A nuculid taxodont. *N. girtyi* Schenck (6a, L; 6b, dorsal; ×2), Middle Pennsylvanian, Oklahoma.

Orthomyalina Newell, Pennsylvanian–Permian (subgenus of *Myalina*). This dysodont is distinguished by its thick shell and rectangular outline. *M. (O.) subquadrata* Shumard (24, R, ×0.5), Upper Pennsylvanian, Kansas.

Palaeoneilo Hall, Devonian–Triassic. A nuculid taxodont. *P. sulcatina* (Conrad) (1, L), Lower Mississippian, Ohio.

Parallelodon Meek, Devonian–Tertiary. An arcid taxodont. *P. obsoletus* (Meek) (10, R), Upper Pennsylvanian, Nebraska.

Pleurophorus King, Devonian–Triassic. A primitive heterodont. *P. oblongus* Meek (16, R, ×2), Upper Pennsylvanian, Nebraska.

Promytilus Newell, Mississippian–Pennsylvanian. A mytilid dysodont. *P. annosus senex* Newell (13, L), Upper Pennsylvanian, Kansas.

Pteria Scopoli, Devonian–Recent. A pectinoid dysodont. *P. longa* (Geinitz) (17, L), Upper Pennsylvanian, Kansas.

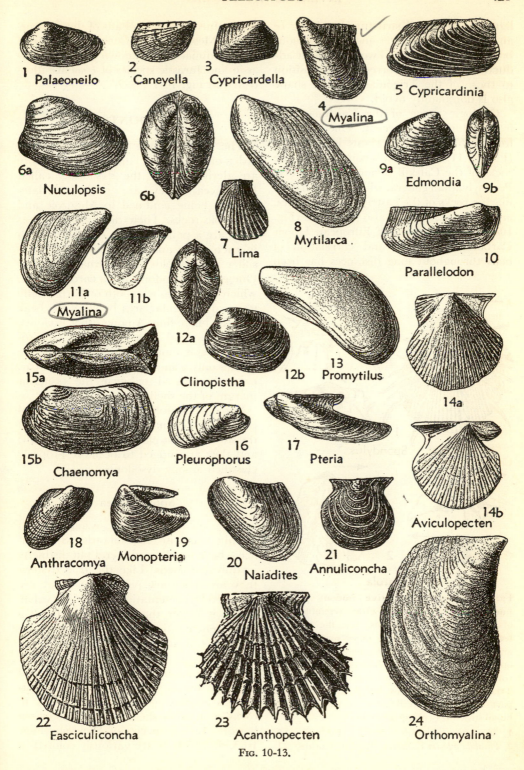

1 Palaeoneilo
2 Caneyella
3 Cypricardella
4 Myalina
5 Cypricardinia
6a Nuculopsis
6b
7 Lima
8 Mytilarca
9a Edmondia
9b
10 Parallelodon
11a Myalina
11b
12a
12b
13 Promytilus
Clinopistha
14a
15a
15b Chaenomya
16 Pleurophorus
17 Pteria
14b Aviculopecten
18 Anthracomya
19 Monopteria
20 Naiadites
21 Annuliconcha
22 Fasciculiconcha
23 Acanthopecten
24 Orthomyalina

Fig. 10-13.

ligament and shell structure consisting of an outer prismatic layer and a conspicuous inner nacreous layer. Two prominent adductor muscle scars correspond in position to those of the Trigoniacea. Shell shape also is dominantly ovoid.

Geologic range of the Cardiniacea is from Silurian to Recent. One of these fresh-water Devonian clams is *Amnigenia* (Fig. 10-11, *23*), which is common in some rocks of the Catskill region in New York. Pennsylvanian genera, which occur in non-marine rocks of the eastern United States, are widely distributed also in Europe (Fig. 10-13, *18*). Zonation of European Upper Carboniferous rocks and correlation of beds throughout long distances has been

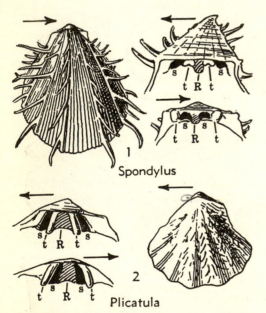

Spondylus

Plicatula

Fig. 10-14. **Representative isodont pelecypods.** The dentition of this group resembles that of some typical schizodonts, but the pairs of teeth on either valve are more nearly equal, and an internal ligament is generally well developed. Gill structure is filibranch. They are confined to post-Paleozoic rocks, ranging from Triassic to Recent. Arrows point to anterior part of shell. (*R*, resilifer; *s*, socket; *t*, tooth; hinge structures enlarged.)

1, *Spondylus regius* Linné, Recent, Pacific, ×0.5.
2, *Plicatula gibbosa* Lamarck, Recent, Florida, ×1.

determined largely on basis of these clams. Many Mesozoic and Cenozoic species of the Cardiniacea are known as fossils. Triassic and Jurassic species of *Unio* are illustrated in Fig. 10-17, *3*, *7*.

ISODONTS

The Isodonta (Fig. 10-14) are a small order of filibranch clams which probably are derivatives of the dysodonts called Pectinacea. They are geological latecomers, being confined to Mesozoic and Cenozoic rocks. Their distinguishing features are a pair of nearly equal-size teeth in each valve and a centrally placed resilifer which holds the internal ligament. One group of isodonts has a cardinal area, which appears as a smooth or horizontally grooved triangular area between the beak and the hinge line. Among other isodonts, such a cardinal area is lacking. The isodonts are monomyarian, for each valve carries only a single adductor muscle scar, centrally located or near the posterior margin of the shell.

Some paleontologists have classed members of the Pectinacea among isodonts, but the shells of these clams have no true teeth; and they are judged to be more properly classified among dysodonts.

Spondylacea

The so-called thorny oysters, which mainly comprise the suborder Spondylacea, have nearly equilateral but distinctly inequivalve shells. Like the true oysters, they are sedentary pelecypods. They are attached by cementation of the right valve, which is larger than the left and has a prominent cardinal area. The valves are radially ribbed, and along some of the ribs are scaly projections or long spines. Rather commonly, the shells are distorted in shape by reason of crowding of individuals as they grow attached firmly to rocks or to other shells on the shallow sea bottom. Living species of *Spondylus* (Figs. 10-3, *5*; 10-14, *1*) are variously colored in brilliant hues, such as orange, yellow, crimson, and violet. Their brightly painted colors and ribbed spinose surface ornamentation make them truly striking. Fossil specimens lack coloration, and many of the spinose projections of the shell are likely to be broken.

The genus *Plicatula* (Figs. 10-14, *2;* 10-25, *2*) is a smaller, more conservative type of shell than *Spondylus*. It also is attached by the umbo of the right valve. The shell is not spinose but is broadly plicate in radial manner. An average-size specimen has a length and height of about 1 in. The Spondylacea range from Triassic to Recent and include approximately 200 known species.

Anomiacea

The shells of anomiacean clams are rounded in outline and mostly very thin. Traces of a prismatic shell layer may be present, but a translucent aragonitic lamellar layer, characterized by pearly luster, is chiefly important. Like the Spondylacea, the valves are nearly equilateral but distinctly inequivalve. The shell lies on its side, with the flat or gently concave right valve lowermost and the moderately convex left valve above it. The shell is attached by a more or less calcified byssus, which extends through a notch or perforation in the lower valve. *Anomia* (Fig. 10-19, *13*) occurs as a fossil in Cretaceous and various Cenozoic deposits, the shell diameter ranging from about 0.5 to 3 in. Modern shells belonging to this genus are common along the Atlantic and southern Pacific coasts of the United States.

DYSODONTS

The order called Dysodonta includes a heterogeneous assemblage of essentially toothless pelecypods, among which two important suborders have filibranch gills (Fig. 10-16) and the rest are characterized by eulamellibranch gill structure (Fig. 10-18). The two filibranch assemblages

are known to extend from Ordovician to Recent, whereas the oldest eulamellibranch dysodonts appear in the Devonian, and the majority of them are post-Paleozoic. Combined shell characters and gill structure set the dysodonts rather clearly apart from other members of the Prionodesmacea, and likewise distinguish them readily from pelecypod groups classed as Teleodesmacea.

Pectinacea

Both among fossils and in modern marine pelecypod assemblages, the Pectinacea are an extremely important group (Fig. 10-16, *6–14*). They include the scallops, most of which have shells of rounded outline, except for projecting anterior and posterior auricles along the short or long straight hinge line. The shell of many genera is equivalve, but that of others is more or less strongly inequivalve. A distinct obliquity of the major shell axis with respect to the hinge line characterizes most genera, but in some Pectinacea this axis is normal to the hinge and the two halves of each valve are nearly identical. Except for slight differences in size and shape of the auricles and the sinuses below them, the shells are equilateral. Very commonly, the surface of the valves is radially costate or plicate. Other shells bear coarse to fine concentric markings or have a perfectly smooth surface. As classified here, the Pectinacea include the group called Pteriacea, which has minor distinguishing features.

Typically, the shell lies flat on one of the valves. At least during early life, it is commonly attached by means of a byssus, which extends from a narrow gap between the valves along the anterior margin not far from the hinge line. Some Pectinacea maintain a sedentary existence throughout life, but others, like modern species of *Pecten* (Fig. 10-1, *3*), become free moving and are able to swim by a pulsatory clapping together of their valves.

The hinge structure of representative genera of Pectinacea is illustrated in Fig. 10-16, *6–14*. Teeth are lacking, although

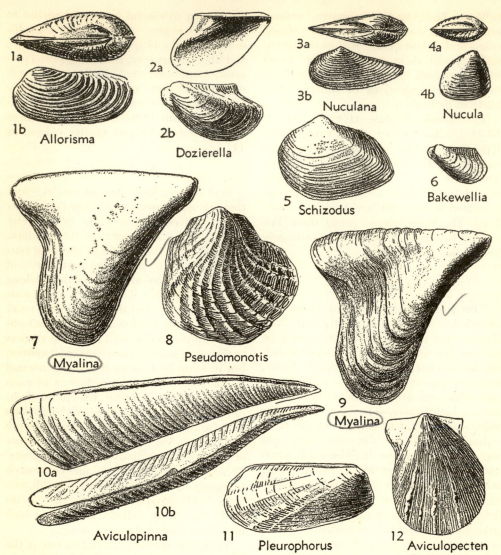

Fig. 10-15. **Representative Permian pelecypods.** General similarity of the Permian and Pennsylvanian pelecypod faunas is noteworthy. Among fossils here illustrated are taxodonts (3, 4), a schizodont (5), dysodonts (2, 6–10, 12), a heterodont (11), and a desmodont (1). Natural size except as indicated otherwise. L, left valve; R, right valve.

Allorisma King, Mississippian–Permian. *A. terminale* Hall (1*a*, dorsal; 1*b*, L; ×0.5), Lower Permian, Kansas.

Aviculopecten McCoy, Silurian–Permian. A pectinoid dysodont. *A. vanvleeti* Beede (12, L), Lower Permian, Oklahoma.

Aviculopinna Meek, Pennsylvanian–Permian. This elongate slender dysodont is a eulamellibranch. The shell commonly is found with the long axis in vertical position, the pointed end down, showing that the clam burrowed in sediment of the sea floor. *A. peracuta* (Shumard) (10*a*, R; 10*b*, dorsal; ×0.5), Lower Permian, Kansas.

Bakewellia King, Pennsylvanian–Permian. A characteristic Permian dysodont. *B. parva* Meek & Hayden (6, L, ×3), Lower Permian, Kansas.

(Continued on next page.)

a few (Fig. 10-16, *14*) have weak radial corrugations next to the hinge line which may be interpreted as a sort of dentition, inasmuch as elevations and grooves on opposite valves fit together. Beneath the beak of some shells (Fig. 10-16, *6*, *12*) is an area indented by narrow, broadly flaring chevron-shaped grooves. This is the ligament area, and the grooves mark places of attachment of the lamellar ligament. The ligament area, which also may be designated as cardinal area, is on the dorsal side of the hinge line; and although the apposed areas of the two valves are only narrowly separated, the ligament is external.

The ligament area and structures of some late Paleozoic and early Mesozoic shells (Fig. 10-16, *7–9*, *13*) differ from those of genera just considered in that a triangular depression below the beak serves for reception of ligament having fibrous structure and constituting a resilium. The depression is a resilifer. The remainder of the smooth ligament area on each side of the resilifer serves for the attachment of the lamellar ligament. Except for these differences, the structure of the hinge area is essentially the same as in older, more simple forms (Fig. 10-16, *6*, *12*).

Euchondria (Fig. 10-16, *10*) has a resilifer on the ligament area of each valve, and the remainder of the area is vertically grooved. The fibrous ligament is fastened to the grooves and the lamellar ligament, in the remaining spaces between the grooves.

Advanced members of the Pectinacea (Fig. 10-16, *11*, *14*) differ from those so far described in that the hinge axis is on the dorsal side of the resilifer, rather than on its ventral side. They have an internal ligament, and when the valves are closed, it is not visible.

Characters of the hinge are important in distinguishing Pectinacea and in classifying them. These clams differ from most others in that the hinge is at one side, for the animal lies on what is anatomically the right valve, and the left valve is uppermost. Right valves are invariably identifiable by the presence of the deep byssal notch or slit below the anterior auricle, by observation of the forward (prosocline) obliquity of growth lines near the beak, and by location of the single adductor muscle scar slightly behind the middle of the valve. Left valves have a very shallow sinus beneath the anterior auricle, and they show characters of the beak and muscle-scar location which may be used for orientation as in the right valves. It is interesting to note that among primitive Pectinacea, the right or lower valve is less convex than the left or upper valve. On the other hand, the valves of some Paleozoic forms and many later ones, including living species, are subequal. Finally, there are many Mesozoic and Cenozoic Pectinacea in which the lower valve is very convex, whereas the upper one is flattened or concave. Yet the immature parts of such shells, near the beak, demonstrate that in youth the lower valve was flattened and the upper more convex.

The oldest known representative of the Pectinacea is *Pterinea* (Figs.; 10-10, *2*;

(Fig. 10-15 continued.)

Dozierella Newell, Permian. A pectinoid dysodont. *D. gouldi* (Beede) (*2a*, *b*, L, ×3), Lower Permian, Oklahoma.

Myalina De Koninck, Mississippian–Permian. A mytilid dysodont. *M. pliopetina* Newell (7, R, ×0.5), *M. copei* Whitfield (9, R, ×0.5); both Lower Permian, Texas.

Nucula Lamarck, Silurian–Recent. Nuculid taxodont. *N. montpelierensis* Girty (4, R, ×2), Lower Permian, Idaho.

Nuculana Link, Silurian–Recent. Nuculid taxo-

dont. *N. bellistriata* (Stevens) (*3a*, dorsal; *3b*, L), Lower Permian, Kansas.

Pleurophorus King, Devonian–Triassic. A primitive heterodont. *P. albequus* Beede (11, R, ×3), Lower Permian, Oklahoma.

Pseudomonotis Beyrich, Pennsylvanian–Permian. A pectinoid dysodont. *P. hawni* (Meek & Hayden) (8, L), Lower Permian, Kansas.

Schizodus King, Devonian–Permian. A schizodont. *S. wheeleri* Swallow (5, L), Lower Permian, Kansas.

10-16, *12*), which occurs in Ordovician rocks. It has an elongate hinge line and well-defined anterior and posterior auricles. The shell has a very distinct forward obliquity (prosocline). The surface is marked by closely spaced concentric lines, or fine radial costae, or both. *Pterinea* differs from most later Pectinacea in that the lower (right) valve is flat and the upper (left) valve, distinctly convex. Shells very similar to *Pterinea* occur in Silurian and Devonian rocks (Figs.; 10-10, *17;* 10-11, *3–5, 11, 21;* 10-16, *6*).

Numerous Mississippian and Permian representatives of the Pectinacea differ from forms so far described in having a resilifer in the mid-portion of the ligament area. They are distinguished by differences in outline of the shell, including auricles, obliquity of the axis, shell structure, and features of surface ornamentation. One of the most important genera is *Aviculopecten*

(Figs. 10-13, *14;* 10-51, *12;* 10-16, *8*), but others (Figs. 10-13, *2, 21–23;* 10-15, *8*) are nearly as numerous and stratigraphically useful. Very oblique shells with prominent posterior auricles occur in the Pennsylvanian (Fig. 10-13, *17, 19*).

Mesozoic representatives of the Pectinacea are common, although not dominant in pelecypod assemblages. The Triassic contains important equivalve pectinacean shells (Fig. 10-17, *1, 2, 5*), and several occur in Jurassic and Cretaceous rocks (Figs. 10-17, *12, 13;* 10-19, *7*).

Cenozoic Pectinacea, which include the highly varied modern types of scallops, are well represented among fossils. The dominant form is *Pecten* and its subgenera (Figs. 10-23, *12;* 10-25, *7, 10*). The shells of *Pecten* are equivalve or have the left valve more convex than the right, and the hinge axis occurs at the dorsal edges of the shell, rather than along a line which is separated

---

FIG. 10-16.   **Representative dysodont pelecypods having filibranch gill structure.** This group, which includes the mytiloid and pectinoid clams, is characterized by complete or nearly complete absence of hinge teeth and generally part-internal part-external ligament. Stratigraphic range is from Ordovician to Recent. Arrows point to anterior part of shell. (*cl,* calcified ligament; *fl,* fibrous ligament; *la,* ligament area; *ll,* laminated ligament; *m,* muscle scar; *R,* resilifer; *t,* tooth). Not to scale.

1, *Mytilus edulis* Linné, Recent, cosmopolitan.

2, *Volsella modiola* (Linné), Recent, Atlantic. 2b is enlarged transverse section at middle of hinge showing ligament structure.

3, *Selenimyalina meliniformis* (Meek & Worthen), Middle Pennsylvanian, Missouri; showing parallel linear ligament grooves on ligament area.

4, *Myalina* (*Myalina*) *goldfussiana* De Koninck, Upper Mississippian, Belgium; showing ligament grooves as in 3.

5, *Myalina* (*Orthomyalina*) *slocomi* Sayre, Upper Pennsylvanian, Kansas; ligament area is wide and the shell thick.

6, *Pterinopecten undosus* Hall, Middle Devonian, New York; linear ligament grooves have the form of widely expanded chevrons.

7, *Fasciculiconcha knighti* Newell, Upper Pennsylvanian, Missouri; a relatively large oblique resilifer divides the smooth ligament area.

8, *Aviculopecten exemplarius* Newell, Upper Pennsyl-

vanian, Kansas; hinge structure much like that of *Fasciculiconcha.*

9, *Pseudomonotis hawni* (Meek & Worthen), Lower Permian, Kansas.

10, *Euchondria subcancellata* Newell, Middle Pennsylvanian, Missouri; ligament area marked by vertical ligament grooves and oblique resilifer beneath beak.

11, *Pernopecten clypeatus* Newell, Upper Pennsylvanian, Kansas; distinguished by small triangular auricles on left valve, projecting above hinge line.

12, *Pterinea demissa* (Conrad), Upper Ordovician, Ohio.

13, *Pteria costa a* Sowerby, Jurassic, Spain.

14, *Pecten* (*Chlamys*) *varius* Linné, Pliocene, Mediterranean region.

15, *Inoceramus cripsi* Mantell, Upper Cretaceous, Austria; numerous ligament pits along ligament area.

16, *Gervillia aviculoides* Sowerby, Jurassic, Spain; vertical ligament grooves.

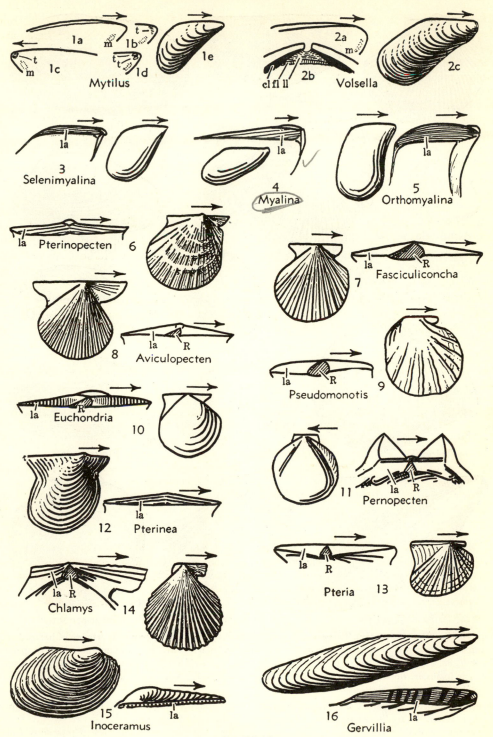

FIG. 10-16.

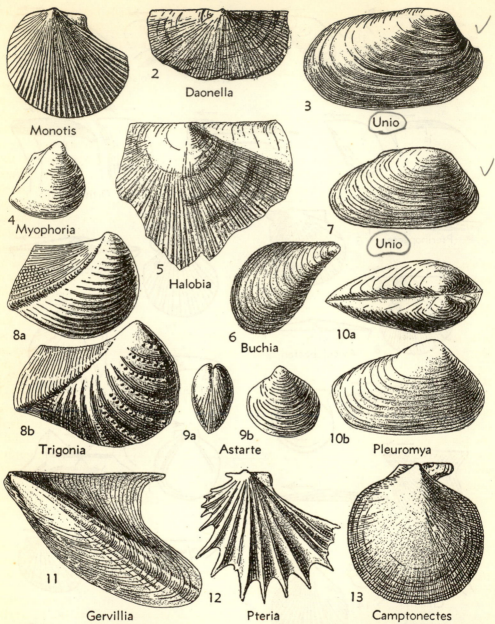

FIG. 10-17. **Representative Triassic and Jurassic pelecypods.** Relative unimportance of heterodonts stamps this assemblage as more akin to ancient than modern forms. Schizodonts include four species (3, 4, 7, 8*a*, *b*), dysodonts seven (1, 2, 5, 6, 11–13), and heterodonts (9) and desmodonts (10) one each. Triassic fossils are illustrated in 1–5 and Jurassic species in 6–13. Natural size except as indicated otherwise. L, left valve; R, right valve.

Astarte Sowerby, Triassic–Recent. A heterodont. *A. packardi* White (9*a*, dorsal; 9*b*, L), Upper Jurassic, Idaho.

Buchia Rouillier, Jurassic–Cretaceous. Mytilid dysodont. *B. piochei* (Gabb) (6, R), Upper Jurassic, California.

Camptonectes Meek, Jurassic–Cretaceous (sub-
(Continued on next page.)

by the ligament area from the beak.

As a group, the Pectinacea are one of the most readily differentiated assemblages of clams. They are a very long ranging division, and throughout their existence, evolutionary change is really not great. The modification represented by their mode of life, lying on one side, is responsible for noteworthy torsion of soft body parts, which distinguishes them from most other clams. Once established, however, their mode of life and general shell characters remained unchanged from Ordovician time to the present.

Mytilacea

Running parallel to the Pectinacea in geological distribution from Ordovician to modern times, and resembling them in having filibranch gill structure, are the Mytilacea. Shells belonging to this group are more varied and certainly less distinctive than the Pectinacea, but they are not less important in stratigraphical paleontology. It is rather difficult to define characters which are common to the whole group, although all are sedentary, rather than active mobile clams. Some are equivalve, living attached, with the valves erect and the plane of commissure approximately in vertical position. Some of the Mytilacea are distinctly inequivalve, and there are differences in shell structure of the two valves, as in Pectinacea. It is reasonable to judge that these shells lay on their sides and that the smaller, flatter right valve was undermost. This deviation from bilateral symmetry denotes specialization which results from adaptation to a particular mode of life. A character that distinguishes most, but not all, Mytilacea is rather strong shell obliquity and the terminal position of the sharp beaks at the anterior extremity of the hinge line.

The type form of the Mytilacea is the modern slipper shell called *Mytilus* (Figs. 10-16, *1;* 10-25, *11*), and from it is derived the adjective mytiliform (Fig. 10-6, *24*) to denote a rather distinctive shell shape. *Mytilus* is a widely distributed pelecypod of the shallow seas, its preferred habitat being from high-water mark to depths of a few fathoms. The shell is attached by a byssus, which extends through a slight antero-ventral gap, and wherever found, it is likely to be extremely abundant, for these pelecypods are highly gregarious. Areas of shallow sea bottom exposed at low tide may be so densely covered by these shells that no free space is left for other organisms. Thousands of growing *Mytilus* individuals to the square foot have been reported. *Mytilus* can tolerate great variations in salinity; and by tightly closing its shell so as to retain a quantity of sea water in the pallial cavity, it can live many hours exposed to the air.

Mytilus has rudimentary teeth just below

(Fig. 10-17 continued.)

genus of *Pecten*). *P. (C.) bellistriatus* Meek (13, R, ×0.5), Upper Jurassic, Wyoming.

Daonella Mosjisovics, Triassic. Pectinoid dysodont. *D. americana* Smith (2, R), Middle Triassic, Nevada.

Gervillia Defrance, Jurassic–Eocene. Dysodont. *G. montanaensis* Meek (11, L), Upper Jurassic, Montana.

Halobia Bronn, Triassic. Pectinoid dysodont. *H. superba* Mojsisovics (5, R), Upper Triassic, Alps.

Monotis Bronn, Triassic. Pectinoid dysodont. *M. subcircularis* (Gabb) (1, L, ×0.5), Upper Triassic, California.

Myophoria Bronn, Triassic. Schizodont. *M. alta* Gabb (4, R, ×2), Upper Triassic, California.

Pleuromya Agassiz, Triassic–Cretaceous. Desmodont. *P. subcompressa* Meek (10a, dorsal; 10b, R), Upper Jurassic, Wyoming.

Pteria Scopoli, Jurassic–Recent. Pectinoid dysodont. *P. submcconnelli* McLearn (12, L, ×0.5), Middle? Jurassic, British Columbia.

Trigonia Bruguière, Jurassic–Recent. Schizodont. *T. americana* Meek (8a, R), Upper Jurassic, Utah. *T. montanaensis* Meek (8b, R), Upper Jurassic, Montana.

Unio Retzius, Triassic–Recent. Schizodont. *U. dockumensis* Simpson (3, R), Upper Triassic, Texas. *U. felchi* White (7, R, ×0.5), Upper Jurassic, Colorado.

the beak, or the shell is toothless (Fig. 10-16, *1*). When the adductor muscles relax, the shell is slightly opened by the very narrow ligament which extends along the dorsal margin where the valves are fastened together. A transverse section of the ligament of *Volsella* (Fig. 10-16, *2b*), a close relative of *Mytilus*, clearly shows its relation to the shell along the hinge line and indicates the construction in which three elements (calcified ligament, fibrous ligament, lamellar ligament) can be distinguished. Comparative study of pelecypod ligaments shows that this feature has considerable importance in classification of the clams, and it permits an understanding of the markings found on ligament areas.

The shell of some mytilaceans, such as *Mytilus*, may be very thin; but in others (many species of *Myalina*, Fig. 10-16, *4*), it is decidedly thick, although by no means so ponderous as in some oysters, pachyodonts, and certain other pelecypods. In *Myalina*, the ostracum of both valves may be largely composed of prismatic calcite; but in some, only the right valve has prismatic structure, the left valve consisting of homogeneous calcareous material.

Two distinct groups of mytilaceans are recognized where the suborder first appears in Ordovician rocks. *Byssonychia*, *Ambonychia* (Fig. 10-10, *5*, *12*), and some others have sharp-pointed terminal beaks, and only a single well-defined adductor muscle is present, located in posterior position. The strongly oblique shells have radially arranged fine to coarse costae. Like *Mytilus*, the valves are equal. The other group is represented by genera (Fig. 10-10, *3*, *6*) which hardly differ in appearance from many nonmytilacean pelecypods. They have anterior and posterior adductor muscles and correspond to *Mytilus* in the nature of the ligament and hinge.

Goniophora (Figs. 10-10, *14;* 10-11, *20*) is one of several mytilaceans occurring in Silurian and Devonian rocks. The shape of the shell and especially the very angular umbonal ridge on each valve are distinguishing external features. Mississippian

and later Paleozoic mytilaceans include *Myalina* and other important forms (Figs. 10-13, *4, 8, 11, 13, 20, 24;* 10-15, *7, 9;* 10-16, *3–5*). The myalinids are especially important among late Paleozoic pelecypods, for they occur abundantly in many formations and are useful as guide fossils. *Naiadites* differs from most other mytilaceans in being adapted to a fresh-water environment, or at least estuarine conditions. It is chiefly found associated with the clams *Anthracomya* and *Carbonicola* in shales above coal beds, and none of the fossils associated with it are normal marine invertebrates.

Mesozoic mytilaceans (Figs. 10-17, *6;* 10-19, *1*) are abundant in some Jurassic and Cretaceous rocks and useful as guide fossils. *Volsella* (Figs. 10-19, *4;* 10-23, *6*), which closely resembles *Mytilus* in shape, occurs in Cretaceous and Paleogene deposits. A fossil *Mytilus* from Neogene rocks is illustrated in Fig. 10-25, *11*.

An important family is the Pernidae, a dysodont assemblage in which some genera resemble *Mytilus* but others are not at all obviously related to it. Examples are *Gervillia* and *Inoceramus* (Figs. 10-16, *15, 16;* 10-17, *11;* 10-19, *15*), of which the first is a common distinctive clam found in Triassic and Jurassic rocks, and the second is a characteristic fossil in Cretaceous deposits. Both genera are characterized by a ligament area which contains numerous depressions or pits arranged normal to the hinge line. These define places of attachment of fibrous ligaments, whereas other parts of the area are covered by lamellar ligament. Hinge teeth are lacking. *Gervillia* has a very oblique outline, and the hinge line may be extended posteriorly as in the manner of an auricle. *Inoceramus* has a thin to moderately thick shell in which the prismatic structure of the ostracum is very prominent. The surface commonly is marked by concentric wrinkles.

Pinnacea

Pelecypods belonging to the suborder Pinnacea, together with three other suborders of the Dysodonta, are characterized by having eulamellibranch gill structure. The eulamellibranchs are more advanced than filibranchs (Fig. 10-7) in that the main gill leaves are connected at regular intervals by junctions that provide more extended circulation of the blood for aeration in the gills. The eulamellibranch gill is evidently more complicated than the filibranch type, and (Fig. 10-29) it characterizes not only an important group of dysodonts, but also heterodonts, pachyodonts, and desmodonts, which have hinge structures indicating specialization.

The Pinnacea have a known range from Devonian to Recent. They are eulamellibranch dysodonts having an elongated form, the shell truncated and somewhat gaping posteriorly, and distinguished by the presence of two adductor muscles, the small one in the anterior part of the shell and the large one in the posterior region. As illustrated by the modern *Atrina* (Fig. 10-18, *2*), the animal is attached by a prominent byssus, which extends through a narrow opening in the antero-ventral part of the shell. Pelecypods of this group are sedentary, living in a fixed position, partly buried in sediment of the sea bottom. Many fossil shells found in rock strata retain their original position of growth, with the long axis of the shell pointing subvertically. Permian and Cretaceous examples of the Pinnacea are shown in Figs. 10-15, *10*, and 10-19, *12*.

Ostreacea

A very important division of the eulamellibranch dysodonts comprises the oys-

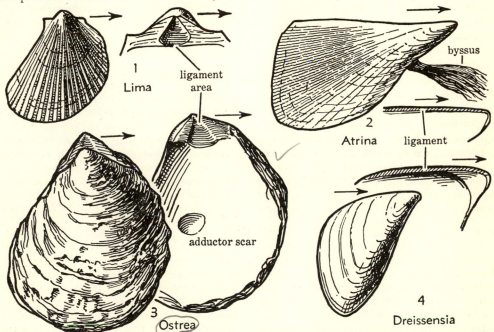

FIG. 10-18. **Representative dysodont pelecypods having eulamellibranch gill structure.** This division of the dysodonts includes especially the oysters, limas, and pinnas, some of which live in fixed position on the sea floor and others are free-moving. They range from Devonian to Recent. Arrows point to anterior part of shell.

1, *Lima pectinoides* Sowerby, Lower Jurassic, Germany, ×1; ligament area enlarged.

2, *Atrina serrata* Sowerby, Recent, Florida, ×0.16; a very thin-shelled member of the Pinnacea.

3, *Ostrea percrassa* Conrad, Miocene, Maryland,

×0.3; the ligament is partly external and partly internal, occupying a resilifer-like depression.

4, *Dreissensia* sp., Recent, Europe, ×2.

ters. The suborder Ostreacea includes not only the genus *Ostrea* (Fig. 10-18, *3*), but several other genera, such as *Exogyra*, *Gryphaea*, and others important in the paleontological record (Fig. 10-19, *2*, *3*, *5*, *6*, *14*). The oysters lack hinge teeth and are characterized by their inequivalve form and attached mode of life. The left valve is lowermost and commonly is cemented to some foreign object, such as a rock or another shell. The right valve is smaller and generally flatter than the left, and thus constitutes a sort of lid. The ligament area, partly external and partly internal, contains a depressed median portion which suggests the resilifer of other pelecypods. The oysters are monomyarian, the single adductor muscle scar being located in the postero-median part of the valve interior (Fig. 10-18, *3*).

The genus *Ostrea* ranges throughout Mesozoic and Cenozoic rocks (Figs. 10-20, *7;* 10-27, *8*). Other genera assigned to the Ostreacea include forms having spirally twisted beaks, and some having exceptionally thick shells (*Exogyra*, Fig. 10-19, *2*, *14; Gryphaea*, Fig. 10-19, *3*). A Cretaceous oyster, *Alectryonia* (Fig. 10-19, *5*, *6*), is characterized by the strongly plicated nature of its shell. Oysters are an abundant and economically important element of shallow-water pelecypod faunas of the present day, and they are able to live in somewhat brackish as well as in normal saline waters.

Limacea

The group of eulamellibranch dysodonts called Limacea includes somewhat inequivalve, strongly inequilateral shells, which resemble those of the Pectinacea. Like the pectens, some of the Limacea are very good swimmers. They are differentiated from Pectinacea in having a eulamellibranch gill structure. Members of this group range from Pennsylvanian to Recent (Figs. 10-13, *7;* 10-20, *12*).

Fɪɢ. 10-19. **Representative Cretaceous pelecypods.** Prominence of ostreacean dysodonts characterizes late Mesozoic clam assemblages. Forms illustrated here include isodonts (10, 13), dysodonts (1–7, 12, 14, 15), and desmodonts (8, 9, 11). Natural size except as indicated otherwise. L, left valve; R, right valve.

Alectryonia Fischer, Cretaceous (subgenus of *Ostrea*). A plicate oyster. *A. quadriplicata* Shumard (5, R); *A. carinata* Lamarck (6, L, ×0.5); both Lower Cretaceous (Washitan), Texas.

Anomia Linné, Jurassic–Recent. Thin-shelled isodont. *A. argenteria* Morton (13*a*, *b*, R), Upper Cretaceous, Tennessee.

Buchia Rouillier, Jurassic–Cretaceous. Mytilid dysodont. *B. terebratuloides* (Lahusen) (1*a*, anterior; 1*b*, R), Lower Cretaceous, California.

Cuspidaria Nardo, Jurassic–Recent. A small desmodont having the highly specialized septibranch type of gill structure. *C. moreauensis* Meek & Hayden (8, R, ×2), Upper Cretaceous, South Dakota. *C. ventricosa* Meek & Hayden (9, L, ×2), Upper Cretaceous, Montana.

Exogyra Say, Jurassic–Cretaceous. An ostreacean dysodont characterized by large spirally twisted left valve and small flat right valve. *E. arietina* Roemer (2*a*, R; 2*b*, L), Lower Cretaceous (Washitan), Texas. *E. costata* Say (14*a*, posterior; 14*b*, L; ×0.5), Upper Cretaceous, Mississippi.

Gryphaea Lamarck, Jurassic–Eocene. Ostreacean dysodont. *G. corrugata* Say (3*a*, *b*, L, ×0.5), Lower Cretaceous (Washitan), Texas.

Inoceramus Sowerby, Jurassic–Cretaceous. Concentrically wrinkled dysodont. *I. labiatus* Schlotheim (15, R, ×0.5), Upper Cretaceous, Kansas.

Paranomia Conrad, Cretaceous. Isodont. *P. scabra* (Morton) (10, L), Upper Cretaceous, New Jersey.

Pholadomya Sowerby, Jurassic–Recent. Desmodont. *P. papyracea* Meek & Hayden (11*a*, dorsal; 11*b*, L), Upper Cretaceous, Kansas.

Pinna Linné, Jurassic–Recent. An elongate dysodont with terminal beaks. *P. laqueata* Conrad (12, R, ×0.5), Upper Cretaceous, New Jersey.

Pteria Scopoli, Devonian–Recent. Pectinoid dysodont. *P. petrosa* (Conrad) (7, R), Upper Cretaceous, Tennessee.

Volsella Scopoli, Devonian–Recent. Mytilid dysodont. *V. multilinigera* (Meek) (4, L), Upper Cretaceous, Colorado.

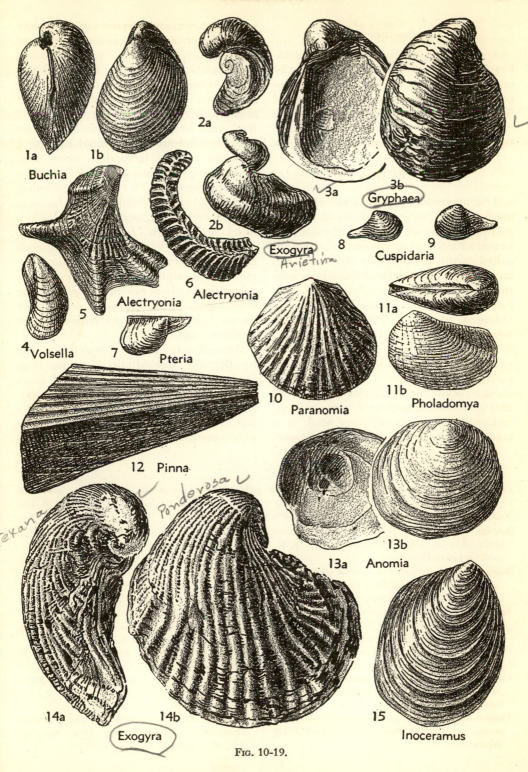

1a 1b
Buchia

2a

2b

3a 3b
Gryphaea

Exogyra
Arietina

8
Cuspidaria 9

5 Alectryonia 6 Alectryonia

4 Volsella 7 Pteria

11a

10 Paranomia 11b Pholadomya

12 Pinna

Texana Ponderosa

13a Anomia 13b

14a 14b
Exogyra

15 Inoceramus

Fig. 10-19.

Dreissensiacea

A minor group of mytiliform eulamellibranch dysodonts is included in this suborder, which is confined to Cenozoic deposits. The shells have a small external ligament and resemble closely those of the Mytilacea, but the group is distinguished by the type of gill structure (Fig. 10-18, *4*).

HETERODONTS

The heterodont group of clams, which are among the most prominent in modern marine faunas, are an order of eulamellibranchs having clearly differentiated teeth called cardinals just below the beaks, and lateral teeth on anterior and posterior portions of the hinge plate (Fig. 10-22). The cardinal teeth are disposed vertically or obliquely to the hinge line, whereas the lateral teeth are approximately parallel to the hinge. The tooth projections on one valve fit into sockets in a corresponding position on the opposite valve. The ligament typically is external and opisthodetic, that is, extending backward from the beak.

The arrangement of hinge teeth is a constant and therefore diagnostic feature, which serves to characterize various genera and families of heterodonts. According to various conventions, the tooth arrangement can be expressed by a formula. Thus, in the system of Steinman, if the presence of a tooth is indicated by the numeral 1 and a socket by 0, and if the succession of teeth and sockets is listed from front to back, the formula for cardinal teeth of a cyrenoid type of heterodont clam (Fig. 10-21) may be indicated as *1 0 1 0 1 0* for a right valve and *0 1 0 1 0 1* for a left valve. The formula for cardinal teeth of a lucinoid type of heterodont, according to this system (Fig. 10-21), is *1 0 1 0* for a right valve and *0 1 0 1* for a left valve. Some specialists prefer this method because of its simplicity and lack of implication as to evolutionary relationships. It is evident, however, that homologous teeth of the cyrenoid and lucinoid groups, as thus indicated, are unidentifiable.

According to another system introduced by Bernard and Munier-Chalmas (Fig. 10-21, *BM*), the central cardinal tooth below the beak of a right valve is designated as *1* and that below the beak of a left valve as *2*. Cardinal teeth of right valves are indicated by odd numerals and

---

FIG. 10-20. **Representative Upper Cretaceous pelecypods.** This group includes four taxodonts (2, 5, 8, 9), a schizodont (3), dysodonts (7, 12), five heterodonts (1, 6, 10, 11), and desmodonts (4, 13). Natural size except as indicated otherwise. L, left valve; R, right valve.

Arctica Schumacher, Jurassic–Recent. Heterodont. *A. ovata* (Meek & Hayden) (1*a*, *b*, R, ×0.5), North Dakota.

Barbatia Gray, Cretaceous–Recent (subgenus of *Arca*). Taxodont. *A.* (*B.*) *micronema* (Meek) (2*a*, *b*, R, ×0.5), Wyoming.

Breviarca Conrad, Cretaceous–Recent. Taxodont. *B. haddonfieldensis* Stephenson (8*a*, R, ×2; 8*b*, R, ×1), New Jersey.

Cardium Linné, Triassic–Recent. Heterodont. *C. eufaulense* Conrad (6*a*, *b*, L), Maryland.

Corbula Bruguière, Triassic–Recent. Inequivalve desmodont. *C. crassiplicata* Gabb (4*a*, L; 4*b*, R, ×4), Maryland.

Crassatellites Krueger, Cretaceous–Recent. Desmodont. *C. vadosus* (Morton) (13*a*, R; 13*b*, L), Maryland.

Cyprimeria Conrad, Cretaceous. Heterodont. *C. alta* Conrad (11*a*, *b*, L, ×0.5), Tennessee.

Idonearca Conrad, Cretaceous (subgenus of *Cucullaea*). Taxodont. *C.* (*I.*) *carolinensis* (Gabb) (5*a*, *b*, R), North Carolina.

Lima Bruguière, Pennsylvanian–Recent. Dysodont. *L. reticulata* Forbes (12, L), Maryland.

Nucula Lamarck, Silurian–Recent. Taxodont. *N. percrassa* Conrad (9*a*, *b*, R), Tennessee.

Ostrea Linné, Triassic–Recent. Dysodont. *O. cretacea* Morton (7*a*, *b*, L), North Carolina.

Trigonia Bruguière, Jurassic–Recent. Schizodont. *T. eufaulensis* Gabb (3*a*, L dorsal; 3*b*, L), North Carolina.

Veniella Stoliczka, Cretaceous–Tertiary. Heterodont. *V. conradi* (Morton) (10*a*, anterior; 10*b*, R, ×0.5), Tennessee.

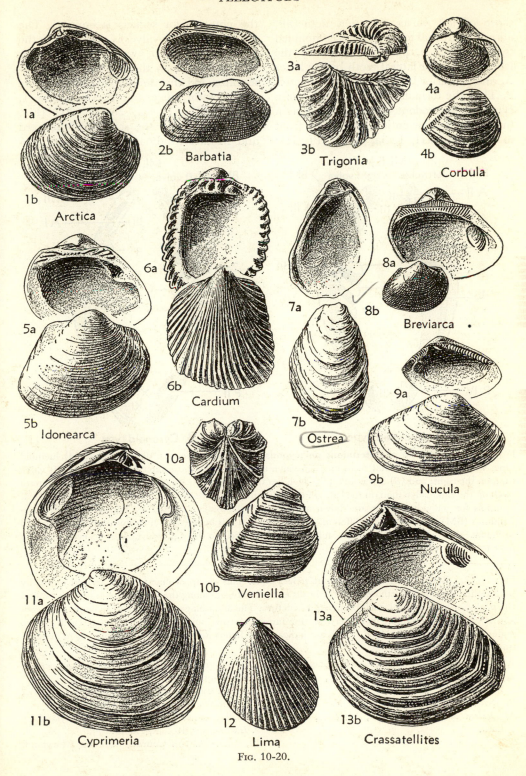

1a
1b
Arctica

2a
2b Barbatia

3a
3b Trigonia

4a
4b Corbula

5a
5b Idonearca

6a
6b Cardium

7a
7b Ostrea

8a
8b Breviarca

9a
9b Nucula

10a
10b Veniella

11a
11b Cyprimeria

12 Lima

13a
13b Crassatellites

Fig. 10-20.

those of left valves by even numerals, anterior cardinal teeth being accompanied by the letter *a*, posterior teeth by the letter *b*, anterior lateral teeth by the letter A, and posterior lateral teeth by the letter P. Thus the formula for a right valve of a cyrenoid heterodont clam in this system (according to convention, listed from front to back) is AIII, I, 3a, 1, 3b, PI, III; the notation for sockets is omitted. The dentition of both valves is expressed in the form of a fraction, with the formula for the right valve in the position of the numerator and that for the left valve as the denominator, for example, AIII, I, 3a, 1, 3b, PI, III/AII, 2a, 2b, 4b, PII (Fig. 10-21). Although the scheme of Bernard and Munier-Chalmas is widely used, be-

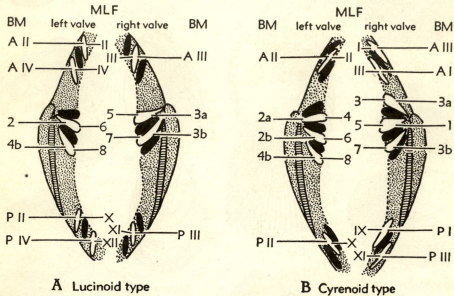

A Lucinoid type   **B** Cyrenoid type

FIG. 10-21.   **Dentition of heterodont pelecypods.** Two main types are distinguished: (*A*) lucinoid type, which has two cardinal teeth on each valve and may have more lateral teeth on the left valve than on the right; and (*B*) cyrenoid type, which has three cardinal teeth on each valve and more lateral teeth on the right valve than on the left. On the diagrams, teeth are shown in white and sockets in black. Symbols for individual teeth are shown according to the system devised by Bernard and Munier-Chalmas (BM) and an entirely different, new one used in this book (MLF). Attention should be called to the fact that in both systems odd numbers refer exclusively to the right valve, whereas even numbers refer to the left valve.

FIG. 10-22.   **Representative heterodont pelecypods.** This group of clams, characterized especially by prominent cardinal teeth beneath the beaks and lateral teeth along the hinge plate in front or behind the beaks, or both, has eulamellibranch gill structure. The assemblage ranges from Devonian to Recent and is especially dominant in modern marine faunas. Arrows point to anterior part of shell. Not to scale; hinge structures enlarged. Teeth are indicated by letters and numbers corresponding to designations given in Fig. 10-21.

1, *Cyrena semistriata* Deshayes, Eocene, England.

2, *Venericardia planicosta* (Lamarck), Eocene, France.

3, *Mercenaria mercenaria* (Linné), Recent, Atlantic.

4, *Dosinia acetabulum* Conrad, Miocene, Maryland.

5, *Cyprimeria alta* Conrad, Upper Cretaceous, Mississippi.

6, *Lucina columbella* Lamarck, Miocene, Austria.

7, *Americardia* sp., Recent, California.

8, *Tellina declivis* Conrad, Miocene, Maryland.

9, *Isocardia* sp., Recent, Atlantic.

10, *Astarte thomasii* Conrad, Miocene, Maryland.

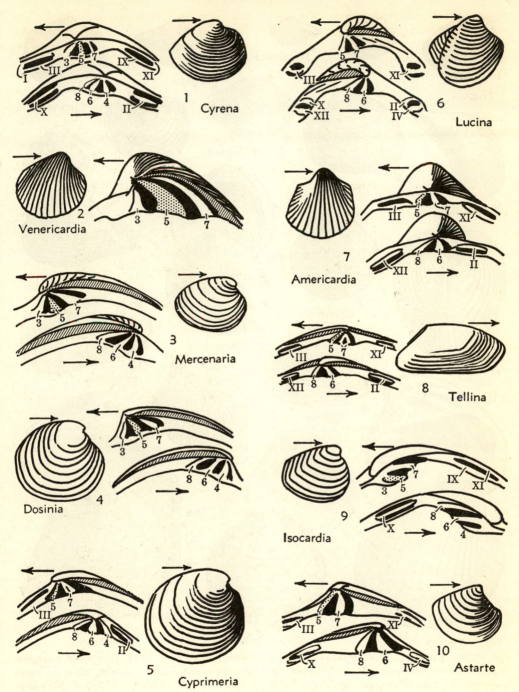

FIG. 10-22.

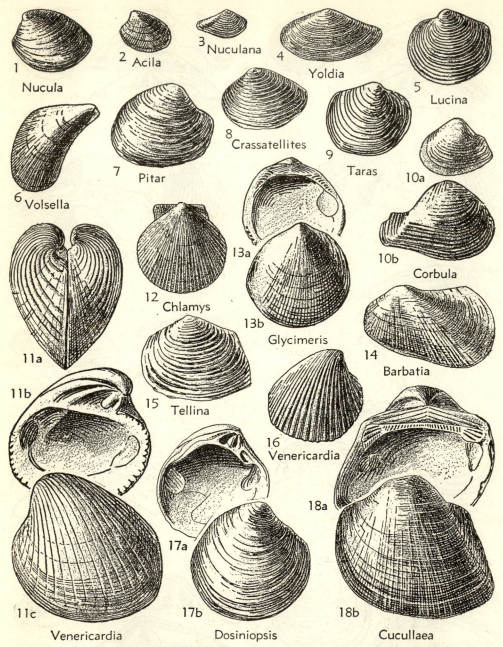

FIG. 10-23. **Representative Paleogene pelecypods.** Heterodonts (5, 7, 9, 11, 15–17) and taxodonts (1–4, 13, 14, 18) predominate in this assemblage; others illustrated are dysodonts (6, 12) and desmodonts (8, 10). Natural size and from Eocene of Maryland except as indicated otherwise. L, left valve; R, right valve.

Acila Adams & Adams, Cretaceous-Recent. A nuculid Taxodont. *A. shumardi* (Dall) (2, L), Oligocene, Washington.

Barbatia Gray, Cretaceous–Recent (subgenus of *Arca*). An arcid taxodont. *A. (B.) morsei* Gabb (14, L, ×2), Eocene, Washington.

(*Continued on next page.*)

cause it is the only proposed method permitting designation of individual teeth, it has numerous defects in addition to being very cumbersome. Accordingly, a practical, much simplified system is introduced in this book. Incorporating the best features of the plan devised by the French paleontologists, the new method avoids duplication of letters and numbers, and it requires no fractional form of statement of the dentition formula. Remembering that all odd numbers denote teeth on the right valve (with corresponding sockets on the left valve), and similarly that even numbers refer to teeth on the left valve, also using Roman and Arabic numerals in the manner of Bernard and Munier-Chalmas, the complete formula for a cyrenoid heterodont may be expressed as I, II, III, 3, 4, 5, 6, 7, 8, IX, X, XI, or briefly as I–III, 3–8, IX–XI. Using a hyphen (-) to represent sockets, the formula for a single cyrenoid right valve is I-III, 3-5-7-, IX-X; or employing parentheses and numbers for notation of sockets, this dentition is expressed as I(II)III, 3(4)5(6)7(8), IX(X)XI.

Cypricardiacea

A primitive generalized type of heterodont, which ranges from Silurian to Recent times, is represented by the Cypricardi-

acea. They have two or three cardinal teeth in each valve, of which the posterior one is nearly parallel to the hinge line. A Silurian representative of this group is illustrated in Fig. 10-10, *15;* Devonian examples in Fig. 10-11, *15, 18;* Mississippian species in Fig. 10-13, *3, 5;* a Pennsylvanian example in Fig. 10-13, *16;* and a Permian form in Fig. 10-15, *11.* Mesozoic pelecypods classed among Cypricardiacea include *Arctica* and *Veniella,* Fig. 10-20, *1, 10.*

Lucinacea

The heterodont group called Lucinacea is more advanced than the Cypricardiacea in having well-differentiated cardinal and lateral teeth, which are well formed but fewer in number than in the third group, called Cyrenacea. Typical lucinoid-type heterodonts (Fig. 10-21) have two lateral teeth on the anterior and posterior portions of the left valve and two cardinal teeth on each valve, the complete formula being II, III, IV, 5, 6, 7, 8, X, XI, XII (Fig. 10-21*A*). Heterodonts of this type range from Silurian to Recent. A Devonian representative of the group is *Paracyclas* (Fig. 10-11, *14);* a Mesozoic example, *Astarte,* is illustrated in Fig. 10-17, *9;* Cenozoic forms include *Lucina* (Fig. 10-23, *5*) and *Tellina* (Fig. 10-25, *3*). The dentition

---

(Fig. 10-23 continued.)

Chlamys Bolten, Triassic–Recent (subgenus of *Pecten*). Dysodont. *P.* (*C.*) *choctavensis* Aldrich (12, L).

Corbula Bruguière, Triassic–Recent. Inequivalve desmodont. *C. aldrichi* Meyer (10*a*, L; 10*b*, R, ×3).

Crassatellites Krueger, Cretaceous–Recent. Desmodont. *C. gabbi* (Safford) (8, L), Paleocene, Texas.

Cucullaea Lamarck, Jurassic–Recent. Arcid taxodont. *C. gigantea* Conrad (18*a*, *b*, L).

Dosiniopsis Conrad, Eocene. Heterodont. *D. lenticularis* (Rogers) (17*a*, L; 17*b*, R).

Glycimeris Costa, Cretaceous–Recent. Arcid taxodont. *G. idonea* (Conrad) (13*a*, *b*, R).

Lucina Bruguière, Triassic–Recent. Heterodont. *L. smithi* Meyer (5, L, ×5).

Nucula Lamarck, Silurian–Recent. Taxodont.

N. ovula Lea (1, L, ×2).

Nuculana Link, Silurian–Recent. Taxodont. *N. parva* (Rogers) (3, R, ×4).

Pitar Roemer, Eocene–Recent. Heterodont. *P. uvasana* (Conrad) (7, R), Eocene, California.

Taras Risso, Eocene–Recent. Heterodont. *T. hopkinsensis* (Clark) (9, R, ×2).

Tellina Linné, Jurassic–Recent. Heterodont. *T. undulifera* Gabb (15, L), Paleocene, California.

Venericardia Lamarck, Cretaceous–Recent. Heterodont. *V. planicosta* Lamarck (11*a*, anterior; 11*b*, *c*, L, ×0.5), Eocene, Texas. *V. smithi* Aldrich (16, R, ×0.5), Paleocene, Texas.

Volsella Scopoli, Devonian–Recent. Mytilid dysodont. *V. alabamensis* (Aldrich) (6, R).

Yoldia Möller, Pennsylvanian–Recent. Nuculid taxodont. *Y. eborea* (Conrad) (4, L), Paleocene, Texas.

of representative examples of the Lucinacea is shown in Fig. 10-22, *6–8, 10*.

Cyrenacea

The most highly differentiated pelecypods included among heterodonts belong to the suborder Cyrenacea. These have a known range which is restricted to Mesozoic and Cenozoic rocks. The group is especially prominent in modern marine pelecypod faunas.

Typical dentition of the Cyrenacea consists of two lateral teeth on anterior and posterior parts of the right valve and three well-developed cardinals on each valve (Fig. 10-21*B*). The hinge structure of representative Cyrenacea is illustrated in Fig. 10-22, *1–5*.

A well-known Cretaceous cyrenoid heterodont is *Cyprimeria* (Fig. 10-20, *11;* 10-22, *5*). This is a moderately large equivalve shell having a nearly circular outline and marked externally by more or less prominent concentric growth lines. *Venericardia* (Figs. 10-22, *2;* 10-23, *11, 16*) is a very widely distributed early Cenozoic genus which includes numerous described species. The chief distinguishing features are the radial flat-topped ribs separated by narrow grooves, and the characters of the dentition. Cenozoic members of the Cyrenacea are numerous (Figs. 10-23, *7, 9, 17;* 10-25, *4, 5, 12, 13*). Pelecypods of this group belong to the so-called normal, active, bottom-dwelling bivalves, which crawl about by means of their foot, the plane of commissure, defined by the junction of the valves, being maintained in an approximately vertical position.

PACHYODONTS

The Pachyodonta are late Mesozoic and Cenozoic specialized derivatives of the Heterodonta, which include some of the most strangely aberrant kinds of pelecypods known. They are adapted to a sedentary mode of life, and like the oysters, one valve lies below and the other one above. A few of the pachyodonts resemble oysters also in having the larger left valve lowermost, the right valve forming a rather flat lid as in *Toucasia* (Fig. 10-24, *5*), or consisting of a somewhat elevated spiral shell as in *Requienia* (Fig. 10-24, *2*). A majority of the pachyodonts are characterized by having a much larger right valve than left; and in this group, the right valve invariably is lowermost. A tendency to develop a snail-like, spirally twisted shell is evident in some genera (Fig. 10-24, *3, 6*), whereas others develop a decidedly coral-like form (Fig. 10-24, *4, 7–9*). The flat left valve may be a mere operculum, but on its inner side are strongly projected, highly specialized, thick teeth (Fig. 10-24, *7b, 9b*). The wall of the lower valve of some pachyodonts is extraordinarily thick; also it may include a complex system of open passageways (Fig. 10-24, *6c, 8c*).

Chamacea

The suborder Chamacea includes some of the least specialized pachyodonts, such as *Chama*, which is not strongly inequivalve and not very different from a typical heterodont (Fig. 10-24, *1*). Others are very strongly modified by a sessile mode of life. The Chamacea range from Jurassic to Recent, and among the forms illustrated in Fig. 10-24, all except the coral-like *Durania* and *Hippurites* (Fig. 10-24, *7, 9*) belong to the Chamacea. Maximum development in most of the described genera is in the Cretaceous. They are gregarious forms and particularly adapted to a reef environment.

Rudistacea

The most remarkably specialized pachyodonts are the Rudistacea, which are confined to rocks of Cretaceous age (Fig. 10-24, *7, 9*). In this group, the tendency toward a spiral form of shell is entirely lost, and the lower valve takes the form of an elongate thick-walled cone. The exterior surface is commonly marked by longitudinal ribs and grooves. Superficial resemblance to a large horn coral is strik-

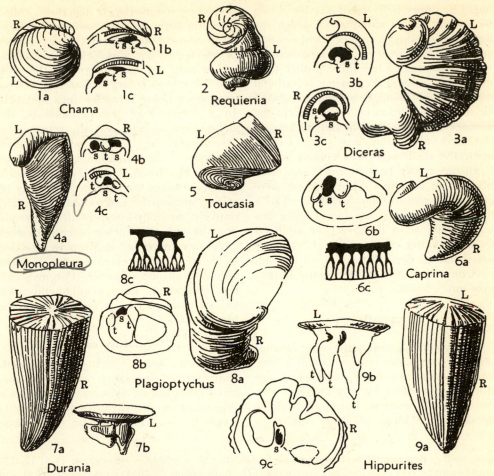

FIG. 10-24. **Representative pachyodont pelecypods.** The pachyodonts comprise specialized sedentary clams which have eulamellibranch gill structure. They tend to become strongly inequivalve, thick-shelled, spirally twisted or coral-like forms. They range from Jurassic to the present. (*L*, left valve; *l*, ligament; *R*, right valve; *s*, socket; *t*, tooth.)

1, *Chama squamosa* Lamarck, Eocene, England, ×0.7.

2, *Requienia ammonia* (Goldfuss), Lower Cretaceous, France, ×0.25.

3, *Diceras arietinum* Lamarck, Jurassic, France, ×0.5.

4, *Monopleura pinguiscula* White, Lower Cretaceous, Texas, ×0.7.

5, *Toucasia texana* (Roemer), Lower Cretaceous, Texas, ×0.7.

6, *Caprina adversa* d'Orbigny, Lower Cretaceous, France, ×0.2; 6*c* shows canals (white) pene-

trating shell (black) as seen in transverse section, enlarged.

7, *Durania cornupastoris* (d'Orbigny), Upper Cretaceous, France, ×0.3. The coral-shaped right valve articulates with a lidlike left valve which bears large hinge teeth, as shown in 7*b*.

8, *Plagioptychus aguilloni* (d'Orbigny), Upper Cretaceous, France; ×0.5; 8*c* shows canals (white) penetrating shell (black) as seen in transverse section, enlarged.

9, *Hippurites gosaviensis* Douvillé, Upper Cretaceous, Austria, ×0.3.

ing. The upper (left) valve is reduced to a nearly flat lidlike structure in which the ligament is shifted to a subcentral position and the elongate teeth are modified for vertical motion of the valve, rather than movement along a hinge. Rotational movement of the upper valve is prevented by the configuration of the teeth and the manner in which they fit into sockets on the fixed valve. Like many of the Chamacea, the shells of genera classed among Rudistacea may be very thick and complicated in structure. Most of them were gregarious reef dwellers which lived in a fixed location from the larval stage onward. They seem to have been confined to warm-water areas of the Cretaceous seas; and in deposits laid down in such environments, they include many important index fossils.

DESMODONTS

Pelecypods assigned to the order Desmodonta are primarily burrowers and borers, and most of them have shells which are distinctly modified by this mode of life. An internal ligament borne by resilifer areas on the hinge plate or carried in projecting spoon-shaped structures, called chondrophores, is characteristic. The posterior edges of the shell typically gape more or less widely for passage of the siphons, which may extend far outside the shell and become so enlarged that they are incapable of being retracted into the shell. In a majority of the desmodonts, the siphons are at least partially retractable; accordingly, the posterior part of the shell interior shows a distinct pallial sinus. Some genera (Fig. 10-26, *1, 2, 6, 7*) have heterodont-type hinge teeth; such forms may be classed among the heterodonts, but alternatively, one may interpret all the desmodonts as derivatives of the heterodont stock which are more or less strongly modified by their burrowing mode of life. With the exception of one suborder (Poromyacea), the desmodonts have a eulamellibranch gill structure, such as characterizes the heterodonts.

Anatinacea

Modern *Anatina* (Fig. 10-26, *7*) is a well-characterized desmodont which lacks special distinguishing features. The anterior projection of the shell is approximately equal to the posterior. The hinge plate

---

FIG. 10-25.   **Representative Neogene pelecypods.** In this assemblage, heterodonts (3–5, 12, 13) predominate; others illustrated include taxodonts (1, 9), dysodonts (7, 10, 11), desmodonts (6, 8), and an isodont (2). Natural size and from Miocene of Maryland except as indicated otherwise. L, left valve; R, right valve.

Acila Adams & Adams, Cretaceous-Recent. Nuculid taxodont. *A. gettysburgensis* Reagen (9, R), Miocene, Washington.

Chione Muhlfeldt, Miocene–Recent (subgenus of *Venus*). Heterodont. *V. (C.) latirata* Conrad (4a, dorsal; 4b, c, R).

Clementia Gray, Eocene–Recent. Heterodont. *C. inoceriformis* (Wagner) (5, L, ×0.5).

Glycimeris Costa, Cretaceous–Recent. Taxodont. *G. subovata* (Say) (1a, b, R, ×0.5), Miocene, Florida.

Lyropecten Conrad, Oligocene–Recent (subgenus of *Pecten*). Dysodont. *P. (L.) estrellanus* (Conrad) (10a, b, L, ×0.5), Miocene-Pliocene, California.

Mactra Linné, Cretaceous–Recent. A desmodont. *M. clathrodon* Lea (6a, b, L).

Mercenaria Schumacher, Miocene–Recent (subgenus of *Venus*). Heterodont. *V. (M.) mercenaria* Linné (13a, L; 13b, R, ×0.5).

Mytilus Linné, Triassic–Recent. Dysodont. *M. conradinus* d'Orbigny (11, R).

Pallium Schumacher, Miocene–Recent (subgenus of *Pecten*). Dysodont. *P. (P.) swifti nutteri* Arnold (7, R, ×0.5), Pliocene, California.

Panope Menard, Jurassic–Recent. Desmodont. *P. generosa* Gould (8, L, ×0.5), Miocene-Pleistocene, California.

Plicatula Lamarck, Triassic–Recent. Isodont. *P. densata* Conrad (2a, b, R), Miocene, Florida.

Taras Risso, Eocene–Recent. Heterodont. *T. acclinis* (Conrad) (12a, b, R).

Tellina Linné, Jurassic–Recent Heterodont. *T. declivis* Conrad (3, R, ×2).

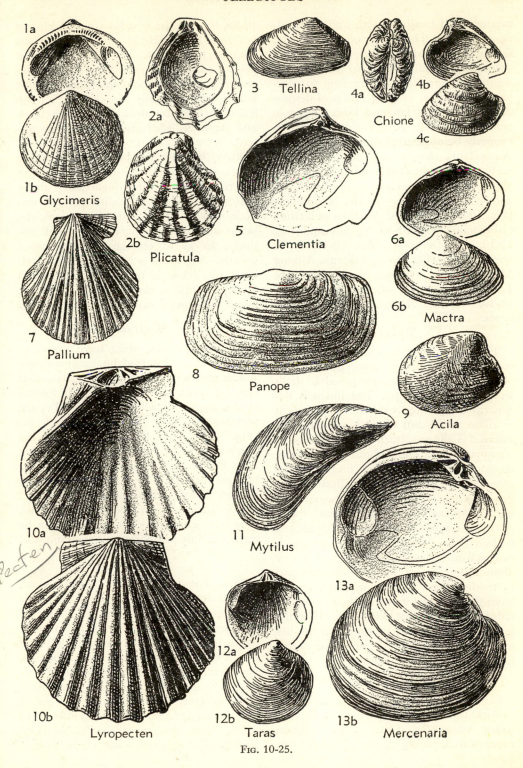

1a
1b Glycimeris
2a
2b Plicatula
3 Tellina
4a
4b
Chione
4c
5 Clementia
6a
6b Mactra
7 Pallium
8 Panope
9 Acila
10a
10b Lyropecten
11 Mytilus
12a
12b Taras
13a
13b Mercenaria
Pecten

FIG. 10-25.

carries small cardinal teeth and a prominent resilifer. The posterior gape of the shell is small. This relatively unspecialized desmodont gives its name to the suborder Anatinacea, which is interpreted to comprise the longest ranging group of desmodonts. If Ordovician and later Paleozoic pelecypods which are assigned to the group properly belong here, differentiation of the desmodonts may be assigned to a very early part of paleontological record—actually before the appearance of the oldest-known representatives of the heterodonts (Fig. 10-29). Paleozoic Anatinacea include Ordovician (Fig. 10-10, *9*), Silurian (Fig. 10-10, *16*), Devonian (Fig. 10-11, *2, 8*), and Permian (Fig. 10-15, *1*) forms. Among Mesozoic and younger Anatinacea are Jurassic (Fig. 10-17, *10*), Cretaceous (Fig. 10-19, *11*), and Neogene (Fig. 10-27, *3*) types.

Myacea

The suborder Myacea comprises burrowing long-siphoned pelecypods, which very commonly have inequivalve shells with a posterior gape. The hinge is degenerate in having weakly developed teeth or none at all, but there is a prominent internal ligament borne by resilifers or chondrophores. The siphons typically are elongate and more or less united. Characters of the genus *Mya* are illustrated in Figs. 10-26, *8;* 10-27, *11*. It is common as a living form and occurs in Tertiary rocks. Another representative of the Myacea is *Corbula* (Figs. 10-23, *10;* 10-26, *4*), which ranges from Triassic to Recent and is very abundant in many Tertiary deposits and in modern seas. It is a small pelecypod having a more or less distinctly inequivalve habit, the right valve being larger than the left. *Corbula* is not a true burrower but lies on its side with the larger valve down, hardly buried in sediment.

Adesmacea

Clams which are specialized for boring in rock and wood are classified in the suborder Adesmacea. One of the common modern rock-boring clams is *Zirphaea* (Fig.

10-26, *9*), and the best known wood borer is *Teredo* (Fig. 10-26, *5*), which ranges from Jurassic to Recent. The valves of these pelecypods have a permanent gape in the posterior region behind the inconspicuous beaks, and generally also a gape in the front part of the shell. The hinge teeth and ligament are much reduced or absent, but a curved projection (myophoric septum) beneath the beaks is a structure developed for muscle attachment. Among the rock borers, various accessory shelly plates are secreted in addition to the two main valves. These plates may cover the beaks and anterior part of the shell entirely, so as to change its appearance greatly. A rodlike accessory plate developed by *Teredo* is termed a pallet (Fig. 10-26, *5*), and variations in its form are found to be useful in distinguishing species.

Mactracea

Burrowing clams, which have a more or less prominent internal ligament and resilifers and which also possess heterodont types of cardinal and lateral teeth, are included in the suborder Mactracea. Typical representatives are *Mactra* and *Schizothaerus* (Fig. 10-26, *1, 2*). Several kinds of living marine clams belong to this group, which has a known distribution from Upper Cretaceous through Cenozoic deposits. They burrow in soft sediments of the sea floor; and as is characteristic of such pelecypods, they have elongate siphons which can be retracted only partly into the shell interior.

Solenacea

Mud-burrowing marine clams, which are characterized by very elongate narrow shells gaping at both ends, are classed as Solenacea. They have a narrow external ligament and small cardinal teeth. The foot is greatly elongated and modified for burrowing, but the siphons are short. Although some Paleozoic pelecypods have been referred to this group, the actual stratigraphic range of the Solenacea seems to be from Cretaceous to Recent (Fig. 10-29). The modern razor clam *Ensis* is

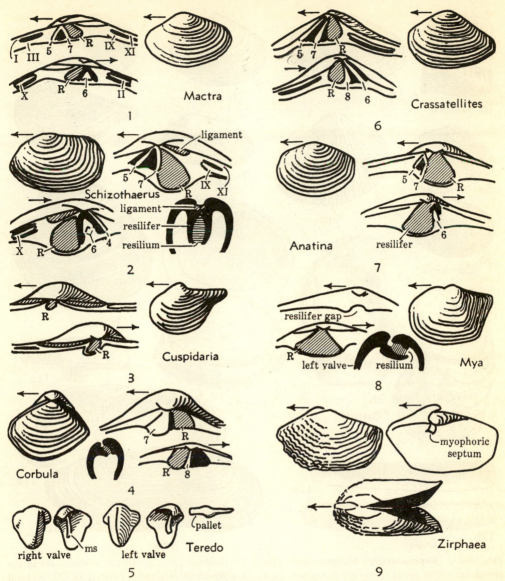

FIG. 10-26. **Representative desmodont pelecypods.** Excepting the Poromyacea (here illustrated by *Cuspidaria*), which have a highly specialized septibranch type of gill structure, the desmodonts have eulamellibranch gills. They are chiefly burrowers. Their stratigraphic range is from Ordovician to Recent. Arrows point to anterior part of shell; *R* indicates resilifer and *ms*, myophoric septum, a structure for muscle attachment; other letters and numbers designate teeth, as shown in Fig. 10-21.

1, *Mactra clathrodon* Lea, Miocene, Maryland.
2, *Schizothaerus* sp., Recent, Pacific.
3, *Cuspidaria cuspidata* Olivi, Miocene, Austria.
4, *Corbula idonea* Conrad, Miocene, Maryland.
5, *Teredo* sp., Recent, Pacific.

6, *Crassatellites turgidulus* (Conrad), Miocene, Maryland.
7, *Anatina* sp., Recent, Pacific.
8, *Mya truncata* (Linné), Recent, Europe.
9, *Zirphaea* sp., Recent, Pacific.

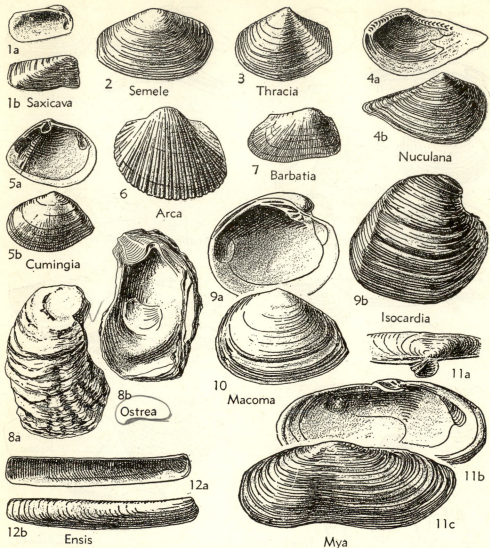

FIG. 10-27. **Representative Neogene pelecypods.** Desmodonts (1–3, 5, 10, 12) predominate in the group here illustrated; others are taxodonts (4, 6, 7), a dysodont (8), and a heterodont (9). Natural size and from Miocene of Maryland except as indicated otherwise. L, left valve; R, right valve.

Arca Linné, Jurassic–Recent. Taxodont. *A. trilineata* Conrad (6, L, ×0.5), Miocene-Pliocene, California.

Barbatia Gray, Cretaceous–Recent (subgenus of *Arca*). Taxodont. *A. (B.) marylandica* Conrad (7, R, ×0.5).

Cumingia Sowerby, Eocene–Recent. Desmodont. *C. medialis* Conrad (5a, b, L).

Ensis Schumacher, Miocene–Recent. Desmodont. *E. directus* (Conrad) (12a, L; 12b, R, ×0.5).

Isocardia Lamarck, Jurassic–Recent. Heterodont. *I. fraterna* Say (9a, L; 9b, R, ×0.5).

Macoma Leach, Miocene–Recent. Desmodont. *M. nasuta* (Conrad) (10, L, ×0.5), Pliocene-Pleistocene, California.

Mya Linné, Miocene–Recent. Desmodont. *M. producta* Conrad (11a, L. dorsal; 11b, c, L, ×0.5).

Nuculana Link, Silurian–Recent. Taxodont. *N. acuta* (Conrad) (4a, b, R), Miocene, Florida.

(Continued on next page.)

represented also by fossils in Miocene and other Tertiary rocks (Fig. 10-27, *12*).

Poromyacea

An offshoot of the desmodonts, which is differentiated from all other clams in having a septibranch type of gill structure (Fig. 10-7, *4*), comprises the assemblage called Poromyacea. These clams have equivalve or nearly equivalve shells, which mostly are rather small. Hinge teeth are small or lacking, and the ligament, which tends also to be reduced, commonly is internal, borne in a resilifer. The group ranges from Jurassic to Recent. *Cuspidaria*, a representative genus, is illustrated in Figs. 10-19, *8, 9*, and 10-26, *3*.

EVOLUTIONARY TRENDS

Several sorts of evolutionary trends are discernible among the pelecypods, viewing this group of mollusks as a whole and taking account of their differentiation in the course of geologic time. The most significant evolutionary changes seem to be in response to mode of life. In many groups, this is obviously expressed both in the characters of soft parts and in the structural features of the shell, including especially the articulation of valves along the hinge, nature and position of the ligament, structure of shell layers, and relative dimensions of the valves. Evolutionary

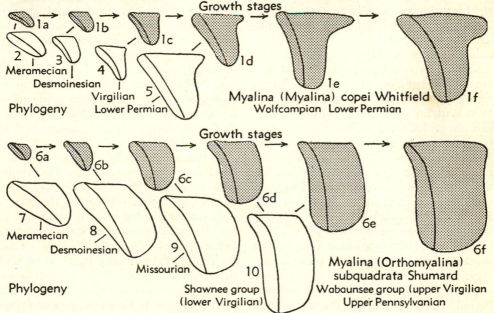

FIG. 10-28. **Evolutionary trends of two late Paleozoic pelecypods.** The successive growth stages (shaded) show gradual decrease in obliquity of the shell, which corresponds to trend in modification of shell shape represented by adults of geologically preceding species (data from Newell). Figures 1a, b, ×0.75; 1c–f, ×0.25; 2, *Myalina* sp., ×0.5; 3, *M. lepta*, ×0.15; 4, *M. miopetina*, ×0.15; 5, *M. pliopetina*, ×0.25; 6a–c, ×0.7; 6d–f, ×0.3; 7, *M. goldfussiana*, ×0.3; 8, *M. lepta*, ×0.45; 9, *M. wyomingensis*, ×0.4; 10, *M. slocomi*, ×0.3.

(*Fig. 10-27 continued.*)

Ostrea Linné, Triassic–Recent. Dysodont. *O. trigonalis* Conrad (*8a, b,* L, ×0.3).

Saxicava Fleuriau, Miocene–Recent. Desmodont. *S. arctica* (Linné) (*1a,* L; *1b,* R).

Semele Schumacher, Eocene–Recent. Desmodont. *S. subovata* (Say) (2, R, ×2).

Thracia Leach, Triassic–Recent. Desmodont. *T. trapezoides* Conrad (3, L, ×0.5), Miocene, Oregon.

trends are also clearly shown by shell shape and thickness, and by development of accessory characters, such as internal buttresses for shell support, myophore plates for muscle attachment, and additional external shell elements developed by some of the borers. The attachment scars of adductor muscles on the shell interior vary in number, relative size, and placement, and thus we may recognize evolutionary trends in the nature of pelecypod musculature. The presence of two subequal adductors, located near the hinge line in anterior and posterior parts of the shell, is a primitive character. By progressive diminution and ultimate disappearance of the anterior adductor, a monomyarian shell is produced, such as characterizes the important dysodont groups of the Pectinacea and Ostreacea.

The nature of the gill structure (Fig. 10-7), observed in various groups of clams, permits recognition of four types which have progressively more complicated structure. With little doubt, this represents a significant evolutionary differentiation, which has far-reaching classificatory importance. Except for the small group of septibranch clams, which are a geologically late invention, the paleontological record indicates that derivation of filibranch gills from the protobranch type, and eulamellibranchs from complication of the filibranch type, are very ancient. They belong to early Paleozoic time, or conceivably, these steps in evolution may partly antedate the beginning of the Cambrian Period. Once established, the main types of pelecypod gill structure have been remarkably stable and persistent (Fig. 10-29). The hypothesis that filibranch gills are independent evolutionary developments in different stocks at different times and, similarly, that eulamellibranch gills of the oysters and other dysodonts, of heterodonts, and of desmodonts constitute separate, multiple development of identical structure is highly implausible. Assumed convergence of this sort is far less reasonable than interpretation of the gill types as genetically significant divergences. In other words, if such changes in gill structure denote simply grades in development, which may appear in any stock at any time, we should reasonably expect to find filibranch derivatives of the Nuculacea belonging in late geologic time; and there should be living branches of the Arcacea, Pectinacea, and Mytilacea, which have eulamellibranch gill structure. No such things are known, although derivatives of them (such as Ostreacea from the Pectinacea) do show such advance.

Because most groups of clams are extraordinarily conservative, it is far more difficult to recognize specific lineages than among such mollusks as ammonoid cephalopods, whose type of suture and other shell structures permit segregation of closely related stocks. A single example of evolutionary change in a group of clams is cited here and illustrated in Fig. 10-28. This relates to late Paleozoic species of *Myalina*, in which progressive changes in shell outline during growth are found to match characters of adult shells occurring in a succession of geologically older strata. An evident evolutionary trend in both of the species illustrated is the tendency toward decreased obliquity of the shell axis; and this is found to be correlated with other progressive changes, such as shell thickness, nature of ligament area, and proportions of the opposed valves. These changes are not shown in the illustrations.

FIG. 10-29. **Geologic distribution of pelecypods.** The main divisions called Prionodesmacea and Teleodesmacea are shown on the left and right sides of the diagram, respectively. Also indicated are subdivisions of each, with graphic differentiation according to gill structure and notation of dominant mode of life. The early geologic appearance of many groups and their very long range characterize pelecypods.

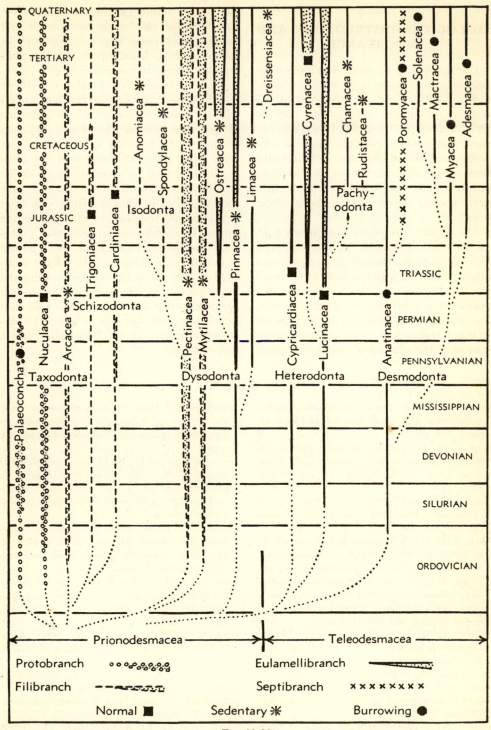

FIG. 10-29.

GEOLOGICAL DISTRIBUTION AND IMPORTANCE

The known geological distribution of main groups of clams is shown graphically in Fig. 10-29, together with types of gill structure, dominant mode of life, and inferred genetic relationship. Especially because knowledge of Paleozoic clams is very inadequate, the phylogenetic relationships shown in the diagram are largely based on morphological comparisons, with attention given also to appearance in the geologic record.

The relative abundance of known genera of clams belonging to the different geologic periods cannot be represented satisfactorily. It is partly indicated, however, by expansion of the lines or patterns representing distribution of the groups. One of the most noteworthy features shown by the diagram is the long geologic range of most of the pelecypod stocks. Although some genera and families of clams have short stratigraphic range, we do not find among pelecypods any major group, as among corals, brachiopods, and cephalopods, which is exclusively lower Paleozoic, any which is confined to middle or late Paleozoic, or any (excepting pachyodonts) which expand greatly in Mesozoic or Cenozoic parts of the column and then disappear. In contrast to such a record, each main group of clams seems to have become adapted to a certain mode of life and to have maintained its existence very successfully in its chosen setting with little alteration in response to changes in the spread of epicontinental seas and multitudinous fluctuations in the local nature of marine environments. They have managed to persist by finding somewhere the living conditions for which they are suited. No good reason is known for thinking that the Pelecypoda are less sensitive to environment than other invertebrates, nor is this evident.

REFERENCES

ARNOLD, R. (1906) *Tertiary and Quaternary pectens of California:* U.S. Geol. Survey Prof. Paper, 47, pp. 1–264, pls. 2–53.

CLARK, B. L. (1925) *Pelecypoda from marine Oligocene of western North America:* Calif. Univ. Pub., Dept. Geol. ser., Bull. 15, pp. 69–136, pls. 8–22.

CLARK, W. B., & MARTIN, G. C. (1901) *Mollusca (Pelecypoda):* Maryland Geol. Survey, Eocene, pp. 160–203, pls. 30–57.

DALL, W. H. (1913) *Pelecypoda:* in Eastman, C. R., & Zittel, K. A., Textbook of palaeontology, Macmillan & Co., Ltd., London, 2d ed., vol. 1, pp. 422–503, figs. 637–836. Gives detailed morphological description and groups clams in three main divisions: Prionodesmacea, Anomalodesmacea, and Teleodesmacea.

DAVIES, A. M. (1935) *Tertiary faunas:* Murby, London (pelecypods), pp. 116–208, figs. 167–282. Describes structural characters, classification, and most important Tertiary types.

GARDNER, J. (1916) *Mollusca (Pelecypoda):* Maryland Geol. Survey, Upper Cretaceous, pp. 511–733, pls. 19–45.

———— (1926–1928) *Molluscan fauna of Alum Bluff group of Florida:* U.S. Geol. Survey, Prof. Paper 142, pp. 1–249, pls. 1–36. Includes description of typical Miocene assemblage of Gulf region.

———— & BOWLES, E. (1939) *Venericardia planicosta group in Gulf province:* Same, Prof. Paper 189, pp. 143–215, pls. 29–46. Discusses evolution and distribution of one of the leading groups of Paleogene clams.

GLENN, L. C. (1904) *Mollusca (Pelecypoda):* Maryland Geol. Survey, Miocene, pp. 274–401, pls. 65–108. A representative early Neogene assemblage of marine clams.

GRANT, U.S., IV, & GALE, H. R. (1931) *Pliocene and Pleistocene marine Mollusca of California:* San Diego Soc. Nat. History Mon. 1, pp. 1–1036, pls. 1–36. Includes clams.

HILL, R. T., & VAUGHAN, T. W. (1898) *Lower Cretaceous Gryphaeas of the Texas region:* U.S. Geol. Survey, Bull. 151, pp. 1–139, pls. 1–31.

KEEN, A. M., & FRIZZELL, D. L. (1939) *Illustrated key to West North American pelecypod genera:* Stanford University Press, Stanford University, pp. 1–28, figs.

MACGINITIE, G. E., & MACGINITIE, N. (1949)

Natural history of marine animals: McGraw-Hill, New York, pp. 1–473, figs. 1–283 (pelecypods, pp. 329–352, figs. 169–197). Excellent description of habits of living clams.

MacNeil, F. S. (1938) *Species and genera of Tertiary Noetinae:* U.S. Geol. Survey Prof. Paper 189, pp. 1–49, pls. 1–6. Describes a group of arcacean taxodonts.

Mansfield, W. C. (1932) *Miocene pelecypods of Choctawhatchee formation of Florida:* Florida Geol. Survey Bull. 8, pp. 1–240, pls. 1–34.

Moret, L. (1948) *Manuel de paléontologie animale:* Masson et Cie., Paris, pp. 1–745, figs. 1–274 (pelecypods, pp. 343–417, figs. 128–154) (French). Excellent general description of fossil clams.

Newell, N. D. (1937) *Late Paleozoic pelecypods, Pectinacea:* Kansas Geol. Survey, vol. 10, pt. 1, pp. 1–123, pls. 1–20, figs. 1–42. An especially important paper on this group of clams.

——— (1942) *Late Paleozoic pelecypods, Mytilacea:* Same, vol. 10, pt. 2, pp. 1–115, pls. 1–15, figs. 1–22. Chiefly describes Pennsylvanian and Permian Myalinidae.

Palmer, K. W. (1927–1929) *The Veneridae of eastern America:* Paleontologia Americana 1 (5), pp. 209–522, pls. 32–76.

Pelseneer, P. (1906) *Mollusca:* in Lankester, E. R., Treatise on zoology, A. & C. Black, London, pt. 5, pp. 1-355, figs. 1–301 (pelecypods, pp. 205–284, figs. 187–251). One of the standard works on clams.

Perry, L. M. (1940) *Marine shells of the southwest coast of Florida:* Bull. Am. Paleontology, vol. 95, pp. 1–260, pls. 1–39. Describes modern marine clams of eastern Gulf region.

Pohl, E. R. (1929) *Devonian of Wisconsin, Lamellibranchiata:* Milwaukee Pub. Mus. Bull. 11, pp. 1–100, pls. 1–14. Treats a typical Devonian clam assemblage.

Robson, G. C. (1929) *Lamellibranchia:* Encyclopaedia Britannica, 14th ed., vol. 13, pp. 617–626, figs. 1–26. A good concise description.

Schenck, H. G. (1934) *Classification of nuculid pelecypods:* Mus. royale histoire nat. Belgique Bull. 10, pp. 1–78, pls. 1–5.

Shimer, H. W., & Shrock, R. R. (1944) *Index fossils of North America:* Wiley, New York, pp. 1–837, pls. 1–303 (pelecypods, pp. 366–433, pls. 144–173). Figures and brief descriptions of many fossil clams; follows classification of Dall (1913).

Stanton, T. W. (1947) *Studies of some Comanche pelecypods and gastropods:* U.S. Geol. Survey Prof. Paper 211, pp. 1–256, pls. 1–67 (pelecypods, pp. 1–52, pls. 1–46). Lower Cretaceous clams of the Gulf and central plains regions.

Stephenson, L. W. (1923) *Upper Cretaceous formations of North Carolina:* North Carolina Geol. Survey Bull. 5, pp. 1–604, pls. 1–102. Includes description of clams.

——— (1941) *The larger invertebrate fossils of the Navarro group of Texas:* Texas Univ. Pub. 4101, pp. 1–641, pls. 1–95, figs. 1–12. Contains descriptions and excellent illustrations of a representative Upper Cretaceous clam assemblage, including very thick-shelled pachyodonts.

Storer, T. I. (1951) *General zoology:* McGraw-Hill, New York, pp. 1–832, figs. 1–551 (pelecypods, pp. 424–431).

Turner, F. E. (1938) *Stratigraphy and Mollusca of Eocene of western Oregon:* Geol. Soc. America Spec. Paper 10, pp. 1–130, pls. 5–22. Pacific coast Paleogene clams.

Ulrich, E. O. (1897) *Lower Silurian [Ordovician] Lambellibranchiata of Minnesota:* Minnesota Geol. Survey, vol. 3, pp. 475–628, pls. 1–72.

Wade, B. (1926) *Fauna of the Ripley formation on Coon Creek, Tennessee:* U.S. Geol. Survey Prof. Paper 137, pp. 1–272, pls. 1–72. Remarkably preserved Upper Cretaceous clams.

CHAPTER 11

Annelids and Other Worms

Before the early nineteenth century all creatures now collectively known as invertebrates, except the insects, were classed as Vermes (worms). Those still called worms are biologically and structurally so highly diverse that they are divided into nine phyla. It is difficult to classify many forms. The worms are animals with an anterior end, or head with sense organs, and a posterior end. They are bilaterally symmetrical and rest or move on a ventral surface.

The only group of worms which has paleontological importance is the Annelida. There is a special reason for studying their fossil remains because the Arthropoda are believed to have been derived from them. The annelids resemble the arthropods in having a segmented body covered by cuticle and in the structure of the nervous system; they differ in having simple unjointed appendages and in lacking specialization of segments in different parts of the body.

Although the worms are represented by fossils in rocks ranging in age from Pre-Cambrian to Recent, the fossil record is a very poor one, and worms are not important as index fossils in any geologic system. The worms are poorly adapted for preservation because only a few of them have hard parts, in the form of jaws or opercula. The fossil record of the worms chiefly consists of trails, tracks, burrows, castings, impressions, tubes, jaws, and opercula. Soft parts of these organisms have been preserved only under unusual conditions of deposition of fine-textured muds in quiet, probably oxygen-poor waters. A large number of worms were preserved under such conditions in the Burgess shale of Middle Cambrian age in British Columbia. The remarkable fossil assemblage in this deposit contains 10 genera of worms representing a number of families, and thus it indicates much differentiation of the worms before Middle Cambrian time.

The Pre-Cambrian evidence of worms consists of trails and burrows. Some Ordovician worms have calcareous tubes, and the genus *Spirorbis*, which is living yet, is a common Paleozoic fossil. Worms are found in many formations of other geologic systems, but mostly they are rare and for this and other reasons unimportant in stratigraphic use.

Scolecodonts, or worm jaws, which are composed of chitinous, horny, or siliceous material, are known from most of the geologic systems. They are common in some Ordovician, Devonian, and Pennsylvanian rocks. Comparisons of Devonian scolecodonts with jaws of living polychaetes indicate that the variation between them is little more than the variations between members of closely related living genera.

MORPHOLOGICAL FEATURES OF ANNELIDS

The annelids, in contrast to the other phyla of worms, are composed of many ringlike essentially similar segments. The segmentation is shown in the external appearance and in the internal structures, such as nerves, muscles, and excretory, circulatory, and reproductive organs. The annelids range in length from a few millimeters to nearly 3 m. (10 ft.) and in diameter to 25 mm. (1 in.). Some other worms are much longer.

The worms live in a variety of environments, in soils, beach sands, marine muds, marine waters to depths of 5,500 m. (18,000 ft.), fresh waters, and in a variety of organs in humans and other animals.

External Features

The elongate body is bilaterally symmetrical and somewhat flattened on the ventral side. A distinct head occurs in some classes of annelids but none in others. The body is divided into as many as 180 segments in the adult. Each segment, except some at each end, bears a bundle of bristle-like setae that project on the lateral surfaces of some of the annelids. The mouth is at the anterior end and the anus at the posterior end. Morphological features of worms are illustrated in Fig. 11-1.

Digestive, Circulatory, and Respiratory Systems

The digestive tract is essentially a straight tube and consists of the mouth, pharynx, which in Polychaeta has horny jaws, a short esophagus with lateral digestive glands, a gizzard (in some forms), and an intestine or stomach intestine, which extends to the anus. The intestine commonly bulges laterally in each segment.

The circulatory system of annelids is closed, with two to five principal vessels extending lengthwise of the body. The circulation results from peristaltic pulsations of the dorsal vessel in the Polychaeta and from contractions of five pairs of "hearts" in the earthworms. The blood consists of a red fluid plasma containing dissolved hemoglobin and colorless corpuscles.

Respiration is effected by capillaries in the appendages (parapodia) and body wall where oxygen is received and carbon dioxide given up.

Excretory and Nervous Systems

A pair of coiled tubular structures, called nephridia, occurs in every segment of the annelids except the first three and last. The nephridia dispose of waste matter.

The annelid nervous system consists of a pair of cerebral ganglia in the anterior or head region and two long connectives joined to the midventral nerve cord, which has a pair of ganglia and lateral nerves in each segment. Nerves extend from the cerebral ganglia to the tentacles, eyes, and other parts.

Reproductive Organs

Each earthworm contains male and female sex organs. Immature sperm cells separate from the testes and mature sperms are discharged from the vesicles during copulation. The sperms are received in the female system in two pairs of seminal receptacles where they are stored until needed to fertilize eggs in cocoons.

The sexes are separate in the Polychaeta, and there are no reproductive organs. Ova or sperms form from cells in the lining of the body cavity (coelom) and pass through the body wall or kidneys (nephridia). Fertilization occurs in sea water. The fertilized egg develops into a ciliated trochophore larva, which floats about and is transformed later into a young worm.

Some worms have great powers of regeneration and may reproduce by fragmentation of the body, each piece growing into an adult form.

GEOLOGIC WORK OF WORMS

Worms modify sediment and the soil mantle by chewing and grinding material and by chemical action within their bodies. The mud and sand of many coastal areas are in continuous passage through the intestinal tracts of worms. The "lobworms," which live on sand flats exposed at low tide, are so numerous that individual castings average 84,400 per acre in the Northumberland, England, area. More than 3,000 tons of material per acre passes through the intestinal tracts of these worms and is changed in the process. The lobworms burrow to a depth of about 2 ft.

Some soils contain more than 50,000

earthworms per acre, and it is estimated that they move 18 tons of material to the surface each year. This material, which has passed through their intestinal tracts, has been changed physically and chemically.

CLASSIFICATION

The worms include a large group of animals which have an anterior end or head with sense organs, a posterior end, are bilaterally symmetrical, and move on a ventral surface. The various kinds of worms differ sufficiently in structural and biological features so they must be divided into several phyla. The classification is based entirely on the soft parts, as follows:

Divisions of Worms

Platyhelminthes (*phylum*), simplest group of worms, which are dorso-ventrally flattened and lack true segmentation. The have a mouth but no anus, skeletal, circulatory, or respiratory system. Length ranges from 0.5 mm. to 12 m. (40 ft.). Recent.

Turbellaria (*class*), free-living flatworms, most of which are found in fresh or salt water or in moist soil. Recent.

Trematoda (*class*), flukes; internal or external parasites, such as the liver fluke, which is a parasite of sheep. Recent.

Cestoidea (*class*), tapeworms; intestinal parasites of vertebrates. Recent.

Nemathelminthes (*phylum*), the unsegmented roundworms, which have slender cylindrical bodies and a complete digestive tract with mouth and anus at opposite ends of the body; circulatory and respiratory organs absent. Recent.

Nematoda (*class*), includes many free-living and parasitic worms, as *Trichina* (pork worm) and many others. Recent.

Gordiacea (*phylum*), "horsehair worms"; extremely long, uniformly cylindrical body, with bluntly rounded anterior end and swollen, coiled posterior end. They live in fresh water. Reported as fossils in insects from rocks of Carboniferous and Tertiary age; also Recent.

FIG. 11-1. Morphologic features of worms. The illustrated elements are explained in the alphabetically arranged list which is cross-indexed to the figure by numbers.

anal cirri (24). Elongate soft sensory processes at posterior end of body.

anus (20). Posterior opening of alimentary canal.

apical organ (18). Cells at apex of trochosphere larva which develop into cerebral ganglia in adults.

basal plate (28). Unpaired plate at base of forceps.

carrier (13). Posterior smooth, slender plates which support forceps.

dental plate (12). Denticulate asymmetrical plates located under forceps (maxillae II).

esophagus (6). Tube leading from pharynx to stomach.

eye (1, 4). Organ of sight.

fans (8). Modified appendages (parapodia) used to draw in water containing oxygen and microscopic organisms.

forceps (27). Pair of toothed jaws, united and supported by a basal pair of carriers (maxillae I).

generative pore (14). Small opening through which genital products are passed.

head (2). Anterior division of body in most worms, containing brain, chief sense organs, and mouth.

mouth (7, 16). Opening for reception of food.

operculum (17). Plate which covers opening of some annelid worms.

paragnaths (26). One or more pairs of minute denticulate distal plates (maxillae IV).

parapodia (22, 25). Lateral lobes on a worm segment, each bearing a cirrus and bundle of bristle-like setae.

pharynx (15). Part of digestive tract between mouth cavity and esophagus.

pre-oral ciliated ring (19). Ring of cilia in front of mouth.

proboscis (3). Prolongation of head.

setae (23). Bundles of chitinous bristles on a parapodium.

somites (9). Segments of body.

stomach (5). Digestive cavity.

tentacles (10). Long flexible tactile processes on head.

tube (21). Hollow burrow in which a worm lives.

unpaired piece (11). Unpaired denticulate piece, located on left side of jaw apparatus.

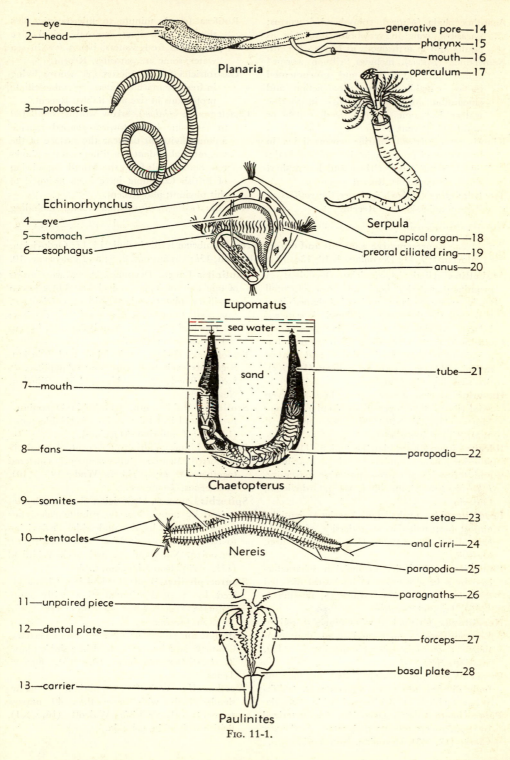

1—eye
2—head
Planaria
generative pore—14
pharynx—15
mouth—16
operculum—17

3—proboscis
Echinorhynchus
4—eye
5—stomach
6—esophagus

Serpula

apical organ—18
preoral ciliated ring—19
anus—20

Eupomatus

sea water
sand
tube—21
7—mouth
8—fans
parapodia—22
Chaetopterus

9—somites
10—tentacles
Nereis
setae—23
anal cirri—24
parapodia—25

11—unpaired piece
12—dental plate
paragnaths—26
forceps—27
basal plate—28
13—carrier
Paulinites
Fig. 11-1.

Acanthocephala (*phylum*), spiny-headed worms; parasites living in vertebrate intestines as adults and in arthropods as larvae. Recent.

Nemertinea (*phylum*), includes "ribbon worms" which have soft, flat, and unsegmented bodies, capable of great contraction and elongation. Most of them live in marine waters, but some occur in fresh waters or moist soil. Recent.

Kinorhyncha (*phylum*), marine worms living in bottom mud or sand, having spinose body, which is made up of rings and has a short retractile proboscis. Recent.

Trochelminthes (*phylum*), includes the rotifers or "wheel animalcules." Recent.

Rotifera (*class*), minute to microscopic (less than 1 mm. long) animals which live mostly in fresh waters, but a few in sea water; some are parasites. Recent.

Gastrotricha (*class*), microscopic worms living in fresh or marine waters; resemble ciliate protozoans in size and habits. Recent.

Chaetognatha (*phylum*), arrow worms; small (20 to 70 mm. long) torpedo-shaped marine animals living at or near the surface of the sea, often abundant constituents of the plankton. Some fossils from the Middle Cambrian of British Columbia have been assigned to this phylum; also Recent.

Annelida (*phylum*), segmented worms, including

FIG. 11-2. **Representative annelids and other worms.** Cambrian fossils include 14–17; Ordovician, 3–6; Silurian 1, 7, 11; Devonian, 2, 12, 13*a*; Pennsylvanian, 13*b*; Cretaceous, 8, 9; and Paleogene, 10.

Arabellites Hinde, Ordovician–Devonian. A prominent anterior hook and row of small teeth; quadrate jaws, and sickle-shaped jaws make up the three types of jaws. *A. oviformis* Eller (7, maxilla I, ×15), Silurian, New York.

Canadia Walcott, Cambrian. Each segment has a pair of parapodia with dorsal and ventral nonjointed setae. *C. setigera* Walcott (15, ×2), British Columbia, Canada.

Hamulus Morton, Cretaceous. Three to seven axial ribs on tube, operculum circular. *H. onyx* Morton (8*a*, side view; 8*b*, operculum view, ×2), Ripley, Tennessee.

Ildraites Eller, Ordovician–Devonian. The wide anterior jaw has a pointed hook on maxillae I and tapers to a narrow posterior extremity. *I. horridus* Eller (3, maxilla I, ×15), Ordovician, Ontario.

Leodicites Eller, Ordovician. A series of denticles almost perpendicular to underside of jaw. *L. buris* Eller (6, maxilla II, ×15), Trenton, Ontario.

Lumbriconereites Eller, Ordovician. Characterized by a large number of backward directed denticles. *L. angustifossus* Eller (5, maxilla I, ×15), Trenton, Quebec.

Nereidavus Grinnel, Ordovician–Devonian. Elongate jaws, a hook, and small blunt denticles. *N. invisibilis* Eller (1, maxilla I, ×15), Silurian, New York.

Ottoia Walcott, Cambrian. Elongate, tapering body divided into many segments. *O. prolifica* Walcott (17, ×1), British Columbia, Canada.

Palaeochaeta Clarke, Devonian. Characterized by numerous segments and setae. *P. devonica* Clarke (12, ×3), Devonian, New York.

Paulinites Lange, Devonian. Armature consists of one pair of long, conical mandibles, seven maxillary plates and a pair of carriers. *P. paranaensis* Lange (2*a*, paragnath; 2*b*, carriers; 2*c*, unpaired piece; 2*d*, dental plate; 2*e*, forcep, ×15), Devonian, Brazil.

Protoscolex Ulrich, Ordovician and Silurian. Segments each bear two rows of papillae. *P. batheri* Ruedemann (11, ×1.5), Silurian, New York.

Scolithus Haldemann, Cambrian–Devonian. Free, cylindrical tube. *S. linearis* (Haldemann) (14, ×1), Cambrian, Maryland.

Serpula Linné, Silurian(?)–Recent. Irregularly contorted calcareous tubes frequently clustered together. *S. pervermiformis* Wade (9, ×10), Cretaceous, Tennessee.

Spirorbis Lamarck, Ordovician–Recent. Characterized by annulations on minute calcareous tubes cemented by flat underside. Ordovician to Recent. *S. gyrus* Clarke & Swartz (13*a*, ×6), Devonian, Maryland, *S. anthracosia* Whitfield (13*b*, ×25), Pennsylvanian, Ohio.

Staurocephalites Hinde, Ordovician. Characterized by a small elongate jaw with many denticles. *S. arcus* Eller (4, maxilla III, ×15), Trenton, Quebec.

Tubulostium Stoliczka, Cretaceous and Tertiary. Characterized by tubes coiled in a single plane. *T. horatianum* Gardner (10, ×3), Eocene, Texas.

Worthenella Walcott, Cambrian. Elongate, slender body with more than 40 narrow segments. *W. cambria* Walcott (16, ×1), British Columbia, Canada.

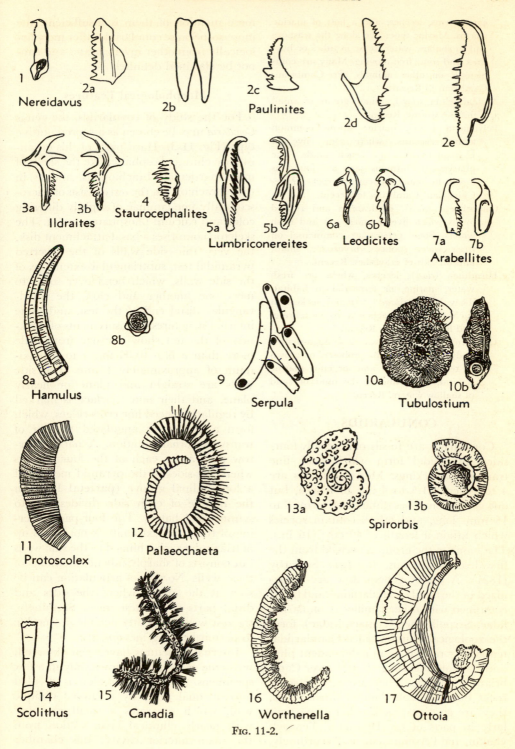

1 Nereidavus 2a 2b 2c Paulinites 2d 2e

3a 3b Ildraites 4 Staurocephalites 5a 5b Lumbriconereites 6a 6b Leodicites 7a 7b Arabellites

8a 8b Hamulus 9 Serpula 10a 10b Tubulostium

11 Protoscolex 12 Palaeochaeta 13a 13b Spirorbis

14 Scolithus 15 Canadia 16 Worthenella 17 Ottoia

Fig. 11-2.

earthworms, leeches, and a host of marine worms. Marine types live along the seashore and in shallow water, some in tubes or burrows and some free moving. Many are commensals on other animals. ?Pre-Cambrian, Cambrian to Recent.

Archiannelida (*class*), minor group of small marine worms. Recent.

Polychaeta (*class*), marine worms common along seacoasts, where some live in U-shaped tubes in beach sands. The pharyngeal chitinous jaws are found as fossils in many systems of rocks. ?Pre-Cambrian, Cambrian to Recent.

Oligochaeta (*class*), earthworms and related forms which live in moist soil and fresh waters. Some genera are commensal on fresh water sponges, bryozoans, snails, and on gills of crayfishes. Recent.

Hirudinea (*class*), leeches, which are fresh water, marine, or terrestrial in habitat. They have enlarged terminal suckers at each end for attachment to vertebrates and for locomotion. Recent.

Gephyrea (*class*), forms with a sausage-shaped body and a retractile proboscis or crown of tentacles at the anterior end; marine habitat, burrowing in the mud or sand of shallow waters. Recent.

CONULARIIDS

Conulariids are fossils of generally four-sided pyramidal form which bear fine transverse markings. Most specimens are 4 to 10 cm. (1.5 to 4 in.) in length, but one genus comprises diminutive forms 8 to 15 mm. long, and others contain species which attain a length of 40 cm. (16 in.). The name of the group is derived from the first-described genus, *Conularia* Sowerby (1821). A dozen genera now are recognized in the family Conulariidae, and these, combined with other families (Conulariellidae, Serpulitidae, ?Tentaculitidae), form the zoological division called Conulariida, tentatively ranked as an independent phylum. Conulariids range from Lower Cambrian to Triassic. They are rare in some rocks but very abundant in others; also, they are stratigraphically useful, particularly in parts of the Devonian, Pennsylvanian, and Permian systems. Accordingly,

mere mention of them is insufficient, the more so because conulariids differ morphologically from other invertebrates and cannot be classified definitely.

Morphological Features

For the study of conulariids, the genus *Conularia* may be chosen as a representative type (Fig. 11-3). Hard parts of this organism are chitinophosphatic, like the shells of most inarticulate brachiopods, and as in these brachiopods, the test consists of microscopically fine laminae of slightly differing color and calcium phosphate content. The skeleton comprises a basal attachment disk, the very thin side walls of the inverted pyramidal test, subtriangular extensions of the side walls, which bend over so as to meet one another and close the quadrangular distal end of the test, and some internal structures. Measurements of thickness of the test show a range from little more than a film (0.06 mm.) to a maximum of approximately 1 mm. The side walls are straight and plane or nearly plane, and their outer surface is marked by regularly spaced fine cross ridges, which form a curved or angulated pattern of transverse ornamentation. A narrow furrow runs along each of the four corners where the sides of the pyramid meet, and a longitudinal groove (**parietal line**) in the middle of each side divides it into symmetrical halves. The four-part covering of the distal extremity is not made up of triangular pieces hinged to the side walls but consists of sharply inbent extensions of these walls. No sign of articulation can be seen at the angles where the sides and distal parts of the test meet. Seemingly, the test was sufficiently flexible during life to permit bending movements.

Internal structures have been observed, but some of them are known only in a few specimens. The apical region is divided into unequal parts by a curved oblique wall, which separates a small chamber with nearly elliptical cross section from the main interior cavity; this chamber extends upward along one side of the test to a point near its mid-length where it terminates (Fig. 11-3, *2*). The function of the chamber is unknown. Deposits of calcium carbonate are found in the apical region of some specimens, mostly in the space outside the small chamber. Opposite the parietal line, in the middle of each side, is an internal keel or septum of varying prominence, and this has been interpreted as a ridge for muscle attachment. A Silurian specimen (Fig. 11-3, *11*) shows Y-shaped inward bifurcations of these septa.

In many specimens of *Conularia*, the narrow pointed extremity is broken off, and commonly such fossils show that the place of fracture or truncation of the test is sealed over by a thin curved diaphragm, convex outward in direction of the lost apex (Fig. 11-3, *10*). This suggests that the breaking loose of the main part of the test from its apex and place of mooring occurred while the animal was alive—not after its death, for such a diaphragm is not seen in specimens which retain their sharp-pointed tip.

Kinds of Conulariids

Typical conulariids, which are those obviously similar in general features to *Conularia*, include approximately 125 described species. They exhibit variation chiefly in the apical angle (mostly 5 to 40 deg. but ranging to 150 deg.); nature and distinctness of the parietal line; spacing, inclination, and matching of transverse ridges on opposite sides of the parietal lines; and surface ornamentation (Fig. 11-3). Nontypical conulariids include several kinds of steep-sided conical shells of chitinophosphatic substance, some of which have transverse markings and attachment disks like those of *Conularia;* these are included in the family Serpulitidae (Fig. 11-3, *15, 16*). Very similar to the last-mentioned group are fossils assigned to the family Tentaculitidae, ranging from Ordovician to Devonian (Fig. 11-3, *13, 14, 17*), which are extraordinarily abundant in some for-

mations and useful as guide fossils. They are classed among conulariids by some paleontologists.

Ecologic Association

Conulariids are found in all sorts of marine deposits, including shales, limestones, and sandstones, where they occur sporadically as associates of sponges, corals, brachiopods, bryozoans, mollusks, echinoderms, and arthropods. Such occurrence indicates adaptation to a "normal" marine environment, even though conulariid individuals generally are not numerous. On the other hand, they are abundant in some special settings where they may predominate over other fossils; such sedimentary environments are those of Ordovician iron-ore beds in Czechoslovakia and phosphate-rich, highly carbonaceous black shales in Pennsylvanian formations of the central United States. Animal remains associated with the conulariids in the black shales consist chiefly of conodonts, inarticulate brachiopods, and fragments of fishes, all of which have hard parts rich in calcium phosphate. Because deposits of this type rest on coal beds, it is reasonably certain that they were formed in very shallow water, although other conditions of the environment are largely speculative. Discovery of flattened but otherwise well-preserved conulariid remains in the Hunsrück (Lower Devonian) shale of western Germany, which almost surely was laid down in deep water, indicates that these animals may have been swimmers after they left the anchorage which they seem to have had during early life. Conulariids are interpreted as nektoplanktonic organisms.

Zoological Affinities

By different authors and at different times, the conulariids have been classified as types of coelenterates, worms, gastropods, cephalopods, and hemichordates; they have been judged to be closely related to brachiopods, bryozoans, and graptolites. This is a wide range, but at least we may be sure that conulariids are not sponges or echinoderms!

Correspondence of the structural pattern of conulariids with that of some coelenterates may signify zoological affinity. This refers especially to the fourfold plan of the body and the similarity of internal branched septa in some conulariids to the organization of hydromedusae and scyphozoans. This suggestion is partly confirmed by evidence that, in early existence, conulariid individuals were attached in fixed position on the sea floor, but later they became free-moving and possibly they were swimmers. Taken all together, these are very weak supports, however, for classification of the conulariids with coelenterates; and when attention is given to the fact that no known coelenterate has hard parts composed of calcium phosphate, one

---

FIG. 11-3. **Conulariids and similar fossils.** Nearly all the fossils shown have very uncertain classificatory status. They may belong to a phylum or phyla not represented by living animals.

1a, b, *Conularia*, restorations showing (1a) individual with sucker-like attachment disk and with the distal extremity closed by infolded extensions of the presumably flexible test; and (1b) distal part of same individual with the triangular flaps of the test raised, providing a doorway to the interior. Sides of the flaps are joined by integument which folds together snugly when the test is closed.

2, *Conularia quadrisulcata* Sowerby, from the Devonian of Morocco, ×1; an internal mold showing at lower right, edge of wall which crosses the apical part of the test obliquely.

3a, b, *Conchopeltis alternata* Walcott, Middle Ordovician, New York, apical and inverted side views, ×0.5; conulariid having a low conical test of distinctly quadripartite form.

4, *Conularia* sp., Middle Ordovician, Ohio, ×12; part of exterior of test showing ornamentation.

5, *Conularia ornata* d'Archiac, Lower Devonian, North Africa, ×6; transverse ridges of side of test.

6, *Mesoconularia ulrichana* Clarke, Devonian, Bolivia; parietal line lacking on middle of side walls.

7, *Conularia laevigata* Morris, Carboniferous, India; shallow furrow in position of the parietal line.

8, *Plectoconularia trentonensis* Hall, Middle Ordovician, New York.

9a–d, Transverse sections of conulariids: (9a) *Metaconularia;* (9b) *Pseudoconularia;* (9c) *Conularina;* (9d) *Conulariella;* all Ordovician.

10, *Metaconularia consobrina* (Barrande), Ordovician, Czechoslovakia; showing rounded diaphragm or cross wall which seals off the test after breaking away from its original anchorage.

11a, b, *Eoconularia loculata* (Wiman), Silurian, Gotland; showing (11a) transverse section of test with bifurcated septa, ×10, and (11b) side view of test, ×6.

12, Transverse section of a jellyfish (*Cratelolophus*) with gonad-bearing inward projections which simulate the septa of *Eoconularia*, although the jellyfish has no hard parts. Suggestion that conulariids are coelenterates rests mainly on such features as this.

13, *Tentaculites gyracanthus* (Eaton), Upper Silurian, New York, ×4; hollow cone, closed at tip, distinguished by pattern of its external rings.

14a, b, *Tentaculites scalariformis* Hall, Middle Devonian, New York, ×1 and portion ×3.

15, *Cornulites cingulatus* Hall, Lower Devonian, New York, enlarged; ringed cone similar to that of *Tentaculites*, interpreted as belonging to annelid worm.

16, *Cornulites corrugatus* (Eaton), Upper Ordovician, Ohio, enlarged.

17, *Styliolina fissurella* (Hall), Middle Devonian, New York, ×1; like *Tentaculites* but lacks external corrugations. Specimens are almost invariably crushed, but the solid apical region retains its form.

18, *Archaeoconularia attenuata* Sinclair, Middle Silurian, Ontario, ×1.

19, *Archaeoconularia slateri* (Reed), Upper Ordovician, Scotland, ×1; distal extremity, showing infolded flaps of test (compare 1a, b).

20a, b, *Conularina triangulata* (Raymond), Middle Ordovician (Chazyan), Quebec, ×2. Conulariids having a triangular cross section are rare.

21, *Climacoconus batteryensis* (Twenhofel), Upper Ordovician (Richmondian), Anticosti, ×1.

(2, 4, 5, from H. & G. Termier; 6–10, 12 from Boucek; others from Sinclair, Knight, and various sources.)

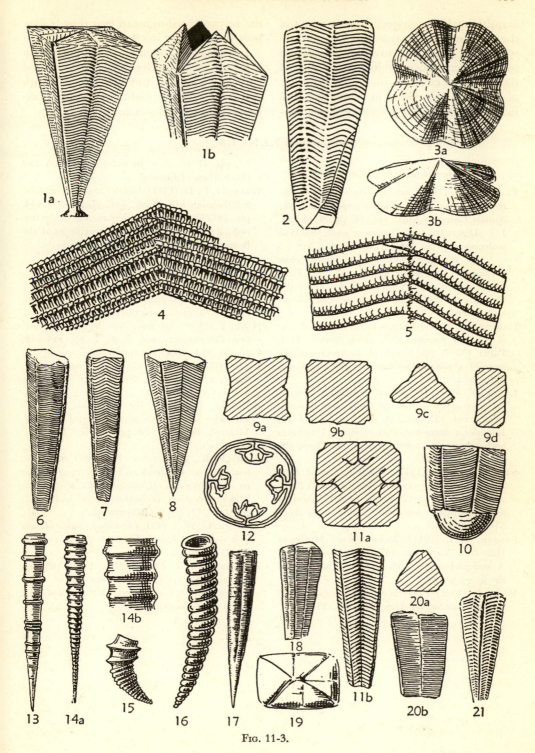

Fig. 11-3.

is quite justified in rejecting assignment of the conulariids to this phylum. Comparison with other groups of invertebrates brings out no resemblances which can be judged significant of close relationship. Similarity of shell structure found in conulariids and inarticulate brachiopods, for example, cannot be construed to mean that they belong near one another in classification.

It seems best to leave the Conulariida in a group by themselves. They are classed here as an extinct phylum grouped with Annelida and the eight other phyla collectively known as worms.

REFERENCES

Annelids and Other Worms

ELLER, E. R. (1940) *New Silurian scolecodonts from the Albian beds of the Niagara gorge, New York:* Annals Carnegie Mus., vol. 28, pp. 9–46, pls. 1–7. Describes and illustrates some Silurian scolecodonts from western New York.

—— (1945) *Scolecodonts from the Trenton series (Ordovician) of Ontario, Quebec, and New York:* Same, vol. 30, pp. 119–212, pls. 1–7. Includes illustrations and descriptions of Ordovician scolecodonts.

GARDNER, J. (1941) *Notes on fossils from the Eocene of the Gulf province:* U.S. Geol. Survey Prof. Paper 193–B, pp. 17–44, pl. 6. Species of the annelid genus *Tubulostium* are described and illustrated.

LANGE, F. W. (1949) *Polychaete annelids from the Devonian of Parana, Brazil:* Bull. Am. Paleontology, vol. 133, no. 134, pp. 1–102, pls. 1–16. Includes a discussion of scolecodonts; illustrates and describes an assemblage of them from the Devonian of Brazil.

STAUFFER, C. R. (1933) *Middle Ordovician Polychaeta from Minnesota:* Geol. Soc. America Bull., vol. 44, pp. 1173–1218, pls. 59–61. Describes and illustrates Ordovician scolecodonts from Minnesota.

STORER, T. I. (1951) *General zoology:* McGraw-Hill, New York, pp. 345–389, 438–460, illus. A comprehensive, well-organized account of the living worms.

WADE, B. (1921) *The fossil annelid Hamulus Morton, and operculate Serpula:* U.S. Nat. Mus. Proc., vol. 59, pp. 41–46, illus. Includes description and illustrations of *Hamulus.*

WALCOTT, C. D. (1911) *Middle Cambrian Annelida:* Smithsonian Misc. Coll., vol. 57, pp. 109–144, pls. 18–23. Describes and illustrates an extraordinary assemblage of fossil worms from the Burgess shale of British Columbia, Canada.

Conulariids

BOUCEK, B. (1939) *Conularida:* Handbuch der Paläozoologie, Borntraeger, Berlin, Bd. 2A, pp. A113–A131, figs. 1–13 (German).

HALL, J. (1879) *Pteropoda:* New York Geol. Survey, Paleontology, vol. 5, pt. 2, pp. 154–216, pls. 31–34A. Describes conulariids.

—— (1888) *Pteropoda, Cephalopoda, and Annelida:* Same, vol. 5, pt. 2, Suppl., pp. 5–24, pls. 114–116A.

KIDERLEN, H. (1937) *Die Conularien: Über Bau und Leben der ersten Scyphozoa:* Neues Jahrb. f. Mineral., Beil.-Bd. 77B, pp. 113–169, figs. 1–47 (German).

RICHTER, R., & RICHTER, E. (1930) *Bemerkenswert erhaltene Conularien und ihre Gattungsgenossen im Hunsrückshiefer (Unterdevon) des Rheinlandes:* Senckenbergiana (Frankfurt a. M.), Bd. 12, pp. 152–171, figs. 1–4 (German).

SINCLAIR, G. W. (1942) *The Chazy Conularida and their congeners:* Annals Carnegie Mus., vol. 29, pp. 219–240, pls. 1–3.

TERMIER, H., & TERMIER, G. (1948) *Position systematique et biologie des conulaires:* Rev. scientifique, vol. 86, fasc. 12, no. 3300, pp. 711–722, figs. 1–25 (French).

CHAPTER 12

Arthropods

Animals called arthropods (*arthro*, joint; *pod*, foot) are invertebrates of highly varied form, distinguished primarily by a segmented organization of the body and the possession of a hardened external covering. To this group belong five main assemblages: Insecta, comprising the multitudinous host of insects; Chelicerata, which include spiders, scorpions, ticks, and mites; Myriapoda, made up of centipedes and thousand-legs (millipedes); Crustacea, containing lobsters, crabs, shrimps, crayfishes, ostracodes, and barnacles; and the important extinct invertebrates called Trilobitomorpha, which consist mainly of the trilobites (Fig. 12-1). Roughly three fourths of all known kinds of animals—invertebrate and vertebrate—belong to the arthropods, which include at least 700,000 species. Many kinds of arthropods, also, are represented by almost incredibly enormous numbers of individuals.

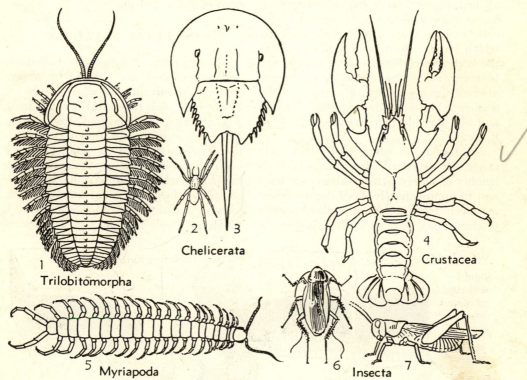

FIG. 12-1. **Representatives of the five main groups of arthropods.** (1) *Triarthrus eatoni*, an Ordovician trilobite, restoration showing appendages, dorsal view. (2) *Lycosa*, a moderately long-legged running spider, living. (3) *Xiphosura polyphemus*, the king crab, living. (4) *Cambarus*, a crayfish, living. (5) *Scolopendra*, a centipede, living. (6) *Periplanata*, a cockroach, living. (7) *Schistocerca*, a common grasshopper, living.

In structural complexity, adaptation to all sorts of environments, and development of a remarkable social organization among some, the arthropods are judged to represent the peak of evolutionary advancement attained by invertebrates. Most of them are exceptionally equipped for defense against their enemies. They are suited to use all sorts of plant and animal substances as food. On the whole, they constitute the only serious competitor of vertebrate animals, including man, in the struggle for existence.

The body segments (**somites** or **metameres**) of arthropods are bilaterally symmetrical, and they are movable on one another except where two or more adjacent somites have become fused. Such union of body parts occurs in the head region of all arthropods, and it may affect median or posterior segments. Each segment typically bears a pair of jointed appendages, which have a stout covering like that of the body. Specialized types of arthropods, however, are characterized by disappearance of the appendages from many body segments. Insects have developed wings and are the only invertebrates capable of flying.

Arthropods are adapted to a very wide range of habitats. They are found to heights of nearly 5 miles above sea level and to approximately equal depths below sea level. They range from polar latitudes to the tropics, various kinds living in the sea, in fresh waters of streams and lakes, in the soil and crevices of rocks, on the land surface, on and in trees and other plants, and flying in the air. Some can live in hypersaline brines. They are distributed throughout continental and oceanic areas of the modern world, and representatives of the phylum occur in rocks of all ages back to Pre-Cambrian.

The size of adult arthropods ranges from 0.1 mm. (0.04 in.) to nearly 3 m. (10 ft.), as represented by the length of some Paleozoic eurypterids, or 4 m. (13 ft.), as measured from extremities of the outspread legs of a giant Japanese crab. One of the late Paleozoic dragonflies attained a wingspread of approximately 75 cm. (30 in.). The great bulk of arthropod individuals, however, has maximum dimensions of less than 1 in.

SKELETAL FEATURES

The soft parts of arthropods, both body and appendages, are encased by a protective chitinous covering which is secreted by the outer skin (epiderm) of the animal. This covering has a layered structure, for its surface consists of a thin waxy waterproof substance, and its relatively thick inner part is made up of alternating laminae of chitin and protein, which are complex compounds of carbon, hydrogen, oxygen, and nitrogen. Such a covering is tough but easily flexible, and it has some

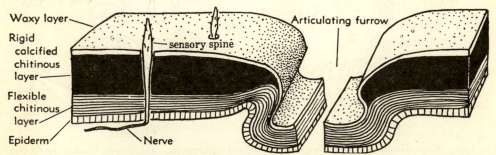

Fig. 12-2. **Diagrammatic section of part of an arthropod exoskeleton.** The chitinous principal layer of the outer covering is stiffened by impregnated mineral substances in areas between joints, whereas the unmineralized chitin at joints remains easily flexible. Commonly such joints are surficially marked as an articulating furrow.

elasticity. By impregnation of the principal layer with mineral salts, mostly calcium carbonate and phosphate, the covering is altered into rigid armor (Fig. 12-2). This strengthening by secretion of mineral matter is not uniformly distributed, for if the whole exterior were stiffened, movement of the appendages and other body parts would be impossible. The deposition of calcium compounds is localized so as to provide stout protection for most of each unit part but leave uncalcified areas between the parts. Thus, rigidly armored segments have free mobility at the joints. The mineralized parts of the arthropod covering are readily capable of fossilization, whereas the flexible unmineralized chitin is ill suited for preservation. This partly explains the common occurrence of disarticulated hard parts of arthropods among fossils.

The rigid covering which has just been described not only serves for protection of the arthropod but furnishes support for its soft tissues and provides places of attachment for its muscles. It is appropriately termed an external skeleton or **exoskeleton.** Mechanically, such a structure seems to be quite as efficient as the internal skeleton of vertebrates, and it has the advantage of being a strong, relatively impermeable cover, which takes the place of a vertebrate's variously constructed outer covering. For life on land, it is necessary to protect the watery internal parts of an animal from drying out rapidly in the air. Thus, land-living vertebrates bear a tough hide which may be further insulated by fur, scales, feathers, or bony plates, whereas arthropods need only their exoskeleton (Fig. 12-3).

Segmentation of Body

Like annelids, which undoubtedly were their ancestors, the arthropods have segmented bodies. The arthropod body segments (metameres or somites) differ basically from those of an annelid in having part of the enclosing walls made rigid by

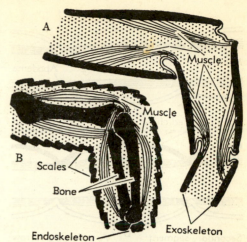

Fig. 12-3. **Comparison of the endoskeletal structure of a vertebrate and exoskeleton of an arthropod.** (*A*) Diagrammatic section of part of the limb of an arthropod showing rigid exoskeletal portions (black), with intervening unmineralized flexible covering at joints. The muscles are attached to the inner surfaces of the exoskeleton so as to move the segments readily. (*B*) Diagrammatic section of part of the limb of a lizard, showing bony endoskeleton, attached muscles, and external cover of heavy scales. These drawings indicate the essential contrast in architectural plan of arthropod invertebrates and the vertebrates.

the addition of mineral matter to the chitin. The hardened parts of the cover belonging to each metamere are termed **sclerites,** and according to their position they comprise a **tergite** (on the dorsal side), a **sternite** (on the ventral side), and among some arthropods, lateral pieces called **pleurites.** Mobility between adjoining sclerites is effected by unmineralized tracts of the flexible chitinous wall, which are slightly offset from the boundaries of the metameres. Thus, the arrangement of successive sclerites constitutes a secondary segmentation which does not correspond exactly to the primary metameres (Fig. 12-4). Inwardly bent parts of the sclerites in the position of the boundaries between primary segments furnish places of muscle attachment (**apodemes**) for movement of the sclerites.

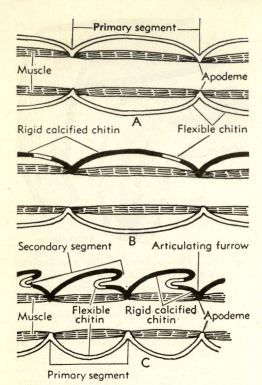

FIG. 12-4. **Diagrammatic longitudinal sections showing comparison of somites (metameres) of an annelid and arthropod.** (*A*) Segments of an annelid; primary divisions of the body are indicated by inflections of the flexible covering, which furnish places of attachment for muscles. (*B*) Diagram showing stiffened portions of the cover which are separated by areas which remain flexible. This represents a condition intermediate between the structure of annelids and arthropods. (*C*) Inflected exoskeleton of an arthropod showing rigid portions (black) and flexible joints (white). This indicates the manner in which secondary segmentation of arthropods is related to primary somites.

Organization of Body Segments

The somites of arthropods can be grouped according to their placement in distinct body regions, but the number and organization of these regions differ in the various divisions of the phylum. All arthropods have a **head,** which is formed by combined anterior somites, generally six or more. Some of these are recognizable in the adults by pairs of modified append-ages, such as antennae, mandibles, and maxillae, which belong to different coalesced somites of the head region; others can be discerned only by a study of embryological development. Behind the head, the body segments are essentially all alike in such arthropods as the centipedes and millipedes (Fig. 12-1, 5), but in insects and some other groups they are well differentiated into a median region, termed **thorax,** and posterior region, called **abdomen** (Fig. 12-1, 1, 7). Some crustaceans have united head and thoracic regions, and this part of the body is then designated as **cephalothorax** (Fig. 12-1, 4). Also, rigid chitinous outgrowths which project over the head and enclose part or all of the thoracic and abdominal regions are seen in some crustaceans. This structure, termed **carapace,** may be bivalve in nature, with hingement along the dorsal edge, as in ostracodes.

Appendages

Each body segment of an arthropod, including those which are fused together in the head or other regions, typically bears a pair of jointed appendages. These are attached to the sides of the somite, between the dorsal and ventral sclerites. The segment of the appendage which directly articulates with the body is moved by muscles attached to the inside of the appendage segment near its extremity and fastened to the inner surface of adjacent body sclerites. Each division of an appendage can be moved by bending at the places of its attachment to adjoining elements, and the movement is effected by muscles inside the appendage which extend across joints so as to be fixed to two skeletal segments.

The number of segments, shape, and relative size of various arthropod appendages vary greatly. A single individual commonly possesses different kinds (Fig. 12-5). Some are slender sensory structures composed of many minute segments; these are **antennae** or **antennules.** Some are short, consisting of a very few rather broad seg-

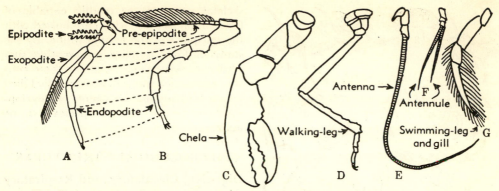

Fig. 12-5. **Appendages of arthropods.** (*A*) Generalized limb of a crustacean, showing branches from upper segments and biramous structure. (*B*) Trilobite limb, showing inferred homology with limb of a crustacean; the endopodite serves for walking, whereas the pre-epipodite is judged to have functioned for respiration and as a swimming organ. (*C*) Pincer-bearing appendage of a lobster. (*D*) Walking leg of an insect. (*E*) Antenna of a lobster. (*F*) Biramous antennules of a lobster. (*G*) Swimming leg of a water beetle.

ments which may bear toothlike projections; these are jaw parts or **mandibles.** Others are limbs, termed **maxillae,** modified for passing food to the mouth. Some are walking legs, like those of most insects, and these may bear claws or thickened pads at their extremities. Some are provided with rows of very closely spaced long bristles, which are used for swimming. Certain appendages are modified to serve as mating organs or for the placement of eggs. The number of paired appendages is very great in some arthropods, such as the millipedes, which carry two pairs to each somite. In others, like the insects, the number is restricted to three pairs of walking or swimming limbs and appendages of the head region. A special type is one having a pincer arrangement at the tip, as in the larger claw of a lobster; this structure, called a **chela,** is characteristic of the major division of the arthropods called Chelicerata (Fig. 12-5*C*).

Among some arthropods, such as the trilobites and crustaceans, appendages may be branched. Attached to a basal segment (**coxa**), which articulates with the body, is an external branch (**exopodite**) and an internal one (**endopodite**); additional branches (**epipodites**) may be present above the exopodite of crustaceans.

GROWTH STAGES

Important characteristics of arthropods, not previously mentioned, relate to their growth stages. Like other invertebrates, except those produced by asexual budding, an arthropod develops from an egg and passes through a ciliated larval stage. Reproduction commonly is sexual, eggs within the body of the female being fertilized by a male, but many kinds develop parthenogenetically from females which have not been fertilized, or from females which produce eggs very long after fertilization. Sexual reproduction by immature individuals (neoteny) and parthenogenetic reproduction at larval stages (paedogenesis) are known among arthropods, especially certain insects.

After the egg hatches, a variable number of pronounced changes in body form (metamorphoses) occurs, and thus appearances of the growing animal at different stages may be radically dissimilar. This is well shown by the familiar changes of butterflies and moths, which reach the winged adult stage only after passing through a period as caterpillars and going into retirement as pupae. Crustaceans similarly traverse a varying number of metamorphic alterations (Fig. 12-6). Fi-

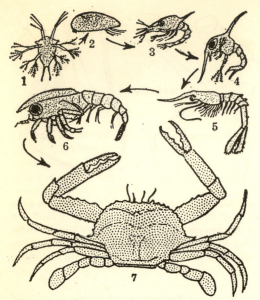

FIG. 12-6. **Stages in development of a crustacean.** Successive larval stages are illustrated in 1–6; the adult is shown in 7. This is the swimming crab, *Portunus*, living. Designations applied to the successive larval stages are: nauplius, cypris, protozoea, zoea, mysis, and megalops. Most arthropods go through a smaller number of metamorphoses.

nally, they attain the form of maturity, but thereafter they continue to increase in size so long as life lasts.

Both the modifications incident to larval development and the size increases during adult life require successive renovations of the chitinous exoskeleton. Parts of this exoskeleton, which are made rigid by impregnation of mineral salts, are inelastic. They cannot accommodate changes of shape or size. Hence, they must be discarded whenever the animal reaches a point in life when the covering is no longer suited to the body. This brings to our notice the phenomenon of molts, or **ecdysis**, which consists of splitting apart the previously worn exoskeleton and discarding it, followed by the secretion of a new, more comfortably fitting outer cover. All arthropods exhibit these changes in varying number. This has importance in paleontological study because the discarded exo-

skeletal parts are as readily capable of fossilization as those which are buried with the soft parts of the animal inside. By the study of fossil remains representing successive growth stages, whether these consist of discarded molts or accidentally killed young individuals, the course of development in the growth of species can be learned.

PHYSIOLOGICAL STRUCTURES

Digestive, Circulatory, and Respiratory Systems

Arthropods have a well-developed digestive system (Fig. 12-7). The mouth, located on the underside of the head, opens into a chitin-lined esophagus (**stomodaeum**), which among higher crustaceans is a somewhat complexly built tract fitted for comminution of food. This leads to the stomach and thence to a long intestine. The anus is on the underside of the most posterior segment, and as in the esophageal region, the chitinous external covering here bends inward to line the terminal part of the digestive tract (**proctodaeum**). Insects, spiders, and some other land-living arthropods have excretory organs (malpighian tubes) which open into the intestine near the anus.

The circulatory system of arthropods differs importantly from that of vertebrates and some other invertebrates in having no arterial and vein passageways. A pulsating heart is located along the dorsal side of the

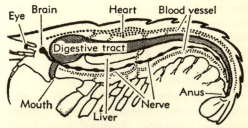

FIG. 12-7. **Longitudinal section of a lobster showing anatomical features.** The digestive tract has an anterior distended portion corresponding to a stomach. The heart is located dorsally. The brain connects with the ventrally located nerve cord and ganglia.

animal. Blood is pumped from an anterior opening of the heart into tissue spaces and large blood cavities in the middle and posterior parts of the body, from which it returns to the heart through paired openings located in each somite. This open sort of circulation is specialized, for primitive arthropods, like annelids, have blood vessels leading from the heart to all main regions of the body.

Most aquatic arthropods breathe by means of gills, which are thin parts of the body wall through which oxygen can reach the interior and carbon dioxide can be expelled. Terrestrial forms, especially insects, have a system of branching tubes which conduct air inward from the body wall so as to oxygenate the tissues directly, in this manner largely replacing the functions of the circulatory system.

Nervous System and Sense Organs

Arthropods have a brain in the dorsal part of the head. It is connected by a ring of nerve fibers around the digestive tract to ganglia and nerves that run backward along the ventral side of the animal (Fig. 12-7). Branches of the nervous system connect with various sense organs which function for seeing, feeling, hearing, and in receiving chemical stimuli.

Two types of eyes are found in arthropods: simple and compound (Fig. 12-8). Many species have both types. The simple eye consists of a transparent lens above a crystalline cone which condenses light on sensitive cells connected by nerve fibers to the brain. Most spiders have eight such simple eyes in a cluster. Compound eyes are composed of many closely packed units, each of which is provided with its own light-condensing apparatus and connected with a retina of light-sensitive cells. The hundreds or thousands of such units which occur in one compound eye are isolated optically from one another so that each receives only a very narrow bundle of parallel rays. The visual image produced by these independent units working together probably is comparable to that

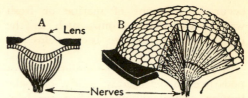

FIG. 12-8. **Eye structure of arthropods.** (*A*) Section of a simple eye showing lens and bordering part of rigid exoskeleton (black); light-sensitive cells and nerve fibers carry impulses to the brain. (*B*) A compound eye sectioned in quadrant. Each element of this eye consists of a lens and conical light-condensing mechanism leading to light-sensitive cells and nerve fibers.

given by a rather coarse half-tone screen, but it is surely an effective mechanism for detecting movement. This is readily demonstrated by experiments with any arthropod having compound eyes.

Organs which are sensitive to touch are first of all the antennae and antennules, which are more or less elongated slender appendages at the front of the head (Fig. 12-5*E*, *F*). In addition, different arthropods have various sorts of sensory bristles, spines, scales, or pits, scattered over the body (Fig. 12-2). These are connected by perforations in the chitinous exoskeleton with nerve fibers. At least some of these structures are sensitive to chemical stimuli, as are the membranes on the feet of various insects. It has been learned that destructive effects of DDT on flies, mosquitoes, and the like are due to penetration of the chemical through the foot membranes so as to paralyze the nervous system. Some arthropods have a flexible membrane stretched across an opening in the exoskeleton and thus can detect sound. Balancing organs which consist of sensory pits containing a hard particle are also known.

MODE OF LIFE

We have already noted that arthropods are found in an extremely wide range of environments. From the viewpoint of paleontological study, however, it is appropriate to focus attention almost wholly on the aquatic forms, because most fossils occur in sediment of marine or fresh waters. Important groups such as the trilobites seem to have been exclusively marine.

A large majority of the aquatic arthropods live on the muddy, sandy, or rocky bottom of water bodies. They are benthonic, but most of them can move about actively. Some dig burrows in which they may stay semipermanently. A few (notably the barnacles) become fixed to some foreign object after their free-swimming larval existence and, largely because of their sessile mode of life, are so modified as hardly to be recognized as arthropods. Many are scavengers, feeding on any dead animal matter they can find; among these are the common lobsters of both the Atlantic and Pacific coasts of North America. Some kill and eat various live animals, including small fishes. Various kinds feed on plants.

Many small arthropods are good swimmers and are found in surface waters of the ocean far from land. Others are floaters or drifters, forming part of the oceanic plankton. These pelagic forms, especially copepods, commonly occur in enormous numbers and are a chief food source of many fishes and of whales. Likewise, the minute bivalved crustaceans called ostracodes, which are very widely distributed as fossils, occur abundantly both in salt water and fresh. Adaptation to excessively saline water is shown by the brine shrimp.

CLASSIFICATION

The problem of classifying the host of varied jointed-leg invertebrates in a manner that expresses their most significant structural similarities and divergences and that fits the probable course of phylogeny in this group is indeed difficult. Such assemblages as the insects, spiders, and crustaceans seem to be quite distinct, but judgment about the affinities of various other arthropods differs. For example, the trilobites, which are very important in paleontological study, have long been treated as a subdivision of the crustaceans, because, like crustaceans, they have antennae and possess biramous limbs. Some recent workers, however, call attention to affinities of trilobites with arachnids and have ranged them in this division of the arthropods. Consensus of the most highly qualified modern opinion places the trilobites in an independent position, joined neither with the crustaceans nor the arachnids, and it is very possible that the stock to which trilobites belong was the source of these other main divisions. A minor group, which is less well known generally than the trilobites, is that of the Onychophora. It has special significance in study of evolution, because of characters which place it halfway between annelids and arthropods. Some zoologists class the onychophores as a class of the arthropods, whereas others treat them as an independent phylum.

The classification which is adopted in this book is shown in the following tabular outline. Divisions below the rank of class are not indicated here, for they are more appropriately treated in later chapters. An asterisk (*) indicates that the group so marked is known only from fossil remains.

Main Divisions of Jointed-leg Invertebrates

Pararthropoda (*phylum*), arthropod-like invertebrates of somewhat uncertain affinities. Cambrian–Recent.

  Onychophora (*class*), similar in form to annelids and myriapods. Cambrian–Recent.

  Tardigrada (*class*), very minute marine and terrestrial forms. Recent.

  Pentastomida (*class*), wormlike parasitic invertebrates, Recent.

Arthropoda (*phylum*), true jointed-leg invertebrates. Cambrian–Recent.

  *Trilobitomorpha (*subphylum*), trilobites and nearly related forms. Cambrian–Permian.

    *Trilobita (*class*), trilobites. Cambrian–Permian.

    *Merostomoidea (*class*). Cambrian–Devonian.

    *Marellomorpha (*class*). Cambrian.

    *Pseudocrustacea (*class*). Cambrian.

    *Arthropleurida (*class*). Upper Carboniferous.

Chelicerata (*subphylum*), merostomes, arachnids, sea spiders, characterized by lack of antennae and possession of some appendages which bear pincers (chelae). Cambrian–Recent.

*Merostomata (*class*), xiphosurans, which include the king crab, and eurypterids. Cambrian–Recent.

Arachnida (*class*), scorpions and spiders. Silurian–Recent.

Pycnogonida (*class*), sea spiders. Devonian–Recent.

Crustacea (*subphylum*), mostly aquatic arthropods which have two pairs of antennae and generally some biramous limbs. Cambrian–Recent.

Branchiopoda (*class*), mainly distinguished by leaflike appendages, include water fleas. Cambrian–Recent.

Ostracoda (*class*), small bivalved forms. Ordovician–Recent.

Copepoda (*class*), minute crustaceans, very abundant in modern seas. Recent.

Branchiura (*class*), resemble copepods, temporarily parasitic on fishes. Recent.

Cirripedia (*class*), barnacles. Ordovician–Recent.

Malacostraca (*class*), lobsters, crabs, shrimps, and many others. ?Cambrian, Ordovician–Recent.

Myriapoda (*subphylum*), elongate, annelid-like body having many similar pairs of appendages. Silurian–Recent.

Diplopoda (*class*), millipedes. Silurian–Recent.

Chilopoda (*class*), centipedes. Carboniferous–Recent.

Insecta (*subphylum*), mostly terrestrial wing-bearing arthropods which typically bear three pairs of walking legs. Devonian–Recent.

Apterygota (*class*), primitively wingless insects. Devonian–Recent.

Pterygota (*class*), typically wing-bearing insects. Devonian–Recent.

GEOLOGICAL DISTRIBUTION

The arthropods are one of the most ancient—and accordingly, we might say most respectable—of all known organic lineages. They make an appearance near the base of Cambrian deposits, where we find structurally complex trilobites. Also, the amazing assemblage of arthropods

found in the Middle Cambrian Burgess shale of western Canada affords a glimpse of the remarkable degree of differentiation which had been attained by these exo-skeleton-bearing invertebrates near the beginning of Paleozoic time. They must have had a long Pre-Cambrian existence, during which they became modified from their annelid ancestors and branched along various evolutionary lines. Fossil arthropods have been reported from Pre-Cambrian rocks, but all such records have doubtful validity. These include chitinous fragments called *Beltina*, from the Belt series of Montana; a supposed onychophore (*Xenusion*) in an erratic block of ?Pre-Cambrian quartzite from Scandinavia; and alleged eurypterid-like fossils (*Protadelaidea*), which actually may be symmetrical concretions, reported from Pre-Cambrian rocks of Australia. Therefore, we can recognize the distribution of arthropods as reaching almost to the base of Paleozoic rocks but not lower.

Cambrian representatives of the arthropods and pararthropods include trilobitomorphs as the most important assemblage,

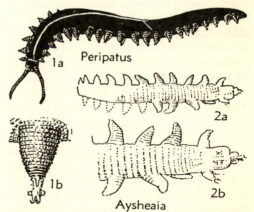

Fig. 12-9. **Recent and fossil onychophores.** (*1a, b*) *Peripatus* is a modern representative of this group which is a connecting link between annelids and arthropods; (*1a*) the animal, ×1; (*1b*) one of the claw-bearing legs, enlarged. (*2a, b*) *Aysheaia pedunculata* Walcott, from the Middle Cambrian Burgess shale of British Columbia; (*2a*) the animal, ×1.3; (*2b*) anterior end. ×2.7.

merostome chelicerates, branchiopod crustaceans, and an onychophore (Fig. 12-10). The Ordovician period was evidently the time of maximum development of the trilobites, and here begins the fossil record of the paleontologically important group of ostracode crustaceans—likewise of cirripeds and malacostracans (lobsters, crabs). The oldest presently known arachnids are scorpions from the upper Silurian strata of England, and myriapods are first represented by a millipede in rocks of the same age. Insects and sea spiders (pycnogonids) begin their record in the upper parts of the Devonian system. Thus, long before the close of Paleozoic time, all main divisions of the arthropods are found to be established.

Subdivisions, which are here treated as classes, subclasses, and orders, have widely varying paleontological distribution and importance. Some are unknown as fossils. Xiphosuran merostomes, among the chelicerate arthropods, are mainly late Paleozoic, whereas the eurypterids are most prominent in Silurian and Devonian rocks; three xiphosuran genera, including the king crab (*Xiphosura*), survive in modern seas. Fossil spiders of several sorts occur in rocks of Pennsylvanian age, and this group is represented by many fossils preserved in Cretaceous and Tertiary amber. Crayfishlike crustaceans occur in Pennsylvanian and Mesozoic strata. Jurassic, Cretaceous, and Tertiary deposits have yielded numerous kinds of lobsters, crabs, and shrimps. Ostracodes are especially abundant and varied in many Paleozoic formations, and they are common in various younger deposits; most kinds are useful in stratigraphical paleontology. Nearly all main stocks of insects are found to be represented in late Paleozoic rocks, but the fossil record of this group becomes progressively more complete in Mesozoic and Cenozoic strata.

PARARTHROPODS

Three minor groups (onychophores, tardigrades, and pentastomids) collectively are termed pararthropods. Among these, the onychophores deserve notice because they are a connecting link between the soft-bodied segmented worms (annelids) and the jointed-leg segmented arthropods provided with an exoskeleton. They include about a dozen living genera, among which *Peripatus* has been most studied. It is a caterpillar-like animal about 3 in. long when fully grown (Fig. 12-9). The body is segmented internally but not externally. Its covering is a thin cuticle of flexible chitin, which contains no hard mineral-impregnated portions. A pair of short stout legs is joined to each body segment except at the front, where three segments are fused together to form the head. Forked claws at the extremities of the legs resemble those of arthropods. The three pairs of appendages attached to the head consist of rather large antennae, short food-handling papillae next to the mouth, and horny jaws, which are effective mandibles for cutting. The modern onychophores are modified in having become adapted to life on land and in giving birth to young alive, but they retain peculiarities derived from distant ancestors of a half-annelid half-arthropod nature. Thus, they approximate closely the intermediate stage in evolution which translated annelid into arthropod.

Aside from a doubtfully identified onychophore (*Xenusion*) from late ?Pre-Cambrian rocks of Scandinavia, the only paleontological record of this group consists of a fossil (*Aysheaia*) from the Middle Cambrian Burgess shale of British Columbia (Fig. 12-9).

Tardigrades and pentastomids are unknown as fossils.

Fɪɢ 12-10. **Geological distribution of arthropods.** All main divisions are represented in Paleozoic rocks, although maximum differentiation, which is attributable chiefly to insects, probably belongs to the present time.

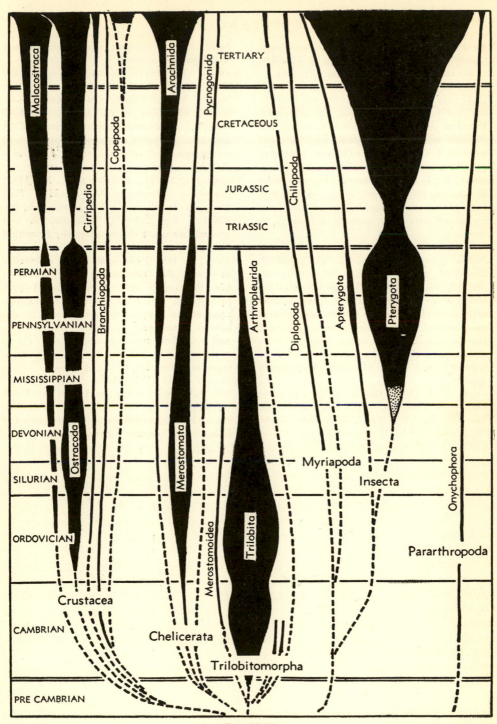

Fig. 12-10.

REFERENCES

BORRADAILE, L. A., *et al.* (1948) *The Invertebrata:* Cambridge, London, 2d ed., pp. 1–725, figs. 1–483 (arthropods in general, pp. 305–316).

BUCHSBAUM, R. (1948) *Animals without backbones:* University of Chicago Press, Chicago, rev. ed., pp. 1–405, illus. (arthropods, pp. 239–298). Excellent general description and figures but contains some misinformation, such as statement (p. 245) that "the head of every arthropod consists of exactly six segments."

CALMAN, W. T. (1929) *Arthropoda:* Encyclopaedia Britannica, 14th ed., vol. 2, pp. 456–459, figs. 1–6. Comparative morphology of the head region of arthropods showing differences in structure.

MACGINITIE, G. E., & MACGINITIE, N. (1949) *Natural history of marine animals:* McGraw-Hill, New York, pp. 1–473, figs. 1–282 (arthropods, pp. 252–326, figs. 116–167). Especially good for observations on life habits of living marine arthropods. Introductory chapters on subjects such as food, sense organs, growth rates, and marine animal habitats supply information on arthropods. Noteworthy is an account of copepods as a source of marine food supply (pp. 44–46).

STORER, T. I. (1951) *General zoology:* McGraw-Hill, New York, pp. 1–832 (arthropods, pp. 461–555, figs. 22.1–22.17, 23.1–23.47, 24.1–24.13) illus. A compact, comprehensive, and well-organized account.

STØRMER, L. (1944) *On the relationships and phylogeny of fossil and recent Arachnomorpha:* Skr. Norske Vidensk.-Akad. Oslo, Mat.-Naturv. Klasse, 1944, no. 5, pp. 1–158, figs. 1–30 (English). A thorough comparative study of the structures of trilobites, arachnids, xiphosurans, eurypterids, and other arthropods; contains a good bibliography.

TWENHOFEL, W. H., & SHROCK, R. R. (1935) *Invertebrate paleontology:* McGraw-Hill, New York, pp. 1–511, figs. 1–175 (arthropods, pp. 406–478, figs. 148–175). Classification and general treatment differs materially from that here given.

VANDEL, A. (1949) *Generalités sur les arthropodes:* in Grassé, P., Traité de zoologie, Masson et Cie., Paris, vol. 6 (pp. 1–979, 3 pls., 871 figs.), pp. 79–158, figs. 1–36 (French). A thorough treatment of morphology accompanied by discussion of ontogeny, phylogeny, classification, and evolution; contains bibliography. This is the most recent and authoritative general account of arthropods.

WOLCOTT, R. H. (1946) *Animal biology:* McGraw-Hill, New York, 3d ed., pp. 1–719, figs. 1–508 (arthropods, pp. 289–357, figs. 201–270). An up-to-date, well-written general account of living forms, but does not treat fossils.

CHAPTER 13

Trilobites

Trilobites are an extinct group of arthropods which occupy a dominant role in the early part of the paleontological record. They are exclusively marine organisms, and especially in some shallow sea deposits of Cambrian and Ordovician age, their remains are both abundant and widely distributed. They are less numerous in Silurian and succeeding Paleozoic formations, but a few genera persisted into Permian time. It is rather surprising to find that one of the most highly organized of all invertebrate phyla comprises the chief element of oldest faunas. The variety and structural complexity of trilobites found near the base of Cambrian rocks surely indicates a very long antecedent existence of animal life, during which the first arthropods became differentiated. Probably their beginning belongs more than 100 million years before earliest Paleozoic time.

The designation of trilobites dates back at least to the first scientific publication on these fossils by Lhwyd (1698) in England. The name refers to the division of the body into three longitudinal segments or lobes, which are very distinct in most specimens. The central segment, running from a point at or near the anterior margin to the posterior extremity, is termed **axis** or **axial lobe** (in many older publications, rhachis), and the segments on its left and right sides are termed **pleural lobes** (Figs. 13-1; 13-2, *9, 39*). In addition, a generally well-defined division of the body into transverse segments is observed: anterior, comprising the head or **cephalon**; median, made up of more or less numerous jointed pieces, which together form the **thorax**; and posterior, called the **pygidium** (Figs. 13-1; 13-2, *1, 3, 6*). The head bears the

eyes on its dorsal side and the mouth on its underside. Head, thorax, and pygidium all carry pairs of jointed limbs. These longitudinal and transverse divisions mark out the main regions of the body and skeleton.

The hard parts of a trilobite, which generally are the only ones preserved as fossils, consist of mineral-impregnated portions of its chitinous covering. These cover the back, or dorsal side, and marginal parts of the lower or ventral side; ventral prolongations of the exoskeleton comprise what is called the **doublure** (Fig. 13-2, *12*). Other skeletal elements on the ventral side are plates next to the mouth, the **hypostome** in front of it, and the **metastome** behind it (Fig. 13-2, *20, 27*). The jointed appendages also are covered by exoskeleton, but only rare fossils retain these parts.

The exoskeleton of trilobites is composed of calcium carbonate and calcium phosphate in varying proportions, up to about

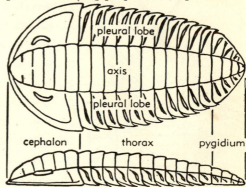

FIG. 13-1. **Diagram of a trilobite showing main divisions of exoskeleton.** The drawing represents a generalized trilobite of primitive aspect in having a completely segmented axis, which extends through the entire length of the animal and has gradually tapered form anteriorly and posteriorly.

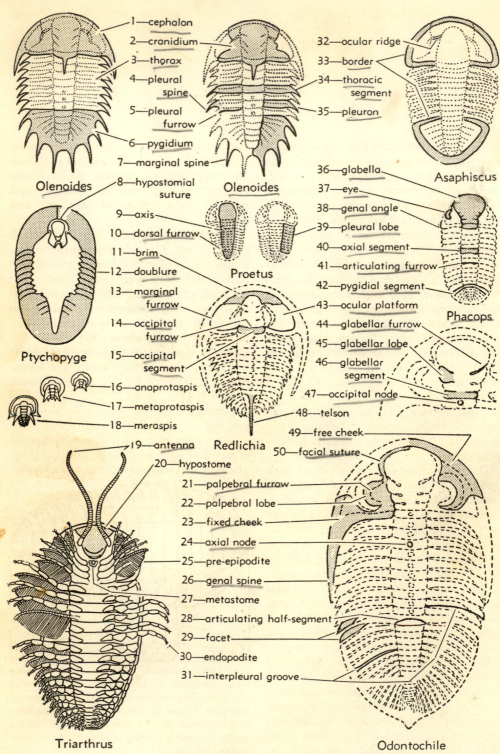

1—cephalon
2—cranidium
3—thorax
4—pleural spine
5—pleural furrow
6—pygidium
7—marginal spine
8—hypostomial suture
9—axis
10—dorsal furrow
11—brim
12—doublure
13—marginal furrow
14—occipital furrow
15—occipital segment
16—anaprotaspis
17—metaprotaspis
18—meraspis
19—antenna
20—hypostome
21—palpebral furrow
22—palpebral lobe
23—fixed cheek
24—axial node
25—pre-epipodite
26—genal spine
27—metastome
28—articulating half-segment
29—facet
30—endopodite
31—interpleural groove
32—ocular ridge
33—border
34—thoracic segment
35—pleuron
36—glabella
37—eye
38—genal angle
39—pleural lobe
40—axial segment
41—articulating furrow
42—pygidial segment
43—ocular platform
44—glabellar furrow
45—glabellar lobe
46—glabellar segment
47—occipital node
48—telson
49—free cheek
50—facial suture

Olenoides

Olenoides

Asaphiscus

Ptychopyge

Proetus

Phacops

Redlichia

Triarthrus

Odontochile

Fig. 13-2.

Fig. 13-2. **Morphological features of trilobites.** The illustrated structures are briefly defined in the following alphabetically arranged list of terms, which is cross-indexed by numbers to the figures.

anaprotaspis (16). Early protaspis stage having no more than six segments, all belonging to cephalon.

antenna (19). Many-segmented sensory appendage attached to front part of head.

articulating furrow (41). Transverse groove between axial segments of thorax.

articulating half segment (28). Arched anterior extension of axial segments of thorax and front of pygidium, which project beneath next-forward segment.

axial node (24). Centrally located tubercle on an axial segment.

axial segment (40). Transverse division of axis of thorax or pygidium.

axis (9). Longitudinal central part of cephalon, thorax, and pygidium, bounded by dorsal furrow.

border (33). Raised peripheral parts of cephalon and pygidium exclusive of axial parts adjoining thorax; adjoins marginal furrow.

brim (11). Part of cranidium bounded anteriorly by marginal furrow and posteriorly by front of glabella and ocular ridges, or lines running from front of eyes to glabella.

cephalon (1). Part of trilobite carapace in front of thorax.

cranidium (2). Central part of cephalon, bounded laterally by facial sutures; among trilobites having marginal sutures, it includes entire dorsal part of cephalon.

dorsal furrow (10). Groove bounding axis; it is located along sides and front of glabella, sides of axial lobe of thorax, and sides and rear of axial lobe of pygidium.

doublure (12). Reflexed portion of carapace along ventral margins of cephalon, pleura, and pygidium.

endopodite (30). Inner branch (walking leg) of biramous paired appendages attached to each post-antennal segment.

eye (37). Visual area containing one or many lenses, located on either side of glabella; generally curved in plan and sloping steeply outward; among proparians and opisthoparians borne by inner margins of free cheeks.

facet (29). Sharply downbent areas along outer front edges of pleura and pygidium; in articular movement providing for impingement on adjacent parts of skeleton.

facial suture (50). Line of junction between cranidium and free cheeks, in different genera

wholly marginal, partly marginal and partly dorsal, or wholly dorsal; sutures crossing dorsal side of cephalon invariably run along upper edge of eye.

fixed cheek (23). Part of cranidium on either side of glabella; the two fixed cheeks comprise all of cranidium exclusive of glabella and may be confluent in front of it.

free cheek (49). Part of cephalon separated by facial suture from cranidium, including part or all of doublure on one side of cephalon; the two free cheeks may be confluent in front of glabella, divided by a suture, or separated by an accessory plate (rostrum).

genal angle (38). Postero-lateral corner of cephalon.

genal spine (26). Backward extension of posterolateral corner of cephalon in form of a spine.

glabella (36). Axial part of cranidium, bounded by dorsal furrow at its front and sides.

glabellar furrow (44). Straight or curved groove extending inward from side of glabella.

glabellar lobe (45). Part of glabella bounded in front or behind, or on both of these sides by short glabellar furrows.

glabellar segment (46). Part of glabella bounded in front and behind by long glabellar furrows.

hypostome (20). Plate on underside of cephalon in front of mouth.

hypostomial suture (8). Line at anterior edge of hypostome where it joins doublure or rostrum.

interpleural groove (31). Transverse furrow between adjoining pleura of thoracic region or crossing a pleural lobe of pygidium.

marginal furrow (13). Groove or abrupt inflection of surface along inner edge of border of cephalon or pygidium.

marginal spine (7). Sharp projection at edge of pygidium.

meraspis (18). Late larval stage characterized by presence of a small pygidium and fewer thoracic segments than in adult stage.

metaprotaspis (17). Late part of protaspis larval stage, characterized by presence of one or more thoracic segments but no pygidium.

metastome (27). Small plate behind mouth.

occipital furrow (14). Transverse groove in front of hindmost glabellar segment.

occipital node (47). Tubercle on mid-portion of occipital segment, in some trilobites observed to have structure of a simple eye.

(Continued on next page.)

30 per cent of the latter in some. Thin sections of well-preserved fossils show canals or minute pores running through the calcareous walls, approximately at right angles to their surface. Also, three more or less distinct layers have been distinguished: an outer thin dark layer, probably pigmented; a middle relatively thick laminated layer; and an inner thin clear layer. The exterior of the carapace commonly is smooth, but it may bear fine to coarse granules or tubercles; a characteristic surface marking, especially on the doublure, is an arrangement of fine, slightly irregular parallel lines (terrace lines), which are corrugations serving to strengthen the test. A considerable part of the undersurface of all trilobites was unprotected by mineralized chitin. Consequently, the covering of this vulnerable region is not preserved in fossils.

The size of adult trilobites ranges from (*Agnostus*) 6 mm. (0.25 in.) to a few giant forms nearly 75 cm. (30 in.) long (*Paradoxides, Isotelus, Uralichas, Terataspis*). Most trilobites, however, are 2 to 7 cm. (0.8 to 2.7 in.) in length and 1 to 3 cm. (0.4 to 1.2 in.) in width.

STRUCTURE OF MAIN SKELETAL DIVISIONS

Cephalon

The head shield, or cephalon, is one of the most important parts of the exoskeleton, because morphologic features of this region furnish the chief basis for classification. Also, because structures of the cephalon are particularly concerned in ontogenetic development, and influenced by evolutionary trends, they merit special study. To the extent that phylogeny is understood, this rests mainly on evidence of cephalic characters. Likewise, comparison of trilobites with other main branches of the phylum Arthropoda and recognition of the Trilobitomorpha as a division correlative with the crustaceans and chelicerates, rather than included with either, takes chief account of the head region.

Facial Sutures. The cephalon of all trilobites, except the group called Agnos-

---

(Fig. 13-2 continued.)

occipital segment (15). Hindmost part of glabella, bounded in front by a complete transverse groove (occipital furrow).

ocular platform (43). Part of fixed cheek behind brim and extending laterally outward from eye.

ocular ridge (32). Narrow elevation extending from front edge of each eye to glabella; lacking in many genera.

palpebral furrow (21). Groove or abrupt inflection of surface along inner edge of palpebral lobe.

palpebral lobe (22). Raised portion of fixed cheek along inner edge of visual area of eye.

pleural furrow (5). Groove extending outward and generally backward from inner front edge of each pleuron; interpreted as a trace of primary segmentation.

pleural lobe (39). Lateral portion of thorax and pygidium.

pleural spine (4). Sharp-pointed extremity of a pleuron.

pleuron (35). Lateral portion of a single thoracic segment.

pre-epipodite (25). Outer and upper branch of paired biramous post-antennal appendages on ventral side.

pygidial segment (42). Transverse division of pygidium representing one of fused body segments composing it; homologous to thoracic segment.

pygidium (6). Posterior part of trilobite carapace, generally formed by fusion of several body segments.

telson (48). Prominent backwardly directed spike borne by an axial segment at or near rear extremity; may or may not be equivalent to true telson of other arthropods.

thoracic segment (34). Transverse division of thorax, consisting of an axial and two pleural portions.

thorax (3). Post-cephalic part of body composed of individually movable segments; region between head and pygidium.

tida, is divided into structurally distinct elements by curved lines of partition called **facial sutures** (*50;* italic numbers in the text describing skeletal features of trilobites refer to Fig. 13-2 unless indicated otherwise). These lines denote the places where the hard exoskeleton splits open at each time of molting (ecdysis) so as to release the imprisoned animal, allowing an increase of size and (during juvenile growth stages) increase of post-cephalic segments before new hard parts are secreted. The facial sutures of most trilobites are partly on the dorsal side of the cephalon, but may be restricted to the lateral and anterior margins. The central, main part of the cephalon, which is bounded laterally and anteriorly by the facial sutures, is termed the **cranidium** (*2*). Among trilobites having marginal position of these sutures, it comprises the whole dorsal part of the cephalon. The parts of the head shield which separate from the cranidium at the time of molting are known as **free cheeks** (*49*), but where facial sutures are confined to the cephalic margins, the free cheeks comprise only parts of the doublure. Trilobites judged to have such marginal sutures as a primitive feature are classed as **protoparian** (*proto,* primitive; *paria,* cheek, Fig. 13-3*A*), whereas those in which the marginal position of the sutures is interpreted to signify evolutionary specialization are called **hypoparian** (*hypo,* under, Fig. 13-3*B*). It is interesting to note that all trilobites having partly dorsal facial sutures show these lines along the crest of the visual area of each eye, but the course of the sutures in front of the eyes and behind them varies considerably. Anteriorly, the sutures may run obliquely outward to the edge of the cephalon and become marginal around the front, or they may extend nearly straight forward; in some, they curve around the front end of the axis so that the sutures on either side are joined together on the dorsal side of the cephalon, rather than along the margin. Posteriorly from the eye, the sutures may bend outward to the lateral margins of the head;

trilobites having this type of sutures are designated as **proparian** (*pro,* in front, Fig. 13-3*C*), because the free cheeks terminate in front of the postero-lateral corners of the cephalon. Also, the sutures may intersect the back edge inside the postero-lateral extremities; such trilobites are termed **opisthoparian** (*opistho,* behind, Fig. 13-3*D*), because the free cheeks include part of the rear margin of the cephalon. Thus, cranidia and free cheeks belonging to different genera may have quite dissimilar outlines. These parts of the ceph-

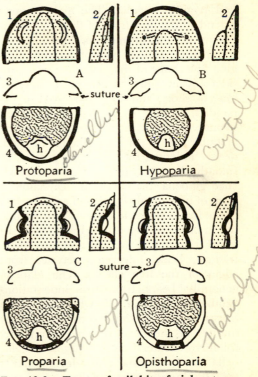

Fɪɢ. 13-3. **Types of trilobite facial sutures.** These sutures mark the place of parting of the exoskeleton when the animal molts. As indicated by the heavy black lines, they are exclusively marginal in (*A*) protoparian and (*B*) most hypoparian trilobites, but partly dorsal in (*C*) proparian and (*D*) opisthoparian trilobites. Among the two latter groups, also, the suture may be located entirely on the dorsal side. (1) Dorsal view of cephalon; (2) side view; (3) transverse profile near the posterior edge of cephalon; (4) ventral view; *h,* hypostome.

alon are found dissociated from one another much more frequently than joined together, just as we should expect.

The chief parts of the cranidium, which can be differentiated in nearly all trilobites, are the **glabella** (*36*) and **fixed cheeks** (*23*).

Glabella. The glabella comprises the axial portion of the cephalon, which almost invariably is clearly differentiated by an elevation above the remainder of the head and by a narrow troughlike indentation along its sides and front. This trough is part of the **dorsal furrow** (*10*), which bounds the axis in the cephalic, thoracic, and pygidial regions. The glabella may extend to the very front of the head, comprising its forwardmost part, or it may terminate anteriorly back of the front margin. The shape of the glabella varies considerably in different genera. It may be long or short, anteriorly narrowed or expanded, straight or curved at the sides, strongly elevated (at maximum, attaining globular form) or indistinctly marked, deeply indented by transverse grooves or smooth. Among trilobites having a shortened glabella, the part of the cephalon in front of it generally consists of an extension of the fixed cheeks, but it may also include the area belonging to the free cheeks. In some trilobites, a median accessory plate (*rostrum*, beak) intervenes between front edges of the free cheeks. The glabella generally bears transverse indentations (**glabellar furrows,** *44*) which constitute traces of the divisions between primary somites of the head region. Primitively, the furrows extend entirely across the glabella so as to define **glabellar segments** (*46*), but in most trilobites the median part of each furrow (except generally the most posterior one) has disappeared, leaving lateral indentations which separate so-called **glabellar lobes** (*45*). The rearmost glabellar furrow, which only rarely is incomplete or indistinct, is known as the **occipital furrow** (*14*); it separates **occipital segment** (*15*) from the remainder of the glabella. This segment may bear a prominent tubercle (**occipital node,** *47*),

which in some trilobites has the structure and function of a simple eye, but in others is prolonged as an occipital spine.

Fixed Cheeks. The fixed cheeks (*23*) include all the cranidium outside the area of the glabella. Marginal tracts of the fixed cheeks generally are somewhat elevated wherever they reach the edge of the cephalon, and these tracts form part of the **border** (*33*) of the head shield. The inner limit of the border is defined by a **marginal furrow** (*13*), which at the rear is confluent with the occipital furrow of the glabella. The anterior part of each fixed cheek is differentiated as an area called **brim** (*11*), and among trilobites having a short glabella, the brim includes the territory lying between the marginal furrow and the dorsal furrow at the front of the glabella. The posterior part of the fixed cheeks is termed the **ocular platform** (*43*), because it contains the eye (in protoparian and hypoparian trilobites) or the elevation called **palpebral lobe** (*palpebra*, eyelid, *22*) which adjoins the visual surface of the eye. On the side toward the glabella, each palpebral lobe is delimited by a groove, the **palpebral furrow** (*21*). Some trilobites have a narrow elevation on the fixed cheeks running from the front of the eye to the glabella; this is known as the **ocular ridge** (*32*). The ocular ridge, or an arbitrarily drawn line across the fixed cheeks at the anterior edge of the eye, is reckoned as the boundary between the brim and the ocular platform. In all except opisthoparian trilobites, the postero-lateral corners of the cephalon are formed by the fixed cheeks; these corners are termed **genal angles** (*gena*, cheek, *38*), or if posteriorly produced, they are called **genal spines** (*26*). In aggregate, the fixed cheeks, like the glabella, exhibit a wide variety of form and structure. Nearly all are useful in distinguishing genera, and they have importance in studying evolutionary trends.

Free Cheeks. The free cheeks (*49*) have fewer elements than other parts of the cephalon. Among proparian and opisthoparian trilobites, they bear part of the border and marginal furrow. The free cheeks of opisthoparians (and of a few hypoparians) also carry the genal angles or genal spines. The lower part of the genal spines of some hypoparians is borne by the free cheeks, whereas the upper part belongs to the fixed cheeks.

Ventral Parts. The ventral side of the cephalon commonly is much less easily studied than the dorsal. The plate called **hypostome** (*20*), which belongs just in front of the mouth, is preserved in position in many fossils, but it is likely to become separated along the suture where it fits against the doublure. The shape of the hypostome varies a good deal in different genera. Accordingly, the nature of this element would be useful in classification if it could be found commonly in association with the exoskeleton to which it belongs. Unfortunately, this is not the case. The **metastome** (*27*) is rarely present.

Five pairs of jointed appendages are attached to the underside of the cephalon in the axial region beneath the glabella. They are found only in exceptionally well-preserved specimens but are more or less completely known in seven genera (*Neolenus, Calymene, Ceraurus, Cryptolithus, Triarthrus, Phacops, Asteropyge*). The front pair of appendages are **antennae** (*19*), made up of many short segments; they are unbranched and are joined to the cephalon at the sides of the hypostome. The other four pairs of appendages are nearly identical in structure to the two-branched limbs carried by all post-cephalic somites; they are joined to the axis of the cephalon behind the mouth. The proximal joint of these limbs is much larger than the others, and larger also than the corresponding segment of thoracic limbs. Accordingly, some authors have judged that this part of the cephalic appendages functioned as maxillae and mandibles in feeding, tearing food, and pushing it along to the mouth. Mainly because these parts lack horny toothlike projections, being almost smooth, competent present-day specialists do not subscribe to this interpretation. The lower or inside branch of each limb is called an **endopodite** (*30*); it has seven nearly even-sized segments and a terminal small claw. The upper or outer branch (called exopodite by many authors because of misinterpreted analogy with the crustacean limb) is named **pre-epipodite** (*25*); it has many short segments which bear a comb-like structure, interpreted to be a gill—not a swimming organ, as formerly held.

Thorax

The thoracic region is distinguished by its unfused segments and by the mobility of their articulation with one another and with the head and tail regions. The median part of the thorax, which is a prolongation of the glabellar portion of the head, comprises part of the axis, and it is defined laterally by the dorsal furrow. The areas on each side are termed **pleural lobes** (*pleura*, rib, *39*). The individual **thoracic segments** (*34*) are wide transversely but short longitudinally. They are composed of an **axial segment** (*40*) and on either side of it a **pleuron** (*35*). The lateral extremities of the pleura may be bluntly rounded or produced as **pleural spines** (*4*). Transverse grooves between adjacent axial segments, which mark the location of movable joints, are designated as **articulating furrows** (*41*), and those between any two adjoining pleura are **interpleural grooves** (*31*). On the dorsal side of each pleuron is a depression which curves backward and outward from the front edge of the pleuron at the dorsal furrow; this is named **pleural furrow** (*5*). Articular movement of the segments is effected by muscles attached to inward projections (**apodemes,** Fig. 13-4*c*, *d*) of the exoskeleton beneath the articulating furrows. An arched forward extension of each axial segment (**articulating half segment,** Fig. 13-4*a*, *c*) reaches beneath the reflexed border of the next preceding segment, and the edges of these two elements are connected by flexible unmineralized chitin (Fig. 12-4). Hence, one segment can move on its neighbors in a hinged manner, and

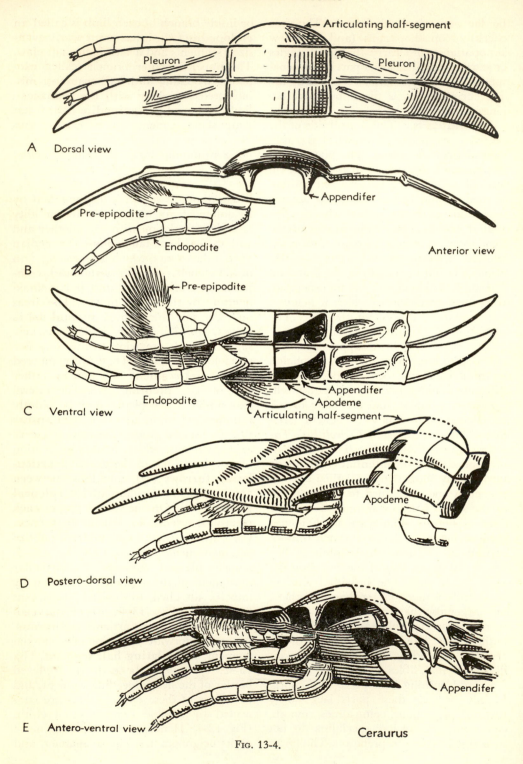

A   Dorsal view

B   Anterior view

C   Ventral view

D   Postero-dorsal view

E   Antero-ventral view

Fig. 13-4.   Ceraurus

the thoracic region may be bent longitudinally so as to bring the pygidium into contact with the underside of the head. Trilobite fossils thus rolled up like a sowbug are known in the Lower Cambrian (Fig. 13-14, *6a–c*) and many younger deposits.

The ventral surface of each thoracic segment bears a pair of biramous appendages, which are almost exactly like the four pairs of post-antennal limbs of the head (*25, 30;* Fig. 13-4). They are moved by muscles attached to the inside of the exoskeleton of the appendage and to ventrally directed prominences (**appendifers,** Fig. 13-4*b, c, e*) on the inside of each axial segment. The pre-epipodites of *Ceraurus* (Fig. 13-4) differ from those of *Triarthrus* (Fig. 13-2) in having gill blades restricted to terminal portions.

Pygidium

The pygidium of a trilobite, like the cephalon, consists of solidly united somites, but unlike the cephalon, in which the number of fused segments is constant, it may be composed of 1 to 30 or more segments. The median portion consists of the posterior part of the axis, bounded laterally and terminally by the dorsal furrow. The axial region reaches to the very tip of the pygidium in some genera, but more commonly it ends somewhere short of this point. Laterally adjoining the axis are **pleural lobes** (*39*), which may bear a **border** (*33*) or not. Subdivisions of the axis and pleural lobes essentially correspond to those of the thorax, but of course they differ in that the pygydial elements are fused. The most anterior axial segment bears an **articulating half segment** (*28*), like those of thoracic segments. This functions to permit movement between the pygidium and the posterior end of the thorax.

The shape of pygidia belonging to different trilobites varies greatly, and the margin may be smoothly rounded or spinose (*7*). A peculiarity observed in some genera is the disparity in the number of axial segments as compared with the interpleural grooves on the pleural lobes. Generally, when these elements do not match, the axial segments exceed somite divisions of the pleural lobes, indicating that, as result of fusion, the latter tend first to become indistinct and disappear. The pygidium of a few trilobites (for example, *Eodiscus*, Fig. 13-15, *3*) may have many distinct axial segments but perfectly smooth pleural lobes. Exceptionally, the reverse of this tendency is observed, as in *Eobronteus* (Fig. 13-18, *5*), which has a short smooth axis adjoined by well-marked interpleural grooves on the large, broadly rounded pleural lobes.

The ventral side of the pygidium carries as many pairs of biramous limbs as there are fused segments, and except for posteriorly diminished size, these appendages exactly correspond to those of the thorax. The anus is located on the underside of the terminal segment.

Fig. 13-4. **Structure of thoracic segments and appendages.** The diagrams represent a partly dissected pair of segments from five viewpoints. An articulating half segment at the front of each axial segment projects forward beneath the recurved posterior border of the next-preceding axial segment. The free edges of these skeletal elements were joined by flexible unmineralized chitin when the animal was alive. The segments were moved by muscles running longitudinally from inward projections (*apodemes*) at the posterior edge of each articulating half segment. The biramous limbs consist of a stout walking leg (*endopodite*), below, and a smaller, gill-bearing branch (*pre-epipodite*), above. Each limb was moved by muscles attached to a projection (*appendifer*) of the apodeme. The pleura are mostly mere porchlike projections of the exoskeleton, which protect the limbs.

INTERNAL ANATOMY

Evidence concerning the soft parts of trilobites is scanty. Markings on the interior of the carapace afford some clues, and a few specimens studied by means of serial sections show traces of longitudinal internal structures interpreted to represent the heart and mud-filled alimentary tract (Fig. 13-5). The inner surface of the fixed cheeks of several genera bear dendritic impressions which radiate from the glabellar region. These are judged to denote placement of some sort of ramified soft structures, possibly connected with the nervous, circulatory, or digestive systems. Along segment divisions of the glabella are pairs of dark spots, which are interpreted as muscle attachments for operation of the cephalic appendages. One group of trilobites (Asaphidae) has revealed the presence of a small round opening near the inner edge of doublures belonging to each thoracic segment and postero-lateral extremities of the cephalon; these structures, termed Panderian organs, have been postulated to denote poison glands or luminescent organs, but their significance really is quite unknown.

GROWTH STAGES

Larval Development

Study of the growth stages of trilobites throws light on the morphologic significance of various adult characters, and is important for an attack on the problems of classification. Also, it bears on the interpretation of evolutionary trends. The fact that each individual trilobite repeatedly molts its exoskeleton, and that cast-off hard parts are as readily capable of fossilization as those borne by the animal at death, greatly facilitate the study of growth stages. Otherwise, we should have to depend on finding the remains of several individuals which died at different stages in development from larval to adult. Whether dealing with molted tests or carapaces left by dead trilobites, it is needful to identify and segregate fossils belonging to the same species, or at least to the same genus. Although all trilobites may be presumed to pass through comparable ontogenetic development, representatives of one genus might differ from those of another as much as in two genera of mammals, for example, a bat and a man. In the past, the finding of suitable materials for research on growth stages of trilobites and accurate sorting of specifically allied fossils have been difficult indeed, but in spite of this, larval stages of some 20 genera have been described. Recent work on limestones containing numerous silicified trilobites has yielded such an abundance of beautifully preserved fossils when the rock is dissolved in acid that unsurpassed materials for investigation of many growth-stage series are available. Very little is yet published, however.

Three stages in the larval development of trilobites are recognized. The first, called **protaspis** (*prot*, primitive; *aspis*, shield), consists of a simple rounded unjointed carapace of minute size (0.25 to 1.25 mm.), which represents the beginning of the cephalon. The earliest protaspis of some genera shows no markings other than a raised axis adjoined by smooth side

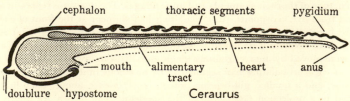

Fig. 13-5. **Longitudinal section of a trilobite through the axis.** Skeletal parts are indicated by solid black. Internal structures, based on interpretation of features revealed by sections of some specimens, are shaded.

regions, but later protaspis molts reveal clearly marked transverse grooves on the axis, and these may extend onto the side areas. They denote the somites which are fused together to form the cephalon. Including a frontmost (pre-antennal) division, which may or may not form part of the axial lobe, the trilobite larva at this very early stage is seen to contain six somites; they are termed pre-antennal, antennal, and first to fourth leg-bearing somites (Fig. 13-6). A protaspis having only these six somite divisions is designated **anaprotaspis** (Fig. 13-2, *16*). The protaspis passes into the **metaprotaspis** sub-

stage when additional segments, representing post-cephalic somites, make an appearance (Fig. 13-2, *17*).

The protaspis stage ends and the **meraspis** stage begins when the first trace of a pygidium develops (Fig. 13-2, *18*). During meraspis larval development, there is an increase of thoracic segments until the number equals that belonging to the adult of the concerned species, and each molt is accompanied by an increase in size. The newly added thoracic segments are inserted at the front margin of the hindmost segment (protopygidium), behind previously formed thoracic segments.

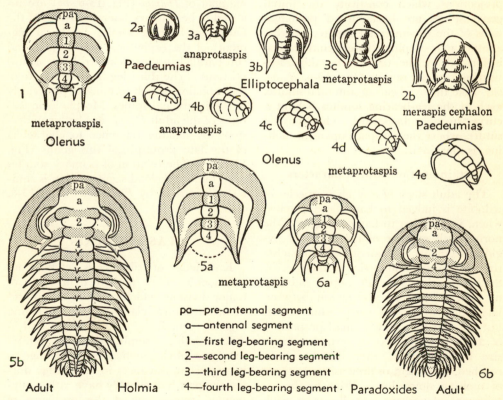

Fig. 13-6. **Growth stages of trilobites.** All except the two lowermost figures show larval forms. Alternate shaded and unshaded segments of some of the drawings are designed to bring out the pattern of primary segmentation, which in the thoracic region obliquely crosses lateral parts of the mechanically induced secondary segmentation (interpreted by Størmer). (1) *Olenus* (×50) is an opisthoparian. (2a, b) *Paedeumias* and (3a–c) *Elliptocephala* (all ×10) show successive growth stages of protoparians. (4a–e) *Olenus* (×25), oblique views of protaspis stages. (5a) *Holmia* protaspis (×15), (5b) adult (×1.3); a protoparian. (6a) *Paradoxides* protaspis (×15), (6b) adult (×0.3); an opisthoparian.

Late larval growth, during which there is merely an increase of size without any significant structural change, is called the **holaspis** stage. It may begin long before the trilobite reaches full adult stature. Also, the adult trilobite may continue to become larger at each time of molting until the individual dies.

The larval stages just described are illustrated by specimens belonging to *Paedeumias* and *Elliptocephala*, which are protoparian trilobites, and *Olenus* and *Paradoxides*, which are opisthoparians (Fig. 13-6). The protaspis of these trilobites shows that the crescentic eyes are connected with the antennal somite. Posterior projections which constitute the initial genal spines are back-curved prolongations of the pre-antennal somite, whereas another pair of spines (**intergenal spines**) is formed by the tips of the third leg-bearing cephalic somite (Fig. 13-6, *2b, 5a, 6a*). Like the eye-bearing antennal somite, each of the leg-bearing somites behind it curve strongly backward. The ecdysial suture, where the carapace opens in molting, is marginal.

Attainment of Adult Characters

The adult stage of any given species of trilobite is judged to have been reached when, in addition to having acquired the normal full number of thoracic and pygidial segments, an individual approaches the average maximum size of the species. Such definition is rather vague. Actually no criteria exist for distinction between late immaturity and early adult status, but the fact that only a small proportion of any species exceeds a certain average size indicates that, after this has been reached, increment of dimensions at times of molting slows down. Prior to the attainment of average size, however, each molt was accompanied by appreciable enlargement.

Secondary Segmentation. A feature of the adult trilobite is a well-articulated thoracic region, which is attached by movable joints to the cephalon and pygidium.

The skeletal pattern associated with these articulations partly conforms to primary segmentation of the body, notably in the axial region, but partly it deviates in crossing primary somites obliquely in the pleural regions. Just as the cephalic somites curve backward on each side of the axis (Fig. 13-6), the primary post-cephalic somites seem to be bent, their borders being marked by the pleural furrows. The development of interpleural separations of the exoskeleton, in line with the transverse articulating furrows of the axial lobe, are reasonably interpreted as a mechanically induced secondary segmentation. Thus, the pleural extremities of the first thoracic segment of *Holmia* (Fig. 13-6, *5b*) are inferred to be primary lateral portions of the fourth leg-bearing somite (occipital segment) of the cephalon. They adjoin and run parallel to the backward bent pleural portions of the third leg-bearing somite of the cephalon, which terminates posteriorly in the intergenal spines. Morphologic features of the adult *Holmia* are better understood by making reference to characters of the late protaspis of this genus (Fig. 13-6, *5a*). The relation of primary somites to secondary skeletal segmentation is similarly illustrated by *Paradoxides* (Fig. 13-6, *6a, b*).

CLASSIFICATION

Knowledge of trilobites is yet far from sufficient to allow positive delineation of major divisions which conform to requirements of natural classification in bringing together genetically related stocks, and in differentiating assemblages which exhibit divergent phylogenetic trends. Strange as it may seem, advances in knowledge during the past half-century have rather augmented than resolved the problems of classification.

We can recognize numerous groups of genera which possess such combinations of characters as to justify firm judgment that they are mutually related, and must have had common ancestry. These are family groups. Also, certain families obviously stand in near association, whereas others are far removed. Evidence furnished by ontogeny is helpful, but geological order of appearance seems untrustworthy as a guide. Considering all things, the most important basis for the approach to a natural classification is the nature of the facial suture, at the margin or on the dorsal side of the cephalon. This may well be judged to have special significance, because of connection of these lines in most genera with the eyes, probably the most important sensory organs of the animal. Grouping of families according to type of facial suture brings together trilobites which seem to be related on other grounds. Classification here adopted, however, does not rest exclusively on this character.

Six orders of trilobites are recognized, as shown in the following tabulation. It should be noted that the sequence in which they are listed does not reflect a progression from the most primitive or archaic to the most specialized. Such an arrangement is not possible, because the trilobites seem to have evolved in a very divergent manner.

Main Divisions of Trilobites

Trilobita (*class*), carapace three-lobed, head formed by six fused somites, of which four bear biramous thorax-type limbs. Cambrian–Permian.

Protoparia (*order*), facial suture marginal, eyes large and crescentic, thoracic segments numerous, pygidium small or rudimentary. Lower Cambrian.

Proparia (*order*), posterior part of facial suture intersects edge of cephalon in front of its postero-lateral extremities. Middle Cambrian–Devonian.

Opisthoparia (*order*), facial suture intersects posterior edge of cephalon. Lower Cambrian–Permian.

Hypoparia (*order*), facial suture marginal, eyes small or lacking, pygidium well-developed but small. Lower Ordovician–Devonian.

Eodiscida (*order*), facial suture marginal or proparian, eyes generally lacking, thoracic segments two or three, pygidium large. Lower and Middle Cambrian.

Agnostida (*order*), facial suture and eyes lacking, thoracic segments two, pygidium large. Lower Cambrian–Upper Ordovician.

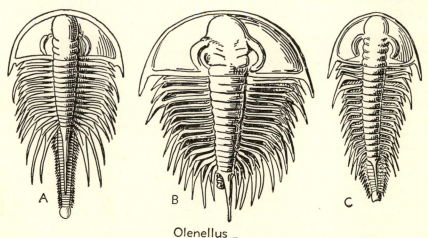

Olenellus

FIG. 13-7. **Three species of Olenellus, a representative protoparian.**

A, O. robsonensis (Burling) (×0.25), Lower Cambrian, western Canada; distinguished by the long spinose projections of pleura just in front of the telson-bearing segment, and the long wormlike posterior part of the thorax.

B, O. thompsoni (Hall) (×0.5), Lower Cambrian, Vermont; this species, type of the genus, was described from specimens lacking small segments beneath and behind the telson, which accordingly was thought to represent a pygidium.

C, O. vermontanus (Hall) (×0.5), Lower Cambrian, common in Appalachian belt; this species is characterized by its narrow, elongate form.

CHARACTERS OF MAIN TRILOBITE GROUPS

Protoparians

The very oldest known trilobites are protoparians, for fossils belonging to this group occur nearest to the base of Cam-

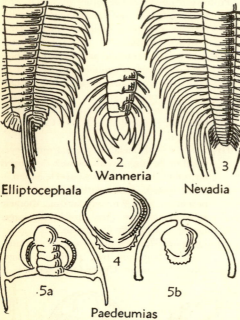

FIG. 13-8. Some structural features of protoparians. The upper three figures are introduced to show thoracic and pygidial characters which differ from those of *Olenellus*. The lower three pertain to the head region.

1, *Elliptocephala asaphoides* Emmons (×0.66), Lower Cambrian, New York; showing posterior spine-bearing axial segments.
2, *Wanneria walcottana* (Wanner) (×2), Lower Cambrian, Pennsylvania; the pygidium seems to consist of a longitudinally divided plate.
3, *Nevadia weeksi* Walcott (×0.66), Lower Cambrian, Nevada; the pygidium is a minute segment preceded by six short pleura-bearing segments.
4, Hypostome of *Olenellus thompsoni* (Hall) (×3).
5a, b, *Paedeumias transitans* Walcott (×2), Lower Cambrian, New York. The cranidium (glabella and fixed cheeks) is shown at left (5a), and the submarginal free cheeks (joined together in front) with attached hypostome are figured at right (5b). The marginal placement of the facial suture is evident.

brian deposits. They are essentially world-wide in distribution. Numerous species representing several genera are described. The chiefly studied forms occur in the eastern and western parts of North America, northwestern Europe, and eastern Asia. Typical examples of the protoparians are species of *Olenellus* (Fig. 13-7). The carapace of this trilobite is larger than average, for complete adult specimens may be as much as 8 in. long.

Cephalon. The head shield of *Olenellus* is semicircular in outline except that the postero-lateral corners are extended to form short or long genal spines. A well-defined border is separated by a strong marginal furrow from the broad, smooth, gently arched main area of the fixed cheeks. Large crescentic eyes adjoin the glabella, which reaches forward to the marginal furrow. The glabella is marked laterally by three to four pairs of furrows. The anterior glabellar lobe, which is at least twice as large as others, is interpreted to represent thoroughly coalesced pre-antennal and antennal somites. Obscure marks on the fixed cheeks of some specimens of *Olenellus* have been thought by a few workers to be remnants of fused facial sutures, but no real basis for such interpretation exists. On the other hand, the existence of a marginal suture separating the dorsal fixed cheeks from a narrow free cheek (doublure) is well established in protoparians such as *Paedeumias* (Fig. 13-8, 5b), very similar to *Olenellus*. The hypostome of *Olenellus* has a pear-shaped outline, the narrow part, which has a toothed margin, being directed backward (Fig. 13-8, 4).

Thorax. The thorax is composed of 18 to 44 or more segments, among which the first 15 (proceeding backward from the head) are wide and the remaining ones narrow. The transversely extended parts of pleura belonging to the wide segments bear prominent pleural furrows, and the pleural extremities have a backwardly directed spinose form. The third thoracic segment is larger than the others and has more prominent pleural spines. The reduction in width which begins at the sixteenth segment is very abrupt, and it is correlated with a striking difference in size, shape, and attitude of pleural portions of the segments, as compared with thoracic segments farther forward. The pleura of the sixteenth and succeeding segments are very short and turn backward less strongly than the anterior ones. Finally, the fifteenth axial segment, unlike any other part of the axis, bears a very prominent, posteriorly directed spine, which overlies and partly conceals the narrow thoracic segments. The body of *Olenellus*, therefore, is divided into two very dissimilar regions. The anterior part of the thorax may be described as trilobite-like, whereas the posterior part is decidedly wormlike (Fig. 13-7).

Pygidium. *Olenellus* has a nearly featureless rudimentary pygidium which consists of a flattened plate of rounded outline having a length equal to that of the next

preceding three or four segments combined. Whether it represents a persistent single-segment larval pygidium of the meraspis stage, or, as in most other trilobites, is made of fused thoracic segments, is uncertain.

The spine borne by the fifteenth axial segment has been interpreted as equivalent to a pygidium and called a **telson** (Fig. 13-2, 48), like the spike tail of the king crab (*Xiphosura*, commonly known as limulus). This seems very plausible when one finds a specimen which has lost the posterior wormlike section of the body, and such specimens are the rule, rather than the exception. Hence, the telson naturally has been misjudged to constitute a special sort of pygidium. The type species of *Olenellus* (*O. thompsoni*, Fig. 13-7B) was not known to have thoracic segments beyond the telson-bearing fifteenth segment until recent years when more complete examples of this species were found. *Olenellus fremonti* (Fig. 13-24, 2), has a spike tail unaccom-

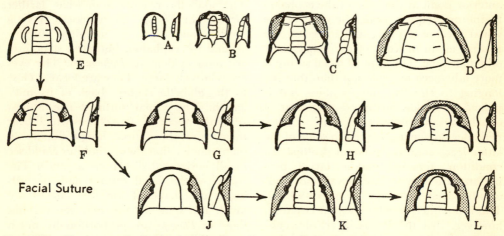

FIG. 13-9. **Characters of trilobite facial sutures.** Except *A–D*, the drawings are wholly diagrammatic.

A–D, Growth stages of *Peltura scarabaeoides* (Wahlenberg), Upper Cambrian, Sweden and Great Britain, showing change from protoparian type of suture in early larva (protaspis) (*A*), to proparian type of suture in late larvae (meraspis) (*B, C*), and opisthoparian type of suture in the adult (*D*). (*After Poulsen.*)

E, Protoparian facial suture, in marginal position. The similar hypoparian type of suture, judged

to be secondarily developed, is not figured.

G–I, Progressively advanced types of proparian sutures, showing tendency of the pre-ocular parts of the sutures to curve inward and become wholly dorsal in position.

J–L, Progressively advanced types of opisthoparian sutures, illustrating the same tendency of the pre-ocular course of sutures as in proparians.

panied by wormlike segments belonging beneath and behind it. Although such segments attached to a specimen of this species have not been seen, probably they existed. Comparison of the posterior portions of various protoparian genera (*Olenellus*, Fig. 13-7*A–C; Holmia*, Fig. 13-6, *5b; Callavia*, Fig. 13-14, *3; Paedeumias*, Fig. 13-14, *4; Elliptocephala*, Fig. 13-8, *1; Wanneria*, Fig. 13-8, *2;* and *Nevadia*, Fig. 13-8, *3*) demonstrates both the rudimentary nature of the pygidium and the incorrectness of interpreting axial spines of these trilobites as pygidial structures.

Appendages. Antennae belonging to *Olenellus* have been observed, but no other appendages of protoparians are known.

Proparians

The proparian trilobites differ from protoparian in having part of the facial suture on the dorsal surface. In front of the glabella, the suture generally is marginal or submarginal, then it crosses from some anterior point at the edge of the cephalon to the eye, curves along the crest of the visual area of the eye, and intersects the margin again in front of the genal angle or genal spine. This type of facial suture is more advanced in evolution than that belonging to *Olenellus* and its allies, but is judged to be less advanced than that of opisthoparian trilobites, because ontogenetic development observed in some of the latter leads from early larval protoparian condition, through middle larval proparian structure, to late larval and adult opisthoparian structure (Fig. 13-9*A–D*). This is not interpreted to mean necessarily, however, that the Proparia are descendants of the Protoparia, because it is altogether possible that unknown ancestors provided with marginal facial sutures are the true progenitors. One good reason for thinking this is that among proparian trilobites we do not find various peculiarities which distinguish all known protoparian genera, and which, even though modified, should be recognizable readily in their descendants.

The Proparia include trilobites ranging in age from Middle Cambrian to Upper Devonian. They are moderate in size, the adults ranging in length from less than 0.5 in. to a maximum of about 8 in. They are mostly conservative in form, having a regularly ovoid outline and smooth carapace, but a few have a bizarre appearance. Some are covered with granules or tubercles and bear spines.

Study of this order may be based on two of the more common, very representative genera, *Dalmanites* and *Phacops*, both of which occur in Silurian and Devonian rocks (Fig. 13-10). We exclude from the group a few Cambrian forms provided with proparian sutures which are judged to be a modified branch of the Eodiscida. These will be noted later.

Cephalon. The most prominent feature of the cephalon of *Dalmanites* and *Phacops* is the bulbous glabella, which has an inverted pear shape and is strongly indented by two or more pairs of furrows. The glabella reaches somewhat farther forward in *Phacops*. Both genera have a border and distinct marginal furrow around the cephalon, but these structures are most evident in *Dalmanites*. The eyes are relatively large. They are located close to the glabella slightly back of its midlength. The visual surfaces rise almost vertically above the free cheeks which bear them, and they curve longitudinally in such manner that these trilobites could see in all horizontal directions at once. The course of the facial sutures and, accordingly, the shape of the free cheeks are much the same in the two genera, but those of *Phacops* are proportionally much smaller. The fixed cheeks have different outlines because of the presence of genal spines in *Dalmanites* and their absence in *Phacops*.

Thorax and Pygidium. The structure of the thoracic region is conservative and very stable in the two genera under discussion. Adults typically have 11 thoracic segments of nearly equal size. The axial lobe is moderately arched and defined by well-incised dorsal furrows. The pleura have nearly straight margins and are rounded or blunt-angled at the extremities. Free mobility of thoracic articulation is indicated by the common occurrence of enrolled specimens among fossils.

The pygidium is similar in shape to the cephalon but somewhat smaller. The axis, which reaches to the posterior border, and pleural lobes bear transverse furrows indicating 10 to 12 pygidial segments in *Phacops* and 10 to 16 in *Dalmanites*. Neither genus has a border on the pygidium, but the posterior extremity of *Dalmanites* is pointed, or produced as a spine.

Paedomorphism. The persistence in adult growth stages of structural features, which in related contemporaneous or antecedent stocks are restricted to larval development, is termed **paedomorphism** (*paedo*, child; *morphism*, form). The Proparia are judged to illustrate this sort of evolutionary retardation, for fully grown individuals, capable of reproduction, have facial sutures like those of contemporary and older opisthoparian larvae.

Distribution. Proparian trilobites are most common in Silurian and Devonian rocks; they disappeared before Mississippian time, whereas Opisthoparia continue nearly to the close of the Permian. Proparians (in addition to *Phacops* and *Dalmanites*) illustrated in this chapter include genera from the Cambrian (Figs. 13-15, *6;* 13-17, *3*), Ordovician (Figs. 13-17, *14, 16;* 13-18, *1–3*), Silurian (Fig. 13-19, *3, 5*), and Devonian (Figs. 13-19, *6, 8, 9, 11;*13-20, *3, 7, 10;* 13-24, *6*).

Opisthoparians

By far the largest number of known kinds of trilobites belong to the order Opisthoparia, characterized mainly by a facial suture which crosses the dorsal side of the cephalon to some point on its rear margin. Genera of this type make an appearance in Lower Cambrian rocks, and the order is found to be represented throughout all of the Paleozoic. The culmination of evolutionary development along several lines is seen in this assemblage. Opisthoparian trilobites attain the largest known size—a length of nearly 2.5 ft. Some show a gradual loss of markings until the whole exoskeleton is featureless, whereas others exhibit strangely modified outlines, eyes elevated on stalks, and long spines on various parts of the carapace. A few have unusually large pygidia, formed by the fusion of many segments.

As a basis for a brief survey of morphologic features of the Opisthoparia, we may refer mainly to some of the genera employed in preceding descriptions—*Triarthrus* (Figs. 12-1, *1;* 13-2), *Olenoides* (Fig. 13-2), *Paradoxides* (Fig. 13-6, *6b*)—and in addition, *Isotelus* (Fig. 13-11).

Cephalon. In outline and characters

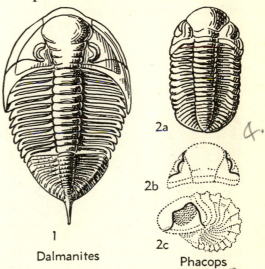

Dalmanites   **Phacops**

FIG. 13-10. **Representative proparian genera.** These two forms are restricted to Silurian and Devonian rocks, occurring both in North America and Europe.

1, *Dalmanites verrucosus* Hall (×0.5), Middle Silurian, Indiana; chiefly distinguished by characters of the cephalon and pygidium.

2a–c, *Phacops rana* Green (×1), Middle Devonian, New York. This trilobite is characterized by relative prominence of its forwardly expanding and protruding glabella, small free cheeks, large eyes, and lack of genal spines. (2a) Dorsal view of complete specimen; (2b) cephalon, showing facial sutures; (2c) side view of enrolled specimen, showing facial suture and eye.

of the glabella, the cephalon of opisthoparian trilobites closely parallels that of
protoparian and proparian forms. Genal
spines may be present or absent. The glabella may be long or short, varied in shape,
and provided with glabellar furrows or
lacking them. These features have importance in distinguishing genera but not
otherwise. *Triarthrus* and *Olenoides*, which
have nearly identical outlines and markings of the glabella, differ in the shape of
the cephalon. *Paradoxides* and *Olenoides*
closely correspond in shape of cephalon
but differ in glabellar features. The nearly
smooth glabella of *Isotelus* is only slightly
raised above the fixed cheeks and is not
bounded by a distinct dorsal furrow. Characters of the eyes differ more or less obviously also among these genera. The most
important structure common to all is the
opisthoparian type of facial suture. Its
course from front to back edges of the

cephalon is most nearly alike in *Paradoxides* and *Olenoides;* the suture curves away
from the glabella in front of and behind
the eye. The suture of *Triarthrus* and *Isotelus* curves toward the glabella in front of
the eye, meeting the cephalic margin (*Triarthrus*) or joining with the opposite suture
(*Isotelus*) in front of the glabella (Figs.
12-1, *1;* 13-11).

The free cheeks of *Isotelus* extend beneath the margins of the cephalon as a
doublure. This ventral part is widest anteriorly, where the cheeks join along a
suture which runs to the margin of the
forked hypostome (Fig. 13-11*D*). Doublure
portions of the free cheeks of *Triarthrus* are
narrow throughout, and they are separated
in front by the doublure extension of the
pre-glabellar fixed-cheek area, which connects with the subtriangular hypostome
(Fig. 13-2, *20*). *Ptychopyge* resembles *Isotelus*
but has broader free-cheek doublures (Fig.
13-2, *8*).

Thorax and Pygidium. No special
structural features of the thoracic and
pygidial regions call for attention. These
parts of the opisthoparians almost exactly
duplicate characters seen in protoparians,
with which the many thoracic segments
and tiny pygidium of *Paradoxides* (Fig.
13-6, *6b*) may be compared. They also
resemble proparians, which have small
pygidia, like *Triarthrus*, and large ones,
like *Olenoides* and *Isotelus*. The loss of markings on the axial and pleural regions of the
pygidium of *Isotelus* points to a trend which
is much more pronounced in some other
opisthoparians.

Appendages. Knowledge of the appendages attached to the underside of the
head, thorax, and pygidium of trilobites
has been obtained mostly from the study of
unusually well-preserved opisthoparians,
especially pyritized specimens of *Triarthrus*
collected from the Utica shale (Middle
Ordovician) of New York. Antennae and
limbs of *Neolenus*, from Middle Cambrian
rocks of western Canada, have also been
found exceptionally perfect. Less complete
information concerning appendages of

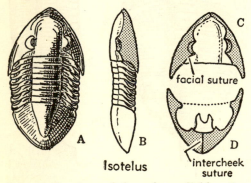

Isotelus

FIG. 13-11. **An opisthoparian trilobite, Isotelus.** This genus, which is restricted to Ordovician rocks, occurs in North America and
northwestern Europe. It is advanced in evolutionary characters, such as absence of glabellar furrows
and pygidial segmentation.

A, Isotelus gigas (Dekay) (×0.5), Middle Ordovician, New York; widespread in the eastern
United States; dorsal view of complete specimen.

B, Side view of same, area of free cheek shaded.

C–D, Dorsal and ventral sides of cephalon, showing free cheeks shaded. The doublure extensions of the free cheeks (*D*) and the intercheek
suture, which runs from the dorsal side of the
cephalon in front of the glabella to the front
edge of the hypostome, are noteworthy.

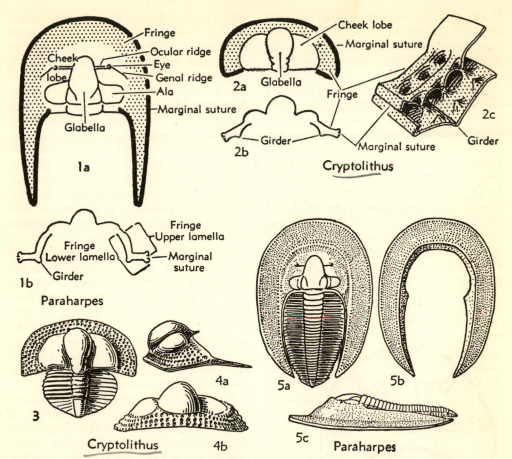

FIG. 13-12. **Characters of hypoparian trilobites.** The two illustrated genera are restricted to Ordovician rocks, but close relatives of *Paraharpes* persist into the Devonian.

1a, b, Diagrams showing structural elements and nomenclature of parts of the cephalon of *Paraharpes*. (1a) Dorsal view of cephalon, the fringe shaded, and position of the facial suture indicated by heavy black line; only the upper lamella of the fringe is visible in this view, its inner part on the side toward the glabella being foreshortened because of its steeply upturned position. (1b) Transverse section of cephalon intersecting the alae; this shows the attitude of the lower lamella of the fringe, which is coextensive in distribution with the upper lamella (shaded area in 1a).

2a–c, Structural features of the cephalon of *Cryptolithus*. (2a) Dorsal view of cephalon showing fringe and facial suture, as in 1a. (2b) Transverse section of cephalon approximately at its mid-length, showing lower lamella of fringe. (2) Diagrammatic oblique view of part of fringe (shaded), sectioned to show

arrangement of pits in the upper and lower lamellae, which meet one another so that perforations in the base of each pit join to make passageways through the fringe (Størmer).

3, 4, *Cryptolithus* Green, Ordovician, North America and Europe. (3) *C. bellulus* (Ulrich) (×4), Upper Ordovician, Ohio; complete carapace, showing absence of eyes on cheek lobes. (4a, b) *C. cryptolithus* Green, Middle Ordovician, New York; lateral and anterior views of cephalon, showing slope of fringe and presence of genal spines.

5a–c, *Paraharpes* Whittington, Upper Ordovician, England. (5a) *P. hornei* (Reed), type of the genus; dorsal view of complete specimen, showing wide posterior prolongations of the fringe, many thoracic segments, and small pygidium. (5b) Ventral view of fringe belonging to *P. hornei*, showing lower lamella. (5c) Side view of *P. hornei*. All ×1.3.

Isotelus and other opisthoparians has been obtained. In general, little difference has been found in the nature of these structures in trilobite genera representing different orders.

Distribution. Opisthoparian trilobites are well known in each geologic system from Cambrian to Permian, but they are much more common in pre-Silurian formations than in younger rocks. Illustrated examples of this order, classed by systems are: Cambrian (Figs. 13-14, *1, 2, 5–13;* 13-15, *5, 7–20;* 13-16, *1–15;* 13-17, *1, 2, 4, 5, 7–11*), Ordovician (Figs. 13-17, *12, 13, 15;* 13-18, *4, 5, 7–11*), Silurian (Fig. 13-19, *1, 4, 10*), Devonian (Fig. 13-20, *4, 5, 8*), Mississippian (Fig. 13-20, *1, 5, 6, 11*), Pennsylvanian (Fig. 13-20, *2, 13*), and Permian (Fig. 13-20, *12, 14*).

Hypoparians

The hypoparian trilobites resemble protoparians in the marginal position of the

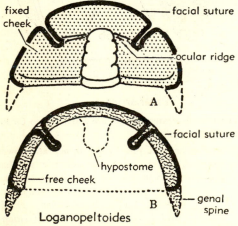

fixed cheek
facial suture
ocular ridge
A
facial suture
hypostome
free cheek
B
genal spine
Loganopeltoides

FIG. 13-13. Origin of the hypoparian facial suture. Diagrammatic figures (×6), based on *Loganopeltoides kindlei* Rasetti, Upper Cambrian of Newfoundland, showing facial suture (heavy black line) and characters of the fixed cheeks (*A*) and free cheeks (*B*), both viewed from the dorsal side. The facial suture is marginal, as in typical hypoparians, except where it crosses the dorsal surface to encircle each eye and (not certainly) to traverse the base of the genal spines, as in species of *Cryptolithus*. By disappearance of the narrow extension of the free cheeks near the eyes, the suture becomes typically hypoparian.

facial suture. Some of them (Harpidae) are like protoparians, also, in the large number of their thoracic segments and diminutive, although better developed, pygidium. Distinction between these orders chiefly rests on the characters of the dorsal side of the cephalon, indicating that hypoparians are specialized trilobites, whereas the protoparians are interpreted as primitive. Peculiarities shown by the two representative hypoparian genera, which are described in the following paragraphs, make evident these dissimilarities. They are of such sort as to indicate that no direct genetic connection exists between Protoparia and Hypoparia.

Hypoparians are also similar to the Agnostida and to most Eodiscida in the absence of dorsally located facial sutures, and for this reason some authors have classed the two latter groups as subdivisions of the Hypoparia. Actually, the agnostids stand rather far removed from hypoparian and other trilobites, and the eodiscids only somewhat less so. They seem best treated as independent orders.

The known stratigraphic range of the hypoparians, from Lower Ordovician to Upper Devonian, suggests that their origin may be sought among Cambrian forms, possibly in a stock such as the opisthoparian Ptychopariidae. The length of described adult hypoparians ranges from 0.15 to 2 in.

Genera chosen for special notice here are the widely distributed, relatively abundant *Cryptolithus* (Middle and Upper Ordovician) and *Paraharpes* (Upper Ordovician) (Fig. 13-12). Another genus, which resembles *Cryptolithus*, is *Reedolithus* (Fig. 13-18, *6*).

Cephalon. The central part of the head shield of *Cryptolithus*, equal to the thorax in width, consists of three finely reticulate or smooth-surfaced bulbous areas. The central one, which has an inverted pear shape, constitutes the glabella; its posterior part bears three pairs of slightly indented furrows. The two lateral areas are termed **cheek lobes;** they are devoid of markings or contain a very small, centrally located eye tubercle. Thin sections of very well-preserved eye tubercles show that it has a single lens. The eye is interpreted as a degenerated equivalent of the normal compound eye of trilobites, a conclusion which is supported by finding that several species of *Cryptolithus* are entirely blind, although an eye tubercle may be present in larval stages. Surrounding the glabella and cheek lobes at the sides and front of the cephalon is a broad pitted band called the **fringe.** The postero-lateral corners of the fringe bear moderately long narrow genal spines in some species, but none at all in others. The marginal facial suture is located on the down-bent edge of the fringe, and it runs all the way around the sides and front of the cephalon from one genal angle to the other, but cutting across the base of genal spines if they are present. Beyond the marginal suture is part of the cephalon which corresponds to the free cheeks of other trilobites. It is equal in width to the fringe, which it underlies, and is termed the lower lamella of the fringe. An angulated part of this lower lamella, which strengthens it, is known as the **girder** (Fig. 13-12, *2b–c*).

The cephalon of *Paraharpes* (and other harpids) generally resembles that of *Cryptolithus*, but the fringe is unusually developed with prolongations reaching back like broadened genal spines nearly or quite to the pygidium (Fig. 13-12, *5a*). The cheek lobes are less convex than in *Cryptolithus*, but adjoining the posterior part of the glabella are localized bulges of the cheeks, termed **alae.** The small round eyes on the front portions of the cheek lobes are of the compound type; each eye is connected with the glabella by a narrow ocular ridge, and extending outward from the eye there may be a less distinct line (genal ridge). The marginal suture extends all the way around the cephalon, including inner and outer sides of the genal-spine prolongations, except at the rear between these prolongations (Fig. 13-12, *1a*).

Thorax and Pygidium. The thoracic segments of *Cryptolithus* and *Paraharpes* are very straight-sided, lacking perceptible backward curvature at the extremities. Only a few (typically six) occur in *Cryptolithus*, whereas *Paraharpes* has 23 to 25. Pleural furrows are relatively broad and shallow.

The pygidium of both described genera is small and is much wider than long. It is composed of a very few fused segments.

Appendages. Among hypoparians, the appendages only of *Cryptolithus* are fairly well known. They correspond in essential plan to those of *Triarthrus*.

Eodiscids

The eodiscids are very diminutive trilobites, less than 0.5 in. in length, which are confined to Lower and Middle Cambrian rocks, so far as known. They are characterized by head and tail shields of nearly equal size, and they possess only two or three thoracic segments. In these respects they resemble agnostids and differ widely from all other trilobites. They are distinguished most readily from the Agnostida by the well-annulated nature of the axis of the pygidium, which indicates fusion of several somites. The pleural lobes of the pygidium, however, are smooth, showing advanced evolutionary change.

Included among the eodiscids are a few trilobites like *Pagetia* (Fig. 13-15, *4*), which corresponds structurally to other genera of the order except in having short proparian-type sutures on the dorsal side of the cephalon, small eyes near the margin of the head, and interpleural grooves more or less clearly marked on the sides of the pygidium. Although these characters suggest a relationship to the order Proparia, *Pagetia* is judged really to belong to the Eodiscida, and there is no good evidence that any Upper Cambrian or younger trilobites were derived from this order.

Agnostids

The chief attributes of agnostid trilobites are their minute size, mostly 0.3 in. or

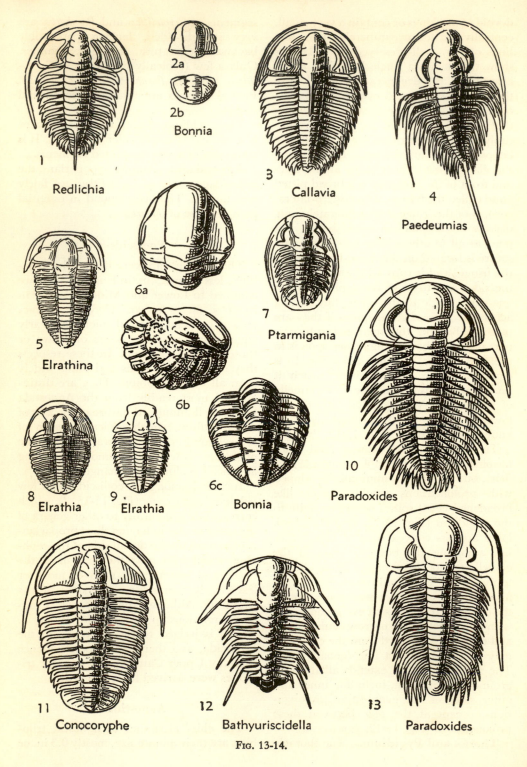

1 Redlichia

2a
2b Bonnia

3 Callavia

4 Paedeumias

5 Elrathina

6a

6b

7 Ptarmigania

8 Elrathia

9 Elrathia

6c Bonnia

10 Paradoxides

11 Conocoryphe

12 Bathyuriscidella

13 Paradoxides

Fig. 13-14.

smaller in length; equal-sized subcircular cephalon and pygidium; lack of cephalic sutures, and rather peculiar simple segmentation of the glabella; and the presence of only two thoracic segments. They are long-ranging as compared with the eodiscids, being distributed from Lower Cambrian to Upper Ordovician, but are most common in some Cambrian formations. So similar are the cephalon and pygidium of some agnostids that difficulty may be encountered in determining which rounded shield belongs in the anterior position and which in the posterior. The agnostids are specialized blind trilobites which stand far removed from others and are not known to have left any descendants. Illustrated representatives of this order are *Hypagnostus* and *Peronopsis* (Fig. 13-15, *1, 2*).

EVOLUTIONARY TRENDS OF TRILOBITES

An initial need in the study of the nature of evolution among trilobites is the ability to recognize characters which may be identified as primitive, and to evaluate others denoting various sorts of special-

---

FIG. 13-14.   **Representative Lower and Middle Cambrian trilobites.** Natural size, except as indicated. Protoparians include 3 and 4; the others are opisthoparians.

Bathyuriscidella Rasetti, Middle Cambrian, North America. Distinguished from *Bathyuriscus* (Fig. 13-16, *8*) mainly by characters of the pygidium and occurs stratigraphically higher. *B. socialis* Rasetti (12, type species of the genus, ×2), Quebec.

Bonnia Walcott, Lower Cambrian, North America. The long quadrate glabella, wide fixed cheeks, moderately large eyes, and character of the pygidium distinguish the genus. *B. fieldensis* Walcott (2a, cranidium; 2b, pygidium), Mt. Whyte formation, British Columbia. *B. bubaris* (Walcott) (6a–c, an enrolled specimen ×3), Lower Cambrian boulders in Levis conglomerate (L. Ordovician), Quebec.

Callavia Matthew, Lower Cambrian, eastern North America and Europe (Atlantic province). Characterized by shape of glabella, which bears a long occipital spine, nature of thorax, and small pygidium. *C. bröggeri* (Walcott) (3, type species of the genus, ×0.25), Newfoundland.

Conocoryphe Hawle & Corda, Middle Cambrian, eastern North America and Europe (Atlantic province). Blind opisthoparian having very narrow free cheeks and a shortened, tapering glabella, which is laterally indented by backwardly curved furrows. *C. sulzeri* (Schlotheim) (11, ×0.5), Czechoslovakia.

Elrathia Walcott, Middle Cambrian, North America. The shortened glabella, broad brim, wide free cheeks, and nature of the pygidium distinguish this genus. *E. kingi* (Meek) (8, a typical example of a very common species),

Wheeler formation, Utah. *E. georgiensis* Resser (9, a specimen lacking free cheeks), Conasauga shale, Georgia.

Elrathina Resser, Middle Cambrian, western North America. Differs from *Elrathia* in its longer glabella, narrower brim, long tapering thorax, and very small pygidium. *E. cordillerae* (Rominger) (5), Stephen formation, British Columbia.

Paedeumias Walcott, Lower Cambrian, North America. The somewhat shortened glabella does not reach the marginal furrow. *P. transitans* Walcott (4), Parker slate, Vermont.

Paradoxides Brongniart, Middle Cambrian, eastern North America and Europe (Atlantic province). The forwardly expanding glabella, which reaches to the marginal furrow, numerous thoracic segments, and small pygidium characterize the genus. It resembles protoparians but differs in its opisthoparian suture. *P. pinus* Holm (10, ×0.5), Middle Cambrian, Sweden. *P. harlani* Green (13, ×0.5), Massachusetts.

Ptarmigania Raymond. Middle Cambrian, western North America. This genus has relatively large crescentic eyes, long glabella, narrow free cheeks which carry exceptionally long genal spines, and moderately large pygidium. *P. rossensis* (Walcott) (7, type of the genus, ×0.5), Ptarmigan formation, British Columbia.

Redlichia Walcott, Lower Cambrian, Asia and Australia. The large crescentic eyes and other characters suggest derivation of this opisthoparian from olenellid ancestors. *R. chinensis* Walcott (1, ×0.5), China.

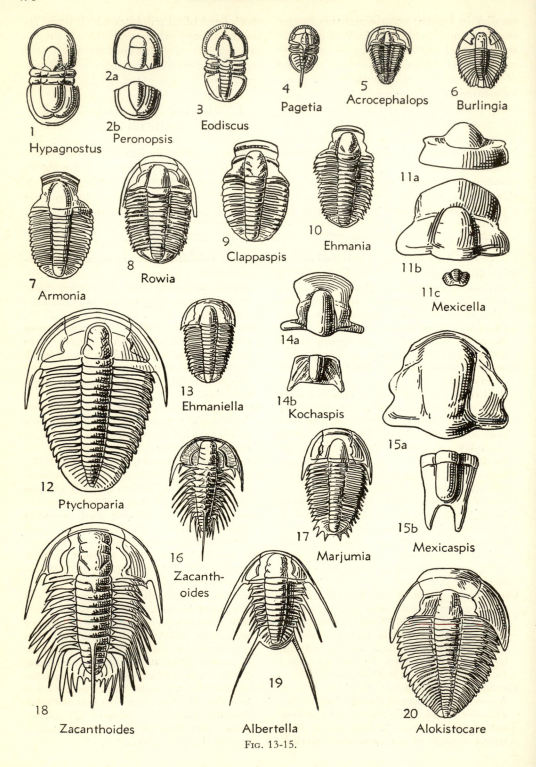

1 Hypagnostus

2a
2b Peronopsis

3 Eodiscus

4 Pagetia

5 Acrocephalops

6 Burlingia

7 Armonia

8 Rowia

9 Clappaspis

10 Ehmania

11a
11b
11c Mexicella

12 Ptychoparia

13 Ehmaniella

14a
14b Kochaspis

15a
15b Mexicaspis

16 Zacanth-oides

17 Marjumia

18 Zacanthoides

19 Albertella

20 Alokistocare

Fig. 13-15.

Fig. 13-15. **Representative Middle Cambrian trilobites.** Natural size except as indicated otherwise. Agnostida include 1 and 2; Eodiscida 3 and 4; Proparia 6; and Opisthoparia the remainder.

Acrocephalops Poulsen, Middle Cambrian, North America. The short, tapering glabella, ocular ridges, slender thorax narrowing posteriorly, and moderately large, furrowed pygidium mark this genus. *A. insignis* (Walcott) (5, ×2), Conasauga shale, Alabama.

Albertella Walcott, Middle Cambrian, western North America. Mainly distinguished by long spines on pygidium, thorax, and head. *A. helena* Walcott (19, type of the genus), a widespread lower Middle Cambrian zone fossil, Montana.

Alokistocare Lorenz, Middle Cambrian, North America. Cephalon has shortened glabella, broad brim, and distinct ocular ridges; thoracic segments numerous, narrowing backward to the very small pygidium. *A. idahoense* Resser (20), Spence formation, Utah.

Armonia Walcott, Middle Cambrian, North America. Very similar to *Elrathia*. *A. elongata* (Walcott) (7, a specimen lacking free cheeks, ×0.5), Maryville limestone, Alabama.

Burlingia Walcott, Middle Cambrian, western North America. Has small quadrate free cheeks on the dorsal side. *B. hectori* Walcott (6, type species of the genus, ×2), Stephen formation, British Columbia.

Clappaspis Deiss, Middle Cambrian, western North America. Wide fixed cheeks bear prominent ocular ridges; brim extends in front of furrowed glabella. *C. typica* Deiss (9, type species of genus, ×2), Pentagon formation, Montana.

Ehmania Resser, Middle Cambrian, North America. Differs from *Clappaspis* in characters of glabella, fixed cheeks, and pygidium. *E. convexa* Deiss (10, specimen lacking free cheeks), Pentagon formation, Montana.

Ehmaniella Resser, Middle Cambrian, North America. Differs from *Ehmania* in cranidium and pygidium. *E. placida* Resser (13), Rogersville shale, Tennessee.

Eodiscus Matthew, Lower and Middle Cambrian, Europe and North America. Short unfurrowed glabella; smooth fixed cheeks and pleural lobes of pygidium; segmented axis on pygidium; three thoracic segments. *E. punctatus* (Salter) (3, ×3), Middle Cambrian, Newfoundland.

Hypagnostus Jaekel, Middle Cambrian, North America, Australia, and Europe. Glabella very short, two thoracic segments. *H. métisensis* Rasetti (1, a complete carapace, ×4), Quebec.

Kochaspis Resser, Lower and Middle Cambrian, North America. Broad curved brim; quadrate

pygidium. *K. coosensis* (Walcott) (14a, b), cranidium, pygidium, ×0.5), Maryville limestone, Alabama.

Marjumia Walcott, Middle Cambrian, western North America. Somewhat like *Ehmaniella* but has spine-bordered pygidium. *M. typa* Walcott (17, type species of genus, ×0.5), Marjum limestone, Utah.

Mexicaspis Lochman, Middle Cambrian, western North America. Glabella reaches marginal furrow; narrow pygidium bearing two posterior projections. *M. stenopyge* Lochman (15a, b, type species of genus, ×2), Sonora.

Mexicella Lochman, Middle Cambrian, western North America. Glabella very short, brim broad, front edge of cranidium turned vertically downward; pygidium disproportionately minute. *M. mexicana* Lochman (11a, b, anterior and dorsal views of cranidium; 11c, pygidium; all ×4), Sonora.

Pagetia Walcott, Middle Cambrian, North America and Australia. Resembles *Eodiscus* in most features but has small eyes and proparian-type facial sutures; free cheeks very small on dorsal side. *P. bootes* Walcott (4, a complete specimen, ×2), Stephen formation, British Columbia.

Peronopsis Corda, Middle Cambrian, North America and Europe. Differs from *Hypagnostus* in larger size and markings of glabella. *P. gaspensis* Rasetti (2a, b, cephalon and pygidium, ×4), Quebec.

Ptychoparia Corda, Middle Cambrian, North America, Europe, and Asia. Short tapering glabella marked by backwardly curved furrows; moderately large, furrowed, smoothly rounded pygidium. *P. striata* Emmrich (12, ×0.5), Czechoslovakia.

Rowia Deiss, Middle Cambrian, western North America. Resembles *Ehmania* but differs in features of cranidium. *R. vulgata* Deiss (8, type species), Pentagon shale, Montana.

Zacanthoides Walcott, Middle Cambrian, North America. Several characters of this genus, such as the large crescentic eyes, presence of intergenal spines, and spinose thoracic segments (including telson-like spines), suggest protoparians, but the opisthoparian facial suture and type of pygidium sharply distinguish it. *Z. typicalis* Walcott (16), Chisholm formation, Nevada. *Z. spinosus* Walcott (18, ×0.5), Spence shale, Utah.

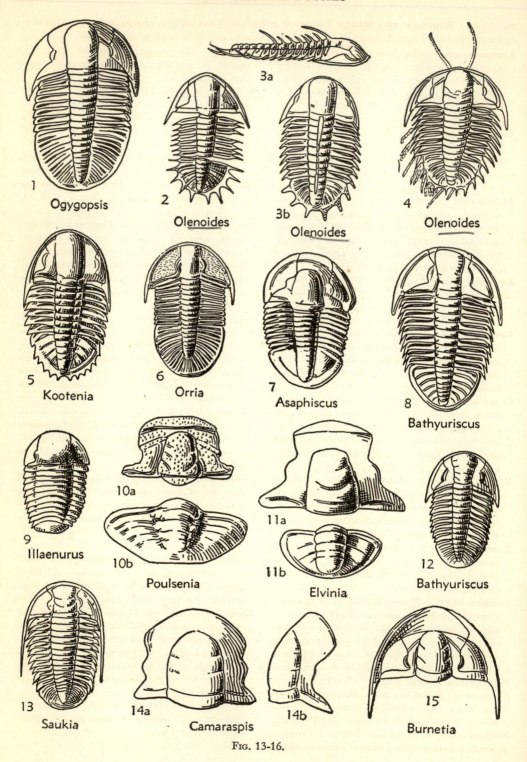

1 Ogygopsis

3a

2 Olenoides

3b Olenoides

4 Olenoides

5 Kootenia

6 Orria

7 Asaphiscus

8 Bathyuriscus

9 Illaenurus

10a

10b Poulsenia

11a

11b Elvinia

12 Bathyuriscus

13 Saukia

14a Camaraspis

14b

15 Burnetia

FIG. 13-16.

ization. A trilobite has numerous structural elements belonging to its exoskeleton, and each of these may exhibit varied sorts of evolutionary trends. In order to learn the actual nature of changes, we must compare corresponding parts of many trilobite genera, at the same time taking account of the placement of the studied forms in taxonomic divisions, and their occurrence in rocks of different geologic age. Knowledge of ontogeny should help us. The task of fitting all evidence together, however, is not easy.

We may begin by trying to construct a hypothetical near-annelid prototype of the trilobites. This should have generalized simple structural elements which may give rise to all actually observed features of these arthropods. With such a form in

mind, we then can review significant divergences in the nature of the main parts of the trilobite carapace: facial sutures, glabella, eyes, thorax, and pygidium.

Primitive Characters

Since the structures of trilobites may be considered primitive in proportion to their approach to the characters of an annelid, distinctness of somite divisions throughout the body, from tip of head to end of tail, is one feature which should denote primitiveness. Others are a long narrow body formed of many somites, and restriction of coalesced somites to the head region. Because trilobites typically possess differentiated axial and pleural lobes, in this respect deviating from an annelid, it is appropriate to picture our very archaic,

---

Fig. 13-16. **Representative Middle and Upper Cambrian trilobites.** Natural size unless indicated otherwise. All are opisthoparians.

Asaphiscus Meek, Middle Cambrian, western North America. Cephalon and pygidium have a wide border; brim extends in front of glabella. *A. wheeleri* Meek (7), Wheeler formation, Utah.

Bathyuriscus Meek, Middle Cambrian, North America. Glabella expanded forward, reaching to marginal furrow; pygidium large, well segmented. *B. formosus* Deiss (8), Meagher formation, Montana. *B. rotundatus* (Rominger) (12), Stephen formation, British Columbia.

Burnetia Walcott, Upper Cambrian, North America. Distinguished by widened frontal part of brim and inward convergence of facial sutures in this region. *B. urania* (Walcott) (15, cephalon, ×0.5), Wilberns formation, Texas.

Camaraspis Ulrich & Resser, Upper Cambrian, North America. Cranidium strongly convex, brim wide. *C. convexa* (Whitfield) (14a, b, dorsal and side views of cranidium, ×2), *Elvinia* zone, Honey Creek formation, Oklahoma.

Elvinia Walcott, Upper Cambrian, North America. Border and brim well developed in front of truncate glabella. *E. roemeri* (Shumard) (11a, b, cranidium and pygidium), Wilberns formation, Texas.

Illaenurus Hall, Upper Cambrian, North America. Axis weakly differentiated, free cheeks wide. *I. quadratus* Hall (9), Trempealeauan, Wisconsin.

Kootenia Walcott, Lower and Middle Cambrian,

North America and Asia. The long, parallel-sided glabella, narrow cephalic border, and relatively large serrate-edged pygidium characterize this genus. *K. dawsoni* (Walcott) (5), Stephen formation, British Columbia.

Ogygopsis Walcott, Middle Cambrian, western North America. Pygidium very large, well-furrowed. *O. klotzi* (Rominger) (1, ×0.5), Stephen formation, British Columbia.

Olenoides Meek, Middle Cambrian, North America. Resembles *Kootenia* but has occipital and thoracic axial spines. *O. superbus* Walcott (2, ×0.5), Marjum formation, Utah. *O. curticei* Walcott (3a, b, side and dorsal views, ×0.5), Conasauga shale, Alabama. *O. serratus* (Rominger) (4), Stephen formation, British Columbia.

Orria Walcott, Middle Cambrian, western North America. Like *Ogygopsis* but has eyes close to glabella and shortened axis on pygidium. *O. elegans* Walcott (6, ×0.5), Marjum formation, Utah.

Poulsenia Resser, Lower and Middle Cambrian, North America and Greenland. *P. granosa* Resser (10a, b, cranidium and pygidium), Middle Cambrian, Wyoming.

Saukia Walcott, Upper Cambrian, North America. Border of cephalon distinct, free cheeks wide, genal spines long. *S. acuta* Ulrich & Resser (13, ×0.5), Trempealeauan, Wisconsin.

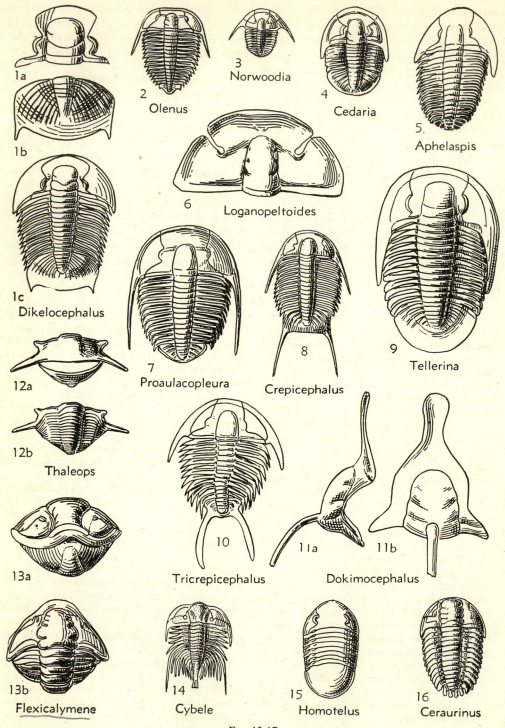

1a

1b

1c
Dikelocephalus

2
Olenus

3
Norwoodia

4
Cedaria

5.
Aphelaspis

6
Loganopeltoides

7
Proaulacopleura

8
Crepicephalus

9
Tellerina

12a

12b
Thaleops

13a

13b
Flexicalymene

10
Tricrepicephalus

11a

11b
Dokimocephalus

14
Cybele

15
Homotelus

16
Ceraurinus

Fig. 13-17.

primitive trilobite as having prominent axial segments, which on each side bear short, relatively insignificant pleura, very much like the posterior thoracic region of *Olenellus*, behind the fifteenth segment (Fig. 13-7*A*).

The cephalic portion of the axial lobe (glabella) should reach to the anterior extremity of the head, and on either side of it there should be well-developed eyes, as in annelids. Probably, but not certainly, the hardened exoskeleton of the head is divided along a marginal facial suture at times of molting. All segments of the body, except an antennae-bearing somite (and pre-antennal one) at the front of the head, are provided with a pair of biramous limbs which have identical structure. The head, or cephalon, of even the most primitive trilobite may be conceived to comprise the forwardmost six somites, of which four bear pairs of biramous appendages, for this

---

Fig. 13-17.   **Representative Upper Cambrian and Ordovician trilobites.** Natural size except as indicated otherwise. Proparians include 3, 14, and 16. All others are opisthoparians, except 6, which is classed with hypoparians (Fig. 13-13).

Aphelaspis Resser, Upper Cambrian, North America. Glabella short, brim wide; pygidium small. *A. hamblenensis* Resser (5), Nolichucky shale, Tennessee.

Cedaria Walcott, Upper Cambrian, North America. A small trilobite; free cheeks wide, pygidium large. *C. minor* (Walcott) (4, ×2), Weeks formation, Utah.

Ceraurinus Barton, Middle and Upper Ordovician, North America and Europe. Glabella strongly lobed; pygidium bordered by six blunt projections. *C. icarus* (Billings) (16, ×0.5), Upper Ordovician (Richmondian), Ohio.

Crepicephalus Owen, Upper Cambrian, North America. Glabella short, free cheeks and pygidium having long spines. *C. iowensis* (Owen) (8, ×0.5), Dresbachian, Iowa.

Cybele Lovén, Ordovician–Silurian, Europe. Distinctively peculiar head, thorax, and pygidium. *C. lovéni* Linnarsson (14, ×0.5), Middle Ordovician, Sweden.

Dikelocephalus Owen, Upper Cambrian, North America and Europe. Broad cephalon, large concave brim; large pygidium has short axis and two short spines at lateral extremities. *D. gracilis* Ulrich & Resser (1*a*, *b*, cranidium, pygidium, ×0.5), Trempealeauan, Wisconsin. *D. oweni* Ulrich & Resser (1*c*, ×0.5), Trempealeauan, Wisconsin.

Dokimocephalus Walcott, Upper Cambrian, North America. Distinguished by long anterior extension of brim. *D. curtus* (Resser) (11*a*, *b*, side and dorsal views of cranidium, ×2), Honey Creek formation (*Elvinia* zone), Oklahoma.

Flexicalymene Shirley, Ordovician–Silurian, North America and Europe. *F. meeki* (Foerste) (13*a–b*, two views of an enrolled specimen), Cincinnatian, Ohio.

Homotelus Raymond, Middle and Upper Ordovician, North America. Resembles *Isotelus* (Fig. 13-11) but lacks concave borders on cephalon and pygidium. *H. obtusus* (Hall) (15, ×0.5), Chazyan, Virginia.

Loganopeltoides Rasetti, Upper Cambrian, North America. Facial suture marginal, except for very narrow proparian-type dorsal portions; classed as a hypoparian (Fig. 13-13). *L. kindlei* Rasetti (6, cranidium, ×6), Newfoundland.

Norwoodia Walcott, Upper Cambrian, North America. Widely flaring genal spines borne by fixed cheeks. *N. gracilis* Walcott (3), Nolichucky shale, Tennessee.

Olenus Dalman, Upper Cambrian–Ordovician, Europe. Glabella short, free cheeks large, distinct ocular ridges; small pygidium. *O. gibbosus* Wahlenberg (2), Upper Cambrian, Sweden.

Proaulacopleura Kobayashi, Upper Cambrian, North America. Like *Olenus* but differs in glabellar and other features. *P. buttsi* Kobayashi (7), Conasauga shale, Alabama.

Tellerina Ulrich & Resser, Upper Cambrian, North America. Resembles *Saukia* but has shortened glabella differently marked by furrows. *T. crassimarginata* (Whitfield) (9, ×0.5), Trempealeauan, Wisconsin.

Thaleops Conrad, Middle Ordovician, North America. Like *Illaenus* but has laterally extended eyes and projecting genal spines. *T. ovata* Conrad (12*a*, *b*, anterior and dorsal views of an enrolled specimen), Trentonian, Ontario.

Tricrepicephalus Kobayashi, Upper Cambrian, North America. Small pygidium bears two large spines. *T. cedarensis* Resser (10), Nolichucky shale, Alabama.

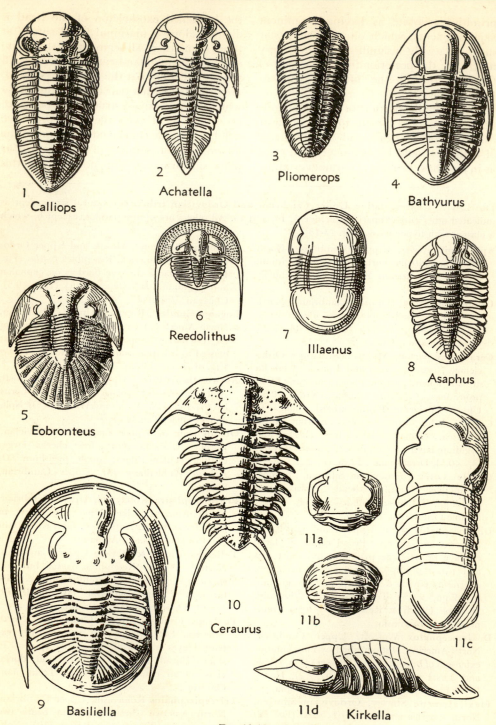

1 Calliops

2 Achatella

3 Pliomerops

4 Bathyurus

5 Eobronteus

6 Reedolithus

7 Illaenus

8 Asaphus

9 Basiliella

10 Ceraurus

11a

11b

11c

11d Kirkella

Fig. 13-18.

seems to be a stable feature of the Trilobita as a whole.

No known fossil approximates the form described other than partially, but we may construe deviations from it as marking evolutionary trends. A very generalized trilobite of primitive aspect, except in having pleural lobes equal to the axis in width and a pygidium composed of four fused segments, is illustrated in Fig. 13-1.

Facial Sutures

A marginal position of the facial suture of the cephalon is judged to be most archaic, because it characterizes the early protaspis larval condition of all trilobites, so far as known. Also, it is seen in adult protoparians, which have other primitive characters, and include the most ancient preserved fossils of the class. Growth series of larvae show the migration of the facial suture first on to the antero-lateral corners of the dorsal side of the cephalon (Fig. 13-9B, F); it runs along the edge of the eye and facilitates freeing of this organ in molting. Enlargement of the dorsal part of the free cheeks follows, but among Proparia the rear part of the suture stays in front of the genal angles (Fig. 13-9G–I). In larval development of an opisthoparian trilobite, the suture is successively marginal, proparian, and opisthoparian, intersecting the back margin of the cephalon (Fig. 13-9A–D). The part of the facial suture in front of the eyes, both in the more primitive proparians and opisthoparians, curves somewhat outward to the front edge of the head shield (Fig. 13-9G, J), but in progressively advanced evolution it curves inward so as eventually to become restricted to the dorsal surface of the cephalon in this part of its course (Fig. 13-9H, I, K, L).

In discussing the characters of hypoparian trilobites, reasons have been given for interpreting this group as specialized. Their marginal facial suture might be interpreted as an archaic structure, which was retained during evolution while other parts of the exoskeleton became modified.

Fig. 13-18. **Representative Ordovician trilobites.** Natural size unless indicated otherwise. Proparians include 1–3, and others are opisthoparians, except 6, which is a hypoparian.

Achatella Delo, Middle and Upper Ordovician, North America. Front lobe of glabella very wide; pygidium triangular. *A. achates* (Billings) (2), Trentonian, Minnesota.

Asaphus Brongniart, Ordovician, Europe and Asia. Glabella expanded forward. *A. expansus* Dalman (8, ×0.5), Middle Ordovician, Norway.

Basiliella Kobayashi, Ordovician, North and South America. Mainly characterized by features of cephalon and pygidium. *B. barrandei* (Hall) (9), Trentonian, New York.

Bathyurus Billings, Ordovician, North America. Resembles *Asaphus* but cephalon has straight-sided glabella and long genal spines. *B. extans* (Hall) (4), Blackriveran, Ontario.

Calliops Delo, Ordovician, North America. Cephalon like that of *Phacops* (Fig. 13-19, 2, 7) but less expanded in front; pygidium subtriangular. *C. callicephala* (Hall) (1), Trentonian, New York.

Ceraurus Green, Ordovician, North America and Europe. Furrowed glabella prominent, eyes small; small pygidium bears two large spines. *C. pleurexanthemus* Green (10), Trentonian, New York. (Compare Figs. 13-4, 13-5.)

Eobronteus Reed, Middle and Upper Ordovician, North America and Europe. *E. lunatus* (Billings) (5), Trentonian, Ontario.

Illaenus Dalman, Ordovician and Silurian, North America and Europe. Axis obsolescent; cephalon and pygidium nearly smooth. *I. davisii* Salter (7, ×0.5), Middle Ordovician, England.

Kirkella Kobayashi, Ordovician, North America. Narrow carapace and the quadrate form of cephalon and pygidium distinguish this from *Isotelus*. *K. vigilans* (Whittington) (11a, b, two views of an enrolled specimen; 11c, d, dorsal and side views, ×3), Canadian, Nevada.

Pliomerops Raymond, Middle Ordovician, North America and Europe. *P. canadensis* (Billings) (3), Chazyan, New York.

Reedolithus Bancroft, Ordovician, Europe. Differs from *Cryptolithus* in having ocular ridges and eyes in adult stage. *R. carinatus* (Angelin) (6, ×3), Middle Ordovician, Sweden.

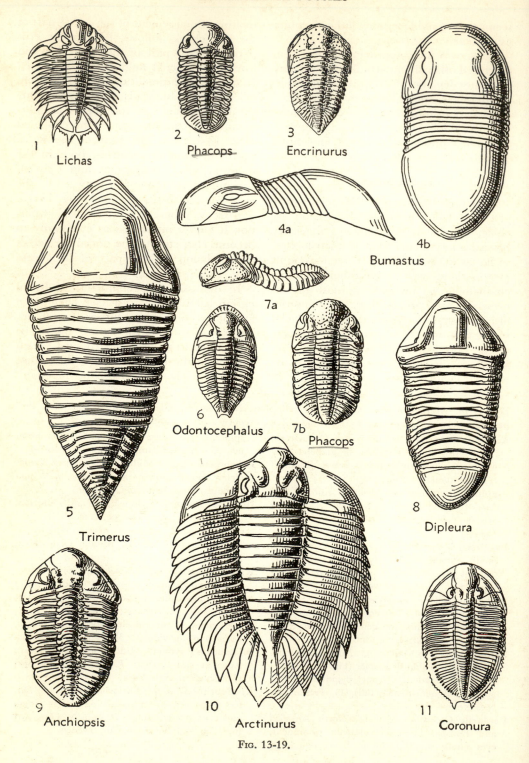

1 Lichas

2 Phacops

3 Encrinurus

4a

4b Bumastus

5 Trimerus

7a

6 Odontocephalus

7b Phacops

8 Dipleura

9 Anchiopsis

10 Arctinurus

11 Coronura

Fig. 13-19.

It is possible, on the other hand, that this suture is "reconstructed" in a marginal position after having gone through a proparian condition, or both proparian and opisthoparian placement on the dorsal side of the cephalon. This is demonstrated, seemingly, by specimens of Upper Cambrian trilobites of hypoparian type, which show closely adjacent portions of the suture running from the front-lateral edge of the cephalon to each eye, the remainder of the suture being marginal. This represents a proparian condition, or opisthoparian, if the free cheeks (which are lost) bear the genal angles or spines (Figs. 13-3*B;* 13-13; 13-17, *6*). Closely similar Ordovician trilobites, which are typical hypoparians, differ from the Cambrian forms just cited in having no part of the facial suture on the dorsal surface. Thus, it seems probable that all Hypoparia have a secondarily developed marginal suture, and in this respect, as well as others, are decidedly specialized.

Glabella

The axial region of the head, composed chiefly of the glabella, ranks next in importance to the facial suture as a generally useful guide in classification. Possibly this reflects dominance in the anatomical significance of the structures lodged here. At all events, the glabella exhibits noteworthy evolutionary modification. Its chief changes are in (1) segmentation; (2) shape, including distinctness of definition; and (3) relative length.

Segmentation. Since larval trilobites more or less clearly indicate the existence of six cephalic somites, the most primitive sort of glabella in an adult should be one that retains a clearly marked sign of all these somites. A glabella transversely crossed by five complete furrows (Fig. 13-21*A*) ideally represents such a condition, and it is seen in some specimens of primitive opisthoparians (*Paradoxides*). Obsolescence of the middle part of a complete

---

FIG. 13-19. **Representative Silurian and Devonian trilobites.** Natural size unless indicated otherwise. Proparians include 2, 3, 5–9, and 11; opisthoparians include 1, 4, and 10.

Anchiopsis Delo, Lower and Middle Devonian, North America. *A. anchiops* (Green) (9, ×0.5), Lower Devonian (Onondaga), New York.

Arctinurus Castelnau, Silurian, North America. The glabella has composite lateral lobes; pygidium distinguished by shortened axis and leaflike appearance of pleural lobes. *A. boltoni* (Bigsby) (10, ×0.5), Middle Silurian (Clintonian), New York.

Bumastus Murchison, Ordovician–Silurian, North America and Europe. Like *Illaenus* but axis entirely obsolete. *B. niagarensis* (Whitfield) (4*a, b,* side and dorsal views), Niagaran, Illinois.

Coronura Hall & Clarke, Devonian, North America. Distinguished mainly by large pygidium composed of many segments and bearing two terminal spines. *C. aspectans* (Conrad) (11, ×0.5), Lower Devonian (Onondaga), New York.

Dipleura Green, Devonian, North America and Europe. Dorsal furrow, defining axial lobe, lacking on thorax and pygidium. *D. dekayi* Green (8, ×0.5), Middle Devonian (Hamilton), New York.

Encrinurus Emmrich, Ordovician-Silurian, North America and Europe. *E. ornatus* Hall & Whitfield (3), Niagaran, New York.

Lichas Dalman, Ordovician, Europe; Silurian, North America and Europe. Composite glabellar lobes; projecting ribs and spines on pygidium. *L. speciosus* Beyrich (1, ×0.5), Silurian, Czechoslovakia.

Odontocephalus Conrad, Lower Devonian, North America. Fluted border in front of glabella; pygidium bears two short terminal projections. *O. aegeria* Hall (6), Onondaga, New York.

Phacops Emmrich, Silurian–Devonian, North America and Europe. Strongly expanded and forwardly bulging glabella. *P. rana* Green (2, 7*a, b,* dorsal and side views), Middle Devonian (Hamilton), New York. (Compare Fig. 13-10, *2a–c.*)

Trimerus Green, Silurian–Devonian, North America and Europe. Like *Dipleura* but has more pointed head and tail. *T. delphinocephalus* Green (5, ×0.5), Middle Silurian (Clintonian), New York.

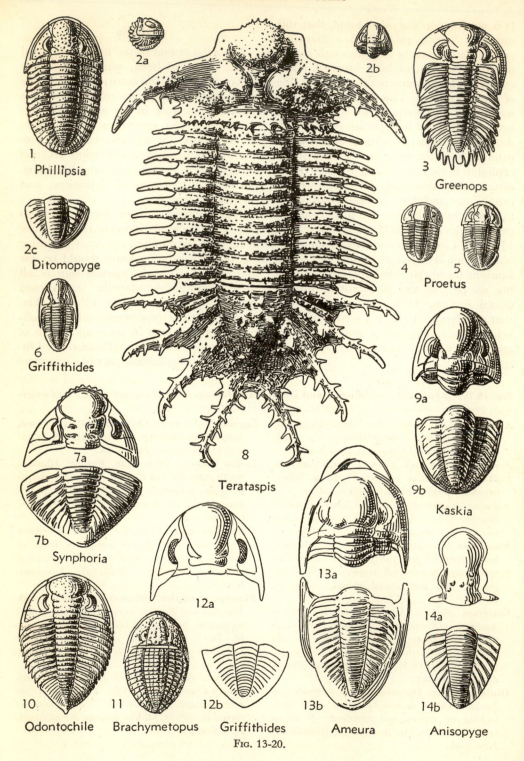

1. Phillipsia

2a

2b

2c Ditomopyge

6 Griffithides

3 Greenops

4   5 Proetus

7a

7b Synphoria

8 Terataspis

9a

9b Kaskia

10 Odontochile

11 Brachymetopus

12a

12b Griffithides

13a

13b Ameura

14a

14b Anisopyge

FIG. 13-20.

glabellar furrow gives rise to a pair of short furrows on the opposite sides of the glabella. Such a change reduces a glabellar segment to a pair of glabellar lobes. Next, the short furrows tend to disappear. These erasures of the external marks of segmentation, characterized first by interruption of complete furrows and secondly by obliteration of remnant side furrows, proceed from front to rear of the glabella. The occipital furrow vanishes last; it is represented at least by short lateral furrows in most trilobites. Some genera, such as *Illaenurus* (Fig. 13-16, *9*), *Homotelus* (Fig. 13-17, *15*), *Illaenus* (Fig. 13-18, *7*), *Kirkella* (Fig. 13-18, *11*), *Bumastus* (Fig. 13-19, *4*), and several others, have lost all trace of glabellar segmentation. Progressive steps in this evolutionary trend are shown diagrammatically in Fig. 13-21*A, G–M*.

A character of short glabellar furrows which calls for notice is the tendency to change direction from directly transverse to backwardly recurved. This may lead to partial or complete isolation of glabellar lobes, as in *Anchiopsis* (Fig. 13-19, *9*), *Coronura* (Fig. 13-19, *11*), *Kaskia* (Fig. 13-20, *9*), and many others. Coalescence of adjacent furrows at the sides of the glabella (Fig. 13-21*N, O*) gives rise to the peculiar appearance of the glabellar region in *Lichas* (Fig. 13-19, *1*), *Arctinurus* (Fig. 13-19, *10*), *Terataspis* (Fig. 13-20, *8*), and various other opisthoparian genera. Obviously, these are specialized trilobites.

Shape. The shape of the glabella, both in horizontal outline and in vertical profile, exhibits wide variation. Change of shape accordingly must be included among evolutionary trends (Fig. 13-21*B–F, P–T*). A most primitive glabella should be one that narrows somewhat evenly in a forward direction and slopes regularly in this direction; in transverse profile it should be only moderately elevated (Figs. 13-1; 13-21*A*). Any sort of deviation from this simple form

---

FIG. 13-20. **Representative Devonian, Mississippian, Pennsylvanian, and Permian trilobites.** Natural size unless indicated otherwise. Except 3, 7, and 10, which are proparians, all are opisthoparians.

Ameura J. M. Weller, Pennsylvanian–Permian, North America. Glabella not reaching to edge of cephalon; pygidium elongate semiellipse composed of many segments. *A. major* (Shumard) (13*a, b*, cephalic and pygidial views of an enrolled specimen), Pennsylvanian, Mid-Continent.

Anisopyge Girty, Permian, North America. Long, low glabella; many axial segments on pygidium but few pleural segments. *A. perannulata* (Shumard) (14*a, b*, cranidium, pygidium), Upper Permian (Guadalupian), western Texas.

Brachymetopus McCoy, Devonian–Permian, Europe, Australia; Mississippian, North America. *B. lodiensis* Meek (11, ×2), Osagian (Waverly), Ohio.

Ditomopyge Newell, Pennsylvanian–Permian, North America and Europe. *D. scitula* (Meek & Worthen) (2*a, b*, views of enrolled specimen, ×1.5; 2*c*, pygidium, ×2), Pennsylvanian, Illinois.

Greenops Delo, Middle and Upper Devonian, North America. *G. boothi* (Green) (3), Middle Devonian (Hamilton), New York.

Griffithides Portlock, Mississippian–Permian, North America, Europe, East Indies. *G.*
longispinus Portlock (6, ×0.5), Mississippian, Ireland. *G. breviceps* Gheyselinck (12*a, b*, cephalon, pygidium, ×2), Permian, Timor.

Kaskia J. M. Weller, Mississippian, North America. Resembles *Griffithides*. *K. chesterensis* Weller & Weller (9*a, b*, views of enrolled specimen, ×2), Upper Mississippian, Mid-Continent.

Odontochile Hawle & Corda, Devonian, North America and Europe. Resembles *Dalmanites*. *O. micrurus* (Green) (10), Lower Devonian, New York.

Phillipsia Portlock, Mississippian–Permian, North America, Europe, East Indies. *P. sampsoni* Vogdes (1), Lower Mississippian, Missouri.

Proetus Steininger, Ordovician–Mississippian, Europe, North America. *P. cuvieri* Steininger (4), Middle Devonian, Germany. *P. haldemani* Hall (5), Middle Devonian, New York.

Synphoria Clarke, Lower Devonian, North America. *S. stemmata* (Clarke) (7*a, b*, ×0.5), Oriskany, New York.

Terataspis Hall, Lower Devonian, North America. Very large, bizarre form. *T. grandis* (Hall) (8, ×0.25), Onondagan, New York.

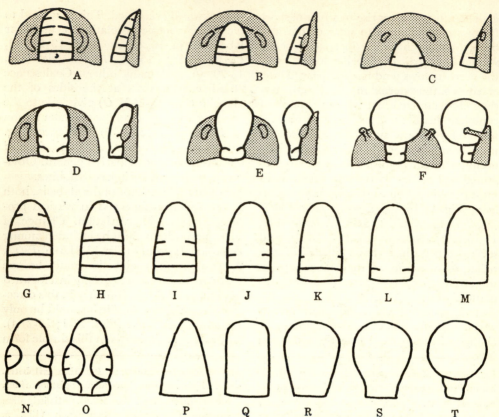

FIG. 13-21. **Evolutionary trends of the glabella.** The diagrams indicate (*A*) the inferred most primitive type of glabella, which is completely segmented, gently tapering forward, and extending to the front of the cephalon; (*B, C*) progressive trends in shortening of the glabella; (*D–F*) lengthening and protrusion of the glabella, both horizontally and vertically; *G–M* steps in disappearance of segmentation; (*N, O*) coalescence of glabellar furrows; (*P–T*) change in outline of the glabella.

may be judged to denote evolutionary change. We find trilobites having pronounced forward convergence of the glabellar margins, straight-sided parallel margins, and abruptly truncate fronts, and variously expanded anterior parts of the glabella; all these indicate shape modification. What may be termed loss of shape is as significant as accentuation in any direction. Illustrating this are several genera in which a weakened dorsal furrow and reduced height of the glabellar region leave nothing to be described as the shape of the glabella. Obviously, such obsolescence distinguishes one type of trend.

Length. The relative length of the glabella is not synonymous with the shape,

as is demonstrated readily by a comparison of specimens such as many here illustrated. Evolution may lead in opposite directions, giving rise to progressive shortening of the glabella and enlargement of the brim, or lengthening of the glabella until it protrudes considerably beyond other anterior parts of the cephalon (Fig. 13-21*D–F*). Many degrees of change along these lines are observed. They aid in the definition of genera and must be taken into account in studying evolution.

Eyes

The eyes of trilobites may be expected to show evolutionary changes, for on the whole, these organs were probably their

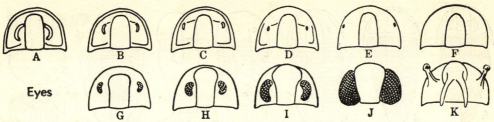

Eyes

FIG. 13-22. Evolutionary trends of the eyes of trilobites. The upper series of figures diagrammatically represents change from a protoparian type of long crescentic eye (*A*) to trilobites having smaller eyes joined to the glabella by ocular ridges (*B–D*), and reduction of the eyes to a state of blindness (*E, F*). The lower diagrams show increasing size of eyes, as observed in various stocks of proparians and opisthoparians (*G–J*), the maximum accentuation of the eyes being seen in such a genus as *Aeglina* (*J*). Aberrant tendencies lead to stalked eyes (*K*).

most important sensory possession. Both simple and compound eyes are recognized. Some trilobites had both kinds, some only one, and some none at all. Opposite directions of evolution of the eyes in different stocks are indicated by (1) reduction of size and ultimate disappearance, and (2) increase of size. A special modification, which is seen in several genera, is elevation of the eyes on stalks. These features are illustrated diagrammatically in Fig. 13-22.

The protoparian trilobites mostly possess well-developed compound eyes, which have a long, narrow, crescent-shaped plan but relatively little elevation above the surface of the fixed cheeks near them. The eyes are structures developed on the backwardly curved pleura of the antennal somite of the head, and the front extremity of the eyes touches or closely adjoins the glabella. The presence of a very large number of individual facets shows that these eyes are highly organized, yet their long, narrow form is considered archaic because it is not matched by the eyes of younger trilobites of other orders.

The ocular ridges of many Cambrian and Ordovician trilobites (but virtually unknown in later forms) may correspond to the anterior portion of the protoparian crescentic eye, even though the Protoparia left no descendants. The comparison merely indicates homology, suggesting that genera characterized by an ocular ridge are derived from ancestors having a visual area restricted to parts of the antennal

pleura located at some distance from the glabella; or alternatively, near-glabellar parts of the pleura, which originally belonged to the eyes, became blind.

All orders of trilobites except the protoparians contain genera in which no eyes are possessed by adults, although they may be present in larval stages. All except the agnostids are represented by genera provided with very small eyes. Reduction in size of the eyes, accompanied by disappearance of the ocular ridge (if present), leads by continued evolution to blindness. This tendency is demonstrated in several groups of related species or genera and is inferred in others.

Increase in the actual and relative size of compound eyes is shown by various proparian and opisthoparian genera. Among these, the visual area of the eye is increased in height and in lateral curvature, so as to become more and more efficient as a visual organ. A peak of evolution in this direction is represented by *Aeglina*, in which the entire cephalon, except the glabella and part of the ventral surface, is occupied by hemispherical, many-faceted eyes (Fig. 13-22*J*).

Thorax

Diametrically opposed views on what constitutes the most primitive type of trilobite thorax have been expressed. One interpretation points to evidence of the annelidan ancestry of trilobites and the occurrence among earliest Cambrian fossils of protoparians having an extremely

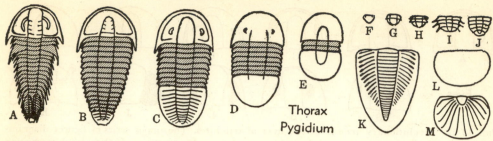

FIG. 13-23. Evolutionary trends of the thorax and pygidium. (*A–E*) Progressive decrease in the number of thoracic segments, correlated with enlargement of the pygidium; also change from spinose pleural extremities to bluntly rounded ones. (*F–J*) Stages in enlargement of the pygidium from a proto-pygidial one-segment structure to the normal, consisting of a number of fused segments. (*K–M*) Types of highly modified pygidia, consisting of very many segments with axial divisions not matching pleural ones (*K*), disappearance of all markings (*L*), and reduction of length of axial lobe (*M*).

large number of thoracic segments. The deduction that primitiveness of the trilo-bite thoracic region consists in a close approach to wormlike attributes—especially the long narrow body and numerous near-identical segments—is wholly logical. Another interpretation calls attention to the characters of trilobite larvae (large head shield, few thoracic segments) and compares these with Lower Cambrian tri-lobites (agnostids, eodiscids) which have only two or three thoracic segments in the adult. These are diminutive, mainly eye-less, larva-like forms, which on the basis of morphology and antiquity can be alleged to denote primitiveness. This second pos-tulate, however, omits some important things. Among larvae cited as correspond-ing to adult agnostids and eodiscids are those of *Olenellus*, which in its adult stage has the most wormlike, abundantly seg-mented thorax of any known trilobite. Also, the agnostids and eodiscids differ from larvae in having a large pygidium which approximately equals the cephalon in size. Undoubtedly it developed by fu-sion of several segments. These seemingly simple forms now are judged to be not at all like the most primitive trilobites.

This discussion leads to the conclusion that genera having very many thoracic segments, like *Olenellus* (which has 45 or more), are primitive. Progressive reduction in the number of these divisions is the normal direction of evolutionary change (Fig. 13-23*A–E*). Shortness of pleura, the presence of strongly incised, obliquely dis-posed pleural furrows, and backwardly curved spinose extremities may likewise be interpreted as primitive. Conversely, widened pleura; shallow, straight, or ob-solescent pleural furrows; and bluntly rounded extremities are advanced char-acters. The weakening of dorsal furrows, so that axial segments become barely dif-ferentiated from pleura, or are not sepa-rable at all in the exterior view of the thorax, is another type of evolutionary development.

Possibly the facility with which a trilo-bite can bend its thorax so as to roll up, thus using the strong dorsal exoskeleton to protect vulnerable ventral parts, is a meas-ure of evolution. Not all trilobites, by any means, seem to have had this ability, for among hundreds of specimens representing many genera none are in an enrolled posi-tion. Some Lower Cambrian opistho-parians (Fig. 13-14, *6*) had acquired the roll-up habit. It seems strange that no protoparian is known to have coiled in this manner.

Pygidium

Evolutionary characters of the pygidium are not difficult to recognize. They include such things as (1) relative size, (2) shape, (3) differentiation of axial and pleural

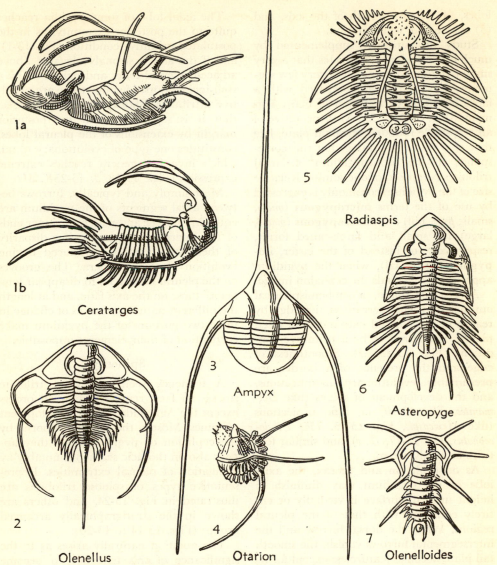

1a

1b
Ceratarges

2
Olenellus

3
Ampyx

4
Otarion

5

Radiaspis

6
Asteropyge

7
Olenelloides

FIG. 13-24. **Evolutionary trend of trilobites toward spinosity.** The tendency to develop spines is seen in representatives of all orders except the Agnostida, but causes for it are largely speculative.

1a, b, *Ceratarges armatus* (Goldfuss), Middle Devonian, Germany. (1a) oblique front view, ×1.5; (1b) side view, ×1. The eyes are elevated on stalks.

2, *Olenellus fremonti* Walcott, Lower Cambrian, Nevada, ×1. This species is distinguished by the remarkable curved extensions of the third thoracic segment and of the genal spines. Beneath the long spike tail (telson), several narrow thoracic segments probably belong, as in other species of *Olenellus*.

3, *Ampyx tetragonus gigas* Linnarsohn, Ordovician, Sweden, ×1. The spinose extensions of the cephalon of this blind trilobite are exceptionally long, and the purpose which they served is highly conjectural.

4, *Otarion ceratophthalmus* (Goldfuss), Middle Devonian, Germany, ×2. This enrolled specimen, in addition to long genal spines, has a still longer curved spine on the sixth axial segment of the thorax.

(Continued on next page.)

lobes, (4) relative length of the axis, and (5) evidence of segmentation.

Study of ontogeny, supplemented by much other evidence, indicates that a very small pygidium, composed of very few segments, is primitive as compared with a large one formed by fusion of many segments (Fig. 13-23F, K). Most primitive of all is a single rounded terminal plate, like that of *Olenellus*, assuming that this corresponds to the protopygidium of larval trilobites composed of one somite. Relative size of the pygidium commonly is expressed by use of the terms **micropygous** (*micro*, small; *pyge*, tail) and **macropygous** (*macro*, large) for small and large-tailed forms, respectively, or instead of the latter, **isopygous** (*iso*, equal), when the pygidium approximately equals the cephalon in size.

As regards shape, a subhemispherical outline, somewhat serrate at the edge by reason of the pointed extremities of fused pleura, is inferred to be a primitive condition (Fig. 13-23G–I). Smoothness of edge and the presence of a border are presumably evolutionary modifications, and the development of spines (like *Dalmanites*, Fig. 13-10, *1*), bifid projections (like *Mexicaspis*, Fig. 13-15, *15b*; *Dikelocephalus*, Fig. 13-17, *1b, c*), and similar features certainly are so.

As on cephalon and thorax, the axial lobe of the pygidium may diminish in height until its surface is virtually or entirely confluent with that of the pleural regions. When the dorsal furrow and the intersegmental furrows vanish, the smooth tail plate becomes featureless except for its broad convexity (Figs. 13-18, *7*; 13-19, *4*, *8*; 13-23L). Of course, this obsolescence of characters defines one of the many evolutionary trends seen in trilobites.

The axial lobe of some pygidia reaches quite to the posterior extremity, as in the postulated primitive condition (Fig. 13-1), and if the axis is strongly elevated above adjacent pleural lobes and delimited by a well-incised dorsal furrow, these are primitive attributes. Shortening of the axis, so that it is separated from the posterior margin by extensions of the pleural lobes, constitutes one type of evolutionary trend, which in a few genera reaches extreme expression (Figs. 13-18, *5*; 13-23K, M).

Most simply and typically, furrows between axial segments of the pygidium are equal in number and match the inner ends of the interpleural grooves, but disparity of these structures is introduced during evolution of many genera. The grooves on the pleural lobes tend to disappear long before those on the axis fade, and at length the different nature and rates of change in these two portions of the pygidium make correlation of their elements impossible.

Spinosity

A tendency seen in several trilobite stocks—in fact, represented in all orders except the Agnostida—is the development of spines. Mostly, the spines are borne by the cephalon or pygidium, but they appear also on thoracic segments, notably by elongation of pleural extremities. Representative types of spinose trilobites are illustrated in Fig. 13-24, and others are shown in the stratigraphically arranged groups (Figs. 13-14 to 13-20).

The question naturally arises as to the significance of this tendency to become spinose. What purpose, if any, can the spines have served? Some of them are remarkably long, more than twice the greatest linear dimension of the skeleton,

---

(*Fig. 13-24 continued.*)

5, *Radiaspis radiata* (Goldfuss), Middle Devonian, Germany, ×2. The spines developed on cephalon, thorax, and pygidium all project laterally.

6, *Asteropyge punctata* (Steininger), Middle Devonian, Germany, ×1. This trilobite resembles

Dalmanites except for its spinose pygidium and posterior part of the thorax.

7, *Olenelloides armatus* Peach, Lower Cambrian, Scotland, ×5. An immature (meraspis) individual characterized by three pairs of cephalic spines.

exclusive of spines (Fig. 13-24, *3*). Did they mainly serve as means of protection from enemies, and if so, what were such enemies? Do they denote overspecialization of some sort, in which skeletal structures became exaggerated in bizarre manner?

Outgrowth of spinous projections from various parts of the body are found to distinguish the free-swimming, partly planktonic larvae of various crustaceans and several other marine animals. The outspread parts add to the resistance against sinking, and spines lying approximately in the plane of the body do not interfere materially with swimming. Hence, trilobites characterized by an unusual number or prominence of spines have been interpreted as evolutionary adaptations to a swimming and floating mode of life, rather than crawling along on the sea bottom. Very possibly this explanation applies accurately to spinose larvae, such as *Olenelloides* (Fig. 13-24, *7*), and perhaps also to adults like the illustrated species of *Ceratarges*, *Radiaspis*, and *Asteropyge* (Fig. 13-24, *1*, *5*, *6*), but it is rather implausible as applied to a blind trilobite like *Ampyx* (Fig. 13-24, *3*). Whatever the true significance of spinosity among these fossils, its existence establishes a distinctive sort of evolutionary change from the conservatively built average form of trilobites, among which progenitors of the spiny forms must belong. We may observe that no Mississippian or younger trilobite exhibits peculiarities of this sort. The last survivors of this class of arthropods are a conservative assemblage, advanced in evolution but not divergent from the mean.

General Conclusions

The survey of the nature of evolutionary trends which has been given omits many things, but it is sufficient to make clear (1) the diversity of skeletal elements which are affected by environment, mode of living, geographic isolation, competition for survival, and other factors causing morphologic change with lapse of time;

and (2) the main directions of modification of these skeletal elements. We must now set down additional observation that (3) evolution of different parts of the trilobite exoskeleton has not been the same in kind, in rate, or in geologic date. This is shown clearly by fossils belonging to the many recognized families distributed among the six orders.

Changes affecting some structures seem to have proceeded rapidly in one group but slowly in another, whereas evolution of other structures in these groups was the reverse—retarded in the first but accelerated in the second. For example, if we compare *Phacops* (Fig. 13-10, *2*) with *Isotelus* (Fig. 13-11), it is not difficult to support the conclusion that in the shape of glabella and the nature of the eyes *Phacops* is the more specialized (advanced), whereas in the type of facial suture, segmentation of the glabella, and characters of thorax and pygidium, *Isotelus* is the more specialized (advanced). *Phacops* is primarily a Devonian genus, and *Isotelus* is restricted to Ordovician rocks. To extend such a comparison so as to include the main characters of all known genera would emphasize the complexities of the evolutionary record, but in the light of present knowledge would not serve to discriminate the order of importance of the voluminous, highly varied data. Therefore, various questions of classification and phylogeny which depend on an accurate understanding of the evolution of the trilobites are not yet answerable unequivocally. On the other hand, it is easy to comprehend the fact that evolution produced an almost endless variety of trilobite characters, which in combinations that differ from others may appropriately be defined to distinguish genera.

GEOLOGICAL DISTRIBUTION

The stratigraphic range of trilobite orders and of the class as a whole has been stated in the table showing classification and in the discussion of the different groups. Distribution of the orders, with an indication of the relative importance of each in different parts of Paleozoic time, is shown graphically in Fig. 13-25. Not yet

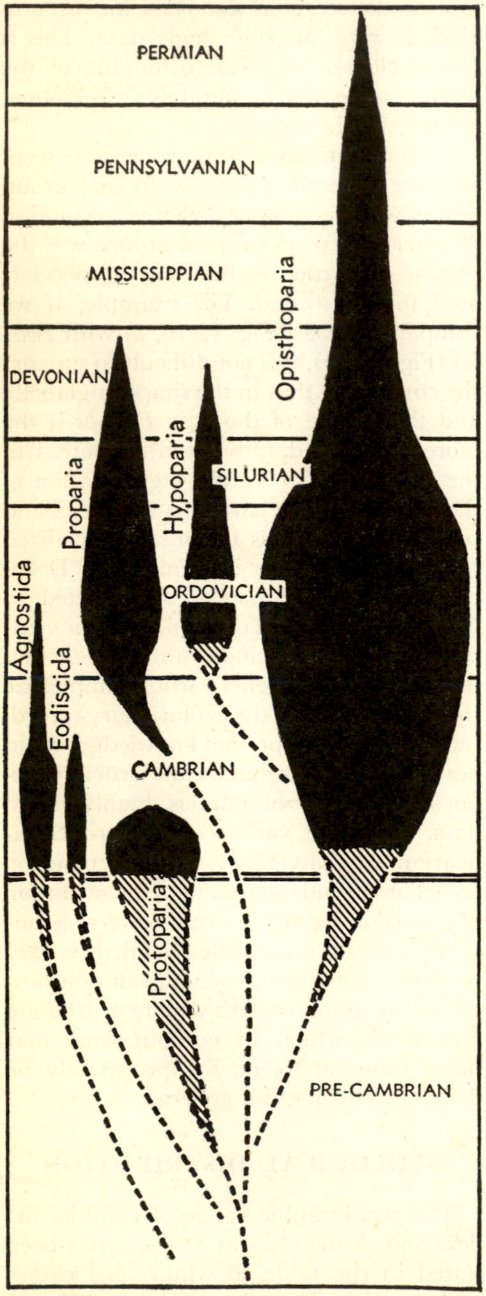

FIG. 13-25. **Geologic distribution of trilobite orders.**

considered are the features of geological distribution which have a bearing on phylogeny and definition of faunal realms, and which relate to the aspect of trilobite assemblages in successive geologic systems.

The first important observation on the geological occurrence of trilobites, which relates to their phylogeny, is the abruptness of their beginning in the fossil record. Protoparians are found nearer to the base of Cambrian rocks than other orders, but within the Lower Cambrian are also many agnostids, eodiscids, and numerous genera of opisthoparians; only proparians and hypoparians are lacking. The four known Lower Cambrian orders are structurally so distinct that none of them reasonably can be interpreted as derived from any other. If they are thus interpreted correctly as independent groups, their common ancestors must belong somewhere in the Pre-Cambrian, perhaps far back. No trilobites have been found in little altered late Pre-Cambrian rocks, such as the Belt and Grand Canyon series, which contain water-laid (presumably marine) limestones and other strata, or the Sinian series of China (which conformably underlies fossiliferous Lower Cambrian). Therefore, the sudden appearance of these fossils in Cambrian deposits seems best explained by assuming that this appearance essentially coincides with the acquisition of fossilizable hard parts by the organisms. If in Pre-Cambrian time they had no exoskeleton, their absence in deposits that otherwise should contain them is explained. A necessary corollary of this hypothesis is that brachiopods and gastropods, as well as trilobites, became capable of preservation at about the same time, for their remains also occur in Lower Cambrian deposits. Further, if it is a fact that the four orders of Lower Cambrian trilobites are mutually independent branches, as seems conclusive, then these several orders also acquired hard parts and became fossilizable at nearly the same time.

The Proparia and subdivisions of the Opisthoparia which first are seen in Mid-

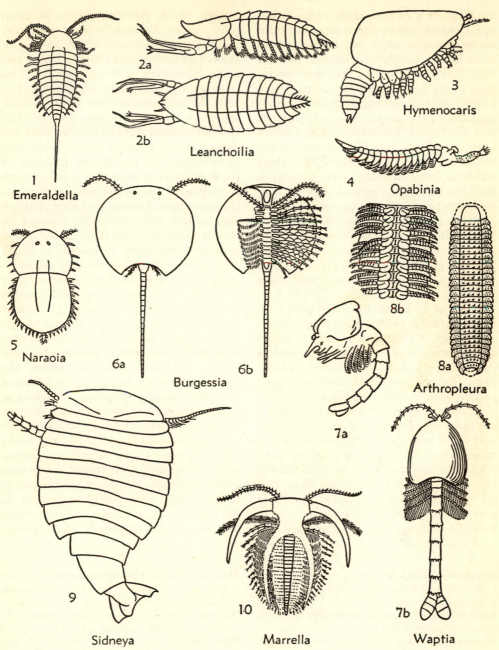

FIG. 13-26. **Minor trilobitomorphs.** All except 8, which is from rocks of Pennsylvanian age in Europe, come from the Middle Cambrian Burgess shale of British Columbia, and all the Cambrian genera (except *Hymenocaris*) are known from nowhere else.

Arthropleura Waterlot, Upper Carboniferous (Pennsylvanian), Europe. Long-bodied freshwater arthropods, primitively constructed of many similar segments; attained length of approximately 5 ft. *A. armata* Jordan (8*a*, dorsal side, ×0.03; 8*b*, part of ventral side, showing biramous limbs, ×0.05), Westphalian, Germany. (*Continued on next page.*)

dle Cambrian rocks, and the order Hypoparia, which makes appearance in Ordovician deposits, may be descendants of some known Cambrian trilobites, but it is logical also to postulate that their beginning in the fossil record represents simply a delay in their becoming capable of preservation. Like cephalopods, bryozoans, and some other invertebrate groups which failed to "learn how" to become fossils until Ordovician time, some or all of these trilobites may actually have existed as shell-less soft-bodied animals during a long period when more "progressive" protoparians and other trilobites were well housed in their various exoskeletons. In any case, trilobites are judged to be a polyphyletic group in which most, if not all, of the main divisions originated before Paleozoic time.

Study of the distribution of genera, both in stratigraphic placement and geographic occurrence, furnishes a means of recognizing deposits of equivalent age and of determining significant variations in Paleozoic shallow seas which were populated by trilobites. Restriction of groups of forms to certain regions constitutes evidence of the existence of barriers of some sort which prevented migration of invertebrates into other areas occupied by shallow seas. These features of distribution define faunal realms or provinces. They are illustrated by the occurrence of various Cambrian genera. For example, *Olenellus* is found in Lower Cambrian deposits of western and eastern North America except in maritime eastern Canada, where equivalent rocks contain *Callavia* (Fig. 13-14, *3*), various agnostids, and other trilobites unknown elsewhere in North America. This latter group, associated with *Holmia* (Fig. 13-6, *5b*) and several other forms, is well-represented in northwestern Europe. The Cordilleran and Appalachian seaways, occupied by the *Olenellus* fauna, are classed as belonging to the Pacific province of Early Cambrian time, whereas maritime Canada and Europe are parts of the contemporaneous Atlantic province. Eastern Asia at this time was partly covered by seas in which *Redlichia* (Fig. 13-14, *1*) and associated trilobites, unknown in North America but reaching into the Mediterranean region of southern Europe, were common. This assemblage distinguishes another province. Likewise, in Middle Cambrian time, *Paradoxides* (Figs. 13-6, *6b;* 13-14, *10, 13*) was fairly abundant in central and northwestern Europe, but unknown in North America outside the eastern Canadian region which belonged in the Atlantic province. Equivalent Pacific province trilobites, which did not invade the Atlantic realm, include *Ogygopsis, Olenoides* (Fig. 13-16, *1–4*), and many other genera. Upper Cambrian, Lower Ordovician, and each higher subdivision of the Paleozoic succession are characterized by somewhat similar variations in distribution of trilobites, which reflect paleogeographic conditions. Many genera

---

(Fig. 13-26 continued.)

Burgessia Walcott. Resembles the modern king crab. *B. bella* Walcott (*6a, b,* dorsal and ventral sides, ×5).

Emeraldella Walcott. A shrimplike form. *E. brocki* Walcott (*1*, ×1).

Hymenocaris Salter, Cambrian, North America and Europe. *H. perfecta* Walcott (*3*, side view, ×1).

Leanchoilia Walcott. A sowbug-type of arthropod provided with specialized anterior appendages. *L. superlata* Walcott (*2a, b,* side and dorsal views, ×0.5).

Marrella Walcott. Very bizarre horned carapace; two pairs of feathery antennae. *M. splendens* Walcott (*10*, dorsal side, ×2).

Naraoia Walcott. Two-segmented carapace. *N. compacta* Walcott (*5*, dorsal side, ×1).

Opabinia Walcott. *O. regalis* Walcott (*4*, side view, ×0.5).

Sidneya Walcott. *S. inexpectans* Walcott (*9*, dorsal side, ×0.5).

Waptia Walcott. Compact carapace in front, long thorax terminating in flattened swimming organ. *W. fieldensis* Walcott (*7a, b,* side and dorsal views, ×1).

are found well represented on two or more continents, among these being forms specially noted in this chapter (*Dalmanites* and *Phacops*, Fig. 13-10; *Isotelus*, Fig. 13-11; and *Cryptolithus*, Fig. 13-12), whereas others have been found only in one continent or part of a continent.

Stratigraphically arranged groups of representative trilobites are illustrated in Figs. 13-14 to 13-20.

MINOR TRILOBITOMORPHS

Mainly because of the discovery of a remarkable assemblage of marine fossils by C. D. Walcott in the Burgess shale, Middle Cambrian, of British Columbia, several kinds of arthropods have been made known. They throw light on the morphologic features of primitive divergent branches of the phylum and merit careful study by specialists concerned with problems of arthropod evolution. Much has been written about them. Partly because genera defined on the basis of these Burgess fossils are known nowhere else in the world, and partly because a discussion of their structure and relationships would require appreciable deviation from the objectives of this book, they are represented here simply by the illustration of selected forms (Fig. 13-26). Their affinity with trilobites is indicated by including them in the subphylum Trilobitomorpha.

The arthropod called *Arthropleura*, shown in Fig. 13-26, *8*, is not from the Cambrian but from rocks of Pennsylvanian age in Europe. It lived in fresh water or on land, for its remains are in nonmarine coal-bearing deposits of England, Belgium, France, Germany, and Czechoslovakia. The largest specimen, which is incomplete, indicates that this creature attained a length of 5 ft. In structure, it is more primitive than a trilobite.

REFERENCES

DEISS, C. (1939) *Cambrian stratigraphy and trilobites of northwestern Montana:* Geol. Soc. America Spec. Paper 18, pp. 1–135, pls. 1–18, figs. 1–7. Defines seven Middle Cambrian trilobite zones; gives descriptions and illustrations of characteristic fossils.

DELO, D. M. (1940) *Phacopid trilobites of North America:* Geol. Soc. America Spec. Paper 29, pp. 1–135, pls. 1–13. Structure and classification of a leading family of proparians, chiefly Devonian.

HOWELL, B. F., & LOCHMAN, C. (1939) *Succession of Late Cambrian faunas in the northern hemisphere:* Jour. Paleontology, vol. 13, pp. 115–122.

LALICKER, C. G. (1935) *Larval stages of Trilobites from the Middle Cambrian of Alabama:* Jour. Paleontology, vol. 9, no. 5, pp. 394–399, pl. 47.

LOCHMAN, C., & DUNCAN, D. (1944) *Early Upper Cambrian faunas of central Montana:* Geol. Soc. America Spec. Paper 54, pp. 1–181, pls. 1–19, fig. 1.

RASETTI, F. (1948) *Cephalic sutures in Loganopeltoides and the origin of the "hypoparian" trilobites:* Jour. Paleontology, vol. 22, pp. 25–29, pl. 7.

RAYMOND, P. E. (1913) *Trilobita:* in Eastman,

C. R., & Zittel, K. A., Macmillan & Co., Ltd., London, Textbook of palaeontology, vol. 1, pp. 692–729, figs. 1337–1412.

———— (1920) *The appendages, anatomy, and relationships of trilobites:* Connecticut Acad. Arts. and Sci., vol. 7, pp. 1–169, pls. 1–11, figs. 1–46. Gives excellent illustrations of appendages of *Triarthrus, Cryptolithus, Ceraurus,* and some other genera.

———— (1925) *Trilobites of lower Middle Ordovician of eastern North America:* Harvard Coll. Mus. Comp. Zoology Bull., vol. 67, pp. 1–180, pls. 1–10.

RESSER, C. E. (1938) *Cambrian system (restricted) of the southern Appalachians:* Geol. Soc. America Spec. Paper 15, pp. 1–140, pls. 1–16. Contains illustrations and brief descriptions of many Lower, Middle, and Upper Cambrian trilobites.

———— & HOWELL, B. F. (1938) *Lower Cambrian Olenellus zone of the Appalachians:* Geol. Soc. America Bull. vol. 49, pp. 195–248, pls. 1–13, fig. 1.

RICHTER, R., & RICHTER, E. (1926) *Die Trilobiten des Oberdevons, Beiträge zur Kenntnis devonischer Trilobiten:* Preuss. geol. Landesanstalt Abh., n.s., vol. 99, pp. 1–314, pls. 1–12 (German). Describes excellently preserved fossils of western Germany.

SHIMER, H. W., & SHROCK, R. R. (1944) *Index fossils of North America:* Wiley, New York, pp. i–ix, 1–837, pls. 1–303 (trilobites, pp. 600–655, pls. 251–276). Illustrates many American species; genera alphabetically arranged.

STØRMER, L. (1930) *Scandinavian Trinucleidae:* Skr. norske vidensk.-Akad. i Oslo, Mat.-Naturv. Kl. 1930, no. 4, pp. 1–111, pls. 1–14, figs. 1–47. Valuable morphologic study on an important group of hypoparians.

——— (1939) *Studies on trilobite morphology, Part 1, The thoracic appendages and their phylogenetic significance:* Norsk geol. Tidskr., vol. 19, pp. 143–273, pls. 1–12, figs. 1–35. Summarizes previous work and reports new observations, with extensive comparison to other arthropods.

——— (1942) *Studies on trilobite morphology, Part 2, The larval development, the segmentation, and the sutures, and their bearing on trilobite classification:* Same, vol. 21, pp. 49–164, pls. 1–2, figs. 1–19. Larval growth stages indicate that position of facial suture in different orders reflects accelerated or retarded dorsal development of the preantennal segment.

——— (1944) *On the relationships and phylogeny of fossil and recent Arachnomorpha:* Skr. norske vidensk.-Akad. i Oslo, Mat.-Naturv. Kl. 1944, no. 5, pp. 1–158, figs. 1–29. Classes trilobites, merostomes, and arachnids together as divisions of an independent phylum, from which crustaceans, insects, and other arthropods are excluded.

——— (1949) *Sous-embranchement des Trilobitomorphes:* in Grassé, P., Traité de zoologie, Masson et Cie., Paris, vol. 6 (pp. 1–979, 3 pls., 871 figs.), pp. 159–216, figs. 1–39 (French). A well-organized, compact description of the trilobites and related Paleozoic arthropods.

SWINNERTON, H. H. (1915) *Suggestions for a revised classification of trilobites:* Geol. Mag., n.s., dec. 6, vol. 2, pp. 487–496, 538–545. Introduces concepts for definition of Protoparia.

——— (1947) *Outlines of paleontology:* E. Arnold & Co., London, 3d ed., pp. 1–393, figs. 1–368 (trilobites, pp. 223–248, figs. 176–198). A good brief description, with emphasis on evolutionary trends.

WALCOTT, C. D. (1908) *Cambrian trilobites:* Smithsonian Misc. Coll., vol. 53, no. 2, pp. 13–52, pls. 1–6. New Middle Cambrian forms from the Cordilleran region.

——— (1910) *Olenellus and other genera of the Mesonacidae:* Same, vol. 53, no. 6, pp. 231–422, pls. 23–44. An important source of information on protoparians.

——— (1910a) *Abrupt appearance of the Cambrian fauna on the North American continent:* Same, vol. 57, no. 1, pp. 1–16, pl. 1, fig. 1. Explains lack of Pre-Cambrian trilobites by assuming that rocks which otherwise might contain them are nonmarine.

——— (1911) *Middle Cambrian Merostomata:* Same, vol. 57, no. 2, pp. 17–40, pls. 2–7. Describes Burgess shale fossils.

——— (1912) *Middle Cambrian Branchiopoda, Malacostraca, Trilobita, and Merostomata:* Same, vol. 57, no. 5, pp. 145–228, pls. 24–34, figs. 8–10. Describes Burgess shale fossils.

——— (1913) *Dikelocephalus and other genera of the Dikelocephalinae:* Same, no. 13, pp. 345–412, pls. 60–70, figs. 13–20.

——— (1916) *Cambrian trilobites:* Same, vol. 64, no. 3, pp. 157–258, pls. 24–38.

——— (1916a) *Cambrian trilobites:* Same, vol. 64, no. 5, pp. 303–456, pls. 45–67.

——— (1918) *Appendages of trilobites:* Same, vol. 67, no. 4, pp. 116–216, pls. 14–42, figs. 1–3.

WARBURG, E. (1925) *The trilobites of the Leptaena limestone in Dalarna:* Geol. Inst. Upsala Bull., vol. 17, pp. 1–446, pls. 1–11, figs. 1–14. Contains extensive discussion of zoological features and describes a large mid-Ordovician trilobite assemblage from Sweden.

WESTERGARD, A. H. (1946) *Agnostidea of the Middle Cambrian of Sweden:* Sveriges geol. Undersökning, ser. C, no. 477, pp. 1–140, pls. 1–16, figs. 1–2. A highly varied assemblage of agnostids.

WHITTINGTON, H. B. (1941) *The Trinucleidae, with special reference to North American genera and species:* Jour. Paleontology, vol. 15, pp. 21–41, pls. 5–6.

——— (1941a) *Silicified Trenton trilobites:* Same, vol. 15, pp. 492–522, pls. 72–75.

——— (1950) *A monograph of the British trilobites of the family Harpidae:* Palaeont. Soc. Mon. 1949, pp. i–ii, 1–55, pls. 1–7, figs. 1–16. A careful modern study of a striking group of hypoparians.

CHAPTER 14

Ostracodes and Other Crustaceans

The Crustacea (*crusta*, crust) are a large group of essentially aquatic arthropods which include the lobsters, crabs, crayfish, shrimps, copepods, prawns, water fleas, and many other less familiar forms (Fig. 12-1). Wood lice and land crabs have become adapted to a land habitat. The main divisions which are recognized in zoological classification, briefly defined in Chap. 12, include the Branchiopoda, Ostracoda, Copepoda, Branchiura, Cirripedia, and Malacostraca. The living Crustacea differ from other groups in having two pairs of antennae in front of the mouth and three pairs of post-oral appendages used as jaws, in breathing by gills or the general body surface, and in being restricted generally to an aquatic mode of life.

CHARACTERS OF CRUSTACEANS

Morphological Features

Crustaceans commonly are covered by an exoskeleton of chitin which may be impregnated with calcium carbonate or calcium phosphate. It consists of about 20 segments which generally are grouped into distinct units—head or cephalon, thorax, and abdomen or pygidium. In some forms the head and thorax are combined as a cephalothorax. During the period of growth, the exoskeleton is shed at frequent intervals, allowing the soft parts of the animal to increase in size. This process of molting is termed **ecdysis.**

The appendages of crustaceans are differentiated for walking, swimming, digging, capturing prey, feeding, breathing, mating, and receiving sensory stimuli (Fig. 12-5). The appendages are typically biramous, consisting of a basal joint (pro-topodite), which bears two branches, an inner, called **endopodite,** and an outer, called **exopodite.** The endopodite, which is jointed, is used for walking. The exopodite may be unsegmented or composed of a varying number of joints. It is used for respiration and swimming. A pair of unbranched many-jointed appendages, termed antennules, is joined to the front of the head.

Growth Stages

The sexes are separate among most Crustaceans, but some parasitic isopods and sessile cirripeds (barnacles) are hermaphrodites. The eggs usually are carried by the female, on the appendages (most decapods), between the valves (some ostracodes and branchiopods), or in a brood pouch (malacostracans).

The crustacean typically hatches from the egg in a form different from the adult, and passes through a number of larval stages. Among species which pass through a complete series of larval stages, the egg hatches as a **nauplius** larva (Fig. 12-6). This is a tiny unsegmented growth form, having three pairs of appendages and a single median eye. The eggs of some crustaceans hatch into a more highly organized **metanauplius** larva which possesses more somites than the nauplius. Accordingly, crustaceans of this type are judged to be more advanced than those which retain a nauplius stage that changes into a metanauplius by molting. Most of the larval forms are free-swimming and live near the surface of the water. A series of molts occurs during the growing period when the shell is shed, and the soft parts grow rapidly before another exoskeleton is developed.

521

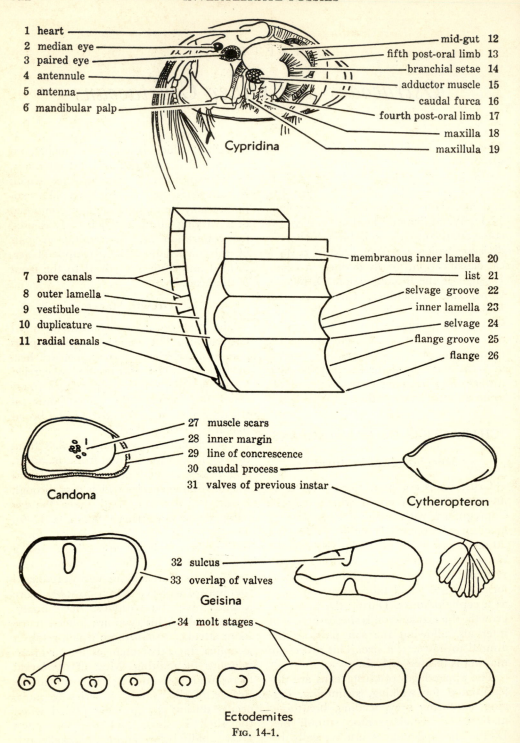

1 heart
2 median eye
3 paired eye
4 antennule
5 antenna
6 mandibular palp

mid-gut 12
fifth post-oral limb 13
branchial setae 14
adductor muscle 15
caudal furca 16
fourth post-oral limb 17
maxilla 18
maxillula 19

Cypridina

membranous inner lamella 20

7 pore canals
8 outer lamella
9 vestibule
10 duplicature
11 radial canals

list 21
selvage groove 22
inner lamella 23
selvage 24
flange groove 25
flange 26

27 muscle scars
28 inner margin
29 line of concrescence
30 caudal process
31 valves of previous instar

Candona

Cytheropteron

32 sulcus
33 overlap of valves

Geisina

34 molt stages

Ectodemites

Fig. 14-1.

Physiological Features

Digestive System.   In most crustaceans the alimentary canal is straight, except at the anterior end where it curves downward to the mouth (Fig. 12-7). It consists of the fore-gut, mid-gut, and hind-gut. The mid-gut is an absorptive and a digestive area, which is enlarged in many forms by tubular extensions that secrete digestive juices and aid in absorption. The fore- and hind-guts are lined by an inturning of the chitinous exoskeleton. The fore-gut is an enlarged area in some forms, such as malacostracans, comprising a sort of stomach where food is prepared for absorption.

Circulatory System. The circulatory system is mainly of the type known as lacunar, in which the blood flows in spaces without distinct walls. The heart, which is small, connects with main arteries that carry the blood to different parts of the body. Some small crustaceans have no circulatory system, the blood being shifted about by movements of the body.

Respiratory System. Respiration of most crustaceans is by means of gills situated on exopodites of the appendages or formed by the lining of the carapace. The gills are flattened, thin-walled, and penetrated by a network of blood vessels. Among many smaller crustaceans, such

---

FIG. 14-1. **Morphology of the Ostracoda and molt stages.** Illustrated features are defined in the alphabetically arranged list, cross-indexed to the figure by numbers.

adductor muscle (15). Principal muscle for opening and closing shell.

antenna (5). Appendage used for swimming or creeping.

antennule (4). Movable, segmented appendage for sensation and swimming.

branchial setae (14). Slender bristle-like breathing organ.

caudal furca (16). Most posterior abdominal segment.

caudal process (30). Posterior pointed projection of shell.

duplicature (10). Calcareous peripheral portion of inner lamella.

fifth post-oral limb (13). Appendage used for cleaning and clasping.

flange (26). Outer margin of shell.

flange groove (25). Furrow between selvage and flange.

fourth post-oral limb (17). Paired appendage used for cleaning and clasping.

heart (1). Organ for circulation of blood.

inner lamella (23). Interior calcareous and membranous layers of shell.

inner margin (28). Limit of duplicature toward shell interior.

line of concrescence (29). Inner limit of junction between duplicature and outer lamella.

list (21). Inward projection of duplicature.

mandibular palp (6). Mouth appendage having a biting or gustatory function.

maxilla (18). Paired appendage used for locomotion.

maxillula (19). Paired appendage used for locomotion.

median eye (2). Single simple eye of nauplius larval stage.

membranous inner lamella (20). Very thin chitinous continuation of inner lamella.

mid-gut (12). Stomach and digestive area.

molt stages (34). Steps in development of an ostracode between any two successive sheddings of its shell.

muscle scars (27). Impressions made by muscles on interior of shell.

outer lamella (8). External calcareous layer constituting most of shell.

overlap of valves (33). Projection of margin of one valve over edge of the other.

paired eye (3). One of the two compound eyes.

pore canals (7). Minute perforations extending through outer lamella approximately at right angles to its surface.

radial canals (11). Passageways through shell extending from line of concrescence to outer margin.

selvage (24). Inward projection of duplicature.

selvage groove (22). Furrow between selvage and list.

sulcus (32). Trough or groove indenting outer surface of shell.

valves of previous instar (31). Valves of previous molt still attached to valves formed later.

vestibule (9). Space between outer lamella and duplicature.

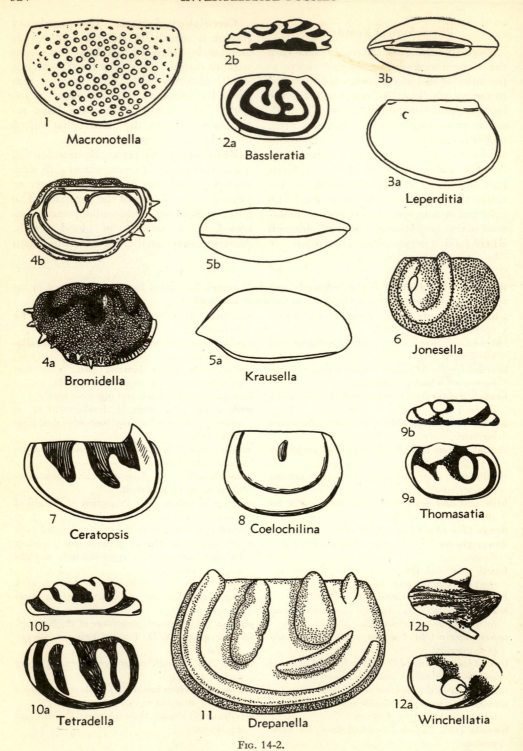

1 Macronotella

2b

2a Bassleratia

3b

3a Leperditia

4b

4a Bromidella

5b

5a Krausella

6 Jonesella

7 Ceratopsis

8 Coelochilina

9b

9a Thomasatia

10b

10a Tetradella

11 Drepanella

12b

12a Winchellatia

Fig. 14-2.

as the copepods, respiration is carried on through the general surface of the body. Enlarged branchial chambers of some land crustaceans are used as lungs. The wood lice have branching tubes like the tracheae of insects and spiders, which are formed from the invaginated integument of their abdominal segments.

Nervous System. The typical nervous system of crustaceans contains a dorsal "brain," united by esophageal connectives to a double ventral nerve cord, which has a pair of ganglia in each somite.

Sense Organs. The eyes of crustaceans are of two types: compound eyes, of which a pair usually is present, and a median eye. The median eye characterizes the nauplius larval stage, and in some groups, such as the copepods, it is the only organ of vision. It persists in most adults along with the paired eyes, but may become vestigial or disappear entirely. It consists of three cups of pigmented matter, each filled with retinal cells connected at their outer ends with nerve fibers. The compound eyes consist of many visual elements, each terminating in a light-condensing apparatus covered by cornea. A single compound eye may have hundreds or thousands of such units.

Organs sensitive to touch include the antennae, antennules, and numerous setae, bristles, and pits, which are variously modified for special functions, one type of sensory setae being associated with the sense of smell.

Certain marine crustaceans of different groups possess the ability of emitting light through secretions of various organs, such as the dermal glands of the ostracodes and the excretory organs of some decapods.

FIG. 14-2. **Representative Ordovician ostracodes.** Ordovician Ostracoda are relatively large in size. Some genera have smooth shells, whereas others have prominent nodes, sulci, and pits. All figures are ×30.

Bassleratia Kay, Ordovician. Two marginal ridges enclose other ridges and nodes. *B. typa* Kay (2a, b, right valve and dorsal views).

Bromidella Harris, Ordovician. The subovate carapace has a straight hinge line, curved dorsal ridge with deep sulcus at its base, marginal rim or "brood pouch," and spinose surface. *B. reticulata* Harris (4a, b, right valve internal and external views), Oklahoma.

Ceratopsis Ulrich, Ordovician–Silurian. Two vertical ridges and a prominent dorsal hornlike process on the carapace. *C. quadrifida* (Jones) (7, right valve), Ordovician, Iowa.

Coelochilina Ulrich & Bassler, Ordovician. The subquadrate carapace has a wide, convex frill, straight hinge, and small median sulcus. *C. aequalis* (Ulrich) (8, right valve), Kentucky.

Drepanella Ulrich, Ordovician–Silurian. The suboblong valves have a sickle-shaped marginal ridge and two or more usually isolated nodes. *D. crassinoda* Ulrich (11, left valve), Ordovician, Kentucky.

Jonesella Ulrich, Ordovician. An anterior U-shaped ridge occurs on the small subovate carapace. *J. crepidiformis* Ulrich (6, left valve), Ohio.

Krausella Ulrich, Ordovician–Silurian. Carapace elongate subovate; a prominent posterior spine on the smaller right valve. *K. arcuata* Ulrich (5a, b, right valve and dorsal views), Ordovician, Oklahoma.

Leperditia Rouault, Ordovician–Carboniferous. Oblong large carapace generally more than 6 mm. long; hinge line short. *L. fabulites* Conrad (3a, b, right valve and dorsal views), Ordovician, Minnesota.

Macronotella Ulrich, Ordovician–Devonian. The equivalve semiovate carapace has a straight hinge line, and surface marked by rounded or reticular pits and a subcentral smooth spot. *M. scofieldi* Ulrich (1, lateral view), Ordovician, Minnesota.

Tetradella Ulrich, Ordovician. Four or fewer curved vertical ridges are ventrally united. *T. ellipsilira* Kay (10a, b, right valve and dorsal views), Iowa.

Thomasatia Kay, Ordovician. Carapace marked by a marginal ridge, an elevated ridge within the borders, and a distinct node just anterior to the middle. *T. falcicosta* Kay (9a, b, right valve and dorsal views), Ontario.

Winchellatia Kay, Ordovician. The oval carapace has a straight hinge, sulcus, rounded anterior lobe, and anterior spine. *W. longispina* Kay (12a, b, left valve and ventral views), Iowa.

Crustaceans as Food

Crustaceans play a very important part in the economy of nature. Most of them are scavengers and swarm in the shallower waters of the sea, feeding on all kinds of vegetable and animal matter. The tiny pelagic copepods, which are so abundant as to color the surface of the sea at times, are one of the greatest sources of food in the ocean. They are the main link in the food chain between the diatoms upon which they feed and the larger animals.

OSTRACODES

Ostracodes are minute bivalved crustaceans which live in both fresh and marine waters (Fig. 14-1). The enclosing shell commonly ranges in length from 0.5 to 4 mm., but some forms reach a length of more than 20 mm. Ostracodes are common in the present seas, and excepting copepods, they are the most abundant type of crustaceans. They were very abundant also during past time, as shown by the profusion of their shells in many formations. They are useful in stratigraphic work and for correlating the formations in wells drilled for petroleum, especially in non-marine beds where Foraminifera and other microfossils are absent. They are distinguished in having the body and appendages completely enclosed in a bivalve shell,

---

Fig. 14-3. **Representative Ordovician ostracodes.** All figures are ×30.

Bellornatia Kay, Ordovician. Carapace subquadrate, marked by nodes and raised ridges. *B. tricollis* Kay (14, left valve), Iowa.

Bollia Jones & Holl, Ordovician–Devonian. A prominent horseshoe-shaped central ridge generally is accompanied by a marginal ridge. *B. subaequata* Ulrich (15, left valve), Ordovician, Iowa.

Ctenobolbina Ulrich, Ordovician–Devonian. The subquadrate carapace has three bulbous lobes separated by two curved sulci. *C. ciliata* (Emmons) (1, right valve), Ordovician, Ohio.

Dicranella Ulrich, Ordovician. A distinct sulcus is bordered on each side by dorsally spinose lobes, and a broad frill occurs at the margin. *D. bicornis* Ulrich (6a, b, left valve and ventral views), Minnesota.

Eukloedenella Ulrich & Bassler, Ordovician–Devonian. The evenly convex carapace bears a median sulcus. *E. richmondensis* Spivey (4, right valve), Ordovician, Iowa.

Euprimitia Ulrich & Bassler, Ordovician–Silurian. A distinct sulcus is located in front of a rounded node; surface pitted. *E. labiosa* (Ulrich) (7, left valve, female), Ordovician, Iowa.

Eurychilina Ulrich, Ordovician. Characterized by postero-dorsal sulcus, broad frill, and in some species a coarsely reticulate or pitted surface. *E. reticulata parvifrons* Kay (13a, b, left valve and ventral views), Wisconsin.

Isochilina Jones, Ordovician–Devonian. Valves subequal, but beveled edge of the right valve overlaps a sloping area of left valve. *I. bulbosa* Harris (3, right valve), Ordovician, Oklahoma.

Leperditella Ulrich, Ordovician–Devonian. Left valve a little larger than the right; commonly a broad shallow depression occurs in the dorsal half. *L. rex* (Coryell & Schenck) (10, left valve), Ordovician, Kentucky.

Milleratia Swartz, Ordovician. Carapace subquadrate bearing a single sulcus and strong asymmetrical dorsal umbos. *M. cincinnatiensis* (Miller) (11, right valve), Iowa.

Primitia Jones & Holl, Ordovician–Permian. The subovate carapace is distinguished by a prominent dorsal sulcus and in some species by a small ventral frill. *P. tumidula* Ulrich (8, left valve), Ordovician, Minnesota.

Primitiella Ulrich, Ordovician–Devonian. The equal valves each bear a broad sulcus and have a finely pitted surface. *P. constricta* Ulrich (5, right valve of female), Ordovician, Minnesota.

Raymondatia Kay, Ordovician. The rectangular-shaped carapace has a marginal ridge and two ridges between depressed portions. *R. goniglypta* Kay (2a, b, left valve and dorsal views), Iowa.

Saccelatia Kay, Ordovician. The small carapace has a straight hinge and ventral swellings. *S. arcuamuralis* Kay (9a, b, left valve and ventral views), Minnesota.

Schmidtella Ulrich, Ordovician–Silurian. The subovate carapace has a depressed area on the ventral slope and is inflated in the dorsal region. *S. crassimarginata* Ulrich (12a, b, left valve and ventral views), Ordovician, Wisconsin.

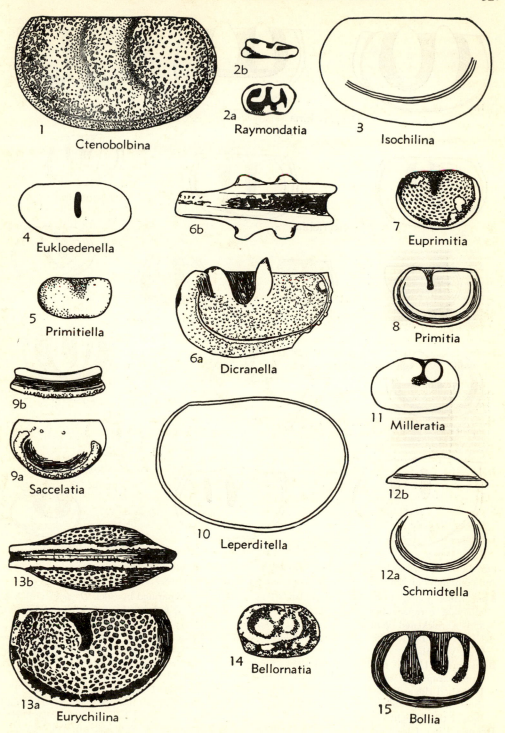

1 Ctenobolbina

2b

2a Raymondatia

3 Isochilina

4 Eukloedenella

6b

7 Euprimitia

5 Primitiella

6a Dicranella

8 Primitia

9b

9a Saccelatia

10 Leperditella

11 Milleratia

12b

12a Schmidtella

13b

13a Eurychilina

14 Bellornatia

15 Bollia

Fig. 14-3.

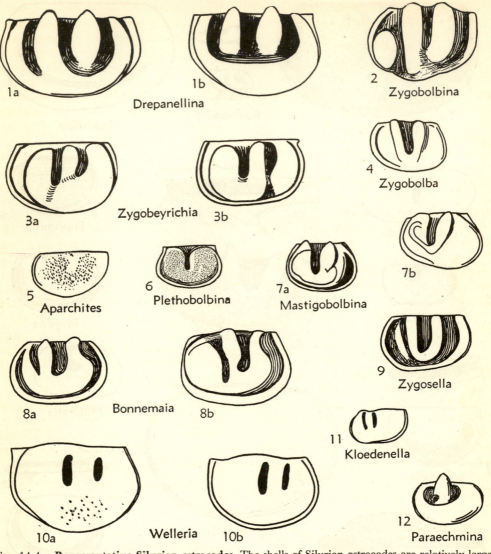

FIG. 14-4. **Representative Silurian ostracodes.** The shells of Silurian ostracodes are relatively large and commonly marked by prominent lobes and sulci. All figures are ×10.

Aparchites Jones, Ordovician–Devonian. The subovate carapace has a straight hinge line and a thickened ventral edge. *A. obliquatus* Ulrich & Bassler (5, left valve), Silurian, West Virginia.

Bonnemaia Ulrich & Bassler, Silurian. This large carapace has the posterior limb of the U-shaped ridge divided in its upper half by a short posterior sulcus. *B. transita* Ulrich & Bassler (8a, b, right valves), Tennessee.

Drepanellina Ulrich & Bassler, Silurian. The anterior lobe is well developed, and the central

lobes are separated by sulci. *D. clarki* Ulrich & Bassler (1a, b, left valves of male and female), Maryland.

Kloedenella Ulrich & Bassler, Silurian–Mississippian. The convex valves have a median and an anterior sulcus. *K. rectangularis* Ulrich & Bassler (11, left valve), Silurian, Maryland.

Mastigobolbina Ulrich & Bassler, Silurian. The trilobate carapace has posterior, anterior, and pyriform median lobes. *M. intermedia* Ulrich & Bassler (7a, b, right valves), Pennsylvania.

Paraechmina Ulrich & Bassler, Silurian–De-

(Continued on next page.)

and in having a smaller number of appendages than any other crustaceans.

Hard Parts

The shell of an ostracode consists of two valves, designated right and left. Generally they are unequal in size. The valves articulate along the dorsal edge, the **hinge margin,** by hinge teeth on one valve and corresponding sockets on the other, or by ridges that fit into grooves. The remainder of the circumference of the shell is the **contact margin.** The two valves may overlap at only the ventral margin, or at the ventral and dorsal margins, or all around, with either the right or left valve overlapping.

The shape of the shell is commonly ovate or reniform, but it may be pointed at one or both ends. The surface may be ornamented by puncta, reticulations, nodes, ridges, swellings, and grooves (sulci). The general surface of the shell is perforated by tiny pores at right angles to the surface, called **pore canals** (7; italic numbers in the text describing ostracodes refer to Fig. 14-1 unless indicated otherwise) from which a hair protrudes in living forms. Also, near the ventral margin are **radial canals** (11) which extend through the shell to its outer margin. The shell is opened and closed by subcentral adductor muscles (15), attachments of which on the interior of the shell are marked by **muscle scars** (27). Each valve has two parts: the calcareous wall composing the externally visible shell, called **outer lamella** (8), and a shelflike inwardly projecting calcareous wall joined to anterior, ventral, and posterior margins of the shell, called **inner lamella** (23). In living ostracodes, a thin chitinous layer covers both sides of the outer lamella. The calcareous peripheral portion of the inner lamella, next to the shell margin, is termed the **duplicature** (10). The free margin of the duplicature, called **flange** (26), is termed **monolamellar** if the intersection of the outer and inner lamellae is a simple turnover, and **bilamellar** if the lower portion of the duplicature is welded to the outer lamella. The **vestibule** (9) is the space between the outer lamella and the duplicature. The inner limit of the welded region of the duplicature, where surfaces of inner and outer lamellae meet, is the **line of concrescence** (29). The **inner margin** (28) is the inner limit of the calcareous inner lamella, but beyond it, a thin chitinous extension (**membranous inner lamella** (20)) may subdivide the shell interior still further. The **selvage** (24) is an inward projection of the duplicature which seals the valves when closed. The selvage forms the outer

---

(Fig. 14-4 continued.)

vonian. The subovate carapace has a prominent dorsal spine. *P. postica* Ulrich & Bassler (12, left valve), Silurian, Maryland.

Plethobolbina Ulrich & Bassler, Silurian. The strongly convex valves have a median sulcus. *P. ornata* Ulrich & Bassler (6, right valve), Pennsylvania.

Welleria Ulrich & Bassler, Silurian–Devonian. There are two sulci and an inflated dorsal area. *W. obliqua* Ulrich & Bassler (10a, b, right valves of male and female), Silurian, West Virginia.

Zygobeyrichia Ulrich & Bassler, Silurian–Devonian. Distinguished by partial or complete obsolescence of the anterior lobe. *Z. regina* Ulrich & Bassler (3a, b, right valves of female and male), Silurian, West Virginia.

Zygobolba Ulrich & Bassler, Ordovician–Devonian. Characterized by a U-shaped ridge with large median sulcus. *Z. decora* (Billings) (4, left valve), Silurian Virginia.

Zygobolbina Ulrich & Bassler, Silurian. The carapace has a U-shaped ridge and bilobed brood pouch. *Z. conradi* Ulrich & Bassler (2, right valve), Maryland.

Zygosella Ulrich & Bassler, Silurian. The carapace has a U-shaped ridge and a brood pouch. *E. mimica* Ulrich & Bassler (9, left valve), Virginia.

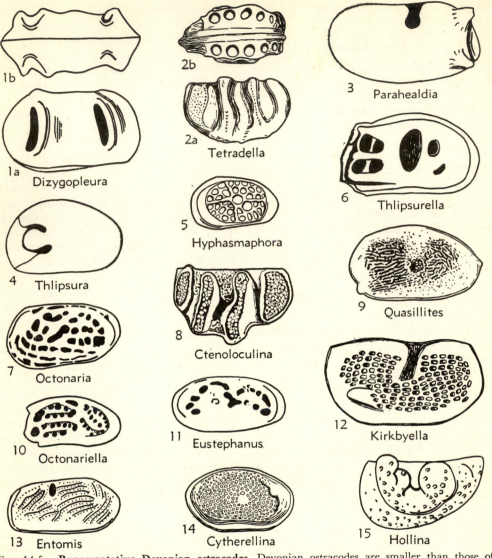

FIG. 14-5. **Representative Devonian ostracodes.** Devonian ostracodes are smaller than those of Ordovician and Silurian deposits. The shells are commonly marked by pits, reticulations, and spines. Lobes and sulci are less prominent. All figures are ×30.

Ctenoloculina Bassler, Devonian. The subquadrate valves have four long subvertical flat lobes and three intervening furrows. *C. acanthophora* Swartz & Oriel (8, left valve), New York.

Cytherellina Jones & Holl, Devonian. The reniform-shaped carapace is covered by funnel-shaped pits. *C. punctulifera* (Hall) (14, left valve), Michigan.

Dizygopleura Ulrich & Bassler, Silurian–Mississippian. Characterized by prominent lobes and long sulci. *D. recta* Roth (1a, b, left

valve and dorsal views), Devonian, Oklahoma.

Entomis Jones, Silurian–Permian. The subovate carapace has a submedian furrow near the ventral edge. *E. rugatulus* Van Pelt (13, left valve), Devonian, Michigan.

Eustephanus Swartz & Swain, Devonian. The elongate subelliptical carapace has pits which lie in curving furrows parallel to the dorsal margin. *E. catastephanes* Swartz & Swain (11, left valve), Pennsylvania.

Hollina Ulrich & Bassler, Devonian. The

(Continued on next page.)

margin in young specimens. Another inward projection frequently develops between the selvage and the inner margin, the **list** (*21*). A groove between the selvage and the flange of the shell is the **flange groove** (*25*), and a groove between the selvage and list is known as the **selvage groove** (*22*).

Orientation of Shell

The orientation of Mesozoic and Cenozoic ostracode shells is well established, because many genera have living representatives and comparisons can be made between fossil and living forms. In general, the pointed end is posterior, the anterior end is highest in side view, and the posterior end commonly is widest in dorsal view. The orientation of Paleozoic ostracodes has been the subject of much discussion and disagreement. It is believed that shell structures formed by direct connection with the body furnish the most reliable clues and, among these, muscle scars are the most significant. Muscle scars can be seen on many fossil shells by placing them in liquids of high refractive index.

Among Recent ostracodes, the central adductor muscle is attached at the center and in front of the center of the shell. Also in many genera the adductor-muscle scars are found to be situated on an internal ridge, which corresponds to a sulcus on the shell exterior. The so-called median sulcus, therefore, should be located in front of the center of the valves. When the shell is oriented in this manner, the lobes, nodes, and spines on the exterior are directed backward, as in living ostracodes. Since the thickest end of living Ostracoda is the posterior where reproductive organs are located, it is reasonable to surmise that the thickest end of most Paleozoic ostracodes also is in the posterior position.

Growth Stages

It is known that certain ostracode species pass through a series of molt stages (*34*), and observations show that ontogenetic development is similar for different species. In some species, there are nine between-molt stages (instars), including the adult. Well-marked differences distinguish shells of the young and old instars.

(Fig. 14-5 continued.)

carapace has a straight hinge, and the anterior lobe is divided into three or four nodes. *H. devoniana* Van Pelt (15, left valve), Michigan.

Hyphasmaphora Van Pelt, Devonian. A large, shallow central pit is surrounded by a coarse reticulated area. *H. textiligera* Van Pelt (5, left valve), Michigan.

Kirkbyella Coryell & Booth, Devonian–Pennsylvanian. Distinguished by a pronounced subcentral sulcus in the dorsal half and a reticulate surface. *K. verticalis* Coryell & Cuskley (12, right valve), Devonian, Oklahoma.

Octonaria Jones, Ordovician–Devonian. The left valve overlaps the right; raised ridges are separated by depressions. *O. loculosa* Ulrich (7, left valve), Devonian, Kentucky.

Octonariella Bassler, Devonian. Each valve bears two prominent anterior spines and pitlike depressions between ridges. *O. bifurcata* Bassler (10, left valve), Tennessee.

Parahealdia Coryell & Cuskley, Devonian. Small subovate carapace marked by a sulcus and two spines at the anterior end. *P. pecorella* Coryell & Cuskley (3, right valve), Oklahoma.

Quasillites Coryell & Malkin, Devonian. Surface marked by fine longitudinal ridges and grooves which bifurcate; there is a median spot. *Q. lobatus* Swartz & Oriel (9, right valve), New York.

Tetradella Ulrich, Ordovician–Devonian. Carapace subquadrate marked by four curved vertical ridges which are ventrally united. *T. cicatricosa* Warthin (2a, b, left valve and ventral views), Devonian, Michigan.

Thlipsura Jones & Holl, Silurian–Devonian. The depressed anterior area contains two furrows extending back on sides of a central high area. *T. furcoides* Bassler (4, left valve), Devonian, Tennessee.

Thlipsurella Swartz, Silurian–Devonian. Characterized by a submedian sulcus and generally two longitudinal pits in the anterior half. *T. putea* Coryell & Cuskley (6, left valve), Devonian, Oklahoma.

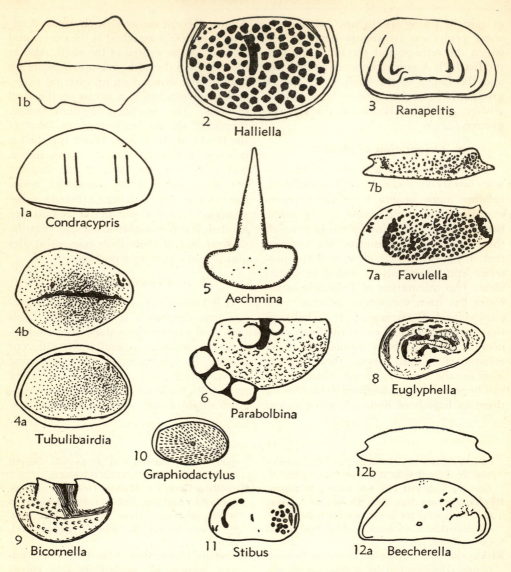

FIG. 14-6. **Representative Devonian ostracodes** (continued). Except as indicated otherwise, all figures are ×30.

Aechmina Jones & Holl, Ordovician–Pennsylvanian. The smooth convex shell has a single dorso-median spine or hornlike process. *A. cuspidata* Jones & Holl (5, right valve), Devonian, Maryland.

Beecherella Ulrich Devonian. The elongate boat-shaped carapace is triangular in cross section, and spines occur at each end. *B. bloomfieldensis* Swartz & Swain (12a, b, right valve and ventral views, ×20), Pennsylvania.

Bicornella Coryell & Cuskley, Devonian. The surface is reticulate and marked by a deep sulcus, on each side of which is a broad swelling that terminates in a spine. *B. tricornis* Coryell & Cuskley (9, right valve), Oklahoma.

Condracypris Roth, Devonian. Characterized by two transverse ridges on each valve. *C. binoda* Roth (1a, b, right valve and dorsal views), Oklahoma.

Euglyphella Warthin, Devonian. Each valve of

(*Continued on next page.*)

Nodes on the young shells are found to disappear entirely in adult shells, or conversely, surface ornamentation, including nodes, tends to increase as growth proceeds. Other changes affect the size and shape of the shell, hinge structures, muscle-scar pattern, and pore-canal systems. The valves of one instar are occasionally found attached to the valves of another instar (31).

Appendages

Ostracodes have fewer appendages than do any other crustaceans, usually only seven pairs. The **antennules** (4) may function as sensory organs, for digging or swimming, or as male clasping organs. The **antennae** (5) are mainly used for locomotion by creeping or swimming. The **mandibles** (6) serve for grasping food and creeping. The remaining limbs, **maxillae** (18), **maxillulae** (19), and thoracic appendages, are varied in structure and use, some being used for locomotion and some for such purposes as cleaning or clasping organs. The abdomen is devoid of appendages.

Physiological Features

Digestive System. The alimentary canal is divided into three parts: esophagus, mid-gut, and rectum. The esophagus of some genera contains chitinous teeth and ridges on the interior, forming a "gastric mill." The mid-gut is quite large.

Circulatory System. No definite blood vessels occur in Ostracoda, and a heart is present in only one group. The blood circulates through channels.

Respiratory System. Among most ostracodes, respiration is accomplished by the general surface of the body and limbs. The appendages keep a current of water circulating through the shell. Definite gills, which occur in a few groups, are lamellar appendages on the dorsal side of the body near the posterior end.

Nervous System. The ventral nerve chain is a single mass in some groups, but is divided into two or three masses in others. The supra-esophageal ganglion is largest in groups which have compound paired eyes.

Sense Organs. Only a few ostracodes have paired compound eyes, and in these the number of visual units varies from 4 to about 50. Most ostracodes have the nauplius single eye, and transparent "eye-spots" on the shell are common.

Reproduction

The sexes are separate, but parthenogenesis is common in many genera. In

---

(Fig. 14-6 continued.)

this subtriangular carapace bears C-shaped ridges and furrows. *E. numismoides* Swartz & Oriel (8, left valve), New York.

Favulella Swartz & Swain, Devonian. The punctate subovate carapace has two well-developed posteriorly directed spines and a submarginal ridge. *F. favulosa* Swartz & Swain (7a, b, left valve and ventral views), Pennsylvania.

Graphiodactylus Roth, Devonian–Pennsylvanian. The larger right valve completely overlaps the left; surface ornamented. *G. catenulatus* Van Pelt (10, left valve), Devonian, Michigan.

Halliella Ulrich, Ordovician–Mississippian. The broad carapace has a prominent sulcus and a very coarsely reticulate surface. *H. pulchra* Bassler (2, left valve), Tennessee.

Parabolbina Swartz, Devonian. The small straight-hinged carapace is marked by a sulcus and spines or frill near the ventral margin. *P. parvinoda* Swartz & Swain (6, left valve), Pennsylvania.

Ranapeltis Bassler, Devonian. Valves have two narrow ridges parallel to ventral margin and turning upward toward dorsal edge. *R. typicalis* Bassler (3, left valve), Tennessee.

Stibus Swartz & Swain, Devonian. The carapace has furrows or rows of pits near the antero-dorsal margins and may bear small pits near the posterior end. *S. kothornostibus* Swartz & Swain (11, left valve), Pennsylvania.

Tubulibairdia Swartz, Devonian. The inequivalve subovoid carapace has pores which in most species pierce the valves. *T. windomensis* Swartz & Oriel (4a, b, right valve and dorsal views), New York.

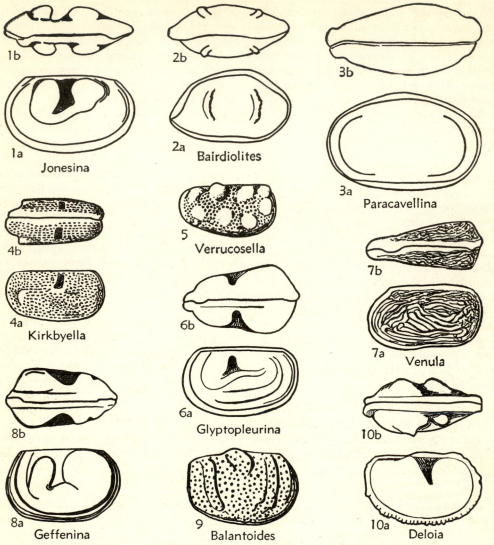

1b

1a Jonesina

2b

2a Bairdiolites

3b

3a Paracavellina

4b

4a Kirkbyella

5 Verrucosella

6b

6a Glyptopleurina

7b

7a Venula

8b

8a Geffenina

9 Balantoides

10b

10a Deloia

FIG. 14-7. **Representative Mississippian ostracodes.** The shells of Mississippian ostracodes are relatively small and commonly sulcate. All figures are ×40.

Bairdiolites Croneis & Gale, Mississippian. Carapace like that of *Bairdia* but has two curved ridges on each valve. *B. procerus* Cooper (2a, b, right valve and dorsal views), Illinois.

Balantoides Morey, Mississippian. The reticulate subquadrate carapace has three lobes which project to or above the hinge. *B. quadrilobatus* Morey (9, left valve), Wyoming.

Deloia Croneis & Thurman, Upper Mississippian. The subovate carapace has serrated flanges and an anterior sulcus. *D. sulcata* Croneis &

Funkhouser (10a, b, right valve and dorsal views), Illinois.

Geffenina Coryell & Sohn, Mississippian. The median sulcus is bordered by swellings. *G. marmerae* Coryell & Sohn (8a, b, right valve and dorsal views), West Virginia.

Glyptopleurina Coryell, Mississippian–Pennsylvanian. The subquadrate carapace has nodes, a median sinus, and inosculating costae. *G. vetula* Cooper (6a, b, left valve and dorsal views), Mississippian, Illinois.

(Continued on page 536.)

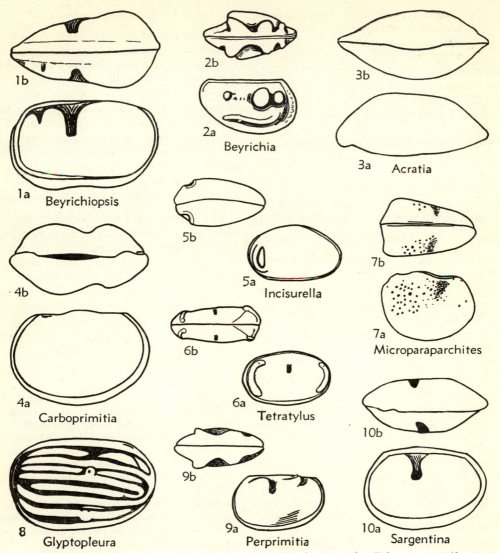

FIG. 14-8. **Representative Mississippian ostracodes** (continued). All figures are ×40.

Acratia Delo, Mississippian. The elongate carapace is lenslike in outline, and the posterior end is produced downward. *A. mucronata* Cooper (3a, b, left valve and ventral views), Illinois.

Beyrichia McCoy, Ordovician–Pennsylvanian. The hinge line is straight, and three rounded lobes occur in the dorsal position. *B. contracta* Cooper (2a, b, right valve and dorsal views), Mississippian, Illinois.

Beyrichiopsis Jones & Kirby, Devonian–Pennsylvanian. The subquadrate carapace has a median sulcus and a small postmedian lobe.

B. brynhildae Coryell & Johnson (1a, b, left valve and dorsal views), Mississippian, Illinois.

Carboprimitia Croneis & Funkhouser, Mississippian. The valves are deeply sulcate, and the right valve overlaps the left. *C. depressa* Croneis & Funkhouser (4a, b, left valve and dorsal views), Illinois.

Glyptopleura Girty, Mississippian–Permian. The subquadrate carapace has inosculating costae and a median pit. *G. alata* Croneis & Funkhouser (8, right valve), Mississippian, Illinois.

Incisurella Cooper, Mississippian. The thick

(Continued on next page.)

some, sexual reproduction occurs at fixed intervals, but in one species of *Cypris* which was kept in an aquarium for 30 years, no males appeared in the colony which was reproducing by parthenogenesis. The male is commonly smaller than the female.

The eggs are carried within the valves of the female above the body or are deposited free in the water or upon plants. When hatched, the larva is already enclosed in a bivalved shell and has three pairs of appendages, like a nauplius stage. The remaining appendages are added between successive molts, and the shell becomes larger with each succeeding molt.

Mode of Life

Ostracodes are common in both fresh and marine waters, generally living on the bottom or on stems of plants. A few groups are planktonic. Many kinds burrow into the mud. Most are omnivorous, feeding on diatoms, algae, tiny animals, and detrital material.

Classification

The Ostracoda are divided into families, subfamilies, genera, and species on the following criteria: (1) general shape, size, and convexity of valves, and position of greatest thickness; (2) position and amount of overlap of the valves; (3) presence of spines, frills, lobes, sulci, and pits on the valves; (4) hinge characteristics, such as reticulations; and (5) structures due to sex differences, as brood pouches.

Geological Distribution and Importance

The stratigraphic distribution of ostracodes is from Lower Ordovician to Recent, and in many formations they are very abundant. Their shells are most common in shales, marls, and limestones. Most genera and species have short vertical ranges and a wide geographic distribution, making them valuable index fossils. They are widely used in identifying formations in
(*Text continued on page 541.*)

(*Fig. 14-7 continued from page 534.*)

Jonesina Ulrich & Bassler, Mississippian–Permian. Carapace ovate, marked by prominent median sulcus and one or more adjacent lobes. *J. insculpta* Croneis & Funkhouser (1*a*, *b*, left valve and dorsal views), Illinois.

Kirkbyella Coryell & Booth, Devonian–Pennsylvanian. The small carapace has a straight hinge line, reticulate surface, and pronounced subcentral sulcus. *K. quadrata* Croneis & Gutke (4*a*, *b*, right valve and dorsal views), Mississippian, Illinois.

Paracavellina Cooper, Mississippian. The cy-therelloid-shaped carapace has a ridge adjacent and parallel to the margins. *P. elliptica* Cooper (3*a*, *b*, right valve and dorsal views), Illinois.

Venula Cooper, Mississippian. The subquadrate carapace has faint anastomosing costae and a small circular pit. *V. striata* (Croneis & Funkhouser) (7*a*, *b*, left valve and dorsal views), Illinois.

Verrucosella Croneis & Gale, Mississippian. The small carapace has a straight hinge line, prominent nodes, and a pitted surface. *V. golcondensis* Croneis & Gale (5, right valve), Illinois.

(*Fig. 14-8 continued.*)

carapace has a suboval shallow area in front of the posterior margin. *I. prima* Cooper (5*a*, *b*, right valve and dorsal views), Illinois.

Microparaparchites Croneis & Gale, Mississippian. This small thick subovate carapace has a globose swelling in the dorso-posterior quarter. *M. spinosus* Croneis & Gale (7*a*, *b*, right valve and dorsal views), Illinois.

Perprimitia Croneis & Gale, Upper Mississippian. The subquadrate carapace has deeply sulcate valves bearing anterior and posterior lobes. *P. turrita* Croneis & Gutke (9*a*, *b*, right valve and dorsal views), Illinois.

Sargentina Coryell & Johnson, Mississippian. Right valve overlaps the left all around, and a depression develops into a sulcus in the dorsal half. *S. allani* Coryell & Johnson (10*a*, *b*, left valve and dorsal views), Illinois.

Tetratylus Cooper, Mississippian. The ovate carapace has low ridges near the ends which have knoblike spines and shallow sulcus. *T. ellipticus* Cooper (6*a*, *b*, left valve and dorsal views), Illinois.

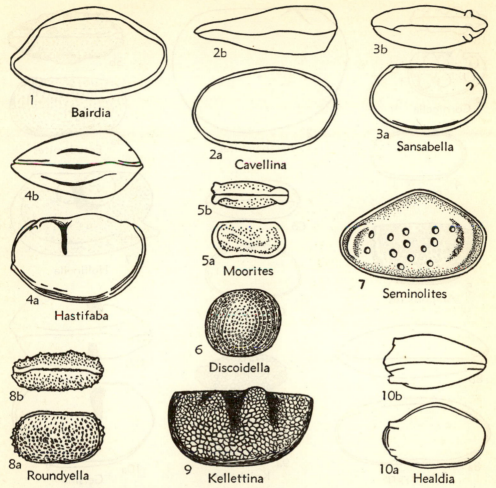

FIG. 14-9. **Representative Pennsylvanian and Permian ostracodes.** Many genera are common to both Pennsylvanian and Permian formations, and the majority are relatively long-ranging. The shells vary from smooth to ornate, and a few are sulcate. All figures are ×35.

Bairdia McCoy, Ordovician–Recent. The carapace is subtriangular or rhomboidal in outline and posteriorly pointed. The left valve overlaps the right throughout. *B. beedei* Ulrich & Bassler (1, right valve), Permian, Kansas.

Cavellina Coryell, Mississippian–Permian. The right valve overlaps the left on all margins; carapace wedge-shaped in dorsal view. *C. nebrascensis* (Geinitz) (2a, b, left valve and dorsal views), Pennsylvanian, Nebraska.

Discoidella Croneis & Gale, Mississippian–Pennsylvanian. This subcircular carapace is strongly convex and reticulated. *D. convexa* Scott & Borger (6, right valve), Pennsylvanian, Illinois.

Hastifaba Cooper, Pennsylvanian. The large, tumid carapace has a deep prominent sulcus forward of mid-length and a greatly inflated posterior. *H. robusta* Cooper (4a, b, left valve and dorsal views), Pennsylvanian, Illinois.

Healdia Roundy, Mississippian–Permian. The subtriangular carapace is marked by a distinct dorso-posterior slope and two backward pointing spines or an elevated vertical ridge near the posterior end. *H. colonyi* Coryell & Booth (10a, b, right valve and dorsal views), Pennsylvanian, Illinois.

Kellettina Swartz, Pennsylvanian–Permian. The straight-hinged reticulated carapace is marked by two large and prominent nodes on the dorsal

(*Continued on page 539.*)

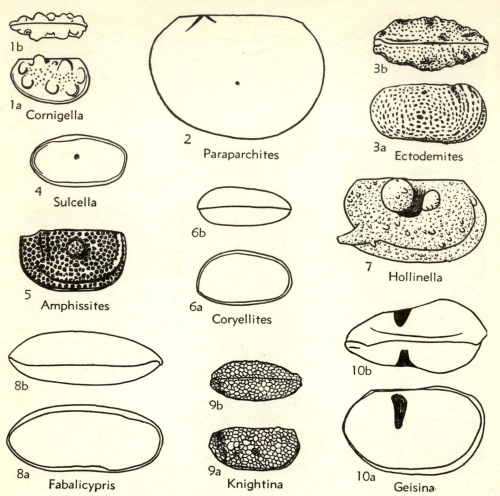

Fig. 14-10. Representative Pennsylvanian and Permian ostracodes (continued). All figures are ×35.

Amphissites Girty, Upper Mississippian–Permian. The subquadrate carapace has a single, centrally located node and two or more carinae near the ends; surface marked by reticulations. *A. centronotus* (Ulrich and Bassler) (5, left valve), Permian, Kansas.

Cornigella Warthin, Pennsylvanian. The minute carapace is characterized by about eight prominent spines on each valve. *C. tuberculospina* (Jones & Kirkby) (1a, b, left valve and dorsal views), Pennsylvanian, Illinois.

Coryellites Kellett, Pennsylvanian. The somewhat tumid carapace is ovate in side view; anterior end broadly rounded. *C. lowelli* Cooper (6a, b, right valve and dorsal view), Pennsylvanian, Illinois.

Ectodemites Cooper, Mississippian–Pennsylva-

nian. The carapace is marked by one or more false keels but without well-defined nodes; surface reticulated. *E. sullivanensis* (Payne) (3a, b, dorsal and right valve views), Pennsylvanian, Illinois.

Fabalicypris Cooper, Mississippian–Pennsylvanian. The tumid smooth carapace is ovate to lenticular in side view. *F. wetumkaensis* Cooper (8a, b, right valve and dorsal views), Pennsylvanian, Illinois.

Geisina Johnson, Pennsylvanian. The thick subquadrate carapace has a prominent and deep sulcus. *G. gallowayi* (Bradfield) (10a, b, left valve and dorsal views), Illinois.

Hollinella Coryell, Devonian–Permian. The rectangular carapace has large anterior and small posterior nodes with a broad, deep sulcus

(*Continued on next page.*)

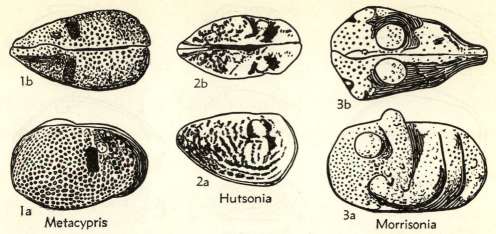

1b

2b

3b

3a

1a

Metacypris

2a

Hutsonia

3a

Morrisonia

FIG. 14-11. **Representative Jurassic ostracodes.** Jurassic ostracodes resemble those found in other Mesozoic and in Cenozoic rocks in shape and ornamentation, muscle-scar patterns, and complex hinge structures.

Hutsonia Swain, Jurassic. The subpyriform carapace is marked by swollen areas, reticulations, and longitudinal ridges; generally it is bisulcate. *H. vulgaris* Swain (2*a, b*, right valve and dorsal views of a male, ×50), Louisiana.

Metacypris Brady, Jurassic–Cretaceous. The subrhomboidal valves are unequally tumid at both extremities; surface pitted in rows. *M.*

pahasapensis (Roth) (1*a, b*, right valve and dorsal views, ×50), Jurassic, South Dakota.

Morrisonia C. C. Branson, Jurassic. The subquadrate carapace is divided by a median sulcus above a crescentric ridge and is marked by strong nodes. *M. wyomingensis* C. C. Branson (3*a, b*, left valve and dorsal views, ×40), Wyoming.

(Fig. 14-9 continued from page 537.)

half. *K. montosa* (Knight) (9, right valve), Pennsylvanian, Missouri.

Moorites Coryell & Billings, Mississippian–Pennsylvanian. The small carapace has a straight hinge line, looplike ridges, and a pitted surface. *M. minutus* (Warthin) (5*a, b*, right valve and dorsal views), Pennsylvanian, Oklahoma.

Roundyella Bradfield, Pennsylvanian–Permian. The ends are equally rounded on this suboblong carapace, and the surface is reticulated *R. simplicissima* (Knight) (8*a, b*, left valve and

dorsal views), Pennsylvanian, Illinois.

Sansabella Roundy, Mississippian–Pennsylvanian. Left valve overlaps smaller right around free margins, and each valve bears a small antero-dorsal spine. *S. laevis* (Warthin) (3*a, b*, right valve and dorsal views), Pennsylvanian, Illinois.

Seminolites Coryell, Pennsylvanian. The subtriangular carapace is marked by several large irregularly distributed circular pits and a curved ridge near each end. *S. truncatus* Coryell (7, right valve), Oklahoma.

(Fig. 14-10 continued.)

between them. *H. digitata* Kellett (7, left valve), Permian, Kansas.

Knightina Kellett, Pennsylvanian–Permian. The small elongate carapace has a prominent anterior shoulder; surface covered with coarse reticulations. *K. hextensis* (Harlton) (9*a, b*, left valve and dorsal views), Pennsylvanian, Illinois.

Paraparchites Ulrich & Bassler, Devonian–

Permian. The subovate carapace has a straight hinge and generally a small spine in the antero-dorsal area. *P. magnus* Kellett (2, left valve), Pennsylvanian, Kansas.

Sulcella Coryell & Sample, Mississippian–Permian. This small cytherelloid carapace is marked by an anterior vertical ridge and a submedian pit. *S. sulcata* Coryell & Sample (4, left valve), Pennsylvanian, Illinois.

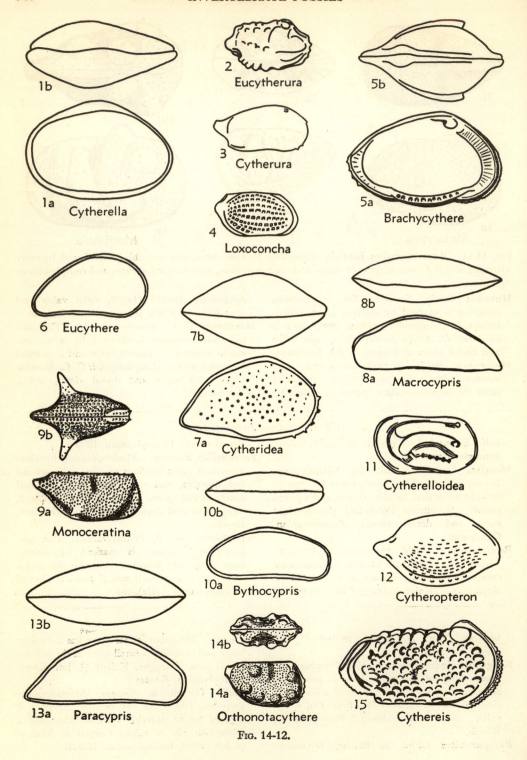

1b

2 Eucytherura

5b

1a Cytherella

3 Cytherura

5a Brachycythere

4 Loxoconcha

6 Eucythere

7b

8b

8a Macrocypris

9b

7a Cytheridea

11 Cytherelloidea

9a Monoceratina

10b

13b

10a Bythocypris

12 Cytheropteron

14b

13a Paracypris

14a Orthonotacythere

15 Cythereis

Fig. 14-12.

(Continued from page 536.)

samples taken from wells drilled for petroleum.

Twelve stratigraphically arranged groups of ostracodes, which include species belonging to many of the most important genera, illustrate the variety of form and surface markings exhibited by these fossils (Figs. 14-2 to 14-13).

OTHER CRUSTACEANS

From the viewpoint of a zoologist, it is anomalous to treat the highly varied host of crustaceans belonging to groups other than ostracodes as a sort of appendage of this class. Lobsters, shrimps, crabs, and numerous other kinds of crustaceans are

---

FIG. 14-12. Representative Cretaceous ostracodes. Cretaceous ostracodes have complex hinge structures, distinctive muscle scar patterns, and some genera are highly ornamented. All figures are ×50.

Brachycythere Alexander, Cretaceous–Recent. The subovate carapace has winglike processes at the ventro-lateral edge of valves. *B. taylorensis* Alexander (5a, b, right valve and dorsal views), Cretaceous, Texas.

Bythocypris Brady, Ordovician–Recent. The elongate compressed carapace is elliptical, and the left valve overlaps the right both dorsally and ventrally. *B. goodlandensis* Alexander (10a, b, right valve and dorsal view), Lower Cretaceous, Texas.

Cythereis Jones, Cretaceous–Recent. The subquadrate carapace has a rounded anterior and a compressed posterior end and an ornamented surface. *C. burlesonensis* Alexander (15, right valve), Lower Cretaceous, Texas.

Cytherella Jones, Paleozoic–Recent. The smooth compressed carapace is ovate in outline and is compressed at both ends. *C. ovata* (Roemer) (1a, b, left valve and dorsal view), Lower Cretaceous, Texas.

Cytherelloidea Alexander, Cretaceous–Tertiary. The subquadrangular carapace is ornamented by raised ridges parallel to the dorsal and ventral margins. *C. subgoodlandensis* Vanderpool (11, left valve), Lower Cretaceous, Oklahoma.

Cytheridea Bosquet, Cretaceous–Recent. The left valve of the subtriangular carapace overlaps the right; surface smooth or sculptured. *C. everetti* Berry (7a, b, right valve and dorsal view), Upper Cretaceous, Texas.

Cytheropteron Sars, Cretaceous–Tertiary. The dorsal margin is arched, the posterior end is pointed, and the ventro-lateral edges are winglike extensions. *C. trinitiensis* (Vanderpool) (12, right valve), Lower Cretaceous, Oklahoma.

Cytherura Sars, Cretaceous–Recent. The elongate subquadrate carapace has a compressed caudal process on the posterior end. *C. texana* Alexander (3, right valve), Upper Cretaceous, Texas.

Eucythere Brady, Cretaceous–Recent. The triangular shell is highest near the anterior end, and the posterior end is subacute. *E. brownstownensis* Alexander (6, right valve), Upper Cretaceous, Texas.

Eucytherura Mueller, Cretaceous–Recent. The small carapace is relatively thick and heavy, short, quadrate. *E. chelodon* (Marsson) (2, right valve), Upper Cretaceous, Texas.

Loxoconcha Sars, Cretaceous–Recent. The posterior end of the rhomboidal carapace bears a blunt caudal process above the center. *L. cretacea* Alexander (4, right valve), Upper Cretaceous, Texas.

Macrocypris Brady, Ordovician–Recent. The elongate, smooth carapace is narrow at both ends; the posterior end is pointed. *M. graysonensis* Alexander (8a, b, left valve and dorsal view), Lower Cretaceous, Texas.

Monoceratina Roth, Devonian–Tertiary. The subquadrate carapace has a subventral backward projecting spine or flap. *M. umbonata* (Williamson) (9a, b, right valve and dorsal view), Upper Cretaceous, Texas.

Orthonotacythere Alexander, Cretaceous–Tertiary. The quadrate carapace has a short caudal process near the postero-dorsal angle, and the surface is marked by several strong tubercles. *O. scrobiculata* Alexander (14a, b, right valve and dorsal view), Upper Cretaceous, Texas.

Paracypris Sars, Cretaceous–Recent. The elongate smooth carapace is pointed posteriorly and has a larger left valve. *P. siliqua* Jones & Hinde (13a, b, right valve and dorsal view), Lower Cretaceous, Texas.

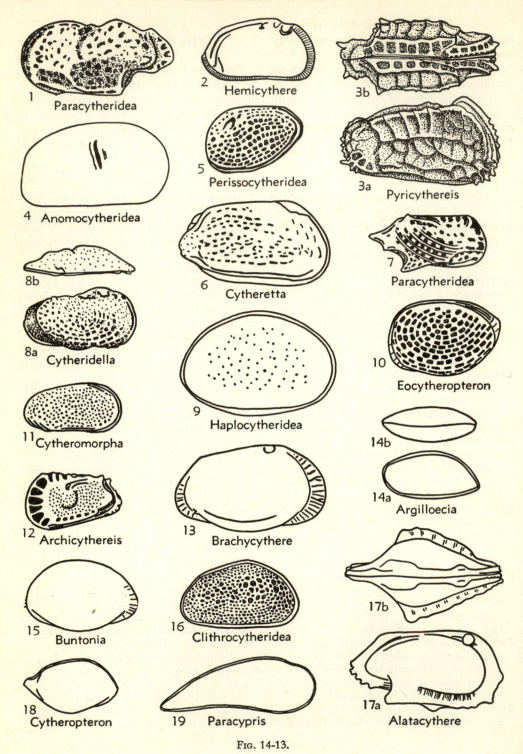

1 Paracytheridea

2 Hemicythere

3b

3a Pyricythereis

4 Anomocytheridea

5 Perissocytheridea

8b

6 Cytheretta

7 Paracytheridea

8a Cytheridella

9 Haplocytheridea

10 Eocytheropteron

11 Cytheromorpha

14b

14a Argilloecia

12 Archicythereis

13 Brachycythere

15 Buntonia

16 Clithrocytheridea

17b

18 Cytheropteron

19 Paracypris

17a Alatacythere

Fig. 14-13.

considerably more advanced in structural organization than ostracodes. Some of them are more ancient, and some are much more abundant at the present day. In the paleontological record, however, representatives of crustacean classes other than the Ostracoda, taken all together, have only a minor part. Accordingly, they are subordinated in our study.

Two of the six classes of crustaceans are

FIG. 14-13. Representative Cenozoic ostracodes. The shells of Cenozoic ostracodes are similar to those of living kinds in shape, ornamentation, complex hinge structures, and muscle scars. Many Tertiary genera persist to Recent time. All figures are ×50.

Alatacythere Murray & Hussey, Eocene–Oligocene. The carapace is triangular in cross section or end view, and has spines or bladelike plates developed along ventral margin. *A. westi* Stephenson (17*a, b*, right valve and dorsal view), Oligocene, Texas.

Anomocytheridea Stephenson, Miocene. The subovate carapace bears anterior and posterior ventral flanges on right valve. *A. inornata* Stephenson (4, right valve), Miocene, Louisiana.

Archicythereis Howe, Oligocene–Miocene. The subrhomboidal carapace is ornamented with spines or reticulate ridges. *A. sylverinica* Howe & Law (12, left valve), Oligocene, Mississippi.

Argilloecia Sars, Cretaceous–Tertiary. This narrow, thin carapace has a rounded anterior end and a pointed posterior end. *A. faba* Alexander (14*a, b*, left valve and dorsal view), Paleocene, Texas.

Brachycythere Alexander, Cretaceous–Recent. The subovate carapace has winglike processes at the ventro-lateral edge of valves. *B. russelli* Howe & Law (13, right valve), Oligocene, Louisiana.

Buntonia Howe, Eocene. The subtriangular carapace has a broadly rounded anterior end and an angulate posterior end. *B. shubutaensis* Howe & Chambers (15, left valve), Jacksonian, Mississippi.

Clithrocytheridea Stephenson, Eocene–Miocene. The carapace is subpyriform to subquadrate in outline, and the surface is smooth, pitted, or variously ornamented. *C. wechesensis* Stephenson (16, right valve), Eocene, Texas.

Cytheretta Muller, Eocene–Recent. The elongate carapace is faintly reticulate to smooth. *C. karlana* Howe & Pyeatt (6, right valve), Miocene, Florida.

Cytheridella Daday, Tertiary–Recent. The subquadrate carapace has a reticulated surface. *C. chambersi* Howe (8*a, b*, right valve and dorsal view), Miocene, Florida.

Cytheromorpha Hirschmann, Eocene–Recent. The elongate ovate carapace has a pitted surface in many species. *C. rosefieldensis* Howe & Law (11, right valve), Oligocene, Louisiana.

Cytheropteron Sars, Cretaceous–Tertiary. The dorsal margin is arched, the posterior end is pointed, and the ventro-lateral edges are winglike extensions. *C. midwayensis* Alexander (18, right valve), Paleocene.

Eocytheropteron Alexander, Cretaceous–Tertiary. The inflated shell has a bluntly pointed posterior end and concentrically arranged reticulations. *E. fiski* Howe & Law (10, left valve), Oligocene, Louisiana.

Haplocytheridea Stephenson, Cretaceous–Recent. The ovate carapace is similar to *Cytheridea* except for the hinge structures. *H. texana* Stephenson (9, right valve), Oligocene (*Marginulina* zone), Texas.

Hemicythere Sars, Cretaceous–Recent. The subquadrate carapace is smooth to slightly reticulate or pitted. *H. dalli* Howe & Brown (2, right valve), Miocene, Florida.

Paracypris Sars, Cretaceous–Recent. The carapace is long and narrow, rounded anteriorly and tapering to an acute angle posteriorly. *P. perapiculata* Alexander (19, right valve), Paleocene, Texas.

Paracytheridea Muller, Cretaceous–Recent. The small, elongate carapace is triangular in dorsal view and has subcentral swellings in the anterior and posterior portions. *P. chipolensis* Howe & Stephenson (1, left valve), Miocene, Florida; *P. byramensis* Howe & Law (7, right valve), Oligocene, Mississippi.

Perissocytheridea Stephenson, Miocene. The subpyriform carapace has a pitted surface and a subacute posterior end. *P. matsoni* (Stephenson) (5, left valve), Louisiana.

Pyricythereis Howe, Tertiary. The elongate carapace is subquadrate in outline, and the surface is ornamented by irregular ridges. *P. simiensis* Le Roy (3*a, b*, right valve and dorsal view), Pliocene, California.

not known to include any fossils. These are the Copepoda and Branchiura. Copepods are minute, extremely abundant marine crustaceans which play an important part in the food economy of near-surface parts of seas throughout the globe, both shallow water and open ocean. They live also in fresh water. Probably they were abundant during past geologic periods (Fig. 12-10), but proof of this is lacking. The branchiures superficially resemble copepods and likewise occur in fresh and salt waters; unlike copepods, they are not abundant in kinds or number of individuals.

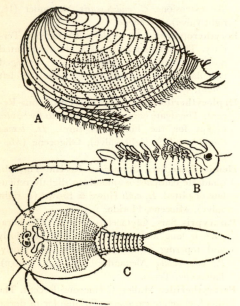

Fig. 14-14. **Living branchiopods.** The parts of these crustaceans which are hidden beneath the bivalve (*A*) or univalve (*C*) shell are indicated by dotted outlines.

A, A conchostracan, *Cyzicus morsei* (×5), common in fresh waters of central and western United States. The two valves are not hinged dorsally, as in ostracodes.

B, An anostracan, *Branchipus vernalis* (×2.5), the fairy shrimp, which lives in ponds of the eastern United States. It is shown in normal position, swimming on its back.

C, A notostracan, *Apus cancriformis* (×2), fresh waters, Europe.

Of the four classes which are represented by fossils, Ostracoda have been described. The others are Branchiopoda, Cirripedia, and Malacostraca.

Branchiopods

The branchiopods (*branchio*, gill; *pod*, foot) are mostly diminutive, rather primitive crustaceans, which include the transparent fairy shrimps of fresh-water ponds; the brine shrimp, able to live in excessively saline lakes and lagoons; the tiny water fleas, so abundant in some lakes and ponds as to color the water; and several other kinds (Fig. 14-14). Less than 5 per cent of modern species of branchiopods live in the sea. Some fossils are judged to denote a marine habitat, but several genera clearly are nonmarine.

The chief distinguishing features of the branchiopods are the large but variable number of their body segments and the flat leaflike nature of their paired thoracic appendages. The head bears compound eyes and antennae. Most, but not all, branchiopods have chitinous hard parts which are impregnated with calcium carbonate. The head and body segments are not individually covered so as to form a complexly built exoskeleton like that of trilobites, but the anterior part of the body is protected by a sheath. This may be a rounded unsegmented structure of flattish form or curved over the back of the animal, or it may consist of two equal symmetrical valves which join together dorsally. The bivalved carapace of a branchiopod differs from that of an ostracode in having no interlocking mechanism for hingement along the line where the valves adjoin. Branchiopod shells have been mistaken frequently for small clam shells. Most fossils are less than 5 mm. (0.2 in.) in length or width, but the largest known branchiopod (*Apus*) attains a length of 65 mm. (2.5 in.) (Fig. 14-14*C*).

The class is divided into four orders, mainly on the nature of the carapace, and of these, three are represented by fossils. The Anostraca (Fig. 14-14*B*), which have no shell, contain an Oligocene genus. The Notostraca (Fig. 14-14*C*), which have an undivided carapace, include forms ranging from Cambrian to Recent, but only a few genera (mostly lower Paleozoic) have been described. The Conchostraca have a bivalved carapace which encloses the whole body (Fig. 14-14*A*). This is the most important group paleontologically, for it includes at least a dozen genera of fossils. Among these, approximately one half are restricted to the Cambrian. Others are common in some late Paleozoic rocks, and a few occur in younger deposits. The most noteworthy genera are *Cyzicus* (formerly called *Estheria*, a pre-occupied name) and *Leaia*, which contain numerous species and are abundant fossils in fresh- and brackish-water deposits, especially in some Pennsylvanian and Permian formations (Figs. 14-14*A*; 14-15*A–C*, *G*). *Cyzicus* is common in various Mesozoic rocks and has been reported also from Pleistocene fresh-water clays in Canada.

Cirripeds

The class called cirripeds includes the barnacles, which in adult life do not resemble other crustaceans at all (Fig. 14-16). With a shelly covering of many plates around them, they grow attached to rocks, shells of other invertebrates, or other foothold on the shallow sea bottom. They incrust the piling of wharves and foul the sides of ships. The name of the group (*cirri*, curl; *ped*, foot) refers to the many curved delicate appendages which are borne by the body segments, used to convey food to the mouth. The identity of barnacles as crustaceans is readily established by study of their anatomy, and especially by investigation of their larval stages. Such examination shows that these animals are not very different from other crustacean stocks when they hatch from an egg and begin existence as free-swimmers. After molting one to three times, with slight changes of form, they develop ostracode-like bivalved shells, and the animal

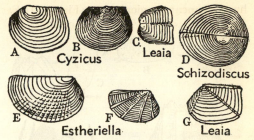

FIG. 14-15. **Fossil branchiopods.** The illustrated forms range from Devonian to Triassic, but genera such as *Cyzicus* (formerly called *Estheria*) persist to the present.

A, B, Cyzicus Audouin, Devonian–Recent, worldwide. (*A*) *C. ortoni* (Clarke) (×5), Upper Pennsylvanian (Conemaugh), Ohio. (*B*) *C. membranacea* (Pacht) (×2.5), Upper Devonian, New York.

C, G, Leaia Jones, Pennsylvanian, North America and Europe. (*G*) *L. leidyi* Jones (×2), Lower Pennsylvanian (Pottsville), Pennsylvania. (*C*) *L. baentschiana* Geinitz (×1.5), Middle Pennsylvanian, western Germany.

D, Schizodiscus Clarke, Devonian, North America. *S. capsa* Clarke (×1.3), Hamilton, New York.

E, F, Estheriella Weiss, Permian–Triassic, Europe. (*E*) *E. costata* Weiss (×4); (*F*) *E. nodocostata* Giebel (×3), both species from Lower Triassic, Germany.

during this stage eventually finds a place of attachment (Fig. 14-16*A–D*). It adheres by parts of the head region, casts off its two-valve shell, undergoes changes of the body, and secretes the calcareous plates which furnish protection throughout the sessile adult life of the barnacle. The upturned fringed limbs of the thorax produce water currents which carry microscopic food particles to the mouth (Fig. 14-16*F*).

Fossil cirripeds are distributed through various Paleozoic, Mesozoic, and Cenozoic rocks from the Ordovician (doubtfully from the Cambrian) onward. Genera are defined on the basis of the form and arrangement of the skeletal parts. Some early Paleozoic barnacles are much taller than wide, having numerous plates arranged in overlapping series (Fig. 14-16*H*). The common modern genus *Balanus*, which occurs as a fossil in many Tertiary deposits, has a truncate conical form (Fig. 14-16*E*; Fig. 14-18, *4a, b*). The sides of the cone

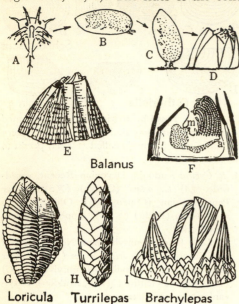

Balanus

Loricula   Turrilepas   Brachylepas

FIG. 14-16. **Living and fossil cirripeds.** The upper group of figures (*A–F*) illustrate the ontogeny and structure of modern *Balanus*, and the lower group (*G–I*) show representative fossils.

A, Larval form (nauplius) of *Balanus* (enlarged), which is very similar to corresponding larvae of other crustaceans.

B, Later larval stage (*cypris*) having form of a bivalved ostracode (not to scale). This stage, like *A*, is free-swimming.

C, Attachment of larva by the head region.

D, Young form of adult after discarding bivalved shell and secreting new calcareous plates, which are attached to a rock or shell.

E, Oblique view of adult specimen of *Balanus*.

F, Section of *Balanus*, showing immovable side and base plates and two movable plates (solid black) moved by muscles; the curved appendages at right produce water movements which carry food to the mouth (*m*); digestive tract stippled, anus (*a*).

G, *Loricula darwini* Woodward (×1), Cretaceous, England; a barnacle growing on a stalk composed of many small plates in series.

H, *Turrilepas wrightianus* De Koninck (×1), Silurian, England; a fossil which resembles a narrow pine cone.

I, *Brachylepas naissanti* (Hébert) (×1), Upper Cretaceous, France.

are formed by six pieces which fit immovably together. Two subtriangular plates which are carried by the back of the barnacle and are not connected to the cone have diagnostic peculiarities that are useful in defining species, but unfortunately, these unjointed plates commonly are missing from fossils.

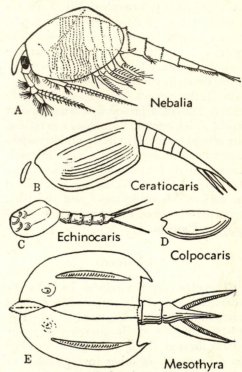

Nebalia

Ceratiocaris

Echinocaris

Colpocaris

Mesothyra

FIG. 14-17. **Living and fossil phyllocarids.** The carapace is not fused to the body in this group of malacostracans.

A, *Nebalia bipes*, a British species which lives along the coast between high- and low-tide marks (enlarged).

B, *Ceratiocaris* McCoy, Ordovician–Silurian, Europe and North America. *C. papilio* Salter (×0.7), Ordovician, Scotland.

C, *Echinocaris* Whitfield, Devonian, North America. *E. socialis* Beecher (×1.3), Upper Devonian (Chemung), New York.

D, *Colpocaris* Meek, Mississippian, North America. *C. elytroides* (Meek) (×0.7), Lower Mississippian, Kentucky.

E, *Mesothyra* Hall & Clarke, Upper Devonian, North America. *M. oceani* Hall (×0.3), Upper Devonian, New York.

Malacostracans

The crustaceans which are best known to almost everyone are malacostracans (Figs. 14-17 to 14-20). These include the crayfishes of fresh-water streams and ponds, the sow bugs or pill bugs which live in damp places under rocks and logs on land, and a host of crabs, lobsters, shrimps, and other arthropods of the sea. The name of the class (*malaco*, soft; *ostracon*, shell) is appropriate for all just after times of molting their external covering, and it suits many which never acquire rigid

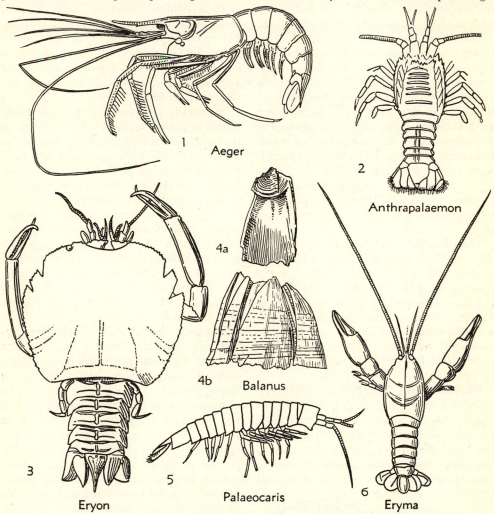

1 Aeger

2 Anthrapalaemon

4a

4b Balanus

3 Eryon

5 Palaeocaris

6 Eryma

FIG. 14-18. **Fossil malacostracans and a cirriped.**

1, *Aeger tipularius* (Schlotheim) (×0.5), Upper Jurassic, Germany; one of the decapods (Natantia) adapted for swimming.

2, *Anthrapalaemon gracilis* Meek & Worthen (×1) Pennsylvanian, Illinois; a percarid.

3, *Eryon propinquus* (Schlotheim) (×0.5), Upper Jurassic, Germany; a lobster-like decapod.

4, *Balanus concavus* Bronn (×0.5), Miocene, Mary-land; a cirriped. (4a) One of the calcareous plates (carina), which is immovably joined to others around the animal. (4b) Side view of complete fossil.

5, *Palaeocaris typus* Meek & Worthen (×3), Pennsylvanian, Illinois; a syncarid.

6, *Eryma leptodactylina* (Germ) (×1), Upper Jurassic, Germany; a true lobster.

hard parts. Many malacostracans, how-ever, like the stone crab, build remarkably strong shells, mainly composed of calcium carbonate. These are well suited for fossilization.

Despite the amazing diversity of form, the fundamental organization of malacostracans is nearly identical. All but a few have a body composed of 20 segments, of which six are fused to form the head, eight comprise the thorax, and six belong to the abdomen; additional is the telson at the hind extremity of the abdomen. One division of the malacostracans (Phyllocarida) is more primitive than others in retaining an additional segment in the abdomen. Inasmuch as essential structures of malacostracan body parts, including appendages, have been described in the introductory part of this chapter, attention may be turned to classification of members belonging to this group and to their geologic record.

Classification. The malacostracans are divisible into five subclasses which are mainly established on the nature of the continuous shell (carapace) that incases the head and part or all of the thoracic region. These subclasses, each of which contains one to five orders, are: Phyllocarida, Syncarida, Hoplocarida, Percarida, and Eucarida. All are represented by fossils.

Phyllocarids. The group called phyllocarids (*phyllo*, leaf; *carid*, shrimp) is distinguished by a relatively large, loose carapace which covers the anterior part of the body but is not joined by fusion to any part of the thorax. Limbs attached to the thoracic segments are all alike, and

their broad leaflike appearance gives the subclass its name. As already noted, the abdomen contains an extra segment. A typical modern member of this group is *Nebalia* (Fig. 14-17*A*).

At least three dozen genera of fossil phyllocarids are described, but only one of these, which occurs in Upper Triassic rocks of Germany, is recorded from post-Paleozoic deposits. The others are distributed from Cambrian to Permian, and they are found in both North America and Europe. A few representative forms are illustrated in Fig. 14-17*B–E*.

Syncarids. A small group of freshwater malacostracans is differentiated in the subclass called Syncarida (*syn*, with). They have no carapace, and the first thoracic segment is fused to the head. Living species are confined to Australia and Tasmania. Five genera from freshwater deposits of Pennsylvanian and Permian age in North America and Europe exhibit features of the body and appendages so closely similar to these modern forms that they are assigned to the same subclass and order. An example is *Palaeocaris*, from the Pennsylvanian of Illinois (Fig. 14-18, *5*).

Hoplocarids. A shallow carapace covering the head and fused to the first three thoracic segments but leaving others uncovered distinguishes the hoplocarids (*hoplo*, armor). The posterior three pairs of thoracic appendages are biramous, but anterior ones are unbranched and those borne by the second thoracic segment are much enlarged, closely resembling the potent weapons of the praying mantis among insects (Fig. 14-19). Five living genera include only marine forms. Although two fossils from rocks of Pennsylvanian age have been placed tentatively in this group, the oldest undoubted hoplocarids occur in Jurassic beds of Europe. They have been found also in Cretaceous and Tertiary strata but are not common fossils.

Percarids. The next to the largest assemblage of malacostracans is termed

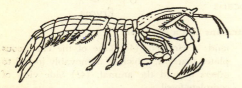

FIG. 14-19. **A modern hoplocarid.** These malacostracans have a carapace which is fused to front segments of the thorax. *Squilla mantis* (×0.5) is a widely distributed marine form.

Percarida (*per*, much). Some have no carapace, whereas in others it is present but fused only with the first two to four thoracic segments. A common feature of crustaceans classed as percarids is the occurrence of plates (oostegites) on the inner side of some or all of the thoracic limbs of females. These form a brood pouch in which the young may be carried. There are other distinguishing features also. Marine, fresh-water, and terrestrial genera are found.

Three of the five orders of percarids are represented by fossils. The Mysidacea, which have a carapace fused to anterior thoracic segments, are found in rocks of Mississippian age in Scotland and of Pennsylvanian age in England and North America. *Anthrapalaemon* (Fig. 14-18, *2*) is a genus of this group. The Isopoda, to which the sow bugs belong, lack a carapace. A Devonian fossil from Ireland doubtfully is assigned to this order. The oldest undoubted isopods occur in the Jurassic, and others are found in Cretaceous and Tertiary rocks, but they are uncommon fossils. The Amphipoda resemble the isopods but have laterally compressed bodies. They live in fresh waters and the tidal zone along the seashore. The oldest definitely determinable representative is from the Baltic amber (Oligocene).

Eucarids. All malacostracans not included in the subclasses just reviewed are assigned to the Eucarida (*eu*, true). They are characterized by a carapace which covers the entire thorax and is fused to all the thoracic segments. Females possess no brood pouch. The group comprises two very unequal orders: the pelagic prawn-like Euphausiacea, comprising a very few genera; and the Decapoda, which contain a multitude of forms, including many genera of crabs, lobsters, shrimp, prawns, and others.

One fossil from the Lower Carboniferous of Scotland is very doubtfully placed in the first-mentioned order, but, as might be supposed, the decapods are represented by very many fossils. None of these, however, is so old as Permian. Lobster-like genera are found in Triassic deposits of Europe and true lobsters in the Jurassic. Many well-preserved specimens have furnished the basis for describing several genera of both. *Eryon* is an example of the lobster-like forms, and *Eryma* is one of the first true lobsters (Fig. 14-18, *3*, *6*). A Triassic and Jurassic group of decapods which closely resembles some modern eucarids (called Natantia, swimmers) is distinguished by the compressed form of the body, the presence of a long projecting beak in front

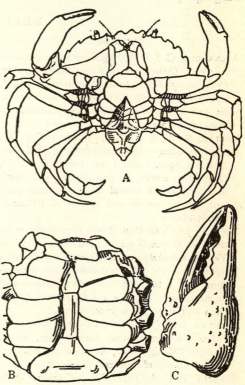

Fig. 14-20. **Living and fossil crabs.** These decapod malacostracans represent the most specialized of all crustaceans.

A, A modern shore crab, *Carcinus*, from the ventral side, the abdomen turned back from its normal position, tucked under the thorax.

B, *Callinectes sapidus* Rathbun (×2), Pleistocene, Maryland; ventral side, showing very narrow abdomen.

C, Chela of *Avitelmessus grapsoideus* Rathbun (×1.3), Upper Cretaceous, Eastern Gulf Coast.

of the eyes, and peculiarities of the limbs. They roughly resemble lobsters, but are quite different. One of these is *Aeger* (Fig. 14-18, *1*).

The crabs (Brachyura, short tails) are differentiated from lobsters by shortening and widening of the body and absence of a tail fan. They are the most highly specialized group of the crustaceans. The abdomen is much shortened and turned forward beneath the thorax, fitting into a depression (Fig. 14-20*A*). Many fossils belonging to this group have been found ranging upward from the Jurassic. They do not appear in Triassic faunas. This group of decapods is most common in some Tertiary deposits; indeed, *Callianassa* and some other crabs are useful zone fossils in part of the Gulf Coast and Pacific Border Tertiary sections. This is an exception, rather than the rule, for generally the remains of malacostracans in different parts of the geologic column have value chiefly as records of then-existent type of life. Mostly, they are not useful, as are ostracodes, in stratigraphic correlation. Two fragments of fossil crabs are illustrated in Fig. 14-20*B*, *C*.

REFERENCES

Ostracodes

ALEXANDER, C. I. (1933) *Shell structure of the ostracode genus Cytheropteron, and fossil species from the Cretaceous of Texas:* Jour. Paleontology, vol. 7, pp. 181–214, pls. 25–27. Contains a short section on shell structure and illustrations and descriptions of Cretaceous species.

BASSLER, R. S., & KELLETT, BETTY (1934) *Bibliographic index of Paleozoic Ostracoda:* Geol. Soc. America Spec. Paper 1, pp. 1–500, figs. 1–24. Contains a section on morphology and classification, one on faunal lists, a bibliography, and lists of genera and species of Paleozoic Ostracoda.

BOLD, W. A. VAN DEN (1946) *Contribution to the study of Ostracoda with special reference to the Tertiary and Cretaceous microfauna of the Caribbean region:* Privately published dissertation, Amsterdam, pp. 1–167, pls. 1–18. Descriptions and illustrations of Tertiary and Cretaceous Ostracoda and a section on morphology.

BONNEMA, J. H. (1932) *Orientation of the carapaces of Paleozoic Ostracoda:* Jour. Paleontology, vol. 6, pp. 288–295, figs. 1–13, A critical discussion of the orientation of Paleozoic Ostracoda.

BRADLEY, P. C. S. (1941) *The shell structure of the Ostracoda and its application to their paleontological investigation:* Annals and Mag. Nat. History, ser. 11, vol. 8, pp. 1–33, figs. 1–18. Includes a discussion of the structure of the shell and descriptions and illustrations of some species.

COOPER, C. L. (1941) *Chester ostracodes of Illinois:* Illinois Geol. Survey, Rept. Inv. 77, pp. 1–101, pls. 1–14. Contains illustrations and descriptions of Mississippian Ostracoda from Illinois.

———— (1945) *Moult stages of the Pennsylvanian ostracode Ectodemites plummeri:* Jour. Paleontology, vol. 19, pp. 368–375, pl. 57. A detailed study of the growth stages of Ostracoda.

———— (1946) *Pennsylvanian ostracodes of Illinois:* Illinois Geol. Survey Bull. 70, pp. 1–177, pls. 1–21.

HESSLAND, I. (1948) *Investigations of the Lower Ordovician of the Siljan District, Sweden. I. Lower Ordovician Ostracods of the Siljan District, Sweden:* Geol. Inst. Upsala Bull., vol. 33, pp. 97–408, pls. 1–26. Includes a discussion of orientation, brood-pouch problem, details of the shell, and descriptions and illustrations of lower Paleozoic species.

HOWE, H. V., & LAW, J. (1936) *Louisiana Vicksburg Oligocene Ostracoda:* Louisiana Dept. Cons. Geol. Bull. 7, pp. 1–96, pls. 1–6. Contains range charts, illustrations, and descriptions of Oligocene Ostracoda from Louisiana.

KELLETT, BETTY (1933) *Ostracodes of the Upper Pennsylvanian and the Lower Permian strata of Kansas: I. The Aparchitidae, Beyrichiidae, Glyptopleuridae, Kloedenellidae, Kirkbyidae, and Youngiellidae:* Jour. Paleontology, vol. 7, pp. 59–108, pls. 13–16. Descriptions and illustrations of Ostracoda of Upper Pennsylvanian and Lower Permian rocks in Kansas.

———— (1934) *II. The genus Bairdia:* Same, vol 8., pp. 120–138, pls. 14–19. Contains illustrations and descriptions of species of *Bairdia* from Kansas.

———— (1935) *III. Bairdiidae (concluded), Cytherellidae, Cypridinidae, Entomoconchidae, Cytheridae and Cypridae:* Same, vol. 9, pp. 132–166, pls. 16–18. Includes range charts for three papers and illustrations and descriptions of Ostracoda from Upper Pennsylvanian and Lower Permian rocks of Kansas.

LE ROY, L. W. (1945) *A contribution to ostracodal ontogeny:* Jour. Paleontology, vol. 19, pp. 81–86, pl. 9. Contains discussion and illustrations of molt stages of Ostracoda.

SWAIN, F. M. (1946) *Upper Jurassic Ostracoda from the Cotton Valley group in northern Louisiana:* Jour. Paleontology, vol. 20, pp. 119–129, pls. 20, 21. Discusses stratigraphy, and illustrates and describes Ostracoda of the Cotton Valley group.

SWARTZ, F. M. (1933) *Dimorphism and orientation in ostracodes of the family Kloedenellidae from the Silurian of Pennsylvania:* Jour. Paleontology, vol. 7, pp. 231–260, pls. 28–30. Contains descriptions and illustrations of Silurian Ostracoda and discussion of orientation and dimorphism.

TRIEBEL, E. (1941) *Zur Morphologie und Ökologie der fossilen Ostracoden mit Beschreibung einiger neuer Gattungen und Arten:* Senckenbergiana, Bd. 23, no. 416, Frankfurt a. M. (German). Includes a discussion on orientation of Ostracoda.

ULRICH, E. O., & BASSLER, R. S. (1923) *Paleozoic Ostracoda: their morphology, classification and occurrence:* Maryland Geol. Survey, Silurian, pp. 271–391, pls. 36–65. Discusses morphology, classification, and occurrence of Ostracoda and gives illustrations and descriptions of Silurian Ostracoda.

Other Crustaceans

BASSLER, R. S. (1913) *Branchiopoda, Ostracoda:* in Eastman, C. R., & Zittel, K. A., Textbook of palaeontology, Macmillan & Co., Ltd., London, 2d ed., vol. 1, pp. 731–742, figs. 1413–1436.

BORRADAILE, L. A., *et al.* (1948) *The Invertebrata:* Cambridge, London, pp. 1–725, figs. 1–483 (crustaceans, pp. 326–417, figs. 222–293).

CALMAN, W. T. (1909) *Crustacea:* in Lankester, E. R., Treatise on zoology, A. & C. Black, London, pp. 1–346, figs. 1–194.

——— (1913) *Cirripedia, Malacostraca:* in Eastman, C. R., & Zittel, K., Textbook of palaeontology, Macmillan & Co., Ltd., London, 2d ed., vol. 1, pp. 742–748, 754–769, figs. 1437–1450, 1463–1495.*fl*

CLARKE, J. M. (1913) *Phyllocarida:* in Eastman, C. R., & Zittel, K., Textbook of palaeontology, Macmillan & Co., Ltd., London, 2d ed., vol. 1, pp. 748–754, figs. 1451–1462.

RATHBUN, M. J. (1908) *Descriptions of fossil crabs from California:* U.S. Nat. Mus. Proc., vol. 35, pp. 341–349, illus.

SHIMER, H. W., & SHROCK, R. R. (1944) *Index fossils of North America:* Wiley, New York, pp. 1–837, pls. 1–303 (crustaceans, pp. 655–701, pls. 277–293, 295–298).

STORER, T. I. (1951) *General zoology:* McGraw-Hill, New York, pp. 1–832, illus. (crustaceans, pp. 461–484, figs. 22-1–22-17).

WOLCOTT, R. H. (1946) *Animal biology:* McGraw-Hill, New York, 3d ed., pp. 1–719, figs. 1–508 (crustaceans, pp. 289–310, figs. 201–220).

CHAPTER 15

Chelicerates, Myriapods, and Insects

A large majority of all known kinds of arthropods belong to divisions treated in this chapter, and this statement is true of fossils, as of living forms. The chelicerates have paleontological importance mainly because they include the eurypterids, but there are also many kinds of fossil xiphosurans, arachnids (scorpions and spiders), and a few pycnogonids. Myriapods, which include the centipedes and millipedes, are represented by fossils, but remains of this group are fragmentary and uncommon. The insects are numerically the greatest division. Kinds of living insects (about 900,000 species) far outnumber all other animals combined (about 220,000 species). It is not generally known that, in addition, kinds of fossil insects discovered to date exceed the combined number of described trilobites, crustaceans, chelicerates, and myriapods.

If this is so, it is appropriate to ask why the study of fossil insects, and likewise myriapods and chelicerates, is rather neglected in courses on invertebrate paleontology. The reasons are not hard to find. Emphasis in paleontological study is not appropriately based on a census of fossil genera and species, but rather on the importance of each group of organisms (1) for correlation of fossil-bearing strata and in determining geochronology, (2) for tracing out the general course of evolutionary differentiation among animals, and (3) for elucidation of principles which apply to organic evolution. Fossil insects, myriapods, and chelicerates qualify for attention in varying degree under the second and third of these criteria, but most of them have little value in considering the first. Also, in order to comprehend significant details, specialized knowledge, such as that

belonging to the field of entomology, is needed. The magnitude of other aspects of paleontological study is a factor also. Accordingly, in trying to meet the demand for reasonable completeness in surveying characters of the fossil arthropod groups, it is advisable to give attention only to the most important features.

CHELICERATES

Distinguishing Features

The arthropods classed as chelicerates differ from trilobites, crustaceans, and other main divisions of the phylum in having a single pair of pre-oral appendages which are provided with pincers (chelae). Their function is primarily for grasping food. The chelicerates are unlike insects, myriapods, crustaceans, and trilobitomorphs, in having no antennae. Another character which sets them apart is the organization of the anterior part of the body in a **prosoma** (cephalothorax), formed by the fusion of pre-oral segments with the first six or seven post-oral segments. Those belonging to the posterior part of the body comprise the **opisthosoma** (abdomen). They may be severally distinct or more or less fused together, and they may be all of similar character or divisible into a forward group (**mesosoma**) of relatively broad segments and a posterior group (**metasoma**) of narrow ones. The terminal segment of the opisthosoma may be a spikelike or oarlike **telson**. These divisions of the body of typical chelicerates are illustrated in Fig. 15-1.

The exoskeleton of chelicerates is composed of chitin, which generally is not impregnated by calcium carbonate or other calcium salts, as among trilobites and crustaceans. Some fossils, however, have hard parts consisting mainly of calcium phosphate. Many spiders, ticks, and mites have a thin and flexible body covering; in others it is tough and leathery; and in still others it is very hard and rigid. The unmineralized exoskeletons of chelicerate arthropods evidently are less readily preserved as fossils than those of trilobites and crustaceans.

The prosoma has six pairs of appendages attached to the underside. The front pair are called **chelicerae** (*chela*, claw; *cera*, horn), and the next following pair, just behind the mouth, are known as **pedipalpi** (*pedi*, foot; *palpus*, feeler). The four remaining pairs are walking legs (Fig. 15-2). Scorpions have chelate (chela-bearing) pedipalpi, whereas in spiders chelae are confined to the chelicerae. The king crab (*Xiphosura*) is equipped with chelae on all prosomal limbs except the posterior pair. Appendages are attached to the abdominal region of some chelicerates but not of others. Where present, they may consist of six pairs of simple walking legs, or, more commonly, six pairs of limbs modified for swimming and as gills.

In size, the chelicerates range from mites less than 0.5 mm. (0.02 in.) in length to giant eurypterids like *Stylonurus* and *Ptery-*

gotus which attained a length of nearly 3 m. (10 ft.).

Merostomes

The division of chelicerates which ranks first in paleontological importance is the class Merostomata (*mero*, part; *stoma*, mouth, signifying well-organized mouth parts). The chief distinguishing feature of the merostomes is their aquatic mode of life, breathing by means of gills. Thus, they are separated readily from arachnids, which comprise the spiders and scorpions.

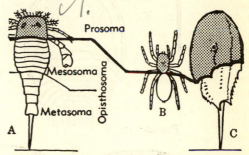

FIG. 15-1.   Divisions of the body of chelicerates. The reason for nomenclature of parts differing from that applied to trilobitomorphs and crustaceans is that segments composing respective main divisions of the body do not correspond exactly. Also, the nature of attached appendages differs in various groups. (*A*) Eurypterid; (*B*) spider; (*C*) xiphosuran (oblique view).

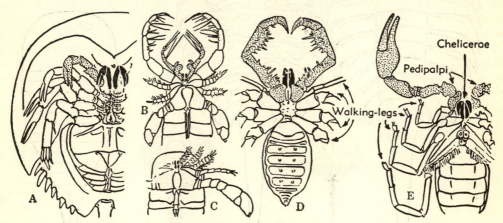

FIG. 15-2.   Appendages of chelicerates. The pair of front appendages (chelicerae), which give the subphylum Chelicerata its name, are indicated by solid black. The next posterior pair (pedipalpi) are differentiated by a stippled pattern. The four pairs of walking legs are attached behind the pedipalpi. (*A*) Xiphosuran; (*B, C*) eurypterids; (*D*) spider; (*E*) scorpion.

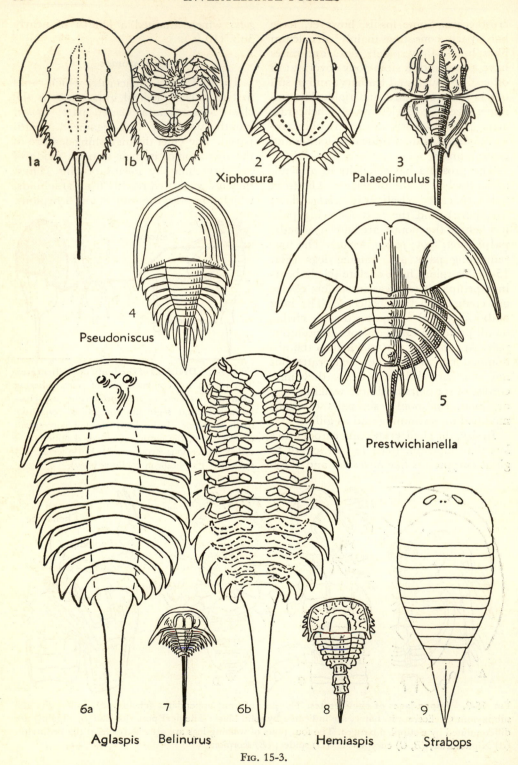

1a 1b Xiphosura 2 3 Palaeolimulus

4 Pseudoniscus

5 Prestwichianella

6a Aglaspis 7 Belinurus 6b 8 Hemiaspis 9 Strabops

Fig. 15-3.

The peculiar marine arthropods called pycnogonids (sea spiders), which also are assigned to the chelicerates, correspond to merostomes in that they are aquatic, but in appearance and structure these groups are very far apart.

The merostomes contain three assemblages which are assigned rank as subclasses: Xiphosura, consisting of the king crab and numerous allied forms; Eurypterida, restricted to Paleozoic time; and Pycnogonida. The first two are represented by a goodly number of fossils, and the third, by two genera.

XIPHOSURANS

The chelicerate group called xiphosurans is mainly characterized by the relatively wide and short main part of the body, which gives it a subrounded outline, and by the prominence of a spiked telson, which resembles a dagger or thick straight sword (*xiphos*, sword; *ura*, tail). A semicircular prosoma articulates with the front of the opisthosoma, which is composed of a small number of movable segments, or of segments fused solidly together, excepting the telson. The first pair of prosomal appendages consists of chelicerae, the others being pedipalpi and walking legs. The opisthosoma also has paired appendages on the underside. In some there is a front rudimentary pair (chilaria), followed by five pairs of swimming limbs which serve also a respiratory function as gills; in others, there are six pairs of simple walking legs, like the five posterior pairs of the prosoma. Most xiphosurans have two well-developed compound eyes, and in addition, many have a median pair of small simple eyes (ocelli). Representatives of the group are distributed from Cambrian to Recent. Their average size is 50 to 75 mm. (2 to 3 in.) from front to back and 40 to 50 mm. (1.5 to 2 in.) across, but maximum length may slightly exceed 60 cm. (2 ft.).

The xiphosurans are divided into three groups which are assigned taxonomic rank as orders. These are the Aglaspida, Synxiphosura, and Limulida. Each division is represented by fossils.

Aglaspids. The most primitive kinds of xiphosurans are included among the 10 now-known genera of aglaspids, of which one occurs in the Middle Cambrian and the rest only in the Upper Cambrian (Fig. 15-3, *6a, b, 9*). They have a phosphatic

FIG. 15-3. **Living and fossil xiphosurans.** These chelicerates are characterized by the relatively compact, rounded outline of the cephalothorax joined with the wide anterior part of the abdomen and the posterior termination of the body in a prominent swordlike telson.

Aglaspis Hall, Upper Cambrian, upper Mississippi Valley, United States. This ancient xiphosuran is distinguished by the 12 movable segments of its opisthosoma and the presence of appendages on their underside. *A. spinifer* Raasch (*6a, b*, dorsal and ventral sides, ×1), Trempealeauan, Wisconsin.

Belinurus König, Devonian, Europe; Pennsylvanian, Europe and North America. A small but typical representative of the Limulida. *B. reginae* Baily (*7*, ×1), Pennsylvanian, Ireland.

Hemiaspis Woodward, Silurian, Europe. A synxiphosuran. *H. limuloides* Woodward (*8*, ×0.7), England.

Palaeolimulus Dunbar, Lower Permian, North America. A Paleozoic member of the Limulida

which is closely similar to modern *Xiphosura*. *P. avitus* Dunbar (*3*, ×3), Kansas.

Prestwichianella Woodward, Pennsylvanian, Europe and North America. The opisthosoma is composed of seven coalesced segments and a telson. *P. danae* (Meek) (*5*, ×1), Illinois.

Pseudoniscus Nieszk, Silurian, Europe and North America. A synxiphosuran. *P. roosevelti* Clarke (*4*, ×2), New York.

Strabops Beecher, Upper Cambrian, North America. An aglaspid which has been misinterpreted as a primitive eurypterid. *S. thacheri* Beecher (*9*, ×0.5), Missouri.

Xiphosura Gronovius, Jurassic–Recent, Europe, North America, Asia. Type form of the Limulida. *X. polyphemus* (Linné) (*1a, b*, dorsal and ventral sides, ×0.2). *X. walchi* Desm. (*2*, ×0.5), Jurassic, Germany.

exoskeleton which is more or less distinctly trilobed longitudinally. The semicircular prosoma bears genal spines in some forms, but these are lacking in others. Moderately large compound eyes, and in some, a pair of small simple eyes, occur near the front on the dorsal side. The abdomen consists of 12 segments, including the swordlike telson. The segments are united by movable joints. A primitive feature of aglaspids in which appendages have been observed is the presence of at least six pairs of unspecialized walking legs on anterior segments of the abdomen; additional pairs occur also, but their structure is not clearly determinable (Fig. 15-3, *6b*). Most complete specimens range in length from 25 to 75 mm. (1 to 3 in.), the largest reaching 220 mm. (8.5 in.).

The aglaspids superficially somewhat resemble trilobites, but they definitely do not belong among the trilobitomorphs. The main evidence of this is their lack of antennae and biramous limbs, and the presence of chelicerae followed by pairs of simple walking legs. Also, the distribution of appendages differs. Whereas trilobites have five pairs (including antennae) attached to the cephalon, the aglaspids, like other merostomes, have six pairs belonging to the prosoma.

The genus *Aglaspis*, from which this order is named, is illustrated in Fig. 15-3, *6a*, *b*. Another fossil of the group is *Strabops* (Fig. 15-3, *9*), which has been misinterpreted as the most primitive discovered eurypterid; it has been figured in various textbooks with wholly conjectural reconstruction of eurypterid-like appendages. This form is judged actually to belong among the aglaspids.

Synxiphosurans. This division differs from aglaspids mainly in the smaller number of abdominal segments and generally in differentiation of the opisthosoma into an anterior broad (mesosomal) region and a posterior narrow (metasomal) region. Known fossils are exclusively Silurian and Devonian in age. Forms shown in Fig. 15-3, *4*, *8*, are representative of the order.

Limulids. A third group of xiphosurans is characterized by the modern limulus or king crab (*Xiphosura*), which has all opisthosomal segments except the telson fused together (Fig. 15-3, *1a*, *b*). Thus, the exoskeleton consists of three main parts: the large semicircular prosoma; the somewhat smaller, subtriangular front part of the opisthosoma; and the long stout telson. These parts are joined by readily movable articulations. On the underside of the prosoma are the chelicerae and five pairs of moderately large pincerbearing walking legs, of which the front pair occupies the position of pedipalpi. The forward part of the abdomen bears closely packed, partly coalesced pairs of bladelike respiratory and swimming appendages. The limulus molts by splitting the carapace of the cephalothorax along its margin, pulling soft parts from the abdomen and appendages, and crawling forward until free. Then a new exoskeleton is secreted.

The oldest discovered fossils which are classifiable as limulids occur in Devonian rocks of Europe and North America. Among three described genera of this age, one (*Belinurus*, Fig. 15-3, *7*) occurs on both sides of the Atlantic and it persists into the Pennsylvanian. Other late Paleozoic genera include *Prestwichianella*, from the Pennsylvanian, and *Palaeolimulus*, from the Permian (Fig. 15-3, *5*, *3*). Other forms make an appearance in Triassic deposits, and the modern genus *Xiphosura* is identified from Jurassic rocks (Fig. 15-3, *2*).

Habitat of Xiphosurans. Some groups of xiphosuran merostomes are marine. Living genera, all of which are limulids, are represented by abundant individuals in shallow waters along the eastern coast of North America, in the western Pacific, and northern Indian Ocean. Jurassic limulids (*Xiphosura*) also lived in shallow seas. Judging by association of aglaspids with trilobites and brachiopods, these Cambrian xiphosurans must also have been marine. One genus of synxiphosurans (*Weinbergia*) is found in marine Devonian rocks of western Germany. All other synxiphosurans, however, seem to have lived in brackish or fresh water, or perhaps some of them were adapted to hypersaline conditions. This conclusion is based on finding their remains in rocks which do not carry a normal marine fauna, and nearly all genera are associated with eurypterids. *Prestwichianella*, *Belinurus*, and probably other fossil limulids lived in fresh or brackish water.

Ontogenetic Recapitulation of Phylogeny. The development of modern *Xiphosura*, common along the Atlantic and Gulf coasts of the United States, parallels in interesting manner the successive adult forms of ancient xiphosurans. An embryonic stage of *Xiphosura* corresponds in some features to the trilobites. Later stages are similar to a Silurian synxiphosuran, and a Pennsylvanian xiphosuran (Fig. 15-4, *1a, b, 2, 3*). The adult stage of present-day *Xiphosura polyphemus* is not very different from *X. walchi* of the Jurassic (Fig. 15-4, *1c, 4*). Thus, the life history of the living species recapitulates the race history of the group to which it belongs. Such parallelism supports the thesis which has come to be known as Haeckel's law, that "ontogeny recapitulates phylogeny."

EURYPTERIDS

Largest among the chelicerates, and indeed, all fossil arthropods (but not living kinds), are the eurypterids (*eury*, broad; *pterid*, winglike, referring to the shape of some appendages). Their remains are distributed in rocks ranging in age from Ordovician to Permian, but are most common in some Silurian and Devonian formations, which evidently are not normal marine deposits. Study of the eurypterids, accordingly, calls both for examination of their structural features and analysis of evidence concerning their mode of life.

Structural Features. The exoskeleton of eurypterids consists of a chitinous covering which completely incases the body and its several appendages. It is divided into numerous parts, corresponding to the region of the prosoma, segments of the opisthosoma, and elements of the jointed limbs, each such part being relatively rigid. As in trilobites, the skeletal parts were joined by thin flexible chitin so as to permit differential movement at the joints. The outer surface of many eurypterid carapaces bears fine tubercles or scalelike markings, but these features are purely superficial.

Description of parts of the eurypterid exoskeleton may be based chiefly on two of the smaller genera, *Eurypterus* and *Hughmilleria*, which exhibit all typical structural features well and which lack peculiarities seen in other forms. Both dorsal and ventral views of these genera are given in Fig. 15-5, *2, 5a, b, 6*.

Taking up first the relatively large prosoma, we note that its front is evenly

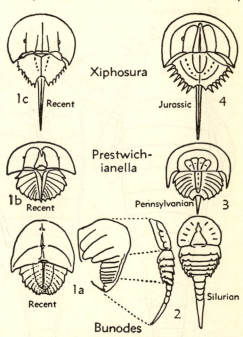

FIG. 15-4. **Comparison of fossil adult xiphosurans with growth stages of modern Xiphosura.** The Silurian genus *Bunodes* is a synxiphosuran; *Prestwichianella* and *Xiphosura* belong to the Limulida. Dorsal and side views of a young larval stage of the living *Xiphosura polyphemus* are shown in 1*a*, a later larval stage of this species is shown in 1*b*, and the adult form in 1*c*; not to scale.

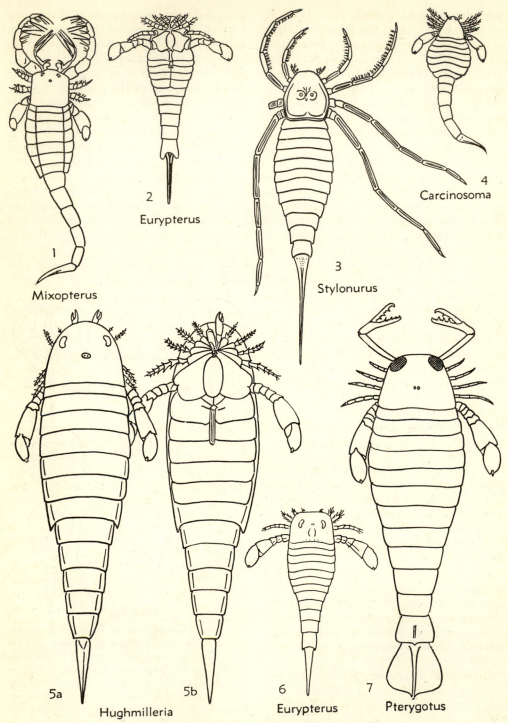

1 Mixopterus
2 Eurypterus
3 Stylonurus
4 Carcinosoma
5a 5b Hughmilleria
6 Eurypterus
7 Pterygotus

Fig. 15-5.

rounded in *Hughmilleria* but decidedly truncate in *Eurypterus*. On the dorsal side is a pair of somewhat widely spaced kidney-shaped compound eyes, and centrally located between them is a closely set pair of simple eyes (ocelli). Owing to the flatly rounded shape of the dorsal side of the cephalothorax and lack of elevation of the eyes, effective vision was mainly upward, not outward. The posterior edge of the prosoma is straight or slightly flexuous where it adjoins the anterior segment of the opisthosoma. Six pairs of appendages occur beneath the prosoma, the frontmost consisting of chelicerae, the next four of short walking legs (including undifferentiated pedipalpi), and the hindmost of large oarlike flippers. The pincer-tipped chelicerae of *Hughmilleria* (Fig. 15-5, 5b) are readily distinguished, but those of *Eurypterus* (Fig. 15-5, 2) are so short that they are not shown clearly in the drawing.

The opisthosoma, or abdomen, of these eurypterids is composed of 13 segments, of which the anterior six (characterized by greater width and presence of gills on the ventral side) are parts of the mesosoma, and the posterior seven constitute the metasoma. The terminal (or seventh) segment of the metasoma is a spikelike telson. Skeletal covering of the mesosomal segments consists of separate dorsal elements (tergites) and ventral elements (sternites), with sutural divisions located at the sides of the body between them. The covering of metasomal segments is not thus divided, for dorsal and ventral parts form a continuous ring. No appendages are attached to any opisthosomal segment, except that male or female genital appendages are joined to the forward edge of the first mesosomal sternite.

Sexual dimorphism among eurypterids is demonstrated by differences in the genital appendages, as well as in secondary characters. The appendage of males is a clasping organ, whereas that of females has a small median extension which fits into the clasping organ.

Other eurypterids are specialized in various ways and are very much larger than *Hughmilleria* and *Eurypterus*. Among

FIG. 15-5. **Representative Silurian and Devonian eurypterids.** This group of chelicerates is distinguished by their elongate, streamlined form, movable articulation of segments behind the cephalothorax, nature of their appendages, and general large size. They are distributed from Ordovician to Permian but are most common in some Silurian and Devonian deposits.

Carcinosoma Claypole, Ordovician–Silurian, North America and Europe. The fore part of the abdomen is widened to form a distinct mesosoma and the rear part narrowed in a tail-like metasoma terminating in curved telson, somewhat resembling the spine of a scorpion. *C. scorpionis* (Grote & Pitt) (4, ×0.1), Silurian, Scotland.

Eurypterus Dekay, Ordovician–Permian, Europe and North America. The two posterior pairs of appendages are not spinose. *E. fischeri* Eichwald (2, ventral side, ×0.2), Upper Silurian, Europe. *E. remipes* Dekay (6, dorsal side, ×0.25), Upper Silurian, New York.

Hughmilleria Sarle, Ordovician–Devonian, North America. Rather small eurypterids which differ from *Eurypterus* in shape of the prosoma and nature of appendages. *H. norvegica* (Kiaer) (5a, b, dorsal and ventral sides, ×1), Upper Silurian, Norway.

Mixopterus Ruedemann, Ordovician–Devonian, North America and Europe. This genus is characterized especially by the very spinose nature of its anterior appendages, as well as by features of the body. *M. kiaeri* Størmer (1, ×0.1), Upper Silurian, Norway.

Pterygotus Agassiz, Ordovician–Devonian, North America, Europe, Australia. Exceptional development of the chelicerae, which project in front as potent grasping organs, the very large marginally placed eyes, shape of the flattened telson, and huge size of some species distinguish this genus. *P. rhenaniae* Jaekel (7, ×0.05), Lower Devonian, Europe.

Stylonurus Page, Ordovician–Mississippian, North America and Europe. The unusually long and slender appendages and long sharp telson are peculiarities of this eurypterid. *S. excelsior* Hall (3, ×0.05), Middle Devonian, New York.

these are *Carcinosoma* (Fig. 15-5, *4*), which has an elongate subtriangular prosoma, rounded mesosoma, and scorpion-like metasoma; *Mixopterus* (Fig. 15-5, *1*), which is specially marked by long spines on the two front pairs of limbs behind the chelicerae; *Stylonurus* (Fig. 15-5, *3*), which has remarkably elongate telson and appendages; and *Pterygotus* (Fig. 15-5, *7*), which is distinguished by unusually formidable chelicerae, large round eyes, and a flattened telson. These variations in form do not mask the evident identity of essential structures belonging to the various genera, and therefore all are recognized readily as eurypterids.

Eurypterids molt during growth, just as do trilobites, crustaceans, and other arthropods. The facial suture, where the exoskeleton splits apart, runs along the margin of the cephalothorax, like that of protoparian trilobites at the edge of the cephalon. The modern limulus (*Xiphosura*) has a similarly placed suture.

Mode of Life. The habitat to which eurypterids were adapted has been much studied. On some points evidence is clear. This group of arthropods did not live side by side with contemporary trilobites, nor are they found associated with corals, bra-

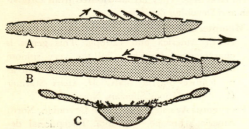

FIG. 15-6. **Adaptation of eurypterids as back swimmers.** These diagrams suggest structural suitability of eurypterids for swimming on their backs, as young xiphosurans and other aquatic arthropods are known to do (Størmer). (*A*) Side profile view of *Hughmilleria* with sternal plates on fore part of abdomen raised; (*B*) same with plates down, the downward movement having pushed the animal ahead (to right); (*C*) front profile showing dorsal side down and outspread appendages used as steering and balancing apparatus.

chiopods, or various other invertebrates which occur commonly in normal marine environments. Therefore, they are not animals of the open shallow seas. Nor are the sedimentary deposits containing eurypterid remains fresh-water formations, laid down in lakes or made by rivers. This is shown by several sorts of observations—physical features of the fossil-bearing strata, stratigraphic relations to adjoining deposits, and specialized nature of associated sparse fossils. Nevertheless, horizontal distribution in rocks, such as the Upper Silurian Bertie water lime and Manlius formation of the New York region, indicates that eurypterids were spread throughout large water bodies. If the water was abnormally saline, as suggested by some features, or if it was very brackish—so low in salt content that normal marine invertebrates could not survive in it—the habitat of eurypterids may find explanation. Such an environment, either unusually high or low in salinity, may exist in semienclosed lagoons, and such a setting seems best to fit all the evidence. A suggestion that eurypterids actually lived in fresh waters of land areas and that discovered fossil remains belong to individuals which had been swept from the mouths of streams and carried by current to the scattered places of burial fails to explain the nonoccurrence of eurypterids mingled with normal marine faunas. Also, it conflicts with the common finding of completely articulated specimens of these arthropods. Transportation of the segmented carapaces should result inevitably in dismemberment.

The streamlined body and oarlike nature of the hindmost pair of appendages, which are seen in various eurypterid genera, support interpretation of these animals as good swimmers. It is very possible also that eurypterids effectively used the movable plates on the ventral side of the mesosoma for propelling the body through the water, just as phyllopods and young xiphosurans are known to do. These swim on their backs, as do some aquatic insects, which leads to the idea that eurypterids may have done likewise, for in this position their outspread flattened oarlike limbs would serve perfectly as stabilizing blades. Thus, such appendages possibly were used chiefly as balancing organs and the tail as a steering apparatus, although only *Pterygotus* seems to have a particularly well-designed rudder (Fig. 15-6). The long, relatively stout legs of *Stylonurus* are adapted for crawling—not at all for swimming—but the so-called walking legs of most other eurypterids seem to be much too diminutive for effective use in dragging around the large body to which they are joined.

Paleontological Importance. On the whole, structural details of the eurypterids are better known than those of almost any fossil arthropods. They are specially interesting from the standpoint of morphologic comparison with other arthropod groups such as trilobites, to which they are obviously, though distantly, related. Study along this line has value for elucidation of evolution and for better understanding of this very remarkable group of extinct animals. Although the eurypterids are far outranked in value to stratigraphical paleontology by many less imposing kinds of fossils, they yield precedence to no other assemblage of invertebrates in arousing the imagination. To find and collect a complete eurypterid specimen, even a small one belonging to a common, well-known species, is exciting. To discover a large eurypterid representing a species new to science is an event. In this connection part of a report by Prof. Johan Kiaer, a Norwegian paleontologist, may be quoted. It was written just after the first specimen of a new species of *Mixopterus* (Fig. 15-5, *7*), about 3 ft. in length—not a huge eurypterid—had been found in Upper Silurian sandstone near Oslo.

I shall never forget the moment when the first excellently preserved specimen of the new giant eurypterid was found. My workmen had lifted up a large slab, and when they turned it over, we suddenly saw the huge animal, with its marvelously shaped feet, stretched out in natural position. There was something so lifelike about it, gleaming darkly in the stone, that we almost expected to see it slowly rise from the bed where it had rested in peace for millions of years and crawl down to the lake that glittered close below us.

Arachnids

The class Arachnida (*arachne*, spider) has been variously interpreted as including all or nearly all of the invertebrates which now are assigned to the subphylum Chelicerata, or as containing only animals which have the essential attributes of spiders. In any case, on fundamental structural grounds the scorpions belong with spiders. It is the present view of leading specialists on this group of arthropods that the arachnids should be defined to exclude forms which do not have the characters of spiders and scorpions. These characters are (1) division of the body into clearly differentiated cephalothorax and abdomen (except in ticks and mites which have them fused); (2) attachment to the underside of the cephalothorax of six pairs of appendages, including chelicerae, pedipalps, and four pairs of legs, but no antennae or mandibles; (3) respiration by means of leaflike closely packed plates containing fine blood vessels (book lungs) or by ramified tubules (tracheae) like those of insects, both signifying that arachnids are air breathers; and (4) essentially terrestrial, solitary mode of living. The internal anatomy is highly organized, as in insects, and sexes almost exclusively are separate. A peculiarity of many, but not all, arachnids is specialization of the chelicerae as poison fangs. Also, most spiders are provided with glands which secrete a fluid that can be extruded through openings on short conical projections (spinnerets) at the rear of the abdomen. This fluid hardens on contact with the air to form the "silk" with which spiders may form webs, build a protective cocoon around eggs, or entangle prey.

Thus, the arachnids are a distinctive group. They differ from other chelicerates and crustaceans in the absence of gills and in being adapted to life on land; from myriapods in body form and number of appendages; and from insects in lacking wings, a separate head and antennae, and in having eight walking legs instead of six.

The exoskeleton of arachnids is a chitinous covering built of lamellae. It varies considerably in thickness and rigidity among different members of the group. The spiders called Ricinulei are veritably armored, whereas nearly the whole body covering of some acarians (ticks) and other arachnids is a soft, slightly leathery layer. The skeletal covering varies also according to parts of the body. Generally it is strongest on the dorsal side of the prosoma. In many spiders it is weak in the abdominal region. During growth, the exoskeleton is molted five to eight times. At each molting, the body increases in size, changes somewhat in form, and commonly acquires new color patterns.

The size of arachnids ranges from mites less than 0.5 mm. (0.02 in.) and spiders 1 mm. (0.04 in.) in length to spiders 90 mm. (3.5 in.) and scorpions 165 mm. (6.5 in.) in length. Most species are not more than 1 in. in maximum dimension.

The arachnids are divided into 4 subclasses and 16 orders, all of which are represented among fossils. The oldest known arachnids are Silurian scorpions. Two orders of true spiders make a first appearance in Devonian rocks of Scotland, and associated with these are the oldest discovered ticks. Nine additional orders of arachnids are represented by many species in deposits of Pennsylvanian age in North America and Europe. Two of the subclasses and five orders are extinct; the other two subclasses, containing eleven orders, comprise modern arachnid faunas. As a whole, scorpions and spiders are adapted to many sorts of environments and are spread throughout the land areas of the globe.

Scorpions. Modern and fossil scorpions so closely correspond in appearance and in main structural features that a description of one serves for the other. We may examine first a typical living genus, *Scorpio*, which has given its name to the order Scorpiones. An average specimen has a length of 2.5 in. Viewed dorsally (Fig. 15-7*A*), divisions of the body and five prominent pairs of appendages are readily apparent. The front half of the body, distinguished by its greater width, consists of a large anterior segment (prosoma) and seven smaller ones, which comprise the mesosomal portion of the opisthosoma. The narrow posterior part of the abdomen, composed of five subequal segments and a terminal needle-sharp curved claw, is the metasoma. All the appendages are joined to the prosoma, none to the opisthosoma. The frontmost pair, overlooked in our casual first inspection, consists of short chelicerae; they are at the anterior extremity of the prosoma between the bases of the most prominent of the scorpion's appendages—two stout pincer-tipped limbs, which actually are pedipalpi. The remaining four pairs of limbs are similar to one another and subequal in size. They correspond to the eight walking legs of spiders. On the underside of the abdomen, two peculiar comblike appendages are attached to the second mesosomal segment; these are a distinctive structure peculiar to scorpions but having functions which are quite unknown.

Fossil scorpions are classed in two suborders: Protoscorpiones (*proto*, primitive), and Euscorpiones (*eu*, true). The first is represented by two genera of Silurian scorpions and by another genus of Pennyslvanian age. One of the Silurian forms is *Palaeophonus* (Fig. 15-7*B*), characterized chiefly by its short-jointed appendages. This ancient scorpion has been considered possibly to be an aquatic, rather than an air-breathing arachnid. Even if the description of the openings of the book lungs on some of the ventral plates of a specimen from Sweden (Gotland) is faulty, the near identity of many structures belonging to the modern and Silurian scorpions indicates that the latter were air breathers.

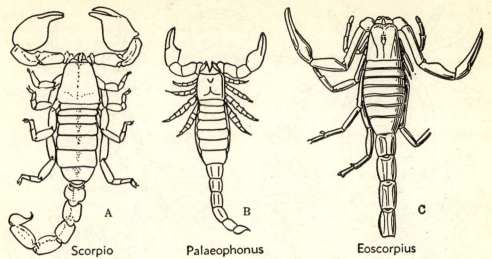

Scorpio   Palaeophonus   Eoscorpius

FIG. 15-7. **Types of living and fossil scorpions.** Structural features of the body and appendages of scorpions correspond closely to those of spiders. Therefore, despite differences in appearance, the two groups are classed together as arachnids.

Eoscorpius Meek & Worthen, Pennsylvanian, North America and Europe. This fossil is characterized by its slender appendages and elongate metasomal segments. *E. dunlopi* (Pocock) (*C*), Scotland.

Palaeophonus Thorell & Lindström, Silurian, Europe and North America. Segments of appendages are very short. *P. nuncius* Thorell & Lindström (*B*), Upper Silurian, Sweden.

Scorpio Linné, Recent, Africa. *S. maurus* Linné (*A*), Algeria.

The Silurian scorpions are known from Sweden, Scotland, and New York.

Eighteen genera of true scorpions have been described from Paleozoic rocks of North America and Europe, of which three occur in Mississippian and the others only in Pennsylvanian deposits. One of the latter is *Eoscorpius* (Fig. 15-7*C*), which has long slender limbs. Most American fossil scorpions have been found in the Mazon Creek beds, Middle Pennsylvanian, of Illinois.

Spiders. Structural characters which are useful for classification of fossil spiders and for study of evolutionary trends in this group of chelicerates consist almost wholly of externally visible features of the exoskeleton. They include elements of both the dorsal and ventral sides. The appendages, particularly the arrangement of their proximal parts on the underside of the cephalothorax, are also important.

Three orders of spiders (Solifugae, Palpigradi, Schizomida) exhibit a segmented structure of the cephalothorax, in which separate anterior, median, and posterior skeletal parts are divided by flexible membrane. All other orders have a fused covering of this part of the body, and although it may carry external furrows denoting the position of segments, there is no parting along these lines when the exoskeleton is molted or when the spider dies. The segmented condition is more primitive, evidently, than the coalesced state; yet only one species of spider having a segmented cephalothorax is known in Paleozoic rocks. Other representatives of the three orders mentioned are almost exclusively Tertiary and Recent. The Pennsylvanian spider shown in Fig. 15-8, *1a*, which seems to have two body divisions in front of the abdomen, really possesses a solidly fused cephalothorax, to which all the appendages are attached, and a small anterior plate (not a segment) which projects over the chelicerae. It belongs to the order Ricinulei.

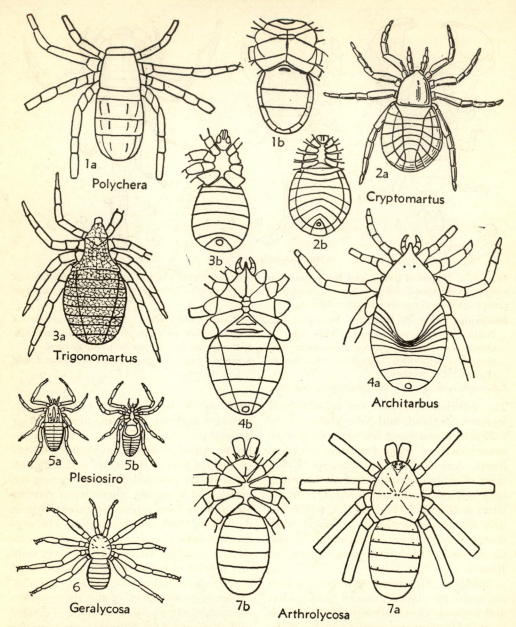

FIG. 15-8. **Representative types of fossil spiders.** The illustrated species are all of Pennsylvanian age. Except figures 6 and 7, each genus belongs to a different order.

Architarbus Scudder, Pennsylvanian, North America and Europe. Order Architarbi, characterized by the broad union of front and rear divisions of the body and shortened abdomen. *A. minor* Petrunkevitch (4*a, b,* dorsal and ventral sides, ×2.5), Mazon Creek, Illinois.

Arthrolycosa Harger, Pennsylvanian, North America and Europe. Belongs to the order Araneae, which includes a large majority of modern spiders. The union of cephalothorax and abdomen is constricted, but less so than in many living forms. *A. danielsi* Petrunkevitch (7*a, b,* dorsal and ventral sides, ×2), Mazon Creek, Illinois.

(*Continued on next page.*)

Abdominal segments of the spiders generally are well defined, both dorsally and ventrally. Only in the order Acari (ticks) is there no trace of segmentation in the adult. The normal complete number of abdominal segments is 12, but in the course of evolution the posterior ones tend successively to disappear. Accordingly, some specialized genera have no more than five abdominal segments. The abdomen of the genus *Polychera* (Fig. 15-8, *1a*, *b*) seems to have only four segments, as viewed from the dorsal side, and three segments, as seen from the ventral side. Two additional ones are concealed beneath the rear part of the cephalothorax, and three tiny segments which form a tail-like appendage belong at the other end of the abdomen. Correlated with the tendency to diminish in numbers is a shortening of the abdominal segments.

Junction of the cephalothorax with the abdomen may be very broad, as in the orders Ricinulei (Fig. 15-8, *1a*, *b*), Anthracomarti (Fig. 15-8, *2a*, *b*), Trigonotarbi (Fig. 15-8, *3a*, *b*), Architarbi (Fig. 15-8, *4a*, *b*), Haptopoda (Fig. 15-8, *5a*, *b*), and a couple of others which have less paleontological importance. In other orders, this union of the two main body parts is constricted in varying degree (Araneae, Fig. 15-8, *6*, *7a*, *b*). Many modern spiders have a decidedly attenuated, wasplike "waist."

Among the 15 main groups of spiders— that is, arachnids exclusive of scorpions— one (Trigonotarbi) ranges from Devonian to Pennsylvanian, and two (Acari, Araneae) from Devonian to Recent. Nine

orders make appearance in rocks of Pennsylvanian age, and of these, three (Architarbi, Haptopoda, Kustarachnae) are confined to this system, one (Anthracomarti) ranges into the Permian, and five (Opiliones, Thelyphonida, Phrynichida, Ricinulei, Solifugae) extend to the present day. The Mesozoic record of spiders is strangely barren, for only one poorly preserved specimen (belonging to the Palpigradi, Jurassic to Recent) is recorded. Two groups (Schizomida, Pseudoscorpiones) of modern arachnids are known as fossils only from Tertiary deposits. The richest assemblage of Tertiary fossil spiders is that from the Baltic region of Europe, preserved in amber; they are nearly all representatives of the modern dominant order (Araneae). Many other, less well-preserved spiders come from the Florissant (Oligocene) beds of central Colorado.

The spiders are a very successful branch of the arthropods which, like insects, became adapted to a terrestrial mode of life. They are less diversified than the insects, but they show almost equally definite evolutionary trends.

Pycnogonids

A paleontologically unimportant but zoologically interesting assemblage of marine arthropods, having uncertain affinities, consists of the so-called sea spiders, or pycnogonids. Because the head, which is provided with a long forward-reaching snout, carries chelicerae and lacks antennae, these creatures are judged to belong to the chelicerates, although they are

---

(Fig. 15-8 continued.)

Cryptomartus Petrunkevitch, Pennsylvanian, North America and Europe. Represents the order Anthracomarti, especially marked by features of the abdomen. *C. hindi* (Pocock) (*2a*, *b*, dorsal and ventral sides, ×1), England.

Geralycosa Kusta, Pennsylvanian, Europe. Order Araneae. *G. fritschii* Kusta (*6*, dorsal side, ×1), Czechoslovakia.

Plesiosiro Pocock, Pennsylvanian, Europe. Order Haptopoda. *P. madeleyi* Pocock (*5a*, *b*, dorsal and ventral sides, ×1), England.

Polychera Scudder, Pennsylvanian, North America and Europe. Represents the Ricinulei, which bear an extra plate in front of the cephalothorax and have no constriction between cephalothorax and abdomen. *P. punctulata* Scudder (*1a*, *b*, dorsal and ventral sides, ×2), Mazon Creek, Illinois.

Trigonomartus Petrunkevitch, Pennsylvanian, North America and Europe. Order Trigonotarbi. *T. pustulatus* (Scudder) (*3a*, *b*, dorsal and ventral sides, ×2), Mazon Creek, Illinois.

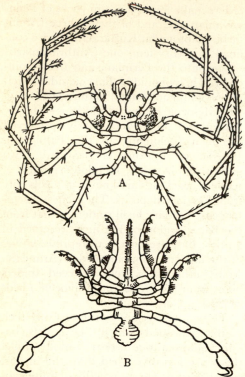

FIG. 15-9. **Recent and fossil pycnogonids.** These strange arthropods, called sea spiders, are aberrant chelicerates. (*A*) A modern form, *Nymphon rubrum* Hodge, approximately ×1. (*B*) A Lower Devonian fossil from western Germany, *Palaeoisopus problematicus* Broili, ×0.25. This form had a median length of 5 in. and width across the base of nearly 9 in.

surely far removed from merostomes and arachnids. Maximum spread of the out-stretched legs of a modern species is about 10 cm. (4 in.), but some fossil specimens are a little larger. Two Paleozoic genera of pycnogonids, both from Lower Devonian rocks of western Germany, constitute the only now-known record of this group in the geologic column. The strange appearance of a representative living member of the group and of one of the fossil forms is indicated by Fig. 15-9.

MYRIAPODS

The jointed-leg invertebrates called myriapods (*myria*, 10,000; *pod*, leg, an

exaggeration allowable as poetic license) have a decidedly wormlike form. All modern species are air breathers, and because the known fossils occur in nonmarine deposits, there is no reason to think that ancient myriapods lived in fresh water or the sea. Some Paleozoic representatives of the subphylum have been interpreted as amphibious, however, and it is probable that the distant arthropod ancestors of this assemblage were marine associates of the trilobites. Actually, we know nothing of evolutionary steps which gave rise to differentiation of the stock. Fossil remains are scanty, and except specimens preserved in amber, they are mostly very imperfect. Because they constitute a minor part of the paleontological record, brief notice of their distinguishing features is sufficient.

Among four recognized classes of myriapods, two are predominant: the Diplopoda, comprising the millipedes (thousand-legs), and the Chilopoda, including various types of centipedes. Both are represented by fossils.

Diplopods

The diplopods are chiefly characterized by their large number of short legs, arranged in double pairs at the lower edges of each segment (Fig. 15-10*A*). These range from 11 to 192 or more in number, and all have virtually identical size and form. There is no differentiation of parts, except that the first three segments behind the head have single pairs of legs, instead of double. The head carries a pair of antennae, and on its upper surface are two clusters of simple eyes, so crowded as to resemble compound eyes. The chitinous exoskeleton is somewhat strengthened by calcium carbonate, which should facilitate fossilization of individuals suitably buried in water-laid fine sediment. These animals molt their test from time to time, like other arthropods.

The oldest discovered fossil millipedes occur in Upper Silurian rocks of Scotland. They are known in the Devonian and Mississippian, but are most numerous in rocks of Pennsylvanian age. Among several genera described from North America and Europe, *Euphoberia* closely resembles some modern millipedes which are provided with longer-than-average legs (Fig. 15-10*B*). Another Pennsylvanian millipede (*Acantherpestes*) attained the unusual length of 20 cm. (8 in.). Few modern kinds, world-wide in distribution, exceed 75 mm. (3 in.).

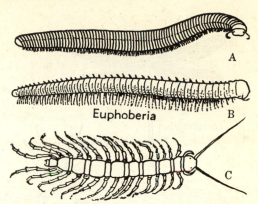

FIG. 15-10. **Living and fossil myriapods.** (*A*) A living diplopod, *Iulus*, ×1. (*B*) *Euphoberia armigera* Meek & Worthen, ×0.7 (restored), Pennsylvanian, Illinois, (*C*) A living chilopod, *Lithobius*, ×1.

Chilopods

The centipedes generally resemble millipedes in their elongate wormlike form and construction of many similar segments, but only one pair of legs is attached to each segment (Fig. 15-10*C*). The legs of the centipede are longer and stouter, and the segments are somewhat flattened dorsoventrally, instead of subcylindrical, as in millipedes. The head bears a pair of antennae and two simple-eye clusters. The first segment behind the head differs from others in that the limbs attached to it are modified as poison claws, or fangs. The total number of segments in various centipede genera ranges from 15 to 175, the largest kinds attaining a length of 20 cm. (8 in.) or more.

Fossils which are classed somewhat uncertainly as centipedes occur in Pennsylvanian rocks of Illinois. The class is definitely represented among early Tertiary fossils by specimens preserved in amber of Oligocene age. Various modern species are distributed throughout the world.

INSECTS

A division of the arthropods, exceeding all others in variety of form, number of species, adaptation to different modes of life, and spread over all land areas of the globe, is called Insecta. The name (*insectus*, cut into) refers to the strongly incised divisions which separate the insect body into three parts: head, thorax, and abdomen. As in other arthropods, these parts are composed of segments, or metameres. Those of the head are so thoroughly fused that no clearly marked external indication of them persists. Appendages attached to the head (antennae and three pairs of mouth parts) denote four primary somites, and embryological study indicates the existence of two more—a total of six. The thorax comprises three segments, each of which bears a pair of legs. Thus, insects have only six post-oral limbs, whereas spiders and scorpions have eight. The abdomen normally contains 10 segments, but some primitive wingless insects have 11 or 12. This part of the body lacks limbs, but other sorts of appendages may be present. Although larvae may lack a skeletal covering, adult insects are provided with more or less rigid chitinous hard parts which enclose all parts of the body and its appendages. Among these appendages are the one or two pairs of wings, which during all or part of the life of most insects are attached to the thorax. The wings are double-layered outgrowths of the insect body wall, the two layers being in contact except along lines called veins, which furnish passageways for nourishment during growth, and which by their thickened structure add to the strength of the wings.

Insects are exclusively air breathers in adult life, but many aquatic larvae obtain oxygen by means of tracheal **gills**. Most

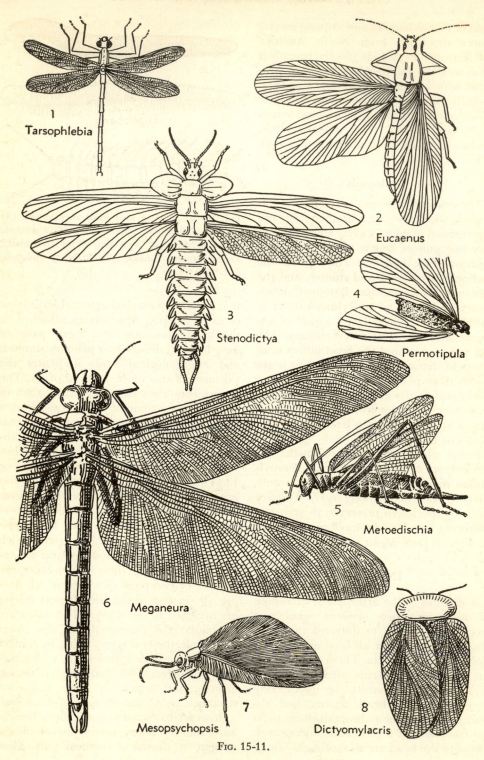

1 Tarsophlebia

2 Eucaenus

3 Stenodictya

4 Permotipula

5 Metoedischia

6 Meganeura

7 Mesopsychopsis

8 Dictyomylacris

Fig. 15-11.

aquatic adult insects, including the few which are adapted to live in salt lakes and the sea, have to come to the surface for air, but there are some which are able to stay submerged indefinitely by transfer of oxygen in the water to air bubbles held at the respiratory openings of the body. The implication of these facts is that the insects are descended from an arthropod stock which had already left the sea and become adapted for life on land.

Like trilobites, crustaceans, and chelicerates, the insects pass through several growth stages between hatching from an egg and becoming an adult. The larval forms of most species differ more or less profoundly from fully grown individuals, as illustrated by the grub which develops into a housefly and the caterpillar which turns into a butterfly. These changes, termed metamorphoses, are accompanied by molting of exoskeletons belonging to the larvae, but once the adult stage is reached, change of shape, increase of size, and molting of hard parts cease. The animal "stays put" until it dies.

The average size of insects is small as compared with most other main groups of arthropods, and without doubt this partly explains their wide distribution, for they are borne by water currents, blown by winds, and transported by birds and mammals for long distances. Some are less than 0.25 mm. (0.01 in.) in full-grown length, and the great majority do not exceed 10 mm. (0.4 in.). Among living insects, the largest are a 15 cm. (6 in.) South American beetle, a tropical moth 23 cm. (9 in.) wide across the wings, and a "walking stick" 28 cm. (11 in.) long. The most

gigantic known insect is a fossil dragonfly (*Meganeura*, Fig. 15-11, *6*) of Pennsylvanian age, which had a wingspread of 60 to 75 cm. (2 to 2.5 ft.).

Classification of the host of known insects, living and extinct, offers many difficulties, although entomologists are generally agreed on most main divisions, including subclasses, orders, and a very large number of families. Two very unequal subclasses are recognized: the primitive wingless insects, called Apterygota (*a*, not; *pterygota*, winged), which are relatively few in number; and the wing-bearing insects (including some degenerate wingless forms), called Pterygota, which include approximately 98 per cent of known species. The first group contains 4 orders and the second about 36 orders.

Primitive Wingless Insects

Two of the four orders of primitive wingless insects are represented by fossils and warrant our notice. The others may be neglected. Especially interesting is the group called Collembola, because they include the oldest known insects, discovered in Middle Devonian rocks (Rhynie chert) of Scotland. Four specimens of a species named *Rhyniella praecursor*, first described in 1926, now are supplemented by a collection of many others, which not only permit identification of structures belonging to the Collembola but reveal that these early representatives of the insects are already specialized in several respects. Indeed, they seem most closely related to fossils from the Cretaceous of Canada. One important aspect of the finding of these apterygote insects in the Devonian is their

---

FIG. 15-11. **Fossil winged insects.** The forms illustrated are mostly late Paleozoic in age.

1, *Tarsophlebia eximia* Hagen, ×1, Jurassic, southern Germany. Dragonfly.

2, *Eucaenus ovalis* Scudder, ×0.5, Pennsylvanian, Illinois. Cockroach.

3, *Stenodictya lobata* Brongniart, ×1, Upper Carboniferous, France. Paleopteran.

4, *Permotipula* sp., ×1, Upper Permian, Queensland. Ancestral dipteran.

5, *Metoedischia* sp., ×1, Permian, Russia. Orthopteran.

6, *Meganeura monyi* Brongniart, ×0.25, Upper Carboniferous, France. Dragonfly.

7, *Mesopsychopsis hospes* Germar, ×2, Jurassic, Germany. Lacewing.

8, *Dictyomylacris poiraulti* Brongniart, ×2, Upper Carboniferous, France. Cockroach.

bearing on the argument that the wingless character of the Collembola and other orders is a result of degeneration, the wings (hypothetical) of their ancestors having been lost. Wingless insects are now known from considerably older rocks than those which contain the first-discovered winged forms.

The Collembola are small, having wedge-shaped bodies, short antennae, and simple eyes. They have only six abdominal segments. Modern kinds have specialized appendages on some of these segments, of which that on the fourth generally is the largest and serves as a springing structure —hence the name springtail, which is applied to this group.

The other group of primitive wingless insects which is found among fossils is that of the Thysanura. A present-day unwelcome representative of the order is the "silverfish," which eats clothing and bindings of books. Fossil thysanurans have been reported from rocks of Pennsylvanian age, but they are doubtfully identified. Well-preserved fossils occur in Tertiary rocks, however.

Winged Insects

The paleontological record of wing-bearing insects, as now known, begins in Pennsylvanian time, when both in North America and Europe numerous kinds of primitive Pterygota had made an appearance. Large numbers of well-preserved specimens have been found in rocks of this age, especially in the Mazon Creek beds (Middle Pennsylvanian, Desmoinesian) of Illinois, Saarbrücken beds (Middle Pennsylvanian, Westphalian) of western Germany, and coal-bearing deposits (Upper Pennsylvanian, Stephanian) of the Commentry region, in France. Somewhat younger, but equally important Paleozoic insect-bearing strata are known in central Kansas, New South Wales, and Russia. Collections from these and other later Paleozoic localities contain many thousands of specimens, which have furnished the basis for a description of hundreds of species distributed among scores of genera. Considering the prevalent unfavorable conditions for the preservation of terrestrial invertebrates such as these insects, the aggregate accumulated information about them is remarkably great.

Paleopterans. A dominant group of Pennsylvanian and Permian insects comprises the paleopterans ("ancient wings"), which include numerous kinds of large insects, 5 to 10 cm. (2 to 4 in.) in average wingspread but attaining a maximum of more than 60 cm. (2 ft.). Some branches of this stock have persisted to the present day.

The paleopterans are primitive in (1) the similarity to one another of thoracic parts and of abdominal parts, (2) possession of two nearly identical wings borne by each side of the thorax, and (3) the permanently outspread attitude of the wings, which could not be folded back above one another over the abdomen, thus resembling the wings of a dragonfly. The relatively small head carries well-developed compound eyes and three centrally located ocelli. The three thoracic segments are subequal, and to each a pair of strong walking legs is attached. The pattern of primary veins which furnish support for the wings and of the tiny cross veinlets differs from that of later, more advanced insects in its generalized plan and the larger number of its elements. A most primitive sort of wing belongs to *Stenodictya* (Fig. 15-11, *3*), for the veinlets are arranged in small polygons; this insect also retains rudiments of a third pair of wings on the first thoracic segment. The distinctly segmented abdomen commonly is characterized by short pleuron-like lateral projections. These structures and the much larger outgrowths of the thorax, which constitute wings, have been interpreted as signs of relationship to the trilobites, but this seems very dubious.

One of the most advanced assemblages of paleopteran insects is made up of ancestral dragonflies. The giant form *Meganeura* (Fig. 15-11, *6*) is a late Pennsylvanian member of this group, and several other kinds occur in Permian rocks of North America, Europe, and Australia. These Paleozoic fossils differ chiefly from Mesozoic and Cenozoic descendants, including modern genera, in the greater complexity of their wing venation. True dragonflies (Odonata) are not known to antedate the Triassic; a Jurassic representative is illustrated in Fig. 15-10, 7.

Blattoids (Cockroaches). A very abundant, widely distributed group of fossil insects consists of the cockroaches and allied forms. These appear in Pennsylvanian deposits, are very common in some Permian, Mesozoic, and Cenozoic formations, and are represented by more than 1,200 living species. Late Paleozoic cockroaches attained a length of 10 cm. (4 in.). Their variety is indicated by the description of approximately 800 Pennsylvanian and Permian species.

The blattoids (Latin, *blattus*, cockroach) are especially characterized by enlargement of the first thoracic segment so that forward-projecting parts largely or entirely conceal the head. Also, they have a rather distinctive pattern of wing veins, which has remained little changed from oldest known kinds down to the present. The wings can be rotated backward in partly overlapping manner, the front pair of wings entirely concealing the hind pair (Fig. 15-11, 2, 8). Occurrence of fossil cockroaches indicates that the habits of these insects in Paleozoic time were little if any different from now, for most abundant remains are associated with accumulations of plant debris such as produced coal beds.

Orthopteroids. Several groups of so-called straight-winged insects, among modern members of which are grasshoppers, locusts, and crickets, are collectively termed orthopteroids (*ortho*, straight; *pteron*, wing). Some of these (Protorthoptera, Plecoptera) are common in the Pennsylvanian and Permian insect faunas. Formerly, true Orthoptera, characterized particularly by the lengthening and en-

largement of the third pair of legs to make effective jumping organs, were not counted among Paleozoic fossils. Partly as the result of new finds in Permian rocks and partly as a consequence of the restudy of Pennsylvanian insects, the orthopteran group, as narrowly defined, is recognized to include contemporaries of the most ancient dragonflies and cockroaches. Some Permian cricket-like insects strikingly resemble living species (Fig. 15-11, 5). The Plecoptera are an orthopteroid assemblage which includes the May flies. They, too, were well differentiated in late Paleozoic time, for seven families containing numerous genera are described from Lower Permian rocks of Kansas and Russia, and two additional families are represented in Upper Permian deposits of Australasia and northern Russia.

Beetles. The wide distribution and abundant kinds (some 250,000 species) of modern beetles suggest that these insects should be represented among fossils, especially when we take note of the thick hard exoskeleton which covers their body and the equally strong forewings. The term Coleoptera (*coleo*, sheath), which is applied to this group, refers to the nature of the beetle forewings. They are not useful in flying but serve as a rigid cover for the rear wings. Except in flight, the convexly rounded forewings fit snugly together above the abdomen.

Coleoptera begin in the Permian—or at least none older have yet been found. Several genera are known from Permian deposits in Russia and additional ones from rocks of the same age in Australia. Some 300 species occur in Mesozoic strata and 2,300 in the Tertiary. Many complete or nearly complete insects occur as fossils, but, as we might expect, many kinds are represented only by dissociated forewings. Shape, vein markings, and similar features distinguish these fragmentary remains.

Neuropteroids. So-called lacewings and allied forms, which are distinguished mainly by the remarkably fine pattern of their wing venation, are placed together under the designation of neuropteroids (*neuro*, vein).

The paleontological record of this group is chiefly Mesozoic and Cenozoic, but beginnings are seen in five Permian families which are defined on the basis of fossils from Kansas and Russia (Fig. 15-11, 7). Especially well-preserved representatives of the group occur in early Tertiary amber from the Baltic region of Europe and in Oligocene lake deposits (Florissant) of Colorado.

Mecopteroids. Four orders of insects, which collectively are termed mecopteroids (*meco*, length, referring to equal length of wings), include butterflies, moths, and some lesser known groups. The flies (Diptera) are placed here because of the nature of the oldest representatives of this stock, which had four equal-size wings having a venation comparable to that of true two-winged flies.

The Lower Permian of Kansas and later Permian in Australia and Russia contain several kinds of insects belonging here, among them being the four-winged ancestral dipteran just mentioned (Fig. 15-11, 4). Others occur in Triassic, Jurassic, and younger deposits. The oldest true flies have been found in Upper Triassic rocks of Queensland. Although mothlike insects, assigned to neuropteroids, occur in Permian and Triassic rocks, no true moths or butterflies older than Eocene are known.

Hymenopteroids. Bees, wasps, ants, and some other groups are classed as hymenopteroids (*hymen*, membrane). The front and hind wings of these insects interlock in flight so as to function as a single pair of wings. Fossils belonging to this group are very abundant in Tertiary deposits of various regions. A few Cretaceous and Jurassic hymenopteroids are known, but none older.

Hemipteroids. Many sorts of bugs of terrestrial and aquatic habitat are included in this assemblage (*hemi*, half, referring to the part stiffened, part membranous nature of the forewings). Among these are several kinds of water bugs, chinch bugs, cicadas, leaf hoppers, and aphids. One of the groups (Homoptera) includes a number of fossil genera from the central Kansas Lower Permian insect-bearing deposits, and others from Upper Permian beds of Russia and Australia. The paleontological record is continued in Triassic forms from Queensland and Jurassic fossils from central Europe, Turkestan, and Siberia.

Other Groups. A number of insect orders which have lesser paleontological importance than those noted briefly above are omitted from notice. All the known fossils have interest for the entomologist who is concerned with problems of the phylogeny of insects, but most of them are beyond the scope of study in general invertebrate paleontology.

REFERENCES

BUCHSBAUM, R. (1948) *Animals without backbones:* University of Chicago Press, Chicago, rev. ed., pp. 1–405, illus. (chelicerates, myriapods, insects, pp. 268–298). Good general descriptions and excellent illustrations.

CARPENTER, F. M. (1930) *A review of our present knowledge of the geological history of the insects:* Psyche, vol. 37, pp. 15–34.

CLARKE, J. M., & RUEDEMANN, R. (1912) *The Eurypterida of New York*, New York State Mus. Mon. 14, pp. 1–627, pls. 1–88, figs. 1–121. A classic work on these fossils.

DENIS, R. (1949) *Sous-classe des Aptérygotes:* in

Grassé, P., Traité de zoologie, Masson et Cie., Paris, vol. 9, pp. 111–275, figs. 1–114 (French). Mainly morphology but refers to fossils.

DESPAX, R., CHOPARD, L., GRASSÉ, P., DENIS, R., & JEANNEL, R. (1949) *Sousclasse des Ptérygotes:* in Grassé, P., Traité de zoologie, Masson et Cie., Paris, pp. 277–1077, figs. 1–752 (French). Describes dragonflies, termites, beetles, orthopterans and lesser groups, with data on fossils.

FAGE, L. (1949) *Classe des Merostomacés:* in Grassé, P., Traité de zoologie, Masson et Cie., Paris, vol. 6, pp. 219–262, figs. 1–46 (French). Xiphosurans and eurypterids.

—— (1949a) *Classe des Pycnogonides:* Same, pp. 906–941, figs. 687–719. (French). Sea spiders.

JEANNEL, R. (1949) *Les insectes: classification et phylogénie; les insectes fossiles; évolution et géonémie:* in Grassé, P., Traité de zoologie, Masson et Cie., Paris, vol. 9, pp. 1–110, figs. 1–109 (French). Comprehensive survey of recent paleo-entomological studies, with discussion of successive insect faunas from oldest to modern, which is much influenced by concepts of continental drift.

MILLOT, J., DAWYDOFF, C., VACHON, M., BERLAND, L., ANDRE, M., WATERLOT, G. (1949) *Classe des Arachnides:* in Grassé, P., Traité de zoologie, Masson et Cie., Paris, vol. 6, pp. 263–905; figs. 47–686 (French). Scorpions and spiders, with summary of fossil forms.

O'CONNELL, M. (1913) *Distribution and occurrence of the eurypterids:* Geol. Soc. America Bull., vol. 24, pp. 499–515.

PETRUNKEVITCH, A. (1913) *A monograph of the terrestrial Palaeozoic Arachnida of North America:* Connecticut Acad. Arts. and Sci. Trans., vol. 18, pp. 1–137, pls. 1–13, figs. 1–88. Morphology, classification, and evolutionary trends of the Paleozoic spiders.

—— (1942) *A study of amber spiders:* Same, vol. 34, pp. 119–464, pls. 1–69. A very technical paper on Araneae but having excellent illustrations of remarkable Baltic-amber early Tertiary fossils.

—— (1945) *Palaeozoic Arachnida of Illinois:* Illinois State Mus. Sci. Papers, vol. 3, no. 2, pp. 1–72, pls. 1–4, figs. 1–34. Mainly deals with Architarbi and Anthracomarti, discussing evolution.

—— (1949) *A study of Paleozoic Arachnida:* Connecticut Acad. Arts and Sci., Trans., vol. 37, pp. 69–315, pls. 1–83. An elaboration and revision of earlier work.

RAASCH, G. O. (1939) *Cambrian Merostomata:* Geol. Soc. America Spec. Paper 19, pp. 1–146, pls. 1–21, figs. 1–14. Chief work on aglaspids.

STORER, T. I. (1951) *General zoology:* McGraw-Hill, New York, pp. 1–832, illus. (chelicerates, myriapods, insects, pp. 485–555). Compact, informative, and comprehensive on main morphologic features and habits.

STØRMER, L. (1934) *Merostomata from the Downtonian sandstone of Ringerike, Norway:* Skr. Norske Vidensk.-Akad. Oslo, Mat.-Naturv. Kl. 1933, no. 10, pp. 1–125, pls. 1–12, figs. 1–39. Describes *Mixopterus* and discusses habitat and evolution of eurypterids.

CHAPTER 16

Echinoderms

The "spiny-skinned" animals called echinoderms (phylum Echinodermata) are highly organized, exclusively marine invertebrates. Their body plan and structural organization is very different from that of any other invertebrate group, so that the best clue to their relationship is furnished by the larvae. Strangely enough, these resemble most closely the larval forms of primitive hemichordates, a group which stands next below the chordates comprising all vertebrate animals.

Living representatives of the Echinodermata include the starfishes, or sea stars (asteroids), brittle stars (ophiuroids), sand dollars and other sea urchins (echinoids), sea cucumbers (holothuroids), sea lilies (stalked crinoids), and feather stars (free crinoids) (Fig. 16-1). A number of additional kinds of echinoderms belong to extinct classes. All have a primitive bilateral symmetry, but this is commonly masked by a strongly developed radial symmetry following a fivefold plan of organization.

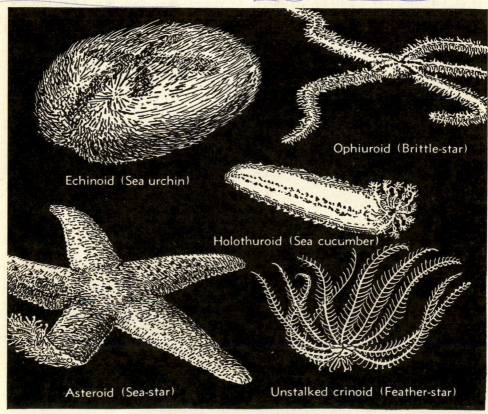

Echinoid (Sea urchin)

Ophiuroid (Brittle-star)

Holothuroid (Sea cucumber)

Asteroid (Sea-star)

Unstalked crinoid (Feather-star)

FIG. 16-1. **Types of echinoderms.**

All echinoderms are characterized by the possession of skeletal elements formed of calcite built up in crystalline form. It is important to observe that the echinoderm skeleton is a product of internal secretion, like the bony skeleton of vertebrates, and that the individual hard parts of this group of invertebrates may increase in size during the life of the organism. This capacity to grow in size applies both to elements of the main body skeleton (**theca** or **calyx**) and those of the appendages (**rays, arms, stem**). The calcareous parts have a characteristic microstructure in which regularly arranged minute passageways permeate the calcite (Fig. 16-2); thus in sections, almost any echinoderm fragment can be identified by its very fine honeycomb structure as belonging to this phylum.

Although a radial plan of body organization is interpreted as a primitive character, associated with animals like sea anemones and corals, which do not move about but live in fixed locations, the body structure of echinoderms is much advanced over that of the coelenterates and other radially symmetrical invertebrates. From the mouth, centrally located on the upper surface (crinoids and other stem-bearing forms), lower surface (asteroids, ophiuroids, and echinoids), or anterior extremity of the body (holothuroids), a long coiled gut extends to the anus, situated between two of the rays (Fig. 16-3) or at the opposite pole. An expansion of the digestive tract in some echinoderms corresponds to a stomach. Surrounding the digestive tube is a distinct body cavity. Within the body cavity and extending to appendages are parts of the nervous system, reproductive parts, and (in most echinoderms) a highly organized water-circulatory system. No heart is present.

Aeration and transfer of nutrient substances to various parts of the body are effected by movement of fluids within the body cavity and, in part, by action of the water-circulatory system. In its simplest form, as seen in primitive, extinct stemmed echinoderms (cystoids, paracrinoids), water circulated through pores in the plates surrounding the body (Fig. 16-3G). In more highly organized echinoderms, such as the starfishes and sea urchins, water is introduced through a special sieve plate (**madreporite,** Fig. 16-4) and by the action of cilia is drawn through a calcified tube (**stone canal,** Fig. 16-4) to a tubular ring which surrounds the gut near the mouth and thence along canals which diverge to each main ray (Fig. 16-3B–E). Innumerable extensions of the radial canals have the form of small rounded muscular sacs (**ampullae**) joined to hollow, thin-walled cylinders which end in sucker-like expansions or have rounded-off ends (**tube feet** or **podia,** Fig. 16-3). By contraction and expansion of the small sacs, water is pressed into the tubular extensions or withdrawn from them, and this serves to lengthen or shorten the small tubes. Many echinoids use the podia in the lower part of the body for crawling and those in the upper part for respiration. The podia of the starfish, which end in sucker disks, are also used for locomotion, and by a collective steady pull, they are effective in opening the shells of bivalves on which the starfish may feed. The radial canal and its many extensions are disposed along tracts called **ambulacra,** and the hard parts of most

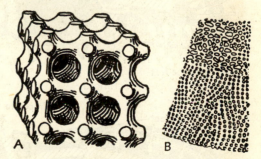

FIG. 16-2. **Microstructure of echinoderm hard parts.** The plates, spines, and other skeletal parts of echinoderms consist of a fine honeycomb of crystalline calcite, each individual plate being a crystallographic unit. The figures show parts of a modern crinoid (*Holopus*) plate, *A*, ×55, *B*, ×5.

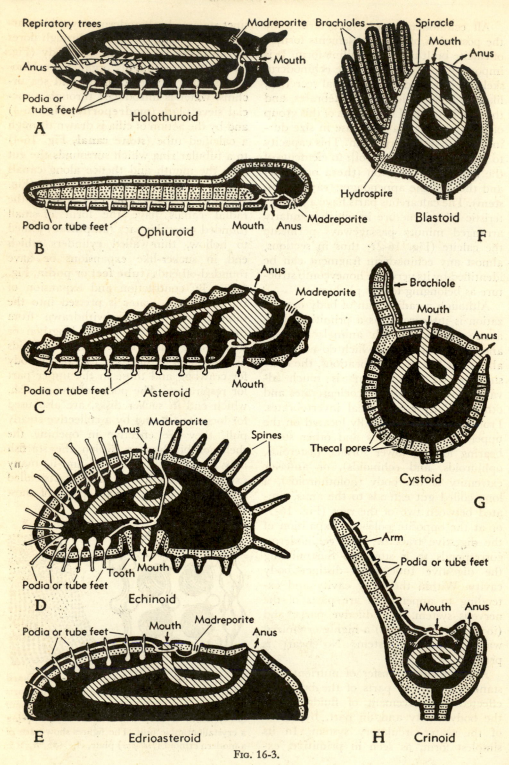

FIG. 16-3.

echinoderms are specially modified along the ambulacral areas.

The water-circulatory system of ancient primitive types of echinoderms does not serve for the purpose of locomotion but for respiration and sensation and for gathering of food particles. Water currents produced by lashing movement of minute cilia converge along ambulacra toward the mouth. Extension of ambulacral areas is developed in a radial manner along the rays.

KINDS OF ECHINODERMS

Echinoderms are divisible into two main groups: those which are more or less permanently fixed in mode of growth (*Pelmatozoa*, mainly stem-bearing animals) and those which are not attached to the sea bottom (*Eleutherozoa*, free-moving animals). Characters which are best suited for further classification of the echinoderms include the general form and organization of the body, the nature of the water-circulatory system, and the structural plan of skeletal elements. This has led to recognition of the following main divisions. Extinct groups are indicated by an asterisk (*).

Main Divisions of the Phylum Echinodermata

Pelmatozoa (*subphylum*), attached forms. Lower Cambrian–Recent.

*Eocrinoidea (*class*), primitive crinoids. Middle Cambrian–Middle Ordovician.

*Paracrinoidea (*class*), primitive crinoids. Middle Ordovician.

*Carpoidea (*class*), laterally compressed forms. Middle Cambrian–Lower Devonian.

*Edrioasteroidea (*class*), discoid forms. Lower Cambrian–Lower Mississippian.

*Cystoidea (*class*) irregular theca bearing pores. Middle Ordovician–Upper Devonian.

*Blastoidea (*class*), symmetrical budlike theca. Middle Ordovician–Upper Permian.

Crinoidea (*class*), sea lilies, feather stars. Lower Ordovician–Recent.

*Inadunata (*subclass*), arms free above lowest ray plates. Lower Ordovician–Triassic.

*Flexibilia (*subclass*), calyx plates movable. Middle Ordovician–Upper Permian.

*Camerata (*subclass*), ray plates rigidly in calyx, Middle Ordovician–Upper Permian.

Articulata (*subclass*), mostly stalked crinoids. Lower Triassic–Recent.

Eleutherozoa (*subphylum*), free-moving forms. ?Middle Cambrian Ordovician–Recent.

Holothuroidea (*class*), sea cucumbers. ?Middle Cambrian, Mississippian–Recent.

Stelleroidea (*class*), starfishes and brittle stars. Ordovician–Recent.

*Somasteroidia (*subclass*), primitive starfishes. Ordovician.

Asteroidia (*subclass*), sea stars. Ordovician–Recent.

Ophiuroidia (*subclass*), brittle stars. Ordovician–Recent.

Echinoidea (*class*), nonstellate eleutherozoans. Ordovician–Recent.

Regularia (*subclass*), regular sea urchins. Ordovician–Recent.

Irregularia (*subclass*), heart urchins and sand dollars. Jurassic–Recent.

*Bothriocidaroidea (*class*), primitive echinoidlike eleutherozoans. Ordovician.

*Ophiocystia (*class*), primitive boxlike eleutherozoans. Ordovician–Devonian.

Some of the groups of echinoderms differentiated in the foregoing list are so dissimilar in appearance that they seem to have nothing in common except the crystalline calcite nature of their hard parts. If, however, representatives of each sort of echinoderm are oriented in like manner, without reference to the normal position of the animal during life, many significant resemblances and correspondences become apparent. In nearly all classes, the mouth is centrally placed on the lower or upper side of the animal, the ambulacra radiating from it. The anus is located generally in an interray area, either on the same side of the body as the mouth (termed the **oral** side), or on the

Fig. 16-3. **Cross sections of echinoderms showing the internal placement of skeletal elements.** The sections are drawn along the antero-posterior plane, the anterior ray at left and posterior interray at right. The digestive and water-circulatory systems are represented diagrammatically.

opposite surface (**aboral**). The position of the opening of the stone canal (madreporite) is constant within most groups (Fig. 16-4).

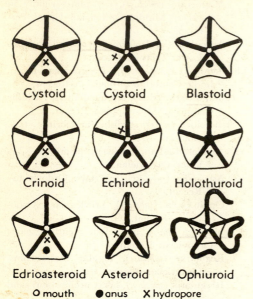

O mouth   ● anus   X hydropore

FIG. 16-4. **Location of the mouth, anus, and madreporite in various echinoderms.** The animals are viewed from the ventral side, mouth in the center; the anus and madreporite mostly occur in interrays (in echinoids, asteroids, and ophiuroids, on the dorsal side).

MODES OF LIFE

No echinoderms are adapted for life in fresh waters, and excepting a few, not even in brackish waters, but they are found distributed in almost every conceivable environment in the normally saline waters of shallow seas and deep oceans. Most holothuroids and many sea urchins prefer a mud bottom. Some obtain their food supply mainly by passing the mud through their alimentary tract, abstracting organic material distributed in the sediment. They function effectively as scavengers, and their action contributes greatly to breaking down organic hard parts on the sea bottom. Other echinoderms prefer sandy or rocky bottoms. Some crinoids and sea urchins live in strong currents and the zone of breaking waves around coral reefs. The

crinoids, except the free-swimming types, are anchored by their stem or by the cemented base of the body. Echinoids hold themselves in place by their spines or podia; some are protected by depressions which they have excavated in hard rocky material.

The habitat of crinoids ranges from very shallow water to abyssal depths. Modern stalked crinoids prefer quiet water, but the occurrence of crinoid remains in Paleozoic and other rocks indicates that these organisms lived profusely in very shallow seas. Free-swimming crinoids are found at all depths in present-day oceans, and many are observed to be highly gregarious.

The sea stars and brittle stars are common inhabitants of the littoral zone of nearly all coasts and live also in moderately deep waters. The brittle stars are the most rapid of all free-moving echinoderms in crawling along the sea bottom. As a rule, both sea stars and echinoids are sluggish animals, remaining immobile for considerable periods of time.

LARVAL DEVELOPMENT

Echinoderms develop from eggs which generally are released into the sea water and fertilized there. The egg passes through successive stages of cell division which produce a minute ciliated hollow sphere (blastula), a two-layered vaselike form (gastrula), and eventually a larva having a mouth, digestive tube, anus, and an enclosed body cavity adjoining the alimentary tract. At this stage, the larvae of nearly all echinoderms are bilaterally symmetrical, free-swimming organisms, which lack calcareous skeletal parts. They have an uncoiled gut, and the body cavity consists essentially of three pairs of pouches. The larva (called *Dipleurula*) of pelmatozoans and some sea stars attaches itself to some foreign object by the anterior part of the body, which elongates into a stalk (Fig. 16-5, *2a, b*). The mouth migrates upward, and the gut becomes twisted. Among pelmatozoans, the larva continues to grow in this attached position (Fig. 16-5, *3a*, *b*), but among starfishes, the larvae shortly break away from their place of attachment, the mouth and anus become closed, and a new mouth and anus break through on the original left and right sides, respectively. This produces an adult axis at right angles to the larval axis. The radial canals develop, and the adult body form begins to appear (Fig. 16-5, *4a*, *b*). Echinoid, holothuroid, and ophiuroid larvae, as well as some belonging to

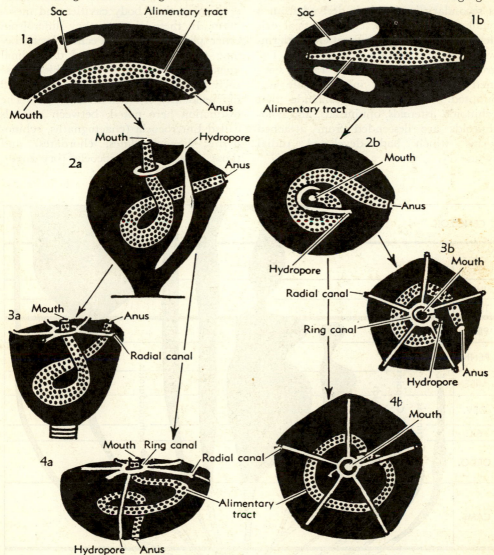

FIG. 16-5. **Larval development of echinoderms.** All echinoderms have an early free-moving larvae stage which shows bilateral symmetry. This is inferred to indicate an ancestral free-moving form called *Dipleurula* (1*a*, 1*b*). A following attached stage of the larva (2*a*, 2*b*) induces a twisted course of the gut and introduces radial symmetry (although actual attachment is by-passed in most eleutherozoans). Structural features of adult stem-bearing echinoderms (3*a*, *b*) and free-moving echinoderms, shown in inverted position (4*a*, *b*), are represented diagrammatically. Figures at left are side views; those at right are top views.

starfishes, lack an attached stage, and many show a more direct development from egg to mature form. This is especially true of young which begin life protected by the parent.

The radial symmetry of all echinoderms is secondary in that it is derived from a bilateral larva. Thus, embryological evidence suggests that ancestral echinoderms were bilateral free-moving animals. The group presumably acquired radial symmetry when a fixed mode of life was adopted. Therefore, we conclude that echinoids, asteroids, ophiuroids, and holothuroids are descended from attached forms which had developed radial symmetry.

RELATIONSHIPS

The echinoderms are closely related to chaetognath worms, the hemichordates, and chordates (vertebrates) in the manner of forming their body cavities and mesoderm. In these respects, they differ from other phyla. Thus, the echinoderms belong to a special branch of the animal kingdom, the chordate line, which arose early in the history of life and developed independently of other invertebrates. The distinction here noted, between animals called Enterocoela (chaetognaths, echinoderms, hemichordates, chordates) and those classed as Schizocoela (bryozoans,

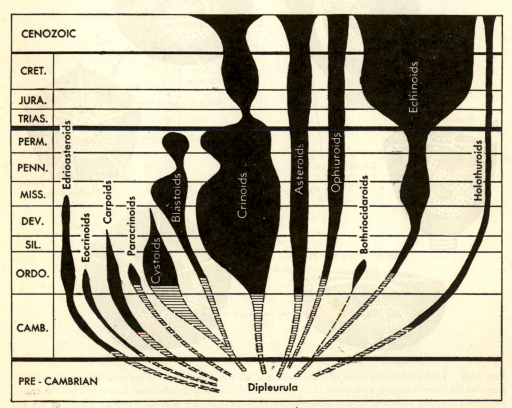

FIG. 16-6. **Geological distribution of echinoderms.** The origin of echinoderms as a distinct group of invertebrates and the beginning of their divergence into widely varied forms are judged surely to have taken place in Pre-Cambrian time. The main development of stem-bearing groups is associated with Paleozoic history, whereas free-moving types are dominant in Mesozoic and Cenozoic time.

brachiopods, mollusks, annelids, arthropods, and some others), is confirmed by features of larval development and chemical nature of body fluids.

GEOLOGICAL DISTRIBUTION

The oldest known echinoderms occur in Lower Cambrian rocks. The structure of these and of forms from Ordovician strata, in which all main divisions of the echinoderms (except holothuroids) are represented, clearly indicates that the origin of this invertebrate stock lies in Pre-Cambrian time. The Paleozoic rocks are characterized by dominance of stem-bearing echinoderms, especially crinoids, and most classes of these did not survive into Mesozoic or Cenozoic parts of geologic history. Although remains of echinoids and other free-moving echinoderms are fairly abundant in some Paleozoic formations, they are most common in post-Paleozoic deposits. These unattached forms are much the most widely varied representatives of the Echinodermata in modern seas. The geological distribution of echinoderms is shown diagrammatically in Fig. 16-6.

REFERENCES

BATHER, F. A. (1900) *The Echinoderma:* in Lankester, E. R., Treatise on zoology, A. & C. Black, London, pt. 3, pp. 1–344, 309 figs. Although 50 years old, this is one of the outstanding books on echinoderms, particularly treating fossil forms. Its usefulness is shown by the demand, which has required numerous reprintings, the latest in 1948.

———— (1937) *Echinoderma:* Encyclopaedia Britannica, 14th ed., vol. 7, pp. 895–904, figs. 1–25. A summary by the foremost British authority on fossil echinoderms, treating the comparative structure, classification, embryological development, occurrence, and evolution of these invertebrates.

BORRADAILE, L. A., *et al.* (1948) *The Invertebrata:* Cambridge, London, pp. 1–725, figs. 1–493 (echinoderms pp. 622–659, figs. 453–461). A widely known text which adequately describes structures of living echinozoans and asterozoans but is otherwise weak, no bibliography.

BUCHSBAUM, RALPH (1938) *Animals without backbones:* University of Chicago Press, Chicago, pp. 1–371, illus; 1948, rev. ed., pp. 1–405, illus. A brief but exceptionally lucid and interesting description of echinoderms is given in this work (pp. 299–311, both editions).

CUÉNOT, L. (1948) *Embranchement des Echinodermes:* in Grassé, P., Traité de zoologie, Masson et Cie, Paris, vol. 6, pp. 1–363, figs. 1–399 (French). A comprehensive, excellently illustrated modern treatise by a leading French worker; good on morphology and embryology of living forms but somewhat weak on paleontology. Contains useful bibliography.

MacGINITIE, G. E., & MacGINITIE, N. (1949) *Natural history of marine animals:* McGraw-Hill, New York, pp. 1–473, figs. 1–282 (echinoderms, pp. 221–251, figs. 84–110). Excellent descriptions of modes of life of starfishes, brittle stars, echinoids, and holothurians along the Pacific Coast.

PARKER, T. J., & HASWELL, W. A. (1910) *A textbook of zoology:* Macmillan & Co., Ltd., London, vol. 1, pp. 1–839, figs. 1–704 (echinoderms pp. 375–438, figs. 301–345). Good description of anatomy.

CHAPTER 17

Primitive Attached Echinoderms

Echinoderms which are grouped together for discussion in this chapter are known only from fossil remains, mainly of early Paleozoic age. They are stem-bearing or attached forms, some of which have a laterally compressed body. Many bear armlike appendages, but in general, the food-gathering and water-circulatory structures are relatively undeveloped (Fig. 17-1). Although a majority are primitive types of echinoderms, curiously specialized organisms are not lacking. On the whole, the group is characterized by the absence of a regular pattern in the arrangement of plates enclosing the body. An exception to this generalization, however, is given by the blastoids, among which 13 or 14 symmetrically arranged plates surround the body.

Classification of the primitive attached echinoderms is based primarily on the structure of the plates of the **theca** surrounding the main portion of the body, particularly in relation to water circulation, arrangement of the plates, and nature of the ambulacra. Knowledge of some of

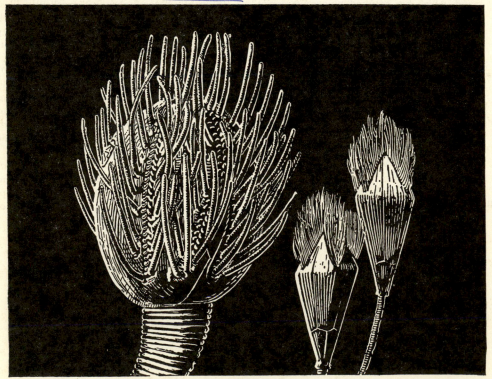

Fig. 17-1. **Restoration of Silurian cystoid and blastoids.** Numerous slender armlike appendages, called brachioles, border pathways (ambulacra) leading to the mouth at the summit of the theca. Attachment is by a stem composed of many discoid plates (columnals).

the groups is decidedly fragmentary, but present information supports the division of this group into six main parts (eocrinoids, paracrinoids, carpoids, edrioasteroids, cystoids, and blastoids), which are treated as classes. They are assigned rank, therefore, correlative with that of the crinoids, starfishes, and echinoids.

EOCRINOIDS

The Eocrinoidea are primitive attached echinoderms having a subglobular or sac-shaped theca, a stem, and free armlike appendages (Fig. 17-2; *1, 3*). They make an appearance in Middle Cambrian rocks and persist to Middle Ordovician.

A typical member of the group, which we may select for study of structural features, is the Cambrian genus *Macrocystella* (Fig. 17-2, *3*). The theca of this stalked echinoderm is composed of four circlets of regularly disposed plates. Each circlet has five plates of similar size and shape, but those of the two intermediate circlets are considerably larger than the ones forming the base and top of the theca. The plates of one circlet alternate in position with those of adjoining circlets. Ridges on the surface of the plates radiate from a central area to the margins so as to meet similar ridges of other plates, thus forming a geometrical pattern. Each of the five topmost plates bears a slender food-gathering appendage, which bifurcates a short distance above the theca; these appendages are termed **brachioles** (Fig. 17-5, *8*) because they correspond in structure to the brachioles of cystoids. They lack branchlets, and they are formed of biserially arranged plates throughout their entire length; ambulacral grooves on the inner (ventral) surface of the brachioles are roofed over by very minute covering plates. Attached to the base of the theca is a tapering stem, composed of relatively wide and short segments in the upper part and of narrow, long segments in the lower part.

The thecal plates of *Macrocystella* and other eocrinoids are solid imperforate bodies of crystalline calcite. In this respect they correspond to crinoids and differ from cystoids. On the other hand, the brachioles of eocrinoids are like those of cystoids and differ from the arms of crinoids. Many crinoids have biserial arms, but the basal part of such arms has a uniserial structure indicating that the biserial arrangement of arm plates in crinoids is an evolutionary modification of an antecedent (and therefore primitive) uniserial plan. The eocrinoids, which have a fairly regular radial symmetry in the arrangement of the thecal plates, may include the ancestral stock of the cystoids and perhaps crinoids, as well as other echinoderms; at any rate, they are judged to stand close to these ancestors.

PARACRINOIDS

Among paracrinoids (Fig. 17-2, *2, 4*), the plates are not arranged in definitely symmetrical manner like eocrinoids, but like many cystoids, the paracrinoids typically have a pore system within the thecal plates. The pores lie beneath a thin solid outer part of the plates (**epistereom**) (Fig. 17-5, *18*) and thus are not visible externally on unweathered specimens. The presence of thecal pores distinguishes paracrinoids from crinoids.

One of the best known paracrinoids is *Comarocystites* (Fig. 17-2, *4*). The theca of this echinoderm is an egg-shaped structure formed of very many irregularly arranged polygonal plates. The exterior surface of these plates is strongly concave (Fig. 17-2, *4a*), and the theca, accordingly, has a peculiar pitted appearance. The inner side of the plates (Fig. 17-2, *4b*) shows arched surfaces extending from the center to the borders. Beneath each arch are parallel pore slits which intersect the plate margin at a right angle so as to connect with similar pores on an adjoining plate. Skeletal appendages of *Comarocystites* consist of the stem and uniserial arms, which bear uniserial branchlets. Some other paracrinoids have food-gathering arms which are recumbent on the surface of the theca.

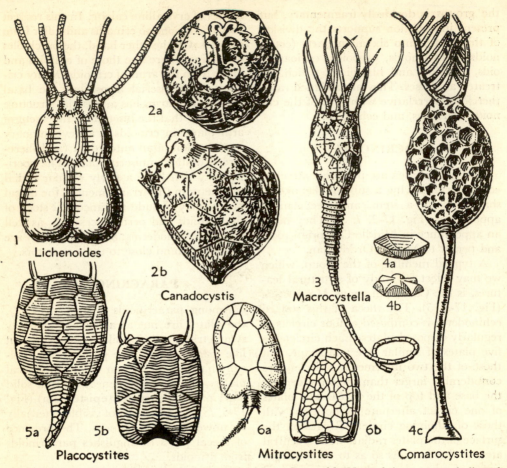

FIG. 17-2. **Representative eocrinoids, paracrinoids, and carpoids.** Natural size except as indicated. All the forms shown are from Cambrian and Ordovician rocks. Eocrinoids include 1 and 3, paracrinoids 2 and 4, and carpoids 5 and 6.

Canadocystis Jaekel, Middle Ordovician. Ventral and side views of a paracrinoid having somewhat irregularly arranged thecal plates. Places of arm attachment are closely grouped around the mouth. *C. emmonsi* (Hudson) (2a, b), Chazyan, New York.

Comarocystites Billings, Middle Ordovician. The plate structure of this interesting paracrinoid is curious in the pronounced thickenings at mid-points of the sutural unions between the plates, coupled with extreme thinness at angles of the plates and in the deeply hollowed central part of the plates. This genus has a fairly regular arrangement of plates near the summit of the calyx but very irregular near the stem. *C. punctatus* Billings (4a, b, oblique views of exterior and interior, respectively, of a thecal plate, enlarged; 4c, a nearly complete indi-

vidual showing ovoid theca, uniserial branched arms, and stem provided with holdfast at the base), Trentonian, Canada.

Lichenoides Barrande, Cambrian. An eocrinoid having five well-developed unbranched biserial brachioles and large bulbous plates (infralaterals) near the base of the theca. *L. priscus* Barrande (1), Middle Cambrian, Czechoslovakia.

Macrocystella Callaway, Cambrian. The symmetrically arranged thecal plates of this eocrinoid bear radial ridges; the brachioles bifurcate, and the stem is very slender toward the base. *M. mariae* Callaway (3), Upper Cambrian, England.

Mitrocystites Barrande, Ordovician–?Devonian. The mouth of this carpoid is located at the

(Continued on next page.)

Paracrinoids are known at present only from Middle Ordovician strata of North America and possibly Europe.

CARPOIDS

The plates of the theca of carpoids are imperforate, and except for a stem, which is peculiarly specialized in the form of a tail-like appendage, there are normally no extensions from the body such as brachioles (Fig. 17-2, 5, 6). A few forms possess well-developed brachiole structures, however.

As shown by *Placocystites* (Fig. 17-2, 5), for example, one side of the body is characterized by large plates which are sym-metrically arranged along a longitudinal axis. This side, which generally is a little more convex than the other, is judged to have lain downward on the sea bottom. The opposite side has a large number of small, irregularly arranged plates, and in this area the anal vent is located. Around the margin are large sturdy plates (**marginals**) which serve as a frame (Fig. 17-5, 21). The mouth is located near the edge opposite the point of stem attachment.

On the whole, this group of echinoderms strikingly resembles some primitive chordates rather than any branch of the cystoids or crinoids. The range of carpoids is from Middle Cambrian to Lower Devonian.

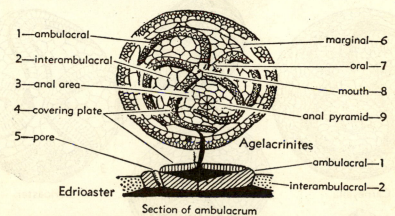

FIG. 17-3. **Terminology applied to edrioasteroids.** This ancient group of primitive attached echinoderms is fastened by the base of the theca rather than a stem. Parts of the skeleton are illustrated and defined.

ambulacral (1). Plate of the paired rows of plates which form food grooves radiating from mouth.

anal area (3). Posterior interambulacrum.

anal pyramid (9). Small cone of plates surrounding anus.

covering plate (4). Ossicle in one of the two rows of plates covering each ambulacrum.

interambulacral (2). Any plate in areas lying between ambulacra.

marginal (6). One of the differentiated plates lying at or near periphery of theca.

mouth (8). Centrally located opening on upper surface of theca which forms inlet to digestive tube.

oral (7). One of the somewhat enlarged plates surrounding or covering mouth.

pore (5). Opening through or between ambulacral plates for a branch of water-circulatory system.

(Fig. 17-2 continued.)

extremity opposite the attachment of the peculiarly specialized stem, which resembles a caudal appendage. *M. mitra* Barrande (6a, b), Lower Ordovician, Czechoslovakia.

Placocystites De Koninck, Silurian–Devonian.

This is a specialized carpoid having highly developed bilateral symmetry. The surface shown in 5a is gently convex and that in 5b concave. *P. forbesianus* De Koninck (5a, b), Middle Silurian, England.

EDRIOASTEROIDS

A very ancient branch of the attached echinoderms, which looks as though it might be an ancestral form of starfish, comprises the edrioasteroids (Fig. 17-3). As typically represented by *Agelacrinites*, they have a saclike or low discoid body. The flexible theca is composed of many small irregularly polygonal or rounded plates. From the mouth (Fig. 17-3, *8*), which is centrally located on the upper surface, five straight or curved unbranched ambulacral areas extend toward the bor-

ders of the theca. In different genera, the pattern of the rays and arrangement of plates along them varies. The anus, provided with a valvular covering, is located in one of the interambulacral areas. An opening for the water-circulatory system is located between the mouth and anus. Some of the edrioasteroids may have moved about on the sea bottom, but most of them were sedentary invertebrates which attached themselves firmly to some object. There are no brachioles or other appendages extending from the body.

This group includes some 24 genera, distributed from Lower Cambrian to Missis-

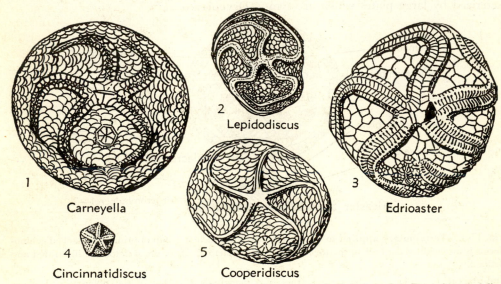

Fig. 17-4. **Representative edrioasteroids.** Ordovician types include 1, 3, and 4; Devonian, 5; Mississippian, 2. Natural size.

Agelacrinites Vanuxem, Devonian–Mississippian. The placement of ambulacra and characters of the interambulacrals and marginals distinguish this genus. *A. hamiltonensis* Vanuxem (Fig. 17-3), Middle Devonian, New York.

Carneyella Foerste, Ordovician. The scalelike plates, especially interambulacrals and marginals, and pattern of the ambulacral areas distinguish this common genus. *C. cincinnatiensis* (Bather) (1), Cincinnatian, Ohio.

Cincinnatidiscus Bassler, Ordovician. This is a diminutive edrioasteroid having straight ambulacra. *C. stellatus* (Hall) (4), Cincinnatian, Ohio.

Cooperidiscus Bassler, Devonian. The narrow, very fine-plated ambulacra all curve to the right. *C. alleganius* (Clarke) (5), Upper Devonian, New York.

Edrioaster Billings, Ordovician. The ambulacra of this genus reach the margin of the theca and are very strongly curved. *E. bigsbyi* (Billings) (3), Trentonian, Canada.

Lepidodiscus Meek & Worthen, Devonian–Mississippian. The very long, slender, strongly curved ambulacra especially distinguish this genus, which is one of the geologically youngest edrioasteroids. *L. squamosus* (Meek & Worthen) (2), Osagian, Indiana.

sippian. A few of them are illustrated in Fig. 17-4.

CYSTOIDS

The primitive stemmed echinoderms called cystoids are a diversified group which is widely distributed in early Paleozoic rocks (Figs. 17-5 to 17-7). Their body outline is spheroidal or saclike. The plates of the theca are generally very numerous and irregularly arranged. Most cystoids exhibit some degree of radial symmetry, and in several it is highly developed. A few, however, resemble carpoids in having dominant bilateral symmetry. Two chief characters of cystoids are the pore structures of thecal plates and the possession of biserial food-gathering appendages which are called **brachioles** (Fig. 17-5, *8*). To some extent, both of these features are seen among blastoids, and accordingly some paleontologists unite cystoids and blastoids in a single class. On the other hand, structural distinctions seem to be sufficient to warrant treatment of the cystoids and blastoids as independent classes of echinoderms. Both are readily distinguished from crinoids by many characters. Relationships of the cystoids to eocrinoids, paracrinoids, and carpoids can be observed, but surely cystoids are not the ancestral stock, as formerly believed, from which all other types of stemmed echinoderms, and possibly the free-moving kinds also, are descended.

The terminology applied to cystoids is indicated and explained in Fig. 17-5.

The cystoids occur chiefly in lower Paleozoic deposits, especially Ordovician and Silurian. The group ranges from Middle Ordovician to Upper Devonian.

Two orders of cystoids are defined on the basis of the arrangements of the thecal pores. These are the pore-rhomb type (Rhombiferida) and the double-pore type (Diploporida). The latter order includes also a few genera in which plates of the theca seem to bear only simple unpaired pores.

Cystoids Bearing Pore Rhombs

Nature of Pore Rhombs. The structures called pore rhombs comprise an arrangement of laterally directed passageways consisting of tubes or grooves in adjoining pairs of thecal plates. The outline of the pore-bearing area is rhombic or diamond-shaped, and each plate of the pair bears one-half of the rhomb (Fig. 17-5, *14*). The tubes or grooves run at right angles to the suture between the plates. They may be open toward the outside throughout their length (**conjunct pore rhomb,** Fig. 17-5, *13*), or they may be covered near the interplate suture which they cross (**disjunct pore rhomb,** Fig. 17-5, *15*). In some pore rhombs of the latter type, only the very ends of the tubes may turn outward so as to connect with the exterior. These openings form aligned rows of round pores, as seen on the thecal plates of *Caryocrinites* (Fig. 17-7, *1C*). Commonly there are openings from the tubes toward the interior of the theca. Also, among some cystoids the tubes are entirely sealed over externally by a thin roof (**epistereom,** Fig. 17-5, *18*); the passageways open to the interior of the theca, so they could permit circulation only of fluids of the body cavity. Clearly, these pores served for circulation of water and doubtless functioned as the means of respiration. Virtually all plates of the theca of many cystoids carry pore rhombs, but some have rhombs only on certain plates of the theca, other plates or parts of plates being imperforate.

One type of pore rhomb (**pectinirhomb,** Fig. 17-5, *16*), is characterized by two sets of short parallel slits which open to the exterior. Commonly, only a few such pectinirhombs are contained in the theca of an individual cystoid, and in various genera they have a constant arrangement as to position (Figs. 17-6, *11;* 17-7, *2, 4, 6*). The borders of the pectinirhomb areas tend to be surrounded by elevated ridges. They are a strikingly distinctive structural feature.

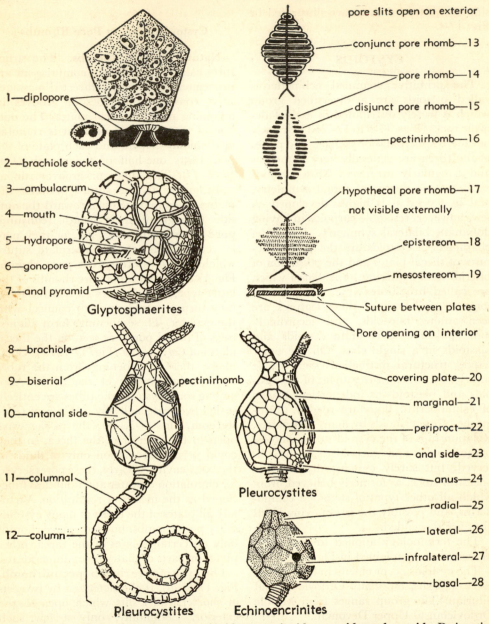

1—diplopore

2—brachiole socket
3—ambulacrum
4—mouth
5—hydropore
6—gonopore
7—anal pyramid
Glyptosphaerites

pore slits open on exterior
conjunct pore rhomb—13
pore rhomb—14
disjunct pore rhomb—15
pectinirhomb—16

hypothecal pore rhomb—17
not visible externally

epistereom—18
mesostereom—19
Suture between plates
Pore opening on interior

8—brachiole
9—biserial
pectinirhomb
10—antanal side
11—columnal
12—column
Pleurocystites

covering plate—20
marginal—21
periproct—22
anal side—23
anus—24
Pleurocystites

radial—25
lateral—26
infralateral—27
basal—28
Echinoencrinites

FIG. 17-5. **Terminology applied to eocrinoids, paracrinoids, carpoids, and cystoids.** Designation of the hard parts belonging to primitive types of echinoderms differs to some extent from that for more advanced groups. The alphabetically arranged list of defined terms is cross-indexed by numbers to the figure.

ambulacrum (pl. ambulacra) (3). Shallow groove on arms or brachioles and on surface of theca, leading to mouth.

anal pyramid (7). Small circlet of somewhat raised plates surrounding or covering anal vent.

anal side (23). In laterally compressed carpoids and cystoids, side containing anal vent.

antanal side (10). In laterally compressed carpoids and cystoids, side opposite to that containing anus.

(Continued on next page.)

The theca of pore-rhomb-type cystoids is oriented by some workers according to a vertical plane passing through the mouth and a special opening (**hydropore**, Fig. 17-5, *5*) located on one of the plates, generally in line between the anus and mouth. The direction toward the mouth along this plane is designated as anterior and the opposite direction posterior. The left and right sides may be differentiated readily with reference to this median plane. Orientation based on the position of the anus and mouth is useful for definition of parts and for comparison with homologous portions of other echinoderms. As pointed out in the study of the eleutherozoans, however, the location of the inlet to the water-circulation system, corresponding to the hydropore of cystoids, is the most constant structural character.

Arrangement of Plates. Typically, pore-rhomb-bearing cystoids have numerous irregularly arranged plates. A few of the Rhombiferida (Figs. 17-6, *3, 9, 11*; 17-7, *1*) are distinguished, however, by a regular arrangement of plates belonging to the theca. In the first two indicated genera, there are four circlets of plates

---

(Fig. 17-5 continued.)

anus (24). External opening for outlet of waste from digestive tract.

basal (*B*, pl. *BB*) (28). In cystoids having comparatively stable, regular plate arrangement, plate of circlet adjoining column.

biserial (9). Arrangement of brachiole plates in interlocking double rows as in eocrinoids and cystoids.

brachiole (8). Free appendage of a cystoid or eocrinoid, which bears a food groove.

brachiole socket (2). Place of attachment of a brachiole.

column (12). The stem, commonly attached to base or one extremity of eocrinoids, paracrinoids, carpoids, and cystoids, but not serving invariably as a means of fixation.

columnal (*C*, pl. *CC*) (11). Individual plate of column.

conjunct pore rhomb (13). Type of pore rhomb in which externally visible slits are continuous across suture between plates bearing rhomb.

covering plate (20). One of the minute plates which roof an ambulacral groove.

diplopore (1). Paired arrangement of thecal pores in some cystoids.

disjunct pore rhomb (15). Type of pore rhomb in which externally visible slits forming parts of rhomb are separated by solid areas of plates.

epistereom (18). Thin surficial calcareous material of plates which completely covers and conceals pore rhombs of some cystoids.

genital pore (6). Same as gonopore.

gonopore (6). Small opening in theca of many cystoids, generally located between mouth and anus, presumably for discharge of eggs and sperm.

hydropore (5). Generally a slitlike opening

adjacent to gonopore, observed in many cystoids; interpreted as inlet for a water-circulatory system.

hypothecal pore rhomb (17). Type of pore rhomb occurring beneath surface of theca, mainly or entirely invisible externally in unweathered cystoids.

infralateral (*IL*, pl. *ILL*) (27). In eocrinoids and some cystoids having four circlets of comparatively regular plates below attachment of brachioles, plate of circlet next above basals.

lateral (*L*, pl. *LL*) (26). In eocrinoids and some cystoids, plate of circlet next above infralaterals.

marginal (21). One of the large, relatively massive plates of carpoids and bilaterally compressed cystoids, which form periphery of flattened theca.

mesostereom (19). Portion of plates of cystoid theca lying below epistereom; may contain pore rhombs.

mouth (4). Opening which serves as inlet for digestive system.

pectinirhomb (16). Type of disjunct pore rhomb; name derived from comblike appearance of each half of rhomb, which commonly is surrounded by a ridge.

periproct (22). Area occupied by numerous, generally small plates adjoining anus; very large in carpoids and some compressed cystoids.

pore rhomb (14). Group of parallel horizontal tubes or slits occupying parts of two adjoining plates, the channels being arranged normal to suture between plates and together generally defining a rhombic area.

radial (*R*, pl. *RR*) (25). In eocrinoids and some regular cystoids, plate of circlet next above laterals.

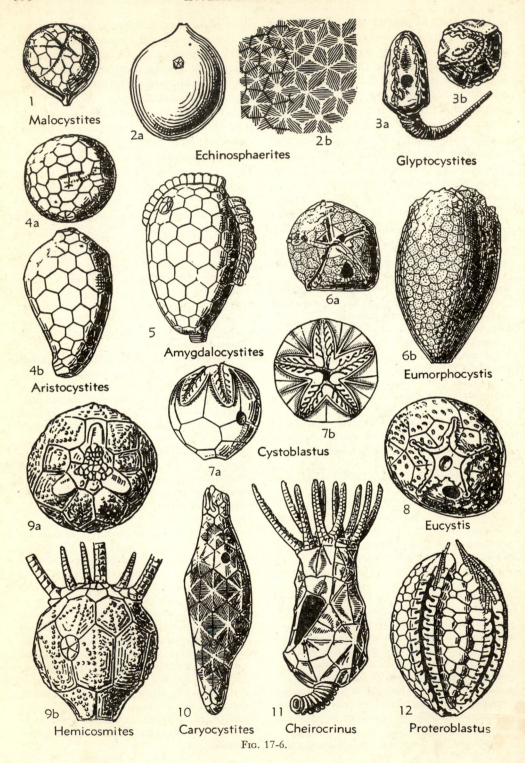

1 Malocystites

2a 2b Echinosphaerites

3a 3b Glyptocystites

4a

4b Aristocystites

5 Amygdalocystites

6a

6b Eumorphocystis

7a 7b Cystoblastus

8 Eucystis

9a

9b Hemicosmites

10 Caryocystites

11 Cheirocrinus

12 Proteroblastus

Fig. 17-6.

below the brachioles, in downward order designated as radials, laterals, infralaterals, and basals. *Hemicosmites* (Fig. 17-6, *9*) and *Caryocrinites* (Fig. 17-7, *1*) each have three circlets of plates below the brachioles; these circlets are called radials, laterals, and infralaterals.

Many-plated Cystoids Having Concealed Pore Rhombs. As an example of pore-rhomb-bearing cystoids which we

may distinguish as primitive, the Middle and Upper Ordovician genus *Echinosphaerites* (Fig. 17-6, *2*) is a good choice for study. This genus is fairly common in the Appalachian region of the United States, where it is recognized as a useful zone fossil, and in northwestern and central Europe, and eastern Asia. Also, it is one of the first cystoids to become known.

An unabraded specimen (as illustrated

FIG. 17-6. **Representative Ordovician cystoids.** These include both pore-rhomb-bearing types (?1–3, 5, 7, 9–11) and genera characterized by diplopores (4, 6, 8, 12). Natural size except as indicated.

Amygdalocystites Billings, Middle Ordovician. The flattened theca bears two unbranched food grooves from which small brachioles arise. *A. florealis* Billings (5), Trentonian, Canada.

Aristocystites Barrande, Ordovician. Brachioles have not been observed on the many-plated theca of this genus, which bears simple and paired pores. *A. bohemicus* Barrande (4a, b), Middle Ordovician, Czechoslovakia.

Caryocystites Buch, Middle Ordovician. The elongate theca somewhat resembles *Echinosphaerites* in structure but has more regular and larger plates. *C. angelini* (Haeckel) (10), Sweden.

Cheirocrinus Eichwald, Middle and Upper Ordovician. The four-circlet theca is surmounted by well-developed brachioles; the anal area is large. *C. insignis* Jaekel (11), Middle Ordovician, Russia.

Cystoblastus Volborth, Middle Ordovician. This regular cystoid has blastoid-like ambulacra *C. leuchtenbergi* Volborth (7a, b), Russia.

Echinosphaerites Wahlenberg, Ordovician. Well-preserved specimens are nearly smooth spheroidal fossils on which the outline of plates may be quite indiscernible. Weathered surfaces reveal many pore rhombs arranged in flower-like pattern, sutures between plates being very indistinct. This valuable guide fossil is extremely abundant in some formations. *E. aurantium* (Gyllenhahl) (2a, an unweathered specimen; 2b, pore rhombs shown by a weathered specimen, plate margins indicated at left, enlarged), Middle Ordovician, Sweden.

Eucystis Angelin, ?Middle Ordovician–Lower Devonian. Five ambulacral grooves radiate from the mouth of this diplopore-bearing

spheroidal genus. *E. raripunctata* Angelin (8), Upper Ordovician, Sweden.

Eumorphocystis Branson & Peck, Middle Ordovician. The very numerous small plates of the elongate theca bear many diplopores. Five ambulacral grooves are well-developed. *E. multiporata* Branson & Peck (6a, b), Chazyan, Oklahoma.

Glyptocystites Billings, Middle Ordovician. Brachiole-bearing ambulacral grooves spread far down over the four circlets of thecal plates. *G. multiporus* Billings (3a, b), Trentonian, Canada.

Glyptosphaerites Müller, Middle Ordovician. The spheroidal theca is composed of many small plates which contain diplopores. The five main ambulacral grooves branch. *G. leuchtenbergi* (Volborth) (Fig. 17-4), Middle Ordovician, Russia.

Hemicosmites Buch, ?Lower, Middle, and ?Upper Ordovician. The thecal plates are arranged in three circlets below the brachioles, which are grouped in three rays. *H. pyriformis* Buch (9a, b), Middle Ordovician, Russia.

Malocystites Billings, Middle Ordovician. Radiating ambulacra are impressed on the summit of the small theca. *M. murchisoni* Billings (1), Chazyan, Canada.

Pleurocystites Billings, Middle Ordovician. This bilaterally compressed cystoid has pectinirhombs and a very large anal area (periproct). *P. filitextus* Billings (Fig. 17-4), Trentonian, Canada.

Proteroblastus Jaekel, Middle Ordovician. Diplopores on the numerous small plates distinguish this form from blastoids. *P. schmidti* (Jaekel) (12), Middle Ordovician, Estonia.

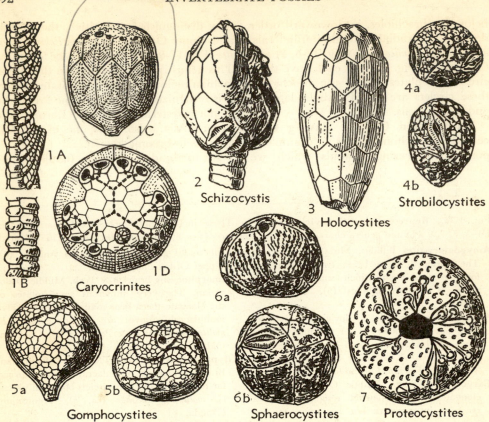

FIG. 17-7. **Representative Silurian and Devonian cystoids.** Natural size, except as indicated. Silurian forms include 1–3, 5, 6, Devonian, 4, 7. All are Rhombiferida except 5 and 7, which are Diploporida.

Caryocrinites Say, Middle Ordovician–Middle Silurian. The regular arrangement of large plates, arrangement of pore rhombs which are marked externally only by openings of the tubes, and numerous strong branching biserial brachioles distinguish this genus. Remains are especially common in some Middle Silurian formations of the United States. *C. ornatus* Say (1*A*, *B*, portions of brachiole, enlarged; 1*C*, side view of theca, ×1; 1*D*, summit of theca, ×2), Niagaran, New York, Indiana.

Gomphocystites Hall, Middle Silurian. The numerous small irregularly arranged plates bear diplopores. Spirally curved ambulacra radiate from the mouth. *G. bownockeri* Foerste (5*a*, *b*), Niagaran, Ohio.

Holocystites Hall, Silurian. The plates of this genus are arranged in several regular series; pores or rhombs not clearly defined. Fifty species are described from the central United States. *H. cylindricus* (Hall) (3), Niagaran, Illinois.

Proteocystites Barrande, Devonian. This is a spheroidal form having prominent diplopores and ambulacra grouped around the large mouth. *P. flavus* Barrande (7), Lower Devonian, Czechoslovakia.

Schizocystis Jaekel, Silurian. The four-circlet theca bears a few prominent pectinirhombs. *S. armata* (Forbes) (2), Middle Silurian, England.

Sphaerocystites Hall, Upper Silurian. Some of the large plates of this spheroidal theca have prominent pectinirhombs. *S. multifasciatus* Hall (6*a*, *b*), Cayugan, Maryland.

Strobilocystites White, Upper Devonian. This genus, one of the last of the cystoids, has regular plates in four circlets, prominent pectinirhombs, and long ambulacra. *S. schucherti* Thomas (4*a*, *b*), Upper Devonian, Iowa.

in Fig. 17-6, *2a*) is a globular object which lacks almost any sign of the many small plates (800 or more in an average-size specimen) which form the theca. The surface is smooth. At the summit of the theca is a small protuberance which indicates the location of the mouth. Faintly impressed markings of ambulacra are seen on the surface of the theca of some specimens; no free brachioles are known. A small pyramidal elevation, composed of five or more plates on the upper third of the theca, is interpreted to mark the location of the anus. Above it and to the right is another special spot, marked on some specimens by three slightly elevated plates, which surround an opening termed the gonopore. A stem was very tiny or it was lacking.

A weathered specimen of *Echinosphaerites* shows the presence of pore rhombs distributed evenly on all the thecal plates (Fig. 17-6, *2b*). The passageways are made visible by the disappearance of the epistereom which in the unweathered condition covered them.

Regular Cystoids Having Disjunct Pore Rhombs. The genus *Caryocrinites* (Fig. 17-7, *1*) represents cystoids of a type which may be considered much more advanced in evolution than *Echinosphaerites*. Pore rhombs are present on all plates of *Caryocrinites*, except at the summit of the theca, but there are far fewer plates than in *Echinosphaerites*, and they are regularly arranged in circlets termed radials, laterals, and infralaterals (Fig. 17-5, *25–27*); also, the pores are of disjunct type, connecting with the exterior by openings which are arranged in lines running to the center of thecal plates. An anal pyramid occurs on the summit of the theca. The biserial brachioles are free appendages which bear closely spaced biserial branchlets (Fig. 17-7, *1A, 1B*). A fairly stout cylindrical stem joins the base of the theca. *Caryocrinites* is a common Middle Silurian fossil.

Pectinirhomb-bearing Cystoids. A somewhat specialized representative of the cystoids which have pectinirhombs is the Ordovician genus *Pleurocystites* (Fig. 17-5). The specialization consists in bilateral flattening of the theca and enlargement of the area of irregular plates adjacent to the anus; otherwise it corresponds in structure to several other genera which lack pore rhombs on a majority of the thecal plates. Certain pairs of plates bear pectinirhombs, and accordingly these structures are definitely localized. Three pectinirhombs occur on each individual. Two stout biserial brachioles are attached to the summit of the theca, and a fairly strong tapering column is joined to its base.

Cystoids Bearing Paired Pores

Another group of cystoids consists of those in which pores extend subvertically through the plates. Commonly, external openings of the pores are arranged in pairs (**diplopores**, Fig. 17-5, *1*). In some genera, the paired pores are thickly distributed over the entire surface of the theca, whereas in others they are more sparingly distributed and may be localized. Many genera belonging to this group have ambulacral grooves on the surface of the theca. Some have well-developed brachioles which extend outward from the vicinity of the mouth. The order Diploporida ranges from Middle Ordovician to Lower Devonian.

Representatives of this group of cystoids are *Glyptosphaerites* (Fig. 17-5) and *Eumorphocystis* (Fig. 17-6, *6*), from Middle Ordovician rocks of Europe and North America, respectively. Both of these genera are characterized by the very large number of their irregularly arranged plates, which lack any definite organization in circlets, and on the summit of the theca are radiating ambulacral grooves which terminate at sockets for the attachment of free brachioles. An anal pyramid and gonopore occur on the posterior side of the theca. Each plate is punctured by several pairs of openings, so-called diplopores, which permit communication of water between the outside and inside of the theca. These paired pores are the chief characteristics of the fossils.

Evolution of Cystoids

Comparative study of the cystoids, including particularly notice of structures belonging to the geologically oldest and those seen most commonly in the geologically youngest representatives of this group, indicates some features of evolutionary change. We must conclude, first, that differentiation of the pore-rhomb- and diplopore-bearing groups of cystoids antedates the appearance of this class of echinoderms in the geological records. When we first see them, in Middle Ordovician rocks, the two groups are clearly separated. Although the presence of thecal pores is a structural character belonging to both, it is not easy to understand how either could be derived from the other. The pore-rhomb arrangement seems, however, to be much more highly developed as a type of respiratory structure than the random perforations of plates represented by the diplopores.

Among pore-rhomb-bearing cystoids, the genera having very numerous, irregularly arranged plates and hypothecal placement of the pores, as in *Echinosphaerites*, must be adjudged primitive as compared with cystoids having fewer, more regularly arranged plates which bear disjunct or conjunct pore rhombs having external openings. Most specialized, as regards thecal pore structures, are the cystoids which have pectinirhombs on a few pairs (minimum three) of adjoining plates. These cystoids have a theca which is mostly composed of solid, imperforate plates. In this respect, they resemble crinoids.

The arrangement of ambulacra as food-gathering grooves affixed to the summit portions of the theca seems to be a decidedly less efficient and more primitive structural plan than free-moving appendages which extend outward and upward from the theca near the mouth. Early cystoids lack well-developed free brachi-

oles, such as are seen in *Caryocrinites* and other advanced forms. Some of the cystoids evolved in a direction which produced close similarity to characteristic structures of the blastoids. Among these cystoids are *Cystoblastus* (Fig. 17-6, *7*) and *Proteroblastus* (Fig. 17-6, *12*); without doubt, the blastoids are derived from cystoidean ancestors. The last survivors among cystoids are the pectinirhomb-bearing forms such as *Strobilocystites* (Fig. 17-7, *4*), from Upper Devonian rocks of Iowa.

BLASTOIDS

The blastoids are grouped in a class distinct from that of the cystoids because of (1) their pronounced symmetry, (2) the uniformity of arrangement of their 13 or 14 main thecal plates, and (3) the development in these echinoderms of specialized ambulacral areas which bear very numerous small free brachioles. A distinctive structure of the blastoids, known as the **hydrospire** (Fig. 17-9, *11*), hangs into the body cavity beneath each ambulacrum. It has some similarity in form and function to the pectinirhombs of rhombiferid cystoids. The detailed nature of the hydrospires, however, and their placement along the ambulacra differ from the organization of any cystoids. Also, no blastoid has openings in the plates corresponding to diplopores.

Except for a few primitive blastoids which have extra thecal plates and various forms which exhibit lateral division of one of the plates adjoining the anus, this group of echinoderms is remarkably uniform in having a theca composed of 13 plates arranged in three circlets.

Structural Features of Pentremites

A blastoid which is specially suitable for study, as representative of the class, is *Pentremites* (Figs. 17-8; 17-9C). This genus is more abundant in the number of described species (about 80) than any other. At various outcrops of Upper Mississippian rocks in Illinois, Indiana, and Kentucky, scores of beautifully preserved specimens of *Pentremites* can be collected. The theca has the shape and approximate dimensions of a good-sized rosebud, but specimens belonging to some unusually robust species may attain a height of 2.5 in. Structural features and terminology applied to *Pentremites* and other blastoids are indicated in Figs. 17-8 and 17-9C.

At the base of the theca are three outwardly flaring **basals,** which bear attachment to the stem. One of the basal plates (**azygous basal**) is only half as large as the other two, and invariably this small plate lies in the antero-right interray (Fig. 17-8 *B, C*). Next above the basals are five evensized, more or less deeply forked plates, called **radials** (Fig. 17-8*A, C*). The ambulacra occupy these forked spaces. The uppermost circlet of plates, distributed around the mouth, consists of relatively small quadrangular **deltoids,** which are placed interradially so that each adjoins a pair of radials (Fig. 17-8*A, C*). At the upper extremity of each deltoid is an opening (**spiracle**) which provides an outlet for the water-circulatory system. In *Pentremites*, the anal opening, joined with one of the spiracles, occurs just above the posterior deltoid; this combined anus and spiracle is readily identifiable because it is the largest opening in any interray. The posterior deltoid of many blastoids contains the anus or is indented by it; in many genera this plate is divided by a horizontal suture into upper and lower elements (**epideltoid, hypodeltoid,** Fig. 17-9, *15, 16*).

The five ambulacra of *Pentremites* are evenly spaced around the mouth opening, which is centrally located on the summit of the theca (Fig. 17-8*A*). Each ambulacral area is occupied by an elongate single plate, which bears many regularly spaced, strongly marked transverse grooves on its exposed surface. Because of its spear-shaped outline, this element has come to be known as a **lancet plate** (Fig. 17-9, *7*). Its middle bears a longitudinal groove

which leads upward to the mouth. Both the transverse and longitudinal grooves of perfectly preserved specimens are roofed over by tiny **covering plates,** a double row of these belonging with each groove (Fig. 17-9, *19*). This series of connected passageways is an important part, although not the whole, of the food-gathering system of *Pentremites*. Yet to receive notice are the

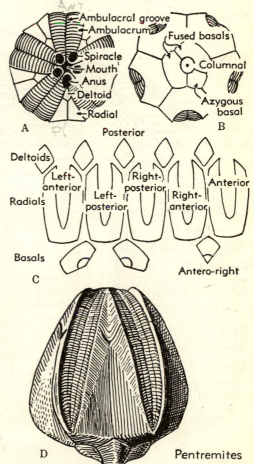

Fig. 17-8. **Structural features of Pentremites.** Ventral view (*A*) and dorsal view (*B*) of theca. A drawing of the 13 individual plates which compose the theca (exclusive of the ambulacra) is shown in *C*; orientation is based on the posterior deltoid, situated just below the large spiracle which serves also as anus. (*D*) Side view (antero-right interray) of a large species, *Pentremites robustus* Lyon, from Upper Mississippian (Chesteran) rocks of Kentucky, natural size.

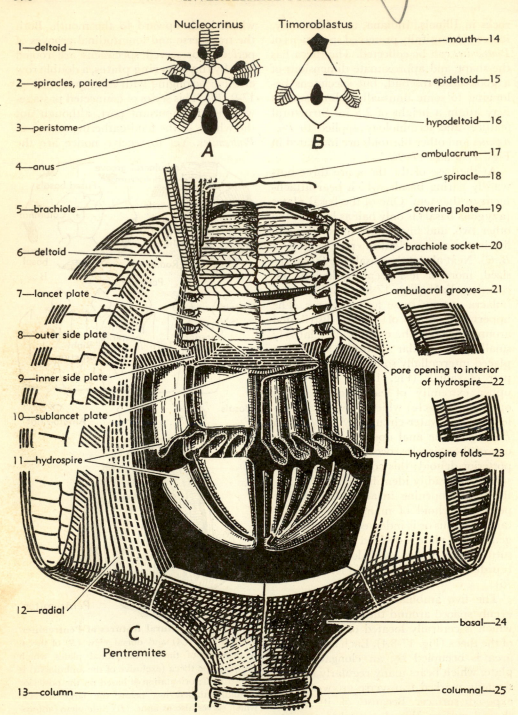

Nucleocrinus

1—deltoid

2—spiracles, paired

3—peristome

4—anus

A

Timoroblastus

mouth—14

epideltoid—15

hypodeltoid—16

B

ambulacrum—17

spiracle—18

5—brachiole

covering plate—19

6—deltoid

brachiole socket—20

7—lancet plate

ambulacral grooves—21

8—outer side plate

9—inner side plate

pore opening to interior
of hydrospire—22

10—sublancet plate

11—hydrospire

hydrospire folds—23

12—radial

basal—24

C

Pentremites

13—column

columnal—25

Fig. 17-9. **Terminology applied to blastoids.** The most commonly used terms employed in the study of blastoids are defined in the alphabetically arranged list which is cross-indexed to the figure by numbers.

(*Continued on next page.*)

very numerous tiny free appendages which rise from the margins of the ambulacra. The appendages, called **brachioles,** are very fragile and therefore are preserved only under exceptional circumstances.

Along the lateral edges of each lancet plate, occupying the space between the lancet plate and the adjoining radials and deltoids, is a double series of small side plates (**outer side plates, inner side plates,** Fig. 17-9, *8, 9*). Each pair of side plates bears a socket for the attachment of a brachiole, and from this socket a shallow depression extends inward to the median furrow of the lancet plate. On the side toward the ambulacra, the brachioles bear a narrow longitudinal groove which is concealed by a double row of minute covering plates. The other plates which compose each brachiole are arranged in a double or single row.

Between the bases of the brachioles and the outer borders of the ambulacral areas are openings between the side plates which lead into a gill-like structure termed the hydrospire (Fig. 17-9, *11, 22, 23*). Each

---

(Fig. 17-9 continued.)

ambulacral grooves (21). Shallow depressions on an ambulacrum or brachiole which serve as passageways for food particles; they lead ultimately to mouth.

ambulacrum (pl. **ambulacra**) (17). Portion of theca traversed by shallow grooves for movement of water which carries food particles; ambulacral areas are bordered by brachioles and may contain pores opening to hydrospires.

anus (4). External opening for outlet of digestive tract, located near summit of theca in posterior interray.

basal (24). One of three plates of circlet adjoining stem.

brachiole (5). Slender food-gathering appendage attached at border of an ambulacrum.

brachiole socket (20). Place of attachment of a brachiole.

column (13). The stem, composed of circular segments.

columnal (25). One of stem segments.

covering plate (19). One of small plates, arranged in an alternating double series, which roofs ambulacral grooves; occurs also on brachioles.

deltoid (1, 6). Generally a small rhomb-shaped interradial plate near summit of theca.

epideltoid (15). Upper part (adjoining mouth) of divided posterior deltoid in some blastoids.

hydrospire (11). Infolded thin-walled respiratory structure beneath border of an ambulacrum or intersecting radial and deltoid plates parallel to an ambulacrum.

hydrospire folds (23). Reflexed portions of hydrospire.

hypodeltoid (16). Lower part (below anus) of a divided posterior deltoid in some blastoids, separated from mouth by an epideltoid.

inner side plate (9). One of a series of small plates along border of an ambulacrum, resting on edge of lancet plate.

lancet plate (7). Elongate spear-shaped plate occupying central area of an ambulacrum; its outer surface is marked by a median longitudinal groove and many transverse grooves, which are concealed by covering plates in a perfectly preserved blastoid.

mouth (14). Opening at summit of theca which is inlet to digestive tract; in perfectly preserved specimens covered by small plates of the peristome.

outer side plate (8). One of the series of small plates along border of an ambulacrum, resting against an inner side plate and bearing a brachiole socket; each lies between pores which open into a hydrospire.

paired spiracles (2). Openings near summit of an ambulacrum, which serve as outlets of hydrospires.

peristome (3). Area of small polygonal plates which cover mouth.

pore (22). Opening at margin of an ambulacrum leading to one of the hydrospires.

radial (12). Plate of circlet next above basals; generally indented strongly by an ambulacral area so as to have a forked shape.

side plate (8, 9). One of the series of inner and outer plates adjoining lancet plate in an ambulacrum.

spiracle (2, 18). Round or slitlike opening near summit of theca which serves as outlet for one or more hydrospires.

sublancet plate (10). Thin plate occurring below lancet plate in some blastoids.

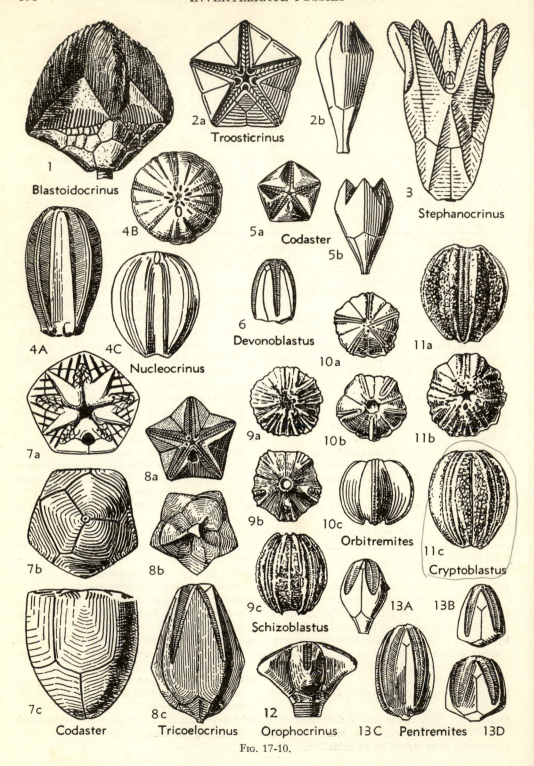

1
Blastoidocrinus

2a Troosticrinus 2b

3 Stephanocrinus

4B

5a Codaster

5b

4A   4C
        Nucleocrinus

6
Devonoblastus

11a

7a

9a   10a

10b   11b

8a

7b   8b

9b   10c

11c
Cryptoblastus

Orbitremites

9c
Schizoblastus

13A   13B

7c   8c
Codaster Tricoelocrinus

12
Orophocrinus 13C Pentremites 13D

Fig. 17-10.

hydrospire may be compared to pleated folds of a curtain hanging downward into the body cavity and arranged so that the axis of the pleats runs parallel to the borders of the ambulacral areas. The folds consist of a very thin sheet of calcite. The outer edge of each hydrospire fold is attached to the two plates (radial and deltoid) which border the ambulacrum. The inner edge, on the side toward the middle of the ambulacrum, joins a thin **sublancet plate** which is attached to the relatively stout lancet plate (Fig. 17-9, *10, 23*). Accordingly, a pair of these relatively complex structures occurs in each ambulacral area. Through the pores, sea water can bathe the interior of the hydrospires, and through the extended thin wall, it operates to aerate the interior body fluids of *Pentremites*. The water moves through a hydrospire to a spiracle at its upper extremity and escapes. Each spiracle connects with two hydrospires, not of the same ambulacrum but of the two different ambulacra nearest the spiracle.

The fairly large pentagonal opening which is seen at the summit of most specimens of *Pentremites*, surrounded by the

---

Fig. 17-10. **Representative Ordovician, Silurian, Devonian, and Mississippian blastoids.** A primitive Ordovician form is shown in 1; Silurian, 2, 3; Devonian, 4–6; Mississippian, 7–13; Natural size, except as indicated.

Blastoidocrinus Billings, Middle Ordovician. This genus, one of the oldest known blastoids, has numerous small plates below the large deltoids, which bear numerous hydrospire slits. *B. carchariaedens* Billings (1), Chazyan, Canada.

Codaster McCoy, Silurian–Pennsylvanian. This genus is primitive in having broadly exposed hydrospire slits and short ambulacra. *C. gracilis* (Wachsmuth) (5a, b), Middle Devonian, Michigan; *C. trilobatus* McCoy (7a–c), Mississippian, England.

Cryptoblastus Etheridge & Carpenter, Mississippian. Resembles *Schizoblastus* but lacks pores along outer edges of ambulacra. *C. melo calcaratus* Cline (11a–c, ×2), Osagian, Missouri.

Devonoblastus Reimann, Middle Devonian. Closely resembles *Pentremites* but has narrower, slightly different ambulacra. *D. leda* (Hall) (6), Middle Devonian, New York.

Nucleocrinus Conrad, Devonian. The very long narrow ambulacra and paired spiracles, combined with differentiated surface markings of the very large deltoids, mark this genus. *N. obovatus* (Barris) (4A), Middle Devonian, Iowa; *N. verneuili* (Troost) (4B, C), Middle Devonian, Kentucky.

Orbitremites Austin & Austin, Mississippian–Permian. Closely resembles *Schizoblastus* in appearance but has only five spiracles, with one of which the anus is joined. *O. norwoodi* (Owen & Shumard) (10a–c), Osagian, Iowa.

Orophocrinus Seebach, Mississippian. This genus resembles *Codaster* but has 10 spiracles and hydrospires. *O. stelliformis* (Owen & Shumard) (12), Osagian, Iowa.

Pentremites Say, Mississippian–Pennsylvanian. The budlike calyx and relatively broad ambulacra characterize this most common of blastoid genera. Five spiracles include one that is enlarged by fusion with the anus. Among approximately 80 described species, all but one is Mississippian. *P. elongatus* Shumard (13C), Osagian, Missouri; *P. conoideus* Hall (13B), Meramecian, Illinois; *P. pyriformis* Say (13A), Chesteran, Illinois; *P. godoni* (Defrance) (13D), Chesteran, Illinois.

Schizoblastus Etheridge & Carpenter, Mississippian–Permian. The melon-shaped calyx, large deltoids, and paired spiracles are features of this genus. *S. sayi bellulus* Cline (9a–c), Osagian, Missouri.

Stephanocrinus Conrad, Middle Silurian. This genus, which possibly may belong among crinoids rather than the blastoids, is distinguished by the strong projections of the radials and very short ambulacra. *S. angulatus* Conrad (3, ×2), Niagaran, New York.

Tricoelocrinus Meek & Worthen, ?Devonian, Mississippian, ?Permian. This blastoid is distinguished by shape of the calyx and structure of the hydrospires. *T. woodmani* (8a–c), Meek & Worthen, Meramecian, Indiana.

Troosticrinus Shumard, Middle Silurian–Upper Mississippian. This tall slender blastoid, very common in some Niagaran rocks, has five unpaired spiracles. *T. reinwardti* (Troost) (2a ventral view, ×2; 2b, side view), Niagaran, Tennessee.

spiracles, is the mouth (Fig. 17-8*A*). Exceptionally preserved specimens show that the mouth is not open directly to the exterior but is roofed over by a group of irregular small plates, which together constitute the **peristome** (as in *Nucleocrinus*, Fig. 17-9, *3*).

Geological Record and Evolutionary Trends of Blastoids

The known geological history of the blastoids extends from Middle Ordovician to Upper Permian. Ordovician, Silurian, and some Devonian and Permian blastoids seem to be distinguished especially by the shortness of the ambulacral areas, which are largely restricted to summit portion of the theca. Other Devonian blastoids and most Mississippian, Pennsylvanian, and Permian species have very long, well-developed ambulacra, which in some reach almost to the base of the theca.

The evolutionary modification of blastoids, on the whole, is not very pronounced in any direction. The class is a widely differentiated, somewhat conservative group of echinoderms, in which stability of the thecal structure is an outstanding character. We find some rather definite trends

which are expressed in the shape of the theca, and we can observe progressive increase in complication of the distinctive blastoid structure which is called the hydrospire.

Among oldest known blastoids, such as *Blastoidocrinus* (Fig. 17-10, *1*) of Middle Ordovician (Chazyan) age, cystoid-like characters are found in the presence of somewhat numerous thecal plates which show only partial symmetry in their arrangement. That *Blastoidocrinus* is a primitive blastoid, rather than a cystoid, is indicated by the absence of diplopores or pore rhombs and by the nature of the five ambulacral areas, which bear very numerous small brachioles.

Silurian blastoids (Fig. 17-10, *2*, *3*) have very regular pentameral symmetry. Their relatively elongate theca is characterized by narrow ambulacra which are confined to the summit.

Some Devonian blastoids (Fig. 17-10, *5*) rather closely resemble Silurian genera in form but have better developed hydrospires. Elongation of the ambulacra and a decidedly ovoid form characterize some (Fig. 17-10, *4*, *6*).

Culmination in variety and numbers of

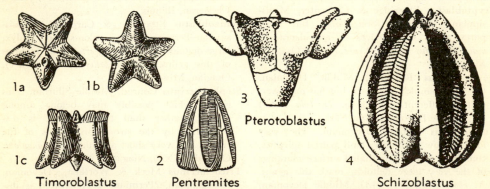

FIG. 17-11. **Representative Pennsylvanian and Permian blastoids.** Natural size.

Pentremites Say, Mississippian–Pennsylvanian. The illustrated species is locally very abundant in Lower Pennsylvanian rocks of Arkansas. *P. angustus* Hambach (2), Morrowan, Arkansas.

Pterotoblastus Wanner, Permian. Bears striking winglike extensions on the upper part of the theca. *P. gracilis* Wanner (3), Upper Permian, East Indies.

Schizoblastus Etheridge & Carpenter, Mississippian–Permian. *S. permiscus* Wanner (4), Permian, Timor.

Timoroblastus Wanner, Permian. The deltoids of this Late Paleozoic blastoid are elevated in a comblike ridge. *T. coronatus* Wanner (1*a–c*), Upper Permian, East Indies.

blastoids is indicated by Mississippian fossils belonging to this class. Associated with *Pentremites* (Fig. 17-10, *13*) are several sorts of pentagonal to stellate genera (Fig. 17-10, *7, 8, 12*), and others characterized by a spheroidal theca (Fig. 17-10, *9–11*).

Pentremites is common in some Early Pennsylvanian (Morrowan) rocks, but blastoids are otherwise almost unknown in this system. A surprising expansion of blastoid stocks is to be recorded mainly on the basis of Permian fossils collected in the Indonesian island of Timor. At least 20 genera (17 confined to Permian) and 67 species of blastoids have been described from this youngest Paleozoic system. Among specializations in form of the theca found here are the winglike projections of the radials (*Pterotoblastus*, Fig. 17-11, *3*) and the comblike prominence of deltoids *Timoroblastus*, Fig. 17-11, *1*). Several Permian genera show a division of the posterior deltoid plate into two (Fig. 17-9, *15, 16*).

Evolution of the hydrospire structures of blastoids is in the direction of an increased number of folds and then in close packing of the folds associated with narrowing of the ambulacra. Species of *Orbitremites* having only one or two hydrospire folds (Fig. 17-12, *5, 2*) represent an early evolutionary stage in the development of this structure. The hydrospire of *Metablastus* (Fig. 17-12, *4*) has four folds. In *Codaster* (Fig. 17-12, *3*) the regular hydrospires are broadly open along the ambulacra, whereas in *Orophocrinus* (Fig. 17-12, *1*) and *Pentremites* (Fig. 17-8) the system of folds has a very narrow opening for the inlet of sea water. The latter condition is clearly the more advanced.

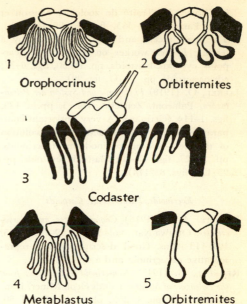

FIG. 17-12. **Cross sections of blastoid hydrospires.** All those illustrated are from Mississippian rocks. The simplest type consists of a single downfold of thin wall (5). The most complex known type (*Neoschisma*), not illustrated here, has 18 downfolds. (1) *Orophocrinus verus* (Cumberland), England; (2) *Orbitremites norwoodi* (Owen & Shumard), Iowa; (3) *Codaster trilobatus* McCoy, England; (4) *Metablastus lineatus* (Shumard), Illinois; (5) *Orbitremites derbiensis* (Sowerby), England.

Primitively, as in the Devonian *Nucleocrinus*, there are separate excurrent openings for each set of hydrospire folds. These consist of the 10 paired spiracles (Fig. 17-9, *2*) at the summit of the theca. Evolution leads to coalescence of the spiracles so as to reduce their number to five, as in *Pentremites*. The coalescence, however, brings together spiracles of different pairs—not the two of any one pair.

REFERENCES

Primitive Attached Echinoderms in General

BASSLER, R. S., & MOODEY, M. W., (1943) *Bibliographic and faunal index of Paleozoic pelmatozoan echinoderms:* Geol. Soc. America Spec. Paper 45, pp. 1–734. Contains outline of classification, faunal lists, and bibliography of genera and species (cystoids, blastoids, edrioasteroids, pp. 3–10, 127–260).

BATHER, F. A. (1929) *Echinoderma:* Encyclopaedia Britannica, 14th ed., vol. 7, pp. 895–904, figs. 1–25. Brief account of carpoids, cystoids, blastoids, and edrioasteroids.

CUÉNOT, L. (1948) *Embranchement des Echinodermes:*

in Grassé, P., Traité de zoologie, Masson et Cie., Paris, pp. 1–363, figs. 1–399 (French). The most recent and best description by a leading French worker on echinoderms (carpoids, cystoids, blastoids, pp. 11–30, figs. 6–37; edrioasteroids, pp. 74–77, fig. 99).

JAEKEL, O. (1918) *Phylogenie und System der Pelmatozoen,* Palaeont. Zeitschr., vol. 3, pp. 1–128, figs. 1–114 (German). A very important comparative study of the structure and evolution of the stem-bearing echinoderms (eocrinoids, pp. 24–27, figs. 17–20; blastoids, cystoids, pp. 93–124, figs. 86–114).

Eocrinoids, Paracrinoids, Carpoids

FOERSTE, A. F. (1916) *Comarocystites and Caryocrinites:* Ottawa Nat., vol. 30, pp. 69–79, 85–93, 101–113, illus. Good description of a representative paracrinoid and a cystoid.

REGNÉLL, G. (1945) *Non-crinoid Pelmatozoa from the Paleozoic of Sweden:* Lunds Geol. Miner. Inst., Medd., 108, pp. 1–255, pls. 1–15, figs. 1–30. A comprehensive comparative study of primitive stemmed echinoderms with description of Swedish forms; contains first-published definition of the paracrinoids (eocrinoids, paracrinoids, carpoids, pp. 35–42, 67–68, 194–197).

Edrioasteroids

BASSLER, R. S. (1935) *Classification of Edrioasteroidea:* Smithsonian Misc. Coll., vol. 93, no. 8, pp. 1–11, pl. 1. Describes several new genera and species.

———— (1936) *New species of American Edrioasteroidea:* Same, vol. 95, no. 6, pp. 1–33, pls. 1–7.

BATHER, F. A. (1900) *The Echinoderma:* in Lankester, E. R., Treatise on zoology, A. & C. Black, London, pp. 1–344, 309 figs. (edrioasteroids, pp. 205–216, figs. 1–8).

———— (1915) *Studies in Edrioasteroidea,* I–IX, Wimbledon, England, reprint of papers published in Geol. Mag., 1898, dec. 4, vol. 5, pp. 543–548; 1899, dec. 4, vol. 6, pp. 94, 134–136; 1900, dec. 4, vol. 7, pp. 193–204, pls. 8–10; 1908, dec. 5, vol. 5, pp. 543–550, pl. 25; 1914, dec. 6, vol. 1, pp. 115–125, 162–171, 193–203, pls. 10–15; 1915, dec. 6, vol. 2, pp. 5–12, 49–60, 211–215, 259–266, 316–322, 393–403, pls. 2–3. A series of important papers on characters of these fossils.

SHIMER, H. W., & SHROCK, R. R. (1944) *Index fossils of North America,* Wiley, New York, pp. 1–837, pls. 1–303 (edrioasteroids, pp. 129–133, pl. 49).

Cystoids

BARRANDE, J. (1887) *Système silurien du centre de la Bohême,* Recherches paléontologiques, Echinodermes, pt. 1, Cystidées: Praha, vol, 7, pp. i-vxii, 1–233, pls. 1–2 (French). A classic work on Paleozoic cystoids of central Europe containing fairly complete bibliography of earlier papers.

BATHER, F. A. (1900) *The Echinoderma;* in Lankester, E. R., Treatise on zoology, A. & C. Black, London, pp. 1–344, 309 figs. One of the best available discussions of general scope on cystoids (pp. 38–77, figs. 1–48).

———— (1913) *Caradocian (Ordovician) Cystidea from Girvan:* Trans. Royal Soc. Edinburgh, vol. 49, pt. 2, pp. 359–529, pls. 1–6, figs. 1–80. First description of several new genera and species of carpoids and cystoids, with discussion of their structure.

BILLINGS, E. (1858) *On the Cystideae of the Lower Silurian (Ordovician) Rocks of Canada:* Canadian Organic Remains, Montreal, dec. 3, pp. 9–74, pls. 1–7. Original publication of several genera and species.

FOERSTE, A. F. (1916) *Comarocystites and Caryocrinites:* Ottawa Nat., vol. 30, pp. 69–79, 85–93, 101–113, illus. Noteworthy for description of brachiole structure of *Caryocrinites.*

JAEKEL, O. (1895) *Organization der Cystoideen,* Deutsch geol. Gesell., Verhandl. 1895, pp. 109–121 (German).

SCHUCHERT, C. (1904) *Siluric and Devonic Cystidea:* Smithsonian Misc. Coll., vol. 27, pp. 201–272, pls. 34–44.

———— (1913) *Cystoidea:* Maryland Geol. Survey, Lower Devonian, pp. 227–248, pls. 32–36.

SHIMER, H. W., & SHROCK, R. R. (1944) *Index fossils of North America,* Wiley, New York, pp. 1–837, pls. 1–303 (cystoids, pp. 123–129, pls. 47–49).

SPRINGER, FRANK (1913) *Cystoidea:* in Zittel, K., & Eastman, C. R., Textbook of palaeontology, Macmillan & Co., Ltd., London, 2d ed., vol. 1, pp. 145–158, figs. 228–249. Includes forms now classified as eocrinoids and carpoids.

Blastoids

BATHER, F. A. (1900) *The Echinoderma:* in Lankester, E. R., Treatise on zoology, A. & C. Black, London, pp. 1–344, 309 figs. (blastoids, pp. 78–93, figs. 1–15).

CLINE, L. M. (1936) *Blastoids of the Osage group, Mississippian, I, Schizoblastus:* Jour. Paleontology, vol. 10, pp. 260–281, pls. 44–45. Morphology of an important Mississippian-Permian genus.

——— (1937) *Blastoids of the Osage group, Mississippian, II, Cryptoblastus:* Same, vol. 11, pp. 634–649, pls. 87–88.

——— (1944) *Class Blastoidea:* in Shimer, H. W., & Shrock, R. R., Index fossils of North America, Wiley, New York, pp. 133–137, pls. 50–51.

CRONEIS, C., & GEIS, H. L. (1940) *Microscopic Pelmatozoa: Part 1, Ontogeny of the Blastoidea:* Jour. Paleontology, vol. 14, pp. 345–355, figs. 1–4. Virtually the only published study on very young blastoids.

ETHERIDGE, R., & CARPENTER, H. (1886) *Catalogue of the Blastoidea in the geological department of the British Museum,* pp. 1–322, pls. 1–20. Contains extensive description of morphology and classification.

HAMBACH, G. (1903) *Revision of the Blastoidea:* St. Louis Acad. Sci. Trans., vol. 13, pp. 1–67, pls. 1–6. A leading work of its time especially for description of American blastoids.

HUDSON, G. H. (1907) *On some Pelmatozoa from the Chazy limestone of New York:* New York State Mus. Bull. 107, pp. 97–152, pls. 1–10. Describes *Blastoidocrinus.*

MOORE, R. C. (1940) *Early growth stages of Carboniferous microcrinoids and blastoids:* Jour. Paleontology, vol. 14, pp. 572–583, figs. 1–3. Notes on homology of certain characters of immature blastoids and microcrinoids.

PECK, R. E. (1938) *Blastoidea of the Chouteau of Missouri:* Missouri Univ. Studies, vol. 13, no. 4, pp. 57–69, pl. 26.

SPRINGER, FRANK (1913) *Blastoidea:* in Zittel, K. A., & Eastman, C. R., Textbook of palaeontology, Macmillan & Co., Ltd., London, 2d ed., vol. 1, pp. 161–172, figs. 252–266.

WANNER, J. (1924) *Die permischen Echinodermen von Timor, Teil 2:* Palaeontologie von Timor, Lief. 14, Abh. 23, Schweizerbart, Stuttgart, pp. 1–81, pls. 1–8 (199–206), figs. 1–31 (German). Describes several new species of *Schizoblastus* and six new genera.

CHAPTER 18

Crinoids

Echinoderms belonging to the class Crinoidea are one of the most complexly organized, highly varied assemblages of all marine invertebrates. Throughout life, beyond an initial free-swimming larval stage, most crinoids are attached to the sea bottom by a stem, and therefore the group constitutes a division of the subphylum Pelmatozoa (*pelmato*, sole or foot). Most living species of crinoids, however, are stemless. These are the feather stars, whereas stem-bearing types, because of fancied resemblance to flowering plants, comprise the sea lilies. Actually, the stemless forms are also fixed by a stem during part of their very early growth, but by breaking away from this anchorage, they become able to crawl or swim about freely.

The crinoids are distinguished especially by their general form and by the structure of their skeleton. A relatively small disk-shaped or globular body, enclosed by armor of symmetrically arranged calcareous plates, bears radially outspread food-gathering appendages, which generally are branched. These appendages, called arms, and the stem are composed of many calcareous segments, joined together in a manner permitting differential movement and providing a degree of flexibility.

Adult crinoids range in size from a few millimeters, including length of stem and arms, to 18 m. (60 ft.) or more; a Cretaceous species had arms at least 120 cm. (4 ft.) long. Most modern free-swimming crinoids are 25 cm. (10 in.) or less in measurement across the outspread arms.

Crinoids have world-wide distribution in present-day oceans, not only in tropical and temperate belts but in frigid waters of the Arctic and Antarctic. They are found at depths ranging from a few feet below sea level to about 4,000 m. (13,000 ft.). The occurrence of fossil crinoids suggests moderately shallow water as the habitat preferred by most species. None are found in fresh-water deposits.

Remains of crinoids are widely distributed in many rock formations ranging in age from Lower Ordovician to Cenozoic, although commonly the hard parts are more or less dissociated. At many places, sedimentary deposits ranging in thickness to 100 ft. or more are composed largely of crinoidal debris, with or without numerous fossils consisting of articulated crinoid hard parts. Study of these remains has paleontological importance because the variety of the fossils is extraordinarily great and the stratigraphic range of individual species almost invariably is very short. Progressive modifications of crinoid structures during geologic time provide rich material for research on the nature of evolutionary trends. A point deserving stress is that completeness of fossil crinoid specimens is not a measure of their worth to paleontology, because a surprisingly large volume of precise and useful knowledge can be gained from the study of crinoid fragments. At present, this is a largely neglected field.

MODERN CRINOIDS

Introduction to the study of fossil crinoids is furnished best by a brief survey of the characters observed in living members of the class. It is fortunate that these are available, for otherwise we would have to interpret crinoid structures by their analogy with parts belonging to very distant echinoderm cousins such as starfishes and sea urchins, which differ greatly from crinoids in many ways. Crinoids are the sole living representatives of the pelmatozoans.

Among the modern crinoids are several genera of stem-bearing forms. These most closely resemble the great majority of fossil crinoids, and accordingly are chosen for examination (Fig. 18-1, *1–3*).

Soft Parts

The soft parts of a relatively simple stalked crinoid (for example, *Ptilocrinus*, Fig. 18-1, *3*) comprise a skin (partly vestigial), alimentary tract, body cavity ((coelom), water-vascular system, nerve tissues, and reproductive organs. Although derived from a bilaterally symmetrical larva (dipleurula, Chap. 16), the anatomical structure of adult crinoids is profoundly modified by the pattern of pentameral radial symmetry, as in other echinoderms.

Skin. A leathery integument surrounds the mouth on the upper (ventral) side of the body and forms a flexible roof (tegmen) over the viscera. It is studded with minute calcareous plates or spicules. Elsewhere, except along furrows on the ventral side of the arms, skin tissue is lacking, but minute pore spaces of the honeycomb-like skeleton contain ectodermal cells, which function in secreting calcium carbonate and in distributing nourishment. Epidermal tissue along the arm grooves bears many cilia and contains numerous nerve fibers.

Ambulacra. The term **ambulacra** is applied to aisle ways along the ventral side of the radially extended appendages of crinoids and crossing parts of the tegmen to the mouth. They are bordered by tube feet and floored by close-spaced cilia,

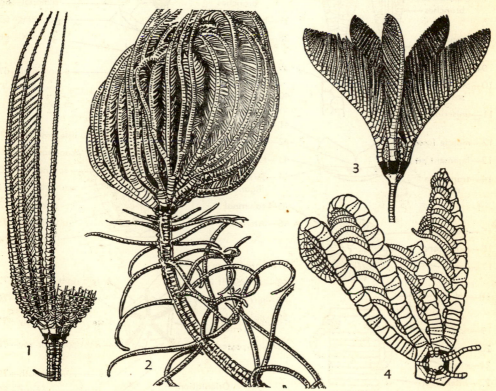

Fig. 18-1. **Modern crinoids.** Stem-bearing crinoids are shown in figures 1 (*Teliocrinus*, ×1), 2 (*Isocrinus*, ×0.5), and 3 (*Ptilocrinus*, ×1). A stemless crinoid (*Comatula*, ×1) is illustrated in figure 4, arms of three of the rays being omitted.

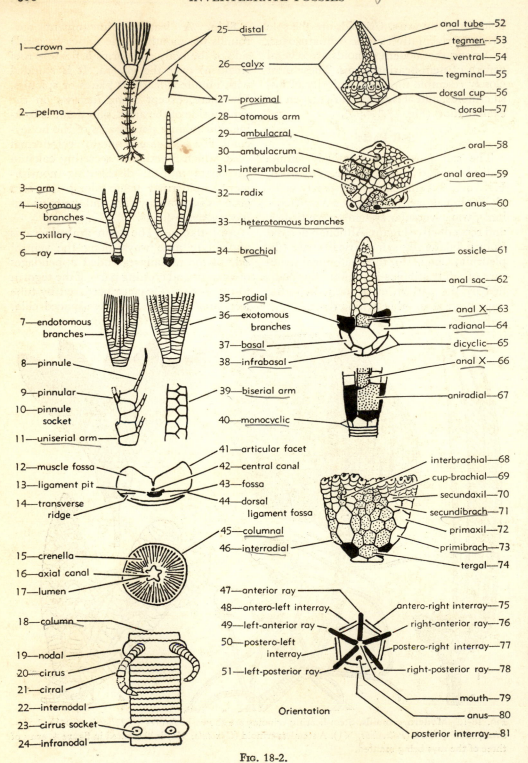

1—crown
2—pelma
3—arm
4—isotomous branches
5—axillary
6—ray
7—endotomous branches
8—pinnule
9—pinnular
10—pinnule socket
11—uniserial arm
12—muscle fossa
13—ligament pit
14—transverse ridge
15—crenella
16—axial canal
17—lumen
18—column
19—nodal
20—cirrus
21—cirral
22—internodal
23—cirrus socket
24—infranodal

25—distal
26—calyx
27—proximal
28—atomous arm
29—ambulacral
30—ambulacrum
31—interambulacral
32—radix
33—heterotomous branches
34—brachial
35—radial
36—exotomous branches
37—basal
38—infrabasal
39—biserial arm
40—monocyclic
41—articular facet
42—central canal
43—fossa
44—dorsal ligament fossa
45—columnal
46—interradial
47—anterior ray
48—antero-left interray
49—left-anterior ray
50—postero-left interray
51—left-posterior ray

Orientation

anal tube—52
tegmen—53
ventral—54
tegminal—55
dorsal cup—56
dorsal—57
oral—58
anal area—59
anus—60
ossicle—61
anal sac—62
anal X—63
radianal—64
dicyclic—65
anal X—66
aniradial—67
interbrachial—68
cup-brachial—69
secundaxil—70
secundibrach—71
primaxil—72
primibrach—73
tergal—74
antero-right interray—75
right-anterior ray—76
postero-right interray—77
right-posterior ray—78
mouth—79
anus—80
posterior interray—81

Fig. 18-2.

FIG. 18-2. **Terminology applied to crinoids.** Alphabetically arranged terms, with accompanying definitions, are cross-indexed to illustrations by numbers. Abbreviations which are used commonly in paleontological literature for some morphological terms are also given.

ambulacral (*Amb*, pl. *Ambb*) (29). Any of the small plates arranged in two rows along sides of ambulacral grooves.

ambulacrum (pl. *ambulacra*) (30). Radial extension of water-circulatory system on ventral surface of tegmen, arms and pinnules converging toward mouth; each part of an ambulacrum is an avenue bordered by podia or tentacles.

anal area (59). Portion of an interambulacral area in which anal tube or vent is located.

anal sac (62). Moderately protuberant to strongly inflated portion of inadunate crinoid tegmen in which pores, presumably for respiration, are commonly intercalated between plates; normally carries anus.

anal tube (52). Constricted pipelike elevation of tegmen in camerate crinoids. Term is applied also to a fleshy conical tube of some modern crinoids; this tube is located on tegmen and bears anus.

anal X (63, 66). Plate between posterior radials, obliquely at left above radianal in most inadunates and flexibles, or obliquely at left above aniradial in disparids.

aniradial (*AR*) (67). Proximal or next to proximal plate of right-posterior ray in disparids; supports anal X obliquely above it at left.

anterior (*Ant*) (47). In position of ray opposite posterior interray, which is normally identified by its greater breadth and content of extra plates.

antero-left (*AntL*) (48). Interray adjoining anterior ray in counterclockwise direction, as viewed from top of dorsal cup.

antero-right (*AntR*) (75). Interray adjoining anterior ray in clockwise direction, as viewed from top of dorsal cup.

anus (60, 80). External opening for discharge of digestive waste, located on side or summit portion of calyx.

arm (3). Branched or unbranched portion of a ray (exclusive of pinnules), extending upward or outward from calyx.

articular facet (41). Surface by which a crinoid plate is united with more or less freedom of movement to an adjoining plate; according to nature of fibers connecting plates, articulation may be muscular or ligamentary (Fig. 18-3).

atomous (28). Unbranched arm.

axial canal (16). Central passageway running through columnals and cirrals.

axillary (*Ax*, pl. *Axx*) (5). Any ray plate at which the ray divides; it adjoins one ray plate proximally (below) and two ray plates distally (above).

basal (*B*, pl. *BB*) (37). Any plate of circlet next below radials, each basal being normally interradial in position.

biserial (39). Arrangement of brachials in an interlocked double series.

brachial (*Br*. pl. *Brr*) (34). Any ray plate (exclusive of pinnulars) above radial; includes free-brachials and cup-brachials.

calyx (26). Dorsal cup and tegmen; hard parts of crinoid exclusive of free arms and pelma.

central canal (42). Small passageway through brachials for nervous system; located just ventral to center of transverse ridge of articular facets.

cirral (*Ci*, pl. *Cii*) (21). Single cirrus segment.

cirrus (pl. *cirri*) (20). Jointed appendage of crinoid stems attached to nodals or to centrodorsal (Fig. 18-34, *4*, *8*) of comatulids; also, rootlike branch at distal extremity (base) of some crinoid columns.

cirrus socket (23). Depression in nodal of crinoid stems or centrodorsal (Fig. 18-34, *4*, *8*) of comatulids marking place of attachment of a cirrus.

column (18). Series of circular, elliptical, or pentagonal segments composing crinoid stem.

columnal (*C*, pl. *CC*) (45). Individual segment of crinoid column.

crenella (15). One of narrow rounded ridges which normally are arranged radially on joint faces of columnals and observed on joint faces of some rigidly united brachials.

crown (1). Crinoid calyx and arms; whole crinoid skeleton, excepting pelma.

cup-brachial (*CBr*, pl. *CBrr*) (69). Any ray plate (except radial) incorporated in dorsal cup; includes cup-primibrach (*CIBr*), cup-secundibrach (*CIIBr*), and so on.

dicyclic (65). Having two circlets of plates below radials.

distal (25). Direction away from center of base of dorsal cup; upward in crown and downward along crinoid column.

dorsal (57). Side away from mouth (aboral), in most crinoids bearing stem attachment.

dorsal cup (56). Plates of crinoid skeleton exclusive of free arms, tegmen, and pelma.

(Continued on next page.)

(Fig. 18-2 continued.)

dorsal ligament fossa (44). Large semicircular depression in muscular articular facets located dorsal to transverse ridge.

endotomous (7). Arm structure in which two main arms of a ray give off branches only on their inner sides.

exotomous (36). Arm structure in which two main arms of a ray give off branches only from their outer sides.

fossa (43). Depressions on articular facets of plates for attachment of muscles and ligaments.

heterotomous (33). Arm structure characterized by inequality of branches.

infrabasal (*IB*, pl. *IBB*) (38). Any plate of lowermost circlet of dorsal cup in dicyclic crinoids, each plate radially disposed.

infranodal (24). Columnal immediately below a nodal.

interambulacral (*iAmb*, pl. *iAmbb*) (31). Any plate of tegmen lying between rows of ambulacrals.

interbrachial (*iBr*, pl. *iBrr*) (68). Any plate occurring between two rows of brachials belonging to one ray; not plates between brachials of adjoining different rays (see interradial).

internodal (22). Normally a columnal between nodals; any columnal lacking cirri.

interradial (*iR*, pl. *iRR*) (46). Any plate of dorsal cup situated between adjacent ray plates.

isotomous (4). Arm structure characterized by equal size of branches.

left-anterior (*LAnt*)(49). Ray next counterclockwise from anterior, as viewed from top of dorsal cup.

left-posterior (*LPost*) (51). Ray next clockwise from posterior interray, as viewed from top of dorsal cup.

ligament pit (13). Centrally placed depression on dorsal side of transverse ridge on articular facet of a ray plate.

lumen (17). Central opening in a columnal or cirral.

monocyclic (40). Having only one circlet of plates (basals) below radials.

mouth (79). Anterior opening of digestive tube, located on or beneath tegmen.

muscle fossa (12). Part of articulating surface of a brachial or radial serving for muscle attachment.

nodal (19). Somewhat enlarged columnal which normally bears cirri.

oral (*O*, pl. *OO*) (58). Interradially disposed, more or less triangular plate belonging to a circlet surrounding mouth.

ossicle (61). Any calcareous segment or plate which forms part of crinoid skeleton.

pelma (2). Crinoid column with all its appendages and anchorage structures.

pinnular (*P*, pl. *PP*) (9). Any segment of a pinnule.

pinnule (8). Slender unbranched jointed branchlets of an arm.

pinnule socket (10). Articular facet on a brachial to which a pinnule is attached.

posterior (*Post*) (81). Interray, normally wider than others, corresponding in position to interambulacrum which carries anus; generally distinguished by presence of extra plates.

postero-left (*PostL*) (50). Interray between left-posterior and left-anterior rays.

postero-right (*PostR*) (77). Interray between right-posterior and right-anterior rays.

primaxil (*IAx*, pl. *IAxx*) (72). Axillary primibrach; the most proximal ray plate having two adjoining brachials on its distal margin; may be either a free-brachial or a cup-brachial.

primibrach (*IBr*, pl. *IBrr*) (73). Any ray plate above radials to and including first axillary; may be either a free-brachial or a cup-brachial.

proximal (27). Direction measured toward contact between base of dorsal cup and topmost columnal; on crinoid stem this direction is upward and in crown, downward.

radial (*R*, pl. *RR*) (35). Proximal plate of each ray, exclusive of radianal and aniradial.

radianal (*RA*) (64). Proximal plate of right-posterior ray in cladid, hybocrinid, and flexible crinoids, but shifted obliquely leftward in many; supports anal X at its left.

radix (32). Rootlike branches at distal extremity of column, for anchorage.

ray (6). A radial combined with all structures which it bears.

right-anterior (*RAnt*) (76). Ray located next clockwise from anterior ray, as viewed from top of dorsal cup.

right-posterior (*RPost*) (78). Ray next counterclockwise from posterior interray, as viewed from top of dorsal cup.

secundaxil (*IIAx*, pl. *IIAxx*) (70). An axillary secundibrach.

secundibrach (*IIBr*, pl. *IIBrr*) (71). Any ray plate forming part of a series next distal to a primaxil, extending to and including a secundaxil; may be either a free-brachial or a cup-brachial.

tegmen (53). Ventral surface of crinoid body covered by a noncalcareous integument or by

(Continued on next page.)

which produce mouthward-moving water currents. The currents conduct microscopic food particles to the digestive tract, and thus a prime function of the ambulacral system is food gathering. In most crinoids, skeletal parts protect the ambulacra.

Alimentary Tract. The mouth and anus of *Ptilocrinus*, like other modern crinoids and most fossil forms, are located not far apart on the ventral surface. A short esophagus leads vertically from the mouth to a stomach, which curves horizontally around the axis of the body, and a short intestine ascends to the anus. Digestive glands adjoin the alimentary tract, but there are no kidneys.

Body Cavity. Fluid-filled space around the digestive system comprises the coelom. Water enters and leaves the coelom by pores leading through the body wall to the exterior. Important structures of the body cavity are an **axial organ** and (surrounding it) the so-called **chambered organ,** which consists of five radially disposed compartments. Both of these organs run longitudinally through the body into the stem. They are associated with nerves which control movements of the animal and connect with genital canals leading to gonads on parts of the arms.

Water-vascular System. Water from the body cavity is admitted into a ring canal surrounding the esophagus near the mouth, and thence into radial canals which extend to the arms. The canals are essential parts of a water-vascular system leading to many short tube feet (podia) located along the arm grooves. These tube feet lack terminal suckers, such as occur in starfishes and echinoids, and there are no water pouches (ampullae) like those of eleutherozoans. The podia are covered by cilia and serve respiratory functions. There is no sieve plate (madreporite) at the entrance to the water-circulatory system of crinoids.

Nervous System. Beside nerve fibers in parts of the skin, nerve cords extend along each arm, a nerve ring surrounds the mouth, and a system of cross-connected nerves forms a somewhat complex pattern on the dorsal side of the body, following the center lines of the skeletal plates in this region. All connect with a "central station" in the chambered organ, which also controls the nerves of the stem. Crinoids are sensitive to touch but lack eyespots or other sense organs.

Reproductive Organs. Sexes of crinoids are separate. Eggs or sperm are formed in genital glands and released directly into surrounding sea water. They are not stored or passed through ducts in the body cavity, as in many other invertebrates.

Hard Parts

The skeleton of *Ptilocrinus* is composed of several thousand individual elements, termed **ossicles** (*61;* italic numbers associated with morphologic terms for crinoid hard parts all refer to Fig. 18-2, unless indicated otherwise). The ossicles are organized to form several structurally different portions of the whole. First, the skeleton is divisible into two main parts: the **crown** (*1*), which comprises all of the crinoid test above the top of the stem; and the **pelma** (*2*), consisting of the stem, all of its appendages, and whatever anchorage structure may be developed. The crown and pelma contain well-differentiated elements, with which we must become acquainted.

Calyx. The crown is made up of skele-

(*Fig. 18-2 continued.*)

ambulacrals and irregularly arranged interambulacrals.

tegminal (55). Plate of the tegmen.

tergal (*T*) (74). Proximal plate of posterior interray in camerate crinoids.

transverse ridge (14). Linear elevation crossing joint face of a muscular articulated brachial or

radial, located just dorsal to central canal and separating large dorsal ligament fossa from ventral areas of facet.

uniserial (11). Arrangement of brachials in a single series.

ventral (54). Adoral, upper surface of crinoid, opposite stem attachment.

tal parts (**calyx,** *26*) enclosing the viscera and radial extensions (**arms,** *3*) of these. The arms bear many small branches (**pinnules,** *8*).

The calyx has two main parts: the teg-

MUSCULAR ARTICULATION

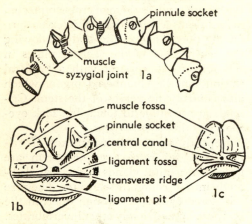

LIGAMENTARY ARTICULATION

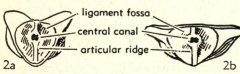

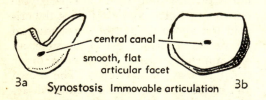

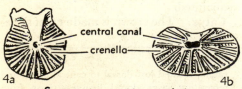

FIG. 18-3. Types of articulation in arm plates of crinoids. These plates all belong to modern crinoids (enlarged, *from Gislén*).

men (*53*), comprising the ventral portion, on the upper side of the body between bases of the arms; and the **dorsal cup** (*56*), which consists of laterally joined plates on the lower side of the body adjacent to the stem, bounded at its upper margin by the bases of the arms. The tegmen of *Ptilocrinus* differs from that of most fossil crinoids in lacking ossicles joined together at their edges.

The dorsal cup is composed of two circlets of plates, both visible in side view (Fig. 18-1, *3*), flaring upward so as to resemble a wineglass. The upper circlet contains five plates of identical size and shape, and because these are in direct line with the arms, they are called **radials** (*35*); they constitute skeletal elements at the beginning points of the **rays** (*6*). The circlet next below the radials, forming the base of the dorsal cup, includes five **basals** (*37*). These have their mid-lines opposite the junctions (**sutures**) between adjoining pairs of radial plates; hence, their position is defined as interradial. In some Recent crinoids, a third circlet of small plates is present in the dorsal cup below the basals, and accordingly these are termed **infrabasals** (*38*). The infrabasals alternate in position with the basals and thus have a radial position, but they are not connected in direct series with the radial plates. The circlet of infrabasals is much more prominent in many fossil crinoids than in living species, most of which have lost this element of the cup.

Arms. *Ptilocrinus* has five arms. They bear many pinnules, but the axial part of each arm is unbranched (**atomous,** *28*; *a*, not; *tomous*, divided). In *Teliocrinus* and *Isocrinus* (Fig. 18-1, *1–2*), each ray bifurcates evenly two or more times, and each of the branches bears many pinnules. Such equal branches are termed **isotomous** (*4*) (*iso*, equal). Other modern and ancient crinoids are characterized by unequal branches (**heterotomous,** *33*) (*hetero*, other or different).

Each individual ossicle forming part of an arm is called a **brachial** (*34; brachium*, arm), and ossicles composing pinnules are named **pinnulars** (*9*). Along ambulacral grooves on the ventral surface of arms and pinnules are small covering plates (ambulacrals) and side plates (adambulacrals).

Articulation between the brachial segments is mostly movable, but there are some immovable joints. The study of living crinoids permits recognition of two types of movable articulation, one having well-developed muscles in addition to ligaments (**muscular articulation,** Fig. 18-3, *1a–c*) and the other provided with ligament fibers only (**ligamentary articulation, synarthry,** Fig. 18-3, *2a, b*). These are distinguished by the nature and arrangement of the depressions on the articular facets of the brachials. Immovable articulation also includes two main types, both ligamentary; smooth facets distinguish one (**synostosis,** Fig. 18-3, *3a, b*) and radially ridged facets (**syzygy,** Fig. 18-3, *4a, b*) the other. The types and distribution of these various sorts of brachial articulation have much importance in classifying genera of modern crinoids, and it is taken into account also in work on fossils. Immovable union of skeletal parts of crinoids may go a step beyond what can be called articulation; this consists of **ankylosis** or fusion, in which calcareous deposits weld the ossicles together, with or without disappearance of an external line (suture) marking the junction. This sort of union is more common in plates of the calyx than among brachials.

Stem. *Ptilocrinus* is attached by means of a stem (**column,** *18*) composed of many superposed disk-shaped ossicles (**columnals,** *45*) of circular outline. Each columnal is perforated centrally by an opening (**axial canal,** *16*, or **lumen,** *17*) for passage of extensions of the chambered and axial organs. Articulating surfaces of these plates are marked by fine radiating ridges (**crenellae,** *15*). The columnals are of uniform size and bear no appendages. Ligamentous tissue holds them together, permitting just enough differential movement at the joints to provide bending of the whole stem.

Some crinoids have less rigidly united columnals and therefore greater mobility of the stem. Several types of stems are illustrated in Fig. 18-4.

The columns of *Teliocrinus* and *Isocrinus* differ from that of *Ptilocrinus* in having a pentagonal cross section, in being composed of columnals not of uniform size, and in possessing laterally directed appendages (Fig. 18-1, *1, 2*). All these characters, combined with features of the crown, are significant grounds for generic differentiation. The larger-than-average columnals, most of which bear whorls of side branches (**cirri,** *20*) define punctuation points or nodes of the stem, and hence are termed **nodals** (*19*); the columnals between any two successive nodals are called **internodals** (*22*). The nodals are seen to be more closely spaced toward the top of the stem near the crown (defined as **proximal,** *27*), and consequently internodals in this region are fewer and thinner than toward the base of the column (defined as **distal,** *25*). This is explained by a study of the manner in which new columnals are formed. Nodals are introduced just below the base of the dorsal cup, one following directly on another. Subsequently, internodals make an appearance between the nodals; they increase in thickness and numbers until the normal full size and complement of these ossicles characteristic of the species are attained. Thus, the youngest part of the stem is at its top and the oldest at its base. Another feature, which is well shown by the stem of *Isocrinus*, is a distinct sort of articulation between nodals and next-lower columnals (termed **infranodals,** *24*); this joint lacks crenellae, and consequently the smooth apposed surfaces of nodal and infranodal are less firmly united than those of other columnals. The stem tends to break apart at these places, during the life of the animal or after death. An *Isocrinus* which becomes loosened from its mooring may be drifted about, dragging remnant portions of its stem. Similarly, the free-swimming stemless crinoids (Fig. 18-1, *4*) break away from their larval stem by a cleavage just below the crown.

Cirri. Lateral appendages of the stem, which characterize many modern and fossil crinoids but which are wholly lacking in others, are parts of the pelma (Fig. 18-4, 7). Although joined to the column, they are distinct from it. Each cirrus is composed of numerous similar ossicles (**cirrals,** 21) which articulate with one another so as to provide considerable mobility. A study of living cirrus-bearing stalked crinoids shows that these appendages tend constantly to be in motion, their function being in part sensory and in part prehensile. They can coil around seaweed or other foreign bodies and cling firmly. Attachment of a cirrus to the nodal which bears it is marked by a facet (**cirrus socket,** 23) having the cross-sectional shape of the cirrus, and the outline of this facet may differ from that of the nodal. For example, pentagonal nodals of *Isocrinus* carry elliptical cirrus sockets on their sides. An axial canal runs through the center of each cirral, and continuation of this passage-

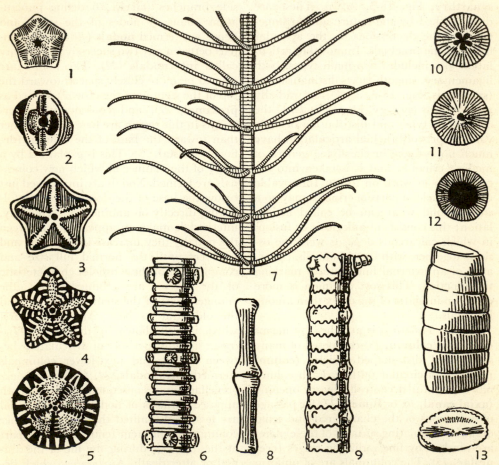

FIG. 18-4. **Characters of crinoid stems.** Recent forms include 7 and 8; the others are fossils. Varied sorts of articular surfaces, outline of stem segments, and shape of axial canal are shown. Cirri and sockets for attachment of those lateral appendages of crinoid stems are seen in figures 6, 7, and 9, which also illustrate characters of nodal and internodal columnals. Elliptical columnals which bear an articulating ridge are shown in 2 and 13; the latter forms a twisted stem. (Not to scale.)

way reaches from the cirrus socket to the axial canal of the nodal. Connection for nerves and vessels carrying nourishment is thus furnished.

Anchorage Structures. Modern stalked crinoids generally are attached by cementation of the distal extremity of the stem to rock or some firm anchorage on the sea bottom, but some possess rootlike terminal branches (**radix,** *32*) which provide fixation by spreading in sediment. Various fossil crinoids exhibit these types of attachment. Some cemented "holdfasts" consist of relatively wide disk-shaped expansions. In addition, there are specialized sorts of anchorage devices (for example, Fig. 18-30, *10*).

Orientation

The dorsal cups of all examples of modern crinoids illustrated in Fig. 18-1 are characterized by diminutive size, compared with the remainder of the crown, and by nearly perfect pentameral symmetry. In these respects, they differ from most species of fossil crinoids, which have proportionally larger cups lacking complete radial symmetry. In these especially, but in living forms too, attention must be given to distinction of individual rays and interrays and this calls for a standard method of orienting the animal.

Crinoids have recognized dorsal and ventral sides, but they lack a front and rear as defined by direction of movement or by location of head and tail. The key to orientation in terms of anterior, posterior, right, and left is furnished by the position of the anus and (generally associated with this) the presence of additional plates in one of the interrays of the dorsal cup. The anus, located in an interray on the tegmen, or in some fossil forms on the side of the cup, is defined as posterior in position, and likewise the extra-plate interray. Then, according to convention, the crinoid is held in its normal living position, with the ventral side upward, and this establishes the sides, respectively, designated as right and left. Each indi-

vidual ray and interray has its own designation (*47–51, 75–78, 81*) in a manner satisfying the needs for precise descriptions.

CLASSIFICATION

Modern crinoids are all included in a subclass called Articulata, which has fossil representatives distributed through Mesozoic and Cenozoic systems. Other groups of crinoids are exclusively Paleozoic, or nearly so, and each of the three subclasses known from rocks of this era range from Ordovician to Permian. The divisions are indicated in the following outline of classification (orders of Articulata except Uintacrinida and Comatulida are newly defined by Dr. H. Sieverts-Doreck).

Main Divisions of Crinoids

Crinoidea (*class*), echinoderms belonging to the subphylum Pelmatozoa, characterized by relatively numerous plates of the calyx, arranged with pentamerous symmetry; radial appendages (arms) prominent, generally branched. Lower Ordovician–Recent.

Inadunata (*subclass*), plates of calyx joined firmly together, mouth concealed by tegmen, dorsal cup lacking interray plates except on posterior side (anals); except in a few primitive forms, arms free above radicals. Lower Ordovician–Upper Permian.

Disparida (*order*), dorsal cup typically composed of radials, basals, and anals, but including brachials in some; arms nonpinnulate. Middle Ordovician–Upper Permian.

Hybocrinida (*order*), like Disparida but have a radianal plate and show tendency to suppress arms. Middle Ordovician–Lower Silurian.

Cladida (*order*), dorsal cup composed of radials, basals, infrabasals, and anals. Lower Ordovician–Upper Permian.

Cyathocrinina (*suborder*), bowl-shaped to globose cup, narrowly rounded radial facets without transverse ridge, arms lacking typical pinnules. Middle Ordovician–Upper Permian.

Dendrocrinina (*suborder*), steeply conical to bowl-shaped cup, narrow or wide radial facets with transverse ridge, anal sac prominent, arms nonpinnulate. Lower Ordovician–Devonian.

Poteriocrinina (*suborder*), cup form diverse, arms pinnulate, otherwise like Dendrocrinina. Middle Ordovician–Upper Permian.

Flexibilia (*subclass*), dorsal cup composed of lower brachials, radials, basals, three infrabasals, and anals, mostly not rigidly united; flexible tegmen bears exposed food grooves and mouth; arms uniserial, nonpinnulate. Middle Ordovician–Upper Permian.

Taxocrinida (*order*), elongate crown with relatively weak calyx. Middle Ordovician–Upper Pennsylvanian.

Sagenocrinida (*order*), subglobular crown with rather firmly united calyx plates. Middle Silurian–Upper Permian.

Camerata (*subclass*), all calyx plates rigidly united, mouth and food grooves concealed beneath strong tegmen, dorsal cup includes brachials and generally also interradials. Middle Ordovician–Upper Permian.

Diplobathrida (*order*), dorsal cup includes infrabasals. Middle Ordovician–Upper Mississippian.

Monobathrida (*order*), dorsal cup lacks infrabasals. Middle Ordovician–Upper Permian.

Tanaocrinina (*suborder*), basal circlet typically with hexagonal outline, composed of three equal plates; radials laterally in contact except at posterior side of cup. Middle Ordovician–Upper Permian.

Glyptocrinina (*suborder*), basal circlet with pentagonal outline, typically composed of two large plates and a small one; radials in contact all around. Middle Ordovician–Upper Permian.

Articulata (*subclass*), dorsal cup mostly reduced greatly, infrabasals five or atrophied; tegmen flexible, with exposed mouth and food grooves; arms uniserial or rarely biserial, pinnulate throughout. Lower Triassic–Recent.

Isocrinida (*order*), stem-bearing with true cirri, cup dicyclic with wide radial facets, arms mostly uniserial. Lower Triassic–Recent.

Millericrinida (*order*), stem-bearing, without true cirri; fixation by radix, cup cryptodicyclic or monocyclic, radial facet wide, arms uniserial. Middle Triassic–Eocene.

Cyrtocrinida (*order*), small compact reef dwellers, mostly with short cirriferous stem or unstalked; monocyclic, processes between radial facets, arms 10. Lower Jurassic–Recent.

Uintacrinida (*order*), unstalked, dicyclic or monocyclic, cup large, thin-plated, with proximal brachials, pinnulars, and interradials incorporated in cup. Upper Cretaceous.

Roveacrinida (*order*), small unstalked, infrabasals atrophied, radial facets narrow, skeleton light, delicate. Upper Triassic–Upper Cretaceous.

Comatulida (*order*), unstalked, infrabasals lacking, basals in form of rosette; centrodorsal generally cirriferous. Lower Jurassic–Recent.

STRUCTURES OF FOSSIL CRINOIDS

Generally speaking, the structural features of fossil crinoids correspond to those described in modern forms, but several differences do exist, especially among ancient stocks which now are extinct. In extending our study, it is preferable to give chief attention to the fossils, adding to observations already made. Discussion is arranged according to much the same outline used in describing examples of modern crinoids, with the addition of sections on the distinguishing characters and evolutionary trends of each main group of fossil crinoids.

Calyx

The calyx of most fossil crinoids constitutes a proportionally larger part of the crown than that of modern crinoids. Exceptionally, the crown of a fossil crinoid is all calyx, no free arms being present, but such a condition marks an extreme. The dorsal cup is invariably formed of two or more circlets of plates joined rigidly or flexibly. The tegmen also is commonly composed of adjoined plates (**tegminals,** *55*), but (except **orals,** *58*) these are not arranged in regular rings. Some fossil species, like modern ones, are judged to have possessed a leathery tegmen in which a varying number of small calcareous plates were scattered; but the tegmen of most fossil crinoids was fairly well built of plates joined together. Thus, the calyx, including the dorsal cup and tegmen, may be preserved as a unit, lacking a trace of arms and stem other than the surfaces (facets) to which these were attached. Especially common examples of such calices are most fossil camerates—the so-called "box crinoids"—because their stoutly constructed tegmen is rigidly joined to the dorsal cup.

Dorsal Cup

On the basis of their structure, dorsal cups of crinoids can be divided into six main groups, which largely reflect the definition of subclasses and orders. Descriptions of the groups which follow are arranged in order of increasing complexity, but this should not be interpreted to mean that they represent successive steps in structural development defining an evolutionary series.

Two-circlet Cups. Type 1. The simplest known type of dorsal cup appears among Paleozoic crinoids which form part of the subclass Inadunata. The cup of these crinoids consists of only two circlets—basals and radials—the plates of which alternate with one another in position (Fig. 18-5). This arrangement of a single ring of plates below the circlet of radials is termed **monocyclic** (Fig. 18-2, 40). The lower circlet bears attachment for the stem, and the upper one supports free arms. Accessory plates or a single plate may occupy part of the posterior interray. Crinoids of this type are included in the orders Disparida and Hybocrinida. Many mod-

ern crinoids, such as *Ptilocrinus* and a majority of the free-swimming types, as well as Mesozoic and Cenozoic fossils, seem to have dorsal cups of the two-circlet type. These crinoids which belong to the subclass Articulata, are highly evolved derivatives of three-circlet-cup ancestors. Accordingly, their simplicity is not a sign of primitiveness.

The posterior side of two-circlet dorsal cups of disparid inadunates contains special structural elements which are important in classification and the study of evolution. Lower plates of the right posterior ray differ from those of other rays in dividing just above the first or second plate so as to form two branches. The branch on the right side leads upward to an arm or arms which begin above the rim of the cup, whereas the branch on the left (although morphologically armlike) constitutes the lower part of a plate series called **anals** (Fig. 18-7, 10–13). The lowest plate of this series is designated as the **anal X** plate (66), and the ray plate below it at the right is called **aniradial** (67). The aniradial is the plate where branching takes place. A radial is present below the aniradial in some genera (Fig. 18-7, 10–12), but in others it is lacking and then the aniradial stands alone as a special sort of radial (Fig. 18-7, 13). Exceptionally, the aniradial and anal X plates occur above the rim of the dorsal cup (Fig. 18-7, 12); in such crinoids, the cup has five even-sized radials.

Two-circlet Cups. Type 2. Mono-

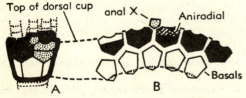

Fig. 18-5. **Two-circlet dorsal cup. Type 1.** (A) Posterior view of cup. (B) Diagram of cup plates. The figures show cup structure of a monocyclic disparid crinoid (such as *Homocrinus*) belonging to the subclass Inadunata. (Radials black; plates of the anal series stippled; X, anal X.)

Fig. 18-6. **Two-circlet dorsal cup, Type 2.** (A) Posterior view of cup. (B) Diagram of cup plates. The figure shows cup structure of a monocyclic hybocrinid (*Hybocrinus*) belonging to the subclass Inadunata. (Radials black; plates of anal series stippled; RA, radianal; X, anal X).

Cladida

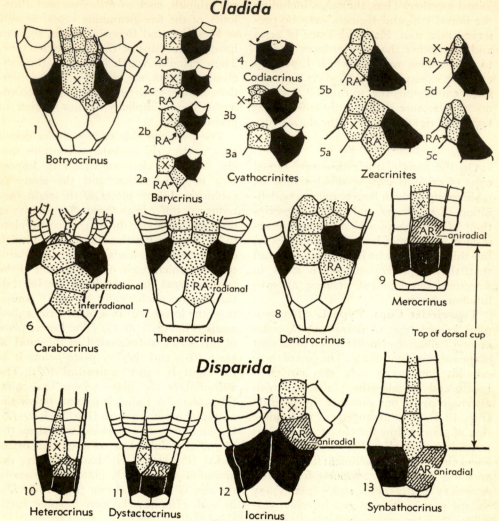

Disparida

FIG. 18-7. **Structure of the posterior part of inadunate crinoid dorsal cups.** The dicyclic inadunates (Cladida) exhibit various structures of relatively advanced evolutionary type, whereas the monocyclic forms (Disparida) have more primitive arrangement of posterior plates. Radial plates are shown in black; anal plates stippled; infrabasals, basals, and brachials without pattern (*AR*, aniradial; infer-radianal, which is the lower of a pair of plates equivalent to radianal; *RA*, radianal; superradianal, upper of pair of plates equivalent to radianal; *X*, anal X plate).

1, Quadrangular *RA* obliquely left below right-posterior radial (*RPostR*) supporting large anal X, which separates posterior radials and is succeeded above by anal-sac plates. This is a characteristic cladid pattern.

2a–d, Series showing disappearance of *RA* by resorption.

3a, b, 4, Series showing reduction in size of anal X accompanied by upward shift out of the dorsal cup.

5a–d, Series showing tendency of anal plates to shift upward.

6, Presumably a very archaic, primitive type of anal-plate structure, which exactly corresponds to that of some monocyclic inadunates (Hybocrinida) except for occurrence of a pair of plates in the position of *RA*.

7, Equivalent to *Carabocrinus* (6) except for substitution of *RA* for inferradianal and superradianal. (*Continued on next page.*)

cyclic crinoids of the order Hybocrinida are inadunates having a structure similar to the cups just described, except that the lowest plate in the right posterior ray is succeeded by anal X and a radial, instead of anal X and a brachial (Figs. 18-6; 18-28, *11*). This radial is smaller than the others, but otherwise entirely like them. The lowest ray plate, which seems to correspond to an aniradial, is here given another name, because it lies beneath a radial rather than above it or taking the place of it. This name is **radianal** (*64;* Fig. 18-6).

Three-circlet Cups. Type 3. Most fossil crinoids of the subclass Inadunata (all of the order Cladida), as well as many living and fossil Articulata, are distinguished by the presence of three circlets of plates between the base of the free arms and the stem attachment (Fig. 18-8). The plates of each circlet alternate in position with those of adjoining circlets. In downward order from the arm bases, the plates are radials, basals, and infrabasals.

Dorsal cups which have two circlets of plates below the radials are termed **dicyclic** (*65*).

The three-circlet-cup inadunates have an extra plate or plates on the posterior side. These are chiefly the radianal and anal X plates. The dorsal cup of a few genera belonging in this group have perfect pentamerous symmetry, some because aniradial and anal X plates are above the edge of the cup (Figs. 18-7, *9;* 18-28, *13*) and others because radianal and anal X plates present in ancestors have vanished in the course of evolution (Figs. 18-33, *9, 10*).

Multiple-circlet Cups. Type 4. Dorsal cups of crinoids which are classed together in the subclass Flexibilia are distinguished by the partly movable articulation of the plates, as contrasted to the rigid sutures between the plates of other cri-

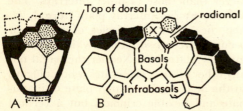

Fig. 18-8. **Three-circlet dorsal cup, Type 3.** (*A*) Posterior view of cup. (*B*) Diagram of cup plates. The drawings represent *Blothrocrinus*, a typical dicyclic inadunate crinoid belonging to the order Cladida. (Radials black; plates of the anal series stippled; *X*, anal X; *RA*, radianal).

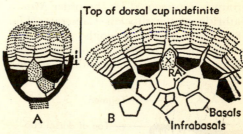

Fig. 18-9. **Multiple-circlet dorsal cup, Type 4.** (*A*) Posterior view of cup. (*B*) Diagram of cup plates (full line) and free-arm plates (dotted). The drawings represent the flexible crinoid, *Lecanocrinus*, which lacks clearly marked separation of dorsal cup and free arms. (Radials black; plates of anal series stippled; *X*, anal X; *RA*, radianal).

(Fig. 18-7 continued.)

8, A basal plate separates *RA* from the infrabasal circlet; *RA* directly below *RPostR*.

9, Dorsal cup perfectly symmetrical, but the crown is not symmetrical, for only the *RPostR* has an axillary plate above it; this is an aniradial (*AR*), bearing anal X at left and a normal arm at right. This dicyclic crinoid duplicates structure seen in the monocyclic *Iocrinus* (12).

10, 11, Monocyclic cups with *AR* above *RPostR* and including cup-brachials above all radials

and *AR*. Anal X and plates above it are morphologically like brachials, and this is judged to be a most primitive plan of posterior-plate arrangement.

12, Like *Merocrinus* (9) except that the cup is monocyclic.

13, Cup with almost perfect pentamerous symmetry of radial circlet (including *AR* as substitute for *RPostR*, or representing fusion of these two plates, as indicated by comparison with cups like 10 and 11).

noids, and by the occurrence of two circlets of plates below the radials (Fig. 18-9). Flexible crinoids are thus dicyclic, and in addition, they are characterized by the presence of only three plates in the infrabasal circlet. Another feature of many of this group is a lack of definite demarcation between plates classifiable as belonging to the dorsal cup and those of free arms (Figs. 18-28, *1;* 18-29, *2, 3, 5, 10*). This lack of sharp separation is accentuated by the occurrence of extra plates between the rays so as to unite them laterally (and in some flexibles, very firmly, Fig. 18-33, *18*). The extra plates are designated as **interradials** (*46*, illustrates them in another type of crinoid).

The posterior interray and right posterior ray of some contain extra plates called radianal and anal X.

Multiple-circlet Cups. Type 5. This group is represented by the order Monobathrida of the subclass Camerata (Fig. 18-10). Camerate crinoids are distinguished by the incorporation of lower ray plates, in addition to the radials, as rigid

elements of the dorsal cup. These lower ray plates of the cup, in linear series above the radials, are termed **cup-brachials** (*69*). One or more bifurcations of the cup-brachial series in each ray may be included in the dorsal cup. Free arms join the calyx at the boundary between the dorsal cup and tegmen, at a variable height above the radials. Beneath the radials is a single circlet of plates, which are basals. Hence, crinoids of this group are monocyclic.

The interray parts of the dorsal cup belonging to most members of this group contain extra plates (interradials) which fill in spaces between the radials and cup-brachials so as to make a strongly built wall enclosing the soft parts of the crinoid. In addition, some monocyclic camerate cups have extra plates between branches of the cup-brachials; these are termed **interbrachials** (*68*). Extra plates belonging to the posterior interray commonly are more numerous, and some of them may be larger than those of other interrays. The lowest posterior interray plate (**tergal,** *74*) may lie above the circlet of radials or occur between two radials, which no other interray plate does. Neither the tergal nor any other plate of camerates is homologous to the anal X or radianal plate of inadunate and flexible crinoids, but a vertical series of posterior interray plates (including the tergal) in some camerates is collectively called anals.

Multiple-circlet Cups. Type 6. This group differs from the last only in having two circlets of plates below the radials, instead of one (Fig. 18-11). The difference in structure of the base of the cup is an important one, however. It is the basis for distinguishing the order Diplobathrida. Most dicyclic camerates have several cup-brachials in each ray, and interradials between the rays; a few have interbrachials also. A tergal is present.

Characters of the Tegmen. Different fossil crinoids show a variety of tegminal features. Many have a nearly flat or gently convex tegmen composed of small plates

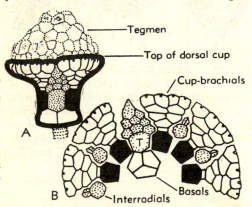

FIG. 18-10. **Multiple-circlet dorsal cup, Type 5.** (*A*) Posterior view of calyx. (*B*) Diagram of cup plates. The drawings show structure of a monocyclic camerate crinoid, *Uperocrinus,* belonging to the order Monobathrida. The boundary between dorsal cup and the stout tegmen is clearly defined at the upper limit of the firmly united cup-brachials; the free arms are attached at the top of the cup, which is equivalent to outer margin of the tegmen. (Radials black; anal series and interradials stippled; *7,* tergal.)

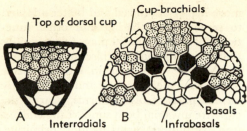

FIG. 18-11. **Multiple-circlet dorsal cup, Type 6.** (*A*) Posterior view of cup. (*B*) Diagram of cup plates. The drawings represent *Ptychocrinus*, a typical dicyclic camerate belonging to the order Diplobathrida. The top of the dorsal cup is clearly marked by the places for attachment of the free arms. (Radials black; anal series and interradials stippled; *T*, tergal.)

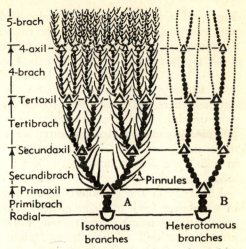

FIG. 18-12. **Types of crinoid arm branching and nomenclatures of ray plates.** Diagrammatically represented are (*A*) equal-size divisions of a ray belonging to a pinnulate crinoid, and (*B*) unequal-size divisions of a ray belonging to a nonpinnulate crinoid. The names of plates are indicated at left.

joined firmly together. They are arranged without pattern. Some low-tegmen crinoids, however, exhibit a well-marked differentiation of plates belonging to this part of the calyx. Five radiating double rows of interlocked plates in position of the **ambulacra** (*30*) are termed **ambulacrals** (*29*); the unevenly disposed plates of the tegmen between these rows comprise **interambulacrals** (*31*). The posterior interambulacrum is sometimes designated as the **anal area** (*59*).

Moderately protuberant or strongly elevated projections of the tegmen are seen in many fossil crinoids. They are known as **anal tubes** (*52*) or **anal sacs** (*62*) because they carry the anus at some point along the side or at the top. The presence of round or slitlike pores between the plates of some anal sacs suggests that these structures also serve a respiratory function. Tegminal plates, including those of anal tubes or sacs, may carry prominent spines.

Ray System

The ray system of fossil crinoids, as of living kinds, is composed of rows of plates, which decrease in size outward from the cup. The number of ray plates belonging to a single crinoid ranges upward to several hundred thousand. The main part of each ray consists of a single arm, or two or more arms formed by bifurcation of the axial

part of the ray. Each bifurcation commonly produces **isotomous branches** (*4;* Fig. 18-12), but **heterotomous branches** (*33;* Fig. 18-12), in which one of the divided arms is distinctly larger than the other, occur in many fossil crinoids. Pinnules may be abundant or lacking in both kinds of branching. Among heterotomously branching rays, two main types are recognized: **endotomous branches** (*7*), in which small arms develop toward the inside of the ray, and **exotomous branches** (*36*), in which small arms are directed toward the outside of the ray.

According to position of the individual arm plates (brachials) with reference to points of branching, they may be classified as first-order brachials (**primibrachs,** *73*), second-order brachials (**secundibrachs,** *71*), and so on (Fig. 18-12). The brachials at the points of bifurcation, called **axillaries** (*5*), differ from others in having a single articular face on the lower or **proximal** (*27*) side and two on the upper or **distal** (*25*) margin for attachment of the two branches which succeed this plate.

These axillary brachials also are classified with respect to their position in the ray as first-order axillary (**primaxil**, *72*), second order (**secundaxil**, *70*), and so forth (Fig. 18-12). Viewed from the external (dorsal) side, axillary plates have a pentangular outline, whereas other brachials mostly appear quadrangular or pentangular. Some brachials are much wider (measured across the branch) than long, others are subequal in width and length, and still others are longer than wide.

Efficiency of the pinnule-bearing arms in gathering food particles clearly must vary with the number and spacing of the pinnules, as well as their length. Figure 18-13 illustrates four types of pinnule arms, which from left to right exhibit increasing number and close spacing of pinnules. The differences in arm structure are mainly those of shape and articulation of the brachials. The structure of some arms is **uniserial** (*11*; Fig. 18-13*A–C*) and that of others **biserial** (*39*; Fig. 18-13*D*). The biserial type of arm plainly seems to be more highly organized and better adapted to serve food-gathering and other functions than the uniserial plan. Observations that crinoids having biserial arms are much more numerous in late Paleozoic than early Paleozoic time and that initially formed parts of biserial arms, nearest the cup, are uniserial both support the conclusion that uniserial arms are more primitive. Some crinoid stocks, such as monocyclic inadunates, flexibles, and most articulates, have only uniserial arms, however.

The nature and arrangement of branching, distribution of types of articulation along the branches, and presence or absence of pinnules are important in the classification of crinoids.

Column

The columns of fossil crinoids, as seen in different genera, may have circular, pentagonal, stelliform, elliptical, quadrangular, or crescentic cross sections. Commonly, each segment is much wider (measured across the stem) than long. The columnals

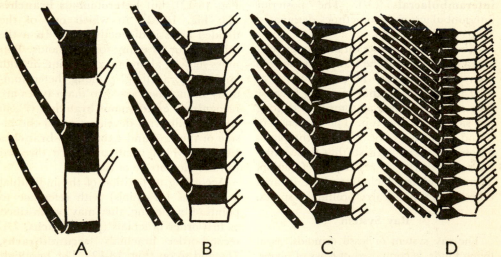

A   B   C   D

Fig. 18-13. **Arrangement of pinnules borne by uniserial and biserial arms.** Each brachial plate normally carries one pinnule, and the direction of branching of the pinnules is the same on each second brachial. Elongate uniserial brachials (*A*) bear widely spaced pinnules. Shortening of uniserial brachials (*B, C*) serves to increase the number of pinnules in unit distance along the arm. Crowding, which results from wedge shapes of the brachials, reaches a maximum in interlocked double series of brachials belonging to biserial arms (*D*). Pinnulate arms are more advanced in evolution than nonpinnulate arms, and pinnulate biserial arms represent culmination of arm development.

of some stems, however, are distinctly longer than wide. Their external faces may be arranged so as to be smoothly confluent or variously bulged and otherwise ornamented, features which are useful for distinction of different types.

Branches at the base of many fossil crinoid stems have the form of roots which evidently serve as holdfasts for the stem in soft mud of the sea bottom. They differ from ordinary cirri in that many of them bifurcate.

A noteworthy feature of many crinoid stems is variation in the appearance of different portions of the column. Thus, the part of the stem closest to the crown may have columnals of different size and shape from those of the middle and distal portions. Nodals tend to be much more closely spaced near the calyx than in parts of the stem distant from it. Some stems and their attached cirri are remarkably specialized, as those of *Myelodactylus*, for example (Fig. 18-29, *15A*, *B*, *21*). Columnals of many species are so distinctive that commonly they can be identified by characters of the external and articular surfaces.

INADUNATE CRINOIDS

The inadunates are an extinct group of Paleozoic crinoids in which the plates of the calyx are joined firmly together, with the mouth concealed by the tegmen, and the arms free above the radials. A few primitive crinoids of this subclass have a dorsal cup in which each ray contains a radial and one or two cup-brachials, all these plates being firmly incorporated in the cup.

The inadunate assemblage includes both monocyclic and dicyclic crinoids, and this difference in the structure is a main basis for discrimination of orders: disparids and hybocrinids (monocyclic) and cladids (dicyclic). More than 1,750 species of Inadunata, classed in 325 genera, are described from Middle Ordovician to Upper Permian formations. Greatest abundance of these crinoids is in Mississippian deposits (Fig. 18-27).

Disparid Inadunates

Crinoids of the disparid group are mostly diminutive forms. Among them are numerous species of microcrinoids with the maximum height of the adult calyx not exceeding 1 mm. Others, like the "pea crinoids" (*Pisocrinus*, Fig. 18-29, *4*), remarkably abundant in some Silurian deposits, and similar Devonian forms (Fig. 18-30, *3*), have calyces 2 to 4 mm. in diameter. This group is of interest mainly because of the light which several members throw on the evolution of plates on the posterior side of the cup; this is important for classification. Most of these crinoids have the simplest type of dorsal-cup structure. The known range of disparids (60 genera) is from Middle Ordovician to Upper Permian.

Evolution of Anal Plates. The posterior side of the cup is identified by the presence of an extra plate or plates belonging to the anal series. Characters shown by several Ordovician disparids clearly prove that the anal series is actually a modified branch of the right-posterior ray. This branch takes on a specialized physiological function and may become modified into an anal sac rising prominently above the borders of the tegmen.

The aniradial plate, located at the proximal extremity of the right-posterior ray or next above it, underlies and supports the anal branch which diverges toward the left (Fig. 18-7, *10–13*). In forms having no ray plate below the aniradial, the plate called aniradial is unquestionably the product of fusing the first two right-posterior ray plates. This fusion is the only essential change in structure of the posterior side of the cup introduced by evolution, but it is important for the study of homologous plates in other crinoids. In the course of evolution, the anal X plate tends to be forced upward out of the cup.

Other Evolutionary Trends. In different families of disparids, the basal circlet shows a tendency toward reduction in the size and number of plates. The end product of this trend, shown by a number of genera, is a single fused disk no larger than the stem in diameter. In no group do remnants of the basals vanish, however.

A similar trend toward reduction of parts affects the rays. Especially, this affects the lowermost (proximal) plates, strengthening and simplifying the cup by fusion of the primitive radial with the next following ray plate. Various genera show different stages in this evolution, some with no fusion, and others with fusion in two rays, three rays, or all five rays. It is an interesting fact that the consolidation is not distributed in a random manner but according to a definite pattern, which yields more or less marked bilateral symmetry. Moreover, the plane of this symmetry is constantly oriented in all families, not as one naturally might suppose, through the anterior ray and posterior

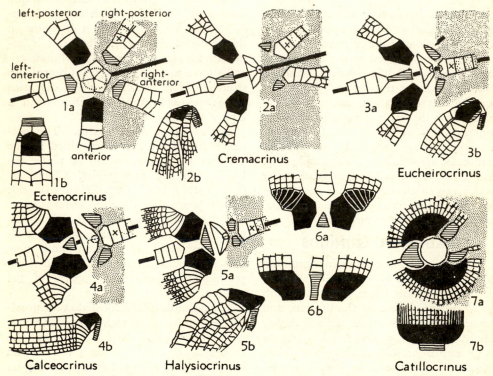

Fig. 18-14. **Bilateral symmetry of specialized monocyclic inadunate crinoids.** The oriented diagrams of plate structure represent three families of Disparida (1, Homocrinidae; 2–5, Calceocrinidae; 7, Allagecrinidae), the posterior interray being located at the top of each diagram and the right side of the crown indicated by shading. (Large radials black, small radials ruled; X, anal X). The plane of bilateral symmetry is observed to extend through the left-anterior ray and postero-right interray.

1a, b, *Ectenocrinus*, Middle and Upper Ordovician, probable ancestor or similar to ancestor of other crinoids shown. (1a) Plate diagram; (1b) anterior side, inverted so as to retain left and right sides in proper position.

2–5, Plate diagrams and anterior views of *Cremacrinus*, Middle Ordovician–Middle Silurian (2a, b); *Eucheirocrinus*, Middle Silurian (3a, b); *Calceocrinus*, Middle Silurian (4a, b); and *Halysiocrinus*, Devonian–Mississippian (5a, b); showing progressive enlargement of two of the rays.

6a, b, Diagram indicating possible derivation of large radial with several directly attached arms (6b) by fusion of radial and axillaries (6a) similar to those of rays in 4a, b and 5a, b.

7a, b, *Catillocrinus*, Mississippian; (7a) plate diagram; (7b) anterior view, not inverted.

interray, but through the left anterior ray and postero-right interray (Fig. 18-14). Explanation is unknown, for analogy with other echinoderms seems to offer no help.

A disparid assemblage of special interest from the standpoint of their strange evolutionary modification consists of the "bent-crown crinoids" (Calceocrinidae) (Figs. 18-28, *6a, b;* 18-29, *14;* 18-31, *18a–c*). Presumably as adaptation to a reef habitat, the cup and arms of these crinoids came to have a recumbent position on the stem, inducing changes which greatly disturbed the arrangement of arms and the anal sac and ultimately produced a crown having remarkable bilateral symmetry. A wide hinge, provided with muscular articulation, was developed between the basals and radials, so that the crown could be flexed through an arc of approximately 90 deg. Representative specimens of this aberrant group of crinoids are distributed from Middle Ordovician to Mississippian, and they clearly exhibit successive steps in remarkable evolutionary changes induced by their adopted mode of life.

Equally strange but specialized in a manner different from calceocrinoids are the members of the family Allagecrinidae (Figs. 18-14, *7a, b;* 18-31, *6;* 18-33, *3, 4*) which differ from normal crinoids in having as many as 32 arms borne directly by a single radial plate. The arms are slender unbranched structures. Enlargement of the multiple-arm-bearing radials and attachment of several independent arms directly to a radial are both explained as a result of fusion in which a succession of axillary brachials is joined to the subjacent radial (Fig. 18-14, *6a, b*).

Hybocrinid Inadunates

The order Hybocrinida includes monocyclic inadunates which have a well-defined radianal instead of an aniradial (Figs. 18-6; 18-15). The radianal underlies the right posterior radial directly or obliquely leftward. The hybocrinids also exhibit a tendency to have fewer than five

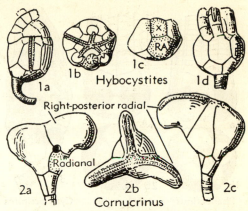

FIG. 18-15. **Types of hybocrinid monocyclic inadunates.** Some genera of this group have five free arms, but the two forms illustrated each have only three. The anal plates of the cup consist of a radianal (*RA*) and anal X (*X*).

1a–d, Hybocystites eldonensis (Parks), Ordovician (Trentonian), Ontario. Left anterior (*1a*), posterior (*1b*), ventral (*1c*), anterior (*1d*); ×1.3.

2a–c, Cornucrinus mirus Regnéll, Middle and Upper Ordovician, Sweden. Posterior (*2a*), ventral (*2b*), right anterior (*2c*), ×1.

free arms and to resemble cystoids in the extension of some ambulacra across the plates of the dorsal cup. This may be a primitive attribute. Lack of free arms does not characterize all genera, but among those having fewer than five such appendages, the same rays are affected (Fig. 18-15). As in bilaterally symmetrical disparids (Fig. 18-14), the anterior and left-posterior radials are dominant, but unlike them, the right-posterior ray is equally prominent. The hybocrinids (six genera) are known from Middle Ordovician to Lower Silurian. Antecedents and descendants are not identified.

Cladid Inadunates

The dicyclic inadunates, called cladids (Fig. 18-7), are far more numerous than the disparids. Their dorsal cup is the three-circlet type, for radials, basals, and infrabasals occur between the bases of free arms and the stem attachment. They are represented by such typical genera as

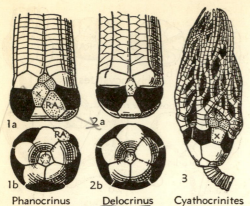

Phanocrinus   Delocrinus   Cyathocrinites

FIG. 18-16. **Representative types of dicyclic inadunate crinoids.** These are posterior and dorsal views of two Mississippian crinoids (1, 3) having uniserial arms, and a Pennsylvanian form having 10 biserial arms (2). (Radials black; plates of the posterior interray stippled; *X*, anal X; *RA*, radianal.)

Phanocrinus, *Delocrinus*, and *Cyathocrinites* (Fig. 18-16) and many others illustrated (Figs. 18-7; 18-28; 18-29; 18-30; 18-31; 18-33). The crowns of some cladids attain a height of at least 12 in. They range from Middle Ordovician to Upper Permian. About 265 genera are described.

Evolution of Cup Shape. The earliest cladids are characterized by very steep-sided or deep bowl-shaped cups (Figs.

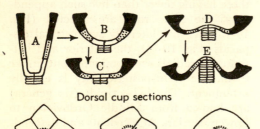

Dorsal cup sections

Infrabasal circlets

FIG. 18-17. **Evolution of cup shape and of the infrabasal circlet among cladid inadunate crinoids.** (*A–E*) Progressive steps in change from a steep-sided cup (primitive) to a shallow discoid cup (advanced) which is deeply concave at the base. (*F–H*) Infrabasal circlets showing reduction in number of plates as result of fusion.

18-7, *1, 6–9;* 18-17*A*). Evolutionary change consists in a gradual decrease in the height of the cup, accompanied by a flattening of the base and the development ultimately of a deep concavity enclosing the proximal part of the stem (Figs. 18-16, *1, 2;* 18-17*B–E*). The most highly modified cups have a discoidal shape, and the summit of the radial plates may actually be lower than the infrabasal circlet; such cups are most common in late Paleozoic strata, especially among Permian inadunates.

Evolution of Infrabasal and Radial Circlets. Infrabasal plates are invariably five in number in all primitive cladids, and they slope steeply upward, being visible in the side view of the cup. Evolution reduces their size and number and shifts their attitude (Fig. 18-17). Late stages are represented by a solidly fused circlet concealed by the stem, or down-flaring plates in the basal concavity of the cup.

The radial plates are one of the most important elements of the cup for purposes of classification and in showing evolutionary modification. Indeed, many cladid genera can be identified on the basis of a single isolated radial plate. The characters of chief significance are the shape and attitude of these plates and the nature of the articular facets at their distal edges. The outer surface of radials in unspecialized cups slopes steeply downward confluently with the basals and infrabasals. The articular surface is subhorizontal and directed upward.

Primitive cladids have narrow, nearly featureless rounded facets, whereas advanced forms have facets as wide as the radial plates, with the surface elaborately differentiated by ridges, grooves, and denticles for the attachment of ligaments and muscles and as fulcrums for articulation. Evolutionary change associated with a depression of the cup produces outward tilting of the radials so that they have an increasingly horizontal attitude. In specialized cladids having a strong basal concavity, the proximal edges of the radials extend nearly to the center of this concavity. The articular facets of these radials commonly slope gently or steeply outward. Finally, various end products of cladid evolution show reversion to very simple structure. One or more radials tend to become atrophied, for their size is much reduced and they are no longer arm bearing (Fig. 18-33, *23*). Eventually, these radials disappear entirely.

Evolution of Anal Plates and Tegmen. The plates of the posterior side of the cup belonging to the anal series are judged to be identical in origin and similar in evolution to those of disparid inadunates. A large majority of all early cladid genera and many of late Paleozoic age have a radianal plate, either in its primitive position directly below the right-posterior radial (Fig. 18-7, *8*), or obliquely at the left below this radial plate, which is the most common position (Figs. 18-7, *1, 2, 5, 7*; 18-16, *1*). Evolutionary change pushes the radianal upward to a position even with the radials or somewhat above them. Eventually this plate disappears, either by continuous upward displacement (Fig. 18-7, *5a–d*) or by resorption (Fig. 18-7, *2a–d*). The anal X plate occurs above the radianal, generally somewhat to the left. The evolutionary course of these plates is like that of the radianal in migrating upward until they vanish from the cup (Fig. 18-7, *3, 4*). In the most advanced condition, all anals have disappeared and the dorsal cup shows perfect pentamerous symmetry (Fig. 18-33, *9, 10*).

An almost diagnostic structural feature of cladid crinoids is the presence of a relatively prominent anal sac (Fig. 18-18). The sac of different genera is highly varied in size and form, but basically it seems to be a product of evolutionary development of the right-posterior ray, in which a main branch of this ray is greatly modified in structure and wholly changed in function. Joined with it are plates of the tegmen. Changes of the tegmen, which are associated with anal-sac structures, include development of prominent interray ridges, making niches into which the arms fit

neatly (Figs. 18-18, *2*; 18-31, *9b*; 18-33, *25*). Some sacs have an expanded summit like a mushroom, with laterally directed spines (Fig. 18-31, *10*). The anal opening has various locations, even on the anterior side of the sac (Fig. 18-31, *9b*).

Evolution of Arms. Arm structure evolves along several lines among the cladids. Undivided (atomous) arms, seen in a few genera, denote primitiveness, but they appear also in Permian genera having perfectly symmetrical cups which signify advanced evolution. Simplification of arms

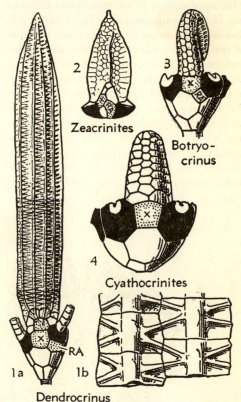

FIG. 18-18. **Structure of the anal sac of dicyclic inadunate crinoids.** The sacs of some of these crinoids exceed the cup in volume. The plates may be arranged in regular series, which reflects derivation of this structure from branching of the right-posterior ray, or they may become very irregular. Commonly there are pores, which presumably served respiratory functions, between the plates. (Radials black; plates of posterior interray stippled; *X*, anal X; *RA*, radianal.)

may also be interpreted as a tendency of late evolutionary change. A few late Paleozoic armless cladids represent extreme specialization in this direction.

A vast majority of these crinoids have arms that branch. Study of Ordovician genera shows that a very primitive type of branching is one composed of uniserially arranged elongate segments having only isotomous divisions. Shortening of brachials, increase in branching, and downward migration of the first bifurcation are early lines of evolutionary advancement. The appearance of heterotomous branching is a sign of specialization. Commonly this is superimposed on isotomous division of the rays, and it leads gradually to pinnulation. Arms composed of biserially arranged brachials bearing pinnules are highly advanced in evolution. Obviously, the change from elongate uniserially arranged brachials (with alternate ones bearing pinnules on the same side of the branch) to short biserially arranged brachials (with all on each side bearing pinnules) greatly increases the food-gathering

efficiency of an arm (Fig. 18-13). This change is an evolutionary trend among cladids. Biserial arm structure is not observed in pre-Mississippian inadunates but is common in late Paleozoic families.

Unique types of ray specialization are illustrated by Middle Silurian crinoids, *Petalocrinus*, in which branches of each ray are fused to form solid petal-like plates (Figs. 18-19, *2;* 18-29, *13*), or in which adjacent branches are united to form a flexible continuous network (Fig. 18-19, *1, 3*). Inasmuch as these specialized crinoids are found in rocks of equivalent age on opposite sides of the Atlantic, their distribution has much significance in relation to the study of paleogeography. Their simultaneous occurrence in North America and Europe is much more plausibly explained by migration than by independent coincident development from different ancestors.

FLEXIBLE CRINOIDS

The crinoids called Flexibilia comprise dicyclic crinoids in which the lower bra-

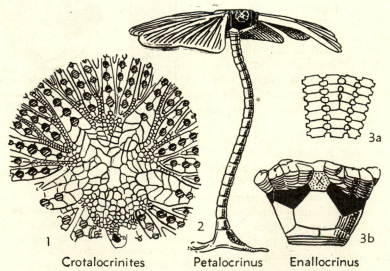

Crotalocrinites   Petalocrinus   Enallocrinus

Fig. 18-19. **Specialized Mid-Silurian dicyclic inadunate crinoids.** The arms of *Crotalocrinites* (1, ventral view) and *Enallocrinus* (3a, b) join one another laterally so as to make a flexible network, whereas those of *Petalocrinus* (2) are five solid petal-like plates having branched ambulacral grooves on their upper surface. *Enallocrinus* is restricted to Europe, but the other two occur both in North America and Europe. (Radials black; plates of posterior interray stippled; *X*, anal X.)

chials are incorporated in the dorsal cup, but not rigidly. They are represented by such fossils as *Talanterocrinus*, which has heterotomous branching of the rays, *Ichthyocrinus*, which has isotomous divisions of the rays (Fig. 18-20), and others illustrated (Figs. 18-28; 18-29; 18-30; 18-31; 18-33). They have multiple-circlet dorsal cups, Type 4, previously described. The lowermost circlet of flexible crinoids consists of three infrabasals—two large plates and a small one—of which the latter invariably occurs in the right-posterior ray. The tegmen is flexible, bearing exposed food grooves and the mouth. The arms are uniserial and nonpinnulate. The stem is round and bears no cirri. The group ranges from Middle Ordovician to Upper Permian.

Collectively, the flexible crinoids are well characterized by a combination of structural peculiarities. Their chief diagnostic feature is the nature of the articulation between the plates of the calyx and the remarkably constant occurrence of three infrabasals. Also important is the absence of an anal sac and the nature of the uniserial nonpinnulate arms, which commonly curl inward distally. There is no well-marked division between the arms and dorsal cup, for the contour of the crown is not broken at the summit of the radials. This group comprises about 50 described genera and more than 300 species. Six genera are taxocrinids and the remainder sagenocrinids. The group ranges from Middle Ordovician to Upper Permian, with peak developments in Middle Silurian and Mississippian time (Fig. 18-27).

Evolutionary Trends. The primitive cup shape is steep-sided and conical, whereas modified late forms are bowl-shaped to flat-based cups; among flexibles, however, a basal concavity is not developed in the cup.

The infrabasal circlet, which persists in all genera except a few Carboniferous and Permian forms, tends to be reduced in size, becoming concealed by the stem, and in a few it is solidly fused or dwindles to disappearance.

Radial plates are not specially differentiated from others except by position,

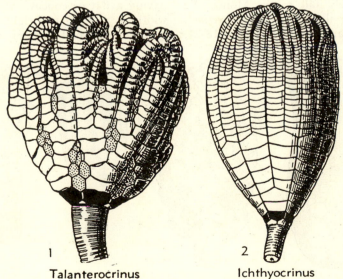

1   2

Talanterocrinus   **Ichthyocrinus**

Fig. 18-20. **Flexible crinoids.** The form at left, which has heterotomous arm branches, is from Pennsylvanian rocks; the one at right, characterized by isotomous branching, is from the Silurian. Both show lack of demarcation between cup and arms and smooth contour of the crown which is common among flexible crinoids. (Radials black; interradial and interbrachial plates stippled.)

but the nature of their articular facets serves readily to separate them from corresponding plates of inadunate and camerate crinoids.

Interradial and interbrachial plates occur in many genera, and they show considerable variation in number and size; a definable evolutionary trend in these elements is not recognized.

Evolution of the posterior interradius, as among inadunates, shows upward migration of radianal and anal plates and their eventual disappearance from the cup. Accordingly, advanced forms have perfect pentamerous symmetry.

The arms branch isotomously in primitive flexible crinoids. Heterotomous division appears in several advanced types, and a few end products show a regressive unbranched condition of the arms.

CAMERATE CRINOIDS

Rigid union of all plates of the crinoid calyx distinguishes the camerates. An example of a dicyclic camerate is the Silurian genus, *Lampterocrinus* (Fig. 18-21, *2*), which has a relatively tall calyx. Monocyclic camerates are illustrated by *Clonocrinus*, *Dimerocrinites*, and *Platycrinites* (Fig. 18-21,

1, 3, 4). Illustrations of many other typical forms are given elsewhere (Figs. 18-23 to 18-25; 18-28 to 18-31). The mouth and food grooves of all camerates are covered over by a rigid tegmen. The arms are uniserial or biserial in structure and bear pinnules. Other important distinguishing features are the firm incorporation of lower ray plates (cup-brachials) above the radials in the dorsal cup and absence of a radianal. Also, no true anal X and associated anal series (having origin as a modified branch of the right-posterior ray, as in inadunates) is recognized. Extra plates do occur on the posterior side of the dorsal cup of many camerates, however, reflecting expansion which is associated with placement of the digestive tract inside the calyx, and some of these are called anals. Because of difference in evolutionary significance, the proximal plate of the posterior interray of camerates is designated as **tergal** (Figs. 18-2, *74;* 18-10; 18-11).

The shape of camerate dorsal cups, which varies greatly, is a main basis for the differentiation of many genera. Evolutionary modification is shown by cups having a flat or concave base. Increased width in relation to height is a definite evolutionary trend, but only a few camer-

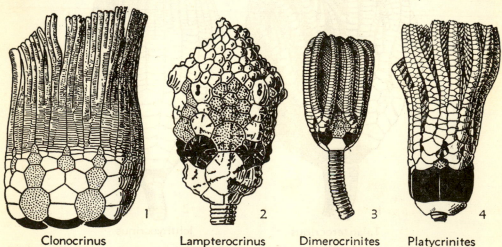

| 1 | 2 | 3 | 4 |
| Clonocrinus | Lampterocrinus | Dimerocrinites | Platycrinites |

FIG. 18-21. **Types of camerate crinoids.** These crinoids are distinguished chiefly by incorporation of lower ray plates in addition to the radials (black) as firmly united elements of the cup. Also, they have a rigid tegmen. *Lampterocrinus* is a dicyclic camerate; the others shown are monocyclic. (Interradial and interbrachial plates stippled.)

ates (like *Marsupiocrinus*, Fig. 18-29, *19*) develop a distinctly discoidal calyx.

Camerata as a whole include approximately 2,500 described species and 160 genera, distributed from Middle Ordovician to Upper Permian. Peak abundance occurred in Mississippian time (Fig. 18-27).

General Evolutionary Trends

Cup Plates. The proximal circlet of camerate crinoids shows a tendency toward reduction in number and size of the constituent plates, but fusion to a single plate is rare. The lowermost circlet becomes concealed by the stem in several genera. The structural attitude of plates in the lowermost circlet changes from upflaring to subhorizontal (downflaring sides of basal concavities commonly being made by other plates).

Radial plates, which are identified only by position, are separated all around by other plates in archaic camerates. Evolutionary upward migration of these intervening plates first brings the radials into contact with one another everywhere, or except at the posterior side. Primitively, the radials and other ray plates bear a prominent ridge, marking the axis of the ray and its branches. In advanced forms, this becomes weak and disappears.

Post-radial ray plates incorporated in the dorsal cup show evolutionary changes of size, shape, and reduction of number until nearly all are eliminated from the cup.

Interradial plates, including those of the posterior interray, and cup plates (interbrachials) between branches of the rays are primitively small, irregular in shape and arrangement, and well depressed below the level of adjacent ray plates (Figs. 18-24*E;* 18-26*D;* 18-28, *4, 7*). Evolutionary change brings increased size, definite polygonal outline, reduction in number, and ultimate upward migration to the summit of the dorsal cup (Figs. 18-24*A–D;* 18-26*A–C, E–G*).

Some camerate crinoids have dorsal cups containing an excessively large number of plates. An example is *Scyphocrinites* (Fig. 18-25), a monocyclic form of Devonian age in which the basals and radials compose a minor bottom part of the whole cup. Although cup-brachials consist only of primibrachs, secundibrachs, and some tertibrachs, their number is large, and plates of numerous long cup-pinnules, interradials, interbrachials, and interpinnulars add considerably to the total. The drawing shows only the lower one third of the crown, which has a height of approximately 65 cm. (25 in.).

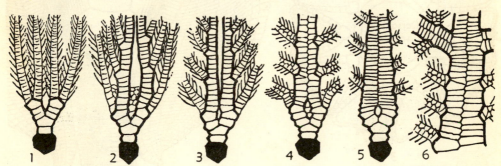

FIG. 18-22. **Evolution of ray structure in a family of monocyclic camerate crinoids.** The diagrams show part of the rays belonging to species of genera in the family Melocrinitidae. Central arms of the ray become larger and more branched, eventually fusing together. (Radials black.)

1, *Alisocrinus*, Middle Silurian, North America.
2, *Promelocrinus radiatus*, Middle Silurian, Gotland.
3, *Ctenocrinus gotlandicus*, Middle Silurian, Gotland.
4, *Ctenocrinus nobilissimus*, Lower Devonian, North America.
5, *Melocrinites splendens*, Upper Devonian, North America.
6, *Melocrinites (Trichotocrinus) harrisi*, Upper Devonian, New York.

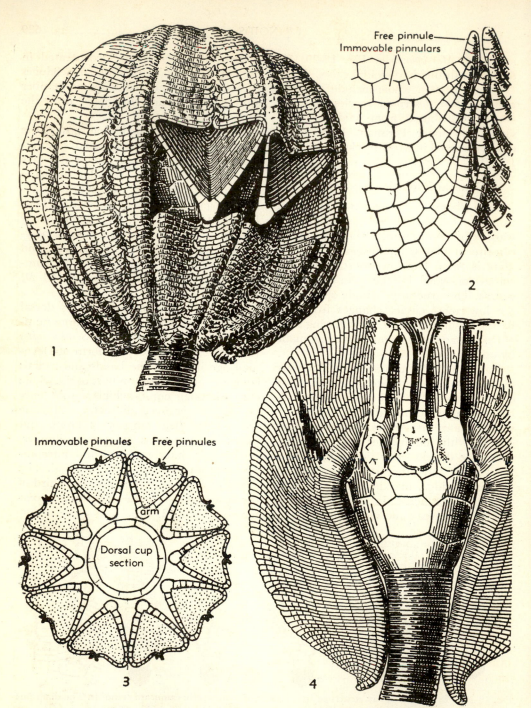

Fig. 18-23.   A specialized camerate crinoid from Silurian rocks of the Baltic region. *Barrandeocrinus sceptrum* Angelin, from Gotland. (1) Side view of crown with two of the arms and their attached pinnules cut away so as to show form of the ambulacral chambers, ×3. (2) Distal part of several pinnules showing interlocking rigidly joined pinnulars and mobile portions, ×10. (3) Diagrammatic cross section of the crown showing ambulacral chambers formed by rigid pinnules, ×2. (4) Side view of dorsal cup and two of the pendent arms, which are virtually immovable, ×3. (1–3, Based on unpublished studies of Prof. G. Ubaghs, Université de Liége; 4, reconstruction by Liljevahl.)

Arm Structure. The arms of camerate crinoids primitively are composed of uniserially arranged segments bearing branchlets classed as pinnules. Biserial arm structure appears very early, however, and characterizes the vast majority of camerates. The biserial arms of camerates differ from the biserial brachioles of cystoids and eocrinoids in having a uniserial proximal portion, which indicates derivation of the biserial pattern from a uniserial one.

Types of arm branching among camerates vary greatly. A few isotomous divisions are seen in oldest forms, whereas advanced types exhibit either a single bifurcation near the base of each ray, or many arms are grouped according to complex structural patterns. Extraordinarily complicated arm structures, which appear in several genera (Figs. 18-22; 18-30, *16;* 18-31, *16*) are remarkable culminations of evolutionary change along certain lines. Paddle-shaped arms (Fig. 18-32, *3*) and aligned pinnule spines are unusual types of specialization.

Tegminal Features. Characters of the tegmen show sporadic divergence in various directions rather than a consistent evolutionary trend. Noteworthy tegminal structures are a prominent central elevation distinguished as an **anal tube** (Fig. 18-32, *12, 13, 21*), lateral protrusion of large forked processes (Fig. 18-32, *20*), formation of compartments for enclosures of the arms (Fig. 18-29, *22*), and appearance of prominent spines (Figs. 18-29, *17;* 18-32, *11, 15*). Simultaneous occurrence in Eurasia and North America of some camerates bearing highly modified structures (Fig. 18-29, *17, 22*) can be explained logically only as a consequent of migration, thus indicating the existence of shallow seaways between the geographically widely separated regions of sedimentation.

Special Adaptations. Camerate cri-

noids, like other divisions of this class, exhibit many sorts of structural modifications which reflect particular modes of life

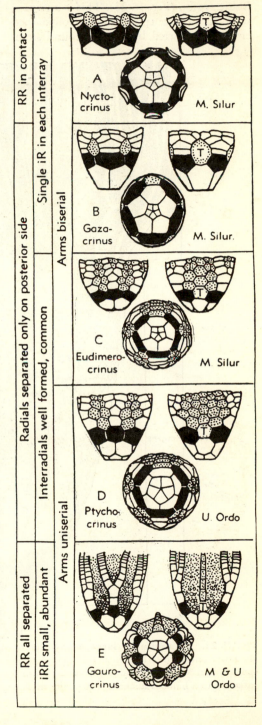

FIG. 18-24. **Evolution of dicyclic camerate crinoids.** Reduction in numbers of interradial plates and gradual elimination of them from the cup are main evolutionary trends. (Radials black; interradial and interbrachial plates stippled; *T*, tergal.)

and also mark evolutionary specialization. A single example is cited here, *Barrandeocrinus*, from Middle Silurian rocks of the Baltic region (Fig. 18-23). This is a crinoid with a nearly spherical, melon-shaped crown, supported by a stout round stem.

No part of the dorsal cup is visible unless some of the arms are broken away, for they not only hang far downward on all sides but fit tightly together. A peculiarity which distinguishes this crinoid is that the arms and pinnules are not flexible, except

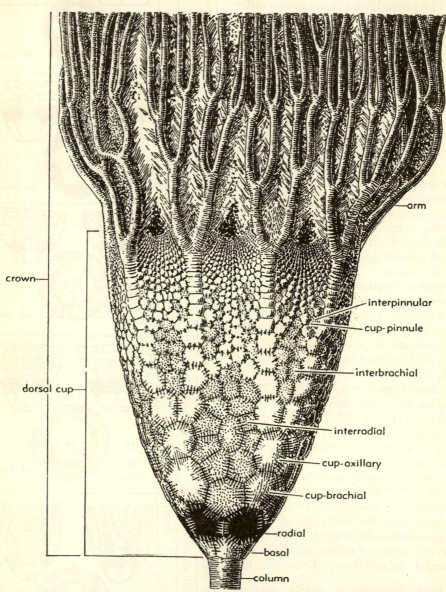

FIG. 18-25. **An unusually large, many-plated monocyclic camerate crinoid.** The drawing shows the lower one-third of the crown of *Scyphocrinites elegans* Zenker (×0.7), from Lower Devonian rocks of southeastern Missouri. Total height of the crown is approximately 25 in. The stem, about 3 ft. long, is attached to a bulbous anchorage structure similar to that shown in Fig. 18-30, 5.

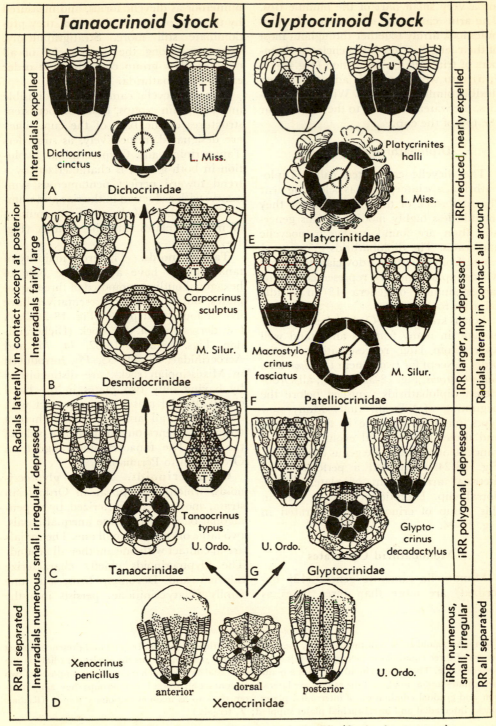

Tanaocrinoid Stock

Glyptocrinoid Stock

Interradials expelled

Interradials fairly large

Interradials numerous, small, irregular, depressed

Radials laterally in contact except at posterior

RR all separated

Dichocrinus cinctus

L. Miss.

A   Dichocrinidae

Carpocrinus sculptus

M. Silur.

B   Desmidocrinidae

Tanaocrinus typus

U. Ordo.

C   Tanaocrinidae

Xenocrinus penicillus

anterior   dorsal   posterior

D   Xenocrinidae

Platycrinites halli

L. Miss.

E   Platycrinitidae

Macrostylocrinus fasciatus

M. Silur.

F   Patelliocrinidae

Glyptocrinus decadactylus

U. Ordo.   U. Ordo.

G   Glyptocrinidae

iRR reduced, nearly expelled

iRR larger, not depressed

iRR polygonal, depressed

iRR numerous, small, irregular

Radials laterally in contact all around

RR all separated

Fig. 18-26.   **Evolution of monocyclic camerate crinoids.** (*Continued on next page.*)

for a very small part of the pinnule tips. The arms cannot be moved, and the pinnules lock firmly together throughout most of their length. They build enclosed chambers of triangular cross section (Fig.18-23, *1, 3*) into which water is admitted where flexible pinnule tips meet. With food particles, the water is carried to the mouth near the top of the cup.

Diplobathrid Camerates

The dicyclic camerates, called Diplobathrida, are interpreted as belonging near the root stock of the Camerata, for they include less highly modified forms generally than are found among monocyclic camerates. The diplobathrids (35 genera) range from Middle Ordovician to Upper Mississippian and are represented by the largest number of genera (15) in Middle Silurian rocks.

Forms known from Ordovician rocks are archaic in many features and stand well apart from later representatives of the order. Except for two genera (*Rhodocrinites, Gilbertsocrinus*, Fig. 18-32, *18, 20*), all Ordovician diplobathrids vanished before the beginning of Silurian time. Many new types made appearance in the Silurian, and these show marked evolutionary advancement. At least one genus (*Nyctocrinus*, Fig. 18-24*A*) attained a perfectly symmetrical arrangement of plates in the dorsal cup. Evolutionary trends seen in this group of crinoids are illustrated in Fig. 18-24.

Monobathrid Camerates

Monocyclic camerate genera (125 described) are more than three times as numerous as dicyclic forms, and measured by the host of described species, they predominate still more. Stratigraphically, they are among the most useful of all crinoids. The group comprises the order called Monobathrida.

The monocyclic camerates are divisible into two main groups, based on persistent structures of the base of the cup. They are designated, respectively, as suborders Tanaocrinina and Glyptocrinina. Evolution in both groups is characterized by a trend toward perfect pentamerous symmetry, with reduction in the size and number of ray plates of the cup and expulsion of interray plates into the tegmen (Fig. 18-26).

Tanaocrinoids. The great majority of tanaocrinoids have a tripartite base of hexagonal outline composed of three equal plates, but primitive representatives have five or four basal plates (Fig. 18-26*C*). A few derivatives of this stock (Dichocrinidae, Figs. 18-26*A*; 18-32, *14, 16;* and Acrocrinidae, Fig. 18-33, *15*), introduced in Mississippian rocks, are distinguished by having only two plates in the basal circlet. Radial plates are in contact with one another all around except where a tergal intervenes on the posterior side of the cup. The tanaocrinoids range from Ordovician to Permian.

Glyptocrinoids. Among glyptocrinoids, which first appear in Ordovician strata, most are characterized by a pentagonal base having three unequal basals, two large ones and a small one. The radials are in contact with one another all around. The glyptocrinoids chiefly characterize Silurian and Devonian beds, but one family (Platycrinitidae) persists into the

---

(*Fig. 18-26 continued.*)

The tanaocrinoid stock is distinguished by the hexagonal outline of the basal circlet and presence in most genera of three equal plates belonging to this circlet; a tergal plate interrupts the circlet of radials on the posterior side. The glyptocrinoid stock has a pentagonal basal circlet which is normally composed of two large plates and a small one; the radials are laterally in contact all around. Both groups show simplification and gradual elimination of interradial plates from the cup as main evolutionary trends. (Radials black; interradial and interbrachial plates stippled; *T*, tergals.)

Permian. Approximately 135 species belonging to the genus *Platycrinites* are described from Mississippian rocks.

Geological Distribution. About a dozen genera of monocyclic camerates are known from Ordovician rocks, all but one occurring in North America. A great expansion of monobathrids is recorded in Middle Silurian deposits (40 additional genera). These include some of the most distinctive and specialized crinoids known, and their appearance in Middle Silurian strata of opposite sides of the Atlantic is strong proof of marine shallow-water connections, probably by way of the Arctic. In Devonian rocks, 30 more genera belonging to this group make appearance, but measured by the number of described species and profusion of individuals, culmination of monocyclic camerate crinoids belongs in Early Mississippian time. Yet, among 25 new genera introduced in the Mississippian, all but six failed to survive into the Late Mississippian. The assemblage is predominantly North American. Few camerate crinoids are known in Pennsylvanian strata (three genera), but survival of this stock into Permian time is shown by occurrence of 12 genera, all described from Timor in the Dutch East Indies, except one from Sicily.

ARTICULATE CRINOIDS

Nearly all post-Paleozoic crinoids are referred to the subclass Articulata (Figs. 18-1; 18-34). The dorsal cup of most articulates is reduced in size and composed only of a few elements—radials, basals, and (actually or potentially) infrabasals. The mouth and food grooves are exposed on the tegmen, which may be studded with small calcareous plates but is invariably a flexible integument. The stratigraphic range of articulates is from Lower Triassic to deposits of the present time (Fig. 18-27).

(*Text continued on page 650.*)

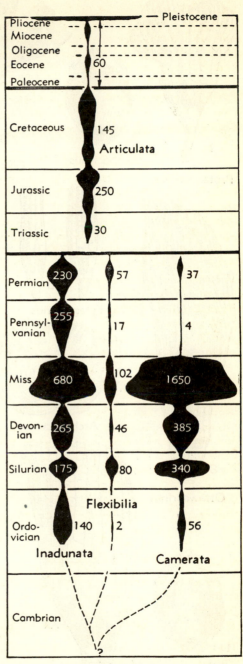

FIG. 18-27. **Geologic distribution of crinoids.** The range of each subclass is plotted separately. The accompanying figures indicate number of described species throughout the world.

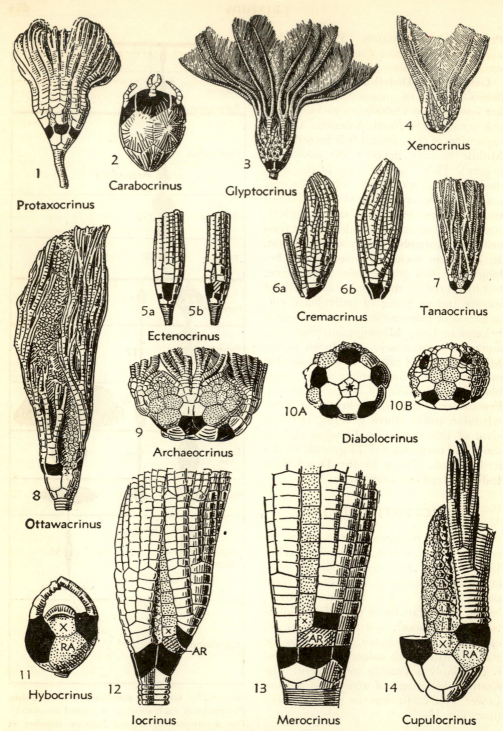

FIG. 18-28. **Representative Ordovician crinoids.** Inadunates are shown in 2, 5, 6, 8, 11–14; the oldest known flexible in 1; dicyclic camerates in 9, 10, and monocyclic camerates in 3, 4, 7. (Radials black, interradials and plates of the anal series stippled; *AR*, aniradial; *RA*, radianal; *X*, anal X).

(Continued on next page.)

(*Fig. 18-28 continued.*)

Archaeocrinus Wachsmuth & Springer, Ordovician–Silurian. An early diplobathrid. *A. desideratus* W. R. Billings (9), Trentonian, Canada.

Carabocrinus Billings, Ordovician. One of the few cladids which has an inferradianal. *C. vancortlandi* Billings (Fig. 18-7, *6*) Trentonian, Canada.

Cremacrinus Ulrich, Ordovician–Silurian. Oldest known type of bent-crown disparids. *C. articulosus* (Billings) (*6a, b*), Trentonian, Kentucky.

Cupulocrinus d'Orbigny, Ordovician. A primitive cladid which shows affinities with flexible crinoids. *C. jewetti kentuckiensis* Springer (14), Trentonian, Kentucky.

Diabolocrinus Wachsmuth & Springer, Ordovician. A diplobathrid. *D. perplexus* Wachsmuth & Springer (10*A*), Chazyan, Tennessee. *D. vesperalis* (White) (10*B*), Chazyan, Tennessee.

Dystactocrinus Ulrich, Ordovician. A primitive disparid. *D. constrictus* (Hall) (Fig. 18-7, *11*), Cincinnatian, Ohio.

Ectenocrinus S. A. Miller, Ordovician. A primitive disparid. E. simplex (Hall) (*5a, b*), Cincinnatian, Kentucky.

Gaurocrinus S. A. Miller, Ordovician. A primitive diplobathrid. *G. nealli* Hall (Fig. 18-24, *E*), Cincinnatian, Ohio.

Glyptocrinus Hall, Ordovician–Silurian. An early monobathrid. *G. decadactylus* Hall (3, and Fig. 18-26*G*), Cincinnatian, Ohio.

Heterocrinus Hall, Ordovician. A primitive disparid. *H. heterodactylus* Hall (Fig. 18-7, *10*), Cincinnatian, Kentucky.

Hybocrinus Billings, Ordovician. A bowl-shaped hybocrinid. *H. tumidus* Billings (11), Trentonian, Kentucky.

Iocrinus Hall, Ordovician. A primitive disparid having no anals in the cup. *I. subcrassus* (Meek & Worthen) (12 and Fig. 18-7, *12*), Cincinnatian, Ohio.

Merocrinus Walcott, Ordovician. A cladid having no anals in the dorsal cup but an aniradial above the cup. *M. typus* Walcott (13, and Fig. 18-7, *9*), Trentonian, New York.

Ottawacrinus W. R. Billings, Ordovician. A primitive cladid distinguished by its arm branching and large anal sac. *O. typus* W. R. Billings (8), Trentonian, Canada.

Protaxocrinus Springer, Ordovician–Silurian. The oldest known flexible crinoid. *P. elegans* (Billings) (1), Trentonian, Canada.

Tanaocrinus Wachsmuth & Springer, Ordovician. A monobathrid. *T. typus* Wachsmuth & Springer (7, and Fig. 18-26*C*), Cincinnatian, Ohio.

Xenocrinus S. A. Miller, Ordovician. A monobathrid. *X. baeri* (Meek) (4, and Fig. 18-26*D*), Cincinnatian, Ohio.

FIG. 18-29. (*See following page.*) **Representative Silurian crinoids.** Inadunates are shown in 1, 4, 6, 13–15, 21; flexibles in 2, 3, 5, 7, 10; dicyclic camerate in 20, and monocyclic camerates in 8, 9, 11, 12, 16–19, 22. (Radials black, interradials and plates of the anal series stippled; *RA*, radianal; *T*, tergal; *X*, anal X).

Botryocrinus Angelin, Silurian–Devonian. A typical cladid. *B. ramosissimus* Angelin (Figs. 18-7, *1*; 18-18, *3*), Middle Silurian, Sweden.

Calceocrinus Hall, Silurian–Mississippian. A bent-crown disparid. *C. bassleri* Springer (14, left-anterior view of crown), Niagaran, Tennessee.

Calliocrinus d'Orbigny, Silurian–Devonian. A monobathrid having a distinctive tegmen. *C. murchisonianus* (Angelin) (17), Middle Silurian, Sweden.

Carpocrinus Müller, Silurian. A moderately advanced monobathrid. *C. sculptus* Springer (Fig. 18-26*B*), Niagaran, Indiana.

Clidochirus Angelin, Silurian–Mississippian. Radianal directly below radial. *C. keyserensis* Springer (3), Cayugan, West Virginia.

Clonocrinus Quenstedt, Silurian–?Devonian. The rotund hollow-based calyx and strongly biserial arms characterize this monobathrid. *C. polydactylus* (McCoy) (Fig. 18-21, *1*), Middle Silurian, England.

Crotalocrinites Austin & Austin, Silurian. A cladid having rays joined laterally in a flexible network. *C. pulcher* (Hisinger) (Fig. 18-19, *1*), Middle Silurian, Sweden.

Dendrocrinus Hall, Ordovician–Silurian. A cladid having primitive position of the radianal but a very prominent anal sac. *D. longidactylus* Hall (Figs. 18-7, *8*; 18-18, *1a, b*), Niagaran, New York.

Dimerocrinites Phillips, Silurian–Devonian. A typically Silurian, simply constructed monobathrid. *D. decadactylus* Phillips (Fig. 18-21, *3*,

(*Continued on page 639.*)

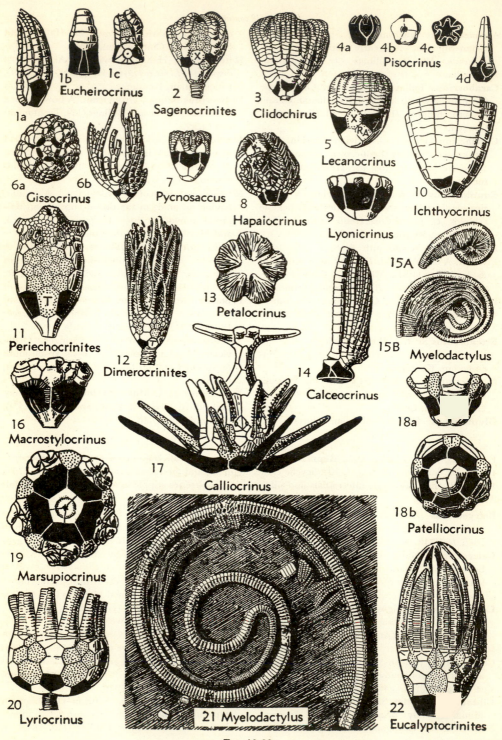

1a 1b 1c Eucheirocrinus
2 Sagenocrinites
3 Clidochirus
4a 4b 4c Pisocrinus 4d
5 Lecanocrinus
6a 6b Gissocrinus
7 Pycnosaccus
8 Hapalocrinus
9 Lyonicrinus
10 Ichthyocrinus
11 Periechocrinites
12 Dimerocrinites
13 Petalocrinus
14 Calceocrinus
15A 15B Myelodactylus
16 Macrostylocrinus
17 Calliocrinus
18a 18b Patelliocrinus
19 Marsupiocrinus
20 Lyriocrinus
21 Myelodactylus
22 Eucalyptocrinites

Fig. 18-29.

(Fig. 18-29 continued.)

posterior), Middle Silurian, England; *D. arborescens* (Talbot) (12, posterior), basal Devonian, New York.

Enallocrinus d'Orbigny, Silurian. A cladid related to *Crotalocrinites*. *E. scriptus* (Hisinger) (Fig. 18-19, *3a, b*), Middle Silurian, Sweden.

Eucalyptocrinites Goldfuss, Silurian–Devonian. The radial partitions between pairs of arms are outgrowths of the tegmen. One of the best known Silurian monobathrids. *E. crassus* (Hall) (22), Niagaran, Indiana.

Eucheirocrinus Meek & Worthen, Ordovician–Silurian. This genus is intermediate in evolutionary features between *Cremacrinus* and *Calceocrinus*. *E. chrysalis* (Hall) (1a, c, right, anterior, and posterior views of crown), Niagaran, New York.

Eudimerocrinus Springer, Silurian. A typical diplobathrid having few interradials. *E. multibrachiatus* Springer (Fig. 18-24C), Niagaran, Tennessee.

Gazacrinus S. A. Miller, Silurian. Interradials are reduced to one between each pair of rays. *G. inornatus* S. A. Miller (Fig. 18-24B), Niagaran, Indiana.

Gissocrinus Angelin, Silurian–Devonian. This cladid has distinctive arms. *G. lyoni* Springer (6a, b, dorsal and posterior views of crown), Niagaran, Kentucky.

Hapalocrinus Jaekel, Silurian–Devonian. A small, almost perfectly symmetrical monobathrid having a few cup plates. *H. gracilis* Springer (8), Niagaran, Tennessee.

Ichthyocrinus Conrad, Silurian–Devonian. The rays are all alike except for presence of a primitive radianal in the right-posterior ray. *I. pyriformis* (Phillips) (Fig. 18-20, *2*), Middle Silurian, England; *I. subangularis* Hall (10, postero-right), Niagaran, Indiana.

Lampterocrinus Roemer, Silurian. A diplobathrid having a high calyx and bulging posterior interray. *L. tennesseensis* Roemer (Fig. 18-21, *2*, posterior), Niagaran, Tennessee.

Lecanocrinus Hall, Silurian–Devonian. Crown small and rotund. *L. macropetalus* Hall (5, posterior), Niagaran, New York.

Lyonicrinus Springer, Silurian. Resembles *Hapalocrinus*. *L. bacca* (Roemer) (9), Niagaran, Tennessee.

Lyriocrinus Hall, Silurian. A diplobathrid having perfect symmetry of the cup. *L. melissa* (Hall) (20), Niagaran, Indiana.

Macrostylocrinus Hall, Silurian–Devonian. This monobathrid has interradial plates showing moderately advanced evolution. *M. fasciatus* (Hall) (Fig. 18-26F), Niagaran, Tennessee; *M. striatus* Hall (16), Niagaran, Indiana.

Marsupiocrinus Morris, Silurian–Devonian. This monobathrid is characterized by its low, very symmetrically constructed calyx. *M. striatissimus* (Springer) (19), Niagaran, Tennessee.

Myelodactylus Hall, Silurian–Devonian. Cirri directed toward one side of the coiled stem protect and conceal the delicate crown of this disparid. *M. keyserensis* Springer (15B), Cayugan, West Virginia; *M. fletcheri* (Salter) (21), Middle Silurian, England; *M. convolutus* Hall (15A), Niagaran, Indiana.

Nyctocrinus Springer, Silurian. A nearly perfectly symmetrical advanced type of diplobathrid. *N. magnitubus* Springer (Fig. 18-24A), Niagaran, Tennessee.

Patelliocrinus Angelin, Silurian. A monobathrid having few interradials and widely separated arm bases. *P. ornatus* Springer (18a, b), Niagaran, Indiana.

Periechocrinites Austin & Austin, Ordovician–Mississippian. A relatively large calyx composed of numerous plates belongs to this monobathrid. *P. tennesseensis* (Hall & Whitfield) (11, posterior), Niagaran, Tennessee.

Petalocrinus Weller & Davidson, Silurian. This cladid has solidly fused petal-like rays. *P. mirabilis* Weller & Davidson (Figs. 18-19, *2*, restoration, one ray removed; 18-29, *13*, ventral view of consolidated rays), Niagaran, Iowa.

Pisocrinus De Koninck, Silurian. The "pea crinoid" is one of the most common and characteristic Silurian disparids. *P. quinquelobus* Bather (4a–c, left-anterior, dorsal, and ventral views of dorsal cup; 4d, left-anterior view of crown), Niagaran, Tennessee.

Pycnosaccus Angelin, Silurian–Devonian. Narrow arms joined to large radials are unusual features among flexible crinoids. *P. tenuibrachiatus* Springer (7, left anterior), Cayugan, West Virginia.

Sagenocrinites Austin & Austin, Silurian. Numerous interradials and short arms mark this genus. *S. americanus* (Springer) (2, posterior), Niagaran, Indiana.

Thenarocrinus Bather, Silurian. The radianal of this cladid interrupts the basal circlet. *T. callipygus* Bather (Fig. 18-7, *6*), Middle Silurian, England.

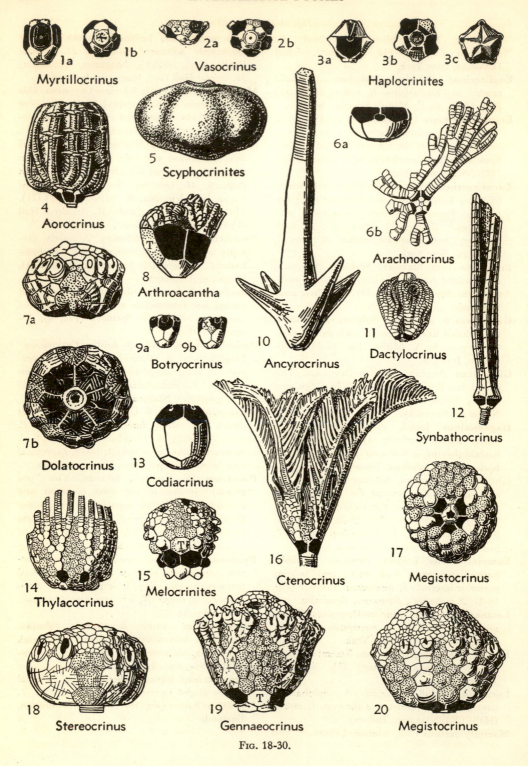

1a 1b Myrtillocrinus

2a 2b Vasocrinus

3a 3b 3c Haplocrinites

4 Aorocrinus

5 Scyphocrinites

6a 6b Arachnocrinus

7a

7b Dolatocrinus

8 Arthroacantha

9a 9b Botryocrinus

10 Ancyrocrinus

11 Dactylocrinus

12 Synbathocrinus

13 Codiacrinus

14 Thylacocrinus

15 Melocrinites

16 Ctenocrinus

17 Megistocrinus

18 Stereocrinus

19 Gennaeocrinus

20 Megistocrinus

Fig. 18-30.

FIG. 18-30. **Representative Devonian crinoids.** Inadunates are shown in 1–3, 6, 9, 10, 12, 13; flexible in 11; dicyclic camerate in 14; and monocyclic camerates in 4, 5, 7, 8, 15–20. Natural size, except as indicated. (Radials black, interradials and plates of the anal series stippled; *T*, tergal.)

Ancyrocrinus Hall, Devonian. This crinoid, which has a crown resembling that of *Botryocrinus*, is distinguished by its peculiar grapnel-like lower extremity of the stem. *A. spinosus* Hall (10), Middle Devonian, Indiana.

Aorocrinus Wachsmuth & Springer, Silurian–Mississippian. The small cup has grouped arms; about 23 described species. *A. armatus* Goldring (4, left-anterior), Middle Devonian, New York.

Arachnocrinus Meek & Worthen, Devonian. A cladid especially characterized by thick isotomous arms. *A. bulbosus* (Hall) (6a, dorsal cup, ×3; 6b, part of crown, dorsal view), Lower Devonian, New York.

Arthroacantha Williams, Devonian. A large tergal interrupts the circlet of radials. *A. punctobrachiata* (Hall) (8, right-posterior), Middle Devonian, New York.

Botryocrinus Angelin, Silurian–Devonian. A cladid distinguished by the nature of its anal plates and arms. *B. crassus* (Whiteaves) (9a, b, right-anterior, posterior), Middle Devonian, New York.

Codiacrinus Schultze, Devonian. A globose cladid which has very small, rounded arm facets on the radials. *C. granulatus* Schultze (13, postero-right), Middle Devonian, Germany.

Ctenocrinus Bronn, Silurian–Devonian. The two main arms of each ray are laterally joined and they give off many branches outward; 20 species. *C. nobilissimus* (Hall) (16, postero-right), Lower Devonian, New York.

Dactylocrinus Quenstedt, Devonian–Mississippian. Two main arms in each ray bear minor branches on the sides facing one another (endotomous). *D. alpena* Springer (11, posterior), Middle Devonian, Michigan.

Dolatocrinus Lyon, Devonian. The somewhat depressed calyx has respiratory pores on the tegmen and a few large plates in each interray of the cup; 43 species. *D. bellulus* Miller & Gurley (7a, b, posterior, dorsal), Middle Devonian, Indiana.

Gennaeocrinus Wachsmuth & Springer, Devonian. Greatest width of cup is at level of the clustered arm bases. *G. eucharis* (Hall)

(19, posterior), Middle Devonian, New York.

Haplocrinites Steininger, Silurian–Mississippian. This small disparid resembles the Silurian *Pisocrinus* but has more regular plates. *H. clio* (Hall) (3a–c, anterior, dorsal, ventral, ×4), Middle Devonian, New York.

Megistocrinus Owen & Shumard, Devonian–Mississippian. Very similar to *Gennaeocrinus*, contains many species, mostly Devonian. *M. nodosus* Barris (20, postero-right); *M. concavus* Wachsmuth (17), both Middle Devonian, Michigan.

Melocrinites Goldfuss, Silurian–Mississippian. Closely resembles *Ctenocrinus*, but main arms of each ray more completely fused; 77 species, nearly all Devonian. *M. nodosus* (Hall) (13, posterior), Middle Devonian, New York.

Myrtillocrinus Sandberger & Sandberger, Devonian. Resembles *Arachnocrinus* but differs in structure of the cup; axial canal in columnals quadripartite. *M. americanus* Hall (1a, b), Lower Devonian, New York.

Scyphocrinites Zenker, Silurian–Lower Devonian. A large crown (Fig. 18-25) is attached to the bulbous chambered structure (perhaps a float) at end of the stem. *S. ulrichi* (Schuchert) (5, lateral view of bulb at distal end of stem, ×0.5), Lower Devonian, Oklahoma.

Stereocrinus Barris, Devonian. Differs only slightly from *Dolatocrinus* in structure of ray plates of the cup. *S. triangulatus* Barris (18, posterior), Middle Devonian, Michigan.

Synbathocrinus Phillips, Devonian–Permian. A disparid having a very tall slender crown, its simple construction indicating advanced evolution. *S. matutinus* Hall (12), Middle Devonian, Iowa.

Thylacocrinus Oehlert, Devonian. A diplobathrid having a deep bowl-shaped cup composed of many plates and showing larger interradial areas. *T. clarkei* Wachsmuth & Springer (14), Middle Devonian, New York.

Vasocrinus Lyon, Devonian. A shallow-cup cladid having subcircular radial facets in a nearly vertical plane. *V. valens* Lyon (2a, b, posterior, dorsal), Lower Devonian, Kentucky.

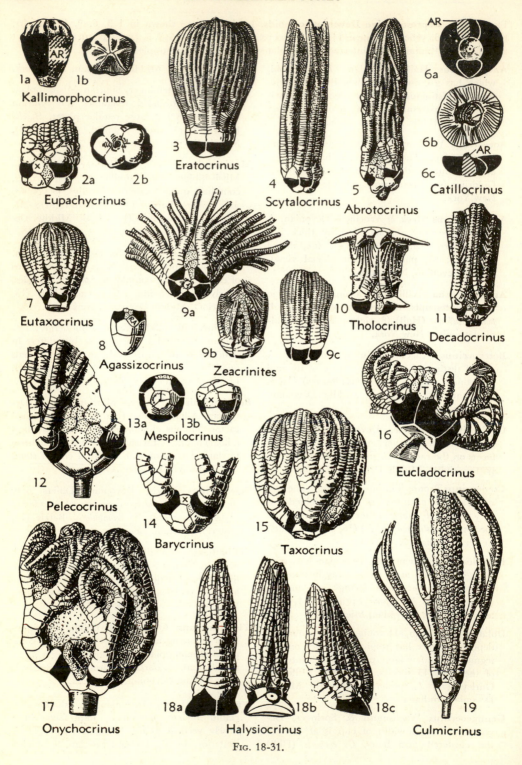

1a 1b
Kallimorphocrinus

2a 2b
Eupachycrinus

3 **Eratocrinus**

4 **Scytalocrinus**

5 **Abrotocrinus**

6a 6b 6c
Catillocrinus

7 **Eutaxocrinus**

8 **Agassizocrinus**

9a 9b 9c
Zeacrinites

10 **Tholocrinus**

11 **Decadocrinus**

12 **Pelecocrinus**

13a 13b
Mespilocrinus

14 **Barycrinus**

15 **Taxocrinus**

16 **Eucladocrinus**

17 **Onychocrinus**

18a 18b 18c
Halysiocrinus

19 **Culmicrinus**

Fig. 18-31.

FIG. 18-31. **Representative Mississippian crinoids.** Inadunates are shown in 1–6, 8–12, 14, 18, 19; flexibles in 7, 13, 15, 17; monocyclic camerate in 16. Natural size, except as indicated. (Radials black, interradials and plates of the anal series stippled; *AR*, aniradial; *RA*, radianal; *T*, tergal; *X*, anal X.)

Abrotocrinus Miller & Gurley, Mississippian. The cup of this cladid has a basal concavity; the slender arms branch two to four times. *A. cymosus* Miller & Gurley (5, posterior), Osagian, Indiana.

Agassizocrinus Owen & Shumard, Mississippian, ?Pennsylvanian, ?Permian. The semiovoid thick-walled cup lacks a stem; a cladid. *A. laevis* (Roemer) (8, posterior), Chesteran, Kentucky.

Barycrinus Wachsmuth, Mississippian. The low bowl-shaped cup has a quadrangular radianal obliquely below the right posterior radial; arms, coarse, rounded. This genus, which contains 37 described species, is a cladid. *B. princeps* Miller & Gurley (14, posterior), Osagian, Indiana.

Catillocrinus Shumard, Mississippian. A robust disparid which has many slender free arms rising directly from the radials. *C. tennesseeae* Shumard (6a–c, dorsal, ventral, right-posterior), Osagian, Kentucky.

Culmicrinus Jaekel, Mississippian. Tall slender arms and anal sac characterize this cladid. *C. elegans* (Wachsmuth & Springer) (19, posterior), Chesteran, Kentucky.

Cyathocrinites Miller, Silurian–Permian. A common bowl-shaped cladid having one anal plate in the cup; the round uniserial arms branch abundantly; nearly 100 described species. *C. multibrachiatus* (Lyon & Casseday) (Fig. 18-6, *3*), Osagian, Indiana.

Decadocrinus Wachsmuth & Springer, Devonian–Mississippian. A 10-armed cladid having projecting wedge-shaped brachials. *D. bellus* (Miller & Gurley) (11, posterior), Osagian, Indiana.

Eratocrinus Kirk, Mississippian. The pear-shaped crown and arm structure are some of the distinguishing features of this cladid. *E. elegans* (Hall) (3, antero-left), Osagian, Iowa.

Eucladocrinus Meek, Mississippian. A platycrinitid which has specialized arms. *E. pleuroviminus* (White) (16, right-posterior), Osagian, Iowa.

Eupachycrinus Meek & Worthen, Mississippian. The cup, which has a deep basal concavity, bears 12 or more biserial arms; a cladid. *E. quattuordecimbrachialis* (Lyon) (2a, b, posterior, dorsal), Chesteran, Kentucky.

Eutaxocrinus Springer, Silurian–Mississippian. A symmetrically constructed taxocrinid which commonly lacks a radianal. *E. fletcheri* (Worthen) (7, right-posterior), Kinderhookian, Iowa.

Halysiocrinus Ulrich, Silurian–Mississippian, ?Permian. Youngest and most advanced of the bent-crown disparid crinoids. *H. nodosus* (Hall) (18a–c, anterior, posterior, left-anterior), Osagian, Indiana.

Kallimorphocrinus J. M. Weller, Mississippian–Permian. A disparid microcrinoid of very simple structure. Besides a basal circlet, commonly fused, it has five radials and five orals. *K. puteatus* (Peck) (1a, b, posterior, ventral, $\times$18), Osagian, Missouri.

Mespilocrinus De Koninck & Lehon, Mississippian. The small rotund crown has a large anal X but no radianal. *M. konincki* Hall (13a, b, dorsal, posterior), Osagian, Iowa.

Onychocrinus Lyon & Casseday, Mississippian. The irregularly heterotomous branching of the rays chiefly distinguishes this robust flexible crinoid. *O. ulrichi* Miller & Gurley (17, posterior), Osagian, Indiana.

Pelecocrinus Kirk, Mississippian. Three anals occur in the cup of this cladid; radial facets slope steeply outward. *P. insignis* Kirk (12, posterior), Osagian, Iowa.

Phanocrinus Kirk, Mississippian. The low cup, concave at the base, bears 10 tall uniserial arms; a cladid. *P. formosus* (Worthen) (Fig. 18-6, 1a, b, posterior, dorsal), Chesteran, Illinois.

Scytalocrinus Wachsmuth & Springer, Mississippian–Pennsylvanian. This cladid resembles *Decadocrinus* except for difference in the brachials. *S. validus* Wachsmuth & Springer (4, anterior), Osagian, Indiana.

Synbathocrinus Phillips, Devonian–Permian. A simple but advanced disparid. *S. robustus* Shumard (Fig. 18-7, *13*), Osagian, Tennessee.

Taxocrinus Phillips, Devonian–Mississippian. Distinguished by ray structure and characters of posterior interray. *T. colletti* White (15, right-posterior), Osagian, Indiana.

Tholocrinus Kirk, Mississippian. The mushroom-shaped sac, bordered by spines at the top, is a distinctive feature of this cladid. *T. spinosus* (Wood) (10, posterior), Chesteran, Kentucky.

Zeacrinites Troost, Mississippian–Pennsylvanian. The discoid cup of this genus has a strong basal concavity, elongate narrow basals, and outward sloping radial facets; a cladid. *Z. magnoliaeformis* (Troost) (Fig. 18-28, *2*; 18-17, 9a–c, posterior, anterior showing sac and anus, anterior), Chesteran, Alabama.

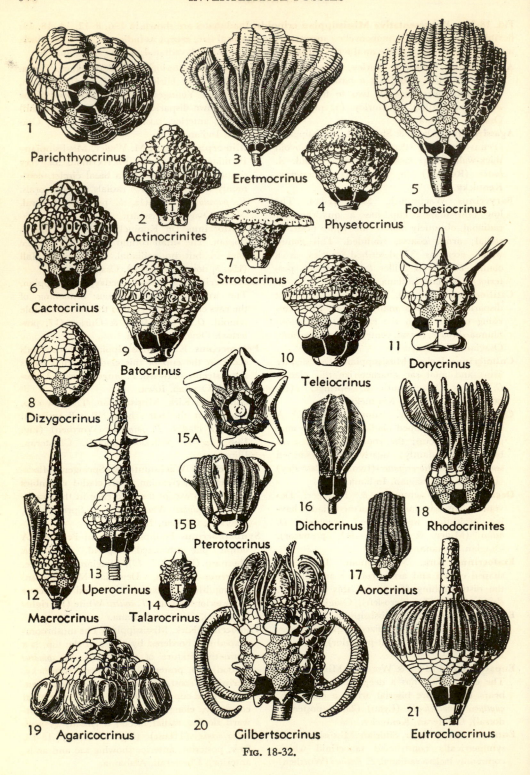

1 Parichthyocrinus
2 Actinocrinites
3 Eretmocrinus
4 Physetocrinus
5 Forbesiocrinus
6 Cactocrinus
7 Strotocrinus
8 Dizygocrinus
9 Batocrinus
10 Teleiocrinus
11 Dorycrinus
12 Macrocrinus
13 Uperocrinus
14 Talarocrinus
15A
15B Pterotocrinus
16 Dichocrinus
17 Aorocrinus
18 Rhodocrinites
19 Agaricocrinus
20 Gilbertsocrinus
21 Eutrochocrinus

Fig. 18-32.

Fig. 18-32. **Representative Mississippian crinoids.** Flexibles are shown in 1, 5; dicyclic camerates in 18, 20; and monocyclic camerates in remaining figures. Natural size. (Radials black, interradials and plates of the anal series stippled; T, tergal.)

Actinocrinites Miller, Mississippian–Permian. A widely distributed monobathrid genus, characterized by strongly grouped arm bases; plates of the cup generally marked by radiating ridges; 92 described species. *A. verrucosus* (Hall) (2, posterior), Osagian, Iowa.

Agaricocrinus Hall, Lower Mississippian. This monobathrid is characterized by its flat or concave base and strongly grouped arms; 41 species. *A. wortheni* Hall (19, anterior), Osagian, Illinois.

Aorocrinus Wachsmuth & Springer, Silurian–Mississippian. The bases of free arms of different rays are not in contact. *A. immaturus* (Wachsmuth & Springer) (17, posterior), Kinderhookian, Iowa.

Batocrinus Casseday, Mississippian. This monobathrid has three equal basals and ungrouped arms; 133 species. *B. grandis* (Lyon & Casseday) (9), Osagian, Indiana.

Cactocrinus Wachsmuth & Springer, Mississippian. A monobathrid like *Actinocrinites* but arms not grouped; the pinnules bear spines which are so spaced as to form aligned rows transverse to the pinnules. *C. multibrachiatus* (Hall) (6, anterior), Osagian, Iowa.

Dichocrinus Münster, Mississippian. A large tergal occurs in line with the radials. This monobathrid has two basals; 52 species. *D. cinctus* Miller & Gurley (Fig. 18-11*A*); *D. inornatus* Wachsmuth & Springer (16; both Kinderhookian, Iowa.

Dizygocrinus Wachsmuth & Springer, Mississippian. A monobathrid characterized by its rotund calyx, tall slender anal tube, and ungrouped arms; 32 species. *D. rotundus* (Yandell & Shumard) (8), Osagian, Illinois.

Dorycrinus Roemer, Mississippian. The tegmen of this monobathrid bears one or more large spines, the arms are grouped, and the base is lobed; 17 species. *D. missouriensis* (Shumard) (11, posterior), Osagian, Missouri.

Eretmocrinus Lyon & Casseday. Mississippian. The paddle-shaped expansion of terminal parts of the arms distinguishes this monobathrid; 27 species. *E. remibrachiatus* (Hall) (3), Osagian, Iowa.

Eutrochocrinus Wachsmuth & Springer, Mississippian. The calyx of this monobathrid is urn-shaped, being expanded at arm bases; arms short, anal tube long. *E. christyi* (Shumard) (21), Osagian, Missouri.

Forbesiocrinus De Koninck & Lehon, Mississippian. The sutures between ray plates are distinctly flexuous, and interradials are numerous. *F. pyriformis* Miller & Gurley (5, antero-right), Osagian, Kentucky.

Gilbertsocrinus Phillips, Devonian–Mississippian. A diplobathrid which is distinguished by peculiar lateral branched appendages of the tegmen. *G. typus* (Hall) (20, antero-left), Osagian, Iowa.

Macrocrinus Wachsmuth & Springer, Mississippian. A small biturbinate monobathrid having 12 to 16 arms. *M. jucundus* (Miller & Gurley) (12, anterior), Osagian, Indiana.

Parichthyocrinus Springer, Mississippian. This genus lacks a radianal and has few interradials. *P. meeki* (Hall) (1, dorsal), Osagian, Indiana.

Physetocrinus Meek & Worthen, Mississippian. Mainly distinguished from *Actinocrinites* by the pattern of cup-brachials and the very small tegminals. *P. ventricosus* (Hall) (4, anterior), Osagian, Iowa.

Pterotocrinus Lyon & Casseday, Upper Mississippian. A monobathrid having prominent winglike expansions of the tegmen; 27 species. *P. bifurcatus* Wetherby (15*A*, dorsal); *P. capitalis* (Lyon) (15*B*, right-posterior); both Chesteran, Kentucky.

Rhodocrinites Miller, Devonian–Mississippian. The globose cup of this important diplobathrid genus is concave at the base; interradials numerous. *R. wortheni* (Hall) (18, right-anterior), Osagian, Missouri.

Strotocrinus Meek & Worthen, Mississippian. The calyx of this monobathrid flares broadly at the arm bases. *S. glyptus* (Hall) (7, posterior), Osagian, Iowa.

Talarocrinus Wachsmuth & Springer, Upper Mississippian. A small thick-plated monobathrid having two basals and an elevated tegmen; 20 species. *T. cornigerus* (Shumard) (14, posterior), Chesteran, Alabama.

Teleiocrinus Wachsmuth & Springer, Mississippian. This genus resembles *Cactocrinus* but has many more arms and cup-brachials. *T. umbrosus* (Hall) (10, anterior), Osagian, Iowa.

Uperocrinus Meek & Worthen, Mississippian. A tall anal tube and 20 grouped arms characterize this monobathrid. *U. nashvillae* (Troost) (13), Osagian, Tennessee.

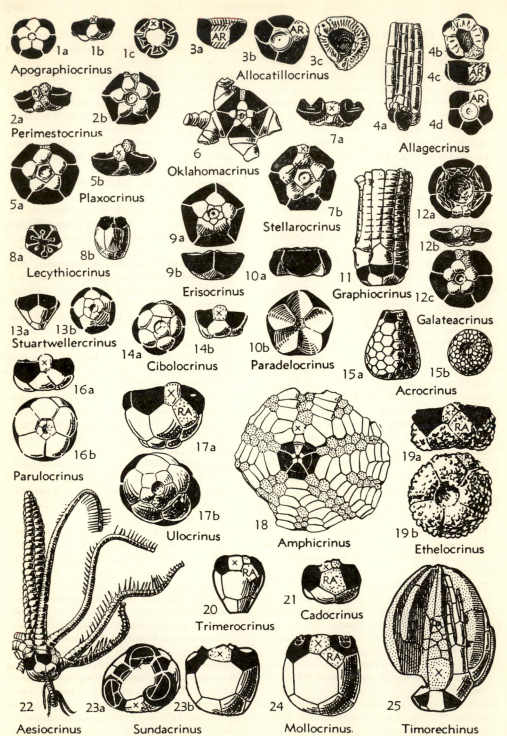

Fig. 18-33. Representative Pennsylvanian and Permian crinoids. Pennsylvanian inadunates are shown in 1–12, 16, 17, 19, 22; flexibles in 14, 18; camerate in 15. Permian inadunates comprise the remainder, 13, 20, 21, 23–25. Natural size, except as indicated. (Radials black, interradials, interbrachials, and plates of the anal series stippled; *AR*, aniradial; *RA*, radianal; *X*, anal X.)

(Fig. 18-33 continued.)

Acrocrinus Yandell, Mississippian–Pennsylvanian. Many plates are intercalated between the two basals and the circlet of radials. *A. pirum* Moore & Plummer (15*a*, *b*, left-posterior, dorsal), Lower Pennsylvanian (Morrowan), Arkansas.

Aesiocrinus Miller & Gurley, Pennsylvanian. The slender round arms and distinctive anal sac are some of the diagnostic marks of this cladid. *A. magnificus* Miller & Gurley (22, left-posterior), Upper Pennsylvanian (Missourian), Missouri.

Allagecrinus Carpenter & Etheridge, Mississippian–Permian. A microdisparid which commonly has two or more arms directly attached to a radial. *A. bassleri* Strimple (4*a*, left-posterior, ×2; 4*b–d*, ventral, posterior, dorsal, ×2.3), Upper Pennsylvanian (Missourian), Oklahoma.

Allocatillocrinus Wanner, Mississippian–Pennsylvanian. This disparid differs from *Allagecrinus* in its larger size and more numerous arms. *A. morrowensis* (Strimple) (3*a–c*, right-posterior, dorsal, ventral, ×1.3), Lower Pennsylvanian (Morrowan), Oklahoma.

Amphicrinus Springer, Mississippian–Pennsylvanian. This genus is characterized by the broad, rather firmly constructed lower part of the calyx; interradials fairly numerous. *A. carbonarius* Springer (18, dorsal), Middle Pennsylvanian (Desmoinesian), Oklahoma.

Apographiocrinus Moore & Plummer, Pennsylvanian–Permian. A low bowl-shaped cup, shallowly concave at the base, has one anal plate and narrow radial facets; arms uniserial. *A. exculptus* Moore & Plummer (1*a–c*, dorsal, posterior, ventral), Middle Pennsylvanian (Desmoinesian), Texas.

Cadocrinus Wanner, Permian. The low globular cup has highly variable anals; anal sac inflated. *C. variabilis* Wanner (21, posterior), Upper Permian, Timor.

Cibolocrinus Weller, Pennsylvanian–Permian. A bowl-shaped cup having one anal and a rotund crown characterizes this genus. *C. tumidus* Moore & Plummer (14*a*, *b*, dorsal, posterior), Lower Pennsylvanian (Morrowan), Arkansas.

Delocrinus Miller & Gurley, Pennsylvanian–Permian. This is one of the most common and characteristic Pennsylvanian cladids. The base of the cup is strongly concave; one anal; arms biserial. *D. verus* Moore & Plummer (Fig. 18-16, 2*a*, *b*, posterior, dorsal), Upper Pennsylvanian (Missourian), Kansas.

Erisocrinus Meek & Worthen, Mississippian–Permian. Resembles *Delocrinus* but has no anal in the cup and no basal concavity. *E. typus* Meek & Worthen (9*a*, *b*, dorsal, posterior), Pennsylvanian, Illinois.

Ethelocrinus Kirk, Pennsylvanian. This cladid has a low rotund cup with basal concavity, two anals in cup, and 16 to 18 biserial arms. *E. texasensis* Moore & Plummer (19*a*, *b*, posterior, dorsal), Lower Pennsylvanian (Morrowan), Texas.

Galateacrinus Moore, Pennsylvanian. The very low discoid cup, horizontal radial facets and three anals in the cup distinguish this cladid. *G. stevensi* Moore (12*a–c*, ventral, posterior, dorsal), Middle Pennsylvanian (Desmoinesian), Oklahoma.

Graphiocrinus De Koninck, Mississippian–Permian. This genus resembles *Delocrinus* but has uniserial arms and lacks a basal concavity of the cup. *G. kingi* Moore & Plummer (11, anterior), Middle Pennsylvanian (Desmoinesian), Texas.

Lecythiocrinus White, Pennsylvanian. The small radial facets and anus on the side of the cup distinguish this genus. *L. adamsi* Worthen (8*a*, *b*, ventral, posterior), Pennsylvanian, Illinois.

Mollocrinus Wanner, Permian. A cladid having a rotund cup and narrow round radial facets sloping outward. *M. poculum* Wanner (24, posterior), Upper Permian, Timor.

Oklahomacrinus Moore, Pennsylvanian. The discoid cup, provided with one narrow anal, has uniserial pendent arms attached to outsloping radial facets. *O. loeblichi* Moore (6, dorsal), Upper Pennsylvanian (Missourian), Oklahoma.

Paradelocrinus Moore & Plummer, Pennsylvanian. Resembles *Erisocrinus* but has a more or less concave base. *P. dubius* (Mather) (10*a*, *b*, anterior, dorsal), Lower Pennsylvanian (Morrowan), Arkansas.

Parulocrinus Moore & Plummer, Pennsylvanian. Resembles *Ethelocrinus* except that the basal concavity is very shallow or lacking and arms are 10. *P. compactus* Moore & Plummer (16*a*, *b*, posterior, dorsal), Upper Pennsylvanian (Missourian), Kansas.

Perimestocrinus Moore & Plummer, Pennsylvanian. The base of the cup has a moderately deep concavity and radial facets slope outward; three anals in the cup. *P. impressus* Moore & Plummer (2*a*, *b*, posterior, dorsal), Middle Pennsylvanian (Desmoinesian), Texas.

(Continued on next page.)

(Fig. 18-33 continued.)

Plaxocrinus Moore & Plummer, Pennsylvanian. Much like *Perimestocrinus* but base of cup and arm structure differ. *P. crassidiscus* (Miller & Gurley) (5a, b, dorsal, posterior), Upper Pennsylvanian (Missourian), Missouri.

Stellarocrinus Strimple, Pennsylvanian. A distinctive cladid having relatively narrow, outsloping radial facets, biserial arms, and one anal plate in the cup. *S. geometricus* (Moore & Plummer) (7a, b, posterior, dorsal), Upper Pennsylvanian (Missourian), Kansas.

Stuartwellercrinus Moore & Plummer, Permian. The cup of this cladid is steeply conical, anal plates excluded from cup. *S. symmetricus* (Weller) (13a, b, posterior, dorsal), Lower Permian, Texas.

Sundacrinus Wanner, Permian. This specialized cladid has only three arm-bearing radials. *S. granulatus* Wanner (23a, b, ventral, posterior), Upper Permian, Timor.

Talanterocrinus Moore & Plummer, Pennsylvanian. Heterotomous branching of the rays is evident. *T. jaekeli* Moore & Plummer (Fig. 18-20, 7), Pennsylvanian, Russia.

Timorechinus Wanner, Permian. This genus is distinguished especially by its remarkable inflated tegmen, which has niches for the arms. *T. mirabilis* Wanner (25, posterior view of calyx, and right-posterior ray), Upper Permian, Timor.

Trimerocrinus Wanner, Permian. Resembles *Cadocrinus* but has higher cup. *T. pumilus* Wanner (20, posterior), Upper Permian, Timor.

Ulocrinus Miller & Gurley, Pennsylvanian–Permian. Cup is deep bowl-shaped, radial facets horizontal and wide, two anals normally in cup. *U. sangamonensis* (Meek & Worthen), (17a, b, posterior, dorsal), Pennsylvanian, Illinois.

Fig. 18-34.   **Representative Mesozoic crinoids.** Among stem-bearing crinoids, Jurassic types are shown in 3, 5, 9, 11, and Cretaceous in 1, 6, 7. Stemless crinoids include, from Jurassic rocks, 2, 8, and from the Cretaceous, 4, 10, 12. Natural size, except as indicated. (Radials black, interradials and interbrachials stippled.)

Bourgueticrinus d'Orbigny, Cretaceous. Crown short, little if any larger in diameter than the elliptical stem. *B. aequalis* d'Orbigny (7a, b, ventral, lateral, ×2), Cretaceous, France.

Eugeniacrinus Miller, Jurassic–Cretaceous. A small specialized articulate having fused radials and prominent projecting axillary primibrachs. *E. caryophyllatus* Schlotheim (3a, interior of primaxil, ×2; 3b, reconstruction of part of crown and stem, ×1.3), Upper Jurassic, France.

Isocrinus von Meyer, Triassic–Tertiary. This genus differs from *Pentacrinites* in arm branching and in having round cirri. *I. knighti* Springer (11a, cup and lower brachials, ×2; 11b, crown, ×1), Upper Jurassic, Wyoming.

Marsupites Miller, Cretaceous. A large-plated stemless crinoid. *M. testudinarius* Schlotheim (10), Upper Cretaceous, England.

Pentacrinites Blumenbach, Triassic–Pliocene. This articulate is characterized especially by its abundant cirri having a diamond-shaped cross section. The stem is pentagonal to strongly stellate. *P. californicus* Clark (9, columnals, ×4), Jurassic, California; *P. fossilis* Blumenbach (5, ×2), Jurassic, Germany.

Rhizocrinus Sars, Jurassic–Recent. The calyx is confluent with the upper part of the stem and about the same in diameter. *R. alabamaensis*

(de Loriol) (6a, columnal; 6b, ventral view of cup; 6c, side view of cup and proximal columnals), Upper Cretaceous, Alabama; *R. cylindricus* Weller (1, ×2), Upper Cretaceous, New Jersey.

Saccocoma L. Agassiz, Jurassic–Cretaceous. A small stemless crinoid having winglike expansions of proximal arm plates, very common in Jurassic lithographic limestone of central Europe. *S. tenella* Goldfuss (2, ×7), Upper Jurassic, Germany.

Semiometra Gislén, Upper Cretaceous. A comatulid having a relatively large central cavity surrounded by prominent radial facets; cirri in two to four whorls attached to centrodorsal. *S. scanica* Gislén (4a, b, ventral, lateral, ×5), Senonian, Denmark.

Solanocrinus Goldfuss. Jurassic–Cretaceous. A comatulid characterized by deep cirrus sockets on the centrodorsal and large central cavity. *S. coliticus* Gislén (8a, lateral view, showing small basals between radials and the large centrodorsal; 8b, ventral view, ×4), Middle Jurassic, England.

Uintacrinus Grinnell, Upper Cretaceous. The rotund calyx of this stemless genus has many small plates; the arms are very long and slender. *U. socialis* Grinnell (12), Upper Cretaceous, Kansas.

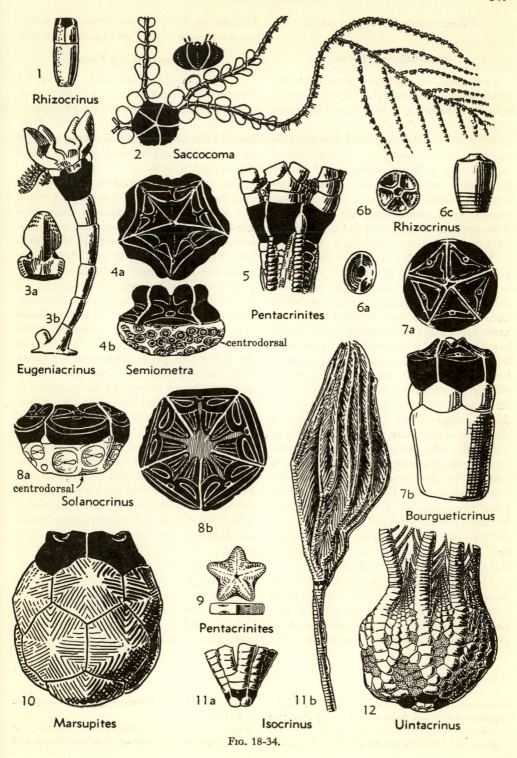

1 Rhizocrinus

2 Saccocoma

3a 3b Eugeniacrinus

4a 4b Semiometra

5 Pentacrinites 4b centrodorsal

6a 6b 6c Rhizocrinus

7a 7b Bourgueticrinus

8a centrodorsal Solanocrinus 8b

9 Pentacrinites

10 Marsupites

11a 11b Isocrinus

12 Uintacrinus

FIG. 18-34.

(*Text continued from page 635.*)

Stem-bearing Forms

Few stem-bearing fossil articulate crinoids are known from the Western Hemisphere, although a host of forms has been described from the Old World. Triassic articulates include long-lived genera (*Pentacrinites* and *Isocrinus*) which are characterized by a large crown and many branching arms, a small, simply constructed dorsal cup, and a strongly pentagonal stem which bears abundant cirri. Also present are crinoids (Millericrinida) having round stems which lack cirri.

Ancestors of the stem-bearing articulates, as well as all others, are judged to belong among the inadunates, but these cannot be forms having highly developed biserial arm structure because nearly all articulates are characterized by uniserial arms. Even the biserial-arm Encrinina from Triassic rocks have long uniserial lower and middle parts of the arms in early representatives, and they have other structural differences from inadunates. Some authors think that at least part of the articulates may be descendants of Flexibilia, with which they agree in uniserial arm structure but not in pinnules (absent in flexibles) or pattern of the infrabasal circlet. The invariable three-plate organization of this circlet among flexible crinoids (except for a few in which infrabasals are fused into a solid disk) cannot be antecedent of the five-plate infrabasal circlet of articulates.

The Lower Jurassic marks the first appearance of several new crinoid genera, most of which range through rocks of this period, and a few continue into the Neogene. Important additions are made in the Cretaceous, but few new genera are found until we come to the host of living crinoids.

Free-swimming Forms

The comatulids structurally correspond closely to the stem-bearing articulates but are distinguished by the abandonment of the stem after an early larval stage. They include the host of free-swimming crinoids (about 90 genera) which dominate crinoid life in modern seas.

The stemless comatulids first become prominent in Jurassic deposits (21 genera, 250 species). Approximately a dozen new fossil genera make appearance in Cretaceous rocks and a few more in the Tertiary. The great majority of living comatulids are not known in the fossil record. Described fossil genera and species of comatulids are based largely on the characters of the structure called **centrodorsal** (Fig. 18-34, *4*, *8*) which commonly bears many cirri. It is formed by solid fusion of the lower cup plates with stem segments. The disconnected fragments of comatulids, which are known to be common in some Mesozoic and Cenozoic formations, probably have value in stratigraphical paleontology.

Based on studies of comparative structure and evidence from ontogeny of the living crinoids, it seems likely that the comatulids are derived from a pentagonal-stemmed crinoid of the type of *Pentacrinites*. The fossil record indicates the dominance of different assemblages during three epochs: (1) Upper Jurassic in which solanocrinids (Fig. 18-34, *8*) were very abundant and nearly restricted; (2) Upper Cretaceous, characterized by conometrids and notocrinids; and (3) Miocene, marked by predominance of palaeantedontids.

REFERENCES

BASSLER, R. S., & MOODEY, M. W. (1943) *Bibliographic and faunal index of Paleozoic pelmatozoan echinoderms:* Geol. Soc. America Spec. Paper 45, pp. 1–734 (crinoids, pp. 10–23, 261–734). Gives references for all described genera and species to 1943.

BATHER, F. A. (1890) *British fossil crinoids: The classification of the Inadunata Fistulata:* Annals and

Mag. Nat. History, ser. 6, vol. 5, pp. 306–334, 373–388, 485–486, pls. 14–15.

——— (1893) *The Crinoidea of Gotland:* Svenska Vetensk. Akad. Handl., Bd. 25, no. 2, pp. 1–200, pls. 1–10. Excellent descriptions and illustrations of inadunate crinoids.

——— (1899) *A phylogenetic classification of the Pelmatozoa:* British Assoc. Rept. 1898, pp. 916–923.

——— (1900) *The Crinoidea:* in Lankester, E. R., Treatise on zoology, A. & C. Black, London, pt. 3, pp. 94–204, text figs. 1–127. A comprehensive comparative treatment of fossil crinoids.

GISLÉN, T. (1924) *Echinoderm studies:* Zoologiska Bidrag från Uppsala, Bd. 9, pp. 1–330, figs. 1–351. A noteworthy work on fossil and recent articulate crinoids.

GOLDRING, W. (1923) *Devonian crinoids of New York:* New York State Mus. Mem. 16, pp. 1–670, pls. 1–60. Chief publication on American Devonian crinoids.

HALL, J. (1858) *Paleontology of Iowa:* Iowa Geol. Survey, vol. 1, pt. 2, pp. 473–724, pls. 1–29. Contains first descriptions of many Mississippian crinoids.

——— (1859) *New species of Crinoidea:* Iowa Geol. Survey, vol. 1, pt. 2, suppl., pp. 1–94, pls. 1–3.

——— (1861) *Descriptions of new species of Crinoidea from the Carboniferous rocks of the Mississippi valley:* Boston Soc. Nat. History Jour., vol. 7, pp. 261–328.

——— (1867) *Account of some new or little known species of fossils from rocks of the age of the Niagara group:* New York State Cabinet Ann. Rept. 20, pp. 305–401 (rev. ed., 1870, pp. 347–438, pls. 2–8).

——— (1882) *Descriptions of the species of fossils found in the Niagara group at Waldron, Indiana:* Indiana Dept. Geology Nat. History Ann. Rept. 11, pp. 217–345, pls. 1–55.

JAEKEL, O. (1918) *Phylogenie und System der Pelmatozoen:* Paläont. Zeitschr., Bd. 3, pp. 1–128, text figs. 1–114 (German). Important study on classification.

MOORE: R. C. (1939) *The use of fragmentary crinoidal remains in stratigraphic paleontology:* Denison Univ., Sci. Lab. Bull., vol. 33 (1938), pp. 165–250, pls. 1–4, text figs. 1–14.

——— (1940) *Relationships of the family Allagecrinidae, with description of new species from Pennsylvanian rocks of Oklahoma and Missouri:* Same, vol. 35, pp. 55–137, pls. 2–3, text figs. 1–14.

——— & LAUDON, L. R. (1943) *Evolution and classification of Paleozoic crinoids:* Geol. Soc. America Spec. Paper 46, pp. 1–153, pls. 1–14,

figs. 1–18. Gives summary of most significant structural characters and conclusions on phylogeny.

——— & ——— (1944) *Class Crinoidea,* in Shimer, H. W., and Shrock, R. R., Index fossils of North America, Wiley, New York, pp. 137–209, pls. 52–79.

——— & PLUMMER, F. B. (1940) *Crinoids from the Upper Carboniferous and Permian strata in Texas:* Texas Univ. Bull. 3945, pp. 1–468, pls. 1–21, text figs. 1–78. Describes several new genera and many species of inadunates.

PECK, R. E. (1936) *Lower Mississippian microcrinoids from the Kinderhook and Osage groups of Missouri:* Jour. Paleontology, vol. 10, pp. 282–293, pls. 46–47.

SCHMIDT, W. E. (1930) *Die Echinodermen des deutsches Unterkarbons:* Preuss. geol. Landesanstalt Jahrb., pp. 1–92, pls. 1–3, text figs. 1–20 (German).

——— (1934) *Die Crinoiden des rheinischen Devons, Teil 1, Die Crinoiden des Hunsrückschiefers:* Preuss. geol. Landesanstalt Abh., neue Folge, Heft 163, pp. 1–149, pls. 1–34 (German).

SPRINGER, F. (1913) *Crinoidea:* in Zittel, K. A., & Eastman, C. R., Textbook of palaeontology, Macmillan & Co., Ltd., London, 2d ed., vol. 1, pp. 173–243, figs. 267–346.

——— (1920) *The Crinoidea Flexibilia:* Smithsonian Inst. Pub. 2501, pp. 1–486, pls. A–C, 1–75, text figs. 1–51. Indispensable monograph on this group of crinoids.

——— (1921) *Dolatocrinus and its allies:* U.S. Nat. Mus. Bull. 115, pp. 1–78, pls. 1–16.

——— (1923) *On the fossil crinoid family Catillocrinidae:* Smithsonian Misc. Coll., vol. 76, no. 3, pp. 1–41, pls. 1–5.

——— (1926) *Unusual fossil crinoids:* U.S. Nat. Mus. Proc., vol. 67, art. 9, pp. 1–137, pls. 1–26.

——— (1926a) *American Silurian crinoids:* Smithsonian Inst. Pub. 2871, pp. 1–239, pls. 1–33. A particularly valuable work.

SUTTON, A. H., & WINKLER, V. D. (1940) *Mississippian Inadunata—Eupachycrinus and related forms:* Jour. Paleontology, vol. 14, pp. 544–567, pls. 66–68.

ULRICH, E. O. (1924) *New classification of the Heterocrinidae:* Canada Geol. Survey Mon. 138, pp. 82–104, text figs. 1–14.

WACHSMUTH, C., & SPRINGER, F. (1897) *The North American Crinoidea Camerata:* Harvard Coll. Mus. Comp. Zoology Mem., vols. 21–22, pp. 1–897, pls. 1–83, text figs. 1–21. Classic study on camerates.

WANNER, J. (1916) *Die permischen Krinoiden von*

Timor, Teil 1: Paläontologie von Timor, Lief. 6, Teil 11, pp. 1–329, pls. 96–114 (1–19), text figs. 1–88 (German). First comprehensive publication on Permian crinoids.

———— (1924) *Die permischen Krinoiden von Timor:* Mijn. nederl. Oost-Indië, Jahrb., Verhandel. 1921, Gedeelte 3, pp. 1–348, pls. 1–22, text figs. 1–61 (German).

———— (1930) *Neue Beiträge zur Kenntnis der permischen Echinodermen von Timor, IV, Flexibilia:* Dienst Mijnbouw nederl.-Indië, Wetensch. Mededeel., no. 14, pp. 1–52, pls. 1–4 (German).

———— (1937) *Neue Beiträge zur Kenntnis der permischen Echinodermen von Timor, VIII-XIII:* **Palaeontographica Suppl., Bd. 4, Abt. 4,** Abschn. 2, pp. 57–212, pls. 5–14, text figs. 1–82, Schweizerbart, Stuttgart (German).

———— (1949) *Neue Beiträge zur Kenntnis der permischen Echinodermen von Timor, XVI, Poteriocrinidae, 4 Teil:* Palaeontographica Suppl., Bd. 4, pp. 1–56, pls. 1–3, Schweizerbart, Stuttgart (German).

WELLER, J. M. (1930) *A group of larviform crinoids from lower Pennsylvanian strata of the Eastern Interior Basin:* Illinois Geol. Survey Rept. Inv. 21, pp. 1–38, pls. 1–2.

WRIGHT, J. (1939) *The Scottish Carboniferous Crinoidea:* Royal Soc. Edinburgh Trans., vol. 60, pt. 1, no. 1, pp. 1–78, pls. 1–12. text figs. 1–86.

CHAPTER 19

Holothuroids

A glimpse of the over-all structure and classification of echinoderms has been provided in Chap. 16. Members of the subphylum Pelmatozoa have been discussed, and we now begin the study of the subphylum Eleutherozoa, members of which are free-roaming, rather than being attached during much or all of their life.

Holothuroids or sea cucumbers seem to bear little resemblance to other echinoderms, but the ontogeny and internal structure show that they are definitely members of this phylum. They are distinguished as a separate class.

MORPHOLOGY

The genus *Holothuria* is illustrated in Fig. 19-1 as an example of holothuroids in general. Italic numbers in the following discussion refer to this figure unless otherwise designated.

Soft Parts

The body of *Holothuria* is elongate, tapering at the ends. It rests on one side, the flattened **sole,** which is accordingly termed **ventral.** The **mouth** (*15*), surrounded by the **tentacles** (*9*), is located at the anterior end, the **anus** (*13*), at the posterior tip. The body is covered by a tough, leathery skin, in which are embedded microscopic calcareous bodies of various types. The skin is unique among echinoderms in lacking cilia. It is underlain by circular transverse muscles, below which lie five strands of longitudinal muscles.

The alimentary or **intestinal tract** (*11*), which is a simple single-looped tube connecting the mouth and anus, is fastened to the body wall by means of a thin membrane (**mesentery,** *10*). Near the anus it is joined by a pair of large ramified structures termed the **respiratory trees** (*12*), features which are without parallel in other classes of living echinoderms. Water is drawn into and expelled from these respiratory organs through the anus and the short intervening section of gut, termed the **cloaca** (*22*). A single **gonad** (*20*) is located alongside the alimentary tract.

The water-vascular system of *Holothuria* consists chiefly of a **ring canal** (*17*) around the esophagus, and five **radial canals** (*4, 21*). These are connected with the ring canal and run the length of the body wall, between the longitudinal and transverse muscles. A large contractile **polian vesicle** (*19*) serves as main reservoir. The radial canals bear numerous branches, each of which terminates in an inner bulblike extension (**ampulla,** *6*) and an elongate outer extension (**papilla,** *3*, or **tube foot,** *7*), which passes to the outside through the body wall. Contraction of the ampulla forces water into the tube foot, and extends it, whereas expansion of the ampulla shrinks the tube foot. During their early life history, holothuroids charge the water-vascular system with sea water, through an opening on the surface of the body. This situation persists through life in other classes of echinoderms, and in some of the sea cucumbers, but most holothuroids lose this outer opening and develop numerous openings of the water-vascular system into the body cavity, to charge the water-vascular system with coelomic fluid. The openings are located on branches of the ring canal which are called **stone canals,** and are generally screened by highly perforated plates termed **internal madreporites** (*18*). The fluid content of the body cavity is maintained by diffusion of water

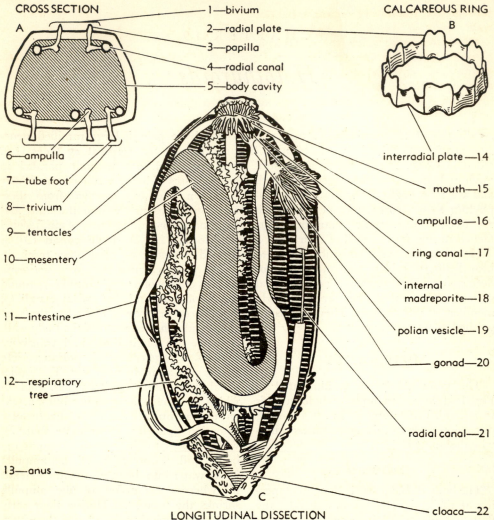

CROSS SECTION

A

1—bivium
2—radial plate
3—papilla
4—radial canal
5—body cavity

6—ampulla
7—tube foot
8—trivium
9—tentacles
10—mesentery
11—intestine
12—respiratory tree
13—anus

LONGITUDINAL DISSECTION

C

CALCAREOUS RING

B

interradial plate—14
mouth—15
ampullae—16
ring canal—17
internal madreporite—18
polian vesicle—19
gonad—20
radial canal—21
cloaca—22

FIG. 19-1. **Structure of sea cucumbers.** The recent genus *Holothuria* is shown in longitudinal dissection (*C*), and cross section (*A*). The skeletal ring surrounding the esophagus is shown in (*B*). Labels of the diagram are cross-indexed by number to the following alphabetical list of terms.

ampullae (6, 16). Contractile bulbs which force fluid into, or withdraw it from, tube feet.

anus (13). Opening of cloaca, serving for expulsion of waste matter and as intake for respiratory organs.

bivium (1). Sides and top of sea cucumber, underlain by two radial canals; contrasted with underside which contains three canals (trivium).

body cavity (5). Internal cavity in which main organs are suspended.

cloaca (22). Last portion of gut, through which waste matter is expelled and fresh water is drawn into respiratory organs.

gonad (20). Single reproductive organ.

internal madreporite (18). Sievelike plate through which coelomic fluid passes into water-vascular system.

interradial plates (14). Five plates in interradial position (between radial canals) among the 10 plates surrounding esophagus.

intestine (11). Digestive tube, forming a simple loop.

mesentery (10). Membrane which attaches gut to body wall.

(Continued on next page.)

through the respiratory organs and body wall.

Of the five radial canals, three (**trivium, 8**) lie on the lower (**ventral**) side of the animal, one extending down the center and two along the edges of the sole. They send out a great many branches bearing tube feet. The latter, which are closely crowded on the sole, possess sucker disks and enable the animal to creep along. The other two radial canals (**bivium, 1**) possess a smaller number of extensions, called **papillae** (*3*). These lack sucker disks and appear merely as pointed protuberances on the surface of the animal. Their function may be partly sensory and partly respiratory. The tentacles which surround the mouth are large modified tube feet, attached to long ampullae (*16*).

A nervous system and well-developed blood-vascular system are present.

Most holothuroids are nearly cylindrical in form, but one pelagic genus is umbrella-shaped, with an upward pointing mouth. Some sea cucumbers do not have one of their sides specialized as a creeping sole. Differentiation and distribution of tube feet vary greatly; the podia may be restricted to rows in the ambulacral areas (above the radial canals) leaving intervening interambulacral areas bare. In *Holothuria*, they are scattered over the entire surface. Some holothuroids possess only normal tube feet in addition to the tentacles, and others lack podia altogether, excepting the tentacles. In one group, the radial canals are lost during ontogeny. Some have lost the respiratory trees.

Skeleton

Most holothuroids possess skeletal elements consisting of small, mainly microscopic ossicles of calcite or iron phosphate. The esophagus is generally encircled by a calcareous ring (*B*), composed of alternating radial and interradial plates (*2, 14*) which may occur singly or in series. The skin contains numerous **spicules** of diverse shape—crosses, plates, anchors, wheels, and others (Fig. 9-2). Some holothuroids possess **anal teeth,** a special group of plates surrounding the anus, and most have spicules within their internal organs. Like other echinoderm plates, these ossicles are single crystal units, and this character serves to distinguish them readily from the remains of other phyla when examined under a polarizing microscope. Some orders bear distinctive ossicles which are useful in identifying species. Many different kinds of spicules occur in one individual, and, to further complicate matters, spicules may change shape during life history. When a sea cucumber dies and its soft tissues decay, the spicules are scattered and may become mixed with skeletal parts of others.

Holothurian spicules are widely distributed in the geologic column, from Mississippian to Recent. Although the association of spicules belonging to any one individual, species, or genus cannot be determined, generic and specific names have been applied to some of the more distinctive types.

(Fig. 19-1 continued.)

mouth (15). Entrance to digestive tract.

papilla (3). Sensory or respiratory extension of water-vascular system.

polian vesicle (19). Large contractile bulb serving to inflate water-vascular system.

radial canals (4, 21). Five longitudinal water vessels which supply fluid to papillae and tube feet.

radial plates (2). Five plates in line with radial canals, among the 10 plates surrounding esophagus.

respiratory trees (12). Large branching organs serving for respiration.

ring canal (17). Water vessel forming a ring around esophagus, and connecting radial canals.

tentacles (9) Greatly extended branches of water-vascular system surrounding mouth.

trivium (8). Ventral part of holothuroid, underlain by three radial canals.

tube foot (7). Prehensile branch of water-vascular system.

LIFE HABITS

Most sea cucumbers creep over the sea floor, or lie in a favorite spot. Some eat mud which may be shoveled into the mouth by the tentacles; others spread

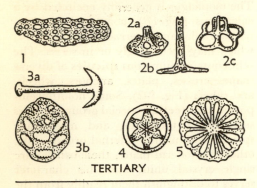

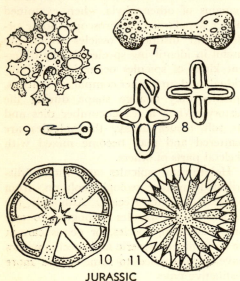

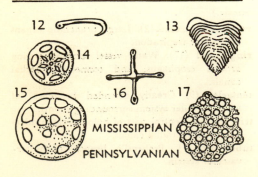

slimy tentacles on the sea floor and trap microscopic organisms; when a sufficient crop of food has accumulated on a tentacle, the appendage is stuffed into the mouth and pulled out again between tightly closed lips, so that mucous and food are stripped off.

Some sea cucumbers (synaptids) are specialized mainly for a burrowing existence; they eat sediment wholesale and are thought to contribute materially to the attrition of sedimentary particles.

A single living genus, *Pelagothuria*, has become adapted to a swimming mode of life. The mouth is surrounded by radiating rays, which are greatly elongate ampullae of the tentacles, between which the thin body wall is stretched as a web. The small body hangs down in the middle, beneath this spread umbrella, which undulates and thus gently propels the animal. Supposed pelagic holothuroids in Middle Cambrian

FIG. 19-2. **Fossil holothuroid spicules.** Representative spicules illustrated are from the Eocene of France, Jurassic (Lias) of Germany, and the Carboniferous of England and North America.

1–5, Eocene spicules, all ×180. (1) Plate of *Priscopedatus normani* Schlumberger. (2a–c) Tables of *Priscopedatus*; (2a) *P. pyramidalis* Schl.; (2b) *P. eiffeli* Schl.; (2c) *P. normani* Schl. (3a, b) *Synapta* spicules; (3a) *S. eocena* Schl., anchor; (3b) *S. truncata* Schl., anchor plate. (4) *Chirodota curriculum* Schl., wheel. (5) *Myriotrochus operculum* Schl., wheel.

6–11, Jurassic spicules. (6, 7) Dendrochirote holothuroid, unnamed, ×130; (6) spicule; (7) rod. (8) Unnamed crosses, ×90. (9) Hook of synaptid genus *Ancistrum* Etheridge, ×30. (10) Wheel of *Chirodota leptalampra* Bartenstein, ×80. (11) Wheel of *Myriotrochus* sp., ×130.

12–17, Carboniferous spicules. (12) *Ancistrum brownwoodensis* Croneis & McCormack, hook, Pennsylvanian, Texas, ×17. (13) Unnamed frond, Upper Carboniferous, England, ×180. (14) *Palaeochirodota plummeri* Croneis & McCormack, wheel of synaptid, Pennsylvanian, Texas, ×17. (15) *Protocaudina traquairi* (Etheridge), Mississippian, Illinois, ×70. (16) Unnamed cross, Pennsylvanian, Texas, ×17. (17) *Ancistrum?* sp., Pennsylvanian, Texas, ×35.

(Burgess) shale are interpreted now as coelenterates.

Holothuroids are generally rather limp but when disturbed, they contract the muscles of the body wall and become turgid. If further irritated, some will contract so forcibly as to explode, either through the body wall or through the anal membrane, spewing forth respiratory trees, intestines, and other internal organs. Some, which specialize in this type of defense, have developed a special eruptive organ. When ejected alone or along with other viscera through the anal region, this expands to form a great mass of sticky threads, in which enemies may become entangled. The evisceration tactics are of no permanent inconvenience, as a new set of internal structures may be generated.

CLASSIFICATION

The major groups of holothuroids are recognized on the basis of soft parts, but there are skeletal differences between some of the orders. One of the late classifications of Recent forms (Cuénot) is as follows:

Main Divisions of Holothuroids

Cucumariida or Dendrochirotida (*order*), relatively conservative forms which may withdraw the head. Dermal ossicles chiefly derived from crosses, lack hooks, anchors, tables, and wheels. A few members are heavily plated on the dorsal side, resembling carpoids.

Molpadida (*order*), lack tube feet, have spicules like Dendrochirotida as well as granules of iron phosphate.

Aspidochirotida (*order*), possess a well-defined sole; dermal spicules are mainly tables, rods, and perforated plates.

Elasipodida (*order*), external madreporite, respiratory trees rudimentary or absent, skeleton may be absent, or, if present, may include wheels. Deep-sea forms.

Synaptida (*order*), radial canals and podia lost, respiratory trees absent, skin liberally studded with dermal ossicles including wheels, anchors, and the perforated anchor-support plates. Burrowing forms.

The classification of fossil holothuroids is in an unsatisfactory state. Nearly all fossil remains are in the form of dissociated spicules. Some of these are sufficiently like certain modern genera to be identified with them. Other kinds have been given new generic and specific names, but their placement within orders generally is not certain. Many distinctive types have been described and illustrated but not named.

GEOLOGICAL DISTRIBUTION

The earliest reported holothuroids are impressions of soft-bodied animals in the Middle Cambrian Burgess shale of British Columbia. Walcott recognized four genera, two pelagic and two benthonic forms. It is now generally believed that the former two are not holothuroids but coelenterates; the assignment of the benthonic forms, such as *Redoubtia* (Fig. 19-3), may be correct. If so, we must conclude that skeletal elements in holothuroids are not rudimentary remains of former armor, but secondary developments, for the wheels, disks, hooks, and other skeletal structures distinctive of the class are not known from rocks older than Mississippian. Impressions of animals in Jurassic limestones of Germany and some other rocks have been interpreted as holothuroids, but such identification is doubtful.

Skeletal elements of holothuroids are locally abundant, but for several reasons they have not been used widely in stratigraphic work. Most of them are much smaller than Foraminifera, and therefore they easily escape notice. Also, they are less widely distributed than foraminifers and ostracodes. Therefore, they have received far less attention than these groups.

FIG. 19-3. **Possible Cambrian holothuroid.** *Redoubtia polypodia* Walcott, from Burgess shale, Middle Cambrian, British Columbia; seems to be the impression of a sea cucumber, ×0.8.

RELATIONSHIPS

The position of holothuroids in the family tree of echinoderms is uncertain. Their larval development is comparatively simple, somewhat like that of pelmatozoans, but lacking an attached stage. It is very unlike the development of asteroids and echinoids, suggesting that the Eleu-

therozoa may be derived from different ancestors. Holothuroids show less radial symmetry than any other living class of echinoderms, and some workers consider this to be a primitive feature inherited from the bilateral ancestors of echinoderms. Others interpret it as a secondary return to bilateral symmetry, analogous to similar trends in the irregular echinoids.

REFERENCES

CRONEIS, C., & McCORMACK, J. T. (1932) *Fossil Holothuroidea:* Jour. Paleontology, vol. 6, pp. 111–148, figs. 1–4, pls. 15–21. Contains skeletal diagnoses of modern holothurian genera, a thorough review of described fossil species, and abundant illustrations of ossicles from fossil and recent forms.

CUÉNOT, L. (1948) *Classe des Holothuries:* In Grassé, P., Traité de zoologie, Masson & Cie., Paris,

vol. 11, pp. 82–120, figs. 101–141 (French). An excellent treatment of structure and classification of modern sea cucumbers.

MacBRIDE, E. W. (1909) *Holothuroidea,* in Harmer, S. F., & Shipley, A. E., Cambridge natural history, Macmillan & Co., Ltd., London, vol. 1, chap. 19, pp. 560–578, figs. 254–262. An old but good and widely available reference on modern sea cucumbers.

CHAPTER 20

Starfishes

The starfishes or stelleroids (class Stelleroidea) differ from other eleutherozoans in their starlike shape. The body is divided into a central disk and radiating arms. The mouth lies in the middle of the lower surface. Starfishes, like other echinoderms, possess a water-vascular system, in which the radial canals extend from near the mouth to the tips of the arms, and bear numerous branches termed tube feet. The skeleton is not a rigid box, but rather is composed of many loosely joined ossicles.

Starfishes are abundant in modern seas, and range from the intertidal zone into abyssal depths. Their remains, which are known from Ordovician to Recent, show great diversity in structure. Despite their long record and rapid evolution, they have been of relatively little value to the stratigrapher, because complete fossil stelleroids are a rarity. The ossicles of the skeleton generally become scattered when the animal dies, and excepting those in Cretaceous rocks, little work has been done on these dissociated fragments. For this reason, most of our knowledge of fossil starfish is derived from fossils found in a few places where exceptionally rapid burial resulted in preservation of many specimens.

The stelleroids are divided into subclasses as shown in the following table.

Classification of Starfishes

Stelleroidea (*class*), star-shaped eleutherozoans. Ordovician–Recent.

Somasteroidia (*subclass*), primitive starfishes having a double row of semicylindrical ossicles partly or completely surrounding the radial canals and rows of rodlike ossicles branching off on either side. Lower Ordovician–Upper Devonian.

Asteroidia (*subclass*), having generally hollow arms which contain large lobes of body

cavity and enclosed organs; radial canals located on outside of skeleton. Ordovician–Recent.

Ophiuroidia (*subclass*), having slender, whiplike arms built around a core of ossicles; radial canals deeply sunken or completely enclosed in skeleton. Ordovician–Recent.

SOMASTEROIDS

The genus *Villebrunaster* furnishes a simple introduction to the study of the somasteroids. It is the most primitive member of the earliest known starfish fauna found in basal Ordovician (Tremadocian) rocks of France. The skeleton of the lower (**oral**) surface is shown in Fig. 20-1 and a diagrammatic cross section of an arm in Fig. 20-4, *1*. The pentagonal body has the shape of a blunt-rayed star. The middle of the oral surface is occupied by a large five-sided opening, which, if it is not the mouth, constitutes a vestibule in front of the mouth. A double series of hemicylindrical ossicles leads from the corners of this opening to the tips of the arms. These are **ambulacral** ossicles which embrace a tunnel, called the **ambulacral channel,** occupied by the radial water vessel (Figs. 20-1; 20-2, *17, 19*). Between the ossicles branches of the water-vascular system led to tube feet, the presence of which is attested by a series of cups located along the sides of the ossicles. These cups are interpreted by analogy with later starfishes as having housed contractile bulbs (ampullae) which operated the tube feet.

The paired rows of ambulacrals split at the mouth to form a **mouth frame.** The two ossicles which constitute the inner corners of the mouth or mouth vestibule, located between the ambulacral areas, are different from the others and project inward (Fig. 20-1). They are known as the **mouth angle plates.** Such plates exist also in advanced starfishes.

Each ambulacral bears a series of rods (**virgalia**), which lead outward toward the periphery of the arms. The oral aspect of the skeleton thus superficially resembles the crown of a crinoid, but the series of virgalia were not free like pinnules. They were connected by skin, forming the oral wall of the starfish.

The aboral (upper) surface of *Villebrunaster* and the other somasteroids was armored only by scattered three- to four-rayed spicules.

The arrangement of virgalia suggests that *Villebrunaster* may be close to the pelmatozoan ancestors of the starfishes.

Furthermore, this genus probably obtained its food in the manner of most pelmatozoans, gathering plankton and other microscopic bits in the grooves between the virgalia. This material was carried to ambulacral areas in the middle of the arms and thence to the mouth. It is conceivable that, in order to accomplish this method of securing food, *Villebrunaster* lived mouth upward (like pelmatozoans and unlike modern starfishes), or it lived mouth downward but with up-curved arms. Advanced types of somasteroids differ from *Villebrunaster* in having longer arms and additional types of plates. The margins of the arms are bounded by a row of marginal plates, and the ambulacrals may be bordered on each side by a row of

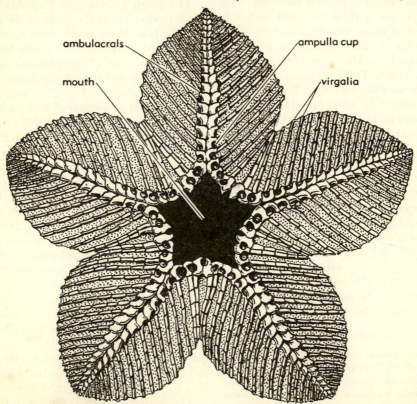

FIG. 20-1.   **Reconstruction of Villebrunaster.** Oral view, ×3. Paired ossicles which extend down the middle of each arm are termed *ambulacrals*. They surround the radial water vessels, and bear *ampulla cups*, excavations for the reception of ampullae. Rows of short rodlike ossicles (*virgalia*) diverging from the ambulacrals are separated by unarmored grooves. *V. thorali* Spencer, Lower Ordovician (Tremadocian), France. (*After Spencer.*)

adambulacrals. Ambulacrals do not close over the radial canals, and thus leave an open ambulacral groove.

Somasteroids have been found in rocks ranging in age from Lower Ordovician to Upper Devonian. It is highly probable that they represent the root stock of the stelleroids, from which asteroids and ophiuroids split off in or before Early Ordovician time.

ASTEROIDS

Members of the subclass Asteroidia are distinguished from somasteroids chiefly by lacking virgalia. They generally retain the large, hollow type of arm and the paired ambulacral ossicles. Among asteroids, the water-vascular system is the chief locomotary mechanism. It bears innumerable prehensile tube feet, the action of which is coordinated by a well-developed nervous system. This type of locomotion is necessarily slow, and various protective devices, including armor, became necessary for survival.

Structure of a Sea Star

We may begin the study of asteroids by reference to the common modern starfish *Asterias*, well known to almost everyone who has visited the seashore. Structures of *Asterias* and other asteroids are shown in Fig. 20-2. Italic numbers in the following description refer to features shown on this figure.

Upper Side. Most specimens of *Asterias* have the size of a dinner plate, but some attain a spread of 1 m. The central part of the body is poorly defined, grading into the five large **arms** (*29*). The upper (**aboral**) side is warty owing to the presence of many **spines** (*54*). These are extensions of plates which form a meshwork in the body wall. The small **anus** (*28*) lies near the center, displaced slightly toward one of the interarm angles. The **madreporite** (*53*) is a highly porous plate through which water is drawn into the water-vascular system. It is located in an angle

between two of the arms. An **ocular spot** at the tip of each arm is sensitive to light.

Lower Side. The **oral** side of the starfish presents a different picture. The **mouth** (*35*) is located in the center. Five deep **ambulacral grooves** (*22, 33*) which extend outward to tips of the arms are thickly crowded with delicate, translucent **tube feet** (*23, 30*) or podia, each ending in a **sucker disk** (*50*). The **radial canal** (*18*) of the water-vascular system lies recessed in the **ambulacral channel** (*19*) which runs down the middle of the ambulacral groove. Ordinarily the grooves are expanded, allowing the tube feet to protrude. In case of danger, they may be contracted, bringing the **marginal spines** (*31*) together so as to form a protective barrier over the delicate podia.

Organs. A tough, leathery, muscular body wall, containing many skeletal plates (**ossicles**) of different sorts, encloses a body cavity filled with organs. This cavity and its enclosed organs are not confined to the central disk, but extend out into the arms. From the mouth an esophagus leads upward into the stomach (*27*), whence five branches lead out to paired, lobate **digestive glands** (*44, 51*) or pyloric caeca, located in each arm. An intestine leads from the stomach to the small anus (*28*), located on the upper surface, but much undigestible matter is ejected through the mouth. The only other large organs in the body cavity are **gonads** (*26*), which lie paired in each arm. The **water-vascular system** is rather similar to that of holothuroids, differing chiefly in the location of the **radial canal** (*18*), which, among starfishes, lies outside the skeleton. A view of the ambulacral areas from the inside shows only the **ampullae** (*47, 52*), which activate the tube feet like rubber syringes. The tube feet extend to the outside through **ambulacral pores** (*48*), which are merely gaps between adjoining plates of the skeleton.

Space between the organs is filled by fluid which carries nutrients and oxygen and serves also to transport waste mate-

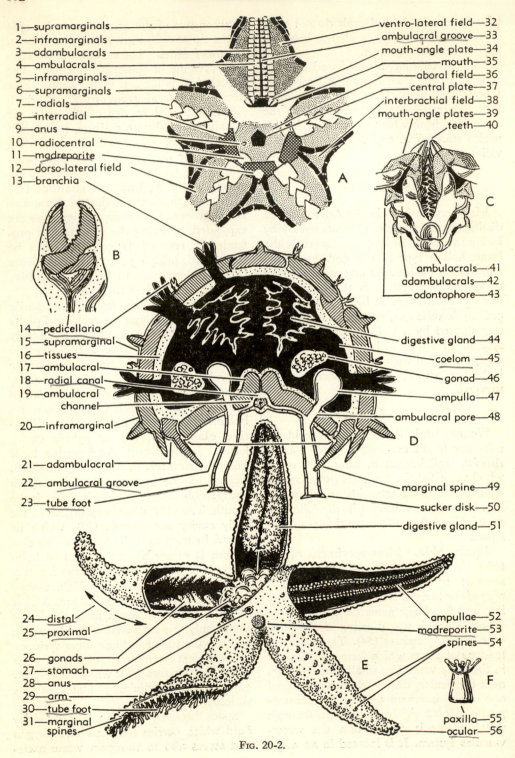

1—supramarginals
2—inframarginals
3—adambulacrals
4—ambulacrals
5—inframarginals
6—supramarginals
7—radials
8—interradial
9—anus
10—radiocentral
11—madreporite
12—dorso-lateral field
13—branchia

ventro-lateral field—32
ambulacral groove—33
mouth-angle plate—34
mouth—35
aboral field—36
central plate—37
interbrachial field—38
mouth-angle plates—39
teeth—40

ambulacrals—41
adambulacrals—42
odontophore—43

14—pedicellaria
15—supramarginal
16—tissues
17—ambulacral
18—radial canal
19—ambulacral channel
20—inframarginal

21—adambulacral
22—ambulacral groove
23—tube foot

digestive gland—44
coelom—45
gonad—46
ampulla—47
ambulacral pore—48

marginal spine—49
sucker disk—50
digestive gland—51

24—distal
25—proximal

26—gonads
27—stomach
28—anus
29—arm
30—tube foot
31—marginal spines

ampullae—52
madreporite—53
spines—54

paxilla—55
ocular—56

FIG. 20-2.

FIG. 20-2. **Structure of asteroids.** Main structural features of asteroids are illustrated and cross-indexed by numbers to the alphabetically arranged list of morphologic terms. (*A*) Organization of plates in advanced forms, upper segment showing oral (under) side, the rest viewed from above. (*B*) Pedicellaria. (*C*) Mouth frame, much enlarged. (*D*) Cross section of arms of *Asterias*, a common modern starfish. (*E*) *Asterias*, partly dissected, viewed from above. (*F*) Paxilla (a starfish spine type).

aboral field (36). Central region of upper (aboral) surface, surrounded by interradial and radiocentral plates.

adambulacrals (3, 21, 42). Plates which form a single row along edges of ambulacral grooves on underside of starfish.

ambulacral channel (19). Linear depression in ambulacral groove containing radial water vessel and nerve.

ambulacral groove (22, 33). Broad, open depression on underside, leading from mouth to tip of each arm.

ambulacral pores (48). Openings between plates, through which tube feet extend from interior into ambulacral groove.

ambulacrals (4, 17, 41). Double row of plates along center of underside of arms.

ampullae (47, 52). Syringe-like bulbs which squeeze water into tube feet or out of them.

anus (9, 28). Opening located excentrically on upper surface.

arm (29). Any radial extension of body.

branchia (13). Branched external pouch of skin, filled with fluids of body cavity, serving for respiration.

central plate (37). Ossicle in center of dorsal surface.

coelom (45). Body cavity.

digestive glands (44, 51). Paired organs lying in each arm.

distal (24). Direction away from center toward tips of arms.

dorso-lateral fields (12). Areas along upper sides of arms, between radial plates and supramarginals; generally only lightly armored.

gonads (26, 46). Paired reproductive organs, located in each arm.

inframarginals (2, 5, 20). Plates arranged in a row bordering adambulacrals.

interbrachial field (38). Unarmored area in interarm angle, bounded by inframarginal and supramarginal plates.

interradials (8). Five plates on upper surface, placed between radials.

madreporite (11, 53). Sievelike intake of water-vascular system, placed interradially on upper surface.

marginal spines (31, 49). Projections of adambulacral plates, which protect tube feet when ambulacral groove is contracted.

mouth (35). Large opening in center of underside, through which food may be drawn into stomach, or stomach may be extruded to digest food outside body.

mouth angle plates (34, 39). Specially modified ambulacral plates shifted into interradial position, forming five jawlike projections on margin of mouth.

ocular (56). Light-sensitive organ at tip of each arm.

odontophore (43). Interradial plate of mouth frame, lying behind mouth-angle plates.

paxilla (55). Type of spine, consisting of short blunt trunk and crown of small spinelets.

pedicellariae (14). Minute pincers for defense.

proximal (25). Direction toward center of starfish.

radial canal (18). Water vessel which supplies tube feet.

radials (7). Plates of upper surface, arranged in row along mid-line of each arm.

radiocentrals (10). Five radial plates nearest center.

spines (54). Calcareous projections of skeleton, some of which are mere extensions of plates, whereas others articulate with plates.

stomach (27). Centrally located organ which secretes powerful digestive juices, and may be extruded through mouth.

sucker disks (50). Prehensile vacuum cups at ends of tube feet.

supramarginals (1, 6, 15). Plates arranged in a row on sides of arms, above inframarginals.

teeth (40). Scaly or spiny ossicles projecting from jaws.

tissues (16). Body wall, consisting of muscles surrounding plates of skeleton.

tube feet (23, 30). Hollow extensions of water-vascular system, used for locomotion and grasping prey.

ventro-lateral fields (32). Lightly armored areas on underside of arms, between adambulacral plates and inframarginals.

rials. Aeration is accomplished through delicate finger-like pouches of skin (dermal **branchiae,** *13,* or papulae) extending out between the plates of the skeleton.

Skeleton. The hard parts of starfishes are of primary importance to the paleontologist as well as to the neontologist, as they have furnished the chief basis for classification. The plates, like those of other echinoderms, are single crystal units of spongy texture. They are covered by skin and pervaded by living cells, and each may grow or be resorbed during ontogeny. Many plates bear spines. *Asterias* bears beaklike structures of microscopic dimensions called **pedicellariae** (*14*), which by their snapping action protect the starfish from attack. Also, it possesses internal spicules of microscopic size, many of which lie in the walls of the tube feet, lending strength to these delicate structures. Others support the digestive tract and internal organs.

The plates and pedicellariae have been used in the classification of Recent species of *Asterias* and other starfishes. Unfortunately, the pedicellariae almost universally are lost on fossil specimens. Several main rows of plates may be recognized in all asteroids, and these are supplemented variously by accessory plates in some groups. The major categories of plates are diagrammatically illustrated in Fig. 20-2*A*.

The ambulacral groove is floored by a paired series of **ambulacrals** (*4, 17*), which ends at the tip of each arm in a single **terminal.** Edges of the ambulacral grooves are formed by plates called **adambulacrals** (*3, 21*), which bear **marginal spines** (*31, 49*). The adambulacrals of most starfishes articulate with adjacent plate rows, so as to permit transverse muscles to contract the ambulacral grooves when in danger. The sides of the arms may bear one conspicuous row of **marginal plates,** or, more commonly, two rows, termed **inframarginals** (*5, 20*) and **supramarginals** (*6, 15*). The inframarginals adjoin the adambulacrals along the arms but are separated from them in the disk so as to leave a lightly armored intervening space termed the **ventro-lateral field** (*32*). Among some Paleozoic starfishes, the inframarginal and supramarginal rows are separated in angles between the arms, leaving flexible **interbrachial fields** (*38*).

The mouth is surrounded by modified ambulacral and adambulacral plates, forming a **mouth frame.** V-shaped **mouth angle plates** (*34, 39*), which project in interambulacral position, bear small spines called **teeth** (*40*). Behind each mouth angle plate lies an ossicle termed **odontophore** (*43*).

On the aboral side of many starfishes lies a **central** (*37*) or dorso-central plate. Generally this is surrounded by the dorsal or **aboral field** (*36*), armored by small ossicles. A row of large plates called **radials** (*7*) or carinals extends down the mid-line of each arm. The most proximal of these commonly is larger and different in shape from the rest, and thus it is singled out as the **primary radial.** In some forms, a smaller plate, called **radiocentral** (*10*), may lie between the primary radial and the central. The primary radials are separated by large plates, termed **interradials** (*8*). The radial rows may be flanked on each side by prominent plates, termed **adradials.** *Asterias,* however, has **dorso-lateral fields** (*12*) comprising the sides which are armored by a latticework of small plates.

The young starfish is disk shaped; in aboral view it may show a central plate surrounded by five interradials and five terminals. The radials and other plates of the arms are inserted during further growth.

Spines and Pedicellariae. Most spines of the aboral surface are mere projections of the skeletal plates. Spines on the underside may be articulated with the supporting plate by means of a ball-and-socket joint and moved by muscles. Many starfishes also possess distinctive spines called **paxillae** (*55*), consisting of a small shaft topped by a rosette or bundle of secondary spines.

Pedicellariae are microscopic calcareous pincers which are an effective means of defense. Large specimens of *Asterias* possess more than a quarter million of them. Six major types of pedicellariae are recognized. They are one of the chief characters on which modern starfishes are classified but are almost unknown in the fossil state.

Symmetry and Regeneration. Most eleutherozoan echinoderms are built on a five-rayed plan and show striking fivefold symmetry. This is true for most of the asteroids, but many species have more than five rays or arms, and some exceed forty in number. Also, four- or six-rayed individuals are not uncommonly found in normally five-armed species.

Many individuals have one or more arms which are shorter than others. This is due to accidental or intended loss, followed by regeneration. Most starfish can regenerate complete arms from the disk, and a few are able to regenerate a complete new starfish from a single arm.

Mode of Life

Asteroids of modern seas fill an ecological niche which is peculiarly their own. They slowly crawl over the sea floor in search of even more slow-moving or sessile prey. Movement is accomplished by the numerous tube feet. These act like little stilts when moving over sand. By means of their suckers, they enable the starfish to cling to rocks in violent surf and to ascend submarine cliffs.

Starfishes are exceedingly voracious. They eat a wide variety of benthonic animals, but clams are generally the chief article of food. Small prey is swallowed whole, but some starfishes are able to extrude the stomach through the mouth for external digestion of objects too large to be introduced into the body. Large clams, for example, may be held with the tube feet and killed and digested within their own shells by the powerful gastric secretions of the asteroid.

Starfishes are among the toughest of echinoderms; they are able to stand environments as rigorous as that of the intertidal zone with its variable salinity and temperature and intermittent exposure to the atmosphere.

Geological Distribution and Evolution

The asteroids were probably developed from somasteroids in Ordovician time. They have persisted to the present day and are an important part of marine faunas. Their plate systems, formerly believed to be inherited from echinoid-like ancestors, are largely absent in the most primitive starfishes. Therefore, they are now interpreted as a secondary acquisition, convergent toward echinoid tests.

As might be expected in a large, actively evolving group, the morphology of starfishes changed greatly during geologic time, and many offshoots came to differ widely from the norm. Some starfishes, such as *Cheiropteraster*, of Devonian age (Fig. 20-6, *4*), lost the starlike form. Others which show a trend toward greater mobility converged toward ophiuroids by withdrawing the body cavity and organs from the arms and pushing the ambulacral ossicles into the middle (Fig. 20-4, *6–8*). A main change is seen in the placement of ampullae. Among Paleozoic asteroids, the ampullae are typically accommodated in depressions (ampulla cups) within the ambulacral ossicles. In later starfishes, the ampullae lie within the body cavity, and the tube feet extend out through ambulacral pores, openings between adjacent ambulacrals.

Various starfishes have deviated from the basic five-rayed plan in the development of many more arms. Such forms are found in Silurian and Devonian rocks and are common today.

OPHIUROIDS

Members of the subclass Ophiuroidia reach great abundance on many sea bottoms. Instead of having a bulky body and thick arms and crawling about on tube feet, the ophiuroids have a well-defined

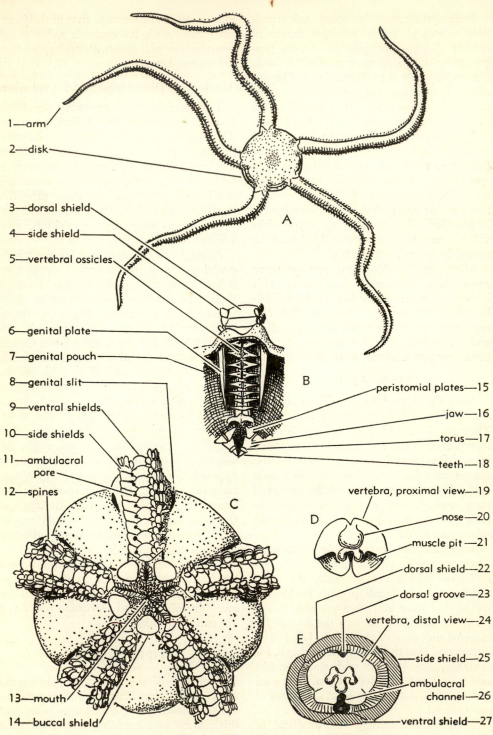

1—arm
2—disk
3—dorsal shield
4—side shield
5—vertebral ossicles
6—genital plate
7—genital pouch
8—genital slit
9—ventral shields
10—side shields
11—ambulacral pore
12—spines
13—mouth
14—buccal shield

A

B

peristomial plates—15
jaw—16
torus—17
teeth—18

C

vertebra, proximal view—19
nose—20
muscle pit—21
dorsal shield—22
dorsal groove—23
vertebra, distal view—24
side shield—25
ambulacral channel—26
ventral shield—27

D

E

FIG. 20-3.

central body disk, surrounded by slender arms which propel the animal by snakelike movements. For this reason, ophiuroids are called "wrigglers" in contrast to asteroids, termed "crawlers."

Structure

The structure of ophiuroids differs widely from that of asteroids. A good example for study is the Recent *Ophioderma panamensis* (Fig. 20-3; italic numbers in the following description refer to this figure). *Ophioderma* is common under rocks in the lower part of the intertidal zone along the Pacific coast. Unlike many other ophiuroids, which merit the term "brittle star" in shedding their arms on slight provocation, *Ophioderma* may be handled without breaking in pieces.

Disk. The central part of the body is disk shaped (*2*). On the upper (**aboral**) side, it is armored only with small granules. This side contains no openings, for an anus is lacking and the inconspicuous slit or plate called **madreporite** is located inter-

radially on the underside. The **mouth** (*13*) is a star-shaped opening in the middle of the under (**oral**) side. It serves for the intake of food and for excretion. The arms are traceable to the periphery of the mouth. Segments of the disk lying between the arms are partitioned off from the body cavity into five **genital pouches** (*7*) or bursae, which open outward by **genital slits** (*8*) located alongside the arms. These pouches function chiefly for respiration, water being drawn in and expelled through the genital slits, either by ciliary currents or by a breathing mechanism. The genital pouches are so named because in some species they have taken on the secondary function of retaining eggs and young. The pouches also serve in excretion, as waste materials diffuse into them from the body cavity and are dumped into them by special roaming cells.

The skeleton of the disk consists of included parts of the arms; the **mouth frame** (*15–18*); **genital plates** (*6*) or scales between the genital pouches and arms;

Fig. 20-3. **Structure of ophiuroids.** Morphologic features of brittle stars are illustrated by *Ophioderma panamensis*, a modern ophiuroid. Structures illustrated are defined in the appended list of terms, which is alphabetically arranged and numerically cross-indexed to the figure. (A) Aboral view, ×0.7. (B) Internal structure, viewed from above, ×3. (C) Oral side, ×3. (D) Arm ossicle, proximal view. (E) Cross section of arm, distal view.

ambulacral channel (26). Canal which houses radial water vessel.

ambulacral pore (11). Passage for tube foot.

arm (1). Flexible extension of body, in ambulacral position, used in locomotion and feeding.

buccal shields (14). Five large plates surrounding mouth.

disk (2). Central part of body, housing main organs.

dorsal groove (23). Canal in aboral (dorsal) part of arms.

dorsal shields (3, 22). Series of plates along aboral (dorsal) side of arms.

genital plates (6). Bladelike plates in disk, lying alongside arms.

genital pouches (7). Interambulacral cavities in disk, serving for respiration and, among some ophiuroids, for brood pouches.

genital slits (8). Openings of genital pouches, through which water is "breathed" in and out.

jaws (16). V-shaped ossicles which are largest members of mouth frame.

mouth (13). Star-shaped opening in center of underside.

muscle pits (21). Cavities in arm ossicles for insertion of powerful arm muscles.

noses (20). Articulating projections on arm ossicles.

peristomial plates (15). Small triangular plates overlying jaws.

side shields (4, 10, 25). Plates covering sides of arms.

spines (12). Small rods or scales attached to side shields of arms.

teeth (18). Small scales on mouth frame.

torus (17). Plate located at jaw tip.

ventral shields (9, 27). Plates which armor underside of arms.

vertebral ossicles (5, 19, 24). Ossicles which form a backbone-like row in core of each arm.

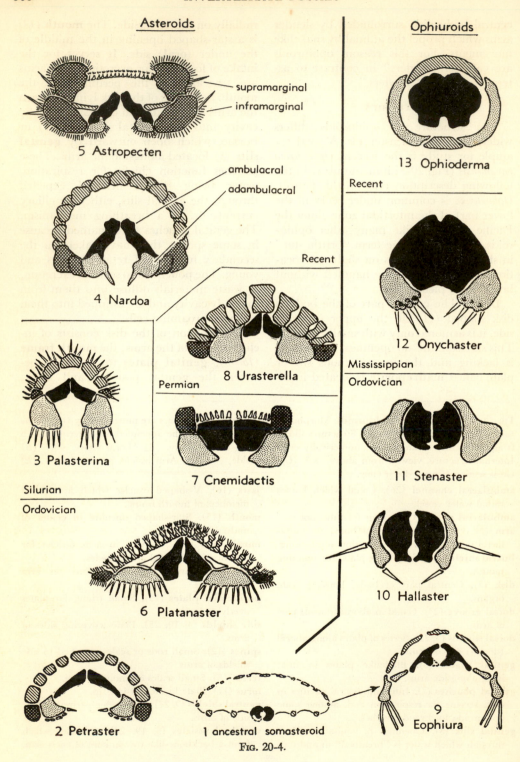

Asteroids

Ophiuroids

5 Astropecten

supramarginal
inframarginal

13 Ophioderma

Recent

4 Nardoa

ambulacral
adambulacral

Recent

8 Urasterella

12 Onychaster

3 Palasterina

Silurian

Ordovician

Permian

7 Cnemidactis

Mississippian

Ordovician

11 Stenaster

6 Platanaster

10 Hallaster

2 Petraster

1 ancestral somasteroid

9 Eophiura

Fig. 20-4.

buccal shields (*14*), large plates surrounding the mouth in interradial position; and many small surficial plates and scales. The mouth frame contains five V-shaped **jaws** (*16*), like the mouth-angle plates of the asteroids. At the tip of each jaw is a small plate termed **torus** (*17*), which carries many small **teeth** (*18*). The jaws have limited motion. The "teeth" of some forms act as strainers, but their function in most is not clear. Flat triangular ossicles above the jaws are termed **peristomial plates** (*15*).

Arms. The arms of ophiuroids differ greatly from those of asteroids. The interior is not hollow but filled with a series of articulating ossicles resembling the vertebrae of a backbone. These **vertebrae** (*5, 19, 24*) articulate with each other by means of ball-and-socket joints. The vertebrae are connected by four sets of powerful longitudinal muscles, especially strong on the ventral side where they are inserted in muscle pits (*21*) on the proximal side of the ossicles (toward the disk). Each vertebra is notched by a dorsal groove (*23*) and bears a deep ventral groove (**ambulacral channel, *26***) containing the radial water vessel, nerve, and blood vessel. Each vertebra starts out as two little ossicles lying side by side, which fuse during ontogeny. This structure is equivalent to a pair of ambulacrals in somasteroids and asteroids.

The vertebrae of *Ophioderma* are completely incased in an outer skeleton consisting of **dorsal shields** (*3, 22*), **ventral shields** (*9, 27*), and **side shields** (*4, 10, 25*). Each side shield bears a row of small scalelike **spines,** which serve to brace the arm against sediment.

The tube feet of ophiuroids do not lie between ambulacral plates, as in asteroids and somasteroids, but arise in pairs from each vertebra, which has two pits on the ventral side for their reception. The watervascular system is less strongly developed than among the asteroids; the tube feet, which project through ambulacral pores (*11*) between ventral and side shields, lack suckers. They serve as respiratory and sensory organs and in handling food.

Mode of Life

Ophioderma lives under rocks. Most other brittle stars also live a rather secretive life, burrowing in soft sediments and extending only the arms to the surface. Some have learned to swim, and some climb around on seaweeds. Locomotion is accomplished mainly by snakelike wriggling of the arms, usually two arms for pulling and three for pushing. When disturbed, most brittle stars cast off parts of their arms, which continue to squirm about; this may detain the pursuer while the intended victim escapes. If hard pressed, all arms and even parts of the disk may be shed, but in time these can be regenerated. Pedicellariae and other organs of defense are lacking.

Many brittle stars feed on detrital bits of organic matter, while others are carnivorous. Large chunks of food may be grasped in a coil of the arm tip and conveyed to the mouth by further coiling; small food particles, however, are transported by being passed along the underside of the arms from tube foot to tube foot.

Geological Distribution and Evolution

The ophiuroids were almost certainly derived from primitive somasteroids in or before Early Ordovician time. An evolutionary sequence leads from the Tremad-

FIG. 20-4. **Arm structure in stelleroids.** Arm cross sections of selected genera show the range of variation in arm structure. Homologous ossicles are shown in similar patterns (labeled in 4 and 5).

1, Ancestral type (somasteroid).
2–5, Normal asteroids having hollow arms.
6–8, Asteroids showing convergence toward ophiuroids.

9–13, Ophiuroids with arms built around a core of vertebral ossicles.

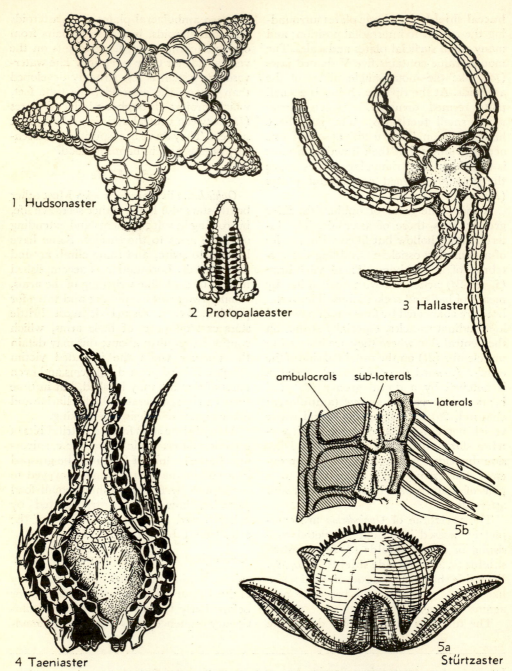

1 Hudsonaster

2 Protopalaeaster

3 Hallaster

ambulacrals   sub-laterals

laterals

5b

4 Taeniaster

5a
Stürtzaster

FIG. 20-5.   **Representative Ordovician and Silurian stelleroids.**

Hallaster Stürtz, Ordovician–Devonian. Ophiuroid. The arms, built around articulating ossicles, mark this as a true ophiuroid. The genus differs from modern brittle stars in having paired vertebral ossicles, and in lacking dorsal and ventral shields. *H. cylindricus* Billings (3, ×0.3), Ordovician (Trentonian), Canada. **Hudsonaster** Stürtz, Middle and Upper Ordovician. Asteroid. A simple asteroid, having *(Continued on next page.)*

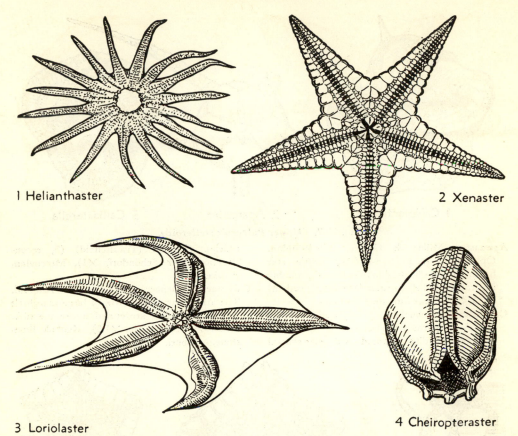

1 Helianthaster

2 Xenaster

3 Loriolaster

4 Cheiropteraster

FIG. 20-6. **Representative Devonian stelleroids.**

Cheiropteraster Stürtz, Devonian. Asteroid. One of the few nonstellate starfishes, having a balloon-shaped body. *C. giganteus* Stürtz (4, ×0.3), Bundenbach slate, Germany.

Helianthaster Roemer, Devonian. Asteroid. This multiarmed form bears only small scales on the aboral side. *H. rhenanus* Roemer (1, ×1), Bundenbach slate, Germany.

Loriolaster Stürtz, Devonian. Asteroid. Like *Cheiropteraster*, this form had a greatly swollen disk lacking armor between the arms; preserved as carbon film in the Bundenbach slate. *L. mirabilis* Stürtz (3, ×0.5).

Xenaster Simonovitch, Lower Devonian. Asteroid. The greatly developed inframarginals form a strong peripheral buttress around this starfish. *X. margaritatus* Simonovitch (2, ×1), Germany.

(Fig. 20-5 continued.)

the aboral surface armored with regularly arranged plates. *H. incomptus* (Meek) (1, ×3), Cincinnatian, Ohio.

Protopalaeaster Hudson, Middle Ordovician–Silurian. Asteroid. Differs from *Hudsonaster* in having differentiated, polygonal inframarginals. *P. matutinatus* (Hall) (2, ×2), Middle Ordovician (Trentonian), New York and Quebec.

Sturtzaster Etheridge, Silurian–Devonian. Asteroid. A primitive asteroid having ambulacrals bordered by sublaterals and laterals. *S. marstoni*

(Salter) (5a, reconstruction after Spencer, ×2; 5b, detail of arm, ×10), Silurian (Ludlovian), England.

Taeniaster Billings, Ordovician–Devonian. Ophiuroid. This tiny brittle star had a bulbous disk and, like most early stelleroids, a very large mouth area. Specimens preserved in life position show that this form lived with the disk buried, extending the up-arched arms to the sea floor. *T. spinosus* Billings (4, ×7), Ordovician Trentonian), Quebec.

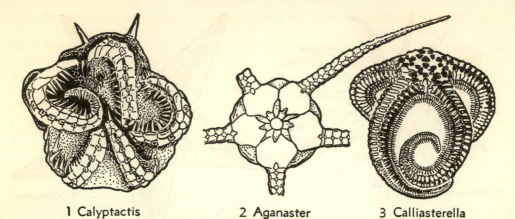

1 Calyptactis   **2 Aganaster**   **3 Calliasterella**

FIG. 20-7. **Upper Paleozoic stelleroids.**

Aganaster Miller & Gurley, Mississippian. Ophiuroid. A modern-looking brittle star having dorsal shields, side shields, and probably ventral shields. *A. gregarius* (Meek & Worthen) (2, ×4), Keokuk limestone, Indiana.

Calliasterella Schuchert, Pennsylvanian. Asteroid. A long-armed starfish having the aboral side of the disk armored with star-shaped ossicles. *C. mira* (Trautschold) (3, reconstruction after Schöndorf, ×1), Moscovian, Russia.

Calyptactis Spencer, Mississippian. Asteroid. The arms bear a single row of large marginals on each side and are curled under the disk. *C. demissus* Miller (1, ×1.5), Keokuk limestone, Missouri.

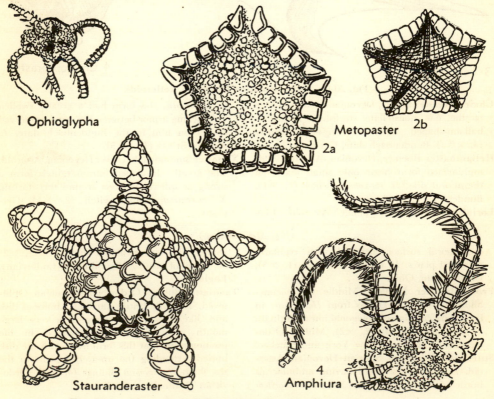

1 Ophioglypha   **Metopaster** **2b**
                                              2a

3
Stauranderaster

4
Amphiura

FIG. 20-8. **Mesozoic and Cenozoic stelleroids.** (*Continued on next page.*)

ocian *Villebrunaster* (somasteroid) to the highly specialized modern type of ophiuroid which first made its appearance in the Devonian. Differentiation of the ophiu-

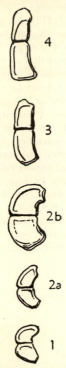

FIG. 20-9. **Evolution of Metopaster.** Scattered ossicles are the most common fossil remains of starfishes. Large ossicles, such as the inframarginals and supramarginals of *Metopaster* and allied genera, are conspicuous and abundant in the Cretaceous chalks of Europe. Careful study of these remains by Spencer has shown that the marginals of *Metopaster* show a distinct evolutionary trend. The marginals illustrated (1–4) are arranged in stratigraphic order as collected

roid stock began with loss of virgalia and with the extension of ambulacral ossicles into the arm cavity, leading through intermediate forms typified by *Eophiura* (Fig. 20-4, *9*) to *Hallaster* (Fig. 20-4, *10*). Additional ossicles were developed alongside the ambulacrals, in a trend paralleling the evolution of asteroids. Early ophiuroids, such as *Eophiura*, show two rows of such plates, termed **infralaterals** and **laterals.** In more advanced forms (*Hallaster* and others) these plates have fused to form the side shields. In early Ordovician time, ball-and-socket joints and longitudinal muscles were concurrently developed between successive ossicles to give the arms mobility and power.

The originally paired ambulacrals are fused in some Devonian and Upper Paleozoic and in all post-Paleozoic ophiuroids. In most Paleozoic forms the radial water vessel is enclosed by the vertebrae, but in later brittle stars it is located in a ventral ambulacral groove.

Early Paleozoic ophiuroids lacked ventral and dorsal shields. These structures first appeared in Devonian time and have become almost universal.

from successive levels in the Upper Cretaceous; they show an evolutionary progression from low curved profile to high straight form.

1, *Metopaster thoracifer*, from zone of *Rhynchonella cuvieri.*
2a, b, *Metopaster parkinsoni;* (2a) from zone of *Micraster coranguinus;* (2b) from zone of *Offaster pilula.*
3, *Metopaster tumidus*, from zone of *Belemnitella mucronata.*
4, *Metopaster mamillatus*, from the Danian.

(Fig. 20-8 continued.)

Amphiura Forbes, Tertiary–Recent. Ophiuroid. The arms of this genus bear comparatively long pointed spines. Some of the Recent forms are hermaphroditic. *A. sanctaecrucis* Arnold, (4, ×0.7), Miocene, California.
Metopaster Sladen, Cretaceous. Asteroid. One of the most common Cretaceous genera, lacking distinct arms and having greatly developed inframarginals and supramarginals. *M. uncatus* Forbes (2a, aboral view; 2b, oral view, ×0.7),

Upper Cretaceous, England. Other species are illustrated in Fig. 20-9.
Ophioglypha Lyman, Jurassic–Recent. Ophiuroid. The aboral surface is armored by small imbricating scales and by five larger radial shields. *O. utahensis* Clark (1, ×2), Jurassic, Utah.
Stauranderaster Sladen & Spencer, Cretaceous. Asteroid. A heavily armored starfish which has expanded arm tips. *S. bulbiferus* Sladen & Spencer (3, ×0.7), Upper Cretaceous, England.

REFERENCES

CUÉNOT, L. (1948) *Les Asterozoaires:* in Grassé, P., Traité de zoologie, Masson & Cie., Paris, vol. 11, pp. 200–270, figs. 236–309 (French). An excellent account of modern starfishes.

MACBRIDE, E. W. (1909) *Echinodermata:* in Harmer, S. F., & Shipley, A. E., Cambridge natural history, Macmillan & Co., Ltd., London, vol. 1, pp. 425–502, figs. 185–222.

RASMUSSEN, H. W. (1950) *Cretaceous Asteroidea and Ophiuroidea with special reference to the species found in Denmark:* Danmarks geol. Undersøgelse, ser. 2, no. 77, pp. 1–134, pls. 1–18, figs. 1–8. Descriptions of Cretaceous starfishes.

SCHUCHERT, C. (1915) *Revision of Paleozoic Stelleroidea with special reference to North American Asteroidea:* U.S. Nat. Mus. Bull. 88, pp. 1–311, pls. 1–33, figs. 1–41. A summary of Paleozoic starfishes, and the standard reference for North American ones, but outdated in classification and by more recent discoveries in Europe.

SLADEN, W. P., & SPENCER, W. K. (1891–1908) *A monograph on the British fossil Echinodermata from the Cretaceous formations, vol. 2, The Aster-* oidea and Ophiuroidea: Paleont. Soc., pp. 1–137, pls. 1–29, figs. 1–34. A standard reference for Cretaceous stelleroids.

SPENCER, W. K. (1913) *The evolution of the Cretaceous Asteroidea:* Royal Soc. London Phil. Trans., ser. B, vol. 204, pp. 99–177, pls. 10–16, figs. A–E. A great deal of biologic and stratigraphic information is here derived from extensive study of scattered starfish ossicles.

——— (1914–1950) *A monograph of the British Paleozoic Asterozoa:* Palaeont. Soc., pp. 1–540, pls. 1–37, figs. 1–348. By far the most extensive work on fossil starfishes, including excellent discussion of general structure, illustrations, and descriptions of all known Paleozoic genera.

——— (1951) *Early Paleozoic starfish:* Royal Soc. London Philos. Trans., ser. B, no. 623, vol. 235, pp. 87–129, pls. 2–9, figs. 1–28. Description of the Lower Ordovician starfish faunas of France and Czechoslovakia and reclassification of Paleozoic starfishes in the light of this new material.

CHAPTER 21

Echinoids

Echinoids are globular, discoidal, heart-shaped, or (rarely) subcylindrical animals, which are armored by a subcutaneous box-like skeleton composed of many plates (Fig. 16-1). This skeleton or test encloses the vital organs, and is studded on the outside with movable spines (Figs. 21-1; 21-3). It is composed of five ambulacral and five interambulacral areas, each made up by two or more rows of plates. The ambulacral plates are perforated for the tube feet of the water-vascular system. The place of echinoids among echinoderms as a whole has been indicated in Chap. 16.

The class Echinoidea is subdivided into two subclasses: the Regularia, in which the anus is located opposite the mouth, within a circle of plates termed the apical system or oculogenital ring (Fig. 21-3, *19*), and the Irregularia, among which the anus has left this central position and has come to lie posterior to the oculogenital ring (Fig. 21-22).

REGULAR ECHINOIDS

The regular echinoids were given their name because they show, with few exceptions, a pronounced radial (fivefold) symmetry, in contrast to the evident bilateral symmetry of the irregular echinoids.

Modern echinoids, like fossil ones, are classified chiefly by means of their skeletal structures, which are amongst the most complex encountered in invertebrates, and

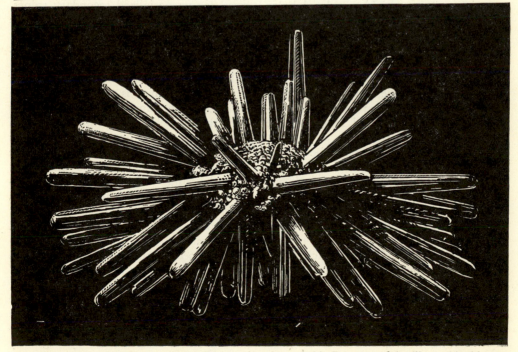

Fig. 21-1. **Slate-pencil sea urchin.** The great spines of this Recent reef-dwelling echinoid were formerly used for writing on slates in school.

have recorded many and diversified evolutionary changes.

Morphology

Let us first examine the skeletal structures of regular echinoids in the light of a selected example, *Stereocidaris*, which is a representative of the family Cidaridae, one of the oldest branches of the regular echinoids. Fossil representatives of this genus are known from the Cretaceous and Eocene of Eurasia, and it flourishes today in the Indo-Pacific region. *Stereocidaris tubifera* Mortensen, a living species, is illustrated in Fig. 21-3 (italic numbers listed after the terms in the following discussion refer to this figure unless otherwise designated).

Stereocidaris tubifera is about the size of an orange, studded with numerous long, slender spines and a multitude of short ones. The color of the skin-covered exterior is creamy white.

Corona. The main part of the skeleton consists of calcareous plates, fused at their margins to form a rigid, nearly globular shell. This shell, an object of exceptional beauty in its symmetry and ornamentation, is termed the **corona** (crown). It contains two large openings at opposite poles: the **peristome** (around the mouth, *36*) on the **oral** (under) side, and the **periproct** (around the anus, *5*) on the **aboral** (upper) side. The equatorial region of the corona is termed the **ambitus.**

The plates which compose the corona are arranged in 10 double rows: five double rows of large **interambulacral plates** (*C*), forming **interambulacral areas** (*24*), alternate with five double rows of small **ambulacral plates** (*B*), forming **ambulacral areas** (*21*).

Interambulacral Plates. Each interambulacral plate bears a shallow smooth pit, the **areole** (*14*), within which is a conical **boss** (*33*) surmounted by a globular **mamelon** (*34*). The boss and its mamelon form a **primary tubercle** (*26*). The mamelon has a dimple in the center, and therefore is termed **perforate.** The areole is surrounded by a ring of small **scrobicular tubercles** (*15*). Still smaller tubercles, grading in size down to very small ones lacking an areole, are scattered over the surface of the plate; the larger ones of these are classed, along with the scrobicular tubercles, as **secondary tubercles,** whereas the nearly microscopic ones are termed **miliary tubercles** (*16*) or granules.

The function of this beautiful beading of the plates is not an ornamental one. Each tubercle, hidden on the living animal by a cover of muscles and skin, bears a spine or modified spine.

Interambulacral Spines. Different types of spines correspond to the various kinds of tubercles. The primary tubercles carry great **primary spines** (*1*), or **radioles,** up to 10 cm. long, which serve for defense and for walking. The scrobicular tubercles carry much shorter, flattened **scrobicular spines** (*2*), which surround the bases of the primary spines in order to protect the muscles by which the latter are moved. The other secondary tubercles serve as bases for unspecialized secondary spines, generally less than 5 mm. long, and the miliary granules carry either tiny spines, or **pedicellariae,** wondrously complex microscopic structures, discussed below.

If we examine one of the primary spines, we find that its base is concave, forming a socket (**acetabulum,** *13*) which articulates with the mamelon of a primary tubercle. Above the acetabulum, the spine expands into a **milled ring** (*30*), to which the muscles are attached. Above the milled ring, the **collar** (*29*) of the spine tapers to the smooth cylindrical **neck** (*28*); that is succeeded by the rough, longitudinally striated **shaft** (*27*), which makes up most of the spine.

A thin section through the shaft reveals a beautiful, complex microstructure (Fig. 21-2). In the center lies an axial zone, which is an irregular meshwork of calcite rods. This is surrounded by perforate **septa** of calcite, which radiate toward the exterior and are connected by regularly spaced cross members. The outermost portion of the spine, termed the **cortex,** is more solid, and extending outward from its surface are delicate little frills and "hairs" of calcite. Like other echinoderm ossicles, this entire complex structure behaves optically as a single crystal unit.

The spine is moved by muscles, which are attached at one end to the milled ring and anchored at the other in the marginal depression of the areole (**scrobicule,** *35*). Young spines are covered with skin, but this cover is worn away from older ones, in which living tissues remain only in the abundant microscopic cavities permeating the spine (Fig. 21-2).

Ambulacral Plates. The ambulacral areas are modified chiefly for the accommodation of the water-vascular system (Chap. 16). In *Stereocidaris,* as well as all other living and most fossil echinoids, the radial water vessels lie inside the test, along mid-lines of the ambulacral areas. Branches lead off on either side to contractile sacs (**ampullae**), whence a pair of tubes pass through the **ambulacral pores** (*8*). These tubes combine on the outside into a single **tube foot** or **podium.** Thus, each ambulacral plate bears two ambulacral pores but only one tube foot. The shell matter separating the two pores of a pair is termed the **wall** (*9*). **Ridges** (*10*) separate the pore pairs of adjacent plates. The pores form double rows, **poriferous zone** (*7*), near the outer margins of the ambulacral plates, leaving an **interporiferous zone** (*6*) in the middle.

The tube feet of the oral side function in grasping and locomotion. Each tube foot is actually tubular, and its end is expanded into a suction cup. Water forced into it by contraction of the ampulla causes the tube foot to become extended. The foot may attach itself to an object by means of its sucker disk, and may then be contracted by means of the muscles in its wall. This results either in a movement of the object toward the echinoid, or, vice versa, in a movement of the echinoid toward the object. The tube feet of the aboral side are modified into bladelike structures

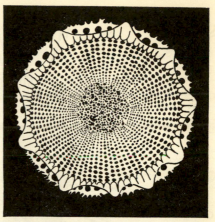

Fig. 21-2. **Spine in cross section.** A transverse thin section of a primary spine of *Stereocidaris japonica* ($\times 30$) shows the three main zones characteristic of cidaroid spines: the irregular meshwork of the core, the radiating spokes of the intermediate region, and the dense cortex or outer layer. Lacy frills and "hairs" project beyond this. (*After Mortensen.*)

which function in respiration. The fluid in the water-vascular system is kept in circulation by the beating of internal cilia. These cilia force fluid out of the ampulla into the tube foot through one of its branches. The fluid is oxygenated by diffusion through the walls of the tube foot, and is circulated back down into the ampulla through the other branch of the tube foot. From the ampulla the oxygen diffuses into the fluids and various organs of the body cavity.

Ambulacral Tubercles and Spines. Each ambulacral plate bears a prominent **marginal tubercle** (*11*) on the boundary between the poriferous and interporiferous zones, and small **secondary tubercles** (*12*) scattered over the latter. The marginal tubercles carry flat **marginal spines** (*3*), similar to the scrobicular spines of the interambulacral plates; these marginal spines may close above the poriferous zones to form a shielding roof over the tube feet. The secondary tubercles support tiny secondary spines.

Oculogenital Ring. At its upper (aboral) extremity, each ambulacral area is

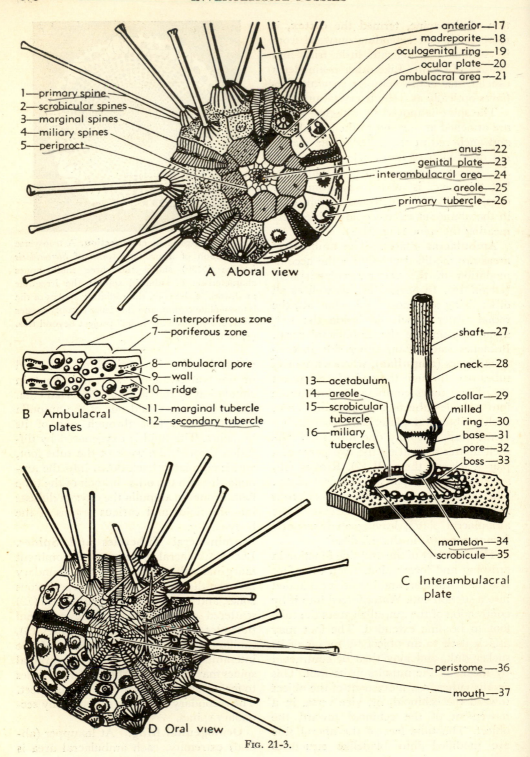

1—primary spine
2—scrobicular spines
3—marginal spines
4—miliary spines
5—periproct

anterior—17
madreporite—18
oculogenital ring—19
ocular plate—20
ambulacral area—21

anus—22
genital plate—23
interambulacral area—24
areole—25
primary tubercle—26

A Aboral view

6— interporiferous zone
7—poriferous zone

8— ambulacral pore
9— wall
10— ridge
11— marginal tubercle
12—secondary tubercle

B Ambulacral plates

13—acetabulum
14—areole
15—scrobicular tubercle
16—miliary tubercles

shaft—27
neck—28
collar—29
milled ring—30
base—31
pore—32
boss—33

mamelon—34
scrobicule—35

C Interambulacral plate

peristome—36

mouth—37

D Oral view

Fig. 21-3.

Fig. 21-3. **Structural characters of regular echinoids.** Morphological features of *Stereocidaris tubifera* are illustrated and defined in the alphabetically arranged list of terms, cross-indexed to the figure by numbers (*A, D,* ×0.7; *B, C,* enlarged). Ambulacral areas shown in black; oculogenital ring shaded.

acetabulum (13). Articulating socket at base of spine.

ambulacral areas (21). Five narrow bands extending from periproct to peristome, composed of a double row of *ambulacral plates.*

ambulacral plates (*B*). Skeletal elements which bear pores for passage of tube feet.

ambulacral pores (8). Perforations in each ambulacral plate, serving as passageways for a tube foot.

anterior (17). Direction or position of interambulacral area at left of madreporite.

anus (22). Outlet of digestive tract, located at center of periproct, opposite mouth.

areole (14, 25). Broad, shallow pit in each interambulacral plate, bearing a large tubercle which serves as base for a primary spine.

base (31). Portion of a spine below milled ring.

boss (33). Cone which supports mamelon of a tubercle.

collar (29). Smooth, tapering portion of a spine, located above milled ring.

genital plate (23). Large plate in each interambulacral area, bordering periproct, and bearing a *genital pore.*

interambulacral areas (24). Five broad bands extending from peristome to periproct, each composed of a double row of large *interambulacral plates.*

interambulacral plate (*C*). Large, imperforate skeletal element composing larger part of interambulacral areas; each plate carries a primary spine.

interporiferous zone (6). Mid-region of each ambulacral area, between pore-bearing margins of ambulacral areas.

madreporite (18). Right anterior genital plate, which has been modified into a sievelike intake for water-vascular system.

mamelon (34). Spheroidal summit of a tubercle.

marginal spines (3). Flat, bladelike spines, occurring singly on ambulacral plates, protecting tube feet.

marginal tubercle (11). Small elevation near an ambulacral pore, for attachment of *marginal spine.*

miliary spines (4). Tiny spines scattered over surface of interambulacral and oculogenital plates.

miliary tubercles (16). Minute elevations which bear miliary spines.

milled ring (30). Flange near base of a spine, serving for attachment of muscles that move spine.

mouth (37). Opening located in center of underside.

neck (28). Smooth cylindrical portion of a primary spine, lying between collar and shaft.

ocular plate (20). Skeletal element adjoining periproct in ambulacral position, and forming, with genital plates, the *oculogenital* ring.

oculogenital ring (19). Circlet of 10 plates (5 ambulacral *oculars,* and 5 interambulacral *genitals,* including one modified as *madreporite*), which surround periproct.

periproct (5). Area surrounding anus and enclosed by oculogenital ring; covered with leathery skin in which small plates are embedded loosely. (Aboral)

peristome (36). Area surrounding mouth, covered with leathery skin studded with small plates. (Oral)

pore (in a primary tubercle) (32). Pit for attachment of a ligament which fastens spine to tubercle.

poriferous zones (7). Pore-bearing outer edges of ambulacral areas, as contrasted with nonperforate strip which extends down middle of each ambulacral area.

primary spines (1). Large movable projections which occur singly on interambulacral plates.

primary tubercles (26). Prominent rounded elevations which bear primary spines.

ridge (10). Area separating adjacent pairs of ambulacral pores.

scrobicular spines (2). Flat spines arranged in a ring around scrobicule of each areole, protecting muscles which move large primary spine.

scrobicular tubercles (15). Small elevations for attachment of *scrobicular spines* encircling areoles.

scrobicule (35). Depressed marginal area of an areole, or any depressed ring around base of a tubercle, serving for attachment of muscles which move spine.

secondary tubercles (12; also 11, 15). Tubercles which carry small (secondary) spines.

shaft (27). Main part of a spine.

wall (9). Shell matter separating two ambulacral pores of a pair.

Insert

Exsert

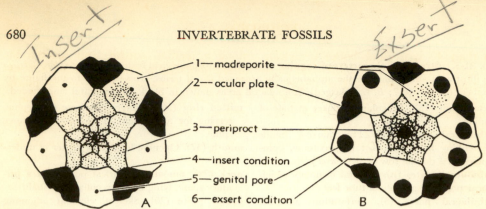

1 — madreporite
2 — ocular plate
3 — periproct
4 — insert condition
5 — genital pore
6 — exsert condition

A   B

FIG. 21-4. **Apical system.** The oculogenital ring, surrounding the periproct (3) consists of ambulacral ocular plates (2) and interambulacral genital plates. The genital plates contain the *genital pores* (5). The left anterior genital plate is also the *madreporite* (1). Apical systems are classified as *insert* (4) if the ocular plates extend to the periproct, *exsert* (6) if they do not. Size of the genital pores marks *A* as male, *B* as female. *B* is anomalous in possessing two right posterior genital plates. (*A*) *Schizocidaris assimilis*, Recent, ×3. (*B*) *Goniocidaris sibogae*, Recent, ×3. (*After Mortensen.*)

terminated by a single **ocular plate** (*20* and Fig. 21-4, *2*) and each interambulacral area by a single **genital plate** (*23*). These 10 plates form the **oculogenital ring** (*19*), which surround the periproct. Each of the ocular plates bears a small ocular pore. Each genital plate is perforated by a **genital pore** (Fig. 21-4, *5*), through which eggs or sperm are discharged. The genital pores of the females are necessarily larger than those of the males, and thus afford a skeletal character for distinction of sexes (Fig. 21-4*A*, *B*). One of the genital plates, termed the **madreporite** (*18* and Fig. 21-4, *1*), is riddled with microscopic pores and acts as a sieve over the intake of the water-vascular system.

Periproct. The oculogenital ring surrounds an area covered by a leathery membrane, in which numerous small irregular plates are loosely embedded. This structure, termed the **periproct** (*5* and Fig. 21-4, *3*), surrounds the **anus** (*22*), which is located at the aboral pole of the animal. The periproct generally is not preserved in fossils.

Peristome. The leathery area on the oral side of the animal is termed the **peristome** (*36*). It is armored by loosely imbricating plates which form a double row in each ambulacral and interambulacral area. They represent coronal plates which

have become detached and have migrated onto the peristome during ontogeny; accordingly, the ambulacral ones are perforate. The peristome surrounds the central, circular **mouth** (*37*). Like the periproct, the peristome generally is not preserved in fossils.

Perignathic Girdle. Inside the test, two plates project toward the interior from the peristomial margin of each interambulacral area (Fig. 21-5). These projections are termed **apophyses**. Collectively, they comprise the **perignathic girdle.** It serves to anchor muscles which raise and lower the complex jaw system.

Jaws. The masticatory apparatus of echinoids (Fig. 21-6) is an intricate mechanical device without parallel among other groups of animals. It consists of no

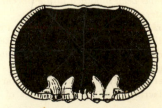

FIG. 21-5. **Perignathic girdle.** Internal projections around the peristome, which serve to anchor the jaw muscles, are termed the perignathic girdle. Among the Cidaridae, the projections of the perignathic girdle lie in the interambulacral areas and are called *apophyses*.

less than 40 individual bones (ossicles), operated by more than 60 individual muscles arranged in seven sets. First described by Aristotle, and bearing a fancied resemblance to a lantern, the structure has come to be known as *Aristotle's lantern.*

The ossicles are arranged in five identical groups. The largest members are the **pyramids** (Fig. 21-6, *6*), hollow, fanglike structures composed of two halves joined along a median suture, and located in interambulacral position. They function as jaws. Each contains a long, rodlike **tooth** (Fig. 21-6, *2*), the lower hard end of which projects like a chisel from the pointed extremity of the pyramid. Hinged at the top, the pyramids may swing their armored tips toward the common center or away from it, thus effecting a gnawing or crushing action. The upper ends of the

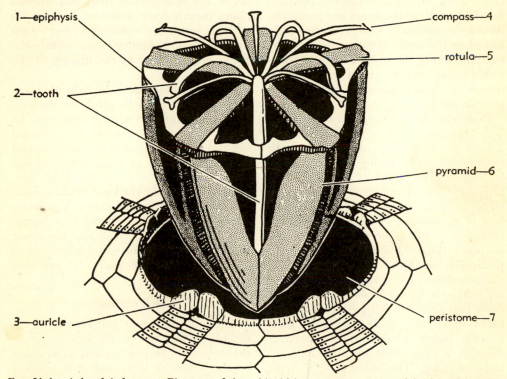

FIG. 21-6. **Aristotle's lantern.** Elements of the echinoid jaw apparatus are explained in the alphabetically arranged list which is cross-indexed by numbers to the figure. Cidaroids and most other echinoids possess lanterns built on the plan illustrated, but in cidaroids the projections around the peristome are interambulacral *apophyses* rather than ambulacral auricles, and adjacent epiphyses of the lantern are not joined.

auricle (3). Internal projection for attachment of lantern muscles, located in each ambulacral area at edge of peristome.

compass (4). Slender arched radial rod in ambulacral position, at top of lantern.

epiphysis (1). Rugged cross member composed of two fused ossicles; occurs in each interambulacrum at top of lantern.

peristome (7). Membranous area around mouth.

pyramid (6). Large beaklike structure in interambulacral position, composed of two fused halves; carries tooth.

rotula (5). Massive radial strut in ambulacral position.

tooth (2). Rod located in each pyramid, having an uncalcified arched upper end but hard, nearly straight lower end; only the chisel-edged tip protrudes from lower end of pyramid.

teeth are soft, uncalcified, and are looped over the tops of the pyramids. As the lower end of a tooth is worn away, more of it solidifies and is fed down through the pyramid.

The pyramids are braced at the top by crossbars called **epiphyses** (Fig. 21-6, *1*). These are not joined in *Stereocidaris*, but among many regular echinoids they form a joined ring around the edge of the lantern. Massive radial spokes termed **rotulae** (Fig. 21-6, *5*), in ambulacral position, interlock with the epiphyses and pyramids. Above the rotulae lie slender, arched rods, forked at the ends; these are termed the **compasses** (Fig. 21-6, *4*). The entire Aristotle's lantern may be raised or lowered by means of muscles leading to the perignathic girdle.

Pedicellariae. Competition for space is as keen over much of the sea floor as it is in a weedy garden, and most animals must beware of becoming a mere pedestal for others. The spines of echinoids may preserve them from being sat upon by large animals, but cannot ward off the host of microscopic larvae and other little creatures scouting for a roost. For this purpose, the echinoids have developed **pedicellariae** (Fig. 21-7), comparable to the avicularia of bryozoans. Pedicellariae are microscopic, sharp-toothed, paired or three fold jaws of calcite, set with flexible necks upon little stalks. Many of them contain poison glands which discharge through a tiny canal opening into the terminal fang. The surface of a living echinoid presents an amazing spectacle under the microscope, for in addition to the various kinds of moving spines, and the motile tube feet, there are countless wicked-looking pedicellariae, lashing back and forth on their stalks, jaws ready to nip any intruder.

The pedicellariae are highly distinctive, and are therefore much used in taxonomy of modern genera and species. They are commonly preserved as fossils, but have largely escaped attention because of their small size. Unfortunately, they rarely remain attached to the tests of fossil sea urchins.

Growth of Test. The plates of echinoids, like those of other echinoderms, are internal and are pervaded by living tissue. Therefore, each plate can grow independently during ontogeny. Growth rings have been found on the individual plates of some species, but in most, the adult carries within it no discernible record of its ontogeny. New plates are added to the corona below the oculogenital ring, and coronal plates may become detached at the peristomial margin, wandering out onto the peristome. Among some echinoids, plates are dissolved at the margin of the peristome.

Internal Skeleton. The Aristotle's lantern and the perignathic girdle are the only conspicuous internal skeletal structures. In addition, countless microscopic internal spicules lie in the walls of the intestine and gonads, as well as in some organs of unknown functions and in the tube feet. These spicules take the form of disks, rods, hooklets, and perforated plates. Vast numbers of them must be present in sedimentary rocks, but their small size has made them unnoticed. Some spicules referred to holothuroids may actually be echinoid remains.

Orientation

The only consistent and readily recognizable asymmetry on the exterior or in the skeleton of *Stereocidaris* and many other regular echinoids is the excentric position of the madreporite, the water intake of the water-vascular system. This madreporite, a specially modified genital plate, is found in all regular echinoids at the top of one of the interambulacral areas, and no good reason exists for concluding that it is not located invariably in the same one. Assuming, then, that the madreporite is constant in position, how are we to know which part of the animal is anterior and which posterior? The movements of the animal are of no help, because it walks with equal facility in any direction. The slight excentricity of the anus among some groups of regular echinoids is also of no assistance, as it varies through an arc of about 35 deg. The key to the riddle is furnished by the members of the bilaterally symmetrical subclass Irregularia and by a few markedly bilateral regulars. Among these, the plane of bilateral symmetry invariably passes through the ambulacral area to the left of the madreporite. Thus, the madreporite lies in the right anterior interambulacral area, as seen from the aboral side. All other echinoids are oriented accordingly (Fig. 21-8).

Several systems have been proposed for the designation of the individual ambu-

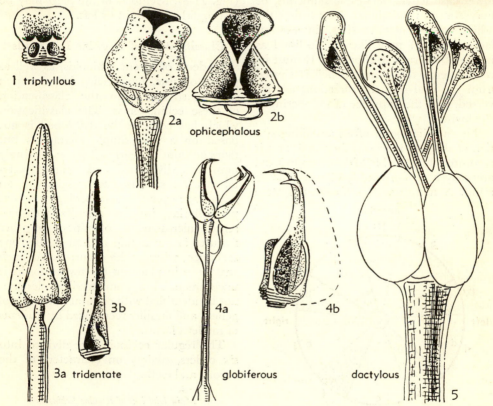

FIG. 21-7. **Pedicellariae of modern echinoids.** Pedicellariae are groups of two, three, or more minute snapping jaws (*valves*), mounted at the tip of small spines scattered over the surface of the test. They function in defense. Among types illustrated and defined below are (1) valve of triphyllous pedicellaria (*Heterocentrotus*, ×130); (2) ophicephalous pedicellaria (*Hygrosoma*, 2a, ×30, 2b, valve ×40); (3) tridentate pedicellaria (*Goniocidaris*, 3a, ×32, 3b, valve, ×40); (4) globiferous pedicellaria (*Allocentrotus*, showing poison glands, 4a, ×32; *Selenechinus*, 4b, valve with poison sac dashed, ×5); (5) dactylous pedicellaria (*Aerosoma*, ×40). (*After Mortensen.*)

dactylous (5). Having a variable number of stalked, spoon-shaped valves.

globiferous (4). Equipped with three sharp-toothed valves connected to poison glands, which in cidaroids lie within each valve but in others are globular sacs partly enclosing valves.

ophicephalous (2). Equipped with three blunt, deeply excavated and strongly ornamented valves; term signifies snake-headed.

tridentate (3). Having three long, slender valves which generally have sharp, finely serrate edges.

triphyllous (1). Having three leaf- or paddle-shaped valves.

lacral and interambulacral areas. The one most widely used, based on the work of Lovén, is illustrated in Fig. 21-8. As seen in aboral view, the ambulacra are labeled with capital letters or Roman numerals, in counterclockwise sequence, the anterior ambulacrum being *C* or *III*. Interambulacral areas are labeled accordingly, by lower-case letters or Arabic numerals, or may simply be designated by listing the adjoining ambulacra; thus, interambulacrum *a* or *1* may likewise be listed as interambulacrum *A–B* or *I–II*. It must be remembered that when a shell is viewed from the oral side the directions are reversed, and the sequence of numbering is clockwise.

The starting point of a numbering system of this sort is readily forgotten, and it is commonly desirable to denote individual areas by a system readily pictured in the mind. A more nearly self-evident system

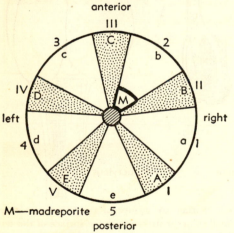

FIG. 21-8. **Orientation of echinoids.** The diagram represents the aboral side of an echinoid with ambulacral areas stippled. Orientation is determined by position of the madreporite (*M*), the anterior-posterior axis being drawn through the ambulacral area (*anterior*) at left of the madreporite, and through the opposite interambulacral area (*posterior*). Numerals or letters have been used to designate the various ambulacral and interambulacral areas as shown on the diagram; the areas may also be designated by reference to their position, as anterior, lateral, right, left, or posterior.

is that in which anterior, posterior, left side, and right side are used as frame of reference. Ambulacral areas comprise the anterior ambulacrum, left and right anteriors, and left and right posteriors. Interambulacral areas may be designated as the posterior interambulacrum, left and right posteriors, and left and right anteriors (Fig. 21-8). Members of the pairs may be distinguished as right or left.

Classification of Regular Echinoids

Students of recent echinoids, as well as paleontologists, have relied largely upon skeletal structures in the classification of these echinoderms. The classification which has come to be developed is not based on a few, simple characters but takes account of nearly all the exceedingly numerous, diversified, and complex skeletal structures. The neozoologist has an advantage over the paleozoologist in that pedicellariae, internal spicules, well-preserved microstructure of spines, and other features are available to him in any one specimen, whereas the occurrence of such structures in place on the test in fossils is uncommon. Thus, many fossil species cannot be identified with certainty in terms of genera and families recognized by students of modern forms.

The regular echinoids are divided into six orders, mainly on characters of the corona and teeth.

Main Divisions of Regular Echinoids

Regularia (*subclass*), periproct located inside oculogenital ring. Ordovician–Recent.

Lepidocentroida (*order*), skeleton flexible, plates with overlapping margins, auricles and apophyses in perignathic girdle, teeth grooved. Ordovician–Recent.

Melonechinoida (*order*), skeleton rigid, two or more rows of thick plates in ambulacral areas and three or more in interambulacral areas. Mississippian.

Cidaroida (*order*), skeleton rigid, 20 rows of plates, only apophyses in perignathic girdle, teeth grooved. Devonian–Recent.

Stirodonta (*order*), skeleton rigid, 20 rows of plates except in primitive forms, auricles and apophyses in perignathic girdle, epiphyses very short, teeth with prominent keel. Triassic–Recent.

Aulodonta (*order*) skeleton rigid in most forms, 20 rows of plates, auricles and apophyses in perignathic girdle, epiphyses short, teeth lack keel or groove. Triassic–Recent.

Camarodonta (*order*), skeleton rigid, 20 rows of plates, only auricles in perignathic girdle, epiphyses very long, teeth keeled. Jurassic–Recent.

Structural Variations of Regular Echinoids

Development of Ambulacral Areas.
Among the earliest well-known echinoids, the Upper Ordovician genera *Aulechinus*, *Ectinechinus*, and *Eothuria*, the radial water vessels are completely enclosed by flanges of the ambulacral plates. In this respect, they resemble the most primitive starfishes. The inner flanges of the plates in succeeding echinoids were reduced to mere ridges along the sides of the radial canals, and later were lost entirely (Fig. 21-9).

The ambulacral pores of *Aulechinus* are not completely enclosed by the ambulacral plates, but are deep embayments of the

Fig. 21-9. **Relation of radial canal and ambulacral plates.** Among the earliest echinoids, such as the Ordovician *Aulechinus*, the radial canals (5), presumably accompanied by radial nerve (3) and blood vessel (4), are completely surrounded by skeletal substance. They are located between the ambulacral plates (7), and are half-embraced by an inner and an outer flangelike part of each plate. The branch canals, leading to the tube feet (1), pierce the inner flanges. The water reservoirs (ampullae, 6) at the base of tube feet, project into the body cavity, while the tube feet pierce the ambulacral plate in one or more ambulacral pores (2). In more advanced echinoids, exemplified by the Permian *Meekechinus*, the inner flanges of the ambulacral plates are reduced to mere ridges along the sides, and the radial canals thus no longer lie in tunnels within the skeleton, but on the inner surface of the plates. In post-Paleozoic echinoids, these vestiges of the inner flanges have disappeared. Nearly all post-Ordovician echinoids possess tube feet which pierce the ambulacral plates in two places, forming two ambulacral pores.

plate margin. This type of structure is intermediate between that of asteroids, in which the tube feet emerge between plates, and echinoids. The tube feet of *Aulechinus* pierce the plates in a single pore, those of *Eothuria* (Fig. 21-16, *1*) in many pores. Some Ordovician and all later Regularia possess double pores.

Ambulacral Plate Rows. The earliest echinoids, such as *Aulechinus* (Fig. 21-16, *2*), possessed two rows of simple plates in each ambulacral area. This is the primitive condition, which has persisted to the present in many echinoids, such as *Stereocidaris*. Among at least two groups of Paleozoic echinoids, the number of ambulacral plate rows was multiplied. The echinoids called Lepidocentroida began to add plate

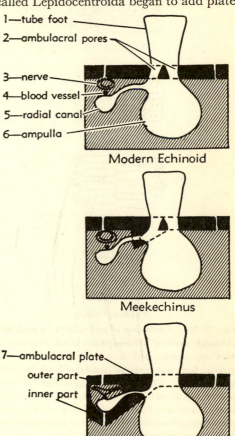

1—tube foot
2—ambulacral pores
3—nerve
4—blood vessel
5—radial canal
6—ampulla

Modern Echinoid

Meekechinus

7—ambulacral plate
outer part
inner part

Aulechinus

rows in Silurian time, by excluding alternate plates from the margins, squeezing them, as it were, into another row. This trend found its climax in the Permian *Meekechinus* (Fig. 21-17, *3*), which possessed 20 rows of plates in each ambulacrum. The Mississippian forms of the order Melonechinoida followed a similar trend (Fig. 21-10).

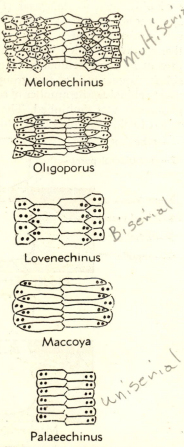

Melonechinus

Multiserial

Oligoporus

Lovenechinus

Biserial

Maccoya

Palaeechinus

Uniserial

Fig. 21-10. **Evolution of ambulacra in melonechinoids during Mississippian time.** *Palaeechinus* McCoy, oldest and simplest member of this group, has two rows of similar plates in each ambulacrum; this makes a single row of pore pairs on each side of the ambulacrum, a condition termed *uniserial*. Gradual restriction of alternate plates toward center and margin of the ambulacra leads through the intermediate *Maccoya* stage to the *biserial* condition illustrated by *Lovenechinus* Jackson. Continuation of this trend produces the *multiserial* arrangement shown in *Melonechinus* Meek & Worthen.

Compound Ambulacral Plates. Among Mesozoic and Cenozoic echinoids, evolutionary development of the ambulacral areas consisted not of monotonously adding new rows of similar plates, but of grouping ambulacrals into different types of compound plates (Fig. 21-11). Two basic styles of these are recognized—**diademoid** and **echinoid.** In diademoid compound plates, the middle one of three component elements (or one below the middle in combinations of four or five) is largest. In the echinoid type, the lowest member is largest.

One of the Jurassic echinothurids, *Pelanechinus*, went a step further in developing supercompound plates by the welding together of several compound plates.

Interambulacral Plate Rows. The interambulacral areas of Ordovician echinoids are armored with irregularly arranged plates, but most later echinoids have interambulacral plates arranged in rows. Many of the Paleozoic echinoids possess more than two rows of plates in each interambulacral area, but only one post-Paleozoic form, the Triassic *Tiarechinus*, is known to have more than two (Fig. 21-18).

Rigidity of Test. The early and mid-Paleozoic echinoids did not have rigid tests. Their plates were thin, overlapping each other like shingles on a roof, and accordingly, the test was flexible. This imbricate, flexible structure has persisted to the present in one family (Echinothuridae, of the order Lepidocentroida), largely limited to the deep sea. Echinoids with rigid tests first appeared in the Mississippian, and dominate in Triassic and later periods.

Oculogenital Ring and Periproct. Oculogenital rings in which the periproct is bounded by both ocular and genital plates are termed **insert** (Fig. 21-4*A*). In many echinoids, the genital plates have crowded the oculars into a peripheral position (Fig. 21-4*B*), a condition termed **exsert.** Some species show an intermediate condition, in which certain oculars are insert and others exsert.

Whereas the anus of *Stereocidaris* and its close relatives is located centrally, it is excentric in many other regular echinoids, being displaced toward either the posterior interambulacrum, the right posterior ambulacrum, or some intermediate position. One or more large plates, called **suranals**, may fill the central and anterior portions of the periproct (Fig. 21-12).

Peristome. The peristomial region of regular echinoids shows a wide range of variation. Although densely covered with ambulacral and interambulacral plates among the cidaroids, it is virtually naked in some groups, and bears five large pairs of **buccal plates** (Fig. 21-13) among others. Interambulacral plates may predominate or be absent.

Many regular echinoids possess special gills, located at the peristomial margin. In such forms, this margin commonly is embayed for the reception of gills, making **gill slits** (Fig. 21-13). The presence and shape of gill slits are important taxonomic characters.

Perignathic Girdle. Whereas cidaroids possess only apophyses, located in interambulacral position, other orders of regular echinoids show strongly developed perignathic processes, termed **auricles** (Fig. 21-6, *3*), located in ambulacral position. Among these groups, the apophyses are small or wanting.

Aristotle's Lantern. Differences in the structure of the jaw apparatus are among the most important features used to differentiate the orders of regular echinoids. The **aulodont** type of lantern, characterized by short epiphyses not joined to each other, may be contrasted with the **camarodont** type of lantern (Fig. 21-14), in which the apophyses are long and joined together. The latter type characterizes the order Camarodonta. The teeth are likewise of taxonomic importance. The cidaroids and lepidocentroids possess grooved teeth, the stirodonts and camarodonts keeled teeth, and the aulodonts unkeeled teeth.

Tubercles and Spines. *Stereocidaris*

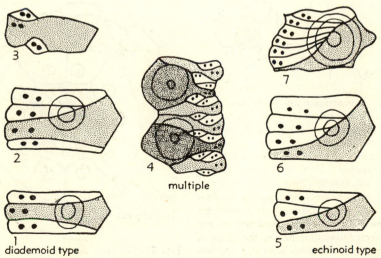

Fig. 21-11. **Compound plates.** Most Mesozoic and Cenozoic regular echinoids possess compound ambulacral plates, formed by fusion of simple plates in which one (stippled) remains larger than others. Two basic types of compound plates are distinguished: the *diademoid* type, in which the lowest member is small (1–3), and the *echinoid* type (5–7) in which the lowest member is large. A special condition in which diademoid compound plates are combined into a multiple compound plate is shown in 4. (3) *Phormosoma*, a Recent echinothurid; (4) *Pelanechinus*, a Jurassic echinothurid; (7) *Strongylocentrotus*, a Recent camarodont; other figures are diagrammatic. (*After Mortensen.*)

and its Mesozoic and Cenozoic relatives exhibit highly differentiated spines. Other echinoids show less diversity, but may carry more than one primary spine on each interambulacral plate. Some fossil and recent urchins are covered by a comparatively uniform blanket of short spines, as with a coarse fur. Others bristle with long, needle-sharp spikes, which may inflict painful injuries on the unwary. The spines may be serrate or beset with barbs. They may have the shape of blunt cylinders or thick clubs (Figs. 21-19, 7; 21-24, 7), which serve to weigh the animal down; this development is found chiefly among surf dwellers which need extra weight to resist the waves. Several recent and fossil forms have spines shaped like paving blocks; these completely hide the surface of the corona from view, forming a thick flexible armor around it. A few bizarre

types of spines are illustrated in Fig. 21-15. Echinoid spines exhibit a widely diverse internal structure, which is useful in classification (Fig. 21-31).

It is evident that the spines have a variety of functions. They protect the shell from physical damage and from attack by enemies, keep the animal from being rolled away by waves and currents, serve as legs to walk on, and in some species serve to carry bits of camouflaging seaweed. Among species which "mother" their offspring, the spines of the parent afford shelter for the young. Some echinoids use their spines to excavate dwelling cavities. No doubt there have been and are numerous other functions.

Diversity in spines necessarily implies diversity in tubercles. Those of cidaroids are perforate, but imperforate mamelons occur in many other groups of echinoids. In many echinoids, the little platform located between the boss and mamelon is conspicuously lobed or **crenulate.** These crenulations interlock with recesses in the

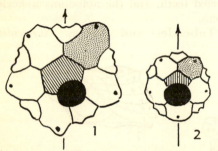

FIG. 21-12. **Excentricity of periproct.** *Polysalenia* (1), a Recent echinoid, shows the saleniid type of asymmetry, in which the periproct has shifted toward the right posterior ambulacral area; the entire oculogenital ring has assumed a bilateral symmetry along the plane passing through the left anterior interambulacrum and the right posterior ambulacrum. *Pseudosalenia zumoffeni* (2), of Cretaceous age, is a stirodont in which the periproct is displaced toward the posterior interambulacrum; hence, the plane of bilateral symmetry coincides with the conventionally defined anterior-posterior axis established by location of the madreporite.

The madreporites are stippled. The large central plate, shaded in the direction of symmetry, is a *suranal* plate. Arrows point forward. (*After Mortensen.*)

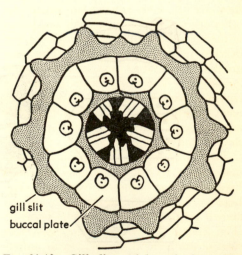

gill slit
buccal plate

FIG. 21-13. **Gill slits and buccal plates.** The genus *Plesiodiadema* illustrates two features common among regular echinoids: embayments (*gill slits*) in the peristomial margin which make room for gills, and a circle of large ambulacral *buccal plates* on the peristome. The stippled area represents the naked peristomial membrane.

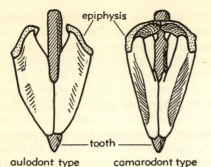

epiphysis

tooth

aulodont type   camarodont type

FIG. 21-14. **Aulodont and camarodont lanterns.** Each figure illustrates a pyramid with associated epiphyses (stippled) and tooth. The *aulodont* type has short, separate epiphyses, whereas *camarodont* type has long epiphyses which join each other.

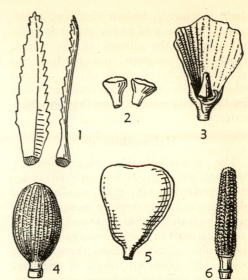

FIG. 21-15. **Diversity in spines.** The cidaroids show exceptionally great diversity of spines, some of which are figured here: (1) *Porocidaris schmidelii* (Münster), Eocene; (2) *Nortonechinus welleri* Thomas, Devonian; (3) *Cyathocidaris erebus* Lambert, Cretaceous; (4) *Balanocidaris glandifera* (Goldfuss), Jurassic; (5) *Balanocidaris pleracantha* (Agassiz), Jurassic; (6) *Plegiocidaris cervicalis* (Agassiz), Jurassic. Other orders of echinoids have convergently produced spines of similar shape.

margin of the socket so as to prevent twisting of the spine.

Lepidocentroids

The order of Regularia called Lepidocentroida (Ordovician–Recent) includes the least specialized echinoids. Probably they are the root stock from which the other orders have been derived. The plates of the corona are not welded together but overlap in a flexible manner. The perignathic girdle is formed by auricles and apophyses. The Aristotle's lantern is equipped with short epiphyses and grooved teeth. The periproct is central. Tubercles are perforate, and the spines are small, simple, and commonly hollow.

Two families are distinguished: Lepidocentridae (Ordovician–Permian) and Echinothuridae (Jurassic–Recent). The Lepidocentridae include the earliest known echinoids, among them *Aulechinus* (Fig. 21-16). These early forms have two rows of ambulacral plates in each ambulacrum and irregularly distributed interambulacral plates. In later members of the family, the rows of interambulacral plates are much reduced, but the ambulacral plate rows are multiplied, culminating in the Permian *Meekechinus* which

shows 20 plate rows at the ambitus of each ambulacrum (Fig. 21-17).

The Echinothuridae, which comprise descendants of the Lepidocentridae, elaborated the ambulacral areas not by multiplying plate rows, but by compounding individual elements into two rows of compound plates in each ambulacrum. The compound plates are formed according to the diademoid plan (Fig. 21-11), and differ from those of most other echinoids in the minute size of the "demi-plates." Oddly enough, the oldest known echinothurid, the Jurassic *Pelanechinus* (Fig. 21-19, *5*), possesses the most complex plates known in any echinoid, namely, compound plates built into supercompound ones.

The echinothurids are the only flexible echinoids found in modern seas, and include the largest known members of the class. Their lack of heavy armor is compensated by the efficacy of the spines, which are poisonous and produce extremely painful wounds. Most modern echinothurids are deep-sea dwellers, but a few brilliantly colored forms live in shallow water.

Melonechinoids

The melonechinoids, exemplified by *Melonechinus* (Fig. 21-17), are an order characterized by very thick, rigidly united plates arranged in three or more columns in the interambulacral areas and in two or more columns in the ambulacra. Each of the genital plates contains three to five pores. The spines are small, of one kind only, and attached to imperforate tubercles. The periproct is centrally located. The Melonechinoida are known only from the Mississippian, within which the five known genera form a remarkable evolutionary series progressing from few to many rows of ambulacral plates (Fig. 21-10).

Cidaroids

The cidaroids, to which the previously discussed *Stereocidaris* belongs, are superficially characterized by having a single large (primary) tubercle and spine located on each of the interambulacral plates (Figs. 21-18, *1;* 21-24, *7, 11*). The unwary student may be misguided by this character because it occurs among other orders of echinoids. To make matters worse, cidaroid-like genera of other orders have commonly been given names ending in *-cidaris*. All true cidaroids possess two rows of simple plates in each ambulacrum. The perignathic girdle consists only of apophyses, a condition not known outside this order. As in the Lepidocentroida, the lantern contains short epiphyses and grooved teeth. The periproct is central. Tubercles are of the perforate type. Spines show a dense outer cortex, a structure which occurs only in cidaroids and stirodonts.

The cidaroids known from Devonian, Mississippian, and Pennsylvanian rocks have more than two plate rows in each interambulacral area; for example, the Devonian *Nortonechinus* (Fig. 21-15, *2*) shows 14, and the Mississippian–Pennsylvanian *Echinocrinus* (Fig. 21-17, *4a, b*) has 4. Permian, Mesozoic, and Cenozoic cidaroids have two rows of simple plates in each ambulacrum. The early Cidaridae possessed flexible skeletons, but rigid forms made their appearance in the Triassic. All Jurassic and later cidaroids are rigid.

As might be expected, the flexible cidaroids of the Paleozoic were not readily preserved intact. Complete tests are exceedingly rare, whereas isolated plates and spines of cidaroids are common fossils in many upper Paleozoic formations. In our present state of knowledge, it is not possible to fit such fragments into recognized genera of echinoids. These largely unstudied remains offer a promising field for investigation; they may be classified temporarily without regard to their true relationships, and thus may prove useful in stratigraphic studies. When discovered in association with tests, many of these distinctive spines and plates can be classified properly.

Stirodonts

The stirodonts (Triassic–Recent) possess rigid, spheroidal tests (Figs. 21-19, *6, 7;* 21-24, *5, 9, 10*). Ambulacra and interambulacra are each composed of two rows of plates, except for the Triassic genus *Tiarechinus* (Fig. 21-18, *2*), which has three rows of interambulacral plates. The order is distinguished from all except Camarodonta by nature of the teeth, which bear a prominent longitudinal keel. It is distinguished from camarodonts by the shortness of epiphyses in the lantern. The perignathic girdle contains both auricles and apophyses. Much of the respiration is effected by special gills, the position of which is recorded in the form of gill slits in the margin of the peristome (Fig. 21-13). Spines of the Stirodonta are built much

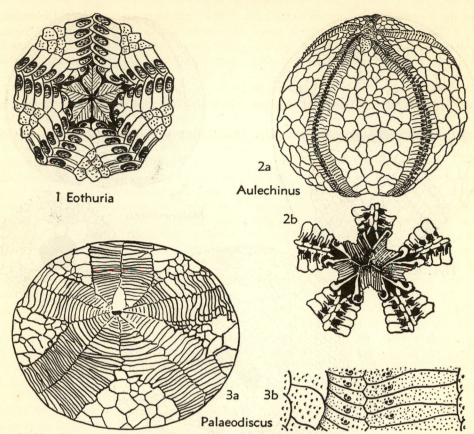

1 Eothuria

2a Aulechinus

2b

3a 3b Palaeodiscus

FIG. 21-16. **Representative Lower Paleozoic echinoids.** The flexible tests of most Ordovician and Silurian echinoids have ambulacral areas composed of two plate rows, interambulacral areas of irregularly overlapping plates. In other respects, the structure shows wide variations. Most of these forms may be classified among the lepidocentroids.

Aulechinus Bather & Spencer, Upper Ordovician, England. In this and associated genera, the radial canals are floored as well as roofed over by flanges of the ambulacral plates (Fig. 21-9). The ambulacral pores are single rather than paired, and are simply deep embayments in the margins of the ambulacral plates. In place of the usual Aristotle's lantern, the mouth of this primitive form is only equipped with five small many-grooved teeth, held in 10 plates which seem to be homologous to the pyramids. *A. grayae* Bather & Spencer (2a, side view, ×1.5; 2b, peristomial region and teeth viewed from inside, ×3).

Eothuria MacBride & Spencer, Upper Ordovician, England. This form resembles *Aulechinus*, but possesses as many as nine closely spaced pores per plate. The mouth is equipped with 10 valves, quite unlike the mouth structures of any other known echinoid. *Eothuria* was first described as a holothuroid, but shows such similarities to *Aulechinus*, and so few to holothuroids, that it is now generally considered to be a primitive sea urchin. *E. beggi* MacBride & Spencer (1, ×2).

Palaeodiscus Salter, Silurian, England, is judged to be intermediate between the Ordovician and the later lepidocentroids. Its radial canals lie in grooves on the inner surface of the ambulacral plates, its tube feet emerging through pore pairs; the mouth contained an Aristotle's lantern. *P. ferox* Salter (3a, peristomial region, ×3; 3b, detail of ambulacral area, ×6).

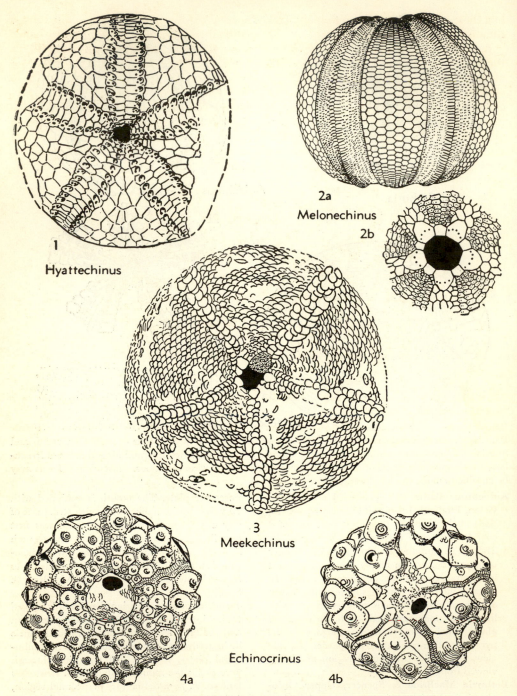

1
Hyattechinus

2a
Melonechinus
2b

3
Meekechinus

Echinocrinus
4a   4b

FIG. 21-17. **Representative Upper Paleozoic echinoids.** Paleozoic echinoids are characterized by flexible tests. Their radial vessels lay on the inner surface of the skeleton, but were generally bordered by skeletal ridges. The mouth contained an Aristotle's lantern. (*Continued on next page.*)

like those of cidaroids, although many lack the distinctive outer cortex layer.

A characteristic but not universal feature of stirodonts is the excentric position of the anus, which is retained within the oculogenital ring but shifted out of the center. The displacement may be in any of three directions: toward the posterior interambulacrum, the right posterior ambulacrum, or a position intermediate between these (Fig. 21-12). The center of the apical system thus vacated may be filled by a specially enlarged periproctal plate, called suranal. Early workers commonly used the position of the periproct as a basis for orientation, considering its excentricity to be either anterior or posterior. Such a procedure places the madreporite in different positions and entails the conclusion that various genital plates have functioned as a madreporite. For reasons already given, this conclusion is rejected. The madre-

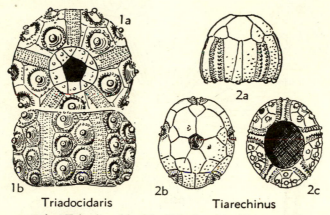

Triadocidaris   Tiarechinus

Fig. 21-18. **Representative Triassic echinoids.** Triassic echinoids are mostly very small and have been found in only a few places.

Tiarechinus Neumayr, Triassic. Stirodont. One of the earliest stirodonts, and the only post-Paleozoic echinoid in which interambulacra are known to be composed of three plate rows. *T. princeps* Neumayr (2a–c, side, aboral, and oral views, ×6), Cassian beds, Alps.

Triadocidaris Döderlein, Triassic. Cidaroid. This echinoid shows traces of the imbricate, flexible structure of its Paleozoic ancestors, but grades into the Triassic genus *Mikrocidaris* Döderlein, which possessed a rigid skeleton. *T. subsimilis* (Münster) (1a, b, side and aboral views, ×1), Triassic, Cassian formation, Alps.

(Fig. 21-17 continued.)

Echinocrinus Agassiz, Mississippian–Pennsylvanian. Cidaroid. The ambulacra contain four rows of plates. *E. rossica* (von Buch) (4a, oral view, 4b, aboral view, ×1), Mississippian, Europe.

Hyattechinus Jackson, Lower Mississippian. Lepidocentroid. This comparatively primitive form, like its early Paleozoic forbears, possessed a double row of plates in each ambulacrum and irregularly plated interambulacra. *H. rarispinus* Jackson (1, oral view, ×1), Waverly group, Ohio.

Meekechinus Jackson, Permian. Lepidocentroid. Last and most highly developed Paleozoic genus of its order. The interambulacra are reduced to narrow bands consisting of three plate rows each, while the ambulacral areas have become greatly enlarged and consist of vast numbers of plates. *M. elegans* Jackson (3, aboral view, ×1.5), Lower Permian, Kansas.

Melonechinus Meek & Worthen, Mississippian. Melonechinoid. The melonechinoids possessed thick skeletons and show a trend toward multiplication of ambulacral plate rows (Fig. 21-10). *Melonechinus* is the most highly developed member of the order. *M. multiporus* (Norwood & Owen) (2a, side view, ×1; 2b, apical area, ×3), Meramecian, Missouri.

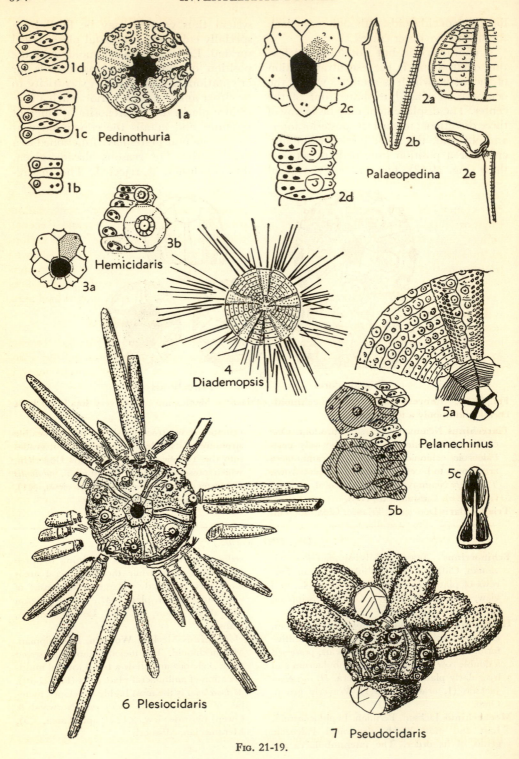

1a
1b
1c
1d Pedinothuria

2a
2b
2c
2d
2e Palaeopedina

3a
3b Hemicidaris

4
Diademopsis

5a
5b
5c Pelanechinus

6 Plesiocidaris

7 Pseudocidaris

Fig. 21-19.

porite has a constant location in the right anterior interambulacrum.

The stirodonts are interpreted as a Mesozoic offshoot of the cidaroids, differing from these chiefly in the development of keeled teeth, auricles, gill slits, and, among advanced forms, displacement of the anus.

Suborders of the stirodont echinoids are named Calycina and Phymosomina. The Calycina (Jurassic–Recent) bear cidaroid-like interambulacral plates and primary spines. One or more suranal plates occupy the center of the apical system. The Phymosomina (Triassic–Recent) differ in being less cidaroid-like and in lacking suranal plates. Both suborders have left a prolific fossil record.

Aulodonts

The aulodonts (Triassic–Recent) possess two columns of plates in each ambulacral and interambulacral area (Fig. 21-19, *1, 2, 4*). The test is rigid in all but the earliest members of the order. Ambu-

lacral plates are simple in primitive forms and diademoid in advanced forms. Gill slits occur on the margin of the peristome. The perignathic girdle is composed of both auricles and apophyses. The lantern bears short epiphyses. Lack of grooves or keels distinguishes the teeth of aulodonts from those of cidaroids, stirodonts, and camarodonts.

Divergent development of spines and tubercles has served as a basis for subdivision of the Stirodonta into three suborders: Pedinina (Triassic–Recent), which possess solid spines and noncrenulate tubercles; Diademina (Jurassic–Recent), which have hollow spines and various sorts of tubercles; and Aspidodiademina (Jurassic–Recent), which are characterized by crenulate tubercles and hollow spines having transversely partitioned axes.

Camarodonts

The camarodonts (Jurassic–Recent) stand morphologically close to the stirodonts. The anus of most camarodonts is

---

Fig. 21-19. **Representative Jurassic regular echinoids.** Most of the modern orders of echinoids became established in the Jurassic.

Diademopsis Desor, Lower Jurassic. Aulodont. The ambulacral plates of this needle-spined form are compound on the oral side, simple on the aboral. *D. heeri* Merian (4, aboral view, ×0.7), Europe.

Hemicidaris Agassiz, Jurassic–Cretaceous. Stirodont. A close relative of *Pseudocidaris* and *Plesiocidaris*. The ambulacral plates are compound according to the diademoid plan (3b, *H. crenularis* Lamarck, enlarged); the apical system has the posterior oculars insert, the anterior ones exsert, and the periproct displaced toward the rear (3a, *H. jauberti* Cotteau, enlarged), Jurassic, France.

Palaeopedina Lambert, Lower Jurassic. Aulodont. This is one of the better-known fossil echinoids, because not only the corona but the jaws and pedicellariae have been described. Nonkeeled teeth and short epiphyses show it to be an aulodont. *B. globulus* Agassiz (2a, side view, ×0.7; 2b lantern, ×3; 2c, apical system, ×3; 2d, oral ambulacral plates, ×3; 2e, pedicellaria, ×40), Hettangian, Europe.

Pedinothuria Gregory, Jurassic. Aulodont. The

peristome of this genus shows deep gill slits. Ambulacral plates are compound, most complex on the oral side, intermediate along the ambitus, and nearly simple on the aboral surface. *P. cidaroides* Gregory (1a, oral aspect, ×2; 1b–d, ambulacral plates from aboral side, ambitus, and oral side, respectively, ×6), Bathonian, Europe.

Pelanechinus Keeping, Jurassic. Lepidocentroid. This is the oldest known post-Paleozoic lepidocentroid, and is unique among echinoids in having combined compound plates into supercompound plates. *P. corallinus* Keeping (5a, oral view, ×0.7; 5b, supercompound plate, ×3; 5c, pedicellaria, ×17), Coral Rag, England.

Plesiocidaris Pomel, Jurassic. Stirodont. The genus is characterized by subcylindrical spines. *P. durandii* Peron & Gauthier, (6, aboral aspect, ×1), Algiers.

Pseudocidaris Étallon, Jurassic–Cretaceous. Stirodont. The primary spines are egg-shaped, perhaps to weigh down the animal in agitated water. *P. mammosa* Agassiz (7, ×1), France.

displaced as in stirodonts, and the central portion of the apical system contains one or more suranal plates. Also, the camarodonts possess keeled teeth. Two structural features which distinguish this order from others are the perignathic girdle, composed of auricles only, and the Aristotle's lantern, in which the epiphyses are so lengthened as to meet (Figs. 21-6, 21-14). All known camarodonts have a rigid skeleton, in which two rows of plates occur in each ambulacrum and interambulacrum. Unique among regular echinoids in having an elliptical test is the Recent family Echinometridae.

The camarodonts are divided into suborders: Orthopsina (Jurassic–Cretaceous), characterized by simple ambulacral plates; Temnopleurina (Cretaceous–Recent), distinguished by compound (diademoid or echinoid) ambulacral plates; and Echinina (Cretaceous–Recent), possessing compound plates of echinoid type. The Camarodonta are the largest group of living regular echinoids, but among fossils they are less important than stirodonts.

IRREGULAR ECHINOIDS

Irregular echinoids differ from members of the Regularia in having the periproct located in the posterior interambulacrum outside the oculogenital ring. The dividing line between the subclasses is somewhat arbitrary because in some Regularia the periproct is just barely within the ring, and in some Irregularia it lies barely outside. The periproct of advanced irregular echinoids, however, has migrated to the margin of the test or even onto the oral surface, and in some it is close to the mouth. The phylogenetic backward displacement of the periproct brought about elimination of the posterior gonad and genital plate, but these structures have been regenerated in some irregular echinoids.

Other characters which distinguish the Irregularia as a whole include the nature of the plates in the test, trend toward bilateral symmetry, features of the ambu-

lacra, specialization of the peristomial region, and divergence along many lines of specialization. Nearly all irregular echinoids possess simple plates only, arranged in double rows within each ambulacrum and interambulacrum. They retain elements of radial structure, but superimposed on this are varying degrees of bilateral symmetry. Unlike the conservative Regularia, the irregulars exhibit marked adaptive radiation. They branched into such unlike end members as the furry, discoidal sand dollars and the bristling, burrowing heart urchins.

Morphological features of the irregular echinoids include elements which have been described previously (Figs. 21-3, 21-4, 21-6).

Study of the Irregularia is taken up by groups because each differs considerably from others. Four such groups are recognized, as shown in the following tabulation.

Main Divisions of Irregular Echinoids

Irregularia (*subclass*), periproct located outside oculogenital ring. Lower Jurassic–Recent.

Holectypoida (*order*), ambulacra not modified into petals, peristome not surrounded by flower-like floscelle, jaws present at least in early ontogeny. Lower Jurassic–Recent.

Cassiduloida (*order*), ambulacra modified into petals on aboral side, mouth surrounded by flower-like floscelle. Lower Jurassic–Recent.

Clypeastroida (*order*), ambulacra petaloid, mouth not surrounded by flower-like floscelle, Aristotle's lantern retained through life but lacking radii. Upper Cretaceous–Recent.

Spatangoida (*order*), ambulacra petaloid, jaw system absent throughout life. Anterior ambulacrum and posterior interambulacrum specialized in more advanced members. Lower Cretaceous–Recent.

Holectypoids

The holectypoids include the least specialized irregular echinoids, that is, forms which most closely resemble the ancestral Regularia (Figs. 21-20, *1, 4;* 21-24, *8*). The dividing line between aulodonts and certain holectypoids is arbitrary. They are

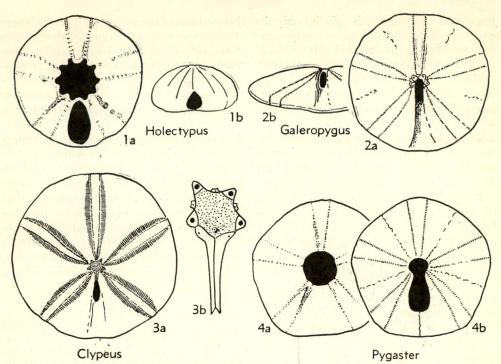

1a Holectypus 1b   2b Galeropygus 2a

3a   3b   4a   4b

Clypeus   Pygaster

FIG. 21-20. **Representative Jurassic irregular echinoids.** Irregular echinoids first appear in the Jurassic and are comparatively primitive in this system.

Clypeus Klein, Jurassic. Cassiduloid. One of the more advanced Jurassic irregulars, this genus has well-developed petals and elongate posterior genital plates which border on the periproct. *C. ploti* Klein (3a, aboral view, ×0.5; 3b, apical system, ×4), Europe.

Galeropygus Cotteau, Jurassic. Cassiduloid. The genus is much more primitive than *Clypeus* in lacking well-defined petals, and having the periproct surrounded on three sides by the oculogenital ring. *G. agariciformis* (Forbes) (2a, b, aboral and rear views, ×0.5), Domerian and Oxfordian, England.

Holectypus Desor, Jurassic–Cretaceous. Holecty-poid). The peristome is extended into gill slits. In some species, such as *H. hemisphaericus* Agassiz (1b, rear view, ×0.7), the periproct is located at the margin; whereas in others, such as *H. depressus* Leske (1a, oral aspect, ×0.7), it is on the oral side and extends close to the mouth. Jurassic, England.

Pygaster Agassiz, Jurassic–Cretaceous. Holecty-poid. One of the most primitive holectypoids, in which the periproct is a keyhole-shaped structure, the front part of which still occupies the middle of the apical system. *P. semisulcatus* Phillips (4a, b, oral and aboral views, ×0.5), Jurassic, England.

small echinoids, possessing a flattened oral side and a hemispherical to conical aboral surface. Except for the posterior displacement of the periproct, the test of most genera retains radial symmetry, becoming elliptical only in very advanced forms. Unlike any other irregulars, holectypoids generally possess compound ambulacral plates. These are restricted to the oral side, and are arranged in groups of three according to the diademoid plan.

The members of the suborder Holectypina (Jurassic–Cretaceous) retained a jaw system throughout life. Their peristomes are surrounded by auricles buttressed by radiating walls, and are scalloped by gill slits. The more advanced Echinoneina (Jurassic–Recent) possess jaws and a perignathic girdle only in early ontogeny, have a transversely elongate peristome, and lack gill slits.

Holectypoids are much the smallest echinoid order in our classification. They were common in Jurassic and Cretaceous faunas, but are represented in modern seas by the single genus *Echinoneus*. Interest attaches to them as connecting links between the regular echinoids and the irregulars ancestral to the orders Cassiduloida, Spatangoida, and Clypeastroida.

The origin of the holectypoids poses an intriguing problem. The primitive Holectypina contain species which are distinct from certain aulodonts only by the irregu-

lar position of the periproct. Clearly, these are descendants of aulodonts. Other species lack such close resemblance and possess crenulate tubercles. Crenulation is considered to be a primitive character, because various crenulate-tubercle echinoids have given rise to smooth-tubercle forms, whereas the reverse is unknown. Therefore, it seems necessary to search for ancestors of crenulate holectypoids among crenulate regulars; the only known group which seems to qualify is classified in the order Stirodonta. We are forced to conclude that holectypoids are polyphyletic for they are judged to include descendants of two distinct orders. These underwent convergent evolution which made them so similar that, for practical reasons, they are classified in a single order.

Cassiduloids

The structures which characterize cassiduloids are illustrated by *Cassidulus* (Fig. 21-21) and several other genera (Figs. 21-20, *2, 3;* 21-24, *2, 6;* 21-28, *3, 5;* 21-29, *5*). Members of the order are differentiated from other echinoids by the possession of a flower-like **floscelle** (Fig. 21-21, *6*),

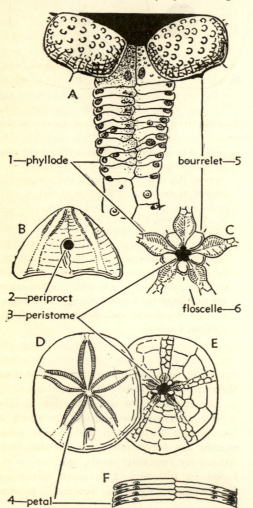

FIG. 21-21. **Structure of cassiduloids.** Special terms which apply to cassiduloids are explained in the alphabetical list below and illustrated by selected representatives: (*A*) *Cassidulus eugeniae* (Agassiz), Recent, portion of an ambulacral area, near the mouth, ×10; (*B–F*) *Cassidulus subconicus* Clark, Upper Cretaceous, Mississippi, (*B*) rear view, ×0.7, (*C*) peristomial region, ×1.5, (*D*) aboral view, ×0.7, (*E*) oral view, ×0.7, (*F*) portion of a petal, ×5.

bourrelets (5). Unpaired humped interambulacral plates adjacent to peristome.
floscelle (6). Flower-like pattern around peristome, formed of interambulacral bourrelets and ambulacral phyllodes.
periproct (2). Opening in test around anus.
peristome (3). Opening in test around mouth.
petals (4). Conspicuous petal-shaped portions of ambulacral areas, on aboral surface, in which the ambulacral pores are slitlike and closely crowded.
phyllodes (1). Depressed oral ends of ambulacral areas, composed of closely crowded plates, and commonly leaf-shaped in outline.

centered around the peristome. It is composed of depressed leaflike areas termed **phyllodes** in the ambulacral segments, separated by bulging interambulacral **bourrelets** (Fig. 21-21, *1*, *5*). Among primitive cassiduloids, close to the holectypoids, these structures are only slightly developed, but among more advanced forms, they are very striking.

The cassiduloids develop a jaw system in early ontogeny, but most of them lose all trace of it before maturity is attained.

On the aboral side of the test, ambulacral plates and pores are modified so as to outline an elliptical area on each ambulacrum. Because the radiating arrangement of the five resulting ellipses suggests a flower, the individual ellipses have been termed **petals** (Fig. 21-21, *4*). The peculiar modification which, above all others, makes the petals stand out as the most conspicuous features of the aboral surface is the development of the ambulacral pores: the outer pore of each pair is transversely elongated, and may be slitlike; in addition, a groove commonly connects the two pores of a pair, a condition termed **conjugate.** Each ambulacral plate carries but one pair of pores, but the plates are short, so that the pore pairs are closely crowded. Most of the aboral tube feet are entirely devoted to respiration. The slitlike pores accommodate broad, blade-shaped tube feet, and the crowding of pores in the petals permits a maximum number in a given space. Thus, the petals are an expression of specialized, efficient respiration on the aboral surface; this, in turn, is correlated with burrowing habits.

The Cassiduloida are divided into two suborders: Conoclypina, characterized by radiating internal partitions, and by lifelong retention of jaws; and Cassidulina, which lack internal walls and lose the jaw system during ontogeny. The most primitive Cassidulina do not possess petals, whereas the more advanced ones do. Some specialized forms have anterior petals which do not match the others. The four

genital plates may be distinct from the central madreporite, a condition termed **tetrabasal;** or may be fused, a condition termed **monobasal.**

Spatangoids

The primitive cassiduloids bridge the gap between holectypoids and the structurally distinctive spatangoids or heart urchins, which are highly specialized for a burrowing mode of life (Figs. 21-22; 21-24, *1*, *3*; 21-26). Some of the spatangoid structures differ so markedly from those of other echinoids that they merit study of a selected representative, the genus *Eupatagus*, known from the Eocene to the present. Italic letters and numbers listed after terms in the following discussion of these echinoids refer to Fig. 21-22, unless indicated otherwise.

Eupatagus (*A*) has a bun-shaped test, which is narrower at the rear than at the front, where it is gently indented by the depressed anterior ambulacrum (*16*). The peristome (*3*) is a transversely elongated opening, well forward of the middle, and the periproct (*6*) is placed vertically at the posterior end of the test.

The most distinctive features of the aboral side of the *Eupatagus* test are the petals (*1*), of which there are only four, the anterior ambulacrum (*16*) being nonpetaloid. Each pore pair of the petaloid area is sunk into an elliptical depression. The petals of many spatangoids are flush with the surface of the test, but in some they are sunken deeply. An Antarctic species uses these depressions as nurseries for its babies.

The interambulacral plates of the aboral side carry large perforate tubercles, which during life support a forest of long, slender, sharp spines, resembling the quills of a porcupine. The apical system consists of a central elongate madreporite and four genital plates. The number of gonads and genital pores (*17*) is reduced in some spatangoids to three or even two.

The oral side of *Eupatagus* and other spatangoids differs remarkably from that

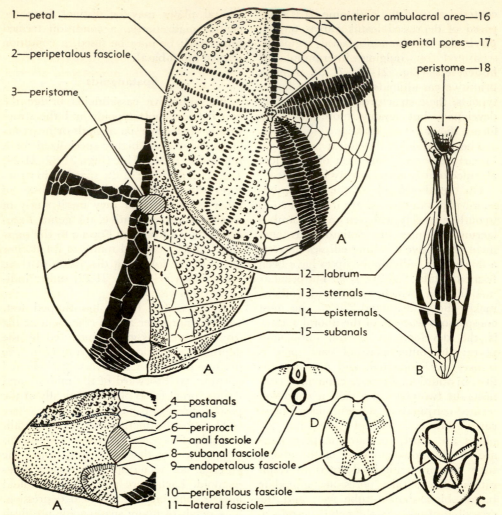

1—petal
2—peripetalous fasciole
3—peristome

anterior ambulacral area—16
genital pores—17
peristome—18

12—labrum
13—sternals
14—episternals
15—subanals

A

B

4—postanals
5—anals
6—periproct
7—anal fasciole
8—subanal fasciole
9—endopetalous fasciole
10—peripetalous fasciole
11—lateral fasciole

A

D

C

FIG. 21-22. **Morphology of spatangoids.** Structures peculiar to spatangoids are defined in the appended alphabetical list, cross-indexed with the numbers on the figure. (*A*) *Eupatagus mooreanus* Pilsbry, Eocene, Florida, aboral, oral, and rear views, ×1; the left side shows the normal appearance of a well-preserved test, exhibiting ornamentation but not much of the plate structure; on the right side, ornamentation is omitted and plates are outlined, with ambulacral areas shown in black. (*B*) *Pourtalesia*, one of the most aberrant spatangoids, a modern deep-sea dweller, oral view, ambulacral plates shown in black, ×3. (*C*) *Linthia*, Paleocene, diagrammatic aboral view to show fascioles, ×1. (*D*) *Echinocardium*, Miocene to Recent, a near-shore heart urchin, diagrammatic aboral and rear views, to show fascioles, ×1. For definitions of terms applicable to echinoids in general, see Fig. 21-3.

anal fasciole (7). Groove encircling periproct.

anals (5). Plates surrounding periproct.

endopetalous fasciole (9). Elliptical groove located within margins of petals.

episternals (14). Pair of interambulacral plates lying between sternals and subanals.

fasciole (2, 7–11). Groove on test lacking larger tubercles and spines; minute spines located

within these grooves are thickly covered with cilia, which move streams of mucus or currents of water for removal of foreign matter from the surface of the animal.

labrum (12). Unpaired interambulacral plate forming rear margin of peristome.

lateral fasciole (11). Groove along side of test.

(*Continued on next page.*)

of other echinoids. The posterior interambulacrum has become highly modified. The single elongate plate next to the mouth is termed the **labrum** (*12*). In *Eupatagus* and some advanced spatangoids, this is completely separated from the remainder of the posterior interambulacral area. It is succeeded by paired **sternal plates** (*13*), followed by a pair of **episternal plates** (*14*). These are barely in contact with the succeeding pair of interambulacral plates, termed **subanals** (15).

The postero-lateral and antero-lateral interambulacra are each composed of a few very large plates covered with flat mamelons, which are set on rhomboidal bosses arranged in oblique rows.

The posterior ambulacral plates of the oral side differ from other ambulacrals in being large, comparatively smooth, and bulging.

The beautiful beading of the test is interrupted by two narrow, depressed bands bearing miliary granules. One of these bands (**peripetalous fasciole**, *2*) forms a ring around the petals; the other (**subanal fasciole**, *8*) includes the subanal plates and parts of the adjacent ambulacral plates beneath the periproct. The fascioles are thickly studded with minute spines which are covered with ciliated tissues. The cilia sweep along a current of mucus, which carries away the sand and mud that settles onto the animal in its burrow. Other spatangoid genera show some other types of fascioles (Fig. 21-22*C*, *D*). **Anal fascioles** (*7*) extend part way around the periproct; **endopetalous fascioles** (*9*) lie within the petals. **Lateral fascioles** (*11*) lead from the peripetalous fasciole to the rear, along the flanks of the test.

The spatangoids lack all trace of a jaw system; their peristomes are scooplike, open toward the front. Forms like *Eupatagus* and *Echinocardium* live in a burrow and obtain food by using extremely elongate tube feet of the nonpetaloid anterior ambulacral area (Fig. 21-26). These reach to the surface and pick up bits of organic matter, drawing them into the burrow and stuffing them into the mouth. Many spatangoids have a floscelle, but this structure is never developed so strongly as among cassiduloids.

The spatangoids include some of the most highly specialized, and therefore structurally "aberrant," echinoids known, such as the modern deep-sea genus *Pourtalesia* (Fig. 21-22*B*), a vase-shaped creature having a flaring peristome located at the anterior end and the periproct near the tapering posterior extremity. It is thought that forms of this type burrow through sediment after the fashion of some holothuroids and the earthworm, namely, by eating their way through.

Clypeastroids

The clypeastroids, known from the Cretaceous on, are the youngest of the orders of echinoids recognized in our classification (Figs. 21-23; 21-24, *4;* 21-28, *1, 4;* 21-29, *3, 4*). Despite their relatively short geological history, they are of considerable importance to paleontologists, for in Cenozoic formations their abundance and rapid evolution have made them stratigraphically important.

The primitive members of the clypeastroids are mostly very small, even microscopic, resembling in structure the more advanced of the holectypoids. *Fibularia*, for example (Figs. 21-24, *4;* 21-28, *4*), is egg-shaped, possesses a persistent Aris-

---

(*Fig. 21-22 continued.*)

peripetalous fasciole (2, 10). Groove which encircles petals.

petal (1). Aboral portion of some or all of ambulacral areas, on which pore pairs are confluent in closely spaced, elongate slits.

postanals (4). Interambulacral plates which lie above periproct.

sternals (13). Pair of plates or single plate in posterior interambulacrum, lying between labrum and episternals.

subanal fasciole (8). Elliptical groove located below periproct.

subanals (15). Pair of interambulacral plates located between episternals and anals.

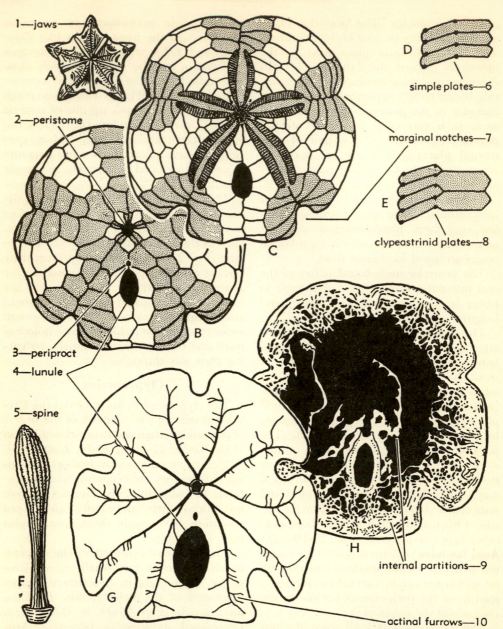

1—jaws

A

2—peristome

3—periproct

4—lunule

5—spine

B

C

D
simple plates—6

marginal notches—7

E
clypeastrinid plates—8

F

G

H
internal partitions—9

actinal furrows—10

FIG. 21-23. **Morphology of clypeastroids.** Structures peculiar to clypeastroids are defined in the accompanying alphabetical list, which is cross-indexed to the figure by numbers. (*A–C*) *Encope tamiamensis* Mansfield, Pliocene, Florida; (*A*) Aristotle's lantern, ×2; (*B, C*) oral and aboral views, ×0.7, ambulacral areas stippled. (*D, E*) Diagrams of plate structure among clypeastroids. (*F–H*) *Encope grandis* Agassiz; (*F*) aboral spine, ×250; (*G*) oral view showing grooves, ×0.7; (*H*) sections of test, showing internal structures, ×0.7. For terms applicable to echinoids in general, see Fig. 21-3.

actinal furrows (10). Branching grooves, generally limited to oral side, along which food particles are carried to mouth.

clypeastrinid plates (8). Structure characteristic of petals of suborder Clypeastrina, in which alternate plates are shortened, and thereby

(Continued on next page.)

totle's lantern, and shows the beginnings of petals.

The more advanced members are generally very much flattened, like *Encope* (Fig. 21-23; italic numbers used in discussing *Encope* refer to this figure). The ambulacra are strongly petaloid on the aboral side. Branching radial grooves, known as **actinal furrows** (*10*), lead to the peristome (*2*), which is small and not surrounded by a floscelle. These grooves serve to transport food particles to the mouth. Although the grooves are branched in *Encope*, they are simple in some other forms. In most advanced genera, they reach over the edge onto the aboral surface. All clypeastroids possess a well-developed Aristotle's lantern (*1*), which lacks radii. The test is covered by a furlike coat of short spines (*5*). In addition to normal tube feet—those of the petals being entirely devoted to respiration—great numbers of smaller secondary tube feet emerge through inconspicuous pores located along the sutures of the plates, and among some forms are found not only on the ambulacral but on interambulacral areas.

All but the most primitive clypeastroids possess **internal partitions** (*9*), which are supports within the tests. These may take the form of radiating ridges or pillars on the inner surface. Among the flat sand dollars, they consist of concentrically arranged or intricately crenulated walls, which divide the interior into narrow, commonly tortuous passages. Such supports, without which the skeletons of the larger sand dollars would be extremely fragile, are generally limited to the peripheral regions of the test, where the gonads are located. In addition, some forms possess pillars or a semicircular wall around the Aristotle's lantern, thus dividing the

interior into three distinct regions: a central pharyngeal area, an intermediate open ring which holds the digestive organs, and the marginal region of tortuous passages filled with gonads.

In some sand dollars, evolution has proceeded beyond the production of a mere flat disk- or shield-shaped test to making marginal indentations (**notches,** 7) and enclosed perforations (**lunules,** 4). Some lunules are developed from notches in early growth stages, whereas others start out as lunules. No doubt such structures, like the internal partitions, serve to strengthen the test; they may serve other, unknown functions also.

The clypeastroids are divided into two suborders: Laganina, which possess normal ambulacra (simple plates, 6), and Clypeastrina, in which the petals are not simple but composed of alternating full plates and "demi-plates," only half as wide as the full plates (clypeastrinid plates, 8). The Clypeastrina are the only known irregular echinoids which show such differentiation of ambulacral plates.

ECOLOGY AND MODE OF LIFE

Echinoids, like other echinoderms, are restricted to marine waters, where they range from the intertidal zone down into abyssal depths.

Males and females are separate, though freak hermaphroditic individuals have been found. Fertilization is external, and the egg generally develops into a free-drifting echinopluteus larva (Fig. 1-15), which settles down somewhere to become an echinoid. Among some species, the larva is not set adrift, but the young are protected by the mother, the baby echinoids finding refuge between her spines until they are able to shift for themselves.

(Fig. 21-23 continued.)

restricted to outer poriferous zones of ambulacral areas.

internal partitions (9). Pillars and anastomosing walls, which buttress interior of clypeastroids.

lunule (4). "Keyhole-like" perforations in tests of many of flat sand dollars.

marginal notches (7). Embayments in margins of sand dollars.

simple plates (6). Ambulacral plate structure in which adjacent plates are equally developed, contrasting with more specialized clypeastrinid type.

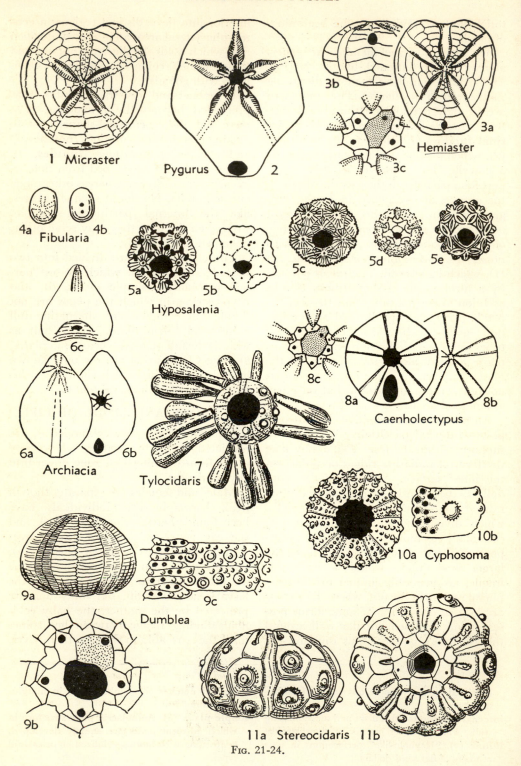

1 Micraster

Pygurus 2

3b

3c

3a

Hemiaster

4a Fibularia 4b

5a 5b

Hyposalenia

5c 5d 5e

6c

6a 6b

Archiacia

7

Tylocidaris

8c

8a Caenholectypus 8b

9a 9c

Dumblea

10a Cyphosoma

10b

9b

11a Stereocidaris 11b

Fig. 21-24.

Regular urchins live upon the sea floor. Some wander about on their spines, as on stilts, while others clamber over submarine cliffs by means of their prehensile tube feet, or nestle in rock cavities (Fig. 21-25). In agitated shallow waters, the regulars are largely restricted to hard bottoms, and show various modifications which permit them to resist the surge of waves and currents. The common shore dweller *Strongylocentrotus*, for example, clings tenaciously to surf-beaten cliffs by its long, tough tube feet equipped with large suckers, while cidaroids may maintain themselves on wave-washed surfaces by means of great clublike spines. Cliff dwellers may take over holes formed inorganically, or made by other organisms; as the echinoids grow

---

FIG. 21-24. **Representative Cretaceous echinoids.** Cretaceous echinoid faunas are rather modern in aspect but differ in general aspect from Recent ones in lacking sand dollars and in the small importance of the order Camarodonta.

Archiacia Agassiz, Cretaceous. Cassiduloid. This peculiarly pear-shaped echinoid has the attenuated apex located far toward the front, and was almost certainly a burrowing form. *A. sandalina* (d'Archiac) (6a–c, aboral, oral, and frontal views, ×0.7), Cenomanian, Mediterranean region.

Caenholectypus Pomel, Cretaceous. Holectypoid. Differs from the Jurassic *Holectypus* in having regained the fifth, posterior genital pore. *C. planatus* (Roemer) (8a, b, oral and aboral views, ×0.7; 8c, apical system, ×4), Middle Cretaceous (Comanchean), Texas.

Cyphosoma Agassiz, Jurassic–Eocene. Stirodont. The ambulacral pore pairs of this genus are arranged in a single row on the oral side, in a double row on the aboral surface. *C. texanum* Roemer (10a, oral aspect, ×0.7; 10b, an enlarged oral ambulacral plate, ×4), Middle Cretaceous (Fredericksburg), Texas.

Dumblea Cragin, Cretaceous. Stirodont. This genus differs from the contemporaneous *Pedinopsis* in having smooth rather than crenulate tubercles. *D. symmetrica* Cragin (9a, side view, ×0.7; 9b, apical system, ×3; 9c, ambulacral plates, ×5), Middle Cretaceous (Washita), Texas.

Fibularia Lamarck, Upper Cretaceous–Recent. Clypeastroid. *Fibularia* is one of the smallest and simplest of clypeastroids, devoid of internal partitions and of ambulacral furrows. The petals are only weakly developed. *F. subglobosa* Goldfuss (4a, b, aboral and oral sides, ×1), Senonian, Europe.

Hemiaster Desor, Cretaceous. Spatangoid. This common spatangoid has depressed petals and a weakly developed peripetalous fasciole. *H. whitei* Clark (3a, b, aboral and rear views, ×1; 3c, apical system, ×5), Middle Cretaceous (Fredericksburg, Washita), Texas.

Hyposalenia Desor, Cretaceous. Stirodont. A saleniid having a large apical system in which the anus is displaced toward the rear or toward the right posterior ambulacral area. Species of *Hyposalenia* are differentiated in part on the ornamentation of the ocular, genital, and suranal plates which form the apical system; several representative apical systems are those of *H. wrighti* (Cotteau) (5a), *H. clathrata* (Cotteau) (5b), *H. heliophora* (Cotteau) (5c), *H. bunburyi* (Forbes) (5e). The entire test of *H. acanthoides* (Desmoulins) is illustrated in 5d, ×1. Cretaceous, Europe.

Micraster Agassiz, Cretaceous–Miocene. Spatangoid. One of the most advanced of Cretaceous spatangoids, *Micraster*, possesses a nonpetaloid anterior ambulacrum and a subanal fasciole. The genus evolved rapidly in the chalk of England, particularly in shape and in the structure of the ambulacral plates, and has therefore served for subdivision of the chalk into a series of *Micraster* zones. *M. cor-testudinarius* Goldfuss (1, aboral side, ×0.7), Upper Cretaceous chalk, Europe.

Pygurus Agassiz, Jurassic–Cretaceous. Cassiduloid. The periproct of this highly advanced cassiduloid is located at the tip of a posterior extension, and the floscelle is exceptionally developed. *P. oviformis* d'Orbigny (2, oral side, ×0.7), from the Cretaceous of France.

Stereocidaris Pomel, Lower Cretaceous–Recent. Cidaroid. The genus *Stereocidaris* has been discussed in the text as model for regular echinoids. *S. sceptrifera* (Mantell) is a fossil representative (11a, b, side and aboral views, ×0.7), Cretaceous, France.

Tylocidaris Pomel, Cretaceous. Cidaroid. The primary spines of this striking echinoid are club shaped. *T. clavigera* König (7, ×0.7), Cretaceous, Europe.

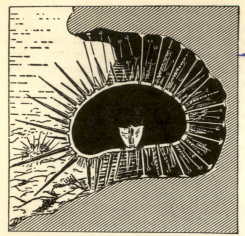

FIG. 21-25. **Regular echinoid anchored in rock cavity.** Small cavities in rocks are favorite roosting places for intertidal Regularia. Various modifications have been developed to permit the animals to cling to the rock in spite of raging surf. Many species, such as the one shown, manage this by bracing themselves by their spines, and by anchoring themselves with their tube feet.

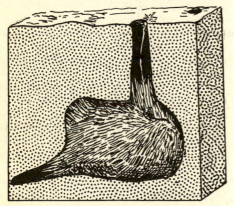

FIG. 21-26. **Spatangoid in its burrow.** Heart urchins, such as *Echinocardium*, burrow in sand. They maintain an open channel to the surface, through which food and fresh water are brought down to the animal.

FIG. 21-27. **Clypeastroid in normal living position.** Most sand dollars live nearly flush with the surface of sediment on the sea bottom.

in size, they may enlarge their domiciles by abrading the walls with teeth and spines.

The diet of the Regularia is a varied one. Some are vegetarians, living on seaweed; others are carnivorous, preying on bryozoans, worms, sponges, or hydroid coelenterates, or stalking larger game, such as clams and crustaceans, which are crushed by means of the Aristotle's lantern. Still others ingest bottom sediment to extract from it the nourishment contained in rotting bits of organic matter.

The irregulars represent divergent lines of adaptation to life on soft bottoms. Most of them are burrowers, living at or under the surface of bottom sediments (Fig. 21-26). This mode of life was made possible by restriction of the respiratory mechanism to the petaloid areas of the aboral side.

The clypeastroid sand dollars live in large colonies on sandy or muddy bottoms. Many occur in the intertidal zone, and the majority are limited to bottoms less than 250 m. deep, but some sand dollars and many of the more primitive, less flattened clypeastroids extend to depths of around 2,000 m. Most sand dollars live flat on the bottom (Fig. 21-27), the aboral surface flush with the sediment or barely covered by a thin veneer of sand. The near-shore dwellers burrow a little more deeply when the tide goes out. Some species do not remain flat but rise on edge when in deeper water. Clypeastroids burrow and move about by means of short furlike spines. Food is obtained from the sediment, by selection of alga-covered sand grains and bits of organic matter, which are passed along the actinal furrows to the mouth.

The spatangoids are adapted to a deeper burrowing existence. The genus *Echinocardium* (Fig. 21-26), for example, lives in burrows 10 to 15 cm. deep. Fresh sea water is introduced, and stale water is expelled through a mucus-lined canal which extends to the sea floor. This canal also permits the long tube feet of the anterior

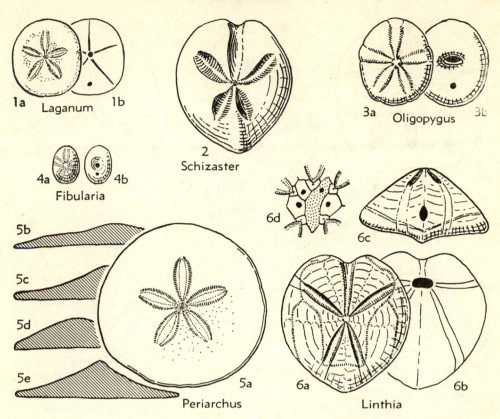

FIG. 21-28. **Representative Paleogene echinoids.** The early Tertiary brought with it the rise of flat, sand-dollar type clypeastroids and great differentiation among spatangoids.

Eupatagus Agassiz, Eocene–Recent. Spatangoid. This genus has been illustrated in Fig. 21-22, *A.*

Fibularia Lamarck, Upper Cretaceous–Recent. Egg-shaped and simple, this echinoid, presumably near the ancestral stock from which the clypeastroids were derived, is small enough to be preserved commonly in well cuttings. *F. vaughani* (Twitchell) (4a, b, aboral and oral views, ×1.5), Eocene (Ocala limestone), Florida.

Laganum Klein, Eocene–Recent. Clypeastroid. A sand dollar possessing an inflated edge and simple actinal furrows. *L. sorigneti* Cotteau (1a, b, aboral and oral views, ×1), Eocene, France.

Linthia Merian, Paleocene–Recent. Spatangoid. *Linthia* has the mouth placed unusually far forward. The petals are depressed, and there are both a peripetalous and a lateral fasciole. *L. tumidula* Clark (6a–c, aboral, oral,

and end views, ×0.7; 6d, apical system, ×3), Paleocene, New Jersey.

Oligopygus de Loriol, Eocene–Oligocene. Cassiduloid. Lacks a floscelle and has a transversely elongated peristome situated in a transverse depression. *O. wetherbyi* de Loriol (3a, b, aboral and oral views, ×1), Eocene (Ocala limestone), Florida.

Periarchus Conrad, Eocene. Clypeastroid. This large sand dollar possesses five genital pores. *P. lyelli* (Conrad) (5a, aboral view, ×0.7), upper Eocene of southeastern United States; a series of "varieties," in part geographic races, are distinguished by various profiles (5b-e).

Schizaster Agassiz, Eocene–Recent. Spatangoid. The petals and anterior ambulacral area are deeply depressed, and the periproct lies under a beak at the posterior end. *S. armiger* Clark (2, aboral view, ×0.7), upper Eocene, southeastern United States.

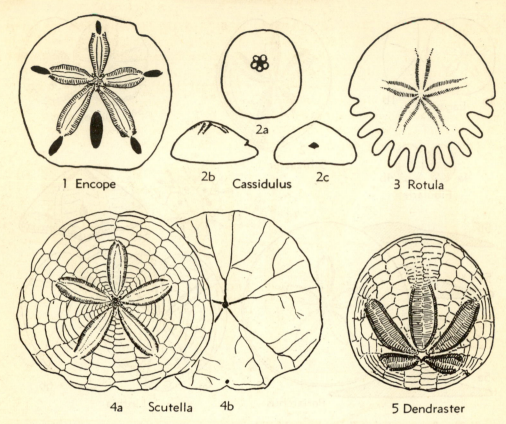

1 Encope   2a   2b   Cassidulus   2c   3 Rotula

4a   Scutella   4b   5 Dendraster

Fig. 21-29. **Representative Neogene echinoids.** Neogene fossil echinoids of the temperate regions show a preponderance of sand-dollar type clypeastroids, a group which did not reach great diversification until this time.

Cassidulus Lamarck, Eocene–Recent. Cassiduloid. This subconical form shows weakly developed petals and a well-marked floscelle. *C. gouldii* (Bouve), Oligocene, southeastern United States (2a–c, oral, side, and rear views, ×0.7).

Dendraster Agassiz, Miocene–Recent. Clypeastroid. The apical system of this distinctive sand dollar is located far behind the center. The ramifying actinal grooves extend over the margin onto the aboral surface. *D. gibbsii* (Remond) (5, aboral view, ×0.7), California.

Echinocardium Gray, Miocene–Recent. Spatangoid. This small spatangoid possesses tapering petals and an endopetalous fasciole. *E. cordatum* Pennant (Fig. 21-22D), Recent, Europe and North America.

Encope Agassiz, Miocene–Recent. Clypeastroid. This "keyhole urchin" shows a large lunule in the posterior interambulacrum. Some species, for example *E. tamiamensis* Mansfield, Pliocene, Florida, have marginal notches in the ambu-

lacral areas (Fig. 21-23A–C); others, such as *E. emarginata* Leske, Pliocene–Recent, North America (1, ×0.7) have five marginal lunules instead.

Pourtalesia Agassiz, Recent. Spatangoid. This aberrant, subcylindrical echinoid has the mouth located at the anterior end, the anus near the posterior; it has lost the petaloid character of the ambulacra. *P. (Echinosigra) paradoxa* Mortensen (Fig. 21-22B, ×3), North Atlantic.

Rotula Klein, Pliocene–Recent. Clypeastroid. The lobation of the posterior margin distinguishes this sand dollar. *R. orbiculus* (Linné) (3, ×0.7), West Africa.

Scutella Lamarck, Eocene–Miocene. Clypeastroid. The periproct is submarginal, and the genital pores number four. *S. leognanensis* Lambert (4, ×0.5), Miocene, France.

Stereocidaris, Cretaceous–Recent. Cidaroid. A Recent representative, *S. tubifera* Mortensen (Fig. 21-3).

ambulacral area to reach the surface. These sensitive tube feet, having finger-like branches at the end, select choice morsels of organic matter and draw them into the burrow. *Echinocardium* burrows chiefly by means of its spines, many of which are shovel-shaped for this purpose. Other spatangoids, such as *Pourtalesia* (Fig. 21-22*B*), may eat their way through sedi-ment. Spatangoids range into the inter-tidal area, but the majority of them live in deeper water, and some species are abyssal.

GEOLOGICAL HISTORY AND IMPORTANCE

The geological distribution of echinoids is summarized in Fig. 21-30. The earliest

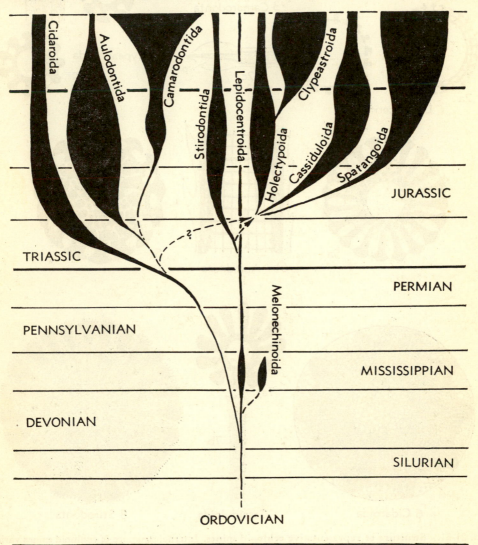

Fig. 21-30. **Geological development of echinoids.** The lepidocentroids seem to be the least modified group. With the melonechinoids and branches shown on the left, these comprise the Regularia. The Irregularia, plotted at upper right, are thought to be polyphyletic, that is, derived from several sources; their ancestors are to be sought in the root stock of aulodonts and camarodonts, as well as among the stirodonts.

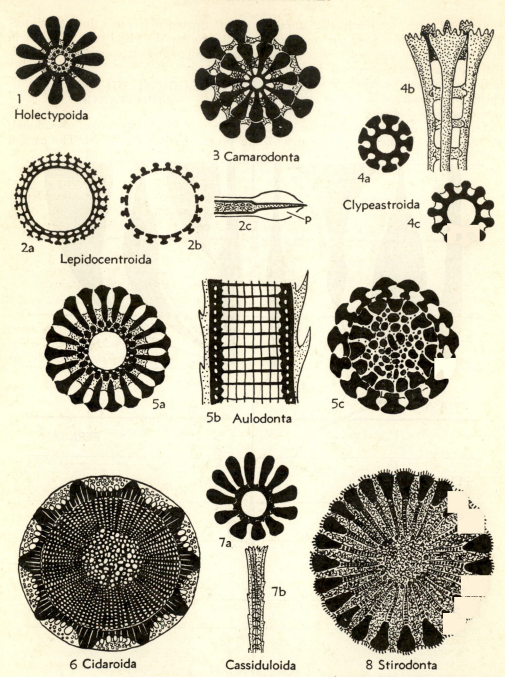

FIG. 21-31. **Structure of representative echinoid spines.** Internal structure of echinoid spines varies widely, and is of taxonomic importance. The spines illustrated are of Recent species belonging to different orders.

1, Holectypoida: *Echinoneus cyclostomus,* cross section of spine, ×100.

2a–c, Lepidocentroida: (2a) *Hemiphormosoma pauci-* *spinum,* oral spine, ×50; (2b) *Phormosoma bursarium,* oral spine, ×50; (2c) diagrammatic (Continued on next page.)

known echinoid is *Myriastiches*, from the Middle Ordovician. From this time on, echinoids seem to have become increasingly diversified, one group after another branching off by adaptation to new environments.

Relationships and Origin

Ordovician echinoids have much in common with Lower Ordovician (Tremadocian) starfishes of the order Somatasteroida. Both groups have rather similar-looking ambulacral plates, which embrace the radial water vessels by means of an outer and an inner flange. The tube feet of some of these early echinoids are not completely enclosed within ambulacral plates, suggesting derivation from ancestors in which, like starfishes, the tube feet lay between ambulacrals. Furthermore, the Ordovician echinoids lacked an Aristotle's lantern, and possessed instead a simple jaw apparatus not unlike that of the starfishes. These similarities point to a close relationship between echinoids and stelleroids. The radial structure and occurrence of a fixed stage in the larvae of many starfishes point to derivation of these classes from fixed (pelmatozoan) ancestors. Recognition of these ancestors awaits the discovery, among pelmatozoans, of ambulacral structures resembling those found in primitive starfishes and echinoids.

Development of Orders

The Ordovician echinoids, though differing from each other in important respects, bear more resemblance to the members of the order Lepidocentroida than to

others, and therefore are included in this group. The lepidocentroids are believed to be closest to the root stock of echinoids. They reached a climax in the Mississippian and are chiefly deep-sea relics today. The cidaroids split off from lepidocentroids in Early Devonian time, if not before, and gave rise to the aulodonts in the late Triassic. The short-lived but rapidly evolving melonechinoids deviated from the lepidocentroids before the Early Mississippian, and the stirodonts came from the same stock in Triassic time. The camarodonts are a Jurassic offshoot of the aulodonts and largely displaced their ancestors in Tertiary time.

The irregular echinoids arose early in the Jurassic period. They may have been derived in part from the aulodonts and in part from the stirodonts. This polyphyletic origin seems best to account for the simplest order of irregulars, the holectypoids, which made their appearance in earliest Jurassic time. The cassiduloids and spatangoids, which immediately follow, probably are derived from these. The comparatively conservative cassiduloids decreased in importance through the Tertiary, while spatangoids continued to diversify. A fourth group of irregulars, the clypeastroids, arose from the holectypoids in the Late Cretaceous. In early Cenozoic time, the clypeastroids developed the sand-dollar type of echinoid, one of the most successful echinoid adaptations in modern seas.

Geological Importance

Echinoids are structurally complex and gave rise to many rapidly evolving stocks.

(Fig. 21-31 continued.)

longitudinal section of a poison spine, having poison gland around tip.
3, Camarodonta: *Temnopleurus michelsoni*, cross section of spine, ×100.
4a–c, Clypeastroida: (4a) *Peronella keiensis*, cross section of primary spine, ×250; (4b) *Peronella pellucida*, side view of miliary spine, ×400; (4c) *Peronella japonica*, cross section of primary spine, ×200.
5a–c, Aulodonta: (5a) *Centrostephanus longispinus*, cross section of primary spine, ×40; (5b)

Plesiodiadema indicum, longitudinal section of primary spine, ×40; (5c) *Aspidodiadema africanum*, cross section of primary spine, ×80.
6, Cidaroida: *Centrocidaris döderleini*, cross section of primary spine, ×120.
7a, b, Cassiduloida: (7a) *Cassidulus caribaearum*, cross section of primary spine, ×40; (7b) same, miliary spine, ×130;
8, Stirodonta: *Arbacia stellata*, cross section of primary spine, ×40.

They are of great interest to students of evolution, and of considerable stratigraphic importance. Their usefulness as guide fossils is limited, however, by their mode of occurrence. A given bed may yield numerous genera and vast numbers of individuals, but most sedimentary rock formations contain few identifiable echinoid remains, or none. This is due partly to close environmental adaptation of the echinoids and partly to conditions affecting preservation. Before the rise of irregulars, few sea urchins were well adapted to life on shallow-water mud bottoms, and even today, large marine areas lack echinoids. Most regular echinoids are rather delicate and therefore are preserved intact only under exceptional conditions. Irregulars are more ruggedly constructed, and because of their burrowing habits, more readily preserved. For these reasons, the Irregularia are of greater stratigraphic importance than their regular cousins. It is probable that regular echinoids have potential stratigraphic usefulness which will be recognized when paleontologic investigations are more widely extended to isolated plates and spines abundant in many formations.

PRIMITIVE ELEUTHEROZOANS OF UNCERTAIN AFFINITIES

Ordovician, Silurian, and Devonian rocks have yielded a number of unique fossil echinoderms, which are specialized for a free-roaming life and therefore grouped properly among eleutherozoans. They do not fit, however, into any of the large classes of this subphylum—Holothuroidea, Stelleroidea, and Echinoidea. Their nonstellate form and heavy boxlike test suggest affinity with echinoids, but other structures differ widely from the echinoid pattern. Consequently they are treated independently.

Bothriocidaroids

One of the best known primitive eleutherozoans is *Bothriocidaris*, from Ordovician rocks of the eastern Baltic area (Fig. 21-32, *2*). It is classed here as the sole

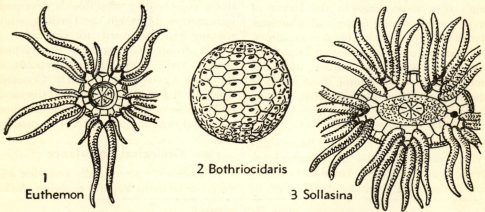

1 Euthemon

2 Bothriocidaris

3 Sollasina

FIG. 21-32. **Primitive eleutherozoans of uncertain affinities.**

1, *Euthemon* Sollas, Silurian, England. The pentagonal test has a large oral opening, equipped with five large jaws. Large-plated tube feet emerge through ambulacral pores located between plates. *E. igerna* Sollas, oral view, ×1.3.

2, *Bothriocidaris* Eichwald, Middle Ordovician, Baltic region. Spheroidal test composed of double rows of perforate ambulacral plates,

and single rows of imperforate interambulacral plates. *B. pahleni* Schmidt, side view, ×2.

3, *Sollasina* Fedotov, Silurian–Devonian, Europe. Resembles *Euthemon*, but has a bilaterally elongate test and a membranous region, armored with small plates around the mouth, *S. woodwardi* (Sollas), oral view, ×1, Silurian, England.

known representative of a class called Bothriocidaroidea.

Morphologic Features. The body is small, globular, and stemless. The mouth and anus are at opposite poles. The surface is armored by rows of thick, rigidly joined plates. Some plates have a pit, within which lie two canals extending through them. The canals served for the passage of tube feet from internal radial water vessels to the outside. The tube feet were armored with calcareous spicules, and have been found preserved in a few specimens of *Bothriocidaris*. Each tube foot penetrated the plate in two branches, which joined on the outside. The perforate plates are organized in double rows, evidently located above the radial water vessels of the water-vascular (or ambulacral) system. They are termed **ambulacral plates,** and the areas covered by them are called **ambulacral areas.** Narrow **interambulacral areas,** each composed of a single row of imperforate **interambulacral plates,** separate the ambulacral areas. Plates of both types bear small dimpled tubercles, which served as bases for short movable spines. The ambulacral areas terminate near the mouth in a single imperforate plate, and near the anus in a single perforate plate. A screen-like ossicle (**madreporite**), which serves as intake for the water-vascular system, is located in one of the ambulacral areas near the anus.

Relationships. *Bothriocidaris* was once held to be the ancestral echinoid. It differs from all echinoids in several important respects, but chiefly in the nature of terminations of ambulacral areas, position of the madreporite, and number of interambulacral plate rows. Also, true echinoids have been discovered in rocks as old as those which have furnished *Bothriocidaris*. These early echinoids show that primitive members of this class possessed flexible skeletons, radial vessels surrounded by skeletal covering, and other features which are quite unlike structures of *Bothriocidaris*. The conclusion is reached, therefore, that this genus is a blind-alley offshoot of early echinoderms (perhaps cystoids), resembling echinoids by reason of convergence rather than close relationship.

Ophiocystids

Another peculiar group of echinoderms from Ordovician, Silurian, and Devonian rocks has been named Ophiocystia. It is treated here as another class of Eleutherozoa.

The body is covered by a rigid skeletal box, composed of calcite plates. The underside shows a centrally located mouth, from which radiate five **ambulacral** regions, each composed of three plate rows, and five **interambulacral** areas, each composed of a single plate row. Ambulacral pores are single, and they lie between ambulacral plates. They give egress to comparatively huge armlike tube feet, which are armored with calcareous scales. These are not uncommon as fossils. The mouth contains a jaw system similar to that of primitive echinoids, but the nature of ambulacral areas and pores points more to the relationship of the Ophiocystia with starfishes. It is possible also that these forms, like *Bothriocidaris*, perhaps were derived independently from pelmatozoan ancestors.

Representative ophiocystids are *Euthemon* and *Sollasina* (Fig. 21-32, *1, 3*).

REFERENCES

COOKE, C. W. (1942) *Cenozoic irregular echinoids of eastern United States:* Jour. Paleontology, vol. 16, pp. 1–62, pls. 1–8.

CLARK, W. B., & TWITCHELL, M. W. (1915) *The Mesozoic and Cenozoic Echinodermata of the United States:* U.S. Geol. Survey Mon. 54, pp. 1–341, pls. 1–100. Although outdated in some details, this is the standard handbook for identification of American fossil echinoids.

JACKSON, R. T. (1907) *Phylogeny of the Echini, with a revision of the Paleozoic species:* Boston Soc. Nat. History Mem. 7, pp. 1–491, pls. 1–76, figs.

1–256. A comprehensive treatment of Paleozoic echinoids known in 1907.

MacBride, E. W. (1909) *Echinodermata: Echinoidea—sea urchins:* in Harmer, S. F., & Shipley, A. E., Cambridge natural history, Macmillan & Co., Ltd., London, pp. 503–559, figs. 223–253.

———— & Spencer, W. K. (1939) *Two new Echinoidea, Aulechinus and Ectinechinus and an adult plated holothurian, Eothuria, from the Upper Ordovician of Girvan, Scotland:* Royal Soc. London Philos. Trans., ser. B, vol. 229, pp. 91–136, pls. 10–17, figs. 1–15. Descriptions of the early echinoids.

Mortensen, Th. (1928–1948) *A monograph of the Echinoidea:* Reitzel, Copenhagen, & Milford, Oxford, New York, (1928) vol. 1, Cidaroida, pp. 1–551, pls. 1–88, figs. 1–173; (1935) vol. 2, Bothriocidaroida, Melonechinoida, Lepidocentroida, and Stirodonta, pp. 1–647, pls. 1–89, figs. 1–377; (1940) vol. 3–1, Aulodonta, with additions to vol. 2 (Lepidocentroida and Stirodonta), pp. 1–370, pls. 1–77, figs. 1–197; (1943) vol. 3–2, Camarodonta, pp. 1–553, pls. 1–56, figs. 1–321; (1943) vol. 3–3, Camarodonta, pp. 1–446, pls. 1–66, figs. 1–215; (1948) vol. 4–1, Holectypoida, Cassiduloida, pp. 1–363, pls. 1–14, figs. 1–326; (1948) vol. 4–2, Clypeastroida, pp. 1–471, pls. 1–72, figs. 1–258. This magnificent series of monographs describes Recent echinoids down to the species level, with fossil forms to genera. It has served as basis for most of this chapter.

Wright, T. (1857–1878) *British fossil Echinodermata of oölitic formations:* Palaeont. Soc. Mon., pp. 1–481, pls. 1–43. A classic monograph on Jurassic sea urchins.

———— (1864–1882) *British fossil Echinodermata from Cretaceous formations:* Same, pp. 1–371, pls. 1–80.

CHAPTER 22

Graptolites and Pterobranchs

Paleontologists have long been acquainted with certain chitinoid fossils, which are preserved most commonly as flattened films of carbon in black shales of Ordovician and Silurian age. Because of the resemblance of the fossils to pencil marks, they were called graptolites (*grapto*, write; *lite*, stone) (Figs. 1-1, 7; 22-1).

Uncompressed graptolite skeletons etched out of limestone and chert reveal a complex structure. Each is composed of many small tubes or cups, which evidently contained soft parts of individual animals. Graptolites are thus seen to be colonial organisms. The tubular cups may not all be of the same type. For example, in the group called dendroids, each colonial skeleton is composed of three distinct types of cups, alternating in regular manner. This indicates specialization for different functions among individual members of the colony.

FIG. 22-1. **Floating communities of graptolite colonies.** The view shows a Mid-Ordovician seascape and water in section, with colonies of *Diplograptus* attached to a float. Such united colonies are termed synrhabdosomes. The individual colonies, clearly visible in the foreground, comprise many zooids, housed in the small cups which are closely spaced along the branches.

RELATION OF GRAPTOLITES AND PTEROBRANCHS

The zoological relationships of graptolites long have been in doubt. Early guesses were based largely on the superficial resemblance of graptolites to some other groups of invertebrates. Thus, the presence of different kinds of individuals in the colonies led some paleontologists to classify them among the coelenterates. Others held that the shape of the colonies, bilateral symmetry of the cups, and occurrence of presumed muscle impressions pointed to a relationship of the graptolites with bryozoans. Studies of exceptionally well-preserved graptolites during the 1930's served to clarify many features of morphology and to establish a firmer basis for comparing these fossils with the hard parts of other organisms. It turned out that, in wall structure, mode of budding, and presence of a system of internal minute tubes (stolons), graptolites are utterly unlike coelenterates and bryozoans but strikingly similar to a small group of little known modern animals, called pterobranchs. The pterobranchs are represented by two genera, both colonial: *Rhabdopleura* (Fig. 22-2), which grows attached, and *Cephalodiscus*, which is free-floating. The existence of pterobranchs was discovered in 1882, but their zoological affinities remained uncertain for many years, until vestiges of a notochord and gill slits as well as a multiple body cavity (coelom) were found in *Cephalodiscus*. The notochord is a longitudinal skeletal rod, which, like gill slits and other structures, characterizes vertebrate animals (phylum Chordata). *Cephalodiscus*, however, is not a true chordate. The pterobranchs, together with burrowing worm-like animals (Enteropneusta) are classed in a subphylum of "half vertebrates," called Hemichordata, and with the ascidians or sea squirts (Tunicata), in the phylum Protochordata. In structure and origin of the coelom, hemichordates, chordates, and echinoderms differ from all other animals having a body cavity. Graptolites now are classified as an extinct group of hemichordates.

The graptolites and pterobranchs include bottom dwellers and free-drifting forms. Graptolites are distributed geologically from Cambrian to Mississippian. They reached their climax in free-floating, rapidly evolving Ordovician and Silurian representatives of the group, and fossil remains of these types are especially useful for stratigraphical zonation and correlation. Fossil pterobranchs are scarce; their known range is Ordovician to Recent.

RHABDOPLEURA, A REPRESENTATIVE PTEROBRANCH

In order better to understand the structures of the geologically important graptolites, the morphology of their closest living relative, *Rhabdopleura* (Fig. 22-2), may be examined first. This animal grows in small encrusting colonies with skeletal parts consisting of chitinoid tubes. The main tubes, called **creeping branches** (Fig. 22-2, *8*), with a diameter of 0.5 mm. or less, are attached along one side to a pebble, shell, or other suitable foundation. They are secreted by specialized **budding zooids** (Fig. 22-2, *12*), similarly attached. The creeping branches are composed of incomplete rings of chitinoid material, joined together at the dorsal side of the tubes by a zigzag suture. A delicate tubule (**stolon**, Fig. 22-2, *10*) extends backward from the budding individual through the creeping branch, which is partitioned off at intervals by **septa** (Fig. 22-2, *11*). Each separate compartment formed by the septa opens into a side branch, which consists either of another creeping branch or an open-ended, partly erect tube (**theca**, Fig. 22-2, *7*). Each theca contains the soft parts of a single individual, which may be extended out of the open end of the tube or withdrawn into the depths of the tube by means of a **contractile stalk** (Fig. 22-2, *5*), linked to the stolon of the creeping branch.

The two types of individuals in a colony

of *Rhabdopleura* thus consist of **budding individuals,** which secrete the creeping branches and other units of the colony, and **feeding individuals,** which gather food and also carry on the function of sexual reproduction, essential for the development of new colonies.

Budding individuals are absent in *Cephalodiscus*, in most species of which males and females are strongly differentiated. Some species contain neuter members of the colonies, neither male nor female;

these have the function of supplying males with food.

Fossil specimens of *Rhabdopleura* have been described from Cretaceous and Paleocene rocks.

CLASSIFICATION OF PROTOCHORDATES

Taking account of structural similarities of the pterobranchs and graptolites, the classification of the phylum Protochordata,

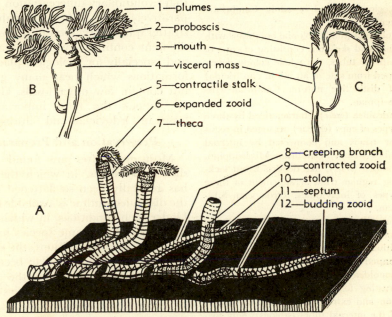

1—plumes
2—proboscis
3—mouth
4—visceral mass
5—contractile stalk
6—expanded zooid
7—theca

8—creeping branch
9—contracted zooid
10—stolon
11—septum
12—budding zooid

Fig. 22-2. **Structure and terminology of Rhabdopleura normanni, a pterobranch.** (*A*) Portion of a colony attached to the substratum, ×20; (*B*) ventral view of a zooid, ×50; (*C*) side view of a zooid, ×50.

budding zooid (12). Animal which secretes creeping branch and gives rise to other zooids by budding.

contracted zooid (9). Animal withdrawn into theca.

contractile stalk (5). Thin muscular connection of zooid with stolon.

creeping branch (8). Portion of skeleton which lies on substratum.

expanded zooid (6). Animal protruded from theca.

mouth (3). Hidden under lower projection of proboscis.

plumes (1). Pair of feathery, tentacle-bearing appendages.

proboscis (2). Hollow lobe lying at anterior end of body.

septum (11). Partition which separates creeping branch into compartments (thecae), each containing a bud or zooid.

stolon (10). Brown chitinoid tube which extends through thecae of creeping branch, and to which all zooids are connected by means of contractile stalks.

theca (7). Compartment occupied by a bud or zooid.

visceral mass (4). Main soft parts of zooid.

which contains these two groups, is summarized in the following tabulation.

Main Divisions of Protochordates

Protochordata (*phylum*), animals which possess a notochord during part of their life history, standing intermediate between invertebrates and true chordates. Their body cavity (coelom) originates in the same manner as that of echinoderms and chordates, in this respect differing from mollusks and all other invertebrates. Cambrian–Recent.

Hemichordata (*subphylum*), notochord restricted to preoral region; three primary segments of coelom retained in adult. Cambrian–Recent.

Graptolithina (*class*), colonial organisms with chitinoid skeletons consisting of rows of cups or tubes, the whole colony developed from side of an initial cup (sicula) of distinctive nature. Cambrian–Mississippian.

Dendroidea (*order*), characterized by three types of cups (thecae) arranged in regular triads and connected by internal tubes (stolons). Cambrian–Mississippian.

Tuboidea (*order*), like dendroids but lacking regular triad arrangement of three types of individuals. Ordovician–Silurian.

Camaroidea (*order*), encrusting colonies composed of two or three types of individuals, the larger with distinctly flask-shaped cups. Ordovician.

Stolonoidea (*order*), encrusting or nonencrusting forms characterized by inflation and extreme irregular convolution of the internal tubes (stolons); only two types of individuals in a colony. Ordovician.

Graptoloidea (*order*), colonies formed of only one type of individual, cups opening into a common internal cavity; stolons lacking. Ordovician–Silurian.

Pterobranchia (*class*), colonial hemichordates with chitinoid hard parts but lacking a sicula; colonies composed of one or two types of individuals; cups opening into a common cavity, or connected by stolons, or entirely separate. Ordovician–Recent.

Enteropneusta (*class*), includes *Balanoglossus* and other burrowing worms. Recent.

Tunicata (*subphylum*), notochord restricted to tail region, coelom disappears in adults. Recent.

GRAPTOLITES

The graptolites are extinct colonial animals, distinguished by the structure of their chitinoid skeletal covering and the arrangement of the tubes or cups occupied by the soft parts of individuals in regular patterns along branches. The world-wide distribution of many genera and the manner in which fossil remains commonly are restricted to narrow zones suggest that the colonies mostly were planktonic organisms, carried by the drift of ocean currents, and that evolutionary differentiation was comparatively rapid. They are valuable for stratigraphic correlation of deposits in the different continents which contain them, and especially for geological placement of formations which bear many graptolites but contain few other fossils. Graptolites are among the most important index fossils in Ordovician and Silurian strata.

Preservation and Preparation

Most graptolites are found in black shales and slates, in which the skeleton has generally been so flattened that only the distended outline is available for study. Under such conditions, in which different specimens of the same species have come to rest in various positions, the specimens may seem to be dissimilar because they present different sides to the observer. Obviously, such material leaves much to be desired, as it does not allow a study of the internal structures; furthermore, the original shape of the external features may not be determinable. Even so, from the early days of paleontological science, fossils of this sort have proved to be stratigraphically useful, since many short-lived strains, classified as genera and species, are sufficiently distinctive to be recognized in the flattened state.

While many genera and species were described and used before 1890, paleontologists had little knowledge of external graptolite morphology until Holm in that year etched undistorted graptolites out of limestones from Sweden. Even less was known about the complex internal structures until a few years later, when Wiman studied similar material in serial sections. Work of this sort has since been carried on elsewhere, and we have come to understand the internal structure of numerous genera, as well as the evolutionary development of some stocks. The most revealing study to date, which has clarified the relationships of the group as a whole, is that of Kozlowski, on fossils from Ordovician cherts in Poland. Kozlowski dissolved the chert matrix by means of hydrofluoric acid. By embedding the freed graptolites in paraffin, he was able, like Wiman, to prepare serial sections with a microtome for the study of structure in minute detail. Despite these and many other works, the internal organization of numerous genera and species remains obscure.

The discovery and study of well-preserved graptolites is a challenge to paleontologists the world over.

Dendroid Graptolites

The dendroid graptolites are so called because of the many-branched treelike mode of growth, which is common to many of their genera. Study of their morphology is undertaken by examining two of the better known genera, *Dictyonema* and *Dendrograptus*.

Occurrence. The Tremadocian rocks of Europe, Schaghticoke shale of North America, and equivalent beds in South America, considered by some to be the topmost Cambrian, but generally classified as basal Ordovician, contain the distinctive dendroid graptolite *Dictyonema flabelliforme* Eichwald. This is a stratigraphically restricted species of a genus which ranges upward into the Mississippian. Because it is so widely distributed and well known, we select this species as an example of dendroid graptolites in general.

Morphologic Features. The colony (**rhabdosome,** *15;* italic numbers and letters in the text describing dendroid graptolites refer to Fig. 22-3) of *Dictyonema flabelliforme* and related species grew in the form of a lacelike bell. In the fossil

state, this bell is almost invariably collapsed into a fan-shaped structure (*E*). The bell consists of two elements: branches (**stipes,** *5*) which diverge from the apex, with division and redivision, and short crossbars (**dissepiments,** *4*), which connect the branches, serving to maintain the bell shape of the entire structure. A third major element of the colony is a thread (**nema,** *2*), by which the bell was suspended. At its distal end, this thread is expanded into a chitinous sheet (**basal disk,** *1*), which served for attachment.

The branches (stipes) housed the individual animals (**zooids**) which formed the colony. Careful investigations of undeformed graptolites etched out of rock matrix have shown that these branches consist of many short overlapping tubes (**thecae,** *6*).

The colony was produced asexually, by budding. The first zooid secreted the first cup or tube (**sicula,** *3*), located at the apex of the later-formed bell. The offspring budded from the first zooid was the builder of the first theca of the first branch. It gave rise to three new zooids, which also secreted thecae, and in turn, one of these budded three additional zooids. The continuation of the budding process formed a single stipe, and the development at intervals of twin buds gave rise to bifurcation of branches.

Wall Structure. The graptolite skeleton consists of a horny material, yellow to brown in transmitted light. In physical and chemical properties, it resembles the hardened chitin which forms the exoskeleton of beetles and other insects. The thecae are thin-walled structures consisting of two main layers (*D*), an inner **fusellar layer** (*12*), and an outer lamellar rind (**cortex,** *13*). The fusellar layer is built up of successive growth bands (**fusellar half rings,** *14*), each of which extends halfway around the theca. The bands of the right side meet those of the left side along zigzig **sutures** (*11*) on the venter and dorsum. The lamellar layer is a secondary feature, laid down over the outside of the wall in successive laminae.

Sicula. The first-formed cup or tube of the colony lies at the apex of the bell. It differs from all succeeding thecae, and is therefore differentiated from them as the sicula (3). The sicula of *Dictyonema flabelliforme* and of most other graptolites is conical in shape. The apex of the cone is attached to the nema (2), while the base is open. The apical portion of the sicula (**prosicula**) is delicate, and is strengthened by a spiral thread or band; the succeeding portion of the sicula (**metasicula**) is built like the other thecae of the skeleton. The zooid of the sicula starts the first stipe by budding a second zooid; this emerges through a hole (**foramen**) in the side of the sicula, and produces the first theca of the first stipe.

Thecae. The plan of thecal development is particularly well shown by undeformed specimens of *Dendrograptus* studied by Kozlowski (Fig. 22-4). *Dendrograptus*

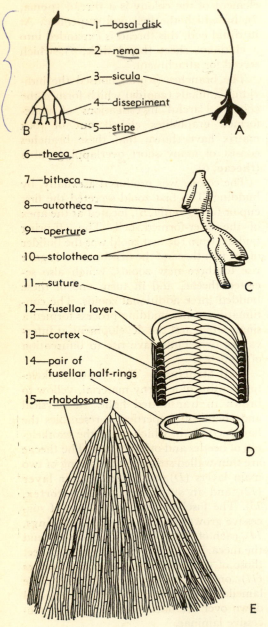

1—basal disk
2—nema
3—sicula
4—dissepiment
5—stipe
6—theca
7—bitheca
8—autotheca
9—aperture
10—stolotheca
11—suture
12—fusellar layer
13—cortex
14—pair of fusellar half-rings
15—rhabdosome

A B C D E

Fig. 22-3. **Structure and terminology of Dictyonema flabelliforme, a representative dendroid.** (*A*) Beginning of a colony, ×3; (*B*) proximal portion of a colony, ×1; (*C*) portion of a branch, ×30; (*D*) wall structure, diagrammatic; (*E*) entire colony, ×0.5.

aperture (9). Opening of one of colonial cups or tubes.

autotheca (8). Largest of three cups or tubes produced by each act of budding (Fig. 22-4, *1*).

basal disk (1). Chitinous patch at end of nema for attachment of colony.

bitheca (7). Small cup or tube accompanying autotheca (Fig. 22-4, *2*).

cortex (13). Outer layer of skeleton, consisting of superimposed laminae.

dissepiment (4). Crossbar of skeleton, uniting adjacent branches.

fusellar layer (12). Inner layer of skeleton, composed of half rings (14).

nema (2). Chitinous threadlike tube which terminates in basal disk at one end and sicula at other.

pair of fusellar half rings (14). Elements of chitinoid skeleton which fit together with zigzag suture.

rhabdosome (15). Whole graptolite colony, developed by budding from a single sicula.

sicula (3). Cup belonging to initial zooid of colony.

stipe (5). Branch of colony, consisting of overlapping thecae.

stolotheca (10). Cup or tube of each set of three thecae from which a succeeding generation of three thecae is budded (Fig. 22-4, *3*).

suture (11). Zigzag union between fusellar half rings.

theca (6). Any cup or tube of colony.

differs from *Dictyonema* in various ways, as in lack of nema and dissepiments. The plan of thecal budding, however, is generally the same for all dendroids, so that we may draw on *Dendrograptus* to supplement the discussion of *Dictyonema flabelliforme*. Growth of the branches is not by simple addition of identical thecae, as might be supposed. The appearance of new individuals, forming a new generation, takes place in groups of three; that is, the normal act of budding gives rise to three thecae, each of which belongs to a distinct type. The three types are named **autotheca** (Fig. 22-4, *1*) (*auto*, self, principal one), **bitheca** (Fig. 22-4, *2*) (*bi*, two, secondary), and **stolotheca** (Fig. 22-4, *3*)

(*stolon*, tube). The autotheca and bitheca open to the exterior, whereas the stolotheca (except at the end of a branch) does not, its end being filled by the base of the next autotheca. Though least conspicuous, the stolotheca is an essential element of the triad, for it housed the **budding zooid** which produced the next generation of zooids. The zooids which occupy the autotheca and bitheca do not give rise to new individuals by budding.

The thecae are connected by delicate tubules (**stolons**, Fig. 22-4, *4*). A stolon extends through the length of each stolotheca. Near the distal end of the stolotheca it divides into three stolons. Of these, a short branch leads to the base of the suc-

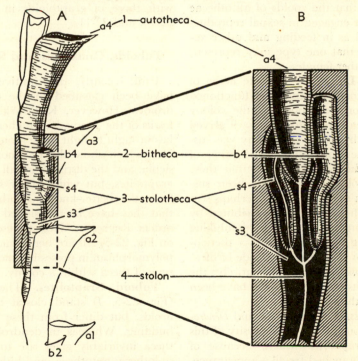

FIG. 22-4. Internal structure of Dendrograptus regularis. (*A*) Portion of a stipe, ×25; (*B*) cutaway view of part included in the shaded rectangle in *A*, ×50. Successive generations of thecae are indicated by numbers: a1, a2, a3, a4, autothecae; b2, b4, bithecae (b3, b5, hidden by autothecae a2 and a4); s3, s4, stolothecae.

autotheca (1). Largest of thecae.
bitheca (2). Smallest of thecae.
stolon (4). Chitinoid tube extending through successive stolothecae of a stipe, and sending off branch stolons to base of each autotheca

and bitheca.
stolotheca (3). Theca which contains stolon and which gives rise, by budding, to triad of offspring thecae: another stolotheca, an autotheca, and a bitheca.

ceeding autotheca, another to the base of the equivalent bitheca, and a third into and through the succeeding stolotheca, where the same organization is repeated.

The function of individuals housed in the stolothecae was evidently that of asexual reproduction in building the colony. The functions of zooids of the autothecae and bithecae are not known; evidently they were specialized for different tasks. By analogy with coelenterates, the suggestion has been made that the gathering and digestion of food for the benefit of the entire colony may have been the function of one group and defense that of the other. The recent proof of the affinities of the graptolites and pterobranchs makes it more likely that the zooids of autothecae and bithecae engaged in sexual reproduction, as well as in feeding and other activities, and that one type may represent males, the other females.

Mode of Life. The delicate nema of *Dictyonema* could hardly have functioned as a stem for the support of the colony above the sea floor. More likely, it served as a thread by which the colony was suspended. The world-wide distribution of *Dictyonema flabelliforme* suggests that these graptolite colonies may have been suspended from drifting objects, perhaps seaweeds, which were widely distributed by ocean currents. This type of hitchhiking upon drifting bodies is termed a pseudoplanktonic or epiplanktonic mode of life.

Dendrograptus lacks a nema entirely; the flat apex of its sicula appears to have been attached to the sea floor.

The discussion of *Dictyonema* and *Dendrograptus* has touched on the diversity of this group of graptolites which, because of their many-branched treelike appearance, are classified as the order Dendroidea. *Dictyonema flabelliforme* is somewhat exceptional in its adaptation to a pseudoplanktonic existence, for most dendroids lack a nema, or they possess a short, thick one, which presumably anchored the colony to the sea floor. Such benthonic mode of existence is not conducive to wide geo-

graphic distribution; hence, it is not surprising that most dendroid species are geographically restricted. The various genera are differentiated chiefly by the shape of the colony and the mode of branching. Some are fan-shaped, others discoidal, still others extended into feathery streamers. Thecal shape also varies widely and is used to distinguish species.

Geological History. The earliest-known dendroid graptolites are of late Cambrian age and the latest ones, Mississippian. In abundance, rapidity of evolution, and geographic distribution they lag far behind their more specialized side branch, the graptoloids. Representative genera of dendroids are illustrated along with those of graptoloids in Figs. 22-9, 22-10, and 22-11.

Tuboids, Camaroids, and Stolonoids

Until recently, all primitive graptolites have been classified in the order Dendroidea. However, Kozlowski's work on fossils of the Tremadocian chert of Poland has brought to light three groups of organisms which show the polymorphism, the sicula, and the distinctive wall structure of graptolites, but which differ so strikingly from the better-known dendroid genera that they have been assigned to separate orders. Representatives of these are shown on Fig. 22-5. All are benthonic, and show polymorphism in possessing more than one type of theca within a colony.

Tuboid Graptolites. The Tuboidea (Fig. 22-5, *1*) stand closest to the dendroids, but differ from these in mode of budding. Whereas in dendroids a stolotheca invariably gives rise to a triad of stolotheca, autotheca, and bitheca, or to a twin pair of such triads, the budding of the tuboids is irregular, so that autothecae and bithecae come to be scattered through the colony at random.

Camaroid Graptolites. The camaroids (Fig. 22-5, *2*) are known only as encrusting colonies, composed of two or three types of thecae. The largest tubes (presumably autothecae) are expanded at the base.

Stolonoid Graptolites. The Stolonoidea (Fig. 22-5, *3*) are represented by encrusting as well as by free-growing colonies, composed of two types of thecae. Their most characteristic feature is the overdevelopment of the stolon system. Whereas the dendroids and the other orders possess slender, straight, tubular stolons, the stolonoid stolons are inflated and convoluted like intestines.

In addition to the remains of graptolites referable to these orders and to the graptoloids, many graptolite faunas contain other chitinoid remains of organisms which lack a sicula and the fusellar structure characteristic of graptolite hard parts. These problematic organisms are considered to be unrelated to graptolites. Some of these classed as hydrozoans are noted in Chap. 4.

Graptoloids

The Tremadocian rocks which contain *Dictyonema flabelliforme* bear other graptolites, sufficiently distinct from the dendroids to be segregated in another order, the Graptoloidea. Intermediate forms (also classified as graptoloids) clearly indicate the development of the graptoloids from *Dictyonema*. The dendroids remained relatively insignificant until their extinction in the Mississippian period. Graptoloids, on the other hand, rapidly expanded in numbers so as to dominate the pelagic faunas of Ordovician and Silurian time, but they vanished without trace before the beginning of the Devonian period.

Structure of Diplograptus. We begin the study of graptoloids by getting acquainted with the structure of *Diplograptus*, a common Middle and Upper Ordovician genus, which ranges into the basal Silu-

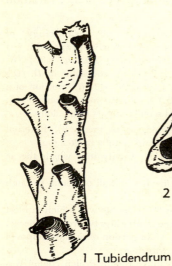
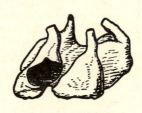
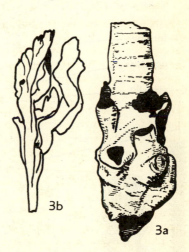

1 Tubidendrum

2 Bithecocamara

3b

3a

Stolonodendrum

Fig. 22-5. **Tuboids, camaroids, and stolonoids.** These three orders are known mainly from the Tremadocian cherts of Poland, but are undoubtedly much more widespread. A representative of each is illustrated.

Bithecocamara Kozlowski (Camaroidea), lower portion of autotheca inflated. *B. vermicollis* Kozlowski (2, ×25), an encrusting type.

Stolonodendrum Kozlowski (Stolonoidea), stolothecae much larger than other thecae and filled with enlarged, convoluted stolons. The exterior is illustrated by *S. ramosum* Kozlowski (3a, ×25), the stolons in 3b (×25).

Tubidendrum Kozlowski (Tuboidea), lacks regular triad arrangement of buds characteristic of dendroids; bithecae generally open in the axil between autotheca and main branch. *T. bulmani* Kozlowski (1, ×2).

rian. It is an advanced, yet not overspecialized member of its order, and has been recorded from all continents.

Much of the terminology of the dendroids is carried over to equivalent features of the graptoloids. Thus, we find that the colony or **rhabdosome** (*10;* italic numbers and letters in the text on morphology of graptoloids refer to Fig. 22-6) of *Diplograptus* began with a **sicula** (*3*) suspended from a **nema** (*7*), that this sicula gave rise to a **theca** (*8*), and so on. Likewise, the wall structure of graptoloids is similar to that of dendroids. *Diplograptus* and all other graptoloids (excepting transitional forms) differ from dendroids, however, in two important respects: they possess only one type of theca within a colony, and they lack stolons.

The nema (commonly also called **virgula**), which is more rigid than in *Dictyonema,* is not merely attached to the apex of the sicula but extends along its dorsal side and projects beyond the aperture as a stout spine termed the **virgella** (*6*). The sicula (*2*) consists of a **prosicula** (*1*) and a **metasicula** (*3*); its aperture is directed downward. The first theca, which grows out of a hole (**foramen,** *4*) in the wall of the metasicula, curves up along the wall of the sicula and opens upward, instead of growing downward. Successively budded thecae do likewise, growing upward along the nema in two rows, back to back. This type of growth, in which the thecae ascend along the nema, is called **scandent.**

The thecae overlap each other to a considerable extent. They vary in shape and may or may not lie in one plane; variation

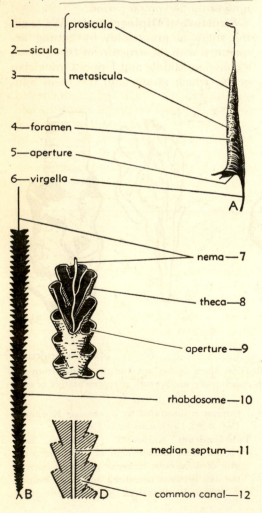

Fig. 22-6. **Structure and terminology of Diplograptus, a representative graptoloid.** (*A*) Sicula of *D. (Orthograptus) gracilis,* ×22; (*B*) carbon film of an entire rhabdosome of *D. (O.) calcaratus incisus* Lapworth, ×2; (*C*) portion of a colony of *D. (O.) gracilis* Roemer, restored, ×8; (*D*) longitudinal section of part of a colony.

aperture (of sicula, 5; of theca, 9). External opening of each cup or tube in skeleton.

common canal (12). Cavity into which thecae open.

foramen (4). Opening in wall of sicula, through which first bud appears.

median septum (11). Double wall which in most members of genus separates two rows of thecae.

metasicula (3). Main portion of sicula, composed of fusellar growth rings.

nema (7). Threadlike rod by which colony was suspended.

prosicula (1). Delicate, juvenile portion of sicula, strengthened by a spiral band.

rhabdosome (10). Skeleton of colony.

sicula (2). Conical skeletal cup of zooid which founded colony.

theca (8). Chitinoid cup or tube which housed zooid.

virgella (6). Spine projecting beyond aperture of sicula.

in these characters has served as the basis for subgeneric divisions.

Studies of undeformed material have shown that each theca of a *Diplograptus* colony opens at the base into a **common canal** (*12*). In some species, this canal opens to both sides of the colony only in the first two thecae; beyond these it splits into two common canals, one for the row of thecae on the left side of the colony, another for the row on the right side; the two canals are separated by a **median septum** (*11*), actually a double wall, which includes the nema. In this case all the thecae budded from the left canal remain on the left side, those from the right canal on the right.

In other species there is a single common canal throughout the colony, and thecae are budded alternately to the right and left. Thus, a colony is produced which may be externally similar to one of the double-canal type but which was constructed by a fundamentally different mode of budding.

Synrhabdosomes. Most specimens of *Diplograptus* occur as single colonies (rhabdosomes) retaining only a portion of that part of the nema which projects beyond the last-formed thecae. However, one occasionally finds radially arranged groups of these fossils, in which the free end of each nema points to the center, the sicular end toward the periphery. A few rare finds have yielded such assemblages showing the nemas centrally united to a basal disk; this is overlain by a delicate film, interpreted as the collapsed remnant of one or several bladder-like floats. Restoration of such an assemblage representing a genus similar to *Diplograptus* is shown in Fig. 22-1. These discoveries have led to the conclusion that *Diplograptus* and similar graptoloids lived in colonies (rhabdosomes) gathered around a central float into a supercolony (synrhabdosome). The supercolonies drifted about as units, but were easily disrupted after death.

Other Graptoloids and Their Evolutionary Trends. Lowermost Ordovician

(Tremadocian) rocks contain graptolites which bridge the gap between the dendroid *Dictyonema* and graptoloids. A few resemble *Dictyonema* to the extent of showing polymorphism, being composed of two or three types of thecae, but most resemble *Diplograptus* in possessing only one type of theca in a colony and in lacking stolons.

The most primitive graptoloids bear 40 or more branches (Fig. 22-7, *1*). Evolutionary development led from these to genera which had 16, 8, 4, 3, 2, and, finally, 1 branch. This trend is illustrated in Fig. 22-7, and may be traced through Figs. 22-9, 22-10, and 22-11.

Another evolutionary trend affects the position of the branches shown in Fig. 22-8 and again in the following ones of this chapter. In primitive position, the branches hung downward from the sicula. From this **pendent** position the branches, during the phylogeny of various stocks, rose through the **declined** or **decurved** positions into a **horizontal** position, and beyond this via the **reclined** or **recurved** into the **scandent** position, illustrated by *Diplograptus*. This trend affected different stocks of graptolites at different times; thus, at least one group reached the scandent stage in the Lower Ordovician, whereas others did not attain it until Mid-Ordovician time.

A third evolutionary trend among graptolites is an increase in the complexity of the individual thecae. Most (though not all) early dendroids possess simple cuplike or tubelike thecae. Likewise, the early graptoloids show simple thecal architecture. Starting from these, various groups developed sigmoidally curved, geniculated, and hooded thecae at different times, as illustrated in Figs. 22-9 to 22-11.

Beside the trends noted, which at one time or another affected nearly all the persistent graptoloid stocks, many groups showed their own peculiar lines of development. Thus, some reduced their skeletons to a meshwork (**clathria**) (Figs. 22-10, *6*; 22-11, *6, 7*). Others developed corkscrew-like colonies (Figs. 22-11, *2, 4*). Some (like

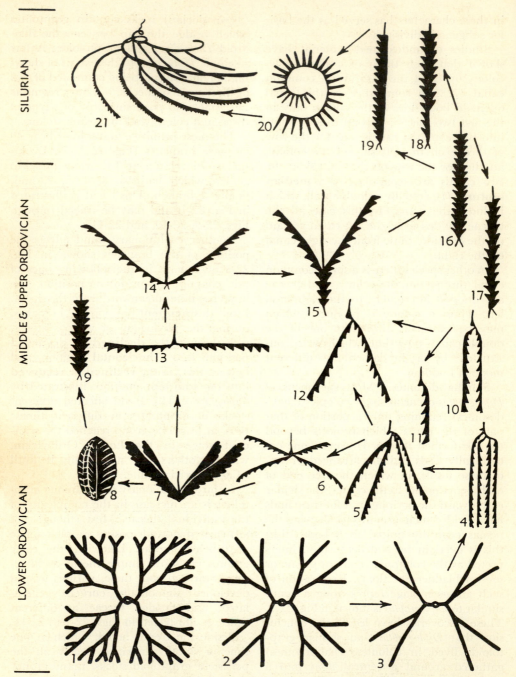

FIG. 22-7. **Evolution in the shape of graptoloid colonies.** The diagram illustrates evolution and stratigraphic distribution of two features of graptolite colonies: number of branches and position of the branches. Some drawings stand for a single genus, whereas others represent several genera. Nature of thecae and other details are conventionalized. Trends are as follows.

(*Continued on next page.*)

Diplograptus, Climacograptus, and *Glossograptus*) lived in supercolonies (Figs. 22-1; 22-10, *4a*), and some Silurian graptoloids, derived from stocks in which the number of branches was reduced to one, again developed branches (Fig. 22-11, *4*). This seeming reversal of an evolutionary trend does not lead to graptolites resembling the ancestral types, because the main branch in the Silurian forms is scandent and coiled.

Each individual zooid in a graptolite colony undergoes a life history or **ontogeny,** reflected in the growth rings of its theca. Likewise, the entire colony shows a life history (called **astogeny**), during which it undergoes change. The thecae formed in the early stages of astogeny, namely, those thecae near the sicula, grow to lesser size than the thecae developed in later stages. But in some genera of graptolites, the astogenic changes are much more striking, since they involve a change in the mode of growth. For example, in Middle Ordovician rocks we commonly find the recurved genus *Dicellograptus* (Fig. 22-10, *1*) and its scandent derivative *Climacograptus* (Fig. 22-10, *2a, b*). The genus *Dicranograptus* (Fig. 22-10, *7*) occupies an intermediate position; it grows according to the scandent plan in the early stages of its life, but changes at a later stage to the reclined or recurved type. Whereas evolution proceeded from a recurved to a scandent stage, the life history of *Dicranograptus,* leading from scandent to recurved, plays the evolutionary record in reverse.

A parallel set of phenomena is to be seen in basal Silurian rocks, in which the biserial *Diplograptus* is associated with its uniserial descendant *Monograptus* (Fig. 22-11, *5a–d*), and the genus *Dimorphograptus* (Fig. 22-11, *3*). *Dimorphograptus*

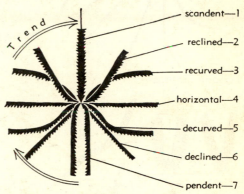

FIG. 22-8. **Evolutionary trends in the position of stipes.** The branches of graptoloids show an evolutionary trend from a position in which they hang downward from nema and sicula, through intermediate positions, to the scandent type of growth, in which the stipes grow upward along the nema.

declined (6). Stipes straight, sloping downward.
decurved (5). Stipes curved, sloping downward.
horizontal (4). Stipes growing out horizontally.
pendent (7). Stipes hanging downward; most primitive position.
reclined (2). Stipes straight, growing upward and outward.
recurved (3). Stipes curved, rising upward and outward.
scandent (1). Stipes growing upward along nema; most advanced position.

(Fig. 22-7 continued)

1–4, Reduction in number of branches from many to four.
4–8, Development of four-branched forms from pendent (4), to declined (5), horizontal (6), recurved (7), and scandent (8). (4–6, *Tetragraptus;* 7, 8, *Phyllograptus.*)
8, 9, By loss of two rows of thecae, production of early biserial forms (9, *Glossograptus*).
4, 10, 11, Loss of branches, resulting in two-stiped (10) and one-stiped (11) forms. (10, *Didymograptus;* 11, *Azygograptus.*)
10, 12–17, Elevation of stipes in two-branched

forms, leading from pendent (10), to declined (12), horizontal (13), reclined (14), partly scandent (15), to scandent (16, 17). (12, 13, *Didymograptus;* 14, *Dicellograptus;* 15, *Dicranograptus;* 16, *Diplograptus;* 17, *Climacograptus.*)
16–18, Loss of one row of thecae, producing a uniserial scandent type (18, *Monograptus*).
19–21, Modifications of the monograptid type, leading to coiled unbranched (20) and coiled branched (21) forms. (20, *Rastrites;* 21, *Cyrtograptus.*)

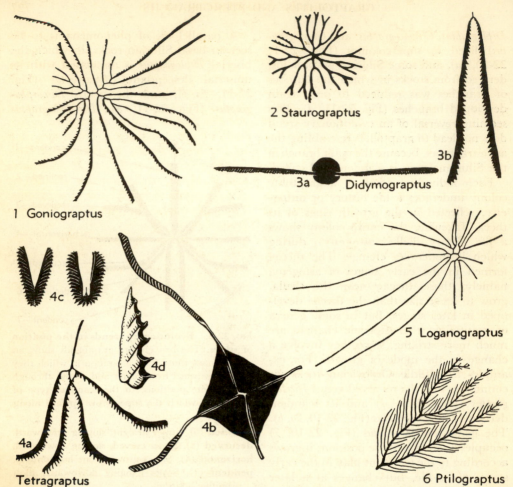

1 Goniograptus

2 Staurograptus

3a **Didymograptus** **3b**

5 Loganograptus

4a 4b 4c 4d

Tetragraptus

6 Ptilograptus

Fig. 22-9. **Representative Lower Ordovician graptolites.** The term Lower Ordovician is here used to include Tremadocian, considered by some as uppermost Cambrian. The oldest fauna is characterized by dendroids (*Dictyonema flabelliforme*). These continue upward but are joined by many-branched graptoloids, which, in ascending the column, largely give way to four- and two-stiped forms; finally, the one-stiped *Azygograptus* appears. Stipe position generally progresses from pendent to horizontal, but *Phyllograptus* carried the evolution of stipe position to a climactic scandent stage. Thecae among graptoloids are simple. All illustrated forms belong to this order except *Ptilograptus*, which is a dendroid.

Didymograptus McCoy, Lower and Middle Ordovician. This two-branched genus varies in growth form from pendent to reclined; thecae of most species are straight. It includes the largest known graptolites, up to 2 m. in length. *D. patulus* Hall (3a, ×0.7), a horizontal species with short nema and large basal disk, Deepkill shale, New York. *D. columbianus* Ruedemann (3b, ×13), a pendent species, Lower Ordovician, British Columbia.

Goniograptus McCoy, Lower Ordovician. A many-stiped graptoloid with distinctive pattern of branching beyond the four second-order stipes; each branch bifurcates into one long undivided branch and a short one which birfurcates also. *G. thoreaui* Ruedemann (1, ×0.7), Deepkill shale, New York.

Loganograptus Hall, Lower Ordovician. Colonies generally show not more than 16 stipes, produced by simple bifurcation. *L. logani* Hall (5, ×0.3), Lower Ordovician, Quebec.

Ptilograptus Hall, Lower Ordovician. The few main branches carry many short, alternate side branches. *P. plumosus* Hall (6, ×1.5), Lower Ordovician, Quebec.

Staurograptus Emmons, basal Ordovician (Tre-

(*Continued on next page.*)

stands structurally between the others; it begins uniserially, but, after having produced a short single row of thecae, switches to a biserial mode of growth. Again, as in *Dicranograptus*, two evolutionary stages are found in a single colony, and these stages follow each other not in the sequence in which they occur during phylogeny, but in reverse order. The colony begins at a phylogenetically advanced stage and reverts later to a more primitive condition. This, a reversal of the usual sequence, is termed **proterogenesis.**

Ecology of Graptoloids. The graptoloids built delicate colonies suspended by a nema. Rhabdosomes of *Diplograptus*, *Glossograptus*, and *Climacograptus* have been found joined around floats into supercolonies. Which other genera of graptoloids possessed floats, and which, lacking such special structures, caught free rides on other drifting bodies, such as seaweeds, remains to be determined. In any case, the structure of the graptoloids, and the wide geographic distribution of their genera and species, have convinced paleontologists that most, if not all, were drifters of one sort or another and were distributed round the globe by currents.

This theory is also supported by the abundant occurrence of graptolites in black, pyritic shales devoid of the remains of benthonic organisms—shales which are judged to have accumulated in stagnant bottom waters low in oxygen and poisoned with hydrogen sulfide, under conditions which were prohibitive to organisms other than anaerobic bacteria. The graptolites evidently did not live in the mud but drifted in the oxygenated surface waters, from which dead colonies or fragments settled into the poisoned waters and onto the anaerobic bottom.

This theory alone does not explain why graptolites are not equally common in more normal types of sediments. Whereas graptolites are exceedingly abundant in the black shales, their occurrence in the lighter colored shales and limestones, rich in benthonic fossils, is sporadic at best. It has been suggested that, if graptolites drifted about, their remains should be equally common in all kinds of sediments. This argument is fallacious on two grounds.

1. The theory that they drifted widely does not imply that they drifted everywhere. Probably they were chiefly dwellers of the open seas and frequently were carried into the major geosynclinal belts which include the chief areas of Ordovician and Silurian black-shale deposits; only rarely may they have reached the remoter parts of the partly landlocked (perhaps commonly reef-girdled) epicontinental seas.

2. It must be remembered that, owing to transportation after death and differential preservation of skeletal remains, distribution of fossil remains is not necessarily the same as that of the living organisms. It is highly probable that dead graptolites which settled to a well-oxygenated bottom teeming with benthonic life were, with few exceptions, demolished to the last theca by multitudes of macroscopic and microscopic scavengers. By contrast, graptolites which settled into poisoned waters and muds of black-shale belts were subjected only to attack of anaerobic bacteria, which evidently were able to break down the less resistant tissues but not the skeleton.

(Fig. 22-9 continued.)

madocian). One of the earliest graptoloids; 40 or more stipes. *S. dichotomus* Emmons (2, ×1), basal Ordovician, New York.

Tetragraptus Salter, Lower Ordovician. All four-branched genera of pendent to reclined habit, and even a partly scandent form leading to *Phyllograptus*, are referred to *Tetragraptus*.

The genus thus represents a polyphyletic assemblage of various stocks. *T. fructicosus* Hall (4a, ×0.7), decurved, basal Ordovician, New York. *T. headi* Hall (4b, ×0.7), horizontal, Lower Ordovician, Quebec. *T. phyllograptoides* Linnarsson (4c, ×1), partly scandent, Sweden; *T. bigsbyi* Hall (4d, ×5), Sweden.

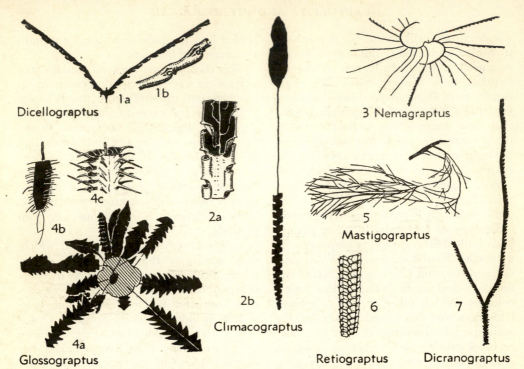

Dicellograptus 1a 1b

3 Nemagraptus

4c

2a

4b

5

Mastigograptus

4a

Glossograptus

2b

Climacograptus

6

7

Retiograptus   Dicranograptus

FIG. 22-10. **Representative Middle and Upper Ordovician graptolites.** Whereas simple thecae were in style in Early Ordovician time, the dominant fashion in the Middle and Late Ordovician included fancy-curved, angled, or spinose thecae. Four-rayed forms had become extinct, and scandent forms became varied and abundant. One side line reduced the skeleton to a thin film strengthened by a meshwork (clathria) of thicker rods. Except *Mastigograptus* (5), all the forms here illustrated are graptoloids.

Climacograptus Hall, Middle and Upper Ordovician. The thecae show a sharp geniculation, emphasized by a flange, producing an angular outline which is distinctive even on specimens reduced to flat films. *C. typicalis* Hall (2a, ×8), Upper Ordovician, North America; *C. parvus* Hall (2b, ×2), Middle Ordovician (Normanskill), New York.

Dicellograptus Hopkinson, Middle and Upper Ordovician. The slender, recurved stipes of this genus bear sigmoidally curved thecae. *D. gurleyi* Lapworth (1a, ×2.5), Normanskill shale, New York; *D. geniculatus* Bulman (1b, part of a stipe, ×10), Middle Ordovician, Sweden.

Dicranograptus Hall, Middle and Upper Ordovician. The initially scandent colonies change to a recurved or reclined habit; thecae are strongly incurved, and in many species spiny. *D. spinifer arkansasensis* Gurley (7, ×0.7), Middle Ordovician (Normanskill), New York.

Glossograptus Emmons, Lower and Middle Ordovician. Differs from *Diplograptus* in bearing rows of spines along sides of the colony. Syn-

rhabdosomes of *G. quadrimucronatus approximatus* Ruedemann (4a, ×3), Mohawkian, New York; *G. ciliatus* Emmons (4b, ×3), Mohawkian, Arkansas; reconstruction (from Bulman) of part of a rhabdosome of the same species from Bolivia (4c, ×15).

Mastigograptus Ruedemann, Upper Cambrian–Upper Ordovician. Generally classed as a dendroid but may belong to another order; one or more main stems branch and rebranch into a multitude of hairlike stipes, which seem to carry two types of thecae. *M. tenuiramus* Walcott (5, ×0.7), Mohawkian, New York.

Nemagraptus Emmons, Middle Ordovician. This distinctive and anomalous genus is an important stratigraphic guide fossil, restricted to a zone in the lower part of the Middle Ordovician (lower Normanskill). The two main stipes give off numerous branches. *N. gracilis* Hall (3, ×1).

Retiograptus Hall, Middle and Upper Ordovician. The very thin thecal walls are supported by a framework of rods (6, diagrammatic, ×2).

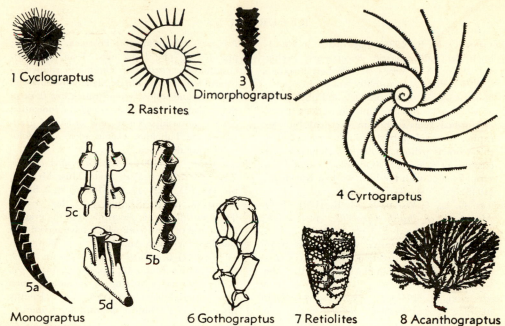

1 Cyclograptus

2 Rastrites

3 Dimorphograptus

4 Cyrtograptus

5c

5b

5a

5d

Monograptus 6 Gothograptus 7 Retiolites 8 Acanthograptus

Fig. 22-11. **Representative Silurian graptolites.** Silurian dendroids are not very distinctive, but the graptoloid fauna contains few of the Ordovician genera, and is dominated by new forms. Trends include (1) reduction of stipes from the two-stiped scandent *Diplograptus* to *Monograptus*, (2) elaboration of thecae, (3) coiling and branching of monograptids, (4) in the descendants of *Retiograptus* a progressive reduction of thecal walls, until only a framework remains. Of the illustrated forms, *Acanthograptus* (8) is a dendroid, *Cyclograptus* (1) a tuboid, and the remainder are graptoloids.

Acanthograptus Spencer, Cambrian–Silurian. A bushy, benthonic dendroid. *A. granti* Spencer (8, ×0.5), Niagaran, Canada.

Cyclograptus Spencer, Silurian. An encrusting member of the order Tuboidea. *C. rotodentatus* Spencer (1, ×0.7), Niagaran, New York.

Cyrtograptus Carruthers, Silurian. A highly developed descendant of *Monograptus*, possessing a main, uniserial stipe coiled in a helical spiral, which gives off numerous branches. *C. kirki* Ruedemann (4, ×0.7, also Fig. 22-7, *21*), Niagaran, Idaho.

Dimorphograptus Lapworth, Lower Silurian. Morphologically intermediate between *Diplograptus* and *Monograptus*, possessing only a single row of thecae in the early portion of the colony, but becoming biserial in the later stages. *D. confertus* Nicholson (3, ×2), Scotland.

Gothograptus Frech, Middle and Upper Silurian. Represents culmination of the skeleton-reducing forms; in some of its species only the framework remains. *G. simplex* Eisenack (6, ×13), Upper Silurian, Baltic region.

Monograptus Geinitz, Silurian. The most common Silurian graptoloid genus. It arose from *Diplograptus* by reduction of thecae to a single row. *M. bohemicus* Barrande (5a, ×3), Niagaran, Oklahoma, common in other parts of the world; *M. dubius* Suess (5b, part of a stipe, restored, ×4), Upper Silurian, Baltic region; *M. lobiferus* McCoy (5c, two restored thecae, ×5), Llandoverian, Wales; *M. convolutus* Hisinger (5d, two restored thecae, ×5), Llandoverian, Wales.

Rastrites Barrande, Lower Silurian. A curved monograptid; the thecae are long, slender, and separated. *R. linnaei* Barrande (2, ×2), Bohemia.

Retiolites Barrande, Lower and Middle Silurian. Descendant of the Ordovician *Retiograptus* from which it may be distinguished by reduction of thecal walls to a reticulate framework within the coarser meshes of the clathria. *R. geinitzianus* Barrande (7, ×7), Sweden.

REFERENCES

BULMAN, O. M. B. (1938) *Graptolithina:* in Handbuch der Paläozoologie, Borntraeger, Berlin, vol. 2D, pp. 1–92, figs. 1–42. The best introduction to a study of graptolites, containing excellent discussions of morphology and phylogeny, as well as descriptions and illustrations of the established genera and an excellent bibliography. Although published in a German series, this work is written in English. Bulman's discussion of dendroid morphology should be supplemented with the later work of Kozlowski (1948).

ELLES, G. L., & WOOD, E. M. R. (1901–1918) *Monograph of British Graptolites:* pts. I–XI, Palaeont. Soc., pp. 1–539, pls. 1–52, figs. 1–359. The most comprehensive work on the graptolites of the British section, which serves as a standard for Ordovician black-shale sections the world over.

KOZLOWSKI, R. (1948) *Les graptolithes et quelques nouveaux groupes d'animaux du Tremadoc de la Pologne:* Palaeontologia Polonica, vol. III, pp. 1–235, pls. 1–42, figs. 1–66 (French). Graptolites etched from chert and studied in transmitted light and in serial sections show wall structure and stolon systems much like those of the hemichordate genus *Rhabdopleura*, indicating close relationship. Three new orders of graptolites, the Tuboidea, Camaroidea, and Stolonoidea are described.

RUEDEMANN, R. (1904–1908) *Graptolites of New York:* Pt. 1, New York State Mus. Mem. 7, pp. 455–803, pls. 1–17, figs. 1–105; Pt. 2, Same, Mem. 11, pp. 1–583, pls. 1–31, figs. 1–484. These are the classic descriptions of the graptolite faunas of New York which have served as a standard for North America.

—— (1947) *Graptolites of North America:* Geol. Soc. America Mem. 19, pp. 1–652, pls. 1–92. A comprehensive treatment, including much new material from the West and a lengthy discussion of the history of graptolite studies.

CHAPTER 23

Conodonts

Conodonts are minute toothlike and plate-like fossils found in rocks ranging in age from Ordovician to Triassic. The largest have a length of slightly more than 2 mm. They are composed of calcium phosphate and are homogeneous throughout. Worm mandibles, called scolecodonts, differ from conodonts in being composed of chitinous and siliceous material. The specific gravity of conodonts is greater than that of quartz, calcite, and dolomite, making it possible to separate them from most sediments by heavy liquids. Because they are not soluble in acetic acid, they may be extracted by the use of this acid from limestones containing them.

Conodonts are good index fossils in many Paleozoic formations. They are common in sandstones and shales of the Ordovician and Mississippian systems and in shales of the Devonian and Pennsylvanian systems. Their small size and resistance to weathering makes them useful in the correlation of surface and subsurface formations.

Morphological characters of conodonts are illustrated in Fig. 23-1 and defined in the accompanying explanation.

SEDIMENTARY ENVIRONMENT

Conodonts occur in greater abundance in shales rich in organic matter than in any other rock type. They are abundant in some black fissile shales, which lack a typical association of marine invertebrates. Also, they are found in sandstones, especially sandstone interbedded with shale, and in some limestones where they are most abundant in thin shaly layers. Conodonts may be concentrated into zones near the top or bottom of lithologic units. To some extent, they seem to represent lag concentrates.

Their wide stratigraphic and geographic distribution, coupled with the mode of occurrence in different types of sedimentary deposits, strongly suggests that the conodonts are remains of pelagic (nektoplanktonic) organisms—not bottom dwellers. This conclusion agrees with and may largely explain the great importance of conodonts in stratigraphic correlation. They are very useful guide fossils. Representative forms from Paleozoic systems are shown in Fig. 23-2.

CLASSIFICATION

The zoological relationships and the function of conodonts is undetermined. They have been considered to belong to the fishes, gastropods, worms, crustaceans, and cephalopods. It is believed that conodonts are parts of a vertebrate animal because they are composed of phosphate of lime, are frequently attached to bonelike material, and have been found in association with remains of vertebrates, as in the Harding sandstone of Ordovician age in Colorado. Probably they should be assigned to the fishes, which have fossil representatives in Ordovician and later times. Conodonts were capable of rejuvenating broken or lost parts, and they may have functioned as internal supports for tissues within the bodies of certain fishes, such as the gills.

Conodonts have two different types of internal structure, fibrous and laminar. The internal structure of the fibrous type is similar to a bundle of fibers. The laminar type seems to be made up of laminated layers. The genera are grouped into suborders and families according to the internal arrangement and other characteristics, as follows.

Divisions of Conodonts

Conodontophoridia (*order*), minute teeth of variable shape, ranging from simple cones to complex plates composed of calcium phosphate and attached to jawlike fragments of similar composition. Ordovician–Triassic.

Neurodontiformes (*suborder*), conodonts made up of bundles of fibers. Ordovician.
Conodontiformes (*suborder*), conodonts composed of laminated layers. Ordovician–Triassic.

Conodont assemblages have been found which represent teeth of various genera in the present classification, all of which may have existed in the body of one individual animal.

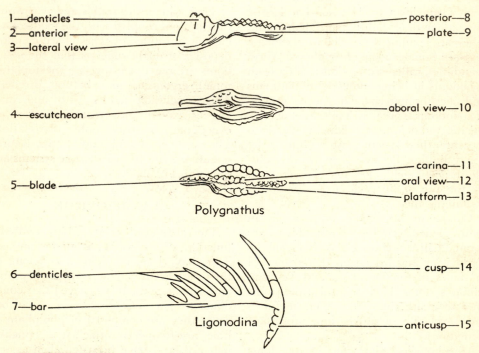

Fig. 23-1. **Morphology of the conodonts.** Parts are defined in the following alphabetically arranged list, cross-indexed to the figure by numbers.

aboral view (10). Attachment side of a conodont blade or bar.
anterior (2). Forward end of a bar or blade, which is convex side of denticles and cones, and direction opposite to inclination of denticles in blades.
anticusp (15). Anterior downward projection.
bar (7). Slender shaft to which discrete denticles are attached, or any conodont with a single large denticle near one end above escutcheon and discrete denticles.
blade (5). Conodont with one large denticle near end or in middle third above escutcheon with fused denticles.
carina (11). Central ridge on oral surface, which

is generally nodose and extended along center of platform.
cusp (14). Large denticle generally above escutcheon.
denticles (1, 6). Small spinelike teeth.
escutcheon (4). Pit or cavity on aboral surface.
lateral view (3). Side view.
oral view (12). Side of a bar or blade to which denticles are attached.
plate (9). Flat edge of a blade.
platform (13). Laterally broadened portions of a blade.
posterior (8). Rear or back end of a bar or blade, which is end toward which denticles are inclined.

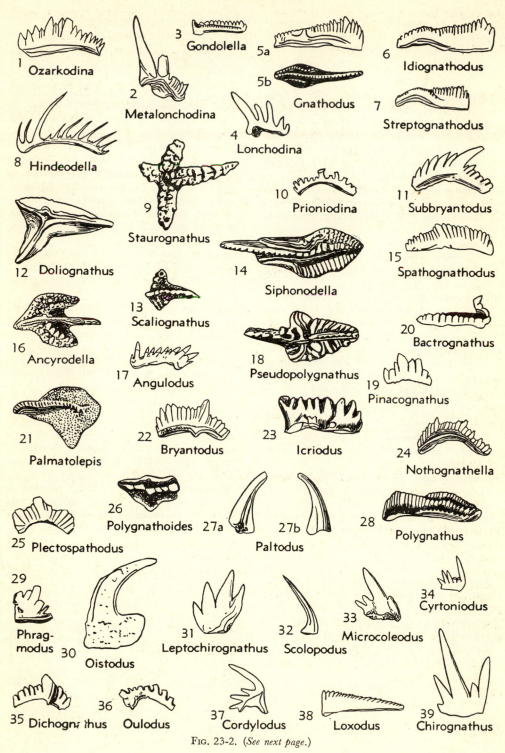

1 Ozarkodina
2 Metalonchodina
3 Gondolella
4 Lonchodina
5a
5b Gnathodus
6 Idiognathodus
7 Streptognathodus
8 Hindeodella
9 Staurognathus
10 Prioniodina
11 Subbryantodus
12 Doliognathus
13 Scaliognathus
14 Siphonodella
15 Spathognathodus
16 Ancyrodella
17 Angulodus
18 Pseudopolygnathus
19 Pinacognathus
20 Bactrognathus
21 Palmatolepis
22 Bryantodus
23 Icriodus
24 Nothognathella
25 Plectospathodus
26 Polygnathoides
27a
27b Paltodus
28 Polygnathus
29 Phragmodus
30 Oistodus
31 Leptochirognathus
32 Scolopodus
33 Microcoleodus
34 Cyrtoniodus
35 Dichognathus
36 Oulodus
37 Cordylodus
38 Loxodus
39 Chirognathus

FIG. 23-2. (*See next page.*)

FIG. 23-2.   **Representative Paleozoic conodonts.** The illustrated genera are stratigraphically grouped by geologic systems. 1–3, 5–7, Pennsylvanian; 4, 8–15, 18–20, Mississippian; 16, 17, 21–24, 28, Devonian; 25–27, Silurian; 29–39, Ordovician.

Ordovician (29–39)

Ordovician conodont faunas contain a large number of simple cones and fibrous types. Many of the bars and blades have prominent cusps.

Chirognathus Branson & Mehl, Ordovician. Conical unequal denticles fused at the bases. *C. varians* Branson & Mehl (39, ×15), Harding sandstone, Colorado.

Cordylodus Pander, Ordovician. Recurved cones with thin, laterally compressed base. *C. intermedius* Furnish (37, ×20), Prairie du Chien, Minnesota.

Cyrtoniodus Stauffer, Ordovician. Laterally compressed bar with backward inclined denticles. *C. complicatus* Stauffer (34, ×20), Bighorn formation, Wyoming.

Dichognathus Branson & Mehl, Middle Ordovician. The bladelike compressed dental units have a large denticle at mid-length. *D. typica* Branson & Mehl (35, ×20), Missouri.

Leptochirognathus Branson & Mehl, Ordovician. Palm-shaped dental units with pointed denticles. *L. quadrata* Branson & Mehl (31, ×20), McLish formation, Oklahoma.

Loxodus Furnish, Ordovician. Bladelike dental units with posteriorly inclined denticles. *L. bransoni* Furnish (38, ×20), Prairie du Chien, Minnesota.

Microcoleodus Branson & Mehl, Middle Ordovician. The deeply excavated cone is flanked by short denticulate bars. *M. typus* Branson & Mehl (33, ×20), Joachim dolomite, Missouri.

Oistodus Pander, Ordovician. Bilaterally symmetrical, laterally compressed simple cones. *O. pandus* Branson & Mehl (30, ×20), Jefferson City dolomite, Missouri.

Oulodus Branson & Mehl, Middle Ordovician. The denticulate bar is twisted and arched. *O. mediocris* Branson & Mehl (36, ×20), Plattin limestone, Missouri.

Phragmodus Brandon & Mehl, Ordovician. Dental units sheathlike. *P. undatus* Branson & Mehl (29, ×20), Viola limestone, Oklahoma.

Scolpodus Pander, Ordovician. Dental units bilaterally symmetrical with fluted faces. *S. quadraplicatus* Branson & Mehl (32, ×20), Prairie du Chien formation, Iowa.

Silurian (25–27)

Silurian conodont faunas contain large numbers of bars and blades, several simple cones, and a few platforms.

Paltodus Pander, Ordovician–Silurian. Simple cones with rounded anterior side and sharp posterior side. *P. unicostatus* Branson & Mehl (27a, b, ×20), Silurian (Bainbridge), Missouri.

Plectospathodus Branson & Mehl, Middle Silurian. A curved bladelike bar set with confluent denticles. *P. flexuosus* Branson & Mehl (25, ×20), Bainbridge formation, Missouri.

Polygnathoides Branson & Mehl, Silurian. A platform type with a median carina of sharp denticles. *P. siluricus* Branson & Mehl (26, ×20), Bainbridge formation, Missouri.

Devonian (16, 17, 21–24, 28)

Devonian conodont faunas include a large variety of bars, blades, and platforms. Many genera are restricted to Devonian rocks such as *Angolodus, Ancyrodella, Icriodus,* and *Nothognathella.*

Ancyrodella Ulrich & Bassler, Devonian. Dental plate heart- or arrow-shaped. *A. lobata* Branson & Mehl (16, ×20), Sweetland Creek shale, Iowa.

Angulodus Huddle, Devonian. Barlike, arched dental units. *A. elongatus* Stauffer (17, ×20), Minnesota.

Bryantodus Ulrich & Bassler, Devonian–Missis-

sippian. Arched blade with a large denticle near mid-length. *B. nitens* Youngquist & Miller (22, ×20), Devonian (Sweetland Creek shale), Iowa.

Icriodus Branson & Mehl, Middle and Upper Devonian. Platform type, dental unit lachrymiform in outline. *I. iowaensis* Youngquist & Peterson (23, ×20), Sheffield formation, Iowa.

Nothognathella Branson & Mehl, Upper Devonian. The bladelike dental units are arched. *N. brevidonta* Youngquist (24, ×20), Devonian, Iowa.

(Continued on next page.)

(Fig. 23-2 continued.)

Palmatolepis Ulrich & Bassler, Devonian. Tongue-shaped dental plate with prominent keel. *P. unicornis* Miller & Youngquist (21, ×20), Sweetland Creek shale, Iowa.

Polygnathus (Hinde), Upper Devonian–Middle Mississippian. A platform type with a long free blade and a transversely concave lanceolate plate. *P. carinifera* Youngquist & Peterson (28, ×20), Devonian (Sheffield formation), Iowa.

Mississippian (4, 8–15, 18–20)

Mississippian conodont faunas contain a large number of bars and blades and many platforms, with several genera (such as *Pinacognathus, Subbryantodus, Bactrognathus, Siphonodella, Pseudopolygnathus, Doliognathus,* and *Scaliognathus*) restricted to Missippian rocks.

Bactrognathus Branson & Mehl, Mississippian. Slender bladelike units with posterior end bent laterally. *B. angularis* Branson & Mehl (20, ×20), Oklahoma.

Doliognathus Branson & Mehl, Mississippian. Platelike dental units with branched carina and keel. *D. lata* Branson & Mehl (12, ×20), Pierson limestone, Missouri.

Hindeodella Ulrich & Bassler, Ordovician–Permian. Arched barlike dental units with large subterminal fang. *H. curvidens* Cooper (8, ×20), Mississippian, Arkansas.

Lonchodina Ulrich & Bassler, Silurian–Permian. An arched bar with irregular denticles. *L. separata* Branson & Mehl (4, ×20), Mississippian, Oklahoma.

Pinacognathus Branson & Mehl, Mississippian. The thick blade has short dental units. *P. subbrevis* (Ulrich & Bassler) (19, ×20), Oklahoma.

Prioniodina Ulrich & Bassler, Devonian–Pennsylvanian. Arched dental units with subequal limb denticles. *P. symmetrica* Branson & Mehl (10, ×20), Mississippian, Missouri.

Pseudopolygnathus Branson & Mehl, Lower Mississippian. The platform is transversely ridged. *P. carinata* Youngquist & Patterson (18, ×20), Prospect Hill sandstone, Iowa.

Scaliognathus Branson & Mehl, Mississippian. Anchor-shaped dental units with posterior transverse bar. *S. anchoralis* Branson & Mehl (13, ×20), Pierson limestone, Missouri.

Siphonodella Branson & Mehl, Lower Mississippian. The platform type dental unit is a transversely concave lanceolate plate. *S. duplicata* (Branson & Mehl) (14, ×20), Prospect Hill sandstone, Iowa.

Spathognathodus Branson & Mehl, Silurian–Permian. Bladelike dental units straight, compressed, and highest near anterior end. *S. abnormis* (Branson & Mehl) (15, ×20), Mississippian (Bushberg sandstone), Missouri.

Staurognathus Branson & Mehl, Mississippian. Barlike dental units with transverse bar. *S. cruciformis* Branson & Mehl (9, ×20), Sycamore formation, Oklahoma.

Subbryantodus Branson & Mehl, Mississippian. Arched blade with one large denticle at apex of arch. *S. camaratus* Cooper (11, ×20), Mississippian, Oklahoma.

Pennsylvanian (1–3, 5–7)

Most of the conodonts found in Pennsylvanian rocks range into the Lower Permian. *Idiognathodus* and *Polygnathodella* are restricted to the Pennsylvanian system. Bars and blades are most numerous, with several platform types present.

Gnathodus Pander, Mississippian–Pennsylvanian. Cuplike dental units with convex oral surface. *G. websteri* Youngquist & Downs (5a–b, ×20), Pennsylvanian, Iowa.

Gondolella Stauffer & Plummer, Pennsylvanian–Permian. Tongue-shaped dental unit with sharply denticulate carina. *G. merrilli* Gunnell (3, ×20), Pennsylvanian, Missouri.

Idiognathodus Gunnell, Pennsylvanian. Platform with oral surface cross ridged. *I. iowaensis* Youngquist & Heezen (6, ×20), Pennsylvanian, Iowa.

Metalonchodina Branson & Mehl, Mississippian–Permian. Arched bar with one exceptionally large denticle. *M. deflecta* Youngquist & Heezen (2, ×20), Pennsylvanian, Iowa.

Ozarkodina Branson & Mehl, Ordovician–Permian. Arched dental units which are thin, bladelike, and with compressed denticles. *O. campbelli* Youngquist & Heezen (1, ×20), Pennsylvanian, Iowa.

Streptognathodus Stauffer & Plummer, Pennsylvanian–Permian. Elongate, medially trenched platform. *S. symmetricus* Youngquist & Heezen (7, ×20), Pennsylvanian, Iowa.

ORIENTATION

The proper orientation of conodonts is a problem since their zoological position is not known. The arbitrary orientation now used by most students is as follows: (1) the points of the denticles are upward or oral, (2) the attachment scar is downward or aboral, (3) the blade end of the platform is forward or anterior, (4) the denticles of both the bars and blades and the simple cones are inclined backward or posterior, and (5) the convex curvature of the longitudinal axis of a blade, bar, or platform is the outer side.

REFERENCES

BRANSON, E. B., & MEHL, M. G. (1933) *Conodont studies Nos.* 1, 2, 3, & 4, Missouri Univ. Studies, vol. 8, nos. 1–4, pp. 1–349, pls. 1–28. Descriptions and illustrations of Ordovician, Silurian, Devonian, and Mississippian conodont faunas of Missouri and Colorado. Hinde's types of conodonts in the British Museum are described and illustrated.

COOPER, C. L. (1939) *Conodonts from a Bushberg-Hannibal horizon in Oklahoma,* Jour. Paleontology, vol. 13, pp. 379–422, pls. 39–47. Descriptions and illustrations · of Mississippian conodonts from southern Oklahoma.

ELLISON, S. P., JR. (1941) *Revision of the Pennsylvanian conodonts,* Jour. Paleontology, vol. 15, pp. 107–143, pls. 20–23. Includes range charts, descriptions, and illustrations of Pennsylvanian conodonts from Missouri and Kansas.

——— (1946) *Conodonts as Paleozoic guide fossils,* Am. Assoc. Petroleum. Geologists Bull., vol. 30, pp. 93–110, 3 figs. Includes an illustrated range chart of conodonts and a discussion of orientation, types, and faunas.

HOLMES, G. B. (1928) *A bibliography of the conodonts with descriptions of early Mississippian species,* U.S. Nat. Mus. Proc., vol. 72, art. 5, pp. 1–38, pls. 1–11. Contains a bibliography of conodont literature to 1926 and a bibliographic list of conodont genera.

STAUFFER, C. R. (1935) *Conodonts of the Glenwood beds,* Geol. Soc. America Bull., vol. 46, pp. 125–168, pls. 9–12. Contains descriptions and illustrations of Ordovician conodonts from Minnesota.

——— (1938) *Conodonts of the Olentangy shale,* Jour. Paleontology, vol. 12, pp. 411–443, pls. 48–53. Illustrations and descriptions of Devonian conodonts in Ohio.

ULRICH, E. O., & BASSLER, R. S. (1926) *A classification of the toothlike fossils, conodonts, with descriptions of American Devonian and Mississippian species,* U.S. Nat. Mus. Proc., vol. 68, art. 12, pp. 1–63, pls. 1–11. Includes a discussion of the classification of conodonts and descriptions and illustrations of Devonian and Mississippian species.

YOUNGQUIST, W., & CULLISON, J. S. (1946) *The conodont fauna of the Ordovician Dutchtown formation of Missouri,* Jour. Paleontology, vol. 20, pp. 579–590, pls. 89, 90. Illustrations and descriptions of Ordovician conodonts from Missouri.

——— & HEEZEN, B. C. (1948) *Some Pennsylvanian conodonts from Iowa,* Jour. Paleontology, vol. 22, pp. 767–773, pl. 118. Descriptions and illustrations of a Pennsylvanian conodont assemblage from Iowa.

Index

A

abdomen, arthropod, 466
aboral field, starfishes, 664
Abrotocrinus, 643
abyssal environment, 20
abyssopelagic environment, 21
Acantherpestes, 567
acanthine septa, 122, 125, 137
Acanthocephala, 456
Acanthocladia, 180
Acanthoclema, 175
Acanthoclymenia, 370
Acanthodesia, 193
Acanthograptus, 731
Acanthopecten, 420
acanthopore, 160, 162, 163
acanthopore ridge, 168
Acari, 565
acceleration, septa, 122, 137
Acephala, 268
acetabulum, 676
Achatella, 505
Acila, 416, 438, 442
acline, 408
Acoelomata, 17
Acratia, 535
Acrocephalops, 499
Acrocrinidae, 634
Acrocrinus, 647
Acrophyllum, 127
Acrothele, 224
Acrotreta, 200, 224
Acrotretacea, 220, 224, 225, 263
Actaeon, 318, 324
Actaeonella, 312
Actaeonina, 329
actinal furrows, 703
Actinaraea, 147
Actiniaria, 104, 152
Actinocamax, 391
Actinoceras, 353, 354
Actinoceroida, 342, 353, 364
actinoceroids, 353, 354, 357, 361
Actinocrinites, 645
actinosiphonate deposits, 342, 350, 352, 361
Actinostroma, 108
Actiplea, 75
acute-angle interarea, 214
adambulacrals, starfishes, 664
adapical, 340
adaptive convergence and divergence, 34
adductor muscles, brachiopods, 201
 ostracodes, 529
 pelecypods, 402
adductor scars, brachiopods, 206, 212 222

adductor scars, pelecypods, 403, 431
Adeona, 193
Adesmacea, 412, 444
adjustor muscle scars, 212
adnation area, 350
adoral, 340
adradials, starfishes, 664
advolute, 288, 347, 354, 370
Aechmina, 532
Aeger, 547, 550
Aeglina, 511
Aerosoma, 683
Aesiocrinus, 647
Aganaster, 672
Agaricocrinus, 645
Agassizocrinus, 643
Agelacrinites, 586
Aglaspida, 555
aglaspids, 555
Aglaspis, 555, 556
Agnostida, 487, 494, 495, 499, 512, 515
Agnostus, 478
Agnotozoa, 17
Agoniatites, 369, 370
Ajacicyathus, 95
ala, brachiopod, 209
 trilobite, 493, 495
alar fossula, 119, 120, 125
alar septum, 119, 120, 123
Alatacythere, 543
alate, 408
Albertan, 37
Albertella, 499
Albian, 36, 383
Alcyonacea, 104
Alcyonaria, 104
alcyonarians, 110, 112
Alcyonidium, 158
Alderina, 190
Alectryonia, 432
Alexander, C. I., 550
Alexandrian, 37
Alisocrinus, 629
alivincular, 408
Allagecrinidae, 622, 623
Allagecrinus, 647
Allocatillocrinus, 647
Allocentrotus, 683
Allomorphina, 50
Allorisma, 424
Alokistocare, 499
alternating reproduction, 23
alternation of generations, 40
alveole, 388
Alveolites, 116
alveolus, 61, 62
amalgamate walls, 159, 162
Amalgamina, 167
Amaltheus, 379

Amaurellina, 317
amber, Baltic, 2, 565, 572
Amberleya, 311, 312
ambitus, 676
Amblysiphonella, 89, 90
Ambocoelia, 256
Ambonychia, 414, 430
ambulacra, 575
 blastoids, 595
 crinoids, 605, 619
 cystoids, 588
ambulacral areas, bothriocidaroids, 713
 echinoids, 676, 685
 ophiocystids, 713
ambulacral channels, 659, 661, 669
ambulacral grooves, 595, 661
ambulacral pores, echinoids, 677, 685
 starfishes, 661, 669
ambulacral spines and tubercles, 677
ambulacrals, bothriocidaroids, 713
 crinoids, 619
 echinoids, 676, 677, 685, 686
 edrioasteroids, 585
 starfishes, 659, 664
Americardia, 436
Ameura, 509
Ammobaculites, 43, 50
Ammodiscus, 35, 44
ammonite suture, 369, 376, 386
Ammonitida, 343
Ammonoidea, 343, 361, 364, 369
ammonoids, 341, 343, 345, 356, 359, 361
 Cretaceous, 369, 378, 382–386
 life habits, 386
Amnicola, 330
Amnigenia, 416, 422
Amphicrinus, 647
amphidetic, 408, 412, 418
Amphidiscophora, 87
amphidisks, 83
Amphigastropoda, 289, 292, 294, 297
Amphigenia, 234
Amphineura, 269–273
Amphiscapha, 308
Amphissa, 326
Amphissites, 538
Amphistegina, 58
Amphitrochus, 311
Amphiura, 673
Amplexocarinia, 135
amplexoid coral, 122, 139
Amplexus, 139
ampulla cups, 660, 665
ampullae, 575
 echinoids, 677, 685
 holothuroids, 653
 starfishes, 661
Ampyx, 513, 515
Amsden, T. W., 153, 264

Amygdalocystites, 591
Anadara, 413
anal area, crinoid, 619
  edrioasteroid, 585
anal cirri, worms, 454
anal fasciole, 701
anal pyramid, 585, 588
anal sac, 619, 625
anal side, 588
anal teeth, 655
anal tube, 619, 631
anal X, 615–618, 621, 625
anals, 615, 621, 625, 627
anaprotaspis, 485
anaptychus, 374
Anascina, 193
Anastrophia, 236
Anatina, 442, 445
Anatinacea, 412, 414, 442
ancestrula, 157, 178
Anchiopsis, 507, 509
anchor plate, 656
anchorage, crinoids, 613
Anchura, 312, 315
Ancilla, 328
Ancistrorhyncha, 240
Ancistrum, 656
Ancyloceras, 383
Ancyrocrinus, 641
Ancyrodella, 736
Andre, M., 573
anemones, 99, 115, 152
Anetoceras, 362
Angulodus, 736
Angulogenerina, 53
Angyomphalus, 304
aniradial, 615, 616, 621
Anisian, 36, 377
anisomyarian, 408
Anisopyge, 509
ankylosis, 611
Annelida, 17, 452, 456
annelids, 452, 466
Annuliconcha, 420
Annulocibicides, 57, 58
annulosiphonate deposits, 348–354, 357
annulus, 339
annulus layer, 339
Anodonta, 401
Anolcites, 377
Anomia, 423, 432
Anomiacea, 411, 423
Anomocytheridea, 543
anomphalous, 290, 308
Anostraca, 544
antanal side, 588
antennae, 466, 481, 533
antennules, 466, 533
anterior, brachiopods, 206, 211
  conodonts, 734
  crinoids, 607
  echinoids, 679, 684
  gastropods, 284
  pelecypods, 403
antetheca, 60, 61
Anthozoa, 100–104
Anthracomarti, 565
Anthracomya, 420, 430

Anthracopupa, 309
Anthrapalaemon, 547, 549
anticusp, 734
Antipatharia, 104
Aorocrinus, 641, 645
Aparchites, 528
apertural face, 42, 48
aperture, bryozoans, 168
  cephalopods, 337, 366, 372
  foraminifers, 42, 44
  gastropods, 280
apex, 205, 283
Aphelaspis, 503
Apheoorthis, 228
Aphetoceras, 355
aphroid coral, 122, 131
apical aperture, 280
apical end, 118, 120
apical line, 389
apical organ, 454
apical system, 680
Aplacophora, 272
Aplousina, 183
apodemes, 465, 481
Apographiocrinus, 647
apophyses, 680
Aporrhais, 315
appendifers, 483
Apterygota, 471, 569
Aptian, 36, 383
aptychus, 374
Aptyxiella, 311
Apus, 544
Aquitanian, 36
Arabellites, 456
Arachnida, 471, 561
arachnids, 553, 561
Arachnocrinus, 641
Arachnophyllum, 124
aragonite, 2, 268, 339, 402
Araneae, 565
Arbacia, 711
Arca, 413, 446
Arcacea, 407–418, 448
Arcestes, 378
Archaeoconularia, 460
Archaeocrinus, 637
Archaeocyathus, 95
Archaeogastropoda, 289, 301
Archaeotrypa, 113, 194
Archiacia, 705
Archiacoceras, 359
Archiannelida, 458
Archicythereis, 543
Archimedes, 171, 177, 182
Architarbi, 565
Architarbus, 564
Architectonica, 320, 324
Archohelia, 151
Arctica, 434, 439
Arctinurus, 507, 509
Arenobulimina, 52
areolae, 193
areole, 676
Argilloecia, 543
Argonauta, 395
Argyrotheca, 205
Aristocystites, 591

Aristotle's lantern, 681, 687
Arkell, W. J., 395
Armonia, 499
arms, crinoids, 604, 610, 625–632
  echinoderms, 575
  squids, 336
  starfishes, 661, 669
Arnold, R., 450
Arthroacantha, 641
Arthroclema, 165
Arthrolycosa, 564
Arthropleura, 519
Arthropleurida, 470
Arthropoda, 17, 452
arthropods, 463, 474
  appendages, 466
  classification, 470
  eye structure, 469
  geological distribution, 471, 472
  habitats, 464, 469
  metamorphoses, 467, 468
  reproduction, 467
  segmentation, 466
  (*See also* Pennsylvanian, Permian, Silurian, Tertiary arthropods)
Arthropomata, 219
articulamentum, 273
articular facet, 607
Articulata, 219–226, 577, 614–617, 635
articulate brachiopods, 198, 201, 207
articulate crinoids, 635
  (*See also* Articulata)
articulating furrow, 481
articulating half segment, 481, 483
articulation, crinoid arms, 610
Artinskian, 37
Asaphidae, 484
Asaphiscus, 501
Asaphus, 505
Aschelminthes, 17
Ascoceras, 347, 351
Ascoceratidae, 351
Ascoceroida, 342, 364
ascoceroids, 342–357
ascocone, 347
ascon type of sponge, 81, 82
Ascophorina, 191
asexual generation, 101, 157, 234
Ashgillian, 37
Aspidobranchia, 301
Aspidochirotida, 657
Aspidodiadema, 711
Aspidodiademina, 695
Astarte, 428, 436, 439
Asterias, 661
Asterocyclina, 72
Asteroidia, 577, 659, 661
asteroids, 574–578, 661, 668–673
Asteropyge, 481, 514, 515
astogeny, 28, 727
Astraeospongia, 88, 91
astreoid coral, 123, 131
Astrhelia, 151
Astrocoenia, 146, 149, 151
Astrocoeniida, 146
Astropecten, 668
astrorhizae, 107
Astrorhizidae, 44, 49

Astylospongia, 88, 93
asymptotic zones, 388
Atactoporella, 165
Athleta, 318, 328
Atokan, 37
atomous arms, 610
Atractites, 391
Atremata, 219–223
Atrina, 431
Atrypa, 212, 217, 251–254
Atrypacea, 220, 251–255, 263
Atrypella, 254
Atrypina, 254
attenuated corals, 123
Aturia, 356, 361, 363
Augustoceras, 359
Aulacoceras, 391
Aulacophyllum, 127, 135
Aulechinus, 685, 689, 691
aulodont, 687
Aulodonta, 685, 711
aulodonts, 695
Aulopora, 115, 131
auloporids, 115
auricle, ammonoids, 374
    echinoids, 681
    pelecypods, 403
auricular sulcus, 403
auriculate, 408
Auriptygma, 300
Austinella, 231
authors, citation of, 12
autopores, 160, 162, 169
autotheca, 721, 722
Auversian, 36
avicularium, 187
Aviculopecten, 420, 424, 426
Aviculopinna, 424
Avitelmessus, 549
Avonia, 250
axial boss, 120, 127
axial canal, 611
axial complex, 120, 127
axial edge, 119, 123
axial lobe, 475
axial organ, 609
axial section, 61
axial segment, 481
axillary, 619
axis, bryozoan, 172
    fusulinid, 60
    gastropod, 284
    trilobite, 475
Aysheaia, 471, 472
Azygograptus, 727, 728
azygous basal, 595

B

Bactrites, 359–362, 388–392
Bactrognathus, 737
Baculites, 386
Bairdia, 537
Bairdiolites, 534
Bajocian, 36
Bakevellia, 424
Balanocidaris, 689
Balanophyllia, 151

Balantoides, 534
Balanus, 545–547
Baltic amber, 2, 565, 572
bar, conodont, 734
Barbarofusus, 326
Barbatia, 413, 434, 438, 446
Barker, R. W., 76
barnacles, 463, 470, 471, 521
Barrande, J., 264, 342, 395, 602
Barrandella, 238
Barrandeoceratida, 342
Barrandeocrinus, 630, 632
Barroisia, 89, 92
Bartonian, 36
Barycrinus, 616, 643
Baryphyllum, 132
basal disk, coral, 117
    graptolite, 719
basal fasciole, 283
basal plate, 454
basals, blastoids, 595
    crinoids, 610, 632
    cystoids, 589
base of gastropod shells, 283
Basiliella, 505
Basommatophora, 290, 330, 331
Bassler, R. S., 153, 195, 550, 601, 650, 738
Bassleratia, 525
Bassleroceratida, 342
Bather, F. A., 581, 601, 650
Bathonian, 36
bathyal, 20
bathypelagic, 21
Bathysiphon, 44
Bathyuriscidella, 497
Bathyuriscus, 501
Bathyurus, 505
Batocrinus, 645
Batostoma, 165
Batostomella, 177
Baylea, 304
Beachia, 234
beak, brachiopods, 199, 207, 209
    pelecypods, 403
beak ridge, 211
Beatricea, 107, 108
Beecher, C. E., 219
Beecherella, 532
beetles, 571
Belemnitella, 391, 392
Belemnoida, 343, 364
belemnoids, 343, 345, 359, 387–391
    Cretaceous, 391–393
Belemnopsis, 391
Belemnosella, 391
Belinurus, 555, 556
Bell, W. C., 264
Bellerophon, 281, 297, 302, 307
Bellerophontacea, 289–294, 297–300, 302–308
Bellifusus, 314
Belliscala, 314
Bellornatia, 526
Beloceras, 369
Beloptera, 391
Belosepia, 391
Beloteuthis, 394

Beltina, 471
Bembexia, 302
benthonic environments, 19
Berenicea, 167
Beretra, 314
Berland, L., 573
Bernard and Munier-Chalmas, 434, 436
Bethanyphyllum, 127
Beyrichia, 535
Beyrichiopsis, 535
biconical form, 288
biconvex shell, 217
Bicornella, 532
Bifarina, 58
Bifurcammina, 44
Bigenerina, 43, 58
bilamellar flange, 529
Bilateria, 17
Billiemia, 315
Billings, E., 602
Billingsastraea, 131
Billingsella, 228
Billingsites, 357
Bimuria, 244
binominal nomenclature, 11
biogenetic law, 28, 40, 557
biologic associations, 19
biozone, 8
bipinnaria, 27
biserial crinoid arms, 620, 631
biserial cystoid brachioles, 589
biserial echinoids, 686
biserial and biserial-uniserial foraminifers, 42, 48
bitheca, 721, 722
Bithecocamara, 723
Bittium, 326
Bittner, A., 264
bivium, 655
Blackriveran, 37
blade, conodont, 734
Blastoidea, 577, 594
Blastoidocrinus, 599, 600
blastoids, 576, 578, 582, 602
blastula, 26, 27
blattoids, 571
Blothrocrinus, 617
Blothrocyathus, 149, 151
Blothrophyllum, 127
body whorl, 283
Böhmers, J. C. A., 395
Bold, W. A. van den, 550
Bolivina, 42, 44, 53, 56
Bolivinoides, 52
Bollia, 526
Boltenella, 314
Bonnema, J. H., 550
Bonnemaia, 528
Bonnia, 497
border, trilobites, 480, 483
Borradaile, L. A., 275, 333, 474, 551, 581
boss, echinoids, 676
Bothriocidaris, 712, 713
Bothriocidaroidea, 577, 713
bothriocidaroids, 712
Botryocrinus, 616, 625, 637, 641
Boucek, B., 462

Bourgueticrinus, 648
bourrelets, 699
Bowles, E., 333, 450
brachia, 197, 199, 247
brachial valve, 201, 206
brachials, 610
brachidia, 213, 255, 259
brachiolaria, 27
brachiole sockets, 589, 597
brachioles, 576, 582, 583, 587, 597
brachiophore, 213
brachiopods, anatomy, 199
  Cambrian, 221–224, 226–228, 237
  classification, 219
  Cretaceous, 234
  Devonian (*see* Devonian brachiopods)
  ecology, 197
  geologic occurrence, 261
  Jurassic, 217, 218
  Mississippian (*see* Mississippian brachiopods)
  morphology, 205–219
  Ordovician (*see* Ordovician brachiopods)
  Pennsylvanian (*see* Pennsylvanian brachiopods)
  Permian (*see* Permian brachiopods)
  Recent, 197, 198, 203, 212
  shell form, 216
  shell growth, 213
  Silurian (*see* Silurian brachiopods)
  Tertiary, 225
  Triassic, 218, 234, 262
Brachiospongia, 87, 91
Brachycythere, 541, 543
Brachylepas, 546
Brachymetopus, 509
Brachyspirifer, 256
Brachythyris, 259
Brachyura, 550
Bradley, P. C. S., 550
Brady, H. B., 76
Bramlette, M. N., 334
branch, bryozoans, 172
branchia, 664
Branchiopoda, 471
branchiopods, 544
Branchipus, 544
Branchiura, 471, 544
Branco, W., 395
Branson, E. B., 264, 738
Breviarca, 413, 434
brevicone, 347, 352
Breviphrentis, 127
Breviphyllum, 124
Brevispirifer, 256
Bridge, J., 334
brim, cephalopods, 350
  trilobites, 480
brittle stars, 574, 667
Bromidella, 525
Brooksella, 109
Bryantodus, 736
Bryozoa, 17, 156
bryozoans, anatomy, 156
  Cambrian, 194
  Cretaceous, 183, 194
  Devonian, 169, 175, 194

bryozoans, ecology, 156
  Eocene, 178, 183, 185, 188, 190
  geologic occurrence, 193
  Jurassic, 178, 187, 193, 194
  Mississippian, 172, 177, 194
  Oligocene, 178, 188, 190
  Ordovician, 161, 165, 173, 194
  Paleocene, 185, 188
  Pennsylvanian, 168, 169, 172, 180, 194
  Permian, 178, 180, 182, 194
  Pleistocene, 186
  Recent, 157, 187, 190
  Silurian, 167, 194
  stony, 194
Bucanopsis, 292
buccal plates, 687
buccal shields, 669
Buccinacea, 314, 320, 322, 326, 328
Buccinofusus, 322, 326
Buccinum, 276, 328
Buchia, 428, 432
Buchiola, 416
Buchsbaum, R., 474, 572, 581
Buckman, S. S., 217, 264
budding zooids, 716
Bugula, 157, 187
Bulimina, 53, 56
Buliminella, 58
Bulimorpha, 304
Bulliopsis, 324
Bulman, O. M. B., 732
Bumastus, 507, 509
Bunodes, 557
Buntonia, 543
Burdigalian, 36
Burgess shale, 109, 329, 452, 471, 519, 657
Burgessia, 518
Burlingia, 499
Burnetia, 501
buttress, 403
Buxtonia, 250, 252
byssal notch, 403
byssal sinus, 403
Byssonychia, 414, 430
byssus, 399, 423, 431
Bythocypris, 541

C

Cactocrinus, 645
Cadocrinus, 647
Caenholectypus, 705
Calapoecia, 114
Calceocrinidae, 622, 623
Calceocrinus, 622, 637
Calceola, 123, 125
calceoloid corals, 118, 123
Calcispongia, 80, 87, 90
calcite, crystalline, 575, 577
calcite shells, 2, 268, 402
calcitization, 5
calcium carbonate, arthropods, 465, 475
  brachiopods, 203
  bryozoans, 157
  coelenterates, 105
  foraminifers, 42

calcium carbonate, sponges, 76
  tintinnids, 87
calcium phosphate, arthropods, 465, 475, 521, 553
  brachiopods, 203
  conulariids, 458, 460
  skeletons, 2
calcium sulfate, 203
Calicantharus, 326
Calipyrgula, 330
Callavia, 490, 497, 518
Callianassa, 550
Calliasterella, 672
Callinectes, 549
Calliocrinus, 637
Calliops, 505
Calliostoma, 312, 324, 326
Callopora, 190
Callovian, 36
callus, 283, 351
Calman, W. T., 474, 551
Calocyclas, 74
calycal pit, 121, 133
calycal platform, 121, 133
Calycina, 695
Calymene, 481
Calyptactis, 672
Calyptraea, 318
Calyptraeacea, 317, 324, 326
Calyptraphorus, 320
calyx, corals, 119, 121, 131
  crinoids, 609, 614
Camaraspis, 501
camarodont, 687
Camarodonta, 685, 690, 696, 711
camarodonts, 695
camaroid graptolites, 722, 723
Camaroidea, 718, 722
Camarotoechia, 240, 242
Cambrian brachiopods, articulates, 226–228, 237
  inarticulates, 221–224
Cambrian bryozoans, 194
Cambrian cephalopods, 347, 356
Cambrian coccoliths, 76
Cambrian conulariids, 458, 460
Cambrian crustaceans, 545, 548
Cambrian echinoderms, 581, 583, 586, 656
Cambrian foraminifers, 58
Cambrian gastropods, amphigastropods, 291, 296–298
  archaeogastropods, 291, 299, 301, 307
  pteropods, 329
Cambrian graptolites, 716, 719
Cambrian pleosponges, 93
Cambrian scyphozoans, 109
Cambrian sponges, 81, 85, 91
Cambrian trilobites, agnostids, 497, 499
  eodiscids, 495, 499
  opisthoparians, 485, 492, 494, 497–503
  proparians, 489–491, 499, 503
  protoparians, 487, 488, 497
Cambrian trilobitomorphs, 518
Cambrian worms, 452, 456, 458
Cambrian xiphosurans, 555
camerae, 340, 348, 351, 366

cameral deposits, 348
Camerata, 15, 577, 614, 618, 629
camerate crinoids, 628
(*See also* Camerata)
Camerella, 236
Camerina, 68, 70
camerinids, 68
Campanian, 36
Campbell, A. S., 78
Campeloma, 322
Camptonectes, 428
Camptostroma, 109, 110
Campylorthis, 212
Canadia, 456
Canadian, 37
Canadocystis, 584
canal, gastropods, 283
Canarium, 315
Cancellaria, 322, 328
cancellate, definition, 408
Cancellina, 66
cancellus, 180
Candona, 522
Caneyella, 420
Caninia, 135, 138, 139
Cantharus, 326
Canu, F., 195
Caprina, 441
Carabocrinus, 616, 637
Caradocian, 37
carapace, 466
Carbonicola, 430
Carboniferous cephalopods, 354
Carboniferous conulariids, 460
Carboniferous crinoids, 627
Carboniferous eucarids, 549
Carboniferous holothuroids, 656
Carboniferous insects, 569
Carboniferous pelecypods, 422
Carboniferous spiders, 565
Carboniferous worms, 454
carbonization, 2
Carboprimitia, 535
Carcinosoma, 559, 560
Carcinus, 549
cardinal area, 405, 412, 422
cardinal extremity, 211
cardinal fossula, 119, 121, 125
cardinal process, 201, 213
cardinal quadrant, 121, 125
cardinal septum, 119, 121, 123
cardinal teeth, 403, 434
cardinalia, 213
Cardiniacea, 411, 419
Cardioeca, 183
Cardiola, 412
Cardium, 434
Caricella, 320
carina, bryozoans, 171
    conodonts, 734
    corals, 121
    gastropods, 283
carinate, 123, 125, 408
Carneyella, 586
Carpenter, F. M., 572
Carpenter, H., 603
Carpocrinus, 633, 637
Carpoidea, 577, 585

carpoids, 585, 602
carrier, worms, 454
Carruthers, R. G., 153
Carybdeida, 104
Caryocrinites, 591–594
Caryocystites, 591
Caryophylliida, 149
Cassadagan, 37
Cassidaria, 318
Cassidulina, 53, 58
Cassidulina, 699
Cassidulinoides, 53
Cassiduloida, 696, 698, 711
cassiduloids, 697, 705–708
Cassidulus, 698, 708, 711
Cassinoceras, 355
Cassiope, 312
cast, 5
Caster, K. E., 97
Catazyga, 254
Catillocrinus, 622, 643
caunopores, 107
Cavellina, 537
Cayugan, 37
Cazenovian, 37
Cedaria, 503
Cellaria, 186
Cenomanian, 36, 384
Cenosphaera, 74
centipedes, 471, 566
central canal, 610
central capsule, 72, 73
central plate, 664
Centroceras, 359
Centrocidaris, 711
centrodorsal, 650
Centroonoceras, 351
Centrostephanus, 711
Cephalodiscus, 716
cephalon, 475, 478, 488–491, 494
Cephalopoda, 269, 270
cephalopods, Cambrian, 347, 356
    Carboniferous, 354
    Cenozoic, 361
    classification, 342
    Cretaceous (*see* Cretaceous cephalo-
        pods)
    Devonian (*see* Devonian cephalopods)
    Eocene, 363, 391, 393
    evolutionary trends, 382, 392
    general, 335
    geologic distribution, 345, 376,
        392
    Jurassic (*see* Jurassic cephalopods)
    largest, 335, 358, 384
    Miocene, 361, 391
    Mississippian (*see* Mississippian ceph-
        alopods)
    Ordovician nautiloids, 347, 357
    Paleocene nautiloids, 361, 363
    Pennsylvanian (*see* Pennsylvanian
        cephalopods)
    Permian (*see* Permian cephalopods)
    Pliocene octopoids, 395
    Recent, 335–337, 391
    Silurian nautiloids, 351, 354, 358
    Tertiary, 356, 363, 393
    Triassic (*see* Triassic cephalopods)

cephalothorax, 466
Ceramella, 175
Ceramoporella, 165
Ceratarges, 513, 515
Ceratiocaris, 546
ceratite suture, 369, 372, 376, 383, 386
ceratites, 377
Ceratobulimina, 56
ceratoid corals, 118, 123
Ceratopea, 292
Ceratopsis, 525
Ceratosida, 89
Ceraurinus, 503
Ceraurus, 481–484, 505
Ceriantharia, 104
cerioid corals, 123, 131
Cerithiacea, 312–314, 320–326
Cerithiopsis, 322
Cerithium, 312, 313
Cestoidea, 454
Chaenomya, 412, 420
Chaetetes, 113, 135
chaetognath worms, 580
Chaetognatha, 17, 456
chain corals, 115
Chama, 440, 441
Chamacea, 407, 411, 440
chamber, foraminifer, 42, 44
chambered organ, 609
Chancelloria, 85, 91
channel, gastropod, 283
Chao, Y. T., 264
Chattian, 36
Chazyan, 37
cheek lobe, 493, 494
Cheilostomata, 161, 183, 186–193
Cheilotrypa, 167
Cheirocrinus, 591
Cheiropteraster, 665, 671
Cheirothyris, 218
chelae, 467, 552
chelicerae, 553
Chelicerata, 463, 467, 471
chelicerates, 552, 553
Chemungian, 37
chevron groove, 405
Chideruan, 37
chilidial plates, 211
chilidium, 211
Chilopoda, 471, 566
chilopods, 567
Chilostomella, 42
Chione, 442
Chirodota, 656
Chirognathus, 736
chitin, 2, 39, 41
    arachnids, 562
    arthropods, 464
    crustaceans, 521
    trilobites, 478
chitinoid, 716
chitinophosphatic, 203
Chitinoteuthidae, 392
Chitinoteuthis, 392
chitons, 269, 271, 272
Chlamys, 426, 439
Chlidonophora, 200

Choia, 85, 93
chomata, 61, 64
chondrophore, 406, **442**
Chonetes, 248, 250
chonetids, 249
Chonetina, 252
Chopard, L., 572
Chordata, 17, 716
chordate line, 580
Choristothyris, 234
Christiania, 244
chromosomes, 24
Cibicides, 53, 58
Cibolocrinus, 647
Cidaridae, 676, 680, 690
Cidaroida, 684, 711
cidaroids, 690, 693, 705
Ciliata, 9
Cimomia, 363
Cincinnatian, 37
Cincinnatidiscus, 586
cirrals, 612
cirri, 611, 612
Cirripedia, 471, 544
cirripeds, 472, 521, 545–547
cirrus sockets, 612
Cirsotrema, 320
Cladida, 613, 616, 617, 623
cladids, 621
 (*See also* Cladida)
Cladochonus, 132
Claibornian, 36
clams, 398
Clappaspis, 499
Clark, B. L., 78, 450
Clark, W. B., 333, 450, 713
Clarke, J. M., 97, 264, 551, 572
classes, 15, 16
classification, animals, 17
 arthropods, 470
 brachiopods, 219
 bryozoans, 159
 cephalopods, 341
 coelenterates, 103, 133, 146
 conodonts, 733
 crinoids, 613
 echinoderms, 577
 echinoids, 684, 696
 foraminifers, 56, 66
 gastropods, 287
 holothuroids, 657
 insects, 569
 mollusks, 269
 ostracodes, 536
 pelecypods, 409
 radiolarians, 75
 spiders, 563, 565
 sponges, 87
 starfishes, 659
 trilobites, 486
 worms, 454
clathria, 725
Clathrodictyon, 108
Clathrodrillia, 326
Clathropora, 167
Clathrospira, 294
Clavulina, 43
clear layer, belemnoids, 387

Cleidophorus, 414
Cleiothyridina, 260
Clementia, 442
Clidochirus, 637
Cliftonian, 37
Climacammina, 48
Climacoconus, 460
Climacograptus, 727, 730
Cline, L. M., 602
Clinopistha, 420
Clintonian, 37
Clio, 276
Clisiophyllum, 123, 129, 139
Clithrocytheridea, 543
cloaca, 79
Clonocrinus, 628, 637
Cloud, P. E., Jr., 264
Clymenidae, 370, 371, 376
clymenids, 370, 376
Clypeastrina, 703
clypeastrinid plates, 703
Clypeastroida, 696, 698, 711
clypeastroids, 701, 702, 705–707
Clypeus, 697
Cnemidactis, 668
Cnidaria, 99
Coblenzian, 37
coccoliths, 76
Cocculina, 304
Cocculinacea, 304, 309
Cochloceras, 378
cockroaches, 571
Codaster, 599, 601
Codiacrinus, 616, 641
Coelenterata, 15, 17
coelenterates, 99
 (*See also* Cretacean, Jurassic, Missis-
 sippian, Ordovician, Recent, Si-
 lurian, Tertiary, Triassic coelen-
 terates)
Coelocaulus, 300
Coelochilina, 525
Coeloptychium, 92
coenelasma, 162, 165
Coenites, 116
coenosteum, 168, 171
Coenothecalia, 104
coiled-biserial and coiled-uniserial, defi-
 nition, 42, 48
Cole, W. S., 78
Coleites, 54
coleoids, 341
Coleoptera, 541
collar, echinoid spines, 676
collar cells, 81
Collembola, 569
Colom, G., 78
Colonammina, 44
colonies, 22
Coloradoan, 36
Colpocaris, 546
Colpomya, 414
Columbites, 376
columella, coral, 121, 129, 137, 145
 gastropod, 283
columellar folds, 283
columellar lip, 283

column, blastoids, 597
 crinoids, 611, 623, 632
 cystoids, 589
columnals, 589, 595, 611, 620
Comarocystites, 583, 584
Comatula, 605
Comatulida, 614
Comleyan, 37
commensalism, 19, 277
commissure, 209, 403
common canal, graptolites, 725
Comophyllia, 147
compass, 682
compensation sac, 191, 193
Composita, 200, 259, 260
compound corals, 118, 121
compressed shell, pelecypods, 408
conacea, 314, 318, 320, 322, 326, 328
concavo-convex brachiopods, 217
concentric shell, pelecypods, 409
Conchidium, 238
conchiolin, 402
Conchopeltis, 460
Conchostraca, 544, 545
Condra, G. E., 77, 195, 265, 396
Condracypris, 532
Conewangoan, 37
Coniacian, 36
conical gastropod shells, 288
conispiral form, 281, 345, 347, 377, 378
conjugate pores, 699
conjunct pore rhomb, 587
connecting ring, 340, 350, 353, 366
Conocardium, 416
Conoclypina, 699
Conocoryphe, 497
Conodontiformes, 734
Conodontophoridia, 734
conodonts, 733, 736, 737
conoidal gastropod shells, 288
consortium, 43
continuous septum, 121, 123, 125
contractile stalk, pterobranchs, 716
Conularia, 458, 460
Conulariella, 460
Conulariellidae, 458
Conulariida, 458
Conulariidae, 458
conulariids, 458–460
Conularina, 460
Conus, 277, 278, 318, 324, 328
convergence, 34
convexi-concave and convexi-plane bra-
 chiopods, 217
convolute, definition, 288, 347, 354, 361
Cooke, C. W., 713
Cooper, C. L., 738
Cooper, G. A., 218, 219, 227, 264, 266,
 550
Cooperidiscus, 586
Cooperoceras, 360
Copepoda, 471, 544
coprolites, 5
coral reefs, 118, **137**
coralla, 121, 129
corallites, 121
corals, alcyonarian, 110
 chain, 115

corals, compound, 118, 121, 131
Cretaceous, 148, 149
cup, 118, 135
Devonian, 127, 129, 131, 135–138
ecology, 117, 135
honeycomb, 115
Jurassic, 147
Mississippian, 132, 137, 138, 141, 142
Ordovician, 113, 114, 118, 135
Pennsylvanian, 113, 135, 138
Permian, 127, 136, 137
rugose, 114, 118, 119, 122–143
scleractinian, 117, 119, 143–152
Silurian, 116, 124, 129, 131, 135, 138
simple, 118
tabulate, 112–116, 131, 132, 135
Tertiary, 151
Triassic, 147
Corbula, 434, 439, 444, 445
Cordylodus, 736
Cornellites, 416
Cornigella, 538
Cornucrinus, 623
Cornulites, 460
Cornuspira, 40, 43, 50
corona, 676
Coronatida, 104, 109
Coronia, 320
Coronura, 507, 509
cortex, echinoid spines, 676
graptolites, 719
Coryellites, 538
Corynella, 93
Corwenia, 132
Coscinopora, 93
Cosmonautilus. 362
costae, 144, 207, 286, 403
costate pelecypods, 409
Costellirostra, 241
Costispirifer, 256
costules, 193
Cotyliscus, 90
counter quadrant, 121, 125
counter septum, 119, 121, 123
counter-lateral septa, 119, 121, 125
covering plates, 585, 589–595
crabs, 25, 463, 471, 550
Cretaceous, 549
Cranaena, 234
Craniacea, 220, 222, 224, 226, 263
Crassatellites, 434, 439, 445
Crassimarginatella, 190
Crateroseris, 147
Cravenoceras, 371
creeping branches, pterobranchs, 716
Cremacrinus, 622, 637
crenella, 610, 611
Crepicephalus, 503
Crepidula, 317, 318, 324, 326
Cretaceous brachiopods, 234
Cretaceous bryozoans, 183, 194
Cretaceous cephalopods, ammonoids, 369, 378, 382–386
belemnoids, 391–393
nautiloids, 356, 361, 362
octopoids, 395
teuthoids, 395

Cretaceous coelenterates, alcyonarians, 110, 112
corals, 148, 149
hydroids, 105
milleporids, 106
sphaeractiniids, 106
Cretaceous crustaceans, cirripeds, 546
crabs, 549
hoplocarids, 548
ostracodes, 541, 543
percarids, 549
Cretaceous echinoderms, crinoids, 648, 650
echinoids, 676, 688, 689, 698, 701, 705
starfishes, 659, 673
Cretaceous gastropods, archaeogastropods, 299, 301, 304, 307–312
mesogastropods, 296, 312–317, 320, 322
neogastropods, 314, 316, 328
opisthobranchs, 312, 314, 316
Cretaceous insects, 569, 572
Cretaceous pelecypods, 399, 402, 416–419, 426, 430–436, 440–444
Cretaceous protozoans, coccoliths, 76
dinoflagellates, 76
foraminifers, 50–55, 70, 72
radiolarians, 74
silicoflagellates, 76
tintinnids, 76
Cretaceous spiders, 472
Cretaceous sponges, 81, 92
Cretaceous worms, 456
cribrate, 48
Cribrillina, 188
Cribrohantkenina, 54, 56
Cribrostomum, 48
Crickmay, C. H., 264
Crinoidea, 15, 577, 604
crinoids, anchorage, 613
bored cups, 332
Carboniferous, 627
Pennsylvanian, 624, 635, 646–648
classification, 613
Cretaceous, 648, 650
cup shape, 624
Devonian, (see Devonian crinoids)
distribution, 604
evolution, 621, 624, 627–633
general, 574–604
geological occurrence, 635
Jurassic, 648, 650
Mississippian (see Mississippian crinoids)
morphology, 607
Ordovician (see Ordovician echinoderms, camerate, flexible, inadunate crinoids)
orientation, 613
size, 604
symmetry, 622
Crioceras, 382
Crisia, 186
Crockford, J., 195
Croixian, 37
Croneis, C., 603, 658
cross, holothuroids, 656
Crotalocrinites, 626, 637

crown, crinoids, 609, 632
crura, 213
crural base, 213
crural lamella, 213
crural process, 213
cruralium, 209
Crurithyris, 259
crus, 213
Crustacea, 463, 471, 521
crustaceans, Cambrian, 545, 548
Cretaceous (see Cretaceous crustaceans)
Devonian (see Devonian crustaceans)
as food, 526
Jurassic, 539, 547–550
limbs, 467
Miocene, 543, 547
Mississippian, 534–536, 546, 549
molts, 468
morphology, 521
ontogeny, 521
Ordovician, 472, 525, 526, 546
physiology, 523
Pleistocene, 545, 544, 549
Recent, 468, 531, 544, 549
Triassic (see Triassic crustaceans)
Cryptoblastus, 599
Cryptolithus, 481, 493–495, 505, 519
Cryptomartus, 565
cryptomphalous, 290
Cryptonella, 234
Cryptophragmus, 107, 108
Cryptostomata, 161, 164–182
Cryptothyrella, 261
crystalline calcite, 577
ctenidia, 279
Ctenobolbina, 526
Ctenocrinus, 629, 641
Ctenodonta, 413–416
Ctenoloculina, 530
Ctenophora, 17, 99
Ctenostomata, 161, 173–175
Cucullaea, 413, 434, 439
Cucumariida, 657
Cuénot, L., 581, 601, 658, 674
Cuisian, 36
Cullison J. S., 738
Culmicrinus, 643
Cumingia, 446
Cumings, E. R., 153, 195
cuniculi, 63
cup corals, 118
cup-axillary, 632
cup-brachial, 618, 628, 632
cup-pinnule, 632
Cupulocrinus, 637
Cushman, J. A., 77
cusp, conodonts, 734
Cuspidaria, 432, 445, 447
cuttlefishes, 393
Cuvier, 16, 268
Cyathocidaris, 689
Cyathocrinina, 613
Cyathocrinites, 616, 624, 625, 643
cyathophylloid corals, 123
Cyathophylloides, 114
Cyathophyllum, 123
cyathotheca, 121

Cybele, 503
Cyclocolposa, 193
Cyclograptus, 731
Cyclolobus, 374, 383
Cyclonema, 294
Cyclonemina, 300
Cyclophoracea, 320, 324
Cyclophorus, 324
Cycloplectoceras, 355
Cyclostomata, 161, 164–166, 175–186
Cyclostomiceras, 355
Cyclotrypa, 180
Cyclozyga, 307
cylindrical corals, 118, 123
Cylindrophyllum, 129
Cymatospira, 309
Cymbalopora, 56
Cyphosoma, 705
Cypraea, 277
Cypraeacea, 317, 318
Cypraedia, 318
Cypraeorbis, 317
Cypricardella, 416, 420
Cypricardiacea, 411, 439
Cypricardinia, 414, 416, 420
Cypridina, 522
Cyprimeria, 434, 436, 440
Cyrena, 436
Cyrenacea, 411, 440
cyrenoid pelecypods, 434, 436
Cyrtia, 256
Cyrtina, 262
Cyrtochoanites, 342, 354
cyrtochoanitic, 342, 349–356
cyrtocone, 347
Cyrtocrinida, 614
Cyrtodonta, 414
Cyrtograptus, 727, 731
Cyrtonella, 291
Cyrtoniodus, 736
Cyrtonotella, 214
Cyrtospirifer, 256
Cystauletes, 89, 91
cystiphragm, 162, 165
Cystiphyllacea, 133, 138
Cystiphylloides, 129, 138, 141
Cystiphyllum, 123, 125, 138
Cystoblastus, 591, 594
Cystoidea, 15, 577, 587
cystoids, 576, 578, 582, 602
   Devonian, 587, 592–594
Cythereis, 541
Cytherella, 541
Cytherellina, 530
Cytherelloidea, 541
Cytheretta, 543
Cytheridea, 541
Cytheridella, 543
Cytheromorpha, 543
Cytheropteron, 522, 541, 543
Cytherura, 541
cytoplasm, 39
Cyzicus, 544, 545

D

dactylethra, 180
Dactylocrinus, 641

dactylopores, 106
dactylous, 683
dactylozooids, 106
Dakotan, 36
Dall, W. H., 333, 450
Dallina, 212
Dalmanella, 227, 232
Dalmanellacea, 220, 231–233, 263
Dalmanites, 490, 491, 514, 519
Danian, 36
Daonella, 429
dark layer, belemnoids, 387
Darwin, 29
Davidson, T., 264
Davidsonia, 203
Davies, A. M., 333, 450
Davis, W. J., 153
Dawydoff, C., 573
Dayia, 255
Decadocrinus, 643
Decapoda, 549
declined graptolites, 727
decurved graptolites, 725
Deerparkian, 37
Deiss, C. F., 196, 519
Dekayella, 161
Delo, D. M., 519
Delocrinus, 624, 647
Deloia, 534
Deltatreta, 218
delthyrial cavity, 211
Delthyris, 256
delthyrium, 207, 211
deltidial plates, 211
deltidium, 211
deltoids, 595
demi-plates, 689, 703
Demospongia, 80, 87, 91
Dendraster, 708
dendrochirote, 656
Dendrochirotida, 657
Dendrocrinina, 613
Dendrocrinus, 616, 625, 637
Dendrograptus, 720–722
dendroid corals, 123, 129
dendroid graptolites, 719, 728, 731
Dendroidea, 718, 722
Dendrophylliida, 151
Denis, R., 572
dental lamellae, 211
dental plates, worms, 454
dental sockets, 213
Dentalina, 50
Dentalium, 274, 318, 324
dentation, septa, 145
denticle, conodonts, 734
dentition, heterodonts, 434, 436
Derbyia, 248
dermal pores, 79
desmas, 83
Desmidocrinidae, 633
Desmodonta, 411, 442
desmodonts, 409, 410, 414, 416, 424,
   428, 432, 434, 438, 442, 445, 446
Desmoinesian, 37
Despax, R., 572
Devonian blastoids, 599, 600

Devonian brachiopods, inarticulates,
   203, 212, 217, 225
  orthids, 231, 233
  pentamerids, 236–238
  rhynchonellids, 239, 241
  spiriferids, 251–262
  strophomenids, 244, 246–251
  terebratulids, 234
Devonian bryozoans, 169, 175, 194
Devonian cephalopods, ammonoids,
   362, 369, 370, 376
  nautiloids, 352, 354, 356, 359–362
Devonian conodonts, 733, 736
Devonian conulariids, 458, 460
Devonian corals, rugose, 127–131, 135–
   138
  tabulate, 115, 129–131
Devonian crinoids, camerates, 629, 632,
   634, 635, 641
  flexibles, 641
  inadunates, 621, 641
Devonian crustaceans, conchostracans,
   545
  ostracodes, 530–533
  percarids, 549
  phyllocarids, 546
Devonian cystoids, 587, 592–594
Devonian echinoids, 689, 690
Devonian edrioasteroids, 586
Devonian eurypterids, 472
Devonian foraminifers, 46
Devonian gastropods, amphigastropods,
   291, 296, 297, 302
  archaeogastropods, 299, 301–304,
   307, 309, 311
Devonian insects, 472, 569
Devonian myriapods, 566
Devonian ophiocystids, 713
Devonian pelecypods, 412, 416, 418,
   419, 423, 426, 430, 431, 439, 444
Devonian pteropods, 329
Devonian pycnogonids, 472, 566
Devonian radiolarians, 74
Devonian receptaculitids, 96
Devonian spiders, 562, 565
Devonian sponges, 88
Devonian starfishes, 661, 665, 671
Devonian stromatoporoids, 108
Devonian tentaculitids, 459, 460
Devonian trilobites, 490–494, 507, 509,
   513
Devonian worms, 452, 456, 460
Devonian xiphosurans, 555, 556, 559
Devonoblastus, 599
dextral, 288
Diabolocrinus, 637
Diacanthopora, 183
Diademina, 695
diademoid, 686
Diademopsis, 695
diaphanotheca, 61
diaphragms, bryozoans, 160, 162, 163,
   168
  cephalopods, 347
Diaphragmus, 250
Dibranchiata, 343, 364
dibranchiates, 337, 341
Dibunophyllum, 135

Dicaelosia, 232
Dicellograptus, 727, 730
Dicellomus, 200, 223
Diceras, 441
Dichocrinidae, 633, 634
Dichocrinus, 633, 645
Dichognathus, 736
Dicranella, 526
Dicranograptus, 727, 729, 730
Dictyoclostus, 217, 218, 250, 252
Dictyoconites, 391, 393
Dictyomylacris, 569
Dictyonema, 719, 720, 728
Dictyonina, 224
dicyclic, 617, 623, 626, 631
diductor muscle scars, 206, 212, 222
Didymograptus, 727, 728
Dielasma, 234
digitations, 286
Dikelocephalus, 503, 514
dilated septa, 123, 139
Dimorecrinites, 628, 637
Dimorphastrea, 148, 149
dimorphism, 22, 340, 559
Dimorphograptus, 727, 731
dimyarian, 409
Dinobolus, 223
dinoflagellates, 76
Dinorthis, 228
Diodora, 320, 324
Diotocardia, 301
dip, gastropod sutures, 283
Diparelasma, 228
Dipleura, 507
dipleurula, 27, 578, 579
Diplhelia, 150, 151
Diploastrea, 148, 149
Diplobathrida, 15, 614, 618, 619, 634
diplobathrids, 634
Diploclema, 167
Diplograptus, 715, 723–729
Diplophyllum, 125
Diplopoda, 471, 566
diplopores, 593
Diploporida, 587, 591, 592
Diplotresis, 183
Diploceras, 383
Dipterophyllum, 125, 133
directive mesenteries, 103, 125
Dirhachopea, 291
Discinacea, 220, 224, 225, 263
Discinisca, 203, 224
Discocyathus, 147
Discocyclina, 70, 72
discoid corals, 118, 123
discoidal gastropod shells, 288
Discoidella, 537
discontinuous septa, 121, 123, 125
Discoporella, 193
Discorbis, 43, 53, 58
Discosoroida, 342, 354, 364
discosoroids, 354
Discotrochus, 151
Discotropites, 378
Discus, 330
disjunct pore rhombs, 587
disk, ophiuroids, 667
Disparida, 613, 615, 616, 621

disparids, 621
dissepiments, bryozoans, 171
corals, 121, 129, 145
graptolites, 719
dissepimentarium, 121, 129
distal, 168
crinoids, 611, 619
starfishes, 663
distortion of fossils, 5
Ditomopyge, 509
divaricate, 409
Diversophyllum, 129
Dizygocrinus, 645
Dizygopleura, 530
Docoglossa, 277, 301
Dokimocephalus, 503
Dolatocrinus, 641
Doleroides, 218, 228
Dolerorthis, 231
Doliognathus, 737
Dollo's law, 34
Dolorthoceras, 357
Domatoceras, 360
d'Orbigny, A. D., 78
Dorothia, 58
dorsal, brachiopods, 207
cephalopods, 339
crinoids, 607
pelecypods, 403
dorsal cup, 610, 615, 632
dorsal furrow, 480
dorsal groove, 667
dorsal shield, 393, 669
dorso-lateral fields, 664
dorso-lateral furrows, 387
dorsum, 368
Dorycrinus, 645
Dosinia, 436
Dosiniopsis, 439
doublure, 475, 481, 488
Douvillé, H., 407
Douvillina, 246
Downtonian, 37
Dozierella, 425
Dreissensia, 431
Dreissensiacea, 411, 434
Drepanella, 525
Drepanellina, 528
Drepanocheilus, 315
Dresbachian, 37
Drillia, 322, 328
Dumblea, 705
Dunbar, C. O., 77, 265, 396
Duncan, D., 519
Duncan, H., 196
Duncan, P. M., 153
duplicature, 529
duplivincular, 409
Duplophyllum, 136
Durania, 440, 441
Durham, J. W., 153, 333
Dybowskiella, 180
Dysodonta, 411, 423
dysodonts, 409, 410, 414, 416, 420, 423–425, 428, 431–434, 438, 442, 446
Dystactocrinus, 616, 637
Dystactospongia, 87, 93

E

Eastman, C. R., 97
Easton, W. H., 153
Eatonia, 212, 241
Ecculiomphalus, 292
ecdysis, 468, 479, 521
Echinina, 696
Echinocardium, 700, 701, 706–709
Echinocaris, 546
Echinoconchus, 252
Echinocrinus, 690, 693
Echinodermata, 15, 17, 574, 577
echinoderms, 574–581
(*See also* Cretaceous, Jurassic, Miocene, Ordovician, Permian, Recent, Silurian, Tertiary, Triassic echinoderms)
Echinoencrinites, 588
echinoid-type ambulacrals, 686
Echinoidea, 15, 577, 675
echinoids, 574–578
classification, 684, 696
Devonian, 689, 690
ecology, 703
Eocene, 67, 689, 699, 707
evolution, 689–703
geological occurrence, 709
irregular, 696, 706
morphology, 676
Oligocene, 708
orientation, 682
Paleocene, 700, 707
Pliocene, 702, 708
regular, 675, 706
terminology, 679–687, 698
Echinometridae, 696
Echinoneina, 697
Echinoneus, 710
echinopluteus, 703
Echinosphaerites, 591, 593, 594
Echinothuridae, 686, 689
echinothurids, 686
Echiuroidea, 17
ecologic adaptations, 16
ecology (*see* specific names of groups)
Ecphora, 322, 326
Ectenocrinus, 622, 637
Ectinochilus, 318
Ectodemites, 522, 538
ectoderm, 100
Ectomaria, 294
ectoplasm, 39
Edenian, 37
edentate, 409
edge zone, 117
Edmondia, 420
Edrioaster, 586
Edrioasteroidea, 577, 586
edrioasteroids, 576, 578, 585, 586, 602
Edwards, H. M., 154
Eggerella, 53
Ehmania, 499
Ehmaniella, 499
Ehrenberg, 16
Eifelian, 37
Eiffelia, 85, 91
Elasipodida, 657

Eleutherozoa, 15, 577, 653, 712
Elias, M. K., 195, 196
Eller, E. R., 462
Elles, G. L., 732
Ellesmeroceroida, 342, 347, 364
ellesmeroceroids, 347, 355, 356, 358
Ellipsonodosaria, 56
Ellipsoscapha, 314
Elliptocephala, 485, 486, 488, 490
Ellis, B. F., 77
Ellison, S. P., Jr., 738
Ellisor, A. C., 77
Elphidium, 50, 53
Elrathia, 497
Elrathina, 497
Elvinia, 501
Elytha, 256
embryonic, definition, 27
embryonic chambers, 72
Emeraldella, 518
Emmonsia, 131
Enallocrinus, 626, 639
Encope, 702, 708
Encrinina, 650
Encrinurus, 507
Endoceras, 352, 353, 356
Endoceroida, 342, 352, 364
endoceroids, 345, 352, 355–358
endocones, 342, 345, 350, 353
endoderm, 100
Endopachys, 150, 151
endopetalous fasciole, 701
endoplasm, 39
endopodites, 467, 481, 521
endosiphotube, 353
endotheca, 144
Endothyra, 42, 46, 58
Endothyranella, 48
endotomous arms, 619
Engonoceras, 384
Ensiphonacea, 412
Ensis, 444, 446
Entalophora, 178, 183, 185
Enteletes, 231, 233
Entelophyllum, 125
Enterocoela, 17, 580
Enterolasma, 127
enteron, 100
Enteropneusta, 716, 718
entocoel, 146
Entomis, 530
Entoprocta, 17, 156
entosepta, 144, 146
environments, adaptation to, 8, 24
  benthonic, 19
  influence of, 31
  land, 21
  marine, 18
  nektoplanktonic, 20
Eoancilla, 316
Eoasianites, 372
Eobronteus, 483, 505
Eocene bryozoans, 178, 183, 185, 188, 190
Eocene cephalopods, 363, 391, 393
Eocene echinoids, 676, 689, 699, 707
Eocene foraminifers, 54–56, 70, 72
Eocene holothuroids, 656

Eocene insects, 572
Eocene ostracodes, 543
Eocene pelecypods, 436, 438, 441
Eocene radiolarians, 74
Eocene worms, 456
Eoconularia, 460
Eocrinoidea, 577, 583
eocrinoids, 583, 602
Eocytheropteron, 543
Eodevonaria, 250
Eodiscida, 487, 494, 495, 499, 512, 515
Eodiscus, 483, 499
Eophiura, 668, 673
Eoscorpius, 563
Eospirifer, 256
Eothalassoceras, 372
Eothuria, 685, 691
Eotomaria, 292
Eotrochus, 304
Eouvigerina, 53
Eozoa, 17
ephebic, definition, 27
epideltoid, 595
epiphysis, 682
epipodite, 467
epirostrum, 389, 392
episeptal deposits, 348
epistereom, 583, 587
episternals, 700
Epistomina, 42, 53
epitheca, 117–121, 144
Epitoniacea, 313–316, 320, 324, 326
Epitonium, 313, 316, 324
Eponidella, 57
Eponides, 53, 56
equilateral, 409
equivalve, 409
Eratocrinus, 643
Eretmocrinus, 645
Eridocampylus, 175
Eridophyllum, 129, 131
Eridotrypa, 167
Erisocrinus, 647
Erkosonea, 185
Eryma, 547, 549
Eryon, 547, 549
escutcheon, 403, 734
Estheria, 545
Estheriella, 545
Ethelocrinus, 647
Etheridge, R., 603
Ethmophyllum, 95
Etroeungtian, 37
Etymothyris, 234
Eucaenus, 569
Eucalyptocrinites, 638
eucarids, 549
Eucheirocrinus, 622, 638
Euchondria, 425, 426
Eucladocrinus, 643
Eucoelomata, 17
Euconospira, 307
Eucyclomphalus, 311
Eucyclus, 311
Eucystis, 591
Eucythere, 541
Eucytherura, 541
Eudimerocrinus, 631, 638

Eugeniacrinus, 648
Euglyphella, 532
Eukloedenella, 526
Eulamellibranchia, 410
eulamellibranchs, 410, 423, 431–436, 441, 448
Eulepidina, 68, 70, 72
Euloxoceras, 360
Eumetazoa, 17
Eumetria, 262
Eumorphocystis, 591
Eunema, 294
Euomphalacea, 292, 299–308, 311
Euomphalus, 299, 307
Euophiceras, 358
Eupachycrinus, 643
Eupatagus, 699, 707
Euphausiacea, 549
Euphoberia, 567
Euplectella, 80
Euprimitia, 526
Euritina, 190
Eurypterida, 555
eurypterids, 472, 552, 553, 557
Eurypterus, 557, 559
Euscorpiones, 562
Eustephanus, 530
Eutaxocrinus, 643
Euthemon, 712
Eutrephoceras, 362
Eutrochocrinus, 645
evolute, 42, 48, 288, 354
evolution, 28, 34
  corals, 138, 146, 151
  crinoids, 621, 624, 627, 629, 631, 633
  fusulinids, 63
  gastropods, 290
  graptolites, 722, 725
  irreversibility in, 34
  rates, 35
  starfishes, 665, 669
  straight-line, 341
  trends, 34
  trilobites, 497, 510
exocoel, 146
Exogyra, 432
exopodites, 467, 521
exosepta, 144, 146
exoskeleton, 465
exotheca, 144
exotomous arms, 619
exsert, 686
external furrow, fusulinids, 60, 61
extinction of animals, 33
extracapsular, 73
extraconical, 288

F

Fabalicypris, 538
facial sutures, 478, 489, 492, 494, 505
facies fossils, 21
Fage, L., 572
Famennian, 37
family, 14, 15
fans, worms, 454
Fardenia, 246

fasciculate corals, 123, 129
Fasciculiconcha, 420, 426
fascioles, 701
fauna, 21
Faviida, 148
Favistella, 114
Favosites, 116, 131
favositids, 115
Favulella, 533
feather stars, 574, 604
Felix, J., 154
Fenestrellina, 167, 171, 175, 177, 180
fenestrules, 171
Fenton, C. L., and M. A., 265
Ferry, H. B. A. T. de, 154
Fibularia, 701, 705, 707
Ficus, 318
Filibranchia, 410
filibranchs, 409, 413, 418, 422, 423, 426, 429, 448
Fimbrispirifer, 256
Fingerlakesian, 37
Finkelnburgia, 218, 227
firmatopores, 180
Fistulipora, 167, 177
fixed cheeks, 480, 494
Flabellammina, 53
Flabellinella, 53
Flabellum, 117, 151
flagella, 39, 81
Flagellata, 39, 87
flagellates, 75
flange, ostracodes, 529
flange groove, 531
Flexibilia, 577, 614, 626, 650
flexible crinoids, 626
Flexicalymene, 503
Floridina, 193
Floridinella, 188
floscelle, 698, 701
Flower, R. H., 342, 395
Foerste, A. F., 396, 602
Foerstephyllum, 114
fold, brachiopods, 207
Foordella, 304
foot, 272, 400
foramen, brachiopods, 205, 211
graptolites, 720, 724
Foraminifera, 39, 60
foraminifers, camerinids, 68
classification, 56, 66
discocyclinids, 72
ecology, 48, 72
Eocene, 54–56, 70, 72
evolution, 63
fusulinids, 60
geologic occurrence, 58, 66
larger, 60
life history, 40
Miocene, 57, 58, 70, 72
miogypsinids, 72
morphology, 44, 60
Oligocene, 57, 70, 72
Ordovician, 44
Pliocene, 58
test, 41
Triassic, 50
Forbesiocrinus, 645

forceps, worms, 454
Forreria, 326
fossa, corals, 145
crinoids, 608
Fossarus, 322
fossils, definition, 1
diversity, 6
value, 7
fossilization requisites, 1
fossula, 121
Franconian and Frasnian divisions, 37
Fredericksburgian division, 36
free cheeks, 479, 480, 494
fresh-water pelecypods, 419, 420, 422, 430
fringe, trilobites, 493, 495
Frizzell, D. L., 450
Fromentel, E. de, 154
Frondicularia, 50
frontal, bryozoans, 191
Fulgerca, 314
Fungia, 148
Fungiida, 148
Furnish, W. M., 396
fusellar half rings, 719
fusellar layer, 719
fusiform, 288
Fusimitra, 320
Fusulina, 62–66
Fusulinella, 62–66
fusulinellid wall structure, 61
fusulinids, 60

G

Galateacrinus, 647
Gale, H. R., 333, 450
Galeropygus, 697
Galloway, J. J., 77, 195
gape, pelecypods, 403, 444
gaping, pelecypods, 409
Gardner, J., 333, 450, 462
Garrett, J. B., Jr., 77
Gastrioceras, 372, 374
Gastrocopta, 330
gastropod borings and trails, 332
Gastropoda, 269, 270
gastropods, amphigastropods, 297–301
anatomy, 277–279
archaeogastropods, 301
classification, 287
ecology, 277
evolution, 290
geologic occurrence, 296–328
mesogastropods, 313
Miocene, 322, 324
morphology, 279–287
neogastropods, 322
opisthobranchs, 328
orientation, 281, 298
prosobranchs, 301–328
pulmonates, 329
shell form, 280, 287
shell structure, 286
types, 276
(*See also* Cambrian, Cretaceous, Devonian, Jurassic, Permian, Recent, Silurian, Triassic gastropods)

gastropores, 106
Gastrotricha, 456
gastrozooids, 106
gastrula, 26, 27
Gaudryina, 42, 53
Gaurocrinus, 637
Gazacrinus, 631, 639
Gedinnian, 37
Geffenina, 534
Geis, H. L., 603
Geisina, 522, 538
Geisonoceras, 359
Gemmellaro, G. G., 265
genal angle, 480
genal ridge, 493, 495
genal spine, 480
gender of names, 11
genera, 9, 11, 15
genes, 24
genetics, 24, 30
genital plates, 667, 680
genital pores, 589, 680
Gennaeocrinus, 641
genotype, 24
genus, 11
geologic occurrence of fossils, 7
(*See also* specific names of groups)
Gephyrea, 458
Geralycosa, 565
gerontic, 27
Gerth, H., 154
Gervillia, 426, 429, 430
Gigantoproductus, 203, 249
Gilbertsocrinus, 634, 645
gill slits, 687
gills, cephalopods, 336, 337
chitons, 273
gastropods, 279
pelecypods, 400, 402, 409
girdle, chitons, 271
girder, trilobites, 493, 495
Girty, G. H., 265
Girtyocoelia, 89, 91
Girtyocoelia, 89, 91
Gislén, T., 651
Gissocrinus, 638
Givetian, 37
glabella, 480, 503, 507
glabellar furrow, 480
glabellar lobe, 480
Glabrocingulum, 307
Glaessner, M. F., 77, 78
glass sponges, 91
Glenn, L. C., 450
globiferous, 683
Globigerina, 50, 58
Globigerinella, 50
Globigerinoides, 50
Globivalvulina, 48
Globorotalia, 50
Globotruncana, 50
Globularia, 318
Glossograptus, 727, 730
Glossorthis, 214
Glycimeris, 413, 439, 442
Glyphostomella, 48
Glyptocrinidae, 633
Glyptocrinina, 614, 634

glyptocrinoids, 633, 634
Glyptocrinus, 633, 637
Glyptocystites, 591
Glyptopleura, 535
Glyptopleurina, 534
Glyptopora, 177
Glyptorthis, 231
Glyptosphaerites, 588, 591
Gnathodus, 737
Goldring, W., 651
Goldringia, 359
Gomphocystites, 592
gonangium, 101, 102
Gondolella, 737
goniatite sutures, 369, 372, 376, 378, 384
Goniatites, 362, 366, 367, 369, 371
Goniobasis, 330
Goniocidaris, 680, 683
Goniograptus, 728
Gonioloboceras, 372
Goniophora, 414, 416, 430
gonopore, 589
Gordiacea, 454
Gorgonacea, 104, 112
Gorgonia, 111
Gosseletina, 304
Gothograptus, 731
Grabau, A. W., 154
Grammysia, 412, 416
Grant, U. S., 333, 450
Graphidula, 316
Graphiocrinus, 647
Graphiodactylus, 533
Graphularia, 111
graptolites, 718
    Cambrian, 716, 719
    classification, 718
    declined, 727
    decurved, 725
    dendroid, 719, 728, 731
    ecology, 722, 729
    evolution, 722, 725
    geologic occurrence, 722
    morphology, 719–729
    Ordovician, 715, 719, 723–728, 730
    Silurian, 723, 731
Graptolithina, 718
Graptoloidea, 718, 723
graptoloids, 723, 728, 730
growth lines, brachiopods, 207
    cephalopods, 339, 366
    corals, 118, 121
    gastropods, 286
    pelecypods, 403
Gryphaea, 432
Gryphochiton, 272
Gshelian division, 37
Guadalupian division, 37
Gümbelina, 53
Gümbelitria, 50
gutter, gastropods, 284
Guttulina, 53
Gymnolaemata, 159, 161
Gymotoceras, 377
Gypidula, 217, 238
Gyraulus, 320
Gyroceras, 347
gyrocone, 347

Gyrodes, 317
Gyroidina, 53, 56, 58

H

Hadrophyllum, 125, 129
Haeckel, E., 16, 39, 78, 557
Haime, J., 154
Haimesiastraea, 151
Haliotis, 284
Hall, J., 97, 154, 265, 412, 651
Hallaster, 668, 670, 673
Halliella, 533
Hallopora, 165, 167
Halobia, 429
Halysiocrinus, 622, 643
Halysites, 115, 116
halysitids, 115
Hambach, G., 603
Hamites, 383
Hamulina, 382
Hamulus, 456
Hanna, G. D., 78
Hantkenina, 54, 56
Hapalocrinus, 639
Haplocrinites, 641
Haplocytheridea, 543
Haplophragmoides, 50
Haploscapha, 399
Hapsiphyllum, 125, 133, **135**
Haptopoda, 565
Harpa, 320, 328
Harpidae, 494
Harris, G. D., 333
Harrisoceras, 358
Hastifaba, 537
Hastites, 391
Haswell, W. A., 581
Hatschek, 16
Healdia, 537
heart urchins, 696
Hebertella, 217, 231
Heezen, B. C., 738
Helderbergian, 37
Helianthaster, 671
Helicoceras, 386
Helicodiscus, 330
Helicotoma, 292
Heliolites, 112, 116
Heliophyllum, 129, **135**
Heliopora, 110
Heliosphaera, 74
Helisoma, 330
Helix, 276
Helvetian, 36
hematitization, 4
Hemiaspis, 555
Hemiaster, 705
Hemichordata, 17, 716, **718**
hemichordates, 580
Hemicidaris, 695
Hemicosmites, 591
Hemicythere, 543
Hemilytoceras, 381
hemiomphalous, 290
hemiperipheral growth, 213
Hemiphormosoma, 710
Hemipronites, 218

hemipteroids, 572
hemisepta, 167
Hemitrypa, 177
Hemizyga, 307
Henbest, L. G., 77
Henderson, J. B., 275
henidium, 211
Hercoglossa, 363
heredity, 24
Heritschia, 127, 136
Hesperorthis, 217, 218
Hessland, I., 550
Heteralosia, 252
Heterelasma, 234
Heterocentrotus, 683
heterochronous homeomorphs, 218
Heterocoelida, 87
Heterocorallia, 104, 142
Heterocrinus, 616, 637
Heterodonta, 411, 434, 440
heterodonts, 409, 410, 414, 416, 420,
    424, 425, 428, 434, 436, 438, 442,
    446
Heterophrentis, 127
Heterophyllia, 142
Heterorthis, 232
Heterostegina, 57, 68, 70
Heterostomella, 50
heterostrophy, 283
heterotomous arms, 610, 619
Hettangian, 36
hexacorals, 104
Hexadoridium, 74
Hexagonaria, 131, 138
Hexagonella, 180
Hexameroceras, 358
Hexaphyllia, 142
Hexasterophorida, 87
Hibolites, 391
Hildoceras, 379
Hill, D., 154
Hill, R. T., 450
Hincksina, 190
Hinde, G. J., 78, 97
Hindeodella, 737
Hindia, 87, 93
hinge line, brachiopods, 199, 207, 210
    pelecypods, 403
hinge plate, brachiopods, 213
    pelecypods, 403
hinge teeth, 211, 403, 422
Hipparionyx, 247
Hipponicacea, 317, 318, 322
Hippurites, 440, 441
Hirudinea, 458
holaspis, 486
holdfast, crinoids, 620
Holectypina, 697, 698
Holectypoida, 696, 710
holectypoids, 696, 697
Holectypus, 697
Hollina, 530
Hollinella, 538
Holm, 718
Holmes, G. B., 738
Holmia, 485, 486, 490, 518
Holochoanites, 342
holochoanitic, 350, 353

Holocystites, 592
Holopea, 294
holoperipheral growth, 213
Holopus, 575
holostomatous, 288
holotheca, 115, 121, 133
Holothuria, 653
Holothuroidea, 15, 577
holothuroids, 574, 576, 578, 653
   classification, 657
   ecology, 656
   geologic occurrence, 655–657
   morphology, 653
   zoological affinities, 658
holotype, 13
Homalophyllum, 125, 127, 133
homeochilidium, 206
homeodeltidium, 207
homeomorphs, 34, 141, 217
Homeospira, 262
Homocoelida, 87
Homocrinidae, 622
Homocrinus, 615
homonyms, 11
Homoptera, 572
Homotelus, 503, 509
honeycomb corals, 115
hood, cephalopods, 337
hook, holothuroids, 656
hoplocarids, 548
horizontal graptoloid stipes, 727
Hormotoma, 294, 301
horn corals, 135, 137
Hornera, 185
Howe, H. V., 550
Howell, B. F., 519
Huang, T. K., 154
Hudson, G. H., 603
Hudsonaster, 670
Hughmilleria, 557, 559, 560
Hungarites, 377
Hustedia, 262
Hutsonia, 539
Huxley, T., 219
hyaline, 42
Hyalospongia, 80, 87, 91
Hyatt, A., 342, 354, 396
Hyattechinus, 693
Hyattidina, 261
Hybocrinida, 613, 615, 623
hybocrinids, 623
Hybocrinus, 615, 637
Hybocystites, 623
Hydnoceras, 88, 91
Hydra, 100
Hydractinia, 105
hydranths, 102
Hydrobia, 331
Hydroida, 104, 105
hydropore, 578, 579, 589
hydrospire, 576, 594, 601
hydrospire folds, 597, 601
hydrotheca, 102, 105
Hydrozoa, 100, 104
Hygrosoma, 683
Hyman, L. H., 97, 153
Hymenocaris, 518
hymenopteroids, 572

Hyolithes, 329
Hypagnostus, 497, 499
Hyperammina, 44
Hyperamminoides, 48
hyperstrophic, 281, 288
Hyphasmophora, 531
hypodeltoid, 575
hyponome, 336, 337
hyponomic sinus, 339, 348
Hypoparia, 487, 494, 515
hypoparian trilobites, 479, 493, 503, 505
Hyposalenia, 705
hyposeptal deposits, 348
hypostege, 193
hypostome, 475, 479, 481, 488
hypostracum, 403, 407
hypothecal pore rhomb, 589
Hypothyridina, 212, 239, 241
Hypselentoma, 307
Hypseloconus, 291

I

Ianthina, 277
Ichthyocrinus, 627, 639
Icriodus, 736
Idiognathodus, 737
Idiotrypa, 167
Idmonea, 185, 186
Idonearca, 413, 434
Ildraites, 456
Illaenurus, 501, 509
Illaenus, 503, 505, 509
immature region, 161, 168
impressed zone, cephalopods, 339
mpunctate shell, brachiopods, 199, 203,
   227, 239, 249
Inadunata, 15, 577, 613, 617, 621
   nadunate crinoids, 621
Inarticulata, 219–226
inarticulate brachiopods, 198, 201, 205,
   221
Incisurella, 535
inductura, 283
inequilateral, 409
inequivalve, 409
inferradianal, 616
nfrabasals, 610, 614, 627
infralaterals, cystoids, 589
   ophiuroids, 673
inframarginals, 664
infranodals, 611
Infusoria, 39
ink gland, 337, 341
inner hinge plate, 210
inner lamella, ostracodes, 529
inner lip, gastropods, 283
inner side plates, blastoids, 597
inner wall, corals, 121, 129
Inoceramus, 402, 426, 430, 432
Insecta, 463, 471, 567
insects, 472, 552, 567, 569, 572
insert, echinoids, 686
integrate wall, 159
Integrina, 167
interambulacral areas, 676, 713
interambulacral spines, 676
interambulacrals, bothriocidaroids, 713

nterambulacrals, crinoids, 619
   echinoids, 676
   edrioasteroids, 585
interarea, brachiopods, 205, 211, 214
interbrachial field, 664
interbrachials, 618, 627, 632
intercheek sutures, 492
intergenal spines, 486
internal madreporite, 653
internal partitions, echinoids, 703
internodals, 611
interpinnulars, 632
interpleural groove, 481
interporiferous zone, echinoids, 677
interradials, crinoids, 618, 627, 629,
   631–633
   starfishes, 664
interseptal ridge, 118, 121
interspaces, bryozoans, 170
intervallum, 94
intracapsular, 73
Intrapora, 175
involute, 42, 48, 288, 354
Iocrinus, 616, 637
Ireland, H. A., 77
Irregularia, 577, 675, 683, 696
irreversibility in evolution, 34
Ischadites, 87, 94
Isocardia, 436, 446
Isochilina, 526
isochronous homeomorphs, 218
Isocrinida, 614
Isocrinus, 605, 611, 612, 648, 650
Isodonta, 411, 422
isodonts, 409, 422, 432, 442
isomyarian, 409
Isonema, 302
isopygous, 514
Isotelus, 478, 491–494, 503, 515, 519
isotomous arms, 610, 619
Itieria, 290

J

Jackson, R. T., 713
Jacksonian, 36
Jaekel, O., 602, 651
jaws, cephalopods, 337
   echinoids, 680
   ophiuroids, 669
Jeannel, R., 572, 573
Jeffords, R. M., 154
Jones, O. A., 154, 265
Jonesella, 525
jugum, 213, 251, 259
Jurassic brachiopods, 217, 218
Jurassic bryozoans, 178, 187, 193, 194
Jurassic cephalopods, ammonoids, 369,
   377, 379–381, 384
   belemnoids, 388–392
   sepioids, 393
   teuthoids, 394
Jurassic chelicerates, spiders, 565
   xiphosurans, 555–557
Jurassic coelenterates, corals, 113, 147
   hydroids, 105
   scyphozoans, 109
   sphaeractiniids, 106

Jurassic crustaceans, eucarids, 547, 549, 550
  hoplocarids, 548
  ostracodes, 539
  percarids, 549
Jurassic echinoderms, crinoids, 648, 650
  echinoids, 686–690, 695, 697
  holothuroids, 656
  starfishes, 668
Jurassic gastropods, archaeogastropods, 299, 301, 304, 307, 309, 311
  mesogastropods, 311, 313, 315, 317, 322, 324
Jurassic insects, 569, 572
Jurassic pelecypods, 419, 422, 426–430, 440–447
Jurassic protozoans, dinoflagellates, 76
  foraminifers, 50
  tintinnids, 76
Jurassic sponges, 81, 93
Juresania, 252

K

Kallimorphocrinus, 643
Karnian, 36, 378
Karreriella, 53
Kaskia, 509
Kazanian, 37
Keen, A. M., 450
Kellett, B., 550
Kellettina, 537
Kerionammina, 44
keriotheca, 61, 62
Kew, W. S. W., 334
Keyseran, 37
Kheraiceras, 381
Kiderlen, H., 462
Kiesling, A., 153
Kimmeridgian, 36
Kinderhookian, 37
King, R. E., 265
King, R. H., 97, 265
Kingena, 234
Kinorhyncha, 456
Kionoceras, 358
Kirkbyella, 531, 536
Kirkella, 505, 509
Kleinpell, R. M., 77
Kloedenella, 528
Knight, J. B., 299, 333
Knightina, 539
Knightites, 297
Kochaspis, 499
Koninck, L. G. De, 333
Kootenia, 501
Kozlowski, R., 219, 243, 265, 719, 720, 732
Krausella, 525
Kuhn, O., 153
Kummel, B., 342, 396
Kungurian, 37
Kustarachnae, 565
Kutorgina, 226
Kyphopyxa, 50

L

Labechia, 107, 108
Labechiida, 104, 107

labrum, 701
Ladinian, 36, 377
Laganina, 703
Laganum, 707
Lagena, 42, 43
Lagenammina, 44
Lagenidae, 49, 60
Lalicker, C. G., 519
Lamarck, J. B. P. A., 16, 29
Lamarckism, 29
Lambeophyllum, 114, 118, 135
lamellae, corals, 121, 127
  ostracodes, 529
lamellar layer, 402, 403
lamellibranchs, 398
Lampterocrinus, 639
lancet plate, 595
Lang, W. D., 154
Lange, F. W., 462
Lankester, E. R., 16
Lapparia, 320
Laramian, 36
larval, 26
lateral fascioles, 701
lateral increase, corals, 131
lateral lobes, 366
lateral saddles, 366
lateral teeth, pelecypods, 403, 434
laterals, cystoids, 589
  ophiuroids, 673
Laterocavea, 183
Latirus, 320
Law, J., 550
Leaia, 545
Leanchoilia, 518
Lecanites, 376
Lecanocrinus, 617, 639
Lecanospira, 292
Lechtricoceras, 358
Leconteiceras, 378
Lecythiocrinus, 647
Leioclema, 180
Leiopteria, 414, 416
Leiosoecia, 183
Leodicites, 456
Leonard, A. B., 333
Leonardian, 37
Leonardophyllum, 136
Leperditella, 526
Leperditia, 525
Lepidocentridae, 689
Lepidocentroida, 684–689, 710
lepidocentroids, 689, 691
Lepidocyclina, 68, 70, 72
Lepidocyclus, 241
Lepidodiscus, 586
Lepidorbitoides, 69
Leptaena, 244, 247
Leptaenisca, 203
Leptellina, 244
Leptobolus, 223
Leptochirognathus, 736
Leptocoelia, 255
Leptocyathus, 151
Leptodesma, 416
Leptodus, 244, 248
Leptoptygma, 307
Leptozyga, 307

LeRoy, L. W., 551
Leuckart, 16
leucon type, sponge, 82
Leuconia, 82
Leucosolenia, 80, 81
Levene, C. M., 266
Levenea, 212
Levifusus, 320, 326
Lhwyd, 475
Liassic, 379
Lichas, 507, 509
Lichenaria, 114
Lichenoides, 584
Liebusella, 56
ligament, 402
ligament area, 403, 431
ligament fossa, 610
ligament groove, 405
ligament pit, 610
Lima, 398, 400, 420, 431, 434
Limacea, 411, 432
Limopsis, 413
Limoptera, 416
Limulida, 555
Lindstroemoceras, 351
line of concrescence, 529
Lingula, 197–200, 203, 221, 222
Lingulacea, 220–223, 263
Lingulella, 223
Lingulepis, 223
Lingulina, 50
Linné, C., 11, 16, 29, 268
Linoproductus, 251, 252
Linthia, 700, 707
Liopeplum, 316, 328
Liospira, 292
Liparoceras, 379
Liroceras, 360
Lirosoma, 322
Listerella, 53
listrium, 207
lithistid sponges, 84
Lithobius, 567
Lithocampe, 74
Lithomelissa, 75
Lithostrotion, 133, 138
Lithostrotionella, 133 , 138
littoral, 20
Littorina, 322, 324, 326
Littorinacea, 311, 320, 324, 326
Littorinopsis, 324
Lituites, 347, 357, 358
lituiticone, 347
Lituola, 53
Lituolidae, 49
Lituotuba, 46
living chamber, cephalopods, 340, 348, 351, 366
Llandeilian, 37
Llandoverian, 37
lobes, cephalopod sutures, 340, 363, 366
lobster, 468
Lochman, C., 519
Lockportian, 37
Loeblich, A. R., Jr., 196
Loganograptus, 728
Loganopeltoides, 494
logarithmic spiral, 339

Loligo, 336, 343
Lonchodina, 737
Longispina, 251
Longstaff, J., 333
Lonsdaleia, 123
lonsdaleoid corals, 123, 127, 139
loop, brachidium, 213
Lophamplexus, 136
lophophore, 156, 197, 200, 202
Lophophyllidium, 135, 137
Lophospira, 294
lorica, 76
Loricata, 272
Loricula, 546
Lorieroceras, 359
Loriolaster, 671
Lovén, 684
Lovenechinus, 686
lower lamella, trilobites, 493
Loxoconcha, 541
Loxodus, 736
Loxonema, 301, 302, 309
Loxonematacea, 300–311
Loxotoma, 301
Lucernariida, 104
Lucina, 436, 439
Lucinacea, 411, 439
lucinoid type, heterodonts, 434, 436
Ludian, 36
Ludlovian, 37
Lumbriconerites, 456
lumen, crinoids, 608
lunaria, 169, 171
Lunatia, 312
lunula, 286
lunules, echinoids, 703
    pelecypods, 403
Lunulicardium, 416
Lutetian, 36
Lyellia, 112, 116
Lymnaea, 331
Lyonicrinus, 639
Lyopomata, 219
Lyriocrinus, 639
Lyriopecten, 416
Lyrodesma, 414, 419
Lyropecten, 442
Lytoceras, 381
Lytospira, 292

M

MacBride, E. W., 658, 674, 714
McCormack, J. T., 658
Maccoya, 686
MacFarlan, A. C., 196
MacGinitie, G. E., and N., 98, 153, 333,
    450, 474, 581
McGuirt, J. H., 196
Macluritacea, 292, 302
Maclurites, 299, 302
McNair, A. H., 196
MacNeil, F. S., 451
Macoma, 399, 446
Macrocephalites, 381
Macrocrinus, 645
Macrocypris, 541
Macrocystella, 583, 584

Macronotella, 525
macropygous, 514
Macrostylocrinus, 633, 639
Mactra, 399, 442, 444, 445
Mactracea, 412, 444
maculae, 167
madreporite, 575, 576, 661, 667, 680,
    713
    internal, 653
Maeandrostia, 89, 91
Maestrichtian, 36
major septa, 119, 121, 125
Malacostraca, 411, 544, 547, 548
Malocystites, 591
mamelons, 676
Mamillopora, 193
mandibles, arthropods, 467
Mangilia, 322
Mansfield, W. C., 333, 451
Manticoceras, 370
mantle, 199, 271, 279, 336, 400
mantle cavity, 336, 402
Marellomorpha, 470
marginal furrow, trilobites, 480
marginal notches, echinoids, 703
marginal spines, echinoids, 677
    starfishes, 661, 664
marginal suture, trilobites, 493
marginal tubercles, 676
marginals, cystoids, 589
    edrioasteroids, 585
    starfishes, 664
Marginella, 324
Marginifera, 252
Marginulina, 50, 53, 58
Marjumia, 499
Marrella, 518
Marsipella, 44
Marssonella, 50
Marsupiocrinus, 629, 639
Marsupites, 648
Martin, G. C., 333, 450
Massilina, 53, 56
massive corals, 123, 131
Mastigobolbina, 528
Mastigograptus, 730
Mastigophora, 39
Mataxa, 316
mature region, 161, 168
maxillae, 467, 533
maxillulae, 533
Maysvillian, 37
Mazzalina, 320, 326
mecopteroids, 572
median lamella, corals, 121, 127
median ridge, brachiopods, 207
median septum, brachiopods, 212
    graptolites, 725
Medionapus, 314
Medlicottia, 371, 374
medusae, 99, 101
Medusina, 109
Meek, F. B., 265
Meekechinus, 685, 689, 693
Meekella, 248
Meekoceras, 376
Meekopora, 180
Meekospira, 309

megacanthopores, 168, 170
Megalomus, 414
megalospheric, 40
Meganeura, 569, 570
megascleres, 83
Megastrophia, 247
Megateuthis, 388
Megistocrinus, 641
Mehl, M. G., 738
Melanella, 324
Meliceritites, 183
Melocrinites, 629, 641
Melocrinitidae, 629
Melonechinoida, 684, 686, 690
melonechinoids, 690, 693
Melonechinus, 686, 690, 693
Membraniporidra, 190
membranous inner lamella, 529
Menetus, 331
Menevian, 37
Meramecian, 37
meraspis, 485, 489
Mercenaria, 436, 442
Meristella, 261
Meristina, 261
Merocrinus, 616, 637
Merostomata, 471, 553
merostomes, 553
Merostomoidea, 470
Merriam, C. W., 333
mesenteries, corals, 102, 117
    holothuroids, 653
Mesoconularia, 460
Mesogastropoda, 289, 313
mesogloea, 100
Mesolobus, 249, 252
mesopores, 160, 162, 163
Mesopsychopsis, 569
mesosoma, 552
mesostereom, 589
mesotheca, 169, 170
Mesothyra, 546
Mesotrypa, 165
Mesozoa, 17
Mespilocrinus, 643
Messina, A. R., 77
Metablastus, 601
Metaconularia, 460
Metacypris, 539
Metalonchodina, 737
metameres, 464
metamesenteries, 144
metamorphoses, 467, 468
metanauplius, 521
Metaplasia, 256
metaprotaspis, 485
metasepta, 119, 121, 123
metasicula, 720, 724
metasoma, 552
metastome, 475, 481
Metazoa, 17
Metoedischia, 569
Metopaster, 673
Metridium, 102, 117
Metriophyllum, 129
Mexicaspis, 499, 514
Mexicella, 499
Michelinia, 135

Michelinoceras, 348, 358–361, **388**
Michelinoceratidae, 348
Michelinoceroida, 342, 348, 364
michelinoceroids, 348, 357–361
Micrabacia, 148
micracanthopores, 168, 170
Micraster, 705
Microcoleodus, 736
microcrinoids, 621
Microcyclus, 129
Micromitra, 225
Microparaparchites, 536
micropygous, 514
microscleres, 83
microspheric, 40
Midwayan, 36
miliary spines, 679
miliary tubercles, 676
Miliolidae, 48, 49, 56
milled ring, 676
Millepora, 106
Milleporida, 104, 105
Miller, A. K., 396
Milleratia, 526
Millerella, 63–66
Millericrinida, 614, 650
millipedes, 463, 471, 566
Millot, J., 573
Mimella, 218
Minilya, 180, 183
minor septa, 119, 121
Miocene cephalopods, 361, 391
Miocene crustaceans, cirripeds, **547**
    ostracodes, 543
Miocene echinoderms, crinoids, 650
    echinoids, 700, 708
    starfishes, 673
Miocene foraminifers, 57, 58, 70, 72
Miocene gastropods, 322, 324
Miocene pelecypods, 436, 442, 445, 446
Miogypsina, 70, 72
miogypsinids, 72
Miogypsinoides, 69, 72
Mississippian blastoids, 594, 595, 599,
    600
Mississippian brachiopods, orthids, 233
    rhynchonellids, 241–243
    spiriferids, 256–262
    strophomenids, 203, 244, 247–251
    terebratulids, 234
Mississippian bryozoans, 172, 177, 194
Mississippian cephalopods, ammonoids,
    366, 369, 371, 376
    belemnoids, 392
    nautiloids, 354, 361
Mississippian chitons, 272
Mississippian coelenterates, corals, 132,
    137–142
    scyphozoans, 110
Mississippian conodonts, 733, 737
Mississippian crinoids, camerates, 629,
    633–635, 643–645
    flexibles, 627, 643, 645
    inadunates, 621, 623, 624, 643
Mississippian crustaceans, ostracodes,
    534–536
    percarids, 549
    phyllocarids, **546**

Mississippian edrioasteroids, 586
Mississippian eleutherozoans, echinoids,
    686, 690, 693
    holothuroids, 655
    starfishes, 668, 672
Mississippian gastropods, amphigastro-
    pods, 296, 297
    archaeogastropods, 299, 301, 304,
    307, 309, 311
Mississippian graptolites, 716, 719
Mississippian myriapods, 566
Mississippian pelecypods, 416, 420, 426,
    430, 439
Mississippian scorpions, 563
Mississippian sponges, 89, 90
Mississippian trilobites, 494, 509
Missourian, 37
Mitrocystites, 584
Mitrodendron, 147
Mixochoanites, 342
mixoperipheral growth, 213
Mixopterus, 559, 561
Modiolopsis, 414
Mohawkian, 37
Mojsisovics, M. E., 396
molds, 5
Mollocrinus, 647
Mollusca, 17, 268
Molluscoidea, 159
mollusks, 270
    (*See also* Ordovician, Pennsylvanian,
    Tertiary mollusks)
Molpadida, 657
molts, arthropods, 468
Monaxonida, 87
monaxons, 83
monobasal, 699
Monobathrida, 15, 614, 618, **634**
monobathrids, 634
Monoceratina, 541
monocyclic, 615, 621, 632
Monodonta, 312
Monogenerina, 42
Monograptus, 727, 731
monolamellar flange, 529
monomyarian, 409, 422, 432
Monopleura, 441
Monopteria, 420
Monopylea, 75
Monotis, 429
Monotocardia, 301, 313, 322
monotypy, 13
Montanan, 36
Montastrea, 148
Montian, 36
monticules, 107, 163, 165
Montlivaltia, 147
Montyoceras, 351
Moodey, M. W., 601, 650
Moore, R. C., 154, 196, 603, 651
Mooreinella, 48
Mooreoceras, 360
Moorites, 539
Morea, 316, 326
Moreman, W. L., 77
Moret, L., 333, 451

morphology, ammonoids, 362–374
    annelids, 452
    arachnids, 561–565
    arthropods, 464–469
        (*See also* specific subdivisions)
    asteroids, 661–664
    blastoids, 594–600
    brachiopods, 199–218
    bryozoans, 156, 161–173, 187–193
    carpoids, 585
    cephalopods, 336
        (*See also* specific subdivisions)
    chelicerates, 552, 553
    chitons, 271–273
    conodonts, 734
    conulariids, 458
    corals, 112, 117–133, 144, 145
    crinoids, 605–621
    cystoids, 587–593
    dibranchiates, 336, 387–394
    echinoderms, 575
        (*See also* specific subdivisions)
    echinoids, 676–688, 696–703
    edrioasteroids, 586
    eocrinoids, 583
    eurypterids, 557–560
    foraminifers, 39–44, 60–63
    gastropods, 277–289
    graptolites, 719–729
    holothuroids, 653–655
    insects, 567–572
    mollusks, 270–271
        (*See also* specific subdivisions)
    myriapods, 566
    nautiloids, 337–356
    ophiuroids, 667–669
    ostracodes, 521–533
    paracrinoids, 583
    pelecypods, 400–406, 434, 439
    pleosponges, 94
    radiolarians, 73
    receptaculitids, 94
    scaphopods, 274
    stromatoporoids, 107
    trilobites, 475–483
    xiphosurans, 555–556
Morrisonia, 539
Morrowan, 37
Mortensen, T., 714
Moscovian, 37
Mourlonia, 301, 304
mouth angle plates, 660, 664
mouth frame, 659, 664, 667
Mucrospirifer, 217, 256
Muensteroceras, 371
Muir-Wood, H. M., 265
Müller-Stoll, H., 396
multicostate, 409
Multicrescis, 183
multispiral, 288
multivincular, 409
mural deposits, 348
mural pores, 61, 62, **115**
Murchisonia, 301, 302
Murex, 318, 326, 328
Muricacea, 314–318, 322–328
muscle fossa, crinoids, 610

muscle scars, briachopods, 207, 211
  ostracodes, 529
  pelecypods, 403
muscles, brachiopods, 200
mussels, 398, 401
mutant and mutation, 24, 31
Mya, 444–446
Myacea, 412, 444
Myalina, 420, 425, 426, 430, 447
Myelodactylus, 620, 639
Myers, E. H., 77
myophore, brachiopods, 213
  pelecypods, 406
Myophoria, 418, 419, 429
myophoric septum, 444, 445
Myriapoda, 463, 471
myriapods, 463, 552, 556, 566, 567
Myriastiches, 711
Myriotrochus, 656
Myrtillocrinus, 641
Mysidacea, 549
Mytilacea, 411, 429, 448
Mytilarca, 420
mytilids, 414, 416, 420, 425, 426
mytiliform, 409
Mytilus, 398, 400, 426, 429, 430, 442

N

nacreous layer, 339
Naef, A., 396
Naiadites, 420, 430
Namurian, 37
Nannobelus, 391
Nanorthis, 228
Naos, 125, 129
Napulus, 314
Naraoia, 518
Nardoa, 668
Nassarius, 326
Natantia, 549
Naticacea, 312–318, 326
Naticopsis, 304
Natland, M. L., 77
natural selection, 31
nauplius, 521
Nautilida, 342, 354, 358, 364
nautilids, 355–362
Nautiloidea, 342, 364
nautiloids, 341–362
Nautilus, 26, 335–337, 343–345, 361, 367
neanic, 27
Nebalia, 546, 548
neck, cephalopods, 350
  echinoid spines, 676
  gastropod shells, 283
nektonic, 10
nektoplanktonic, 19
nema, 719, 724
Nemagraptus, 730
Nemathelminthes, 454
Nematifera, 183
nematocysts, 100, 103
Nematoda, 454
Nemertina, 17, 456
Neocomian, 36, 382
Neogastropoda, 289, 314, 322
Neogene pelecypods, 416, 418, 430, 442–446

Neolenus, 481, 492
Neolobites, 384
Neoshumardites, 369
Neospirifer, 259
neoteny, 28, 138, 467
Neotremata, 219–224
Neozaphrentis, 133
Nephrolepidina, 68, 70, 72
nepionic, 27
Neptunea, 326
Nereidavus, 456
Nerinea, 311, 312, 317
Nerineacea, 311, 312, 317, 318
Nerita, 311
Neritacea, 302, 304, 308, 313
neritic, 20
neritopelagic, 21
Neurodontiformes, 734
neuropteroids, 571
Nevadia, 488, 490
Neverita, 317, 326
Newell, N. D., 451
Newsomella, 414
Niagaran, 37
Nicholson, H. A., 153
Nicholsenella, 167
Nidulites, 87, 97
Nidulitida, 89
Nipponites, 372, 386, 387
Nisusia, 228
nodals, crinoids, 611
Nodosaria, 42, 43
Nonion, 53, 56
Nonionella, 53, 56
Norian, 36, 378
Normannites, 374, 380
Nortonechinus, 689, 690
Norwoodia, 503
noses, ophiuroids, 667
Nothognathella, 736
Notostraca, 545
notostracans, Cambrian, 544, 545
notothyrium, 207, 211
Nucella, 326
Nucleocrinus, 596, 599, 600
nucleus, gastropod shells, 283
  radiolarians, 73
Nucula, 409, 413, 420, 425, 434, 439
Nuculacea, 411, 414, 418
Nuculana, 413, 416, 425, 439, 446
nuculids, 416, 425
Nuculina, 413
Nuculoidea, 416
Nuculopsis, 420
Nudirostra, 242
Nummulites, 68
Nummulitidae, 68
Nyctocrinus, 631, 634, 639
Nyctopora, 114
Nymphon, 566

O

obconical, 288
Obelia, 101, 103
Obolella, 225
obtuse-angle interarea, 214
obverse side, bryozoans, 171

occipital node, 480
occipital segment, 480
oceanopelagic, 21
Ochetosella, 188
O'Connell, M., 573
Octonaria, 531
Octonariella, 531
Octopoida, 343, 364
octopoids, 395
Octopus, 335, 341
ocular plates, echinoids, 680
ocular platforms, trilobiter, 480
ocular ridges, trilobites, 480, 493, 494
ocular spots, starfishes, 661
oculogenital ring, 677, 686
Odontocephalus, 507
Odontochile, 509
Odontofusus, 315, 326
odontophores, 664
Oecoptychius, 374, 381
Ogygopsis, 501, 518
Oistodus, 736
Oklahomacrinus, 6 47
Okulitch, V. J., 98, 154
Olenellus, 487–490, 503, 512, 513, 518
Olenoides, 491, 492, 501, 504, 515, 518
Oligocene bryozoans, 178, 188, 190
Oligocene echinoids, 708
Oligocene foraminifers, 57, 70, 72
Oligocene insects, 572
Oligocene myriapods, 567
Oligocene ostracodes, 543
Oligocene pelecypods, 438
Oligocene percarids, 549
Oligocene spiders, 565
Oligochaeta, 458
Oligoporus, 686
Oligopygus, 707
Oliva, 324, 328
Olivella, 326, 328
Omospira, 294
Omphalotrochus, 309
Oncoceroida, 342, 352, 364
oncoceroids, 352, 357–361
Onesquethawan, 37
ontogeny, 26, 28
  ammonoids, 366
  arthropods, 467
  brachiopods, 204
  cephalopods, 368, 372, 389
  crustaceans, 521
  echinoderms, 578
  graptolites, 727
  ostracodes, 531
  trilobites, 484, 489
Onychaster, 668
Onychites, 389
Onychocrinus, 643
Onychophora, 470
oostegites, 549
Opabinia, 518
Opalia, 326
Operculina, 43
Operculinoides, 68, 70
operculum, bryozoans, 175, 190
  gastropods, 276, 285
  worms, 454
ophicephalous, 683

Ophileta, 292
Ophioceras, 357
Ophiocystoidea, 577, 713
ophiocystoids, 713
Ophioderma, 667, 668
Ophioglypha, 673
Ophiuroidia, 577, 659, 665
ophiuroids, 574–578, 667–670, 673
Öpikina, 244
Opiliones, 565
Opisthobranchia, 289, 312–318, 324, 328
opisthocline, 409
opisthodetic, 409, 418, 434
opisthogyral, 409, 416
Opisthoparia, 487, 491, 515
opisthoparian trilobites, 479, 489, 491, 497–509
opisthosoma, 552
Oppelia, 380
oral side, 577
orals, 585, 614
orbicular pelecypods, 409
Orbiculoidea, 200, 217
Orbitoides, 69
orbitoidids, 68
Orbitremites, 599, 601
Orbulina, 50
orders, 15, 16
Ordovician brachiopods, inarticulates, 222–226
  orthids, 212, 215, 218, 227–232
  pentamerids, 236, 237
  rhynchonellids, 239–241
  spiriferids, 251, 259
  strophomenids, 243–247
  triplesiids, 237, 239
Ordovician bryozoans, 161, 165, 173, 194
Ordovician coelenterates, corals, 113, 114, 118, 135
  hydroids, 105
  scyphozoans, 110
  stromatoporoids, 108
Ordovician conodonts, 733, 736
Ordovician conulariids, 460
Ordovician crustaceans, cirripeds, 472
  ostracodes, 472, 525, 526
  phyllocarids, 546
Ordovician echinoderms, blastoids, 599, 600
  bothriocidaroids, 712, 713
  camerate crinoids, 629, 631, 633–637
  carpoids, 584
  cystoids, 587, 591, 593
  echinoids, 685, 686, 691, 711
  edrioasteroids, 586
  eocrinoids, 583
  flexible crinoids, 627, 635–637
  inadunate crinoids, 621–623, 626, 635–637
  ophiocystoids, 713
  paracrinoids, 584
  starfishes, 659, 660, 665, 668, 670
Ordovician foraminifers, 44
Ordovician graptolites, 715, 719, 723–728, 730

Ordovician mollusks, cephalopods, 347–357
  chitons, 272
  gastropods, 292–297, 299, 301, 307, 309
  pelecypods, 414–419, 423, 426, 430, 444
  pteropods, 329
  tentaculitids, 459
Ordovician receptaculitids, 96
Ordovician sponges, 87
Ordovician trilobites, agnostids, 497
  hypoparians, 493–495, 503–507
  opisthoparians, 463, 492–494, 503–505
  proparians, 491, 503–505
Ordovician worms, 452, 456, 460
orientation, brachiopods, 204
  conodonts, 738
  crinoids, 613
  echinoids, 682
  gastropod shells, 281, 298
  pelecypods, 403
Ornopsis, 315
Orophocrinus, 599, 601
Orospira, 292
Orria, 501
Orthacea, 216–220, 227–231, 263
Orthambonites, 228
Orthida, 204, 218, 220, 227–233
Orthis, 212
Orthoceras, 348
Orthochoanites, 342
orthochoanitic, 349–351
orthocones, 347
Orthodesma, 414
orthogenesis, 34
Orthograptus, 724
Orthomyalina, 420, 426, 447
Orthonema, 307
Orthonota, 416
Orthonotacythere, 541
Orthopsina, 696
Orthoptera, 571
orthopteroids, 571
Orthorhynchula, 241
orthostrophic, 281, 288
Orthovertella, 48
Osagian, 37
oscula, 79
osphradium, 279
ossicles, 609, 659, 669, 681
Ostracoda, 291
ostracodes, 463, 470, 471, 521–543
  appendages, 533
  classification, 536
  Eocene, 543
  geologic occurrence, 536
  molt stages, 523
  morphology, 523, 529
  Oligocene, 543
  ontogeny, 531
  orientation, 531
  physiology and reproduction, 533
  Pliocene, 543
  size, 526
ostracum, 339, 402, 430
Ostrea, 400, 431, 432, **447**

Ostreacea, 411, 431, 448
Otarion, 513
Otionella, 188
Ottawacrinus, 637
Ottoia, 456
Oulodus, 736
outer hinge plate, 210
outer lamella, ostracodes, 529
outer lip, gastropod shells, 283
outer side plates, blastoids, 597
ovicells, 157
Ovoceras, 359
ovoid gastropod shells, 289
Owenella, 291
Oxfordian, 36
Oxycerites, 380
Oxynoticeras, 379
Oxytropidoceras, 383
oysters, 399, 400, 407, 411, 422, **431**
Ozarkispira, 292
Ozarkodina, 737

P

Pachyodonta, 411, 440
pachyodonts, 409, 440, 441
Pachyphyllum, 131, 138
Pachyteichisma, 93
Pachyteuthis, 391
Paeckelmann, W., 266
Paedeumias, 485–490, 497
paedogenesis, 138, 467
paedomorphism, 28, 491
Pagetia, 495, 499
paired spiracles, 597
Paladmete, 316
Palaeastraea, 147
Palaeechinus, 686
Palaeoalveolites, 114
Palaeoacris, 547, 548
Palaeochaeta, 456
Palaeochirodota, 656
Palaeoconcha, 411
palaeoconchs, 412–416, **420**
Palaeoctopus, 395
Palaeodiscus, 691
Palaeoisopus, 566
Palaeolimulus, 555, 556
Palaeoneilo, 416, 420
Palaeopedina, 695
Palaeophonus, 562, 563
Palaeosmilia, 137
Palaeotremata, 219, 220, 226, 263
Palaeotuba, 105
Palasterina, 668
Paleocene bryozoans, 185, 188
Paleocene cephalopods, 361, 363
Paleocene echinoids, 700, 707
Paleocene ostracodes, 543
Paleogene pelecypods, 416, 418, 438
Paleogene worms, 456
paleopterans, 570
pali, 145, 146
palingenesis, 28, 371
palintrope, 210, 214
pallet, 444
pallial line, 406
pallial markings, 199, 207, 211, **212**
pallial sinus, 406, 442

Pallium, 442
Palmatolepis, 737
Palmer, K. V., 333, 451
Palmula, 53
palpebral furrows, 480
palpebral lobes, 480
Palpigradi, 563, 565
Paltodus, 736
Paludina, 277
palus, 145, 146
panderian organs, 484
Panope, 442
papillae, 653
Parabolbina, 533
Paracavellina, 536
parachomata, 63
Paracrinoidea, 577, 583
paraerinoids, 583, 602
Paracyclas, 416, 439
Paracypris, 541, 543
Paracytheridea, 543
Paradasyceras, 380
Paradelocrinus, 647
Paradoxides, 478, 485, 486, 491, 492, 497, 507, 518
Paraechmina, 528
Parafusulina, 62, 63, 66
paragnaths, 454
Paraharpes, 493–495
Parahealdia, 531
parallelism, 34
Parallelodon, 413, 416, 420
Paranomia, 432
Paraparchites, 539
parapodia, 453, 454
Parapopanocaras, 377
Pararca, 416
Pararthropoda, 470
pararthropods, 472
Paraschwagerina, 62
parasitism, 19
Parasmilia, 148
Parasipirifer, 256
paratheca, 144
paratypes, 13
Parawocklumeria, 370
Parazoa, 17
Parichthyocrinus, 645
paries, 94
parietal line, 458
parietal lip, 283
parietes, 94
parivincular, 409
Parker, F. L., 78
Parker, T. J., 581
Parkeria, 106
Parks, W. A., 153
Parmorthis, 232
Parulocrinus, 647
Parvitonnia, 315
Patella, 277, 304
Patellacea, 304, 309
patellate, 118, 123
patelliform, 289
Patelliocrinidae, 633
Patelliocrinus, 639
Paterinacea, 220, 224, 225, 263
paucispiral, 289

Paulinites, 456
Paurorhyncha, 241
paxillae, 664
Peck, R. E., 603, 651
Pecten, 10, 398, 400, 402, 423, 426
Pectinacea, 407, 411, 422, 423, 448
Pectinibranchia, 301, 313, 322
pectinirhombs, 587
pectinoid, 414, 416, 420, 424, 426
pedicellariae, 663, 664, 676, 682
pedicle, 197, 199, 200
   muscle scar, 206
   valve, 200, 207
Pedinina, 695
Pedinothuria, 695
pedipalpi, 553
pelagic, 19
Pelagothuria, 656
Pelanechinus, 686–689, 695
Pelecocrinus, 643
Pelecypoda, 269, 270
pelecypods, 398
   anatomy, 400
   classification, 409
   ecology, 398-400, 407
   Eocene, 436, 438, 441
   evolution, 447
   geologic occurrence, 448, 450
   Jurassic, 419, 422, 426–430, 440-447
   largest, 399
   life habits, 398–400, 406, 429
   morphology, 406
   Neogene, 416, 418, 430, 442-446
   ontogeny, 407
   Paleogene, 416, 418, 438
   perforate, 409
   Pliocene, 446
   reproduction, 407
   (*See also* Cretaceous, Devonian, Miocene, Mississippian, Ordovician, Pennsylvanian, Permian, Recent, Silurian, Triassic pelecypods)
pelma, 609
Pelmatozoa, 577, 604
Pelseneer, P., 275, 334, 451
Peltura, 489
pen, squid, 337, 394
pendent graptolites, 725
Pennatula, 111
Pennatulacea, 104, 110
Penniretepora, 180
Pentacrinites, 648, 650
Pentagonia, 261
Pentameracea, 220, 237, 238, 263
Pentamerella, 238
Pentamerida, 212, 220, 236–238
Pentamerus, 237, 239
Pentastomida, 470
Pentremites, 594–596, 599–601
Pennsylvanian arthropods, conchostracans, 545
   insects, 569, 571
   malacostracans, 547–549
   myriapods, 567
   ostracodes, 537–539
   scorpions, 562
   spiders, 472

Pennsylvania arthropods, trilobites, 494, 509
   trilobitomorphs, 517
   xiphosurans, 555, 557
Pennsylvanian brachiopods, inarticulates, 223
   rhynchonellids, 241–243
   spiriferids, 259, 262
   strophomenids, 203, 218, 244, 248, 249, 252
   terebratulids, 234
Pennsylvanian bryozoans, 168, 169, 172, 180, 194
Pennsylvanian conodonts, 733, 737
Pennsylvanian conulariids, 458
Pennsylvanian corals, 113, 135, 138
Pennsylvanian echinoderms, blastoids, 600, 601
   crinoids, 624, 635, 646–648
   echinoids, 690
   holothuroids, 656
Pennsylvanian fusulinids, 64
Pennsylvanian mollusks, cephalopods, 360, 361, 372, 376
   gastropods, 296–301, 306–311, 324, 331
   pelecypods, 416, 419–422, 426, 432, 439
   scaphopods, 274
Pennsylvanian sponges, 90, 91
Pennsylvanian worms, 452
percarids, 548, 549
Percival, E., 204, 266
perforate pelecypods, 409
perforate septa, corals, 123, 125
Periarchus, 707
periderm, 101, 102
Periechocrinites, 639
periembryonic chambers, 68, 72
Perigastrella, 190
perignathic girdle, 680, 687
Perimestocrinus, 647
periostracum, 286, 402
Peripatus, 471, 472
peripetalous fascioles, 701
peripheral edge, corals, 119, 123
periphery, foraminifers, 42
   gastropods, 283
periproct, 589, 676, 680, 686
Peripylea, 75
perispatia, 352–354
Perisphinctes, 381
Perissocytheridea, 543
peristome, blastoids, 600
   bryozoans, 169, 171, 190
   echinoids, 676, 680, 681, 687
   gastropods, 283
peristomial plates, 669
peristomice, 190
peritheca, 144
Peritrochia, 369
Permian arthropods, conchostracans, 545
   insects, 569-572
   malacostracans, 548
   ostracodes, 537–539
   spiders, 565
   trilobites, 494, 509
   xiphosurans, 555

Permian brachiopods, rhynchonellids, 241–243
spiriferids, 259, 262
strophomenids, 244, 247–249, 252
terebratulids, 234
Permian bryozoans, 178, 180, 182, 194
Permian cephalopods, ammonoids, 368, 369, 372, 374, 377
belemnoids, 389, 391, 392
nautiloids, 350, 360, 361
Permian conulariids, 458
Permian corals, 127, 136, 137
Permian echinoderms, blastoids, 600, 601
crinoids, 621, 627, 629, 634, 635, 646–648
echinoids, 685, 686, 693
starfishes, 668
Permian fusulinids, 66
Permian gastropods, amphigastropods, 296, 297
archaeogastropods, 299, 301, 304, 307–309, 311
mesogastropods, 324
Permian pelecypods, 416, 424, 426, 431, 439, 444
Permian sponges, 91
permineralization, 4
Permotipula, 569
Perner, J., 334
Pernidae, 430
Pernopecten, 426
Peronella, 711
Peronopsis, 497, 499
Perprimitia, 536
Perrinites, 368, 371, 374, 383
Perry, L. M., 334, 451
Pervinquiéria, 383
Petalocrinus, 626, 639
Petalopora, 185
petals, echinoids, 699, 701
Petraster, 668
petrifaction, 4
Petrocrania, 225
Petrunkevitch, A., 573
Peytoia, 109
phaceloid corals, 123, 129
Phacops, 481, 490, 491, 507, 515, 519
Phaenopora, 167
phaneromphalous, 290
Phanocrinus, 624, 643
Pharetrones, 89
Pharkidonotus, 309
phenotypes, 24
Philhedra, 225
Phillipsia, 509
Philoxene, 302
Phlycticeras, 381
Pholadella, 416
Pholadomya, 432
Pholas, 399
Phormosoma, 687, 710
Phoronida, 17
phosphatic shells, 2
Phragmoceras, 358
phragmocone, 343, 348, 388, 393
Phragmodus, 736
Phragmolites, 292

Phragmoteuthis, 391
Phrynichida, 565
phyla, 15, 16
Phylactolaemata, 159, 161
Phyllocarida, 548
phyllocarids, 546, 548
Phylloceras, 380, 382, 386
phyllodes, 699
Phyllograptus, 727, 728
Phyllopachyceras, 382
Phylloporina, 165
phyllotheca, 121, 137
phylogeny, 28
Phymatopleura, 307
Phymosomina, 695
Physa, 276, 331
Physetocrinus, 645
Pictetia, 383
Piestochilus, 316
Pilsbry, H. A., 275, 334
Pinacognathus, 737
Pinna, 432
Pinnacea, 411, 430
pinnulars, 611
pinnule sockets, 610
pinnules, 610, 620, 630
Pionodema, 218, 232
Pisocrinus, 621, 639
Pitar, 439
Placenticeras, 386
Placocystites, 585
Plaesiomys, 212, 231
Plagioptychus, 441
plane of commissure, 403
planispiral, 42, 48, 281, 289
planktonic environment, 19
plano-convex brachiopods, 217
Planorbula, 331
Planularia, 56
plate, conodonts, 734
platform, brachiopods, 207, 222
conodonts, 734
Platyceras, 304
Platycrinites, 628, 633
Platycrinitidae, 633, 634
platyhelminthes, 17, 454
Platyorthis, 233
Platyostoma, 302
Platyrachella, 256
Platystrophia, 228, 231
Platytrochus, 151
Plaxocrinus, 648
Plecoptera, 571
Plectoconcha, 234
Plectoconularia, 460
Plectodonta, 247
Plectogyra, 58
Plectospathodus, 736
Plectronoceras, 347
Plegiocidaris, 689
Pleistocene bryozoans, 186
Pleistocene crustaceans, 545, 549
Pleospongea, 89, 93
pleospongus, 93
Plesiocidaris, 695
Plesiodiadema, 688, 711
Plesiosiro, 565
Plethobolbina, 529

Plethospira, 294
pleural angle, gastropods, 283
pleural furrows, 481
pleural lobes, 475, 481, 483
pleural spines, 481
pleurites, 465
Pleurocoela, 290, 328
Pleurocora, 148
Pleurocystites, 588, 591, 593
Pleurodictyum, 129
Pleuromya, 429
pleuron, 481
Pleuronea, 185
Pleurophorus, 420, 425
Pleurotomaria, 298, 304, 309, 311
Pleurotomariacea, 291–294, 299–308, 311, 320, 324
plicae, 403, 409
plications, 207, 210
Plicatula, 422, 442
Pliensbachian, 36
Pliocene cephalopods, 395
Pliocene echinoids, 702, 708
Pliocene foraminifers, 58
Pliocene ostracodes, 543
Pliocene pelecypods, 446
Pliolepidina, 70, 72
Pliomerops, 505
Plocezyga, 307
plocoid corals, 123, 131
Plocostoma, 309
Plummer, F. B., 396
podia, 575, 677
Pohl, E. R., 451
Poleumita, 301
polian vesicles, 653
Polinices, 315, 317, 318, 324
polyaxons, 83
Polychaeta, 453, 458
Polychera, 565
Polydiexodina, 62, 63, 66
Polygnathoides, 736
Polygnathus, 373
Polygyra, 331
Polylepidina, 68, 70, 72
Polymorphinidae, 49, 60
polymorphism, 22, 163, 725
polyphyletic, 133
Polypora, 177, 180, 183
polyps, 99, 100, 104
Polysalenia, 688
Polyzoa, 156
Pomatiopsis, 331
Pontian, 36
Ponton, G. M., 77
Popanoceras, 377
populations, 10, 24
porcelaneous layer, 339
Porcellia, 304
pore rhombs, 587
pores, blastoids, 597
echinoid spines, 679
edrioasteroids, 585
pleospongus, 93
Porifera, 15, 17, 79, 87
poriferous zones, echinoids, 677
Porocidaris, 689
Porodiscus, 75

Poromyacea, 410, 412, 442, 445, 447
Porpites, 125
Portlandian, 36
postanals, 701
posterior, brachiopods, 207, 210
  conodonts, 734
  crinoids, 608
  echinoids, 684
  gastropods, 285
  pelecypods, 403
posterior extremity, brachiopods, 210
posterior margin, brachiopods, 210
Poterioceras, 361
Poteriocrinina, 614
Poulsenia, 501
Pourtalès plan, 144
Pourtalesia, 700, 701, 708, 709
Prasopora, 165
Pre-Cambrian arthropods, 471, 472
Pre-Cambrian radiolarians, 74
Pre-Cambrian sponges, 81, 91
Pre-Cambrian worms, 452, 458
pre-epipodites, 481
Prestwichianella, 555–557
Priapuloidea, 17
primary radials, starfishes, 664
primary spines, echinoids, 676
primary tubercles, echinoids, 676
primaxils, 620
primibrachs, 619
Primitia, 526
Primitiella, 526
Prionoidina, 737
Prionodesmacea, 411, 423, 448
Priscochiton, 272
Priscopedatus, 656
prismatic layer, 402
Prismodictya, 88, 91
Proaulacopleura, 503
proboscis, gastropods, 278
Procerithiopsis, 309
Proclydonautilus, 362
proctodaeum, 468
prodissoconchs, 403, 407
Prodromites, 371
produced pelecypod, 409
Productacea, 218, 220, 243, 247, 253, 263
Productella, 251
productids, 249
Productorthis, 218
Proetus, 509
Profusulinella, 62, 64, 66
Prolecanites, 371, 372
proloculus, 40, 42, 44, 60
Promelocrinus, 629
Promytilus, 420
Pronorites, 371, 372
pro-ostracum, 388
proparea, 207
Proparia, 487, 490, 495, 505, 515
proparians, 479, 489, 490, 499, 503–509
Properrinites, 369, 371, 374, 383
Proplanulites, 369
Prorichtofenia, 249, 252
prosicula, 720, 724
prosiphonate, 371
Prosobranchia, 289, 301

prosocline, 409, 425
prosogyral, 409
prosoma, 552
Protadelaidea, 471
protaspis, 484, 489
Protaxocrinus, 637
protegulum, 203, 213
Proteocystites, 592
Proteonina, 53
Proteroblastus, 591, 594
proterogenesis, 142, 729
Protista, 39
Protobranchia, 410
protobranchs, 409, 413, 414, 448
Protocaudina, 656
Protochordata, 716, 718
protochordates, classification, 717
protoconchs, 339, 366, 389
protomesenteries, 144
Protopalaeaster, 671
Protoparia, 487, 488, 515
protoparians, 479, 487, 488, 497
protopygidium, 485
Protorthoptera, 571
Protoscolex, 456
Protoscorpiones, 562
protosepta, 119, 121, 123
Protospongia, 85, 91
Prototreta, 225
Protozoa, 17, 39
protractor muscle scars, 207, 222
Protremata, 219
proximal, 168, 611, 619, 663
Psammophax, 46
Psammosphaera, 44
Pseudobelus, 391
pseudocardinals, 420
pseudoceratites, 384
Pseudocidaris, 695
Pseudoclavulina, 53
Pseudocoelomata, 17
Pseudoconularia, 460
Pseudocrustacea, 470
pseudodeltidium, 211
Pseudogastrioceras, 372, 374
Pseudogaudryinella, 53
Pseudoglandulina, 53
pseudolaterals, 420
Pseudoliva, 318
Pseudomonotis, 425, 426
Pseudonerinea, 313
Pseudoniscus, 555
Pseudophragmina, 72
pseudoplanispiral, 281
pseudoplanktonic, 722
pseudopodia, 39, 40, 73
Pseudopolygnathus, 737
pseudopunctate, 199, 203, 243
Pseudorbitoides, 70, 72
Pseudorthoceras, 349, 361
Pseudorthoceratidae, 349, 360
pseudorthoceroids, 350
Pseudosalenia, 688
Pseudoschwagerina, 62, 66
Pseudoscorpiones, 565
pseudosepta, 112, 169, 171
pseudospondylium, 212
Pseudostaffella, 64

Psilosolen, 186
Ptarmigania, 497
Pteria, 400, 420, 426, 429, 432
Pteriacea, 423
Pterinea, 414, 425, 426
Pterinopecten, 426
Pterobranchia, 718
pterobranchs, 716
Pterocera, 317
Pterocorys, 75
Pteropoda, 290, 328, 329
Pterotoblastus, 600, 601
Pterotocrinus, 645
Pterygia, 322
Pterygota, 471, 569
Pterygotus, 553, 559–561
Ptilocrinus, 605, 609–611, 615
Ptilograptus, 728
Ptychoceras, 382
Ptychocrinus, 619, 631
Ptychomphalus, 311
Ptychoparia, 499
Ptychopariidae, 494
Ptychophyllum, 125
Ptychopyge, 492
Ptychosalpinx, 322
Ptychospirina, 302
Pugnoides, 241
Pullenia, 42
Pulmonata, 290, 308, 329
puncta, 199, 241
punctate shells, 199, 203, 227, 234, 241, 259
Punctospiracea, 220, 259, 262, 263
Punctospirifer, 262
Pupa, 332
pupaeform, 289
Pupilla, 331
Pupoides, 331
Purpuroidea, 311
Pustulina, 255
Pycnogonida, 471, 555
pycnogonids, 472, 555, 565, 566
Pycnoidocyathus, 95
Pycnosaccus, 639
Pygaster, 697
pygidium, 475, 483, 489, 490, 492, 512
Pygurus, 705
pyramid, echinoids, 681
pyramidal corals, 118, 123
Pyramidella, 317, 322
Pyramidellacea, 317, 318, 322, 324
Pyrgo, 53
Pyricythereis, 543
pyritization, 4
Pyropsis, 315

Q

Quasillites, 531
Quinqueloculina, 48, 53

R

Raasch, G. O., 573
Rachiglossa, 277, 294
radial canals, echinoderms, 579, 653, 661, 685
  ostracodes, 529

radials, blastoids, 595
   crinoids, 610, 624, 627–633
   cystoids, 589
   holothuroids, 655
   starfishes, 664
radianal, 616, 617, 623, 625
Radiaspis, 514, 515
Radiata, 17
radiocentral, 664
Radiolaria, 72
radiolarian ooze, 72
radiole, 676
radix, crinoids, 613
radula, cephalopods, 337
   chitons, 272, 277
   gastropods, 277, 294, 301
   scaphopods, 274
Rafinesquina, 217, 244
ramp, 283
Ranapeltis, 533
Raphistoma, 292
Raphistomina, 292
Rasetti, F., 519
Rasmussen, H. W., 674
Rastrites, 727, 731
Rathbun, M. J., 551
Raymond, P. E., 266, 519
Raymondatia, 526
Rayonnoceras, 361
rays, 575, 610, 619, 629
Recent brachiopods, inarticulates, 197, 203
   rhynchonellids, 203
   terebratulids, 198, 212
Recent bryozoans, 157, 187, 190
Recent cephalopods, 335–337, 391
Recent chelicerates, pycnogonids, 566
   spiders, 463, 553, 565
   xiphosurans, 556
Recent coelenterates, alcyonarians, 111
   hydroids, 105
   milleporids, 106
   zoantharians, 117
Recent crustaceans, branchiopods, 544
   crabs, 468, 549
   ostracodes, 531
   phyllocarids, 546, 548
Recent echinoderms, crinoids, 604
   echinoids, 676, 683, 685, 687, 698, 700, 708, 710
   holothuroids, 656
   starfishes, 664, 668
Recent foraminifers, 58
Recent gastropods, archaeogastropods, 301, 304, 307, 311
   mesogastropods, 278, 313, 315, 317, 322, 324
   neogastropods, 276, 328
   pteropods, 276
   pulmonates, 276
Recent myriapods, 463, 567
Recent onychophores, 471
Recent pelecypods, 399, 412, 419, 422, 436, 445
Recent worms, 454
Receptaculites, 87, 94, 96
Receptaculitida, 89
receptaculitids, 94, 96

reclined graptolites, 727
Rectocibicides, 57, 58
Rectocornuspira, 48
Rectogümbelina, 53
recrystallization, 4
recurrent faunas, 21
recurved graptolite stipes, 725
Redlichia, 497, 518
Redoubtia, 657
Reedolithus, 494, 505
reefs, 110, 118, 137, 442
Reeside, J. B., 396
reflex-angle interareas, 214
regeneration, 131, 665
Regnéll, G., 602
Regularia, 577, 675, 684, 696
Reineckia, 381
rejuvenescence, 131
Remera, 315
Rensselaeria, 234
replacement, 4
reproduction, alternating and asexual, 23
   sexual, 23, 40, 101
Reptolunulites, 190
Requienia, 440, 441
resilifer, 406, 422, 425, 426, 442, 445
resilium, 403, 425
respiratory trees, 576, 653
Resser, C. E., 519
Resserella, 232
resupinate brachiopods, 217
Reticulariina, 262
Retiograptus, 730
Retiolites, 731
retractor muscle scars, 207, 222
Retrorsirostra, 214, 231
retrosiphonate, 371
reverse side, bryozoans, 171
Rhabdammina, 43, 44
Rhabdomeson, 180, 183
Rhabdopleura, 716
rhabdosome, 719, 724, 725
Rhacophyllites, 380
Rhaetian, 36, 378
Rhineoderma, 304
Rhinidictya, 165
Rhipidium, 239
Rhipidoglossa, 277, 292, 301
Rhipidogyra, 148
Rhipidomella, 233
Rhizocrinus, 648
Rhizopoda, 39
Rhizostomida, 104, 109
Rhizostomites, 109
Rhodocrinites, 634, 645
Rhombiferida, 587, 591, 592
Rhombocladia, 180
Rhombopora, 168, 170
Rhopalodictyon, 75
rhopaloid septa, 123, 137, 139
Rhopalonaria, 173
Rhynchonellacea, 220, 239–243, 263
Rhynchonellida, 204, 212, 220, 237–243
Rhynchopora, 241, 243
Rhynchoporacea, 220, 241–243, 263
Rhynchospirina, 262
Rhynchotrema, 241

Rhynchotreta, 241
Rhyniella, 569
Rhytimya, 414
Richmondian, 37
Richter, R., and E., 462, 519
Ricinulei, 563
ridge, echinoids, 677
right-angle interareas, 214
ring canal, 579, 653
Rissoa, 322
Rissoacea, 320, 322
Rissoina, 322
Robinson, W. I., 154
Robson, G. C., 275, 451
Robulus, 43, 53, 56
Roman, F., 396
Rominger, C. L., 154
Romingeria, 131
rostrate, 409
Rostrospiracea, 218, 220, 259–261, 263
rostrum, 343, 388, 393, 480
Rotalia, 50, 53
rotator muscle scars, 207, 222
Rotifera, 456
Rotulc, 708
rotulae, 682
Roundyella, 539
Roveacrinida, 614
Rowia, 499
Rudistacea, 407, 411, 440
Ruedemann, R., 572, 732
rugae, 210
Rugosa, 104, 114–143
Rupelian, 36
Rustella, 226
Rutoceratida, 342

S

Saccammina, 44
Saccamminidae, 44
Saccelatia, 526
Saccocoma, 648
Sacoglossa, 290
saddles, 340, 363, 366
   lateral and ventral, 366
Sagenocrinida, 614
Sagenocrinites, 639
sagittal section, 61
Sahni, M. R., 266
St. Joseph, J. K. S., 266
Sakmarian, 37
Salinan, 37
salinity, 18
Salmon, E. S., 266
Salpingoteuthis, 391
sand dollars, 696, 706–708
Sanford, W. G., 154
Sansabella, 539
Santonian, 36
Saracenaria, 56
Sarcodina, 39
Sargana, 315
Sargentina, 536
Sarmatian, 36
Saturnalis, 75
Saukia, 501, 503
Savage, T. E., 266

Saxicava, 447
Scalaripora, 175
Scalaspira, 322
Scaliognathus, 737
Scalitina, 302
scallops, 398, 411, 426
scandent graptolites, 724
Scaphella, 324
Scaphites, 386
Scaphopoda, 269, 270, 273, 318, 324
Scenella, 291
Schackoina, 53
Schenck, H. G., 451
Schindewolf, O. H., 154, 396
Schistoceras, 372
Schistochoanites, 342
Schizambon, 200, 225
Schizaster, 707
Schizoblastus, 599, 600
Schizocidaris, 680
Schizocoela, 17, 580
Schizocoralla, 104, 113
Schizocrania, 225
Schizocystis, 592
Schizodiscus, 545
Schizodonta, 411, 418
schizodonts, 409, 414–416, 420, 424, 425, 428, 434
Schizodus, 418, 419, 425
Schizolopha, 301
Schizomida, 563, 565
Schizopea, 292
Schizophorella, 218
Schizophoria, 218, 233
Schizorthosecos, 188
Schizostoma, 304
Schizothaerus, 444, 445
Schloenbachia, 384
Schlotheimia, 380
Schmidt, W. E., 651
Schmidtella, 526
Schuchert, C., 98, 218, 266, 267, 274, 602
Schuchertella, 247
Schuchertoceras, 351
Schwagerina, 62–64, 66
schwagerinid wall structure, 61, 62
scientific names, 11
Scleractinia, 104, 143
sclerites, 465
Scobinella, 320
Scofield, W. H., 334
scolecodonts, 452, 456
scolecoid corals, 118, 123
Scolithus, 456
Scolpodus, 736
Scorpio, 562, 563
Scorpiones, 562
scorpions, 463, 471, 472, 562, 563
Scott, G., 396
scrobicular spines, 676
scrobicular tubercles, 676
Scrobicularia, 399
scrobicules, 676
Scutella, 708
Scyphocrinites, 629, 632, 641
Scyphozoa, 104
scyphozoans, 109

Scythian, 36, 376
Scytalocrinus, 643
sea cucumbers, 574, 653
sea lilies, 574, 604
sea spiders, 471
sea stars, 661
sea urchins, 574
secondary tubercles, 676, 677
secundaxils, 620
secundibrachs, 619
Selenechinus, 683
Selenimyalina, 426
selenizone, 286
selliform, 68
selvage, ostracodes, 529
selvage groove, 531
Semaeostomida, 104, 110
Semele, 447
Seminolites, 539
Semiometra, 539
Semitextularia, 46
Senonian, 36
Sepia, 343, 393
sepia, 337, 393
Sepioida, 343, 364
sepioids, 391, 393
septa, acanthine, 122, 125, 137
  cephalopods, 340, 348, 362, 366
  corals, 112, 117, 119, 121, 145
  echinoid spines, 676
  fusulinids, 61, 62
  pterobranchs, 716
septa acceleration, 122, 137
septa cycles, 145
septal fluting, 60, 367
septal grooves, 118, 121
septal necks, 340, 348, 353, 366
septal pores, 60, 61
Septastrea, 151
Septibranchia, 410
septibranchs, 410, 432, 445, 448
septotheca, 144
septula, 63, 64
septum (*see* septa)
  of truncation, 351
Serpula, 456
Serpulitidae, 458
sessile, 19
sexual reproduction, 23, 40, 101
shaft, cardinal process, 213
  echinoid spine, **676**
Shamattawaceras, 357
Shansiella, 309
Sharpeiceras, 384
shelf, gastropods, **283**
Sherborn, C. D., 78
Shimer, H. W., 153, 334, 451, 520, 551, 602
shoulder, gastropod shells, **283**
shrimps, 463, 471
Shrock, R. R., 153, 334, 451, 474, 520, 551, 602
Shumardella, 243
Shumardites, 369, 372
sicula, 719, 720, 724
side plates, blastoids, 597
side shields, 669
Sidneya, 518

Sieberella, 239
Sieverts-Doreck, H., 613
Sigmomorphina, 56
silica, skeletons, 2, 41, 73, 76, 87
silicification, 5
silicoflagellates, 76
Silurian arthropods, cirripeds, 546
  eurypterids, 472
  myriapods, 566
  ostracodes, 528, 529
  scorpions, 472, 562
  trilobites, 491, 494, 507
  xiphosurans, 555–559
Silurian brachiopods, inarticulates, 222, 223
  orthids, 231, 232
  pentamerids, 212, 238, 239
  rhynchonellids, 239–241
  spiriferids, 251, 254–256, 261, 262
  strophomenids, 244, 246–248
  terebratulids, 234
Silurian bryozoans, 167, 194
Silurian cephalopods, 351, 354, 358
Silurian coelenterates, corals, 116, 124, 129, 131, 135, 138
  hydroids, 105
  scyphozoans, 110
  stromatoporoids, 108
Silurian conodonts, 736
Silurian conulariids, 460
Silurian echinoderms, blastoids, 582, 599, 600
  camerate crinoids, 630–635, 637, 639
  cystoids, 582, 587, 592, 593
  echinoids, 686, 691
  flexible crinoids, 627, 637, 639
  inadunate crinoids, 621, 623, 626, 637, 639
  ophiocystoids, 712, 713
  starfishes, 665, 668, 671
Silurian foraminifers, 44
Silurian gastropods, 296–301, 304, 307, 309, 329
Silurian graptolites, 723, 731
Silurian pelecypods, 412, 414, 419, 422, 426, 430, 439, 444
Silurian sponges, 88
Silurian tentaculitids, 460
Silurian worms, 456
simple corals, 118, 121
Simpson, G. B., 196
Sinclair, G. W., 462
Sinemurian, 36
sinistral, 289
Sinum, 317, 318
Sinuopea, 291
sinus, 284
Sinuspira, 301
Siphodictyum, 183
Siphogenerina, 58
Siphogeneroides, 50
siphon, cephalopods, 340
  gastropods, 279
  pelecypods, 399, 400, 402
siphonal caecum, 339
siphonal deposits, 350
siphonal notch, 283
Siphonalia, 322

Siphonina, 56
Siphonodella, 737
siphonofossula, 121
Siphonophorida, 104
Siphonophrentis, 125, 127
siphonostomatous, 289
Siphonotreta, 225
Siphonotretacea, 220, 224, 225, 263
siphuncle, 340, 350, 362, 366, 371, 387
Sipunculoidea, 17
skeletal composition, arenaceous, 41
    calcareous, 2, 42, 76, 87
    chitinous, 2, 41
    phosphatic, 2
    siliceous, 2, 41, 73, 76, 87
    strontium sulfate, 73
Skiddavian, 37
Skinner, J. W., 77
Sladen, W. P., 674
slit, gastropod shells, 280, 284
slit band, 284
Sloss, L. L., 155
Smith, J. P., 396
Smith, S., 154, 155
Sochkineophyllum, 137
socket, pelecypods, 403, 422
Solanocrinus, 648
Solariella, 318, 320
Solemya, 409, 412, 416
Solenacea, 411, 444
Soleniscus, 307
Solenocheilus, 361
Solenochilida, 342
Solenopsis, 412
Soliclymenia, 370
Solifugae, 563, 565
solitary corals, 118
Sollasina, 712
Solnhofen limestone, 110
Solvan, 37
Somasteroidia, 577, 659
somasteroids, 659, 668
somites, 464
Sorosphaera, 46
Sowerbyella, 244
Sowerbyites, 244
Spatangoida, 696, 698
spatangoids, 699, 700, 705, 706
Spath, L. F., 396
Spathognathodus, 737
species, 9, 15
    subdivisions of, 13
Spencer, W. K., 674, 714
Sphaeractinia, 106
Sphaeractiniida, 104, 106
Sphaerirhynchia, 241
Sphaerocystites, 592
Sphaerospongia, 96
Sphenodiscus, 386
Sphenotus, 416
spicules, echinoid, 684
    holothuroid, 655
    pennatulacean, 110
    sponge, 79, 83, 84
spiders, 463, 471, 472, 562, 563, 565
spines, brachiopod, 210
    bryozoan, 171
    starfishes, 661, 664, 669

Spinocyrtia, 256
spiracles, 595
spiral, logarithmic, 339
spiral angle, 283
spiralia, 213, 251
spiramen, 191, 193
spire, 283
Spirifer, 216, 255, 259
Spiriferacea, 220, 255–259, 263
Spiriferella, 259
Spiriferida, 202, 204, 212, 218, 220, 249
Spirillina, 41
Spiroceras, 380
Spiroplectammina, 43, 50, 53
Spiropora, 178
Spirorbis, 452
spirotheca, 60, 61
Spirula, 391, 393
Spirulirostra, 391
spitz, 94
Spondylacea, 411, 422
spondylium, 212
Spondylospira, 262
Spondylus, 400, 422
spongelike organisms, 89, 93
sponges, calcareous, 80, 89
    Cambrian, 81, 85, 91
    classification, 87
    Cretaceous, 81, 92
    Devonian, 88
    geologic occurrence, 81, 87
    Jurassic, 81, 93
    Mississippian, 89, 90
    morphology, 79
    Pennsylvanian, 90, 91
    siliceous, 80, 91
    Silurian, 88
spongocoel, 79
Springer, F., 602, 603, 651
Sporozoa, 39
squids, 335, 336, 392
Squilla, 548
Stacheoceras, 374
Stainbrook, M. A., 266
Stanton, T. W., 334, 451
starfishes, 27, 574
    classification, 659
    Devonian, 661, 665, 671
    ecology, 665, 669
    evolution, 665, 669
    geologic occurrence, 669
    morphology, 659
statistical analysis, 11
Stauffer, C. R., 462, 738
Stauranderaster, 673
Stauriacea, 133, 137
Staurocephalites, 456
Staurognathus, 737
Staurograptus, 728
Staurolonche, 75
Steganoporella, 190
Stegerhynchus, 241
Stegnammina, 46
Stegocoelia, 304
steinkern, 5, 286
Steinman, 434
Stellarocrinus, 648
Stelleroidea, 15, 577, 659

stelleroids, 659
    (See also starfishes)
Stellispongia, 93
stems, 575, 611, 612
Stenaster, 668
Stenodictya, 569, 670
Stenopoceras, 361
Stenopora, 183
Stenoscisma, 241, 243
Stensiöina, 50
Stephanocrinus, 599
Stephanophyllia, 151
Stephenson, L. W., 334, 451
Stereocidaris, 676, 705, 708
Stereocorypha, 135
Stereocrinus, 641
stereozone, 121, 127, 139
sternals, 701
sternites, 465, 559
Stewart, G. A., 155
Stibus, 533
Stichocados, 183
Stictoporella, 165
Stigmatella, 165, 167
stipes, graptolites, 719
Stirodonta, 684, 690, 698, 711
stirodonts, 690, 695
Stolonifera, 104, 112
Stolonodendrum, 723
stolonoid graptolites, 723
Stolonoidea, 718, 723
stolons, alcyonarian, 110, 112
    bryozoan, 173
    foraminifer, 68
    graptolite, 719
    hydrozoan, 101
    pterobranch, 716
stolotheca, 721, 722
Stomatopora, 175
stomodaeum, 102, 468
stone canal, 575, 653
stony bryozoans, 194
Storer, T. I., 451, 462, 474, 551, 573
Størmer, L., 474, 485, 520, 573
Strabops, 555, 556
straight-angle interareas, 214
Straparolus, 299, 302, 304, 308
stratigraphic use of fossils, 7
Streblites, 382
Streblotrypa, 183
Streptelasma, 114, 118, 135
Streptelasmacea, 133, 135
Streptognathodus, 737
Streptoneura, 301
Striatopora, 116, 135
Stricklandinia, 212
Strobeus, 302
Strobilocystites, 592, 594
Stromatocerium, 107, 108
Stromatopora, 108
Stromatoporella, 107, 108
Stromatoporida, 104, 107
Strombacea, 312, 315, 318, 320
Strombus, 315
Strongylocentrotus, 687, 705
strontium sulfate, 73
Strophecdonta, 247
Strophomena, 212, 217, 244

Strophomenacea, 220, 243–247, 263
Strophomenida, 204, 212, 218–220, 243
Strophonella, 247
Strophostylus, 302
Strotocrinus, 645
Strotogyra, 148
struggle for existence, 31
Stuartwellercrinus, 648
Stumm, E. C., 155
Stürtzaster, 671
Stylaster, 106
stylasterids, 104, 105
Stylina, 146, 148
Styliolina, 460
Stylommatophora, 290, 330, 332
Stylonurus, 553, 559–561
Stylopoma, 193
Stylosphaera, 75
subanal fasciole, 701
subanals, 701
Subbryantodus, 737
subfamilies, 14
subgenera, 14
sublancet plate, 599
suborthochoanitic, 349–352
subspecies, 13
Subulitacea, 294, 300–311
Subulites, 294, 309
Succinea, 331
Suecoceras, 355
Sulcella, 539
Sulcoretepora, 169, 170
sulcus, 207, 529
Sumatrina, 66
Sundacrinus, 648
superfamilies, 14
superradianal, 616
supramarginals, 664
suranals, 687, 693
Surcula, 322
Sutton, A. H., 266, 651
sutures, ammonoid, 362, 366, 368, 369
  crinoid, 610
  foraminifer, 42, 44
  gastropod, 283
  nautiloid, 339, 348, 356
Swain, F. M., 551
Swartz, F. M., 551
Swinnerton, H. H., 153, 520
Sycetta, 81, 82
Sycon, 80–82
Sycones, 89
symbiosis, 19, 73, 173
Synapta, 656
synapticula, 144
synapticulotheca, 144, 151
Synaptida, 657
Synaptophyllum, 125, 127, 129, 131
synarthry, 611
Synbathocrinus, 616, 641, 643
syncarids, 548
synonyms, 12
Synphoria, 509
synrhabdosome, 715, 725
Syntrophiacea, 220, 236, 263
Syntrophopsis, 236
syntypes, 13

Synxiphosura, 555, 556
Syringopora, 115, 116
syringoporids, 115
syzygial joints, 610
syzygy, 611

T

tabella, 121, 127
tabula, 106, 112, 121, 125, 145
tabularium, 121
Tabulata, 104, 112
tabulate corals, 112, 114–116, 131–135
Tabulipora, 180
Taeniaster, 671
Taenioglossa, 277, 294, 313
Taeniopora, 175
Taghanican, 37
Talanterocrinus, 627, 648
Talantodiscus, 311
Talarocrinus, 645
Talpaspongia, 91
Tanaocrinidae, 633
Tanaocrinina, 614, 634
tanaocrinoids, 633, 634
Tanaocrinus, 633, 637
tangential sections, 61
Taras, 439, 442
Tardigrada, 470
Tarphyceratida, 342
Tarsophlebia, 569
tautonymy, 13
Taxocrinida, 614
Taxocrinus, 643
Taxodonta, 411, 414
taxodonts, 409, 413–420, 424, 425, 434, 438, 446
tectorium, 61
tectum, 61, 62
teeth, echinoids, 681
  gastropod shell, 286
  starfishes, 664, 669
tegmen, 605, 610, 618, 625, 631
tegmentum, 273
teginal, 614, 619
Tegulorhyncha, 203
Teichert, C., 396
Teleiocrinus, 645
Teleodesmacea, 411, 448
Telestacea, 104, 112
Teliocrinus, 605, 611
Tellerina, 503
Tellina, 439, 442
Telotremata, 219
telson, 489, 552
Temnopleurina, 696
Temnopleurus, 711
temperature, 18, 53
tentacles, coelenterates, 100, 117
  bryozoans, 156
  gastropods, 278
  holothuroids, 653
  nautiloids, 337
  squids, 336
  worms, 454
Tentaculites, 460
Tentaculitidae, 458
tentaculitids, 459, 460

Tenticospirifer, 214, 256
Terataspis, 478, 509
Terebra, 320, 324, 328
Terebratulida, 212, 218, 220, 234–236, 263
Terebratulina, 198
Terebrirostra, 198
Teredo, 444, 445
tergal, 618, 619, 628, 631, 634
tergites, 465, 559
tergopores, 180
Termier, H., and G., 462
terminal aperture, 42, 48
terminals, 664
terminology, anthozoans, 103
  blastoids, 595–597
  brachiopods, 206–210
  bryozoans, 162, 163, 168–172, 191
  cephalopods, 336–354, 365–367, 387
  chelicerates, 353
  chitons, 273
  conulariids, 458
  corals, 117, 120–123, 145
  crinoids, 606–610
  cystoids, 588–589
  edrioasteroids, 585
  foraminifers, 42, 60–62
  gastropods, 284–289
  hydrozoans, 100, 101
  milleporids, 106
  ostracodes, 522–523
  pelecypods, 404–406, 408, 409
  pleosponges, 94
  radiolarians, 73
  scaphopods, 274
  sponges, 82–84
  trilobites, 476–478, 493
  worms, 454, 455
Tertiary arthropods, cirripeds, 545, 547
  crabs, 550
  hoplocarids, 548
  insects, 570–572
  myriapods, 567
  percarids, 549
  spiders, 472, 563, 565
Tertiary brachiopods, 225
Tertiary coelenterates, alcyonarians, 110, 112
  corals, 151
  hydroids, 105
  milleporids, 106
Tertiary echinoderms, crinoids, 650
  holothuroids, 656
  starfishes, 673
Tertiary mollusks, archaeogastropods, 301, 304, 307, 309, 311
  cephalopods, 356, 363, 393
  mesogastropods, 313, 315, 317–327
  neogastropods, 318–328
  opisthobranchs, 318, 322–325
  pelecypods, 438, 442–446
  scaphopods, 318, 324
Tertiary protozoans, coccoliths, 76
  dinoflagellates, 76
  foraminifers, 54–59, 68–72
  radiolarians, 74, 75
  silicoflagellates, 76
tertiary septa, 119, 121

Tertiary sponges, **81**
Tertiary worms, **454**
Tessardoma, 190
Testacella, 276
tetrabasal, 699
Tetrabranchiata, 341
Tetracoralla, 118
tetracorals, 104
Tetractinella, 218
Tetradella, 531, 535
Tetradium, 113, 114
Tetragraptus, 727, **729**
Tetralobula, 236
Tetranota, 294
Tetrataxis, 48
Tetratylus, 536
Tetraxonida, 89
tetraxons, 83
Teuthoida, 343, 364
teuthoids, 394, 395
Texanites, 386
Textularia, 41–44, 50, 53, **56**
Thaleops, 503
Thallocoralla, 104, **113**
Thamnasteria, 148
thamnasterioid corals, 123, **131**
Thamniscus, 180
Thanetian, 36
theca, coelenterates, 145
 echinoderms, 575, 582
 graptolites, 719, 720, **724**
 pterobranchs, 716
thecal pores, 576
thecarium, 118, 120, **121**
Thecia, 116
Thecosmilia, 148
Thelyphonida, 565
Thenarocrinus, 616, **639**
Theodiscus, 75
Theodossia, 256
thin sections, **159**
Thlipsura, 531
Thlipsurella, 531
Tholocrinus, 643
Tholosina, 44
Thomas, H. D., **154**
Thomas, I., 266
Thomasatia, 525
Thompson, M. L., 78
Thomson, J. A., 219, 266
thoracic segments, 481
thorax, 475, 481, 488–492, 511
Thracia, 447
Thurammina, 44
Thylacocrinus, 641
Thysanura, 570
Tiarasmilia, 148
Tiarechinus, 686, 690, **693**
Timanites, 370
Timor, 601, 635
Timorechinus, 648
Timorites, 374, 383
Timoroblastus, 596, 600, **601**
tintinnids, 76
Tirolites, 377
Tissotia, 386
Titusvillia, 89, **91**
Toarcian, **36**

Tongrian, 36
Tonnacea, 315, 317, 318
Tonolowayan, 37
Tornatellaea, 317, 318
Tortonian, 36
torus, 669
Toucasia, 440, 441
Tournaisian, 37
Toxiglossa, 277, 294
trabeculae, 117, 121, 125, 145
Trachyceras, 377
Trachydermon, 277
Trachydomia, 309
Trachylida, 104
Trachypora, 131
trail, 210, 247
transverse ridge, 610
Tremadocian, 37, 659, 719
Tremanotus, 301
tremata, 286
Trematis, 225
Trematoda, 454
Trematospira, 262
Trempealeauan, 37
Trentonian, 37
Trepospira, 307
Trepostomata, 161–167, 175–182
Triadocidaris, 693
Triarthrus, 463, 481–483, 491, 492
Triassic brachiopods, 218, 234, 262
Triassic cephalopods, ammonoids, 369, 376–378, 384
 belemnoids, 389, 391
 nautiloids, 349, 356, 361
Triassic coelenterates, alcyonarians, 110
 corals, 147
Triassic conulariids, 458
Triassic crustaceans, conchostracans, 545
 eucarids, 549
 phyllocarids, 548
 xiphosurans, 556
Triassic echinoderms, crinoids, 635
 echinoids, 686, 690, 693
Triassic foraminifers, 50
Triassic gastropods, amphigastropods, 297
 archaeogastropods, 299, 301, 304, 307, 309, 311
 mesogastropods, 313, 315, 324
Triassic insects, 572
Triassic pelecypods, 419, 422, 423, 426, 428, 429, 444
triaxons, 83
Trichotocrinus, 629
Tricoelocrinus, 599
Tricrepicephalus, 503
Tridacna, 399, 407
tridentate, 683
Triebel, E., 551
trigonal, 409
Trigonia, 418, 419, 429, **434**
Trigoniacea, 411, 419
Trigonoglossa, 223
Trigonomartus, 565
Trigonopora, 188, 193
Trigonotarbi, 565
Triloculina, 50, 53

Trilobita, 470, 487
trilobites, appendages, 481, 490, 492, 495
 Cambrian (*see* Cambrian trilobites)
 classification, 486
 Devonian, 490–494, 507, 509, 513
 evolution, 497
 eyes, 510
 geologic occurrence, 491, 494, 515
 hypoparian, 479, 493, 503, 505
 Mississippian, 494, 509
 morphology, 467, 477
 ontogeny, 484, 485, 489
 Ordovician (*see* Ordovician trilobites)
 primitive characters, 501
 secondary segmentation, 486
 size, 478
 spinosity, 513, 514
 zoological affinities, 470
Trilobitomorpha, 463, 470, **478**
Trimerella, 223
Trimerellacea, 220–223, 263
Trimerocrinus, 648
Trimerus, 507
Trinitian, 36
triphyllous, 683
Tripilidium, 75
Triplesia, 239
Triplesiida, 220, 237, 239, **263**
Tripleuroceras, 359
Triplophyllites, 125, 233
Tripylea, 75
triserial, triserial-biserial, and triserial-uniserial, definition, 42, 48
Tritaxia, 50
Triticites, 60, 62–66
Tritoechia, 200, 227
Tritonalia, 324, 326
Tritonoatractus, 320
trivial names, 11
trivium, 655
Trochacea, 307, 311, 312, **318–326**
Trochactaeon, 313
Trochammina, 53
Trochelminthes, 456
trochiform, 289
Trochifusus, 317
Trochocyathus, 148, 151
trochoid, 42, 48, 118, 123, 347, 358
Trochonema, 292, 307
Trochonemnatacea, 292, 294, 300–304, 307, 308, 311, 312
Trochopora, 188
Trochosmilia, 148
Trochus, 277, 307
Troosticrinus, 599
Tropidodiscus, 302
Tropidoleptus, 231, **233**
Tropites, 378
trough, 207
Trueman, A. E., 386, 396
Trupetostroma, 108
Tryblidiacea, 289–291, 296, **297**
Tryblidium, 296, 298
Tryoniella, 317
Tryplasma, 125
tube feet, 575, 653, 661, 669, 677, 685
Tubidendrum, 723

Tubipora, 111
tuboid graptolites, **722, 723, 731**
Tuboidea, 718, 722
Tubucellaria, 186, 188
tubular, 42
Tubulibairdia, 533
Tubulostium, 456
Tunicata, 716, 718
tunnel, 61
Turbellaria, 454
turbinate, 118, 123, **289**
Turbinolia, 151
Turbonopsis, 302
Turcica, 326
Turner, F. E., 451
Turonian, 36, 386
turreted, 289
Turrilepas, 546
Turrilites, 384, 386
Turris, 324, 328
Turritella, 313, 315, 320, 324, 326, 328
Tschernyschew, T., 266
Twenhofel, W. H., 474
Twitchell, M. W., **713**
Tylocidaris, 705
Tylostoma, 313, 317
types, 10, 12

U

Uddenites, 371, 372, **376**
Uhligerites, 382
Uintacrinida, 614
Uintacrinus, 648
Ulocrinus, 648
Ulrich, E. O., 196, 266, 334, 396, 451, 551, 651, 738
umbilical lobes, 366
umbilical perforations, 339
umbilical plugs, 339
umbilical shoulders, 289, 339, 366, 378
umbilicus, 42, 48, 283, 339, 362
umbo, 207, 210, 403
umbonal cavities, 210
Unio, 419, 422, 429
Unionidae, 407
uniserial, crinoids, 620, 631
    echinoids, 686
    foraminifers, 42, 48
univalve, 398
Unklesbay, A. G., 396
Uperocrinus, 618, 645
upper lamella, trilobites, 493
Uralian, 37
Uralichas, 478
Urasterella, 668
Urosalpinx, 322
Ussuria, 376
Uvigerina, 53, 58

V

Vachon, M., **573**
vacuoles, 180
vagile benthos, 19
Vaginulina, 50
Valcourea, 217, 228
Valcouroceras, 352, 357

Vallonia, 331
Valvata, 278, 322, 324, 330
Valvatacea, 320, 324, 330
valves, 197, 273, 406
Valvulina, 53
Valvulineria, 58
Vandel, A., 474
Vanuxemia, 414
variations, 31
variety, 13
varix, 286, 366
Vasocrinus, 641
Vaughan, A., 155
Vaughan, T. W., 78, 153, 155, 450
Vauxia, 85, 91
Velella, 22
Vellamo, 214
Velumella, 190
Venericardia, 436, 440
Veniella, 434, 439
venter, 339, 366, 368
Ventilabrella, 53
ventral, 207, 339, 403, 609
ventral lobe, 366
ventral saddle, 366
ventral shield, 669
Ventriculites, 93
ventro-lateral fields, 664
ventro-lateral furrows, 387
Venula, 536
Venus, 398, 442
Verbeekina, 64, 68
Vermes, 452
Vermetus, 277
Vermiceras, 380
Verneuiliana, 43
Verrucosella, 536
vertebrae, ophiuroids, 669
Vertigo, 331
vestibules, 167, 529
vibracula, 193
Villebrunaster, 659, 660
Vinella, 175
virgalia, 660
virgella, 724
Virgilian, 37
virgula, 724
Virgulina, 53, 58
Virgulinella, 57
Viséan, 37
Viviparus, 322
Vogt, 16
Vokes, H. E., 334
Volborthella, 347, 356
Volsella, 426, 430, 432, 439
Volutacea, 314, 316, 318, 320, 322, 326, 328
Volutoderma, 315, 328
Volutomorpha, 317, 328
Volvula, 322
Vulvulina, 58

W

Waagenoceras, 374, 383
Waagenoconcha, 249, 252
Waagenophyllum, 137
Wachsmuth, C., 651

Wade, B., 334, 451, 462
Walcott, C. D., 98, 153, 267, 462, 520, 652
wall, echinoids, 677
Wang, H. C., 155
Wang, Y., 267
Wanner, J., 603
Wanneria, 488, 490
Waptia, 518
Warburg, E., 520
Warthia, 309
Washitan, 36
water vascular system, 661
Waterlot, G., 573
Waters, J. A., 77
Waucoban, 37
Webbinella, 44
Webbinelloidea, 46
Wedekind, R., 397
Wedekindellina, 62, 64
Weinbergia, 556
Weller, J. M., 98, 652
Weller, S., 267
Wellerella, 243
Welleria, 529
Wells, J. W., 155
Wenlockian, 37
Wenz, W., 334
Westergard, A. H., 520
Wewokella, 91
wheels, holothuroids, 656
Whiteavesia, 414
Whitfield, R. P., 265, 334
Whitfieldella, 261
Whittington, H. B., 520
whorls, 283, 337
Wilcoxian, 36
Wiman, 718
Winchell, N. H., 98, 267
Winchellatia, 525
Winkler, V. D., 651
Wolcott, R. H., 474, 551
Wolfcampian, 37
Wood, E. M. R., 732
Woodring, W. P., 334
worms, 452, 454–460
Worthenella, 456
Worthenia, 307
Worthenopora, 177
Wright, J., 652
Wright, T., 714

X

Xenaspis, 369
Xenaster, 671
xenidium, 211
Xenocrinidae, 633
Xenocrinus, 633, 637
Xenusion, 471, 472
Xiphosura, 463, 472, 489, 553–557, 560
xiphosurans, 555, 556, 559

Y

Yabeina, 66
Yoldia, 439
Yonge, C. M., 334

Youngquist, W., 738
Ypresian, 36

Z

Zacanthoides, 499
Zaphrenthis, 125, 127, 135
Zaphrentoides, 141
Zeacrinites, 616, 625, 643
zigzag septa, 123

Zirphaea, 440, 445
Zittel, K. A., 97
Zoantharia, 104, 115
Zoanthidia, 104
zoaria, 161
zones, 8
zooecia, 160, 161
zooids, 156
  budding, 716

zoological nomenclature, rules, 11, 12
zooxanthellae, 73
Zygobeyrichia, 529
Zygobolba, 529
Zygobolbina, 529
Zygopleura, 311
Zygosella, 529
Zygospira, 255